ENGINEERING HEAT AND MASS TRANSFER

ENGINEERING HEAT AND MASS TRANSFER

BY

MAHESH M. RATHORE

Energy Auditor and Chartered Engineer,
Professor and Head, Mechanical Engineering
SNJB's K.B. Jain
College of Engineering, Chandwad
Maharashtra, India

(An Imprint of Laxmi Publications Pvt. Ltd.)
An ISO 9001:2008 Company

BENGALURU • CHENNAI • COCHIN • GUWAHATI • HYDERABAD
JALANDHAR • KOLKATA • LUCKNOW • MUMBAI • RANCHI • NEW DELHI
BOSTON (USA) • ACCRA (GHANA) • NAIROBI (KENYA)

ENGINEERING HEAT AND MASS TRANSFER

Printed and bound in India

Typeset at Goswami Associates

First Edition: 2004; Second Edition: 2006; Third Edition: 2015; Reprint: 2016

ISBN 978-81-318-0613-5

PUBLISHED IN INDIA BY

UNIVERSITY SCIENCE PRESS
(An Imprint of Laxmi Publications Pvt. Ltd.)
An ISO 9001:2008 Company
113, GOLDEN HOUSE, DARYAGANJ,
NEW DELHI - 110002, INDIA
Telephone : 91-11-4353 2500, 4353 2501
Fax : 91-11-2325 2572, 4353 2528
www.laxmipublications.com info@laxmipublications.com

Branches

Bangalore	080-26 75 69 30	
Chennai	044-24 34 47 26,	24 35 95 07
Cochin	0484-237 70 04,	405 13 03
Guwahati	0361-254 36 69,	251 38 81
Hyderabad	040-27 55 53 83,	27 55 53 93
Jalandhar	0181-222 12 72	
Kolkata	033-22 27 43 84	
Lucknow	0522-220 99 16	
Mumbai	022-24 91 54 15,	24 92 78 69
Ranchi	0651-220 44 64	

C—10720/015/08
Printed at:

Contents

8. EXTERNAL FLOW 266–292

9. INTERNAL FLOW 293–332

10. NATURAL CONVECTION 333–371

11. CONDENSATION AND BOILING 372–401

12. THERMAL RADIATION: PROPERTIES AND PROCESSES 402–433

13. RADIATION EXCHANGE BETWEEN SURFACES 434–485

14. HEAT EXCHANGERS 486–553

15. MASS TRANSFER 554–591

16. EXPERIMENTS IN ENGINEERING HEAT TRANSFER 592–619

APPENDIX 621–645

INDEX 647–651

Preface to the Third Edition

The first edition of Comprehensive Engineering Heat Transfer was published in 2000. It was written principally to cater syllabi of Pune and North Maharashtra Universities.

The second revised and enlarged edition was published in year 2005, in which I had tried to incorporate the relevance of heat and mass transfer applicable to Mechanical, Chemical, Aerospace, Civil Engineering, Computer Science, Information Technology, Biotechnology, Pharmacology, and Alternative Energy generation.

Confronted with economic realities, many colleges and universities have set clear priorities. In recognition of its value and applications to society, investment in engineering education has increased. Pedagogically, there is reintroduced emphasis on the fundamental principles that are the foundation for lifelong learning. The important and sometimes dominant role of heat transfer in many applications, particularly in conventional as well as in alternative energy generation and concomitant environmental effects, has reaffirmed its relevance.

In preparing third edition, I have attempted to incorporate recent heat transfer research at a level that is appropriate for an undergraduate student. I have strived to include new examples and problems that motivate students with interesting applications, but whose solutions are based firmly on fundamental principles. We have remained true to the pedagogical approach of previous editions by retaining a rigorous and systematic methodology for problem solving. I have tried to continue the tradition of providing a text that will serve as a valuable, everyday resource for students and practicing engineers throughout their careers.

Approach and Organization

Previous editions of the text have adhered to four learning objectives:

1. The student should adopt the meaning of the terminology and physical principles associated with heat transfer.
2. The student should be able to describe relevant transport phenomena for any process or system involving heat transfer.
3. The student should be able to use requisite inputs for calculating heat transfer rates and/or material temperatures.
4. The student should be able to develop representative models of processes and systems and draw conclusions concerning process/system design or performance from the attendant analysis.

Moreover, as in previous editions, specific learning objectives for each chapter are clarified, as are means by which achievement of the objectives may be assessed. The summary and glossary at the end of each chapter highlight key terminology and concepts developed in the chapter and poses questions designed to test and enhance student comprehension.

What's New in the Third Edition?

In order to reduce the volume and cost of book, it is prepared in two columns and two colours to make it attractive and interesting. The constructive criticisms and suggestion sent by users have been amalgamated. Answer(s) to almost all problems presented for practice at the end of each chapter are provided.

Chapter-by-Chapter Content Changes

Chapter 1, Concepts and Mechanisms of Heat Flow is re-written and modified to accentuate the significance of heat transfer in various contemporary applications. It has also been improved to elaborate the complementary nature of heat transfer and thermodynamics. The economic thickness of insulation is augmented with a new section with the help of cost and year of service. Two more sections on heat transfer in boiling and condensation and mass transfer are included at the end of chapter.

Chapter 2, deals with Conduction-Basic equations and their applications with the help of boundary conditions. In this edition, the boundary conditions are elaborated with extensive graphical support. The radiation and interface boundary conditions are incorporated.

Chapters 3, 4 and 5 have undergone extensive revision and some examples are reorganized in order to give them justification. Some parallel illustrations are withdrawn from these Chapters.

Chapter 6, Transient Conduction was substantially modified in the previous edition and has been augmented in this edition with a streamlined presentation of the methods. The multi-dimensional, semi-infinite body transient heat transfer has been restructured.

Chapter 7, Principles of Convection includes clarification of how temperature-dependent properties should be evaluated when calculating the convection heat transfer coefficient. Specifically, presentation of the similarity solution for flow over a flat plate has been simplified.

Chapter 8, External Flow has been updated and reduced in length. New results for flow over noncircular cylinders have been added, replacing the correlations of previous editions. The discussion of flow across circular tubes has been reduced, abolishing redundancy without sacrificing content.

Chapter 9, Internal Flow; entry length correlations have been updated and rearranged.

Chapter 10, Free Convection include a new correlation for free convection from flat plates, replacing a correlation from previous editions. The discussion of boundary layer effects has been modified.

Aspects of Boiling and Condensation have been updated to incorporate recent advances in, for example, external condensation on finned tubes in *Chapter 11,* Condensation phenomena and heat transfer rates are explained. The coverage of forced convection condensation and related enhancement techniques has been expanded.

The concepts of emissive power, irradiation, radiosity, radiation function and net radiative flux are presented in *Chapter 12,* Radiation: Properties and Processes. The coverage of environmental radiation has undergone substantial revision, with the inclusion of separate discussions of solar radiation, the atmospheric radiation balance, and terrestrial solar irradiation. Concern for the potential impact of anthropogenic activity on the temperature of the earth is addressed.

Chapter 13, Radiation Exchange Between Surfaces highlights the difference between geometrical surfaces and radiative surfaces, a key concept that is often difficult for students to appreciate. Increased coverage of radiation exchange between diffused grey surfaces, included in older èditions of the text, has been retained. In doing so, radiation exchange between differentially small surfaces is briefly introduced.

The content of *Chapter 14,* Heat Exchangers is experiencing a resurgence in interest due to the critical role such devices play in conventional and alternative energy generation technologies. Much of the coverage of compact heat exchangers included in the previous edition was limited to a specific heat exchanger. Although general coverage of compact heat exchangers have been retained.

Chapter 15, Mass Transfer has been entirely revised extensively from the previous edition. General mass diffusion equation and boundary conditions are restructured. Concept of solubility, permeability, mass diffusion with and without homogeneous chemical reaction, steady state diffusion through a plane membrane, water vapour migration have been incorporated in new sections.

Chapter 16, Experiments in Engineering Heat Transfer have been updated in the interest of students to cater curriculum. Appendix is added as it was in previous edition.

—Author

Acknowledgements

We wish to acknowledge and thank many of our colleagues in the heat transfer community. In particular, we would like to express our appreciation to Prof. (Dr.) R. M. Warkhedkar, Govt. Engg. College Karad (M.S.), Prof. S. B. Patil of MET's IoE. Nashik, Prof. H. R. Thakare of SNJB's CoE. Chandwad, Prof. D. H. Dubey of B.S. Deore CoE. Dhule, Prof. P. A. Deshmukh of R.S. CoE Pune and many friends, students, users whose constructive suggestions helped me to improve the text and to bring out this edition.

I would like to extend my gratitude to administration and executive management of SNJB's Late Sau. K. B. Jain College of engineering, Neminagar, Chandwad (M.S.), India, for providing me facilities, moral support and cherish cooperation during the preparation of this manuscript.

I also take the opportunity to express my heartiest thanks to Mr. Saurabh Gupta, Managing Director, Laxmi Publications (P) Ltd. New Delhi, who adapted my desire to bring this volume in two columns and two colours, even after three proof readings of manuscript.

In closing, I am deeply grateful to my spouse and children, Dr. Ankit, and Pratik for their endless love and patience.

A human creation can never be perfect. Some mistakes might have crept in the text. My effort in writing the book will be rewarded, if readers send their constructive suggestions and objective criticism with a view to improve the usefulness of the book.

—Author

Nomenclature

A	Area normal to heat transfer, m^2
A_c	Cross-sectional area, m^2
A_{fin}	Fin surface area, m^2
A_p	Profile area, m^2
A_s	Surface area, m^2
$A_{un\ fin}$	Area or bare (un-finned) surface, m^2
$A_{no\ fin}$	Area of surface without fin, m^2
a	Acceleration, m/s^2
Bi	Biot number
Bo	Bond number
C	Heat capacity rate, W/K, specific heat of solids, J/kg. K
C_A	Molar concentration of species A in a mixture
C_D	Drag coefficient
C_p	Specific heat of liquids, J/kg. K
C_f	Coefficient of friction
C_o	Condensation number
D, d	Diameter, m
D_{AB}	Mass diffusivity, m^2/s
D_h	Hydraulic diameter, m
E	Emissive power, W/m^2
E′	Rate of energy, W
F	Force, N
Fo	Fourier number
F_{i-j}	Radiation view factor
f	Friction factor
$f_{0-\lambda}$	Blackbody radiation function
G	Irradiation, W/m^2
Gr	Grashof number
Gz	Graetz number
g	Acceleration due to gravity, m/s^2
g_o	Uniform heat generation per unit volume, W/m^3
H	Height, m
h, h_c	Convection heat transfer coefficient, $W/m^2.K$
h	Plank's constant
h_{fg}	Specific enthalpy of vaporization, J/kg
h_{rad}, h_r	Radiation heat transfer coefficient, $W/m^2.K$
I	Electrical current, A, radiation intensity, $W/m^2.sr$
I_{sc}	Solar constant, W/m^2
i	Electric current density, A/m^2
J	Radiosity, W/m^2
J_A	Diffusion flux, $kgmol/m^2.s$
Ja	Jacob number
K_c	Mass transfer coefficient, m/s
k	Thermal conductivity, W/m.K
L	Length, thickness, fin height, m
L_c	Characteristic length, corrected lenght, m
Le	Lewis number
LMTD	Log mean temperature difference, °C
M	Molecular number, kg/kgmol
m	Mass, kg
$\dot{m}$	Mass flow rate, kg/s
m_f	Mass fraction
N	Number of tubes
N_{fin}	Number of fins
Nu	Nusselt number
NTU	Number of transfer units
P	Perimeter, m, power, W
Pe	Peclet number
Pr	Prandtl number
P	Pressure, N/m^2
Q	Heat transfer rate, W
q	Heat flux, W/m^2
R	Specific gas constant, J/kg.K
R_u	Universal gas constant = 8314, J/kgmol. K
Ra	Rayleigh number
R_{cont}	Contact resistance, K/W
Re	Reynolds number

R_e	Electrical resistance, ohms
R_f	Fouling resistance, $m^2.K/W$
R_{th}	Thermal resistance, K/W
r_{cr}	Critical radius, m
r_i	Inner radius, m
r_o	Outer radius, m
r, θ, z	Cylindrical coordinates
r, θ, ϕ	Spherical coordinates
S	Shape factor for two dimensional heat conduction, m
Sc	Schmidt number
S_c	Solar constant
Sh	Sherwood number
St	Stanton number
T	Temperature, °C or K
t	Time, s
U	Overall heat transfer coefficient, $W/m^2.K$
u, v	Mass average velocity components, m/s
V	Volume of solid, m^3
u	Fluid velocity, m/s
v	Specific volume, m^3/kg
w	Depth, width, m
x, y, z	Cartesian coordinates
x	Local distance, m
x_{cr}	Critical distance, m
x_e	Hydrodynamic entry length, m
y_i	Mole fraction of ith species

Greek Letters

α	Thermal diffusivity, m^2/s
β	Volumetric expansion coefficient, K^{-1}
δ	Thickness of velocity boundary layer, characteristic length, mm
δ_{th}	Thickness of thermal boundary layer, mm
ε	Emissivity of the radiating surface, effectiveness of heat exchanger
ε_{fin}	Fin effectiveness
ϵ_H	Turbulent diffusivity for heat transfer
ϵ_M	Turbulent diffusivity for momentum
η_f, η_{fin}	Fin efficiency
η_{total}	Total fin efficiency
θ	Temperature difference, °C
κ_B	Boltzmann constant
λ	Wavelength, μm
μ	Dynamic viscosity, kg/m.s
ν	Kinematic viscosity, momentum diffusivity, m^2/s, frequency
ρ	Mass density, kg/m^3, reflectivity of the radiating surface, electrical resistively, Ω-m
ρ_i	Mass concentration of ith species in a mixture
σ	Stefan Boltzmann constant, $W/m^2.K^4$, surface tension, N/m
τ	Shear stress, N/m^2, transmissivity
ω	Solid angle, sr

Subscripts

b	blackbody
c	Cross-sectional area
cr	Critical
$cond$	Conduction
$conv$	Convection
D	diameter
e	excess, emission
f	fluid properties
fg	Phase transformation
g	saturated vapour
H	heat transfer
h	hydraulic, hot
i	inner surface, initial condition, tube inlet condition, incident radiation
L	Based on characteristic length
l	saturated liquid
lm	log mean condition
M	momentum transfer
m	mean value
max	maximum
o	centre or mid-plane condition, outer, tube outlet condition
R	radiaton surface
rad	radiation
s	surface condition
sat	saturated condition
sky	sky condition
sur	Surroundings condition
th	thermal
α	absorbed
v	vapour condition
x	local condition
λ	spectral
∞	free stream condition
ρ	reflected
τ	transmitted

UNITS AND DIMENSIONS

BASE UNITS:

QUANTITY	*UNITS*	*DIMENSION*
Length	metre	m
Mass	kilogram	kg
Time	second	s
Electric current	ampere	A
Temperature	kelvin	K

SUPPLEMENTARY UNITS:

Plane angle	radian	rad
Solid angle	steradian	sr

DERIVED UNITS:

Acceleration	a	metre per second squared	m/s^2
Angular acceleration	ω	radian per second squared	rad/s^2
Angular velocity	θ	radian per second	rad/s
Area	A	square metre	m^2
Electric potential difference	V	Volt	W/A
Electric resistance	Re	Ohm	Ω
Energy	E	joule	J or N.m
Entropy	s	joule per kelvin	J/K
Force	F	newton	Nor kg.m/s^2
Frequency	υ	hertz	Hz or 1/s
Heat energy	Q	joule	J or N.m
Magnetomotive force	emf	ampere	A
Power	P	watt	W or J/s
Radiation Intensity	I	watt per steradian	W/sr
Specific heat	C_P	joule per kilogram-kelvin	J/kg-K
Stress	σ	pascal	N/m^2
Thermal conductivity	k	watt per metre-kelvin	W/m–K
Velocity	u	meter per second	m/s
Volume	V	cubic metre	m^3
Work	W	Joule	J or N.m

SYMBOLS

Greek Alphabets

A	α	Alpha	I	ι	Iota	P	ρ	Rho
B	β	Beta	K	κ	Kappa	Σ	σ	Sigma
Γ	γ	Gamma	Λ	λ	Lambda	T	τ	Tau
D	δ	Delta	M	μ	Mu	Y	υ	Upsilon
E	ε	Epsilon	N	ν	Nu	Φ	φ	Phi
Z	ζ	Zeta	Ξ	ξ	Xi	X	χ	Chi
H	η	Eta	O	o	Omicorn	Ψ	ψ	Psi
Θ	θ	Theta	Π	π	Pi	Ω	ω	Omega
	∃	there exists		∀	for all			

Metric Weights and Measures

LENGTH

10 millimetres	= 1 centimetre
10 centimetres	= 1 decimetre
10 decimetres	= 1 metre
10 metres	= 1 dekametre
10 dekametres	= 1 hectometre
10 hectometres	= 1 kilometre

VOLUME

1000 cubic centimetres	= 1 centigram
1000 cubic decimetres	= 1 cubic metre

WEIGHT

10 milligrams	= 1 centigram
10 centigrams	= 1 decigram
10 decigrams	= 1 gram
10 grams	= 1 dekagram
10 dekagrams	= 1 hectogram
10 hectograms	= 1 kilogram
100 kilograms	= 1 quintal
10 quintals	= 1 metric ton (tonne)

CAPACITY

10 millilitres	= 1 centilitre
10 centilitres	= 1 decilitre
10 decilitres	= 1 litre
10 litres	= 1 dekalitre
10 dekalitres	= 1 hectolitre
10 hectolitres	= 1 kilolitre

AREA

100 square metres	= 1 are
100 ares	= 1 hectare
100 hectares	= 1 square kilometre

ABBREVIATIONS

kilometre	km	tonne	t
metre	m		
centimetre	cm	kilogram	kg
millimetre	mm	gram	g
kilolitre	kl	are	a
litre	l	hectare	ha
millilitre	ml	centiare	ca

Concepts and Mechanisms of Heat Flow

1

Objective of this chapter is to:

- give an introduction to heat transfer rate, heat flux,
- elaborate three modes of heat transfer—conduction, convection and thermal radiation,
- offer an introduction of physical laws of heat transfer,
- enlighten thermal conductivity, R value, thermal conductors and insulators.

The science of *Thermodynamics* deals with the *amount* of heat transfer as system undergoes a process from one equilibrium state to another, without any information concerning the nature of interaction or the time rate at which it occurs. *Heat Transfer* is a branch of thermal science which deals with analysis of rate of heat transfer and temperature distribution taking place in a system as well as the nature of heat transfer. The design of boilers, condensers, evaporators, heaters, refrigerators, and heat exchangers, requires considerations of the amount of heat to be transmitted as well as the rate at which heat is to be transferred. The successful operation of equipment component such as turbine blades, walls of combustion chambers, etc. depends on the cooling rate, in order to avoid their metallurgical failure. A heat transfer analysis must also be accounted in the design of electronic components, electric machines, transformers, and bearings to avoid the overheating and damage of equipment.

1.1. WHAT IS HEAT TRANSFER ?

Its simple answer is the definition of heat or heat energy.

Heat is a form of energy in transit due to temperature difference. Heat transfer is transmission of energy from one region to another region as a result of temperature difference between them. Whenever there exists a temperature difference in mediums or within a media, heat transfer must occur.

The amount of heat transferred per unit time is called **heat transfer rate** and is denoted by Q. The heat transfer rate has unit J/s which is equivalent to Watt.

When the rate of heat transfer Q is available, then total amount of heat energy transferred ΔU during a time interval Δt can be obtained from

$$\Delta U = \int_0^{\Delta t} Q dt = Q\Delta t \text{ (joule)} \quad ...(1.1)$$

The rate of heat transfer per unit area normal to direction of heat flow is called **heat flux** and is expressed as;

$$q = \frac{Q}{A} \text{ (W/m}^2\text{)} \quad ...(1.2)$$

Steady and Unsteady State Heat Transfer

For analysis of heat transfer problems, two types of heat transfer are considered—steady state and unsteady state. In case of *steady state heat transfer*, the temperature at any location on the system does not vary with time. The temperature is function of space coordinates only, but it is independent of time. Mathematically, for rectangular coordinate system ;

$$T = f(x, y, z) \quad ...(1.3)$$

During steady state conditions, the heat transfer rate is constant and there is no change of internal energy of the system. For example, the heat transfer in coolers, heat exchangers, heat transfer from large furnaces, etc.

In *unsteady state heat transfer*, the temperature varies with time as well as position. The temperature is a function of time and space coordinates. Mathematically, for rectangular coordinates ;

$$T = f(x, y, z, t) \quad ...(1.4)$$

During unsteady state or transient heat transfer, rate of heat transfer varies with time due to change in internal energy of the system. Most of the actual heat transfer processes are unsteady in nature , but some of them are considered in steady state to simplify them. For example, heat transfer from hot coffee left in a room, cooling and heating process, etc. are transient processes.

The heat transfer may be one, two or three directional, depends upon the configuration of the system considered.

1.2. MODES OF HEAT TRANSFER

When the temperature gradient exists in a medium, which may be solid, liquid, or gas, heat transfer occurred is called *conduction*. In contrast, the *convection* refers to heat transfer that will occur between a surface and a moving medium, when they are at different temperatures. The third mode of heat transfer is *thermal radiation*. All surfaces at finite temperature emit energy in the form of electromagnetic waves. The thermal radiation can also occur in absence of any medium.

1.3. PHYSICAL MECHANISM OF MODES OF HEAT TRANSFER

1.3.1. Conduction

The conduction occurs usually in the stationary mediums. It is the mode of heat transfer in which energy exchange takes place from a region of high temperature to that of low temperature by direct molecular interactions and by the drift of electrons. The heat conduction may be viewed as the transfer of energy from more energetic molecules to adjacent less energetic molecules of a substance. When a fast moving molecules from a region of high temperature collide with slower moving molecules from a region of lower temperature, the heat energy transfer takes place between them. The low energy molecules absorb energy and thus their temperature is increased and the temperature of high energy molecules is lowered.

The conduction heat transfer in liquids and gases occurs due to *collisions* and *diffusion* of molecules during their random motion. However, the nature is much more complex.

The temperature gradient is the potential for heat conduction. If a body in any phase exists a temperature gradient, will definitely have the conduction heat transfer.

1.3.2. Convection

The convection is a mode of heat transfer in which the energy is transported by moving fluid particles. The convection heat transfer comprises two mechanisms. First is transfer of energy due to random molecular motion (diffusion) and second is the energy transfer by bulk or macroscopic motion of the fluid (advection). The molecules of fluid are moving collectively or as aggregates thus carry energy from high temperature region to low temperature region. Therefore, the faster the fluid motion, the greater the convection heat transfer.

Convection heat transfer may be classified according to nature of fluid flow.

If the fluid motion is artifically induced by a pump, fan or a blower, that forces the fluid over a surface to flow as shown in Fig. 1.1(*a*), the heat transfer is said to be by the *forced convection.*

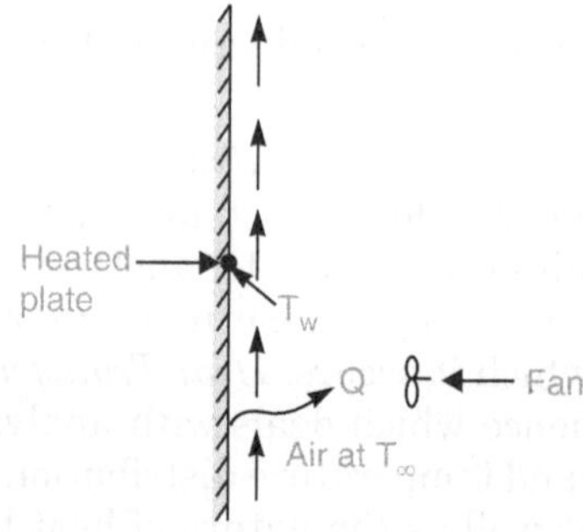

Fig. 1.1. (*a*) Forced convection of air ($T_w > T_\infty$)

If the fluid motion is set-up by buoyancy effects, resulting from density difference caused by temperature difference in the fluid as shown in Fig. 1.1(*b*), the heat transfer is said to be by the *free or natural convection.* For example, a hot plate vertically suspended in stagnant cool air, causes a motion in air layer adjacent to the plate surface because of temperature difference in air gives rise to density gradient which in turn sets-up the air motion.

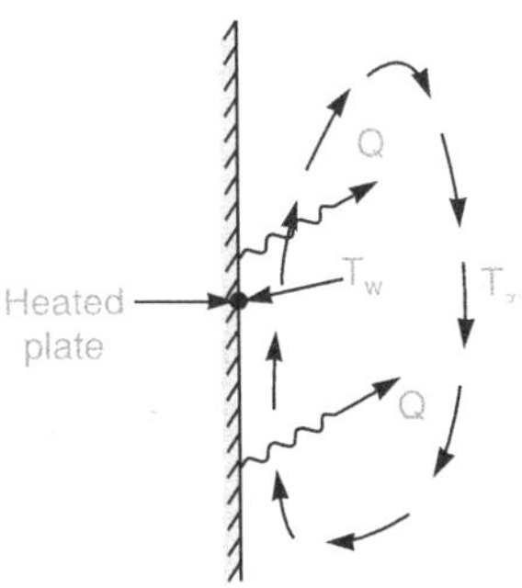

Fig. 1.1. (*b*) Natural or free convection of air ($T_w > T_\infty$)

1.3.3. Radiation

Thermal radiation is the energy emitted by a substance because of its temperature. The radiation energy emitted by a body is transmitted in the space in the form of electromagnetic waves according to Maxwell wave theory or in the form of discrete photons according to Max Plank's theory. Both concepts have been used in analysis of radiation heat transfer. Regardless of the form of substance (solid, liquid or gas) the emission of energy is due to change in electron configuration of the constituent molecules. While the transfer of energy by conduction or convection requires the presence of material medium, radiation does not. In fact, the radiation heat transfer is more efficient in vacuum. Thermal radiation occurs in the region of wavelengths 0.1 μm to 100 μm on electromagnetic spectrum.

1.4. LAWS OF HEAT TRANSFER

Like all subjects in physical science, some fundamental and subsidiary laws are also used in heat transfer analysis.

The fundamental laws, which are used in broad area of applications are :

1. The law of conservation of mass,
2. Newton's second law of motion,
3. First and second laws of thermodynamics.

The subsidiary laws, which are based on experimental facts are :

4. Fourier law of heat conduction,
5. Newton's law of cooling,
6. Stefan Boltzmann law for thermal radiation,
7. Equation of state.

1.4.1. Law of Conservation of Mass : Continuity Equation

It states that the mass of an incompressible fluid system is constant in absence of nuclear reaction.

It is the conservation of mass equation for steady state incompressible fluid flow. In general, the mass flow rate ($\dot{m}$) is expressed as ;

$$\dot{m} = \rho u A_c \qquad \text{...(1.5)}$$

where, A_c = cross-sectional area of flow (m^2),

ρ = fluid density (kg/m^3),

u = fluid velocity (m/s).

The volume of a fluid flowing through a pipe or duct per unit time is called ***volume flow rate or discharge rate,*** denoted by $\dot{V}$ and is expressed as ;

$$\dot{V} = u A_c = \frac{\dot{m}}{\rho} \qquad \text{...(1.6)}$$

1.4.2. Newton's Second Law of Motion

It states that the rate of change of momentum in any direction is always equal to sum of all external forces acting on the body in such direction.

The momentum = mass flow in particular direction × directional velocity

$$\text{The rate of change of momentum} = \frac{d(mv)}{dt}$$

Newton's second law of motion :

$$\Sigma Fx = \frac{d(mv)}{dt} \qquad \text{...(1.7)}$$

1.4.3. Laws of Thermodynamics

In heat transfer analysis, the first and second laws of thermodynamics are useful. The *first law of thermodynamics* states that the energy can neither be created nor be destroyed, only its form can be changed. In fact its quantity remains constant in either form.

The *second law of thermodynamics* states that the energy cannot be upgraded, or heat energy cannot flow from a body at lower temperature to a body at higher temperature. In other words, the second law of thermodynamics talks about the quality of energy, not of quantity like first law of thermodynamics.

1.4.4. Fourier Law of Heat Conduction

Whenever, a temperature gradient exists in a body, there is an energy transfer from the high temperature region to low temperature region by conduction. The Fourier law states that *the rate of heat conduction per unit area (heat flux) is directly proportional to the temperature gradient.*

$$\frac{Q}{A} \propto \frac{dT}{dx}$$

or
$$\frac{Q}{A} = -k\frac{dT}{dx} \quad \text{or} \quad Q = -kA\frac{dT}{dx} \qquad \text{...(1.8)}$$

where, Q = rate of heat transfer in W,

A = heat transfer area in m^2 ; normal to direction of heat flow,

$\frac{dT}{dx}$ = temperature gradient in °C/m ; slope of temperature curve on T–x diagram,

k = constant of proportionality, called the *thermal conductivity* of material in W/m·°C or W/m.K.

The minus sign is inserted to make the natural heat flow, a positive quantity. According to the second law of thermodynamics, the heat energy always flows in the direction of decreasing temperature, thus the temperature gradient dT/dx becomes negative.

A simple case of one dimensional steady state heat flow through a plane wall is shown in Fig. 1.2. For constant thermal conductivity k and heat transfer area A, equation (1.8) can be written as ;

$$\frac{Q}{A}\,dx = -\,k dT$$

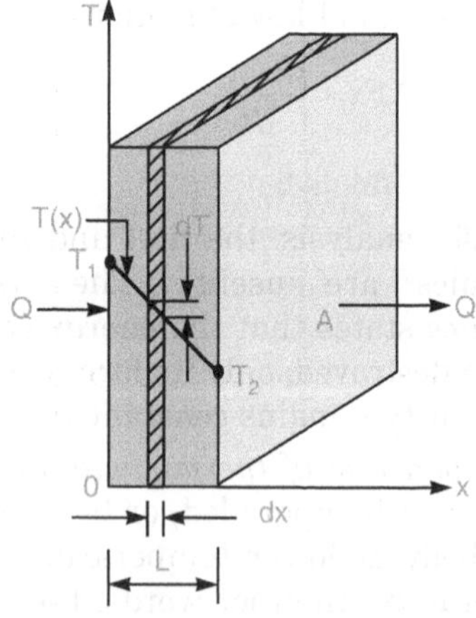

Fig. 1.2. One dimensional steady state conduction through a plane wall

Integrating above equation within the limits as ;

$$\frac{Q}{A}\int_0^L dx = -\,k\int_{T_1}^{T_2} dT$$

or $$\frac{Q}{A}L = -\,k(T_2 - T_1)$$

or $$Q = k\,A\,\frac{(T_1 - T_2)}{L} \qquad \text{...(1.9)}$$

Comparing with equation (1.8), the temperature gradient is linear and is given by

$$\frac{dT}{dx} = -\,\frac{(T_1 - T_2)}{L} \qquad \text{...(1.10)}$$

Example 1.1. *The wall of a furnace is constructed from 15 cm thick fire brick having constant thermal conductivity of 1.6 W/m.K. The two sides of the wall are maintained at 1400 K and 1100 K, respectively. What is the rate of heat loss through the wall which is 50 cm × 3 m on a side?*

Solution

Given : A furnace wall with

T_1 = 1400 K, T_2 = 1100 K

A = 50 cm × 3 m = 0.5 × 3 = 1.5 m^2

k = 1.6 W/m.K

L = 15 cm = 0.15 m.

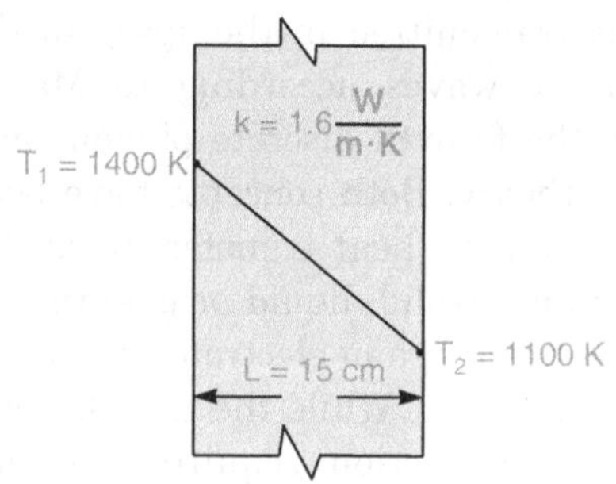

Fig. 1.3. Schematic of furnace wall

To find : Heat loss through the wall.

Assumptions :

1. Steady state conditions.
2. One dimensional heat conduction through the wall.
3. Constant properties.

Analysis : According to Fourier law of heat conduction, equation (1.9)

$$Q = kA\,\frac{(T_1 - T_2)}{L}$$

Using numerical values

$$Q = \frac{(1.6\ \text{W/m.K}) \times (1.5\ \text{m}^2) \times (1400\ \text{K} - 1100\ \text{K})}{(0.15\ \text{m})}$$

= **4800 W. Ans.**

Example 1.2. *To determine thermal conductivity of hydrogen, a hollow tube with a heating wire concentric to the tube is often used. Essentially the gas between the wire and the wall is a hollow cylinder and an electric current passing through the wire acts as a heat source. Determine thermal conductivity of the gas, using following data :*

T_1 = wire temperature = 200°C,

T_2 = tube wall temperature = 150°C,
I = current in the wire = 0.5 A,
V = voltage drop over 0.3 m section of wire = 3.6 V,
r_2 = tube radius = 0.125 cm,
r_1 = wire radius = 0.0025 cm,
L = length of the wire = 0.3 m.

Solution

Given : Hollow cylinder of hydrogen gas with

T_1 = 200°C, T_2 = 150°C, I = 0.5 A,
V = 3.6 V r_2 = 0.125 cm,
r_1 = 0.0025 cm, L = 0.3 m.

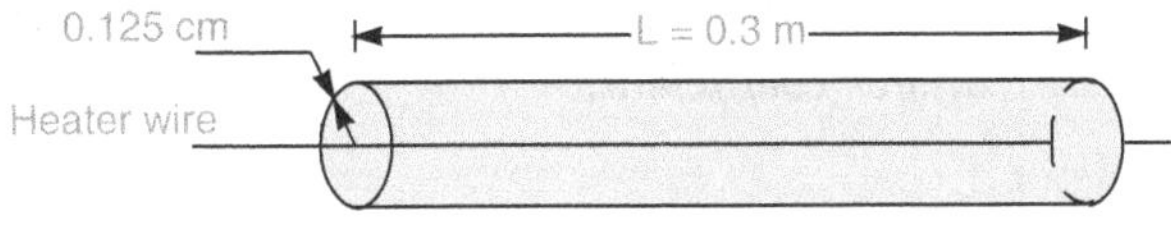

Fig. 1.4. Hydrogen filled tube with concentric heating wire

To find : Thermal conductivity of hydrogen gas.

Assumptions :

1. Steady state heat conduction.
2. Heat conduction in radial direction only.
3. Constant properties.

Analysis : The Fourier law of heat conduction for radial system is given as ;

$$\frac{Q}{A} = -k\frac{dT}{dr}$$

where, $A = 2\pi r L$

and $Q = VI = (3.6 \text{ V}) \times (0.5 \text{ A}) = 1.8 \text{ W}$

Hence $$\frac{Q}{2\pi L}\int_{r_1}^{r_2}\frac{dr}{r} = -k\int_{T_1}^{T_2} dT$$

or $$\frac{Q}{2\pi L}\ln\left(\frac{r_2}{r_1}\right) = -k(T_2 - T_1)$$

or $$k = \frac{Q\ln(r_2/r_1)}{2\pi L\,(T_1 - T_2)}$$

Substituting numerical values,

$$k = \frac{(1.8 \text{ W}) \times \ln(0.125/0.0025)}{2\pi \times (0.3 \text{ m}) \times (200°\text{C} - 150°\text{C})}$$

$= \mathbf{0.075\ W/m.K.}$ Ans.

1.4.5. Newton's Law of Cooling

It is the fundamental law for heat convection and it states that the rate of heat transfer is directly proportional to the temperature difference between a surface and a fluid, or mathematically

$$\frac{Q}{A} \propto (T_s - T_\infty)$$

or $$q = \frac{Q}{A} = h(T_s - T_\infty)$$

or $$Q = hA(T_s - T_\infty) \quad ...(1.11)$$

where, T_s = surface temperature, °C,
T_∞ = fluid temperature, °C,
A = surface area for convection heat transfer, m^2,
h = constant of proportionality, is called

the *heat transfer coefficient.* It is measured in W/(m^2.K) or W/(m^2.°C).

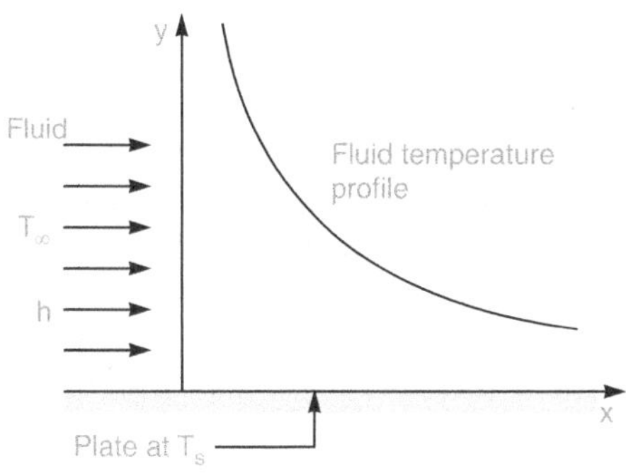

Fig. 1.5. Temperature profile in convection heat transfer

The value of heat transfer coefficient depends on the properties of fluid as well as fluid flow conditions. Fig. 1.5 shows a temperature profile in convection heat transfer. Table 1.1 shows typical values of heat transfer coefficient for some fluid flow conditions.

TABLE 1.1. Typical values of heat transfer coefficient *h*

Fluid flow condition	*h (W/m².K)*
Air (1 bar, free convection)	6 – 30
Air (1 bar, forced convection)	10 – 200
Water (free convection)	500 – 1000
Water (forced convection)	600 – 8000
Vapourisation of water	2500 – 1,00,000
Condensation of steam	4000 – 25,000

Example 1.3. *Hot air at 150°C flows over a flat plate maintained at 50°C. The forced convection heat transfer coefficient is 75 W/m²·K. Calculate the heat gain rate by the plate through an area of 2 m².*

Solution

Given : Flow of hot air over a flat plate

T_∞ = 150°C, T_s = 50°C
h = 75 W/m^2.K, A = 2 m^2.

To find : Heat transfer rate by air to plate.

Assumptions :

(*i*) Steady state conditions,
(*ii*) Constant properties,

(*iii*) Heat is transferred by forced convection only.

Analysis : According to Newton's law of cooling

Fig. 1.6. Flow over a flat plate

$$Q = hA_s(T_\infty - T_s)$$

$$= (75\ \text{W/m}^2.\text{K}) \times (2\ \text{m}^2) \times (150 - 50)\ (°\text{C or K})$$

$$= 15 \times 10^3\ \text{W} = \mathbf{15\ kW}.$$ **Ans.**

Example 1.4. *A refrigerator stands in a room, where air temperature is 21°C. The surface temperature on the outside of the refrigerator is 16°C. The sides are 30 mm thick and has an equivalent thermal conductivity of 0.10 W/m.K. The heat transfer coefficient on the outside is 10 W/m².K. Assume one dimensional conduction through the sides, calculate the net heat flow rate and the inside surface temperature of the refrigerator.*

Solution

Given : Heat transfer from a refrigerator wall.

$T_\infty = 21°\text{C}$, $T_{s,o} = 16°\text{C}$,

$L = 30\ \text{mm} = 0.03\ \text{m}$,

$k = 0.10\ \text{W/m.K}$, $h = 10\ \text{W/m}^2.\text{K}$.

To find : (*i*) Net heat flow rate, and

(*ii*) Inside surface temperature of refrigerator.

Assumptions :

1. Steady state conditions.
2. 1 m² surface area normal to heat transfer.
3. Constant properties.

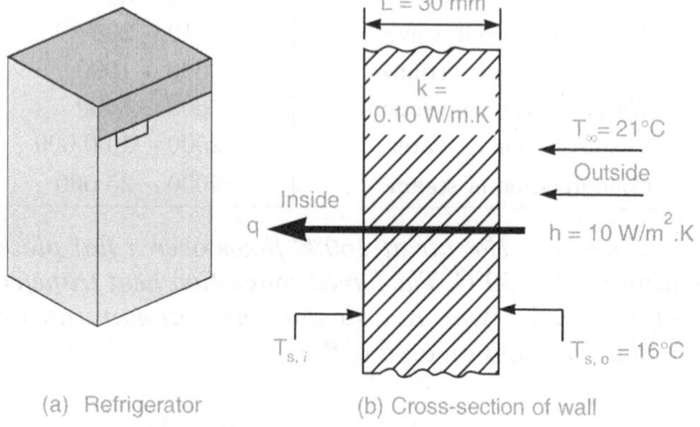

Fig. 1.7

Analysis : (*i*) The convective heat flux is given as;

$$q = \frac{Q}{A} = h(T_\infty - T_{s,o})$$

$$= (10\ \text{W/m}^2.\text{K}) \times (21°\text{C} - 16°\text{C})$$

$$= \mathbf{50\ W/m^2}.$$ **Ans.**

(*ii*) It is also the net heat flux conducted through the wall, therefore for plane wall

$$q = \frac{Q}{A} = \frac{k(T_{s,o} - T_{s,i})}{L}$$

or $$(50\ \text{W/m}^2) = \frac{(0.10\ \text{W/m.K}) \times (16°\text{C} - T_{s,i})}{(0.03\ \text{m})}$$

or $$T_{s,i} = 16°\text{C} - \frac{(50\ \text{W/m}^2) \times (0.03\ \text{m})}{(0.10\ \text{W/m.K})}$$

$$= \mathbf{1°C}.$$ **Ans.**

Example 1.5. *A hot plate is exposed to an environment at 100°C. The temperature profile of the environment fluid is given as T(°C) = 60 + 40 y + 0.1 y². The thermal conductivity of the plate material is 40 W/m.K. Calculate the heat transfer coefficient.*

Solution

Given : A hot plate exposed to an environment

$$T = (60 + 40\,y + 0.1\,y^2)°\text{C}$$

$$k = 40\ \text{W/m.K}$$

$$T_\infty = 100°\text{C}$$

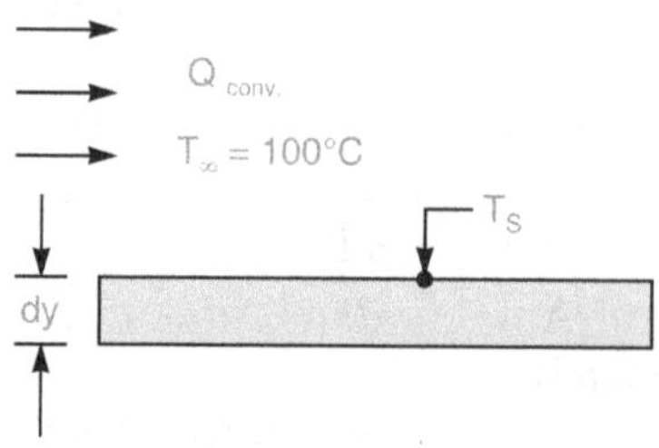

Fig. 1.8. Schematic for example 1.5

To find : Heat transfer coefficient.

Analysis : Assuming steady state conditions, the energy balance for the plate.

Heat conduction through plate = Heat convection from the plate

or $$Q_{cond} = Q_{conv}$$

i.e., $$-kA\left.\frac{dT}{dy}\right|_{y=0} = hA\,(T_s - T_\infty) \qquad ...(i)$$

where, T_s = temperature of the plate surface,

i.e., at $y = 0$,

$$T_s = 60 + 40 \times 0 + 0.1 \times (0)^2 = 60°\text{C}$$

and $$\left.\frac{dT}{dy}\right|_{y=0} = \Big[40 + 0.1 \times 2y\Big]_{y=0} = 40°\text{C/m}$$

Using numerical values in eqn. (*i*)

Then

$$(-40\ \text{W/m.K}) \times (40°\text{C/m}) = h(60°\text{C} - 100°\text{C})$$

or $$h = \frac{(-40 \text{ W/m.K}) \times (40°\text{C/m})}{-40°\text{C}}$$

$$= \mathbf{40 \text{ W/m}^2.\text{K}}.$$ Ans.

1.4.6. The Stefan Boltzmann Law of Thermal Radiation

It states that the rate of the radiation heat transfer per unit area from a black surface is directly proportional to fourth power of the absolute temperature of the surface and is given by

$$\frac{Q}{A} \propto T_s^4 \quad \text{or} \quad \frac{Q}{A} = \sigma T_s^4 \qquad ...(1.12)$$

where, T_s = absolute temperature of the surface, K

σ = constant of proportionality, is called the *Stefan Boltzmann constant* and has a value of 5.67×10^{-8} W/m^2.K^4.

The heat flux emitted by a real surface is less than that of black surface and is given by

$$\frac{Q}{A} = \sigma \varepsilon (T_s^4) \text{ (W/m}^2) \qquad ...(1.13)$$

where, ε = a radiative property of the surfaces and is called the *emissivity*.

The net rate of radiation heat exchange between a real surface and its surroundings is

$$\frac{Q}{A} = \sigma \varepsilon (T_s^4 - T_\infty^4) \qquad ...(1.14)$$

where, T_∞ = surrounding temperature, K

T_s = surface temperature, K

The three other radiation laws, Plank's law, Wein's displacement law and Kirchhoff 's law are also used in radiation heat transfer.

1.5. COMBINED CONVECTIVE AND RADIATION HEAT TRANSFER

In many engineering applications, if the surface temperature is high enough then the heat transfer from a surface may take place simultaneously by convection and radiation to the surroundings.

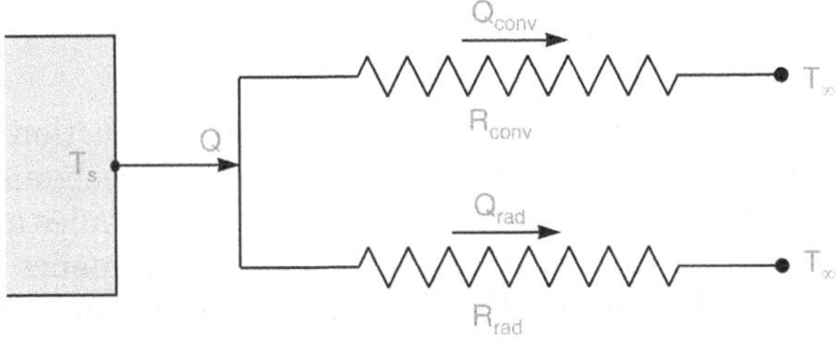

Fig. 1.9. Schematic for convection and radiation resistances at the surface

Consider a surface of emissivity ε is maintained at temperature T_s and exchanges energy by convection and radiation with its surroundings at temperature T_∞ as shown in Fig. 1.9. The rate of heat loss from the surface by combined mechanisms of convection and radiation can be expressed as :

$$Q = h_c A(T_s - T_\infty) + \varepsilon \sigma A(T_s^4 - T_\infty^4) \qquad ...(1.15)$$

Introducing the radiation or surface heat transfer coefficient (h_r) in very similar to convection heat transfer coefficient :

$$Q = h_r A(T_s - T_\infty) \qquad ...(1.16)$$

Equating with equation (1.14),

$$A \sigma \varepsilon (T_s^4 - T_\infty^4) = h_r A(T_s - T_\infty)$$

we get $$h_r = \frac{\varepsilon \sigma (T_s^4 - T_\infty^4)}{(T_s - T_\infty)}$$

$$= \varepsilon \sigma (T_s^2 - T_\infty^2)(T_s + T_\infty) \qquad ...(1.17)$$

If $T_\infty << T_s$, the result is linearised as ;

$$h_r = \varepsilon \sigma T_s^3 \qquad ...(1.18)$$

Using the radiation heat transfer coefficient, the expression can be written as ;

$$Q = h_c A(T_s - T_\infty) + h_r A(T_s - T_\infty) \qquad ...(1.19)$$

or $$Q = (h_c + h_r) A (T_s - T_\infty) \qquad ...(1.20)$$

where h_c and h_r are the convection and radiation heat transfer coefficients, respectively.

Example 1.6. *A black surface is positioned in a vacuum container so that it absorbs incident solar radiant energy at the rate of 950 W/m^2. If the surface conducts no heat to its surroundings, determine its equilibrium temperature.*

Solution

Given : A black surface absorbs solar energy in vacuum

$$q = 950 \text{ W/m}^2.$$

To find : The equilibrium temperature of the surface.

Assumptions :

1. The surface is perfectly black.
2. No heat loss by conduction and convection.
3. Stefan Boltzmann's constant,

$$\sigma = 5.67 \times 10^{-8} \text{ W/m}^2.\text{K}^4.$$

Analysis : The radiant heat flux for a black surface can be expressed as ;

$$q = \sigma T_s^4$$

or $$T_s = \left(\frac{q}{\sigma}\right)^{1/4}$$

$$= \left[\frac{950 \text{ W/m}^2}{5.67 \times 10^{-8} \text{ W/m}^2.\text{K}^4}\right]^{1/4}$$

$$= 359.78 \text{ K}$$

The equilibrium temperature of black surface will be,

$$T_s = \mathbf{86.78°C.} \quad \text{Ans.}$$

Example 1.7. *A black body at 30°C is heated to 100°C. Calculate the increase in its emissive power.*

Solution

Given : A black body emission

$T_1 = 30 + 273 = 303$ K,

$T_2 = 100 + 273 = 373$ K.

To find : The increase in emissive power.

Assumptions :

1. Stefan Boltzmann's constant,

$\sigma = 5.67 \times 10^{-8}$ W/m².K⁴.

2. No heat loss by conduction and convection.

Analysis : The radiant heat flux or emissive power for a black surface can be expressed as ;

$$E_b = \sigma T_s^4$$

Hence the increase in emissive power of a black body can be calculated as ;

$$E_{b_2} - E_{b_1} = \sigma(T_2^4 - T_1^4)$$

or $$E_{b_2} - E_{b_1} = (5.67 \times 10^{-8} \text{ W/m}^2.\text{K}^4) \times [(373 \text{ K})^4 - (303 \text{ K})^4]$$

$$= \mathbf{619.62 \ W/m^2.} \quad \text{Ans.}$$

Example 1.8. *The surface temperature of a central heating radiator is 60°C. What is the net black body radiation heat transfer unit surface area between the radiator and its surroundings at 20° ?*

Take $\sigma = 5.67 \times 10^{-8}$ W/m²·K⁴

Solution

Given : Central heating radiator as black body with

$T_s = 60°C = 333$ K, $T_\infty = 20°C = 293$ K.

To find : Radiation heat transfer

Analysis : The black body radiation heat transfer rate per unit area between radiator surface and its surroundings is expressed as ;

$$q = \frac{Q}{A} = \sigma(T_s^4 - T_\infty^4)$$

$$= (5.67 \times 10^{-8} \text{ W/m}^2.\text{K}^4) \times [(333 \text{ K})^4 - (293 \text{ K})^4]$$

$$= 5.67 \times 10^{-8} \times 4.9263 \times 10^9$$

$$= \mathbf{279.32 \ W/m^2.} \quad \text{Ans.}$$

1.5.1. Equation of State

It is the relation between the properties of an ideal gas. The perfect gas law is used in convection heat transfer, which is :

$$\frac{p}{\rho} = RT \qquad \text{...(1.21)}$$

where, p = gas pressure in kN/m²,

ρ = gas density in kg/m³,

R = specific gas constant in kJ/kg.K, (= 0.287 kJ/kg.K for air)

T = absolute temperature of gas in K.

1.6. THERMAL CONDUCTIVITY

The *thermal conductivity* is the property of materials and is defined as the ability of the materials to conduct the heat through it. By inspection of equation (1.9), thermal conductivity can be interpreted as the *rate of heat transfer through a unit thickness of material per unit area per unit temperature difference.* The thermal conductivity of a material is a measure of how fast heat will flow in that material. The large value of thermal conductivity indicates that the material is a good heat conductor and low value indicates that the material is a poor heat conductor or an insulator.

The thermal conductivity is measured in watts per metre per degree Celsius or Watt per metre per kelvin, when heat flow rate is expressed in watts. The thermal conductivity of a substance is highest in solid phase and lowest in gaseous phase. Fig. 1.10 shows typical range of thermal conductivity of various materials at 20°C.

The value of thermal conductivity depends upon the manner in which energy is transferred. The pure metals allow faster transmission of heat energy through the vibrations of the crystal lattices. Therefore, a metal in pure state has the maximum thermal conductivity and is a good conductor of heat. The thermal conductivity decreases with increasing amount of impurities in the metals. Most non-metals are poor conductor of heat transfer, therefore, have low values of thermal conductivity and are called the *thermal insulators.*

In gases, the faster the molecules move, the faster, they will transport energy. Therefore, the thermal conductivity of gases depends on the square root of absolute temperature. The thermal conductivities of some typical gases are given in Table 1.2.

The physical mechanism of heat conduction in liquids is also same as in gases, however, the mechanism is slightly complex due to close spacing of molecules and molecular attraction force exerts a strong influence on energy exchange in collision process. The thermal conductivity of liquids usually lies between those of solids and gases. The value of thermal conductivity for some standard liquids is also given in Table 1.2.

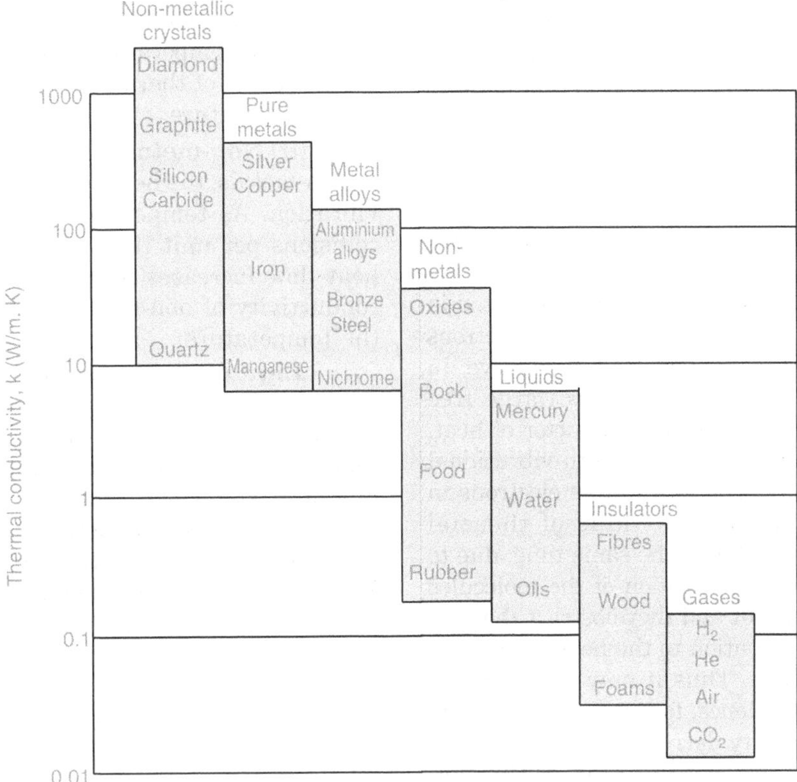

Fig. 1.10. Typical range of thermal conductivity of various materials at room temperature

TABLE 1.2. Typical values of thermal conductivities at 20°C

Material	*Thermal conductivity, k (W/m.K)*
Metals :	
Diamond	2300
Silver	429
Copper (pure)	401
Gold	317
Aluminium (pure)	237
Iron (pure)	73
Carbon steel, 1% C	43
Non-metallic solids :	
Window glass	0.780
Brick	0.720
Asbestos	0.149
Cork	0.045
Glass wool	0.038
Liquids :	
Water	0.556
Ethylene glycol	0.249
Ammonia	0.54
Gases :	
Helium	0.152
Air	0.024
Steam	0.0206
Carbon dioxide	0.0146

1.6.1. Variation in Thermal Conductivity

The temperature is a measure of kinetic energies of molecules of a substance. Thus the thermal conductivity of materials varies with the temperature. It may also change with pressure in fluids.

Effect of Temperature

Solids. The solids are classified into two groups : (*i*) Metals and (*ii*) Non-metals.

(*i*) **Metals.** The heat energy may be conducted in the metals by two mechanisms : migration of free electrons and lattice vibrations. These two effects are additive. In general, the presence of the *electron gas* (large free electrons) in metals, makes it a good conductor of heat, but the conduction also takes place due to vibrational energy in lattice structure. The flow of free electrons in metal results in an increase in value of thermal conductivity several times. But at the same time, due to increase in temperature, the vibration of the molecules in the metals becomes violent and they obstruct the flow of free electrons and contribution to the heat conduction by free electrons decreases. Thus it may result in net decrease in the heat flow. Hence, for most of metals, the value of thermal conductivity decreases as temperature increases. Fig. 1.11 shows the variaton of thermal conductivity with temperature for few solids.

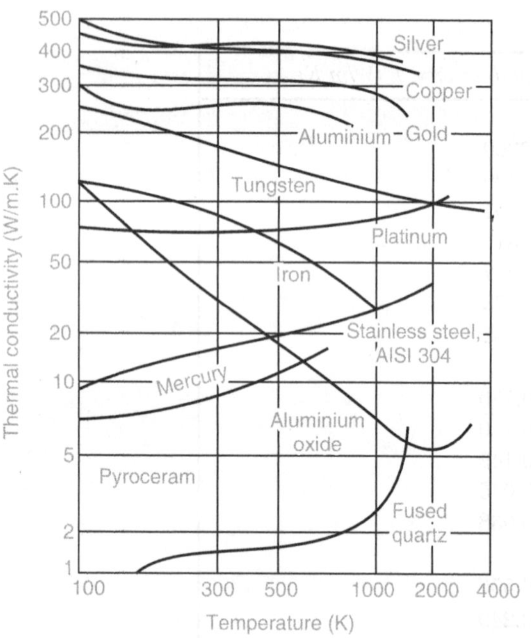

Fig. 1.11. Effect of temperature on thermal conductivity of selected solids

The thermal conductivity of mercury increases with increase in temperature is an exceptional case of metals.

The thermal conductivity of alloys generally increases as temperature increases. Fig. 1.11 also shows the variation of thermal conductivity of stainless steel with temperature.

(*ii*) **Non-metals.** Due to absence of free electrons in non-metals, the heat conduction is only due to lattice vibration. As temperature increases, the number of collisions per unit time increases ; hence, the rate of heat flow increases in non-metals. Thus the thermal conductivity of non-metals increases with increase in the temperature.

Liquids. For most of the liquids, the thermal conductivity decreases with increase in the temperature. But water and glycerine are the exceptional cases. The thermal conductivity of liquids is independent of pressure. As a general rule, the thermal conductivity of liquids decreases with increasing molecular weight. The value of thermal conductivity of liquids are taken from Table 1.2 as function of temperature in saturated state and plotted in Fig. 1.12.

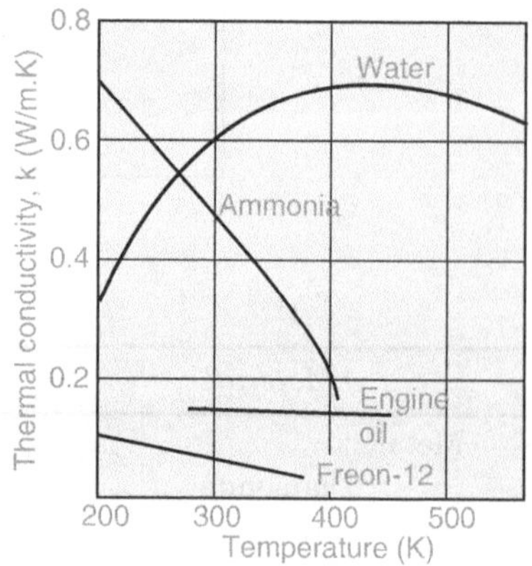

Fig. 1.12. Effect of temperature on thermal conductivity of selected

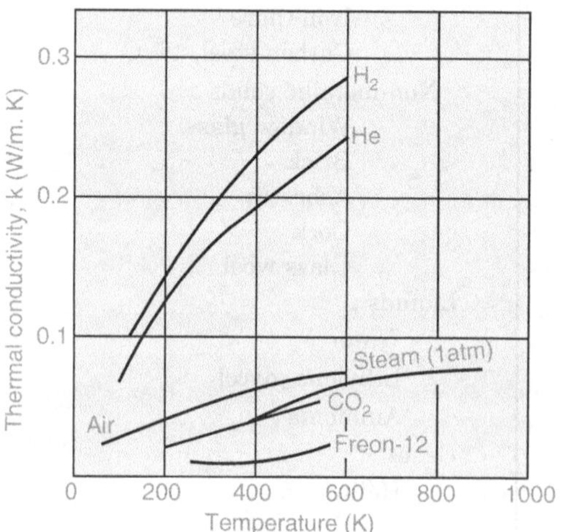

Fig. 1.13. Effect of temperature on thermal conductivity of selected gases at normal pressure

Gases. For the gases, the molecules are in continuous random motion. As temperature increases, velocities of molecules become higher than in some lower temperature region. The molecules move from high temperature region to a region of low temperature and give up its energy through *collisions* to lower energy molecules. Thus the thermal conductivity of gases increases with increase in temperature and it is proportional to square root of the absolute temperature. It is also affected by change in pressure and humidity.

1.6.2. Determination of Thermal Conductivity

The thermal conductivity, k can be defined by Fourier law, equation (1.8)

$$k = -\frac{(Q/A)}{(dT/dx)} \quad \text{...(1.22)}$$

This equation is used for determination of thermal conductivity of a material. A layer of solid material of thickness L and area A is heated from one side by an electric resistance heater as shown in Fig. 1.14. If the outer surface of heater is perfectly insulated, then all the heat generated by resistance heater will be transferred through the exposed layer of material. When steady state condition is reached, the temperature of two surfaces of material T_1 and T_2 are measured and thermal conductivity of material is determined by relation

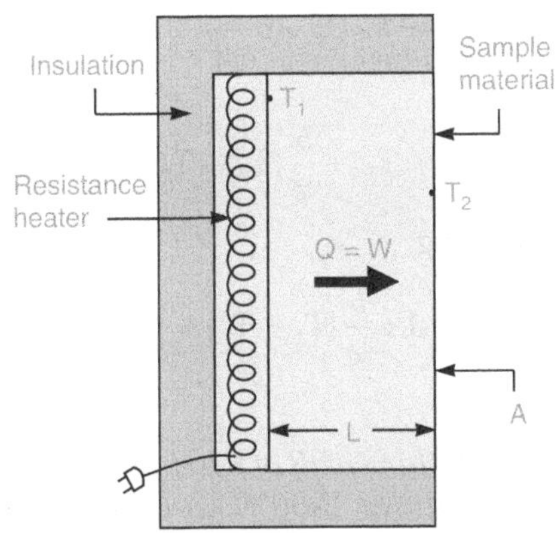

Fig. 1.14. Experimental set-up for determination of thermal conductivity

$$k = \frac{Q\,L}{A(T_1 - T_2)} \quad \text{...(1.23)}$$

1.6.3. Variable Thermal Conductivity

For many materials, the thermal conductivity can be approximated as a linear function of temperature over limited range and it is expressed as ;

$$k = k_0(1 + \alpha T) \quad \text{...(1.24)}$$

where, k_0 is the value of thermal conductivity at some reference temperature and α is an empirical constant, whose value depends on material. The value of constant α may be negative or positive as shown in Fig. 1.15, depending upon whether thermal conductivity increases or decreases with rise in temperature. The constant α is usually positive for non-metals and insulating materials and negative for most of the metals.

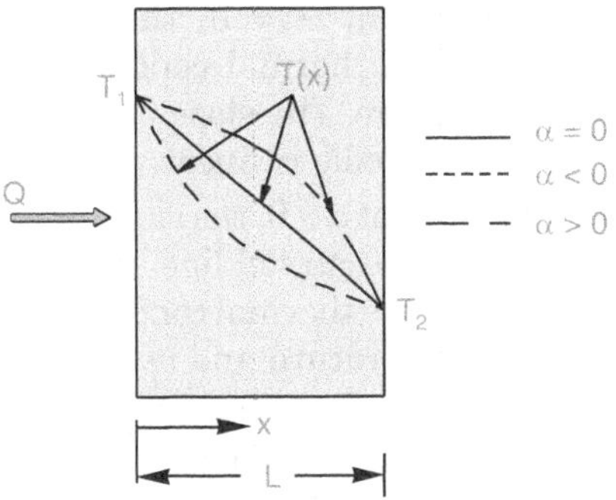

Fig. 1.15. Variation of thermal conductivity with temperature

With a variable thermal conductivity, Fourier law of heat conduction through a plane wall can be expressed as ;

$$\frac{Q}{A} = -k\frac{dT}{dx} = -k_0(1 + \alpha T)\frac{dT}{dx} \quad \text{...(1.25)}$$

or

$$\frac{Q}{A}\,dx = -k_0(1 + \alpha T)\,dT$$

Integrating both sides within the boundary conditions :

$$\frac{Q}{A}\int_0^L dx = -k_0\int_{T_1}^{T_2}(1 + \alpha T)\,dT$$

or

$$\frac{Q}{A}(L - 0) = -k_0\left[T + \alpha\frac{T^2}{2}\right]_{T_1}^{T_2}$$

or

$$Q = -\frac{k_0 A}{L}\left[(T_2 - T_1) + \alpha\frac{(T_2^2 - T_1^2)}{2}\right]$$

Rearranging

$$Q = \frac{k_0 A(T_1 - T_2)}{L}\left[1 + \frac{\alpha}{2}(T_1 + T_2)\right]$$

or

$$Q = \frac{k_m A(T_1 - T_2)}{L}$$

where,

$$k_m = k_0\left[1 + \frac{\alpha}{2}(T_1 + T_2)\right] \quad \text{...(1.26)}$$

The quantity k_m represents the mean value of thermal conductivity evaluated at arithmetic mean temperature $\left(\frac{T_1 + T_2}{2}\right)$. The term $\frac{dT}{dx}$ in eqn. (1.25) represents the slope of the temperature profile for the conducting medium. Mathematically, the slope is :

$$\frac{dT}{dx} = -\left[\frac{Q}{k_0(1+\alpha T)A}\right] \quad ...(1.27)$$

(*i*) When constant $\alpha = 0$, the equation (1.24) reduces to $k = k_0$ and thermal conductivity does not change with temperature. The slope of curve is constant and the temperature profile is linear.

(*ii*) When constant $\alpha > 0$, the slope or temperature profile follows a positive curved line along the material thickness. Therefore, the thermal conductivity increases with increase in temperature and *vise versa.*

(*iii*) When constant $\alpha < 0$, the slope or temperature profile follows a negative curved line along the material thickness. Therefore, the thermal conductivity decreases with increase in temperature and *vise versa.*

1.7. ISOTROPIC MATERIAL AND ANISOTROPIC MATERIAL

If the thermal conductivity of a material does not vary with change in direction or its value is same in all directions then material is called the *isotropic material.* If the thermal conductivity of the material depends on the direction of the heat flow, the material is called *anisotropic material.* There are some materials in which thermal conductivity depends upon directions. For example, the thermal conductivity of wood in the direction of grains is different from that in the transverse direction. So wood is an anisotropic material.

Example 1.9. *The thermal conductivity of a plane wall varies as :*

$$k = k_0(1 + bT + cT^2)$$

If the wall thickness is L and surface temperatures are maintained at T_1 and T_2, show that the heat flux q through the wall is given by :

$$q = \frac{k_0(T_1 - T_2)}{L}\left\{1 + \frac{b}{2}[T_1 + T_2] + \frac{c}{3}[T_1^2 + T_1T_2 + T_2^2]\right\}$$

Solution

Given : The relation for variable thermal conductivity as ;

$$k = k_0(1 + bT + cT^2).$$

Assumptions :

1. k_0 is constant.
2. Steady state conditions.
3. One dimensional heat conduction.

Analysis : According to Fourier law of heat conduction

$$\frac{Q}{A} = q = -k\frac{dT}{dx}$$

or
$$q = -k_0(1 + bT + cT^2)\frac{dT}{dx}$$

or
$$q\,dx = -k_0(1 + bT + cT^2)\,dT$$

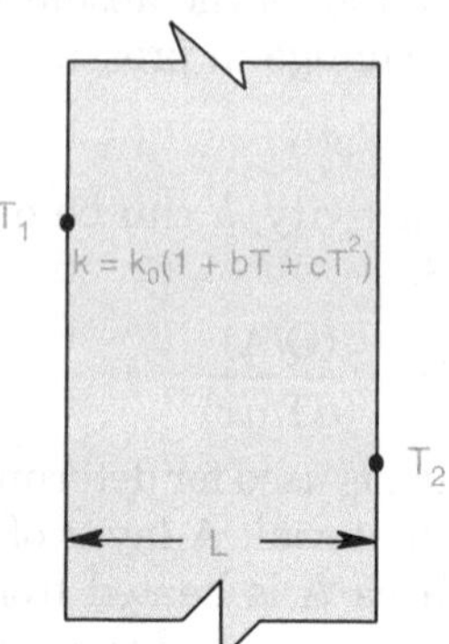

Fig. 1.16. Schematic for example 1.9

Integrating both sides,

$$q\int_0^L dx = -k_0\int_{T_1}^{T_2}(1 + bT + cT^2)\,dT$$

or
$$q(L-0) = -k_0\left[T + b\frac{T^2}{2} + c\frac{T^3}{3}\right]_{T_1}^{T_2}$$

or
$$q = -\frac{k_0}{L}\left\{(T_2 - T_1) + \frac{b}{2}(T_2^2 - T_1^2) + \frac{c}{3}(T_2^3 - T_1^3)\right\}$$

Rearranging

$$q = \frac{k_0(T_1 - T_2)}{L}\left\{1 + \frac{b}{2}[T_1 + T_2] + \frac{c}{3}[T_1^2 + T_1T_2 + T_2^2]\right\}.$$

Proved.

Example 1.10. *A plane wall of fireclay brick, 25 cm thick is having temperatures 1350°C and 50°C on two sides. The thermal conductivity of fireclay varies as ;*

$$k = 0.838\,(1 + 0.0007\,T),$$

where T is in degree celcius.

Calculate the heat loss per square metre through the wall.

Solution

Given : A plane wall with variable thermal conductivity

$L = 25$ cm $= 0.25$ m, $T_1 = 1350$°C,
$T_2 = 50$°C
$k = 0.838(1 + 0.0007\,T)$.

To find : Heat loss per sq m through the wall.

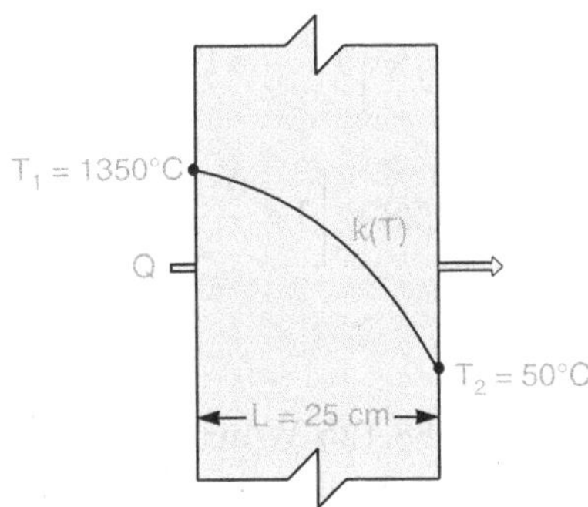

Fig. 1.17. Schematic of plane wall of example 1.10

Assumptions :

1. Steady state heat conduction.
2. One dimensional heat conduction.

Analysis : According to Fourier law of heat conduction

$$\frac{Q}{A} = -k\frac{dT}{dx}$$

or
$$\frac{Q}{A}dx = -0.838 \times (1 + 0.0007\ T)\ dT$$

Integrating both sides within the boundary conditions :

$$\frac{Q}{A}\int_0^L dx = -0.838 \times \int_{T_1}^{T_2} (1 + 0.0007\ T)\ dT$$

or
$$\frac{Q}{A}(L - 0) = -0.838 \times \left[T + 0.0007\left(\frac{T^2}{2}\right)\right]_{T_1}^{T_2}$$

or
$$\frac{Q}{A} = -\frac{0.838}{L} \times \left\{(T_2 - T_1) + \frac{0.0007}{2} \times (T_2^2 - T_1^2)\right\}$$

Using numerical values

$$\frac{Q}{A} = -\frac{0.838}{0.25} \times [(50 - 1350) + 0.00035 \times (50^2 - 1350^2)]$$

$$= \mathbf{6492.82\ W.}$$ **Ans.**

Example 1.11. *A metal piece, 50 cm long is in the form of a sector of a circle of radius 10 cm and includes an angle of π/2. The thermal conductivity of the metal piece varies as ;*

$$k = k_0(1 + \alpha T)$$

where, k_0 = *111.63 W/m.K and* $\alpha = -1 \times 10^{-4}$ *W/m.K*^{2}.

Calculate the heat transfer rate, when the two ends of the metal piece are maintained at temperatures of 100°C and 20°C. Assume heat flow takes place in axial direction only.

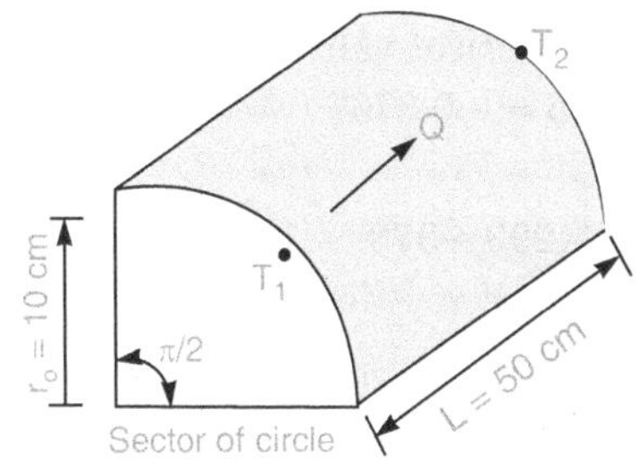

Fig. 1.18. Schematic for example 1.11

Solution

Given : A metal piece in the form of a sector of a circle

$$L = 50\ cm = 0.5\ m, \quad r_o = 10\ cm = 0.1\ m,$$
$$\theta = \pi/2 = 90°, \quad k = 111.63\ (1 - 1 \times 10^{-4}\ T)$$
$$T_1 = 100°C = 373\ K, T_2 = 20°C = 293\ K.$$

To find : Heat loss from the cylindrical sector in axial direction.

Assumptions :

1. Steady state heat conduction.
2. Heat transfer in axial direction only.
3. In expression of thermal conductivity, the scale of temperature is not mentioned thus assuming kelvin scale for temperature.

Analysis : The area of cylindrical piece :

$$A = \frac{\pi r_o^2}{4} = \frac{\pi(0.1\ m)^2}{4} = 7.854 \times 10^{-3}\ m^2$$

Fourier law,

$$\frac{Q}{A} = -k\frac{dT}{dx}$$

or
$$Q\ dx = -111.63 \times 7.854 \times 10^{-3} \times (1 - 10^{-4}\ T)\ dT$$

Integrating both sides within boundary conditions

$$Q\int_0^L dx = -0.8767 \times \int_{T_1}^{T_2} (1 - 1 \times 10^{-4}\ T)\ dT$$

or
$$Q\ L = -0.8767 \times \left[T - 1 \times 10^{-4}\left(\frac{T^2}{2}\right)\right]_{T_1}^{T_2}$$

$$= -0.8767 \times \left[(T_2 - T_1) - 1 \times 10^{-4} \times \left(\frac{T_2^2 - T_1^2}{2}\right)\right]$$

Using numerical values :

$$Q \times 0.5 = -0.8767$$

$$\times \left[(293 - 373) - 1 \times 10^{-4} \times \left(\frac{293^2 - 373^2}{2}\right)\right]$$

or $$Q = -\frac{0.8767 \times (-80 + 2.664)}{0.5}$$

$$= \mathbf{135.6\ W.}\ \text{Ans.}$$

Example 1.12. *A pipe carrying pressurised water is located in a 1.2 m thick brick wall, whose surfaces are held at constant temperatures of 200°C and 50°C, respectively. It is required to locate the pipe in the wall where temperature should not exceed 125°C. Find how far from the hot surface the pipe should be imbeded ? The thermal conductivity of wall material (brick) varies with temperature as ;*

$$k = 0.28\ (1 + 0.035\ T)\ W/m.K, \text{ where } T \text{ is in } °C.$$

and $k = k_0(1 + \alpha T)$.

Solution

Given : A pressurised water pipe is located in a brick wall as shown in Fig. 1.19

Thickness of wall,

$$L = 1.2\ m$$

Temperature of wall surfaces,

$$T_1 = 200°C, \quad T_2 = 50°C$$

Allowed pipe temperature,

$$T_3 = 125°C$$

Thermal conductivity,

$$k = 0.28\ (1 + 0.035\ T)\ W/m.K.$$

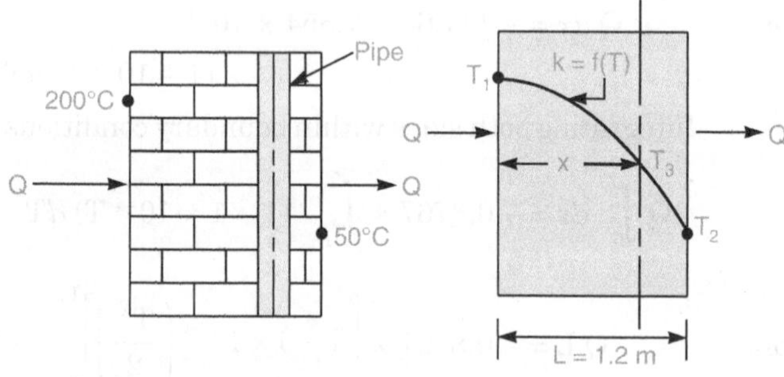

Fig. 1.19. Schematic for example 1.12

To find : The location of pipe in the wall.

Analysis : The steady state heat transfer rate per unit area, through wall.

$$\frac{Q}{A} = -k\frac{dT}{dx} = -k_0(1 + \alpha T)\frac{dT}{dx}$$

Integrating and rearranging, we get

$$\frac{Q}{A} = k_0 \times \left[1 + \frac{\alpha}{2} \times (T_1 + T_2)\right] \frac{(T_1 - T_2)}{L}$$

$$= 0.28 \times \left[1 + \frac{0.035}{2} \times (200 + 50)\right] \times \frac{(200 - 50)}{1.2}$$

$$= 188.125\ W/m^2.$$

Further, if x is the distance from the hot surface in the wall, where temperature, $T_3 = 125°C$, then heat flow rate can be expressed by replacing T_2 by T_3 and L by x ;

$$\frac{Q}{A} = 188.125$$

$$= 0.28 \times \left[1 + \frac{0.035}{2} \times (200 + 125)\right] \times \frac{(200 - 125)}{x}$$

$$= \frac{140.44}{x} \quad \text{or} \quad x = \frac{140.44}{188.125}$$

$$= \mathbf{0.75\ m\ from\ left.}\ \text{Ans.}$$

1.8. INSULATION MATERIALS

There are many situations in engineering applications, when the heat flow rate from a system has to be reduced. Such cases include lagging on heat pipes, a thermos bottle, ruberisation on electrical cables, etc. An industrial furnace is provided with innermost layer with refractory bricks, one or two layers of insulating bricks, and outer layer with ordinary bricks. All these layers form a thermal insulation to the furnace. Thus a *thermal insulation* is a material or combination of materials, which is mainly used to minimise the heat flow to or from a system.

Thermal insulations are put on the surfaces exposed to certain environment. They offer a strong thermal resistance in the path of heat flow. An insulation reduces the total heat transfer rate from a system. It does not only minimise conduction rate, but also minimises convection and radiation heat transfer rates.

Thermal insulation material must have a very low value of thermal conductivity. It should also be chemically inert, dimensionally stable, and easily available in form, suitable for application on the surfaces. It should also be cheap and light. In most of the cases, the thermal insulators are manufactured by mixing fibre, powders or flakes of insulating materials with air. The air is trapped inside small cavities of solids. The same effect can also be produced by filling the space across which heat flow is to be minimised with small solid particles and trapping the air between them. A gas has very poor thermal conductivity. Commercial insulators are ceramics (*e.g.,* insulating bricks), rock-wool, gypsum and polymeric (expanded polyurethane, expanded polystyrene etc.) materials. These materials are highly porous, or have a high void volume filled with an inert gas.

The heat transfer through an insulation is by conduction through the solid material, and by conduction and convection through the air space as well as by radiation as shown in Fig. 1.20. Such insulation materials are characterised by an *apparent thermal conductivity* k_{eff}. It is an effective value that accounts for all mechanisms of heat transfer and it should not change with temperature, pressure and moisture contents. Fig. 1.21 shows the ranges of effective thermal conductivity for evacuated and non evacuated insulations.

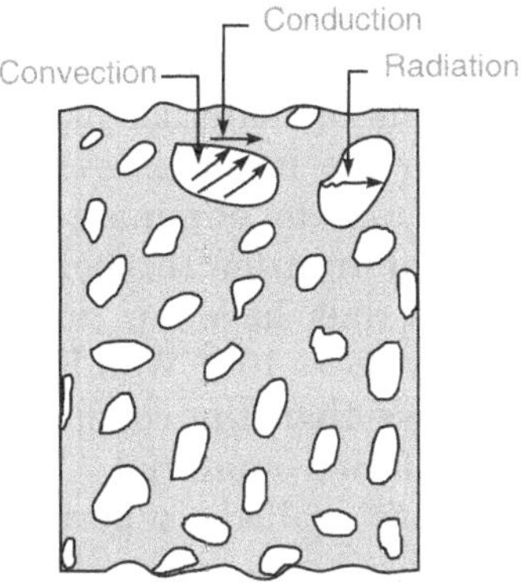

Fig. 1.20. Heat transfer through an insulating material

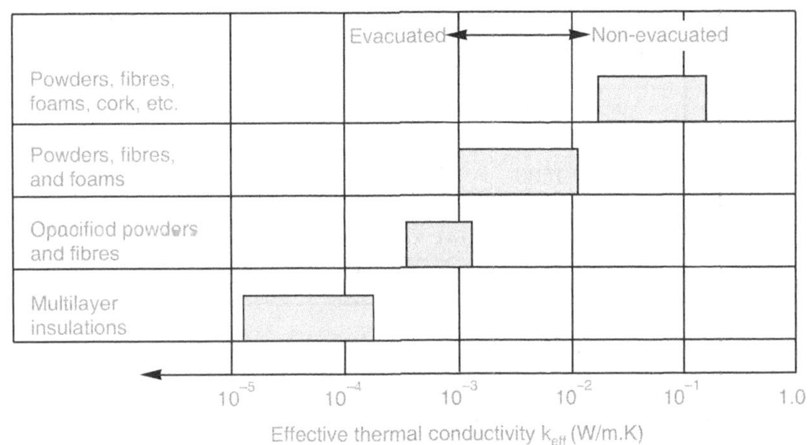

Fig. 1.21. Range of thermal conductivities of thermal insulations

Insulating materials are classified into three categories :

1. **Fibrous.** The fibrous insulations are obtained by mixing small particles or flakes of low density materials with air. The material is poured into small gaps as loose fill or formed into boards or blankets. Fibrous materials have very high porosity. Mineral wool is a common fibrous material for application at temperature below 700°C and fibrous glass is used for temperature below 200°C. For temperature range 700°C to 1700°C, the refractory fibre, such as alumina or silica is used.

2. **Cellular.** Cellular insulations are closed or open cell materials that are usually in the form of extended flexible or rigid boards. These can easily be formed or sprayed in the place to achieve desired geometrical shapes. These materials have low density, low heat capacity, and good compressive strength. Some cellular materials are polyurethane, expanded polystyrene, cellular glass, and cellular silica, etc.

3. **Granular.** Granular insulations consist of small flakes or particles of inorganic materials, bonded into some common shapes or used as powders. For example, asbestos, perlite powder, diatomaceous silica and vermiculite.

1.8.1. Superinsulators

The insulators with extremely low apparent thermal conductivity (about one thousand of that of air), called *superinsulators* and are obtained by using layers of highly reflective sheets separated by glass fibres in an evacuated space. Like a thermos bottle in which the space between the two surfaces is evacuated to suppress conduction and convection and inner surface is coated with reflective layer to prevent the radiation heat transfer, the heat transfer between two surfaces can also be reduced by placing highly reflective sheets. The radiation heat transfer is inversely proportional to the number of such reflective sheets placed between the surfaces. Very effective insulations are obtained by using closely packed layers of highly reflective thin metal sheets such as aluminium foils separated by fibres made of insulating materials like glass. Further, the space between the layers is evacuated to form a very strong vacuum to eliminate the conduction and convection heat transfer through the air space. The resulting materials have an apparent thermal conductivity below 2×10^{-5} W/m.K, which is one thousand times less than the conductivity of air or any common insulating material. These specially built insulators are called superinsulators. The superinsulators are used in space applications and cryogenics.

1.8.2. Selection of Insulating Materials

The selection and design of a suitable insulation depends upon following factors :

1. Thermal conductivity,
2. Density of material,
3. Upperlimit of operating temperature,
4. Structural rigidity,
5. Degradation rate,
6. Chemical stability,
7. Cost, *i.e.,* economic thickness of insulation.

The range of thermal conductivities for common temperature insulating materials is given in Table 1.3.

TABLE 1.3. Some insulating materials with their range of operating temperatures

Insulating materials	*Max. operating temperature (K)*	*Density (kg/m^3)*	*Thermal conductivity (W/m.K)*
Asbestos fibre	420	190–300	0.078–0.098
Cellular glass	700	145	0.046–0.079
Diatomaceous	1145	345	0.92–0.104
Alumina silica fibre	1530	48	0.071–0.15
Magnesia 85%	590	185	0.051–0.061
Polystyrene	350	16–56	0.023–0.040
Mineral fibres	922–1255	430–560	0.071–0.137

1.8.3. The R-Value of Insulation

For building materials, the effectiveness of insulation is characterised by a term called R-value. The *R-value* is the thermal resistance of material for a unit area and it is the ratio of thickness and effective thermal conductivity of the material. That is :

$$\text{R-value} = \frac{\text{Thickness}}{\text{Effective thermal conductivity}} = \frac{L}{k_{eff}} \quad \text{...(1.28)}$$

The R-value is generally given in English unit h.ft^2. °F/Btu.

For example, R-value of 6″ thick glass fibre insulation (k_{eff} = 0.025 Btu/h.ft^2.°F) is designated as R-20 insulation by builders, *i.e.,*

$$\text{R-value} = \frac{L}{k_{eff}} = \frac{6'' \times (1/12)\text{ ft}}{0.025\text{ Btu/h.ft}^2.°\text{F}}$$

$$= 20 \text{ h.ft}^2.\ °\text{F/Btu.}$$

1.8.4. Economic Thickness of Insulation

The energy crisis of 1970s had a tremendous impact on energy awareness and energy conservation. Since then, new and more effective materials are developed and use of insulation is considerably increased. The walls and roofs of our house are also applied with some insulation like plaster of paris, etc., to minimise the heat transfer rate with its surroundings.

When an insulation layer is put on a heating system, the heat loss from the system reduces. The cost of insulation material adds the system cost, while cost of reduction in heat loss reduces the operating cost. Therefore, the cost of insulation material is subsidised by saving in energy cost. *Insulation pays for itself from the energy it saves.*

The insulating material has certain period of service. Over the service life of the insulation material, the thickness of insulation at which the sum of cost of insulation and cost of heat loss is minimum as shown in Fig. 1.22, is referred to as *economic thickness of insulation.* If the thickness of insulation is more than the economic thickness, the cost of insulation will not compensate the energy it saves. Thickness of insulation is controlled by density of material. As density of material increases, the required thickness of insulation decreases for same thermal effect. If the layer after layer of insulation is applied over a surface, the heat transfer reduces gradually. The inner layer saves more heat than outer one. A limiting thickness of insulation balances the cost of energy saved and cost of insulation itself. This particular critical layer is called **optimum insulation thickness.**

The optimum thickness of insulation can be obtained by plotting a graph of value of heat loss and cost of insulation against the thickness of insulation.

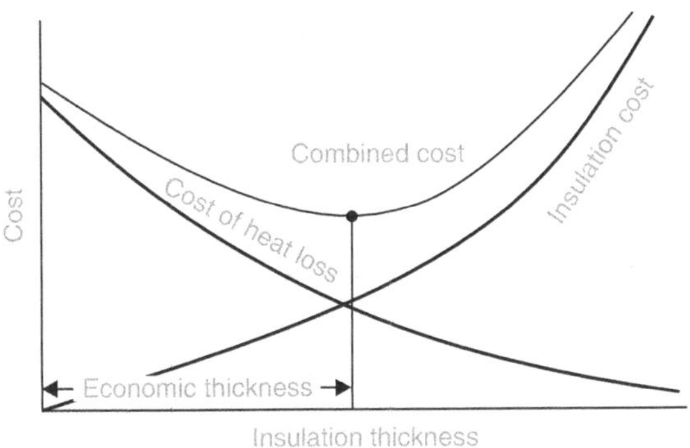

Fig. 1.22. Economic thickness of insulation

1.9. THERMAL DIFFUSIVITY

Thermal diffusivity is an important thermophysical property. It is the ratio of thermal conductivity k of the medium to **heat capacity** ρC. It is denoted by α, and measured in m^2/s.

$$\alpha = \frac{k}{\rho C} \qquad \text{...(1.29)}$$

The thermal conductivity k indicates how well a material can conduct heat and the heat capacity ρC represents how much energy a material can store per unit volume. Therefore, the *thermal diffusivity* of a material is viewed as the ratio of the *heat conducted* through the material to the *heat stored* per unit volume. In other words, the thermal diffusivity of a material is associated with the propagation of heat energy into the medium during the change of temperature with time. The higher the thermal diffusivity, faster the propagation of heat into the medium.

Example 1.13. *The inside temperature of a furnace wall (k = 1.35 W/m.K), 200 mm thick, is 1400°C. The heat transfer coefficient at the outside surface is a function of temperature difference and is given by*

$$h = 7.85 + 0.08\,\Delta T \ (W/m^2.K)$$

where ΔT is the temperature difference between outside wall surface and surroundings. Determine the rate of heat transfer per unit area, if the surrounding temperature is 40°C.

Solution

Given : A furnace wall exposed to convection environment on one side.

$$k = 1.35 \text{ W/m.K}$$

$$L = 200 \text{ mm} = 0.2 \text{ m}$$

$$T_1 = 1400°C$$

$$T_\infty = 40°C$$

$$h = 7.85 + 0.08\,(\Delta T)\ (W/m^2.K)$$

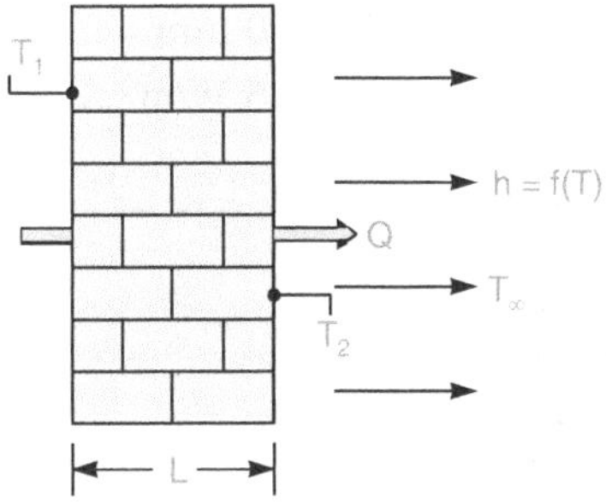

Fig. 1.23. Schematic of a furnace wall

To find : Heat flux.

Analysis : Steady state heat transfer rate per unit area

$$q = \frac{k(T_1 - T_2)}{L} = h(T_2 - T_\infty)$$

or $$\frac{1.35 \times (1400 - T_2)}{0.2} = [7.85 + 0.08\,(T_2 - 40)] \times (T_2 - 40)$$

or $$9450 - 6.75\,T_2 = 7.85\,T_2 - 314 + 0.08\,(T_2 - 40)^2$$

$$= 7.85\,T_2 - 314 + 0.08 \times (T_2^2 - 80\,T_2 + 1600)$$

$$= 7.85\,T_2 - 314 + 0.08\,T_2^2 - 6.4\,T_2 + 128$$

Rearranging, $$0.08\,T_2^2 + 8.2\,T_2 - 9636 = 0$$

or $$T_2 = \frac{-8.2 + \sqrt{(8.2)^2 + 4 \times 0.08 \times 9636}}{2 \times 0.08}$$

$$= 299.57°C$$

The heat flow rate per unit area

$$q = \frac{k}{L}(T_1 - T_2)$$

$$= \frac{1.35}{0.2} \times (1400 - 299.57)$$

= **7427.88 W. Ans.**

Example 1.14. *An uninsulated steam pipe is passed through a room in which air and walls are at 25°C. The outer diameter of the pipe is 50 mm and surface temperature and emissivity are 500 K and 0.8, respectively. If the free convection heat transfer coefficient is 15 W/m².K, what is the rate of heat loss from the surface per unit length of pipe ?*

Solution

Given : An uninsulated pipe exposed to room air

$T_\infty = T_w = 25°C = 298\text{ K}$

$T_s = 500\text{ K}$, $D = 50\text{ mm} = 0.05\text{ m}$

$\varepsilon = 0.8$, $h = 15\text{ W/m}^2\text{.K}$.

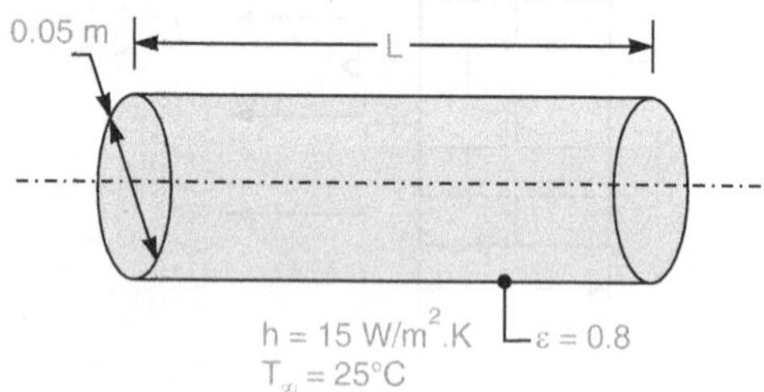

Fig. 1.24. Schematic of example 1.14

To find : Heat loss per unit length of pipe.

Assumptions :

1. Steady state conditions.
2. Heat loss by radiation and convection only.
3. Stefan Boltzmann constant, $\sigma = 5.67 \times 10^{-8}$ W/m².K⁴.
4. Constant properties.

Analysis :

(*i*) Heat loss from the pipe by convection is given by

$$Q_{conv} = hA_s(T_s - T_\infty) = h(\pi DL)(T_s - T_\infty)$$

or
$$\frac{Q_{conv}}{L} = 15 \times (\pi \times 0.05) \times (500 - 298)$$
$$= 476\text{ W/m}$$

(*ii*) Heat loss per unit length of pipe by radiation is given by

$$\frac{Q_{rad}}{L} = \sigma\varepsilon(\pi D)(T_s^4 - T_\infty^4)$$
$$= 5.67 \times 10^{-8} \times 0.8 \times (\pi \times 0.05) \times (500^4 - 298^4)$$
$$= 389.13\text{ W/m}$$

Total heat loss from pipe surface per unit length :

$$\frac{Q}{L} = \frac{Q_{conv}}{L} + \frac{Q_{rad}}{L}$$
$$= 476 + 389.13$$
$$= \mathbf{865.13\ W/m.}\ \text{Ans.}$$

Example 1.15. *A horizontal plate (k = 30 W/m.K) 600 mm × 900 mm × 30 mm is maintained at 300°C. The air at 30°C flows over the plate. If the convection coefficient of air over the plate is 22 W/m².K and 250 W heat is lost from the plate by radiation. Calculate the bottom surface temperature of the plate.* (P.U., Dec. 2008)

Solution

Given : A horizontal plate as shown in Fig. 1.25.

$k = 30\text{ W/m.K}$, $A = 600\text{ mm} \times 900\text{ mm}$

$T_s = 300°C$, $T_\infty = 30°C$

$h = 22\text{ W/m}^2\text{.K}$, $L = 30\text{ mm} = 0.03\text{ m}$

$Q_{rad} = 250\text{ W}$.

Fig. 1.25. Schematic of horizontal plate, conducting, convecting and radiating heat

To find : Temperature of bottom surface of the plate.

Assumptions :

1. Steady state conditions.
2. One dimensional heat conduction in the plate.
3. Constant properties.

Analysis : The surface area of the plate

$$A = 600\text{ mm} \times 900\text{ mm}$$
$$= 5.40 \times 10^5\text{ mm}^2 = 0.54\text{ m}^2$$

Making the energy balance for the plate :

Rate of heat conduction = Rate of heat convection + Rate of heat radiation

or
$$Q_{cond} = Q_{conv} + Q_{rad}$$

or
$$\frac{kA(T_i - T_s)}{L} = hA(T_s - T_\infty) + 250$$

Using numerical values :

$$\frac{30 \times 0.54 \times (T_i - 300)}{0.03} = 22 \times 0.54 \times (300 - 30) + 250$$

or
$$T_i - 300 = \frac{0.03}{30 \times 0.54} \times (3207.6 + 250)$$

or
$$T_i = 300 + 6.40 = \mathbf{306.4°C.}\ \text{Ans.}$$

Example 1.16. *A black metal plate (k = 25 W/m.K) at 300°C is exposed to surrounding air at 30°C. It convects and radiates heat to surroundings. If the convection coefficient is 25 W/m².K, what is the temperature gradient in the plate ?*

Solution

Given : An iron plate convects and radiates heat to surroundings.

$$k = 25 \text{ W/m.K}$$
$$T_s = 300°C = 573 \text{ K}$$
$$T_\infty = 30°C = 303 \text{ K}$$
$$h = 25 \text{ W/m}^2\text{.K.}$$

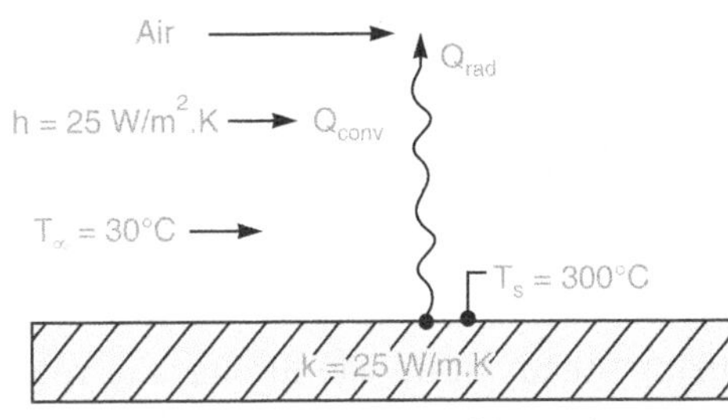

Fig. 1.26. Schematic of iron plate

To find : Temperature gradient in the plate.

Assumptions :

1. Steady state conditions.
2. Black metal plate is black body for radiation.
3. Stefan Boltzmann constant, $\sigma = 5.67 \times 10^{-8}$ W/m².K⁴.

Analysis : The energy balance for the metal plate is given as ;

Heat conducted through the plate
= Heat convection from the surface
+ Heat radiated from the surface

i.e., $$Q_{cond} = Q_{conv} + Q_{rad}$$

or $$-kA\frac{dT}{dx} = hA(T_s - T_\infty) + \sigma A(T_s^4 - T_\infty^4)$$

Using numerical values

$$-25 \times \frac{dT}{dx} = 25 \times (573 - 303) + 5.67 \times 10^{-8} \times (573^4 - 303^4)$$
$$= 25 \times 270 + 5.67 \times 10^{-8} \times 9.937 \times 10^{10}$$
$$= 6750 + 5634.34 = 12384.34$$

or $$\frac{dT}{dx} = -\frac{12384.34}{25} = \mathbf{-495.37°C/m.} \quad \text{Ans.}$$

Example 1.17. *A metal plate with dimension 5 m× 3 m with negligible thickness has a surface temperature of 300°C. One side of it looses heat to the surroundings air at 30°C. The heat transfer coefficient between plate surface and air is 20 W/m².K. The emissivity of the plate surface is 0.8. Calculate.*

(*i*) *Rate of heat loss by convection.*

(*ii*) *Rate of heat loss by radiation.*

(*iii*) *Combined convection and radiation heat transfer coefficient.* (P.U., May 2009)

Solution

Given : A thin metal plate surface exposed to convection and radiation environment.

$$A = 5 \text{ m} \times 3 \text{ m} = 15 \text{ m}^3$$
$$T_s = 300°C = 573 \text{ K}$$
$$h = 20 \text{ W/m}^2\text{ . K,}$$
$$T_\infty = 30°C = 303 \text{ K} \quad \varepsilon = 0.8$$

To find :

(*i*) Rate of heat transfer by convection.

(*ii*) Radiation heat transfer rate.

(*iii*) Combined heat transfer coefficient.

Assumption : The Stefan Boltzmann's constant

$$\sigma = 5.67 \times 10^{-8} \text{ W/m}^2\text{.K}^4.$$

Analysis :

(*i*) Convection heat transfer rate from the plate surface

$$Q_{conv} = hA(T_s - T_\infty)$$
$$= 20 \times 15 \times (300 - 30)$$
$$= \mathbf{81{,}000 \text{ W.}} \quad \text{Ans.}$$

(*ii*) Radiation heat transfer rate from the plate

$$Q_{rad} = \varepsilon\sigma A(T_s^4 - T_\infty^4)$$
$$= 0.8 \times 5.67 \times 10^{-8} \times 15 \times (573^4 - 303^4)$$
$$= \mathbf{67{,}612 \text{ W.}} \quad \text{Ans.}$$

(*iii*) Combined convection and radiation heat transfer coefficient

Total heat transfer rate by convection and radiation

$$Q = Q_{conv} + Q_{rad}$$
$$= 81{,}000 + 67{,}612$$
$$= 148{,}612 \text{ W}$$

It can be expressed as :

$$Q = h_{comb} A(T_s - T_\infty)$$

or $$148{,}612 = h_{comb} \times 15 \times (300 - 30)$$

or $$h_{comb} = \mathbf{36{,}694 \text{ W/m}^2.} \quad \text{Ans.}$$

1.10. HEAT TRANSFER IN BOILING AND CONDENSATION

The boiling and condensation are the phase change phenomena with heat transfer. During boiling, evaporation and vaporization, the liquid absorbs latent heat, gets converted to vapour. Reverse process occurs in condensation, where the vapour gets converted into liquid by rejecting its latent heat to some cooling medium. In all these cases of heat transfer with phase change, the temperature remains constant during the process. Heat transfer in boiling and condensation is characterized by very high values of heat transfer coefficient at constant temperature and therefore, these processes of heat transfer are preferred in actual practices. Boiling process takes place in steam generators, distillation columns and evaporators, while the condensation process occurs in the condensers, during formation of dew, etc.

The mathematical treatment to actual mechanism of boiling and condensation is very complicated; therefore, empirical relations are used to calculate the heat transfer coefficient and heat flux. Many useful empirical relations are presented in chapter 11 for estimation of various parameters during boiling and condensation processes.

1.11. MASS TRANSFER

Mass transfer is defined as movement of mass due to concentration difference in a mixture. The concentration difference is the driving potential for the mass transfer. Mass transfer occurs in many processes, such as absorption, evaporation, adsorption, desorption, solvent extraction, humidification and drying. In many practical applications, heat transfer processes occur simultaneously with mass transfer processes and the principles of mass transfer are very similar to those of heat transfer, therefore, the analogy between heat and mass transfer can easily be established.

1.12. SUMMARY

Heat is a form of energy, which transfers due to temperature difference. The heat transfer is a branch of thermodynamics, which deals with analysis of rate of heat transfer, temperature distribution and the nature of heat transfer taking place in a system.

During steady state conditions, the rate of heat transfer is always constant and the temperature at any location does not change with time. In unsteady state, the temperature changes with time and position, thus the rate of heat transfer varies with time.

TABLE 1.4. Summary of heat transfer rate processes

Mode	*Mechanism*	*Governing equation*	*Transport property*
Conduction	Exchange of energy due to direct molecular interactions	$q(\text{W/m}^2) = -k \dfrac{dT}{dx}$	Thermal conductivity k(W/m.K)
Convection	Diffusion of energy due to random molecular motion plus energy due to bulk motion (advection)	$q(\text{W/m}^2) = h(T_s - T_\infty)$	Heat transfer coefficient h(W/m^2.K)
Radiation	Energy transfer by electromagnetic waves	$q(\text{W/m}^2) = \varepsilon\,\sigma(T_s^4 - T_{sur}^4)$ or $Q(\text{W}) = h_r A(T_s - T_{sur})$	Emissivity ε, or radiation heat transfer coeff. h_r

TABLE 1.5. Glossary of heat transfer terms

Terms	*Interpretation*
Heat energy	A form of energy in transit.
Conduction	Energy transfer into the medium due to existence of temperature gradient.
Convection	Energy transportation by moving fluid particles from hot region to cold region.
Radiation	Emission of energy in the form of electromagnetic waves by the surface.
Emissivity	A property of the radiating surface.
Thermal conductivity	Ability of the materials, which allows the heat conduction through them.
Heat transfer coefficient	A property of the fluid environment associated with heat convection.
Mass transfer	Movement of mass due to concentration difference in a mixture.

The thermal insulation is a material or combination of materials, which is mainly used to minimise the heat flow to or from a system.

Thermal diffusivity is the ratio of thermal conductivity k of the medium and heat capacity ρC. It is denoted by α, and measured in m^2/s, *i.e.*,

$$\alpha = \frac{k}{\rho C}$$

REVIEW QUESTIONS

1. How does the heat transfer differ from the thermodynamics ?
2. How does transient heat transfer differ from steady state conduction ?
3. What is heat flux ? How it relates heat transfer rate ?
4. What are different modes of heat transfer ? Explain their potential for occurrence.
5. How does the heat conduction differ from convection ?
6. Prove that convection is not fundamentally different mode of heat transfer. It consists of conduction from the surface to the adjacent layer + energy transfer due to mass transfer + conduction to the adjacent fluid layer.
7. What are the laws of heat transfer ?
8. State Fourier law of heat conduction and by using it derive an expression for steady state heat conduction through a plane wall of thickness L maintains its two surfaces at temperatures T_1 and T_2, respectively.
9. Identify the mode(s) of heat transfer in the following cases :
   (*i*) Heat transfer from a room heater,
   (*ii*) Hot plate exposed to atmosphere,
   (*iii*) Heat loss from thermos flask,
   (*iv*) Cooling of a scooter engine,
   (*v*) Heat loss from automobile radiator,
   (*vi*) Heat transfer from sun to a living room.
10. Define thermal conductivity and explain its significance in heat transfer.
11. How does the thermal conductivity of liquids and gases vary with temperature ?
12. What are the thermal insulators ?
13. How is the thermal conductivity of plane wall determined experimentally ? Explain.
14. Define isotropic and anisotropic materials.
15. Define apparent thermal conductivity.
16. Explain superinsulators. How do they differ from ordinary insulations ?
17. What is the R-value of an insulation ? How is it determined ?
18. How does the R-value of an insulation differ from its thermal resistance ?
19. What is the physical significance of thermal diffusivity ?

PROBLEMS

1. Determine the heat flow across a plane wall of 10 cm thickness with a thermal conductivity of 8.5 W/m.K. When the surface temperatures are steady and at 200°C and 50°C, the wall area is 2 m^2. Also find the temperature gradient in flow direction.
   [Ans. 25500 W, 1500°C/m]
2. Consider a plane wall 20 cm thick. The inner surface is kept 400°C, and the outer surface is exposed to an environment at 800°C with a heat transfer coefficient of 12 W/(m^2.K). If the temperature of the outer surface is 685°C, calculate the thermal conductivity of the wall. [Ans. 0.968 W/m.K]
3. A glass window 60 cm × 60 cm is 16 mm thick. If its inside and outside surface temperatures are 20°C and – 20°C respectively, determine the conduction heat transfer rate through the window. Take thermal conductivity of glass as 0.78 W/m.K. [Ans. 702 W]
4. The wall of a house 7 m wide, 6 m high is made from 0.3 m thick brick (k = 0.6 W/m.K). The surface temperature on inside of wall is 26°C and that on outside is 16°C. Find the heat flux and total heat loss through the wall. [Ans. 20.0 W/m^2, 840 W]
5. Determine steady state heat transfer rate per unit area through a 3.8 cm thick homogeneous wall with its two faces maintained at uniform temperature of 35°C and 25°C. Thermal conductivity of wall material is 0.19 W/m.K. [Ans. 50 W/m^2]
6. Calculate rate of heat flow for a red brick wall 5 m long × 4 m high × 0.25 m thick. Temperature of inner surface is 110°C and that of outer surface is 40°C. Thermal conductivity of red brick is 0.7 W/m.K. Also calculate the temperature at 20 cm away from the inner surface of the wall. [Ans. 3920 W, 54°C]
7. A brass condenser tube [k = 115 W/(m.K)] with an outside diameter of 2 cm and a thickness of 0.2 cm is used to condense steam on its outer surface at 50°C with a heat transfer coefficient of 2000 W/(m^2.K). Cooling water at 20°C with a heat transfer coefficient of 5000 W/(m^2.K) flow inside.
   (*a*) Determine the heat flow rate from the steam to the cooling water per meter of length of the tube.
   (*b*) What would be the heat transfer rate per metre of length of the tube if the outer and the inner surfaces of the tube were at 50°C and 20°C, respectively ? Compare this result with (*a*) and explain the reason for the difference between the two results. [Ans. (*a*) 2450 W/m, (*b*) 97.4 kW/m]

8. A 20 mm dia copper pipe is used to carry heated water. The external surface of the pipe is exposed to convection environment at 20°C, with a heat transfer coefficient of 6 W/m^2.K. The pipe surface temperature is 80°C. Assume black body radiation and calculate heat loss by convection and radiation.

[Ans. 22.6 W/m, 29.1 W/m]

9. Determine heat transfer rate through a spherical copper shell of thermal conductivity of 386 W/m.K, inner radius of 20 mm and outer radius of 60 mm. The inner surface and outer surface temperatures are 200°C and 100°C, respectively. [Ans. 14551.87 W]

10. A hollow sphere (k = 20 W/m.K) of inside radius 30 mm and outside radius 50 mm is electrically heated at its inner surface at a constant rate of 10^5 W/m^2. The outer surface is exposed to a fluid at 30°C with heat transfer coefficient of 170 W/m^2.K. Calculate inner and outer surface temperature of sphere [Ans. 301.75°C and 241.75°C]

11. Consider a furnace wall [k = 1 W/(m.K)] with the inside surface at 1000°C and the outside surface at 400°C. If the heat flow through the wall should not exceed 2000 W/m^2, what is the minimum wall thickness L ? [Ans. 30 cm]

12. Determine the heat transfer rate by convection over a surface of 1 m^2, if a surface at 100°C is exposed to a fluid at 40°C with convection coefficient of 25 W/m^2.K.

[Ans. 1500 W]

13. An electrically heated plate dissipates by convection at a rate of 8000 W/m^2 into the ambient air at 25°C. If the surface of the hot plate is at 125°C, calculate the heat transfer coefficient for convention between the plate and air. [Ans. 80 W/m^2.K]

14. A 25 cm diameter sphere at 120°C is suspended in air at 20°C. If the natural heat transfer coefficient is 15 W/m^2.K, determine the heat loss from the sphere.

[Ans. 294.5 W]

15. A flat plate of length 1 m and width 0.5 m is placed in air stream at 30°C blowing parallel to it. The convective heat transfer coefficient is 30 W/m^2.K. Calculate the heat transfer rate, if the plate is maintained at a temperature of 300°C.

[Ans. 4.05 kW]

16. A surface is at 200°C is exposed to surroundings at 60°C and convects and radiates heat to the surroundings. Calculate the heat transfer rate from surface to surroundings, if the convection coefficient is 80 W/m^2.K. Consider the black bodies for radiation heat transfer.

Take $\sigma = 5.67 \times 10^{-8}$ W/m^2.K^4.

[Ans. 13.34 kW/m^2]

17. A cube shaped solid 20 cm side having density 2500 kg/m^3, specific heat 520 J/kg.K has a uniform heat generation rate of 100 kW/m^3. If the heat received over each surface is 40 W, determine the time rate of temperature change of solid.

[Ans. 0.1°C/s]

18. A solid with thermal conductivity 25 W/m.K, has a temperature gradient of – 5°C/cm. Determine the steady state heat flux. If the heat is exchanged by radiation from the surface (black) to the surroundings at 30°C, determine the surface temperature. [Ans. – 12500 W/m^2, 418.6°C]

19. Two black bodies exchange radiation heat, are maintained at 1500°C and 150°C respectively. Calculate the radiation heat flux due to radiation between them. [Ans. 558.48 kW/m^2]

20. A pipe of outer diameter 10 cm and inner diameter 8 cm, whose thermal conductivity is expressed as k = (5 + 0.01 T) W/m°C, where T is expressed in °C. The inside and the outside surfaces are maintained at 100°C, and 20°C, respectively. What is the heat loss for 2 m long pipe ? [Ans. 25229.2 W]

21. In a solar flat plate heater, some of the heat is absorbed by a fluid while the remaining heat is lost by convection, bottom surface is insulated. The fraction absorbed is known as efficiency of the collector. If the flux incident has a value of 1100 W/m^2 at collection temperature of 60°C. Determine the collector efficiency when it is exposed to surroundings at 32°C with convection coefficient of 15 W/m^2.K. Also find the collector efficiency, if collection temperature is 45°C. [Ans. 61.8%, 82.2%]

22. Calculate the heat transfer by radiation from the surface of a 60 mm dia spherical lamp (black body) at temperature of 80°C into an ambient at 20°C.

[Ans. 5.2 W]

23. A commercial heat flux meter uses thermocouple junctions to measure the temperature difference across a thin layer of vermiculite (k = 0.059 W/m.K), that is 0.0005 m thick. What is heat flux when the temperature difference is 3°C ? [Ans. 354 W/m^2]

24. The inside and outside surfaces of hollow sphere of radii r_1 and r_2 are maintained at constant temperature T_1 and T_2, respectively. The thermal conductivity of sphere material varies with temperature as $k(T) = k_0(1 + \alpha T + \beta T^2)$.

Prove that the heat flow rate Q through the sphere is given by

$$Q = \frac{4\pi k_0 r_1 r_2}{r_2 - r_1}(T_1 - T_2)$$

$$\left[1 + \frac{\alpha}{2}(T_1 + T_2) + \frac{\beta}{3}(T_1^2 + T_1T_2 + T_2^2)\right]$$

25. A solid (k = 38 W/m.K) is having temperature gradient of 350°C/m. Determine the steady state heat flux. If the heat is exchanged by radiation from a surface (black) to the surrounding at 30°C, determine the surface temperature of solid.

[Ans. 13300 W/m^2, 429.1°C]

MULTIPLE CHOICE QUESTIONS

1. Transfer of heat energy takes place in accordance with
   (*a*) Zeroth law of thermodynamics
   (*b*) First law of thermodynamics
   (*c*) Second law of thermodynamics
   (*d*) Third law of thermodynamics
2. Heat energy can be considered as :
   (*a*) Form of energy
   (*b*) Form of energy in transit
   (*c*) Internal energy.
   (*d*) All of the above.
3. The assumption in the Fourier law $Q = -kA(dT/dx)$,
   (*a*) Constant value of thermal conductivity
   (*b*) Constant and uniform temperature at the surface of wall.
   (*c*) Steady state one dimensional flow
   (*d*) Only (*b*) and (*c*)
   (*e*) All of the above
4. Temperature difference between two sides of a wall can be increased by
   (*a*) Increasing the heat flow rate
   (*b*) Decreasing thermal conductivity of material
   (*c*) Either (*a*) or (*b*)
   (*d*) Both (*a*) and (*b*)
5. A slab 50 cm thick is made of fire brick (k = 1.5 W/m.K). For same heat transfer and same temperature drop, what will be the wall thickness of material having thermal conductivity 0.75?
   (*a*) 0.05 m (*b*) 0.1 m
   (*c*) 0.2 m (*d*) 0.25 m
6. Arrange thermal conductivity of materials in ascending order. Copper, steel, brick and aluminium
   (*a*) Copper steel, brick aluminium
   (*b*) Brick, aluminium, copper, steel
   (*c*) Brick, steel, aluminium, copper
   (*d*) Steel, copper, brick, aluminium
7. The thermal conductivity of a material varies with
   (*a*) Area (*b*) Thickness
   (*c*) Temperature (*d*) None of above
8. For same thickness, the temperature drop in an insulation material is :
   (*a*) Equal to that of conductor
   (*b*) Less than that of conductor
   (*c*) Greater than that of conductor
   (*d*) None of above
9. Thermal conductivity of powderly and porous materials
   (*a*) Decreases with increasing temperature
   (*b*) Increases with increasing temperature
   (*c*) Is independent of temperature change, and
   (*d*) None of the above
10. Low temperature insulating material is :
    (*a*) Asbestos (*b*) Glass wool
    (*c*) Magnesia (*d*) Diatomaceous earth
11. When a fan is switched on in a class room of 50 students, comfort level of students increases due to
    (*a*) Decrease in temperature of room
    (*b*) Increase in heat transfer coefficient in room
    (*c*) Both (*a*) and (*b*) (*d*) None of the above
12. What should be the convection heat flux, if heat transfer coefficient is 40 W/m^2. K and the temperature difference between surface and fluid is 200°C?
    (*a*) 7.6 kW/m^2 (*b*) 10 kW/m^2
    (*c*) 8 kW/m^2 (*d*) 8.5 kW/m^2
13. A thin flat plate is hanging freely in air at 27°C. Solar radiation falls in one of its side at the rate of 500 W/m^2. For maintaining the temperature of plate constant at 32°C, what is the value of heat transfer coefficient?
    (*a*) 25 W/m^2.K (*b*) 50 W/m^2.K
    (*c*) 100 W/m^2.K (*d*) 200 W/m^2.K
14. The radiation heat transfer rate per unit area between two black bodies at temperature 900° and 40° (in kW/m^2) is :
    (*a*) 37.2 (*b*) 10.7
    (*c*) 107 (*d*) 1070
15. The emissivity of real surfaces is always
    (*a*) Equal to unit
    (*b*) Less than unity
    (*c*) Greater than unity, and
    (*d*) Less than or greater than unity

Answers

1. (*c*)	2. (*b*)	3. (*e*)	4. (*d*)
5. (*d*)	6. (*c*)	7. (*c*)	8. (*c*)
9. (*b*)	10. (*d*)	11. (*b*)	12. (*c*)
13. (*c*)	14. (*c*)	15. (*b*)	

Conduction—Basic Equations

2

2.1. Generalised One Dimensional Heat Conduction Equation. 2.2. Three Dimensional Heat Conduction Equation—For the cartesian coordinates—Three dimensional heat conduction equation in cylindrical coordinates—Three dimensional heat conduction equation in spherical coordinates. **2.3. Initial and Boundary Conditions**—Prescribed temperature boundary conditions—Prescribed heat flux boundary conditions—Convection boundary conditions : Surface energy balance—Radiation boundary condition—Interface boundary condition. **2.4. Summary**—*Review Questions—Problems.*

The objective of this chapter is to provide a good understanding of the heat conduction equations and boundary conditions for the use in mathematical formulation of heat conduction problems.

2.1. GENERALISED ONE DIMENSIONAL HEAT CONDUCTION EQUATION

For the thermal analysis of the bodies having shapes such as slab, rectangle, the cartesian coordinates are used, while for cylindrical and spherical bodies, the polar and spherical coordinate systems are used.

In this section, we derive one dimensional, time dependent generalised heat conduction equation which may be obtained in either coordinate system.

Considering one dimensional element as shown in Fig. 2.1.

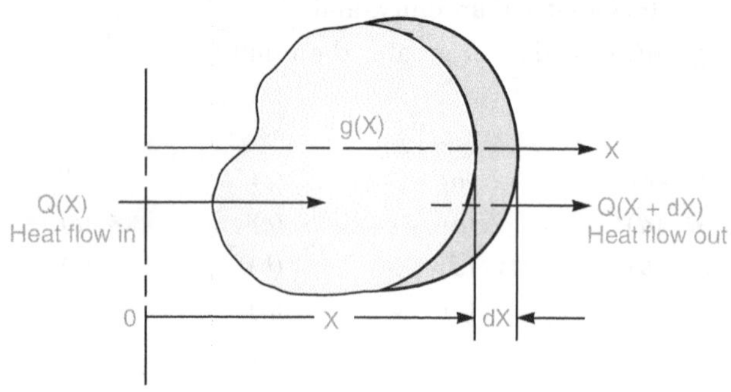

Fig. 2.1. Element for one dimension heat conduction equation

The element having

Heat conduction rate into the element = Q(X)

Heat conduction rate from the element

$$= Q(X + dX)$$

Net rate of heat conduction into the element

$$Q_{net} = Q(X) - Q(X + dX)$$

If the heat is generated within the element due to resistance heating, chemical or nuclear reactions, etc., and the rate of volumetric heat generation is g (W/m^3).

Then rate of energy generation, $Q_{gen} = g\,(A dX)$

Due to unequal heat transfer rates to and from the element, its internal energy will change. The rate of change of internal energy,

$$\frac{\Delta E}{\partial t} = mC\frac{\partial T}{\partial t} = (\rho A dX)C\frac{\partial T}{\partial t} \qquad ...(2.1)$$

where, T = F(X, t), temperature of element as function of time and direction, °C,

g = G(X, t), the function of time and direction, W/m^3,

k = K(X), the function of direction, W/m.K,

C = specific heat of the material (solid having only one specific heat), J/kg.K,

m = mass of the element = (ρ AdX), kg,

A = area of element normal to the heat transfer, m^2,

ρ = density of the material, kg/m^3,

t = time, s,

dX = directional thickness of element, m.

Making the energy balance on the element.

Net rate of heat gain by conduction + rate of energy generation

= The net rate of change of internal energy.

$$Q_{net} + Q_{gen} = \frac{\Delta E}{\partial t}$$

or $$\{Q(X) - Q(X + dX)\} + g\,A\,dX = \rho CA\,dX\,\frac{\partial T}{\partial t} \quad ...(2.2)$$

According to *Taylor's series*

$$Q(X + dX) = Q(X) + \frac{\partial\,Q(X)}{\partial X}\,dX + \frac{\partial^2 Q(X)}{\partial X^2}\frac{dX^2}{2!} + \frac{\partial^3 Q(X)}{\partial X^3}\frac{dX^3}{3!} + ...$$

If the control volume is considered small enough, then the higher powers of dX such as dX^2, dX^3 etc., are negligibly small, therefore, neglected from above equation and it reduces to

$$Q(X + dX) = Q(X) + \frac{\partial\,Q(X)}{\partial X}\,dX \quad ...(2.3)$$

Substituting this equation in eqn. (2.2), we get

$$-\frac{\partial Q(X)}{\partial X}\,dX + g\,A\,dX = \rho\,C\,A\,dX\,\frac{\partial T}{\partial t} \quad ...(2.4)$$

Substituting $Q(X) = -\,kA\,\frac{\partial T}{\partial X}$

$$\text{Then, } -\frac{\partial}{\partial X}\left\{-\,kA\,\frac{\partial T}{\partial X}\right\} dX + g\,A\,dX = \rho\,C\,A\,dX\,\frac{\partial T}{\partial t}$$

Rearranging above, we get

$$\frac{1}{A}\frac{\partial}{\partial X}\left\{kA\,\frac{\partial T}{\partial X}\right\} + g = \rho\,C\,\frac{\partial T}{\partial t} \quad ...(2.5)$$

It is general one dimensional time dependent differential heat conduction equation with heat generation and directional dependent k.

If the conducting material is isotropic, its thermal conductivity is independent of direction, it is treated as constant quantity, then

$$\frac{1}{A}\frac{\partial}{\partial X}\left\{A\,\frac{\partial T}{\partial X}\right\} + \frac{g}{k} = \frac{\rho\,C}{k}\frac{\partial T}{\partial t} = \frac{1}{\alpha}\frac{\partial T}{\partial t} \quad ...(2.6)$$

where, $\frac{k}{\rho C} = \alpha$ is the *thermal diffusivity,* a property of material.

The above eqn. (2.6) is in general coordinate system. It is one dimensional time dependent differential equation for heat conduction with constant thermal conductivity. It is known as *unidirectional governing equation for heat conduction.*

If there is no internal heat generation within the material, the above equation reduces to :

$$\frac{1}{A}\frac{\partial}{\partial X}\left\{A\,\frac{\partial T}{\partial X}\right\} = \frac{1}{\alpha}\frac{\partial T}{\partial t} \quad ...(2.7)$$

It is known as *unidirectional Fourier equation.*

We may write this equation in particular coordinate system by introducing proper area A and directional thickness dX as described below.

Rectangular (Cartesian) Coordinate System

For rectangular coordinate system,

X = x, directional variable,

A = heat transfer area, does not vary with x direction but remains constant.

Therefore, the eqn. (2.5) reduces to :

$$\frac{\partial}{\partial x}\left\{\frac{\partial T}{\partial x}\right\} + \frac{g}{k} = \frac{1}{\alpha}\frac{\partial T}{\partial t} \quad ...(2.8)$$

It is one dimensional time dependent heat conduction equation in rectangular coordinate system. It is used for the analysis of plane wall (slab), with and without heat generation for one dimensional steady state as well as in transient heat conduction.

Cylindrical Coordinate System

For cylindrical coordinate system,

X = r, directional variable,

A = heat transfer area, varies with radius;

= $2\pi r$L, for the cylinder element of radius r and length L.

Using in the eqn. (2.5), we get

$$\frac{1}{r}\frac{\partial}{\partial r}\left\{r\,\frac{\partial T}{\partial r}\right\} + \frac{g}{k} = \frac{1}{\alpha}\frac{\partial T}{\partial t} \quad ...(2.9)$$

It is one dimensional time dependent heat conduction equation in cylindrical coordinate system.

Spherical Coordinate System

For spherical coordinate system :

X = r, directional variable

A = heat transfer area varies with radius

= $4\pi r^2$, for the spherical element of radius r.

Using in the eqn. (2.5), we get

$$\frac{1}{r^2}\frac{\partial}{\partial r}\left\{r^2\,\frac{\partial T}{\partial r}\right\} + \frac{g}{k} = \frac{1}{\alpha}\frac{\partial T}{\partial t} \quad ...(2.10)$$

It is one dimensional time dependent heat conduction equation in spherical coordinate system.

In compact form,

$$\frac{1}{X^n}\frac{\partial}{\partial X}\left\{X^n\frac{\partial T}{\partial X}\right\}+\frac{g}{k}=\frac{1}{\alpha}\frac{\partial T}{\partial t} \quad ...(2.11)$$

where, $n = 0$ and $X = x$ for cartesian coordinate system,
$n = 1$ and $X = r$ for cylindrical coordinate system,
$n = 2$ and $X = r$ for cylindrical coordinate system.

Steady State Conditions

For steady state heat conduction, the temperature at each point within the solid does not vary with time, but it decreases in direction of heat flow (steady means no change with time).

Hence on right hand side of eqns. (2.6) to (2.11)

$$\frac{\partial T}{\partial t}=0 \quad \text{and} \quad T = f(X) \text{ only}$$

Then the one dimensional governing eqn. (2.11) reduces to

$$\frac{1}{X^n}\frac{d}{dX}\left\{X^n\frac{dT}{dX}\right\}+\frac{g}{k}=0 \quad ...(2.12)$$

It is known as unidirectional *Poisson equation.* It can also be written as :

$$\frac{1}{A}\frac{d}{dX}\left\{A\frac{dT}{dX}\right\}+\frac{g}{k}=0 \quad ...(2.13)$$

where area A is constant for plane wall but it is variable for cylinder and sphere.

In cartesian coordinate,

$$\frac{d}{dx}\left(\frac{dT}{dx}\right)+\frac{g}{k}=0 \quad ...(2.14)$$

It is known as unidirectional Poisson equation in the cartesian coordinate.

In cylindrical coordinate,

$$\frac{1}{r}\frac{d}{dr}\left\{r\frac{dT}{dr}\right\}+\frac{g}{k}=0 \quad ...(2.15)$$

It is known as unidirectional Poisson equation in the cylindrical coordinate.

In spherical coordinate,

$$\frac{1}{r^2}\frac{d}{dr}\left\{r^2\frac{dT}{dr}\right\}+\frac{g}{k}=0 \quad ...(2.16)$$

It is known as unidirectional Poisson equation in the spherical coordinate.

If the ***heat is not generated within the solid*** then eqn. (2.12) is reduced to *unidirectional Laplace equation,*

$$\frac{d}{dX}\left\{X^n\frac{dT}{dx}\right\}=0 \quad ...(2.17)$$

In cartesian coordinate,

$$\frac{d}{dx}\left\{\frac{dT}{dx}\right\}=0 \quad ...(2.18)$$

In cylindrical coordinate,

$$\frac{d}{dr}\left\{r\frac{dT}{dr}\right\}=0 \quad ...(2.19)$$

In spherical coordinate,

$$\frac{d}{dr}\left\{r^2\frac{dT}{dr}\right\}=0 \quad ...(2.20)$$

2.2. THREE DIMENSIONAL HEAT CONDUCTION EQUATION

The eqn. (2.6) is the generalized one dimensional time dependent heat conduction equation. By similar approach, the above equation can be extended in the three dimensions.

2.2.1. For the Cartesian Coordinates

Consider a differential volume element with thicknesses *dx, dy* and *dz* in *x, y* and *z* directions, respectively. The rate of incoming and outgoing energy by conduction in respective direction is as shown in Fig. 2.2.

The volume of the element $V = dx\,dy\,dz$

Net rate of heat conduction into the element in *x, y* and *z* directions

$$Q_{net} = Q_x + Q_y + Q_z - Q_{x+dx} - Q_{y+dy} - Q_{z+dz} \quad ...(i)$$

If the heat is generated into the element at the rate of g(W/m^3), then volumetric heat generation rate.

$$Q_{gen} = g\,dx\,dy\,dz \quad ...(ii)$$

The rate of change of internal energy of the differential volume

$$\frac{\Delta E}{\partial t} = mC\frac{\partial T}{\partial t} = (\rho\,dx\,dy\,dz)\,C\frac{\partial T}{\partial t} \quad ...(iii)$$

Making the energy balance on the element by using quantities from eqns. (*i*), (*ii*) and (*iii*)

Net rate of heat gain by conduction + rate of energy generation in the element

= The net rate of change of internal energy

$$[Q_x + Q_y + Q_z - Q_{x+dx} - Q_{y+dy} - Q_{z+dz}] + g\,dx\,dy\,dz$$

$$= \rho\,C\frac{\partial T}{\partial t}\,dx\,dy\,dz \quad ...(iv)$$

Using Taylor's series approximation and rearranging, we get

$$-\frac{\partial}{\partial x}\{Q_x\,dx\} - \frac{\partial}{\partial y}\{Q_y\,dy\} - \frac{\partial}{\partial z}\{Q_z\,dz\} + g\,dx\,dy\,dz = \rho\,C\,\frac{\partial T}{\partial t}\,dx\,dy\,dz \qquad ...(2.21)$$

where the heat conduction quantities in each direction are shown in Fig. 2.2.

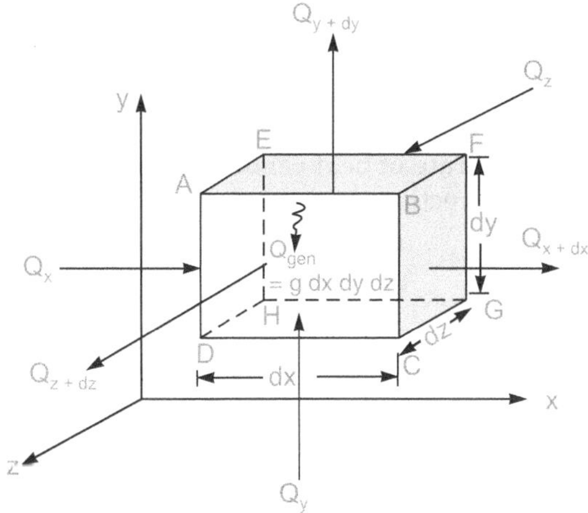

Fig. 2.2. Three dimensional element in cartesian coordinate

$$Q_x = -k_x\,dy\,dz\,\frac{\partial T}{\partial x}$$

$$Q_y = -k_y\,dx\,dz\,\frac{\partial T}{\partial y}$$

$$Q_z = -k_z\,dx\,dy\,\frac{\partial T}{\partial z}$$

Substituting in eqn. (2.21) and rearranging, we get

$$\frac{\partial}{\partial x}\left\{k_x\frac{\partial T}{\partial x}\right\} + \frac{\partial}{\partial y}\left\{k_y\frac{\partial T}{\partial y}\right\} + \frac{\partial}{\partial z}\left\{k_z\frac{\partial T}{\partial z}\right\} + g = \rho\,C\,\frac{\partial T}{\partial t} \qquad ...(2.22)$$

It is three dimensional time dependent, differential heat conduction equation with heat generation and direction dependent k.

The functional relations for used parameter are :

$$T = A(x, y, z, t)$$
$$g = B(x, y, z, t)$$
$$k = D(x, y, z)$$

where A, B, D are some functions, and

C = specific heat of the material, J/kg.K

ρ = density of the material, kg/m^3,

t = time, s

dx, dy, dz = thicknesses of element in x, y and z directions, respectively, m

k_x, k_y, k_z = thermal conductivities in x, y, z directions, respectively, W/m.K.

g = heat generation rate per unit volume, W/m^3.

The above eqn. (2.22) is three dimensional differential equation for ***unsteady state heat conduction for anisotropic material.***

1. If the thermal conductivity of the material is constant in all directions, *i.e.*, for ***isotropic material,***

$k_x = k_y = k_z = k$ (constant value of thermal conductivity)

Eqn. (2.22) reduces to,

$$\frac{\partial^2 T}{\partial x^2} + \frac{\partial^2 T}{\partial y^2} + \frac{\partial^2 T}{\partial z^2} + \frac{g}{k} = \frac{1}{\alpha}\frac{\partial T}{\partial t} \qquad ...(2.23)$$

where, $\alpha = \frac{k}{\rho C}$ = Thermal diffusivity of the material.

The eqn. (2.23) is the three dimensional differential equation for the *transient heat conduction* with constant thermal conductivity. It is also known as governing equation for heat conduction.

2. If there is ***no internal heat generation*** within the material (*i.e.*, $g = 0$), the governing equation reduces to the ***Fourier equation as :***

$$\frac{\partial^2 T}{\partial x^2} + \frac{\partial^2 T}{\partial y^2} + \frac{\partial^2 T}{\partial z^2} = \frac{1}{\alpha}\frac{\partial T}{\partial t} \qquad ...(2.24)$$

3. ***For steady state conditions,*** $\frac{\partial T}{\partial t} = 0$

The eqn. (2.23) becomes

$$\frac{\partial^2 T}{\partial x^2} + \frac{\partial^2 T}{\partial y^2} + \frac{\partial^2 T}{\partial z^2} + \frac{g}{k} = 0 \qquad ...(2.25)$$

The eqn. (2.25) is the three dimensional differential equation for steady state heat conduction with constant thermal conductivity. It is also called the ***Poisson equation.***

4. ***If the solid has no heat generation,***

$$g = 0$$

The eqn. (2.25) reduces to

$$\frac{\partial^2 T}{\partial x^2} + \frac{\partial^2 T}{\partial y^2} + \frac{\partial^2 T}{\partial z^2} = 0 \qquad ...(2.26)$$

The eqn. (2.26) is the three dimensional differential equation for steady state heat conduction without heat generation, with constant thermal conductivity. It is also known as ***Laplace equation.***

2.2.2. Three Dimensional Heat Conduction Equation in Cylindrical Coordinates

Consider a cylindrical differential volume element of isotropic material (k, is constant in all directions). Its thicknesses are dr, $rd\theta$, and dz in r, θ and z directions, respectively as shown in Fig. 2.3.

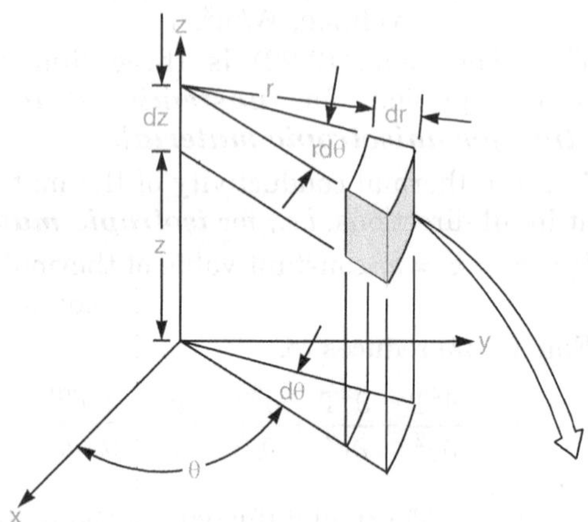

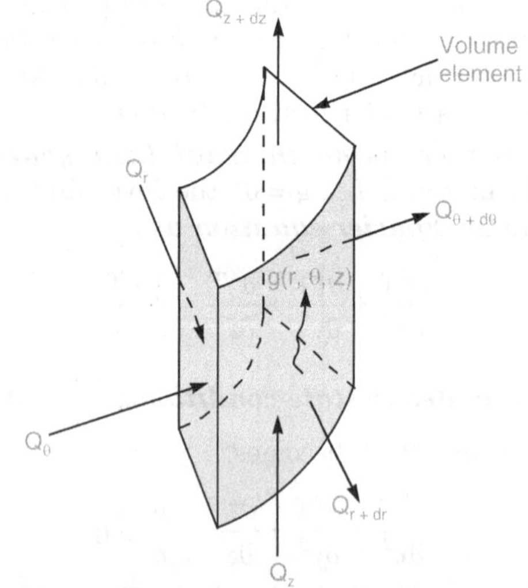

Fig. 2.3. Differential element for cylindrical coordinate system

The volume of elements $V = rd\theta\, dr\, dz$ ...(2.27)

Heat conduction rate into the element in r direction *i.e.,* $\theta - z$ plane

$$Q_r = -k\,(rd\theta\, dz)\,\frac{\partial T}{\partial r} \quad ...(i)$$

Heat conduction rate into the element in θ direction *i.e.,* $r - z$ plane

$$Q_\theta = -k\,(dr\, dz)\,\frac{\partial T}{r\partial\theta} \quad ...(ii)$$

and heat conduction rate into the element in z direction *i.e.,* $r - \theta$ plane

$$Q_z = -k\,(rd\theta\, dr)\,\frac{\partial T}{\partial z} \quad ...(iii)$$

Net rate of heat conduction out the element in r, θ and z directions, respectively.

$$Q_{r+dr} + Q_{\theta+d\theta} + Q_{z+dz} \quad ...(iv)$$

Using Taylor's series approximation

$$Q_{r+dr} = Q_r + \frac{\partial}{\partial r}(Q_r)\,dr \quad ...(v)$$

$$Q_{\theta+d\theta} = Q_\theta + \frac{\partial}{r\partial\theta}(Q_\theta)\,rd\theta \quad ...(vi)$$

and

$$Q_{z+dz} = Q_z + \frac{\partial}{\partial z}(Q_z)\,dz \quad ...(vii)$$

The net rate of heat conduction into the element in r, θ and z directions

$$Q_{net} = (Q_r + Q_\theta + Q_z) - (Q_{r+dr} + Q_{\theta+d\theta} + Q_{z+dz})$$

Using eqns. (v), (vi) and (vii), we get

$$Q_{net} = -\left\{\frac{\partial}{\partial r}(Q_r)\,dr + \frac{\partial}{r\partial\theta}(Q_\theta)\,rd\theta + \frac{\partial}{\partial z}(Q_z)\,dz\right\} \quad ...(2.28)$$

Using eqns. (i), (ii), (iii), we get

$$= \frac{\partial}{\partial r}\left(kr\frac{\partial T}{\partial r}\right)dr\,d\theta\,dz + \frac{\partial}{r\partial\theta}\left(k\frac{\partial T}{r\partial\theta}\right)rd\theta\,dr\,dz + \frac{\partial}{\partial z}\left(k\frac{\partial T}{\partial z}\right)dz\,rd\theta\,dr$$

For an isotropic material k = constant, then

$$Q_{net} = k\left[\frac{1}{r}\frac{\partial}{\partial r}\left(r\frac{\partial T}{\partial r}\right) + \frac{\partial}{r\partial\theta}\left(\frac{\partial T}{r\partial\theta}\right) + \frac{\partial}{\partial z}\left(\frac{\partial T}{\partial z}\right)\right]dr\,rd\theta\,dz \quad ...(2.29)$$

If the heat is generated into the element at the rate of g(W/m^3), then volumetric heat generation rate :

$$Q_{gen} = g\,V = g\,(dr\,rd\theta\,dz) \quad ...(2.30)$$

Due to these heat transfer rates into the element, the internal energy of the element may change. The rate of change of internal energy of the differential volume element is :

$$\frac{\Delta E}{\partial t} = mC\,\frac{\partial T}{\partial t} = (\rho\,dr\,rd\theta\,dz)\,C\,\frac{\partial T}{\partial t} \quad ...(2.31)$$

Making the energy balance on the differential element :

Net rate of heat gain by conduction + Rate of energy generation

= Net rate of change of internal energy

Using the quantities from eqns. (2.29), (2.30) and (2.31) respectively, we get

$$k\left[\frac{1}{r}\frac{\partial}{\partial r}\left(r\frac{\partial T}{\partial r}\right)+\frac{\partial}{r\partial\theta}\left(\frac{\partial T}{r\partial\theta}\right)+\frac{\partial}{\partial z}\left(\frac{\partial T}{\partial z}\right)\right]+g = \rho C\frac{\partial T}{\partial t} \qquad ...(2.32)$$

or $$\frac{\partial^2 T}{\partial r^2}+\frac{1}{r}\frac{\partial T}{\partial r}+\frac{1}{r^2}\frac{\partial^2 T}{\partial\theta^2}+\frac{\partial^2 T}{\partial z^2}+\frac{g}{k}=\frac{\rho C}{k}\frac{\partial T}{\partial t}$$

or $$\frac{\partial^2 T}{\partial r^2}+\frac{1}{r}\frac{\partial T}{\partial r}+\frac{1}{r^2}\frac{\partial^2 T}{\partial\theta^2}+\frac{\partial^2 T}{\partial z^2}+\frac{g}{k}=\frac{1}{\alpha}\frac{\partial T}{\partial t} \qquad ...(2.33)$$

It is the general heat conduction equation in cylindrical coordinates.

Note: The eqn. (2.33) can also be obtained by transformation from rectangular coordinates using

$x = r\cos\theta,\ y = r\sin\theta$ and $z = z$

The steady state one dimensional heat conduction equation in radial direction takes the form

$$\frac{\partial^2 T}{\partial r^2}+\frac{1}{r}\frac{\partial T}{\partial r}+\frac{g}{k}=0$$

or $$\frac{1}{r}\frac{\partial}{\partial r}\left(r\frac{\partial T}{\partial r}\right)+\frac{g}{k}=0$$

It is the *Poisson equation* derived earlier by eqn. (2.15). If no heat is generated within the body, then above equation is reduced to :

$$\frac{\partial}{\partial r}\left(r\frac{\partial T}{\partial r}\right)=0$$

2.2.3. Three Dimensional Heat Conduction Equation in Spherical Coordinates

Consider a three dimensional spherical differential element of isotropic material. The sides of the element are dr, $rd\theta$ and $r\sin\theta\, d\phi$ in r, θ and ϕ directions, respectively.

Volume of element,

$$V = dr \times rd\theta \times r\sin\theta\, d\phi$$

The rate of heat conduction into the element in r direction, *i.e.*, $\theta - \phi$ plane ;

$$Q_r = -k\,(rd\theta \times r\sin\theta\, d\phi)\frac{\partial T}{\partial r} \qquad ...(i)$$

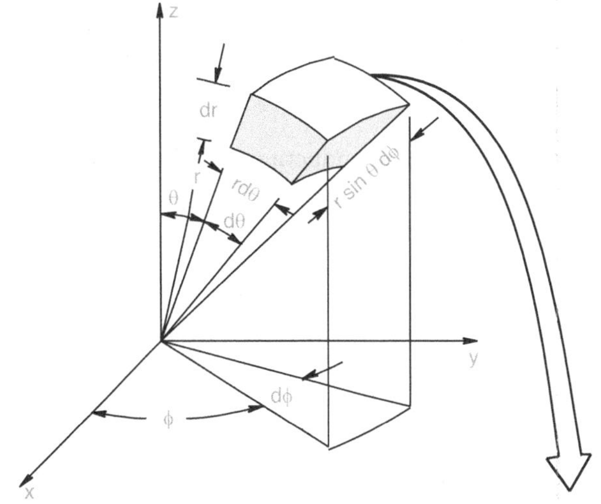

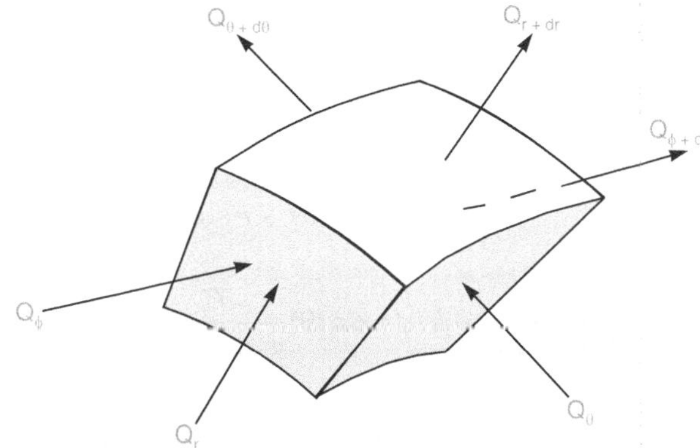

Fig. 2.4. Volume element for spherical coordinate system

The rate of heat conduction into the element in θ direction, *i.e.*, $r - \phi$ plane ;

$$Q_\theta = -k\,(dr \times r\sin\theta\, d\phi)\frac{\partial T}{r\partial\theta} \qquad ...(ii)$$

The rate of heat conduction into the element in ϕ direction, *i.e.*, $r - \theta$ plane ;

$$Q_\phi = -k\,(dr \times rd\theta)\frac{\partial T}{r\sin\theta\, d\phi} \qquad ...(iii)$$

The net rate of heat conduction out the element from r, θ and ϕ directions, respectively.

$$Q_{r+dr} + Q_{\theta+d\theta} + Q_{\phi+d\phi} \qquad ...(iv)$$

Using Taylor's series approximation :

$$Q_{r+dr} = Q_r + \frac{\partial}{\partial r}(Q_r)\,dr \qquad ...(v)$$

$$Q_{\theta+d\theta} = Q_\theta + \frac{\partial}{r\partial\theta}(Q_\theta)\,rd\theta \qquad ...(vi)$$

$$Q_{\phi+d\phi} = Q_\phi + \frac{\partial}{r \sin\theta\, d\phi}(Q_\phi)\, r \sin\theta\, d\phi \qquad ...(vii)$$

Net rate of heat conduction into the element in r, θ and ϕ directions :

$$Q_{net} = (Q_r + Q_\theta + Q_\phi) - (Q_{r+dr} + Q_{\theta+d\theta} + Q_{\phi+d\phi})$$

Using eqns. (v), (vi) and (vii), we get

$$Q_{net} = -\left[\frac{\partial}{\partial r}(Q_r)\,dr + \frac{\partial}{r\partial\theta}(Q_\theta)\, r d\theta + \frac{\partial}{r \sin\theta\, d\phi}(Q_\phi)\, r \sin\theta\, d\phi\right]$$

Using eqns. (i), (ii) and (iii), we have

$$Q_{net} = -\left[\frac{\partial}{\partial r}\left\{-k\,(rd\theta \times r \sin\theta\, d\phi)\frac{\partial T}{\partial r}\right\}dr + \frac{\partial}{r\partial\theta}\left\{-k\,(dr \times r\sin\theta\, d\phi)\frac{\partial T}{r\partial\theta}\right\} rd\theta + \frac{\partial}{r\sin\theta\, d\phi}\left\{-k\,(dr \times rd\theta)\frac{\partial T}{r\sin\theta\, d\phi}\right\} r\sin\theta\, d\phi\right]$$

$$= k\left[\frac{\partial}{\partial r}\left(r^2 \frac{\partial T}{\partial r}\right) dr \sin\theta\, d\theta\, d\phi + \frac{\partial}{r\partial\theta}\left(\sin\theta \frac{\partial T}{r\partial\theta}\right) r\, dr\, d\phi\, rd\theta + \frac{\partial}{r\sin\theta\,\partial\phi}\left(\frac{\partial T}{r\sin\theta\,\partial\phi}\right) dr\, rd\theta\, r\sin\theta\, d\phi\right.$$

or

$$Q_{net} = k\left[\frac{1}{r^2}\frac{\partial}{\partial r}\left(r^2\frac{\partial T}{\partial r}\right) + \frac{1}{r^2 \sin\theta}\frac{\partial}{\partial\theta}\left(\sin\theta\frac{\partial T}{\partial\theta}\right) + \frac{1}{r^2\sin^2\theta}\frac{\partial}{\partial\phi}\left(\frac{\partial T}{\partial\phi}\right)\right] r^2 \sin\theta\, d\theta\, d\phi\, dr \qquad ...(2.34)$$

If heat is generated within the element at the rate of g (W/m^3), then the volumetric heat generation rate :

$$Q_{gen} = g\, dr\, rd\theta\, r\sin\theta\, d\phi$$
$$= g\, r^2 \sin\theta\, d\theta\, d\phi\, dr \qquad ...(2.35)$$

Due to these heat transfer rates into the element, the internal energy of the element may change. The rate of change of internal energy of the element is :

$$\frac{\Delta E}{\partial t} = mC\frac{\partial T}{\partial r} = \rho\,(dr\, rd\theta\, r\sin\theta\, d\phi)\, C\frac{\partial T}{\partial t}$$

$$= \rho C\,(r^2 \sin\theta\, d\theta\, d\phi\, dr)\frac{\partial T}{\partial t} \qquad ...(2.36)$$

Making the energy balance on the element :

Net rate of heat gain by conduction + Rate of energy generation

= Rate of change of internal energy.

Using the quantities from eqns. (2.34), (2.35) and (2.36), respectively :

$$k\left[\frac{1}{r^2}\frac{\partial}{\partial r}\left(r^2\frac{\partial T}{\partial r}\right) + \frac{1}{r^2\sin\theta}\frac{\partial}{\partial\theta}\left(\sin\theta\frac{\partial T}{\partial\theta}\right) + \frac{1}{r^2\sin^2\theta}\frac{\partial}{\partial\phi}\left(\frac{\partial T}{\partial\phi}\right)\right] \times r^2\sin\theta\, d\theta\, d\phi\, dr + g\, r^2\sin\theta\, d\theta\, d\phi\, dr$$

$$= \rho C\,(r^2\sin\theta\, d\theta\, d\phi\, dr)\frac{\partial T}{\partial t}$$

or

$$\frac{1}{r^2}\frac{\partial}{\partial r}\left(r^2\frac{\partial T}{\partial r}\right) + \frac{1}{r^2\sin\theta}\frac{\partial}{\partial\theta}\left(\sin\theta\frac{\partial T}{\partial\theta}\right) + \frac{1}{r^2\sin^2\theta}\frac{\partial^2 T}{\partial\phi^2} + \frac{g}{k}$$

$$= \frac{\rho C}{k}\frac{\partial T}{\partial t} = \frac{1}{\alpha}\frac{\partial T}{\partial t} \qquad ...(2.37)$$

It is a the general heat conduction equation in spherical coordinates.

In absence of any heat generation, the steady state one dimensional heat conduction equation in r direction, the eqn. (2.37) reduces to :

$$\frac{1}{r^2}\frac{\partial}{\partial r}\left(r^2\frac{\partial T}{\partial r}\right) = 0$$

It is a unidirectional *Laplace equation*, derived earlier by eqn. (2.20).

Note: The eqn. (2.37), the general heat conduction equation in spherical coordinates can also be transformed from Cartesian coordinates by using

$$x = r\sin\theta\cos\phi$$
$$y = r\sin\theta\sin\phi$$
$$z = r\cos\theta.$$

2.3. INITIAL AND BOUNDARY CONDITIONS

To determine temperature distribution in a medium, it is necessary to solve the general heat conduction equation. However, such solution depends on physical conditions existing at the boundaries of the medium and if the situation is time dependent (unsteady), some initial conditions are needed.

The mathematical expressions of thermal conditions at the boundaries of an object are called **boundary conditions**. The boundary conditions are several common physical effects, which are simply expressed in mathematical form.

The temperature at any point on the medium at a specified time also depends on the condition of the medium. The ***initial condition*** at the beginning of the heat conduction process is a mathematical expression for the temperatue distribution of the medium initially *i.e.,* $t = 0$.

Since the general heat conduction equation is second order differential equation in spatial coordinates, in any direction at least two thermal conditions are needed at the boundary surfaces. Because the equation is first order in time, only one initial condition must be specified.

Following boundary conditions commonly appeared in heat transfer are discussed below.

2.3.1. Prescribed Temperature Boundary Conditions

For a plane wall of thickness L, whose left face ($x = 0$) is maintained at uniform temperature of T_1 and right face at $x = L$ at uniform temperature of T_2 as shown in Fig. 2.5. Then the boundary conditions at two faces are written as :

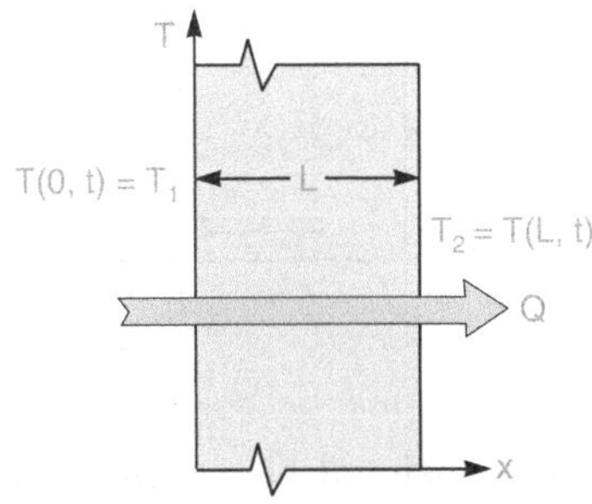

Fig. 2.5. Prescribed temperature boundary conditions

At $x = 0, \quad t = 0 \quad T(x, t) = T_1$ ...[2.38(*a*)]

and $x = L, \quad t = 0 \quad T(x, t) = T_2$ ...[2.38(*b*)]

2.3.2. Prescribed Heat Flux Boundary Conditions

Sometimes, the rate of heat transfer to a boundary is constant. For example, an electrically heated surface, the rate of heat supply (capacity of heater) is constant. Such conditions are called prescribed heat flux boundary condition as illustrated in Fig. 2.6.

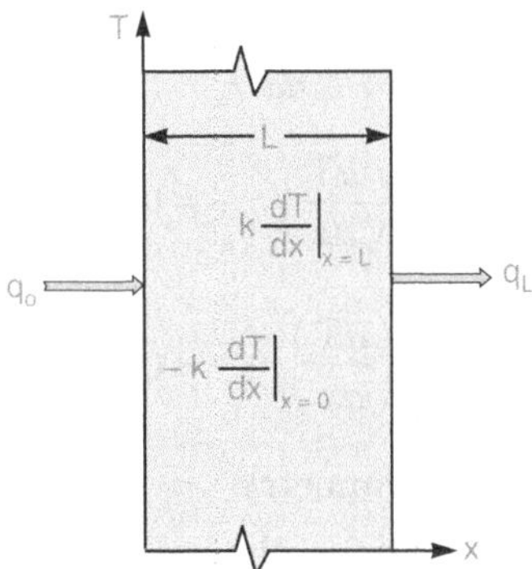

Fig. 2.6. Prescribed heat flux boundary conditions

Heat flux is given by :

$$q_x = -k\left(\frac{dT}{dx}\right)_x$$

Suppose, at $x = 0$, $q_x = q_o$ = (the left face)

and at $x = L$, $q_x = q_L$ (right face)

where,
$$q_o = -k\left\{\frac{dT}{dx}\right\}_{x=0} \quad \text{...[2.39(a)]}$$

and
$$q_L = -k\left\{\frac{dT}{dx}\right\}_{x=L} \quad \text{...[2.39(b)]}$$

If the direction of heat flux at the right face is opposites *i.e.,* towards face, then q_L should be considered negative.

There are two special cases of prescribed heat flux boundary condition.

(*i*) Insulated boundary

In some engineering applications, the system boundary is insulated in order to minimise the heat loss from the system. Although this insulation is not perfect, but in thermal analysis, the heat loss is assumed negligible from the boundary with thermal insulation.

i.e.,
$$q_x = 0 = -k\left(\frac{dT}{dx}\right)_x \quad \text{...(2.40)}$$

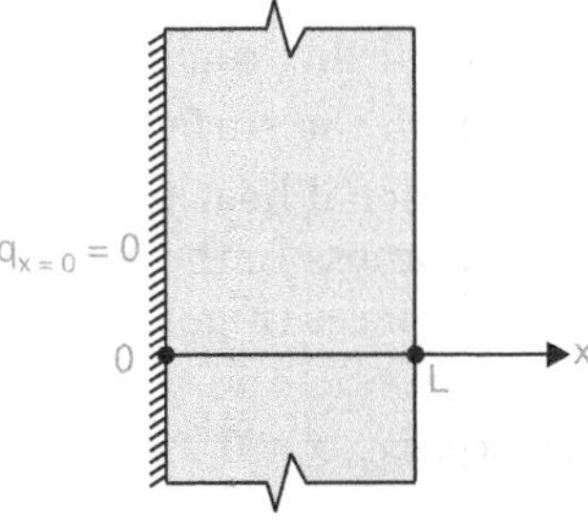

Fig. 2.7. Plan wall with left face insulated

or $$\left(\frac{dT}{dx}\right)_x = 0$$

For left face $$\left(\frac{dT}{dx}\right)_{x=0} = 0$$

For right face $$\left(\frac{dT}{dx}\right)_{x=L} = 0$$

(*ii*) Thermal symmetry

In many situations, the boundary conditions imposed on two sides of plane wall, or solid cylinder or solid sphere are identical, then heat flow from the centre to two sides is also identical and centre of the plane is treated as plane of symmetry. This plane is equivalent to insulated boundary.

Centre plane of solid cylinder or sphere as shown in Fig. 2.8

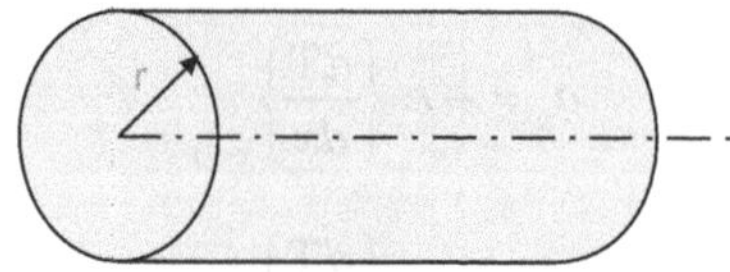

Fig. 2.8. A solid cylinder exposed to convection environment at its outer surface

$$\left(\frac{dT}{dr}\right)_{r=0} = 0 \qquad ...[2.41(a)]$$

So for centre plane of plane wall of thickness L ;

$$\left(\frac{dT}{dr}\right)_{x=L/2} = 0 \qquad ...[2.41(b)]$$

2.3.3. Convection Boundary Conditions : Surface Energy Balance

In most practical applications, the heat dissipates by convection with a known value of heat transfer coefficient h at one or both boundary surfaces. The energy balance at any boundary surface can be written as :

Convection flux from the fluid to the surface

= Heat flux conducted into the body from the surface

For one dimensional heat transfer in x direction of a plane wall of thickness L, the convection boundary conditions on both surfaces (Fig. 2.9) can be expressed as :

At left surface, *i.e.*, $x = 0$

$$h_1(T_{\infty 1} - T_{x=0}) = -k\left\{\frac{dT}{dx}\right\}_{x=0} \qquad ...(2.42)$$

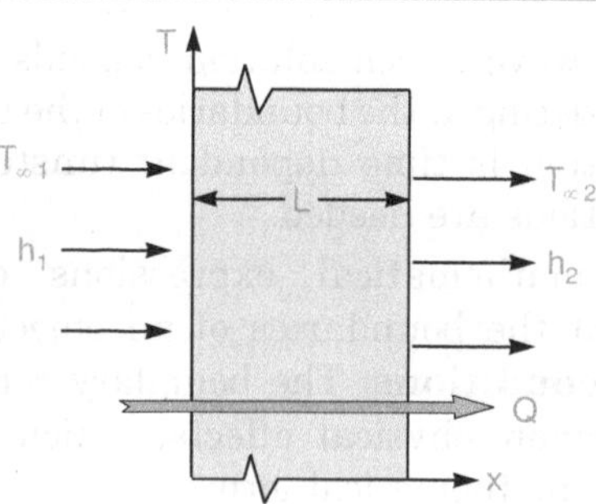

Fig. 2.9. Plane wall exposed to convection boundaries at both sides

At the right surface, *i.e.*, $x = L$

$$-k\left\{\frac{dT}{dx}\right\}_{x=L} = h_2(T_{x=L} - T_{\infty 2}) \qquad ...(2.43)$$

These are the convection boundary conditions at the faces of the plane wall. *Similarly the boundary conditions can be written for cylinders and spheres.*

2.3.4. Radiation Boundary Condition

In some practical cases, for example, in space and cryogenic applications, the outer surface is surrounded by evacuated space in order to minimize conduction and convection heat transfer. In such cases, only radiation heat transfer can take place from surface and surroundings and boundary conditions are specified as :

Heat conduction to surface = radiation heat transfer from the surface to surrounding

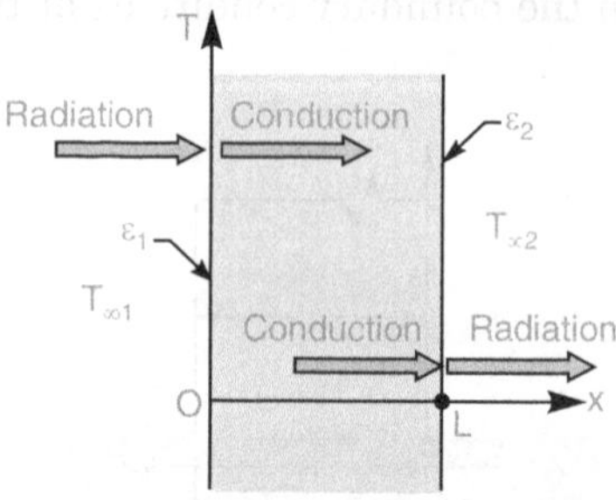

Fig. 2.10. Radiation boundary conditions on both surfaces

For one dimensional heat conduction in a plane wall of thickness L and thermal conductivity k, the radiation boundary conditions on both surfaces can be expressed as shown in Fig. 2.10.

At left surface

$$-k\left(\frac{dT}{dx}\right)_{x=0} = \sigma\varepsilon_1(T^4_{\infty 1} - T^4_{x=0}) \qquad ...[2.44(a)]$$

At right surface

$$-k\left(\frac{dT}{dx}\right)_{x=L} = \sigma\varepsilon_2(T^4_{x=L} - T^4_{\infty 2}) \qquad ...[2.44(b)]$$

Where ε_1 and ε_2 are the emissivities of left and right boundary surfaces, respectively. $\sigma = 5.67 \times 10^{-8}$ W/m^2.K^4 is the Stefan Bolzmann constant.

Note: The temperature in radiation calculations must be used in kelvin (K) (not in °C).

2.3.5. Interface Boundary Condition

When one or more layers in perfect contact made a composite wall, then both body will have same temperature at interface, because flow rate will be same through both layers. (Fig. 2.11)

$$-k_A\left(\frac{dT}{dx}\right)_A = -k_B\left(\frac{dT}{dx}\right)_B$$

and $$T_A(x) = T_B(x) \quad ...(2.45)$$

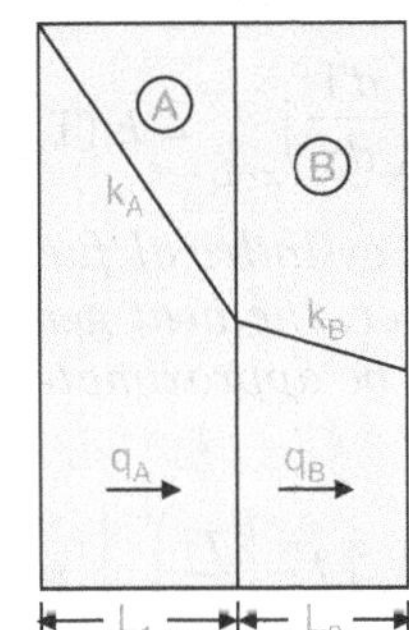

Fig. 2.11. Boundary condition at interface of two layers

These are some boundary conditions at the faces of the plane wall, these boundary conditions can also be written on surfaces of cylinders and spheres.

The boundary conditions explained above do not cover all possible boundary conditions, that may be imposed on the surfaces. However, in other situations, the boundary conditions can be designed by applying the energy balance at the surface that is :

Rate of heat entering in = Rate of heat going out.

Example 2.1. *The temperatures on two sides of a 25 mm thick steel plate with constant thermal conductivity having uniform heat generation are at 180°C and 120°C. Develop a mathematical formulation of one dimensional steady state heat conduction in the plate.*

Solution

Given : A steel plate with constant thermal conductivity and uniform heat generation

$$L = 25 \text{ mm} = 0.025 \text{ m}$$

$$T_1 = 180°C, \quad T_2 = 120°C$$

Mathematical Formulation :

Recognition of Problem :

1. Constant thermal conductivity.
2. Specified temperatures at two faces of plate.
3. Uniform heat generation rate in the plate.

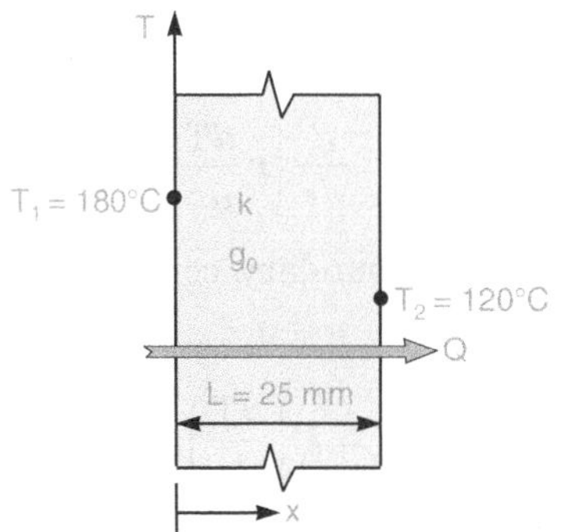

Fig. 2.12. Schematic for example 2.1

The governing one dimensional steady state heat conduction equation with heat generation in cartesian coordinates

$$\frac{d}{dx}\left(\frac{dT}{dx}\right) + \frac{g(x)}{k} = 0$$

Assuming uniform heat generation at g_0 W/m^3, then

$$\frac{d}{dx}\left(\frac{dT}{dx}\right) = -\frac{g_0}{k}$$

Subjected to boundary conditions as shown in Fig. 2.12.

At left face *i.e.,* $x = 0$, $T = T_1 = 180°C$

At right face *i.e.,* $x = L = 0.025$ m, $T = T_2 = 120°C$

where $T = f(x)$ only.

Example 2.2. *Develop the mathematical formulation of one dimensional steady state heat conduction for hollow cylinder with constant thermal conductivity k. The heat is supplied into the cylinder at inner surface at $r = r_1$ at a rate of q W/m^2 and heat is dissipated by convection from the surface at $r = r_2$ into an ambient at temperature T_∞ with heat transfer coefficient h.*

Solution

Given :

1. Steady state heat conduction in radial direction.
2. Constant properties.
3. No energy generation.

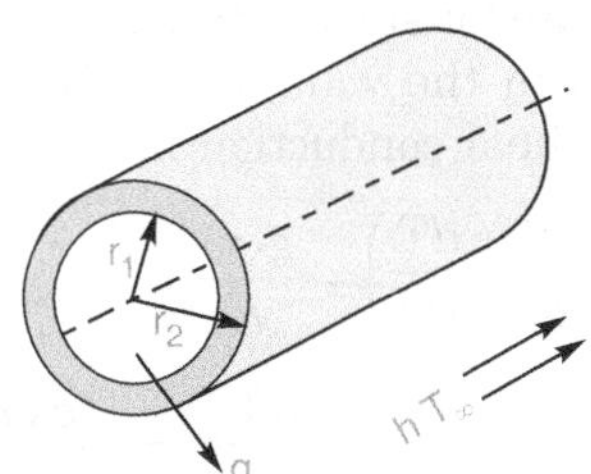

Fig. 2.13. Schematic for example 2.2

Mathematical Formulation :

The governing heat conduction equation in steady state without heat generation for a cylinder

$$\frac{d}{dr}\left\{r\frac{dT}{dr}\right\} = 0$$

Subjected to boundary conditions as shown in Fig. 2.13

$$\text{At } r = r_1, \quad -k\left(\frac{dT}{dr}\right)_{r=r_1} = q$$

and

$$\text{At } r = r_2, \quad -k\left(\frac{dT}{dr}\right)_{r=r_2} = h(T_{r=r_2} - T_\infty)$$

where $T = F(r)$.

Example 2.3. *The volumetric heat generation in a plane wall is given by*

$$g(x) = g_0\, e^{-\beta x}\ (W/m^3)$$

where g_0 and β are constants. The left face of the wall is insulated, while right face dissipates heat by convection into an ambient air at T_∞. Formulate the problem mathematically.

Solution.

Given : For a plane wall with volumetric heat generation as,

$$g(x) = g_0\, e^{-\beta x}$$

1. Boundary conditions.
2. Constant thermal conductivity of wall.

Mathematical Formulation :

Recognition of Problem :

1. Heat is generation as a function of x in the wall.
2. No time dependent quantity is given.
3. Boundary conditions at two faces.
4. No information regarding status of thermal conductivity thus assuming it as constant. These conditions indicate for steady state heat conduction with heat generation in the wall. The differential equation for steady state heat conduction in x direction is :

$$\frac{d}{dx}\left(\frac{dT}{dx}\right) + \frac{g(x)}{k} = 0$$

or

$$\frac{d}{dx}\left(\frac{dT}{dx}\right) = -\frac{g_0\, e^{-\beta x}}{k}$$

Applying boundary conditions.

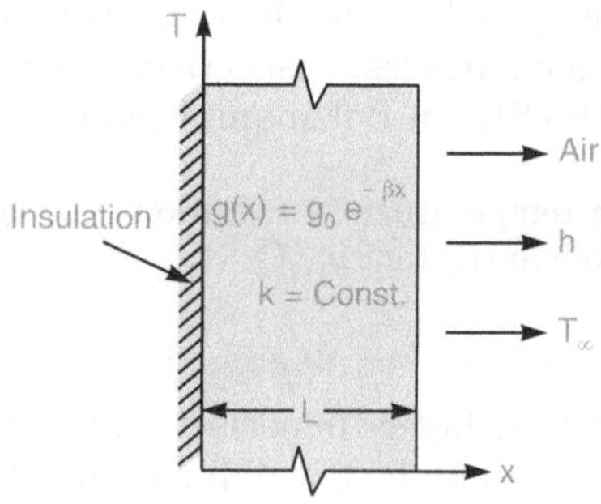

Fig. 2.14. Schematic for example 2.3

(*i*) For insulated surface, at $x = 0$

$$\left(\frac{dT}{dx}\right)_{x=0} = 0$$

(*ii*) For convection heat transfer from right face

$$-k\left(\frac{dT}{dx}\right)_{x=L} = h\,(T_{x=L} - T_\infty).$$

Example 2.4. *In a cylindrical fuel element for a gas cooled nuclear reactor, the heat generation rate within the fuel element can be approximated as :*

$$g(r) = g_0\left[1 - \left(\frac{r}{r_o}\right)^2\right]\ W/m^3$$

where r_o is outer radius of fuel element and g_0 is a constant. The outer surface is maintained at a uniform temperature T_o.

Develop a mathematical formulation assuming one dimensional heat flow.

Solution

Given :

(*i*) A cylindrical fuel element with heat generation

$$g(r) = g_0\left[1 - \left(\frac{r}{r_o}\right)^2\right]\ \text{W/m}^3$$

(*ii*) Outer surface at uniform temperature T_o.

Mathematical Formulation :

Recognition of Problem :

(*i*) Heat is generated in the fuel element.

(*ii*) No time dependent quantity is given.

(*iii*) The outer surface of fuel element is maintained at uniform temperature T_o.

(*iv*) Heat conduction in one dimension.

These conditions indicate for one dimensional steady state heat conduction with heat generation.

Its governing equation in radial direction is given by eqn. (2.15)

$$\frac{1}{r}\frac{d}{dr}\left(r\frac{dT}{dr}\right)+\frac{g(r)}{k}=0$$

or

$$\frac{d}{dr}\left(r\frac{dT}{dr}\right)=-\frac{g_0}{k}\,r\left[1-\left(\frac{r}{r_o}\right)^2\right]$$

Subjected to boundary conditions

(*i*) At $r = r_o$, $T = T_o$

(*ii*) For solid rod in steady state, the temperature gradient at centre is always zero due to symmetry

i.e., at $r = 0, \frac{dT}{dr} = 0$

where $T = f(r)$.

Example 2.5. *The temperature distribution across a wall, 1 m thick at a certain instant of time is given as :*

$$T(x) = 900 - 300\,x - 50\,x^2$$

where T is in degree Celsius and x in metres.

The uniform heat generation of 1000 W/m³ is present in wall of area 10 m² having the properties $\rho = 1600$ *kg/m³*, $k = 40$ *W/m.K and C = 4 kJ/kg.K*

(a) Determine the rate of heat transfer entering the wall (x = 0) and leaving the wall (x = 1 m).

(b) Determine the rate of change of internal energy of the wall.

(c) Determine the time rate of temperature change at x = 0, 0.5 m.

Solution

Given : Temperature distribution across a wall $T(x) = 900 - 300\,x - 50\,x^2$

$g_0 = 1000$ W/m³, A = 10 m²,

L = 1 m,

$\rho = 1600$ kg/m³, $k = 40$ W/m.K,

C = 4 kJ/kg = 4000 J/kg K

To find :

(*a*) (*i*) The rate of heat transfer at left face ($x = 0$)

(*ii*) The rate of heat transfer at right face ($x = L$).

(*b*) The rate of change of internal energy.

(*c*) The time rate of temperature change at $x = 0$ and 0.5 m.

Assumptions :

1. One dimensional conduction in x direction.
2. Medium with constant properties.
3. Uniform internal heat generation at the rate of g_0 W/m³.

Analysis :

(*a*) Since the temperature distribution is given as :

$$T(x) = 900 - 300x - 50x^2$$

and temperature gradient

$$\frac{dT}{dx} = -300 - 100x \text{ (°C/m or K/m)}$$

(*i*) Using boundary condition of prescribed heat flux entering the left face of the wall :

$$q_{x=0} = -k\left\{\frac{dT}{dx}\right\}_{x=0}$$

$$= -k\,(-300 - 100\,x)_{x=0}$$

or

$$q_{x=0} = -(40 \text{ W/m.K}) \times (-300 \text{ K/m})$$

$$= 12{,}000 \text{ W/m}^2$$

The heat entering the left face

$= A\,q_{x=0} = 10 \times 12{,}000$

$= 1{,}20{,}000$ W = **120 kW. Ans.**

(*ii*) Similarly using temperature gradient, the heat flux at the right face :

$$q_{x=L} = -k\left\{\frac{dT}{dx}\right\}_{x=L}$$

$$= -k\,(-300 - 100\,x)_{x=L}$$

or

$$q_{x=L} = -40 \times (-300 - 100 \times 1)$$

$$= 16{,}000 \text{ W/m}^2$$

The heat leaving the right face

$= A\,q_{x=L} = 1{,}60{,}000$ W = **160 kW. Ans.**

(*b*) The rate of change of internal energy

= Rate of heat entering the left face

+ Rate of heat generation

– Rate of heat leaving right face

$= Q_{x=0} + g_0\,A\,L - Q_{x=L}$

= 120 kW + 1(kW/m³) × 10(m²)

× 1(m) – 160 kW = **– 30 kW. Ans.**

(*c*) The rate of change of temperature in the wall can be calculated by using eqn. (2.8)

$$\frac{\partial}{\partial x}\left\{\frac{\partial T}{\partial x}\right\}+\frac{g_0}{k}=\frac{1}{\alpha}\frac{\partial T}{\partial t}$$

or

$$\frac{dT}{dt}=\alpha\left[\frac{d}{dx}\left\{\frac{dT}{dx}\right\}+\frac{g_0}{k}\right]$$

where, $\alpha = \frac{k}{\rho C} = \frac{40 \text{ W/mK}}{1600(\text{kg/m}^3) \times 4000 \text{ (J/kg.K)}}$

$= 6.25 \times 10^{-6} \text{ m}^2\text{/s}$

and $\frac{d}{dx}\{-300 - 100x\} + \frac{g_0}{k} = (-100) + \frac{1000}{40}$

$= -75°\text{C/m}^2$

Hence,

$$\frac{dT}{dt} = 6.25 \times 10^{-6} \times (-75)$$

$$= \mathbf{-4.6875 \times 10^{-4}°C/s.} \quad \textbf{Ans.}$$

The change of temperature is independent of position. **Ans.**

Example 2.6. *At a certain time, the temperature distribution in a long cylindrical tube with an inner radius of 250 mm and outside radius of 400 mm is given by*

$$T(r) = 750 + 1000\, r - 5000\, r^2 \ (°C)$$

where r in metres. Thermal conductivity and thermal diffusivity of the tube material are 58 W/m.K and 0.004 m²/h, respectively. Calculate :

(i) Rate of heat flow at inside and outside surfaces per unit length,

(ii) Rate of heat storage per unit length, and

(iii) Rate of change of temperature at inner and outer surfaces.

Solution

Given : Temperature distribution in hollow cylinder :

$T(r) = 750 + 1000\, r - 5000\, r^2$ (°C)

$k = 58$ W/m.K,

$\alpha = 0.004$ m²/h

$r_1 = 250$ mm $= 0.25$ m,

$r_2 = 400$ mm $= 0.4$ m

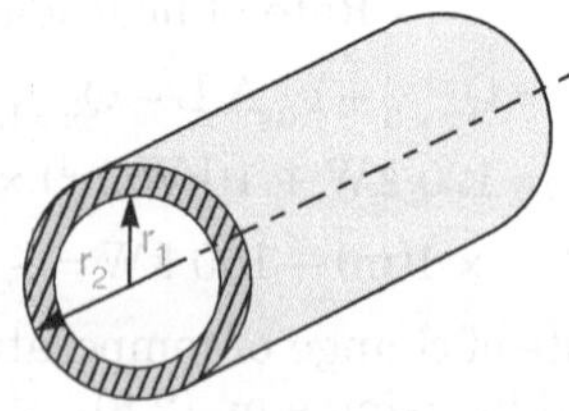

Fig. 2.15. Schematic of cylindrical tube

To find :

(*i*) Rate of heat flow per metre length at $r = r_1$ and $r = r_2$.

(*ii*) Rate of heat storage per unit length,

(*iii*) Rate of change of temperature at $r = r_1$ and $r = r_2$.

Assumptions :

(*i*) No heat generation within the element.

(*ii*) Heat flow in radial direction only.

(*iii*) Constant properties.

Analysis :

(*i*) For given temperature distribution in cylinder, the temperature gradient at any radius r :

$$\frac{dT}{dr} = 1000 - 10{,}000\, r$$

Rate of heat transfer at inside surface ($r = r_1$)

$$Q_{r=r_1} = -kA\left(\frac{dT}{dr}\right)_{r=r_1}$$

$$= -k\, 2\pi r_1 L\, [1000 - 10{,}000\, r]_{r=r_1}$$

or $$\left(\frac{Q}{L}\right)_{r=r_1} = -58 \times 2\pi \times 0.25 \times [1000 - 10000 \times 0.25]$$

$$= \mathbf{13.66 \times 10^4\ W/m.} \quad \textbf{Ans.}$$

(in radial outward direction)

Rate of heat flow at outer surface ($r = r_2$) :

$$Q_{r=r_2} = -kA\left(\frac{dT}{dr}\right)_{r=r_2}$$

$$= -k\, 2\pi r_2 L\, [1000 - 10{,}000 \times r]_{r=r_2}$$

or $$\left(\frac{Q}{L}\right)_{r=r_2} = -58 \times 2\pi \times 0.4 \times [1000 - 10000 \times 0.4]$$

$$= \mathbf{4.37 \times 10^5\ W/m.} \quad \textbf{Ans.}$$

(in radial outward direction)

(*ii*) Rate of heat storage per unit length

$$= \left(\frac{Q}{L}\right)_{r=r_1} - \left(\frac{Q}{L}\right)_{r=r_2}$$

$$= 13.66 \times 10^4 - 4.37 \times 10^5$$

$$= \mathbf{-3.0 \times 10^5\ W/m.} \quad \textbf{Ans.}$$

(It is decrease rate of internal energy)

(*iii*) Rate of change of temperature at inner and outer surfaces

One dimensional Fourier equation in radial coordinate

$$\frac{1}{r}\frac{d}{dr}\left(r\frac{dT}{dr}\right) = \frac{1}{\alpha}\frac{dT}{dt}$$

or

$$\frac{dT}{dt} = \frac{\alpha}{r}\frac{d}{dr}\left(r\frac{dT}{dr}\right)$$

For given temperature distribution

$$\frac{d}{dr}\left(r\frac{dT}{dr}\right) = \frac{d}{dr}\,[1000\,r - 10{,}000\,r^2]$$

$$= 1000 - 20{,}000\,r$$

and at inner surface

$$\left(\frac{dT}{dt}\right)_{r=r_1} = \frac{\alpha}{r_1}\,[1000 - 20{,}000\,r_1]$$

$$= \frac{0.004}{0.25}\,[1000 - 20{,}000 \times 0.25]$$

$$= \mathbf{-\ 64°C/h.} \text{ (decrease) Ans.}$$

At outer surface

$$\left(\frac{dT}{dt}\right)_{r=r_2} = \frac{0.004}{0.4}\,[1000 - 20{,}000 \times 0.4]$$

$$= \mathbf{-\ 70°C/h.} \text{ (decrease) Ans.}$$

Example 2.7. *The temperature distribution in a plate of thickness 20 mm is given by T(°C) = 6x² + 10x + 4. Assume no heat generation in the plate, calculate heat flux on two sides of the plate. Also calculate rate of temperature change with respect to time, if k = 300 W/m.K, ρ = 580 kg/m³ and C = 420 J/kg.K.*

Solution

Given : Temperature distribution in the plate as :

$$T(x) = 6x^2 + 10x + 4 \text{ (°C)}$$

$$L = 20 \text{ mm} = 0.02 \text{ m}$$

$$k = 300 \text{ W/m.K}$$

$$\rho = 580 \text{ kg/m}^3$$

$$C = 420 \text{ J/kg.K.}$$

To find : (*i*) Heat flux on two sides of the plate

(*ii*) $\frac{dT}{dt}$, rate of temperature change with time.

Assumption : No heat generation in the plate.

Analysis : The one dimensional governing heat conduction equation without heat generation in cartesian coordinate

$$\frac{\partial}{\partial x}\left(\frac{\partial T}{\partial x}\right) = \frac{1}{\alpha}\frac{\partial T}{\partial t} = \frac{\rho C}{k}\frac{\partial T}{\partial t}$$

The temperature gradient from temperature distribution

$$\frac{\partial T}{\partial x} = \frac{dT}{dx} = 12x + 10$$

$$\frac{\partial}{\partial x}\left(\frac{\partial T}{\partial x}\right) = \frac{\partial^2 T}{\partial x^2} = \frac{d^2T}{dx^2} = 12$$

Using above equation with the numerical values

$$12 = \frac{580 \times 420}{300} \times \frac{\partial T}{\partial t}$$

or

$$\frac{\partial T}{\partial t} = \frac{dT}{dt} = \mathbf{0.147\ °C/s.} \text{ Ans.}$$

The heat flux is given by

$$q_x = -k\,\frac{dT(x)}{dx}$$

Using $\frac{dT}{dx}$, we get

$$q_x = -k\,(12x + 10)$$

At left face, $x = 0$

$$q_{x=0} = -300 \times (12 \times 0 + 10)$$

$$= \mathbf{-\ 3000\ W/m^2.} \text{ Ans.}$$

At the right face,

$$x = L = 0.02 \text{ m}$$

$$q_{x=L} = -300 \times (12 \times 0.02 + 10)$$

$$= \mathbf{-\ 3072\ W/m^2.} \text{ Ans.}$$

Example 2.8. *A cylindrical nuclear fuel rod of 50 mm diameter has uniform heat generation of 5 × 10⁷ W/m³. Under steady state conditions, the temperature distribution in the rod is given by*

$$T(r) = 800 - 4.2 \times 10^5\,r^2,$$

where T in deg. celsius and r in metres. The fuel rod properties are :

$$k = 30 \text{ W/m.K}, \; \rho = 1100 \text{ kg/m}^3$$

and

$$C = 800 \text{ J/kg . K}$$

(a) What is the rate of heat transfer per unit length of rod at its centre and outer surface?

(b) If reactor power is suddenly increased to 2 × 10⁸ W/m³, what is the initial time rate of temperature change at its centre and its outer surface?

Solution

Given : A cylindrical nuclear fuel rod with uniform heat generation.

$$g_0 = 5 \times 10^7 \text{ W/m}^3,$$

$$r_o = 25 \text{ mm} = 0.025 \text{ m}$$

The temperature distribution in the rod

$$T(r) = 800 - 4.2 \times 10^5\, r^2$$

and properties

$$k = 30 \text{ W/m.K},$$

$$\rho = 1100 \text{ kg/m}^3,$$

$$C = 800 \text{ J/kg.K}$$

To find :

(*i*) Rate of heat transfer per unit length of rod at its centre and outer surface.

(*ii*) Initial rate of temperature change at centre and outer surface of the rod, when reactor power is suddenly raised to 2×10^8 W/m^3.

Assumptions :

1. Heat generation rate is uniform throughout the nuclear rod.

2. Constant properties.

Analysis : The temperature distribution in the nuclear fuel rod is given by

$$T(r) = 800 - 4.2 \times 10^5\, r^2$$

Its first order derivative with respect to r is :

$$\frac{dT}{dr} = -8.4 \times 10^5\, r \qquad \ldots(i)$$

and second order derivative w.r.t. r is :

$$\frac{d^2T}{dr^2} = -8.4 \times 10^5 \qquad \ldots(ii)$$

(*i*) The heat transfer rate per unit length in the rod is :

$$\frac{Q}{L} = -k\, 2\pi \left(r \frac{dT}{dr} \right)_r$$

At centre of the rod ($r = 0$)

$$\left(\frac{Q}{L} \right)_{r=0} = -30 \times 2\pi \times [0 \times (-8.4 \times 10^5 \times 0)]$$

$$= \mathbf{0.} \quad \text{Ans.}$$

At outer surface of the rod ($r_o = 0.025$ m)

$$\left(\frac{Q}{L} \right)_{r=r_o} = -30 \times 2\pi \times [0.025 \times (-8.4 \times 10^5 \times 0.025)]$$

$$= \mathbf{98960.2 \text{ W/m.}} \quad \text{Ans.}$$

(*ii*) For initial rate of cooling, using eqn (2.9) with uniform volumetric heat generation g_0,

$$\frac{1}{r} \frac{\partial}{\partial r} \left(r \frac{\partial T}{\partial r} \right) + \frac{g_0}{k} = \frac{1}{\alpha} \frac{\partial T}{\partial t} = \frac{\rho C}{k} \frac{\partial T}{\partial t}$$

or

$$\frac{\partial T}{\partial t} = \frac{k}{\rho C} \left[\frac{1}{r} \frac{\partial}{\partial r} \left(r \frac{\partial T}{\partial r} \right) + \frac{g_0}{k} \right]$$

$$= \frac{k}{\rho C} \left[\frac{\partial^2 T}{\partial r^2} + \frac{1}{r} \frac{\partial T}{\partial r} + \frac{g_0}{k} \right]$$

Using eqn. (*i*) and (*ii*) then

$$\frac{\partial T}{\partial t} = \frac{dT}{dt} = \frac{30}{1100 \times 800} \times \left[-8.4 \times 10^5 + \frac{1}{r}(-8.4 \times 10^5\, r) + \frac{2 \times 10^8}{30} \right]$$

At centre ($r = 0$)

$$\left(\frac{dT}{dt} \right)_{t=0} = 3.409 \times 10^{-5} \times \left[-8.4 \times 10^5 + \frac{2 \times 10^8}{30} \right] = \mathbf{198.63°C/s.} \quad \textbf{Ans.}$$

At outer surface ($r_o = 0.025$ m)

$$\left(\frac{dT}{dt} \right)_{t=0} = 3.409 \times 10^{-5} \times \left[-8.4 \times 10^5 - 8.4 \times 10^{-5} + \frac{2 \times 10^8}{30} \right]$$

$$= \mathbf{170°C/s.} \quad \text{Ans.}$$

Example 2.9. *A long conducting rod of diameter D and electrical resistance per unit length R_e, is initially in thermal equilibrium with the ambient air and its surroundings. The equilibrium is disturbed, when an electric current I is passed through the rod. Develop an expression that could be used to compute the variation of rod temperature during passage of electric current. Consider all possible types of heat transfer.*

(N.M.U., May 2000)

Solution

Considering a rod exposed to convection and radiation environment. The energy transfers are :

Q_g = energy generation rate
= (current)^2 × (resistance per metre) × length of the conductor
= $I^2R_e.L$,

$\frac{\Delta E}{dt}$ = rate of change of internal energy in the rod
= mass × specific heat × rate of temperature change with time

$$= \rho VC \frac{dT}{dt} = \rho C \left(\frac{\pi D^2 L}{4}\right) \frac{dT}{dt}$$

Q_{out} = energy discharge rate by convection and radiation :

$$= h(\pi DL)(T - T_\infty) + \varepsilon\sigma(\pi DL)(T^4 - T_\infty^4)$$

At anytime, the energy balance on the control volume

$$Q_g - Q_{out} = \frac{\Delta E}{dt}$$

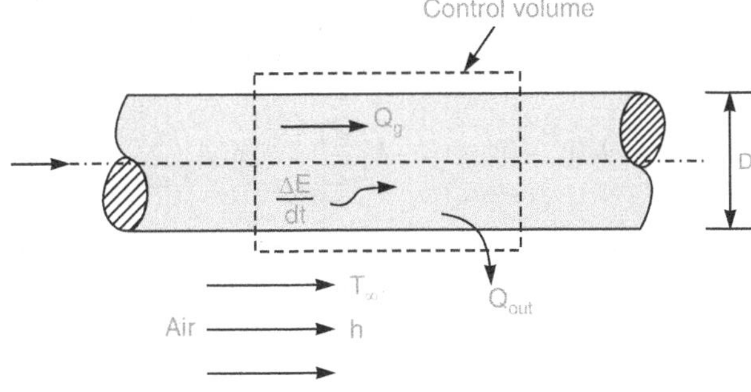

Fig. 2.16. Schematic for example 2.9

or $I^2R_eL - h(\pi DL)(T - T_\infty) - \varepsilon\sigma(\pi DL)(T^4 - T_\infty^4)$

$$= \rho C\left(\frac{\pi}{4} D^2 L\right)\frac{dT}{dt}$$

or $I^2R_e - \pi Dh(T - T_\infty) - \varepsilon\sigma\pi D(T^4 - T_\infty^4)$

$$= \rho C\left(\frac{\pi}{4} D^2\right)\frac{dT}{dt}$$

or $\frac{dT}{dt}$

$$= \frac{4\{I^2R_e - \pi Dh(T - T_\infty) - \varepsilon\sigma\pi D(T^4 - T_\infty^4)\}}{\rho C(\pi D^2)}$$ **Ans.**

2.4. SUMMARY

1. The generalised one dimensional heat conduction equation for isotropic material can be expressed as :

$$\frac{1}{A}\frac{\partial}{\partial X}\left\{A\frac{\partial T}{\partial X}\right\} + \frac{g}{k} = \frac{\rho C}{k}\frac{\partial T}{\partial t}$$

2. The generalised one dimensional heat conduction equation in cartesian coordinate system :

$$\frac{\partial}{\partial x}\left\{\frac{\partial T}{\partial x}\right\} + \frac{g}{k} = \frac{1}{\alpha}\frac{\partial T}{\partial t}$$

3. The generalised one dimensional heat conduction equation in cylindrical coordinate system :

$$\frac{1}{r}\frac{\partial}{\partial r}\left\{r\frac{\partial T}{\partial r}\right\} + \frac{g}{k} = \frac{1}{\alpha}\frac{\partial T}{\partial t}$$

4. The generalised one dimensional heat conduction equation in spherical coordinate system :

$$\frac{1}{r^2}\frac{\partial}{\partial r}\left\{r^2\frac{\partial T}{\partial r}\right\} + \frac{g}{k} = \frac{1}{\alpha}\frac{\partial T}{\partial t}$$

5. For steady state conditions :

$$\frac{\partial T}{\partial t} = 0$$

and the generalised differential equation reduces to Poisson equation

$$\frac{1}{A}\frac{\partial}{\partial X}\left\{\delta A\frac{\partial T}{\partial X}\right\} + \frac{g}{k} = 0$$

6. For steady state and without heat generation (one dimensional Laplace equation) :

$$\frac{\partial}{\partial X}\left\{A\frac{\partial T}{\partial X}\right\} = 0.$$

The three dimensional governing heat conduction equation for isotropic materials in cartesian coordinate system is :

$$\frac{\partial^2 T}{\partial x^2} + \frac{\partial^2 T}{\partial y^2} + \frac{\partial^2 T}{\partial z^2} + \frac{g}{k} = \frac{1}{\alpha}\frac{\partial T}{\partial t}$$

Cylindrical coordinate system

$$\frac{\partial T}{\partial r^2} + \frac{1}{r}\frac{\partial T}{\partial r} + \frac{1}{r^2}\frac{\partial T}{\partial \theta^2} + \frac{\partial T}{\partial z^2} + \frac{g}{k} = \frac{1}{\alpha}\frac{\partial T}{\partial t}$$

Spherical coordinate system

$$\frac{1}{r^2}\frac{\partial}{\partial r}\left(r^2\frac{\partial T}{\partial r}\right) + \frac{1}{r^2\sin\theta}\frac{\partial}{\partial\theta}\left\{\sin\theta\frac{\partial T}{\partial\theta}\right\} + \frac{1}{r^2\sin^2\theta}\frac{\partial^2 T}{\partial\phi^2} + \frac{g}{k} = \frac{1}{\alpha}\frac{\partial T}{\partial t}$$

REVIEW QUESTIONS

1. Derive one dimensional time dependent heat conduction equation with internal heat generation and variable thermal conductivity in cartesian coordinate system.
2. Write an energy balance for a differential volume element in r direction, derive one dimensional time dependent heat conduction equation with internal heat generation and constant thermal conductivity.
3. Simplify the three dimensional heat conduction equation in cartesian coordinates to obtain one dimensional steady state heat conduction with heat generation and constant thermal conductivity.
4. Derive an expression for one dimensional time dependent heat conduction with internal heat generation and constant thermal conductivity in cartesian coordinate system. Reduce it as :
   (*i*) Poisson equation,
   (*ii*) Fourier equation,
   (*iii*) Laplace equation.
5. Derive generalized one dimensional heat conduction equation and deduce it for
   (*i*) Cartesian coordinate in x direction,
   (*ii*) Cylindrical coordinate in r variable,
   (*iii*) Spherical coordinate in r variable.
6. A plane wall of thickness L is subjected to a heat flux q_0 at its left surface, while its right surface dissipates heat by convection with a heat transfer coefficient h into an ambient at T_∞. Write the boundary conditions at the two surfaces of the wall.
7. A spherical shell is electrically heated at the rate of q_1 (W/m^2) at its inner surface at radius r_1, and its outer surface dissipates heat by convection with heat transfer coefficient h into an ambient at T_∞. Write boundary conditions at two surfaces of shell.
8. A copper bar of radius r = R, is heated by the passage of an electric current. It dissipates heat by convection from its outer surface with convection coefficient h into an ambient at T_∞. Write boundary condition for its outer surface.
9. A plane wall of thickness L is insulated at its left face, while its right face dissipates heat by convection with convection coefficient h into an ambient at T_∞. Write boundary conditions at two faces of the wall.
10. A long hollow cylinder has its inner radius r_1 and outer radius r_2. It is insulated at its inner surface and its outer surface is maintained at constant temperature T_s. Write boundary conditions.

PROBLEMS

1. The thermal conductivity k, the density ρ, and the specific heat C of steel are 61 W/m.K, 7865 kg/m^3, and 0.46 kJ/kg.K, respectively. Calculate the thermal diffusivity of the material. [Ans. 1.686×10^{-5} m^2/s]
2. The thermal conductivity k, the density ρ, and specific heat C of an aluminium plate are 160 W/m.K, 2790 kg/m^3 and 0.88 kJ/kg.K respectively. Calculate the thermal diffusivity of the material.
   [Ans. 6.516×10^{-5} m^2/s]
3. Consider a plate fuel element of thickness L for a water cooled nuclear reactor. The energy is generated in the fuel element at the rate of $g = g_0 \cos(x)$ W/m^3. The thermal conductivity of the material is constant. Write steady state heat conduction equation governing the temperature distribution in the fuel element.
4. A copper bar of radius r_o is suddenly heated by passage of an electric current, which generates heat in the rod at the rate of $g_0 e^{-\alpha t}$. The thermal conductivity of the rod varies with radius, $k = k(r)$. Write the transient heat conduction equation governing the temperature distribution in the rod.
5. Consider a plate of thickness L. The boundary surface at $x = 0$ is subjected to forced convection with heat transfer coefficient h into an ambient at temperature T_∞. The boundary surface at x = L is insulated. Write the boundary conditions for both the surfaces.
   $$\left[\text{Ans. } h(T_\infty - T_{x=0}) = -k\left(\frac{dT}{dx}\right)_{x=0} \text{ and } \left(\frac{dT}{dx}\right)_{x=L} = 0\right]$$
6. One of surface of a marble slab (k = 2 W/m.K) is maintained at 300°C, while other boundary surface is subjected to constant heat flux of 5000 W/m^2. Write the boundary conditions.
   $$\left[\text{Ans. } T_{x=0} = 300°\text{C and } \left(\frac{dT}{dx}\right)_{x=L} = 2500°\text{C/m}\right]$$
7. Energy is generated at a constant rate g_0 W/m^3 in a copper rod of radius r_o by passage of an electric current. The heat dissipation is by convection at boundary surface at $r = r_o$ into an ambient air at temperature T_∞ with the heat transfer coefficient h. Develop the mathematical formulation for steady state conditions.
   $$\left[\begin{aligned}&\text{Ans. at } r = 0, \frac{dT}{dr} = 0 \text{ and} \\ &\text{at } r = r_o,\ -k\left(\frac{dT}{dr}\right)_{r=r_o} = h\,(T_{r=r_o} - T_\infty)\end{aligned}\right]$$

8. A tomato with diameter D and thermal conductivity k, initially at uniform temperature T_i is suddenly dropped into boiling water at T_∞ with very large convection coefficient. Develop a mathematical formulation of the problem for determining the temperature distribution within the tomato.

$$\left[\text{Ans. } T_{(x,0)} = T_i \text{ and } -k\left(\frac{dT}{dr}\right)_{r=r_o} = h\,(T_{r=r_o} - T_\infty)\right]$$

9. A spherical shell has an inside radius r_1, an outside radius r_2 and thermal conductivity k, the inside surface is heated at a rate of q W/m², while the outside surface dissipates heat by convection with heat transfer coefficient h into an ambient T_∞. Develop mathematical formulation for determining the temperature distribution within the body.

$$\left[\begin{array}{l}\text{Ans. at } r = r_1 ;\ q = -k\left(\dfrac{dT}{dr}\right)_{r=r_1} \text{ and} \\ \text{at } r = r_2 ;\ -k\left(\dfrac{dT}{dr}\right)_{r=r_2} = h\,(T_{r=r_2} - T_\infty)\end{array}\right]$$

10. A long, rectangular copper bar of thickness L is maintained at a temperature T_0 at its lower surface throughout the bar. Suddenly an electric current is passed through the bar and its upper surface is exposed to an air stream at T_∞, with convection coefficient h, while its bottom surface continues to be maintained at T_0. Obtain differential heat conduction equation and write initial and boundary conditions in order to obtain the temperature distribution as a function of position and time.

$$\left[\begin{array}{l}\text{Ans. } \dfrac{\partial^2 T}{\partial x^2} + \dfrac{g}{k} = \dfrac{1}{\alpha}\dfrac{\partial T}{\partial t},\ T_{(x,0)} = T_0,\ T_{(0,t)} = T_0 \\ \text{and } -k\left(\dfrac{\partial T}{\partial x}\right)_{x=L} = h(T_{x=L} - T_\infty)\end{array}\right]$$

11. Steam at 200°C flows through a pipe. The inner and outer radii of pipe are 8 cm and 8.5 cm, respectively. The outer surface of the pipe is heavily insulated. If the convection heat transfer coefficient at the inner surface of the pipe is 65 W/m².K, express the boundary conditions at inner and outer surfaces of the pipe.

$$\left[\begin{array}{l}\text{Ans. at } r = r_1,\ -k\left(\dfrac{dT}{dr}\right)_{r=r_1} = h\,[T_\infty - T_{r=r_1}] \\ \text{and at } r = r_2,\ \left(\dfrac{dT}{dr}\right)_{r=r_2} = 0\end{array}\right]$$

12. A spherical metal ball of radius r_o, initially at 600°C is allowed to cool in an ambient at 38°C. The heat transfer coefficient on outer surface of the ball is 15 W/m².K and emissivity of outer surface of ball is 0.6. Thermal conductivity of the ball material is 30 W/m.K. Express initial and boundary conditions for cooling process of the ball.

[Ans. Initial condition = $T_{(r,0)} = T_i = 600°C$

Boundary condition at centre $\left(\dfrac{dT}{dr}\right)_{(0,t)} = 0$

Boundary condition at outer surface $-k\left(\dfrac{dT}{dr}\right)_{(r_o,t)} = h$

$[T_{(r_o)} - T_\infty] + \varepsilon\sigma\,[T^4_{(r_o)} - T^4_\infty]]$

Steady State Conduction Without Heat Generation

3

Objective of this chapter is to:

- obtain steady state temperature distribution without heat generation in slab, hollow cylinders and spheres.
- obtain heat conduction rate from differential heat conduction equation without heat generation in solids.
- study concept of thermal resistance in series and parallel.
- study of concept of contact resistance.
- study concept of critical thickness of insulation on cylinders and spheres.

The steady state temperature distribution within the solid without heat generation is governed by differential equation.

$$\frac{d}{dX}\left(X^n \frac{dT(X)}{dX}\right) = 0 \qquad ...(3.1)$$

where, $X = x$ and $n = 0$ for cartesian coordinates,

$X = r$ and $n = 1$ for cylindrical coordinates,

$X = r$ and $n = 2$ for spherical coordinates.

The solution of eqn. (3.1) for the solid subjected to boundary conditions at both surfaces gives the temperature distribution in the body. Knowing the temperature distribution, the heat flux $q(X)$ anywhere in the solid can be obtained from Fourier law,

$$q(X) = -k\frac{dT(X)}{dX} \qquad ...(3.2)$$

and heat flow $\qquad Q = A\,q(X) \qquad ...(3.3)$

3.1. PLANE WALL

Consider a plane wall of thickness L, its left face at $x = 0$, is at a temperature T_1 and right face at $x = L$ is at temperature T_2. The wall has no heat generation and its thermal conductivity k is assumed constant.

Rewriting the governing differential eqn. (2.18) for plane wall.

$$\frac{d}{dx}\left\{\frac{dT}{dx}\right\} = 0 \qquad ...(3.4)$$

Integrating, we get, slope

$$\frac{dT(x)}{dx} = C_1 \qquad ...(3.5)$$

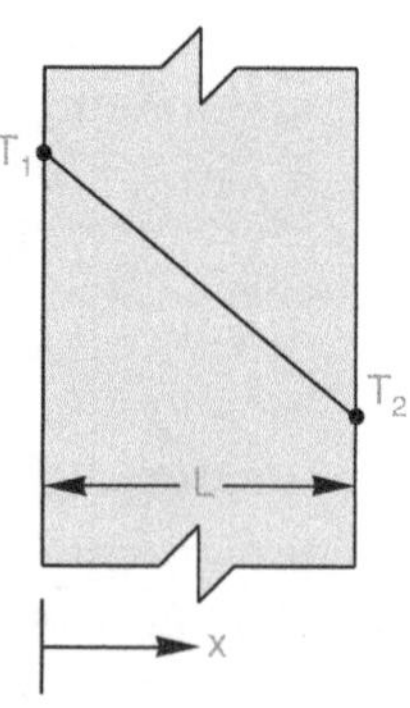

Fig. 3.1. Plane wall

Integrating again, we get equation of straight line

$$T(x) = C_1 x + C_2 \qquad ...(3.6)$$

where C_1 and C_2 are constant of integrations and are evaluated with use of boundary conditions.

The boundary conditions :

$$T(x) = T_1 \text{ at } x = 0$$

and $$T(x) = T_2 \text{ at } x = L$$

Using first boundary condition in eqn. (3.6), we get

$$C_2 = T_1$$

and using second boundary condition in eqn. (3.6) ;

$$T_2 = C_1 L + T_1$$

or $$C_1 = \frac{T_2 - T_1}{L}$$

Then the temperature distribution in the plane wall is given by

$$T(x) = (T_2 - T_1)\frac{x}{L} + T_1 \quad ...(3.7)$$

This is the temperature distribution $T(x)$ in a plane wall. It is a linear function of x as shown in Fig. 3.1.

Differentiation eqn. (3.7) with respect to x, we get, slope,

$$\frac{dT(x)}{dx} = \frac{T_2 - T_1}{L} \quad ...(3.8)$$

The heat flux from the Fourier law,

$$q(x) = -k\frac{dT(x)}{dx}$$

$$= -k\frac{T_2 - T_1}{L} = \frac{k(T_1 - T_2)}{L} \quad ...(3.9)$$

The total heat flow rate Q, through an area A normal to direction of heat flow,

$$Q = kA\left\{\frac{T_1 - T_2}{L}\right\} \quad ...(3.10)$$

3.2. ELECTRICAL ANALOGY OF HEAT TRANSFER RATE THROUGH A PLANE WALL

The eqn. (3.10) for rate of heat conduction through a plane wall (Fig. 3.1) can be rearranged as

$$Q = \frac{T_1 - T_2}{\frac{L}{kA}} = \frac{\Delta T}{\frac{L}{kA}} \quad ...(3.11)$$

In this equation, the temperature difference, ΔT on two sides of the wall is driving potential, that causes flow of heat. The term L/kA is the quantity, which opposes the heat flow in the material and it is equivalent to a thermal resistance R_{th} of wall. It is also called *conduction resistance* of wall. Then eqn. (3.11) can be rearranged as

$$Q = \frac{\Delta T}{R_{th}} = \frac{\Delta T}{R_{wall}} (W) \quad ...(3.12)$$

where, $$R_{th} = R_{wall} = \frac{L}{kA} \ (°C/W \text{ or } K/W) \quad ...(3.13)$$

There is an analogy between a heat flow system and an electric current flow system, where current I is expressed as

$$\text{current, } I = \frac{V_1 - V_2}{R_e} = \frac{\text{Potential difference}}{\text{Electrical Resistance}} \quad ...(3.14)$$

where R_e is *electric resistance* and it is expressed as

$$R_e = \frac{\rho L}{A_c}$$

$$= \frac{\text{Resistivity of the material} \times \text{Conductor length}}{\text{Cross-section area of conductor}} \quad ...(3.15)$$

and the potential difference or voltage difference across the resistance is $V_1 - V_2$.

Comparison of eqn. (3.12) with eqn. (3.14) indicates that the rate of heat transfer Q, through a layer analogous to electric current I, thermal resistance R_{th} analogous to an electric resistance R_e and the temperature difference ΔT analogous to voltage difference ΔV. Such comparison is referred as **electrical analogy** of rate of heat transfer through a plane wall.

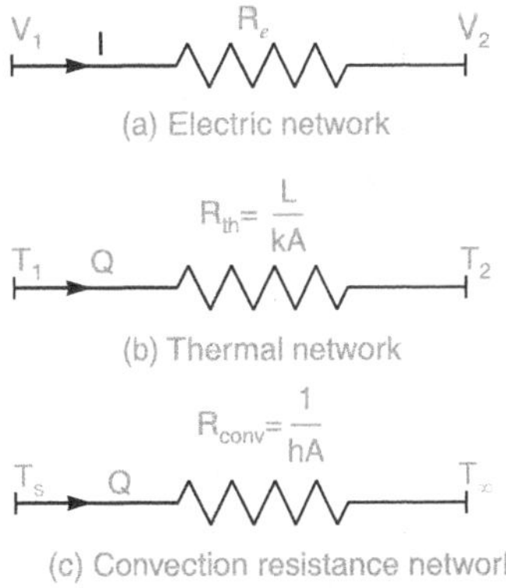

Fig. 3.2. Analogy between thermal and electrical resistance concepts

The analogous quantities in the expression are,

$$\Delta V \Rightarrow \Delta T,$$
$$I \Rightarrow Q,$$
$$R_e \Rightarrow \frac{L}{kA}$$

Similarly the convection heat transfer given by eqn. (1.11) can be rearranged as

$$Q = \frac{T_s - T_\infty}{\frac{1}{hA}} = \frac{\Delta T}{R_{conv}} \quad ...(3.16)$$

where, $$R_{conv} = \frac{1}{hA} \ (°C/W \text{ or } K/W) \quad ...(3.17)$$

The R_{conv} is a thermal resistance acts between the surface and its surroundings against the convection heat transfer, thus it is called **convection resistance** or film resistance or thermal resistance for convection.

Thermal conductance K_c is defined as the reciprocal of thermal resistance and is expressed as

$$K_c = \frac{kA}{L} \quad ...(3.18)$$

It is equal to the rate of heat transfer through a solid of area A and thickness L per degree temperature difference.

Consider a plane wall of thickness L, exposed on its both sides to two different environments at temperatures T_{∞_1}, and T_{∞_2} with heat transfer coefficients h_1 and h_2, respectively as shown in Fig. 3.3.

The steady state heat transfer rate through the wall, when its two surfaces are maintained at constant temperatures T_1 and T_2 can be expressed as

$$Q_1 = \frac{kA\,(T_1 - T_2)}{L}$$

Fig. 3.3. Plane wall subjected to convection boundaries

When the left face and right face involve convection heat transfer due to temperature difference between surface and surroundings.

The convection heat transfer rate at the left face exposed to environment at T_{∞_1}

$$Q_2 = h_1A(T_{\infty_1} - T_1) \quad ...[3.18\ (a)]$$

The convection heat transfer rate at the right face exposed to environment at T_{∞_2},

$$Q_3 = h_2A(T_2 - T_{\infty_2}) \quad ...(3.19)$$

In steady state conditions, the heat transfer rate remains constant ;

Thus $Q_1 = Q_2 = Q_3 = Q$ (say)

Then $Q = h_1A(T_{\infty_1} - T_1)$

$$= \frac{kA(T_1 - T_2)}{L} = h_2\,A(T_2 - T_{\infty_2}) \quad ...(3.20)$$

Fig. 3.4. Equivalent thermal network for plane wall of Fig. 3.3.

It can be modified as

$$Q = \frac{T_{\infty_1} - T_1}{R_{conv,\,1}} = \frac{T_1 - T_2}{R_{wall}} = \frac{T_2 - T_{\infty_2}}{R_{conv,\,2}} \quad ...(3.21)$$

where, $R_{conv,\,1} = \frac{1}{h_1A}$, $R_{wall} = \frac{L}{kA}$, $R_{conv,\,2} = \frac{1}{h_2A}$.

The thermal circuit representation provides a useful tool for the analysis of heat transfer problems. The equivalent thermal resistance for a plane wall with convection on both sides is shown in Fig. 3.4

Since these three resistances are in series, therefore, the total thermal resistance ΣR_{th} is sum of the series resistances as in electrical network

$$\Sigma R_{th} = R_{conv,\,1} + R_{wall} + R_{conv,\,2}$$

or

$$\Sigma R_{th} = R_{total} = \frac{1}{h_1A} + \frac{L}{kA} + \frac{1}{h_2A} \quad ...(3.22)$$

and overall temperature difference, $(\Delta T)_{overall} = T_{\infty_1} - T_{\infty_2}$.

Therefore, heat current,

$$Q = \frac{T_{\infty_1} - T_{\infty_2}}{R_{total}} = \frac{(\Delta T)_{overall}}{\Sigma R_{th}} \quad ...(3.23)$$

3.3. MULTILAYER PLANE WALL

The concept of thermal circuit may also be extended for composite wall. Such wall may involve any number of series and parallel thermal resistances due to layers of different materials.

3.3.1. Plane Slabs in Series

Consider a composite wall with three layers in series and convection heat transfer on both boundary surfaces as shown in Fig. 3.5.

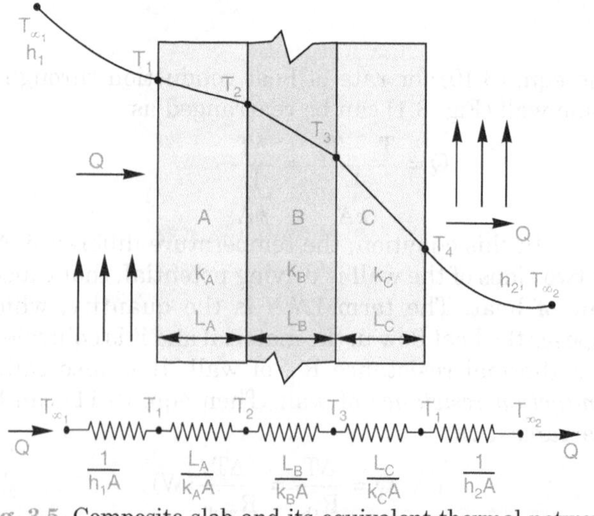

Fig. 3.5. Composite slab and its equivalent thermal network

Heat transfer rate can be expressed as :

$$Q = \frac{(\Delta T)_{overall}}{\Sigma R_{th}}$$

where, $(\Delta T)_{overall} = T_{\infty_1} - T_{\infty_2}$

$$\Sigma R_{th} = \frac{1}{h_1 A} + \frac{L_A}{k_A A} + \frac{L_B}{k_B A} + \frac{L_C}{k_C A} + \frac{1}{h_2 A} \quad ...(3.24)$$

where, A = area normal to heat transfer.

$$\text{Then,} \quad Q = \frac{T_{\infty_1} - T_{\infty_2}}{\frac{1}{h_1 A} + \frac{L_A}{k_A A} + \frac{L_B}{k_B A} + \frac{L_C}{k_C A} + \frac{1}{h_2 A}} \quad ...(3.25)$$

Alternatively, the heat transfer rate associated with each layer in composite wall can be expressed as :

$$Q = \frac{T_{\infty_1} - T_1}{\frac{1}{h_1 A}} = \frac{T_1 - T_2}{\frac{L_A}{k_A A}} = \frac{T_2 - T_3}{\frac{L_B}{k_B A}}$$

$$= \frac{T_3 - T_4}{\frac{L_C}{k_C A}} = \frac{T_4 - T_{\infty_2}}{\frac{1}{h_2 A}} \quad ...(3.26)$$

3.3.2. Heat Conduction Through Parallel Slabs

Conduction heat transfer can also occur through a wall section with two different materials in parallel as shown in Fig. 3.6 (*a*). The material A has thermal conductivity k_A and heat transfer area A_A, while the material B has thermal conductivity k_B and heat transfer area A_B. The temperature over left and right faces are uniform at T_1 and T_2, respectively. The equivalent thermal circuit is shown in Fig. 3.6 (*b*).

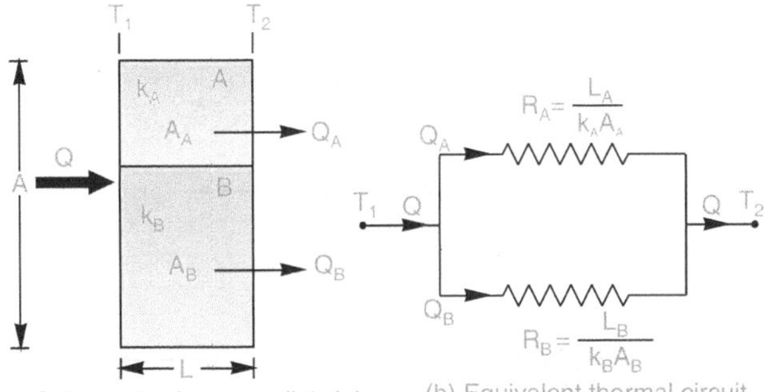

(a) Schematic of two parallel slabs (b) Equivalent thermal circuit

Fig. 3.6

Since the heat is conducted through two different paths between the same temperature difference, the total rate of heat transfer is sum of heat flow through areas A_A and A_B.

$$Q = Q_A + Q_B \quad ...(3.27)$$

$$= \frac{T_1 - T_2}{\frac{L}{k_A A_A}} + \frac{T_1 - T_2}{\frac{L}{k_B A_B}}$$

$$= (T_1 - T_2)\left\{\frac{1}{\frac{L}{k_A A_A}} + \frac{1}{\frac{L}{k_B A_B}}\right\}$$

Using $R_A = \frac{L}{k_A A_A}$ and $R_B = \frac{L}{k_B A_B}$,

$$\text{Then} \quad Q = (T_1 - T_2)\left(\frac{1}{R_A} + \frac{1}{R_B}\right) = \frac{T_1 - T_2}{R_{eq}} \quad ...(3.28)$$

where, $\frac{1}{R_{eq}} = \frac{1}{R_A} + \frac{1}{R_B}$

or $$R_{eq} = \frac{1}{\frac{1}{R_A} + \frac{1}{R_B}} = \frac{R_A R_B}{R_A + R_B} \quad ...(3.29)$$

It is called the equivalent resistance of parallel resistances.

3.3.3. Composite Wall in Series and Parallel

Consider a composite wall with series and parallel configurations as shown in Fig. 3.7

A_E – heat transfer area of layer E

A_F = heat transfer area of layer F

and $A = A_E + A_F$

The equivalent resistance of parallel resistances in Fig. 3.7 (*b*) can be calculated as :

$$\frac{1}{R_{eq}} = \frac{1}{R_E} + \frac{1}{R_F} = \frac{k_E A_E}{L_E} + \frac{k_F A_F}{L_F} \quad ...(3.30)$$

$$R_{eq} = \frac{1}{\frac{k_E A_E}{L_E} + \frac{k_F A_F}{L_F}}$$

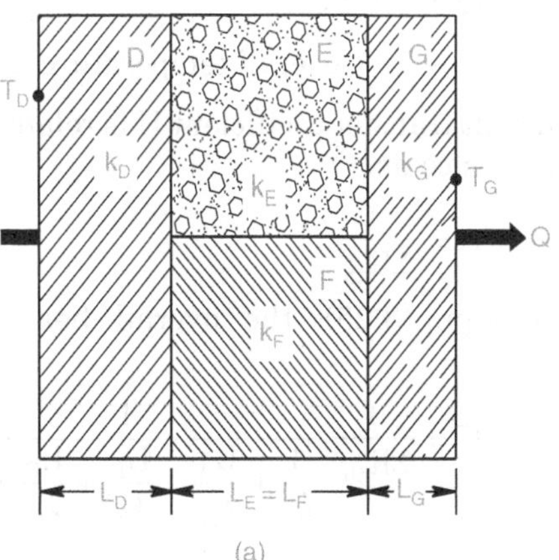

(a)

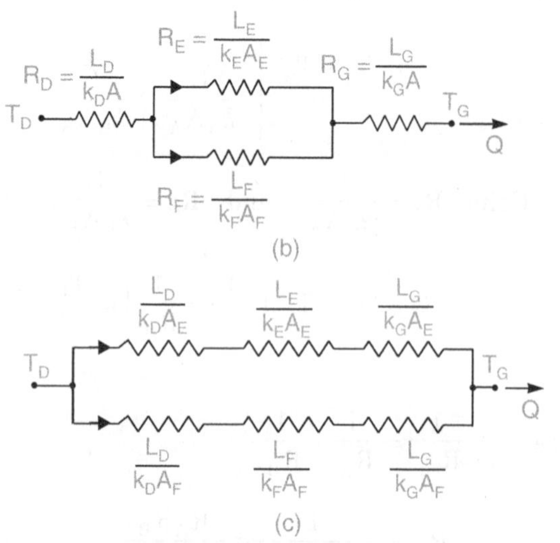

Fig. 3.7. Equivalent thermal circuit for series and parallel composite wall

and total thermal resistance from Fig. 3.7(*b*)

$$R_{total} = \Sigma R_{th} = R_D + R_{eq} + R_G$$

$$R_{total} = \frac{L_D}{k_D A} + \frac{1}{\frac{k_E A_E}{L_E} + \frac{k_F A_F}{L_F}} + \frac{L_G}{k_G A} \quad ...(3.31)$$

The total resistance can also be obtained from Fig. 3.7(*c*)

$$\frac{1}{\Sigma R_{th}} = \frac{1}{\frac{L_D}{k_D A_E} + \frac{L_E}{k_E A_E} + \frac{L_G}{k_G A_E}} + \frac{1}{\frac{L_D}{k_D A_F} + \frac{L_F}{k_F A_F} + \frac{L_G}{k_G A_F}} \quad ...(3.32)$$

3.3.4. Overall Heat Transfer Coefficient

For composite systems, it is more appropriate to work with overall heat transfer coefficient U, which is defined by equation

$$Q = UA(\Delta T) \quad ...(3.33)$$

where, $\Delta T = T_{\infty_1} - T_{\infty_2}$, overall temperature difference.

The overall heat transfer coefficient U is related to total thermal resistance as

$$UA = \frac{1}{\Sigma R_{th}} \quad ...(3.34)$$

From eqn. (3.25), the overall heat transfer coefficient

$$U = \frac{1}{\Sigma R_{th}} \frac{1}{\frac{1}{h_1} + \frac{L_A}{k_A} + \frac{L_B}{k_B} + \frac{L_C}{k_C} + \frac{1}{h_2}} \quad ...(3.35)$$

Example 3.1. *The walls of a house, 4 m high, 5 m wide and 0.3 m thick are made from brick with thermal conductivity of 0.9 W/m.K. The temperature of air inside the house is 20°C and outside air is at –10°C. There is a heat transfer coefficient of 10 W/m².K on the inside wall and 30 W/m².K on the outside wall. Calculate the inside and outside wall temperatures, heat flux and total heat transfer rate through the wall.* (N.M.U., May 2007)

Solution

Given : A wall of house as shown in Fig. 3.8

$A = 4 \text{ m} \times 5 \text{ m} = 20 \text{ m}^2$, $h_1 = 10 \text{ W/m}^2.\text{K}$

$L = 0.3 \text{ m}$, $k = 0.9 \text{ W/m.K}$, $h_2 = 30 \text{ W/m}^2.\text{K}$

$T_{\infty_1} = 20°\text{C}$, $T_{\infty_2} = -10°\text{C}$

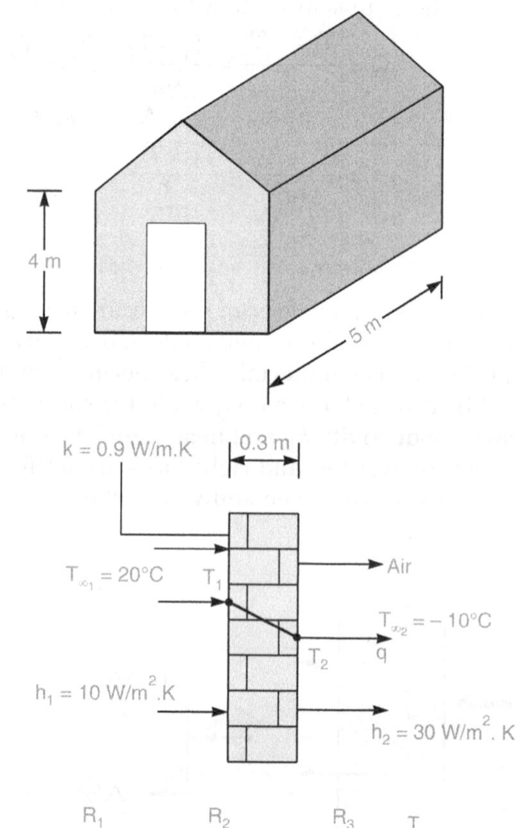

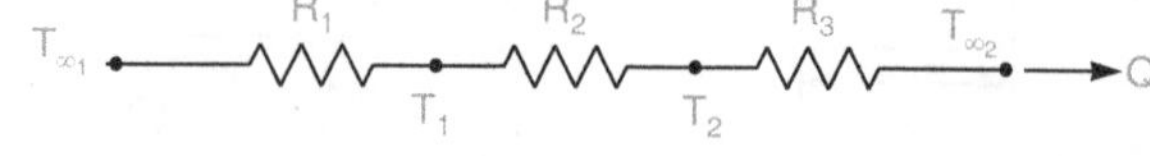

Fig. 3.8. Schematic and thermal network

To find :

(*i*) Inside and outside wall temperatures T_1 and T_2,

(*ii*) Heat flux, and

(*iii*) Total heat transfer rate.

Assumptions :

1. Steady state heat conduction in one direction only.

2. Constant properties.

3. Negligible radiation heat transfer.

Analysis : The heat transfer rate through the house wall, by using eqn. (3.23)

$$Q = \frac{\Delta T}{\Sigma R_{th}} = \frac{T_{\infty_1} - T_{\infty_2}}{R_1 + R_2 + R_3}$$

where

$$R_1 = \frac{1}{h_1 A} = \frac{1}{10 \times 20} = 0.005 \text{ K/W}$$

$$R_2 = \frac{L}{kA} = \frac{0.3}{0.9 \times 20} = 0.0167 \text{ K/W}$$

$$R_3 = \frac{1}{h_2 A} = \frac{1}{30 \times 20} = 1.67 \times 10^{-3} \text{ K/W}$$

(*i*) Heat transfer rate through wall

$$Q = \frac{20 - (-10)}{0.005 + 0.0167 + 1.67 \times 10^{-3}}$$

$$= \mathbf{1285.7 \text{ W}}. \quad \text{Ans.}$$

(*ii*) Heat flux

or

$$q = \frac{Q}{A} = \frac{1285.7}{20} = \mathbf{64.28 \text{ W/m}^2}. \quad \text{Ans.}$$

(*iii*) Inside and outside wall surface temperatures: Steady state convective heat transfer rate is given by eqn. (3.16)

$$Q = \frac{T_s - T_\infty}{R_{conv}}$$

At inside surface

$$Q = \frac{T_{\infty_1} - T_1}{R_1}$$

or

$$T_1 = T_{\infty_1} - QR_1 = 20 - 1285.7 \times 0.005$$

$$= \mathbf{13.57°C}. \quad \text{Ans.}$$

At outside surface

$$Q = \frac{T_2 - T_{\infty_2}}{R_3}$$

or

$$T_2 = QR_3 + T_{\infty_2}$$

$$= 1285.7 \times 1.67 \times 10^{-3} + (-10)$$

$$= \mathbf{-7.85°C}. \quad \text{Ans.}$$

Example 3.2. *A thermopane window consists of two 5 mm thick glass (k = 0.78 W/m.K) sheets separated by 10 mm stagnant air gap (k = 0.025 W/m.K). The convection heat transfer coefficient for inner and outside air are 10 W/m².K and 50 W/m².K, respectively.*

(*a*) *Determine the rate of heat loss per m² of the glass surface for a temperature difference of 60°C between the inside and outside air.*

(*b*) *Compare the result with the heat loss, if the window had only a single sheet of glass of thickness 5 mm instead of thermopane.*

(*c*) *Compare the result with the heat flow, if window has no stagnant air (i.e., a sheet of glass, 10 mm thick).*

Solution

Given : A thermopane glass window :

$L_1 = L_3 = 5$ mm $= 0.005$ m,

$L_2 = 10$ mm $= 0.01$ m,

$k_1 = k_3 = 0.78$ W/m.K,

$k_2 = 0.025$ W/m.K

$h_1 = 10$ W/m².K,

$h_2 = 50$ W/m².K

$\Delta T = T_{\infty_1} - T_{\infty_2} = 60°C$.

To find :

(*a*) Heat flow rate through the glass per m².

(*b*) Heat flow rate when window has only a glass sheet of 5 mm thick.

(*c*) Heat flow rate when window has only a glass sheet of 10 mm thick.

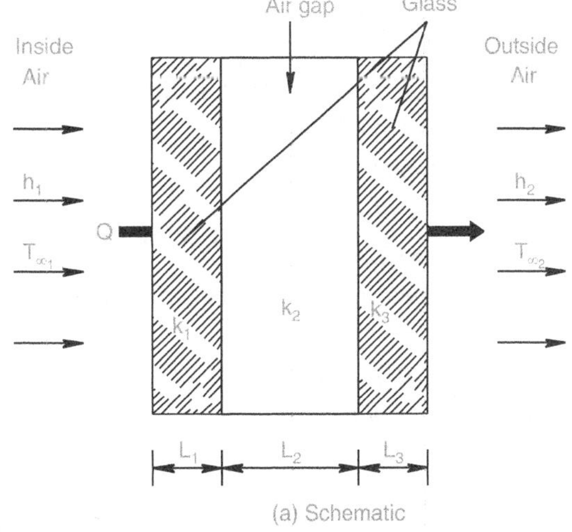

(a) Schematic

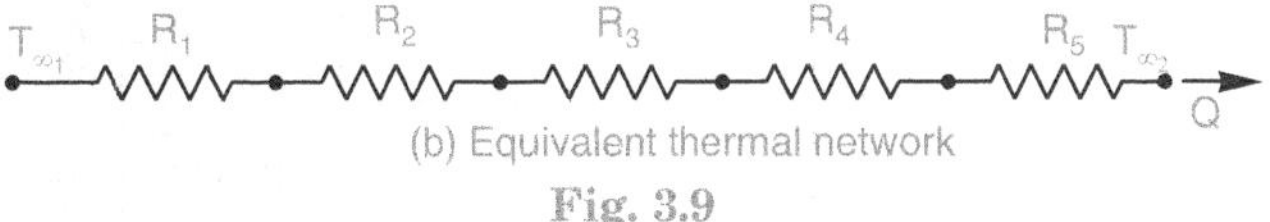

(b) Equivalent thermal network

Fig. 3.9

Assumptions :

(*i*) One dimensional steady state heat flow.

(*ii*) Constant properties.

Analysis : The various specific thermal resistances (for 1 m²) are shown and calculated as :

$$R_1 = \frac{1}{h_1 A} = \frac{1}{10 \times 1} = 0.10 \text{ K/W}$$

$$R_2 = R_4 = \frac{L_1}{k_1 A} = \frac{0.005}{0.78 \times 1} = 0.00641 \text{ K/W}$$

$$R_3 = \frac{L_2}{k_2 A} = \frac{0.01}{0.025 \times 1} = 0.40 \text{ K/W}$$

$$R_5 = \frac{1}{h_2 A} = \frac{1}{50 \times 1} = 0.020 \text{ K/W}$$

(*a*) The heat flow rate through the thermopane window is given by

$$Q = \frac{\Delta T}{\Sigma R_{th}} = \frac{T_{\infty_1} - T_{\infty_2}}{R_1 + R_2 + R_3 + R_4 + R_5}$$

$$= \frac{60}{0.1 + 2 \times 0.00641 + 0.4 + 0.020}$$

$$= \frac{60}{0.53282} = \textbf{112.60 W/m}^2. \quad \text{Ans.}$$

(*b*) If the window has a single sheet of glass of 5 mm thick, the total thermal resistance

$$\Sigma R_{th} = R_1 + R_2 + R_5$$

$$= 0.1 + 0.00641 + 0.02 = 0.12641 \text{ K/W}$$

$$Q = \frac{60}{0.12641} = \textbf{474.65 W/m}^2. \quad \text{Ans.}$$

The heat loss is about four times that of previous case.

(*c*) If the window has a glass sheet of 10 mm thick only, then total resistance

$$\Sigma R_{th} = R_1 + 2\,(R_2) + R_5$$

$$= 0.1 + 2 \times 0.00641 + 0.02$$

$$= 0.13282 \text{ K/W}$$

The heat loss per m^2 ;

$$Q = \frac{60}{0.13282} = \textbf{451.74 W/m}^2. \quad \text{Ans.}$$

Heat loss does not decrease appreciably by increasing the glass thickness.

Example 3.3. *A wall is constructed of several layers. The first layer consists of brick (k = 0.66 W/m.K), 25 cm thick, the second layer 2.5 cm thick mortar (k = 0.7 W/m.K), the third layer 10 cm thick limestone (k = 0.66 W/m.K) and outer layer of 1.25 cm thick plaster (k = 0.7 W/m.K). The heat transfer coefficients on interior and exterior of the wall fluid layers are 5.8 W/m².K and 11.6 W/m².K, respectively. Find :*

(i) Overall heat transfer coefficient,

(ii) Overall thermal resistance per m²,

(iii) Rate of heat transfer per m², if the interior of the room is at 26°C while outer air is at – 7°C,

(iv) Temperature at the junction between mortar and limestone. (P.U., Dec. 2009)

Solution

Given : A multilayer composite exposed to different atmospheres on both boundary surfaces.

L_1 = 25 cm = 0.25 m, k_1 = 0.66 W/m.K,
L_2 = 2.5 cm = 0.025 m, k_2 = 0.7 W/m.K,
L_3 = 10 cm = 0.1 m, k_3 = 0.66 W/m.K,
L_4 = 1.25 cm = 0.0125 m, k_4 = 0.7 W/m.K,
h_1 = 5.8 W/m².K, h_2 = 11.6 W/m².K,
T_{∞_1} = 26°C, T_{∞_2} = – 7°C.

To find :

(*i*) Overall heat Transfer coefficient,

(*ii*) Total thermal resistance, ΣR_{th},

(*iii*) The heat flow rate through composite, Q,

(*iv*) T_3, temperature at the interface of mortar and limestone.

Assumptions :

(*i*) Steady state heat conduction rate through 1 m² area.

(*ii*) Heat transfer in one direction only.

(*iii*) Constant properties.

(*iv*) No thermal contact resistance at interfaces.

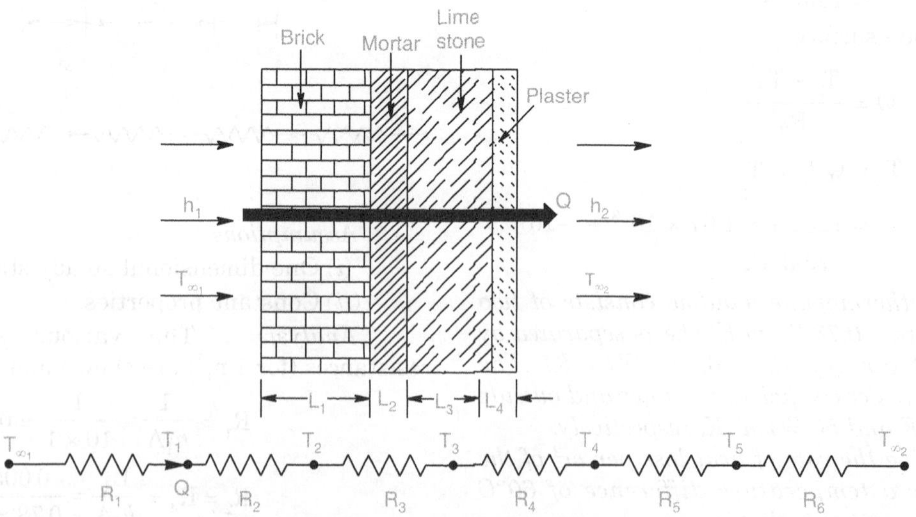

Fig. 3.10. Multilayer wall and its equivalent thermal circuit

Analysis : The schematic and equivalent thermal resistances for composite wall are shown in Fig. 3.10 :

where,
$$R_1 = \frac{1}{h_1 A} = \frac{1}{5.8 \times 1} = 0.1724 \text{ K/W}$$
$$R_2 = \frac{L_1}{k_1 A} = \frac{0.25}{0.66 \times 1} = 0.378 \text{ K/W}$$
$$R_3 = \frac{L_2}{k_2 A} = \frac{0.025}{0.7 \times 1} = 0.0357 \text{ K/W}$$
$$R_4 = \frac{L_3}{k_3 A} = \frac{0.1}{0.66 \times 1} = 0.1515 \text{ K/W}$$
$$R_5 = \frac{L_4}{k_4 A} = \frac{0.0125}{0.7 \times 1} = 0.0178 \text{ K/W}$$
$$R_6 = \frac{1}{h_2 A} = \frac{1}{11.6 \times 1} = 0.0862 \text{ K/W}$$

(*i*) Overall heat transfer coefficient,
$$U = \frac{1}{A \Sigma R_{th}} = \frac{1}{A(R_1 + R_2 + R_3 + R_4 + R_5 + R_6)}$$
$$U = \frac{1}{1 \times (0.1724 + 0.378 + 0.0357 + 0.1515 + 0.0178 + 0.0862)}$$
$$= \frac{1}{0.8424}$$
or $U = \mathbf{1.187\ W/m^2.K.}$ **Ans.**

(*ii*) The overall thermal resistances
$$\Sigma R_{th} = \frac{1}{UA} = \frac{1}{1.187 \times 1}$$
$= \mathbf{0.8424\ K/W.}$ **Ans.**

(*iii*) The heat flow rate through the composite per m^2 ;
$$Q = \frac{T_{\infty_1} - T_{\infty_2}}{\Sigma R_{th}} = \frac{26 - (-7)}{0.8424} = \frac{33}{0.8424}$$
$= \mathbf{39.17\ W/m^2.}$ **Ans.**

(*iv*) The temperature at the interface of mortar and limestone can be calculated as :
$$Q = \frac{T_{\infty_1} - T_3}{R_1 + R_2 + R_3}$$
or $T_3 = T_{\infty_1} - 39.17 \times (0.1724 + 0.378 + 0.0357) = \mathbf{3°C.}$ **Ans.**

Example 3.4. *The composite wall of an oven consists of three materials, two of them are of known thermal conductivity, $k_A = 20$ W/m.K and $k_C = 50$ W/m.K and known thickness $L_A = 0.3$ m and $L_C = 0.15$ m. The third material B, which is sandwiched between material A and C is of known thickness, $L_B = 0.15$ m, but of unknown thermal conductivity k_B.*

Under steady state operating conditions, the measurement reveals an outer surface temperature of material C is 20°C and inner surface of A is 600°C and oven air temperature is 800°C. The inside convection coefficient is 25 W/m^2.K. What is the value of k_B ?

(P.U., May 1997)

Solution

Given : A composite wall of an oven with

$k_A = 20$ W/m.K, $k_C = 50$ W/m.K
$L_A = 0.3$ m, $L_C = 0.15$ m
$L_B = 0.15$ m, $T_i = 600°C$
$T_o = 20°C$, $T_\infty = 800°C$
$h_i = 25$ W/m^2.K

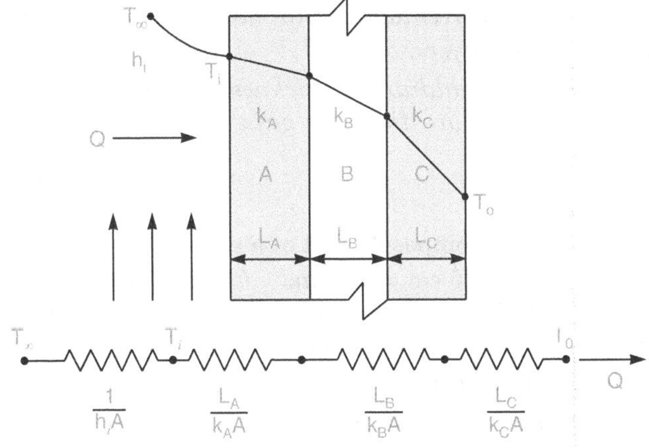

Fig. 3.11. Schematic and equivalent resistances

To find : The thermal conductivity k_B.

Assumptions :

(*i*) Steady state heat conduction in axial direction only.

(*ii*) Constant properties.

Analysis : The heat transfer rate per unit area in the slab can be calculated by considering convection at inner side.
$$\frac{Q}{A} = h_i (T_\infty - T_i) = 25 \times (800 - 600)$$
$$= 5000 \text{ W/m}^2$$

Further this heat is conducted through composite wall, therefore ;
$$\frac{Q}{A} = \frac{T_i - T_o}{\frac{L_A}{k_A} + \frac{L_B}{k_B} + \frac{L_C}{k_C}}$$
or
$$5000 = \frac{600 - 20}{\frac{0.3}{20} + \frac{0.15}{k_B} + \frac{0.15}{50}}$$

or $0.018 + \frac{0.15}{k_B} = 0.116$

or k_B = **1.53 W/m.K.** Ans.

Example 3.5. *The wall of a cold storage consists of three layers: an outer layer of ordinary bricks, 25 cm thick, a middle layer of cork, 10 cm thick and an inner layer of cement, 6 cm thick. The thermal conductivities of the materials are 0.7, 0.043 and 0.72 W/m.K, respectively. The temperature of the outer surface of the wall is 30°C and that of inner is – 15°C. Calculate :*

(*a*) *Steady state rate of heat gain per unit area,*

(*b*) *Temperature at the interfaces of composite wall,*

(*c*) *The percentage of total heat resistance offered by individual layers, and*

(*d*) *What additional thickness of cork should be provided to reduce the heat gain 30% less than the present value ?*

Solution

Given : Composite wall of a cold storage :

L_1 = 25 cm = 0.25 m, k_1 = 0.7 W/m.K
L_2 = 10 cm = 0.1 m, k_2 = 0.043 W/m.K
L_3 = 6 cm = 0.06 m, k_3 = 0.72 W/m.K
T_1 = 30°C, T_4 = – 15°C.

To find :

(*a*) Heat flux, q,

(*b*) Temperatures, T_2, T_3,

(*c*) Percentage resistance offered by individual layers,

(*d*) Thickness of cork layer for 30% heat reduction of existing value.

Analysis : (*a*) Individual thermal resistances per m^2 in the network

$$R_1 = \frac{L_1}{k_1} = \frac{0.25}{0.7} = 0.357 \text{ m}^2.\text{K/W}$$

$$R_2 = \frac{L_2}{k_2} = \frac{0.1}{0.043} = 2.326 \text{ m}^2.\text{K/W}$$

$$R_3 = \frac{L_3}{k_3} = \frac{0.06}{0.72} = 0.083 \text{ m}^2.\text{K/W}$$

All resistances are in series, thus total thermal resistance

$$\Sigma R_{th} = R_1 + R_2 + R_3 = 0.357 + 2.326 + 0.083 = 2.766 \text{ m}^2.\text{ K/W}$$

Rate of heat gain per m^2

$$q = \frac{\Delta T}{\Sigma R_{th}} = \frac{T_1 - T_4}{\Sigma R_{th}} = \frac{30-(-15)}{2.766} = \frac{45}{2.766}$$

= **16.27 W/m².** Ans.

(*b*) Temperature at the interfaces :

(*i*) Heat flux across brick layer

$$q = \frac{T_1 - T_2}{R_1}$$

or $T_2 = T_1 - qR_1 = 30 - 16.27 \times 0.357$

= **24.19°C.** Ans.

It is the temperature at interface between brick and cork layers.

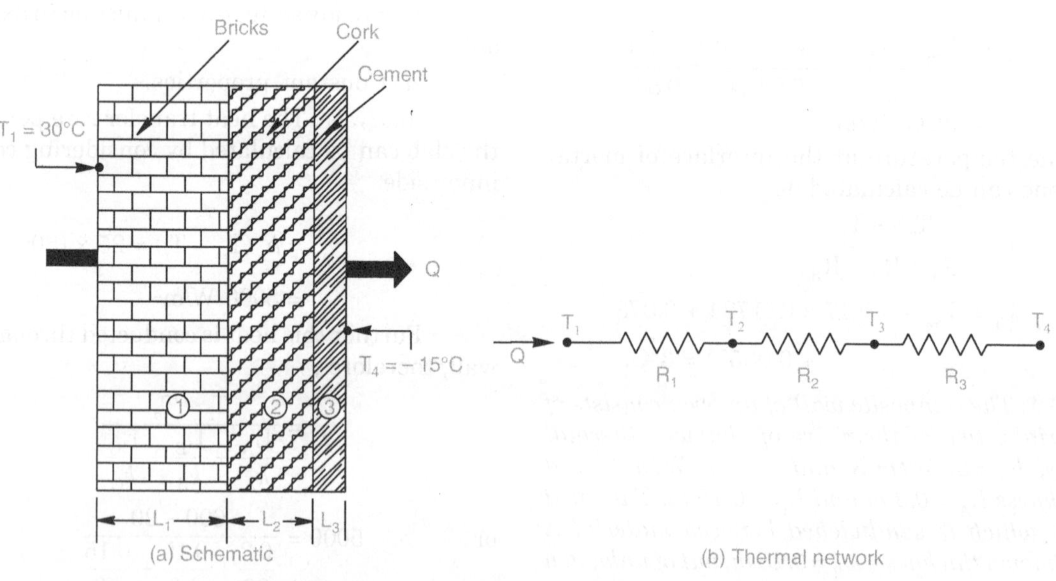

Fig. 3.12. Schematic and thermal network

(*ii*) Heat flux across cork layer

$$q = \frac{T_2 - T_3}{R_2}$$

or $T_3 = T_2 - qR_2 = 24.19 - 16.27 \times 2.326$

$= -$ **13.65°C.** Ans.

It is the temperature at interface between cork and cement layers.

(*c*) Percentage thermal resistance offered by an individual *i*th layer

$$= \frac{R_i}{\Sigma R_{th}} \times 100$$

% resistance offered by brick layer

$$= \frac{0.357}{2.766} \times 100 = 12.9\%$$

% resistance offered by cork layer

$$= \frac{2.326}{2.766} \times 100 = 84.1\%$$

% resistance offered by cement layer

$$= \frac{0.083}{2.766} \times 100 = \mathbf{3.0\%}.$$ Ans.

(*d*) Desired rate of heat flow = 30% less of present value = 0.7 × present value

$$q_1 = 0.7 \times 16.27 = 11.39 \text{ W/m}^2$$

$\therefore$ $$q_1 = \frac{\Delta T}{\Sigma R_{th_1}}$$

or $$11.39 = \frac{30 - (-15)}{\Sigma R_{th_1}}$$

or $$\Sigma R_{th_1} = \frac{45}{11.39} = 3.95 \text{ m}^2\text{.K/W}$$

Additional resistance to be offered by cork

$$= \Sigma R_{th_1} - \Sigma R_{th}$$

$$R_2' = 3.95 - 2.766 = 1.184 \text{ m}^2\text{.K/W}$$

Further $$R_2' = \frac{L_2'}{k_2}$$

or $$L_2' = R_2' k_2 = 1.184 \times 0.043$$

$= 0.051$ m $=$ **5.1 cm.** Ans.

It is an additional thickness of cork to be provided.

Example 3.6. *The door of an industrial furnace is 2 m × 4 m in surface area and is to be insulated to reduce the heat loss to not more than 1200 W/m². The interior and exterior walls of the door are 10 mm and 7 mm thick steel sheets (k = 25 W/m.K). Between these two sheets, a suitable thickness of insulation material is to be placed. The effective gas temperature inside the furnace is 1200°C and the overall heat transfer coefficient between the gas and door is 20 W/m².K. The heat transfer coefficient outside the door is 5 W/m².C. The surrounding air temperature is 20°C. Select suitable insulation material and its size.*

Solution

Given : A door of an industrial furnace as shown below :

Size = 2 m × 4 m

$q = 1200 \text{ W/m}^2$

$L_1 = 10 \text{ mm} = 0.01 \text{ m}, k_1 = 25 \text{ W/m.K}$

$L_3 = 7 \text{ mm} = 0.007 \text{ m}, k_3 = 25 \text{ W/m.K}$

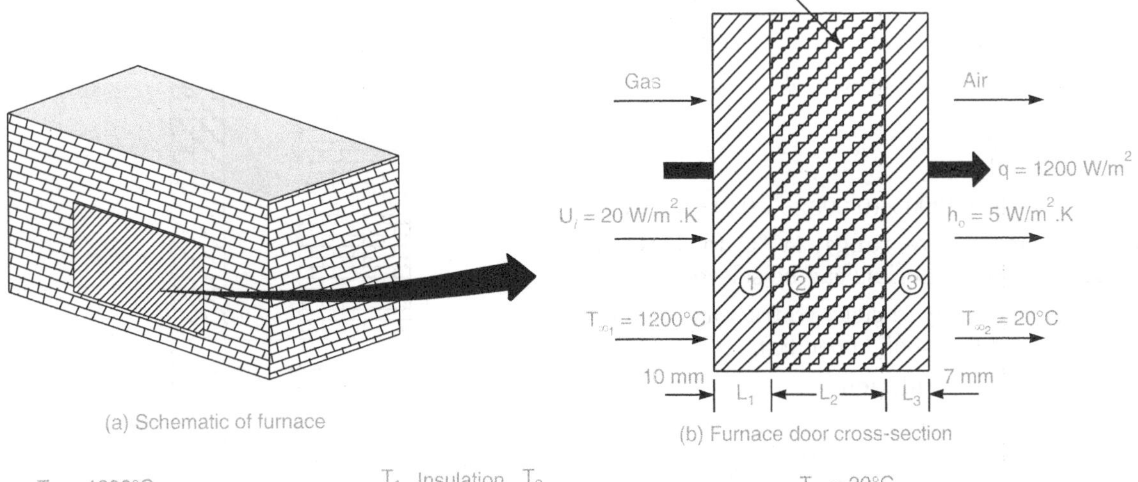

(a) Schematic of furnace

(b) Furnace door cross-section

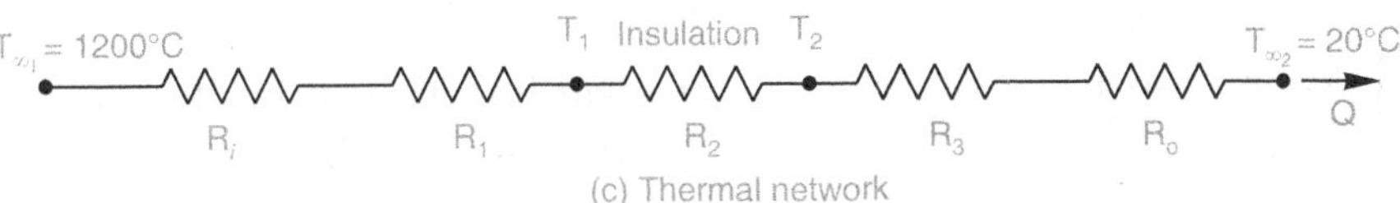

(c) Thermal network

Fig. 3.13. Schematic and thermal network

To find :

(*a*) Material of insulation, and

(*b*) Thickness of insulation.

Assumptions :

(*i*) Steady state conditions.

(*ii*) 1 m^2 area of the door for analysis.

Analysis : The thermal circuit for heat transfer through furnace door is shown in Fig. 3.13.

The individual thermal resistances in the network ;

$$R_i = \frac{1}{U_i} = \frac{1}{20} = 0.05 \text{ m}^2.\text{K/W}$$

$$R_1 = \frac{L_1}{k_1} = \frac{0.01}{25} = 0.0004 \text{ m}^2.\text{K/W}$$

$$R_3 = \frac{L_3}{k_3} = \frac{0.007}{25} = 0.00028 \text{ m}^2.\text{K/W}$$

$$R_o = \frac{1}{h_0} = \frac{1}{5} = 0.2 \text{ m}^2.\text{K/W}$$

The total material resistance

$$\Sigma R_{th,\,m} = R_1 + R_3 = 0.0004 + 0.00028$$
$$= 0.00068 \text{ m}^2.\text{K/W}.$$

The material resistance is negligible compared to other three resistances *i.e.*, R_i, R_2 and R_o. The temperature T_1 on inner side of insulation

$$\frac{Q}{A} = q = \frac{T_{\infty_1} - T_1}{R_i + R_1}$$

or $$1200 = \frac{1200 - T_1}{0.05 + 0.0004}$$

or $$T_1 = 1200 - 1200 \times 0.0504 = 1139.5°\text{C}$$

Temperature T_2 on outer side of insulation

$$q = \frac{T_2 - T_{\infty_2}}{R_3 + R_o}$$

or $$T_2 = 20 + 1200 \times (0.00028 + 0.2) = 260.3°\text{C}$$

For the temperature range 240°C to 1140°C, the following isulation materials can be selected.

(*a*) The fire clay brick (refractory brick, k = 1.09 W/m.K) (Table A-3 of Appendix),

The thickness of insulation

$$q = \frac{(\Delta T)}{L_2/k} \quad \text{or} \quad L_2 = \frac{k(\Delta T)}{q}$$

$$L_2 = \frac{1.09 \times (1139.5 - 260.3)}{1200}$$

or L_2 = **0.8 m.** Ans.

The thickness is too large and door would be heavy also.

(*b*) For same temperature range, silica fibre (k = 0.115 W/m.K), a light material can also be selected (Table A-3 of Appendix),

Thickness

$$L_2 = \frac{k(\Delta T)}{q} = \frac{0.115 \times (1139.5 - 260.3)}{1200}$$

= 0.084 m = **8.5 cm.** Ans.

It is more appropriate choice as compared to above solution.

Example 3.7. *An exterior wall of a house consists of a 10.16 cm layer of common brick having thermal conductivity 0.7 W/m.K. It is followed by a 3.8 cm layer of gypsum plaster with thermal conductivity of 0.48 W/m.K. What thickness of loosely packed rockwool insulation (k = 0.065 W/m.K) should be added to reduce the heat loss through the wall by 80% ?*

(P.U., May 1992)

Solution

Given : L_1 = 10.16 cm = 0.1016 m

k_1 = 0.7 W/m.K

L_2 = 3.8 cm = 0.038 m

k_2 = 0.48 W/m.K

k_3 = 0.065 W/m.K

$q_2 = 0.2\, q_1$.

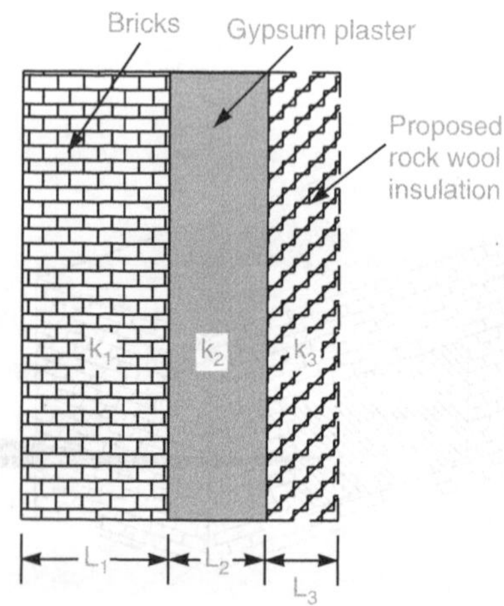

Fig. 3.14. Exterior wall of a house

To find :

(*i*) Heat loss without insulation.

(*ii*) Heat loss with insulation.

(*iii*) Thickness of rockwool insulation.

Assumptions :

(*i*) Steady state heat conduction.

(*ii*) Heat conduction in one direction only.

(*iii*) Constant properties.

(*iv*) No contact thermal resistance at the interfaces.

Analysis : (*i*) Considering ΔT is the temperature difference across the composite wall, then heat flow per unit area or heat flux.

$$\frac{Q_1}{A} = q_1 = \frac{\Delta T}{\frac{L_1}{k_1} + \frac{L_2}{k_2}} = \frac{\Delta T}{\frac{0.1016}{0.7} + \frac{0.038}{0.48}}$$

$$= 4.458\ \Delta T$$

(*ii*) After addition of insulation, heat loss is reduced by 80%, therefore, permissible heat flux will only be 20 per cent of q_1

$$q_2 = 0.2 \times q_1 = 0.2 \times 4.458\ \Delta T = 0.8916\ \Delta T$$

where, q_2 = heat flux with insulation and it can be expressed as :

$$(iii)\ q_2 = \frac{\Delta T}{\frac{0.1016}{0.7} + \frac{0.038}{0.48} + \frac{L_3}{0.065}}$$

or
$$0.224 + \frac{L_3}{0.065} = 1.12$$

or
$$L_3 = 0.0583 \text{ m} = 5.83 \text{ cm},$$

thickness of rockwool insulation. **Ans.**

Example 3.8. *Two M.S. (k = 52 W/m.K) circular rods I and II are interconnected by a sphere III as shown in Fig. 3.15. The respective cross-sectional areas of rods are A_I = 12.5 cm² and A_{II} = 6.25 cm². The system is well insulated except for end faces of rods. Under steady state conditions following data are known.*

T_{∞_1} = 77°C, T_{∞_2} = 3°C,

h_1 = 25 W/m².K, T_1 = 60°C,

T_3 = 10°C,

Temperature T_3 is measured at a point 7.5 cm from the right face of rod II. Find the heat transfer coefficient h_2. (P.U., May 1997)

Solution

Given : Two M.S. Rods connected by a sphere as shown in Fig. 3.15 below :

k = 52 W/m.K, L = 7.5 cm

A_1 = 12.5 cm², A_2 = 6.5 cm²

T_{∞_1} = 77°C, T_{∞_2} = 3°C

h_1 = 25 W/m².K.

To find : Heat transfer coefficient at right face h_2.

Assumptions :

(*i*) One dimensional heat flow.

(*ii*) No contact thermal resistance at interfaces.

(*iii*) Constant properties.

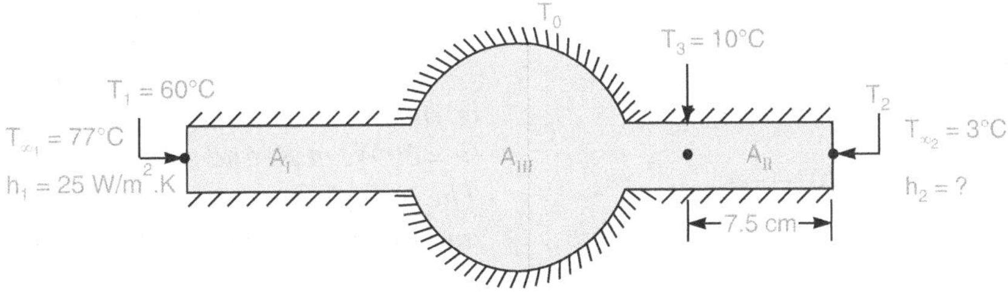

Fig. 3.15. Schematic

Analysis : The heat flow at the left face

$$Q = h_1 A_1 (T_{\infty_1} - T_1)$$

$$= 25 \times (12.5 \times 10^{-4})(77 - 60)$$

$$= 0.53125 \text{ W}.$$

Since system is insulated on its lateral surfaces, therefore, in steady state, same heat will flow at right end of the rod. Applying electrical analogy for heat flow, in the right side of rod and its ambient

$$Q = \frac{T_3 - T_{\infty_2}}{\frac{L}{kA_{II}} + \frac{1}{h_2 A_{II}}}$$

or
$$0.53125 = \frac{10 - 3}{\frac{0.075}{52 \times (6.5 \times 10^{-4})} + \frac{1}{h_2 \times (6.5 \times 10^{-4})}}$$

or
$$\frac{0.075}{0.0338} + \frac{1}{6.5 \times 10^{-4} h_2} = \frac{7}{0.53125} = 13.18$$

or
$$h_2 = \frac{1}{6.5 \times 10^{-4} \times (13.18 - 2.22)}$$

$$= \mathbf{140.42\ W/m^2.K.}$$ **Ans.**

Example 3.9. *The temperature of the inner side of a furnace wall is 640°C and that of on other side is 240°C and it is exposed to an atmosphere at 40°C. In order to reduce the heat loss from the furnace, its wall thickness is increased by 100%.*

Calculate the percentage decrease in the heat loss due to increase in wall thickness. Assume no change in properties except temperature. (P.U., Dec. 2008)

Solution

Given : Furnace wall with

$L_1 = L$, $L_2 = 2L_1$

$T_1 = 640°C$, $T_2 = 240°C$

$T_\infty = 40°C$.

To find :

(*i*) Heat flow with original wall thickness.

(*ii*) Heat flow with change in the wall thickness.

(*iii*) Percentage change in heat flow.

Assumptions :

(*i*) One dimensional steady state heat flow.

(*ii*) No contact thermal resistance at interfaces.

(*iii*) Constant properties.

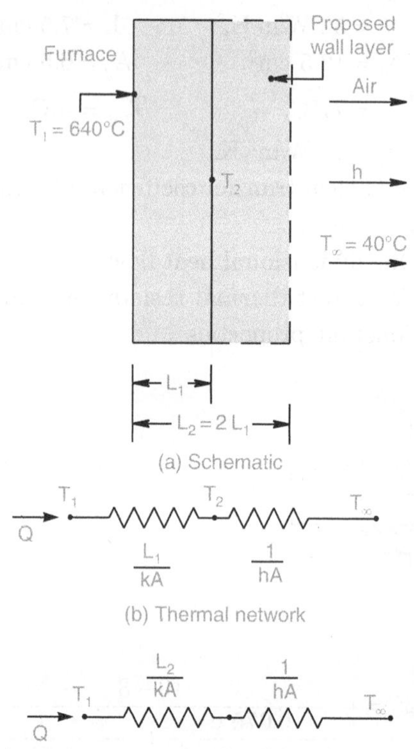

Fig. 3.16. Schematic and thermal network

Analysis : From thermal network, Fig. 3.16 (*b*) we have

$$Q_1 = \frac{kA}{L_1}(T_1 - T_2) = hA(T_2 - T_\infty)$$

or $\frac{k}{L_1} \times (640 - 240) = h\,(240 - 40)$

or $400 \times \frac{k}{L_1} = 200\,h$ or $L_1 = \frac{2k}{h}$ ...(*i*)

Also
$$Q_1 = \frac{T_1 - T_\infty}{\frac{L_1}{kA} + \frac{1}{hA}} = \frac{T_1 - T_\infty}{\frac{2k}{h\,kA} + \frac{1}{hA}}$$

$$= \frac{(T_1 - T_\infty)}{\frac{3}{hA}} = 0.333\,h\,A(T_1 - T_\infty) \quad ...(ii)$$

When wall thickness is increased by 100%, then refer Fig. 3.16 (*c*) thermal network, we have

$$Q_2 = \frac{T_1 - T_\infty}{\frac{L_2}{kA} + \frac{1}{hA}} = \frac{T_1 - T_\infty}{\frac{2L_1}{kA} + \frac{1}{hA}}$$

$$= \frac{T_1 - T_\infty}{\frac{2 \times 2k}{h\,kA} + \frac{1}{hA}} = 0.2\,hA\,(T_1 - T_\infty) \quad ...(iii)$$

Change in heat flow

$$Q_1 - Q_2 = 0.333\,hA(T_1 - T_\infty) - 0.2\,hA(T_1 - T_\infty)$$
$$= 0.1333\,hA(T_1 - T_\infty)$$

% decrease in heat flow

$$= \frac{0.1333\,hA\,(T_1 - T_\infty)}{0.333\,hA\,(T_1 - T_\infty)} \times 100$$

= **40%. Ans.**

Example 3.10. *A square plate heater (size : 15 cm × 15 cm) is inserted between two slabs. Slab A is 2 cm thick (k = 50 W/m.K) and slab B is 1 cm thick (k = 0.2 W/m.K). The outside heat transfer coefficient on both sides of A and B are 200 and 50 W/m².K, respectively. The temperature of surrounding air is 25°C. If the rating of the heater is 1 kW, find :*

(i) Maximum temperature in the system.

(ii) Outer surface temperature of two slabs.

Draw equivalent electrical circuit of the system. (P.U., Nov. 2008)

Solution

Given :

$A = 15\text{ cm} \times 15\text{ cm}$
$= 225\text{ cm}^2 = 225 \times 10^{-4}\text{ m}^2$

$L_A = 2\text{ cm} = 0.02\text{ m}$, $L_B = 1\text{ cm} = 0.01\text{ m}$

$k_A = 50\text{ W/m.K}$, $k_B = 0.2\text{ W/m.K}$

$h_A = 200\text{ W/m}^2\text{.K}$, $h_B = 50\text{ W/m}^2\text{.K}$

$T_\infty = 25°C$, $Q = 1\text{ kW} = 1000\text{ W}$.

To find :

(*i*) Maximum temperature of the system.

(*ii*) Outer surface temperature on both slabs.

Assumptions :

(*i*) Steady state heat flow in one direction only.

(*ii*) Constant thermal conductivity.

(*iii*) No contact thermal resistance at interfaces.

Analysis : (*i*) Since heater is inserted between two slabs, therefore, two arrangements for equivalent thermal circuit are shown in Fig. 3.17(*b*) and (*c*). On left side of arrangement (*b*)

$$R_A = \frac{L_A}{k_A A} = \frac{0.02}{50 \times (225 \times 10^{-4})} = 0.0177 \text{ K/W}$$

$$R_{conv,1} = \frac{1}{h_A A} = \frac{1}{200 \times (225 \times 10^{-4})} = 0.22 \text{ K/W}$$

And total series resistance on left side :

$$R_{series1} = R_{conv,1} + R_A = 0.0177 + 0.22 = 0.2377 \text{ K/W}$$

On right side of heater

$$R_B = \frac{L_B}{k_B A} = \frac{0.01}{0.2 \times (225 \times 10^{-4})} = 2.22 \text{ K/W}$$

$$R_{conv,2} = \frac{1}{h_B A} = \frac{1}{50 \times (225 \times 10^{-4})} = 0.88 \text{ K/W}$$

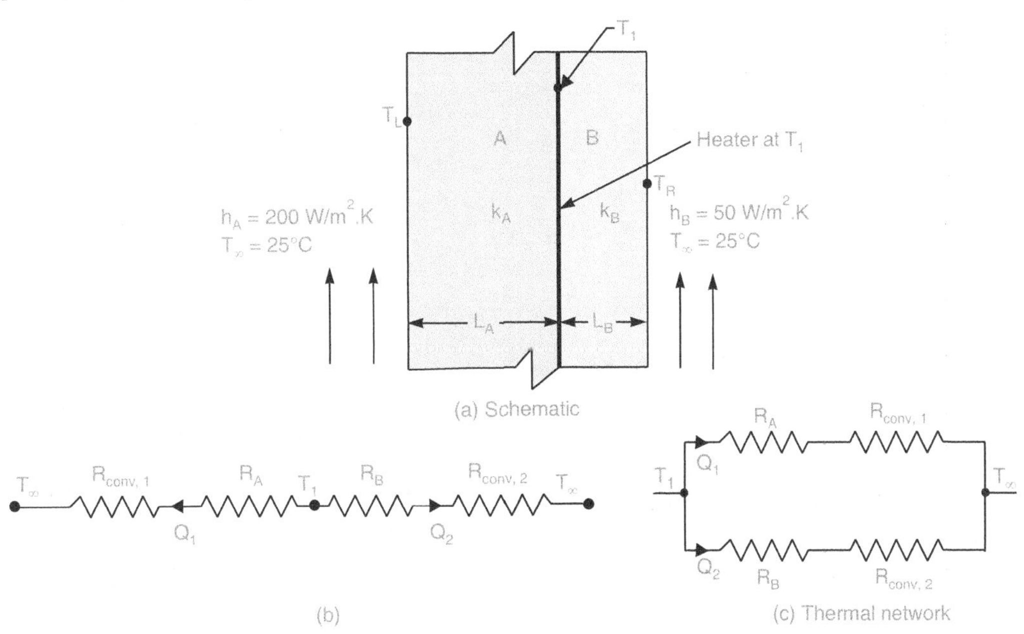

Fig. 3.17. Schematic and thermal network

And total series resistance on right side

$$R_{series2} = R_B + R_{conv,2} = 2.22 + 0.88 = 3.1 \text{ K/W}$$

The equivalent resistance of two sides as in arrangement (*c*)

or $$\frac{1}{R_{eq}} = \frac{1}{R_{series1}} + \frac{1}{R_{series2}} = \frac{1}{0.2377} + \frac{1}{3.1}$$

$$R_{eq} = 0.22077 \text{ K/W}$$

Further heat flow from heater to surroundings

$$Q = 1000 = \frac{T_1 - T_\infty}{R_{eq}} = \frac{T_1 - 25}{0.22077}$$

or $T_1 =$ **245.77°C. Ans.**

(*ii*) Heat flow towards left side

$$Q_1 = \frac{T_1 - T_\infty}{R_{series1}} = \frac{245.77 - 25}{0.2377} = 928.77 \text{ W}$$

And this heat will also flow through left face of the wall

$$Q_1 = \frac{T_1 - T_L}{R_A}$$

or $928.77 = \dfrac{245.77 - T_L}{0.0177}$ or $T_L =$ **229.33°C. Ans.**

Hence, outer surface temperature of left slab is 229.33°C

Heat flow towards right side of the wall

$$Q_2 = \frac{T_1 - T_\infty}{R_{series2}} = \frac{245.77 - 25}{3.1} = 71.21 \text{ W}$$

and $Q_2 = 71.21 = \dfrac{T_1 - T_R}{R_B}$ or $T_R = 87.6°C$

Hence the right side outer surface temperature = **87.6°C. Ans.**

Example 3.11. *The large furnace wall consists of 250 mm thick common brick layer (k = 0.65 W/m.K), lined on inside with 300 mm thick layer of magnesite bricks (k = 11.5 W/m.K). The inner side of the furnace is exposed to hot gases at 1400°C with convective heat transfer coefficient of 17.5 W/m².K, and radiative heat transfer coefficient of 23.2 W/m².K. The temperature of surrounding air is 30°C with convective heat transfer coefficient of 7.5 W/m².K and radiation heat transfer coefficient of 11.5 W/m².K. Calculate :*

(i) Rate of heat transfer through the wall per unit area.

(ii) Maximum temperature to which the common brick is subjected.

Solution

Given : Furnace wall

L_1 = 300 mm = 0.3 m, L_2 = 250 mm = 0.25 m

k_1 = 11.5 W/m.K, k_2 = 0.65 W/m.K

T_{∞_1} = 1400°C, T_{∞_2} = 30°C

h_{c_1} = 17.5 W/m².K, h_{c_2} = 7.5 W/m².K

h_{r_1} = 23.2 W/m².K, h_{r_2} = 11.5 W/m².K.

To find :

(*i*) Rate of heat transfer per unit area.

(*ii*) Maximum temperature T_2, at interface.

Assumptions :

(*i*) Steady state one dimensional heat flow.

(*ii*) No contact resistance at interface.

(*iii*) Constant properties.

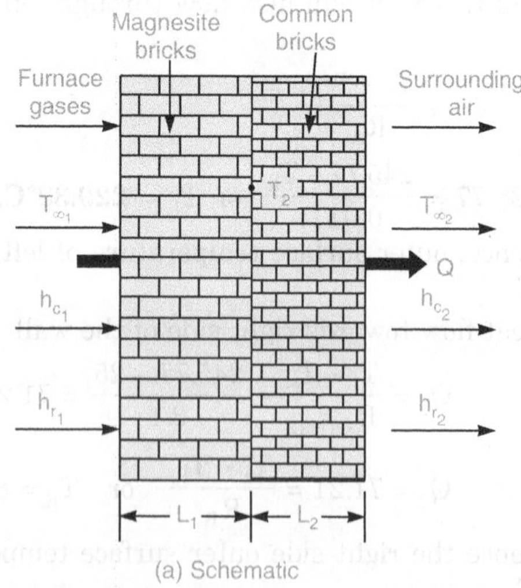

(a) Schematic

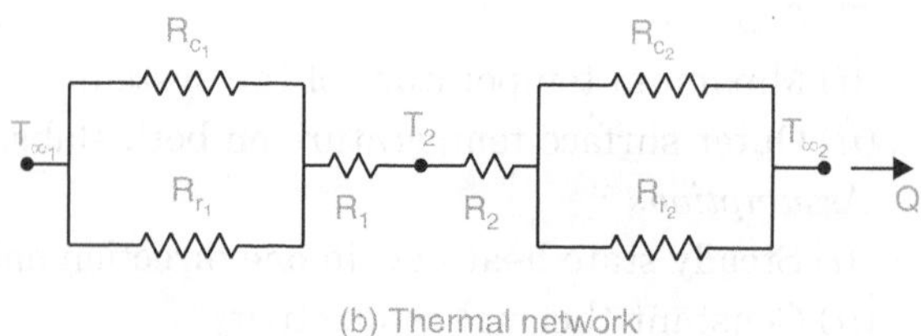

(b) Thermal network

Fig. 3.18. Schematic and thermal network for furnace wall

Analysis : (*i*) The individual thermal resistances per m² in thermal network ;

$$R_{c_1} = \frac{1}{h_{c_1}} = \frac{1}{17.5} = 0.0571 \text{ m}^2\text{.K/W}$$

$$R_{r_1} = \frac{1}{h_{r_1}} = \frac{1}{23.2} = 0.0431 \text{ m}^2\text{.K/W}$$

$$R_1 = \frac{L_1}{k_1} = \frac{0.3}{11.5} = 0.026 \text{ m}^2\text{.K/W}$$

$$R_2 = \frac{L_2}{k_2} = \frac{0.25}{0.65} = 0.3846 \text{ m}^2\text{.K/W}$$

$$R_{c_2} = \frac{1}{h_{c_2}} = \frac{1}{7.5} = 0.1333 \text{ m}^2\text{.K/W}$$

$$R_{r_2} = \frac{1}{h_{r_2}} = \frac{1}{11.5} = 0.087 \text{ m}^2\text{.K/W}$$

The resistances R_{c_1} and R_{r_1} are in parallel, their equivalent resistance

$$\frac{1}{R_{eq_1}} = \frac{1}{R_{c_1}} + \frac{1}{R_{r_1}} = \frac{1}{0.0571} + \frac{1}{0.0431} = 40.7$$

$$R_{eq_1} = \frac{1}{40.7} = 0.0245 \text{ m}^2\text{.K/W}$$

The resistances R_{c_2} and R_{r_2} are also in parallel, their equivalent resistance,

$$\frac{1}{R_{eq_2}} = \frac{1}{R_{c_2}} + \frac{1}{R_{r_2}} = \frac{1}{0.1333} + \frac{1}{0.087} = 19$$

$$R_{eq_2} = \frac{1}{19} = 0.05263 \text{ m}^2\text{.K/W}$$

Now R_{eq_1}, R_1, R_2 and R_{eq_2} are in series, and their sum.

$$\Sigma R_{th} = R_{eq_1} + R_1 + R_2 + R_{eq_2}$$
$$= 0.0245 + 0.026 + 0.3846 + 0.05263$$
$$= 0.4877 \text{ m}^2\text{.K/W}$$

The heat flow rate per unit area

$$q = \frac{T_{\infty_1} - T_{\infty_2}}{\Sigma R_{th}} = \frac{1400 - 30}{0.4877}$$

$$= \mathbf{2808.92\ W/m^2}.\ \text{Ans.}$$

(*ii*) Maximum temperature T_2, to which common brick layer is subjected can be obtained by considering resistance R_{eq_1} and R_1

$$q = \frac{T_{\infty_1} - T_2}{R_{eq_1} + R_1}$$

or $$2808.92 = \frac{1400 - T_2}{0.0245 + 0.026}$$

or $$T_2 = 1400 - 141.85 = \mathbf{1258.15°C}.\ \text{Ans.}$$

Example 3.12. *A wall 30 cm thick has size 5 m × 3 m made of red bricks (k = 0.35 W/m.K). It is covered on both sides by the layers of plaster 2 cm thick (k = 0.6 W/m.K). The wall has a window of size 1 m × 2 m. The window door is made of glass, 12 mm thick having thermal conductivity 1.2 W/m.K. Estimate the rate of heat flow through the wall. Inner and outer surface temperatures are 10°C and 40°C, respectively.* (P.U., Dec. 1995)

Solution

Given : A wall with a window with

For wall $L_w = 30\ \text{cm} = 0.3\ \text{m}$,

$A_w = 5 \times 3 = 15\ \text{m}^2$

$k_w = 0.35\ \text{W/m.K}$,

$L_P = 2\ \text{cm} = 0.02\ \text{m}$

$k_P = 0.6\ \text{W/m.K}$,

For glass : $A_g = 1 \times 2 = 2\ \text{m}^2$,

$L_g = 12\ \text{mm} = 0.012\ \text{m}$

$k_g = 1.2\ \text{W/m.K}$,

$T_1 = 10°C$

$T_2 = 40°C$.

To find : The rate of heat transfer through composite.

Assumptions :

(*i*) Steady state heat flow in one direction only.

(*ii*) Constant thermal conductivity.

(*iii*) No contact thermal resistance at interfaces.

Analysis : Since, wall has a window, therefore, the effective wall area,

$$A_e = A_w - A_g = 15 - 2 = 13\ \text{m}^2$$

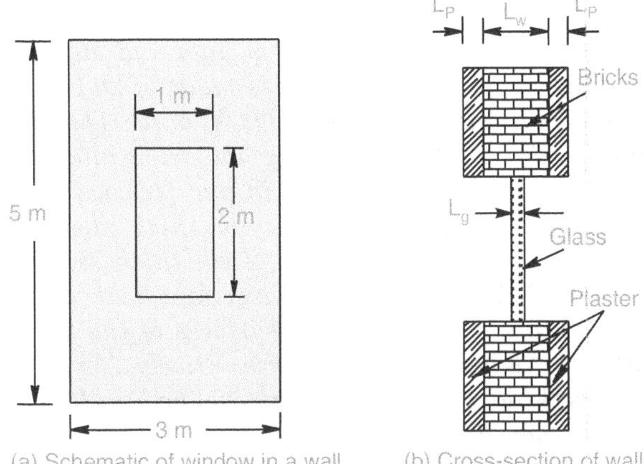

(a) Schematic of window in a wall (b) Cross-section of wall

(c) Equivalent thermal network

Fig. 3.19. Schematic and thermal network

Now referring thermal circuit Fig. 3.19(*c*), the three resistances are in series, therefore, the total wall resistance,

$$R_w = R_1 + R_2 + R_3 = \frac{L_P}{k_P A_e} + \frac{L_w}{k_w A_e} + \frac{L_P}{k_P A_e}$$

or $$R_w = \frac{1}{13} \times \left(\frac{0.02}{0.6} + \frac{0.3}{0.35} + \frac{0.02}{0.6}\right) = 0.0710\ \text{K/W}$$

The resistance of window glass :

$$R_g = \frac{L_g}{k_g A_g} = \frac{0.012}{1.2 \times 2} = 5 \times 10^{-3}\ \text{K/W}$$

The window glass resistance acts in parallel to wall resistances. Therefore, the equivalent resistance :

$$\frac{1}{R_{eq}} = \frac{1}{R_w} + \frac{1}{R_g}$$

$$R_{eq} = \frac{1}{\dfrac{1}{0.0710} + \dfrac{1}{5 \times 10^{-3}}} = 4.67 \times 10^{-3}\ \text{K/W}$$

Now heat flow rate through composite

$$Q = \frac{T_1 - T_2}{R_{eq}} = \frac{40 - 10}{4.67 \times 10^{-3}}$$

$= \mathbf{6422.5\ W.}$ Ans.

Example 3.13. *A 5 m wide, 4 m high and 40 m long klin, used to cure concrete pipe is made of 20 cm thick concrete wall and ceiling (k = 0.9 W/m.K). The klin is maintained at 40°C by injecting hot steam into it. The two ends of the klin 4 m × 5 m in size are covered by a door which is made of a 3 mm thick steel sheet (k = 22 W/m.K) covered with 2 cm thick styrofoam (k = 0.033 W/m.K). The convection heat transfer coefficient on inner and outer surfaces of the klin are 3000 W/m².K and 25 W/m².K, respectively. Neglect any heat loss through the floor, determine the heat loss from the klin when the ambient air is at – 4°C.*

Solution

Given : A klin used for curing of concrete pipes

Width, $w = 5$ m

Height, $H = 4$ m

Length, $z = 40$ m

Thickness, $L = 20$ cm

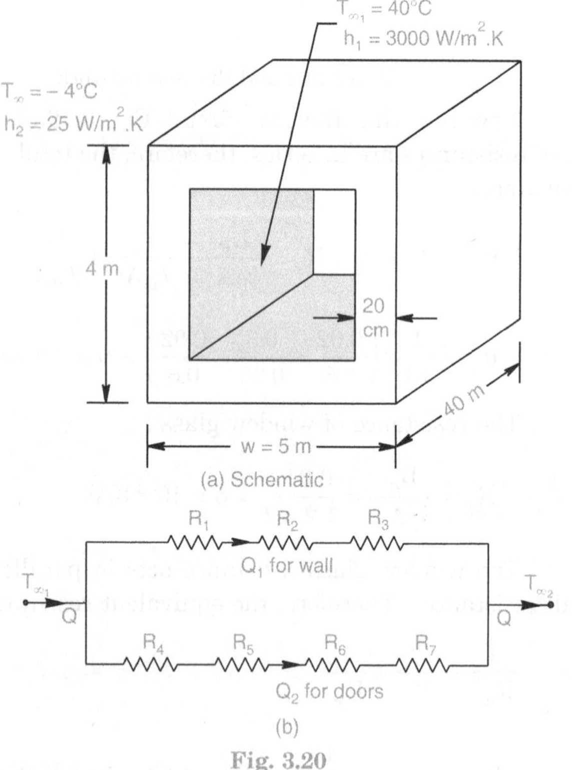

Fig. 3.20

For door $k = 0.9$ W/m.K of klin

size = 4 m × 5 m

$L_1 = 3$ mm $= 0.003$ m

$L_2 = 2$ cm $= 0.02$ m

$k_1 = 22$ W/m.K

$k_2 = 0.033$ W/m.K

$h_1 = 3000$ W/m².K

$h_2 = 25$ W/m².K

$T_{\infty_1} = 40$°C

$T_{\infty_2} = -4$°C

To find : Rate of heat transfer from klin to surroundings.

Assumptions :

(*i*) Steady state conditions.

(*ii*) No heat transfer from edges and corner of klins.

(*iii*) Heat transfer area corresponds to inner dimensions.

Analysis : For heat transfer from klin to its surroundings, through left, right, and top sides and front and back door, the thermal network is shown in Fig. 3.20 (*b*).

Since left, right and top wall construction is same, thus area of heat transfer corresponds to inner dimensions.

$$A_1 = [\underbrace{2 \times (4\text{ m} - 2 \times 0.2\text{ m})}_{\text{Left and right sides}} + \underbrace{(5\text{ m} - 2 \times 0.2\text{ m})}_{\text{Top side}}] \times \underbrace{40\text{ m}}_{\text{Length}}$$

$= 11.8$ m × 40 m $= 472$ m²

Area corresponds to inner dimension of doors

$$A_2 = 2 \times (4\text{ m} - 2 \times 0.2\text{ m}) \times (5\text{ m} - 2 \times 0.2\text{ m}) = 33.12\text{ m}^2$$

The individual thermal resistance of network

For walls, $R_1 = \frac{1}{h_1 A_1} = \frac{1}{3000 \times 472}$

$= 7.062 \times 10^{-7}$ K/W

$R_2 = \frac{L}{kA_1} = \frac{0.2}{0.9 \times 472}$

$= 4.70 \times 10^{-4}$ K/W

$R_3 = \frac{1}{h_2 A_1} = \frac{1}{25 \times 472}$

$= 8.474 \times 10^{-5}$ K/W

For doors, $R_4 = \frac{1}{h_1 A_2} = \frac{1}{3000 \times 33.12}$

$= 1.006 \times 10^{-5}$ K/W

$R_5 = \frac{L_1}{k_1 A_2} = \frac{0.003}{22 \times 33.12}$

$= 4.117 \times 10^{-6}$ K/W

$R_6 = \frac{L_2}{k_2 A_2} = \frac{0.02}{0.033 \times 33.12}$

$= 0.0183$ K/W

$$R_7 = \frac{1}{h_2 A_2} = \frac{1}{25 \times 33.12} = 0.0012 \text{ K/W}$$

The resistances R_1, R_2 and R_3 for wall are in series, its total resistance

$$\Sigma R_{s_1} = R_1 + R_2 + R_3$$
$$= 7.062 \times 10^{-7} + 4.70 \times 10^{-4} + 8.474 \times 10^{-5}$$
$$= 5.554 \times 10^{-4} \text{ K/W}$$

Further, the resistances R_4, R_5, R_6 and R_7 for door are also in series, its total resistance

$$\Sigma R_{s_2} = R_4 + R_5 + R_6 + R_7$$
$$= 1.006 \times 10^{-5} + 4.117 \times 10^{-6} + 0.0183 + 0.0012$$
$$= 0.0195 \text{ K/W}$$

The resistances ΣR_{s_1} and ΣR_{s_2} are parallel to each other, its equivalent resistance

$$\frac{1}{R_{eq}} = \frac{1}{\Sigma R_{s_1}} + \frac{1}{\Sigma R_{s_2}}$$
$$= \frac{1}{5.554 \times 10^{-4}} + \frac{1}{0.0195} = 1851.63$$

or
$$R_{eq} = \frac{1}{1851.63} = 5.40 \times 10^{-4} \text{ K/W}$$

The heat flow rate through klin walls and doors

$$Q = \frac{T_{\infty_1} - T_{\infty_2}}{R_{eq}} = \frac{40 - (-4)}{5.40 \times 10^{-4}} = \frac{44}{5.40 \times 10^{-4}}$$
$$= 81.47 \times 10^3 \text{ W} = \mathbf{81.47 \text{ kW}}. \text{ Ans.}$$

Example 3.14. *A 3 m high and 5 m wide wall consists of 16 cm × 22 cm cross-section horizontal bricks (k = 0.72 W/m.K) separated by 3 cm thick mortar layers (k = 0.25 W/m.K). The brick wall also consists of 2 cm thick plaster (k = 0.22 W/m.K) layers on each side of brick and 3 cm thick rigid foam (k = 0.026 W/m.K) on the inner side of the wall as shown in Fig. 3.21(a). The indoor and outdoor temperatures are 55 and 25°C and convection heat transfer coefficients on inner and outer sides 10 W/m².K and 25 W/m².K, respectively. Assume one dimensional heat transfer and disregard radiation, determine the rate of heat transfer through the wall.*

(A.U., July 1999)

Solution

Given : A composite plane wall

Wall size : 3 m × 5 m

Brick size : 16 cm × 22 cm

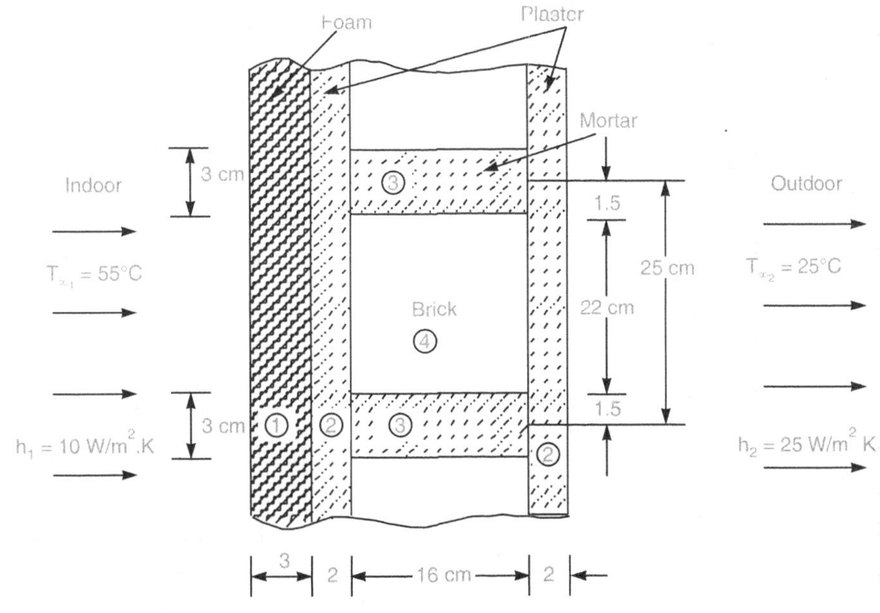

(a) Schematic

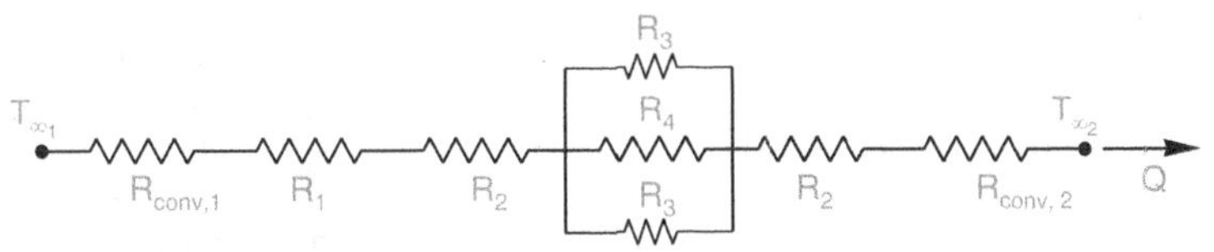

(b) Equivalent thermal network

Fig. 3.21

$L_1 = 3$ cm $= 0.03$ m, $L_2 = 2$ cm $= 0.02$ m

$L_3 = 16$ cm $= 0.16$ m, $L_4 = 16$ cm $= 0.16$ m

$k_1 = 0.026$ W/m².K, $k_2 = 0.22$ W/m.K

$k_3 = 0.25$ W/m.K, $k_4 = 0.72$ W/m.K

$T_{\infty_1} = 55°C$, $T_{\infty_2} = 25°C$,

$h_1 = 10$ W/m².K, $h_2 = 25$ W/m².K.

To find : Rate heat transfer through wall.

Assumptions :

(*i*) Steady state conditions.

(*ii*) No contact resistance at interfaces.

(*iii*) Wall size 3 m × 5 m is too large, but construction repeats itself after every 25 cm distance in vertical direction, therefore, for analysis, considering a portion of wall 0.25 m high and 1 m deep as representation of entire wall.

Analysis : Wall area,

$$A_{wall} = 3 \text{ m} \times 5 \text{ m} = 15 \text{ m}^2$$

Area under consideration,

$$A_1 = 0.25 \text{ m} \times 1 \text{ m} = 0.25 \text{ m}^2$$

Central Area of mortar,

$$A_2 = A_{mortar} = 0.015 \times 1 = 0.015 \text{ m}^2$$

Area of brick,

$$A_3 = A_{brick} = 0.22 \times 1 = 0.22 \text{ m}^2$$

The individual thermal resistance of thermal network ;

$$R_{conv,\,1} = \frac{1}{h_1 A_1} = \frac{1}{10 \times 0.25} = 0.4 \text{ K/W}$$

$$R_1 = R_{foam} = \frac{L_1}{k_1 A_1} = \frac{0.03}{0.026 \times 0.25} = 4.6 \text{ K/W}$$

$$R_2 = R_{plaster} = \frac{L_2}{k_2 A_1} = \frac{0.02}{0.22 \times 0.25} = 0.36 \text{ K/W}$$

$$R_3 = R_{mortar} = \frac{L_3}{k_3 A_2} = \frac{0.16}{0.25 \times 0.015} = 42.67 \text{ K/W}$$

$$R_4 = R_{brick} = \frac{L_3}{k_4 A_3} = \frac{0.16}{0.72 \times 0.22} = 1.01 \text{ K/W}$$

$$R_{conv,\,2} = \frac{1}{h_2 A} = \frac{1}{25 \times 0.25} = 0.16 \text{ K/W}$$

The three resistance R_3, R_4 and R_3 are parallel to each other and their equivalent resistance

$$\frac{1}{R_{eq}} = \frac{1}{R_3} + \frac{1}{R_4} + \frac{1}{R_3} = \frac{1}{42.67} + \frac{1}{1.01} + \frac{1}{42.67}$$

$$= 1.036 \text{ W/K}$$

or

$$R_{eq} = \frac{1}{1.036} = 0.964 \text{ K/W}$$

Now the equivalent resistance is in series with other resistances and total resistance

$$\Sigma R_{th} = R_{conv,\,1} + R_1 + R_2 + R_{eq} + R_2 + R_{conv,\,2}$$

$$= 0.4 + 4.6 + 0.36 + 0.964 + 0.36 + 0.16$$

$$= 6.844 \text{ K/W}$$

Then the heat transfer rate through the portion (0.25 m²) is

$$Q = \frac{T_{\infty_1} - T_{\infty_2}}{\Sigma R_{th}} = \frac{55 - 25}{6.844} = 4.383 \text{ W}$$

For heat transfer rate from total surface area of wall

$$= Q \times \frac{\text{Area of wall}}{\text{Area considered}}$$

$$= 4.383 \times \frac{15}{0.25} = \mathbf{262\ W.} \quad \text{Ans.}$$

Example 3.15. *Consider a 5 m high and 8 m long and 0.22 m thick wall whose representation is shown in Fig. 3.22(a). The thermal conductivity of various materials used are $k_A = k_F = 2$, $k_B = 8$, $k_C = 20$, $k_D = 15$, and $k_E = 35$ W/m.K. The left surface of the wall is maintained at uniform temperatures of 300°C. The right surface is exposed to convection environment at 50°C with $h = 20$ W/m².K. Determine (a) one dimensional heat transfer rate through the wall, (b) temperature at the point where section B, D and E meet, and (c) temperature drop across the section F.* (V.T.U., May 2001)

Solution

Given : A composite wall

$$A_{wall} = 5 \text{ m} \times 8 \text{ m} = 40 \text{ m}^2$$

For representative cross-section of wall

$k_A = k_F = 2$ W/m.K,

$k_B = 8$ W/m.K,

$k_C = 20$ W/m.K,

$k_D = 15$ W/m.K

$k_E = 35$ W/m.K,

$L_A = 1$ cm $= 0.01$ m,

$L_B = L_C = 5$ cm $= 0.05$ m

$L_D = L_E = 10$ cm $= 0.1$ m,

$L_F = 6$ cm $= 0.06$ m

$z = 6$ cm $+ 6$ cm $= 12$ cm $= 0.12$ m,

$z_B = z_C = 4$ cm $= 0.04$ m

$w = 100$ cm $= 1$ m,

$z_D = z_E = 6$ cm $= 0.06$ m,

$T_1 = 300°C$,

$T_\infty = 50°C$,

$h = 20$ W/m².K.

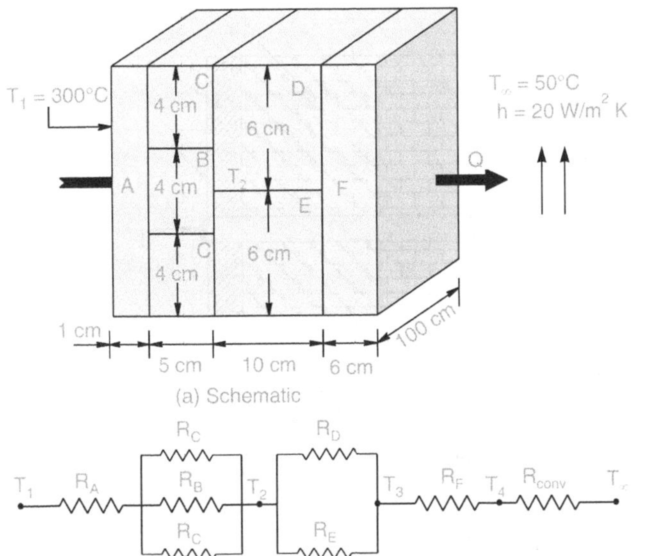

Fig. 3.22. Representation of wall 5 m × 8 m × 0.22 m in size

To find :

(*i*) Heat transfer rate through composite wall,

(*ii*) Temperature T_2 at the point where section B, D and E meet, and

(*iii*) Temperature drop across the section F.

Assumptions :

(*i*) Steady state heat conduction.

(*ii*) No contact resistance at interfaces.

(*iii*) No radiation heat transfer in the system.

Analysis : (*i*) Equivalent thermal network for given composite wall is shown in Fig. 3.22 (*b*).

The cross-section area of representative portion of wall

$$A_1 = w \times z = (1 \text{ m}) \times (0.12 \text{ m}) = 0.12 \text{ m}^2$$

Area for section B and C,

$$A_2 = w \times z_B = 1 \text{ m} \times 0.04 = 0.04 \text{ m}^2$$

Area for section B and D,

$$A_3 = w \times z_D = 1 \text{ m} \times 0.06 \text{ m} = 0.06 \text{ m}^2$$

Individual thermal resistances of thermal network

$$R_A = \frac{L_A}{k_A A_1} = \frac{0.01}{2 \times 0.12} = 0.04167 \text{ K/W}$$

$$R_B = \frac{L_B}{k_B A_2} = \frac{0.05}{8 \times 0.04} = 0.15625 \text{ K/W}$$

$$R_C = \frac{L_C}{k_C A_2} = \frac{0.05}{20 \times 0.04} = 0.0625 \text{ K/W}$$

$$R_D = \frac{L_D}{k_D A_3} = \frac{0.1}{15 \times 0.06} = 0.111 \text{ K/W}$$

$$R_E = \frac{L_E}{k_E A_3} = \frac{0.1}{35 \times 0.06} = 0.0476 \text{ K/W}$$

$$R_F = \frac{L_F}{k_F A_1} = \frac{0.06}{2 \times 0.12} = 0.25 \text{ K/W}$$

$$R_{conv} = \frac{1}{hA_1} = \frac{1}{20 \times 0.12} = 0.4167 \text{ K/W}$$

The resistance R_B and R_C are parallel as shown in thermal network Fig. 3.32(*b*), their equivalent resistance

$$\frac{1}{R_{eq_1}} = \frac{1}{R_C} + \frac{1}{R_B} + \frac{1}{R_C}$$

$$= \frac{1}{0.0625} + \frac{1}{0.15625} + \frac{1}{0.0625}$$

$$R_{eq_1} = \frac{1}{38.4} = 0.0260 \text{ K/W}$$

Further, the resistance R_D and R_E are parallel as shown and their equivalent resistance :

$$\frac{1}{R_{eq_2}} = \frac{1}{R_D} + \frac{1}{R_E}$$

$$= \frac{1}{0.111} + \frac{1}{0.0476} = 30.017 \text{ W/K}$$

$$\therefore \quad R_{eq_2} = \frac{1}{30.017} = 0.0333 \text{ K/W}$$

Now, resistances R_A, R_{eq_1}, R_{eq_2}, R_F and R_{conv} are in series as shown in Fig. 3.22(*c*)

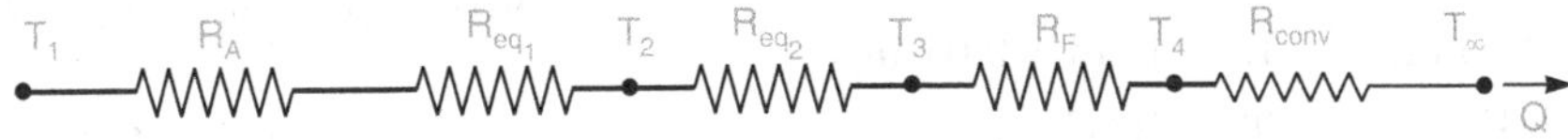

Fig. 3.22. (*c*) Modified thermal network

Total thermal resistance

$$\Sigma R_{th} = R_A + R_{eq_1} + R_{eq_2} + R_F + R_{conv}$$

$$= 0.04167 + 0.0260 + 0.0333 + 0.25 + 0.4167$$

$$= 0.7677 \text{ K/W}$$

Heat flow rate in representative section of wall

$$Q = \frac{T_s - T_\infty}{\Sigma R_{th}} = \frac{300 - 50}{0.7677} = 325.65 \text{ W}$$

Heat transfer rate from total surface of the wall

$$= Q \times \frac{\text{Wall area}}{\text{Representative area}}$$

$$= Q \times \frac{A_{wall}}{A_1} = 325.65 \times \frac{40}{0.12}$$

$$= 108.55 \times 10^3 \text{ W} = \mathbf{108.55 \text{ kW}}. \quad \text{Ans.}$$

(*ii*) Temperature T_2 :

Considering the heat flow through resistances R_A and R_{eq_1},

$$Q = \frac{T_s - T_2}{R_A + R_{eq_1}}$$

or
$$T_2 = T_s - Q \times (R_A + R_{eq_1})$$

$$= 300 - 325.65 \times (0.04167 + 0.0260)$$

$$= 300 - 22.03 \approx \mathbf{278°C}. \quad \text{Ans.}$$

(*iii*) Temperature drop across section F :

The heat flow through section F can be expressed as :

$$Q = \frac{T_3 - T_4}{R_F} \quad \text{or} \quad T_3 - T_4 = Q\, R_F$$

$$= 325.65 \times 0.25 = \mathbf{81.41°C}. \quad \text{Ans.}$$

Example 3.16. *A composite insulating wall has three layers of material held together by 3 cm diameter aluminium rivet per 0.1 m² of surface. The layers of material consists of 10 cm thick brick with hot surface at 200°C, 1 cm thick wood with cold surface at 10°C. These two layers are interposed by third layer of insulating material 25 cm thick. The conductivity of the materials are :*

$k_{brick} = 0.93$ W/m.K, $k_{insulation} = 0.12$ W/m.K

$k_{wood} = 0.175$ W/m.K, $k_{Aluminium} = 204$ W/m.K

Assuming one dimensional heat flow. Calculate the percentage increase in heat transfer rate due to rivets.

(N.M.U., May 1998)

Solution

Given : The schematic is shown in Fig. 3.23 (*a*)

$A_w = 0.1 \text{ m}^2$, $d = 3 \text{ cm} = 0.03 \text{ m}$,

$L_1 = 10 \text{ cm} = 0.1 \text{ m}$, $T_i = 200°C$

$L_2 = 25 \text{ cm} = 0.25 \text{ m}$, $L_3 = 1 \text{ cm} = 0.01 \text{ m}$,

$k_1 = 0.93$ W/m.K, $T_o = 10°C$

$k_2 = 0.12$ W/m.K, $k_3 = 0.175$ W/m.K,

$k_4 = 204$ W/m.K.

To find :

(*i*) Heat flow rate without rivet.

(*ii*) Heat flow rate with rivet.

(*iii*) Percentage increase in heat flow due to rivet.

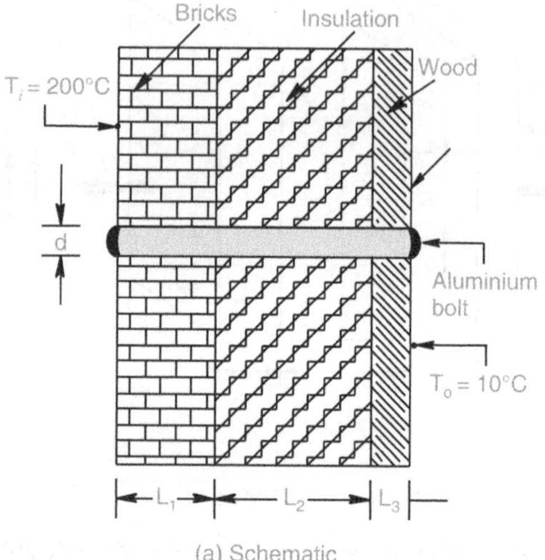

(a) Schematic

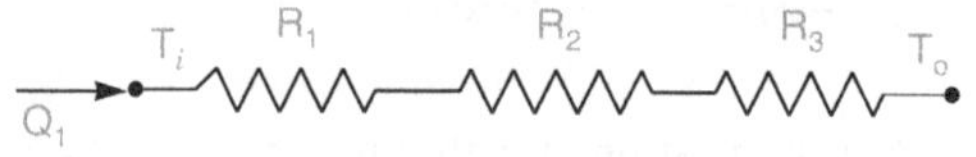

(b) Thermal network without rivet

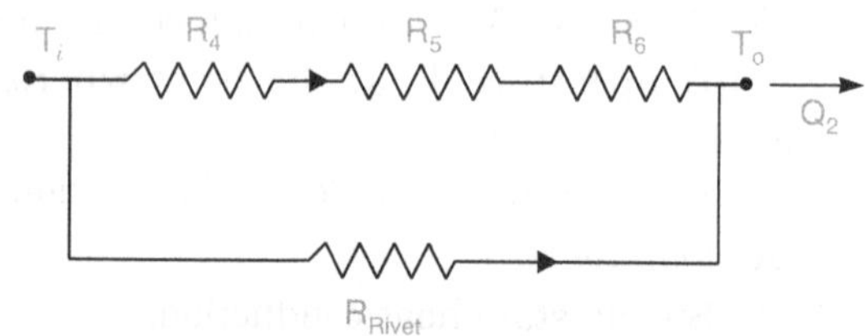

(c) Thermal network with rivet

Fig. 3.23

Assumptions :

(*i*) Steady state heat flow in one direction only.

(*ii*) Constant thermal conductivity.

(*iii*) No contact thermal resistances at interfaces.

Analysis : (*i*) The resistances acting in path of heat flow without rivets as shown in 3.23 (*b*) :

$$\Sigma R_{series} = \frac{1}{A_w}\left[\frac{L_1}{k_1} + \frac{L_2}{k_2} + \frac{L_3}{k_3}\right]$$

$$= \frac{1}{0.1}\left[\frac{0.1}{0.93} + \frac{0.25}{0.12} + \frac{0.01}{0.175}\right] = 22.48 \text{ K/W}$$

Heat transfer rate without rivet,

$$Q_1 = \frac{\Delta T}{\Sigma R_{series}} = \frac{200 - 10}{22.48} = \mathbf{8.45 \text{ W}}. \quad \text{Ans.}$$

(*ii*) The heat transfer rate with rivet : Thermal network in Fig. 3.23 (*c*)

The area of rivet through which heat is conducted (*i.e.,* cross-sectional area)

$$A_{Rivet} = (\pi/4)\,(d_{Rivet})^2 = (\pi/4) \times (0.03)^2$$
$$= 7.068 \times 10^{-4}\ m^2$$

$$R_{Rivet} = \frac{L_{Rivet}}{A_{Rivet} k_{Rivet}} = \frac{L_1 + L_2 + L_3}{A_{Rivet} k_4}$$

$$= \frac{0.1 + 0.25 + 0.01}{7.068 \times 10^{-4} \times 204} = 2.496\ K/W$$

With consideration of rivet, the net effective area of the wall

$$A_e = A_w - A_{Rivet} = 0.1 - 7.068 \times 10^{-4}$$
$$= 0.099293\ m^2$$

Now the wall resistance ;

$$\Sigma R_w = \frac{1}{A_e}\left[\frac{L_1}{k_1} + \frac{L_2}{k_2} + \frac{L_3}{k_3}\right]$$

$$= \frac{1}{0.099293}\left[\frac{0.1}{0.93} + \frac{0.25}{0.12} + \frac{0.01}{0.175}\right]$$

$$= 22.64\ K/W$$

and $$\frac{1}{R_{eq}} = \frac{1}{\Sigma R_w} + \frac{1}{R_{Rivet}} = \frac{1}{22.64} + \frac{1}{2.496}$$

$$R_{eq} = 2.245\ K/W$$

Further, heat flow through composite wall

$$Q_2 = \frac{T_i - T_o}{R_{eq}} = \frac{200 - 10}{2.245} = \mathbf{84.6\ W.}$$ Ans.

(*iii*) The percentage increase in heat flow due to rivet

$$\frac{Q_2 - Q_1}{Q_1} = \frac{84.6 - 8.45}{8.45} \approx \mathbf{9 \ \text{or}\ 900\%.}$$ Ans.

Example 3.17. *The following data refers to wall of an industrial furnace :*

Temperature of gases in the furnace = 1700°C

Temperature of air outside the furnace = 35°C

Combined convective and radiative heat transfer coefficient of the furnace gases = 50 W/m²$.K. Heat transfer coefficient of surrounding air = 10 W/m^2.K.

The inner wall of the furnace is made of refractory bricks [k = 0.28 (1 + 0.000833 T) W/m.°C], 250 mm thick and it is followed by diatomaceous brick layer, [k = 0.113 (1 + 0.000206 T) W/m.°C].

Calculate the thickness of diatomaceous brick layer, so the heat loss to surrounding air should not exceed 900 W/m^2.

Solution

Given : Wall of an industrial furnace as shown in Fig. 3.24.

T_{∞_1} = 1700°C, h_1 = 50 W/m^2.K

T_{∞_2} = 35°C, h_2 = 10 W/m^2.K

L_1 = 250 mm = 0.25 m

$k_1 = 0.28\,(1 + 8.33 \times 10^{-4}\ T)$ W/m°C

$k_2 = 0.113(1 + 2.06 \times 10^{-4}\ T)$ W/m°C.

q = 900 W/m^2.

To find : Thickness of diatomaceous brick layer.

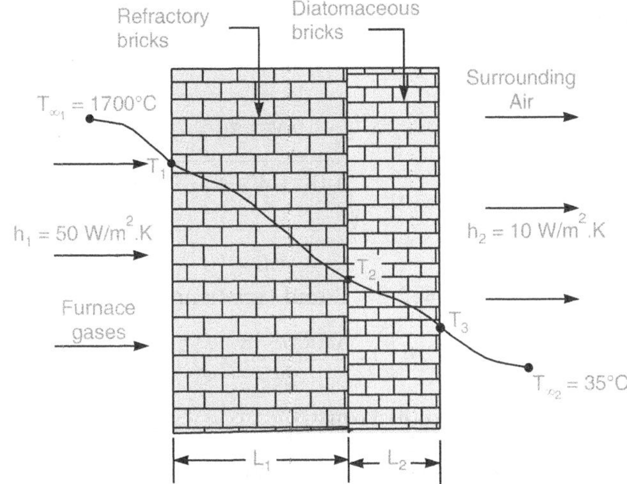

Fig. 3.24. Schematic of furnace wall

Assumptions :

(*i*) No contact resistance at interface.

(*ii*) Steady state one dimensional heat flow.

Analysis : The average thermal conductivity of each layer can be expressed by eqn. (1.20) as :

$$k_1 = 0.28\left[1 + 8.33 \times 10^{-4}\left(\frac{T_1 + T_2}{2}\right)\right]\ W/m.°C$$

$$k_2 = 0.113\left[1 + 2.06 \times 10^{-4}\left(\frac{T_2 + T_3}{2}\right)\right]\ W/m.°C$$

where T_1, T_2 and T_3 are temperature of surfaces as shown in Fig. 3.24 and for steady state heat transfer, T_1 and T_3 are calculated as :

$$q = h_1\,(T_{\infty_1} - T_1)$$

or $$T_1 = T_{\infty_1} - \frac{q}{h_1} = 1700 - \frac{900}{50} = 1682°C$$

and $$q = h_2\,(T_3 - T_{\infty_2})$$

or $$T_3 = \frac{q}{h_2} + T_{\infty_2} = \frac{900}{10} + 35 = 125°C$$

Now for refractory brick layer

$$q = \frac{k_1 (T_1 - T_2)}{L_1}$$

$$= 0.28 \times \left[1 + 8.33 \times 10^{-4}\left(\frac{T_1 + T_2}{2}\right)\right] \times \left(\frac{T_1 - T_2}{L_1}\right)$$

or $$\frac{900}{0.28} = [1 + 4.165 \times 10^{-4} (1682 + T_2)] \frac{(1682 - T_2)}{0.25}$$

or $$\frac{900 \times 0.25}{0.28} = (1682 - T_2) + 4.165 \times 10^{-4} \times (1682^2 - T_2^2)$$

$$803.57 = 1682 - T_2 + 4.165 \times 10^{-4} \times (2829124 - T_2^2)$$

$$= 1682 - T_2 + 1178.33 - 4.165 \times 10^{-4}\, T_2^2$$

Rearranging,

$$4.165 \times 10^{-4}\, T_2^2 + T_2 - 2056.76 = 0$$

or $$T_2 = \frac{-1.0 \pm \sqrt{1.0^2 + 4 \times 4.165 \times 10^{-4} \times 2056.76}}{2 \times 4.165 \times 10^{-4}}$$

$= 1325.25°C$ and (ignoring –ve sign)

Then $$k_1 = 0.28\left[1 + \frac{0.000833}{2} \times (1682 + 1325.25)\right]$$

$= 0.630$ W/m°C

and $$k_2 = 0.113\left[1 + 0.000206 \times \left(\frac{1325.25 + 125}{2}\right)\right]$$

$= 0.130$ W/m°C

Further heat transfer through diatomaceous bricks layer

$$q = \frac{k_2 (T_2 - T_3)}{L_2}$$

or $$L_2 = \frac{0.130 \times (1325.25 - 125)}{900}$$

$= 0.17337 =$ **173.37 mm.** Ans.

The thickness of diatomaceous brick layer is 173.37 mm.

3.4. THERMAL CONTACT RESISTANCE

In composite system, the contact resistance develops when two surfaces in contact do not fit tightly due to significant surface asperities. The direct contact between the solid surfaces takes place at a limited number of spots and the voids between them are usually filled with air or surrounding fluid as shown in Fig. 3.25. Heat transfer is therefore, due to conduction across the actual contact spots and layer of fluid filling the voids, Fig. 3.25(*b*). The convection and radiation heat transfer in thin layer of fluid are negligible. The thermal conductivity of the fluid is very less than that of solid, the interface acts as a strong resistance to heat flow, this resistance is called as **contact resistance.** This resistance is primarily function of surface roughness, the pressure holding the two surfaces in contact, the interference fluid and interface temperature. The contact resistance acts as parallel resistance that due to contact spots and air voids as shown in Fig. 3.25(*d*).

If the heat flux through the two solid surfaces in contact is q and the temperature difference across the contact (fluid gap) is ΔT $(= T_{c_1} - T_{c_2})$ as shown in Fig. 3.25(*e*), the contact resistance is defined by

$$R_{contact} = \frac{T_{c_1} - T_{c_2}}{q} \text{ (m}^2\text{.K/W)} \qquad ...(3.36)$$

$$= \frac{\text{Temperature drop across contact surfaces}}{\text{Heat flux}}$$

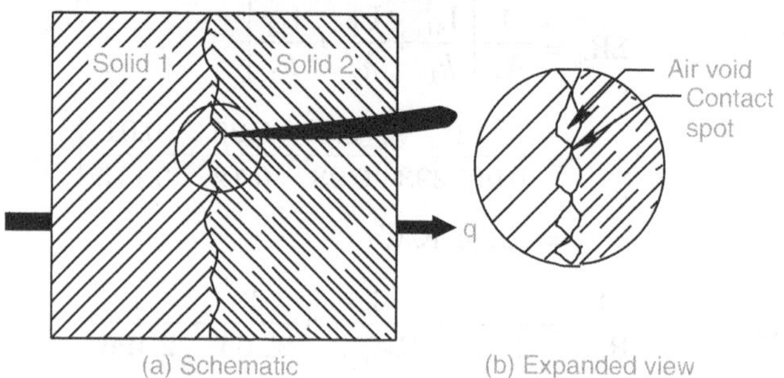

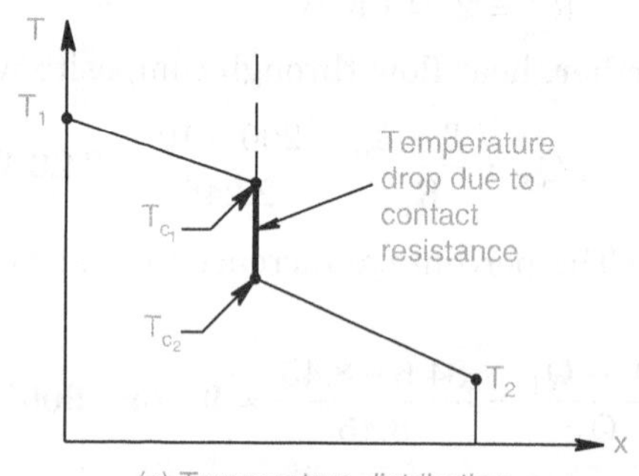

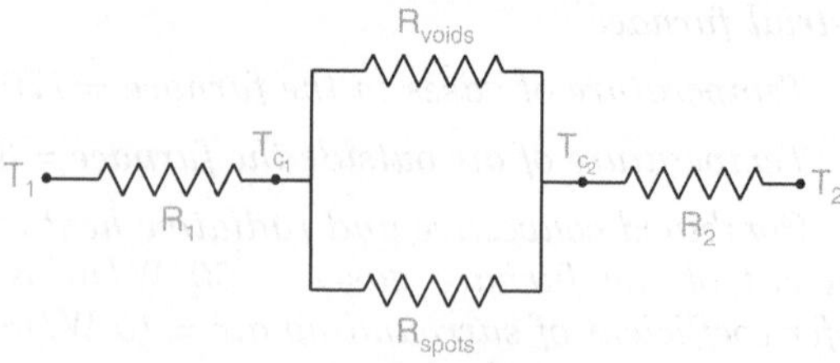

Fig. 3.25. Contact resistance between two solids

An increase in contact pressure can reduce the contact resistance drastically. The Table 3.1 gives some representative values, that illustrate the effect of pressure on thermal contact resistance for some material.

TABLE 3.1. Typical values of thermal contact resistance

Material	*Thermal contact resistance ($m^2.K/W$)*	
	Contact pressure, 1 bar	*Contact pressure, 100 bar*
Aluminium	0.00015–0.0005	0.00002–0.00004
Copper	0.0001–0.001	0.00001–0.00005
Magnesium	0.00015–0.00035	0.00002–0.00004
Stainless steel	0.0006–0.0025	0.00007–0.0004

The properties of interfacial fluid also affects the contact resistance as given in Table 3.2. A viscous fluid like glycerine on the interface can reduce the contact resistance 10 times with respect to air at a given pressure. A thermally conducting liquid is called a **thermal grease.** A high conducting pastes like silicon oil are applied between the contact surfaces before they are pressed against each other and these are usually used to mount the electronic components to heat sink.

TABLE 3.2. Thermal contact resistance for aluminium-aluminium interface with different interfacial fluids having 10 μm surface roughness under 1 bar contact pressure

Interfacial fluid	*Contact resistance ($m^2.K/W$)*
Air	2.75×10^{-4}
Helium	1.05×10^{-4}
Hydrogen	0.72×10^{-4}
Silicon oil	0.525×10^{-4}
Glycerin	0.265×10^{-4}

Example 3.18. *Two large aluminium plates (k = 240 W/m.K), each 1 cm thick with 10 μm surface roughness are placed in contact under pressure of 1 bar in air (k = 0.026 W/m.K). The temperature at inside and outside surfaces are 400°C and 150°C. Calculate (a) the heat flux, and (b) temperature drop due to contact resistance.*

Solution

Given : Two large aluminium plates in contact.

k = 240 W/m.K, L = 1 cm = 0.01 m

$L_a = 10\ \mu m = 10 \times 10^{-6}$ m, T_1 = 400°C

T_2 = 150°C, k_a = 0.026 W/m.K.

To find :

(*a*) Heat flux

(*b*) Temperature drop $(T_3 - T_4)$ due to contact resistance.

Analysis : (*a*) The individual thermal resistance per 1 m^2 area

$$R_1 = R_3 = \frac{L}{k} = \frac{0.01}{240} = 4.167 \times 10^{-5}\ m^2.K/W$$

From Table 3.1, contact resistance $R_{contact}$ is 2.75×10^{-4} m^2.K/W

i.e., $R_2 = R_{contact} = 2.75 \times 10^{-4}\ m^2.K/W$

The heat flux

$$q = \frac{\Delta T}{\Sigma R_{th}} = \frac{T_1 - T_2}{R_1 + R_2 + R_3}$$

or

$$q = \frac{400 - 150}{4.167 \times 10^{-5} + 2.75 \times 10^{-4} + 4.167 \times 10^{-5}}$$

$$= \mathbf{2.79 \times 10^4\ W/m^2.}\ \textbf{Ans.}$$

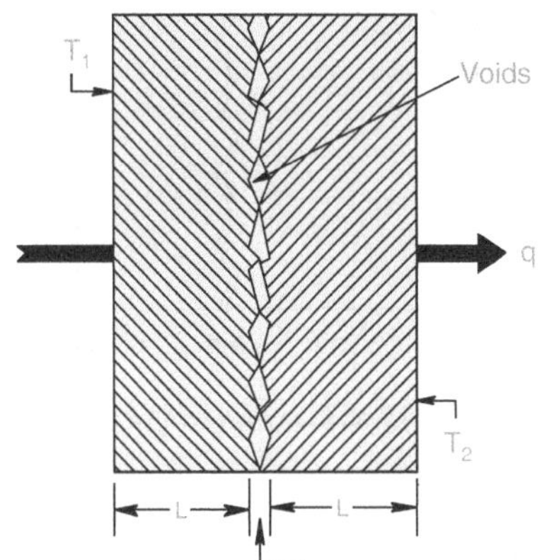

(a) Schematic of aluminium plates in contact

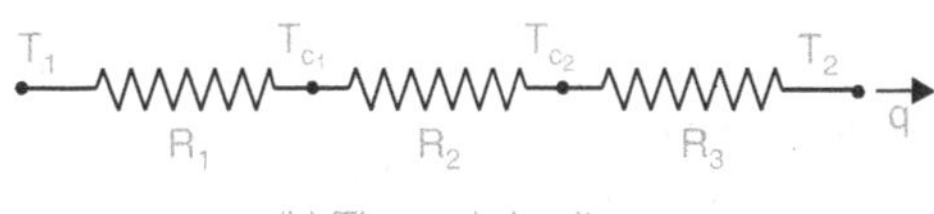

(b) Thermal circuit

Fig. 3.26

(*b*) The temperature drop across contact resistance :

Heat flux across contact resistance is given as :

$$q = \frac{T_{c_1} - T_{c_2}}{R_{contact}}$$

or $(T_{c_1} - T_{c_2}) = q\,R_{contact} = 2.79 \times 10^4 \times 2.75 \times 10^{-4}$

$$= \mathbf{7.67°C.}\ \textbf{Ans.}$$

Example 3.19. *A plane composite slab with unit cross-sectional area is made up material A, (100 mm thick, k_A = 60 W/m.K) and material B (10 mm thick, k_B = 2 W/m.K). Thermal contact resistance at the interface is 0.003 m^2.K/W. The temperature of open side of slab A is 300°C and that of open side of slab B is 50°C. Calculate:*

(i) Rate of heat flow through the slab.

(ii) Temperature on both sides of interface.

(P.U., May 2009)

Solution

Given : A composite wall with contact resistance as shown in Fig. 3.27.

$L_A = 100$ mm = 0.1 m $\quad L_B = 10$ mm = 0.01 m

$k_A = 60$ W/m.K $\quad k_B = 2$ W/m.K

$T_1 = 300°C \quad T_4 = 50°C$

$R_{cont} = 0.003$ m²K/W

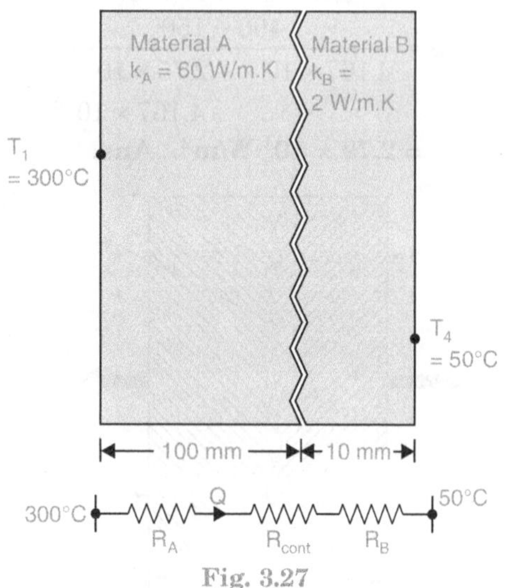

Fig. 3.27

To find :

(*a*) Heat transfer rate through slab.

(*b*) Temperature on both sides interface.

Analysis: Assuming 1 m² slab area, and calculating material resistance.

Thermal resistance of material A

$$R_A = \frac{L_A}{k_A} = \frac{0.1}{60}$$
$$= 1.667 \times 10^{-3} \text{ m}^2\text{.K/W}$$

Thermal resistance of material B ;

$$R_B = \frac{L_B}{k_B} = \frac{0.01}{2} = 0.005 \text{ m}^2\text{.K/W}$$

Total thermal resistance in the circuit

$$\sum R_{th} = R_A + R_{cont} + R_B$$
$$= 1.667 \times 10^{-3} + 0.003 + 0.005$$
$$= 9.667 \times 10^{-3} \text{ m}^2\text{.K/W}$$

(*a*) Heat flow rate

$$Q = \frac{\Delta T}{\sum R_{th}} = \frac{300 - 50}{9.667 \times 10^{-3}}$$
$$= \mathbf{25{,}862\ W.} \quad \text{Ans.}$$

(*b*) Temperature on interfaces.

The heat flow rate in material A is given by

$$Q = \frac{T_1 - T_2}{R_A}$$

or
$$T_2 = T_1 - Q R_A$$
$$= 300 - 25{,}862 \times 1.667 \times 10^{-3}$$
$$= \mathbf{256.9°C.} \quad \text{Ans.}$$

Heat transfer rate in material B

$$Q = \frac{T_3 - T_4}{R_B}$$

or
$$T_3 = Q R_B + T_4$$
$$= 25{,}862 \times 0.005 + 50 = \mathbf{179.3°C.} \quad \text{Ans.}$$

Example 3.20. *A furnace wall is made of three layers. First layer is of insulation (k = 0.6 W/m.K), 12 cm thick. Its face is exposed to gases at 870°C with convection coefficient of 110 W/m².K. It is covered with (backed with), a 10 cm thick layer of fire brick (k = 0.8 W/m.K) with a contact resistance of 2.6 × 10⁻⁴ m².K/W between first and second layer. The third layer is a plate of 10 cm thickness (k = 4 W/m.K) with a contact resistance between second and third layer of 1.5 × 10⁻⁴ m².K/W. The plate is exposed to air at 30°C with convection coefficient of 15 W/m².K. Determine the heat flow rate and overall heat transfer coefficient.* (V.T.U., July 2002)

Solution

Given : A composite wall

$k_1 = 0.6$ W/m.K $\quad L_1 = 12$ cm = 0.12 m

$T_{\infty_1} = 870°C \quad h_1 = 110$ W/m².K

$L_2 = 10$ cm = 0.1 m $\quad k_2 = 0.8$ W/m.K

$R_{c_1} = 2.6 \times 10^{-4}$ m².K/W $\quad L_3 = 10$ cm = 0.1 m

$k_3 = 4$ W/m.K $\quad R_{c_2} = 1.5 \times 10^{-4}$ m².K/W

$T_{\infty_2} = 30°C \quad h_2 = 15$ W/m².K.

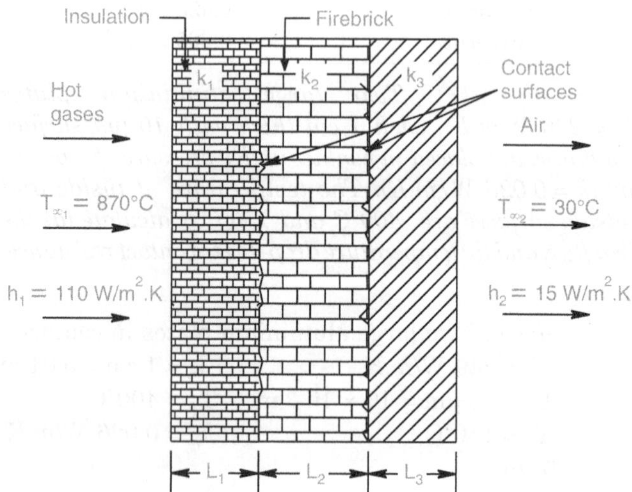

Fig. 3.28. Schematic of furnace wall

To find :

(*i*) Heat flow rate through composite wall, and

(*ii*) Overall heat transfer coefficient.

Assumptions :

(*i*) Steady state one dimensional heat conduction in the wall.

(*ii*) 1 m^2 area of wall surface.

(*iii*) Constant properties.

Analysis : (*i*) The equivalent thermal network of composite wall is shown below :

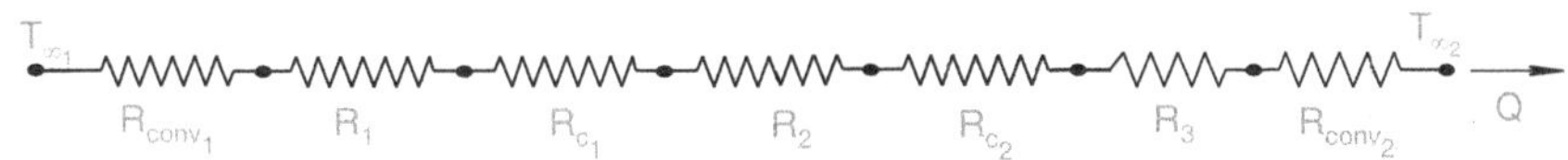

Fig. 3.28 (*a*)

All resistance are in series, total thermal resistance

$$\Sigma R_{th} = R_{conv_1} + R_1 + R_{c_1} + R_2 + R_{c_2} + R_3 + R_{conv_2}$$

Calculating each resistance individually

$$R_{conv_1} = \frac{1}{h_1 A} = \frac{1}{110 \times 1} = 9.090 \times 10^{-3} \text{ m}^2.\text{K/W}$$

$$R_{conv_2} = \frac{1}{h_2 A} = \frac{1}{15} = 0.0667 \text{ m}^2.\text{K/W}$$

$$R_1 = \frac{L_1}{k_1 A} = \frac{0.12}{0.6 \times 1} = 0.2 \text{ m}^2.\text{K/W}$$

$$R_2 = \frac{L_2}{k_2 A} = \frac{0.1}{0.8 \times 1} = 0.125 \text{ m}^2.\text{K/W}$$

$$R_3 = \frac{L_3}{k_3 A} = \frac{0.1}{4 \times 1} = 0.025 \text{ m}^2.\text{K/W}$$

Then

$$\Sigma R_{th} = 9.090 \times 10^{-3} + 0.2 + 2.6 \times 10^{-4} + 0.125 + 1.5 \times 10^{-4} + 0.025 + 0.0667$$

$$= 0.4262 \text{ m}^2.\text{K/W}$$

The heat flow rate can be obtained from thermal circuit as :

$$Q = \frac{T_{\infty_1} - T_{\infty_2}}{\Sigma R_{th}} = \frac{870 - 30}{0.4262}$$

$$= \mathbf{1971\ W/m^2}. \quad \textbf{Ans.}$$

(*ii*) The overall heat transfer is expressed as :

$$U = \frac{1}{A\ \Sigma R_{th}} = \frac{1}{1 \times 0.4262}$$

$$= \mathbf{2.346\ W/m^2.K}. \quad \textbf{Ans.}$$

or it can also be calculated from heat transfer rate,

$$Q = UA\ (\Delta T)_{overall}$$

or

$$U = \frac{Q}{A \times (T_{\infty_1} - T_{\infty_2})} = \frac{1971}{1 \times (870 - 30)}$$

$$= \mathbf{2.346\ W/m^2K}. \quad \textbf{Ans.}$$

Example 3.21. *A layer of 5 cm refractory brick (k = 2 W/m.K) is to be placed between two 4 mm thick steel (k = 40 W/m.K) plates. The both faces of brick adjacent to the plates have rough solid to solid contact over 20% of the area, where the average height of asperities is 1 mm. The outer surface temperature of steel plates are 400°C and 100°C, respectively.*

(*i*) *Find the rate of heat flow per unit area and assume that the cavity area is filled with air (k = 0.02 W/m.K).*

(*ii*) *Find the rate of heat flow, if the faces of brick are smooth and have solid to solid perfect contact over entire area.* (P.U., Dec. 2008)

Solution

Given : The schematic is shown in Fig. 3.29(*a*).

L_1 = 4 mm = 0.004 m (Steel plate),

L_2 = 1 mm = 0.001 m (Air gap),

L_3 = 5 cm = 0.05 m (brick layer),

k_1 = 40 W/m.K,

k_2 = 0.02 W/m.K,

k_3 = 2 W/m.K,

T_1 = 400°C,

T_2 = 100°C.

Contact area = 20%

To find :

(*i*) Heat flow rate with contact resistance.

(*ii*) Heat flow rate without contact resistance.

Assumptions :

(*i*) Steady state one direction heat flow.

(*ii*) Constant properties.

(*iii*) Heat transfer area as 1 m^2.

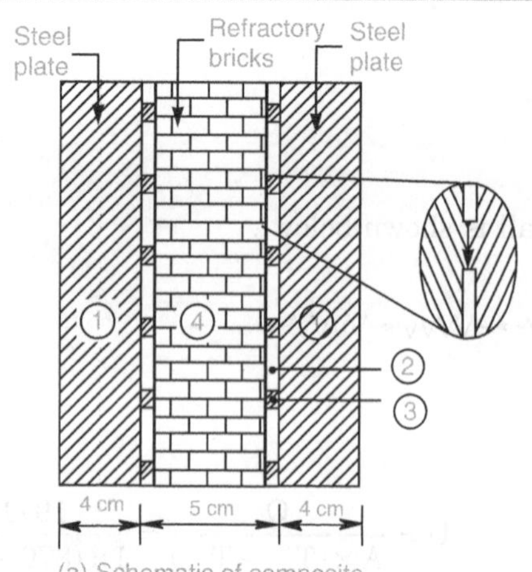

(a) Schematic of composite

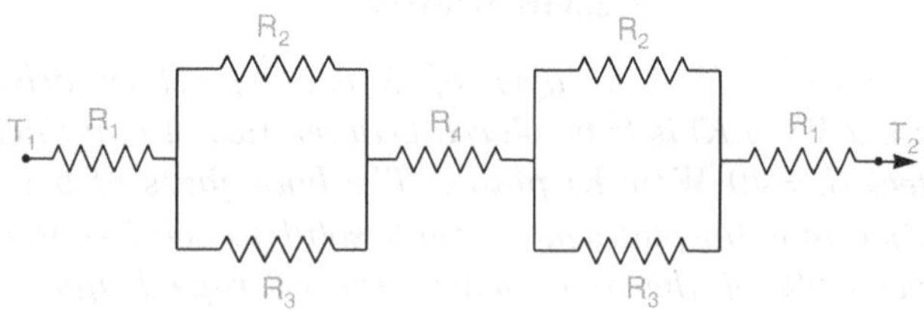

(b) Equivalent thermal network

Fig. 3.29

Analysis : (*i*) When cavities are filled with air. The heat flow rate through the plates :

$$Q = \frac{\Delta T}{\Sigma R_{th}} = \frac{T_1 - T_2}{2R_1 + 2R_{contact} + R_4}$$

For steel plate :

$$R_1 = \frac{L_1}{k_1 A} = \frac{0.004}{40 \times 1} = 1 \times 10^{-4} \text{ K/W}$$

Since the contact area is 20% (*i.e.*, $A_2 = 0.2 \text{ m}^2$), hence area of air pockets is 80% (*i.e.*, $A_1 = 0.8 \text{ m}^2$)

Air resistance,

$$R_2 = \frac{L_2}{k_2 A_1} = \frac{1 \times 10^{-3}}{0.02 \times 0.8} = 0.0625 \text{ K/W}$$

For brick contact,

$$R_3 = \frac{L_2}{k_3 A_2} = \frac{1 \times 10^{-3}}{2 \times 0.2} = 0.0025 \text{ K/W}$$

Since the refractory bricks having 1 mm air pockets on both sides, hence effective thickness of refractory layer, L_4 = 50 cm – 2 × 1 mm = 48 mm = 0.048 m.

For refractory brick,

$$R_4 = \frac{L_4}{k_3 A} = \frac{0.048 \text{ m}}{2 \times 1} = 0.024 \text{ K/W}$$

Equivalence of parallel resistances R_2 and R_3 :

$$\frac{1}{R_{eq}} = \frac{1}{R_2} + \frac{1}{R_3} = \frac{1}{0.0625} + \frac{1}{0.0025}$$

$$R_{eq} = 2.403 \times 10^{-3} \text{ K/W}$$

The total resistance,

$$\Sigma R = 2R_1 + 2R_{eq} + R_4$$
$$= 2 \times 1 \times 10^{-4} + 2 \times 2.403 \times 10^{-3} + 0.024$$
$$= 0.029 \text{ K/W}$$

The heat flow rate with contact resistance :

$$Q_1 = \frac{400 - 100}{0.029} = \mathbf{10342 \ W.} \quad \text{Ans.}$$

(*ii*) The heat transfer rate without contact resistance

$$Q_2 = \frac{T_1 - T_2}{2R_1 + R_{brick}}$$

The resistance of brick layer without air pockets,

$$R_{brick} = \frac{L_{brick}}{k_{brick} A} = \frac{0.05}{2 \times 1} = 0.025 \text{ K/W}$$

The heat flow rate :

$$Q_2 = \frac{400 - 100}{2 \times 1 \times 10^{-4} + 0.025}$$
$$= \mathbf{11904.7 \ W.} \quad \text{Ans.}$$

3.5. LONG HOLLOW CYLINDER

Consider a long hollow cylinder as shown in Fig. 3.30. The inner surface at $r = r_1$ is kept at temperature T_1 and outer surface at $r = r_2$ is kept at temperature T_2. There is no heat generation and the thermal conductivity of the solid is kept constant.

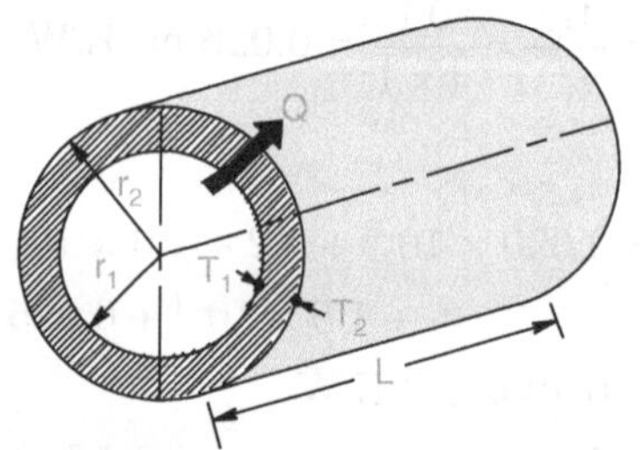

(a) Hollow cylinder with specified temperatures

(b) Equivalent thermal resistances

Fig. 3.30

Rewriting eqn. (2.18) for cylinder,

$$\frac{d}{dr}\left\{r \frac{dT}{dr}\right\} = 0 \qquad \text{...(3.37)}$$

where, T = T(*r*) the function of *r* direction.

Integrating with respect to *r*, we get

$$\frac{r\,dT}{dr} = C_1 \quad \text{or} \quad \frac{dT}{dr} = \frac{C_1}{r} \qquad \text{...(3.38)}$$

Integrating again, we get

$$T(r) = C_1 \ln(r) + C_2 \quad \text{...(3.39)}$$

Subjected to boundary conditions for evaluation of constants C_1 and C_2,

$$T(r) = T_1 \text{ at } r = r_1 \quad \text{and} \quad T(r) = T_2 \text{ at } r = r_2,$$

Using in equation (3.39), we get

$$T_1 = C_1 \ln (r_1) + C_2 \quad \text{...[3.40 (a)]}$$

$$T_2 = C_1 \ln (r_2) + C_2 \quad \text{...[3.40 (b)]}$$

Solving, we get

$$C_1 = \frac{T_2 - T_1}{\ln\left(\dfrac{r_2}{r_1}\right)}$$

and $$C_2 = T_1 - \frac{T_2 - T_1}{\ln\left(\dfrac{r_2}{r_1}\right)} \ln (r_1)$$

Substituting these constants in eqn. (3.39),

$$T(r) = \frac{T_2 - T_1}{\ln\left(\dfrac{r_2}{r_1}\right)} \ln (r) + T_1 - \frac{T_2 - T_1}{\ln\left(\dfrac{r_2}{r_1}\right)} \ln (r_1)$$

Rearranging, we get

$$\frac{T(r) - T_1}{T_2 - T_1} = \frac{\ln (r) - \ln (r_1)}{\ln\left(\dfrac{r_2}{r_1}\right)}$$

or $$\frac{T(r) - T_1}{T_2 - T_1} = \frac{\ln (r/r_1)}{\ln (r_2/r_1)} \quad \text{...(3.41)}$$

Differentiating with respect to r, we get slope

$$\frac{dT(r)}{dr} = \frac{(T_2 - T_1)}{\ln (r_2/r_1)} \times \frac{r_1}{r} \times \frac{1}{r_1}$$

$$= \frac{T_2 - T_1}{r \ln (r_2/r_1)} \quad \text{...(3.42)}$$

The heat flux :

$$q(r) = -k\frac{dT(r)}{dr} = -\frac{k}{r}\left(\frac{T_2 - T_1}{\ln (r_2/r_1}\right)$$

$$= \frac{k}{r}\left(\frac{T_1 - T_2}{\ln (r_2/r_1)}\right) \quad \text{...(3.43)}$$

The total heat transfer rate,

$$Q = Aq$$

or $$Q = \frac{2\pi r L\, k(T_1 - T_2)}{r \ln (r_2/r_1)} = \frac{2\pi L\, k(T_1 - T_2)}{\ln (r_2/r_1)} \quad \text{...(3.44)}$$

where L is the length of cylinder and $A = 2\pi r L$, area of cylinder at radius r.

3.5.1. Electrical Analogy for Hollow Cylinder

Like plane wall, the electrical analogy can also be extended for hollow cylinder without heat generation.

Rearranging eqn. (3.44)

$$Q = \frac{T_1 - T_2}{\dfrac{\ln (r_2/r_1)}{2\pi L\, k}} = \frac{\Delta T}{R_{cyl}} \quad \text{...(3.45)}$$

where, $R_{cyl} = \dfrac{\ln (r_2/r_1)}{2\pi L\, k}$ = thermal resistance to heat flow through hollow cylinder as shown in Fig. 3.30 (*b*).

r_1 = inner radius,

r_2 = outer radius.

Now consider steady state one dimensional heat flow through a cylindrical layer that is exposed to convection on both sides to fluids at T_{∞_1} and T_{∞_2} with heat transfer coefficients h_1 and h_2, respectively as shown in Fig. 3.31. The thermal resistance network in this case consists of one conduction and two convection resistances in series, and the rate of heat transfer is expressed as :

$$Q = \frac{T_{\infty_1} - T_{\infty_2}}{\Sigma R_{th}}$$

where $\Sigma R_{th} = R_{conv,\,1} + R_{cyl} + R_{conv,\,2}$

$$= \frac{1}{(2\pi r_1 L)h_1} + \frac{\ln (r_2/r_1)}{2\pi L\, k} + \frac{1}{(2\pi r_2 L)\, h_2} \quad \text{...(3.46)}$$

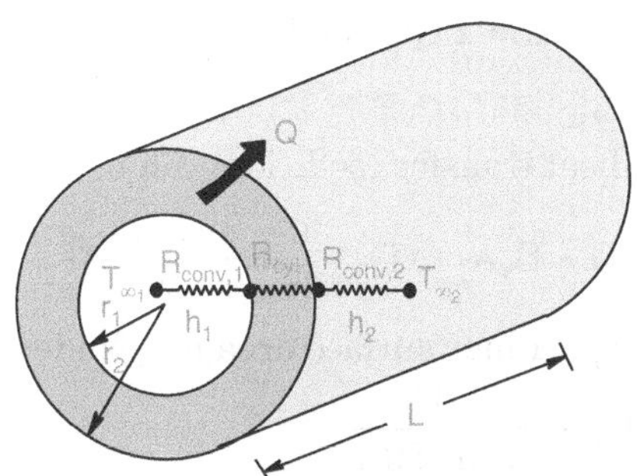

Fig. 3.31. Thermal resistance network for hollow cylinder subjected to convection heat transfer at inner and outer surfaces

3.5.2. Multilayer Hollow Cylinders

Now consider a composite system of three hollow cylinders as shown in Fig. 3.32 (thermal conductivities k_A, k_B, and k_C, respectively) with convection on inner and outer surfaces. Recalling the treatment given to composite wall, the total thermal resistance :

$$\Sigma R_{th} = R_{conv,\,1} + R_1 + R_2 + R_3 + R_{conv,\,2}$$

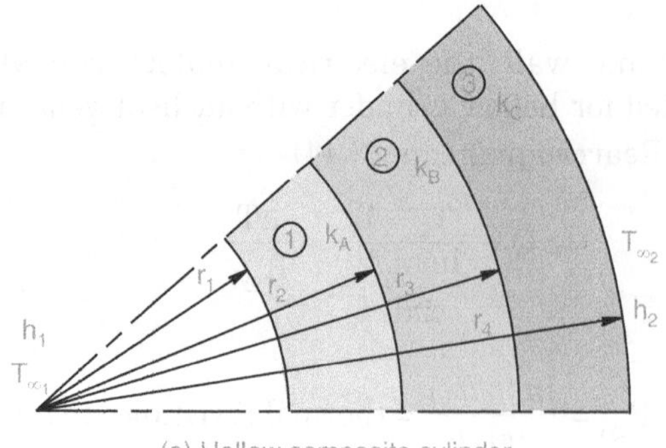

(a) Hollow composite cylinder

T_{∞_1} — $R_{conv,1}$ — T_1 — R_1 — T_2 — R_2 — T_3 — R_3 — T_4 — $R_{conv,2}$ — T_{∞_2}

(b) Equivalent thermal network

Fig. 3.32

or
$$\Sigma R_{th} = \frac{1}{2\pi r_1 L h_1} + \frac{\ln(r_2/r_1)}{2\pi L k_1} + \frac{\ln(r_3/r_2)}{2\pi L k_2} + \frac{\ln(r_4/r_3)}{2\pi L k_3} + \frac{1}{2\pi r_4 L h_2} \quad ...(3.47)$$

$$(\Delta T)_{overall} = T_{\infty_1} - T_{\infty_2}$$

Thus,
$$Q = \frac{T_{\infty_1} - T_{\infty_2}}{\dfrac{1}{h_1 A_1} + \dfrac{\ln(r_2/r_1)}{2\pi L\, k_A} + \dfrac{\ln(r_3/r_2)}{2\pi L\, k_B} + \dfrac{\ln(r_4/r_3)}{2\pi L\, k_C} + \dfrac{1}{h_2 A_2}} \quad ...(3.48)$$

where, $A_1 = 2\pi\, r_1 L$ and $A_2 = 2\pi\, r_4 L$

3.5.3. Overall Heat Transfer Coefficient

An overall heat transfer coefficient can be evaluated as :

$$Q = U_1 A_1\, (\Delta T)_{overall} = \frac{(\Delta T)_{overall}}{\Sigma R_{th}}$$

where A_1 = Inner surface area of cylinder = $2\pi r_1 L$

and
$$U_1 = \frac{1}{2\pi r_1 L\, \Sigma R_{th}}$$

$$U_1 = \frac{1}{2\pi r_1 L \left[\dfrac{1}{2\pi r_1 L\, h_1} + \dfrac{\ln(r_2/r_1)}{2\pi L\, k_A} + \dfrac{\ln(r_3/r_2)}{2\pi L\, k_B} + \dfrac{\ln(r_4/r_3)}{2\pi L\, k_C} + \dfrac{1}{2\pi r_4 L\, h_2} \right]}$$

or
$$U_1 = \frac{1}{\dfrac{1}{h_1} + \dfrac{r_1 \ln(r_2/r_1)}{k_A} + \dfrac{r_1 \ln(r_3/r_2)}{k_B} + \dfrac{r_1 \ln(r_4/r_3)}{k_C} + \dfrac{r_1}{r_4 h_2}} \quad ...(3.49)$$

where, U_1 = overall heat transfer coefficient based on inner surface area A_1 (= $2\pi r_1 L$). It may also be defined in terms of any of the intermediate areas providing

$$U_1A_1 = U_2A_2 = U_3A_3 = U_4A_4 \quad ...(3.50)$$

The specific forms of U_2, U_3 and U_4 may be evaluated from eqn. (3.47)

3.5.4. Log Mean Area

Rewriting eqn. (3.45) in the form for cylinder shown in Fig. 3.33(*a*);

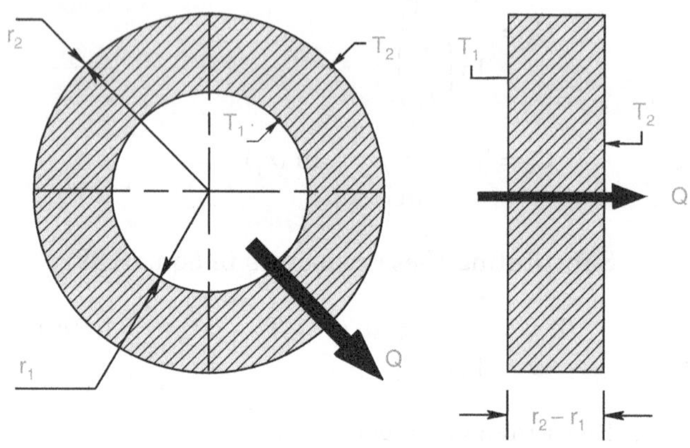

(a) Cylinder (b) Equivalent slab

Fig. 3.33

$$Q = \frac{T_1 - T_2}{R_{cyl}}$$

where
$$R_{cyl} = \frac{\ln(r_2/r_1)}{2\pi L\, k} \quad ...(3.51)$$

Rearranging this equation as

$$R_{cyl} = \frac{(r_2 - r_1)}{(r_2 - r_1)} \times \frac{\ln\left[\dfrac{2\pi r_2 L}{2\pi r_1 L}\right]}{2\pi L\, k}$$

$$= \frac{(r_2 - r_1)\ln(A_2/A_1)}{(A_2 - A_1)\,k}$$

or
$$R_{cyl} = \frac{(r_2 - r_1)}{A_m k} \quad ...(3.52)$$

where,
$$A_m = \frac{A_2 - A_1}{\ln(A_2/A_1)} \quad ...(3.53)$$

$A_2 = 2\pi r_2 L$ = area of outer surface of cylinder,

$A_1 = 2\pi r_1 L$ = area of inner surface of cylinder,

A_m = logarithmic mean area or log mean area of cylinder,

$r_2 - r_1$ = thickness of cylinder,

R_{cyl} = thermal resistance of cylinder.

Now the heat flow for hollow cylinder can be written as :

$$Q = \frac{A_m k(T_1 - T_2)}{r_2 - r_1} \quad ...(3.54)$$

This approach can be used to transform a cylinder into a equivalent slab of thickness $(r_2 - r_1)$ as shown in Fig. 3.33(*b*).

Example 3.22. *A long hollow cylinder (k = 50 W/m.K) has an inner radius of 10 cm, and outer radius of 20 cm. The inner surface is heated uniformly at constant rate of 1.16 × 10⁵ W/m² and outer surface is maintained at 30°C. Calculate the temperature of inner surface.*

Solution

Given : A long hollow cylinder heated uniformly at its inner surface :

$r_1 = 10$ cm = 0.1 m, $r_2 = 20$ cm = 0.2 m

$k = 50$ W/m.K, $q_1 = 1.16 \times 10^5$ W/m²

$T_2 = 30°C$.

To find : Inner surface temperature.

Assumptions :

(*i*) Steady state heat conduction in radial direction only.

(*ii*) 1 m length of cylinder for analysis.

Analysis : The heat transfer rate through the cylinder

$$Q = A(r)\, q(r) = (2\pi r_1 L)\, q_1 \quad ...(i)$$

Further, the heat conduction rate through the cylinder can be expressed as :

$$Q = \frac{2\pi k\, L(T_1 - T_2)}{\ln(r_2/r_1)} \quad ...(ii)$$

For steady state combining these two equations, we get

$$(2\pi r_1 L)\, q_1 = \frac{2\pi k\, L(T_1 - T_2)}{\ln(r_2/r_1)}$$

Substituting the numerical values to obtain T_1

$$(2\pi \times 0.1\, L) \times 1.16 \times 10^5 = \frac{2\pi \times 50\, L(T_1 - 30)}{\ln(0.2/0.1)}$$

It gives $T_1 = 160.8 + 30 =$ **190.8°C. Ans.**

Example 3.23. *A hollow cylinder with inner radius 30 mm and outer radius 50 mm is heated at the inner surface at a rate of 10⁵ W/m² and dissipated heat by convection from outer surface into a fluid at 80°C with heat transfer coefficient of 400 W/m².K. There is no energy generation and thermal conductivity of the material is constant at 15 W/m.K. Calculate the temperatures of inside and outside surfaces of the cylinder.*

Solution

Given : A hollow cylinder is heated at its inner surface

$r_1 = 30$ mm, $r_2 = 50$ mm

$q = 10^5$ W/m², $T_\infty = 80°C$

$h = 400$ W/m².K, $k = 15$ W/m.K.

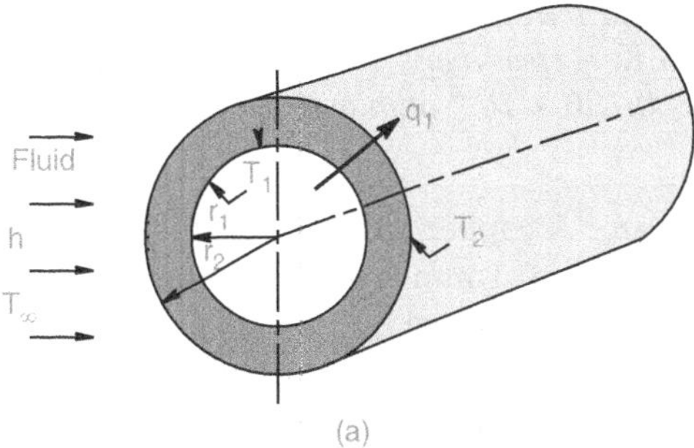

(a)

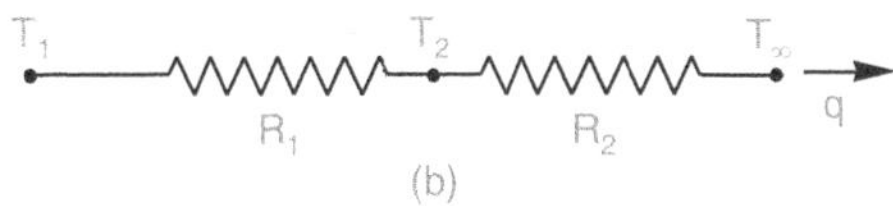

(b)

Fig. 3.34

To find : Temperatures of inner and outer surfaces of the cylinder.

Analysis : The individual thermal resistance

$$R_1 = \frac{\ln(r_2/r_1)}{2\pi L\, k} = \frac{\ln(50/30)}{2\pi L \times 15} = \frac{0.0340}{2\pi L}$$

$$R_2 = \frac{1}{2\pi r_2 L\, h} = \frac{1}{(2\pi L) \times 0.05 \times 400} = \frac{0.05}{2\pi L}$$

These resistances are in series, total thermal resistance

$$\Sigma R_{th} = R_1 + R_2 = \frac{0.0340}{2\pi L} + \frac{0.05}{2\pi L} = \frac{0.084}{2\pi L}$$

The heat flow rate through the network

$$Q = q\,(2\pi r_1 L) = \frac{\Delta T}{\Sigma R_{th}} = \frac{(T_1 - T_\infty) \times 2\pi L}{0.084}$$

Equating 2nd and 4th terms

$$10^5 \times (2\pi L) \times 0.03 = \frac{(T_1 - 80) \times 2\pi L}{0.084}$$

or $T_1 - 80 = 252°C$ or $T_1 = 252 + 80$

= **332°C. Ans.**

The temperature of inner surface is 332°C.

Further, heat convection rate

$$Q = q\,(2\pi r_1 L) = h(2\pi\, r_2 L)\,(T_2 - T_\infty)$$

or $10^5 \times 0.03 = 400 \times 0.05 \times (T_2 - 80)$

or $$T_2 = \frac{3000}{20} + 80 = \mathbf{230°C.}$$ **Ans.**

The temperature of outer surface is 230°C.

Example 3.24. *A copper wire 0.1 mm in diameter is insulated with plastic to an outer diameter of 0.3 mm and is exposed to an environment at 40°C. The heat transfer coefficient from the outer surface of the plastic to surroundings is 8.75 W/m².K. What is the maximum steady current in amperes, that this wire can carry without heating any part of plastic above 95°C ? The thermal conductivities of plastic and copper are 0.35 and 384 W/m.K, respectively. The electrical resistivity of the copper is 0.196 × 10⁻⁵ ohm.cm.*

Solution

Given : A copper wire plastic insulation

$d_1 = 0.1 \text{ mm} = 0.1 \times 10^{-3}$ m,

$r_1 = 5 \times 10^{-4}$ m

$d_2 = 0.3 \text{ mm} = 0.3 \times 10^{-3}$ m,

$r_2 = 15 \times 10^{-4}$ m

$T_\infty = 40°C$,

$T_s = 95°C$

$h = 8.75 \text{ W/m}^2 \text{ K}$,

$k_{plastic} = 0.35$ W/m.K

$k_{cu} = 384$ W/m.K,

$\rho_{cu} = 0.196 \times 10^{-5} \ \Omega\text{.cm} = 1.96 \times 10^{-8} \ \Omega\text{.m}$.

To find : The maximum steady current flow in wire.

Assumptions :

(*i*) Steady state heat conduction in radial direction only.

(*ii*) Constant properties.

(*iii*) No contact resistance.

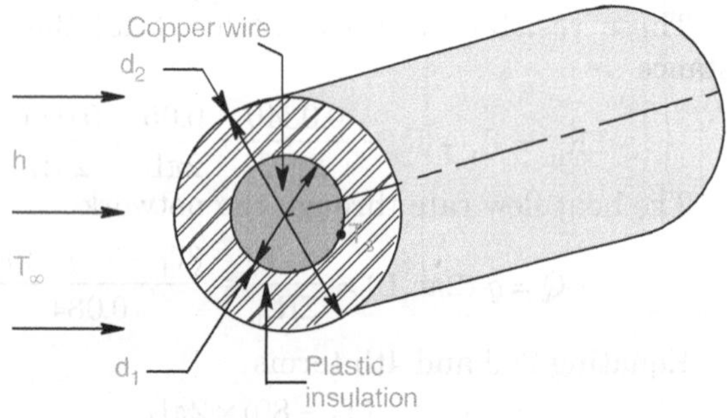

Fig. 3.35

Analysis : The heat flow rate from the wire to surroundings can be calculated by electrical analogy.

$$Q = \frac{2\pi L (T_s - T_\infty)}{\dfrac{\ln (r_2/r_1)}{k_{plastic}} + \dfrac{1}{r_2 h}}$$

$$= \frac{2\pi L (95 - 40)}{\dfrac{\ln (15/5)}{0.35} + \dfrac{1}{15 \times 10^{-4} \times 8.75}}$$

$$= 4.356L \text{ (W)}$$

When steady current flows through copper wire

$$Q = I^2 R_e = I^2 \frac{\rho_{cu} L}{A_c} = I^2 \frac{\rho_{cu} L}{(\pi/4) d_1^2}$$

or

$$I^2 = \frac{\pi}{4} \times \frac{Q d_1^2}{\rho_{cu} L}$$

$$= \frac{\pi}{4} \times \frac{4.356 \text{ L} \times (0.1 \times 10^{-2})^2}{1.96 \times 10^{-8} \times \text{L}} = 174.53$$

$I =$ **13.21 Amp.** **Ans.**

Example 3.25. *A steel tube (k = 45 W/m.K) of outside diameter 7.6 cm, and thickness 1.3 cm, is covered with an insulating material (k = 0.2 W/m.K) of thickness 2 cm. A hot gas at 330°C, with convection coefficient of 200 W/m².K, is flowing inside the tube. The outer surface of the insulation is exposed to ambient air at 30°C, with convection coefficient of 50 W/m².K.*

Calculate :

(i) Heat loss to air from the 5 m long tube,

(ii) The temperature drop due to thermal resistances of the hot gases, steel tube, the insulation layer and the outside air.

Solution

Given : A steel pipe covered with an insulation layer :

$d_1 = (7.6 - 2 \times 1.3) \text{ cm} = 5.0$ cm,

$r_1 = 2.5 \text{ cm} = 0.025$ m,

$r_2 = 3.8 \text{ cm} = 0.038$ m,

$r_3 = (3.8 + 2.0) \text{ cm} = 0.058$ m,

$L = 5$ m

$k_1 = 45$ W/m.K,

$k_2 = 0.2$ W/m.K

$h_1 = 200 \text{ W/m}^2\text{.K}$,

$h_2 = 50 \text{ W/m}^2\text{.K}$

$T_{\infty_1} = 330°C$,

$T_{\infty_2} = 30°C$

To find :

(*i*) The heat loss from 5 m length of the tube.

(*ii*) Temperature drop due to thermal resistances of hot gases, steel tube, insulation layer and the outside air.

Assumptions :

(*i*) Steady state heat conduction in radial direction only.

(*ii*) No contact resistance at interface.

(*iii*) Constant properties.

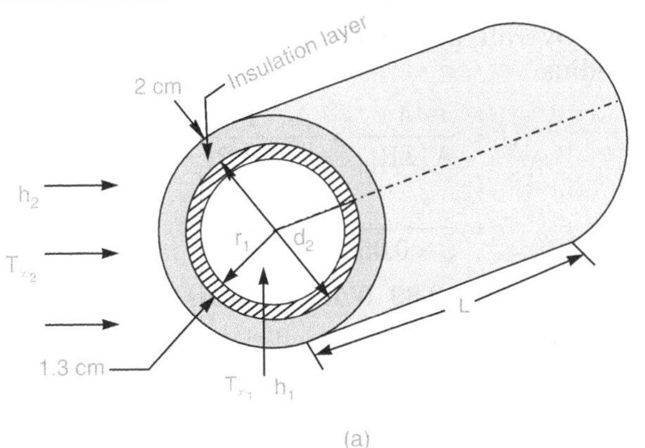

(a)

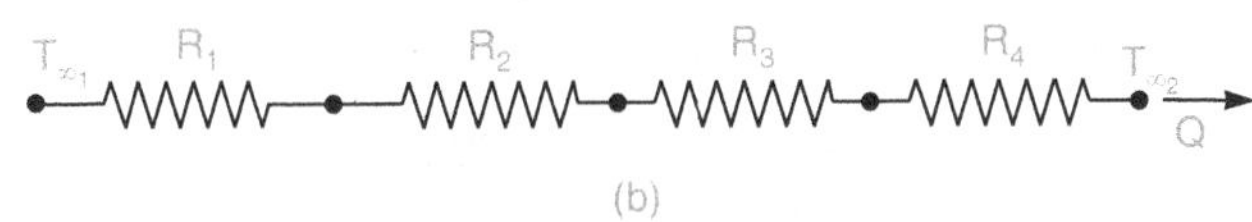

(b)

Fig. 3.36

Analysis : (*i*) Applying electrical analogy for the radial heat flow through the tube :

$$Q = \frac{T_{\infty_1} - T_{\infty_2}}{\Sigma R_{th}}$$

where, $A_1 = 2\pi r_1 L = 2\pi \times 0.025 \times 5 = 0.785 \text{ m}^2$

$A_2 = 2\pi r_3 L = 2\pi \times 0.058 \times 5 = 1.822 \text{ m}^2$

The various thermal resistances

$$R_1 = \frac{1}{h_1 A_1} = \frac{1}{200 \times 0.785} = 6.37 \times 10^{-3} \text{ K/W}$$

$$R_2 = \frac{\ln (r_2/r_1)}{2\pi L k_1} = \frac{\ln (0.038/0.025)}{2\pi \times 5 \times 45}$$

$$= 2.96 \times 10^{-4} \text{ K/W}$$

$$R_3 = \frac{\ln (r_3/r_2)}{2\pi L k_2} = \frac{\ln (0.058/0.038)}{2\pi \times 5 \times 0.2} = 0.0673 \text{ K/W}$$

$$R_4 = \frac{1}{h_4 A_2} = \frac{1}{50 \times 1.822} = 0.01098 \text{ K/W}$$

All resistances are in series.

Thus, the total resistance,

$$\Sigma R_{th} = R_1 + R_2 + R_3 + R_4$$

$$\Sigma R_{th} = 6.37 \times 10^{-3} + 2.96 \times 10^{-4} + 0.0673 + 0.01098$$

$$= 84.94 \times 10^{-3} \text{ K/W}$$

The total heat loss,

$$Q = \frac{330 - 30}{84.84 \times 10^{-3}} = \mathbf{3531.8 \text{ W.}} \quad \text{Ans.}$$

(*ii*) The temperature drop can be calculated by relation :

$$\Delta T_i = Q \times R_i$$

$$\Delta T_1 = Q \times R_1 = 3531.8 \text{ W} \times 6.37 \times 10^{-3} \text{ K/W}$$

$$= 22.5°\text{C}$$

$$\Delta T_2 = Q \times R_2 = 3531.8 \text{ W} \times 2.96 \times 10^{-4} \text{ K/W}$$

$$= 1.045°\text{C}$$

$$\Delta T_3 = Q \times R_3 = 3531.8 \text{ W} \times 0.0673 \text{ K/W}$$

$$= 237.68°\text{C}$$

$$\Delta T_4 = Q \times R_4 = 3531.8 \text{ W} \times 0.01098 \text{ K/W}$$

$$= \mathbf{38.77°C.} \quad \text{Ans.}$$

Example 3.26. *A steam pipe of 5 cm inside diameter and 6.5 cm outside diameter is covered with a 2.75 cm radial thickness of high temperature insulation (k = 1.1 W/m.K). The surface heat transfer coefficient for inside and outside surfaces are 4650 W/m².K and 11.5 W/m².K, respectively. The thermal conductivity of the pipe material is 45 W/m.K. If the steam temperature is 200°C and ambient air temperature is 25°C, determine :*

(*i*) *Heat loss per metre length of pipe.*

(*ii*) *Temperature at the interface.*

(*iii*) *Overall heat transfer coefficient.*

(V.T.U., July 2002)

Solution

Given : A steam pipe covered with high temperature insulation :

$d_1 = 5$ cm or $r_1 = 0.025$ m

$d_2 = 6.5$ cm $r_2 = 0.0325$ m

or $k_1 = 45$ W/m.K $k_2 = 1.1$ W/m.K

$r_3 = 0.0325 + 0.0275 = 0.06$ m

$h_1 = 4650$ W/m².K, $h_2 = 11.5$ W/m².K

$T_{\infty_1} = 200°C$, $T_{\infty_2} = 25°C$

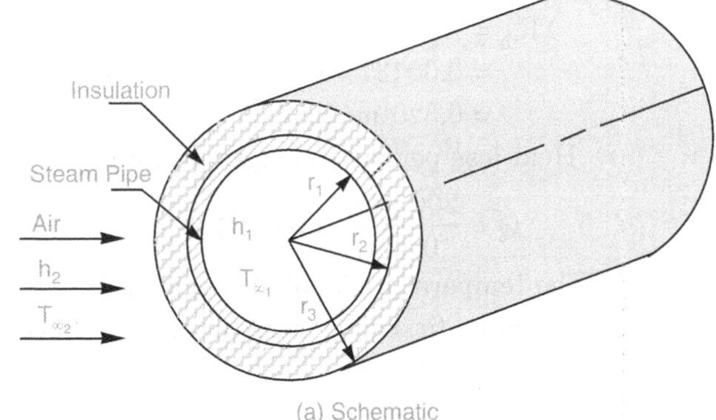

(a) Schematic

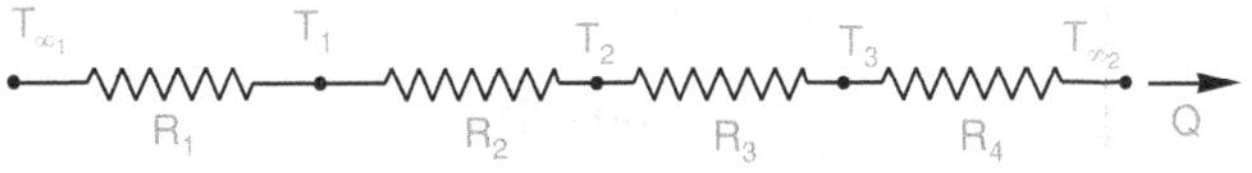

(b) Thermal network

Fig. 3.37

To find :

(*i*) Heat loss per metre length of the pipe.

(*ii*) Temperature T_2 at the interface.

(*iii*) Overall heat transfer coefficient.

Assumptions :

(*i*) Steady state conditions.

(*ii*) Heat transfer in radial direction only.

(*iii*) 1 m length of the pipe.

(*iv*) Constant properties.

(*v*) No contact resistance at the interface.

Analysis : Using the electrical analogy for radial heat flow through composite cylinder.

$$Q = \frac{T_{\infty_1} - T_{\infty_2}}{\Sigma R_{th}}$$

where $R_1 = \dfrac{1}{h_1 A_i} = \dfrac{1}{2\pi r_1 L h_1}$

$$= \frac{1}{2\pi \times 0.025 \times 1 \times 4650}$$

$= 0.00137 \text{ m}^2.\text{K/W}$

$$R_2 = \frac{\ln\left(\frac{r_2}{r_1}\right)}{2\pi L k_1} = \frac{\ln\left(\frac{0.0325}{0.025}\right)}{2\pi \times 1 \times 45} = 9.28 \times 10^{-4} \text{ m}^2.\text{K/W}$$

$$R_3 = \frac{\ln\left(\frac{r_3}{r_2}\right)}{2\pi L k_2} = \frac{\ln\left(\frac{0.06}{0.0325}\right)}{2\pi \times 1 \times 1.1} = 0.088 \text{ m}^2.\text{K/W}$$

$$R_4 = \frac{1}{h_2 A_o} = \frac{1}{2\pi r_3 L \times h_2} = \frac{1}{2\pi \times 0.06 \times 1 \times 11.5}$$

$= 0.23 \text{ m}^2.\text{K/W}$

All resistances are in series, thus

$$\Sigma R_{th} = R_1 + R_2 + R_3 + R_4$$

$= 0.00137 + 9.28 \times 10^{-4} + 0.088 + 0.23$

$= 0.320 \text{ m}^2.\text{K/W}$

(*i*) Heat loss per metre length of the pipe

$$Q = \frac{200 - 25}{0.320} = \textbf{545.25 W/m.} \quad \text{Ans.}$$

(*ii*) The temperature T_2 at the interface :

Considering first two resistances of thermal network, then heat flow rate

$$Q = \frac{T_{\infty_1} - T_2}{R_1 + R_2}$$

or $$545.25 = \frac{200 - T_2}{0.00137 + 9.28 \times 10^{-4}}$$

or $200 - T_2 = 545.25 \times 2.298 \times 10^{-3} = 1.253$

or $T_2 = 200 - 1.138 =$ **198.75°C.** Ans.

(*iii*) Overall heat transfer coefficient based on inner radius

$$U_i = \frac{1}{A_i \Sigma R_{th}} = \frac{1}{\pi d_1 L \Sigma R_{th}}$$

$$= \frac{1}{\pi \times 0.05 \times 1 \times 0.320}$$

$=$ **19.89 W/m².K. Ans.**

Overall heat transfer coefficient based on outer radius

$$U_o = \frac{1}{A_o \Sigma R_{th}} = \frac{1}{2\pi r_3 L \Sigma R_{th}}$$

$$= \frac{1}{2\pi \times 0.06 \times 1 \times 0.320}$$

$=$ **8.29 W/m².K.** Ans.

Example 3.27. *A pipe line (d_i = 160 mm, d_o = 170 mm) is covered with a layer of insulation, 100 mm, with variable thermal conductivity as k = 0.062 (1 + 0.000363T) W/m°C. Calculate the heat loss per metre length of the pipe for temperature of outer pipe surface as 300°C and outer insulation layer as 50°C.*

Solution

Given : An insulation layer on a pipe line

d_i = 160 mm, d_o = 170 mm

r_2 = 85 mm = 0.085 m

r_3 = 85 mm + 100 mm = 0.185 m

For insulation

$k = 0.062 \times (1 + 0.000363\,T)$ W/m°C

$= k_0 (1 + \alpha T)$

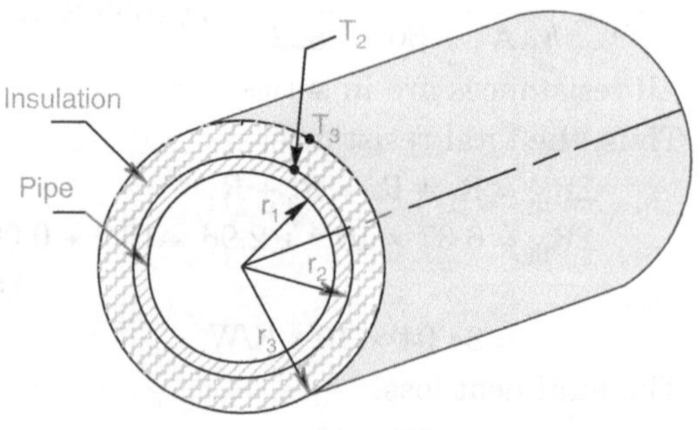

Fig. 3.38

To find : Heat loss rate from pipe per metre length.

Analysis : Heat transfer rate through insulation cylinder

$$Q = -kA \frac{dT}{dr}$$

or $$Q\,dr = -k_0 (1 + \alpha T)(2\pi r L)\, dT$$

or $$\frac{Q}{2\pi L}\int_{r_2}^{r_3}\frac{dr}{r} = -k_0\int_{T_2}^{T_3}(1+\alpha T)\,dT$$

or $$\frac{Q}{2\pi L}\ln\left(\frac{r_3}{r_2}\right) = -k_0\left[(T_3 - T_2)+\frac{\alpha}{2}(T_3^{\,2} - T_2^{\,2})\right]$$

$$= k_0(T_2 - T_3)\left\{1+\frac{\alpha}{2}(T_2+T_3)\right\}$$

or $$Q = \frac{2\pi L k_0(T_2 - T_3)}{\ln(r_3/r_2)}\left\{1+\frac{\alpha}{2}(T_2+T_3)\right\}$$

Using numerical values

$$Q = \frac{2\pi\times 1\times 0.062\times(300-50)}{\ln(0.185/0.085)}\times\left\{1+\frac{0.000363}{2}\times(300+50)\right\}$$

$$= \mathbf{133.18\ W/m.}$$ **Ans.**

Example 3.28. *A 3 cm outer diameter steam pipe is to be covered with two layers of insulation each having thickness of 2.5 cm. The average thermal conductivity of one material is five times of the other. Determine the percentage decrease in heat transfer, if better insulating material is kept next to pipe surface than it is as outer layer. Assume that the outside and inside temperatures are fixed.* (P.U., May 1996)

Solution

Given : A steam pipe covered with two layers of insulation ;

$d_1 = 3$ cm or $r_1 = 1.5$ cm

$r_2 = 1.5$ cm + 2.5 cm = 4 cm

$r_3 = 4$ cm + 2.5 cm = 6.5 cm

$k_1 = 5k_2$

To find : Percentage reduction in heat loss, if better order of insulation is placed.

Assumptions :

(*i*) Steady state heat conduction in radial direction only.

(*ii*) No contact resistance at interfaces.

(*iii*) k_1 and k_2 are the thermal conductivities for the two layers of insulation. We consider $k_1 = 5\,k_2$ *i.e.*, k_2 is good insulator.

Analysis : The steady state heat transfer rate is expressed as :

$$Q = \frac{2\pi L\,\Delta T}{\dfrac{\ln\left(\dfrac{r_2}{r_1}\right)}{k_1}+\dfrac{\ln\left(\dfrac{r_3}{r_4}\right)}{k_2}}$$

when good insulating material is kept next to pipe.

$$Q_1 = \frac{2\pi L(T_1 - T_3)}{\dfrac{\ln(4/1.5)}{k_2}+\dfrac{\ln(6.5/4)}{5k_2}} = 1.856\pi L(\Delta T)k_2$$

Now if poor insulating material (of higher thermal conductivity) is kept next to pipe surface

$$Q_2 = \frac{2\pi L(\Delta T)}{\dfrac{\ln(4/1.5)}{5k_2}+\dfrac{\ln(6.5/4)}{k_2}} = 2.928\pi L(\Delta T)k_2$$

Percentage reduction in heat flow

$$\frac{Q_2 - Q_1}{Q_2}\times 100$$

$$\frac{2.928 - 1.856}{2.928}\,100 = \mathbf{36.6\%.}$$ **Ans.**

Example 3.29. *A steam pipe, 10 cm in outer diameter is covered with two layers of insulation material each 2.5 cm thick, one having thermal conductivity thrice the other. Show that the effective thermal conductivity of two layers is approximately 15% less when better insulation material is placed as inside layer, than when it is on the outside.* (P.U., May 1998)

Solution

Given : Two layers of insulation on a steam pipe.

$d_1 = 10$ cm, $r_1 = 5$ cm

$r_2 = 5$ cm + 2.5 cm = 7.5 cm

$r_3 = 7.5$ cm + 2.5 cm = 10 cm

$k_1 = 3k_2$

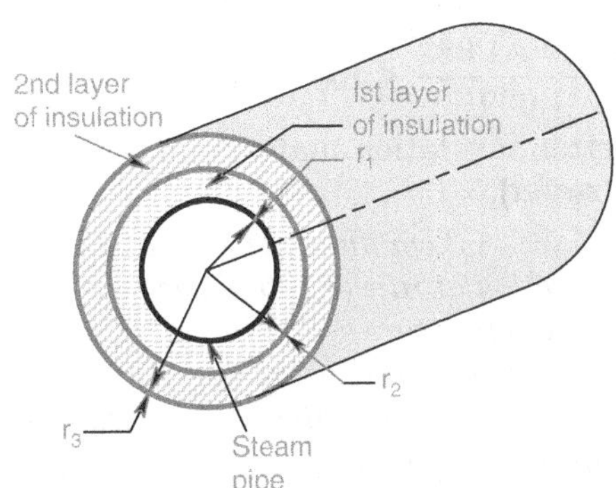

Fig. 3.39

Analysis : The steady state heat transfer rate through composite cylinder.

$$Q = \frac{\Delta T}{\dfrac{\ln\left(\dfrac{r_2}{r_1}\right)}{2\pi L\,k_1}+\dfrac{\ln\left(\dfrac{r_3}{r_2}\right)}{2\pi L\,k_2}} = \frac{2\pi L(\Delta T)}{\dfrac{\ln\left(\dfrac{r_2}{r_1}\right)}{k_1}+\dfrac{\ln\left(\dfrac{r_3}{r_2}\right)}{k_2}}$$

When better insulating matter (k_2) is placed as inside layer,

$$Q_1 = \frac{2\pi L(\Delta T)}{\frac{\ln\left(\frac{7.5}{5}\right)}{k_2} + \frac{\ln\left(\frac{10}{7.5}\right)}{3k_2}} = 1.99 \times [2\pi L\, k_2\, (\Delta T)]$$

The effective thermal conductivity of two layers insulation in this arrangement

$$Q_1 = \frac{2\pi L k_i\, (\Delta T)}{\ln\left(\frac{10}{5}\right)}$$

or $\quad 1.99 \times 2\pi L\, k_2\, (\Delta T) = \dfrac{2\pi L\, k_i\, (\Delta T)}{0.6931}$

or $\quad k_i = 1.382\, k_2 \qquad \ldots(i)$

When effective insulation layer (k_2) is placed as outside layer.

$$Q_2 = \frac{2\pi L\, (\Delta T)}{\frac{\ln\left(\frac{r_2}{r_1}\right)}{3k_2} + \frac{\ln\left(\frac{r_3}{r_2}\right)}{k_2}} = \frac{2\pi L\, k_2\, (\Delta T)}{\frac{\ln\left(\frac{7.5}{5}\right)}{3} + \ln\left(\frac{10}{7.5}\right)}$$

$$= 2.365 \times [2\pi L\, k_2\, (\Delta T)]$$

Effective thermal conductivity of two layer in this arrangement.

$$2.365\, [2\pi L\, k_2\, (\Delta T)] = \frac{2\pi L\, k_o\, (\Delta T)}{\ln\left(\frac{10}{5}\right)}$$

or $\quad k_o = 1.639\, k_2$

% change in effective thermal conductivity

$$\frac{1.639 - 1.382}{1.639} \times 100 = 15.7\%$$

less, if better insulation material is placed as inside layer. **Proved.**

Example 3.30. *A 3 cm diameter pipe at 120°C is losing heat by convection at rate of 120 W per metre length. The surrounding temperature is 30°C. It is required to reduce the heat loss to a minimum value by providing insulation. The following insulation materials are available :*

Insulation 1 : Quantity = 3.15×10^{-3} m^3 per metre length of pipe

Thermal conductivity, k_1 = 5 W/m.K.

Insulation 2 : Quantity = 4×10^{-3} m^3 per metre length

Thermal conductivity, k_2 = 1 W/m.K.

Examine the better insulating layer relative to pipe and determine the percentage change in heat transfer from that arrangement.

Solution

Given : Two insulation layer on a heat pipe.

d = 3 cm,

r_1 = 1.5 cm = 0.015 m

T_1 = 120°C

T_∞ = 30°C

Q = 120 W/m

k_1 = 5 W/m.K,

$V_1 = 3.15 \times 10^{-3}$ m^3/m

k_2 = 1 W/m.K,

$V_2 = 4 \times 10^{-3}$ m^3/m.

To find :

(*i*) Better arrangement of insulation.

(*ii*) Percentage change in heat loss with better arrangement.

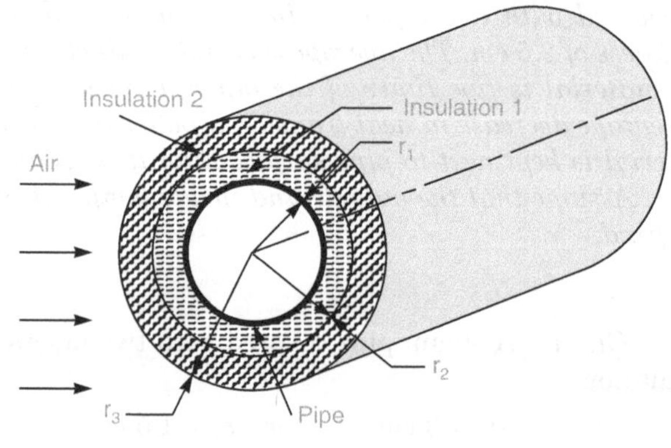

Fig. 3.40

Analysis : (*i*) When heat is transferred from pipe to surroundings by convection,

$$Q = hA(T_1 - T_\infty) = \frac{T_1 - T_\infty}{R_{conv}}$$

The resistance of convection

$$R_{conv} = \frac{T_1 - T_\infty}{Q} = \frac{120 - 30}{120} = 0.75 \text{ K/W}$$

For a pipe with two layer of insulation

$$\Sigma R_{th} = R_{conv} + R_1 + R_2$$

$$= 0.75 + \frac{\ln\left(\frac{r_2}{r_1}\right)}{2\pi L k_1} + \frac{\ln\left(\frac{r_3}{r_2}\right)}{2\pi L k_2}.$$

Arrangement 1 : Insulation material 1 is placed as inside layer of insulation

$$V_1 = \pi\,(r_2^2 - r_1^2)L$$

or $\quad 3.15 \times 10^{-3} = \pi(r_2^2 - 0.015^2) \times 1$

or $\quad r_2^2 = 1.2276 \times 10^{-3}$ m^2

or $r_2 = 0.035$ m

and $V_2 = \pi (r_3^2 - r_2^2) \times L$

$$r_3^2 = \frac{4 \times 10^{-3}}{\pi} + 0.035^2 = 2.498 \times 10^{-3} \text{ m}^2$$

or $r_3 = 0.05$ m

Now, $$\Sigma R_{th} = 0.75 + \frac{\ln\left(\frac{0.035}{0.015}\right)}{2\pi \times 1 \times 5} + \frac{\ln\left(\frac{0.05}{0.035}\right)}{2\pi \times 1 \times 1}$$

$$= 0.75 + 0.0269 + 0.0568 = 0.834 \text{ K/W}.$$

Heat loss rate

$$Q_1 = \frac{\Delta T}{\Sigma R_{th}} = \frac{120 - 30}{0.834} = 107.96 \text{ W}$$

Arrangement 2 : The insulation 2 is placed next to pipe surface and insulation 1 as outer layer.

$$V_2 = \pi (r_2^2 - r_1^2)L$$

or $$r_2^2 = \frac{4 \times 10^{-3}}{\pi} + (0.015)^2$$

$$= 1.498 \times 10^{-3} \text{ m}^2$$

or $r_2 = 0.0387$ m

and $V_1 = \pi (r_3^2 - r_2^2) \times L$

or $$r_3^2 = \frac{3.15 \times 10^{-3}}{\pi} + (0.0387)^2$$

$$= 2.500 \times 10^{-3} \text{ m}^2$$

or $r_3 = 0.05$ m

$$\Sigma R_{th} = 0.75 + \frac{\ln\left(\frac{0.0387}{0.015}\right)}{2\pi \times 1 \times 1} + \frac{\ln\left(\frac{0.050}{0.0387}\right)}{2\pi L \times 5}$$

$$= 0.75 + 0.150 + 0.0081$$

$$= 0.909 \text{ K/W m}^2$$

Heat loss rate

$$Q_2 = \frac{120 - 30}{0.909} = 99.0 \text{ W}.$$

Comment : The second arrangement is more effective.

(*ii*) The percentage decrease in heat loss

$$= \frac{107.96 - 99.0}{107.96} \times 100 = \mathbf{8.2\%}. \quad \text{Ans.}$$

Example 3.31. *A long cylindrical rod of radius 12 cm, consists of nuclear reacting material (k = 2 W/m.K) generating 30 kW/m³ uniformly throughout its volume. The rod is encapsulated within another cylinder (k = 5 W/m.K) whose outer radius is 24 cm and surface is surrounded by air at 30°C with heat transfer coefficient of 20 W/m².K. Find the temperature at the interface between the two cylinders and at the outer surface.*

(P.U., Dec. 2009)

Solution

Given : A long cylindrical nuclear reacting material with uniform heat generation ;

$r_1 = 12$ cm $= 0.12$ m,
$k_1 = 2$ W/m.K
$g_o = 30 \times 10^3$ W/m³,
$r_2 = 24$ cm $= 0.24$ m
$k_2 = 5$ W/m.K,
$T_\infty = 30°C$
$h = 20$ W/m².K.

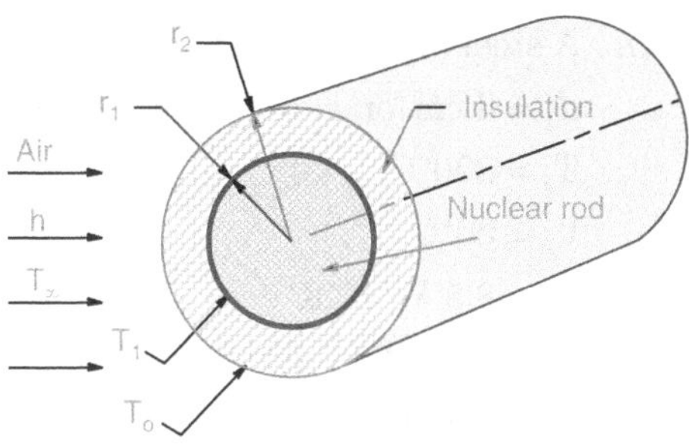

Fig. 3.41. Schematic of cylinder consists of nuclear fuel, covered with insulation

To find : Temperatures T_1 and T_o.

Assumptions :

(*i*) Constant properties.

(*ii*) No contact resistance.

(*iii*) Steady state heat conduction in radial direction.

(*iv*) 1 m length of the cylinder for analysis.

Analysis : Since the heat is generated uniformly throughout the volume of nuclear reacting material cylinder, hence total heat generation rate per m length ;

$$Q_g = g_o \pi r^2 L = 30 \times 10^3 \times \pi \times (0.12)^2 \times 1$$
$$= 1357.16 \text{ W/m}$$

In steady state conditions, this heat will be convected from outer cylinder surface therefore ;

$$Q_g = Q_{out} = 2\pi r_2 L h(T_o - T_\infty)$$

or $$1357.16 = 2 \times \pi \times 0.24 \times 1 \times 20 \times (T_o - 30)$$

or $$T_o = 30 + 45 = \mathbf{75°C}. \quad \text{Ans.}$$

Further, this heat will also be conducted through insulation cylinder, hence

$$Q_g = \frac{2\pi L k_2 (T_1 - T_o)}{\ln (r_2/r_1)}$$

or $$T_1 = \frac{Q_g \ln(r_2/r_1)}{2\pi L k_2} + T_o$$

$$= \frac{1357.16 \times \ln(0.24/0.12)}{2\pi \times 1 \times 5} + 75$$

$$= \mathbf{105°C.}$$ **Ans.**

Example 3.32. *A steel pipe 3 cm in diameter has its outer surface at 200°C, is placed in air at 30°C with heat transfer coefficient of 8.5 W/m².K. It is proposed to add insulation (k = 0.07 W/m.K) on its outer surface to reduce the heat loss by 40%. Estimate the thickness of insulation required, if pipe temperature and heat transfer coefficient remain unchanged.*

Solution

Given : A steel pipe proposed for insulation layer

d_1 = 3 cm or r_1 = 1.5 cm = 0.015 m

T_s = 200°C,

T_∞ = 30°C

h = 8.5 W/m².K

k_{ins} = 0.07 W/m².K

$Q_1 = 0.6\, Q_b$.

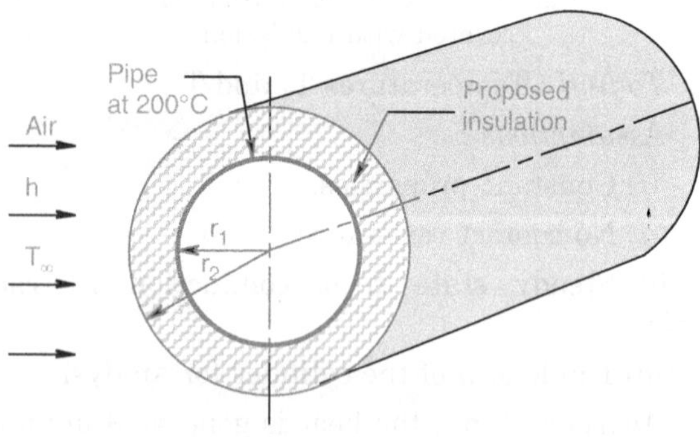

(a) Schematic of pipe with insulation

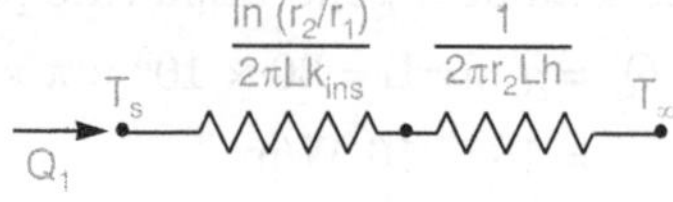

(b) Equivalent thermal network

Fig. 3.42

To find : The thickness of insulation.

Assumptions :

(*i*) Steady state heat conduction in radial direction only.

(*ii*) No contact resistance, when insulation is placed on steel pipe.

(*iii*) Constant properties.

Analysis : The heat loss rate per metre of bare pipe surface

$$\frac{Q_b}{L} = 2\pi r_1 h (T_s - T_\infty)$$
$$= 2\pi \times 0.015 \text{ m} \times 8.5 \times (200 - 30)$$
$$= 136.18 \text{ W/m}$$

After addition of insulation, the heat loss is reduced by 40%. Hence allowable heat loss is 60% only. Hence allowable heat loss,

$$Q_1 = 0.6\, Q_b = 0.6 \times 136.18 = 81.71 \text{ W/m}$$

Now from thermal network ;

Heat loss per metre length of pipe

$$\frac{Q_1}{L} = \frac{2\pi(\Delta T)}{\frac{\ln(r_2/r_1)}{k_{ins}} + \frac{1}{r_2 h}}$$

$$= \frac{2\pi(200 - 30)}{\frac{\ln(r_2/0.015)}{0.07} + \frac{1}{8.5 r_2}}$$

$$= 81.71 \text{ W/m}$$

or $$\frac{\ln(r_2/0.015)}{0.07} + \frac{1}{8.5 r_2} = 13.072$$

or $$\frac{\ln(r_2)}{0.07} + \frac{1}{8.5 r_2} = 13.072 + \frac{\ln(0.015)}{0.07}$$

$$= 13.072 - 59.995 = -46.923$$

or $$\frac{\ln(r_2)}{0.07} + \frac{1}{8.5 r_2} + 46.923 = 0$$

It is a transcedental equation and can be solved by numerical methods, we get

$$r_2 = 2.79 \times 10^{-2} \text{ m} = 2.79 \text{ cm}$$

So required thickness of insulation

$$= 2.79 - 1.5 = \mathbf{1.29 \text{ cm.}}$$ **Ans.**

Example 3.33. *Air at 90°C flows in a copper tube (k = 384 W/m.K) of 4 cm inner diameter and with 0.6 cm thick walls which are heated from the outside by water at 125°C. A scale of 0.3 cm thick is deposited on outer surface of the tube whose thermal conductivity is 1.75 W/m.K. The air and water side heat transfer coefficients are 221 and 3605 W/m².K, respectively. Find (a) overall heat transfer coefficient on the outside area basis (b) water to air heat transfer (c) temperature drop across the scale deposit.*

Solution

Given :

T_{∞_1} = 90°C,

T_{∞_2} = 125°C,

d_1 = 4 cm or r_1 = 2 cm,

r_2 = 2 cm + 0.6 cm = 2.6 cm

r_3 = 2.6 cm + 0.3 cm = 2.9 cm

k_1 = 384 W/m.K,

k_2 = 1.75 W/m.K

h_o = 3605 W/m².K,

h_i = 221 W/m².K

To find :

(*a*) Overall heat transfer coefficient based on outer surface.

(*b*) Heat transfer rate.

(*c*) Temperature drop across scale deposit.

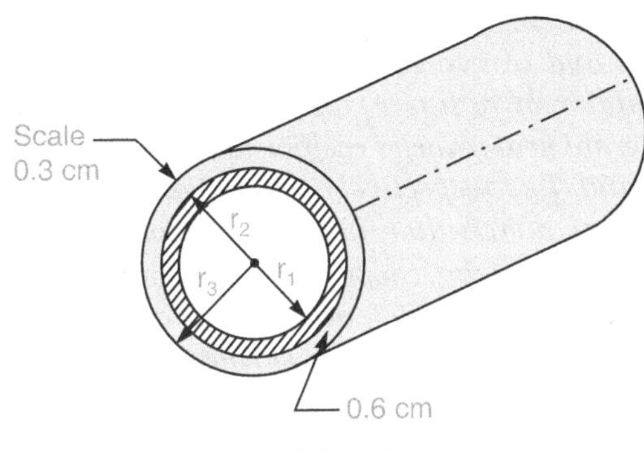

(a) Schematic

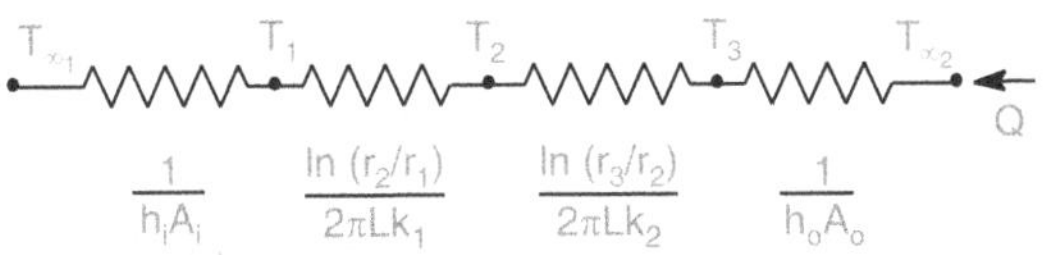

(b) Thermal network

Fig. 3.43

Assumptions :

(*i*) Steady state heat conduction in radial direction.

(*ii*) No contact resistance.

(*iii*) Pipe length is 1 metre.

Analysis : (*a*) Overall heat transfer coefficient based on outer area :

$$U_o = \frac{1}{\frac{r_3}{r_1 h_i} + \frac{r_3}{k_1}\ln(r_2/r_1) + \frac{r_3}{k_2}\ln(r_3/r_2) + \frac{1}{h_o}}$$

$$U_o = \frac{1}{\frac{0.029}{0.02 \times 221} + \frac{0.029}{384} \times \ln\left(\frac{2.6}{2}\right) + \frac{0.029}{1.75} \times \ln\left(\frac{2.9}{2.6}\right) + \frac{1}{3605}}$$

$$= \mathbf{115.37\ W/m^2.K.} \quad \text{Ans.}$$

(*b*) The heat transfer rate per metre length ;

$$Q = U_o A_o (\Delta T)_{overall}$$

$$= U_o \times (2\pi r_3 L)(T_{\infty_2} - T_{\infty_1})$$

$$= 115.37 \times (2\pi \times 0.029 \times 1) \times (125 - 90)$$

$$= \mathbf{735.76\ W/m.} \quad \text{Ans.}$$

(*c*) Temperature drop across the scale deposition :

$$Q = \frac{2\pi L k_2 (T_2 - T_3)}{\ln(r_3/r_2)}$$

$$T_2 - T_3 = \frac{Q \ln(r_3/r_2)}{2\pi L k_2} = \frac{735.76 \times \ln(2.9/2.6)}{2\pi \times 1 \times 1.75}$$

$$= \mathbf{7.3°C.} \quad \text{Ans.}$$

Example 3.34. *A steel pipe, 30 cm in outer diameter, carries steam and its surface temperature is 250°C. It is exposed to ambient air at 30°C. The heat is lost by convection and radiation. The convective heat transfer coefficient is 22 W/m².K. Calculate the heat loss from 1 m length of pipe.*

If a layer of insulation (k = 0.36 W/m.K), 75 mm thick is applied on the pipe in order to minimise the heat loss. The cost of heat is ₹ 200 per 10⁶ kJ. The cost of insulation is ₹8000 per m length. The unit is in operation for 2000 h/year. The cost of capital should be recovered in two years. Check the economical merits of insulation. Neglect radiation heat transfer after addition of insulation.

Solution

Given : An insulation on steam pipe

(*i*) d_1 = 30 cm, r_1 = 15 cm = 0.15 m

T_1 = 250°C = 523 K

T_∞ = 30°C = 303 K

h_c = 22 W/m².K

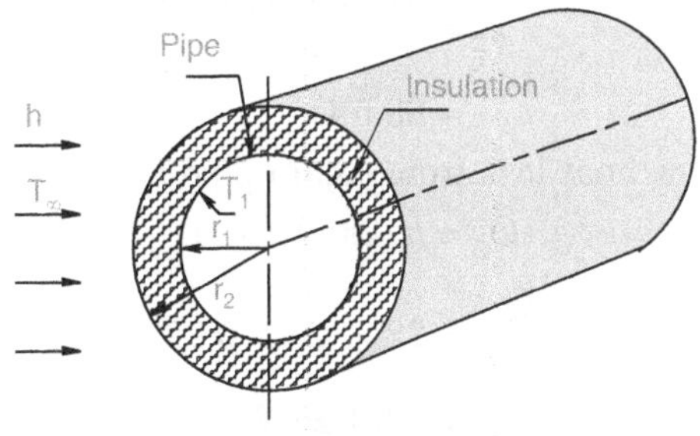

Fig. 3.44. Schematic

(*ii*) k = 0.36 W/m.K

$r_2 - r_1$ = 75 mm = 75 × 10⁻³ m

r_2 = 0.225 m

Cost of heat : ₹ 200 per 10⁶ kJ

Cost of insulation : ₹ 8000 per metre

Operation hours in 2 years = 2000 × 2 = 4000 *h*

To find :

(*i*) The heat loss/m from bare pipe.

(*ii*) Economical merits of insulation.

Assumptions :

(*i*) Steady state heat loss in radial direction only.

(*ii*) No contact resistance.

(*iii*) Bare pipe surface as black surface.

(*iv*) Stefan Boltzmann constant

$$\sigma = 5.67 \times 10^{-8} \text{ W/m}^2 \text{ K}^4.$$

Analysis : (*i*) Heat loss/m from bare pipe = heat loss by convection and radiation.

$$Q_1 = h_c(2\pi r_1 L)(T_1 - T_\infty) + \sigma(2\pi r_1 L)(T_1^4 - T_\infty^4)$$

or
$$\frac{Q_1}{L} = 22 \times (2\pi \times 0.15) \times (523 - 303) + 5.67 \times 10^{-8} \times (2\pi \times 0.15) \times (523^4 - 303^4)$$

$$= 4561.6 + 3547.7 = \mathbf{8109.33\ W.} \quad \text{Ans.}$$

(*ii*) When insulation is added, then outer surface temperature of insulation (T_2) is unknown, and at outer surface

$$Q_{cond} = Q_{conv} + \text{negligible radiation}$$

$$\frac{2\pi L k(T_1 - T_2)}{\ln\left(\frac{r_2}{r_1}\right)} = h_c(2\pi r_2 L)(T_2 - T_\infty)$$

$$\frac{0.36 \times (250 - T_2)}{\ln\left(\frac{0.225}{0.15}\right)} = 22 \times (0.225) \times (T_2 - 30)$$

or $0.179 \times (250 - T_2) = T_2 - 30$

or
$$T_2 = \frac{250 \times 0.179 + 30}{1.179}$$

$$= 63.5°\text{C}$$

Now heat loss from steam pipe per metre length

$$Q_2 = h_c(2\pi r_2 L) \times (T_2 - T_\infty)$$

or
$$\frac{Q_2}{L} = 22 \times (2\pi \times 0.225) \times (63.5 - 30)$$

$$= 1041.25 \text{ W}$$

Saving in heat loss

$$= \frac{Q_1}{L} - \frac{Q_2}{L} = 8109.33 - 1041.25$$

$$= 7068.07 \text{ W}$$

Cost of heat saved per year

= Rate of heat saving × Life of insulation

$$\times \frac{\text{Cost of insulation}}{\text{Amount of heat}}$$

$$= \frac{7068.07\left(\frac{J}{s}\right) \times 3600\left(\frac{s}{h}\right) \times 4000\text{ h} \times ₹\,200}{(10^6 \times 10^3\text{ J})}$$

= ₹ 20356 in two years

Saving in amount = 20356 – 8000 = ₹ 12356

Hence it is economical, because for 1 m length, the cost of insulation is ₹ 8000 for service of two years. **Ans.**

Example 3.35. (*a*) *A cable of radius r_1 and resistance $R_e(\Omega/m)$ and carrying a current I(A) is surrounded by an insulator of radius r_2 and thermal conductivity k. The external heat transfer coefficient and air temperature are h_o and T_∞, respectively. Derive an expression for temperature distribution in the insulator.*

(*b*) *A 1 mm dia. copper wire of resistance 0.02 Ω/m is surrounded by a 2.3 mm dia. plastic coating of k = 0.2 W/m.K. The outside surface of the coating is cooled by air, where the convective heat transfer coefficient is 16 W/m².K. Determine the maximum current, if the surface to air temperature difference is to be limited to 35°C. What is the temperature of the copper wire, if ambient temperature is 25°C ?*

Solution

(*a*) *Given :* The insulation system on an electrical cable.

To find : An expression for temperature distribution in the insulation layer.

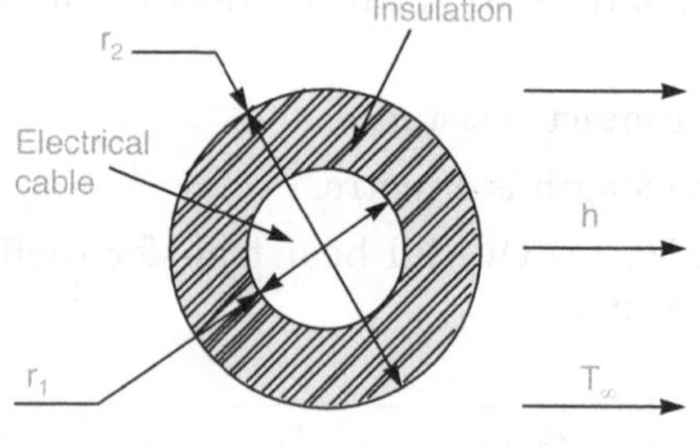

Fig. 3.45

Assumptions :

(*i*) Steady state heat conduction.

(*ii*) Heat conduction in radial direction only.

(*iii*) 1 m length of the cable and insulation.

Analysis : (*a*) The temperature distribution in cylinder, eqn. (3.30)

$$T(r) = C_1 \log_e r + C_2 \quad \ldots(i)$$

With boundary conditions,

(*i*) At $r = r_1$, $\frac{Q}{L} = I^2R_e$

$$= -2\pi r_1 k \left(\frac{dT}{dr}\right)_{r=r_1} \quad ...(ii)$$

Differentiating eqn. (*i*) with respect to *r*

$$\left(\frac{dT}{dr}\right)_{r=r_1} = \frac{C_1}{r_1}$$

Using in eqn. (*ii*),

$$I^2R_e = -2\pi r_1 k \frac{C_1}{r_1} \quad \text{or} \quad C_1 = -\frac{I^2R_e}{2\pi k}$$

Then $\quad T(r) = -\frac{I^2R_e}{2\pi k}\ln(r) + C_2$

(*ii*) At $r = r_2$, $\frac{Q}{L} = I^2R_e$

$$= h_o(2\pi r_2)\,(T_{r=r_2} - T_\infty)$$

or
$$I^2R_e = h_o(2\pi r_2)\left\{-\frac{I^2R_e}{2\pi k}\ln(r_2) + C_2 - T_\infty\right\}$$

or
$$C_2 = \frac{I^2R_e}{2\pi r_2 h_o} + \frac{I^2R_e}{2\pi k}\ln(r_2) + T_\infty$$

Using the values of C_1 and C_2 in eqn. (*i*), we get

$$T(r) = -\frac{I^2R_e}{2\pi k}\ln(r) + \frac{I^2R_e}{2\pi r_2 h_o} + \frac{I^2R_e}{2\pi k}\ln(r_2) + T_\infty$$

$$= \frac{I^2R_e}{2\pi k}\ln\left(\frac{r_2}{r}\right) + \frac{I^2R_e}{2\pi r_2 h_o} + T_\infty$$

or
$$\mathbf{T(r) = \frac{I^2R_e}{2\pi}\left\{\frac{\ln\left(\frac{r_2}{r}\right)}{k} + \frac{1}{r_2 h_o}\right\} + T_\infty.} \quad \text{Ans.} \quad ...(iii)$$

(*b*) *Given :* $d_1 = 1$ mm, $r_1 = 0.5$ mm
$d_2 = 2.3$ mm, $r_2 = 1.15$ mm,
$k = 0.2$ W/m.K
$h = 16$ W/m^2.K,
$R_e = 0.02$ Ω/m,
$T_s - T_\infty = 35$°C,
$T_\infty = 25$°C

To find :

(*i*) Maximum current carrying conductor.

(*ii*) Temperature of copper wire.

Analysis : (*i*) Rearranging eqn. (*iii*),

$$I^2 = \frac{2\pi\,(T - T_\infty)}{R_e\left\{\frac{\ln(r_2/r)}{k} + \frac{1}{r_2 h_o}\right\}}$$

At $r = r_2$, (surface of insulation)

$$I^2 = \frac{2\pi \times 35}{0.02 \times \left\{\frac{\ln(1.15/1.15)}{0.2} + \frac{1}{1.15\times10^{-3}\times16}\right\}}$$

$$= \frac{70\pi}{0.02 \times 54.35} = 202.32$$

or $\quad$ $I = $ **14.22 A.** Ans.

(*ii*) Temperature of copper wire *i.e.*, at $r = r_1$.

Using eqn. (*iii*)

$$T_1 = \frac{(14.22)^2 \times 0.02}{2\pi} \times \left\{\frac{\ln\left(\frac{1.15}{0.5}\right)}{0.2} + \frac{1}{1.15\times10^{-3}\times16}\right\} + 25$$

$$= 0.6436 \times (4.164 + 54.35) + 25 = \mathbf{62.66°C.} \quad \text{Ans.}$$

3.6. CRITICAL THICKNESS OF INSULATION ON CYLINDERS

It is our general perception that the addition of insulation on a surface minimizes the heat loss rate. For a plane wall, thicker the insulation layer, lower the heat transfer rate. It is because of constant heat transfer area and addition of insulation always increases total thermal resistance in the path of heat flow, without affecting the convection resistance.

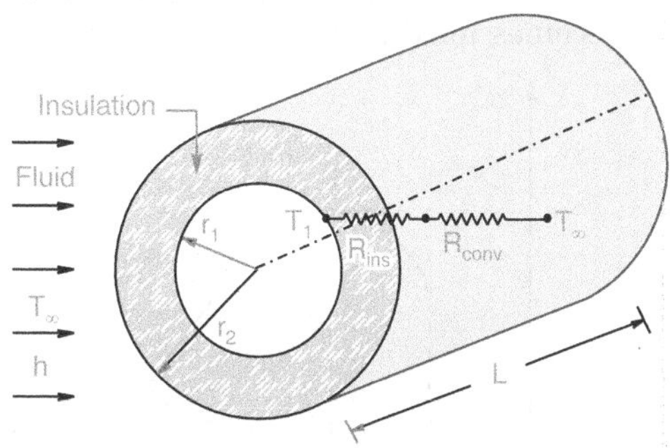

Fig. 3.46. Insulated cylinder exposed to ambient

For cylindrical pipes and spheres exposed to convection environment, the addition of insulation, however, is a different matter. The addition of insulation increases the conduction resistance but decreases the convection resistance due to increase in surface area exposed to environment. The heat transfer rate from these bodies may decrease or increase depending on the effects of dominating resistance.

Consider a layer of insulation of thermal conductivity *k*, applied on a circular pipe of radius r_1 as shown in Fig. 3.46. The inner surface temperature of insulation is T_1 and outer surface is exposed to environment at T_∞ with heat transfer coefficient h.

From thermal network, the heat transfer rate :

$$Q = \frac{T_1 - T_\infty}{R_{ins} + R_{conv}} = \frac{(T_1 - T_\infty)}{\frac{\ln(r_2/r_1)}{2\pi L k} + \frac{1}{(2\pi r_2 L)h}}$$

$$= \frac{2\pi L(T_1 - T_\infty)}{\frac{\ln(r_2/r_1)}{k} + \frac{1}{r_2 h}} \quad ...(3.55)$$

In order to determine outer radius of insulation, which will maximise the heat transfer rate, differentiating above equation with respect to r_2 and equating it to zero.

$$\frac{dQ}{dr_2} = 0 = \frac{2\pi L(T_1 - T_\infty)\left[\frac{1}{k}\left(\frac{r_1}{r_2}\right)\left(\frac{1}{r_1}\right) - \left(\frac{1}{r_2^2 h}\right)\right]}{\left[\frac{\ln(r_2/r_1)}{k} + \frac{1}{r_2 h}\right]^2}$$

or $\frac{1}{r_2 k} - \frac{1}{r_2^2 h} = 0$

or $r_2 = r_{cr,\ cylinder} = \frac{k}{h}$ ...(3.56)

where, r_{cr} = Critical radius of insulation.

Addition of insulation increases the heat loss upto certain radius of cylinder that is called ***critical radius of insulation.*** This thickness of insulation layer is called the **critical thickness of insulation,** at which heat loss becomes maximum.

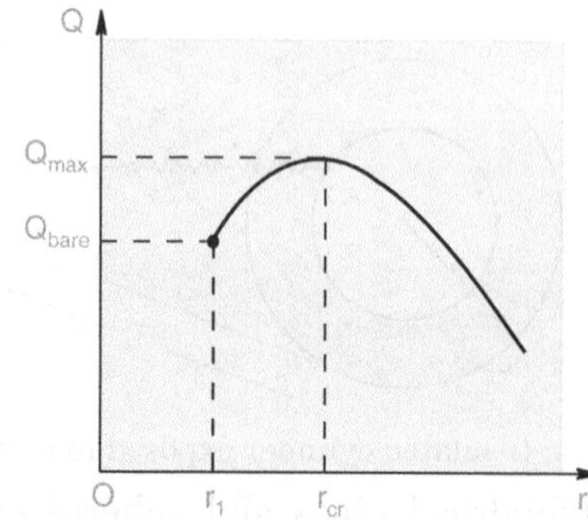

Fig. 3.47. Effect of insulation thickness on heat transfer rate

If the outer radius is greater than the critical radius r_{cr}, any addition of insulation on the pipe surface, decreases the heat loss as one expects. But if the radius is less than the critical radius as in small diameter tubes, cables or wires, the heat loss will increase with addition of insulation upto critical radius of insulation, where the heat loss becomes maximum and heat loss begins to decrease with addition of insulation beyond the critical radius as shown in Fig. 3.47. Therefore, the insulation thickness on heat pipes is always kept greater than critical thickness in order to minimise the heat loss.

The electrical wires or cables, carrying current are insulated with rubber, PVC or some polymer insulation to provide safety against some grounded surface. When current flows through the conductor, the heat is generated at the rate of

$$Q = I^2 R_e \quad ...(3.57)$$

In order to keep the wire or cable temperature steady and within safety limit, this generated heat must be dissipated to the surroundings at the same rate at which it is generated. Therefore, in electrical wires or cables, the thickness of insulation is always kept at critical thickness $(r_{cr} - r_1)$ in order to maximise heat loss.

3.6.1. Effect of Thermal Resistances

Actually with addition of insulation, the exposure area increases. This can be explained in other way with the help of material and surface resistances as shown in Fig. 3.48.

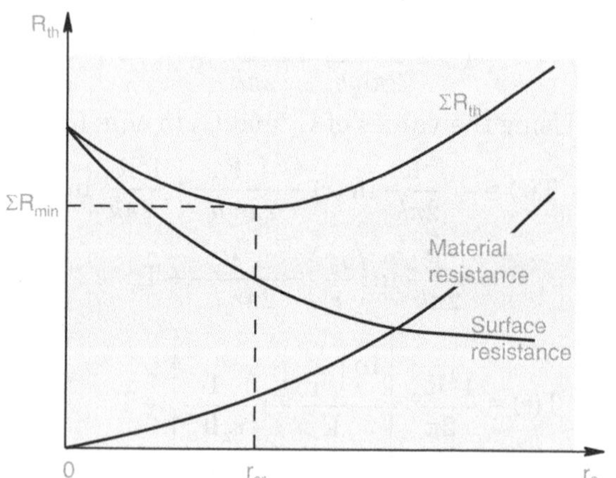

Fig. 3.48. Effect of resistance on heat transfer

When the outer radius of insulation is increased, it increases the material resistance $R_{ins} = \frac{\ln(r_2/r_1)}{2\pi r_2 Lk}$, while outer surface resistance $R_{conv} = \frac{1}{2\pi r_2 Lh}$ decreases. Until r_2 is smaller than the critical radius r_{cr}, the surface resistance decreases at faster rate than increase in material resistance. Hence, the net resistance decreases causing the heat flow to increase. But when r_2 becomes more than critical radius r_{cr}, the material resistance increases at faster rate, resulting into increase in net resistance, causing the heat flow to decrease.

Example 3.36. *An electrical wire, 2 mm in diameter is covered with a 2.5 mm thick layer of plastic insulation (k = 0.5 W/m.K) to reduce the heat loss. Heat is dissipated* *from the outer surface of insulation to surrounding air at 25°C by convection with heat transfer coefficient of 10 W/m².K. The wire is maintained at constant temperature of 120°C. Estimate the rate of heat dissipation from the wire per unit length with and without insulation. Calculate the thickness of insulation when the heat dissipation rate is maximum. What is maximum value of heat dissipation ?*

Solution

Given : $d_1 = 2$ mm, $r_1 = 1$ mm

$r_2 = 1$ mm + 2.5 mm = 3.5 mm

$k = 0.5$ W/m.K

$h = 10$ W/m².K

$T_1 = 120°C$

$T_\infty = 25°C$.

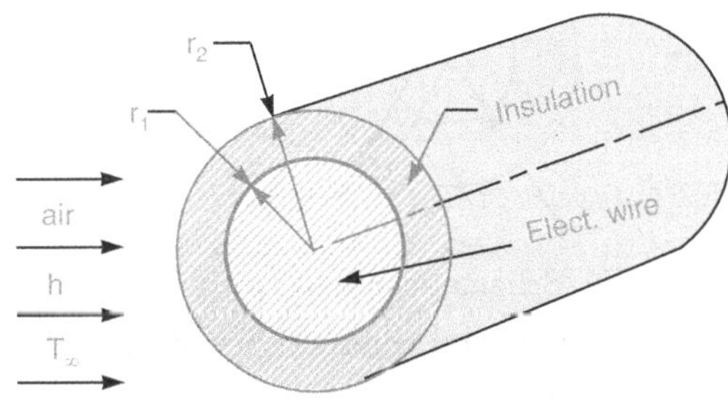

Fig. 3.49

To find :

(*i*) Rate of heat dissipation from the wire per m, without insulation ;

(*ii*) Rate of heat dissipation from wire per m, with insulation ;

(*iii*) Critical thickness of insulation ; and

(*iv*) Maximum heat dissipation rate.

Analysis : (*i*) Heat dissipation rate from bare wire per m, (without insulation)

$$Q_1 = hA(T_1 - T_\infty) = h(\pi d_1 L)(T_1 - T_\infty)$$

or
$$\frac{Q_1}{L} = (10 \text{ W/m}^2.\text{K}) \times (\pi \times 2 \times 10^{-3} \text{ m}) \times (120 - 25)\,(°\text{C}) = \mathbf{5.97\ W/m.} \quad \text{Ans.}$$

(*ii*) Heat dissipation rate from the insulated wire per metre length.

$$\frac{Q_2}{L} = \frac{2\pi(T_1 - T_\infty)}{\frac{\ln(r_2/r_1)}{k} + \frac{1}{hr_2}} = \frac{2\pi(120-25)}{\frac{\ln\left(\frac{3.5}{1}\right)}{0.5} + \frac{1}{10 \times 3.5 \times 10^{-3}}}$$

$$= \frac{596.9}{2.505 + 28.57} = \mathbf{19.2\ W/m.} \quad \text{Ans.}$$

Comment : The addition of insulation increases the heat dissipation from the wire by a factor 6.44.

(*iii*) Critical thickness of insulation :

$$\text{Critical radius, } r_{cr} = \frac{k}{h} = \frac{0.5}{10} = 0.05 \text{ m} = 50 \text{ mm}$$

Critical thickness of insulation

$$= r_{cr} - r_1 = 50 \text{ mm} - 1 \text{ mm} = \mathbf{49\ mm.} \quad \text{Ans.}$$

(*iv*) Maximum heat dissipation rate :

$$\frac{Q_{max}}{L} = \frac{2\pi(T_1 - T_\infty)}{\frac{\ln(r_{cr}/r_1)}{k} + \frac{1}{h\,r_{cr}}}$$

$$= \frac{2\pi(120-25)}{\frac{\ln\left(\frac{50}{1}\right)}{0.5} + \frac{1}{10 \times (50 \times 10^{-3})}} = \frac{596.9}{9.824}$$

$$= \mathbf{60.75\ W/m.} \quad \text{Ans.}$$

Example 3.37. *An electric cable of 20 mm diameter is insulated with rubber, which is exposed to atmosphere at 30°C. Calculate the most economical thickness of rubber insulation (k = 0.175 W/m.K). When cable surface temperature with and without insulation is at 70°C. Also calculate the percentage increase in heat dissipation and current carrying capacity when most economical thickness is provided. Take heat transfer coefficient, h = 9.3 W/m².K.*

Solution

Given : An electric cable insulated with rubber

$d_1 = 20$ mm,

$r_1 = 10$ mm = 0.01 m

$T_\infty = 30°C$,

$T_s = 70°C$

$k = 0.175$ W/m.K,

$h = 9.3$ W/m².K.

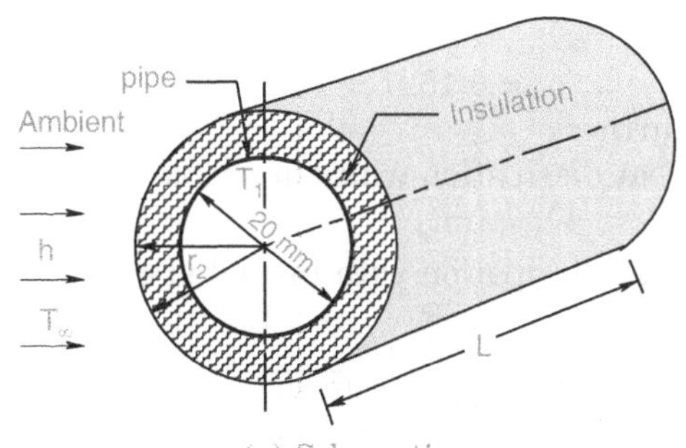

(a) Schematic

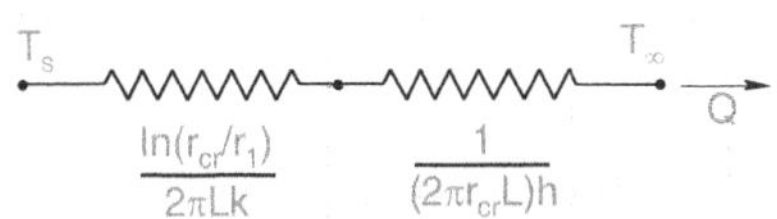

(b) Thermal network

Fig. 3.50

To find :

(*i*) The critical thickness of insulation.

(*ii*) Percentage increase in heat dissipation and current carrying capacity with critical thickness insulation.

Assumptions :

(*i*) Steady state conduction in radial direction only.

(*ii*) No contact resistance at interface.

Analysis : (*i*) The critical radius of insulation

$$r_{cr} = \frac{k}{h} = \frac{0.175}{9.3} = 0.188 \text{ m}$$
$$= 18.81 \text{ mm}$$

Then critical thickness of insulation

$$= r_{cr} - r_1 = 18.81 - 10$$
$$= \mathbf{8.81 \text{ mm.}} \quad \textbf{Ans.}$$

(*ii*) Heat dissipation rate per metre length from bare surface of pipe

$$\frac{Q_1}{L} = 2\pi r_1 h(T_s - T_\infty)$$
$$= 2\pi \times 0.01 \times 9.3 \times (70 - 30)$$
$$= 23.37 \text{ W/m}$$

The heat dissipation rate with critical thickness of insulation can be calculated by electrical analogy.

$$\frac{Q_2}{L} = \frac{\Delta T}{\dfrac{\ln(r_{cr}/r_1)}{2\pi k} + \dfrac{1}{2\pi r_{cr}\, h}}$$
$$= \frac{2\pi\,(70-30)}{\dfrac{\ln(18.81/10)}{0.175} + \dfrac{1}{18.81 \times 10^{-3} \times 9.3}}$$
$$= 26.94 \text{ W/m}$$

The percentage increase in the heat dissipation rate

$$= \frac{Q_2 - Q_1}{Q_1} = \frac{26.94 - 23.37}{23.37}$$
$$= 0.1527 = \mathbf{15.275\%.} \quad \textbf{Ans.}$$

Further,

Heat dissipation with bare cable

$$Q_1 = I_1^2 R_e$$

Heat dissipation with insulated cable

$$Q_2 = I_2^{\,2} R_e$$

or

$$\frac{I_2}{I_1} = \sqrt{\frac{Q_2}{Q_1}} = \sqrt{\frac{26.94}{23.37}} = 1.0733$$

Hence, percentage increase in current carrying capacity

$$\frac{I_2 - I_1}{I_1} \times 100 = (1.0733 - 1) \times 100$$
$$= \mathbf{7.33\%.} \quad \textbf{Ans.}$$

Example 3.38. *An electric cable of 12 mm diameter is insulated to increase the current capacity. Due to insulation the current carrying capacity is increased by 15% without increasing cable surface temperature above 70°C. The environmental temperature is 30°C. Assume that the heat transfer coefficient from the bare or insulated cable is 14 W/m^2.K. Calculate the conductivity of insulating material.*

Solution

Given : An electric cable insulated to increase the current carrying capacity

$$d = 12 \text{ mm}, \qquad T_s = 70°\text{C},$$
$$T_\infty = 30°\text{C}, \qquad h = 14 \text{ W/m}^2.\text{K}.$$

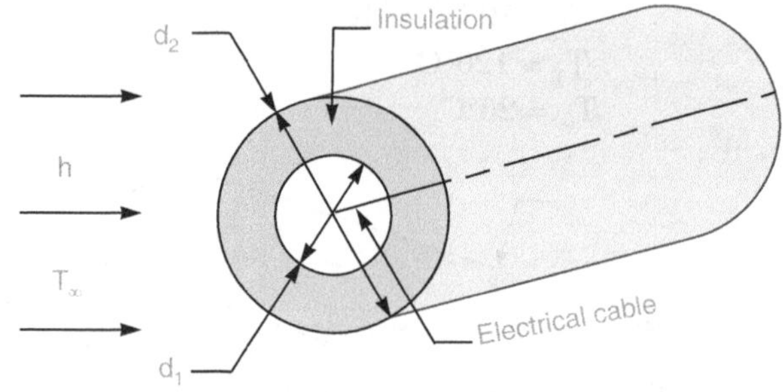

Fig. 3.51. Schematic of cable with insulation

To find : Thermal conductivity of insulation.

Assumptions :

(*i*) Steady state heat transfer in radial direction.

(*ii*) Heat loss from the cable or insulation surface by convection only.

(*iii*) 1 m long electrical cable.

Analysis : The heat dissipation rate per metre length of the bare cable

$$Q_1 = hA\,(\Delta T) = (\pi d \text{ L})\, h\,(T_s - T_\infty)$$
$$= (\pi \times 12 \times 10^{-3} \times 1) \times 14 \times (70 - 30)$$
$$= 21.11 \text{ W}$$

The heat dissipation rate with insulation will be maximum when,

$$r_2 = r_{cr} = \frac{k}{h} \quad \text{or} \quad r_2 h = k$$

Using for maximum heat dissipation rate from insulated cable :

$$Q_2 = \frac{T_s - T_\infty}{\dfrac{\ln(r_2/r_1)}{2\pi L k} + \dfrac{1}{(2\pi r_2 L)h}}$$
$$= \frac{2\pi \times 1 \times (70-30)}{\dfrac{\ln(k/hr_1)}{k} + \dfrac{1}{k}} = \frac{251.32\,k}{1 + \ln(11.90\,k)}$$

Now,

$$\frac{I_2^2 R_e}{I_1^2 R_e} = \frac{Q_2}{Q_1}$$

$$= \frac{251.32\,k}{1 + \ln(11.90\,k)} \times \frac{1}{21.11}$$

$$= \frac{11.90\,k}{1 + \ln(11.90\,k)}$$

But, $I_2 = 1.15 I_1$

or $$\frac{I_2^2}{I_1^2} = 1.3225 = \frac{11.90k}{1 + \ln(11.90k)}$$

Considering 2nd and 3rd terms of equation, we get

$1 + \ln(11.90k) = 9k$ or $\ln(11.9k) - 9k + 1 = 0$

By trial and error

Let $k = 0.1$ W/m.K, then,

$1.19 - 0.9 + 1 \neq 0$

Let $k = 0.2$, then

$0.8671 - 1.9422 + 1 = (-0.075) \neq 0.031$

Let $k = 0.2158$,

satisfy the equation, therefore,

Thermal conductivity of insulating material is **0.2158 W/m.K. Ans.**

Example 3.39. *A current of 1000 A is flowing through a long copper conductor (k = 390 W/m.K), 25 mm in diameter, having its electric resistivity of 1.08 μΩ cm. This rod is insulated to a radius of 17.5 mm with fibrous cotton (k = 0.058 W/m.K), which is further covered by a layer of plastic (k = 0.42 W/m.K) and then it is exposed to surrounding air at 20°C with a heat transfer coefficient of 20.5 W/m².K. Calculate :*

(i) thickness of plastic layer, which gives minimum temperature in a cotton insulation.

(ii) the temperature of copper rod and maximum temperature in the plastic layer for above condition.

Solution

Given : Two insulation layers on a copper conductor ;

$I = 1000$ A,
$k = 390$ W/m.K
$d_1 = 25$ mm $= 25 \times 10^{-3}$ m,
$r_1 = 12.5 \times 10^{-3}$ m
$\rho = 1.08 \times 10^{-8}$ Ω-m
$r_2 = 17.5$ mm $= 17.5 \times 10^{-3}$ m
$k_1 = 0.058$ W/m.K,
$k_2 = 0.42$ W/m.K
$T_\infty = 20$°C,
$h = 20.5$ W/m².K.

To find :

(*i*) Thickness of plastic corresponds to minimum temperature in a cotton insulation.

(*ii*) Temperature of copper rod and maximum temperature in plastic layer.

Assumptions :

(*i*) Steady state conditions.

(*ii*) Heat transfer in radial direction only.

(*iii*) No contact resistance at interfaces.

(*iv*) 1 m length of copper conductor.

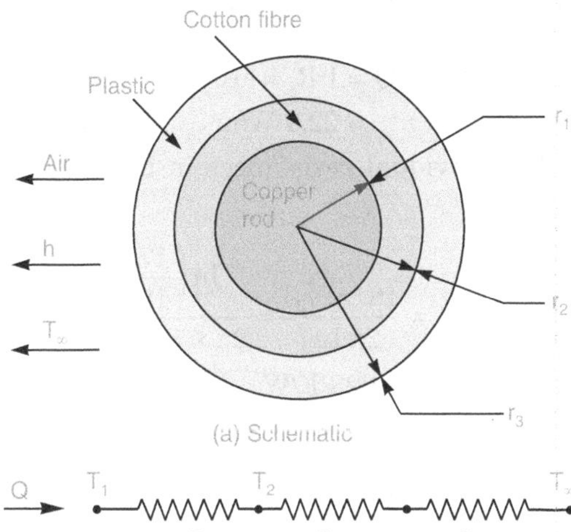

(a) Schematic

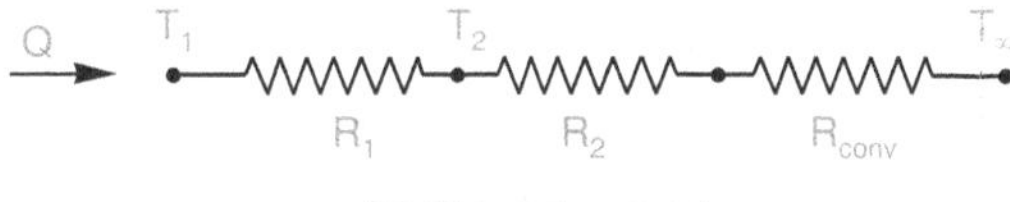

(b) Thermal network

Fig. 3.52

Analysis : (*i*) Thickness of plastic layer for minimum temperature in cotton fibre insulation :

For given system the heat transfer rate is given as :

$$Q = \frac{\Delta T}{\Sigma R_{th}} = \frac{\Delta T}{\dfrac{\ln(r_2/r_1)}{2\pi L k_1} + \dfrac{\ln(r_3/r_2)}{2\pi L k_2} + \dfrac{1}{(2\pi r_3 L)h}}$$

For maximum heat transfer rate through plastic layer, which give minimum temperature of cotton insulation.

Differentiating above equation w.r.t. r_3

$$\frac{dQ}{dr_3} = 0 = \Delta T \left[\frac{0 + \dfrac{1}{2\pi L\, k_2}\left(\dfrac{r_2}{r_3}\right) \times \left(\dfrac{1}{r_2}\right) - \dfrac{1}{2\pi L h}\left(\dfrac{1}{r_3^2}\right)}{\left\{\dfrac{\ln(r_2/r_1)}{2\pi L k_1} + \dfrac{\ln(r_3/r_2)}{2\pi L k_2} + \dfrac{1}{2\pi r_3 L h}\right\}^2} \right]$$

or $$\frac{1}{r_3 k_2} = \frac{1}{r_3^2\, h} \quad \text{or} \quad r_3 = \frac{k_2}{h}$$

(condition of critical radius of insulation)

or $r_3 = \frac{0.42}{20.5} = 0.02048 \text{ m} = 20.48 \text{ mm}$

Thickness of plastic layer

$= r_3 - r_2 = 20.48 - 17.5 = \mathbf{2.98 \text{ mm}}$. **Ans.**

(*ii*) Temperature of copper rod and maximum temperature in plastic layer :

(*a*) Temperature of copper rod :

The resistance of 1 m copper conductor

$$R_e = \frac{\rho L}{A_c} = \frac{\rho L}{(\pi/4)\, d_1^2} = \frac{1.08 \times 10^{-8} \times 1}{(\pi/4) \times (25 \times 10^{-3})^2}$$

$$= 2.2 \times 10^{-5} \ \Omega/\text{m}$$

Heat generation rate per metre, in copper conductor

$$Q = I^2 R_e = (1000)^2 \times 2.2 \times 10^{-5}$$
$$= 22.0 \text{ W/m}$$

The individual resistances in thermal network (Fig. 3.52)

$$R_1 = \frac{\ln (r_2/r_1)}{2\pi L k_1} = \frac{\ln\left(\frac{17.5}{12.5}\right)}{2\pi \times 1 \times 0.058}$$
$$= 0.923 \text{ K/W}$$

$$R_2 = \frac{\ln (r_3/r_2)}{2\pi L k_2} = \frac{\ln\left(\frac{20.48}{17.5}\right)}{2\pi \times 1 \times 0.42}$$
$$= 0.0595 \text{ K/W}$$

$$R_{conv} = \frac{1}{2\pi r_3 L h}$$

$$= \frac{1}{2\pi \times 20.48 \times 10^{-3} \times 1 \times 20.5}$$

$$= 0.379 \text{ K/W}$$

The heat flow rate

$$Q = \frac{\Delta T}{\Sigma R_{th}} = \frac{(T_1 - T_\infty)}{R_1 + R_2 + R_{conv}}$$

or $T_1 - 20 = 22 \times (0.923 + 0.0595 + 0.379)$

or $T_1 = 20 + 29.95 = \mathbf{49.95°C}$. **Ans.**

(*b*) Maximum temperature in plastic layer :

It will occur at cotton fibre insulation and plastic layer interface, say it is T_2. From thermal network

$$Q = \frac{T_1 - T_2}{R_1} \quad \text{or} \quad 49.95 - T_2 = 22 \times 0.923$$

or $T_2 = 49.95 - 21.077 = \mathbf{28.87°C}$. **Ans.**

3.7. HOLLOW SPHERE

Consider a hollow sphere of inner radius r_1 and outer radius r_2 as shown in Fig. 3.53. Its inner and outer surfaces are maintained at uniform temperatures T_1 and T_2, respectively. There is no heat generation in the solid and thermal conductivity, k is assumed constant.

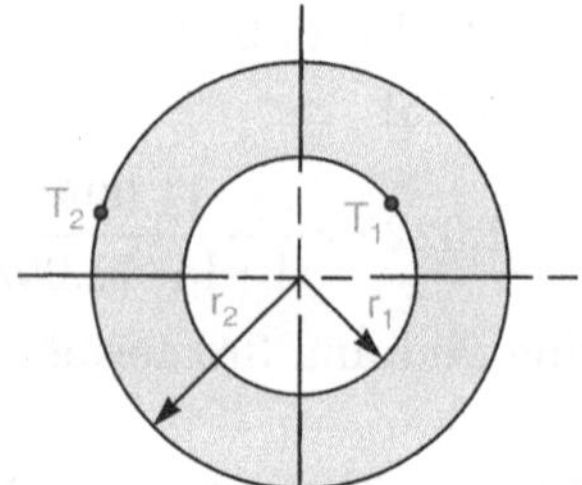

Fig. 3.53. Sphere with specified temperature

The governing differential equation in spherical coordinates eqn. (2.20).

$$\frac{d}{dr}\left[\frac{r^2 dT(r)}{dr}\right] = 0$$

Integrating with respect to r,

$$\left[\frac{r^2 dT(r)}{dr}\right] = C_1$$

or
$$\left[\frac{dT(r)}{dr}\right] = \frac{C_1}{r^2} \quad \text{...(3.58)}$$

Integrating again, we get

$$T(r) = -\frac{C_1}{r} + C_2 \quad \text{...(3.59)}$$

Subjected to boundary conditions,

$$T(r) = T_1 \quad \text{at} \quad r = r_1$$

and
$$T(r) = T_2 \quad \text{at} \quad r = r_2$$

Using, we get two simultaneous equations as,

$$T_1 = -\frac{C_1}{r_1} + C_2 \quad \text{...}(i)$$

$$T_2 = -\frac{C_1}{r_2} + C_2 \quad \text{...}(ii)$$

Subtracting eqn. (*ii*) from (*i*), we get

$$T_1 - T_2 = C_1\left[\frac{1}{r_2} - \frac{1}{r_1}\right]$$

or
$$C_1 = -\frac{r_1 r_2 (T_1 - T_2)}{r_2 - r_1}$$

and
$$C_2 = \frac{r_2 T_2 - r_1 T_1}{r_2 - r_1}$$

Substituting the values of C_1 and C_2 in eqn. (3.59)

$$T(r) = \frac{r_1 r_2}{r}\left(\frac{T_1 - T_2}{r_2 - r_1}\right) + \frac{r_2 T_2 - r_1 T_1}{r_2 - r_1} \quad \text{...(3.60)}$$

Differentiating eqn. (3.60) with respect to r, we get

$$\frac{dT(r)}{dr} = -\frac{1}{r^2}\frac{(r_1 r_2)(T_1 - T_2)}{(r_2 - r_1)}$$

The heat flux

$$q(r) = -k\frac{dT(r)}{dr}$$

$$= -k\left[-\frac{1}{r^2}(r_1 r_2)\frac{(T_1 - T_2)}{(r_2 - r_1)}\right]$$

$$q(r) = \frac{k}{r^2}(r_1 r_2)\frac{(T_1 - T_2)}{(r_2 - r_1)} \quad \text{...(3.61)}$$

The heat transfer rate :

$$Q = q(r)\,A = q(r)\,(4\pi r^2)$$

$$= \frac{4\pi\, r_1 r_2\, k(T_1 - T_2)}{(r_2 - r_1)} \quad \text{...(3.62)}$$

3.7.1. Electrical Analogy for Hollow Sphere

Rearranging the equation (3.62)

$$Q = \frac{(T_1 - T_2)}{\dfrac{(r_2 - r_1)}{4\pi\, r_1 r_2 k}} \quad \text{...(3.63)}$$

T_1 —R_{sph}— T_2 → Q

Fig. 3.54. Equivalent thermal network

It can be written in the form,

$$Q = \frac{T_1 - T_2}{R_{sph}} \quad \text{...(3.64)}$$

where $$R_{sph} = \frac{r_2 - r_1}{4\pi\, r_1 r_2 k} \quad \text{...(3.65)}$$

Where R_{sph} is called the thermal resistance to heat flow for a hollow sphere. The equivalent thermal network is shown in Fig. 3.54. If convection heat transfer is involved at the boundary surfaces as shown in Fig. 3.55 then heat flow,

$$Q = \frac{T_{\infty_1} - T_{\infty_2}}{R_{conv,1} + R_{sph} + R_{conv,2}}$$

$$= \frac{T_{\infty_1} - T_{\infty_2}}{\dfrac{1}{4\pi r_1^2 h_1} + \dfrac{(r_2 - r_1)}{4\pi r_1 r_2 k} + \dfrac{1}{4\pi r_2^2 h_2}} \quad \text{...(3.66)}$$

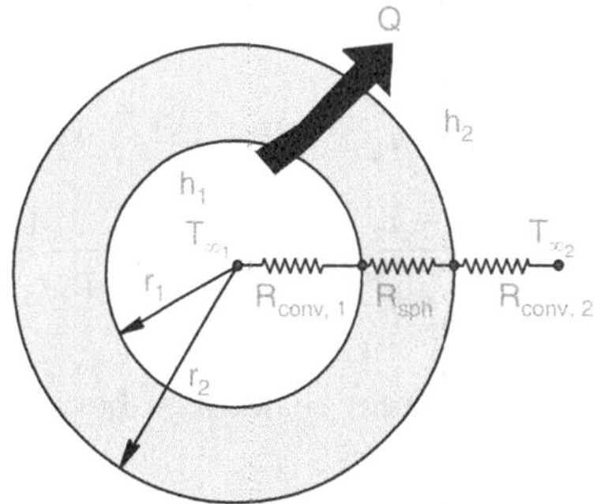

Fig. 3.55. Thermal resistance network for a hollow sphere subjected to convection heat transfer at inner and outer surfaces

where h_1 and h_2 represent convection coefficient at inner and outer surfaces of hollow sphere, while T_{∞_1} and T_{∞_2} are temperatures of ambient on two sides.

3.7.2. Multilayer Hollow Sphere

The radial heat flow Q through a multilayer sphere as shown in Fig. 3.56 (*a*) can be obtained by using thermal resistance concept to each layer

$$Q = \frac{T_{\infty_1} - T_1}{R_{conv,1}} = \frac{T_1 - T_2}{R_1} = \frac{T_2 - T_3}{R_2} = \frac{T_3 - T_{\infty_2}}{R_{conv,2}} \quad \text{...(3.67)}$$

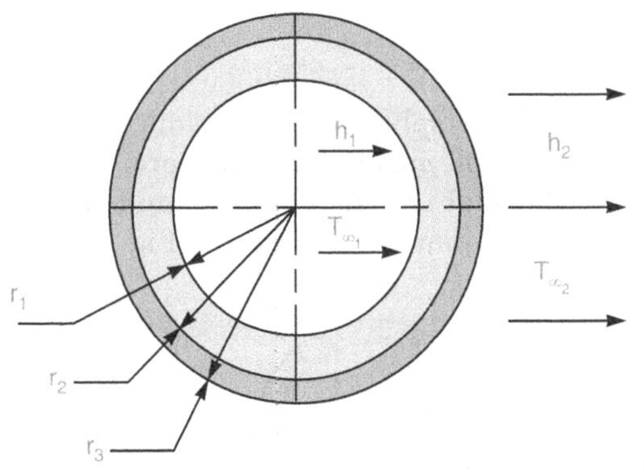

(a) Composite sphere

T_{∞_1} —$R_{conv,1}$— T_1 —R_1— T_2 —R_2— T_3 —$R_{conv,2}$— T_{∞_2} → Q

(b) Equivalent thermal resistances

Fig. 3.56

Since all thermal resistances are in series as shown in Fig. 3.56(*b*) ;

Therefore,

$$Q = \frac{T_{\infty_1} - T_{\infty_2}}{R_{conv,1} + R_1 + R_2 + R_{conv,2}} = \frac{(\Delta T)_{overall}}{\Sigma R_{th}} \quad \text{...(3.68)}$$

where various resistances are

$$R_{conv,1} = \frac{1}{4\pi r_1^2 h_1}, \qquad R_1 = \frac{r_2 - r_1}{4\pi r_1 r_2 k_1}$$

$$R_2 = \frac{r_3 - r_2}{4\pi r_2 r_3 k_2}, \quad R_{conv,2} = \frac{1}{4\pi r_3^2 h_2} \quad ...(3.69)$$

and $\Sigma R_{th} = R_{conv,1} + R_1 + R_2 + R_{conv,2}$.

It is total thermal resistance between the temperatures T_{∞_1} and T_{∞_2}.

3.7.3. Overall Heat Transfer Coefficient

The heat flow rate Q through multilayer sphere can be expressed in form :

$$Q = U_1A_1 (\Delta T) = U_2A_2 (\Delta T) \quad ...(3.70)$$

where U is overall heat transfer coefficient and can be expressed as :

$$U_2A_2 = U_1A_1 = \frac{1}{\Sigma R_{th}} \quad ...(3.71)$$

or $$U_2 = \frac{1}{A_2 \Sigma R_{th}} \quad \text{and} \quad U_1 = \frac{1}{A_1 \Sigma R_{th}}$$

where U_2 is the overall heat transfer coefficient based on outer surface

U_1 is the overall heat transfer coefficient based on inner surface.

3.7.4. Critical Radius of Insulation on Sphere

Consider a hollow sphere of outer radius r_1 is covered with a layer of insulation (of outer radius r_2) of constant thermal conductivity k and exposed to convection environment as shown in Fig. 3.57.

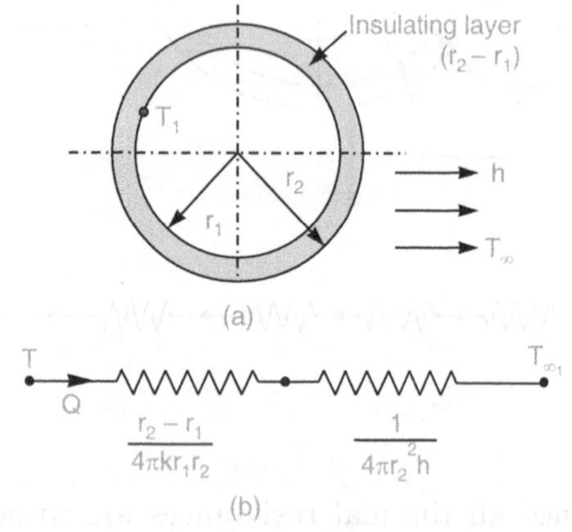

Fig. 3.57. Sphere with insulation

The heat flow rate can be expressed as :

$$Q = \frac{T_1 - T_\infty}{\frac{(r_2 - r_1)}{4\pi r_1 r_2 k} + \frac{1}{4\pi r_2^2 h}} \quad ...(3.72)$$

For maximum heat transfer rate,

$$\frac{dQ}{dr_2} = 0$$

we get, $$r_{cr,\,sphere} = \frac{2k}{h} \quad ...(3.73)$$

where $r_{cr,\,sphere}$ is the critical radius of insulation. The heat transfer rate would be maximum with this radius of insulation on spheres.

Example 3.40. *A spherical thin walled metallic container is used to store liquid nitrogen at 77 K. The container has a diameter of 0.5 m and is covered with an evacuated reflective insulation system composed of silica powder (k = 0.0017 W/m.K). The insulation is 25 mm thick and its outer surface is exposed to ambient air at 300 K. The convective coefficient is known to be 20 W/m².K. The latent heat of vaporization and density of liquid nitrogen are 2 × 10⁵ J/kg and 804 kg/m³, respectively.*

(i) What is the rate of heat transfer to the liquid nitrogen ?

(ii) What is the rate of liquid boil off ?

(N.M.U., May 2009; P.U., Dec. 2002)

Solution

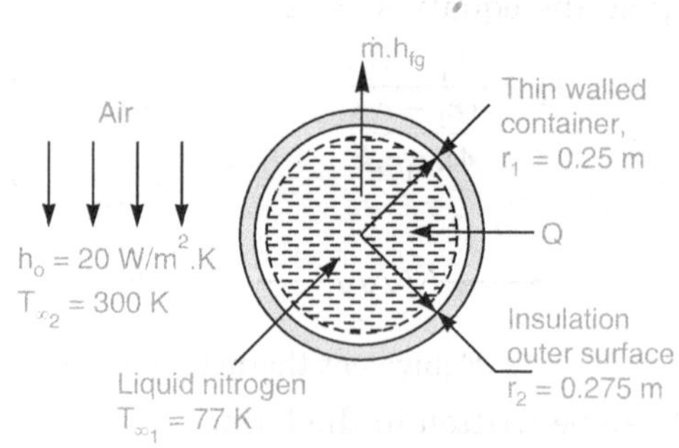

Fig. 3.58

Given : A spherical metallic container filled with liquid nitrogen

$d_1 = 0.5$ m or $r_1 = 0.25$ m

$h_{fg} = 2 \times 10^5$ J/kg,

$r_2 = 0.25$ m + 25 mm = 0.275 m

$k = 0.0017$ W/m.K,

$h_o = 20$ W/m².K

ρ = 804 kg/m³,

T_{∞_1} = 77 K,

T_{∞_2} = 300 K.

To find :

(*i*) The heat transfer rate to the nitrogen.

(*ii*) The mass rate of nitrogen boil off.

Assumptions :

(*i*) Steady state conditions. One dimensional heat transfer in radial direction.

(*ii*) Negligible resistance between container wall and liquid nitrogen.

(*iii*) Constant properties.

(*iv*) Negligible radiation heat loss.

Analysis : (*i*) The thermal circuit involves a conduc-tion and convection resistance in series, therefore

$$R_{sph} = \frac{r_2 - r_1}{4\pi r_1 r_2 k}$$

$$= \frac{0.275 - 0.25}{4\pi \times 0.25 \times 0.275 \times 0.0017}$$

$$= 17.022 \text{ K/W}$$

$$R_{conv} = \frac{1}{4\pi r_2^2 h_o} = \frac{1}{4\pi \times (0.275)^2 \times 20}$$

$$= 0.052 \text{ K/W}$$

Total thermal resistance of series resistances ;

$$\Sigma R_{th} = R_{sph} + R_{conv}$$

$$= 17.022 + 0.52 = 17.074 \text{ K/W}$$

The rate of heat transfer to liquid nitrogen :

$$Q = \frac{T_{\infty_2} - T_{\infty_1}}{\Sigma R_{th}} = \frac{300 - 77}{17.074}$$

$$= \mathbf{13.06\ W.} \quad \text{Ans.}$$

(*ii*) The heat loss to nitrogen will also cause evaporation.

Thus, $Q = \dot{m} h_{fg}$

or $$\dot{m} = \frac{Q}{h_{fg}} = \frac{13.06}{2 \times 10^5}$$

$$= 6.53 \times 10^{-5} \text{ kg/s}$$

$$= \mathbf{0.235\ kg/hr.} \quad \text{Ans.}$$

or on volumetric basis

$$\dot{V} = \frac{\dot{m}}{\rho} = \frac{0.235}{804} = 2.923 \times 10^{-4} \text{ m}^3\text{/hr}$$

$$\approx \mathbf{7\ lit/day.} \quad \text{Ans.}$$

Example 3.41. *A hollow sphere of inside radius 30 mm and outside radius 50 mm is electrically heated at its inner surface at a constant rate of 10^5 W/m². The outer surface is exposed to a fluid at 30°C, with heat transfer coefficient of 170 W/m².K. The thermal conductivity of the material is 20 W/m.K. Calculate inner and outer surface temperatures.*

Solution

Given : A hollow sphere

r_1 = 30 mm = 0.03 m

r_2 = 50 mm = 0.05 m

q = 10^5 W/m², T_∞ = 30°C

h = 170 W/m².K, k = 20 W/m.K.

To find : Inner and outer surface temperatures of sphere.

Analysis : For hollow sphere, the individual thermal resistance R_{sph} and R_{conv}

$$R_{sph} = \frac{r_2 - r_1}{4\pi r_1 r_2 k} = \frac{0.05 - 0.03}{4\pi \times 0.03 \times 0.05 \times 20}$$

$$= 0.053 \text{ K/W}$$

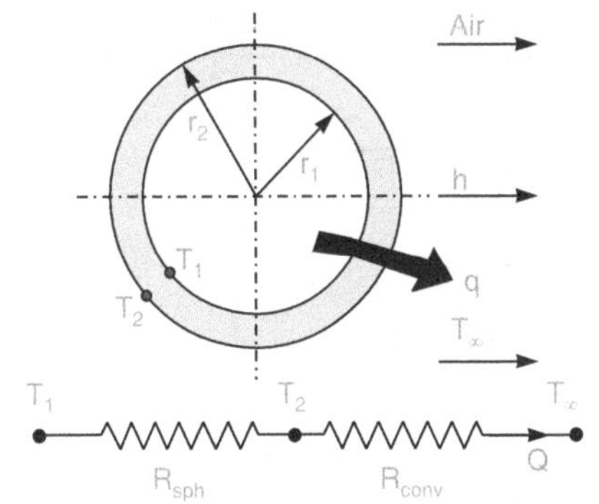

Fig. 3.59

$$R_{conv} = \frac{1}{4\pi r_2^2 h} = \frac{1}{4\pi \times (0.05)^2 \times 170}$$

$$= 0.187 \text{ K/W}$$

These resistances are in series, therefore,

$$\Sigma R_{th} = R_{sph} + R_{conv}$$

$$= 0.053 + 0.187 = 0.24 \text{ K/W}$$

The heat flow rate

$$Q = q\,(4\pi r_1^2) = 10^5 \times 4\pi \times (0.03)^2$$

$$= 1130.97 \text{ W}$$

Further, the heat flow rate by using thermal network ;

$$Q = \frac{T_1 - T_\infty}{\Sigma R_{th}}$$

or $$T_1 = Q\,\Sigma R_{th} + T_\infty$$
$$= 1130.97 \times 0.24 + 30$$
$$= \mathbf{301.5°C.} \quad \text{Ans.}$$

Again for convection heat flow

$$Q = h(4\pi r_2^2)(T_2 - T_\infty)$$

or $$1130.97 = 170 \times 4\pi \times (0.05)^2 (T_2 - 30)$$

or $$T_2 = 211.76 + 30 = \mathbf{241.76°C.} \quad \text{Ans.}$$

Example 3.42. *A hollow spherical form is used to determine thermal conductivity of an insulating material. The inner diameter is 50 mm and outer diameter is 100 mm. A 40 W heater is placed inside and under steady state conditions, the temperature at 32 and 40 mm radii were found to be 100°C and 70°C, respectively. Determine the thermal conductivity of the material. Also calculate the outside temperature of sphere. If surrounding air is at 30°C, calculate convection heat transfer coefficient over the surface.*

Solution

Given : A hollow sphere as shown in Fig. 3.60.

$d_1 = 50$ mm,	$r_1 = 25$ mm $= 0.025$ m
$d_2 = 100$ mm,	$r_2 = 50$ mm $= 0.05$ m
$r_3 = 32$ mm $= 0.032$ m	$r_4 = 40$ mm $= 0.04$ m
$Q = 40$ W	$T_3 = 100°C$
$T_4 = 70°C$	$T_\infty = 30°C$.

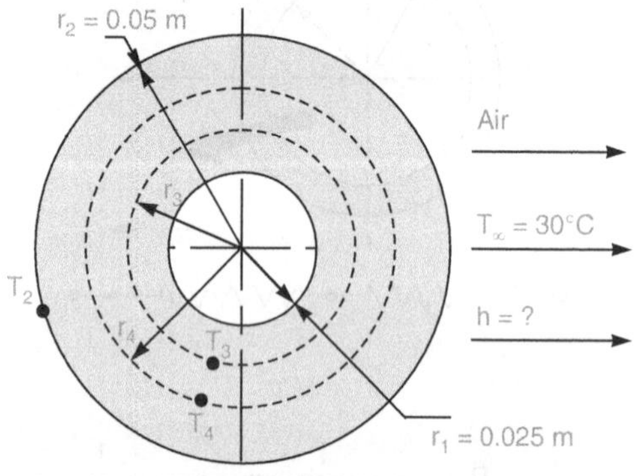

Fig. 3.60

To find :

(*i*) Thermal conductivity of insulating material.

(*ii*) Outside temperature (T_2) of sphere.

(*iii*) Convection heat transfer coefficient.

Analysis : (*i*) Under steady state conditions, the heat transfer rate through hollow sphere

$$Q = \frac{4\pi r_3 r_4 k (T_3 - T_4)}{r_4 - r_3}$$

or $$k = \frac{Q(r_4 - r_3)}{4\pi r_3 r_4 (T_3 - T_4)}$$

$$= \frac{40 \times (0.04 - 0.032)}{4\pi \times 0.032 \times 0.04 \times (100 - 70)}$$

$$= \mathbf{0.663\ W/m.K.} \quad \text{Ans.}$$

(*ii*) Outside temperature of sphere,

$$Q = \frac{4\pi r_4 r_2 k (T_4 - T_2)}{r_2 - r_4}$$

or $$T_2 = T_4 - \frac{Q(r_2 - r_4)}{4\pi r_2 r_4 \times k}$$

$$= 70 - \frac{40 \times (0.05 - 0.04)}{4\pi \times 0.04 \times 0.05 \times 0.663}$$

$$= \mathbf{46°C.} \quad \text{Ans.}$$

(*iii*) Convection heat transfer coefficient

$$Q = h(4\pi r_2^2)(T_2 - T_\infty)$$

or $$h = \frac{Q}{4\pi r_2^2 (T_2 - T_\infty)}$$

$$= \frac{40}{4\pi \times (0.05)^2 \times (46 - 30)}$$

$$= \mathbf{79.6\ W/m^2.K.} \quad \text{Ans.}$$

Example 3.43. *The inside and outside surfaces of a hollow sphere of radii r_1 and r_2 are maintained at constant temperatures T_1 and T_2, respectively. The thermal conductivity of insulating material varies with temperature as*

$$k = k_o(1 + \alpha T + \beta T^2)$$

where k_o is constant.

Derive an expression for heat flow through the sphere. (P.U., Nov. 1998)

Solution

Analysis : Consider an elemental spherical ring of thickness dr at radius, r as shown in Fig. 3.61. The temperature difference across this ring is dT, then Fourier law

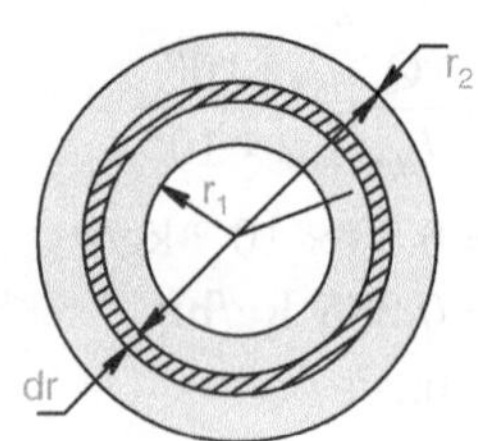

Fig. 3.61

$$Q = -kA\frac{dT}{dr}$$

where $\quad A = 4\pi r^2$

$k = k_o (1 + \alpha T + \beta T^2)$

Using $\quad Q = -k_o (1 + \alpha T + \beta T^2)\, 4\pi r^2 \dfrac{dT}{dr}$

or $\quad \dfrac{Q}{4\pi}\dfrac{dr}{r^2} = -k_o (1 + \alpha T + \beta T^2)\, dT$

Integrating both sides within limits

$$\frac{Q}{4\pi}\int_{r_1}^{r_2}\frac{dr}{r^2} = -k_o\int_{T_1}^{T_2}(1+\alpha T+\beta T^2)\,dT$$

or $$\frac{Q}{4\pi}\left[-\frac{1}{r}\right]_{r_1}^{r_2} = -k_o\left[T+\alpha\frac{T^2}{2}+\beta\frac{T^3}{3}\right]_{T_1}^{T_2}$$

or $$\frac{Q}{4\pi}\left[\frac{1}{r_1}-\frac{1}{r_2}\right] = -k_o\left[(T_2-T_1)+\frac{\alpha}{2}(T_2^2-T_1^2)+\frac{\beta}{3}(T_2^3-T_1^3)\right]$$

or $$\frac{Q}{4\pi}\left(\frac{r_2-r_1}{r_1r_2}\right) = -k_o(T_2-T_1)\left[1+\frac{\alpha}{2}(T_1+T_2)+\frac{\beta}{3}(T_1^2+T_1T_2+T_2^2)\right]$$

or $$Q = \frac{4\pi r_1r_2}{r_2-r_1}k_o(T_1-T_2)\times\left[1+\frac{\alpha}{2}(T_1+T_2)+\frac{\beta}{3}(T_1^2+T_1T_2+T_2^2)\right]$$

It is the required expression. **Ans.**

Example 3.44. *A 3 m ID spherical tank made of 2 cm thick stainless steel (k = 15 W m.K) is used to store iced water at T_{∞_1} = 0°C. The tank is located in a large room maintained at 22°C. The outer surface of the tank is black and heat is convected and radiated on the outer surface of the tank. The convection heat transfer coefficient at inner and outer surfaces of the tank are 80 W/m².K and 10 W/m².K, respectively. The radiation heat transfer coefficient is 5.34 W/m².K. Determine (a) rate of heat transfer to iced water, (b) amount of ice melts during a 24 hours period.*

The latent heat of fusion for ice at atmospheric pressure is 333.7 kJ/kg.

Solution

Given : A spherical tank

$d_1 = 3$ m, $\quad r_1 = 1.5$ m

thickness, $\quad t = 2$ cm

$k = 15$ W/m.K

$T_{\infty_1} = 0$°C, $\quad T_{\infty_2} = 22$°C

$h_1 = 80$ W/m².K, $\quad h_{rad} = 5.34$ W/m².K

$h_2 = 10$ W/m².K, $\quad h_{fg} = 333.7$ kJ/kg.

To find :

(*a*) Rate of heat transfer to iced water

(*b*) Amount of ice melts in 24 hours.

Assumptions :

(*i*) Steady state conditions.

(*ii*) Heat transfer in radical direction only.

(*iii*) Constant properties.

Analysis : (*a*) The outer radius of spherical tank

$d_2 = d_1 + 2t = 3 \text{ m} + 2 \times 0.02 = 3.04$ m

$\therefore \quad r_2 = 1.52$ m

Inner surface area of tank,

$A_1 = \pi d_1^2 = \pi \times (3 \text{ m})^2 = 28.27 \text{ m}^2$

Outer surface area of tank,

$A_2 = \pi d_2^2 = \pi \times (3.04 \text{ m})^2$

$= 29.03 \text{ m}^2$

The individual thermal resistances ;

$$R_{conv, 1} = \frac{1}{h_1A_1} = \frac{1}{80\times 28.27} = 4.421 \times 10^{-4} \text{ K/W}$$

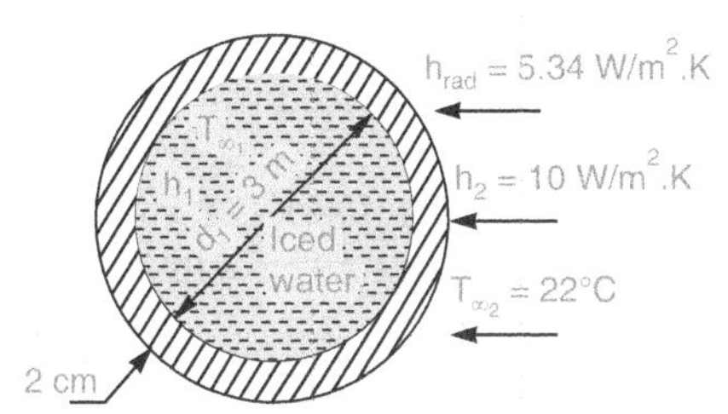

(a) Schematic of spherical tank located in a room

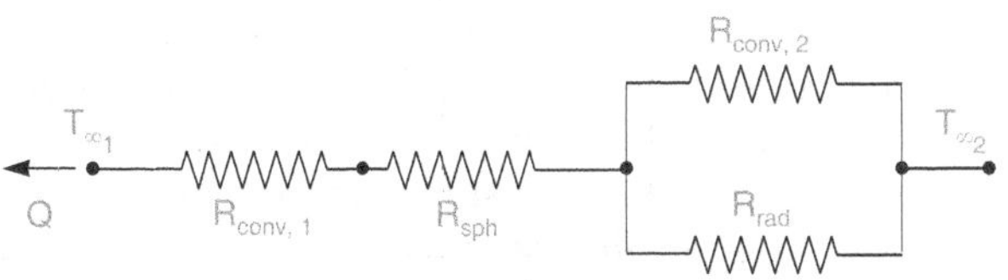

(b) Equivalent thermal network

Fig. 3.62

$$R_{sph} = \frac{r_2-r_1}{4\pi r_1r_2k} = \frac{1.52-1.5}{4\pi\times 1.52\times 1.5\times 15}$$

$= 4.653 \times 10^{-5}$ K/W

$$R_{conv, 2} = \frac{1}{h_2A_2} = \frac{1}{10\times 29.03}$$

$= 3.444 \times 10^{-3}$ K/W

$$R_{rad} = \frac{1}{h_{rad} \times A_2} = \frac{1}{5.34 \times 29.03}$$
$$= 6.45 \times 10^{-3} \text{ K/W}$$

Equivalent resistance of parallel resistances $R_{conv,\,2}$ and R_{rad}

$$\frac{1}{R_{eq}} = \frac{1}{R_{conv,\,2}} + \frac{1}{R_{rad}}$$

$$= \frac{1}{3.444 \times 10^{-3}} + \frac{1}{6.45 \times 10^{-3}} = 445.3$$

or $$R_{eq} = \frac{1}{445.3} = 2.245 \times 10^{-3} \text{ K/W}$$

Now all resistances are in series and total resistance

$$\Sigma R_{th} = R_{conv,\,1} + R_{sph} + R_{eq}$$
$$= 4.421 \times 10^{-4} + 4.653 \times 10^{-5} + 2.245 \times 10^{-3}$$
$$= 2.734 \times 10^{-3} \text{ K/W}$$

The heat transfer rate to iced water

$$Q = \frac{T_{\infty 2} - T_{\infty 1}}{\Sigma R_{th}} = \frac{22 - 0}{2.734 \times 10^{-3}}$$
$$= \textbf{8047.1 W.} \quad \text{Ans.}$$

(*b*) The amount of ice melts to water during a period of 24 hours :

Total heat transfer to iced water during 24 hours period

$$U = (8047.1 \text{ J/s}) \times (24 \times 3600 \text{ s})$$
$$= 695.27 \times 10^6 \text{ J} = 695270 \text{ kJ}$$

and amount of ice melts

$$m_{ice} = \frac{U}{h_{fg}} = \frac{695270}{333.7} = \textbf{2083.51 kg.} \quad \text{Ans.}$$

Example 3.45. *A 7 m diameter vertical klin has a hemispherical dome at its top. The dome is made from 30 cm thick layer of chrome brick (k = 1.16 W/m.K). During an operation, its inside surface temperature is 900°C and outer surface is exposed to surrounding air at 30°C with heat transfer coefficient of 15 W/m².K. Calculate the outside surface temperature of dome and the heat loss from the klin.*

Compare this heat loss with that would result from a flat dome made of same material and klin operating under identical conditions.

Solution

Given : A dome of a klin with

$$d_1 = 7 \text{ m}, \quad r_1 = 3.5 \text{ m}$$
$$r_2 = r_1 + 30 \text{ cm} = 3.8 \text{ m}$$
$$k = 1.16 \text{ W/m.K}$$
$$T_1 = 900°C$$
$$T_\infty = 30°C$$
$$h = 15 \text{ W/m}^2\text{.K.}$$

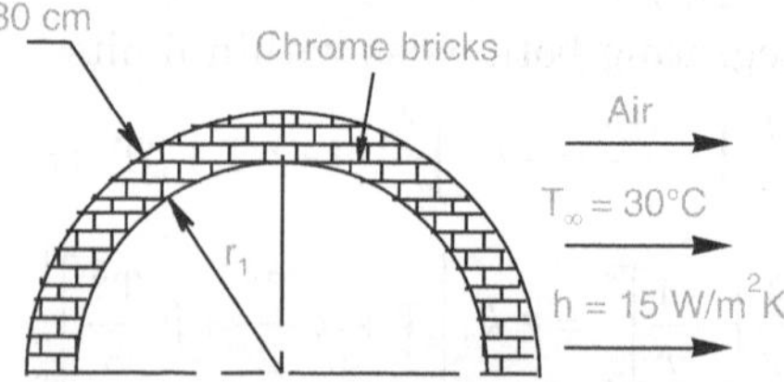

Fig. 3.63. Hemispherical top of klin.

To find :

(*i*) Outside dome temperature,

(*ii*) Heat loss from hemispherical dome, and

(*iii*) Heat loss from flat dome, and percentage change in heat loss.

Analysis : (*i*) Under steady state conditions, heat loss from hemispherical dome.

$$Q_1 = \frac{1}{2} \times \frac{4\pi r_1 r_2 k (T_1 - T_2)}{r_2 - r_1}$$
$$= h(2\pi r_2^2)(T_2 - T_\infty)$$

or $$\frac{r_1 k (T_1 - T_2)}{r_2 - r_1} = h r_2 (T_2 - T_\infty)$$

Using numerical values

$$\frac{3.5 \times 1.16 \times (900 - T_2)}{3.8 - 3.5}$$
$$= 15 \times 3.8 \times (T_2 - 30)$$

or $$213.68 - 0.237\, T_2 = T_2 - 30$$

or $$T_2 = \frac{213.68 + 30}{1.237} = \textbf{196.92°C.} \quad \text{Ans.}$$

(*ii*) Heat loss from hemisphere dome

$$Q_1 = \frac{1}{2} \times \frac{4\pi r_1 r_2 k (T_1 - T_2)}{r_2 - r_1}$$
$$= \frac{2\pi \times 3.5 \times 3.8 \times 1.16 \times (900 - 196.92)}{3.8 - 3.5}$$
$$= 227.18 \times 10^3 \text{ W} = \textbf{227.18 kW.} \quad \text{Ans.}$$

(*iii*) For flat top dome, L = 30 cm = 0.3 m

$$Q_2 = \frac{kA(T_1 - T_2)}{L} = hA(T_2 - T_\infty)$$

$$= \frac{1.16 \times (900 - T_2)}{0.3} = 15 \times (T_2 - 30)$$

or $T_2 = 208.3°C$

and $$Q_2 = \frac{1.16 \times (\pi \times 3.5^2) \times (900 - 208.3)}{0.3}$$

$$= 102.93 \times 10^3 \text{ W} = 102.93 \text{ kW}$$

Reduction in heat loss due to flat top dome

$$\frac{227.18 - 102.93}{227.18} \times 100$$

$$= \mathbf{54.69\%}.$$ Ans.

Example 3.46. *A cylindrical liquid oxygen tank has a diameter of 1.4 m, 7 m long and has hemispherical ends. The boiling point of liquid oxygen is –182°C and its latent heat of evaporation is 214 kJ / kg. The tank is insulated in order to reduce the heat transfer to the tank in such a way that in steady state, the rate of oxygen boil-off should not exceed 14 kg / h. Calculate the thermal conductivity of insulating material, if its 8 cm thick layer of insulation is applied and its outside surface is maintained at 30°C.*

Solution

Given : A cylindrical tank covered with hemispherical end as shown in Fig. 3.64.

$d_1 = 1.4$ m, $r_1 = 0.7$ m

$r_2 = 0.7 \text{ m} + 0.08 = 0.78$ m

$\dot{m} = 14$ kg/h, $h_{fg} = 214$ kJ/kg

$T_i = -182°C$, $T_o = 30°C$,

$L_1 = 7$ m.

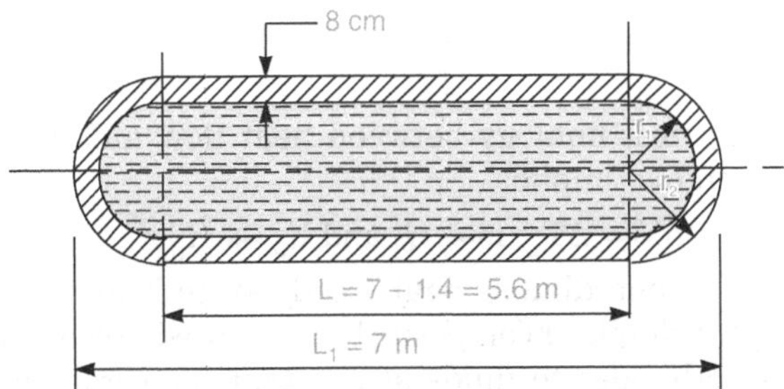

Fig. 3.64. A cylindrical tank covered with spherical ends

To find : Thermal conductivity (k) of insulating material.

Assumption : No convection at inner side of tank and tank inside surface temperature is at – 182°C.

Analysis : The rate of heat transfer to oxygen in steady state

$$Q = \dot{m} \times h_{fg} = (14 \text{ kg/h}) \times (214 \text{ kJ/kg})$$
$$= 2996 \text{ kJ/h} = 832.2 \text{ W}$$

Length of the cylindrical portion of the tank

= Total length – 2 × radius of hemispherical ends

$= L_1 - 2r_1 = 7 \text{ m} - 2 \times 0.7 \text{ m} = 5.6$ m.

This heat will be received from cylinder wall and two hemispherical ends (1 sphere), therefore

$$Q = Q_{cyl} + Q_{sphere}$$

$$= \frac{2\pi L k (T_o - T_i)}{\ln\left(\frac{r_2}{r_1}\right)} + \frac{4\pi r_1 r_2 k (T_o - T_i)}{r_2 - r_1}$$

Using numerical values

$$832.2 = k\{30 - (-182)\}\left\{\frac{2\pi \times 5.6}{\ln\left(\frac{0.78}{0.7}\right)} + \frac{4\pi \times 0.7 \times 0.78}{0.78 - 0.7}\right\}$$

or $832.2 = 212k \times [325.15 + 85.76] = 87113\, k$

or $$k = \frac{832.2}{87113} = \mathbf{0.0095\ W/m.K.}$$ Ans.

Example 3.47. *The two insulation materials are purchased in powder form as A and B with thermal conductivities 0.005 and 0.035 W / m.K, respectively. These materials was to apply over a 40 cm dia. sphere as inner layer 4 cm thick and outer layer 5 cm thick, respectively. But due to lapse of attention, the material B was applied as first layer and subsequently material A as outer layer.*

Estimate its effect on conduction heat transfer.

(M.U., May 2001)

Solution

Given : Two layer insulation of a sphere

$k_A = 0.005$ W/m.K

$k_B = 0.035$ W/m.K

$d_1 = 40$ cm, $r_1 = 20$ cm $= 0.2$ m

$r_2 = 20 \text{ cm} + 4 \text{ cm} = 24 \text{ cm} = 0.24$ m

$r_3 = r_2 + 5 \text{ cm} = 29 \text{ cm} = 0.29$ m.

To find : Effect of wrong arrangement of insulation.

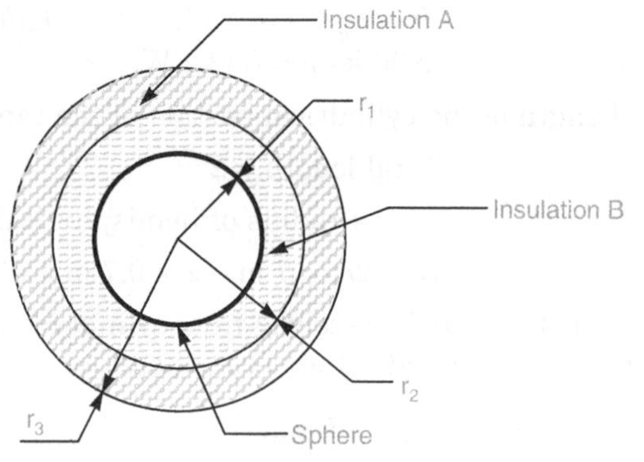

(a) Schematic of insulation on sphere

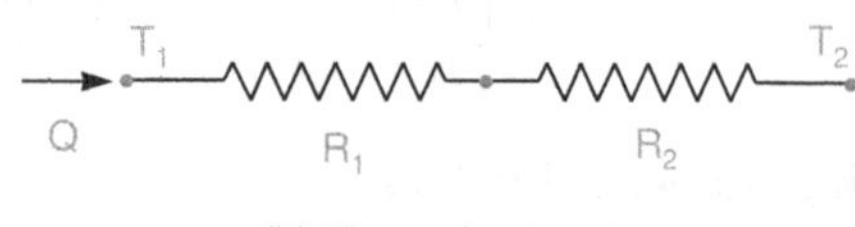

(b) Thermal network

Fig. 3.65

Analysis : The heat flow rate through composite sphere is expressed as :

$$Q = \frac{\Delta T}{R_1 + R_2}$$

where, $R_1 = \frac{r_2 - r_1}{4\pi k_1 r_1 r_2}$ and $R_2 = \frac{r_3 - r_2}{4\pi r_2 r_3 k_2}$

For proposed case

$$k_1 = k_A, \quad k_2 = k_B$$

$$R_1 = \frac{0.24 - 0.2}{4\pi \times 0.2 \times 0.24 \times 0.005}$$

$$= 13.263 \text{ K/W}$$

$$R_2 = \frac{0.29 - 0.24}{4\pi \times 0.24 \times 0.29 \times 0.035}$$

$$= 1.633 \text{ K/W}$$

Then heat loss rate

$$Q_1 = \frac{\Delta T}{13.263 + 1.633} = 0.0671\ (\Delta T) \text{ W}$$

When material get interchanged, then radii will also change.

Volume of material A,

$$V_A = \frac{4\pi}{3}(r_2^3 - r_1^3) = \frac{4\pi}{3} \times (0.24^3 - 0.2^3)$$

$$= 0.0244 \text{ m}^3$$

Volume of material B,

$$V_B = \frac{4\pi}{3}(0.29^3 - 0.24^3) = 0.04425 \text{ m}^3$$

New radius of material B as r_{2B}

$$V_B = \frac{4\pi}{3}(r_{2B}^3 - r_1^3)$$

or $$r_{2B}^3 = \frac{0.04425 \times 3}{4\pi} + (0.2)^3 = 0.018565$$

or $$r_{2B} = 0.2648 \text{ m}$$

and $$r_{3A}^3 = \frac{0.0244 \times 3}{4\pi} + (0.2648)^3 = 0.02439$$

or $$r_{3A} = 0.29 \text{ m.}$$

Now, $$R_1' = \frac{0.2648 - 0.2}{4\pi \times 0.2 \times 0.2648 \times 0.035}$$

$$= 2.782 \text{ K/W}$$

$$R_2' = \frac{0.29 - 0.2648}{4\pi \times 0.2648 \times 0.29 \times 0.005}$$

$$= 5.225 \text{ K/W}$$

and heat loss rate

$$Q_2 = \frac{\Delta T}{R_1' + R_2'} = \frac{\Delta T}{2.782 + 5.225}$$

$$= 0.125\ (\Delta T) \text{ W}$$

Rate of heat loss increases with wrong arrangement of insulation. Percentage increase in heat transfer.

$$\frac{0.125 - 0.0671}{0.0671} \times 100 = \mathbf{86.29\%}. \quad \textbf{Ans.}$$

3.8. SUMMARY

The electrical analogy between heat flow and current flow systems, implies

$$\Delta V \rightarrow \Delta T \quad I \rightarrow Q, \text{ and } R_e \rightarrow R_{th}$$

The one dimensional steady state heat transfer through a simple or composite body exposed to convection on its both sides to fluids at constant temperature T_{∞_1} and T_{∞_2} can be expressed as :

$$Q = \frac{T_{\infty_1} - T_{\infty_2}}{\Sigma R_{th}}$$

where, ΣR_{th} is total thermal resistance between two fluids. The elementary thermal resistance relations can be expressed as follows :

Conduction resistance of wall,

$$R_{wall} = \frac{L}{kA}$$

Conduction resistance of hollow cylinder,

$$R_{cyl} = \frac{\ln (r_2/r_1)}{2\pi Lk}$$

Conduction resistance of hollow sphere,

$$R_{sph} = \frac{r_2 - r_1}{4\pi r_1 r_2 k}$$

Convection resistance,

$$R_{conv} = \frac{1}{hA}$$

Radiation resistance,

$$R_{rad} = \frac{1}{h_r A}$$

The total of resistances in series can be obtained as :

$$\Sigma R_{th} = R_1 + R_2 + R_3 + \ldots\ldots$$

If resistances are parallel to each other, its equivalent resistance is calculated as :

$$\frac{1}{R_{eq}} = \frac{1}{R_1} + \frac{1}{R_2} \quad \text{or} \quad R_{eq} = \frac{R_1 R_2}{R_1 + R_2}$$

The contact resistance at any interface can be calculated as :

$$R_{contact} = \frac{\text{Temp. drop across contact surfaces}}{\text{Heat flux}} = \frac{T_{c_1} - T_{c_2}}{q}$$

The temperature drop across any layer, $\Delta T_i = QR_i$.

Addition of insulation on cylinders and spheres, will increase the rate of heat transfer upto critical radius defined as :

For cylinder $\quad r_{cr} = \frac{k}{h}$

For sphere $\quad r_{cr} = \frac{2k}{h}$

For minimisation of heat loss with insulation ;

$$r_{insulation} >> r_{cr}.$$

REVIEW QUESTIONS

1. Show that the temperature profile for the heat conduction through a wall of constant thermal conductivity is a straight line.
2. Prove that the thermal resistance offered by a hollow long cylinder of constant thermal conductivity is given by

$$R_{cyl} = \frac{\ln (r_2/r_1)}{2\pi Lk}$$

3. Show that the resistance offered by a hollow sphere of radii r_1, r_2 and constant thermal conductivity is given by

$$R_{sph} = \frac{r_2 - r_1}{4\pi r_1 r_2 k}$$

4. What do you mean by mean thermal conductivity ? Derive an expression for mean thermal conductivity of a hollow cylinder where

$$k = k_0(1 + \alpha T)$$

5. What do you mean by critical radius of insulation ? Explain it concept with help of material and surface resistances.
6. A steam pipe is insulated to reduce the heat loss. However, the measurement reveals that the rate of heat loss has increased instead of decreasing. Can you comment why ?
7. Discuss overall heat transfer coefficient. Obtain an expression for overall heat transfer coefficient based inner diameter of a hollow cylinder ?
8. Discuss the effect of contact resistance on heat transfer and temperature distribution.
9. Discuss critical radius and economical thickness of insulation on cylinders.
10. Derive an expression for log mean area for hollow cylinders.

PROBLEMS

1. The wall of a building consists of 10 cm of brick [k = 0.69 W/(m°C)], 1.25 cm of Celotex [k = 0.048 W/(m°C)], 8 cm of glass wool [k = 0.038 W/(m°C)], and 1.25 cm of asbestos cement board [k = 0.74 W/(m°C)]. If the outside surface of the brick is at 5°C and the inside surface of the cement board is at 20°C. Calculate the heat flow rate per square metre of wall surface.
[Ans. – 5.94 W/m^2]
2. An iron plate 2.5 cm thick [k = 62 W/(m°C)] is in contact with asbestos insulation 1 cm thick [k = 0.2 W/(m°C)] on one side and exposed to hot gas with a heat transfer coefficient of 200 W/(m^2.°C) on the other surface. If the outer surface of the asbestos is exposed to cool air with a heat transfer coefficient of 40 W/(m^2.°C), calculate the overall heat transfer coefficient U and the heat flow rate across the composite wall per square metre of the surface for a ΔT of 200°C between the hot gas and cool air. [Ans. 12.43 W/(m^2.°C), 2.48 kW/m^2]
3. A container made of 2 cm thick iron plate [k = 62 W/(m°C)] is insulated with a 1 cm thick asbestos layer [k = 0.1 W/(m°C)]. If the inner surface of the iron plate is exposed to hot gas at 530°C with a heat transfer coefficient of 100 W/(m^2.°C) and the outer surface of the asbestos is in contact with cool air at 30°C with a heat transfer coefficient of 20 W/(m^2.°C), calculate (a) the heat flow rate across the layers per

square metre of the surface area, and (*b*) the interface temperature between the layers.

[Ans. (*a*) 3119 W/(m^2.°C), (*b*) 497.8°C]

4. A composite slab is made of 75 mm thick layer of material with thermal conductivity of 0.15 W/m.K and 0.597 m thick layer of material of thermal conductivity of 1.7 W/m.K. The inner surface is maintained at 1000°C while the outer surface was exposed to convection air at 30°C with convection coefficient of 27 W/m^2.K. The heat flow was measured as 1 kW as against the calculated value of 1.092 kW. It is presumed that this may be due to contact resistance. Determine the contact resistance and the temperature drop at the interface.

[Ans. 0.082 K/W, 82°C]

5. A double glazed window is made of 2 glass panes of 6 mm thick each with an airgap of 6 mm between them. Assuming that the layer is stagnant and only conduction is involved. Determine the thermal resistance and the overall heat transfer coefficient, if the inside surface is exposed to convection with h = 1.5 W/m^2.K. Compare the values with that of a single glass of 12 mm thickness. The conductivity of the glass = 1.4 W/m.K and that for air is 0.025 W/m.K.

[Ans. 0.915 m^2.K/W, 1.092 W/m^2.K, 0.67 m^2.K/W]

6. A composite slab is made of 3 layers of thicknesses of 28 cm, 10 cm and 15 cm with thermal conductivities of 1.7, k_B and 9.5 W/m.K. The outside surface is exposed to air at 20°C with convection coefficient of 15 W/m^2.K and the inside surface is exposed to gases at 1200°C with convection coefficient of 28 W/m^2.K and the inside surface is at 1080°C. Determine the unknown thermal conductivity, all surface temperatures, resistances of each layer and the overall heat transfer coefficient. Compare the temperature gradients in the three layers.

[Ans. K_B = 1.46 W/m.K, 526.6°C, 297°C, 244°C, R = 0.035, 0.164, 0.068, 0.0158, 0.067 m^2.K/W, 2.84 W/m^2.K]

7. A 2 kW heater element of area 0.04 m^2 is protected on the backside with insulation 50 mm thick of k = 1.4 W/m.K and on the front side by a plate 10 mm thick with thermal conductivity of 45 W/m.K. The backside is exposed to air at 5°C with convection coefficient of 10 W/m^2.K and the front is exposed to air at 15°C with convection coefficient including radiation of 250 W/m^2.K. Determine the heater element temperature and the heatflow into the room under steady conditions. [Ans. 190°C, 1753.1 W]

8. To reduce frosting, it is desired to keep the outside surface of a glazed window at 4°C. The outside air is at – 10°C and the convection coefficient is 60 W/m^2.K. In order to maintain the conditions, a uniform heat flux is provided at the inner surface, which is in contact with room air at 22°C with convection coefficient of 12 W/m^2.K. The glass is 7 mm thick and has a thermal conductivity of 1.4 W/m.K. Determine the heating required per m^2 area. [Ans. 203.77 W/m^2]

9. A proposed self cleaning oven design involves use of a composite window separating the oven cavity from the room air. The composite consists of two high temperature plastics. The thickness of plastic exposed to interior of the oven is twice than that of the face exposed to room air. The thermal conductivity of interior plastic is 0.15 W/m.K and that of outer is 0.08 W/m.K. During the self cleaning process, the oven enclosed air temperature is maintained at 430°C and the room air temperature is at 30°C. The inside convection and radiation and outside convection coefficients are the same and equal to 25 W/m^2.K. Calculate the minimum window thickness required to ensure maximum temperature of 55°C at outer surface of window. [Ans. 0.0673 mm]

10. A composite wall separates combustion gases at 2600°C from a liquid coolant at 100°C, with gas and liquid-side convection coefficients of 50 and 1000 W/m^2.K. The wall is composed of a layer of beryllium oxide (k = 272 W/m.K) on the gas side 10 mm thick and a slab of stainless steel (AISI 304) (k = 15 W/m.K) on the liquid side 20 mm thick. The contact resistance between the oxide and the steel is 0.05 m^2.K/W. What is the heat loss per unit surface area of the composite ? Sketch the temperature distribution from the gas to the liquid. [Ans. 34544.6 W/m^2]

11. A wall is constructed of two layers of 1 cm thick plaster board (k = 0.17 W/m.K) placed 12 cm apart. The space between these two is filled with fibre glass insulation (k = 0.035 W/m.K). Determine (*a*) thermal resistance of the wall (*b*) R value of insulation (*c*) heat transfer rate for temperature difference of 150°C.

[Ans. (*a*) 3.546 m^2.K/W, (*b*) 3.546 m^2.K/W, (*c*) 42.3 W/m^2]

12. The wall of a refrigerator is constructed to fibre glass insulation (k = 0.035 W/m.K), sandwiched between the two layers of 1 mm thick steel sheet (k = 15.1 W/m.K). The refrigerator space is maintained at 3°C, while the average kitchen temperature is 25°C. The average inner and outer heat transfer coefficient are 4 W/m^2.K and 9 W/m^2.K, respectively. It is observed that the condensation occurs at the outer surface of refrigerator when its outer surface temperature drops below 20°C. Calculate the minimum thickness of insulation needed to avoid condensation on the outer surface.

[Ans. 13.22 mm]

13. A 4 m high and 6 m wide wall consists of 18 cm × 30 cm cross-section horizontal bricks (k = 0.72 W/m.K) separated by 3 cm thick plaster layer (k = 0.22 W/m.K). There are also 2 cm thick plaster layers on each side of wall and 2 cm thick rigid foam (k = 0.026 W/m.K) on the inner side of the wall. The indoor and outdoor temperatures are 22°C and – 4°C and convection heat transfer coefficients are 10 W/m^2.K and 20 W/m^2.K respectively. Assuming one dimensional steady state conditions. Calculate heat transfer rate the composite wall. [Ans. 421.0 W]

14. A 10 cm thick wall is to be constructed with 2.5 m long wood studs (k = 0.11 W/m.K), that has a cross-section of 10 cm × 10 cm. At some point the builder has run out of those stud and started using pairs of 2.5 m long wood stud with cross section of 5 cm × 10 cm nailed to each other instead. The manganese steel nails (k = 50 W/m.K) are 10 cm long and have dia of 0.4 cm. A total 50 nails are used to connect the two studs. The temperature difference between inner and outer wall is 15°C. Assuming negligible thermal contact resistance between two layers. Calculate heat transfer rate (*a*) through a solid stud, (*b*) through stud pair with equal length and width nailed to each other, (*c*) also, determine the effective thermal conductivity of the nailed stud pair.

[Ans. (*a*) 4.125 W, (*b*) 481.6 W, (*c*) 12.84 W/m.K]

15. Consider a ski jacket is made of five layers of 0.1 mm thick synthetic fabric (k = 0.13 W/m.K) with 1.5 mm thick air space (k = 0.026 W/m.K) between the layers. The surface area of the jacket is 1.1 m^2. Calculate the rate of heat loss through the jacket for an average temperature difference between inner surface of jacket and surrounding air (h = 25 W/m^2.K) is 33°C. [Ans. 230.2 W]

16. Consider a steel pipe [k = 10 W/(m°C)], with an inside radius of 5 cm and an outside radius of 10 cm. The outer surface is to be insulated with fibre glass insulation [k = 0.05 W/(m°C)] to reduce the heat flow rate through the pipe wall by 50%. Determine the thickness of the fibre glass. [Ans. 0.05 cm]

17. A metal pipe with an outside diameter (OD) of 12 cm is covered with an insulation material [k = 0.07 W/(m°C)] of 2.5 cm thick. If the outer pipe wall is at 100°C and the outer surface of the insulation is at 20°C, find the heat loss from the pipe per metre length. [Ans. 101 W/m length]

18. Consider two stainless steel slabs [k = 20 W/(m°C)] with a thickness of 1 cm and 1.5 cm that are pressed together with a pressure of 20 atm. The surfaces have roughness of about 0.76 m μm. The outside surface of the blocks are at 100°C and 150°C. Calculate the heat flow rate across the slabs and the temperature drop at the interface. [Ans. 3.7°C]

19. A metal pipe of 10 cm OD is covered with a 2 cm thick insulation [k = 0.07 W/(m°C)]. The heat loss from the pipe is 100 W per metre of length when the pipe surface is at 100°C. What is the temperature of the outer surface of the insulation ? [Ans. 23.5°C]

20. A 5 cm OD and 0.5 cm thick copper pipe [k = 386 W/(m°C)] has hot gas flowing inside at a temperature of 200°C with a heat transfer coefficient of 30 W/(m^2.°C). The outer surface dissipates heat by convection into the ambient air at 20°C with a heat transfer coefficient of 15 W/(m^2.°C). Determine the heat loss from the pipe per metre of length. [Ans. 261 W/m length]

21. A 6 cm OD, 2 cm thick copper hollow sphere [k = 386 W/(m°C)] is uniformly heated at the inner surface at a rate of 150 W/m^2. The outer surface is cooled with air at 20°C with a heat transfer coefficient of 10 W/(m^2.°C). Calculate the temperature of the outer surface. [Ans. 21.7°C]

22. A steel tube [k = 15 W/(m°C)], with an outside diameter of 7.6 cm and a thickness of 1.3 cm is covered with an insulation material [k = 0.2 W/(m°C)] 2 cm thick. A hot gas at 330°C with a heat transfer coefficient of 400 W/m^2.°C flows inside the tube. The outer surface insulation is exposed to cooler air at 30°C with a heat transfer coefficient of 60 W/(m^2.°C). Calculate the heat loss from the tube to the air for a 10 m length of the tube. [Ans. 7453 W]

23. Consider a brass tube [k = 115 W/(m°C)], with an outside radius of 4 cm and a thickness of 0.5 cm. The inside surface of the tube is kept at uniform temperature, and outside surface is covered with two layers of insulation each 1 cm thick, with thermal conductivities of 0.1 W/(m°C) and 0.05 W/(m°C) respectively. Calculate the overall heat transfer coefficient based on the outside surface area of the outer insulation. [Ans. 2.83 W/m^2.°C]

24. Steam at 320°C flow in CI pipe (k = 80 W/m.K) ID = 5 cm, OD = 5.5 cm. The pipe is covered with 3 cm thick glass wool insulation [k = 0.05 W/m.K]. Heat is lost to surroundings at 5°C by natural convection and radiation, with a combined heat transfer coefficient of 18 W/m^2.K. The heat transfer coefficient at inner surface of the pipe is 60 W/m^2.K. Determine the rate of heat loss from steam per metre length of pipe. Also calculate the temperature drop across pipe and insulation. [Ans. Q = 120.7 W/m, ΔT_{pipe} = 0.02°C, $\Delta T_{insulation}$ = 284°C]

25. A steel tube (k = 15 W/m.K) with 5 cm inner diameter and 7.6 cm outer diameter is covered with an insulation (k = 0.2 W/m.K), 0.2 cm thick. A hot gas at 330°C flows through tube with h_i = 400 W/m^2.K. The outer surface of the insulation is exposed to air at 30°C with h_o = 60 W/m^2.K. Calculate (*a*) the rate of heat loss from 10 m long tube, (*b*) temperature drop resulting from each of thermal resistances.

[Ans. (*a*) 23.533 kW, (*b*) 37.45°C, 10.47°C, 96°C and 156.06°C]

26. The inner and outer radii of a hollow cylinder are 5 cm and 10 cm respectively. The inside surface is maintained at 300°C, while outside surface at 100°C. The thermal conductivity of the material varies with temperature as [k = 0.5(1 + 0.001 T) W/m.K], where T in °C. Calculate the heat flow rate per metre length of the cylinder. [Ans. 1087.77 W/m]

27. A copper rod, 6 mm in diameter is heated by flow of an electric current. The surface of the rod is maintained at 200°C, while it dissipates heat by convection, with h = 150 W/m^2.K into an ambient at 25°C. If the rod is covered with 2 mm thick coating

(k = 0.75 W/m.K), will the heat loss from the rod decrease or increase ?

[Ans. r_{cr} = 5 mm, heat loss increases]

28. A pipe of 60 mm dia. carries oil at 230°C, with heat transfer coefficient of 250 W/m^2.K. The pipe is insulated with a material (k = 0.06 W/m.K). On the outer surface of insulation, a plastic coating (k = 1.6 W/m.K) 1 mm thick is applied. Calculate the radial thickness of insulation, which will reduce the outside temperature of coating to 50°C, when the ambient air temperature is 20°C with convective and radiative heat transfer coefficients of 6 and 8 W/m^2.K, respectively. Calculate the percentage reduction in the heat loss with this thickness of insulation. [Ans. 20 mm, 75%]

29. The steam at 320°C flows in a steel tube (k = 15 W/m.K) with inner and outer radii as 2.5 cm and 2.75 cm respectively. The pipe is covered with 3 cm thick glass wool insulation (k = 0.038 W/m.K). The heat is lost to surrounding air at 20°C by natural convection and radiation with a combined heat transfer coefficient of 15 W/m^2.K. The heat transfer coefficient at inside surface of pipe is 80 W/m^2.K. Calculate the heat loss from the steam per unit length of the pipe. Also calculate the temperature drop across the pipe shell and insulation. [Ans. 14.23 W/m]

30. The steam at 300°C is flowing through a steel pipe (k = 15.1 W/m.K) with inner and outer radii as 4 cm and 4.4 cm, respectively. The pipe is insulated with fibre glass (k = 0.035 W/m.K) and it is exposed in an ambient at 15°C. The heat transfer coefficients on inner and outer side of the pipe is 150 and 25 W/m^2.K respectively. Calculate the rate of heat loss from steam per metre length of the pipe. What is the error involved in neglecting the thermal resistance of steel pipe in calculation ? [Ans. 115.1 W]

31. A 2 cm dia. electrical cable at 45°C is covered by 0.5 mm thick plastic insulation (k = 0.13 W/m.K). The wire is exposed to an ambient at 10°C with h = 12 W/m^2.K. Investigate if the plastic insulation on the cable will help or hurt the heat transfer from the wire ?

[Ans. r_{cr} = 10.83 mm, it helps]

32. A typical domestic central heating installation utilises 50 m long, 15 mm outer diameter copper pipe (k = 400 W/m.K) with 1 mm wall thickness. It is used to convey water at 70°C. Calculate the heat loss from this pipe with a 15 mm radial thickness of insulation (k = 0.05 W/m.K) and compare it to the value without insulation. The ambient air temperature is 15°C and internal and external heat transfer coefficients are 100 W/m^2.K and 8 W/m^2.K. [Ans. 600 W, 1037 W]

33. A tube with outer diameter of 2 cm is maintained at uniform temperature and is covered with an insulating layer (k = 0.18 W/m.K) in order to reduce the heat loss. Heat is dissipated from the outer surface of insulation with h = 12 W/m^2.K into an ambient at constant temperature. Determine the critical thickness of insulation. Calculate the ratio of heat loss from the tube with insulation to heat loss without insulation for

(*i*) the thickness of insulation equal to the critical thickness, and

(*ii*) the thickness of insulation 2.5 cm thicker than the critical thickness.

[Ans. 0.5 cm, (*i*) 1.067, (*ii*) 0.851]

34. A hollow sphere [k = 15 W/(m°C)], with an outside diameter of 8 cm and a thickness of 2 cm is covered with an insulation material [k = 0.2 W/(m°C)] 2 cm thick. Inside the sphere energy is generated at a rate of 3×10^5 W/m^3. The temperature of the interface between the outer surface of the sphere and the insulation is measured to be 300°C. Calculate the outside surface temperature of the insulation material.

[Ans. 266.7°C]

35. Consider a steel sphere [k = 10 W/(m°C)], with an inside radius of 5 cm and an outside radius of 10 cm. The outer surface is to be insulated with fibre glass insulation [k = 0.05 W/(m°C)] to reduce the heat flow rate through the sphere by 50%. Determine the thickness of the fibre glass. [Ans. 0.05 cm]

36. A hollow steel sphere (k = 10 W/m.K) has an inside radius of 10 cm and outside radius of 20 cm. The inside surface is maintained at a uniform temperature of 230°C, while its outside surface dissipates heat by convection with h = 20 W/m.K, into an ambient air at 30°C. Calculate the thickness of asbestos insulation (k = 0.5 W/m.K) required to reduce the heat loss by 50%. [Ans. 5.8 cm]

37. Estimate the rate of evaporation of liquid oxygen from a spherical container with 1.8 m diameter, covered with an insulation of asbestos, 30 cm thick. The temperature of inner and outer sphere surfaces are –183°C and 0°C respectively. The boiling point of the oxygen is –183°C and its latent heat of evaporation is 212.5 kJ/kg. The thermal conductivity of the insulation is 0.157 and 0.125 W/m.K at 0°C and –185°C respectively. Assume that the thermal conductivity of the insulation varies as $k = k_0 + (k_1 - k_0)\left[\dfrac{T - T_0}{T_1 - T_0}\right]$.

[Ans. 0.0055 kg/s]

38. An electrically heated sphere of 6 cm is exposed to an ambient at 25°C with h = 20 W/m^2.K. The surface of the sphere is maintained at 125°C. Calculate the rate of heat loss (*a*) when the sphere is uninsulated.

[Ans. 22.2 W and 44.4 W]

39. A 5 mm dia. spherical ball at 50°C is covered by 1 mm thick plastic insulation (k = 0.13 W/m.K). The wall is exposed to an ambient at 15°C with h = 20 W/m^2°C. Investigate, if the plastic insulation on the ball will help or hurt heat transfer from the ball ?

[Ans. r_{cr} = 13 mm, it helps]

40. A hollow spherical form is used to determine the conductivity of a material. The inner diameter is 20 cm and the outer diameter is 50 cm. A 30 W heater is

placed inside and under steady conditions, the temperatures at 15 and 20 cm radii were found to be 80 and 60°C. Determine the thermal conductivity of the material. Also find the outside temperature. If the surrounding is at 30°C, determine the convection heat transfer coefficient over the surface. Plot the temperature along the radius.

[Ans. k = 0.198 W/m.K, T_o = 48°C, h = 2.12 W/m^2.K]

41. A spherical container holding a cryogenic fluid at – 140°C and having an outer diameter of 0.4 m is insulated with three layers each of 50 mm thick insulations of k_1 = 0.02, k_2 = 0.06 and k_3 = 0.16 W/m.K (starting from inside). The outside is exposed to air at 30°C with h = 15 W/m^2.K. Determine the heat gain and the various surface temperatures.

[Ans. 33.05W, – 8.5°C, 20.71°C, 28.54°C]

42. A spherical electronic device of 10 mm dia generates 1 W. It is exposed to air at 20°C with a convection coefficient of 20 W/m^2.K. Find the surface temperature. The heat transfer consultant advises to enclose it in a glass like material of k = 1.4 W/m.K, to a thickness to obtain 50°C surface temperature. Calculate the thickness of enclosure.

[Ans. T_s = 179.15°C, t = 50 mm]

43. A layer of 50 mm thick firebrick (k = 0.72 W/m.K) is placed between two 8 mm thick steel plates (k = 22 W/m.K). The faces of brick adjacent to the plates are rough, having solid to solid contact over only 30% of the total area, with the average height of asperties being 0.8 mm. If the surface temperature of steel plates are 100 and 400°C, calculate the rate of heat flow per unit area. Assume cavity area is filled with air (k = 0.026 W/m.K) [Ans. 4024 W/m^2]

44. Find steady flow heat flux through a composite slab made of two materials A and B. Thermal conductivity of two materials vary linearly with temperature as :

$$k_A = 0.4\,(1 + 0.008\ T)$$
$$k_B = 0.5\,(1 + 0.001\ T)$$

where T is temperature in deg. Celsius

The thickness : L_A = 10 cm, L_B = 5 cm.

The innerside temperature of slab A is 600°C and outside temperature of slab B is 30°C.

[Ans. q = 4.27 × 10^3 W/m^2]

45. A steam pipe is covered with two layers of insulation, first layer being 3 cm thick and second 5 cm. The pipe is made of steel (k = 58 W/m.K) having ID of 160 mm and OD of 170 mm. The inside and outside film coefficients are 30 and 5.8 W/m^2.K, respectively. Calculate the heat lost per metre of pipe, if the steam temperature is 300°C and air temperature is 50°C. The thermal conductivity of two insulating materials are 0.17 and 0.093 W/m.K, respectively.

[Ans. 220.5 W/m]

46. A pipe, 4 cm in outer diameter is maintained at uniform temperature at T_1 and is covered with an insulation (k = 0.20 W/m.K) in order to reduce the heat loss. The heat is dissipated from outer surface of insulation into an ambient at T_∞ , with heat transfer coefficient of 8 W/m^2.K. Determine the thickness of insulation at which the heat dissipation rate would be the maximum. Calculate the ratio of the heat loss from the outer surface of insulated pipe and that of from bare pipe for

(*a*) thickness of insulation equal to critical thickness.

(*b*) the thickness of insulation is 2 cm thicker than the critical thickness.

[Ans. 5 mm (*a*) 1.022, (*b*) 0.911]

47. A steel pipe (k = 35 W/m.K), with inner radius 25 mm and outer radius 30 mm is insulated with 85% magnesia insulation (k = 0.055 W/m.K). The temperature at the interface between pipe and insulation is 300°C, while the temperature on outside surface of insulation must not exceed 70°C, with permissible heat loss of 700 W/m. Calculate :

(*i*) The minimum thickness of insulation, and

(*ii*) The temperature of inside surface of pipe.

[Ans. (*i*) 8 mm, (*ii*) 300.58°C]

48. A copper pipe carrying the refrigerant at – 20°C is 10 mm in outer diameter and is exposed to ambient at 25°C with convective coefficient of 50 W/m^2.K. It is proposed to apply the insulation of material having thermal conductivity of 0.5 W/m.K. Determine the thickness beyond which the heat gain will be reduced. Calculate the heat losses for 2.5 mm, 7.5 mm and 15 mm thick layer of insulation over 1 m length.

[Ans. 5 mm, 81.3 W/m, 82.37 W/m and 74.95 W/m]

49. An electric cable, 8 mm in diameter is covered by plastic sheathing (k = 0.18 W/m.K). The surface temperature of cable was observed as 50°C when it is exposed to air at 20°C with convective coefficient of 12.0 W/m^2.K. Calculate :

(*i*) the thickness of insulation to keep the wire as cool as possible, and

(*ii*) surface temperature of insulated cable, if the intensity of current flowing the conductor remains unchanged. [Ans. (*i*) 11 mm, (*ii*) 38.57°C]

Steady State Conduction with Heat Generation

4

4.1. The Plane Wall—Specified temperatures on both sides—Plane wall without heat generation—Plane wall with insulated and convective boundaries—Plane wall exposed to convection environment on its both boundaries—The maximum temperature in the wall. 4.2. The Cylinder—Solid cylinder with specified surface temperature—Solid cylinder exposed to convection environment. 4.3. Hollow Cylinder with Heat Generation and Specified Surface Temperatures—Hollow cylinder insulated at its inner surface—The location of maximum temperature in the cylinder—4.4. The Sphere—Solid sphere with convective boundary—Solid sphere with specified surface temperature—4.5. Summary—*Review Questions—Problems—References and Suggested Reading.*

Most of the engineering applications involve heat generation in the solids, such as nuclear reactors, resistance heaters etc. In this chapter, we will consider one dimensional steady state heat conduction with heat generation and determination of temperature distribution and heat flow in solids of simple shapes such as plane wall, a long cylinder and a sphere. Such type of problems cannot be solved with electrical analogy concept presented in previous chapter.

4.1. THE PLANE WALL

The one dimensional heat conduction eqn. (2.14) with $n = 0$ and $X = x$

$$\frac{d}{dx}\left\{\frac{dT}{dx}\right\} + \frac{g(x)}{k} = 0$$

or

$$\frac{d}{dx}\left\{\frac{dT}{dx}\right\} = -\frac{g(x)}{k} \quad ...(4.1)$$

The temperature distribution in the plane wall can be determined by solving the above heat conduction equation with prescribed boundary conditions.

Assuming thermal conductivity k and heat generation rate $[g(x) = g_o]$ are constants. Integrating above equation with respect to x, we get

$$\frac{dT}{dx} = -\frac{g_o x}{k} + C_1 \quad ...(4.2)$$

Integrating again w.r.t. to x, we get

$$T(x) = -\frac{g_o x^2}{2k} + C_1 x + C_2 \quad ...(4.3)$$

where C_1 and C_2 are constants of integration and can be determined according to boundary conditions. The solution of eqn. (4.3) gives temperature distribution and heat transfer in a plane wall.

4.1.1. Specified Temperatures on Both Sides

Consider a plane wall of thickness L, its left face at $x = 0$ is maintained at temperature T_1 and right face at $x = L$ is at temperature T_2 *i.e.,*

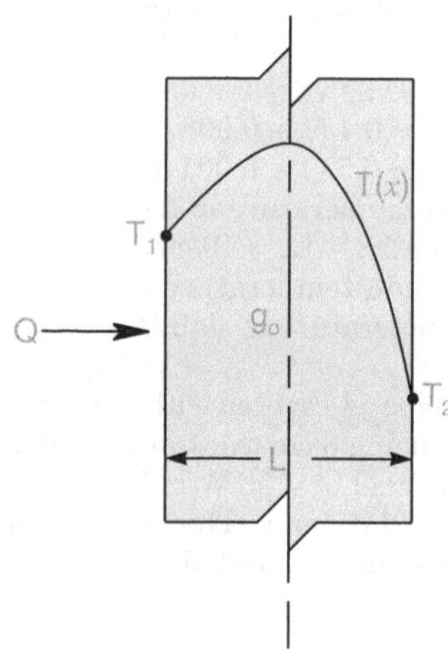

Fig. 4.1. Specified temperature on both faces

(*i*) The boundary condition at left face

At $x = 0$; $T(x) = T_1$

Substituting in eqn. (4.3), we get

$$T_1 = -\frac{g_o(0)^2}{2k} + C_1(0) + C_2$$

It gives $C_2 = T_1$

(*ii*) Boundary condition at right face,

At $x = L$; $T(x) = T_2$

Then eqn. (4.3) becomes

$$T_2 = -\frac{g_o L^2}{2k} + C_1 L + T_1$$

or
$$C_1 L = T_2 - T_1 + \frac{g_o L^2}{2k}$$

It gives

$$C_1 = \frac{T_2 - T_1}{L} + \frac{g_o L}{2k} \quad ...(4.4)$$

Substituting the values of C_1 and C_2 in eqn. (4.3), we get

$$T(x) = -\frac{g_o x^2}{2k} + \left\{\frac{T_2 - T_1}{L} + \frac{g_o L}{2k}\right\} x + T_1$$

or
$$T(x) = -\frac{g_o x^2}{2k} + \frac{x}{L}(T_2 - T_1) + \frac{g_o L x}{2k} + T_1$$

or
$$T(x) = \frac{g_o x}{2k}(L - x) + \frac{x}{L}(T_2 - T_1) + T_1 \quad ...(4.5)$$

It is the equation for the temperature distribution in a plane wall with uniform heat generation.The equation is quadratic thus the temperature distribution is parabolic in nature as shown in Fig. 4.1.

4.1.2. Plane Wall Without Heat Generation

If the wall experiences no heat generation, then $g_o = 0$ and eqn. (4.5) reduces to

$$T(x) = \frac{x}{L}(T_2 - T_1) + T_1 \quad ...(4.6)$$

and slope
$$\frac{dT}{dx} = \frac{T_2 - T_1}{L} \quad ...(4.7)$$

Fig. 4.2. Wall without heat generation

The eqn. (4.6) represents one dimensional steady state, temperature distribution in plane wall without heat generation. Temperature distribution in plane wall without heat generation is shown in Fig. 4.2.

The heat transfer rate in the slab without heat generation

$$Q = -kA\frac{dT}{dx}$$

$$= -kA\frac{(T_2 - T_1)}{L} = \frac{kA(T_1 - T_2)}{L} \quad ...(4.8)$$

eqn. (4.8) is already obtained earlier as eqn. (1.9) with Fourier law of heat conduction

4.1.3. Plane Wall With Insulated and Convective Boundaries

Consider a plane wall of thickness L, with heat generation g_o. The boundary at $x = 0$ is insulated and that $x = L$ dissipates heat by convection with heat transfer coefficient h into a fluid (ambient) at temperature T_∞.

The boundary conditions :

(*i*) At $x = 0$; $Q = 0$

or
$$-kA\left\{\frac{dT}{dx}\right\}_{x=0} = 0$$

Here neither thermal conductivity k, nor area of plane wall A may be zero, therefore,

$$\left(\frac{dT}{dx}\right)_{x=0} = 0 \quad ...(4.9)$$

(*ii*) At $x = L$

Heat conduction to right face = Heat convection from right face

$$-kA\left\{\frac{dT}{dx}\right\}_{x=L} = hA(T_{x=L} - T_\infty)$$

or
$$-k\left\{\frac{dT}{dx}\right\}_{x=L} = h(T_{x=L} - T_\infty) \quad ...(4.10)$$

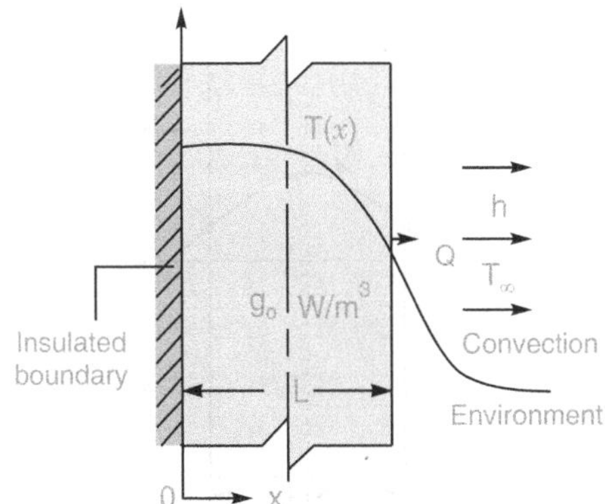

Fig. 4.3. A plane wall insulated on one face and exposed to convection environment on other face

Substituting first boundary condition in eqn. (4.2)

$$\left\{\frac{dT}{dx}\right\}_{x=0} = -\frac{g_o(0)}{k} + C_1 = 0$$

It gives $C_1 = 0$

Using eqns. (4.2) and (4.3) with second boundary condition, we have

$$-k\left\{-\frac{g_o L}{k}\right\} = h\left\{-\frac{g_o L^2}{2k} + 0 + C_2 - T_\infty\right\}$$

or
$$\frac{g_o L}{h} = -\frac{g_o L^2}{2k} + C_2 - T_\infty$$

It gives $C_2 = \frac{g_o L^2}{2k} + \frac{g_o L}{h} + T_\infty$ ...(4.11)

Substituting the value of C_1 and C_2 in eqn. (4.3), we get

$$T(x) = \frac{g_o(L^2 - x^2)}{2k} + \frac{g_o L}{h} + T_\infty \quad ...(4.12)$$

and the slope

$$\frac{dT(x)}{dx} = -\frac{g_o x}{k} \quad ...(4.13)$$

and heat transfer rate at any section

$$Q(x) = -kA\frac{dT}{dx} = -kA\left\{-\frac{g_o x}{k}\right\} = g_o Ax \quad ...(4.14)$$

Note: The eqns. (4.2) and (4.3) are used to obtain temperature distribution and heat transfer in a plane wall according to prescribed boundary conditions. With other combinations of boundary conditions, the above two equations can easily be worked out.

4.1.4. Plane Wall Exposed to Convection Environment on its Both Boundaries

The plane wall exposed to convection environment at T_∞ and heat transfer coefficient h on its both sides is shown in Fig. 4.4.

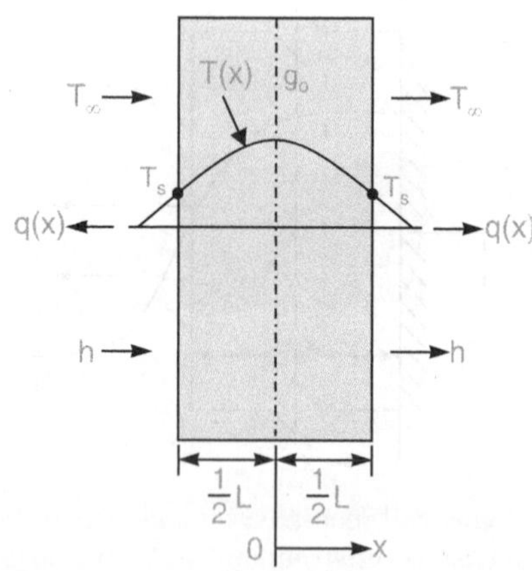

Fig. 4.4. Plane wall with convective boundaries

Due to symmetry in temperature field, its half portion can be analysed with boundary conditions as

At $x = 0$, $\frac{dT}{dx} = 0$

(Wall temperature is maximum, thus temperature gradient is zero at the centre)

and at any boundary surface *i.e.*, at $x = \frac{1}{2}L$.

Heat conduction to the surface

= Heat convection from the surface

or
$$-kA\left[\frac{dT}{dx}\right]_{x=L/2} = hA\,[T_{x=L/2} - T_\infty]$$

or
$$-k\left(\frac{dT}{dx}\right)_{x=L/2} = h\,(T_{x=L/2} - T_\infty)$$

Using temperature gradient in plane wall given by eqn. (4.2), and applying first boundary condition

$$\frac{dT}{dx} = \left[-\frac{g_o x}{k} + C_1\right]_{x=0} = 0$$

It gives $C_1 = 0$.

Using second boundary condition with $C_1 = 0$

$$-k\left[-\frac{g_o x}{k}\right]_{x=L/2} = h\left[\left\{-\frac{g_o x^2}{2k} + C_2\right\}_{x=L/2} - T_\infty\right]$$

or
$$\frac{g_o L}{2h} = -\frac{g_o L^2}{8k} + C_2 - T_\infty$$

It gives $C_2 = \frac{g_o L^2}{8k} + \frac{g_o L}{2h} + T_\infty$ ...(4.15)

and the temperature distribution in the wall

$$T(x) = -\frac{g_o x^2}{2k} + \frac{g_o L^2}{8k} + \frac{g_o L}{2h} + T_\infty \quad ...(4.16)$$

The maximum temperature in the wall *i.e.*, at $x = 0$

$$T_{max} = \frac{g_o L^2}{8k} + \frac{g_o L}{2h} + T_\infty \quad ...(4.17)$$

Note: The students can also perform the exercise for same problem by measuring x from left face of the wall and using same boundary conditions at

$$x = \frac{1}{2}L,\ \frac{dT}{dx} = 0 \quad \text{and} \quad x = L,\ -k\left(\frac{dT}{dx}\right) = h(T - T_\infty).$$

4.1.5. The Maximum Temperature in the Wall

The location of maximum temperature in the wall can be obtained by equating eqn. (4.2) to zero, which gives the location x_{cr} of maximum temperature. Using this value of x_{cr} in eqn. (4.3), the maximum temperature can be obtained.

For the plane wall shown in Fig. 4.3, the slope dT/dx given by eqn. (4.13) is equated to zero, which gives the location of maximum temperature as $x_{cr} = 0$ and maximum temperature is expressed as

$$T_{max} = T(x_{cr} = 0) = \frac{g_o L^2}{2k} + \frac{g_o L}{h} + T_\infty \quad ...(4.18)$$

Example 4.1. *A plane wall (k = 45 W/m.K), 10 cm thick, has heat generation at a uniform rate of 8 × 10⁶ W/m³. The two sides of the wall are maintained at 180°C and 120°C. Neglect end effects; calculate*

(i) temperature distribution across the plate,

(ii) position and magnitude of maximum temperature,

(iii) the heat flow rate from each surface of the wall.

Solution

Given : A wall with uniform heat generation

k = 45 W/m.K, L = 10 cm = 0.1 m

$g_o = 8 \times 10^6$ W/m³

T_1 = 180°C, T_2 = 120°C.

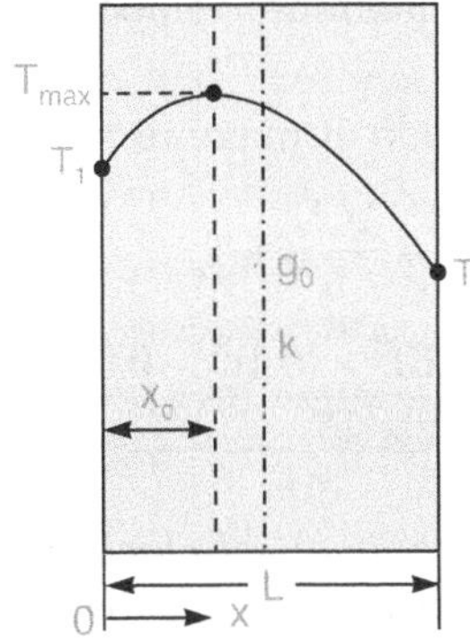

Fig. 4.5. A plane wall

To find :

(*i*) Temperature distribution.

(*ii*) Position and magnitude of maximum temperature.

(*iii*) Heat flow rate from each face.

Analysis : (*i*) Temperature distribution in the plane wall

$$T(x) = -\frac{g_o x^2}{2k} + C_1 x + C_2 \quad ...(i)$$

Subjected to boundary conditions

At $x = 0$, $T = T_1$

It gives $C_2 = T_1$

At $x = L$, $T = T_2$

$$T_2 = -\frac{g_o L^2}{2k} + C_1 L + T_1$$

or

$$C_1 = \frac{(T_2 - T_1)}{L} + \frac{g_o L}{2k}$$

Then temperature distribution takes the form

$$T(x) = -\frac{g_o x^2}{2k} + \frac{g_o L\, x}{2k} + (T_2 - T_1)\frac{x}{L} + T_1 \quad ...(ii)$$

Using numerical values,

$$T(x) = -\frac{8 \times 10^6 \times x^2}{2 \times 45} + \frac{8 \times 10^6 \times 0.1 x}{2 \times 45} + (120 - 180)\frac{x}{0.1} + 180$$

$$= \mathbf{-88888.89x^2 + 8288.89x + 180} \quad ...(iii)$$

It is the required expression. **Ans.**

(*ii*) Position and magnitude of maximum temperature :

Differentiating eqn. (*iii*) w.r.t. x, and equating it to zero.

$$\frac{dT}{dx} = -2 \times 88888.89x + 8288.89 = 0$$

It gives x_{cr} = 0.0466 m

and $T_{max} = -88888.89 \times (0.0466)^2 + 8288.89 \times (0.0466) + 180$

= **373.02°C. Ans.**

(*iii*) Heat flow rate from each face :

Temperature gradient

$$\frac{dT}{dx} = -2 \times 88888.89\, x + 8288.89$$

$$= -177777.78\, x + 8288.89$$

(*a*) Heat flow rate per m² at left face (x = 0)

$$q_{x=0} = -k\left(\frac{dT}{dx}\right)_{x=0} = -45 \times 8288.89$$

= **– 373 × 10³ W/m²**

= **373 kW/m² (towards left out). Ans.**

(*b*) Heat flow rate per m² at right face

$$q_{x=L} = -k\left(\frac{dT}{dx}\right)_{x=L}$$

= – 45 × [– 177777.78 × 0.1 + 8288.89]

= **427 × 10³ W/m²**

≈ **427 kW/m² (From right face). Ans.**

Example 4.2. *A plane wall of thickness 0.1 m and thermal conductivity 25 W/m.K having uniform volumetric heat generation of 0.3 MW/m³, is insulated on one side, while other side is exposed to a fluid at 92°C. The convection heat transfer coefficient between plane wall and fluid is 500 W/m².K. Determine the maximum temperature in the plane wall.*

Solution

Given : A plane wall with heat generation and insulated on its one face

$L = 0.1, \quad k = 25$ W/m.K,

$g_o = 0.3$ MW/m³ $= 0.3 \times 10^6$ W/m³

$T_\infty = 92°C, \quad h = 500$ W/m².K

The boundary conditions

(*i*) At $x = 0$, $\dfrac{dT}{dx} = 0$ (For insulated boundary)

(*ii*) At $x = L$, $-k\left\{\dfrac{dT}{dx}\right\}_{x=L} = h\,(T_{x=L} - T_\infty)$ (For convective boundary)

To find : The maximum temperature in the wall.

Assumptions :

(*i*) One dimensional steady state conduction in x direction.

(*ii*) Uniform heat generation rate $g(x) = g_o$ (constant).

(*iii*) Constant properties.

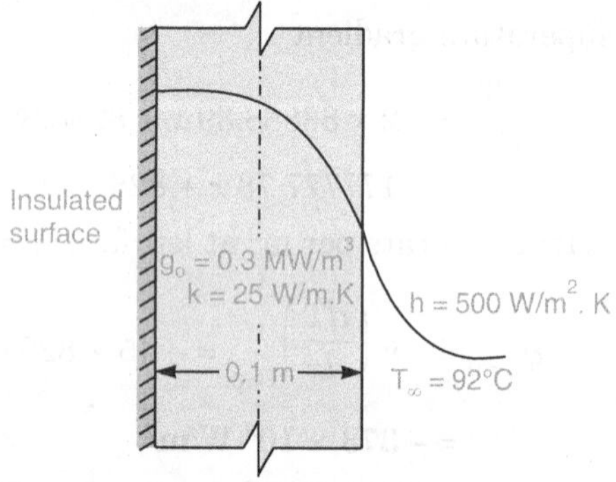

Fig. 4.6. Plane wall for example 4.2

Analysis : The temperature distribution in the plane wall is given by

$$T(x) = -\frac{g_o x^2}{2k} + C_1 x + C_2 \qquad ...(i)$$

and

$$\frac{dT(x)}{dx} = -\frac{g_o x}{k} + C_1$$

At $x = 0$;

$$\frac{dT}{dx} = 0$$

It gives $C_1 = 0$

At $x = L$;

$$-k\left\{-\frac{g_o L}{k}\right\} = h\left\{-\frac{g_o L^2}{2k} + C_2 - T_\infty\right\}$$

It gives $C_2 = \dfrac{g_o L}{h} + \dfrac{g_o L^2}{2k} + T_\infty$

And the temperature distribution as obtained by eqn. (4.12) ;

$$T(x) = \frac{g_o(L^2 - x^2)}{2k} + \frac{g_o L}{h} + T_\infty \qquad ...(ii)$$

The location of maximum temperature :

Differentiating eqn. (*ii*) w.r.t. x and equating it to zero

$$\frac{dT}{dx} = -\frac{g_o x_{cr}}{k} = 0$$

which gives, $x_{cr} = 0$

Hence the maximum temperature will occur at the left face

The magnitude of maximum temperature

$$T_{max} = \frac{g_o L^2}{2k} + \frac{g_o L}{h} + T_\infty$$

or $$T_{max} = \frac{0.3 \times 10^6 \times (0.1)^2}{2 \times 25} + \frac{0.3 \times 10^6 \times (0.1)}{500} + 92$$

$= \mathbf{212°C}$. **Ans.**

Example 4.3. *An amount of chicken in the form of a rectangular block, 25 mm thick is roasted in a microwave heating system. The centre temperature of the chicken block is 100°C, when surrounding temperature is 30°C. The heat transfer coefficient between the chicken block and air is 15 W/m².K. The thermal conductivity of the chicken can be taken as 1 W/m.K. Calculate microwave heating capacity during steady state operation.*

Solution

Given : A rectangular block (slab) of chicken

$L = 25$ mm $= 0.025$ m,

$k = 1$ W/m.K

$h = 15$ W/m².K,

$T_c = 100°C, \quad T_\infty = 30°C.$

To find : Heating capacity of microwave heater.

Analysis : The centre line temperature will be maximum temperature in the slab, and it is given by eqn. (4.17)

$$T_{max} = T_c = \frac{gL}{2h} + \frac{gL^2}{8k} + T_\infty$$

$$= \frac{g}{2}\left(\frac{L}{h} + \frac{L^2}{4k}\right) + T_\infty$$

$$100°C = \frac{g}{2}\left[\frac{(0.025\text{ m})}{(15\text{ W/m}^2.\text{K})} + \frac{(0.025\text{ m})^2}{4 \times (1\text{ W/m.K})}\right] + 30°C$$

$$2 \times (100 - 30) = g \times 1.823 \times 10^{-3}$$

or g = **76800 W/m³ = 76.8 kW/m³.** Ans.

Example 4.4. *The steady state temperature distribution in a 0.3 m thick plane wall is given by*

$$T(x) = 600 + 2500x - 12{,}000x^2$$

where T is in °C and x in metres measured from left surface of the wall. One dimensional steady state heat conduction occurs in the wall along x direction. The thermal conductivity of the wall material is 23.5 W/m.K.

(i) What are the surface temperatures and average temperature of the wall ?

(ii) Calculate the maximum temperature in the wall and its location.

(iii) Calculate the heat fluxes at its surfaces.

(iv) Do you think that there is any heat generation in the wall ? If so, what is the average volumetric rate of heat generation ?

Solution

Given : Temperature distribution in a plane wall as

$$T(x) = 600 + 2500x - 12{,}000x^2$$

with $k = 23.5$ W/m.K, $L = 0.3$ m.

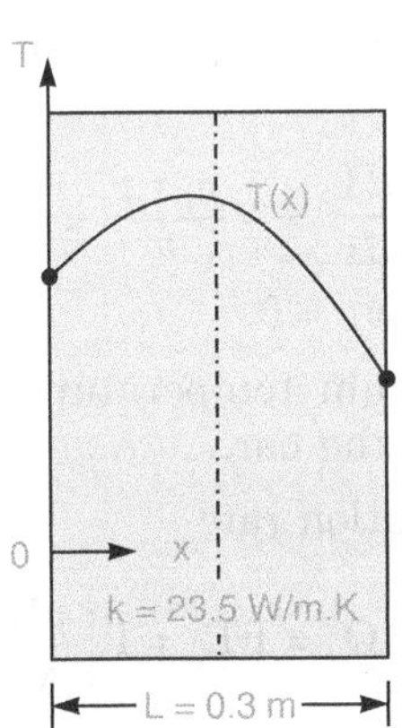

Fig. 4.7. Schematic for example 4.4

To find :

(*i*) Surface temperatures, and average temperature of wall,

(*ii*) Location and magnitude of maximum temperature,

(*iii*) Heat fluxes at the surfaces, and

(*iv*) Heat generation in wall, if any.

Analysis : (*i*) Temperature at the left surface *i.e.,* $x = 0$

$$T_1 = 600 + 2500 \times (0) - 12{,}000 \times (0)^2$$

= **600°C.** Ans.

Temperature at the right face *i.e.,* $x = L = 0.3$ m.

$$T_2 = 600 + 2500 \times (0.3) - 12000 \times (0.3)^2$$

= **270°C.** Ans.

Average temperature of the wall

$$T_{av} = \frac{1}{L}\int_0^L T(x)\,dx$$

$$= \frac{1}{0.3} \times \int_0^{0.3} (600 + 2500x - 12000x^2)\,dx$$

$$= \frac{1}{0.3}\left[600 \times 0.3 + 2500 \times \frac{(0.3)^2}{2} - 12000 \times \frac{(0.3)^3}{3}\right]$$

= **615°C.** Ans.

(*ii*) Location of maximum temperature

$$\frac{dT}{dx} = 2500 - 24000x_{cr} = 0$$

or $x_{cr} = \dfrac{2500}{24000}$ = **0.104 m** from left. Ans.

Maximum temperature *i.e.,* at $x = x_{cr}$

$$T_{max} = 600 + 2500 \times (0.104) - 12000 \times (0.104)^2$$

= **730.2°C.** Ans.

(*iii*) Heat fluxes at any face

$$q(x) = -k\frac{dT}{dx}$$

where $\dfrac{dT}{dx} = 2500 - 24000x$

Heat flux at left face, ($x = 0$)

$$q_{x=0} = -k\left(\frac{dT}{dx}\right)_{x=0}$$

$$= -23.5 \times [2500 - 24000 \times (0)]$$

= **– 58,750 W/m²** (towards left out).

Ans.

Heat flux at right face, ($x = 0.3$ m)

$$q_{x=L} = -23.5 \times [2500 - 24000 \times (0.3)]$$

= **110450 W/m²** (towards right out).

Ans.

(*iv*) Since heat is coming out both the surfaces of wall. This means, there must be heat source within the wall, the amount of heat generation

$$q_g = 110450 - (-58750)$$
$$= \mathbf{169200\ W/m^2}.\ \text{Ans.}$$

Example 4.5. *A rectangular copper bar 80 mm × 6 mm in cross-section (k = 370 W/m.K) is insulated at top, bottom and left faces. It is observed that when a current of 8000 A is passed through the conductor, the bare face has a constant temperature of 50°C. If the resistivity of the copper is 2 × 10⁻⁸ Ωm, calculate*

(*i*) *The maximum temperature in bar and its location,*

(*ii*) *Temperature at the centre of the bar.*

Solution

Given : A rectangular copper bar insulated at its three faces as shown in Fig. 4.8

$$A_c = 80\text{ mm} \times 6\text{ mm}$$
$$T_2 = T_{x=L} = 50°\text{C}$$
$$L = 6\text{ mm} = 0.006\text{ m}$$
$$\left(\frac{dT}{dx}\right)_{x=0} = 0$$
$$k = 370\text{ W/m.K}$$
$$I = 8000\text{ A}$$
$$\rho = 2 \times 10^{-8}\ \Omega\text{m}.$$

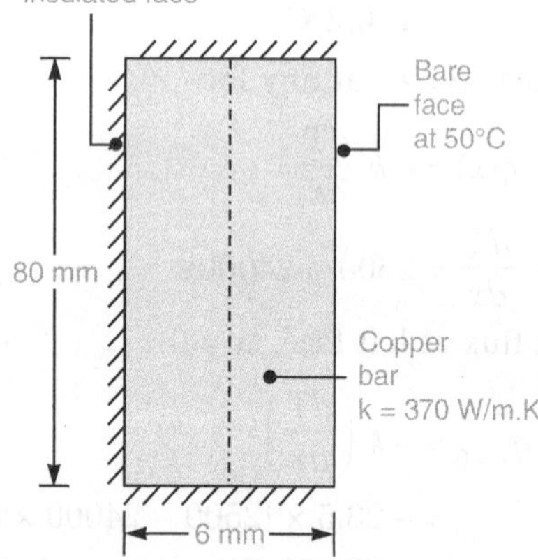

Fig. 4.8. A wall covered on three sides

To find : (*i*) T_{max} and its location

(*ii*) Temperature at centre.

Assumptions :

(*i*) Steady state conditions,

(*ii*) Left face is perfectly insulated,

(*iii*) Heat transfer in axial direction only.

Analysis : The temperature distribution in a plane wall with uniform heat generation rate g_o ;

$$T(x) = -\frac{g_o}{2k}x^2 + C_1 x + C_2$$

and
$$\frac{dT}{dx} = -\frac{g_o}{k}x + C_1$$

Using boundary condition at left (insulated) face *i.e.,*

At $x = 0$,

$$\frac{dT}{dx} = 0$$

$$\left(\frac{dT}{dx}\right)_{x=0} = 0 = -\frac{g_o \times 0}{k} + C_1$$

It gives $C_1 = 0$

With boundary condition at $x = L$, $T = T_2$

$$T_2 = -\frac{g_o L^2}{2k} + C_2$$

It gives
$$C_2 = T_2 + \frac{g_o L^2}{2k}$$

and the temperature distribution in copper bar

$$T(x) = \frac{g_o}{2k}(L^2 - x^2) + T_2$$

The location of maximum temperature can be obtained by differentiating above equation w.r.t. x and equating it to zero.

$$\frac{dT}{dx} = -\frac{g_o x_{cr}}{k} = 0$$

It gives $x_{cr} = 0$.

The maximum temperature will occur at left (insulated) face of the bar. **Ans.**

Heat generation rate

$$Q_g = I^2 R_e = I^2\left(\frac{\rho L}{A_c}\right)$$

and heat generation per unit volume

$$g_o = \frac{Q_g}{V} = \frac{I^2}{A_c L}\left(\frac{\rho L}{A_c}\right) = \left(\frac{I}{A_c}\right)^2 \times \rho$$

$$= \left(\frac{8000}{0.08 \times 0.006}\right)^2 \times 2 \times 10^{-8}$$

$$= 5.55 \times 10^6\ \text{W/m}^3.$$

and the maximum temperature (at $x = 0$)

$$T_{max} = \frac{5.55 \times 10^6}{2 \times 370} \times (0.006)^2 + 50$$
$$= 0.27 + 50 = \mathbf{50.27°C.} \quad \text{Ans.}$$

Temperature at the centre of bar, (at $x = 0.003$ m) ;

$$T_c = \frac{5.55 \times 10^6}{2 \times 370} [(0.006)^2 - (0.003)^2] + 50$$
$$= 7500 \times 2.7 \times 10^{-5} + 50$$
$$= \mathbf{50.2°C.} \quad \text{Ans.}$$

Example 4.6. *Two large steel plates at temperature of 150°C and 100°C are separated by a copper rod ($k = 390$ W/m.K), 300 mm long and 25 mm in diameter. The rod is welded to each plate. The rod is insulated on its lateral surface, so heat can only flow axially. The current flows through the rod, generating heat energy at the rate of 150 W. Find the maximum temperature in the rod and heat flux at ends of the rod.*

Solution

Given : The two large plates separated by a copper rod.

$T_1 = 150°C$, $T_2 = 100°C$

$k = 390$ W/m.K, $L = 300$ mm $= 0.3$ m

$d = 25$ mm $= 0.025$ m, $Q_g = 150$ W.

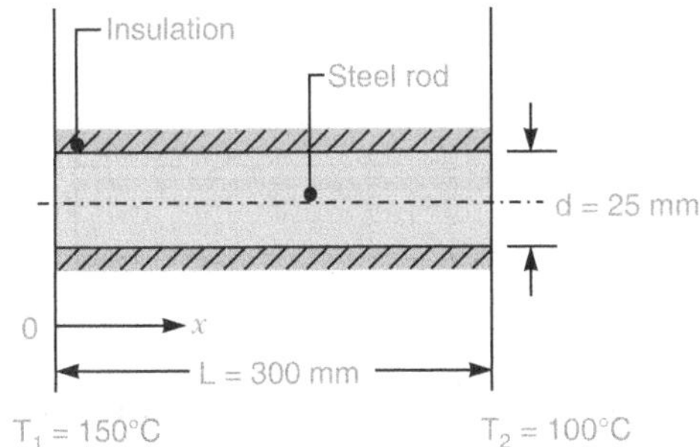

Fig. 4.9. Schematic for example 4.6

To find :

(*i*) Maximum temperature in the rod, and

(*ii*) Heat flux at two ends of the rod.

Analysis : (*i*) The rod is insulated on its lateral surface, therefore, the heat flows axially, the temperature distribution in the rod can be expressed as

$$T(x) = -\frac{g_o x^2}{2k} + C_1 x + C_2$$

where g_o is the uniform volumetric heat generation rate per unit volume.

The boundary conditions

At $x = 0$, $T = T_1 = 150°C$

At $x = L = 300$ mm, $T = T_2 = 100°C$

and
$$g_o = \frac{Q_g}{V} = \frac{Q_g}{\frac{\pi}{4} d^2 \times L} = \frac{150}{\frac{\pi}{4} \times (0.025)^2 \times (0.3)}$$
$$= 1.018 \times 10^6 \text{ W/m}^3$$

Using first boundary condition, we get

$$C_2 = T_1 = 150°C$$

and at right end *i.e.*, $x = L$

$$100 = -\frac{1.018 \times 10^6}{2 \times 390} \times (0.3)^2 + C_1 \times 0.3 + 150$$

or
$$C_1 = \frac{100 + 117.53 - 150}{0.3} = 225.01°C/m$$

$$\therefore \quad T(x) = -1305.88x^2 + 225.01x + 150$$

For location of maximum temperature

$$\frac{dT}{dx} = -2 \times 1305.88x_c + 225.01 = 0$$

or
$$x_{cr} = 0.08618 \text{ m} = 86.18 \text{ mm from left}$$

The maximum temperature in the rod

$$T_{max} = -1305.88 \times (0.08618)^2 + 225.01 \times 0.08618 + 150$$
$$= -9.7 + 19.39 + 150 = \mathbf{159.7°C.} \quad \text{Ans.}$$

(*ii*) Heat flux towards left end

$$q_{x=0} = k\left(\frac{dT}{dx}\right)_{x=0}$$
$$= k \times [-2 \times 1305.88x + 225.01]_{x=0}$$
$$= 390 \times 225.01 = \mathbf{87753.9 \text{ W/m}^2}$$
$$\approx \mathbf{87.754 \text{ kW/m}^2.} \quad \text{Ans.}$$

and
$$q_{x=L} = -k\left(\frac{dT}{dx}\right)_{x=L}$$
$$= -390 \times [-2 \times 1305.88 \times (0.3) + 225.01]$$
$$= \mathbf{217822.02 \text{ W/m}^2}$$
$$\approx \mathbf{217.822 \text{ kW/m}^2.} \quad \text{Ans.}$$

Check : Total heat transfer rate,

$$Q = \left(\frac{\pi}{4}d^2\right)(q_{x=0} + q_{x=L})$$
$$= (\pi/4) \times (0.025)^2 \times (87753.9 + 217822.02)$$
$$= 150 \text{ W}.$$

Example 4.7. *The heat generation rate in a plane wall, insulated at its left face and maintained at a uniform temperature T_2 on right face is given as*

$$g(x) = g_o e^{-\gamma x} \text{ W/m}^3$$

where g_o and γ are constants and x is measured from left face.

Develop an expression for temperature distribution in the plane wall, and deduce the expression for temperature of the insulated surface.

Solution

Given : The heat generation rate in the wall as

$$g(x) = g_o e^{-\gamma x} \text{ W/m}^3$$

Subjected to boundary conditions

At $x = 0$, (insulated face), $\frac{dT}{dx} = 0$

At $x = L$, (specified temperature), $T = T_2$.

To find :

(*i*) An expression for temperature distribution, T(x), and

(*ii*) Temperature of insulated surface.

Analysis : (*i*) Governing differential equation in steady state

$$\frac{d^2T}{dx^2} + \frac{g(x)}{k} = 0$$

or
$$\frac{d^2T}{dx^2} = -\frac{g_o e^{-\gamma x}}{k}$$

Integrating with respect to x,

$$\frac{dT}{dx} = -\frac{g_o e^{-\gamma x}}{k} \frac{1}{(-\gamma)} + C_1$$

where C_1 is constant of integration.

Using first boundary condition *i.e.,* at $x = 0$,

$$\frac{dT}{dx} = 0$$

$$\left(\frac{dT}{dx}\right)_{x=0} = \frac{g_o}{k} \frac{e^{-\gamma \times 0}}{\gamma} + C_1 = 0$$

It gives $C_1 = -\frac{g_o}{k\gamma}$

$$\therefore \quad \frac{dT}{dx} = \frac{g_o e^{-\gamma x}}{k\gamma} - \frac{g_o}{k\gamma} = \frac{g_o}{k\gamma}[e^{-\gamma x} - 1]$$

Integrating again with respect to x

$$T(x) = \frac{g_o}{k\gamma}\left[\frac{e^{-\gamma x}}{-\gamma} - x\right] + C_2$$

where C_2 is constant of integration.

Using second boundary condition, *i.e.,* at $x = L$, $T = T_2$

$$T_2 = \frac{g_o}{k\gamma}\left[\frac{e^{-\gamma L}}{-\gamma} - L\right] + C_2$$

It gives
$$C_2 = T_2 + \frac{g_o}{k\gamma}\left[\frac{e^{-\gamma L}}{\gamma} + L\right]$$

$$\therefore \quad T(x) = \frac{g_o}{k\gamma}\left[\frac{e^{-\gamma x}}{-\gamma} - x\right] + \frac{g_o}{k\gamma}\left[\frac{e^{-\gamma L}}{\gamma} + L\right] + T_2$$

or
$$T(x) = \frac{\mathbf{g_o}}{\mathbf{k\gamma^2}}\left[\mathbf{e^{-\gamma L} - e^{-\gamma x}}\right] + \frac{\mathbf{g_o}}{\mathbf{k\gamma}}[\mathbf{L - x}] + \mathbf{T_2}$$

It is the required expression. **Ans.**

(*ii*) Temperature at insulated surface *i.e.,* $x = 0$

$$T_{x=0} = \frac{\mathbf{g_o}}{\mathbf{k\gamma^2}}\left[\mathbf{e^{-\gamma L} - 1}\right] + \frac{\mathbf{g_o L}}{\mathbf{k\gamma}} + \mathbf{T_2}. \quad \textbf{Ans.}$$

Example 4.8. *Two ends of circular rod of length 2L, perfectly insulated on its lateral surface are held at same temperature T_0. The left half of rod has uniform heat generation at the rate of g_o W/m³, while right half portion has no heat generation. Thermal conductivity of the rod material is constant (independent of temperature). In steady state conditions*

(*a*) *Develop the expressions for the temperature distribution in the left and right portion of the end.*

(*b*) *Find the location of maximum temperature.*

Solution

Given :

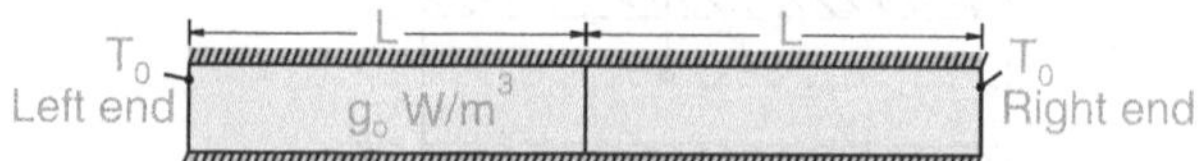

Fig. 4.10. Schematic for example 4.8

Analysis : (*a*) The governing equation in left half of the rod.

$$\frac{d^2T_L(x)}{dx^2} + \frac{g_o}{k} = 0$$

and temperature distribution in left portion of rod

$$T_L(x) = -\frac{g_o x^2}{2k} + C_1 x + C_2 \quad \ldots(i)$$

The boundary condition

At $x = 0$, $T_L(x) = T_0$,

It gives $C_2 = T_0$

Hence
$$T_L(x) = -\frac{g_o x^2}{2k} + C_1 x + T_0 \quad \ldots(ii)$$

The governing differential equation in right portion of the rod is reduced to

$$\frac{d^2T_R(x)}{dx^2} = 0$$

and temperature distribution,

$$T_R(x) = C_3x + C_4 \quad ...(iii)$$

With boundary condition

At $x = 2L$, $T_R(x) = T_0$

It gives $C_4 = T_0 - 2LC_3$

Hence $T_R(x) = C_3(x - 2L) + T_0 \quad ...(iv)$

Due to symmetry, the temperature at mid-point

At $x = L$, $T_R(x) = T_L(x)$

Therefore, equating eqns. (*ii*) and (*iv*),

$$\left[-\frac{g_o x^2}{2k} + C_1x + T_0\right]_{x=L} = \left[C_3(x - 2L) + T_0\right]_{x=L}$$

$$-\frac{g_o L^2}{2k} + C_1L + T_0 = -C_3L + T_0$$

or $$C_1 + C_3 = \frac{g_o L}{2k} \quad ...(v)$$

Also at section $x = L$.

Heat transfer rate from left = Heat flow rate to right

$$Q_L = Q_R$$

or $$\left[\frac{dT_L(x)}{dx}\right]_{x=L} = \left[\frac{dT_R(x)}{dx}\right]_{x=L}$$

or $$-\frac{g_o L}{k} + C_1 = C_3$$

or $$C_1 - C_3 = \frac{g_o L}{k} \quad ...(vi)$$

Adding eqns. (*v*) and (*vi*), we get

$$2C_1 = \frac{3g_o L}{2k}$$

It gives $$C_1 = \frac{3g_o L}{4k}$$

Subtracting eqns. (*v*) and (*vi*), we get

$$2C_3 = -\frac{g_o L}{2k}$$

or $$C_3 = -\frac{g_o L}{4k}$$

Substituting the values of C_1 and C_3 in eqns. (*ii*) and (*iv*), respectively, we get

$$\mathbf{T_L(x) = -\frac{g_o x^2}{2k} + \frac{3g_o L}{4k}x + T_0} \quad ...(vii)$$

and $$\mathbf{T_R(x) = -\frac{g_o L}{4k}(x - 2L) + T_0}. \quad \textbf{Ans.}$$

(*b*) The location of maximum temperature, the left portion will have maximum temperature, therefore, differentiating equation (*vii*) w.r.t. x and equating it to zero.

$$\frac{dT_L(x)}{dx} = -\frac{g_o x}{k} + \frac{3g_o L}{4k} = 0$$

or $$x = \frac{3}{4}\mathbf{L}. \quad \textbf{Ans.}$$

Example 4.9. *A plane wall is composite of two materials A and B. The wall of material A has a uniform heat generation of 2.5 × 10⁶ W/m³. Its thermal conductivity is 110 W/m.K and it is 60 mm thick. The wall of material B has no heat generation and its thermal conductivity is 150 W/m.K and its thickness is 20 mm. The inner surface of material A is well insulated, while the outer surface of material of B is cooled by water stream at 30°C with convection coefficient of 1000 W/m².K. For steady state conditions :*

(*i*) *Sketch the temperature distribution in the composite wall.*

(*ii*) *Determine the temperatures of insulated surface of A and cooled surface of B.*

Solution

Given : A composite wall of material A and B ;

$g_A = 2.5 \times 10^6$ W/m³

$k_A = 110$ W/m.K

$L_A = 60$ mm $= 0.06$ m

$g_B = 0$

$k_B = 150$ W/m.K

$L_B = 20$ mm $= 0.02$ m

$T_\infty = 30$°C

$h = 1000$ W/m².K

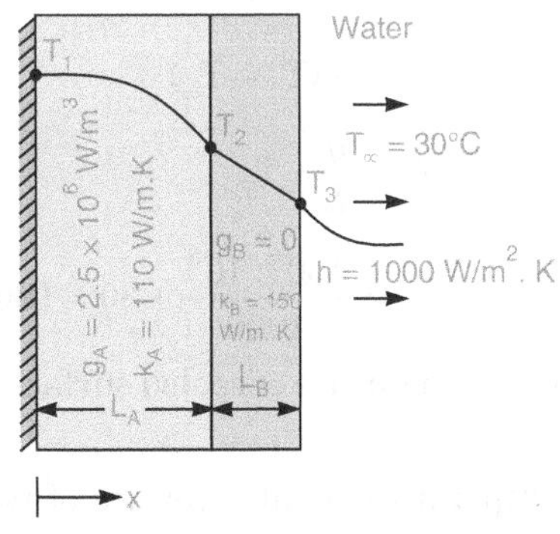

Fig. 4.11. Schematic of composite wall for example 4.19

To find :

(*i*) Temperature distribution in the composite.

(*ii*) Temperature of insulated surface of A and cooled surface of material B.

Assumptions :

(*i*) Steady state heat conduction in axial direction only,

(*ii*) Negligible contact resistance at interface.

(*iii*) Constant properties.

Analysis : (*i*) (*a*) The temperature distribution in material A is given as

$$T_A(x) = -\frac{g_A x^2}{2k_A} + C_1 x + C_2$$

It is *parabolic temperature distribution* in material A as shown in Fig. 4.11 and it is subjected to boundary conditions

At $x = 0$, the slope $\frac{dT}{dx} = 0$

and at $x = L_A$, $T_A(x) = T_2$

(*b*) The temperature distribution in material B is given as

$$T_B(x) = C_3 x + C_4$$

It is a *linear distribution* between temperatures T_2 and T_3.

(*c*) Large gradient near wall B due to water cooling.

(*ii*) The heat flux in wall material A can be calculated as

$$q = g_A L_A = 2.5 \times 10^6 \times 0.06$$

$$= 150 \times 10^3 \text{ W/m}^2$$

Since inner surface of material A is well insulated and hence under steady state, this heat must be dissipated from outer surface of material B to water stream

Thus $q = h\,(T_3 - T_\infty)$

or $$T_3 = \frac{q}{h} + T_\infty$$

$$= \frac{150 \times 10^3}{1000} + 30 = \mathbf{180°C}$$

It is the temperature of cooled surface of material B. **Ans.**

The temperature T_2 at interface of two material can be calculated as

$$q = \frac{k_B\,(T_2 - T_3)}{L_B}$$

or $$T_2 = \frac{qL_B}{k_B} + T_3$$

$$= \frac{150 \times 10^3 \times 0.02}{150} + 180 = 200°C$$

Now temperature distribution in material A

$$T_A(x) = -\frac{g_A x^2}{2k_A} + C_1 x + C_2$$

Subjected boundary conditions :

At $x = 0$, $\frac{dT}{dx} = 0$

It gives $C_1 = 0$

and at $x = L_A$, $T = T_2$

or $$T_2 = -\frac{g_A L_A^2}{2k_A} + C_2$$

It gives $$C_2 = T_2 + \frac{g_A L_A^2}{2k_A}$$

Then $$T_A(x) = \frac{g_A}{2k_A}(L_A^2 - x^2) + T_2$$

$$= \frac{2.5 \times 10^6}{2 \times 110}(0.06^2 - x^2) + 200$$

$$= 11363.63 \times (0.06^2 - x^2) + 200.$$

The inner surface temperature of material A, $(x = 0)$

$$T_1 = 11363.63 \times (0.06)^2 + 200$$

$$= \mathbf{240.9°C.} \text{ Ans.}$$

Example 4.10. *A plane wall is a composite of three materials A, B, and C. The wall of material A has a heat generation at the rate of 2 × 10⁶ W/m³. The thermal conductivity of wall A is 190 W/m.K, while its thickness is 50 mm.*

The wall materials of B and C do not have heat generation with

$$k_B = 150 \text{ W/m.K}, \quad L_B = 30 \text{ mm}.$$

$$k_C = 50 \text{ W/m.K.}, \quad L_C = 15 \text{ mm}.$$

The inner surface of material A is well insulated, while outer surface of material C is cooled by water stream at $T_\infty = 50°C$ with convection coefficient $h = 2000$ W/m².K.

(i) Sketch the temperature distribution in the composite under steady state conditions.

(ii) Determine the temperature of insulated surface and cooled surface. (N.M.U., Nov. 1996)

Solution

Given :

$g_A = 2 \times 10^6$ W/m³,
$L_A = 50$ mm = 0.05 m
$k_A = 190$ W/m.K.,
$g_B = g_C = 0$
$k_B = 150$ W/m.K,
$L_B = 30$ mm = 0.03 m
$k_C = 50$ W/m.K,
$L_C = 15$ mm = 0.015 m
$T_\infty = 50°C$,
$h = 2000$ W/m².K.

To find :

(*i*) Sketch the temperature distribution in the composite wall.

(*ii*) The temperature of cooled surface.

(*iii*) Temperature of insulated surface.

Assumptions :

(*i*) Steady state heat conduction in axial direction only.

(*ii*) Negligible contact resistance at interfaces.

(*iii*) Inner surface of material A is adiabatic.

(*iv*) Constant properties.

Analysis : (*i*) The temperature distribution in composite wall.

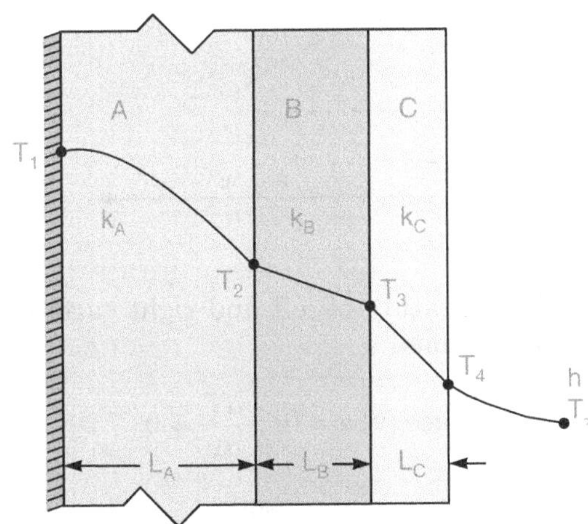

Fig. 4.12. (*a*) Schematic and temperature distribution in the composite

(*a*) The wall material A has a *parabolic distribution,* since its temperature distribution is given by

$$T_A(x) = -\frac{g_A x^2}{2k} + C_1 x + C_2 \qquad ...(i)$$

(*b*) Since inner surface of material A is insulated *i.e.,* slope $\frac{dT}{dx} = 0$ at the inner surface.

(*c*) The material B and C will have the *linear slope*, [Fig. 4.12(*a*)] since their temperature distribution can be expressed as

$$T(x) = C_3 x + C_4. \qquad ...(ii)$$

(*d*) The slope changes as $k_B/k_C = 3$ at the interface of materials B and C.

(*e*) Large gradient near the wall surface of C due to water cooling.

The temperature distribution is shown in Fig. 4.12(*a*).

(*ii*) Temperature of cooled (right) surface :

The heat flux in wall A can be calculated as

$$q = g_A L_A = 2 \times 10^6 \times 0.05$$
$$= 1 \times 10^5 \text{ W/m}^2.$$

Since the inner side of material A is insulated, hence under steady state conditions, this heat must be dissipated from outer surface of material C to water stream.

Therefore, $q = h(T_4 - T_\infty)$

It gives $T_4 = \frac{1 \times 10^5}{2000} + 50 =$ **100°C. Ans.**

(*iii*) Temperature of insulated surface :

The temperature at the interface of wall A and B can be calculated by resistance analogy as :

$$q = \frac{T_2 - T_4}{\frac{L_B}{k_B} + \frac{L_C}{k_C}}$$

or
$$1 \times 10^5 = \frac{T_2 - 100}{\frac{0.03}{150} + \frac{0.015}{50}}$$

or
$$T_2 = 1 \times 10^5 \times 5 \times 10^{-4} + 100 = 150°C$$

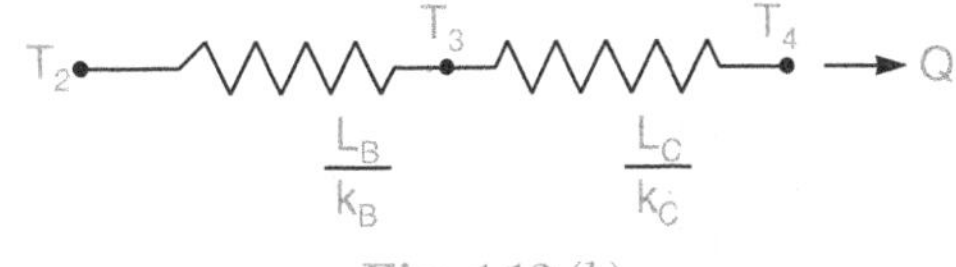

Fig. 4.12 (*b*)

Now considering eqn. (*i*) for temperature distribution $T_A(x)$, with boundary conditions.

The boundary condition at left face of wall A

At $x = 0$; $\frac{dT}{dx} = 0$

It gives $C_1 = 0$

The boundary condition at right face of wall A

At $x = L_A$, $T = T_2 = 150°C$

Using in eqn. (*i*),

$$150 = -\frac{2 \times 10^6 \times (0.05)^2}{2 \times 190} + C_2$$

It gives $\quad C_2 = 163.15°C$

and the temperature distribution in wall A is

$$T_A(x) = -\frac{2 \times 10^6 x^2}{2 \times 190} + 163.15$$

The temperature at the insulated face of wall A at $x = 0$ $\quad T(x = 0) = $ **163.15°C. Ans.**

Example 4.11. *Air inside a chamber at 50°C is heated convectively with convective coefficient of 20 W/m².K by a 200 mm thick wall having thermal conductivity of 4 W/m.K. It has uniform heat generation of 1000 W/m³. Its other side is exposed to an ambient 25°C with convection coefficient of 5 W/m².K. In order to prevent the heat loss from the outside surface of the wall, a very thin strip heater is placed on the outer wall to provide a uniform heat flux* q_o.

(*a*) *Sketch the temperature distribution in the wall on T–x coordinates for condition, where no heat generated within the wall is lost to outside of the chamber.*

(*b*) *What are the temperatures of inside and outside surfaces for condition of part (a).*

(*c*) *Determine the value of heat flux* q_o *that must be supplied by the strip heater, so that the heat generated within the wall is transferred to inside of the chamber.*

(*d*) *If the heat generated within the wall is switched off, while the heat flux* q_o *to the strip heater remains constant. What would be the steady state temperature of outer wall surface?*

Solution

Given: A wall with internal heat generation and insulated outer surface as shown in Fig. 4.13(*a*).

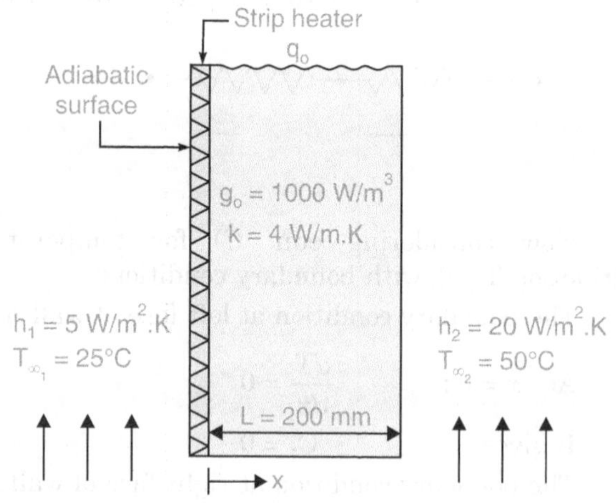

Fig. 4.13(*a*). Schematic for example 4.11

To find:

(*a*) Sketch of temperature distribution in the wall.

(*b*) Temperature of wall at its inside and outside surfaces.

(*c*) Value of heat flux required to maintain adiabatc condition at the left face.

(*d*) Temperature of left surface, when $g_o = 0$ and q_o is kept constant as in above case.

Assumptions:

1. Steady-state one-dimensional heat conduction.

2. Heat generated in the strip heater does not contribute to heat generation in the wall.

3. Constant properties

Analysis: (*a*) The one-dimensional heat conduction with internal heat generation is given by

$$T(x) = -\frac{g_o x^2}{2k} + C_1 x + C_2 \qquad ...(i)$$

and the slope $\frac{dT}{dx} = 0$ for left insulated surface. The temperature distribution is parabolic in the wall as shown in Fig. 4.13(*b*).

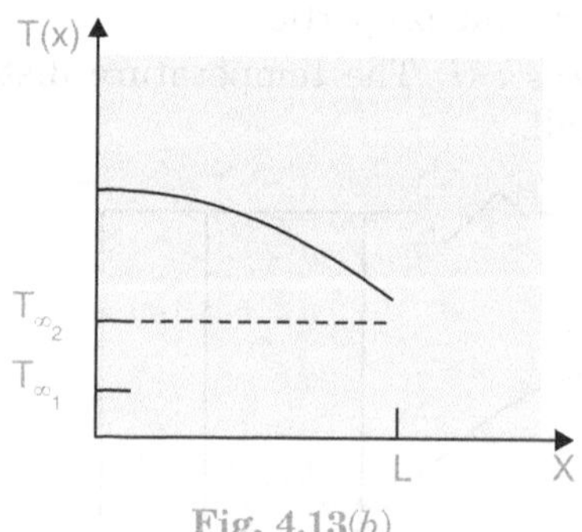

Fig. 4.13(*b*)

(*b*) Temperature of left and right surfaces. The boundary conditions

(*i*) At left face (at $x = 0$) ; $\frac{dT}{dx} = 0$

(*ii*) At right face ; $-k\left(\frac{dT}{dx}\right)_{x=L} = h[T_{x=L} - T_{\infty_2}]$

The differential of eqn. (*i*)

$$\frac{dT}{dx} = -\frac{g_o x}{k} + C_1 \qquad ...(ii)$$

Using first boundary condition in eqn. (*ii*), we get

$$C_1 = 0$$

Using second boundary condition

$$-k\left(-\frac{g_o\text{L}}{k}\right) = h_2\left[-\frac{g_o\text{L}^2}{2k} + \text{C}_2 - \text{T}_{\infty_2}\right]$$

It gives $\quad \text{C}_2 = \dfrac{g_o\text{L}^2}{2k} + \dfrac{g_o\text{L}}{h_2} + \text{T}_{\infty_2}$

Then the temperature distribution in the wall is given by

$$\text{T}(x) = \frac{g_o\,(\text{L}^2 - x^2)}{2k} + \frac{g_o\text{L}}{h_2} + \text{T}_{\infty_2} \quad \ldots(iii)$$

Using the numerical values in eqn. (*iii*), it results into

$$\text{T}(x) = -\,125x^2 + 65°\text{C}$$

The temperature at left surface

$\text{T}(x = 0) = \text{T}_1 =$ **65°C. Ans.**

The temperature at right surface

$\text{T}\,(x = \text{L}) = \text{T}_2 = -\,125 \times (0.2)^2 + 65$

$= $ **60°C. Ans.**

(*c*) Heat flux supplied to strip heater is equal to heat convection rate per unit area to outside surroundings, therefore,

$$q_o = h_1(\text{T}_1 \quad \text{T}_{\infty_1})$$

$= 5 \times (65 - 25) =$ **200 W. Ans.**

(*d*) When heat generation is switched off, the situation can be represented by thermal network as shown in Fig. 4.13(*c*)

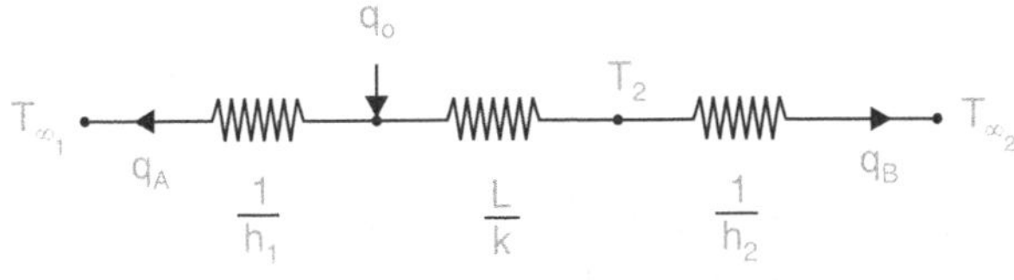

Fig. 4.13(*c*)

and the heat flux generated is sum of heat flux flow on two sides of the wall

$$q_o = q_\text{A} + q_\text{B} = \frac{\text{T}_1 - \text{T}_{\infty_1}}{\dfrac{1}{h_1}} + \frac{(\text{T}_1 + \text{T}_{\infty_2})}{\dfrac{\text{L}}{k} + \dfrac{1}{h_2}}$$

or $\quad 200 = \dfrac{\text{T}_1 - 25}{\dfrac{1}{5}} + \dfrac{\text{T}_1 - 50}{\dfrac{0.2}{4} + \dfrac{1}{20}}$

$= 15\,\text{T}_1 - 625$

or $\quad \text{T}_1 =$ **55°C. Ans.**

4.2. THE CYLINDER

For steady state heat conduction in the cylinders, rewriting the eqn. (2.15),

$$\frac{1}{r}\frac{d}{dr}\left\{r\frac{d\text{T}}{dr}\right\} + \frac{g(r)}{k} = 0$$

where k is treated constant and using $g(r) = g_o$, (uniform heat generation), then above

Expression may be written as

$$\frac{d}{dr}\left\{r\frac{d\text{T}}{dr}\right\} = -\frac{g_o r}{k} \quad \ldots(4.19)$$

Integrating with respect to r on both sides, we get

$$r\frac{d\text{T}}{dr} = -\frac{g_o r^2}{2k} + \text{C}_1$$

or $$\frac{d\text{T}}{dr} = -\frac{g_o r}{2k} + \frac{\text{C}_1}{r} \quad \ldots(4.20)$$

Integrating again, with respect to r, we get

$$\text{T}(r) = -\frac{g_o r^2}{4k} + \text{C}_1\ln(r) + \text{C}_2 \quad \ldots(4.21)$$

where C_1 and C_2 are constants of integration and can be evaluated according to boundary conditions.

4.2.1. Solid Cylinder with Specified Surface Temperature

Consider a long solid cylinder with uniform heat generation g_o as shown in Fig. 4.14. Its outer surface is maintained at temperature T_s. Then boundary conditions (B.C.) are :

(*i*) Due to symmetry of the solid, the centre line temperature of the solid cylinder must be constant and thus temperature gradient must be zero.

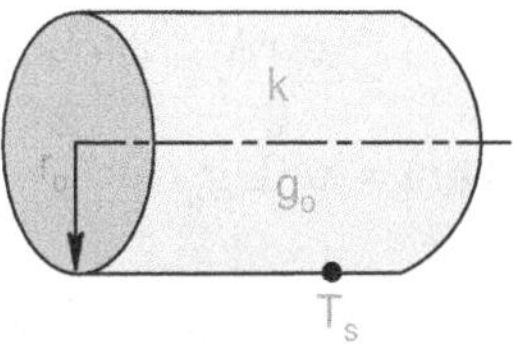

Fig. 4.14. Solid cylinder

At $r = 0$; $\quad \left\{\dfrac{d\text{T}}{dr}\right\} = 0$

or $$\left\{\frac{d\text{T}}{dr}\right\}_{r=0} = 0 \quad \ldots[4.22(a)]$$

(*ii*) At $r = r_o$, $\quad \text{T} = \text{T}_s \quad \ldots[4.22(b)]$

Substituting first B.C. into eqn. (4.20), we get

$$\left\{\frac{d\text{T}}{dr}\right\}_{r=0} = -\frac{g_o(0)}{2k} + \frac{\text{C}_1}{0} = 0$$

It gives $\quad C_1 = 0$

Now using second B.C. into eqn. (4.21), we get

$$T_s = -\frac{g_o r_o^2}{4k} + C_2$$

It gives $\quad C_2 = T_s + \frac{g_o r_o^2}{4k}$

Therefore, the temperature distribution in solid cylinder is

$$T(r) = \frac{g_o}{4k}(r_o^2 - r^2) + T_s$$

or
$$T(r) - T_s = \frac{g_o r_o^2}{4k}\left\{1 - \frac{r^2}{r_o^2}\right\} \quad ...(4.23)$$

Temperature at centre of cylinder is

$$T_c - T_s = \frac{g_o r_o^2}{4k} \quad ...(4.24)$$

Dividing eqn. (4.23) by eqn. (4.24), we get temperature distribution in non-dimensional form.

$$\frac{T(r) - T_s}{T_c - T_s} = 1 - \frac{r^2}{r_o^2} \quad ...(4.25)$$

Heat transfer rate :

The heat transfer rate Q(r) at any radius in the cylinder can be evaluated by using eqn. (4.23), in Fourier equation.

$$Q(r) = -kA\frac{dT}{dr}$$

where, $A = 2\pi rL$

and
$$\frac{dT}{dr} = -\frac{g_o r}{2k} \quad ...(4.26)$$

Then
$$Q(r) = -kA\left(-\frac{g_o r}{2k}\right)$$

$$= \frac{g_o rA}{2} = \frac{2\pi rLrg_o}{2}$$

or
$$Q(r) = \pi r^2 L g_o \quad ...(4.27)$$

4.2.2. Solid Cylinder Exposed to Convection Environment

Consider a long solid cylinder with uniform heat generation g_o W/m³. Its outer surface is exposed to an ambient at T_∞ with heat transfer coefficient h. The temperature distribution is given by eqn. (4.21)

$$T(r) = -\frac{g_o r^2}{4k} + C_1 \ln(r) + C_2 \quad ...(i)$$

and temperature gradient

$$\frac{dT}{dr} = -\frac{g_o r}{2k} + \frac{C_1}{r}$$

Boundary conditions imposed on cylinder

at $r = 0$,
$$\frac{dT}{dr} = 0$$

(for solid cylinder due to symmetry)

It gives $\quad C_1 = 0$

At outer surface *i.e.*, at $r = r_o$.

Rate of heat conduction to outer surface = Rate of heat convection from outer surface

$$-kA\left(\frac{dT}{dr}\right)_{r=r_o} = hA(T_{r=r_o} - T_\infty) \quad ...(4.28)$$

or
$$-k\left[-\frac{g_o r}{2k}\right]_{r=r_o} = h\left[\left\{-\frac{g_o r^2}{4k} + C_2\right\}_{r=r_o} - T_\infty\right]$$

or
$$\frac{g_o r_o}{2h} = -\frac{g_o r_o^2}{4k} - T_\infty + C_2$$

It gives $\quad C_2 = \frac{g_o r_o^2}{4k} + \frac{g_o r_o}{2h} + T_\infty \quad ...(ii)$

Using in eqn. (i), we get temperature distribution as

$$T(r) = \frac{g_o}{4k}(r_o^2 - r^2) + \frac{g_o r_o}{2h} + T_\infty$$

or
$$T(r) - T_\infty = \frac{g_o}{4k}(r_o^2 - r^2) + \frac{g_o r_o}{2h} \quad ...(4.29)$$

The surface temperature

$$T_s = T_\infty + \frac{g_o r_o}{2h} \quad ...(4.30)$$

The maximum temperature occurs at the centre and the temperature at the centre *i.e.*, at $r = 0$

$$T_c - T_\infty = \frac{g_o r_o^2}{4k} + \frac{g_o r_o}{2h} \quad ...(4.31)$$

4.3. HOLLOW CYLINDER WITH HEAT GENERATION AND SPECIFIED SURFACE TEMPERATURES

Consider a hollow cylinder as shown in Fig. 4.15, subjected to boundary conditions :

At $r = r_i$, $\quad T = T_1$

and at $r = r_o$, $\quad T = T_2$

Eqn. (4.21) for temperature distribution gives

$$T_1 = -\frac{g_o r_i^2}{4k} + C_1 \ln(r_i) + C_2 \quad ...(i)$$

$$T_2 = -\frac{g_o r_o^2}{4k} + C_1 \ln(r_o) + C_2 \quad ...(ii)$$

Solving for constants C_1 and C_2, eqn. (i) – eqn. (ii)

$$T_1 - T_2 = \frac{g_o}{4k}(r_o^2 - r_i^2) + C_1 \ln\left(\frac{r_i}{r_o}\right)$$

or
$$C_1 = \frac{(T_1 - T_2) - \frac{g_o}{4k}(r_o^2 - r_i^2)}{\ln(r_i/r_o)}$$

or $$C_1 = \frac{1}{\ln(r_o/r_i)} \left\{ \frac{g_o}{4k}(r_o^2 - r_i^2) + (T_2 - T_1) \right\}$$

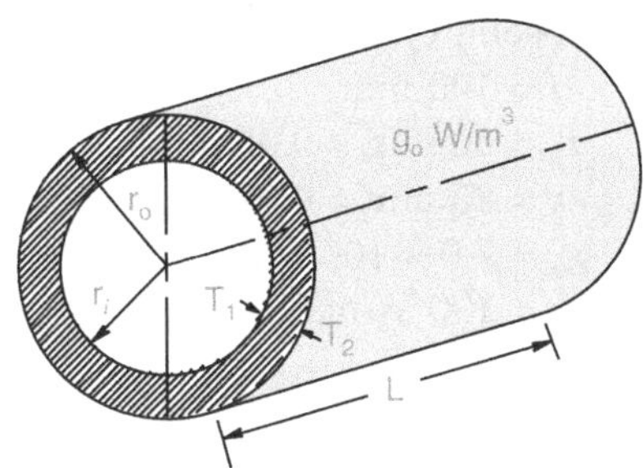

Fig. 4.15. Hollow cylinder with specified temperatures

Using in eqn. (*ii*), we get

$$C_2 = T_2 + \frac{g_o r_o^2}{4k} - \frac{\ln(r_o)}{\ln(r_o/r_i)} \times \left[\frac{g_o}{4k}(r_o^2 - r_i^2) + (T_2 - T_1) \right]$$

$$= \frac{g_o r_o^2}{4k} + \frac{\ln(1/r_o)}{\ln(r_o/r_i)} \times \left[\frac{g_o}{4k}(r_o^2 - r_i^2) + (T_2 - T_1) \right] + T_2$$

Substituting C_1 and C_2 in eqn. (4.21), we get temperature distribution as

$$T(r) = \frac{g_o}{4k}(r_o^2 - r^2) + \frac{\ln(r/r_o)}{\ln(r_o/r_i)} \times \left[\frac{g_o}{4k}(r_o^2 - r_i^2) + (T_2 - T_1) \right] + T_2 \quad ...(4.32)$$

4.3.1. Hollow Cylinder Insulated at its Inner Surface

Consider a hollow cylinder, insulated at its inner surface ($r = r_i$) has internal heat generation and dissipates heat from its outer surface ($r = r_o$) to convection environment at temperature T_∞ with convection coefficient h as shown in Fig. 4.16.

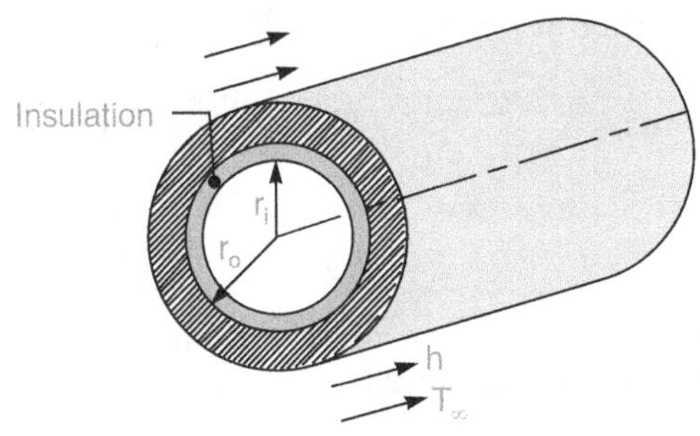

Fig. 4.16. Hollow cylinder insulated on its inner surface

The boundary conditions are :

(1) At $r = r_i$, $$\left\{ \frac{dT}{dr} \right\}_{r=r_i} = 0$$

(2) At $r = r_o$, $$-k\left\{ \frac{dT}{dr} \right\}_{r=r_o} = h\{T(r_o) - T_\infty\}$$

The relation for temperature distribution in cylinder is given by eqn. (4.21)

$$T(r) = -\frac{g_o r^2}{4k} + C_1 \ln(r) + C_2$$

and $$\frac{dT}{dr} = -\frac{g_o r}{2k} + \frac{C_1}{r}$$

Using boundary condition at $r = r_i$

$$\left\{ \frac{dT}{dr} \right\} = -\frac{g_o r_i}{2k} + \frac{C_1}{r_i} = 0 \quad ...(4.33)$$

It gives, $$C_1 = \frac{g_o r_i^2}{2k} \quad ...(4.34)$$

Now, $$\frac{dT}{dr} = -\frac{g_o r}{2k} + \frac{g_o r_i^2}{2kr}$$

Using it with boundary condition at $r = r_o$

$$-k\left\{ -\frac{g_o r}{2k} + \frac{g_o r_i^2}{2kr} \right\}_{r=r_o} = h\left\{ -\frac{g_o r^2}{4k} + \frac{g_o r_i^2}{2k}\ln(r) + C_2 - T_\infty \right\}_{r=r_o}$$

or $$\frac{g_o r_o}{2h} - \frac{g_o r_i^2}{2h\, r_o} = -\frac{g_o r_o^2}{4k} + \frac{g_o r_i^2}{2k}\ln(r_o) + C_2 - T_\infty$$

or $$C_2 = \frac{g_o r_o^2}{4k} - \frac{g_o r_i^2}{2r_o h} + \frac{g_o r_o}{2h} - \frac{g_o r_i^2}{2k}\ln(r_o) + T_\infty \quad ...(4.35)$$

Introducing the C_1 and C_2 in eqn. (4.21)

$$T(r) = -\frac{g_o r^2}{4k} + \frac{g_o r_i^2}{2k}\ln(r) + \frac{g_o r_o^2}{4k} - \frac{g_o r_i^2}{2r_o h} - \frac{g_o r_i^2}{2k}\ln(r_o) + \frac{g_o r_o}{2h} + T_\infty$$

or $$T(r) - T_\infty = \frac{g_o r_o^2}{4k}\left\{ 1 - \frac{r^2}{r_o^2} \right\} + \frac{g_o r_i^2}{2k}\ln\left(\frac{r}{r_o} \right) + \frac{g_o r_o}{2h}\left\{ 1 - \frac{r_i^2}{r_o^2} \right\} \quad ...(4.36)$$

Note : These are some cases, we have discussed as examples, but the heat conduction problems with heat generation are worked out according to prescribed boundary condition within the problem. As the boundary condition changes, the equations for temperature distribution and heat transfer rate take some new form. Therefore, the resulting equations for temperature distribution obtained above cannot be used as standard relations. The students are advised to proceed always with basic equations.

4.3.2. The Location of Maximum Temperature in the Cylinder

The location of maximum temperature in cylinder of given boundary conditions can be obtained by applying condition of maxima *i.e.,* differentiating the relation for temperature distribution T(*r*) with respect to directional coordinate, *r* and equating it to zero.

For cylinder, $$\frac{dT(r)}{dr} = 0 \quad ...(4.37)$$

We get the location r_{cr}, where temperature would be maximum. Hence the maximum temperature T_{max} can be obtained by using this value of r_{cr} in equation of temperature distribution T(*r*).

Example 4.12. *A 2 kW resistance heater wire (k = 15 W/m.K) has its diameter 4 mm and length 0.5 m, is used to boil water. If the outer surface of the wire is 105°C. Calculate the centre line temperature of wire.*

Solution

Given : Q_g = 2 kW = 2000 W,
k = 15 W/m.K
d = 4 mm or r = 2 mm = 0.002 m,
L = 0.5 m
T_s = 105°C.

To find : Centre temperature of water.

Analysis : The uniform heat generation per unit volume

$$g_o = \frac{Q_g}{V} = \frac{Q_g}{\pi r^2 L}$$
$$= \frac{2000 \text{ W}}{\pi \times (0.002 \text{ m})^2 \times (0.5 \text{ m})}$$
$$= 0.318 \times 10^9 \text{ W/m}^3$$

Then the centre line temperature of solid cylinder, eqn. (4.27)

$$T_c = T_s + \frac{g_o r_o^2}{4k}$$
$$= 105°\text{C} + \frac{(0.318 \times 10^9 \text{ W/m}^3) \times (0.002 \text{ m})^2}{4 \times (15 \text{ W/m.K})}$$
$$= \mathbf{126.2°C.} \quad \text{Ans.}$$

Example 4.13. *A solid cylinder, 100 mm in diameter generating heat at a uniform rate of 7 × 10⁶ W/m³. The thermal conductivity of solid is 190 W/m.K and its surface temperature is maintained at 100°C. Calculate*

(i) Temperature at the centre of cylinder.

(ii) Temperature at the distance 25 mm from the centre.

(iii) Temperature gradient at 25 mm radius.

(iv) Heat flux at the surface.

Solution

Given : A solid cylinder with heat generation
d = 100 mm
r_o = 50 mm = 0.05 m
r = 25 mm = 0.025 m
$g_o = 7.0 \times 10^6$ W/m³
k = 190 W/m.K
T_s = 100°C.

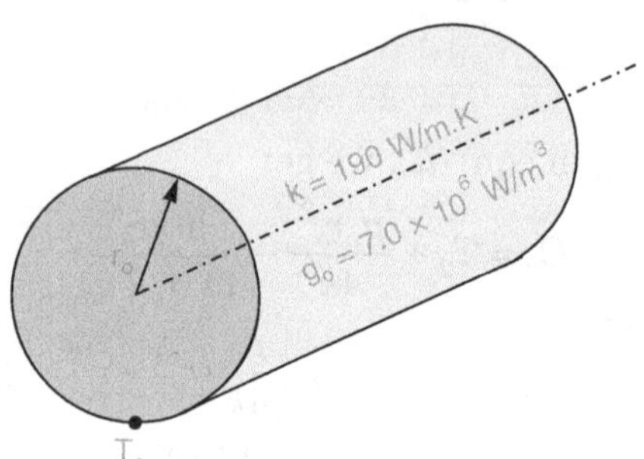

Fig. 4.17. Solid cylinder

To find :

(*i*) Centre line temperature of cylinder.

(*ii*) Temperature at mid radius (r = 0.025 m)

(*iii*) Temperature gradient at r = 0.025 m.

(*iv*) Heat flux at outer surface.

Analysis : The solid cylinder is with specified surface temperature; the temperature distribution in cylinder is given by

$$T(r) = -\frac{g_o r^2}{4k} + C_1 \ln(r) + C_2$$

Subjected to boundary conditions

At $r = 0$, $\quad \frac{dT}{dr} = 0$

It gives $\quad C_1 = 0$

Further at $r = r_o$, $\quad T = T_s$

Thus $$T_s = -\frac{g_o r_o^2}{4k} + C_2$$

or $$C_2 = T_s + \frac{g_o r_o^2}{4k}$$

This temperature distributions obtained is as already given by eqn. (4.23)

$$T(r) = \frac{g_o}{4k}(r_o^2 - r^2) + T_s \quad ...(i)$$

(*i*) Temperature at the centre, ($r = 0$)

$$T_c = \frac{g_o}{4k} r_o^2 + T_s$$

$$= \frac{7.0 \times 10^6 \times (0.05)^2}{4 \times 190} + 100$$

$$= \mathbf{123°C.} \quad \text{Ans.}$$

(*ii*) Temperature at a distance of 25 mm from centre

$$T = \frac{7.0 \times 10^6}{4 \times 190} [(0.05)^2 - (0.025)^2] + 100$$

$$= \mathbf{117.27°C.} \quad \text{Ans.}$$

(*iii*) Temperature gradient at radius of 25 mm

Differentiating eqn. (*i*) w.r.t. r

$$\frac{dT}{dr} = -\frac{g_o r}{2k}$$

at $r = 0.025$ m

$$\left.\frac{dT}{dr}\right|_{r=0.025\text{ m}} = -\frac{7 \times 10^6 \times 0.025}{2 \times 190}$$

$$= \mathbf{-\ 460.5°C/m.} \quad \text{Ans.}$$

(*iv*) Heat flux at outer surface :

The temperature gradient at $r = r_o$

$$\left[\frac{dT}{dr}\right]_{r=r_o} = -\frac{g_o r_o}{2k} = -\frac{7 \times 10^6 \times (0.05)}{2 \times 190}$$

$$= 921.05°C/m$$

$$\text{Heat flux } q_{r=r_o} = -k\left[\frac{dT}{dr}\right]_{r=r_o}$$

$$= -190 \times 921.05$$

$$= 175 \times 10^3 \text{ W/m}^2$$

$$= \mathbf{175\ kW/m^2.} \quad \text{Ans.}$$

Example 4.14. *A long rod of radius 50 cm with thermal conductivity of 10 W/m.K contains radioactive material, which generates heat uniformly within the cylinder at a rate of 0.3 × 10⁵ W/m³. The rod is cooled by convection from its cylindrical surface at T_∞ = 50°C with a heat transfer coefficient of 60 W/m².K. Determine the temperature at the centre and outer surface of the cylindrical rod.* (J.N.T.U., May 2004)

Solution

Given : A nuclear reactor in form of a long, solid cylinder

$$r_o = 50 \text{ cm} = 0.5 \text{ m}$$
$$k = 10 \text{ W/m.K}$$
$$g_o = 0.3 \times 10^5 \text{ W/m}^3$$
$$T_\infty = 50°C$$
$$h = 60 \text{ W/m}^2.K$$

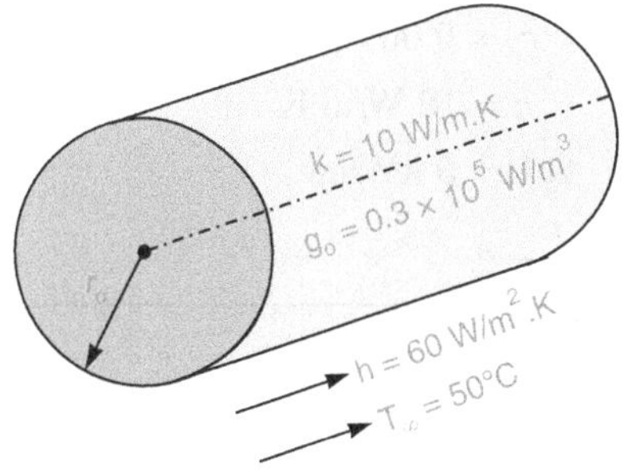

Fig. 4.18. Nuclear fuel rod exposed to coolant

To find :

(*i*) Centre temperature of the rod.

(*ii*) Temperature at the outer surface of the rod.

Assumptions :

(*i*) Steady state conditions.

(*ii*) Heat conduction in radial direction only.

(*iii*) Constant properties.

Analysis : (*i*) Temperature at the centre :

Using eqn. (4.32) for solid cylinder exposed to convection environment

$$T(r) - T_\infty = \frac{g_o(r_o^2 - r^2)}{4k} + \frac{g_o r_o}{2h}$$

At centre, *i.e.*, $r = 0$

$$T_c = T_\infty + \frac{g_o r_o^2}{4k} + \frac{g_o r_o}{2h}$$

$$= 50 + \frac{0.3 \times 10^5 \times (0.5)^2}{4 \times 10} + \frac{0.3 \times 10^5 \times 0.5}{2 \times 60}$$

$$= 50 + 187.5 + 125 = \mathbf{362.5°C.} \quad \text{Ans.}$$

(*ii*) Temperature at the outer surface.

At $r = r_o$

$$T_s = T_\infty + \frac{g_o r_o}{2h}$$

$$= 50 + \frac{0.3 \times 10^5 \times 0.5}{2 \times 60}$$

$$= \mathbf{175°C.} \quad \text{Ans.}$$

Example 4.15. *Heat is generated in a 2 mm diameter electric resistance wire (k = 10 W/m.K) uniformly at the rate of 5 kW/m length. Calculate the temperature difference between centre line and the surface of the wire.*

Solution

Given : An electric resistance wire

$$d = 2 \text{ mm} = 0.002 \text{ m}$$

or $r_o = 0.001$ m

$k = 10$ W/m.K

$Q/L = 5$ kW/m.

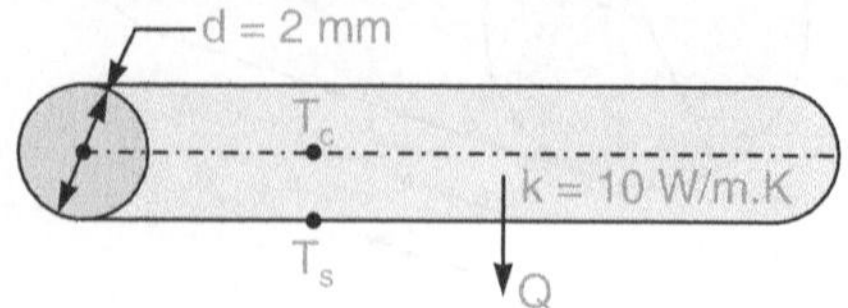

Fig. 4.19. Schematic for example 4.15

To find : Temperature difference between centre line and surface of the wire.

Analysis : The wire is treated as long solid cylinder, and the temperature distribution is given by

$$T(r) = -\frac{g_o r^2}{4k} + C_1 \ln(r) + C_2$$

Subjected to boundary conditions

At $r = 0$, $\frac{dT}{dr} = 0$ (for solid cylinder due to symmetry)

At $r = r_o$, $T = T_s$ (because $T_s - T_c$ is to be calculated)

with $$g_o = \frac{Q}{L} \times \frac{1}{A_c} = \frac{5 \times 10^3}{\frac{\pi}{4} \times (0.002)^2}$$

$$= 1.591 \times 10^9 \text{ W/m}^3$$

Using boundary conditions, we get temperature distribution as in eqn. (4.26)

$$\therefore \quad T(r) - T_s = \frac{g_o}{4k}(r_o^2 - r^2)$$

And at centre *i.e.*, $r = 0$

$$T_c - T_s = \frac{g_o}{4k} r_o^2$$

$$= \frac{1.591 \times 10^9 \times (0.001)^2}{4 \times 10}$$

$$= \mathbf{39.7°C.} \quad \text{Ans.}$$

Example 4.16. *A stainless steel rod 20 mm diameter is carrying an electric current of 1000 Amp. The thermal and electrical conductivities are 20 W/m.K and 1.25 × 10⁴ (Ω cm)⁻¹. What is the temperature at the centre of the rod, if its surface temperature should not exceed 400°C ?*

Solution

Given : $I = 1000$ Amp,

$k = 20$ W/m.K

$$k_e = \frac{1}{\rho} = 1.25 \times 10^4 \ (\Omega \text{ cm})^{-1},$$

$T_s = 400°C$

$d = 20 \text{ mm} = 20 \times 10^{-3}$ m

or $r_o = 10 \times 10^{-3}$ m

Boundary conditions,

(*i*) At $r = 0$, $\frac{dT}{dr} = 0$ $\because$ Solid cylinder, due to symmetry

(*ii*) At $r = r_o$, $T = T_s = 400°C$.

To find : Centre temperature of the rod.

Assumption : Steady state heat conduction in radial direction only.

Analysis : Since stainless steel rod can be treated as solid cylinder and its surface temperature is specified, thus eqn. (4.27) can be used for temperature distribution at the centre

$$T_c - T_s = \frac{g_o r_o^2}{4k}$$

where $$g_o = \frac{Q_g}{V} = \frac{I^2 R_e}{A_c L}$$

$$= \frac{I^2}{A_c L} \times \left\{\frac{\rho L}{A_c}\right\} = \left(\frac{I}{A_c}\right)^2 \times \left(\frac{1}{k_e}\right)$$

$$= \left\{\frac{1000}{(\pi/4) \times (20 \times 10^{-3})^2}\right\}^2 \times \left(\frac{1 \times 10^{-2} \ \Omega\text{m}}{1.25 \times 10^4}\right)$$

$$= 8.1 \times 10^6 \text{ W/m}^3$$

The temperature at the centre of the rod

$$T_c = 400 + \frac{8.1 \times 10^6 \times (10 \times 10^{-3})^2}{4 \times 20}$$

$$= 400 + 10.13 = 410.13°C$$

or $T_c = T_{max} = \mathbf{410.13°C.}$ Ans.

Example 4.17. *A nichrome wire having a resistivity of 110 μΩ cm is to be used as heating element. The wire diameter is 2 mm and other design features are*

Current, I = 25 A,

Ambient temperature, $T_\infty = 20°C$

$k_{nichrome} = 17.5$ *W/m.K,*

Convection coefficient, $h = 46.5$ *W/m².K*

Calculate the heat loss from one metre long heater and also the temperature at the surface and centre line of nichrome wire. (N.M.U., Dec. 2002)

Solution

Given : A nichrome wire heating element

$\rho = 110\ \mu\Omega\ \text{cm} = 110 \times 10^{-8}\ \Omega\text{m}$,

$d = 2\ \text{mm} = 2 \times 10^{-3}$ m,

or $r_o = 1 \times 10^{-3}$ m

$I = 25$ A,

$h = 46.5$ W/m^2.K

$k = 17.5$ W/m.K,

$T_\infty = 20°C$,

$L = 1$ m.

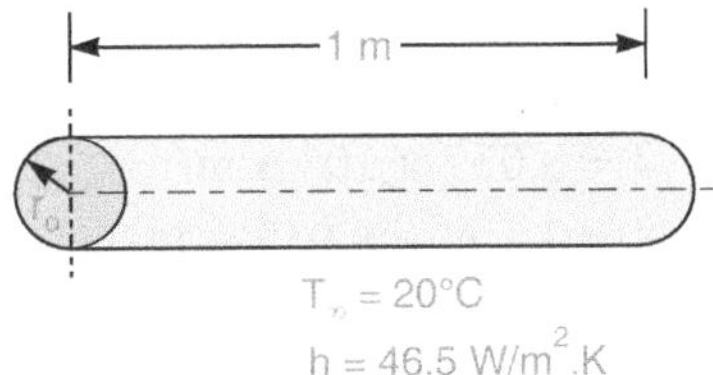

Fig. 4.20. Nichrome wire

To find :

(*i*) Heat flow rate from 1 m long wire.

(*ii*) Temperature at the surface of wire.

(*iii*) Temperature at the centre line of wire.

Assumptions :

(*i*) Steady state conditions.

(*ii*) Heat transfer in radial direction only.

(*iii*) Constant properties.

Analysis : (*i*) Heat flow rate : The resistance per metre length of wire

$$R_e = \frac{\rho L}{A_c} = \frac{(110 \times 10^{-8}\ \Omega\text{m}) \times (1\,\text{m})}{(\pi/4) \times (2 \times 10^{-3}\ \text{m})^2}$$

$$= 0.35\ \Omega/\text{m}$$

The heat generated

$$Q_g = I^2 R_e = (25\ \text{A})^2 \times (0.35\ \Omega/\text{m})$$

$$= 218.83\ \text{W/m}.$$

Under steady state conditions, heat generation rate is always equal to heat dissipation rate from the wire

$$Q_g = Q = \mathbf{218.83\ W/m.}\quad \textbf{Ans.}$$

(*ii*) Temperature at the surface of wire :

The temperature distribution in the wire (a long solid cylinder) can be expressed as

$$T(r) = -\frac{g_o r^2}{4k} + C_1 \ln(r) + C_2 \qquad ...(i)$$

Subjected to boundary conditions

At the centre of wire,

$$\left\{\frac{dT}{dr}\right\}_{r=0} = 0$$

and at $r = r_o$

$$-k\left(\frac{dT}{dr}\right)_{r=r_o} = h\,(T_{r=r_o} - T_\infty)$$

We get the temperature distribution in the wire as obtained by eqn. (4.32)

$$T(r) - T_\infty = \frac{g_o}{4k}(r_o^2 - r^2) + \frac{g_o r_o}{2h}$$

where $$g_o = \frac{Q_g}{V} = \frac{218.83}{(\pi/4) \times (2 \times 10^{-3})^2 \times 1}$$

$$= 69.655 \times 10^6\ \text{W/m}^3$$

Substituting the values for temperature distribution,

$$T(r) - 20 = \frac{69.655 \times 10^6}{4 \times 17.5} \times [(1 \times 10^{-3})^2 - r^2]$$

$$+ \frac{69.655 \times 10^6 \times (1 \times 10^{-3})}{2 \times 46.5}$$

$$T(r) = 995071.42\,[(1 \times 10^{-3})^2 - r^2] + 769$$

Temperature at the surface ($r = r_o$)

$$T_s = \mathbf{769°C.}\quad \textbf{Ans.}$$

(*iii*) Temperature at the centre ($r = 0$)

$$T_c = 995071.42\,(1 \times 10^{-3})^2 + 769$$

$$= \mathbf{770°C.}\quad \textbf{Ans.}$$

Example 4.18. *(a) Prove that the maximum temperature at the centre of wire, carrying electrical current is given by relation*

$$T_{max} = T_s + \frac{J^2}{4kk_e} r_o^2$$

where T_s *= surface temperature,*

J *= current density,*

k *= thermal conductivity of wire material,*

k_e *= electrical conductivity,*

r_o *= radius of wire.*

(b) A 3 mm dia. copper wire 10 m long is carrying electric current and has a surface temperature of 30°C.

The thermal and electrical conductivities of copper are 390 W/m.K and 5.15 × 10⁷ (Ωm)⁻¹, respectively. Calculate the voltage drop, if the temperature rise at the wire axis must not exceed 18°C.

(N.M.U., Dec. 2002 ; M.U., May 1998)

Solution

(*a*) The temperature distribution in a wire (a long solid cylinder) is given by eqn. (4.24)

$$T(r) = -\frac{g_o r^2}{4k} + C_1 \ln(r) + C_2 \qquad ...(i)$$

Wire is a solid cylinder, thus

At $r = 0$, $\frac{dT}{dr} = 0$

It gives, $C_1 = 0$

Then $T(r) = -\frac{g_o r^2}{4k} + C_2$

Given that at $r = r_o$, $T = T_s$

$$\therefore \quad T_s = -\frac{g_o r_o^2}{4k} + C_2$$

or

$$C_2 = T_s + \frac{g_o r_o}{4k}$$

Then temperature distribution

$$T(r) = \frac{g_o (r_o^2 - r^2)}{4k} + T_s \qquad ...(ii)$$

For maximum temperature occurs in the wire at $r = 0$

$$T_{max} = \frac{g_o r_o^2}{4k} + T_s \qquad ...(iii)$$

The heat generation in the wire,

$$Q_g = I^2 R_e = I^2 \frac{\rho L}{A_c} = \frac{I^2 L}{A_c} \times \frac{1}{k_e}$$

Heat generation per unit volume,

$$g_o = \frac{Q_g}{V} = \left(\frac{I^2 L}{A_c k_e}\right) \times \frac{1}{A_c L}$$

$$= \left(\frac{I}{A_c}\right)^2 \times \frac{1}{k_e} = \frac{J^2}{k_e}$$

where J = I/A, current density.

Using in eqn. (*iii*), we get

$$T_{max} = \frac{J^2 r_o^2}{4kk_e} + T_s \quad \textbf{(Proved)} \qquad ...(iv)$$

(*b*) Given $d = 3 \text{ mm} = 3 \times 10^{-3}$ m,

or $r_o = 0.0015$ m

$L = 10$ m,

$T_s = 30°C$

$k = 390$ W/m.K

$k_e = 5.15 \times 10^7\ (\Omega m)^{-1}$

$\Delta T = T_{max} - T_s = 18°C.$

To find : Voltage drop across the wire.

Analysis : The current density, eqn. (*iv*)

$$J^2 = (T_{max} - T_s) \times \frac{4kk_e}{r_o^2}$$

$$= \frac{18 \times 4 \times 390 \times 5.15 \times 10^7}{(0.0015)^2}$$

$$= 6.427 \times 10^{17}$$

or $J = 8.017 \times 10^8$ A/m²

$I = J \times A_c = 8.017 \times 10^8 \times (\pi/4) \times (0.003)^2$

$= 5666.87$ A

Voltage drop

$$\Delta V = IR_e = I\frac{L}{A_c k_e}$$

$$= \frac{(5666.87) \times 10}{(\pi/4) \times (0.003)^2 \times 5.15 \times 10^7}$$

$= \mathbf{155.67\ V.}$ **Ans.**

Example 4.19. *Calculate the maximum current that a 2 mm bare aluminium (k = 210 W/m.K) wire can carry without exceeding a temperature of 225°C, when exposed in an ambient at 25°C with heat transfer coefficient of 10 W/m².K. Take electrical resistance of aluminium wire as 0.037 Ω/m.*

Solution

Given : An electric wire exposed to ambient air

$d = 2 \text{ mm} = 0.002$ m,

or $r_o = 0.001$ m

$k = 210$ W/m.K,

$T_{max} = 225°C$

$T_\infty = 25°C$,

$h = 10$ W/m².K,

$R_e = 0.037$ Ω/m.

To find : Maximum current carrying capacity of conductor.

Analysis : Wire is a solid cylinder and exposed to convection environment thus the temperature distribution is given by eqn. (4.33)

$$T(r) = \frac{g_o}{4k}(r_o^2 - r^2) + \frac{g_o r_o}{2h} + T_\infty$$

Location of maximum temperature

$$\frac{dT}{dr} = -\frac{g_o r_{cr}}{2k} = 0$$

or $r_{cr} = 0$

Then for T_{max},

$$225 = \frac{g_o}{4 \times 210} \times (0.001)^2 + \frac{g_o \times 0.001}{2 \times 10} + 25$$

or $200 = g_o (1.19 \times 10^{-9} + 5 \times 10^{-5})$

or $g_o = 4 \times 10^6$ W/m³

Heat flow rate.

$$Q = g_o \times V = g_o \times \left(\frac{\pi}{4} d^2 \times L\right)$$

$$= 4 \times 10^6 \times (\pi/4) \times (0.002)^2 \times 1$$

$$= 12.56 \text{ W/m}$$

and $Q = I^2 R_e$

or $I^2 = \frac{12.56}{0.037} = 339.62$

or **I = 18.42 A.** Ans.

Example 4.20. *A thin hollow stainless steel tube with ID = 7.6 mm and O.D. = 8 mm is heated with a current 250 A intensity. The outer surface of the tube is insulated and all the heat generated in the tube wall is transferred through its inner surface. The specific resistance and thermal conductivity of steel are 85 μΩ cm and 18.6 W/m.K, respectively.*

Calculate :

(i) Volumetric rate of heat generation in the tube.

(ii) Temperature drop across the wall.

Solution

Given : A stainless steel, hollow tube with

$d_1 = 7.6$ mm, or $r_1 = 3.8 \times 10^{-3}$ m

$d_2 = 8$ mm or $r_2 = 4 \times 10^{-3}$ m

$I = 250$ A, $\rho = 85 \times 10^{-8}$ Ωm

$k = 18.6$ W/m.K

Boundary condition

At $r = r_2$, $\frac{dT}{dr} = 0$

To find :

(*i*) Volumetric heat generation rate in the tube.

(*ii*) Temperature drop across the tube wall.

Assumptions :

(*i*) Steady state conditions with uniform heat generation g_o W/m³.

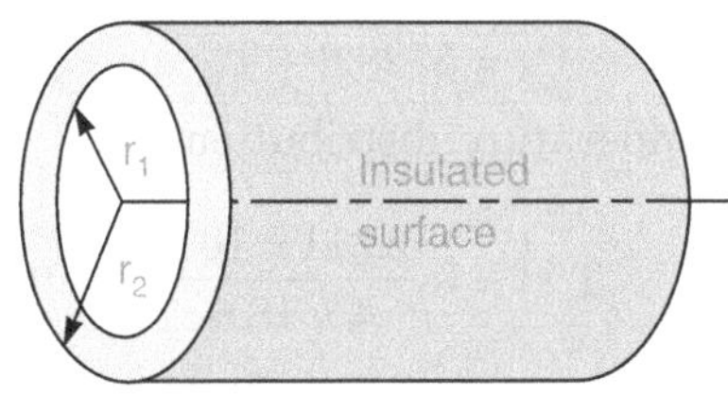

Fig. 4.21. Hollow cylinder, insulated at its outer surface

(*ii*) Heat conduction in radial direction only.

(*iii*) Constant properties.

Analysis : (*i*) The heat generation in the tube wall

$$Q_g = I^2 R_e = I^2 \frac{\rho L}{(\pi/4)(d_2^2 - d_1^2)}$$

$$= \frac{(250)^2 \times 85 \times 10^{-8} \times L}{(\pi/4) \times (8^2 - 7.6^2) \times 10^{-6}}$$

$$= 10.839 \times 10^3 \text{ L W/m.}$$

The volumetric heat generation rate

$$g_o = \frac{Q_g}{V} = \frac{Q_g}{\left(\frac{\pi}{4}\right)(d_2^2 - d_1^2) \times L}$$

$$= \frac{10.839 \times 10^3 L}{\left(\frac{\pi}{4}\right) \times (8^2 - 7.6^2) \times 10^{-6} \times L}$$

$$= \mathbf{2.211 \times 10^9 \ W/m^3.} \quad \text{Ans.}$$

(*ii*) The temperature distribution in the tube

$$T(r) = -\frac{g_o r^2}{4k} + C_1 \ln(r) + C_2$$

(*a*) At $r = r_2$; $\frac{dT}{dr} = 0$

or $$\left\{\frac{dT}{dr}\right\}_{r=r_2} = -\frac{g_o r_2}{2k} + \frac{C_1}{r_2} = 0$$

It gives

$$C_1 = \frac{g_o r_2^2}{2k} = \frac{2.211 \times 10^9 \times (4 \times 10^{-3})^2}{2 \times 18.6} = 951$$

Then $T(r) = -\frac{g_o r^2}{4k} + 951 \times \ln(r) + C_2$

(*b*) At $r = r_1$,

$T = T_1$ (say)

$$T_1 = -\frac{g_o r_1^2}{4k} + 951 \times \ln(r_1) + C_2$$

or $$C_2 = \frac{2.211 \times 10^9 \times (3.8 \times 10^{-3})^2}{4 \times 18.6} - 951 \times \ln(3.8 \times 10^{-3}) + T_1$$

It gives $C_2 = 5728.8 + T_1$

The temperature distribution

$$T(r) = -\frac{2.211\times 10^9}{4\times 18.6} r^2 + 951 \times \ln(r) + 5728.8 + T_1$$

$$T(r) - T_1 = -29.718 \times 10^6 r^2 + 951 \times \ln(r) + 5728.8$$

Temperature drop $(T_2 - T_1)$ across the wall

$$T_{r=r_2} - T_1 = -29.718\times 10^6 \times (4\times 10^{-3})^2 + 951 \times \ln(4\times 10^{-3}) + 5728.8$$

$$= \mathbf{2.40°C.} \quad \text{Ans.}$$

Example 4.21. *A chemical reaction takes place in a packed bed (k = 0.6 W/m.K) between two coaxial cylinders with radii 15 mm and 45 mm. The inner surface is at 580°C and is insulated. Assuming the reaction rate of 0.55 MW/m³ in reactor volume. Calculate the temperature at the outer surface of the reactor.*

(P.U., Nov. 2001)

Solution

Given : A chemical reactor in form of hollow cylinder

$$r_1 = 15 \text{ mm} = 0.015 \text{ m}$$

$$r_2 = 45 \text{ mm} = 0.045 \text{ m}$$

$$k = 0.6 \text{ W/m.K}$$

$$T_1 = 580°C$$

$$\left(\frac{dT}{dr}\right)_{r=r_1} = 0 \text{ (insulated inner surface)}$$

$$g_o = 0.55 \text{ MW/m}^3 = 0.55 \times 10^6 \text{ W/m}^3.$$

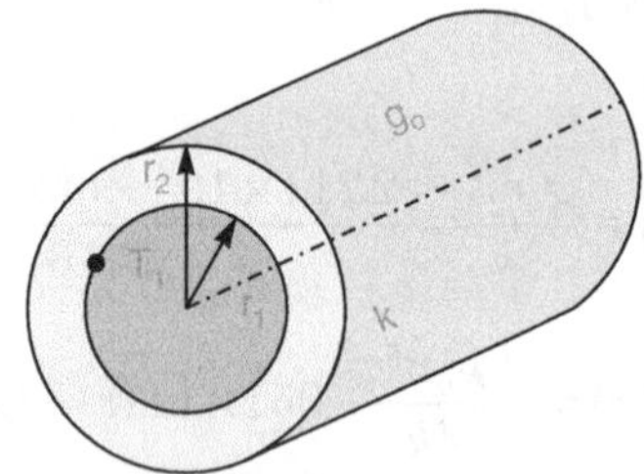

Fig. 4.22. Two coaxial cylinders, with insulated inner surface

To find : Outer surface temperature of the reactor.

Analysis : The temperature distribution in the cylinder

$$T(r) = -\frac{g_o r^2}{4k} + C_1 \ln(r) + C_2$$

Subjected to boundary conditions

At $r = r_1$, $\frac{dT}{dr} = 0$

and at $r = r_1$, $T = T_1$

Using first condition

$$\left.\frac{dT}{dr}\right|_{r=r_1} = \left[-\frac{g_o r}{2k} + \frac{C_1}{r}\right]_{r=r_1} = 0$$

or

$$C_1 = \frac{g_o r_1^2}{2k}$$

and

$$T(r) = -\frac{g_o r^2}{4k} + \frac{g_o r_1^2}{2k}\ln(r) + C_2$$

Using second boundary condition, at $r = r_1$, $T = T_1$

$$T_1 = -\frac{g_o r_1^2}{4k} + \frac{g_o r_1^2}{2k}\ln(r_1) + C_2$$

It gives $C_2 = T_1 + \frac{g_o r_1^2}{4k} - \frac{g_o r_1^2}{2k}\ln(r_1)$

Then temperature distribution becomes

$$T(r) = \frac{g_o}{4k}(r_1^2 - r^2) + \frac{g_o r_1^2}{2k}\ln\left(\frac{r}{r_1}\right) + T_1$$

The temperature at outer surface *i.e.*, $r = r_2$

$$T_2 = \frac{g_o}{4k}(r_1^2 - r_2^2) + \frac{g_o r_1^2}{2k}\ln\left(\frac{r_2}{r_1}\right) + T_1$$

$$= \frac{0.55\times 10^6}{4\times 0.6}(0.015^2 - 0.045^2) + \frac{0.55\times 10^6}{2\times 0.6}\times(0.015)^2 \ln\left(\frac{45}{15}\right) + 580$$

$$= -412.5 + 113.29 + 580$$

$$= \mathbf{280.79°C.} \quad \text{Ans.}$$

Example 4.22. *(a) A cable of radius r_1 and resistance R_e(Ω/m) and carrying a current I (A) is surrounded by an insulator of radius r_2 and thermal conductivity k. The external heat transfer coefficient and air temperature are h and T_∞, respectively. Derive an expression for the temperature distribution in the insulator.*

(b) A 1 mm dia. copper wire of resistance 0.02 Ω/m is surrounded by a 2.3 mm dia. Plastic coating of k = 0.2 W/m.K. The outside surface of the coating is cooled by air, where the convective heat transfer coefficient is 16 W/m².K. Determine the maximum current that the wire can carry, if the surface to air temperature difference should not exceed 35°C. What is the temperature rise of copper wire above ambient temperature ? (Jiwaji Univ., Dec. 2001)

Solution

(*a*) *Given* : An electrical cable with insulation.

To find : Temperature distribution in an insulator.

Analysis : The temperature distribution in the insulator (a hollow cylinder) is given by eqn. (4.24).

$$T(r) = -\frac{g_o r^2}{4k} + C_1 \ln(r) + C_2$$

But there is no heat generation in the insulator, thus

$$T(r) = C_1 \ln(r) + C_2 \quad ...(i)$$

and

$$\frac{dT}{dr} = \frac{C_1}{r} \quad ...(ii)$$

Subjected to boundary conditions

(*i*) At $r = r_1$

Heat generated in cable/m = Heat conducted into insulator/m

i.e.,
$$I^2R_e = -kA\left(\frac{dT}{dr}\right) = -k(2\pi r_1 \times 1)\left(\frac{dT}{dr}\right)$$

or
$$I^2R_e = -k(2\pi r_1)\left(\frac{C_1}{r_1}\right)$$

It gives
$$C_1 = -\frac{I^2R_e}{2\pi k} \quad ...(iii)$$

(*ii*) At $r = r_2$

Heat generated in the cable/m or heat conducted through insulator/m

= Heat convected into ambient/m

i.e.,
$$I^2R_e = hA(T_{r=r_2} - T_\infty)$$
$$= (2\pi r_2 \times 1)\, h\, (T_{r=r_2} - T_\infty)$$

or
$$I^2R_e = 2\pi r_2 h\left[-\frac{I^2R_e}{2\pi k}\ln(r_2) + C_2 - T_\infty\right]$$

It gives
$$C_2 = \frac{I^2R_e}{2\pi r_2 h} + \frac{I^2R_e}{2\pi k}\ln(r_2) + T_\infty \quad ...(iv)$$

Substituting C_1 and C_2 in eqn. (*i*),

$$T(r) = -\frac{I^2R_e}{2\pi k}\ln(r) + \frac{I^2R_e}{2\pi k}\ln(r_2) + \frac{I^2R_e}{2\pi r_2 h} + T_\infty$$

$$= \frac{I^2R_e}{2\pi k}\ln\left(\frac{r_2}{r}\right) + \frac{I^2R_e}{2\pi r_2 h} + T_\infty$$

$$= \left(\frac{I^2R_e}{2\pi}\right)\left\{\frac{\ln(r_2/r)}{k} + \frac{1}{r_2 h}\right\} + T_\infty \quad ...(v)$$

It is required expression for temperature distribution in the insulator. **Ans.**

(*b*) *Given* :

$d_1 = 1$ mm

or $r_1 = 0.5$ mm $= 0.0005$ m

$d_2 = 2.3$ mm,

or $r_2 = 1.15$ mm $= 0.00115$ m

$R_e = 0.02\ \Omega$/m

$k = 0.2$ W/m.K

$h = 16$ W/m^2.K,

$T_2 - T_\infty = 35$°C.

To find :

(*i*) Maximum current in the wire,

(*ii*) Temperature rise of copper wire above ambient temperature *i.e.*, $(T_1 - T_\infty)$.

Analysis : (*i*) Maximum current in the wire :

Using eqn. (*v*) with $r = r_2$, (at outer surface of insulator)

$$T_2 = \left(\frac{I^2R_e}{2\pi}\right)\left\{\frac{1}{r_2 h}\right\} + T_\infty$$

or
$$T_2 - T_\infty = \frac{I^2R_e}{2\pi r_2 h}$$

or
$$I^2 = \frac{2\pi r_2 h (T_2 - T_\infty)}{R_e}$$

$$= \frac{2\pi \times (0.00115\text{ m}) \times (16\text{ W/m}^2.\text{K}) \times (35°\text{C})}{(0.02\ \Omega/\text{m})}$$

$$= 202.318$$

or $I = $ **14.22 A**

(Maximum current in wire). **Ans.**

(*ii*) Temperature rise of copper wire above ambient temperature :

At outer surface of copper wire *i.e.*, $r = r_1$, $T = T_1$

$$T_1 = \frac{I^2R_e}{2\pi}\left\{\frac{\ln(r_2/r_1)}{k} + \frac{1}{r_2 h}\right\} + T_\infty$$

or
$$T_1 - T_\infty = \frac{(14.22)^2 \times 0.02}{2\pi} \times \left[\frac{\ln\left(\frac{0.00115}{0.0005}\right)}{0.2} + \frac{1}{0.00115 \times 16}\right]$$

= **37.68°C. Ans.**

Example 4.23. *A long hollow cylinder has inner and outer radii as 5 cm and 15 cm respectively. It generates the heat at the rate of 1 kW/m³. The thermal conductivity of cylinder material is 0.5 W/m.K. If the maximum temperature occurs at radius of 10 cm and temperature of outer surface is 50°C. Find*

(i) Temperature at inner surface,

(ii) Maximum temperature in the cylinder.

(P.U., Nov. 1993)

Solution

Given : $r_1 = 5$ cm = 0.05 m,

$r_2 = 15$ cm = 0.15 m

$r_{cr} = 10$ cm = 0.1 m,

$g_o = 1$ kW/m³ = 1000 W/m³

$k = 0.5$ W/m.K,

$T(r = r_2) = 50°C$.

To find :

(*i*) Temperature at inner surface,

(*ii*) Maximum temperature in the cylinder.

Assumptions :

(*i*) Steady state heat conduction in radial direction only.

(*ii*) Uniform heat generation per unit volume, g_o W/m³.

(*iii*) Constant properties.

Analysis : The temperature distribution in the cylinder

$$T(r) = -\frac{g_o r^2}{4k} + C_1 \ln(r) + C_2 \quad ...(i)$$

With slope $\dfrac{dT(r)}{dr} = -\dfrac{g_o r}{2k} + \dfrac{C_1}{r}$ ...(*ii*)

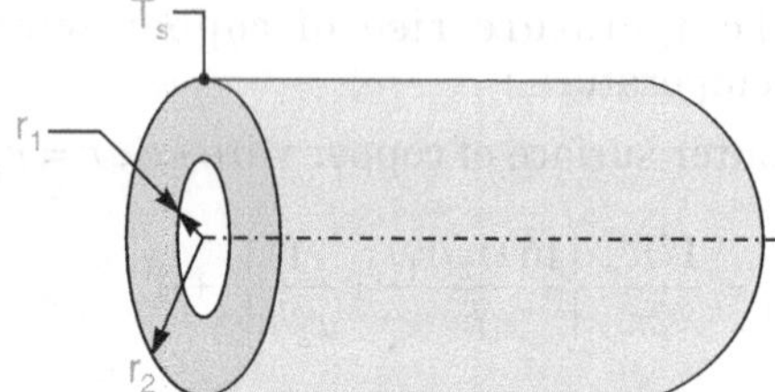

Fig. 4.23. Hollow cylinder

Since, the location of maximum temperature is given, therefore, applying condition of maxima, *i.e.*,

$$\frac{dT}{dr} = -\frac{g_o r_{cr}}{2k} + \frac{C_1}{r_{cr}} = 0$$

or $$-\frac{1000 \times 0.1}{2 \times 0.5} + \frac{C_1}{0.1} = 0$$

It gives $C_1 = 10$

Further at $r = 0.15$ m, T = 50°C, applying, we get

$$50 = -\frac{1000 \times (0.15)^2}{4 \times 0.5} + 10 \times \ln(0.15) + C_2$$

It gives $C_2 = 80.22$

Substituting C_1 and C_2 in eqn. (*i*) for temperature distribution

$$T(r) = -\frac{1000 r^2}{4 \times 0.5} + 10 \ln(r) + 80.22$$

or $$T(r) = -500 r^2 + 10 \ln(r) + 80.22$$

(*i*) The temperature at inner surface *i.e.*, at $r_1 = 0.05$ m

$$T_{r=r_1} = -500 \times (0.05)^2 + 10 \times \ln(0.05) + 80.22$$

= **49°C. Ans.**

(*ii*) Maximum temperature in the cylinder *i.e.*, at $r = 0.1$ m

$$T_{max} = -500 \times (0.1)^2 + 10 \ln(0.1) + 80.22$$

= **52.2°C. Ans.**

Example 4.24. *In a cylindrical fuel element for a gas cooled nuclear reactor, the energy generation can be approximated by*

$$g(r) = g_o \left\{1 - \left(\frac{r}{r_o}\right)^2\right\} W/m^3$$

where r_o is the radius of the fuel element and g_o is constant. The outer surface is maintained at constant temperature T_s.

(a) For the radius of 1 cm, thermal conductivity of 10 W/m.K and $g_o = 1.6 \times 10^8$ W/m³. Calculate the temperature drop from centre line to surface.

(b) If the heat removal rate from the outer surface of nuclear reactor 1.6×10^5 W/m², what would be the temperature drop from centre to surface?

Solution

Given : A nuclear fuel rod with heat generation rate of

$$g(r) = g_o \left\{1 - \left(\frac{r}{r_o}\right)^2\right\} \text{W/m}^3$$

(*a*) $g_o = 1.6 \times 10^8$ W/m³, $r_o = 1$ cm = 0.01 m,

$k = 10$ W/m.K.

(*b*) $q = 1.6 \times 10^5$ W/m².

To find :

(*i*) Temperature drop from centre to surface for given $g_o = 1.6 \times 10^8$ W/m³.

(*ii*) Temperature drop from centre to surface for $q = 1.6 \times 10^5$ W/m^2.

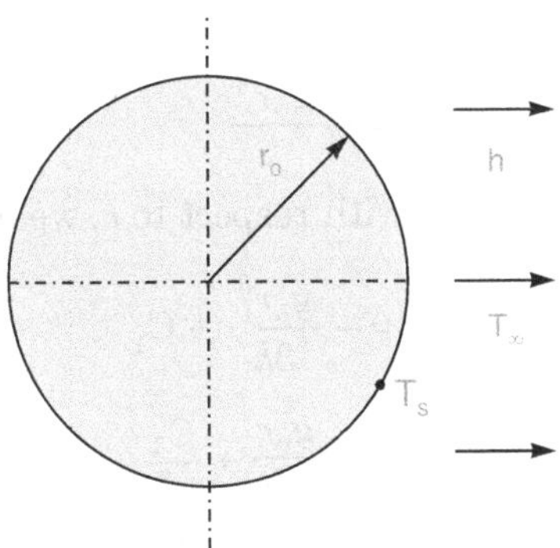

Fig. 4.24. A solid nuclear fuel rod as a solid cylinder

Assumptions :

(*i*) Steady state conditions with constant properties.

(*ii*) Heat transfer in radial direction only.

Analysis : (*a*) The governing differential equation for cylinder

$$\frac{1}{r}\frac{d}{dr}\left\{r\frac{dT}{dr}\right\} + \frac{g(r)}{k} = 0$$

Using $\qquad g(r) = g_o\,[1 - (r/r_o)^2]$

Then
$$\frac{d}{dr}\left\{r\frac{dT}{dr}\right\} = -\frac{g_o r}{k}\left[1 - \left(\frac{r}{r_o}\right)^2\right]$$

or
$$\frac{d}{dr}\left\{r\frac{dT}{dr}\right\} = -\frac{g_o}{k}\left[r - \frac{r^3}{r_o^2}\right]$$

Integrating with respect to r

$$r\left\{\frac{dT}{dr}\right\} = -\frac{g_o}{k}\left\{\frac{r^2}{2} - \frac{r^4}{4r_o^2}\right\} + C_1$$

or
$$\frac{dT}{dr} = -\frac{g_o}{k}\left\{\frac{r}{2} - \frac{r^3}{4r_o^2}\right\} + \frac{C_1}{r}$$

Integrating again, we get

$$T(r) = -\frac{g_o}{k}\left\{\frac{r^2}{4} - \frac{r^4}{16r_o^2}\right\} + C_1\ln(r) + C_2$$

Subject to boundary conditions

(*i*) At the centre, $r = 0$; $\quad \frac{dT}{dr} = 0,$

It gives $\qquad C_1 = 0$

Now, the temperature distribution becomes,

$$T(r) = -\frac{g_o}{k}\left\{\frac{r^2}{4} - \frac{r^4}{16r_o^2}\right\} + C_2$$

Using the boundary condition at the surface,

(*ii*) At $r = r_o$; $\quad T = T_s$

Then,
$$T_s = -\frac{g_o}{k}\left\{\frac{r_o^2}{4} - \frac{r_o^4}{16r_o^2}\right\} + C_2$$

It gives
$$C_2 = T_s + \frac{3}{16}\frac{g_o r_o^2}{k}$$

Substituting C_1 and C_2, the temperature distribution becomes

$$T(r) = -\frac{g_o}{k}\left\{\frac{r^2}{4} - \frac{r^4}{16r_o^2}\right\} + \frac{3}{16}\frac{g_o r_o^2}{k} + T_s$$

The maximum temperature occurs at the centre (at $r = 0$), we get

$$T_{max} = T_c = \frac{3}{16}\frac{g_o r_o^2}{k} + T_s$$

Temperature drop from centre line to surface

$$T_c - T_s = \frac{3}{16}\frac{g_o r_o^2}{k}$$

Using numerical values in above equation

$$T_c - T_s = \frac{3}{16} \times \frac{1.6 \times 10^8 \times (0.01)^2}{10}$$

= 300°C. Ans.

(*b*) For heat removal rate $q = 1.6 \times 10^5$ W/m^2

The heat removal rate can be expressed as

$$Q = 2\pi r_o\,L\,q = \int_0^{r_o} g(r)\,(2\pi r L)\,dr$$

$$r_o q = \int_0^{r_o} g_o\left[1 - \left(\frac{r}{r_o}\right)^2\right] r dr$$

$$= g_o \int_0^{r_o}\left[r - \frac{r^3}{r_o^2}\right] dr$$

or
$$r_o q = g_o\left[\frac{r_o^2}{2} - \frac{r_o^4}{4r_o^2}\right] = \frac{g_o r_o^2}{4}$$

or
$$q = \frac{g_o r_o}{4}$$

or
$$g_o = \frac{4q}{r_o} = \frac{4 \times 1.6 \times 10^5}{0.01} = 6.4 \times 10^7 \text{ W/m}^3$$

Then temperature drop from centre to surface

$$T_c - T_s = \frac{3}{16}\frac{g_o r_o^2}{k}$$

$$= \frac{3}{16} \times \frac{6.4 \times 10^7 \times (0.01)^2}{10}$$

= 120°C. Ans.

Example 4.25. *Consider a thorium (k = 54 W/m.K) fuel rod, 20 mm in diameter, has a thin aluminium (k = 237 W/m.K) cladding 2 mm in thickness. The aluminium looses its mechanical strength above temperature 427°C. If the rod is exposed to a fluid at 90°C with h = 6000 W/m². K. Is the system safe, if the heat generation rate in the thorium rod is 4 × 10⁸ W/m³ ?*

Solution

Given : A nuclear fuel rod contains thorium

d = 20 mm, r_o = 0.01 m

$g_o = 4 \times 10^8$ W/m³, r_{cl} = 0.012 m

T_∞ = 90°C, h = 6000 W/m².K

k_{cl} = 237 W/m.K $T_{s,f}$ = 427°C

k_f = 54 W/m.K

To find : Feasibility of system for its safety.

Analysis : The total amount of heat generated per unit length in the rod

$$Q = g_o (\pi r_o^2 \times 1 \text{ m})$$
$$= 4 \times 10^8 \times \pi \times (0.01)^2 \times 1$$
$$= 125.663 \times 10^3 \text{ W/m}$$

This heat must be dissipated across the aluminium cladding, therefore, for surface temperature of fuel rod

$$Q = \frac{2\pi L(T_{s,f} - T_\infty)}{\frac{\ln\left(\frac{r_{cl}}{r_o}\right)}{k_{cl}} + \frac{1}{r_{cl} h}}$$

$$125.663 \times 10^3 = \frac{2\pi \times 1 \times (T_{s,f} - 90)}{\frac{\ln\left(\frac{0.012}{0.01}\right)}{237} + \frac{1}{0.012 \times 6000}}$$

or
$$T_{s,f} = 90 + \frac{125.663 \times 10^3}{2\pi} \times 0.01466$$
$$= 383.1°C$$

It is the surface temperature of fuel rod, on which the aluminium coating is applied, it is well below the working temperature of aluminium thus it is safe. **Ans.**

4.4. THE SPHERE

The one dimensional steady state temperature distribution T(r) in a sphere in which energy is generated at a rate of $g(r)$ W/m³ is given by Poisson's equation (2.16)

$$\frac{1}{r^2}\frac{d}{dr}\left[r^2 \frac{dT(r)}{dr}\right] + \frac{g(r)}{k} = 0$$

Rearranging the above equation, assuming uniform heat generation g_o and constant thermal conductivity k ;

$$\frac{d}{dr}\left[r^2 \frac{dT(r)}{dr}\right] = -\frac{g_o r^2}{k}$$

Integrating it with respect to r, we get

$$r^2 \frac{dT(r)}{dr} = -\frac{g_o r^3}{3k} + C_1$$

or
$$\frac{dT(r)}{dr} = -\frac{g_o r}{3k} + \frac{C_1}{r^2} \quad ...(4.38)$$

Integrating further, we get

$$T(r) = -\frac{g_o r^2}{6k} - \frac{C_1}{r} + C_2 \quad ...(4.39)$$

where C_1 and C_2 are constants of integration and are evaluated according to boundary conditions.

4.4.1. Solid Sphere with Convective Boundary

Consider a solid sphere exposed to an ambient at temperature, T_∞ with convection coefficient, h as shown in Fig. 4.25. The boundary condition at the centre can be defined as in solid cylinder as :

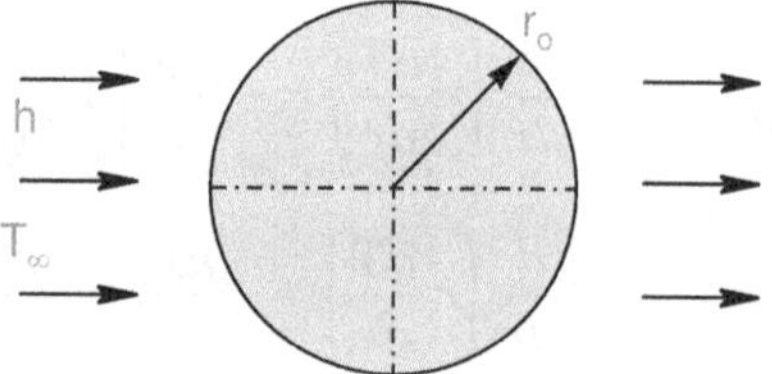

Fig. 4.25. Solid sphere with convection environment

At $r = 0$; T(r) = finite

Thus
$$\frac{dT(r)}{dr} = 0$$

Substituting in equation (4.38), we get

$$C_1 = 0$$

Hence
$$\frac{dT(r)}{dr} = -\frac{g_o r}{3k} \quad ...(4.40)$$

And the boundary condition at outer surface :

At $r = r_o$;
$$-k\left\{\frac{dT(r)}{dr}\right\} = h[T(r) - T_\infty]$$

Substituting the values of $\frac{dT(r)}{dr}$ and T(r) at $r = r_o$ from the equations (4.40) and (4.39), respectively, we get

$$-k\left\{-\frac{g_o r_o}{3k}\right\} = h\left\{-\frac{g_o r_o^2}{6k} + C_2 - T_\infty\right\}$$

It gives $C_2 = \frac{g_o r_o}{3h} + \frac{g_o r_o^2}{6k} + T_\infty$ ...(4.41)

Using the values of C_1 and C_2 in eqn. (4.39), we get temperature distribution in solid sphere, exposed to convection environment at outer surface

$$T(r) = \frac{g_o}{6k}(r_o^2 - r^2) + \frac{g_o r_o}{3h} + T_\infty \quad ...(4.42)$$

4.4.2. Solid Sphere with Specified Surface Temperature

If the outer surface of a solid sphere is subjected to constant temperature T_s as shown in Fig. 4.26, then boundary conditions become.

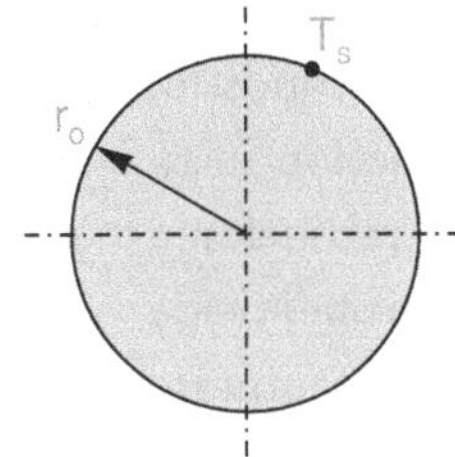

Fig. 4.26. Solid sphere with specified temperature at outer surface

(*i*) At centre, $r = 0$; $\frac{dT}{dr} = 0$

and (*ii*) At surface $r = r_o$; $T = T_s$

The first boundary condition gives

$$C_1 = 0$$

and with second boundary condition in eqn. (4.39)

$$T_s = -\frac{g_o r_o^2}{6k} + C_2$$

It gives $C_2 = T_s + \frac{g_o r_o^2}{6k}$

Substituting C_1 and C_2 in eqn. (4.39), we get

$$T(r) = \frac{g_o}{6k}(r_o^2 - r^2) + T_s \quad ...(4.43)$$

The other boundary conditions may be used as explained in case of cylinder and plane wall.

Once the temperature distribution $T(r)$ is known, the heat flux $q(r)$, anywhere in the sphere can be determined as

$$q(r) = -k\frac{dT(r)}{dr} \text{ W/m}^2. \quad ...(4.44)$$

Example 4.26. *The inner and outer surface of hollow sphere are maintained at temperature T_1 and T_2, respectively. The inner and outer radii are r_1 and r_2, respectively. The thermal conductivity k of the sphere material is constant.*

(*i*) *Develop an expression for one dimensional steady state temperature distribution in the sphere.*

(*ii*) *Develop an expression for radial heat flow rate through the hollow sphere.*

(*iii*) *Develop an expression for thermal resistance of hollow sphere.*

Solution

(*i*) The governing differential equation (Poisson equation),

$$\frac{1}{r^2}\frac{d}{dr}\left[r^2\frac{dT(r)}{dr}\right] + \frac{g(r)}{k} = 0$$

Here no heat generated in the solid *i.e.*,

$$g(r) = 0$$

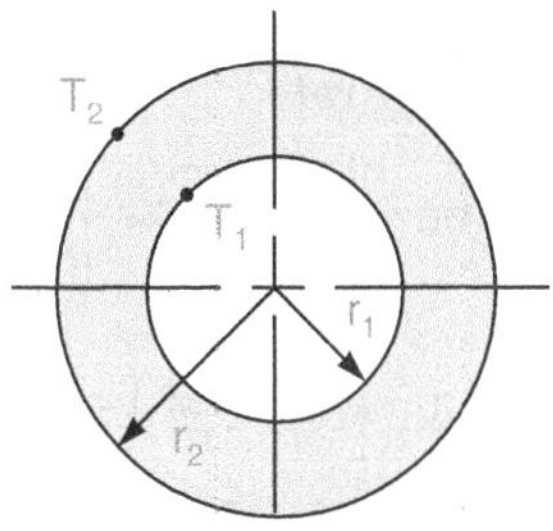

Fig. 4.27. A hollow sphere subjected fixed temperatures T_1 at inner surface and T_2 at outer surface

Then the above equation is reduced to

$$\frac{1}{r^2}\frac{d}{dr}\left[r^2\frac{dT(r)}{dr}\right] = 0 \quad ...(i)$$

Subjected to the boundary conditions

(*i*) At $r = r_1$, $T = T_1$

(*ii*) At $r = r_2$, $T = T_2$

The first and second integration of differential eqn. (*i*) gives

$$\frac{dT}{dr} = \frac{C_1}{r^2}$$

$$T(r) = -\frac{C_1}{r} + C_2$$

Applying the boundary conditions

$$T_1 = -\frac{C_1}{r_1} + C_2$$

$$T_2 = -\frac{C_1}{r_2} + C_2$$

Solution of these two simultaneous equations leads to

$$C_1 = -\frac{r_1 r_2}{r_2 - r_1}(T_1 - T_2)$$

$$C_2 = \frac{r_2 T_2 - r_1 T_2}{r_2 - r_1}$$

Then the temperature distribution T(r) becomes,

$$T(r) = \frac{r_1 r_2 (T_1 - T_2)}{r(r_2 - r_1)} + \frac{r_2 T_2 - r_1 T_1}{r_2 - r_1}$$

$$= \frac{r_1 r_2 T_1 - r_1 r_2 T_2 + r r_2 T_2 - r r_1 T_1}{r(r_2 - r_1)}$$

$$= \frac{r_1 (r_2 - r) T_1 + r_2 (r - r_1) T_2}{r(r_2 - r_1)}$$

or $$T(r) = \frac{\mathbf{r_1}\ \mathbf{(r_2 - r)}}{\mathbf{r}\ \mathbf{(r_2 - r_1)}} \mathbf{T_1} + \frac{\mathbf{r_2}\ \mathbf{(r - r_1)}}{\mathbf{r}\ \mathbf{(r_2 - r_1)}} \mathbf{T_2}.$$

Ans.

(*ii*) The heat flow rate

$$Q = 4\pi r^2 \left\{ -k \left[\frac{dT(r)}{dr} \right] \right\}$$

$$= 4\pi r^2 \left[-k \times \frac{C_1}{r^2} \right] = -\mathbf{4\pi k C_1}.$$ Ans.

(*iii*) $$Q = -4\pi k \left[-\frac{r_1 r_2 (T_1 - T_2)}{r_2 - r_1} \right]$$

or $$Q = \frac{\mathbf{4\pi k\ r_1 r_2 (T_1 - T_2)}}{\mathbf{r_2 - r_1}}.$$ Ans.

Example 4.27. *A solid sphere (k = 39 W/m.K) 10 cm in diameter generates heat at a uniform rate of 5 × 10⁶ W/m³. The outer surface of sphere is exposed to an ambient at 50°C with heat transfer coefficient of 400 W/m².K. Calculate :*

(*i*) *Maximum temperature in solid and its location*

(*ii*) *Temperature at the radius of 3 cm.*

Solution

Given : A solid sphere with heat generation and exposed to an ambient

$k = 39$ W/m.K

$d = 10$ cm, $r_o = 5$ cm $= 0.05$ m

$g_o = 5 \times 10^6$ W/m³

$T_\infty = 50$°C

$h = 400$ W/m².K

$r = 0.03$ m

To find :

(*i*) Location and magnitude of maximum temperature in the sphere.

(*ii*) Temperature at radius 3 cm.

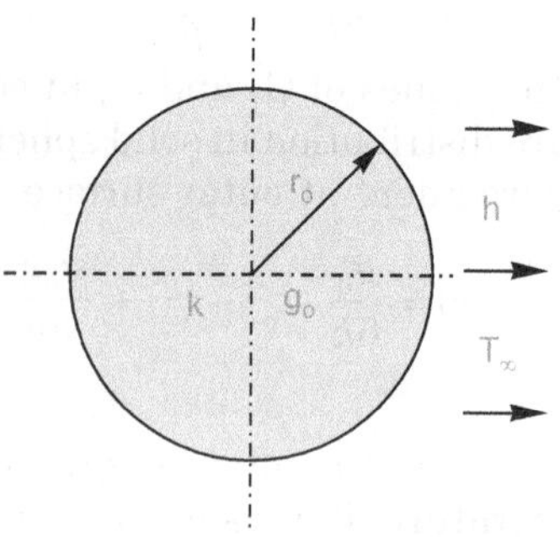

Fig. 4.28. Solid sphere for example 4.27

Assumptions :

(*i*) Steady state conditions.

(*ii*) Heat conduction in radial direction only.

(*iii*) Negligible radiation.

(*iv*) Constant properties.

Analysis : (*i*) The temperature distribution in solid sphere exposed to convective boundary is given by eqn. (4.52)

$$T(r) = \frac{g_o}{6k}(r_o^2 - r^2) + \frac{g_o r_o}{3h} + T_\infty$$

Differentiating w.r.t. r and equating to zero, we get $r_{cr} = 0$

Thus the maximum temperature occurs at the centre. Ans.

Maximum temperature

$$T_{max} = T_c = \frac{5 \times 10^6 \times (0.05)^2}{6 \times 39} + \frac{5 \times 10^6 \times (0.05)}{3 \times 400} + 50$$

$= \mathbf{311.75°C.}$ Ans.

(*ii*) Temperature at radius 0.03 m

$$T = \frac{5 \times 10^6}{6 \times 39}(0.05^2 - 0.03^2) + \frac{5 \times 10^6}{3 \times 400} \times 0.05 + 50$$

$= \mathbf{292.52°C.}$ Ans.

Example 4.28. *During the ripening process of oranges, the energy released is estimated as 563 W/m³. If the orange is assumed to be homogeneous sphere with k = 0.15 W/m.K. Compute the temperature at the centre of orange and the heat flow from the outer surface. Assume a diameter of 8 cm and outer surface temperature of 2°C.* (P.U., Dec. 2009)

Solution

Given : Orange as sphere

$d = 8 \text{ cm} = 0.08 \text{ m}$

or $r = 4 \times 10^{-2}$

$g_o = 563 \text{ W/m}^3$,

$k = 0.15 \text{ W/m.K}$

$T_s = 2°C$.

Fig. 4.29. Orange

To find :

(*i*) Centre temperature of orange.

(*ii*) Heat flow rate from outer surface of orange

Assumptions :

(*i*) Steady state conduction in radial direction only.

(*ii*) Uniform heat generation as g_o, W/m³.

(*iii*) Constant properties.

Analysis : (*i*) The temperature distribution in a sphere with uniform heat generation is given by eqn. (4.49)

$$T(r) = -\frac{g_o r^2}{6k} - \frac{C_1}{r} + C_2$$

where C_1 and C_2 are constants of integration and can be obtained from boundary conditions.

(*a*) The boundary condition at the centre

At $r = 0$, $\quad \frac{dT}{dr} = 0$

It gives $\quad C_1 = 0$

(*b*) The boundary condition at the surface

At $r = r_o$, $T = T_s$

$$T_s = -\frac{g_o r_o^2}{6k} + C_2$$

or

$$C_2 = T_s + \frac{g_o r_o^2}{6k}$$

Substituting the values in order to evaluate C_2

or

$$C_2 = 2 + \frac{563 \times (4 \times 10^{-2})^2}{6 \times 0.15} = 2 + 1 = 3$$

Using the values of C_1 and C_2 for temperature distribution in orange

$$T(r) = -\frac{g_o r^2}{6k} + 3$$

And temperature at the centre ($r = 0$)

$T_c = T(r = 0) = $ **3°C.** Ans.

(*ii*) Heat flow

$$Q = \int_0^{r_o} g_o A(r)\, dr = g_o\, 4\pi \int_0^{r_o} r^2\, dr$$

$$= \frac{4\pi g_o}{3}(r_o^3 - 0)$$

$$Q = \frac{4\pi \times 563}{3} \times (4 \times 10^{-2})^3$$

$= $ **0.1509 W.** Ans.

Example 4.29. *A solid sphere of radius r_o is generating heat uniformly at the rate of g_o W/m³. The thermal conductivity of solid is given by*

$$k = k_o (1 + \alpha T)$$

Assuming surrounding temperature as T_∞ and heat transfer coefficient as h, prove that the temperature distribution in sphere is given by

$$T = -\frac{1}{\alpha} \pm \sqrt{\frac{g_o r_o^2}{3\alpha k_o}\left[1 - \left(\frac{r}{r_o}\right)^2\right] + \left(\frac{g_o r_o}{3h} + T_\infty + \frac{1}{\alpha}\right)^2}$$

(P.U.P., May 1996)

Solution

Given : (*i*) The relation for thermal conductivity of solid sphere

$$k = k_o (1 + \alpha T)$$

(*ii*) Heat generation rate $= g_o$

(*iii*) Heat transfer coefficient $= h$

(*iv*) Ambient temperature $= T_\infty$.

Boundary conditions :

At $r = 0$, $\quad \frac{dT}{dr} = 0$

At $r = r_o$, $\quad -k\left(\frac{dT}{dr}\right)_{r=r_o} = h\,[T_{r=r_o} - T_\infty]$.

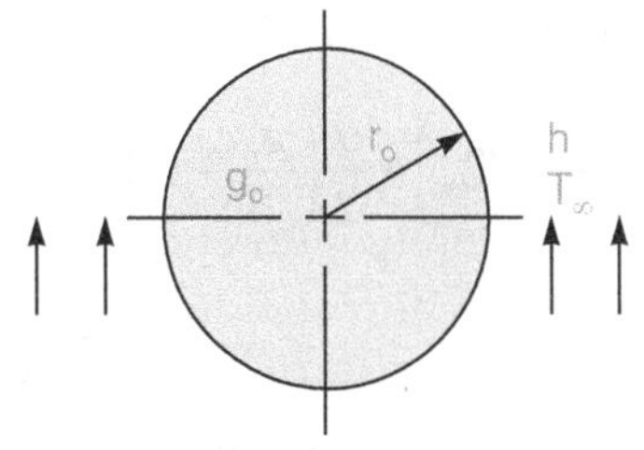

Fig. 4.30

Analysis : The governing differential equation with variable thermal conductivity and constant heat generation rate g_o can be written as

$$\frac{1}{r^2}\frac{d}{dr}\left(r^2 k \frac{dT}{dr}\right) + g_o = 0$$

or $$\frac{d}{dr}\left(r^2 k \frac{dT}{dr}\right) = -g_o r^2$$

Integrating with respect to r, we get

$$r^2 k \frac{dT}{dr} = -\frac{g_o r^3}{3} + C_1$$

Applying boundary condition at the centre *i.e.*,

At $r = 0$; $\dfrac{dT}{dr} = 0$

It gives, $C_1 = 0$

with using $C_1 = 0$, we get

$$r^2 k \frac{dT}{dr} = -\frac{g_o r^3}{3}$$

or $$k \frac{dT}{dr} = -\frac{g_o r}{3} \quad ...(i)$$

Using $k = k_o(1 + \alpha T)$

$$k_o(1 + \alpha T)\frac{dT}{dr} = -\frac{g_o r}{3}$$

Rearranging $k_o(1 + \alpha T)dT = -\dfrac{g_o r dr}{3}$

Integrating both sides, we get

$$k_o\left(T + \frac{\alpha T^2}{2}\right) = -\frac{g_o r^2}{6} + C_2 \quad ...(ii)$$

Now using the boundary condition

At $r = r_o$; $-k\left(\dfrac{dT}{dr}\right)_{r=r_o} = h(T_{r=r_o} - T_\infty)$

Using eqn. (*i*) we get $\dfrac{g_o r_o}{3} = h(T_{r=r_o} - T_\infty)$

or $$T_{r=r_o} = \frac{g_o r_o}{3h} + T_\infty \quad ...(iii)$$

Substituting the value of $T_{r=r_o}$ in eqn. (*ii*), at $r = r_o$

$$k_o\left\{\left(\frac{g_o r_o}{3h} + T_\infty\right) + \frac{\alpha}{2}\left(\frac{g_o r_o}{3h} + T_\infty\right)^2\right\} = -\frac{g_o r_o^2}{6} + C_2$$

or $$C_2 = k_o\left\{\left(\frac{g_o r_o}{3h} + T_\infty\right) + \frac{\alpha}{2}\left(\frac{g_o r_o}{3h} + T_\infty\right)^2\right\} + \frac{g_o r_o^2}{6} \quad ...(iv)$$

Substituting the value of C_2 in eqn. (*ii*)

$$k_o\left\{T + \frac{\alpha T^2}{2}\right\} = -\frac{g_o r^2}{6} + \frac{g_o r_o^2}{6} + k_o\left\{\left(\frac{g_o r_o}{3h} + T_\infty\right) + \frac{\alpha}{2}\left(\frac{g_o r_o}{3h} + T_\infty\right)^2\right\}$$

or $$T + \frac{\alpha T^2}{2} = \frac{g_o r_o^2}{6k_o}\left\{1 - \left(\frac{r}{r_o}\right)^2\right\} + \left\{\left(\frac{g_o r_o}{3h} + T_\infty\right) + \frac{\alpha}{2}\left(\frac{g_o r_o}{3h} + T_\infty\right)^2\right\}$$

or $$T^2 + \frac{2T}{\alpha} = \frac{g_o r_o^2}{3k_o\alpha}\left[1 - \left(\frac{r}{r_o}\right)^2\right] + \left\{\frac{2}{\alpha}\left(\frac{g_o r_o}{3h} + T_\infty\right) + \left(\frac{g_o r_o}{3h} + T_\infty\right)^2 + \frac{1}{\alpha^2} - \frac{1}{\alpha^2}\right\}$$

$$T^2 + \frac{2T}{\alpha} = \frac{g_o r_o^2}{3k_o\alpha}\left\{1 - \left(\frac{r}{r_o}\right)^2\right\} + \left\{\left(\frac{g_o r_o}{3h} + T_\infty + \frac{1}{\alpha}\right)^2 - \frac{1}{\alpha^2}\right\}$$

$$T^2 + \frac{2T}{\alpha} - \frac{g_o r_o^2}{3k_o\alpha}\left\{1 - \left(\frac{r}{r_o}\right)^2\right\} - \left(\frac{g_o r_o}{3h} + T_\infty + \frac{1}{\alpha}\right)^2 + \frac{1}{\alpha^2} = 0$$

It is quadratic equation and its root for T are

$$T = \frac{-\dfrac{2}{\alpha} \pm \sqrt{\dfrac{4T^2}{\alpha^2} + \dfrac{4g_o r_o^2}{3\alpha k_o}\left\{1 - \left(\dfrac{r}{r_o}\right)^2\right\} - \dfrac{4}{\alpha^2} + 4\left(\dfrac{g_o r_o}{3h} + T_\infty + \dfrac{1}{\alpha}\right)^2}}{2 \times 1}$$

or $$\mathbf{T = -\frac{1}{\alpha} \pm \sqrt{\frac{g_o r_o^2}{3\alpha k_o}\left\{1 - \left(\frac{r}{r_o}\right)^2\right\} + \left(\frac{g_o r_o}{3h} + T_\infty + \frac{1}{\alpha^2}\right)^2}}$$

Proved.

Example 4.30. *A hollow sphere of 12 cm inner diameter and 21 cm outer diameter is made of a material (k = 30 W/m.K), in which heat is generated uniformly at a rate of 5 × 10⁶ W/m³. The inside surface is insulated and outside surface is maintained at 360°C. Calculate the maximum temperature in the solid.*

Solution

Given : A hollow sphere with uniform heat generation

$d_1 = 12$ cm or $r_1 = 6$ cm $= 0.06$ m

$d_2 = 21$ cm or $r_2 = 10.5$ cm $= 0.105$ m

$k = 30$ W/m.K, $g_o = 5 \times 10^6$ W/m³

$$\left(\frac{dT}{dr}\right)_{r=r_1} = 0, \qquad T_2 = 360°C.$$

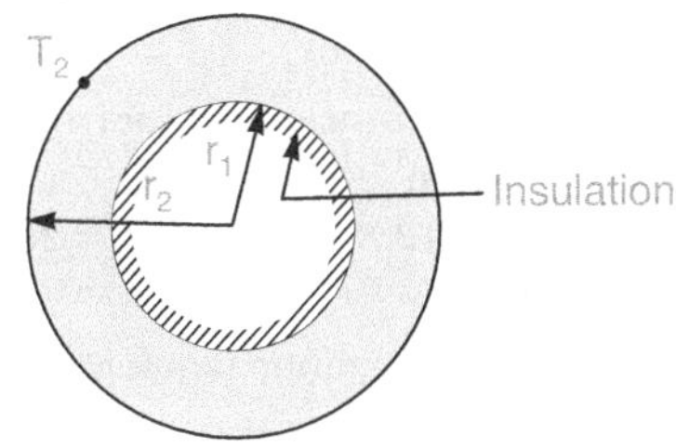

Fig. 4.31

To find : The maximum temperature in sphere.

Analysis : The temperature distribution in a sphere with uniform heat generation

$$T(r) = -\frac{g_o r^2}{6k} - \frac{C_1}{r} + C_2$$

Subjected to boundary conditions

(*i*) At $r = r_1$; $\quad \frac{dT}{dr} = 0$

(*ii*) At $r = r_2$, ; $\quad T = T_2$

Using first condition

$$\left(\frac{dT}{dr}\right)_{r=r_1} = \left[-\frac{g_o r}{3k} + \frac{C_1}{r^2}\right]_{r=r_1} = 0$$

It gives $\quad C_1 = \frac{g_o r_1^3}{3k}$

And at $r = r_2$;

$$T_2 = -\frac{g_o r_2^2}{6k} - \frac{g_o r_1^3}{3k r_2} + C_2$$

It gives $\quad C_2 = T_2 + \frac{g_o r_2^2}{6k} + \frac{g_o r_1^3}{3k r_2}$

Then temperature distribution in hollow sphere

$$T(r) = -\frac{g_o r^2}{6k} - \frac{1}{r} \times \frac{g_o r_1^3}{3k} + \frac{g_o r_2^2}{6k} + \frac{g_o r_1^3}{3k r_2} + T_2$$

$$= \frac{g_o}{6k}(r_2^{\,2} - r^2) + \frac{g_o r_1^3}{3k}\left[\frac{1}{r_2} - \frac{1}{r}\right] + T_2$$

The position of maximum temperature in the solid can be obtained by

$$\frac{dT}{dr} = 0$$

or $\quad \frac{dT}{dr} = -\frac{2g_o r_{cr}}{6k} + \frac{g_o r_1^3}{3k}\left(\frac{1}{r_{cr}^{\,2}}\right) = 0$

It gives $r_{cr} = r_1$(at inner surface)

Then the maximum temperature in the sphere

$$T_{max} = \frac{g_o}{6k}(r_2^2 - r_1^2) + \frac{g_o r_1^3}{3k}\left(\frac{1}{r_2} - \frac{1}{r_1}\right) + T_2$$

$$= \frac{5 \times 10^6}{6 \times 30}(0.105^2 - 0.06^2)$$

$$+ \frac{5 \times 10^6 \times (0.06)^3}{3 \times 30}\left(\frac{1}{0.105} - \frac{1}{0.06}\right) + 360$$

$$= 206.25 - 85.71 + 360$$

$= \mathbf{480.53°C.}$ Ans.

4.5. SUMMARY

The general one dimensional heat conduction equation with heat generation in steady state conditions in cartesian coordinate is

$$\frac{d}{dx}\left(\frac{dT}{dx}\right) + \frac{g(x)}{k} = 0$$

It is also called one dimensional Poisson equation. If heat generation rate $g(x)$ (= g_o) is uniform, then its solution for temperature distribution in the plane wall is

$$T(x) = -\frac{g_o x^2}{2k} + C_1 x + C_2$$

For cylindrical coordinate system, the one dimensional Poisson equation is

$$\frac{1}{r}\frac{d}{dr}\left(r\frac{dT}{dr}\right) + \frac{g(r)}{k} = 0$$

The temperature distribution with uniform heat generation g_o is

$$T(r) = -\frac{g_o r^2}{4k} + C_1 \ln(r) + C_2$$

For spherical coordinate system, the one dimensional Poisson equation is

$$\frac{1}{r^2}\frac{d}{dr}\left(r^2\frac{dT}{dr}\right) + \frac{g(r)}{k} = 0$$

The temperature distribution for $g(r) = g_o$

$$T(r) = -\frac{g_o r^2}{6k} - \frac{C_1}{r} + C_2$$

The C_1 and C_2 are constants of integration and are evaluated according to boundary conditions imposed on the solid.

The location of maximum temperature in any solid can be obtained by applying the condition of maxima *i.e.,*

$$\frac{dT}{dr} \text{ or } \frac{dT}{dx} = 0$$

It gives location of maximum temperature x_{cr} or r_{cr} and

$$T_{max} = T(r_{cr} \text{ or } x_{cr}).$$

REVIEW QUESTIONS

1. What is heat generation ? What do you mean by uniform heat generation ? Give some examples.
2. Consider uniform heat generation in a cylinder and a sphere of equal radius made of same material in the same environment. Which geometry will have a higher temperature at its centre ? Why ?
3. Show that for a plane wall of thickness 2L with uniform heat generation g_o per unit volume, the temperature at the mid plane is given by

$$T_c = \frac{g_o L^2}{2k} T_s$$

where T_s = surface temperature on either side.

4. Develop an expression for the steady state temperature distribution in slab of thickness L, when the boundary surface at $x = 0$ is kept insulated and the boundary surface at $x = L$ is kept at zero temperature. The thermal conductivity of the wall k is constant and within the wall energy is generated at the rate of $g(x) = g_o x^2$ W/m^3. [Ans. $g_o L^4/12k$]
5. Show that the maximum temperature in a cylindrical rod with heat generation g_o W/m^3 is given by

$$\frac{T_{max}}{T_\infty} = 1 + \frac{g_o r_o}{4hT_\infty}\left(2 + \frac{hr_o}{k}\right)$$

where r_o = outer radius of cylinder, T_∞ = ambient temperature, h = convection heat transfer coefficient.

6. Derive an expression for temperature distribution during steady state heat conduction in a solid sphere with internal heat generation and exposed to convection environment.
7. Heat is generated uniformly at the rate of g_o W/m^3, in a fuel rod of nuclear reactor. The rod has a long hollow cylindrical shape with its inner and outer surface temperatures of T_1 and T_2, respectively. Derive an expression for temperature distribution.
8. A nuclear reactor fuel element consists of a solid cylindrical rod of radius r_o and thermal conductivity k_f. The fuel rod is cladded with a material having thermal conductivity k_{cl} and its outer radius is r_{cl}. Consider steady state conditions, with uniform heat generation within the fuel as g_o W/m^3, outer surface is exposed to coolant at temperature T_∞ with convective heat transfer coefficient of h.
   (a) Derive an expression for temperature distribution $T_f(r)$ and $T_{cl}(r)$ in the fuel rod and cladding, respectively.
   (b) Consider a uranium oxide fuel rod for which k_f = 2W/m.K, r_o = 6 mm, k_{cl} = 25 W/m.K, r_{cl} = 9 mm, $g_o = 2 \times 10^8$ W/m^3, h = 2000 W/m^2.K, and T_∞ = 300 K,

   What would be the maximum temperature in the fuel rod ? [Ans. 1031.2°C]
9. A hollow cylindrical conductor of constant thermal conductivity k, with inside radius r_1, outside radius r_2 is perfectly insulated at its outside radius and held at temperature T_1 by a coolant at inside surface. Electrical energy is generated within the conductor at a uniform rate of g_o W/m^3. If the steady state conditions prevail and temperature distribution is radial only, derive an expression for temperature as a function of radius, r.
10. The heat generation rate per unit volume in a long cylinder of radius r_o is given as

$$g(r) = a + br$$

where a and b are constants and r is any radius. The cylinder is exposed to a medium temperature T_∞ with heat transfer coefficient h. Derive an expression for steady state temperature distribution in the solid cylinder.

PROBLEMS

1. Consider a slab 0.1 m thick, with its left face insulated and the right boundary surface dissipates heat by convection with a heat transfer coefficient of 200 W/m^2.K into an ambient air at 150°C. The thermal conductivity of the wall is 10 W/m.K and within the wall the energy is generated at a constant rate of 10^6 W/m^3. Determine the boundary surface temperatures. [Ans. 1150°C, 650°C]
2. A long cylindrical rod of radius 5 cm and k = 20 W/m.K contains radioactive material which generated energy uniformly within the cylinder at a constant rate of 2×10^5 W/m^3. The rod is cooled by convection from its cylindrical surface into an ambient at 20°C with h = 50 W/m^2.K. Determine the temperature at the centre and the outer surface of this cylindrical rod. [Ans. 126.3°C, 120°C]
3. An electric resistance wire of radius 1 mm with thermal conductivity of 25 W/m.K is heated by passage of an electric current, which generates heat within the wire at the constant rate of 2×10^9 W/m^3. Determine the centre line temperature

rise above the surface temperature of the wire, if its outer surface is maintained at constant temperature. [Ans. 20°C]

4. An electric current of 500 A flows through a stainless steel conductor of 5 mm diameter, that has an electric resistance $R_e = 5 \times 10^{-4}$ Ω/m. Energy is generated as result of passage of electric current and that is dissipated by convection into an ambient at 0°C with convection coefficient of 50 W/m^2. K. The thermal conductivity of the conductor is 60 W/m.K. Calculate the centre and surface temperature of the cable. [Ans. 159.32°C,159.15°C]

5. Heat is generated at the constant rate of 2×10^8 W/m^3 in a copper sphere (k = 386 W/m.K) of 1 cm radius. The sphere is cooled by convection from its outer surface into an ambient at 10°C with a convection coefficient of 2000 W/m^2.K. Determine the surface and centre temperature of the sphere. [Ans. 343.3°C, 352°C]

6. Consider a composite wall consists of three layers A, B and C. The outer surfaces are exposed to a fluid at 25°C with convection coefficient of 1000 W/m^2. K. The middle wall layer B experiences uniform heat generation g_B W/m^3, while there is no heat generation in wall layer A and C. The temperature at inner surface of layer A and at outer surface of C are 260°C and 210°C, respectively and thicknesses and thermal conductivity of the three layers are :

   k_A = 25 W/m.K L_A = 30 mm
   k_B = 15 W/m.K L_B = 60 mm
   k_C = 50 W/m.K L_C = 20 mm

   (*a*) Assume negligible contact resistance at the interfaces, determine the volumetric heat generation g_B.
   (*b*) Sketch the temperature distribution in the composite. [Ans. (*a*) $g_B = 3.083 \times 10^6$ W/m^3]

7. A 1.2 m thick slab of poured concrete (k = 1.148 W/m.K) with both of side surfaces maintained at a temperature of 20°C. During its curing, the chemical energy is released at a rate of 80 W/m^3. Presuming that the temperature does not vary with time, calculate the maximum temperature of the concrete.

   What maximum thickness of concrete can be poured without causing temperature gradient to exceed 8.5°C per metre any where in the slab ? [Ans. 29.73°C, 0.314 m]

8. A semiconductor material (k = 2 W/m.K) of electrical resistivity ρ = 2×10^5 Ωm is used to fabricate a cylindrical rod 10 mm in diameter and 40 mm long. The longitudinal surface of the rod is well insulated while the ends are maintained at temperatures of 100°C and 0°C. If the rod carries a current of 10 A, what is the mid point temperature? What is the heat transfer rate at the each end of the rod ? [Ans. 82.03°C, – 0.116 W, 0.992 W]

9. A chemical reaction is being carried out at constant pressure in a packed bed between two coaxial cylinders with radii 1.14 cm and 1.27 cm. The entire inner wall is at a uniform temperature of 500°C and there is almost no heat transfer through this surface.

   The reaction releases heat at a uniform rate of 560 kW/m^3 throughout the reactor. The effective thermal conductivity of the packed bed may be taken as 0.525 W/m.K. What is the temperature of the outer wall ? [Ans. 499°C]

10. In an experiment to determine the thermal conductivity of an insulating material, a thick layer of same is provided over a long copper tube inside which is, a current carrying conductor. The temperature at two points A and B inside the insulation layer at radial distances of 2 cm and 6 cm from the centre of the tube are measured as 115°C and 42°C respectively. If the current flowing through the electric conductor is 10 A and its resistance is 2 Ω/m, determine the thermal conductivity of the insulating material. Neglect the heat loss from the end faces. [Ans. k = 0.48 W/m.K]

11. A copper cable (k = 395 W/m.K) 30 mm in diameter carries a current of 300 A, when exposed to air at 30°C with convection coefficient of 20 W/m^2.K. The cable resistance is 5×10^{-3} Ω/m. Determine the surface and centre temperature of the cable. [Ans. T_s = 268.73°C, T_c = 268.82°C]

12. It is proposed to heat the window glass planes in a living space at 26°C. A company offers resistance embedded glasses with uniform heat generation. The outside is at –15°C, and the convection coefficient on the outside is 20 W/m^2.K. The pane is 8 mm thick and has a conductivity of 1.4 W/m.K. What should be heat generation rate if the inside surface temperature is equal to the room temperature ? [Ans. g_o = 97 kW/m^3]

13. A 3 mm diameter stainless wire, one metre long has a voltage of 100 V impressed on it. The outer surface of the wire is maintained at 100°C. Calculate the centre temperature of the wire. Take ρ = 10 μΩ-cm and k = 20 W/m.K.

    If the heated wire is submerged in a fluid maintained at 50°C find the heat transfer coefficient on the surface of the wire for above given conditions. [Ans. 128°C and 149.55 W/m^2.K]

14. In construction of a bridge concrete columns cylindrical in shape and 0.75 m in diameter are erected by pouring concrete in a short time. The hydration of concrete results in uniform heat generation of 0.8 W/kg. The outside surface temperature is 55°C. Thermal conductivity of concrete = 0.95 W/mk.

    Density of concrete = 2305 kg/m^3.

Determine temperature at the centre of the cylinder and at a distance of 0.2 m, 0.3 m, and 0.35 m from the centre. *(N.M.U., May 2004)*

[Ans. 123.24°C, 103.83°C, 79.57°C, 63.79°C]

15. The heat is generated uniformly in a steel plate (k = 20 W/m.K), 1 cm thick is at the rate of 500 MW/m^3. If the two sides of the plate are maintained at 100°C and 200°C, respectively. Calculate the temperature at the centre of the plate. Also calculate the location and magnitude of maximum temperature in the plate.

[Ans. 462.5°C, 5.4 mm from left, 464.5°C]

16. A plane wall (k = 12 W/m.K) 7.5 cm thick generates heat internally at the rate of 10^5 W/m^3. One side of the wall is insulated and other side is exposed to an environment at 90°C with h = 500 W/m^2.K. Calculate the maximum temperature in the wall.

[Ans. T_{max} = 128.4°C]

17. A nuclear reactor has a flat plate fuel element 10 mm thick. The element is cladded on its both faces with aluminium plate 2 mm in thickness. The rate of heat generation with the fuel element is 4 × 10^4 W/kg of uranium. Calculate the temperature at the outer surface of the aluminium and at the interface of the uranium-aluminium and at the centre of the fuel element. The coolant circulates around the aluminium cladding is at 120°C with heat transfer coefficient of 28000 W/m^2.K.

Take

Thermal conductivity of the uranium = 24.4 W/m.K.,

Thermal conductivity of aluminium = 206 W/m.K.

[Ans. 150°C, 223.4°C, 611°C]

18. A 10 kW heater using nichrome wire (k = 17.5 W/m.K) is to be designed. The maximum operating temperature is 1650 K, other design criteria are

h = 850 W/m^2.K $\quad T_\infty$ = 370 K

ρ = 110 μΩ cm. Power available = 12 V

What size of wire is required, if the heater is 0.6 m long ? [Ans. 2.6 mm]

19. An electrical conductor of copper with a diameter of 1 mm is covered with a plastic insulation of thickness 1 mm. The temperature of its surroundings is 20°C. Find the maximum current carried by conductor so that no part of plastic is above 80°C

k_{copper} = 400 W/m.K, $k_{plastic}$ = 0.5 W/m.K, h = 8 W/m^2.K, specific electric resistance of copper = 3 × 10^{-8} ohm-m. *(N.M.U., May 2004)*

[Ans. I = 10.76 A]

20. A copper cable (k = 380 W/m.K) 25 mm in diameter has an electrical resistance of 0.005 Ω/m and it is used to carry an electrical current of 250 Amps. The cable is exposed to an ambient air at 20°C and the associate convection current is 25 W/m^2.K. What are surface and centre line temperature of copper cable ?

[Ans. T_s = 179.15°C, T_c = 179.22°C]

21. A long cylindrical fuel element 25 mm in diameter in a nuclear reactor has energy generation at a uniform rate of 7 × 10^8 W/m^3. It is wrapped in a thin aluminium cladding (k = 237 W/m.K). The coolant is circulated around it at uniform temperature of 95°C with h = 7000 W/m^2.K. Is this proposal satisfactory ?

Take melting point of nuclear fuel as 1750°C and working temperature of aluminium up to 450°C.

$k_{fuel,\ rod}$ = 54 W/m.K.

[Ans. T_s = 407.5°C, which is safe even with thin aluminium cladding]

22. An electric heater of 30 kW is to be designed from a steel wire (k = 15 W/m.K) having an electrical resistivity of 1 × 10^{-8} Ωm. The operating temperature of the steel should not be more than 1200°C. The minimum expected value of heat transfer coefficient at outer surface of wire is 1500 W/m^2.K and the maximum ambient temperature is 60°C. Show that the expected temperature drop between the centre and surface of the wire is independent of the wire diameter. Then calculate the wire diameter and current required. [Ans. 6.49 mm, 9962.89 A]

23. An internally cooled copper conductor of 2 cm outer radius and 0.75 cm inner radius carries a current density (I/A_c) of 5000 Amp/cm^2. A constant temperature of 70°C is maintained at inner surface and there is no heat transfer through insulation surrounding copper conductor. Derive an expression for temperature distribution through copper conductor.

Calculate the maximum temperature of copper conductor and the radius at which it occurs. Also calculate the internal heat transfer rate and check this equals the total energy generated in the conductor.

Take

For copper, k = 380 W/m.K

ρ = 2 × 10^{-8} Ω cm

[Ans. 845°C, at r = 2 cm, – 5368 W/m]

24. A hollow cylindrical conductor (k = 17.5 W/m.K) with r_1 = 0.6 cm and r_2 = 0.75 cm is insulated at its outer surface, while its inner surface is maintained at 37.5°C by circulating cooling fluid. The electrical resistance per metre is 2.5 × 10^2 ohms. Calculate the maximum allowable current, if the temperature is not to exceed 48.5°C anywhere in the conductor.

[Ans. 317.4 Amp.]

25. A nuclear fuel element is in the form of a hollow cylinder insulated at inner surface. Its inner and outer radii are 5 cm and 10 cm, respectively. The outer surface gives heat to the fluid at 50°C, where the unit surface conductance is 100 W/m^2.K. The thermal conductivity of the material is 50 W/m^2.K. Find the rate of heat generation so that the maximum temperature in the system will not exceed 200°C. [Ans. 379.57 kW/m^3]

26. A solid sphere of radius 5 cm and thermal conductivity of 20 W/m.K is heated uniformly throughout its volume at the rate of 2×10^6 W/m^3, and heat is dissipated by convection to ambient air at 25°C with convection coefficient of 100 W/m^2.K. Determine the steady state temperature at the centre and the outer surface of the sphere. [Ans. 400°C, 358.33°C]

27. A metal rod 6 mm in diameter and 1 m long runs between two large bus bars. The rod is insulated on its lateral surface against the flow of heat and electric current. The bus bars are at 20°C. What is the maximum current, the rod can carry if its temperature is not to exceed 150°C at any point? Assume resistivity of rod material is 1.7×10^{-6} Ω cm and thermal conductivity is 300 W/m.K. [Ans. 121.12 Amp]

28. In a thick infinite slab of thickness 20 cm, the temperature of the fluid on one side is 30°C and 20°C on other side. The heat transfer coefficient on the hot side is 20 W/m^2.K and on the cold side is 40 W/m^2.K, the conductivity of the slab material is 20 W/m.K. The heat generated in the slab at a uniform rate of 5 kW/m^3 ;

(*i*) Derive an expression for temperature distribution in the slab.

(*ii*) Find the maximum temperature in the slab and its location.

(*iii*) Find the temperature at the centre of the slab and at its two surfaces.

(*iv*) Compute the heat transferred from each surface.

[Ans. (*i*) $T(x) = -\dfrac{g_o x^2}{2k} + 11.76x + 41.76$, (*ii*) 42°C, (*iii*) 41.63°C, 41.76°C, 39.11°C, (*iv*) – 235.2 W/m^2 and 764.8 W/m^2]

REFERENCES AND SUGGESTED READING

1. H.S. Carslaw and J.C. Jaeger, *"Conduction of Heat Transfer in Solids"*, Oxford University Press, London, 1986.
2. F Krieth and M.S. Bohn, *"Principles of Heat Transfer"*, 5th ed., PWS Pub. Company, 1997.
3. P.J. Schneider, *"Conduction Heat Transfer"*, Addison-Wesley, Cambrige, MA, 1955.
4. V.S. Arpaci, *"Conduction Heat Transfer"*, 2/e, Addison-Wesley, Reading, MA, 1966.
5. J.P. Holman, *"Heat Transfer"*, 7th ed. McGraw Hill, New York, 1990.
6. F.P. Incropera and D.P. DeWitt, *"Introduction to Heat Transfer"*, 2nd ed., John Wiley and Sons, 1990.
7. M.N. Ozisik, *"Heat Transfer—A Basic Approach"*, McGraw Hill, New York, 1985.
8. B.V. Karlekar, and R.M. Desmond, *"Heat Transfer"* Prentice Hall of India, New Delhi, 1989.
9. Vijay Gupta, *"Elements of Heat and Mass Transfer"*, New Age (I), New Delhi, 1995.
10. Vedat S. Arpaci, *"Conduction Heat Transfer"*, Addison-Wesley Publishing Company, New York, 1966.
11. Donatello Annaratone, *"Engineering Heat Transfer"*, Springer Heidelberg Dordrecht London New York, 2010.
12. J.R. Welty, C.E. Wicks, R.E. Wilson, G.L. Rorrer, *"Fundamentals of Momentum, Heat and Mass Transfer"*, 5th Edition, John Wiley and Sons, Inc., 2008.
13. William S. Janna, *"Engineering Heat Transfer"*, 2nd edition, CRC Press, New York, 2000.

Heat Transfer from Extended Surfaces

5

The term 'extended surface' is commonly used in reference to a solid that experiences energy transfer by conduction and convection between its boundary and surroundings. A temperature gradient in x direction sustains heat transfer by conduction internally, at the same time, there is heat dissipation by convection into an ambient at T_∞ from its surface at temperature T_s, given as

$$Q = h\,A_s\,(T_s - T_\infty)$$

where h = convection heat transfer coefficient,

and A_s = heat transfer area of a surface.

When the temperatures T_s and T_∞ are fixed by design considerations, there are only two ways to increase the heat transfer rate : (*i*) to increase the convection coefficient h, or (*ii*) to increase the surface area A. In the situations, in which an increase in h is not practical or economical, the heat transfer rate can be improved by increasing surface area.

For heat transfer from a hot liquid to a gas, through a wall, the value of heat transfer coefficient on the gas side is usually very less compared to that for liquid side ($h_{gas} << h_{liquid}$). To compensate low heat transfer coefficient, the surface area on the gas side may be extended for a given temperature difference between surface and its surroundings. These extended *surfaces* are called *fins. The fins are normally thin strips of highly conducting metals such as aluminium, copper, brass etc.* The fins enhance the heat transfer rate from a surface by exposing larger surface area to convection.

5.1. TYPES OF FINS

The fins are designed and manufactured in many shapes and forms, some of them are shown in Fig. 5.1 and Fig. 5.2.

Longitudinal Fin : It is a *straight rectangular fin* attached to a plane wall. It may be of uniform cross-sectional area, or its cross-sectional area may vary along its length to form a *triangular, parabolic* or *trapezoidal* shape.

Annular Fin : An *annular fin* is a fin that is circumferentially attached to a cylinder and its cross-section varies with radius from centre line of cylinder.

Spine : In contrast, a *pin fin,* or *spine* is an extended surface of circular cross-section whose diameter is much smaller than its length. The pin fins may also be of uniform or non-uniform cross-section.

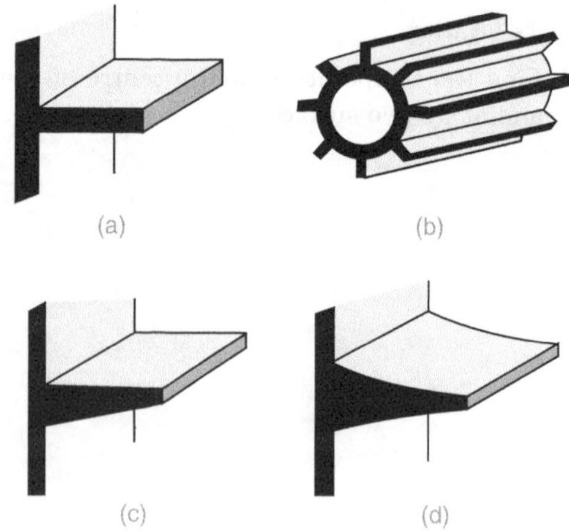

Fig. 5.1. (*a*) Longitudinal fins ; (*b*) Cylindrical tube with fins of rectangular profile ; (*c*) Longitudinal fin of trapezoidal profile ; (*d*) Longitudinal fin of parabolic profile.

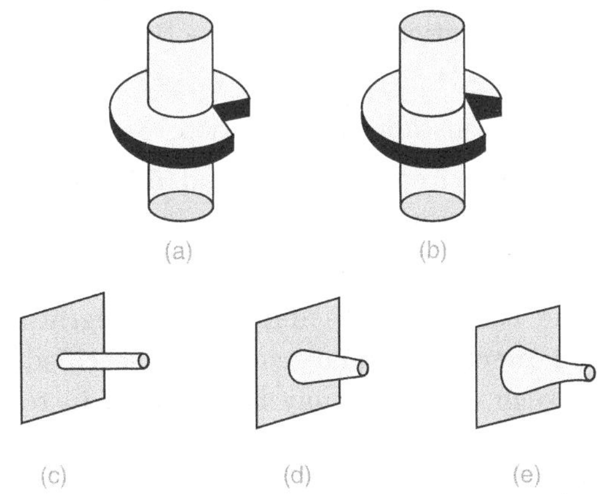

Fig. 5.2. Annular fins and spines ; (*a*) Cylindrical tube with annular fin of rectangular profile ; (*b*) Cylindrical tube with annular fin of truncated conical profile ; (*c*) Cylindrical pin fin ; (*d*) Truncated conical spine ; (*e*) Parabolic spine.

5.2. FIN SELECTION AND APPLICATIONS

Generally, the fins are used on the surfaces where the heat transfer coefficient is very low. For example, in a car radiator the outer surface of the tubes is finned because the heat transfer coefficient for air at the outer surface is much smaller than that of water flow inside the tubes. Similarly, the electrical transformers and the motors in which the generated heat is dissipated to air by providing fins on its outer surface. The fins are also provided on cylinder and cylinder head of an air cooled I.C. engine and large variety of the heat exchangers.

The selection of fins is made on the basis of thermal performance and cost. The selection of suitable fin geometry requires a compromise among the cost, weight, available space, pressure drop of heat transfer fluid and heat transfer characteristics of the extended surface.

It should be noted that the fins of triangular and parabolic profiles contain less material and are more efficient than the fins of rectangular profiles and thus are more suitable for applications that require minimum weight such as space applications.

5.3. GOVERNING EQUATION

Consider the surface of a plane wall at temperature T_0, exposed to an ambient at T_∞. Let us consider a fin has a constant cross-section area A_c, and length L, attached to the surface as shown in Fig. 5.3. In order to determine temperature distribution and heat flow through a fin, the energy balance on the fin is required.

To derive an equation for temperature distribution, we make an energy balance for a differential element of a fin made of a material of uniform thermal conductivity k.

The rate of heat conduction into the element

= The rate of heat conduction out the element + The rate of heat convection from the element surface

The rate of heat conduction into the element is given by

$$Q(x) = -kA_c \frac{dT(x)}{dx} \quad ...(5.1)$$

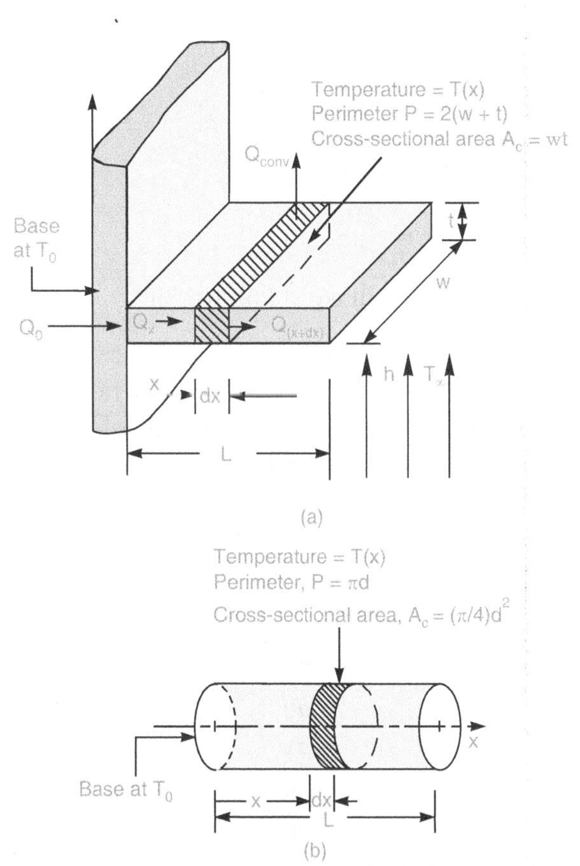

Fig. 5.3. Nomenclature for the derivation of one dimensional governing fin equation

The rate of heat conduction out the element

$$Q(x + dx) = Q(x) + \frac{d}{dx}[Q(x)]\, dx \quad ...(5.2)$$

The rate of heat convection from the element surface of perimeter P ;

$$Q_{conv} = h dA_s [T(x) - T_\infty] \quad ...(5.3)$$

where dA_s is surface area of differential element.

Substituting these quantities in energy balance ;

$$Q(x) = Q(x) + \frac{d}{dx}[Q(x)]\,dx + hdA_s\,[T(x) - T_\infty]$$

$$0 = \frac{d}{dx}\left[-kA_c\frac{dT}{dx}\right]dx + hdA_s\,[T(x) - T_\infty]$$

or

$$\frac{d}{dx}\left(A_c\frac{dT}{dx}\right) - \frac{hdA_s}{kdx}[T(x) - T_\infty] = 0 \quad ...(5.4)$$

Assuming the cross-section area A_c, perimeter P, heat transfer coefficient h, thermal conductivity of fin material k as constants. Using element surface area $dA_s = Pdx$, then

$$\frac{d}{dx}\left(\frac{dT(x)}{dx}\right) - \frac{hPdx}{kA_c dx}[T(x) - T_\infty] = 0$$

or

$$\frac{d^2T(x)}{dx^2} - \frac{hP}{kA_c}[T(x) - T_\infty] = 0 \quad ...(5.5)$$

Let $\theta(x) = T(x) - T_\infty$ ...(5.6)

and

$$m^2 = \frac{dP}{kA_c} \quad ...(5.7)$$

Substituting, in eqn. (5.5), we get

$$\frac{d^2\theta(x)}{dx^2} - m^2\theta = 0 \quad ...(5.8)$$

It is a linear homogeneous, second order differential equation. It can be solved by using operator D,

$$(D^2 - m^2)\,\theta = 0$$

or $(D - m)(D + m)\,\theta = 0$

Either $(D - m)\,\theta = 0$

or $(D + m)\,\theta = 0$

If $(D + m)\,\theta = 0$

$$\frac{d\theta}{dx} = -m\theta$$

or

$$\frac{d\theta}{\theta} = -mdx$$

On integration,

$$\ln\theta = -mx + A$$

or

$$\theta = e^{-mx}e^{A} = C_1e^{-mx} \quad ...(5.9)$$

where C_1 is a constant of integration.

If $(D - m)\,\theta = 0$

$$\frac{d\theta}{dx} = m\theta$$

or

$$\frac{d\theta}{\theta} = mdx$$

On integration, $\ln\theta = mx + B$

or

$$\theta = e^{mx}e^{B} = C_2\,e^{mx} \quad ...(5.10)$$

where C_2 is a constant of integration.

Therefore, from eqns. (5.9) and (5.10), the general solution to temperature distribution is

$$\theta(x) = C_1e^{-mx} + C_2\,e^{mx} \quad ...(5.11)$$

where the constants C_1 and C_2 of integration are evaluated from the boundary conditions specified for the fin. One such condition may be specified in terms of temperature T_0 at the base of fin *i.e.*,

At $x = 0$ $\quad \theta_0 = T_0 - T_\infty$

Substituting, we get

$$\theta_0 = C_1 + C_2 \quad ...(5.12)$$

The second boundary condition, specified at the fin tip, the free end of fin, that may correspond to any of four different physical situations given below :

Case 1. The fin is very long, and the temperature at the fin tip approaches that of the surrounding fluid.

Case 2. The finite long fin and with negligible heat loss from fin tip.

Case 3. Finite long fin with convection heat loss from its fin tip.

Case 4. The finite long fin with specified temperature at its fin tip.

Case 1. The boundary condition at fin tip for very long fin is shown in Fig. 5.4.

At $x \to \infty$; $\theta(x) = T(x) - T_\infty \to 0$

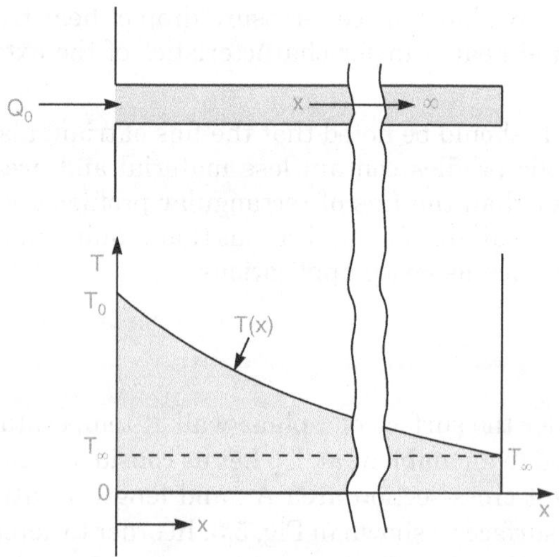

Fig. 5.4. Infinite long fin

Substituting in eqn. (5.11), we get

$$C_2 = 0$$

Using it in the eqn. (5.12), we get

$$C_1 = \theta_0$$

and $$\theta(x) = T(x) - T_\infty = \theta_0 \, e^{-mx}$$

Thus, the temperature distribution in infinite long fin yields to

$$\frac{\theta(x)}{\theta_0} = \frac{T(x) - T_\infty}{T_0 - T_\infty} = e^{-mx} \quad \text{...(5.13)}$$

The dependence of dimensionless temperature $\frac{T - T_\infty}{T_0 - T_\infty}$ along the fin length for different values of m is shown in Fig. 5.5. The plot indicates :

(*i*) as the value of m increases the dimensionless temperature falls, the fin temperature drops.

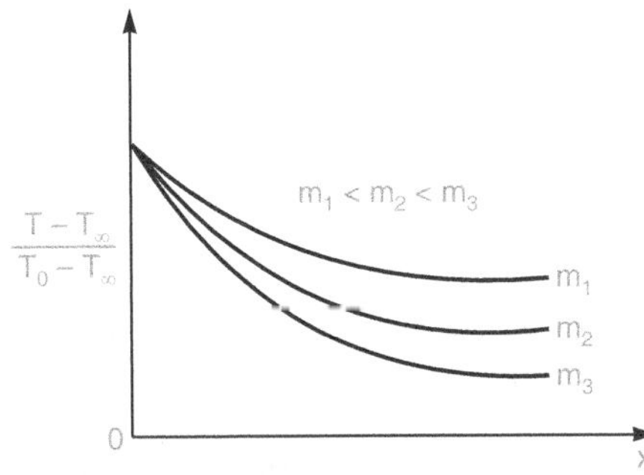

Fig. 5.5. Temperature distribution in an infinite long fin

(*ii*) as the length of fin approaches infinity, all the curves approach.

$$\frac{T - T_\infty}{T_0 - T_\infty} = 0 \text{ (asymptotically)}$$

The Heat transfer rate :

The total heat transfer rate by fin = Heat conduction rate into the fin at its base ($x = 0$)

$$Q_{fin} = Q_{x=0}$$

$$= -kA_c \left[\frac{dT}{dx}\right]_{x=0} = -kA_c \left[\frac{d\theta}{dx}\right]_{x=0}$$

$$= -kA_c \left[(T_0 - T_\infty)\frac{d}{dx}(e^{-mx})\right]_{x=0}$$

$$= -kA_c (T_0 - T_\infty)(-m)\, e^{-m\times 0}$$

$$= kA_c\, m(T_0 - T_\infty)$$

$$= \sqrt{hPk\,A_c}\;(T_0 - T_\infty) \quad \text{...(5.14)}$$

The heat flow rate from a fin can also be obtained by calculating the convection heat transfer from the fin surface to the surrounding fluid

$$Q_{fin} = \int_0^\infty h(Pdx)\,(T_{(x)} - T_\infty)$$

or $$Q_{fin} = hP \int_0^\infty (T_0 - T_\infty)\, e^{-mx}\, dx \quad \text{...(5.15)}$$

$$= hP\,(T_0 - T_\infty) \int_0^\infty e^{-mx}\, dx$$

$$= hP\,(T_0 - T_\infty) \times \left(-\frac{1}{m}\right)(e^{-m\times\infty} - e^{-m\times 0})$$

$$= hP\,(T_0 - T_\infty) \times \frac{(-1)}{-m}$$

$$= \sqrt{hPkA}\;(T_0 - T_\infty)$$

same as eqn. (5.14).

Case 2. A fin is usually very thin and long enough so that the heat loss from the fin tip may be assumed negligible. The boundary condition for such finite long fin, at its tip is :

$$\left[\frac{d\theta}{dx}\right]_{x=L} = 0$$

(Tip may be treated as adiabatic)

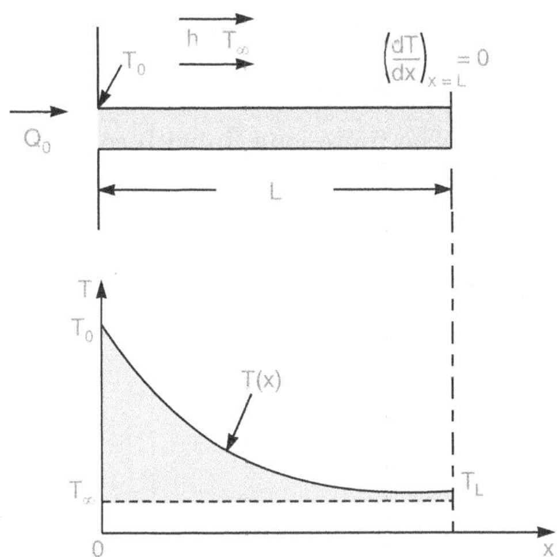

Fig. 5.6. Finite long fin insulated at free end

Substituting in eqn. (5.11), we get

$$\frac{d}{dx}(C_1 e^{-mx} + C_2 e^{mx})_{x=L} = 0$$

or $$C_1 e^{-mL} - C_2 e^{mL} = 0 \quad \text{...(5.16)}$$

Substituting $C_1 = C_2 \dfrac{e^{mL}}{e^{-mL}}$ in eqn. (5.12), we get

$$\theta_0 = C_2 \frac{e^{mL}}{e^{-mL}} + C_2 = C_2 \left[\frac{e^{mL} + e^{-mL}}{e^{-mL}}\right]$$

It gives $C_2 = \dfrac{\theta_0 e^{-mL}}{(e^{mL} + e^{-mL})}$ ...(5.17)

and
$$C_1 = \theta_0 - C_2 = \theta_0 \left[1 - \frac{e^{-mL}}{e^{mL} + e^{-mL}}\right]$$

It gives $C_1 = \dfrac{\theta_0 e^{mL}}{e^{mL} + e^{-mL}}$ ...(5.18)

Using the value of C_1 and C_2 in eqn. (5.11), we get

$$\frac{\theta(x)}{\theta_0} = \frac{T(x) - T_\infty}{T_0 - T_\infty} = \frac{e^{m(L-x)} + e^{-m(L-x)}}{e^{mL} + e^{-mL}} \quad ...(5.19)$$

In terms of hyperbolic functions

$$\frac{\theta(x)}{\theta_0} = \frac{T(x) - T_\infty}{T_0 - T_\infty} = \frac{\cosh\{m(L-x)\}}{\cosh mL} \quad ...(5.20)$$

It is the equation for *the temperature distribution* in finite long fin, insulated at its free end.

The fin heat transfer rate from the fin base :

$$Q_{fin} = Q_{x=0} = -kA_c \left[\frac{d\theta}{dx}\right]_{x=0}$$

or
$$Q_{fin} = -kA_c (T_0 - T_\infty) \frac{\sinh(mL)}{\cosh(mL)} (-m)$$
$$= \sqrt{hPkA_c}\,(T_0 - T_\infty) \tanh(mL) \quad ...(5.21)$$

Case 3. The finite long fin with convection heat loss from its free end,

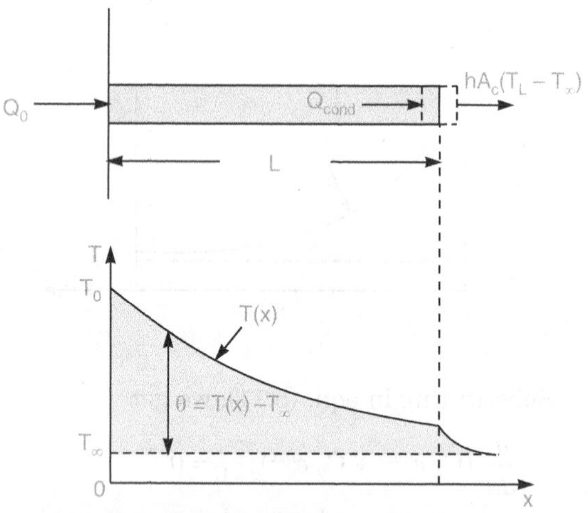

Fig. 5.7. Finite long fin with convection heat transfer from its free end

The boundary condition can be obtained by energy balance at fin tip.

$$-kA_c \left[\frac{dT}{dx}\right]_{x=L} = hA_c[T_{(x=L)} - T_\infty]$$

or
$$-k\left[\frac{d\theta}{dx}\right]_{x=L} = h\theta_{x=L} \quad ...(5.22)$$

Substituting the values from eqn. (5.11), we get
$$-km[-C_1e^{-mL} + C_2e^{mL}] = h[C_1e^{-mL} + C_2e^{mL}]$$
Rearranging as,
$$C_1e^{-mL} - C_2e^{mL} = \frac{h}{mk}[C_1e^{-mL} + C_2e^{mL}]$$

Substituting C_1 from eqn. (5.12), in terms of C_2 and θ_0,
$$\theta_0e^{-mL} - C_2e^{-mL} - C_2e^{mL} = \frac{h}{mk}[\theta_0e^{-mL} - C_2e^{-mL} + C_2e^{mL}]$$

or
$$C_2\left\{e^{mL} + e^{-mL} + \frac{h}{mk}[e^{mL} - e^{-mL}]\right\} = \theta_0\left[e^{-mL} - \frac{h}{mk}e^{-mL}\right]$$

or
$$C_2 = \frac{\theta_0\left[e^{-mL} - \frac{h}{mk}e^{-mL}\right]}{e^{mL} + e^{-mL} + \frac{h}{mk}[e^{mL} - e^{-mL}]} \quad ...(5.23)$$

Similarly, we get
$$C_1 = \frac{\theta_0\left[e^{mL} + \frac{h}{mk}e^{mL}\right]}{e^{mL} + e^{-mL} + \frac{h}{mk}[e^{mL} - e^{-mL}]} \quad ...(5.24)$$

Substituting C_1 and C_2 in eqn. (5.11), we get
$$\frac{\theta(x)}{\theta_0} = \frac{e^{m(L-x)} + \frac{h}{mk}e^{m(L-x)} + e^{-m(L-x)} - \frac{h}{mk}e^{-m(L-x)}}{e^{mL} + e^{-mL} + \frac{h}{mk}[e^{mL} - e^{-mL}]}$$

It may be rearranged and expressed in terms of hyperbolic functions as
$$\frac{\theta(x)}{\theta_0} = \frac{T(x) - T_\infty}{T_0 - T_\infty} = \frac{\cosh\{m(L-x)\} + \left(\frac{h}{mk}\right)\sinh\{m(L-x)\}}{\cosh(mL) + \left(\frac{h}{mk}\right)\sinh(mL)} \quad ...(5.25)$$

The total heat transfer from fin
$$Q_0 = Q_{fin} = -kA_c\left[\frac{d\theta}{dx}\right]_{x=0}$$
$$= \sqrt{hPkA_c}\,(T_0 - T_\infty)\frac{\sinh(mL) + (h/mk)\cosh(mL)}{\cosh(mL) + (h/mk)\sinh(mL)} \quad ...(5.26)$$

Case 4. The boundary condition for finite long fin with specified temperature at its free end.

At $x = L$; $\theta_L = T_L - T_\infty$

Substituting in eqn. (5.11)

$$\theta_L = C_1 e^{-mL} + C_2 e^{mL} \qquad ...(5.27)$$

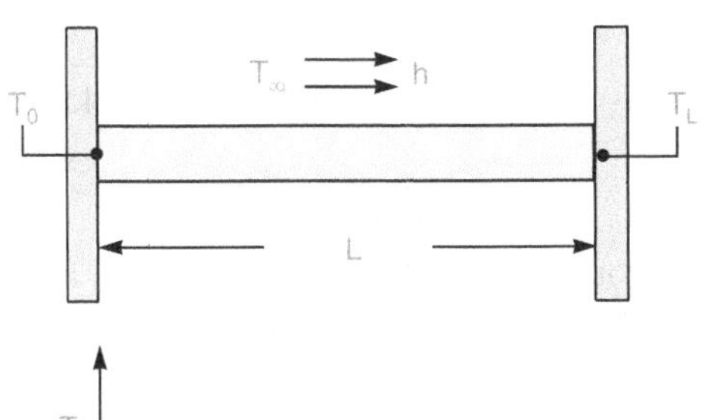

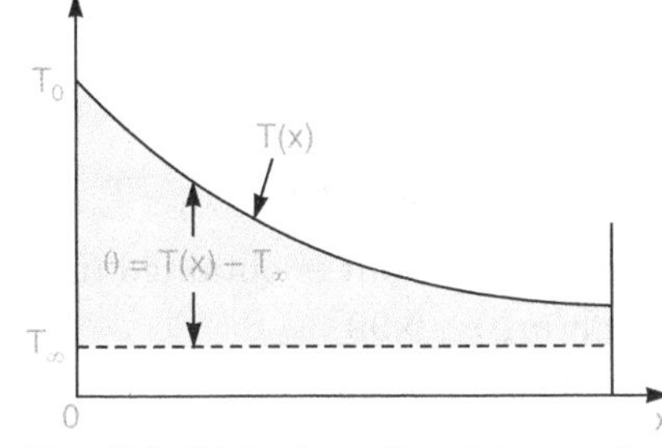

Fig. 5.8. Finite long fin with specified temperature at its free end

Using the C_1 in terms of C_2 and θ_0 from eqn. (5.12)

$$\theta_L = (\theta_0 - C_2)e^{-mL} + C_2 e^{mL}$$
$$- \theta_0 e^{-mL} + C_2[e^{mL} - e^{mL}]$$

It gives $\quad C_2 = \dfrac{\theta_L - \theta_0 e^{-mL}}{e^{mL} - e^{-mL}}$

And $\quad C_1 = \theta_0 - \dfrac{\theta_L - \theta_0 e^{-mL}}{e^{mL} - e^{-mL}}$

It gives $\quad C_1 = \dfrac{\theta_0 e^{mL} - \theta_L}{e^{mL} - e^{-mL}}$

Substituting C_1 and C_2 in eqn. (5.11), we get

$$\theta(x) = \frac{\theta_0\,[e^{m(L-x)} - e^{-m(L-x)}] + \theta_L\,(e^{mx} - e^{-mx})}{e^{mL} - e^{-mL}} \qquad ...(5.28)$$

or in terms of hyperbolic functions ;

$$\frac{\theta(x)}{\theta_0} = \frac{T - T_\infty}{T_0 - T_\infty}$$

$$= \frac{\sinh m\,\{(L - x)\} + \left(\dfrac{T_L - T_\infty}{T_0 - T_\infty}\right)\sinh mx}{\sinh (mL)} \qquad ...(5.29)$$

The heat conduction to fin at the base surface :

The heat conduction rate at any section is given by

$$Q = -kA_c \frac{dT}{dx}$$

At base of fin, *i.e.*, at $x = 0$

$$Q_{x=0} = -kA_c\left[\frac{dT}{dx}\right]_{x=0} = -kA_c\left[\frac{d\theta}{dx}\right]_{x=0}$$

Using eqn. (5.29) and differentiating it, with respect to x,

$$Q_{x=0} = -kA_c\,(T_0 - T_\infty) \times \frac{d}{dx}\left[\frac{\sinh m(L-x) + \dfrac{T_L - T_\infty}{T_0 - T_\infty}\sinh mx}{\sinh mL}\right]_{x=0}$$

$$= -\frac{kA_c\,(T_0 - T_\infty)}{\sinh mL} \times \left[\cosh m(L-x)(-m) + \frac{T_L - T_\infty}{T_0 - T_\infty}\cosh mx\,(m)\right]_{x=0}$$

$$= -\frac{kA_c \times (-m)\,(T_0 - T_\infty)}{\sinh mL}\left[\cosh mL - \frac{T_L - T_\infty}{T_0 - T_\infty}\right]$$

$$\text{or } Q_{x=0} = \sqrt{hPkA_c}\,(T_0 - T_\infty)\left[\frac{\cosh mL - \left(\dfrac{T_L - T_\infty}{T_0 - T_\infty}\right)}{\sinh mL}\right] \qquad ...(5.30)$$

The heat leaving the base surface at $x = 0$ will not be totally convected to surrounding medium, but some heat will also reach to the surface at $x = L$.

The rate of heat dissipation to surrounding fluid can be worked out from relation

$$Q_{fin} = \int_0^L hP\,dx\,(T - T_\infty)$$

Using eqn. (5.29), we get

$$Q_{fin} = hP\,(T_0 - T_\infty) \times \int_0^L \left(\frac{\sinh m\,(L-x) + \dfrac{T_L - T_\infty}{T_0 - T_\infty}\sinh mx}{\sinh mL}\right) dx$$

$$= \frac{hP(T_0 - T_\infty)}{\sinh mL}\int_0^L \left(\sinh m(L-x) + \frac{T_L - T_\infty}{T_0 - T_\infty}\sinh mx\right) dx$$

$$= \frac{hP\,(T_0 - T_\infty)}{\sinh mL}\left[-\frac{\cosh m(L-x)}{m} + \left(\frac{T_L - T_\infty}{T_0 - T_\infty}\right)\frac{\cosh mx}{m}\right]_0^L$$

$$= \frac{hP\,(T_0 - T_\infty)}{m \sinh mL} \times \left[\left(\frac{T_L - T_\infty}{T_0 - T_\infty}\right)(\cosh mL - 1) - (1 - \cosh mL)\right]$$

$$= \sqrt{hPkA_c}$$

$$\times \left[\frac{(T_L - T_\infty)(\cosh mL - 1) + (T_0 - T_\infty)(\cosh mL - 1)}{\sinh mL}\right]$$

or $$Q_{fin} = \sqrt{hPkA_c}\ [(T_L - T_\infty) + (T_0 - T_\infty)]\ \frac{(\cosh mL - 1)}{\sinh mL} \quad ...(5.31)$$

The heat conducted to other surface at $x = L$

$$Q_{x = L} = Q_{x = 0} - Q_{fin} \quad ...(5.32)$$

For fins of non-uniform cross-section, the solution is quite complex. The solution for triangular and parabolic fins are presented in section 5.5, but the approximate solution to these fins in graphical form is presented in Fig. 5.26. The interested students can also refer Schneider [1] and Arpaci [2].

Example 5.1. *A very long 25 mm diameter copper (k = 380 W/m.K) rod extends from a surface at 120°C. The temperature of surrounding air is 25°C and the heat transfer coefficient over the rod is 10 W/m².K. Calculate:*

(i) Heat loss from the rod,

(ii) How long the rod should be in order to be considered infinite ? (Shivaji University, Nov. 2002)

Solution

Given : A very long (infinite long) copper rod as a fin :

d = 25 mm = 0.025 m, k = 380 W/m.K

T_0 = 120°C, T_∞ = 25°C

h = 10 W/m².K

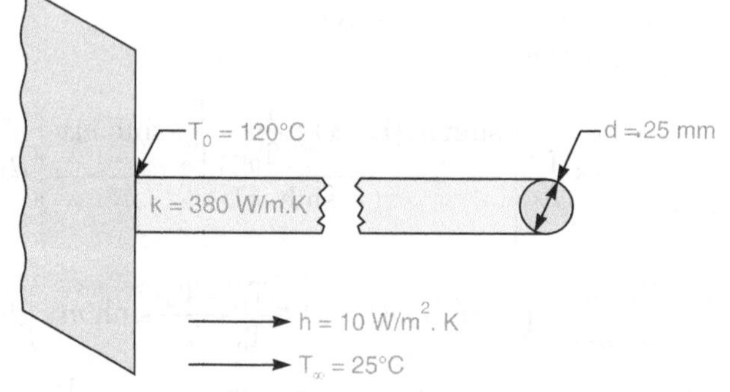

Fig. 5.9. A very long rod extends from a surface.

To find :

(i) The heat loss from infinite long fin.

(ii) Length of rod as infinite long fin.

Analysis : (i) The heat loss from infinite long fin, eqn. (5.14)

$$Q_{\text{infinite fin}} = \sqrt{hPkA_c}\ (T_0 - T_\infty)$$

where $$P = \pi d = \pi \times (0.025 \text{ m}) = 0.0785 \text{ m}$$

$$A_c = \frac{\pi}{4} d^2 = \left(\frac{\pi}{4}\right) \times (0.025 \text{ m})^2$$

$$= 4.908 \times 10^{-4} \text{ m}^2$$

$$\therefore Q_{\text{infinite fin}} = \sqrt{10 \times 0.0785 \times 380 \times 4.908 \times 10^{-4}} \times (120 - 25)$$

$$= \mathbf{36.35\ W.}$$ **Ans.**

(ii) From an infinite log fin $T_L = T_\infty$, and no heat transfer from its free end. Therefore,

$$Q_{\text{infinite fin}} = Q_{\text{insulated tip fin}}$$

Thus $$\sqrt{hPkA_c}\ (T_0 - T_\infty) = \sqrt{hPkA_c}\ (T_0 - T_\infty) \tanh mL$$

The equivalent result is obtained if

$$\tanh(mL) \geq 0.99$$

$$mL = 2.646$$

Here $$m = \sqrt{\frac{hP}{kA_c}}$$

$$= \sqrt{\frac{10 \times 0.0785}{380 \times 4.908 \times 10^{-4}}}$$

$$= 2.052$$

$$\therefore \quad L = \frac{2.646}{2.052} = \mathbf{1.29\ m.}$$ **Ans.**

Example 5.2. *One end of a long rod 3 cm in diameter is inserted into a furnace with the outer end projecting into the outside air. Once the steady state is reached the temperature of the rod is measured at two points, 15 cm apart and found to be 140°C and 100°C, when the atmospheric air is at 30°C with convection coefficient of 20 W/m².K. Calculate the thermal conductivity of the rod material.* (P.U., May 1992)

Solution

Given : One end of the long rod inserted into a furnace ;

d = 3 cm = 0.03 m, L = 15 cm = 0.15 m,

T_0 = 140°C, T_∞ = 30°C,

T_L = 100°C, h = 20 W/m².K.

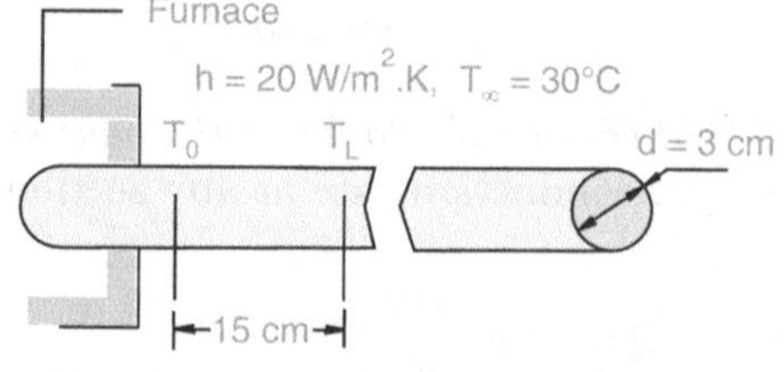

Fig. 5.10. Schematic for example 5.2

To find : The thermal conductivity of the rod material.

Assumptions :

(*i*) Steady state conditions.

(*ii*) One dimensional conduction along the rod.

(*iii*) Constant properties.

(*iv*) No internal heat generation.

(*v*) Infinite long fin.

Analysis : For infinite long fin, the temperature distribution is given by eqn. (5.13)

$$\frac{T(x) - T_\infty}{T_0 - T_\infty} = e^{-mx}$$

The starting point,

$$\text{at } x = 0, \qquad T_0 = 140°C$$

and $$\text{at } x = L = 0.15 \text{ m}, \qquad T_L = 100°C$$

Thus $$\frac{100 - 30}{140 - 30} = e^{-m \times 0.15}$$

It gives $$m = 3.013$$

We have $$m = \sqrt{\frac{hP}{kA_c}}$$

Thus $$\sqrt{\frac{20 \times (\pi \times 0.03)}{k \times \{(\pi/4) \times (0.03)^2\}}} = 3.013$$

It gives k = **293.74 W/m.K. Ans.**

Example 5.3. *The two long rods A and B, equivalent in every respect except that one is fabricated from material of known thermal conductivity of k_A while other of material of unknown thermal conductivity k_B, are attached to a surface of fixed temperature T_0, and are exposed to a fluid at T_∞, with convection coefficient h. These rods are instrumented with thermocouples to measure the temperature at a fixed distance x_1 from the heat source. If the standard material is of aluminium k_A = 200 W/m.K and measurements reveal T_A = 75°C and T_B = 60°C at x_1 when T_0 is 100°C and T_∞ is 25°C. What is the thermal conductivity of the test material ?*

(N.M.U., May 1999)

Solution

Given : Two long similar rods.

$$k_A = 200 \text{ W/m.K}, \qquad x = x_1,$$
$$T_A = 75°C, \qquad T_0 = 100°C,$$
$$T_B = 60°C, \qquad T_\infty = 25°C.$$

To find : The thermal conductivity of the test material B.

Assumptions :

(*i*) Steady state conditions.

(*ii*) One dimensional conduction along the rod.

(*iii*) Constant properties.

(*iv*) No internal heat generation.

(*v*) Infinite long fin.

Analysis : For infinite long fin, the temperature distribution is given by eqn. (5.13)

$$\frac{T(x) - T_\infty}{T_0 - T_\infty} = e^{-mx}$$

For rod A, at x_1

$$\frac{T_A - T_\infty}{T_0 - T_\infty} = e^{-m_A x_1}$$

or $$m_A x_1 = -\ln\left[\frac{75 - 25}{100 - 25}\right] = 0.405 \qquad ...(i)$$

Similarly for rod B at x_1

$$m_B x_1 = -\ln\left[\frac{T_B - T_\infty}{T_0 - T_\infty}\right]$$

$$= -\ln\left[\frac{60 - 25}{100 - 25}\right] = 0.762 \qquad ...(ii)$$

Dividing eqn. (*i*) by eqn. (*ii*)

$$\frac{m_A}{m_B} = \sqrt{\frac{k_B}{k_A}} = \frac{0.405}{0.762}$$

or $$k_B = 200 \times \left[\frac{0.405}{0.762}\right]^2$$

$$= \mathbf{56.5 \text{ W/m.K.}} \quad \textbf{Ans.}$$

Example 5.4. *It is required to heat the oil to 300°C for frying purpose. A long laddle is used in a frying pan. The section of the laddle is 5 mm × 18 mm. The surrounding air is at 30°C. The thermal conductivity of the material is 205 W/m.K. If the temperature at a distance of 380 mm from the oil should not exceed 40°C, determine convective heat transfer coefficient.*

(N.M.U., Dec. 2002)

Solution

Given : The long handle of a laddle as shown in Fig. 5.11.

$$T_0 = 300°C$$
$$A_c = 5 \text{ mm} \times 18 \text{ mm} = 90 \text{ mm}^2$$
$$= 90 \times 10^{-6} \text{ m}^2$$
$$P = 2(w + t) = 2 \times (18 + 5)$$
$$= 46 \text{ mm} = 0.046 \text{ m}$$
$$T_\infty = 30°C$$
$$k = 205 \text{ W/m.K}$$
$$x = 380 \text{ mm} = 0.38 \text{ m}$$
$$T(x) = 40°C.$$

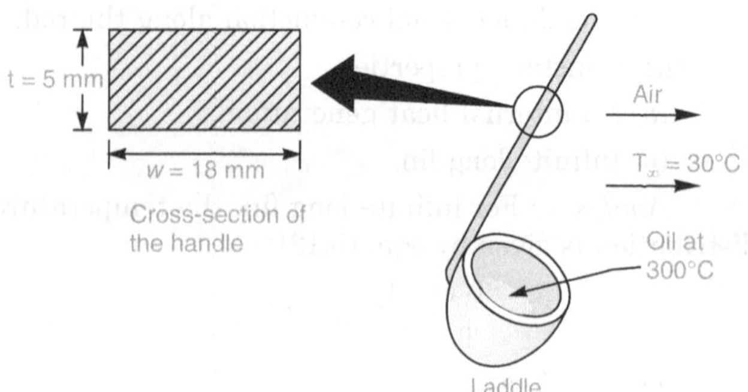

Fig. 5.11. Schematic for example 5.4

To find : The heat transfer coefficient.

Assumptions :

(*i*) Steady state conditions.

(*ii*) Long handle is treated as infinite long fin.

(*iii*) Oil temperature as base temperature of fin.

(*iv*) Constant properties.

Analysis : As the long handle may be treated as an infinite long fin, the temperature distribution is given by eqn. (5.13)

$$\frac{T(x) - T_\infty}{T_0 - T_\infty} = e^{-mx}$$

or $$\frac{40 - 30}{300 - 30} = e^{-0.38\,m}$$

or $$\ln\left(\frac{10}{270}\right) = -\,0.38\,m$$

It gives $m = 8.673$

and m is expressed as

$$m = \sqrt{\frac{hP}{kA_c}}$$

It gives $$h = \frac{m^2 kA_c}{P}$$

$$= \frac{(8.673)^2 \times 205 \times 90 \times 10^{-6}}{0.046}$$

$$= \mathbf{30.17\ W/m^2.K.}$$ **Ans.**

Example 5.5. *Two long rods of the same diameter, one made of brass (k = 85 W/m.K) and the other made of copper (k = 375 W/m.K) have one of their ends inserted into a furnace. Both the rods are exposed to same environment. At a distance of 105 mm away from the furnace, the temperature of brass rod is 120°C. At what distance from the furnace, the same temperature would be reached in the copper rod ?* (I.E.S., 1993)

Solution

Given : Two similar long rods as infinite long fins.

k_1 = 85 W/m.K

k_2 = 375 W/m.K

x_1 = 105 mm = 0.105 m

T_1 = 120°C.

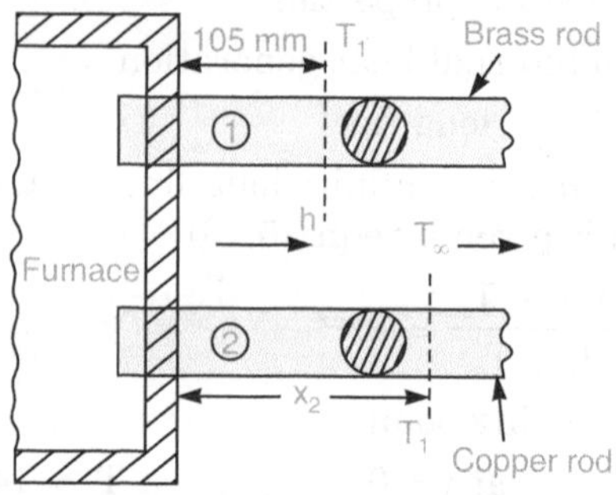

Fig. 5.12. Schematic for example 5.5

To find : The distance x_2 from furnace of copper rod, where temperature of 120°C will reach.

Analysis : The long rods are treated as infinite long fins. For infinite long fin, the temperature distribution

$$\frac{T - T_\infty}{T_0 - T_\infty} = e^{-mx}$$

For brass rod $$\frac{T_1 - T_\infty}{T_0 - T_\infty} = e^{-m_1 x_1} \quad ...(i)$$

For copper rod $$\frac{T_1 - T_\infty}{T_0 - T_\infty} = e^{-m_2 x_2} \quad ...(ii)$$

Equating eqns. (*i*) and (*ii*), we get

$$m_1 \times 105 = m_2\, x_2$$

or $$x_2 = \frac{m_1}{m_2} \times 105$$

$$= \frac{\sqrt{hP/k_1 A_c}}{\sqrt{hP/k_2 A_c}} \times 105 = \sqrt{\frac{k_2}{k_1}} \times 105$$

$$= \sqrt{\frac{375}{85}} \times 105 = \mathbf{220.5\ mm.}$$ **Ans.**

Example 5.6. *Three rods of copper, aluminium and stainless steel are coated with wax all around and are dipped vertically in a water bath at 85°C. The length of each rod projecting outside the bath is 300 mm. Diameter of each rod is 20 mm and length is 400 mm. Convective heat transfer coefficient at the surface of each rod is 11 W/m².K.*

Thermal conductivity of

(*i*) *Copper rod = 380 W/mK*

(*ii*) *Aluminium rod = 206 W/mK*

(*iii*) *Steel rod = 17 W/mK*

Calculate the ratio of lenghts of rod up to which wax melting occurs due to transfer of heat.

(N.M.U., May 2004)

Solution

Given : The rods identical in length and cross-section, coated with wax

$T_0 = 85°C$, $L = 300$ mm,
$d = 20$ mm, $h = 11$ W/m^2.K,
$k_{cu} = 380$ W/m.K, $k_{Al} = 206$ W/m.K,
$k_{st} = 17$ W/m.K.

To find : Ratio of lengths of rods at which wax melts.

Assumptions :

(*i*) Steady state condition.

(*ii*) Since diameter of rods is small as compared to length, thus treating infinite long fin.

(*iii*) Constant properties.

Analysis : For long rod (infinite long fins), the temperature distribution is given as

$$\frac{T(x) - T_\infty}{T_0 - T_\infty} = e^{-mx}$$

For melting of wax along the rod of three different materials, let us assume x_1, x_2 and x_3 are lengths for copper, aluminium and steel rods, respectively up to which wax melts at temperature T (say). Then

For copper rod, $$\frac{T - T_\infty}{T_0 - T_\infty} = e^{-m_1 x_1} \quad ...(i)$$

For aluminium rod,

$$\frac{T - T_\infty}{T_0 - T_\infty} = e^{-m_2 x_2} \quad ...(ii)$$

For steel rod, $$\frac{T - T_\infty}{T_0 - T_\infty} = e^{-m_3 x_3} \quad ...(iii)$$

Since L.H.S. and base e on R.H.S. are same in all three equations thus

$$m_1 x_1 = m_2 x_2 = m_3 x_3$$

or $$\sqrt{\frac{hP}{k_{cu}A_c}}\, x_1 = \sqrt{\frac{hP}{k_{Al}A_c}}\, x_2 = \sqrt{\frac{hP}{k_{st}A_c}}\, x_3$$

or $$\frac{x_1}{\sqrt{k_{cu}}} = \frac{x_2}{\sqrt{k_{Al}}} = \frac{x_3}{\sqrt{k_{st}}}$$

Thus $$\frac{x_1}{x_3} = \sqrt{\frac{k_{cu}}{k_{st}}} = \sqrt{\frac{380}{17}} = 4.727$$

and $$\frac{x_2}{x_3} = \sqrt{\frac{k_{Al}}{k_{st}}} = \sqrt{\frac{206}{17}} = 3.481$$

Thus $x_1 : x_2 : x_3$ = **4.727 : 3.481 : 1.** Ans.

Example 5.7. *An electric motor is to be connected by a horizontal steel shaft (k = 42.56 W/m.K), 25 mm in diameter to an impeller of a pump, circulating liquid metal at a temperature of 540°C. If the temperature of electric motor is limited to a maximum value of 52°C with the ambient air at 27°C and heat transfer coefficient of 40.7 W/m².K, what length of shaft should be specified between the motor and pump ?*

Solution

Given : A horizontal steel shaft as a fin.

$k = 42.56$ W/m.K, $d = 25$ mm $= 0.025$ m,
$T_0 = 540°C$, $T_L = 52°C$,
$T_\infty = 27°C$, $h = 40.7$ W/m^2.K.

To find : The length of steel shaft

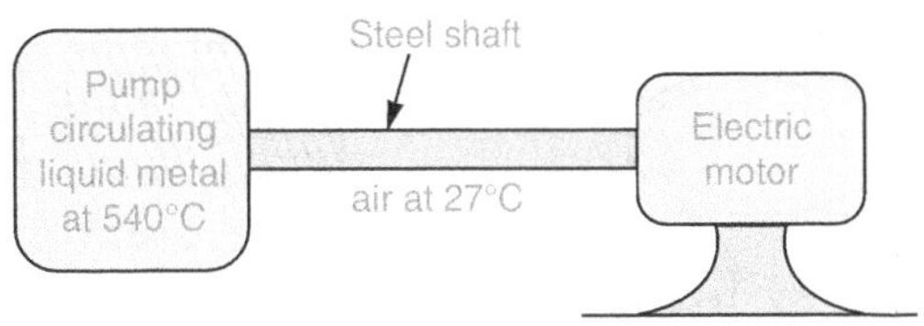

Fig. 5.13. Schematic for example 5.7

Assumptions :

(*i*) Steady state conditions

(*ii*) Since one end of shaft is connected to electric motor, thus assuming, no heat loss from the fin tip.

(*iii*) Constant properties.

Analysis : The diameter of fin is very small and hence treating fin of insulated tip.

$$A_c = \frac{\pi}{4} d^2 = \frac{\pi}{4} \times (0.025)^2 = 4.90 \times 10^{-4} \text{ m}^2$$

$$P = \pi d = 0.025\,\pi \text{ m}$$

$$m = \sqrt{\frac{hP}{kA_c}} = \sqrt{\frac{40.7 \times 0.025\,\pi}{42.56 \times 4.90 \times 10^{-4}}}$$

$$= 12.37 \text{ m}$$

The temperature distribution in the fin at $x = L$

$$\frac{T_L - T_\infty}{T_0 - T_\infty} = \frac{1}{\cosh mL}$$

$$\frac{52 - 27}{540 - 27} = \frac{1}{\cosh mL}$$

or $$\cosh mL = 20.52$$

or $$mL = 3.714$$

or $$L = 0.3 \text{ m} = \mathbf{30\ cm.}$$ Ans.

Example 5.8. *Consider a stainless steel spoon (k = 15.1 W/m.K), partially immersed in the boiling water at 95°C in a kitchen at 25°C. The handle of the spoon has a cross-section 0.2 cm × 1 cm and it extends 18 cm in the air from the free surface of the water. If the heat transfer coefficient on the exposed surface of the spoon is 15 W/m²*.*K, calculate the temperature difference across the exposed surface of the spoon handle. State your assumptions, if any.* (Shivaji University, 2008)

Solution

Given : The handle of a stainless steel spoon :

k = 15.1 W/m.K, T_0 = 95°C,
T_∞ = 25°C, A_c = 0.2 cm × 1 cm,
L = 18 cm = 0.18 m, h = 15 W/m².K.

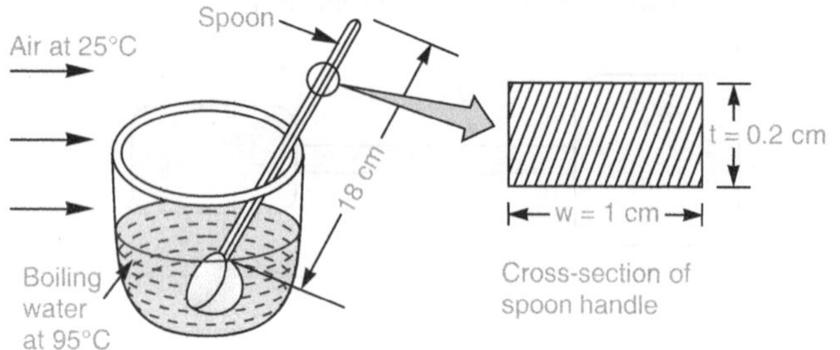

Fig. 5.14. Schematic for example 5.8

To find : The temperature difference across exposed surface of handle.

Assumptions :

1. Steady state conditions.
2. The handle of spoon is thin and heat loss from its free end be negligible.
3. No heat radiation.
4. Constant cross-section of handle.
5. Constant properties.

Analysis : The temperature distribution for insulated tip fin is given by eqn. (5.20)

$$\frac{T - T_\infty}{T_0 - T_\infty} = \frac{\cosh m(L - x)}{\cosh mL}$$

For temperature difference across the exposed surface of spoon handle *i.e.*, x = L ;

$$\frac{T_L - T_\infty}{T_0 - T_\infty} = \frac{1}{\cosh mL}$$

where $m = \sqrt{\dfrac{hP}{kA_c}}$

$$P = 2(w + t) = 2(1 \text{ cm} + 0.2 \text{ cm})$$
$$= 2.4 \text{ cm} = 0.024 \text{ m}$$
$$A_c = wt = 1 \text{ cm} \times 0.2 \text{ cm}$$
$$= 0.2 \text{ cm}^2 = 0.2 \times 10^{-4} \text{ m}^2$$

$$\therefore \quad m = \sqrt{\frac{15 \times 0.024}{15.1 \times 0.2 \times 10^{-4}}} = 34.52 \text{ m}^{-1}$$

$$mL = 34.52 \times 0.18 = 6.21$$
$$\cosh mL = 250.0$$

Then $T_L - T_\infty = \dfrac{95 - 25}{250} = 0.2799 \approx 0.28$°C

or T_L = 25 + 0.28 = 25.28°C,

Thus $T_0 - T_L$ = 95 – 25.28 = **69.72°C.** Ans.

Example 5.9. *The handle of a saucepan, 30 cm long and 2 cm in diameter is partially immersed in boiling water at 100°C. The average unit conductance over the handle surface is 7.35 W/m²*.*K in the kitchen air at 24°C. The cook is likely to grasp the last 10 cm of the handle and hence, the temperature of this portion should not exceed 32°C. What should be the material conductivity of handle ? The handle may be treated as a fin of insulated tip.* (N.M.U., May 2002)

Solution

Given : The handle of a saucepan :

L = 30 cm = 0.3 m, d = 2 cm = 0.02 m,
h = 7.35 W/m².K,
x = L – 10 cm = 20 cm = 0.2 m,
T_0 = 100°C, T_∞ = 24°C,
T(x) = 32°C.

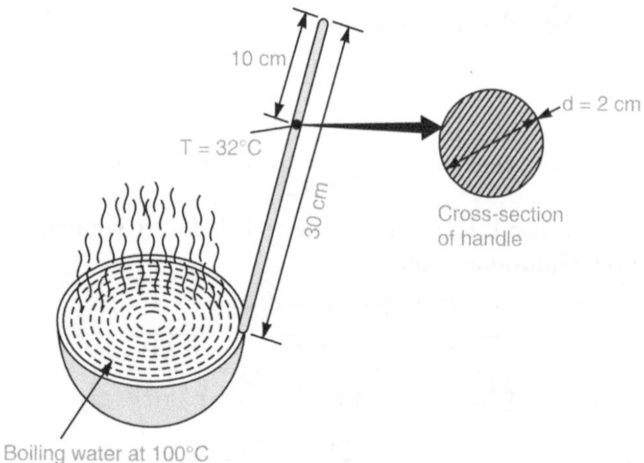

Fig. 5.15. Schematic for example 5.9

To find : Thermal conductivity of handle material.

Analysis : Since handle is treated as a fin of insulated tip, hence, the temperature distribution in the fin.

$$\frac{T(x) - T_\infty}{T_0 - T_\infty} = \frac{\cosh m(L - x)}{\cosh mL}$$

Using numerical values

$$\frac{32-24}{100-24} = \frac{\cosh m(0.3-0.2)}{\cosh m(0.3)}$$

or $$0.10526 = \frac{\cosh(0.1\,m)}{\cosh(0.3\,m)}$$

By trial and error : $m = 11.71$

For a circular handle

$$A_c = \frac{\pi}{4}d^2 = \frac{\pi}{4} \times (0.02)^2$$

$$P = \pi d = (\pi \times 0.02)$$

and $$m = \sqrt{\frac{hP}{kA_c}}$$

or $$(11.71)^2 = \frac{7.35 \times (\pi \times 0.02)}{k \times (\pi/4) \times (0.02)^2}$$

It gives $$k = \frac{0.4618}{(11.71)^2 \times 3.141 \times 10^{-4}}$$

$$= \mathbf{10.72\ W/m.K.} \quad \text{Ans.}$$

It must be stainless steel.

Example 5.10. *A steel fin (k = 54 W/m.K) with a cross-section of an equilateral triangle, 5 mm in side is 80 mm long. It is attached to a plane wall maintained at 400°C. The ambient air temperature is 50°C and unit surface conductance is 90 W/m².K. Calculate the heat dissipation rate from the rod.*

Solution

Given : A finite long fin with an equilateral triangle cross-section.

side $a = 5$ mm, L = 80 mm,

$k = 54$ W/m.K, $T_0 = 400$°C,

$T_\infty = 50$°C, $h = 90$ W/m².K.

Fig. 5.16. Schematic of a triangular fin

To find : The heat dissipation rate from the fin.

Assumptions :

1. Steady state conditions.

2. Length of fin is very large in comparison to its cross-section, thus treating fin as infinite long fin.

Analysis : For triangular fin

Cross-sectional area,

$$A_c = \frac{1}{2}\text{ base} \times \text{height} = \frac{1}{2}a \times \frac{\sqrt{3}}{2}a$$

$$= \frac{\sqrt{3}}{4} \times (5 \times 10^{-3})^2 = 1.0825 \times 10^{-5}\ \text{m}^2$$

Perimeter,

$$P = 3a = 3 \times 5 \times 10^{-3} = 0.015\ \text{m}$$

$$m = \sqrt{\frac{hP}{kA_c}} = \sqrt{\frac{90 \times 0.015}{54 \times 1.0825 \times 10^{-5}}}$$

$$= 48.06\ \text{m}^{-1}$$

The heat transfer rate from infinite long fin

$$Q_{fin} = \sqrt{hPkA_c}\,(T_0 - T_\infty)$$

$$= \sqrt{90 \times 0.015 \times 54 \times 1.0825 \times 10^{-5}} \times (400-50)$$

$$= \mathbf{9.82\ W.} \quad \text{Ans.}$$

Note: The students can also solve this example by assuming insulated tip fin or with corrected length approximation. The result does not change in either case.

Example 5.11. *Calculate the temperature distribution, temperature at the middle and rate of heat flow at the root of a turbine blade with 80 mm long, 600 mm² in cross-section and 150 mm in perimeter. The blade is made of stainless steel (k = 23.3 W/m.K) and is exposed to steam at 1000°C, while its root is maintained at 600°C. The heat transfer coefficient between the blade surface and steam is 500 W/m².K.*

Solution

Given : A turbine blade as a fin

$A_c = 600$ mm², L = 80 mm,

P = 150 mm, $k = 23.3$ W/m.K,

$T_0 = 600$°C, $T_\infty = 1000$°C,

$h = 500$ W/m².K.

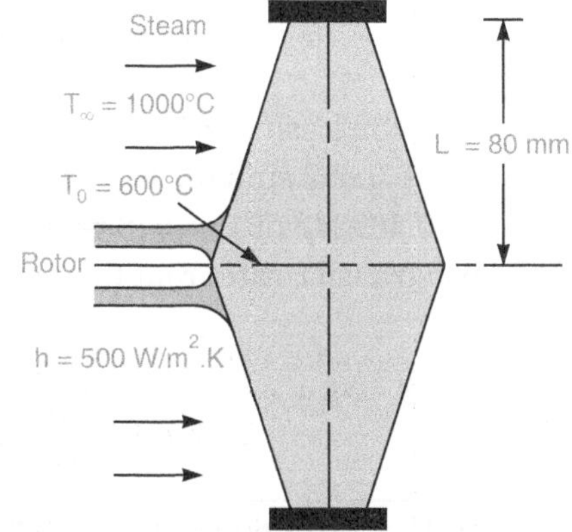

Fig. 5.17. Schematic for example 5.11

To find :

(*i*) Temperature distribution in the turbine blade.

(*ii*) Temperature at the middle of blade, and

(*iii*) The heat flow at root of the blade.

Assumptions :

1. Steady state conditions.

2. Finite long fin with convection heat transfer from its tip.

3. Constant properties.

Analysis : (*i*) For given turbine blade

$A_c = 600 \text{ mm}^2 = 600 \times 10^{-6} \text{ m}^2$

$P = 150 \text{ mm} = 0.15 \text{ m}$

$L = 80 \text{ mm} = 0.08 \text{ m}$

$$m = \sqrt{\frac{hP}{kA_c}} = \sqrt{\frac{500 \times 0.15}{23.3 \times 600 \times 10^{-6}}} = 73.245$$

$$mL = 73.245 \times 0.08 = 5.86$$

$$\frac{h}{mk} = \frac{500}{73.245 \times 23.3} = 3.662$$

The temperature distribution in finite long fin with convective tip

$$\frac{T - T_\infty}{T_0 - T_\infty} = \frac{\cosh m(L - x) + \frac{h}{mk} \sinh m(L - x)}{\cosh mL + \frac{h}{mk} \sinh mL}$$

$$\frac{T(x) - 1000}{600 - 1000} = \frac{\cosh [73.245 \times (0.08 - x)] + 3.662 \sinh [73.245 \times (0.08 - x)]}{\cosh (5.86) + 3.662 \times \sinh (5.86)}$$

or $$T(x) - 1000 = \frac{-400 \times [\cosh (5.86 - 73.245x) + 3.662 \sinh (5.86 - 73.245x)]}{817.534}$$

$$\mathbf{T(x) - 1000 = -0.489 [\cosh (5.86 - 73.245\, x) + 3.662 \sinh (5.86 - 73.245\, x)]}$$

It is the required expression for temperature distribution. **Ans.**

(*ii*) Temperature at the middle of blade *i.e.*, $x = 40 \text{ mm} = 0.04 \text{ m}$

$$T_{middle} = 1000 - 0.489 [\cosh (5.86 - 73.245 \times 0.04) + 3.662 \sinh (5.86 - 73.245 \times 0.04)]$$

$$= 1000 - 0.489 [43.591] = \mathbf{978.68°C.} \quad \textbf{Ans.}$$

(*iii*) The heat transfer rate from fin

$$Q_{fin} = \sqrt{hPkA_c}\,(T_0 - T_\infty) \times \frac{\sinh mL + \frac{h}{mk} \cosh mL}{\cosh mL + \frac{h}{mk} \sinh mL}$$

$$= \sqrt{500 \times 0.15 \times 23.3 \times 600 \times 10^{-6}} \times (600 - 1000) \times \frac{\sinh (5.86) + 3.662 \times \cosh (5.86)}{\cosh (5.86) + 3.662 \times \sinh (5.86)}$$

$$= \mathbf{-409.58 \text{ W.}} \quad \textbf{Ans.}$$

(Heat transfer towards centre)

Example 5.12. *The handle of a ladle used for pouring molten metal at 327°C is 30 cm long and is made of 2.5 cm × 1.5 cm mild steel bar stock (k = 43 W/m.K). In order to reduce the grip temperature it is proposed to make a hollow handle of mild steel plate 0.15 cm thick to the same rectangular shape. If the surface heat transfer coefficient is 14.5 W/m².K and the ambient temperature is at 27°C, estimate the reduction in the temperature of grip. Neglect the heat transfer from inner surface of the hollow shape.* (N.M.U., Nov. 1999)

Solution

Given : A handle of ladle as a finite long fin

$T_0 = 327°C$, $\quad T_\infty = 27°C$,

$k = 43$ W/m.K, $\quad h = 14.5$ W/m².K,

$L = 30$ cm, $\quad w = 2.5$ cm,

$H = 1.5$ cm,

$A_c = 2.5 \times 1.5 \text{ cm}^2$ (solid handle),

$t = 0.15$ cm thick.

To find : The reduction in temperature of grip.

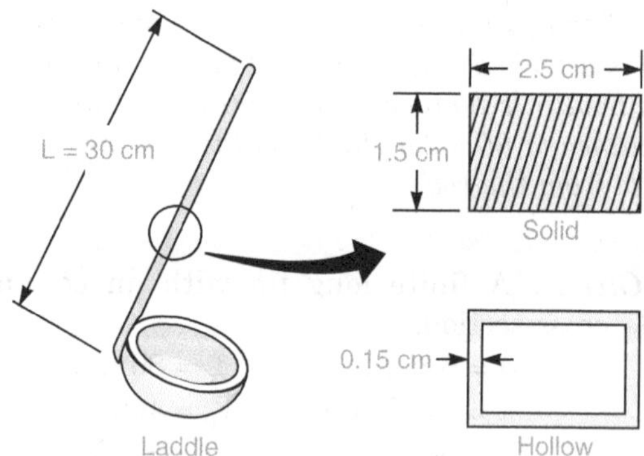

Fig. 5.18. Cross-section of handle

Assumptions :

1. Steady state one dimensional conduction along the handle.

2. Constant properties.

3. No internal heat generation.

4. Assuming no heat loss from end face of handle due to grip cover.

Analysis : **Case 1.** Solid handle

$$A_c = wH = 2.5 \times 10^{-2} \times 1.5 \times 10^{-2} = 3.75 \times 10^{-4} \text{ m}^2$$

$$P = 2(w + H) = 2(2.5 \times 10^{-2} + 1.5 \times 10^{-2}) = 0.08 \text{ m}$$

$$m = \sqrt{\frac{hP}{kA_c}} = \sqrt{\frac{14.5 \times 0.08}{43 \times 3.75 \times 10^{-4}}}$$

$$= 8.48 \text{ m}^{-1}$$

$$mL = 8.48 \times 0.3 = 2.544$$

The temperature at the grip of handle can be calculated by relation

$$\frac{T_{(x=L)} - T_\infty}{T_0 - T_\infty} = \frac{1}{\cosh(mL)}$$

$$\frac{T_L - 27}{327 - 27} = \frac{1}{\cosh(2.544)} = 0.1561$$

or $$T_L = 27 + 300 \times 0.1561 = 73.84°C$$

Case 2. When steel plate is used in a hollow handle.

$$A_{c,H} = 2Ht + 2(w - 2t)t$$

$$= 2 \times (1.5 \times 10^{-2} \times 0.15 \times 10^{-2}) + 2 \times (2.5 - 2 \times 0.15) \times 10^{-2} \times 0.15 \times 10^{-2}$$

$$= 1.11 \times 10^{-4} \text{ m}^2$$

$P = 0.08$ m and $L = 30.0$ cm

$$m = \sqrt{\frac{hP}{kA_{c,H}}} = \sqrt{\frac{14.5 \times 0.08}{43 \times 1.11 \times 10^{-4}}} = 15.59 \text{ m}^{-1}$$

$$mL = 15.59 \times 0.3 = 4.677$$

The temperature at the grip handle

$$\frac{T_{L,H} - 27}{327 - 27} = \frac{1}{\cosh(4.677)} = 0.0186$$

$$T_{L,H} = 32.58$$

The reduction in temperature

$$= T_L - T_{L,H}$$

$$= 73.84 - 32.58 = \mathbf{41.25°C}. \text{ Ans.}$$

Example 5.13. *Two ends of a fin of the cross-sectional area 2 cm², perimeter 2 cm, 100 cm long are maintained at 127°C and 227°C, respectively. It losses heat from the surface due to natural convection to surroundings at 27°C with heat transfer coefficient of 5 W/m².K. Thermal conductivity of fin material is 45 W/m.K. Find the minimum temperature in the fin and its location. Also calculate the heat conducted from each end.*

(P.U. and N.M.U.)

Solution

Given : Finite long fin with specified temperature at two ends.

$A_c = 2$ cm², $P = 2$ cm,
$L = 100$ cm, $T_0 = 127°C$,
$T_L = 227°C$, $T_\infty = 27°C$,
$h = 5$ W/m².K, $k = 45$ W/m.K.

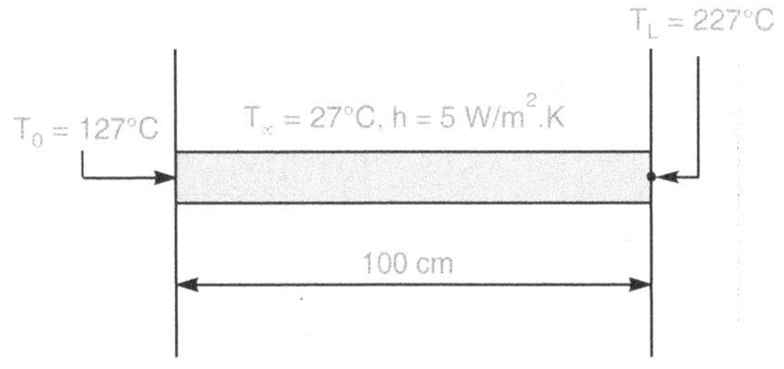

Fig. 5.19. Schematic for example 5.13

To find :

(*i*) The minimum temperature.

(*ii*) Heat loss at two ends.

Analysis : (*i*) In steady state the temperature distribution in the finite long fin,

$$\frac{T - T_\infty}{T_0 - T_\infty} = \frac{\frac{T_L - T_\infty}{T_0 - T_\infty}\sinh(mx) + \sinh\{m(L-x)\}}{\sinh(mL)} \quad ...(i)$$

where, $$m = \sqrt{\frac{hP}{kA_c}} = \sqrt{\frac{5 \times 2 \times 10^{-2}}{45 \times 2 \times 10^{-4}}} = 3.33$$

For location of minimum temperature, differentiating equation (*i*) w.r.t. x and equating it to zero.

$$\frac{dT}{dx} = (T_0 - T_\infty) \times \left\{\frac{\frac{T_L - T_\infty}{T_0 - T_\infty}\cosh mx(m) + \cosh[m(L-x)](-m)}{\sinh(mL)}\right\} = 0$$

or $$(T_L - T_\infty)\cosh mx = (T_0 - T_\infty)\cosh m(L - x)$$

or $$(227 - 27) \times \cosh(3.33x) = (127 - 27) \times \cosh[3.33(1 - x)]$$

or $$2 \times \left[\frac{e^{3.33x} + e^{-3.33x}}{2}\right] = \left[\frac{e^{3.33(1-x)} + e^{-3.33(1-x)}}{2}\right]$$

or $$2e^{3.33x} + 2e^{-3.33x} = e^{3.33} \times e^{-3.33x} + e^{-3.33} \times e^{3.33x}$$

or $$25.94\, e^{-3.33x} - 1.964 e^{3.333x} = 0$$

It gives, $x = 0.3875$ m

The minimum temperature

$$\frac{T(x) - 27}{127 - 27} = \frac{\frac{227 - 27}{127 - 27} \times \sinh(3.33 \times 0.3875) + \sinh[3.33(1 - 0.3875)]}{\sinh(3.333 \times 1)} = 0.5109$$

or $$T(x) = 27 + 100 \times 0.5109 = \mathbf{78.1°C}. \text{ Ans.}$$

Alternatively : The minimum temperature can also be calculated by assuming two fins with insulated tip.

From left face, $$T(x) = \frac{(T_0 - T_\infty)}{\cosh mL} + T_\infty$$

$$\left\{\because \frac{T(x) - T_\infty}{T_0 - T_\infty} = \frac{1}{\cosh mL}\right\}$$

From right face,

$$T(L - x) = \frac{T_L - T_\infty}{\cosh m(L - x)} + T_\infty$$

Equating $T(x) = T(L - x)$, we get

$$x = 0.3875 \text{ m.}$$

$$T(x) = T(L - x) = \mathbf{78.1°C.}$$ **Ans.**

(*ii*) The heat conducted from left end

$$Q_L = \sqrt{hPkA_c}\,(T_0 - T_\infty) \tanh (mx)$$

$$= \sqrt{5 \times 2 \times 10^{-2} \times 45 \times 2 \times 10^{-4}}$$

$$\times (127 - 27) \times \tanh (3.33 \times 0.3875)$$

$$= \mathbf{2.578\ W.}$$ **Ans.**

The heat loss from right end

$$Q_R = \sqrt{hPkA_c}\,(T_L - T_\infty) \tanh m\,(L - x)$$

$$= \sqrt{5 \times 2 \times 10^{-2} \times 45 \times 2 \times 10^{-4}}$$

$$\times (227 - 27) \times \tanh [3.33 \times (1 - 0.3875)]$$

$$= \mathbf{5.8\ W.}$$ **Ans.**

Example 5.14. *The both ends of a 6 mm diameter copper rod (U-shaped) having k = 330 W/m.K are rigidly connected to a vertical wall as shown in Fig. 5.20 (a). The wall temperature is constant at 100°C. The developed length of the rod is 50 cm and is exposed to air at 30°C. The combined convective and radiative heat transfer coefficient is 30 W/m².K. Calculate :*

(i) The temperature at the centre of the rod.

(ii) Heat transfer by the rod.

(P.U., Nov. 1996 ; N.M.U., Nov. 2000)

Solution

Given : U-shaped circular fin.

$d = 6$ mm, $L = 50$ cm,

$k = 330$ W/m.K, $T_0 = 100°C$,

$T_\infty = 30°C$, $T_L = 100°C$,

$h = 30$ W/m².K.

To find :

(*i*) The temperature at the centre of the fin.

(*ii*) The heat transfer rate.

Assumption : Steady state heat conduction along the rod.

Analysis : Considering finite long fin with specified temperature at its free end as shown in Fig. 5.20 (*a*).

$$A_c = \frac{\pi}{4} d^2 = \frac{\pi}{4} \times (6 \times 10^{-3})^2$$

$$= 2.827 \times 10^{-5} \text{ m}^2$$

$$P = \pi d = \pi \times 6 \times 10^{-3} = 0.0188 \text{ m}$$

$$m = \sqrt{\frac{hP}{kA_c}} = \sqrt{\frac{30 \times 0.0188}{330 \times 2.827 \times 10^{-5}}}$$

$$= 7.785 \text{ m}$$

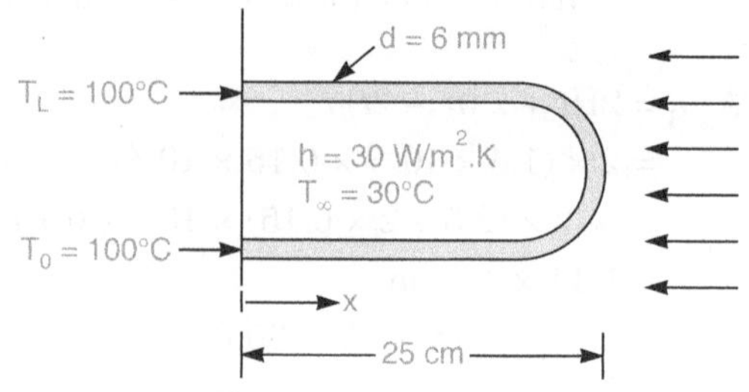

Fig. 5.20. (*a*) Schematic for example 5.14

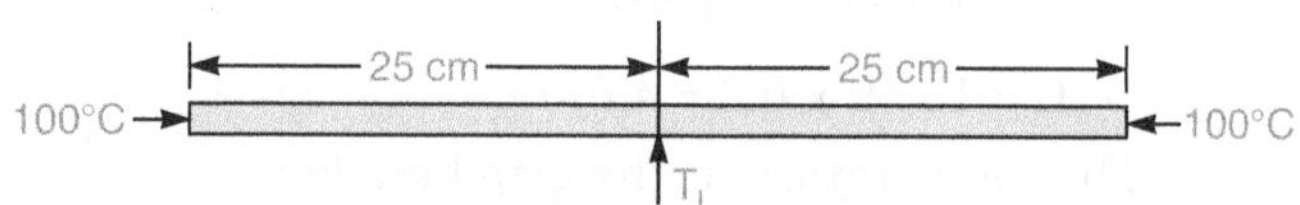

Fig. 5.20. (*b*) Alternative arrangement

(*i*) The temperature distribution in the fin

$$\frac{T(x) - T_\infty}{T_0 - T_\infty} = \frac{\left[\frac{\theta_L}{\theta_0}\right] \sinh (mx) + \sinh \{m(L - x)\}}{\sinh (mL)}$$

The temperature at the centre of fin *i.e.*, At $x = 25$ cm,

Due to symmetry $\theta_L = \theta_0$

$$\text{Thus} \quad \frac{T_c - 30}{100 - 30} = \frac{\sinh (7.785 \times 0.25) + \sinh \{7.785 (0.5 - 0.25)\}}{\sinh (7.785 \times 0.5)}$$

$$= 0.28$$

or $$T_c = 30 + 70 \times 0.28$$

$$= \mathbf{49.6°C.}$$ **Ans.**

Alternatively : At the centre of fin, temperature would be constant due to symmetry.

The temperature of U-shaped fin can also be obtained by considering two insulated tip fins, 25 cm long each, as shown in Fig. 5.20 (*b*).

$$\frac{T_c - T_\infty}{T_0 - T_\infty} = \frac{1}{\cosh(mL)}$$

$$T_c = \mathbf{49.6°C.} \quad \text{Ans.}$$

(*ii*) The heat transfer rate from the fin :

Considering two fins of 25 cm long each with insulated tip.

$$Q = 2\sqrt{hPkA_c}\,(T_0 - T_\infty)\tanh(mL)$$

$$= 2\sqrt{30 \times 0.0188 \times 330 \times 2.827 \times 10^{-5}} \times (100 - 30) \times \tanh(7.785 \times 0.25)$$

$$= \mathbf{9.76\ W.} \quad \text{Ans.}$$

Example 5.15. *One end of a copper rod (k = 380 W/m.K), 300 mm long is connected to a wall which is maintained at 300°C. The other end is firmly connected to other wall at 100°C. The air is blown across the rod so that the heat transfer coefficient of 20 W/m²*.*K is maintained. The diameter of the rod is 15 mm and temperature of air is 40°C. Determine :*

(i) Net heat transfer rate to air,

(ii) The heat conducted to other end which is at 100°C. (M.U., Nov. 1999)

Solution

Given : A copper rod as fin with specified temperature at its two ends.

$k = 380$ W, $\quad L = 300$ mm $= 0.3$ m,

$T_0 = 300°C$, $\quad T_L = 100°C$,

$h = 20$ W/m².K, $\quad d = 15$ mm $= 0.015$ m,

$T_\infty = 40°C$.

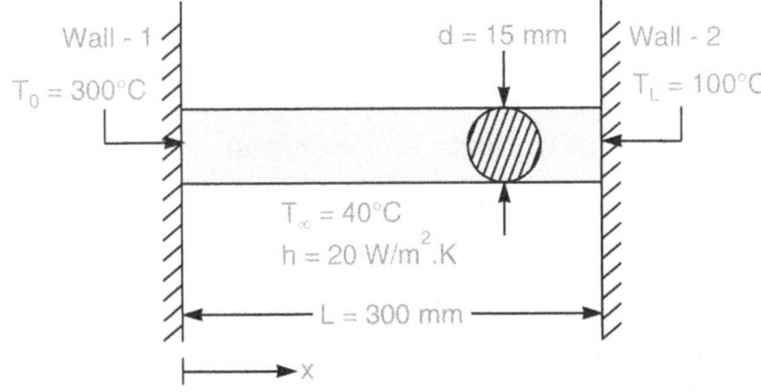

Fig. 5.21. Schematic for example 5.15

To find :

(*i*) Net heat transfer rate to air,

(*ii*) Heat conduction rate to wall at 100°C.

Analysis : For the copper rod as a fin

$$A_c = \frac{\pi}{4}d^2 = \frac{\pi}{4} \times (0.015)^2$$

$$= 1.767 \times 10^{-4}\ \text{m}^2$$

$$P = \pi d = 0.015\,\pi\ \text{m}$$

$$m = \sqrt{\frac{hP}{kA_c}} = \sqrt{\frac{20 \times 0.015\,\pi}{380 \times 1.767 \times 10^{-4}}}$$

$$= 3.746\ \text{m}^{-1}$$

$$mL = 3.746 \times 0.3 = 1.124$$

(*i*) The heat dissipated to air from fin is given by eqn. (5.31)

$$Q_{fin} = \sqrt{hPkA}\,[(T_L - T_\infty) + (T_0 - T_\infty)]\left\{\frac{\cosh mL - 1}{\sinh mL}\right\}$$

$$= \sqrt{20 \times 0.015\pi \times 380 \times 1.767 \times 10^{-4}} \times [(100 - 40) + (300 - 40)]\,\frac{\cosh(1.124) - 1}{\sinh(1.124)}$$

$$= 0.2515 \times 320 \times 0.5094 = \mathbf{41.0\ W.} \quad \text{Ans.}$$

(*ii*) The heat conduction rate to fin from surface at $x = 0$, eqn. (5.30)

$$Q_{x=0} = \sqrt{hPkA_c}\,(T_0 - T_\infty) \times \frac{\cosh mL - \left(\frac{T_L - T_\infty}{T_0 - T_\infty}\right)}{\sinh mL}$$

$$Q_{x=0} = \sqrt{20 \times 0.015\pi \times 380 \times 1.767 \times 10^{-4}} \times (300 - 40) \times \frac{\cosh(1.124) - \left(\frac{100 - 40}{300 - 40}\right)}{\sinh(1.124)}$$

$$= 0.2515 \times 260 \times 1.0684 = 69.88\ \text{W.}$$

The heat reaches to other end at 100°C

$$= Q_{x=0} - Q_{fin} = 69.88 - 41.0$$

$$= \mathbf{28.87\ W.} \quad \text{Ans.}$$

5.4. FIN PERFORMANCE

The fins are used to increase the heat transfer rate from a surface by increasing the effective surface area. The use of fins on a surface cannot be recommended unless the increase in heat transfer justifies the added cost and complexity associated with fins. However, the fin itself puts a conduction resistance to heat transfer from original surface. For this reason, there is no assurance that the heat transfer rate will be increased through the use of fins. A parameter called as *fin effectiveness*, justifies the use of fins, if its value is greater than unity.

5.4.1. Fin Effectiveness

The *fin effectiveness* is defined as the *ratio of the fin heat transfer rate to that of which would occur from the surface on which the fin was attached*, therefore,

$$\varepsilon_{fin} = \frac{\text{Heat transfer rate with fin from base area}}{\text{Heat transfer rate from base area without fin}} = \frac{Q_{fin}}{Q_{no\ fin}} \quad \text{...(5.33)}$$

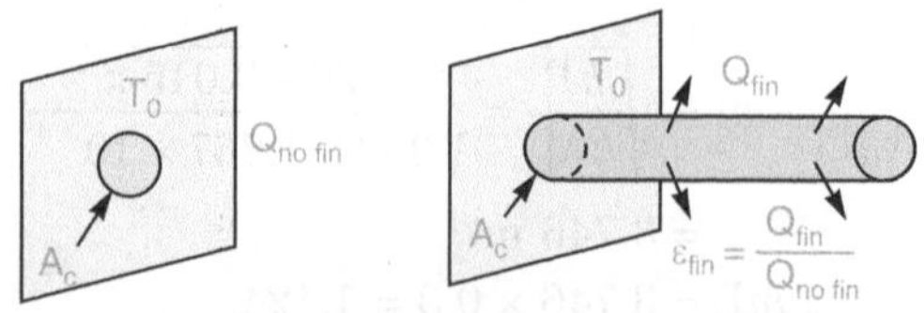

Fig. 5.22. Effectiveness of a fin

Some observations with use of fins are :

(*i*) An effectiveness $\varepsilon_{fin} = 1$, indicates that the addition of fins to the surface does not affect the heat transfer rate at all.

(*ii*) An effectiveness $\varepsilon_{fin} < 1$, indicates that the fin actually acts as insulation and decreasing the heat transfer rate from the surface. It may occur, if fin of low thermal conductivity materials is used.

(*iii*) An effectiveness $\varepsilon_{fin} > 1$, indicates that fins are increasing the heat transfer rate from the surface.

However, the use of fins cannot be justified unless ε_{fin} is more than 5. For an infinite long fin of uniform cross-section, using eqn. (5.14) to obtain the fin effectiveness.

$$\varepsilon_{fin} = \frac{Q_{fin}}{Q_{no\ fin}} = \frac{\sqrt{hPkA_c}\,(T_0 - T_\infty)}{hA_c\,(T_0 - T_\infty)}$$

$$\varepsilon_{fin} = \left(\frac{kP}{hA_c}\right)^{1/2} \qquad ...(5.34)$$

where A_c is cross-sectional area of fin.

We can conclude from the fin effectiveness eqn. (5.34) for consideration in design and selection of the fins :

1. The fin effectiveness can be increased by choice of a material of *higher thermal conductivity*. Therefore, the fins are usually made from metals, with copper, aluminium and iron. But aluminium is the best choice due to its low cost and weight and its resistance to corrosion.
2. The fin effectiveness is also enhanced by increasing the ratio of *perimeter to cross-sectional area of the fin* (P/A_c). For this-reason, aluminium or copper thin fins, or slender pin fins, closely spaced are preferred in most engineering applications.
3. The use of fins can be better justified under conditions for which the *convection heat transfer coefficient is small*. Therefore, fins are placed on a surface on gas side, where the heat transfer is by natural convection instead of forced convection.

5.4.2. Fin Efficiency

Consider a fin of uniform cross-sectional area, made of constant thermal conductivity material as shown in Fig. 5.23. The heat flows from the surface to the fin by conduction and from the fin to the surrounding medium by convection. The convection from the fin surface causes gradually temperature drop along the fin length.

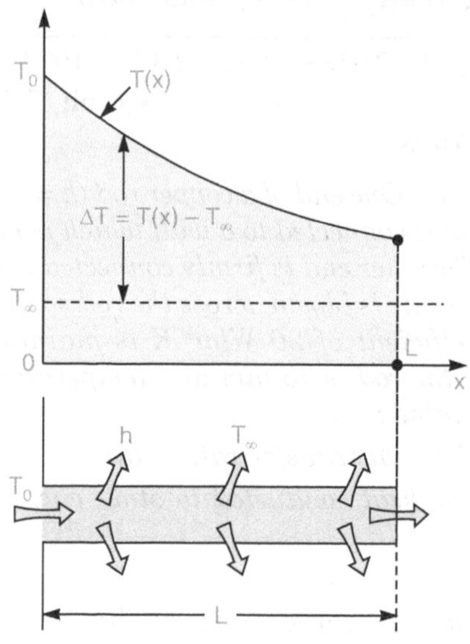

Fig. 5.23. Temperature of a fin drops gradually along the fin

In the limiting case of zero thermal resistance ($k \to \infty$), the temperature of fin along its length will be uniform at the base temperature T_0. The heat transfer from such an ideal fin will be maximum and can be expressed as

$$Q_{ideal} = hA_{fin}\,(T_0 - T_\infty) \qquad ...(5.35)$$

The area A_{fin} represent total surface area of the fins. For a fin of uniform cross-section,

$$A_{fin} = A_{lateral} + A_{tip} = A_{lateral} + A_c$$

Usually $A_c << A_{lateral}$

$\therefore$ $A_{fin} \approx A_{lateral} \approx PL$

However, in actual practice, the temperature drops along the fin length and the resulting heat dissipation from the fin will be less due to decreasing temperature difference $T(x) - T_\infty$ towards the fin tip. A parameter called the *fin efficiency*, evaluates the thermal performance of a fin and is defined as

$$\eta_{fin} = \frac{\text{Actual heat transfer rate from fin}}{\text{Ideal heat transfer rate from fin, if entire fin surface were at fin base temperature } T_0}$$

$$= \frac{Q_{fin}}{Q_{ideal}} \qquad ...(5.36)$$

If the fin efficiency is known, the heat transfer rate Q_{fin} through the fin is determined as

$$Q_{fin} = \eta_{fin}\, Q_{ideal} = \eta_{fin}\, hA_{fin}\,(T_0 - T_\infty)$$

or

$$Q_{fin} \approx \eta_{fin}\, h\, PL\,(T_0 - T_\infty) \quad ...(5.37)$$

The fin efficiency for earlier defined fin geometrics are given below :

(*i*) The fin efficiency of infinite long fin :

$$\eta_{fin} = \frac{\sqrt{hPkA_c}\,(T_0 - T_\infty)}{hPL\,(T_0 - T_\infty)} = \sqrt{\frac{kA_c}{hP}} \times \frac{1}{L} = \frac{1}{mL} \quad ...(5.38)$$

(*ii*) The fin efficiency for insulated fin tip :

$$\eta_{fin} = \frac{\sqrt{hPkA_c}\,(T_0 - T_\infty)\tanh(mL)}{hPL\,(T_0 - T_\infty)}$$

$$= \frac{\tanh(mL)}{mL} \quad ...(5.39)$$

(*iii*) Similarly for finite long fin with convection heat transfer at the fin tip :

$$\eta_{fin} = \frac{\sinh(mL) + \dfrac{h}{mk}\cosh(mL)}{mL} \quad ...(5.40)$$

The fin efficiency relations are developed for fins of various profiles. These are plotted in Fig. 5.24 for fins of plane surfaces with insulated tip.

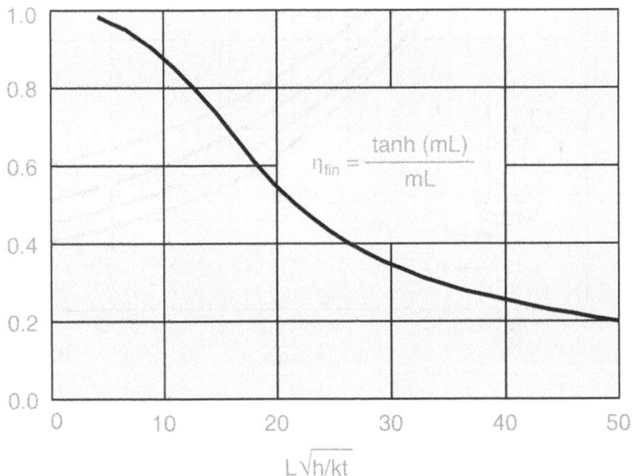

Fig. 5.24. Fin efficiency for insulated tip fin

5.4.3. Overall Fin Effectiveness

The *overall fin effectiveness* for a finned surface can be defined as

$$\varepsilon_{fin\ overall} = \frac{\text{Total heat transfer rate from the finned surface}}{\text{Heat transfer rate from the base surface if there were no fins}}$$

or

$$\varepsilon_{fin\ overall} = \frac{Q_{total}}{Q_{no\ fins}} = \frac{h(A_{unfin} + \eta_{fin}\, A_{fin})(T_0 - T_\infty)}{h\, A_{no\ fin}\,(T_0 - T_\infty)} \quad ...(5.41)$$

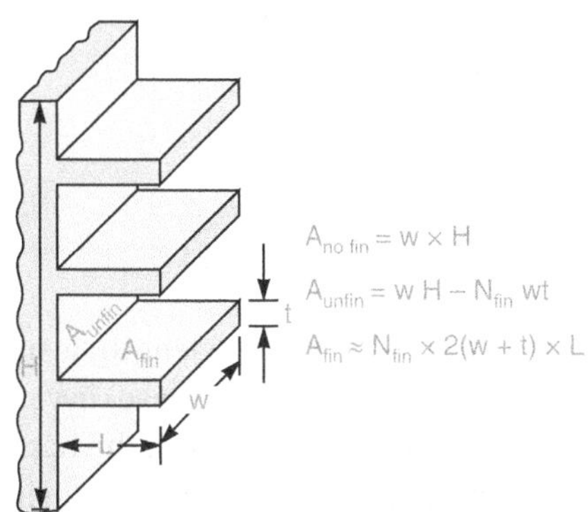

Fig. 5.25. Various surface areas associated with rectangular fins

where $A_{no\ fin}$ = area of the surface, when there are no fins,

A_{fin} = total surface area of all fins,

A_{unfin} = area of the unfinned portion of the surface,

η_{fin} = fin efficiency.

The overall fin effectiveness depends on the *fin density* (number of fins per unit length), as well as the effectiveness of individual fins. The overall effectiveness is a better measure of the performance of a finned surface than the effectiveness of the individual fins.

5.4.4. Area Weighted Fin Efficiency

In practical applications, a finned heat transfer surface is composed of the fin surfaces and the unfinned portion. Therefore, total heat transfer

$$Q_{total\ fin} = Q_{fin} + Q_{unfin}$$
$$= \eta_{fin}\, hA_{fin}\,(T_0 - T_\infty) + (A_{total} - A_{fin})\, h(T_0 - T_\infty) \quad ...(5.42)$$

where A_{fin} = heat transfer (surface) area of all (N_{fin}) fins

A_{total} = total heat transfer area = $A_{fin} + A_{unfin}$

We can also define *area weighted fin efficiency* or *total fin efficiency* which evaluates the thermal performance of the finned surface and defined as

$$\eta_{total\ fin} = \frac{\text{The total heat transfer rate from finned surface}}{\text{The heat which would be transferred if the total area was maintained at base temperature } T_0}$$

$$\eta_{fin} = \frac{\eta_{fin}\, h A_{fin}(T_0 - T_\infty) + (A_{total} - A_{fin})\, h\,(T_0 - T_\infty)}{h\, A_{total}\,(T_0 - T_\infty)}$$

$$= 1 - \frac{A_{fin}}{A_{total}}(1 - \eta_{fin}) \qquad ...(5.43)$$

5.5. APPROXIMATE SOLUTION OF FIN: CONCEPT OF CORRECTED FIN LENGTH

The solution in case of fin of finite length, losing heat by convection from its free end (actual case) is very tedious. In order to avoid complex calculations, the heat loss from the fin tip can be approximated by increasing the fin length by δ (friction length) and assuming the fin of insulated tip.

In this approach, the solution for finite long fin with convection heat loss from its tip can be approximated by the expressions for finite long fin, insulated at its tip, when length of fin L is replaced by corrected length L_c. The resulting error from this approximation will be less than 8%, when

$$\frac{ht}{2k} \le 1/4 \qquad ...(5.44)$$

Corrected length, $L_c = L + \delta$ ...(5.45)

where δ is called *friction length* of fin and is calculated as

$$\delta = \frac{A_c}{P} \qquad ...(5.46)$$

For rectangular or square fin

$$A_c = w \times t$$
$$P \approx 2w$$

then,
$$\delta = \frac{w \times t}{2w} = \frac{t}{2}$$

For circular fin, pin fin,

$$\delta = \frac{(\pi/4)d^2}{\pi d} = \frac{d}{4} = \frac{r}{2}$$

It is often convenient to use the profile area of fin A_p

For a rectangular fin, $A_p = Lt$

For a triangular fin, $A_p = Lt/2$

For a parabolic fin, $A_p = Lt/3$

The fin efficiency relations are developed for fins of various profiles and are plotted in Fig. 5.26 for fins on a plane surface and of three common profiles: rectangular fin of uniform cross-section, and triangular and parabolic fins of non-uniform cross-section. Fig. 5.27 presents the efficiency for annular or circular fins of rectangular profile. These results are presented in terms of corrected length L_c and profile area. The maximum heat transfer rate from rectangular, triangular and parabolic fins is

$$Q_{max} = hPL_c\,(T_0 - T_\infty) \qquad ...(5.47)$$

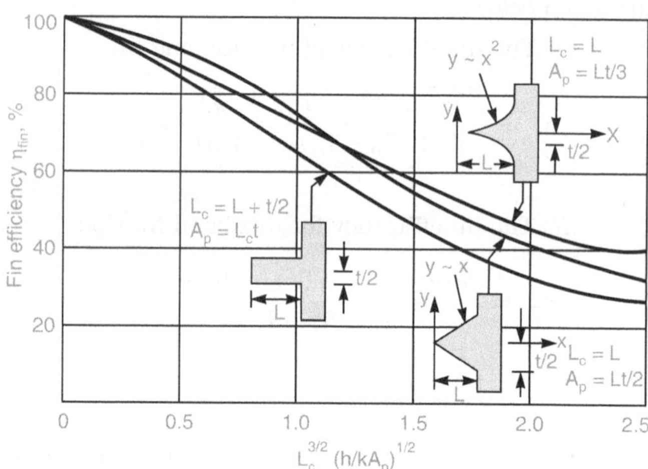

Fig. 5.26. Efficiency of straight fins (rectangular, triangular and parabolic profiles) (from Gardner)

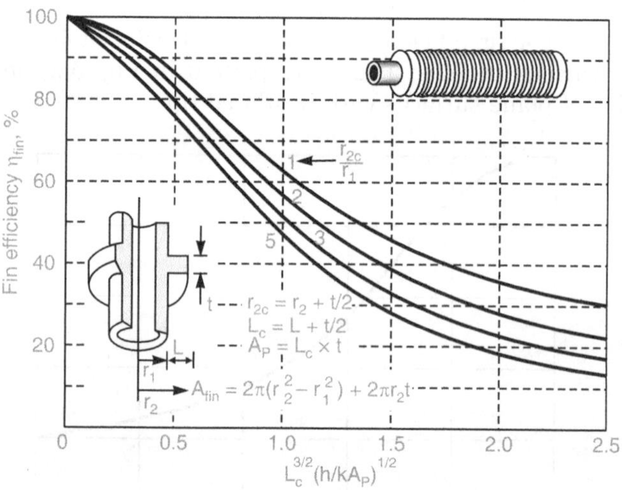

Fig. 5.27. Efficiency of circular fins of rectangular profile (from Gardner)

For circumferential fins of constant thickness

$$Q_{max} = 2\pi h\,(r_{2c}^2 - r_1^2)\,(T_0 - T_\infty) \qquad ...(5.48)$$

The triangular and parabolic fins contains less material and are more efficient than the fins of rectangular profile, and thus they are more suitable for applications that require minimum weight such as space applications.

Example 5.16. *An aluminium alloy fin (k = 200 W/m.K), 3.5 mm thick and 2.5 cm long protrudes from a wall. The base is at 420°C and ambient air temperature is 30°C. The heat transfer coefficient may be taken as 11 W/m².K. Find the heat loss and fin efficiency, if the heat loss from fin tip is negligible.*

Solution

Given : A rectangular fin

$L = 2.5$ cm, $T_\infty = 30°C$,

$t = 3.5$ mm, $h = 11$ W/m².K,

$T_0 = 420°C$, $k = 200$ W/m.K.

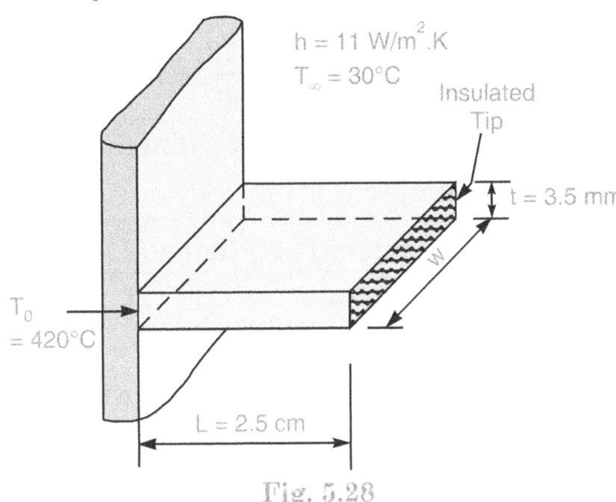

Fig. 5.28

To find :

(*i*) Heat transfer rate from fin surface.

(*ii*) Fin efficiency.

Assumptions :

1. Steady state conditions.
2. One dimensional conduction along the rod.
3. Constant properties.
4. No internal heat generation.
5. The width of fin is 1 m.

Analysis : (*i*) The heat transfer rate

$$Q_{fin} = \sqrt{hPkA_c}\,(T_0 - T_\infty)\tanh(mL)$$

where, $$mL = L\sqrt{\frac{hP}{kA_c}}$$

and $$P = 2w = 2 \text{ m}$$

$$A_c = w \times t = 1 \times 3.5 \times 10^{-3} \text{ m}^2$$

$$mL = 2.5 \times 10^{-2} \times \sqrt{\frac{11 \times 2}{200 \times 3.5 \times 10^{-3}}} = 0.14$$

$$Q_{fin} = \sqrt{11 \times 2 \times 200 \times 3.5 \times 10^{-3}} \times (420 - 30) \times \tanh(0.14)$$

$= \mathbf{213\ W.}$ **Ans.**

(*ii*) The fin efficiency

$$\eta_{fin} = \frac{\tanh mL}{mL} = \frac{\tanh(0.14)}{0.14}$$

$= \mathbf{99.35\%.}$ **Ans.**

Example 5.17. *It is better to use 10 fins of 5 cm length than 5 fins of 10 cm length. State and prove correctness of the statement. Take properties as follows :*

Diameter of fin = 10 mm

Thermal conductivity = 45 W/m.K

Heat transfer coefficient = 95 W/m².K.

(P.U., May 2012)

Solution

Given : Two arrangement of fins with

$d = 10$ mm,

$k = 45$ W/m.K, $h = 95$ W/m².K

Arrangement 1 : $L_1 = 5$ cm, $N_{fin} = 10$

Arrangement 2 : $L_2 = 10$ cm, $N_{fin} = 5$

To find : Better arrangement.

Analysis : The length of fin in either case is much larger than its diameter, therefore, assuming negligible heat loss from fin tip. For a fin

$$A_c = \frac{\pi}{4}d^2 = \frac{\pi}{4} \times (0.01)^2 = 78.54 \times 10^{-6} \text{ m}^2$$

$$P = \pi d = \pi \times 0.01 = 0.01\,\pi \text{ m}$$

$$m = \sqrt{\frac{hP}{kA_c}} = \sqrt{\frac{95 \times 0.01\pi}{45 \times 78.54 \times 10^{-6}}} = 29.06$$

For arrangement 1 :

$$mL_1 = 29.06 \times 0.05 = 1.453$$

$$\tanh(mL_1) = 0.896$$

The fin efficiency

$$\eta_{fin} = \frac{\tanh mL_1}{mL_1} = \frac{0.896}{1.453} = 61.68\%$$

The heat transfer rate from 10 fins

$$Q_{fin} = N_{fin}\sqrt{hPkA_c}\,(T_0 - T_\infty)\tanh(mL_1)$$

$$= 10 \times \sqrt{95 \times 0.01 \times \pi \times 45 \times 78.54 \times 10^{-6}} \times (T_0 - T_\infty) \times 0.896$$

$$= 9.2\,(T_0 - T_\infty) \text{ Watts.}$$

For arrangement 2 : 5 fins 10 cm long

$$mL_2 = 29.06 \times 0.1 = 2.906$$

$$\tanh(mL_2) = 0.994$$

$$\eta_{fin} = \frac{\tanh mL}{mL} = \frac{0.994}{2.906} = 34.2\%$$

and heat transfer rate with this arrangement

$$Q_{fin} = 5 \times \sqrt{95 \times 0.01\pi \times 45 \times 78.54 \times 10^{-6}} \times (T_0 - T_\infty) \times 0.994$$

$$= 5.10\,(T_0 - T_\infty) \text{ Watts.}$$

Comment : The fin efficiency and heat transfer rate, both are much less in second arrangement of fins, *i.e.,* 5 fin, 10 cm long in comparison to arrangement of 10 fins, 5 cm long. The heat dissipation decreases along the length of the fin. The short fins are always effective. Therefore, *large number of short fins should be installed on a surface to increase the heat transfer rate.*

Example 5.18. *Three identical straight fins, 10 mm in diameter and 120 mm long are exposed to an ambient with convective heat transfer coefficient of 32 W/m².K. Compare their efficiency and relative heat flow performance. The three fin materials and their thermal conductivities are :*

Copper : 380 W/m.K
Aluminium : 210 W/m.K
Mild steel : 45 W/m.K.

Solution

Given : Three identical circular fins

d = 10 mm, L = 120 mm,
h = 32 W/m².K, k_1 = 380 W/m.K,
k_2 = 210 W/m.K, k_3 = 45 W/m.K.

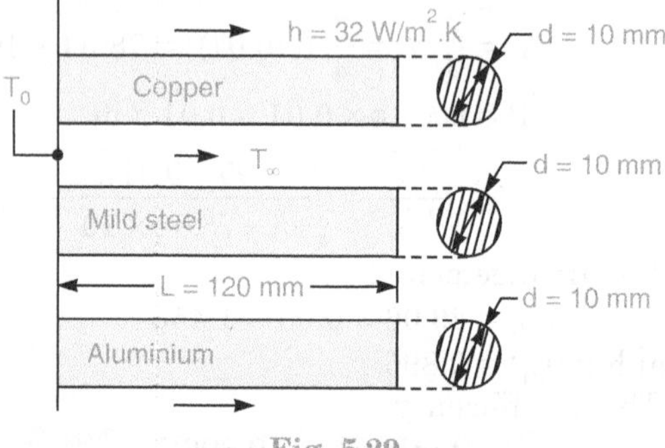

Fig. 5.29

To find :

(*i*) Efficiency of each fin

(*ii*) Relative heat dissipation rate.

Analysis : (*i*) The corrected length of fin

$$L_c = L + \frac{d}{4} = 120 \text{ mm} + 2.5 \text{ mm}$$
$$= 122.5 \text{ m}$$
$$P = \pi d = \pi \times 0.01 = 0.01\,\pi$$
$$A_c = \frac{\pi}{4} \times d^2 = \frac{\pi}{4} \times (0.01)^2$$
$$= 78.53 \times 10^{-6} \text{ m}^2$$
$$m = \sqrt{\frac{hP}{kA_c}} = \sqrt{\frac{32 \times 0.01\pi}{k \times 78.53 \times 10^{-6}}} = \frac{113.13}{\sqrt{k}}$$

For copper fin $m_1 = \frac{113.13}{\sqrt{380}} = 5.804$

$$m_1 L_c = 5.803 \times 0.1225 = 0.711$$

Fin efficiency

$$\eta_{fin} = \frac{\tanh(mL_c)}{mL_c} = \frac{\tanh(0.711)}{0.711}$$
$$= 0.86 = 86\%$$

(*ii*) Heat dissipation rate for copper fin

$$Q_{fin} = \sqrt{hPkA_c}\,(T_0 - T_\infty) \tanh mL_c$$
$$= \sqrt{32 \times 0.01\pi \times 380 \times 78.53 \times 10^{-6}}$$
$$\times (T_0 - T_\infty) \tanh(0.711)$$
$$= 0.106\,(T_0 - T_\infty) \text{ Watts}$$

The calculated values of all materials are tabulated below.

Material	*m*	*mL*	*tanh m*	η_{fin}	Q_{fin} *for* $(T_0 - T_\infty)$	*Relative heat dissipation with respect to copper fin*
Copper	5.804	0.711	0.611	86%	0.106	100
Aluminium	7.807	0.956	0.742	77.6%	0.0956	90.2%
Mild steel	16.865	2.066	0.968	46.8%	0.0577	54.5%

The heat dissipation rate by mild steel fin is least, the fin efficiency of mild steel fin is also low, due to its low value of thermal conductivity. The efficiency of copper fin is highest. The fin efficiency with same material can also improved, if the fins are made short.

Example 5.19. *An electronic semiconductor device has a rating of 60 mW. In order to keep its proper operation, the inside temperature should not exceed 70°C. The device can dissipate about 20 mW of heat on its own when placed in an environment at 40°C with heat transfer coefficient of 12.5 W/m².K. To avoid overheating of the device, it is proposed to install aluminium (k = 190 W/m.K) square fins 0.6 mm side, 10 mm long, to provide additional cooling. Find the number of fins required. Assume no heat loss from the tip of fins.* (P.U., May 1989)

Solution

Given : Square fins with insulated tip

Q_{unfin} = 20 mW, Q_{total} = 60 mW,
T_0 = 70°C, T_∞ = 40°C,
h = 12.5 W/m².K, k = 190 W/m.K,
t = 0.6 mm, L = 10 mm.

To find : The number of fins.

Assumptions :

1. Steady state conditions, one dimensional heat conduction along the rod.
2. Constant properties.
3. No internal heat generation.

Analysis : The net heat to be dissipated by fins

$$Q_{fin} = Q_{total} - Q_{unfin} = 60 - 20$$
$$= 40 \text{ mW} = 40 \times 10^{-3} \text{ W}$$

Cross-section area of square fin

$$A_c = (0.6 \times 10^{-3})^2 = 3.6 \times 10^{-7} \text{ m}^2$$

Perimeter of square fin

$$P = 4t = 4 \times 0.6 \times 10^{-3}$$
$$= 2.4 \times 10^{-3} \text{ m}$$

$$mL = L\sqrt{\frac{hP}{kA_c}}$$

$$= 10 \times 10^{-3} \times \sqrt{\frac{12.5 \times 2.4 \times 10^{-3}}{190 \times 3.6 \times 10^{-7}}}$$

$$= 0.209$$

The heat transfer rate ;

$$Q_{fin} = N_{fin}\sqrt{hPkA_c}\,(T_0 - T_\infty) \tanh mL$$

Using the values,

$$40 \times 10^{-3} = N_{fin}\sqrt{12.5 \times 2.4 \times 10^{-3} \times 190 \times 3.6 \times 10^{-7}} \times (70 - 40) \times \tanh(0.209)$$

or Number of fins

$$N_{fin} = \frac{40 \times 10^{-3}}{8.853 \times 10^{-3}} = \mathbf{4.518 \approx 5 \text{ fins.}}$$ Ans.

Example 5.20. *A 1 m long, 5 cm diameter, cylinder placed in an atmosphere of 40°C is provided with 12 longitudinal straight fins (k = 75 W/m.K), 0.75 mm thick. The fins protrudes 2.5 cm from the cylinder surface. The heat transfer coefficient is 23.3 W/m².K. Calculate the rate of heat transfer, if the surface temperature of cylinder is at 150°C.* (P.U., Dec. 2008)

Solution

Given : A cylinder with longitudinal fins

$w = 1$ m, $d = 5$ cm $= 0.05$ m,

$T_\infty = 40°C$, $N_{fin} = 12$,

$h = 23.3$ W/m².K,

$t = 0.75 \text{ mm} = 0.75 \times 10^{-3}$ m,

$L = 2.5 \text{ cm} = 2.5 \times 10^{-2}$ m,

$T_0 = 150°C$, $k = 75$ W/m.K.

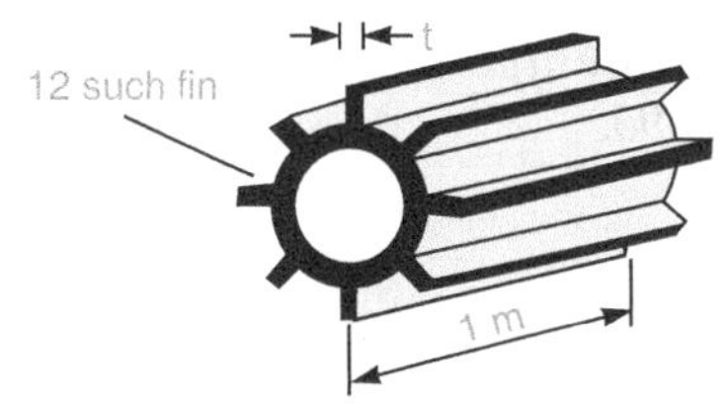

Fig. 5.30. Schematic for example 5.20

To find : The heat transfer rate from surface.

Analysis : The fin is of finite length, it will dissipate heat by convection from its tip. Therefore, using corrected length.

$$L_c = L + \frac{t}{2} = 2.5 \times 10^{-2} + \frac{0.75 \times 10^{-3}}{2}$$
$$= 0.025375 \text{ m}$$
$$P = 2w = 2 \text{ m}$$
$$A_c = w \times t$$
$$= 1 \times 0.75 \times 10^{-3} = 0.75 \times 10^{-3} \text{ m}^2$$

$$mL_c = L_c\sqrt{\frac{hP}{kA_c}}$$

$$= 0.025375 \times \sqrt{\frac{23.3 \times 2}{75 \times 0.75 \times 10^{-3}}} = 0.73$$

The heat transfer rate from fins surface

$$Q_{fin} = N_{fin}\sqrt{hPkA_c}\,(T_0 - T_\infty)\tanh(mL_c)$$
$$= 12 \times \sqrt{23.3 \times 2 \times 75 \times 0.75 \times 10^{-3}} \times (150 - 40) \times \tanh(0.73)$$
$$= 1332 \text{ W}$$

The heat transfer rate from unfinned (base) area

$$Q_{unfin} = hA_{unfin}(T_0 - T_\infty)$$
$$= h(\pi dw - 12 \times A_c)(T_0 - T_\infty)$$
$$= 23.3 \times (\pi \times 0.05 \times 1 - 12 \times 0.75 \times 10^{-3}) \times (150 - 40)$$
$$= 379.5 \text{ W}$$

Hence the total heat transfer

$$= Q_{unfin} + Q_{fin}$$
$$= 379.5 + 1332 = \mathbf{1711.5 \text{ W.}}$$ Ans.

Example 5.21. *The cylinder barrel of a motorcycle is constructed of aluminium alloy (k = 186 W/m.K), 0.15 m high and 50 mm in diameter. Under typical operating conditions, the outer surface of the cylinder is at a temperature of 500 K and is exposed to the ambient air at 300 K with convection coefficient of 50 W/m².K. Annular fins of rectangular profiles are typically added to increase the heat transfer rate to the surroundings. Assume that the five such fins, 6 mm thick, 20 mm long and equally spaced are added. What is the increase in* *heat transfer rate due to addition of fins ? Take fin efficiency as 0.95.*

(Anna University, April 1999 ; N.M.U., Nov. 1996)

Solution

Given : Annular fins of rectangular profile, around a cylinder

k = 186 W/m.K, H = 0.15 m,

d = 50 mm or r_1 = 25 mm,

T_0 = 500 K, T_∞ = 300 K,

h = 50 W/m^2.K, N_{fin} = 5,

t = 6 mm, $L = r_2 - r_1$ = 20 mm,

η_{fin} = 0.95.

To find : Increase in heat transfer rate due to addition of fins on cylinder surface.

Schematic :

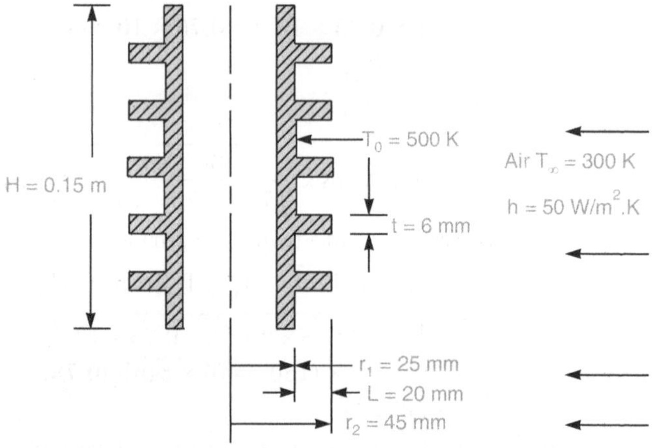

Fig. 5.31. Schematic of motorcycle barrel

Assumptions :

(*i*) Steady state one dimensional conduction in radial direction only.

(*ii*) Constant properties.

(*iii*) No internal heat generation.

(*iv*) Uniform convection coefficient over entire outer surface.

(*v*) Negligible radiation exchange from fin surface.

Analysis : The finite long fins are used on cylinder, thus corrected radius

$$r_{2c} = r_2 + t/2 = 45 + 3 = 48 \text{ mm} = 0.048 \text{ m}$$

The fins are exposed on two sides to air, the surface area of all fins

$$A_{fin} = 2\pi (r_{2c}^2 - r_1^2) \times N_{fin}$$
$$= 2\pi \times (0.048^2 - 0.025^2) \times 5 = 0.0527 \text{ m}^2$$

The total heat transfer rate from fins surface

$$Q_{fin} = \eta_{fin} \times h\, A_{fin} (T_0 - T_\infty)$$
$$= 0.95 \times (50 \text{ W/m}^2.\text{K}) \times (0.0527 \text{ m}^2) \times (500 - 300)(\text{K})$$
$$= 500.65 \text{ W}$$

The heat transfer rate from unfinned portion of cylinder :

Unfinned area,

$$A_{unfin} = \text{Cylinder surface area} - \text{No. of fins} \times \text{Cross-section area of a fin}$$
$$= 2\pi r_1 H - N_{fin} \times 2\pi r_1 t$$
$$= 2\pi \times 0.025 \times (0.15 \text{ m} - 5 \times 0.006 \text{ m})$$
$$= 0.01885 \text{ m}^2$$

The heat transfer rate from unfinned portion

$$Q_{unfin} = hA_{unfin} (T_0 - T_\infty)$$
$$= (50 \text{ W/m}^2.\text{K}) \times (0.01885 \text{ m}^2) \times (500 - 300)(\text{K})$$
$$= 188.5 \text{ W}$$

Total heat transfer rate from finned surface

$$Q_{total\ fin} = Q_{fin} + Q_{unfin}$$
$$= 500.65 + 188.5 = 689.15 \text{ W}$$

Heat transfer rate from corresponding cylinder without any fin

$$Q_{no\ fin} = hA_{no\ fin} (T_0 - T_\infty)$$
$$= h(2\pi r_1 H)(T_0 - T_\infty)$$
$$= (50 \text{ W/m}^2.\text{K}) \times (2\pi \times 0.025 \text{ m} \times 0.15 \text{ m}) \times (500 - 300)(\text{K})$$
$$= 235.62$$

Increase in the heat transfer rate

$$= Q_{total\ fin} - Q_{no\ fin}$$
$$= 689.15 - 235.62 = \mathbf{453.53\ W.}$$ **Ans.**

Comment : The overall effectiveness of finned surface

$$\varepsilon_{fin} = \frac{Q_{total\ fin}}{Q_{no\ fin}} = 2.925$$

% increase in heat transfer

$$= \frac{689.15 - 235.62}{235.62} = \mathbf{192.5\%.}$$ **Ans.**

Example 5.22. *Steam in a heating system flows through tubes whose outer diameter is 3 cm and whose walls are maintained at a temperature of 120°C. Circular aluminium fins [k = 180 W/(m.K)] of outer diameter 6 cm and constant thickness t = 2 mm are attached to the tube, as shown in Fig. 5.32. The space between the fins is 3 mm, and thus there are 200 fins per metre length of the tube. Heat is transferred to the surrounding air at T_∞ = 25°C,* *with a combined heat transfer coefficient of $h = 60\ W/(m^2.K)$. Determine the increase in heat transfer rate from the tube per metre of its length as a result of adding fins. Also calculate fin effectiveness.*

Solution

Given : A tube with annular fins :

$d_1 = 3$ cm, $r_1 = 1.5$ cm $= 0.015$ m,
$d_2 = 6$ cm, $r_2 = 3$ cm $= 0.03$ m,
$T_0 = 120°C$, $k = 180$ W/m.K,
$t = 2$ mm $= 0.002$ m, S = 3 mm,
$N_{fin} = 200$ /m, $T_\infty = 25°C$,
$h = 60\ W/m^2.K$, H = 1 m.

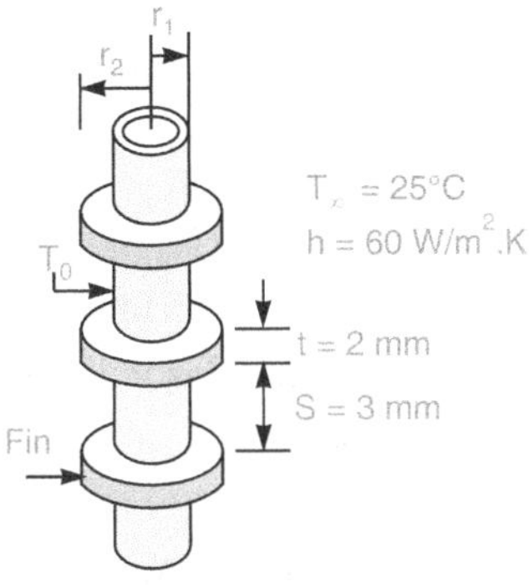

Fig. 5.32

To find : (*i*) Increase in heat transfer rate from the tube per metre length.

(*ii*) Fin effectiveness.

Analysis : (*i*) The finite long fins, with convection at its free end, thus using corrected length of fin

$$r_{2c} = r_2 + t/2 = 0.03 + 0.002/2 = 0.031 \text{ m}$$

$$L = r_2 - r_1 = 0.03 - 0.015 = 0.015 \text{ m}$$

$$L_c = L + t/2 = 0.015 + \frac{0.002}{2} = 0.016 \text{ m}$$

$$A_p = L_c t = 0.016 \times 2 \times 10^{-3} = 3.2 \times 10^{-5} \text{ m}^2$$

$$L_c^{3/2} (h/kA_p)^{1/2} = (0.016)^{3/2} \times \left(\frac{60}{180 \times 3.2 \times 10^{-5}}\right)^{1/2} = 0.2065$$

$$\frac{r_{2c}}{r_1} = \frac{0.031}{0.015} = 2.067$$

From Fig. 5.27, the efficiency of annular fin

$$\eta_{fin} = 0.95$$

Surface area of fins

$$= N_{fin} \times 2\pi (r_{2c}^2 - r_1^2)$$
$$= 200 \times 2\pi \times (0.031^2 - 0.015^2)$$
$$= 0.925 \text{ m}^2$$

The heat transfer rate from fins

$$Q_{fin} = \eta_{fin} \times Q_{ideal} = \eta_{fin} \times h A_{fin} (T_0 - T_\infty)$$
$$= 0.95 \times 60 \times 0.925 \times (120 - 25)$$
$$= 5008.25 \text{ W}$$

Unfinned area of the tube

$$A_{unfin} = \text{Pipe surface area} - \text{No. of fin} \times \text{Cross-section area of a fin}$$
$$= 2\pi r_1 H - N_{fin} \times 2\pi r_1 t$$
$$= 2\pi r_1 (H - N_{fin} t)$$
$$= 2\pi \times 0.015 \times (1 \text{ m} - 200 \times 0.002)$$
$$= 0.0565 \text{ m}^2$$

Heat transfer rate from unfinned portion of tube

$$Q_{unfin} = h A_{unfin} (T_0 - T_\infty)$$
$$= 60 \times 0.0565 \times (120 - 25)$$
$$= 322.32 \text{ W}$$

The total heat transfer rate from finned tube

$$Q_{total\ fin} = Q_{fin} + Q_{unfin}$$
$$= 5008.25 + 322.32$$
$$= 5330.57 \text{ W}$$

In case, if there are no fins on pipe, then,

Area of bare pipe surface,

$$A_{no\ fin} = 2\pi r_1 H = 2\pi \times 0.015 \times 1 = 0.0942 \text{ m}^2$$

Heat transfer rate from tube bare area

$$Q_{no\ fin} = h A_{no\ fin} (T_0 - T_\infty)$$
$$= 60 \times 0.0942 \times (120 - 25)$$
$$= 537.21 \text{ W}$$

Therefore, increase in heat transfer rate from 1 m long tube as a result of addition of fins

$$= Q_{total\ fin} - Q_{no\ fin}$$
$$= 5330.57 - 537.21$$

$= \mathbf{4793.35\ W.}$ **Ans.**

(*ii*) Fin effectiveness.

$$\varepsilon_{fin} = \frac{Q_{total\ fin}}{Q_{no\ fin}} = \frac{5330.57}{537.21} = 9.1$$

% increase in heat transfer

$= \frac{(5330.57 - 537.21)}{537.21} = \mathbf{892.2\%.}$ **Ans.**

Example 5.23. *Copper plates fins of rectangular cross-section, 1 mm thick, 10 mm long and thermal conductivity as 380 W/m.K are attached to a plane wall maintained at a temperature of 230°C. The fins dissipate heat by convection into an ambient at 30°C with a heat* *transfer coefficient of 40 W/m².K. Fins are spaced at 8 mm. Assume negligible heat loss from the fin tip.*

Calculate :

(i) Fin efficiency,

(ii) Area weighted fin efficiency,

(iii) The total heat transfer rate per m² of plane wall surface,

(iv) The heat transfer rate from the plane wall if there were no fins attached. (N.M.U., Nov. 1994)

Solution

Given : Finite long fin with insulated tip :

$t = 1$ mm, $T_0 = 230°C$,

$L = 10$ mm, $T_\infty = 30°C$,

$k = 380$ W/m.K, $h = 40$ W/m².K,

$S = 8$ mm.

To find :

(*i*) η_{fin}, fin efficiency,

(*ii*) η_{total}, area weighted fin efficiency,

(*iii*) Heat transfer rate from finned surface,

(*iv*) Heat transfer rate from base surface.

Assumptions :

1. Steady state one dimensional heat conduction.
2. Constant properties.
3. No internal heat generation.
4. The plate size of 1 m × 1 m.

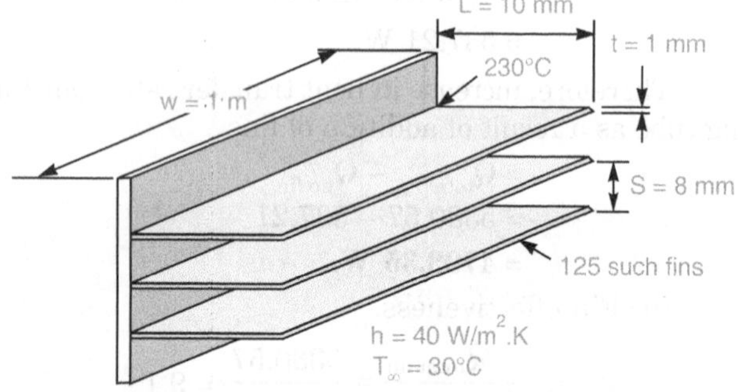

Fig. 5.33. A section of the plane wall

Analysis : (*i*) Fin efficiency :

$$A_c = w \times t = (1\text{ m}) \times (1 \times 10^{-3}\text{ m}) = 1 \times 10^{-3}\text{ m}^2$$

$$P \approx 2w = 2 \times 1\text{ m} = 2\text{ m}$$

$$mL = \sqrt{\frac{hP}{kA_c}} \times L = \sqrt{\frac{40 \times 2}{380 \times 1 \times 10^{-3}}} \times 10 \times 10^{-3} = 0.145$$

The fin efficiency for insulated tip fin is given by

$$\eta_{fin} = \frac{\tanh mL}{mL} = \frac{\tanh(0.145)}{0.145} = 0.993$$

$$= \mathbf{99.3\%.} \quad \textbf{Ans.}$$

(*ii*) Area weighted fin efficiency :

Since the considered plate is 1 m × 1 m in size and fin spacing is 8 mm. The number of fins

$$N_{fin} = \frac{1000\text{ mm}}{8\text{ mm}} = 125\text{ fins/metre}$$

The total fin surface area,

$$A_{fin} = \text{No. of fins } (N_{fin}) \times \text{Surface area of a fin (PL)}$$

$$A_{fin} = 125 \times (2\text{ sides} \times 1\text{ m} \times 0.01\text{ m}) = 2.5\text{ m}^2$$

The unfinned portion,

$$A_{unfin} = \text{Plate area} - \text{Area where fins are attached } (N_{fin} \times A_c)$$

$$= 1\text{ m}^2 - 125 \times 1 \times 10^{-3} = 0.875\text{ m}^2$$

Total heat transfer area,

$$A_{total} = A_{fin} + A_{unfin}$$

$$A_{total} = 2.5 + 0.875 = 3.375\text{ m}^2$$

The area weighted fin efficiency or total fin efficiency

$$\eta_{total} = 1 - \frac{A_{fin}}{A_{total}}(1 - \eta_{fin})$$

$$= 1 - \frac{2.5}{3.375}(1 - 0.993)$$

$$= 0.9948 = \mathbf{99.48\%.} \quad \textbf{Ans.}$$

(*iii*) Total heat transfer rate per m² of plane wall surface

$$Q_{total} = Q_{fin} + Q_{unfin}$$

$$= N_{fin} \times \sqrt{hPkA_c}\,(T_0 - T_\infty)\tanh(mL) + h\,A_{unfin}(T_0 - T_\infty)$$

$$= 125 \times \sqrt{40 \times 2 \times 380 \times 1 \times 10^{-3}} \times (230 - 30) \times \tan(0.145) + 40 \times 0.875 \times (230 - 30)$$

$$= 26848\text{ W/m}^2 = \mathbf{26.84\text{ kW/m}^2}. \quad \textbf{Ans.}$$

Alternatively,

$$Q_{total} = Q_{fin} + Q_{unfin}$$
$$= \eta_{fin}\, Q_{ideal} + Q_{unfin}$$
$$= \eta_{fin} h\, A_{fin}\,(T_0 - T_\infty) + hA_{unfin}(T_0 - T_\infty)$$
$$= (\eta_{fin}A_{fin} + A_{unfin})\, h(T_0 - T_\infty)$$
$$= (0.993 \times 2.5 + 0.875) \times 40 \times (230 - 30)$$
$$= \mathbf{26.86\ kW/m^2}.$$ **Ans.**

(*iv*) The heat transfer if there were no fins attached

$$Q_{no\ fin} = hA_{no\ fin}(\Delta T)$$
$$= (40\ W/m^2.K) \times (1\ m^2) \times (230 - 30)(K)$$
$$= 8000\ K/m^2 = \mathbf{8\ kW/m^2}.$$ **Ans.**

Example 5.24. *An aluminium heat sink for electronics components has a base of length 50 mm and width 70 mm. The eight aluminium (k = 180 W/m.K) fins are attached in such a way that their width is 70 mm. The fins are 12 mm long, and 3 mm thick. The fins cooled by air at 25°C with a convective heat transfer coefficient of h = 10 W/m².K. Assuming that the same value of heat transfer coefficient acts on the tip of the fins as along the rest of the external surface, determine :*

(*i*) *the heat flow through the heat sink for a base temperature of 50°C,*

(*ii*) *the fin effectiveness,*

(*iii*) *the fin efficiency,*

(*iv*) *the length of the fin such that the heat flow is 95% of the heat flow for an infinite long fin,*

(*v*) *the percentage increase in heat transfer with fins.* (P.U., May 2008)

Solution

Given : Finite long fins

$T_0 = 50°C$, $T_\infty = 25°C$,
$h = 10\ W/m^2.K$, $N_{fin} = 8$ fins,
$k = 180\ W/m.K$, $t = 3$ mm,
$L = 12$ mm, $w = 70$ mm,
$H = 50$ mm.

To find :

(*i*) Total heat flow from the finned surface, Q_{total}.

(*ii*) Fin effectiveness, ε_{fin}.

(*iii*) Fin efficiency, η_{fin}.

(*iv*) The length of the fin, so that the heat flow is 95% of the infinite long fin.

(*v*) Percentage increase in heat transfer with fins.

Assumptions :

1. Steady state conditions.
2. Constant properties.
3. Uniform spacing of fins.

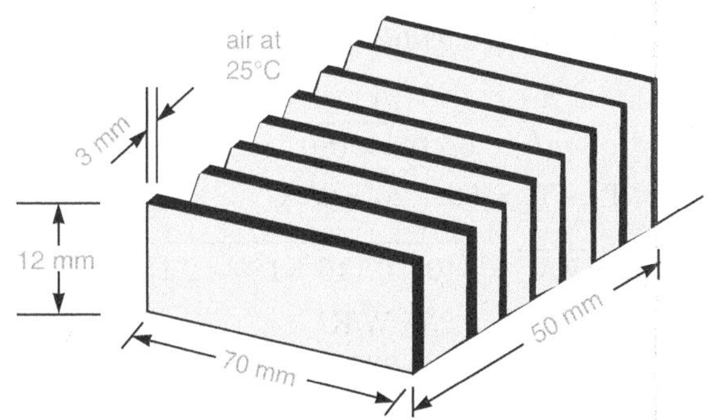

Fig. 5.34. Aluminium heat sink

Analysis : (*i*) The total heat flow from the heat sink = Heat transfer from unfinned portion of the base + Heat flow through the fins

$$Q_{total} = Q_{unfin} + Q_{fin}$$

Heat transfer rate from unfinned portion of base

$$Q_{unfin} = hA_{unfin}\,(T_0 - T_\infty)$$

where $A_{unfin} = w\,(H - N_{fin}t)$

$$= (70 \times 10^{-3}\ m) \times (50 - 8 \times 3) \times 10^{-3}$$
$$= 1.82 \times 10^{-3}\ m^2$$

Then

$$Q_{unfin} = 10 \times 1.82 \times 10^{-3} \times (50 - 25)$$
$$= 0.455\ W$$

Heat transfer rate through fins (of finite long convecting heat from their tips)

$$Q_{fin} = N_{fin}\sqrt{hPkA_c}\,(T_0 - T_\infty) \times \frac{\sinh mL + \dfrac{h}{mk}\cosh mL}{\cosh mL + \dfrac{h}{mk}\sinh mL}$$

where, $A_c = w \times t$

$$= (70 \times 10^{-3}) \times (3 \times 10^{-3})$$
$$= 2.1 \times 10^{-4}\ m^2$$
$$P = 2(w + t) = 2 \times (70 + 3) \times 10^{-3}$$
$$= 0.146\ m$$
$$m = \sqrt{\frac{hP}{kA_c}}$$
$$= \sqrt{\frac{10 \times 0.146}{180 \times 2.1 \times 10^{-4}}} = 6.2148\ m^{-1}$$
$$mL = 6.2148 \times (12 \times 10^{-3}) = 0.074578$$
$$\cosh mL = 1.00278$$

$\sinh mL = 0.07465$

$$\frac{h}{mk} = \frac{10}{6.2148 \times 180} = 8.94 \times 10^{-3}$$

$$T_0 - T_\infty = 50 - 25 = 25°C$$

and $\sqrt{hPkA_c} = \sqrt{10 \times 0.146 \times 180 \times 2.1 \times 10^{-4}}$

$= 0.235$ W/K

$$\therefore \quad Q_{fin} = 8 \times 0.235 \times 25 \times \frac{(0.07465 + 8.94 \times 10^{-3} \times 1.00278)}{(1.00278 + 8.94 \times 10^{-3} \times 0.07465)}$$

$= 3.916$ W

Total heat transfer

$Q_{total} = 0.455 + 3.916 =$ **4.4 W.** Ans.

(*ii*) Effectiveness of the fin

$$\varepsilon_{fin} = \frac{\text{Heat transfer rate with fin}}{\text{Heat transfer rate from the surface on which fin was attached } (w \times t)}$$

$$= \frac{Q_{\text{fin for 1 fin}}}{h\, A_c\, (T_0 - T_\infty)}$$

$$= \frac{(3.916/8)}{10 \times 2.1 \times 10^{-4} \times (50 - 25)}$$

= **9.32.** Ans.

(*iii*) Fin efficiency

$$\eta_{fin} = \frac{\text{Actual heat transfer rate from fin surface}}{\text{Heat transfer rate from the fin, if its entire surface is maintained at base temperature}}$$

$$= \frac{Q_{\text{fin for 1 fin}}}{hA_{fin}\, (T_0 - T_\infty)}$$

where, $A_{fin} = PL + A_c = 0.146 \times 12 \times 10^{-3} + 2.1 \times 10^{-4}$

$= 1.962 \times 10^{-3}$ m

$$\therefore \quad \eta_{fin} = \frac{(3.916/8)}{10 \times 1.962 \times 10^{-3} \times (50 - 25)}$$

$= 0.998 =$ **99.8%.** Ans.

(*iv*) For heat flow within 95% of infinite long fin

$$\frac{\sinh mL + \left\{\frac{h}{mk}\cosh mL\right\}}{\cosh mL + \left\{\frac{h}{mk}\sinh mL\right\}} \geq 0.95$$

$\because \quad \frac{h}{mk} <<< 1.0$, therefore, above equation approximated as

$\tanh mL \geq 0.95$

$mL \geq 1.83$

or $\quad L \geq 0.295$

= **295 mm.** Ans.

(*v*) Heat transfer from heat sink, if there were no fins attached

$$Q_{\text{no fin}} = hA_{\text{no fin}}\, (T_0 - T_\infty)$$

where $\quad A_{\text{no fin}} = (70 \times 10^{-3}) \times (50 \times 10^{-3})$

$= 3.5 \times 10^{-3}$ m^2

$\therefore \quad Q_{\text{no fin}} = 10 \times 3.5 \times 10^{-3} \times (50 - 25)$

$= 0.875$ W

% increase in heat transfer after fin addition

$$= \frac{Q_{total} - Q_{\text{no fin}}}{Q_{\text{no fin}}} = \frac{4.4 - 0.875}{0.875} \times 100$$

= **405%.** Ans.

Example 5.25. *A hot surface at 100°C is to be cooled by attaching 3 cm long, 0.25 cm diameter aluminium fins (k = 237 W/m.K) to it, with a centre to centre distance of 0.6 cm. The temperature of surrounding air is 30°C and heat transfer coefficient on surface is 35 W/m².K. Calculate the rate of heat transfer from the surface for a 1 m × 1 m section of the plate. Also determine the overall effectiveness of the fins.*

Solution

Given : A hot surface attached with pin fins.

$T_0 = 100°C$, $\quad L = 3$ cm $= 0.03$ m,

$d = 0.25$ cm $= 0.25 \times 10^{-2}$ m,

$k = 237$ W/m.K, $\quad S = 0.6$ cm,

$T_\infty = 30°C$, $\quad h = 35$ W/m^2.K.

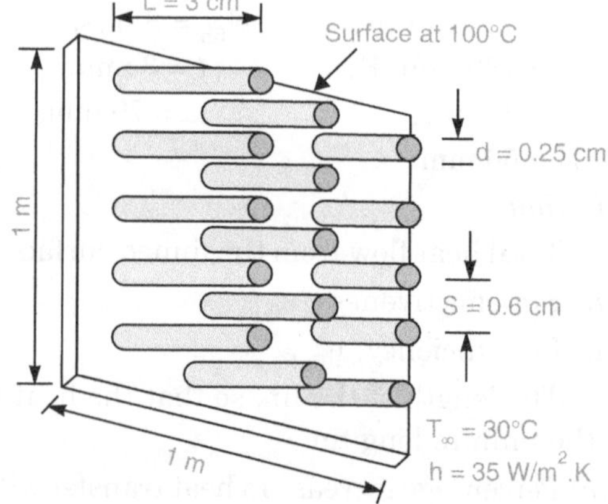

Fig. 5.35. Schematic for example 5.25

To find :

(*i*) Heat transfer rate from 1 m^2 finned surface,

(*ii*) Overall fin effectiveness.

Assumptions :

(*i*) Steady state conditions.

(*ii*) Finite long fin, but L >> d, hence assuming insulated tip.

(*iii*) Uniform heat transfer coefficient on fins as well as on base surface.

(*iv*) Constant properties.

Analysis : (*i*) Heat transfer rate from finned surface :

$$\text{No. of fins in a row} = \frac{w}{S} = \frac{100 \text{ cm}}{0.6 \text{ cm}} = 166.67 \approx 167$$

Similarly the no. of fins in the column

$$= \frac{100}{0.6} = 167$$

These fins are in matrix of $n \times n$, thus

Total no. of fins in

$$N_{fin} = 167 \times 167 = 27{,}889 \text{ fins/m}^2$$

For a fin

$$A_c = \frac{\pi}{4} d^2 = \frac{\pi}{4} \times (0.25 \times 10^{-2} \text{ m})^2$$
$$= 4.9087 \times 10^{-6} \text{ m}^2$$
$$P = \pi d = \pi \times 0.25 \times 10^{-2}$$
$$= 7.8539 \times 10^{-3} \text{ m}$$

$$m = \sqrt{\frac{hP}{kA_c}} = \sqrt{\frac{35 \times 7.8539 \times 10^{-3}}{237 \times 4.9087 \times 10^{-6}}}$$
$$= 15.37$$

The heat transfer rate from all fins

$$Q_{fin} = N_{fin} \sqrt{hPkA_c}\ (T_0 - T_\infty) \tanh mL$$
$$= 27{,}889 \times \sqrt{35 \times 7.8539 \times 10^{-3} \times 237 \times 4.9087 \times 10^{-6}}$$
$$\times (100 - 30) \times \tanh (15.37 \times 0.03)$$
$$= 15.046 \times 10^3 \text{ W}$$

The heat transfer from unfinned portion

$$A_{unfin} = 1 \text{ m}^2 - N_{fin} \times A_c$$
$$= 1 \text{ m}^2 - 27{,}889 \times 4.9087 \times 10^{-6}$$
$$= 0.863 \text{ m}^2$$
$$Q_{unfin} = h\, A_{unfin} \times (T_0 - T_\infty)$$
$$= 35 \times 0.863 \times (100 - 30)$$
$$= 2114.6 \text{ W}$$

Total heat transfer from finner surface/m^2

$$Q_{total\ fin} = Q_{fin} + Q_{unfin}$$
$$= 15.046 \times 10^3 + 2114.6$$
$$= \mathbf{17.16 \times 10^3\ W.} \quad \text{Ans.}$$

(*ii*) Overall effectiveness

$$\varepsilon_{overall} = \frac{Q_{total\ fin}}{Q_{no\ fin}} = \frac{Q_{total\ fin}}{h\,(1\text{ m}^2)\,(T_0 - T_\infty)}$$
$$= \frac{17.16 \times 10^3}{35 \times 1 \times (100 - 35)} = \mathbf{7.0.} \quad \text{Ans.}$$

Example 5.26. *In a transfer type heat exchanger, heat is transferred from hot water at 90°C on one side of the metal partition wall to cold air at 25°C on the other side. Thickness of the metal wall is 1 cm and its conductivity is 20 W/m.K. If the metal wall is 1 m², find the rate of heat transfer if heat transfer coefficient on water and air side are 100 and 10 W/m².K, respectively.*

It is proposed to increase the heat transfer rate by providing fins on one side. On which side the fins should be provided to get maximum heat transfer rate ? If 500 fins of 6 mm diameter and 30 mm long are provided. Find the maximum heat transfer rate achieved. Assume that the fins have insulated ends.

Solution

Given : Partition wall of a heat exchanger.

$T_w = 90°C$, $T_a = 25°C$,
$\delta = 1 \text{ cm} = 0.01 \text{ m}$, $k = 20 \text{ W/m.K}$,
$h_w = 100 \text{ W/m}^2\text{.K}$,
$h_a = 10 \text{ W/m}^2\text{.K}$, $N_{fin} = 500$,
$d = 6 \text{ mm} = 6 \times 10^{-3} \text{ m}$, $A = 1 \text{ m}^2$,
$L = 30 \text{ mm} = 0.03 \text{ m}$.

To find :

(*i*) Heat transfer rate without fins.

(*ii*) Justified of fins attachment side.

(*iii*) Heat transfer rate when fins attached on surface.

Assumptions :

1. Steady state conditions.
2. One dimensional heat transfer.
3. Constant properties.

Analysis : (*i*) Heat transfer rate when one side is exposed to water and other side to air, (No fins on any surface) using electrical analogy.

$$Q_{\text{no fin}} = \frac{(\Delta T)_{\text{overall}}}{\dfrac{1}{h_w A} + \dfrac{\delta}{kA} + \dfrac{1}{h_a A}}$$

or
$$= \frac{90 - 25}{\dfrac{1}{100 \times 1} + \dfrac{0.01}{20 \times 1} + \dfrac{1}{10 \times 1}}$$

$$= \mathbf{588.23\ W.} \quad \text{Ans.}$$

(*ii*) Fins are always provided on side of wall, where heat transfer coefficient is low, hence, use of fins on air side will maximise the heat transfer rate. Ans.

(*iii*) When fins are provided on air side :

$$A_c = \left(\frac{\pi}{4}\right) d^2 = \left(\frac{\pi}{4}\right) \times (6.0 \times 10^{-3})^2$$
$$= 2.827 \times 10^{-5}\ \text{m}^2$$
$$P = \pi d = \pi \times 6 \times 10^{-3} = 0.0188\ \text{m}$$
$$mL = \sqrt{\frac{hP}{kA_c}}\ L$$
$$= \sqrt{\frac{10 \times 0.0188}{20 \times 2.827 \times 10^{-5}}} \times 0.03 = 0.547$$

The fin efficiency

$$\eta_{\text{fin}} = \frac{\tanh mL}{mL} = \frac{\tanh (0.547)}{0.547} = 0.9107$$

The surface area of fins

$$A_{\text{fin}} = N_{\text{fin}} PL$$
$$= 500 \times 0.0188 \times 0.03$$
$$= 0.2827\ \text{m}^2$$

Unfinned area of wall

$$A_{\text{unfin}} = 1\ \text{m}^2 - \text{area occupied by fins}$$
$$= 1\ \text{m}^2 - N_{\text{fin}} A_c$$
$$= 1\ \text{m}^2 - 500 \times 2.827 \times 10^{-5}\ \text{m}^2$$
$$= 0.985$$

Total heat transfer rate of fins

$$Q_{\text{total}} = Q_{\text{unfin}} + \eta_{\text{fin}} Q_{\text{ideal}}$$
$$= h_a A_{\text{unfin}} (T_0 - T_a) + \eta_{\text{fin}} h_a A_{\text{fin}} (T_0 - T_\infty)$$
$$= (A_{\text{unfin}} + \eta_{\text{fin}} A_{\text{fin}})\, h_a\, (T_0 - T_\infty)$$

or
$$Q_{\text{total}} = \frac{T_0 - T_\infty}{\dfrac{1}{(A_{\text{unfin}} + \eta_{\text{fin}} A_{\text{fin}}) h_a}}$$

Since all resistances are still in series,
Hence

$$Q = \frac{(\Delta T)_{\text{overall}}}{\Sigma R_{\text{th}}}$$

or
$$Q = \frac{T_w - T_a}{\dfrac{1}{h_w A} + \dfrac{\delta}{kA} + \dfrac{1}{(A_{\text{unfin}} + \eta_{\text{fin}} A_{\text{fin}}) h_a}}$$

$$= \frac{90 - 25}{\dfrac{1}{100 \times 1} + \dfrac{0.01}{20 \times 1} + \dfrac{1}{(0.985 + 0.9107 \times 0.2827) \times 10}}$$

$$= \mathbf{714.83\ W.} \quad \text{Ans.}$$

Example 5.27. *Air and water are flowing on two sides of a mild steel wall (k = 52 W/m.K). The heat transfer coefficients on air and water sides are,*

$$h_a = 11.4\ \text{W/m}^2.\text{K and } h_w = 256\ \text{W/m}^2.\text{K}.$$

It is proposed to increase the heat transfer rate by adding rectangular mild steel fins of the following features :

Fin thickness = 0.13 cm, Fin height = 2.5 cm,

Fin spacing = 1.3 cm.

What percentage increase in the heat transfer rate can be realised by adding fins on

(i) Air side only.

(ii) Water side only.

(iii) Fins on both sides.

Solution

Given : The mild steel wall with proposed finite long rectangular fins

$k = 52$ W/m.K, $h_a = 11.4$ W/m^2.K,

$h_w = 256$ W/m^2.K, $t = 0.13$ cm,

$L = 2.5$ cm, $S = 1.3$ cm.

To find : The percentage increase in the heat transfer rate, when

(*i*) Fins are added on air side.

(*ii*) Fins are added on water side.

(*iii*) Fins are added on both sides.

Assumptions :

1. Steady state conditions.
2. Negligible thermal resistance offered by wall thickness.
3. No radiation from fins.
4. No internal heat generation.
5. Wall size 1 m × 1 m or $A_w = 1$ m^2.

Analysis : The heat transfer rate through mild steel wall, without fins

$$Q_{\text{no fin}} = \frac{\Delta T}{\dfrac{1}{h_a A_w} + \dfrac{1}{h_w A_w}}$$

$$= \frac{\Delta T}{\frac{1}{11.4 \times 1} + \frac{1}{256 \times 1}} = 10.91\ \Delta T$$

For fins and finned surface :

$$A_c = w \times t$$
$$= 1 \times 0.13 \times 10^{-2}$$
$$= 1.3 \times 10^{-3}\ m^2$$
$$P = 2w = 2\ m,$$
$$L_c = L + (t/2)$$
$$= 2.5 + (0.13/2) = 2.565\ cm.$$

The number of fins/m,

$$N_{fin} = \frac{1\ m}{S} = \frac{100\ cm}{1.3\ cm} = 77\ fins/m$$

Fins surface area,

$$A_{fin} = N_{fin} P L_c$$
$$= 77 \times 2 \times 2.565 \times 10^{-2}$$
$$= 3.95\ m^2/m^2 \text{ of wall}$$

Unfinned (bare) area,

$$A_{unfin} = 1\ m^2 - N_{fin}\ A_c$$
$$= 1 - 77 \times 1.3 \times 10^{-3}$$
$$= 0.9\ m^2/m^2 \text{ of wall area}$$

(*i*) Heat transfer rate with fins on air side.

$$m_a = \sqrt{\frac{h_a P}{k A_c}} = \sqrt{\frac{11.4 \times 2}{52 \times 0.13 \times 10^{-2}}}$$
$$= 18.37\ m^{-1}$$
$$\eta_{fin,\,a} = \frac{\tanh(m_a L_c)}{m_a L_c}$$
$$= \frac{\tanh(18.37 \times 2.565 \times 10^{-2})}{18.37 \times 2.565 \times 10^{-2}} = 0.932$$

Since two resistances act parallel on air side, hence equivalent resistance,

$$\frac{1}{R_{a\ finned}} = \frac{1}{R_1} + \frac{1}{R_2}$$
$$= h_a\ A_{unfin} + h_a\ \eta_{fin,\,a}\ A_{fin}$$

or
$$R_{a\ finned} = \frac{1}{h_a(A_{unfin} + \eta_{fin,\,a}\ A_{fin})}$$

Then

$$Q_{fin,\,a} = \frac{\Delta T}{R_{water} + R_{a\ finned}}$$
$$= \frac{\Delta T}{\frac{1}{h_w A_w} + \frac{1}{h_a(A_{unfin} + \eta_{fin,\,a}\ A_{fin})}}$$
$$= \frac{\Delta T}{\frac{1}{256 \times 1} + \frac{1}{11.4 \times (0.9 + 0.932 \times 3.95)}}$$
$$= 43.38\ \Delta T$$

The percentage increase in heat transfer rate with fins on air side

$$= \frac{Q_{fin,\,a} - Q_{no\ fin}}{Q_{no\ fin}} \times 100$$
$$= \frac{43.38 - 10.91}{10.91} \times 100$$

= **298%. Ans.**

(*ii*) The heat transfer with fins on water side (two parallel resistances on water side)

$$m_w = \sqrt{\frac{h_w P}{k A_c}} = \sqrt{\frac{256 \times 2}{52 \times 1.3 \times 10^{-3}}}$$
$$= 87.03\ m^{-1}$$
$$\eta_{fin,\,w} = \frac{\tanh(m_w L_c)}{m_w L_c}$$
$$= \frac{\tanh(87.03 \times 2.565 \times 10^{-2})}{87.03 \times 2.565 \times 10^{-2}}$$
$$= 0.438$$
$$Q_{fin,\,w} = \frac{\Delta T}{R_{air} + R_{w\ finned}}$$
$$= \frac{\Delta T}{\frac{1}{h_a A_w} + \frac{1}{h_w(A_{unfin} + \eta_{fin,\,w}\ A_{fin})}}$$
$$= \frac{\Delta T}{\frac{1}{11.4 \times 1} + \frac{1}{256 \times (0.9 + 0.438 \times 3.95)}}$$
$$= 11.21\ \Delta T$$

The percentage increase in heat transfer

$$= \frac{Q_{fin,\,w} - Q_w}{Q_w} \times 100$$
$$= \frac{11.21 - 10.91}{10.91} \times 100$$

= **2.75%. Ans.**

(*iii*) Heat transfer with fins on both sides

$$Q_{fin,\ both} = \frac{\Delta T}{\frac{1}{h_a(A_{unfin} + \eta_{fin,\,a}\ A_{fin})} + \frac{1}{h_w(A_{unfin} + \eta_{fin,\,w}\ A_{fin})}}$$

$$= \frac{\Delta T}{\dfrac{1}{11.4 \times (0.9 + 0.932 \times 3.95)} + \dfrac{1}{256 \times (0.9 + 0.438 \times 3.95)}}$$

$= 48.46\ \Delta T$

The percentage increase in heat transfer

$$= \frac{48.46 - 10.91}{10.91} \times 100 = \mathbf{345\%}. \quad \text{Ans.}$$

Comment :

(*i*) The addition of fins on air side is very effective due to its low value of heat transfer coefficient.

(*ii*) The addition of fins on water sides is not effective thus cannot be justified.

(*iii*) The addition of fins on both sides has only marginal effect due to addition of material resistance on both sides. It is not practicable.

Example 5.28. *A composite fin consists of a cylindrical rod (k = 15 W/m.K) 3 mm in diameter and 100 mm long. It is uniformly covered with another material (k = 45 W/m.K) forming outer diameter 10 mm and 100 mm long. It is exposed into an ambient with h = 12 W/m².K.*

(i) Derive an expression for the efficiency of this fin and its value for given data.

(ii) Calculate effectiveness of the composite fin.

Assume heat coduction in axial direction only and tip of fin as insulated.

Solution

Given : A composite fin as shown in Fig. 5.36.

$k_1 = 15$ W/m.K, $k_2 = 45$ W/m.K,

$d_1 = 3$ mm, $d_2 = 10$ mm,

$L = 100$ mm, $h = 12$ W/m².K.

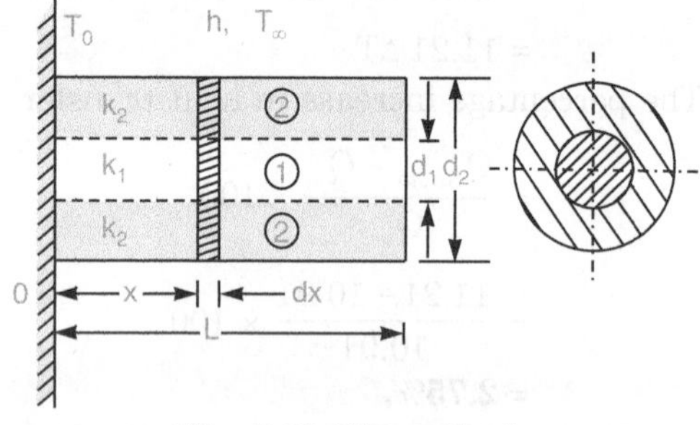

Fig. 5.36. Bimetallic fin

To find :

(*i*) An expression for fin efficiency and its calculation.

(*ii*) Fin effectiveness.

Assumptions :

(*i*) Steady state conduction in axial direction only.

(*ii*) Even though the fin of composite material, the temperature at any cross-section in the fin is uniform.

(*iii*) No contact resistance at interface of two materials.

(*iv*) No heat generation within the fin.

(*v*) Temperature at free end of fin approaches to T_∞.

Analysis : (*i*) Consider a differential element of composite fin of thickness dx at a distance x from the base.

Heat conducted into element by both fins

$$Q_{1x} + Q_{2x}$$

Heat conducted out the element

$$= Q_{1(x+dx)} + Q_{2\,(x+dx)}$$

Heat dissipation by convection from outer surface element

$$Q_{conv} = h\,(Pdx)\,(T - T_\infty),$$

where $P = \pi d_2 = \pi \times 0.01 = 0.01\pi$ m

For steady state conditions, the energy balance yields

$$Q_{1x} + Q_{2x} = Q_{1(x+dx)} + Q_{2(x+dx)} + h\,(Pdx)\,(T - T_\infty)$$

or
$$Q_{1x} + Q_{2x} = Q_{1x} + \frac{d}{dx}(Q_{1x})\,dx + Q_{2x} + \frac{d}{dx}(Q_{2x})\,dx + hP\,dx\,(T - T_\infty)$$

or
$$\frac{d}{dx}\left(-k_1 A_1 \frac{dT}{dx}\right)dx + \frac{d}{dx}\left(-k_2 A_2 \frac{dT}{dx}\right)dx + hP\,dx\,(T - T_\infty) = 0$$

Since the cross-section areas and thermal conductivities are constant. Thus,

$$-k_1 A_1 \frac{d^2T}{dx^2} - k_2 A_2 \frac{d^2T}{dx^2} + hP\,(T - T_\infty) = 0$$

or
$$(k_1 A_1 + k_2 A_2)\frac{d^2T}{dx^2} - hP\,(T - T_\infty) = 0$$

Introducing $\theta = T - T_\infty$, then

$$\frac{d^2\theta}{dx^2} - \frac{hP}{k_1 A_1 + k_2 A_2}\,\theta = 0$$

Using
$$m^2 = \frac{hP}{k_1 A_1 + k_2 A_2}$$

Then
$$\frac{d^2\theta}{dx^2} - m^2\theta = 0$$

It is the second order differential equation for composite fin and its solution is

$$\theta = C_1 e^{-mx} + C_2 e^{mx}.$$

Cross-sectional areas :

$$A_1 = \frac{\pi}{4} d_1^2 = \frac{\pi}{4} \times (0.003)^2$$
$$= 7.0686 \times 10^{-6} \text{ m}^2$$

$$A_2 = \frac{\pi}{4} (d_2^2 - d_1^2) = \frac{\pi}{4} \times (0.01^2 - 0.003^2)$$
$$= 7.147 \times 10^{-5} \text{ m}^2$$

For insulated tip fin, the efficiency is given by

$$\eta_{fin} = \frac{\tanh mL}{mL}$$

where

$$m = \sqrt{\frac{hP}{k_1 A_1 + k_2 A_2}}$$
$$= \sqrt{\frac{12 \times 0.01\pi}{15 \times 7.0686 \times 10^{-6} + 45 \times 7.147 \times 10^{-5}}}$$
$$= 10.652$$

$$\eta_{fin} = \frac{\tanh (10.652 \times 0.1)}{10.652 \times 0.1}$$
$$= 0.7394 = \mathbf{73.94\%}. \quad \text{Ans.}$$

(*ii*) Fin effectiveness

$$\varepsilon_{fin} = \frac{Q_{fin}}{Q_{no\ fin}}$$

$$Q_{fin} = Q_{inner\ fin} + Q_{outer\ fin}$$

$$Q_{inner\ fin} = k_1 A_1 \frac{(T_0 - T_\infty)}{L}$$
$$= \frac{15 \times 7.0686 \times 10^{-6} (T_0 - T_\infty)}{0.1}$$
$$= 0.00106 (T_0 - T_\infty)$$

$$Q_{outer\ fin} = \sqrt{hPk_2A_2}\ (T_0 - T_\infty) \tanh mL$$

where

$$m = \sqrt{\frac{hP}{kA_2}}$$
$$= \sqrt{\frac{12 \times 0.01\pi}{45 \times 7.147 \times 10^{-5}}} = 10.826$$

Then

$$Q_{outer\ fin} = \sqrt{12 \times 0.01\pi \times 45 \times 7.147 \times 10^{-5}} \times (T_0 - T_\infty) \tanh (10.826 \times 0.1)$$
$$= 0.0276(T_0 - T_\infty)$$

$$Q_{fin} = 0.00106 (T_0 - T_\infty) + 0.0276 (T_0 - T_\infty)$$
$$= 0.0285 (T_0 - T_\infty)$$

The heat transfer rate from corresponding base surface, without any fin.

$$Q_{no\ fin} = hA_{no\ fin} (T_0 - T_\infty)$$
$$= h \times (\pi/4) \times d_2^2 \times (T_0 - T_\infty)$$
$$= 12 \times (\pi/4) \times (0.01)^2 \times (T_0 - T_\infty)$$
$$= 0.0009425 (T_0 - T_\infty)$$

$$\varepsilon_{fin} = \frac{0.0285}{0.0009425} = \mathbf{30.24}. \quad \text{Ans.}$$

5.6. ERROR IN TEMPERATURE MEASUREMENT BY THERMOMETERS

The temperature of fluid flowing through a duct is measured by thermometer, placed in thermometer pocket as shown in Fig. 5.37. The thermometer pocket or thermometer well is a small tube welded radially into the duct or pipe. The pocket is filled with some liquid of low specific heat and thermometer is dipped in this liquid. The heat is transferred to the fluid in the pocket and its temperature is recorded by thermometer.

As the wall of duct or pipe is at the temperature less than that of fluid flowing within, the heat will flow from bottom of the pocket towards the wall of duct or pipe. Therefore, the temperature measured by thermometer will not be the true temperature. The error included can be calculated by assuming pocket as a fin (spine) protruded from the wall of the duct in which the fluid is flowing.

Considering duct wall at temperature T_0, temperature recorded by thermometer T_L, flowing fluid temperature T_∞, convection coefficient of h, pocket thermal conductivity k, diameter d and its thickness t. Then, for such spine (at x = L).

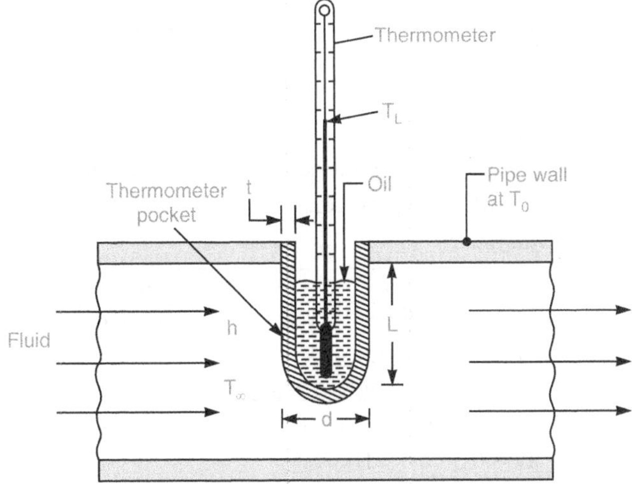

Fig. 5.37. Thermometer in a thermometer pocket

$$\frac{T_L - T_\infty}{T_0 - T_\infty} = \frac{1}{\cosh mL + (h/mk) \sinh mL} \quad ...(5.49)$$

The quantity (h/mk) sinh mL is very small, thus can be neglected and the temperature distribution is approximated as

$$\frac{T_L - T_\infty}{T_0 - T_\infty} = \frac{1}{\cosh mL}$$

where $$m = \sqrt{\frac{hP}{kA_c}} = \sqrt{\frac{h\pi d}{k\pi dt}} = \sqrt{\frac{h}{kt}}$$

$$\text{The error } (T_\infty - T_L) \propto \frac{1}{\cosh (mL)} \propto \frac{1}{L}\sqrt{\frac{kt}{h}}$$

The error in temperature measurement by thermometer can be reduced by :

(*i*) Choosing the thermometer pocket material of moderate thermal conductivity such as steel.

(*ii*) Keeping thermometer pocket thickness t as small as possible.

(*iii*) Keeping the length of the thermometer pocket large.

(*iv*) Maintaining the heat transfer coefficient large.

Example 5.29. *The temperature of hot gas flowing through a pipe is measured by a mercury thermometer inserted in an oil well made of steel (k = 40 W/m.K). The thermometer reads the temperature at the end of the well which is lower than the gas temperature due to transfer of heat along the well. Calculate percentage error in temperature measurement, if thermometer reads 150°C. The temperature of the pipe wall is 80°C. The well is 10 cm long, 2 mm thick. Take h = 40 W/m².K.*

Solution

Given : Thermometer well as spine

k = 40 W/m.K, T_L = 150°C,

T_0 = 80°C, L = 10 cm,

t = 2 mm, h = 40 W/m².K.

To find : The percentage error in measured temperature.

Assumptions :

1. Steady state heat conduction along spine.
2. Insulated spine tip.

Analysis :

$$m = \sqrt{\frac{hP}{kA_c}} = \sqrt{\frac{h}{kt}} = \sqrt{\frac{40}{40 \times 2 \times 10^{-3}}}$$

$$= 22.36 \text{ m}^{-1}$$

$$mL = 22.36 \times 0.1 = 2.236$$

The temperature at x = L

$$\frac{T_L - T_\infty}{T_0 - T_\infty} = \frac{1}{\cosh mL}$$

$$\frac{150 - T_\infty}{80 - T_\infty} = \frac{1}{\cosh (2.236)} = 0.211$$

or $$150 - T_\infty = 16.9 - 0.211\, T_\infty$$

or $$0.789\, T_\infty = 133.09$$

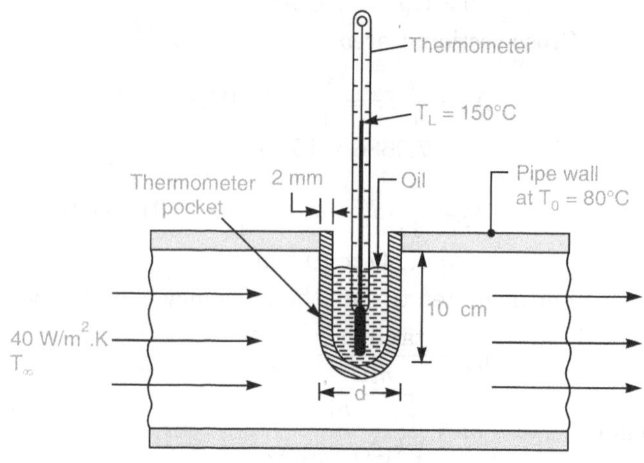

Fig. 5.38. Schematic for example 5.29

True temperature, T_∞ = 168.68°C

$$\text{Percentage error} = \frac{168.68 - 150}{168.68} \times 100$$

$$= \mathbf{11.01\%}. \text{ Ans.}$$

Example 5.30. *A thermometer pocket, 2.2 cm in diameter, 0.5 mm thick is made of steel (k = 27 W/m.K) and it is used to measure the temperature of steam flowing through a pipe. Calculate the minimum length of the pocket so that the error is less than 0.5% of applied temperature difference. Take steam at 250°C and h = 98 W/m².K.*

Solution

Given : Thermometer well as hollow spine.

k = 27 W/m.K, T_∞ = 250°C,

h = 98 W/m².K, d = 2.2 cm,

t = 0.5 mm,

error = 0.5% of applied temperature difference.

To find : Length of thermometer pocket.

Assumptions :

1. Steady state heat conduction along spine.
2. Insulated tip spine.

Analysis : The given error

$$= \frac{0.5}{100} \times \text{Applied temperature difference}$$

or $$\frac{T_L - T_\infty}{T_0 - T_\infty} = \frac{0.5}{100} = \frac{1}{\cosh mL}$$

or $$\cosh mL = 200 \quad \text{or} \quad mL = 6$$

or $$L\sqrt{\frac{h}{kt}} = 6 \quad \text{or} \quad L\sqrt{\frac{98}{27 \times 0.5 \times 10^{-3}}} = 6$$

or $$L = \mathbf{70.42 \text{ mm}}. \text{ Ans.}$$

Example 5.31. *The steam at 300°C is passing through a steel tube. A thermometer pocket of steel (k = 45 W/m.K) of inside diameter 14 mm, and 1 mm thick is used to measure the temperature. Calculate the length of thermometer pocket needed to measure the temperature within 1.8% permissible error. The diameter of steam tube is 95 mm. Take heat transfer coefficient as 93 W/m²*.*K and tube wall temperature as 100°C.*

Solution

Given : The temperature measurement by thermometer in a pocket.

$T_\infty = 300°C$, $k = 45$ W/m.K,

$d_i = 14$ mm, $t = 1$ mm,

$\varepsilon = 1.8\%$, $T_0 = 100°C$,

$h = 93$ W/m^2.K.

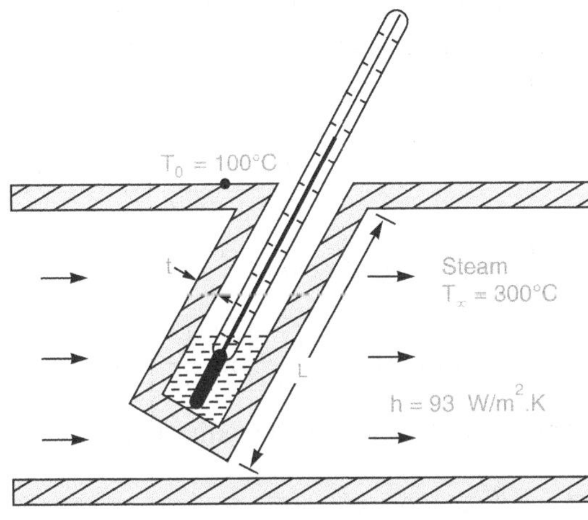

Fig. 5.39. Schematic for example 5.31

To find : The length of the pocket.

Analysis : Thermometer pocket is treated as fin of insulated tip. The temperature at x = L is given by

$$\frac{T_L - T_\infty}{T_0 - T_\infty} = \frac{1}{\cosh mL}$$

where $m = \sqrt{\dfrac{hP}{kA_c}}$

$d_o = d_i + 2t = 14 + 2 \times 1$

$= 16 \text{ mm} = 0.016 \text{ m}$

$P = \pi d_o$

$= \pi \times 0.016 = 0.0502 \text{ m}$

$A_c = \pi/4\ (d_o^2 - d_1^2)$

$= \pi/4 \times (0.016^2 - 0.014^2)$

$= 4.712 \times 10^{-5}$

$$m = \sqrt{\frac{93 \times 0.0502}{45 \times 4.712 \times 10^{-5}}} = 46.9 \text{ m}^{-1}$$

The permissible error = 1.8%

$T_\infty - T_L = 0.018\ T_\infty$

or $T_L = (1 - 0.018)\ T_\infty$

$= 0.982\ T_\infty = 0.982 \times 300$

$= 294.6$

And hence

$$\frac{294.6 - 300}{100 - 300} = \frac{1}{\cosh mL}$$

or $\cosh mL = 37.037$ or $mL = 4.305$

or $L = \dfrac{4.305}{46.9} = 0.0917$ m

$=$ **91.78 mm. Ans.**

Example 5.32. *The temperature of a gas stream is measured by using two thermocouples attached to a tube of perimeter 50 mm and cross-sectional area 25 mm². The tube 250 mm long and is mounted normal to the duct wall. If the thermocouples are attached to the tube at 125 mm and 250 mm from the duct wall and indicate the tube wall temperatures of 350°C and 390°C respectively. Calculate the gas temperature and the duct wall temperature.*

The heat transfer coefficient between tube wall and gas stream is 5 W/m².K and thermal conductivity of tube material is 45 W/m.K. Neglect any heat transfer into exposed end of tube.

Solution

Given : Temperature measurement of a gas stream.

$P = 50$ mm, $A_c = 25$ mm^2,

$L = 250 \text{ mm} = 0.25$ m,

$x_1 = 125$ mm, $T_1 = 350°C$,

$x_2 = 250$ mm, $T_2 = 390°C$,

$h = 5$ W/m^2.K, $k = 45$ W/m.K.

To find :

(*i*) Gas stream temperature

(*ii*) Duct wall temperature.

Analysis : The thermometer tube is assumed as insulated tip fin, and temperature distribution is given by

$$\frac{T - T_\infty}{T_0 - T_\infty} = \frac{\cosh mL(L - x)}{\cosh mL}$$

where $m = \sqrt{\dfrac{hP}{kA_c}} = \sqrt{\dfrac{5 \times 50 \times 10^{-3}}{45 \times 25 \times 10^{-6}}}$

$= 14.90 \text{ m}^{-1}$

Using the data at $x = x_1$, $T = T_1$

$$\frac{350 - T_\infty}{T_0 - T_\infty} = \frac{\cosh [14.90 \times (0.25 - 0.125)]}{\cosh (14.90 \times 0.25)} = 0.158$$

or $\quad 350 - T_\infty = 0.158\, T_0 - 0.158\, T_\infty$

or $\quad 0.158\, T_0 = 350 - 0.8412\, T_\infty$

or $\quad T_0 = 6.297 \times (350 - 0.8412\, T_\infty)$ ...(*i*)

Using data at $x = x_2 = L$, $T = T_2$

$$\frac{390 - T_\infty}{T_0 - T_\infty} = \frac{1}{\cosh mL} = \frac{1}{\cosh (14.90 \times 0.25)} = 0.0481$$

or $\quad 0.0481\, T_0 = 390 - (1 - 0.0481)\, T_\infty$

or $\quad 0.0481 \times 6.297 \times (350 - 0.8412\, T_\infty) = 390 - 0.952\, T_\infty$

or $\quad 0.6972\, T_\infty = 284$

Gas stream temperature

T_∞ = **407.33°C.** Ans.

and duct wall temperature

$T_0 = 6.297\,(350 - 0.8412 \times 407.33)$

= **46.3°C.** Ans.

5.7. DESIGN CONSIDERATIONS FOR FINS

The following factors should be considered for optimum design of fins :

1. Cost.
2. Manufacturing difficulties.
3. Pressure drop caused by fluid friction on fins.
4. Space consideration *e.g.*, length of fins.
5. Weight consideration.

A design of fins should be considered *ideal* (optimum), when the fins require less cost and easy to manufacture. They offer minimum resistance to fluid flow and are light in weight.

5.7.1. Space Considerations : Condition for use of Fins

An important consideration in the design of finned surfaces is the selection of proper fin length L. Normally, it is understood that, longer the fin, the larger the heat transfer area and thus the higher the rate of heat dissipation from the fin surface. But at the same time, with long fins, the weight, cost, and fluid friction increase. Therefore, increasing the length beyond a certain value cannot be justified unless the added benefits outweigh the increased cost. Further, the fin efficiency decreases with increasing fin length because decrease in temperature along the fin length. The fin length that causes the fin efficiency to drop below 60% cannot be justified economically and should not be used.

With regard to limiting condition of fin length, when heat transfer does not increase with an increase in the length of fin can be recognised by

$$\frac{dQ_{fin}}{dL} = 0$$

For the fins loosing heat by convection at its tip, the rate of heat transfer is given by eqn. (5.26)

$$Q_{fin} = \sqrt{hPkA_c}\,(T_0 - T_\infty) \times \frac{\sinh (mL) + \frac{h}{mk} \cosh (mL)}{\cosh mL + \frac{h}{mk} \sinh mL}$$

$$= \sqrt{hPkA_c}\,(T_0 - T_\infty) \times \frac{\tanh (mL) + \frac{h}{mk}}{1 + \frac{h}{mk} \tanh (mL)} \quad ...(5.50)$$

Treating k, h, P, A_c $(T_0 - T_\infty)$ and m as constant quantities and differentiating above equation with respect to fin length L and equating it to zero;

$$\frac{dQ_{fin}}{dL} = \sqrt{hPkA_c}\,(T_0 - T_\infty) \times \frac{d}{dL}\left[\frac{\tanh (mL) + h/mk}{1 + (h/mk) \tan mL}\right] = 0$$

or

$$\left[1 + \frac{h}{mk} \tanh (mL)\right] \times m \sec^2 h\,(mL) - \left[\tanh mL + \left(\frac{h}{mk}\right)\right] \times \frac{h}{mk}\, m \sec^2 h\, mL = 0$$

The simplification of this equation leads to

$$1 - \frac{h^2}{m^2k^2} = 0 \quad \text{or} \quad mk = h$$

or

$$\sqrt{\frac{kP}{hA_c}} = 1 \quad \text{or} \quad \frac{kP}{hA_c} = 1 \quad ...(5.51)$$

Introducing $P \approx 2w$, and $A_c = wt$

$$\frac{2k}{ht} = 1 \quad \text{or} \quad \frac{1}{h} = \frac{t/2}{k} \quad ...(5.52)$$

The term $1/h$ represents an external (convection) thermal resistance and $\frac{t/2}{k}$ represents internal (conduction) thermal resistance of a plane wall of thickness one half of a fin thickness.

The ratio of conduction resistance to convection resistance is known as the Biot number, explained in Chapter 6, that is

$$Bi = \frac{h(t/2)}{k} \quad ...(5.53)$$

We can draw the following conclusion with the help of eqns. (5.52) and (5.53)

1. After attachment of fins to a surface, if external thermal resistance is equal to internal thermal resistance as in eqn. (5.52), *i.e.,*

$$\frac{1}{h} = \frac{t/2}{k}$$

or $$\text{Bi} = 1 \quad i.e., \quad \frac{h}{mk} = 1$$

Then eqn. (5.50) for fin heat transfer rate yields to

$$Q_{fin} = hA_c (T_0 - T_\infty)$$

which represents the heat transfer rate from primary (root) surface without any fin. It suggests that as long as h/mk = 1, the heat transfer rate from the primary surface will not change by attaching fins as shown in Fig. 5.40(a).

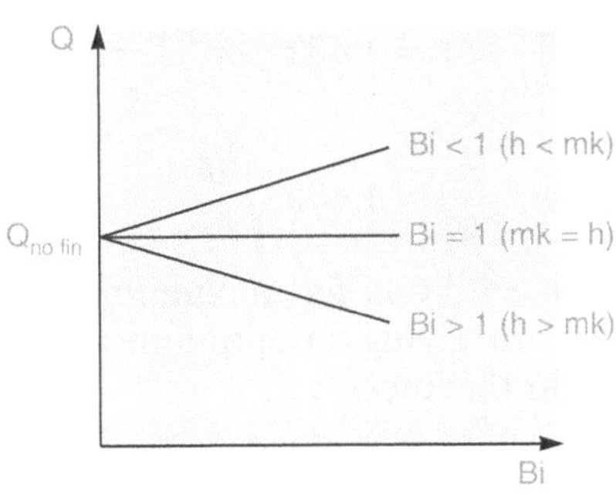

Fig. 5.40. (a) The fin heat transfer rate as a function of Biot number.

2. When internal resistance of fin is greater than the external resistance *i.e.,*

$$\frac{t/2}{k} > \frac{1}{h} \quad \text{or} \quad \text{Bi} > 1 \quad \text{or} \quad \frac{h}{mk} > 1$$

The addition of fins (secondary surface) on the primary surface will reduce the heat transfer rate or the fins will act as insulating medium on the surface. It may happen when value of h is very high as for flowing liquids, and during change of phase. Therefore, the fins are not used on liquid side and on evaporating and condensing surfaces.

3. When internal resistance of fin is less than the external resistance *i.e.,*

$$\frac{t/2}{k} < \frac{1}{h} \quad \text{or} \quad \text{Bi} < 1 \quad \text{or} \quad \frac{h}{mk} < 1$$

The use of fins will increase the heat transfer from the primary surface. In actual practice, the use of fins can only be justified, when the parameter $\frac{2k}{ht}$ has a value equal to or exceeding five.

$$\frac{2k}{ht} \geq 5 \text{ or } \text{Bi} \leq \frac{1}{5} \qquad \ldots(5.54)$$

Further, it should be noted that the rate of heat dissipation beyond a certain length of fin is quite less as compared with the rate of heat dissipation with increase in parameter $2k/ht$. Therefore, the use of shorter fins of higher conducting materials is more effective than longer fins. However, as the fins become shorter, the heat flow becomes two dimensional and therefore, result differs from that obtained from eqn. (5.51). Fig. 5.40(b) shows the variation of heat dissipation rate Q_{fin} with respect to fin length L for given values of parameter $\frac{2k}{ht}$. It indicates that as $\frac{2k}{ht} \to 1$, the $\frac{dQ_{fin}}{dL} \to 0$ and the fin becomes ineffective.

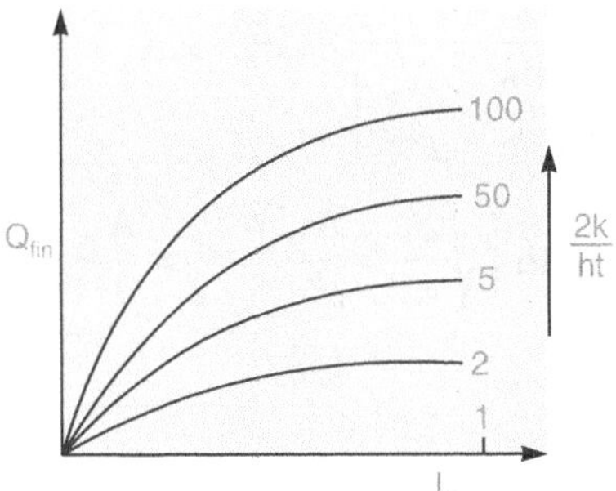

Fig. 5.40. (b) Variation of heat transfer rate with fin length

5.7.2. Weight Consideration

The weight of the fin is very important, when designing the fins for automobiles and aircrafts. In such problems, the maximum heat transfer rate is required with least amount of weight of heat exhanger. For a given weight, the maximum heat dissipation is required.

$$\text{Weight of one fin} = \text{Length} \times \text{width} \times \text{thickness} \times \text{density of material}$$
$$= L \times w \times t \times \rho$$

For given dimensions, the length L of the fin is fixed, whereas the width w and thickness t of the fin are optimised to get maximum heat flow.

If the heat loss from the tip is neglected

$$Q_{fin} = \sqrt{hPkA_c}\,(T_0 - T_\infty) \tanh mL$$
$$= mkA_c (T_0 - T_\infty) \tanh mL$$

Introducing $m = \sqrt{\frac{2h}{kt}}$ and $A_p = Lt$, profile area

Then

$$Q_{fin} = \sqrt{\frac{2h}{kt}} \times k \times (wt)(T_0 - T_\infty) \tanh\left(\sqrt{\frac{2h}{kt}} \times \frac{A_p}{t}\right)$$

or $$Q_{fin} = (2kh)^{1/2} t^{1/2} w\,(T_0 - T_\infty) \tanh\left(\sqrt{\frac{2h}{k}} \times \frac{A_p}{t^{3/2}}\right) \qquad \ldots(5.55)$$

For given profile area A_p, the heat transfer, Q will be maximum, when

$$\frac{dQ_{fin}}{dt} = (2kh)^{1/2}\, w(T_0 - T_\infty)$$

$$\times \frac{d}{dt}\left\{t^{1/2} \tanh\left(\sqrt{\frac{2h}{k}} \times \frac{A_p}{t^{3/2}}\right)\right\} = 0$$

Differentiating by parts

$$\left(\frac{1}{2}t^{-1/2}\right) \tanh\left(\sqrt{\frac{2h}{k}} \times \frac{A_p}{t^{3/2}}\right)$$

$$+ \frac{t^{1/2}}{\cosh^2\left(\sqrt{\frac{2h}{k}} \times \frac{A_p}{t^{3/2}}\right)} \times \sqrt{\frac{2h}{k}} \times A_p \times \left(-\frac{3}{2}t^{-5/2}\right) = 0$$

or $$\frac{1}{2} \times \frac{1}{\sqrt{t}} \tanh\left(\sqrt{\frac{2h}{k}} \times \frac{A_p}{t^{3/2}}\right) - \frac{3}{2} \times \frac{A_p}{t^2}$$

$$\times \sqrt{\frac{2h}{k}} \times \frac{1}{\cosh^2\left(\sqrt{\frac{2h}{k}} \times \frac{A_p}{t^{3/2}}\right)} = 0$$

or $$\tanh\left(\sqrt{\frac{2h}{k}} \times \frac{A_p}{t^{3/2}}\right)$$

$$= \frac{3A_p}{t^{3/2}} \times \sqrt{\frac{2h}{k}} \times \frac{1}{\cosh^2\left(\sqrt{\frac{2h}{k}} \times \frac{A_p}{t^{3/2}}\right)} \quad ...(5.56)$$

Using $A_p = Lt$ and $m = \sqrt{\frac{2h}{kt}}$, we get

$$\tanh mL = 3\, mL \times \frac{1}{\cosh^2 (mL)}$$

or $$\frac{\sinh mL}{\cosh mL} = \frac{3mL}{\cosh^2 (mL)}$$

or $$\sinh (mL) \cosh mL = 3\, mL \quad ....(5.57)$$

or $$\frac{\sinh (2mL)}{2} = 3\, mL$$

or $$\frac{e^{2mL} - e^{-2mL}}{2} = 6\, mL \quad ...(5.58)$$

or $$\left(1 + 2mL + \frac{4(mL)^2}{2!} + \frac{8(mL)^3}{3!} + \frac{16(mL)^4}{4!} + ... +\right)$$

$$- \left(1 - 2(mL) + \frac{4(mL)^2}{2!} - \frac{8(mL)^3}{3!} + \frac{16(mL)^4}{4!} - ... +\right) = 12\, mL$$

or $$4\, mL + \frac{16(mL)^3}{3!} + \frac{64(mL)^5}{5!} + ... = 12\, mL$$

or $$4mL + \frac{16}{6}(mL)^3 + \frac{64}{120}(mL)^5 + ... = 12\, mL$$

Discarding higher order terms, we have

$$1 + \frac{2}{3}(mL)^2 + \frac{2}{15}(mL)^4 = 3$$

or $$15 + 10(mL)^2 + 2(mL)^4 = 45$$

or $$(mL)^4 + 5(mL)^2 - 15 = 0$$

or $$(mL)^2 = \frac{-5 \pm \sqrt{5^2 + 4 \times 1 \times 15}}{2 \times 1} = 2.1$$

or $$mL = 1.452$$

or $$\sqrt{\frac{2h}{kt}} \times L = 1.452 \quad \text{or} \quad L = 1.452\sqrt{\frac{kt}{2h}}$$

or $$\frac{L}{t/2} = 1.452\sqrt{\frac{2k}{ht}} \quad ...(5.59)$$

It is the condition for maximum heat flow for a given weight of fin, giving the optimum ratio of fin height and half the fin thickness.

Further for insulated fin tip, the temperature ratio at its tip is given by eqn. (5.20)

$$\frac{T - T_\infty}{T_0 - T_\infty} = \frac{\theta}{\theta_0} = \frac{1}{\cosh mL}$$

or $$\theta = \frac{\theta_0}{\cosh (1.452)} = 0.453\, \theta_0 \quad ...(5.60)$$

and fin effectiveness for insulated tip fin

$$\varepsilon_{\text{fin insulated tip}} = \frac{Q_{fin}}{Q_{no\ fin}}$$

$$= \frac{\sqrt{hPkA_c}\, \theta_0 \tanh mL}{h(wt)\theta_0}$$

$$= \left(\frac{2k}{ht}\right)^{1/2} \tanh (mL)$$

$$= \left(\frac{2k}{ht}\right)^{1/2} \times \tanh (1.452)$$

$$\varepsilon_{\text{fin, insulated tip}} = 0.896\left(\frac{2k}{ht}\right)^{1/2} \quad ...(5.61)$$

This equation determines the heat flow increase through a wall as a result of addition of fin.

Example 5.33. *The 4 mm thick fins of mild steel are used to transfer heat from water to air. Decide the utility of fin on either side. The heat transfer coefficient of air is 80 W/m^2.K, while that for water is 5600 W/m^2.K. Take thermal conductivity of mild steel as 45 W/m.K.*

Solution

Given : The mild steel fin fins :

$t = 4$ mm $= 4 \times 10^{-3}$ m, $h_1 = 80$ W/m^2.K,

$h_2 = 5600$ W/m^2.K, $k = 45$ W/m.K.

To find : The utility of fin on either side.

Analysis : The condition for use of fins is $\dfrac{2k}{ht} \geq 5$

(*i*) For air, $h_1 = 80$ W/m^2.K

$$\frac{2 \times 45}{80 \times 4 \times 10^{-3}} = 281.25 \geq 5$$

The fins can be used on air side, they will enhance the heat transfer rate on this side. **Ans.**

(*ii*) For water, $h_2 = 5600$ W/m^2.K

$$\frac{2k}{ht} = \frac{2 \times 45}{5600 \times 4 \times 10^{-3}} = 4.01$$

which is less than 5 and hence the use of fins on water side will not serve the purpose, they are installed for. **Ans.**

Example 5.34. *A steel fin having 8 mm diameter and 100 mm long is exposed to boiling water having convective heat transfer coefficient of 4000 W/m^2.K. The thermal conductivity of steel can be taken as 17 W/m.K. Show by calculation how much heat dissipation is achieved and is it advisable to use the fin ? How the heat dissipation performance change, if a material with thermal conductivity of 45 W/m.K is used ? All other conditions are same.*

Solution

Given : A finite long fin

$d = 8$ mm $= 0.008$ m, L = 100 mm = 0.1 m,

$h = 4000$ W/m^2.K, $k_1 = 17$ W/m.K,

$k_2 = 45$ W/m.K.

To find :

(*i*) Validity of fin attachment.

(*ii*) Change in performance if $k_2 = 45$ W/m.K, instead of $k_1 = 17$ W/m.K.

Assumptions :

(*i*) The diameter of the fin is very less compared to its length, thus treating infinite long fin.

(*ii*) The boiling water temperature is 100°C.

(*iii*) Constant properties.

Analysis : The corrected length of fin

$$P = \pi d = \pi \times (0.008)\ m$$

$$A_c = \frac{\pi}{4} d^2 = \frac{\pi}{4} \times (0.008 \text{ m})^2$$

$$m = \sqrt{\frac{hP}{kA_c}} = \sqrt{\frac{4000 \times \pi \times (0.008)}{17 \times (\pi/4) \times (0.008)^2}} = 343$$

The heat transfer rate by fin

$$Q_{fin} = \sqrt{hPkA_c}\ (T_0 - T_\infty)$$

$$= \sqrt{4000 \times \pi \times (0.008) \times 17 \times \frac{\pi}{4} \times (0.008)^2}\ \times (T_0 - T_\infty)$$

$$= 0.293\ (T_0 - T_\infty)\ W$$

The heat transfer rate from root surface, before fin attachment

$$Q_{no\ fin} = hA_c\ (T_0 - T_\infty)$$

$$= 4000 \times (\pi/4) \times (0.008)^2 \times (T_0 - T_\infty)$$

$$= 0.201\ (T_0 - T_\infty)\ W$$

% increase in heat transfer

$$\frac{Q_{fin} - Q_{no\ fin}}{Q_{no\ fin}} \times 100 = \frac{0.293 - 0.201}{0.201} \times 100 = 45.8\%$$

The use of fin increases the heat dissipation by 45.8% but use of fin is not justified, because condition for used

$$\frac{2k}{ht} = \frac{2k}{h(d/4)} = \frac{\mathbf{2 \times 17}}{\mathbf{4000 \times 0.002}}\ \mathbf{\leq 5 \text{ is not satisfied.}}$$ **Ans.**

If material of thermal conductivity, $k_2 = 45$ W/m.K used, then

$$m = \sqrt{\frac{hP}{kA_c}} = \sqrt{\frac{4h}{kd}} = \sqrt{\frac{4 \times 4000}{45 \times 0.008}} = 210.8$$

$$Q_{fin} = \sqrt{4000 \times \pi \times 0.008 \times 45 \times (\pi/4) \times (0.008)^2} \times (T_0 - T_\infty)$$

$$= 0.476\ (T_0 - T_\infty)\ W$$

% increase in heat transfer

$$= \frac{0.476 - 0.201}{0.201} \times 100$$

$$= \mathbf{137.2\%.}$$ **Ans.**

and $$\frac{2k}{ht} = \frac{2 \times 45}{4000 \times 0.002} = 11.25 > 5$$

Thus, the use of fins of mild steel is justified on the surface.

5.8. SUMMARY

A fin is normally a thin strip of metal. The finned surfaces are commonly used to increase the heat dissipation rate. Fins increase the exposure area, thereby the heat dissipation rate by convection. The addition of fins is justified when the value of heat transfer coefficient is low. The temperature distribution and heat transfer for fins of uniform cross-section are tabulated below.

TABLE 5.1. Temperature distribution and heat loss for fins of uniform cross-section

Case	*Tip condition at x = L*	*Temperature distribution* θ/θ_0 =	*Fin heat transfer* Q_{fin} =
A	Infinite long fin	e^{-mx}	M
B	Insulated tip	$\dfrac{\cosh m(L-x)}{\cosh mL}$	M tanh mL
C	Convection heat transfer at tip	$\dfrac{\cosh m(L-x) + (h/mk)\sinh m(L-x)}{\cosh mL + (h/mk)\sinh mL}$	$M\dfrac{\sinh mL + (h/mk)\cosh mL}{\cosh mL + (h/mk)\sinh mL}$
D	Prescribed temperature	$\dfrac{(\theta_L/\theta_0)\sinh mx + \sinh m(L-x)}{\sinh mL}$	$M(\theta_L/\theta_0 + 1)\dfrac{\cosh m(L-1)}{\sinh mL}$
where, $m^2 = hP/(kA_c)$		$\theta_0 = T_0 - T_\infty$, $\theta = T - T_\infty$	$M = \sqrt{hPkA_c}\,(T_0 - T_\infty)$

Thermometer pocket accommodates thermometer, in a small tube welded radially to duct. The thermometer pocket is considered as a fin with insulated tip and recorded temperature is approximated as

$$\frac{T_L - T_\infty}{T_0 - T_\infty} = \frac{1}{\cosh mL}$$

Installation of fin is justified by a term called fin effectiveness, defined as

$$\varepsilon_{fin} = \frac{Q_{fin}}{Q_{no\ fin}} = \frac{Q_{fin}}{hA_c(T_0 - T_\infty)}$$

$$= \frac{\text{The heat transfer rate with fin from base area } A_c}{\text{Heat transfer rate without fin from the surface of area } A_c}$$

The overall effectiveness for the finned surface is defined as

$$\varepsilon_{fin\ overall} = \frac{\text{Heat transfer rate from finned surface}}{\text{Heat transfer rate from same surface, if there were no fins}}$$

$$= \frac{Q_{total\ fin}}{Q_{no\ fin}} = \frac{Q_{fin} + Q_{unfin}}{Q_{no\ fin}}$$

The fin efficiency is used to evaluate the thermal performance of a fin. It is defined as

$$\eta_{fin} = \frac{Q_{fin}}{Q_{ideal}} = \frac{\text{Actual heat transfer rate from a fin}}{\text{Ideal heat transfer rate from a fin, if entire fin surface was at base temperature } T_0}$$

If fin efficiency is known, the heat transfer rate Q_{fin} from a fin can be obtained as

$$Q_{fin} = \eta_{fin} \times Q_{ideal} = \eta_{fin}\, hA_{fin}(T_0 - T_\infty)$$

The total fin efficiency evaluates the thermal performance of the finned surface. It is expressed as

$$\eta_{total\ fin} = \frac{\text{Total heat transfer rate from the finned surface}}{\text{The heat transfer rate which would be possible, if total finned surface were maintained at base temperature } T_0}$$

$$= \frac{Q_{fin} + Q_{unfin}}{h\,A_{total}(T_0 - T_\infty)}$$

where $A_{total} = A_{fin} + A_{unfin}$

The analysis of fin with convection at its tip is tidious. The solution to such fins can be approximated by expressions for finite long insulated tip fin by considering corrected length L_c of fin

$$L_c = L + \frac{A_c}{P}$$

The use of fin is justified when

$$\frac{2k}{ht} \geq 5$$

REVIEW QUESTIONS

1. Why extended surfaces are most commonly used ?
2. A fin attached to a surface shows an effectiveness of 0.9. Do you think the heat transfer rate from the surface has increased or decreased with addition of fins ? Comment.
3. Define fin effectiveness. When the use of fins is not justified.
4. Explain the criteria of selection of fins.
5. What is the difference between fin effectiveness and fin efficiency ?
6. How does overall effectiveness of a finned surface differ from the effectiveness of a single fin ?
7. If a thin and long fin, insulated at its tip is used, show that the heat transfer from the fin is given by

$$Q_{fin} = \sqrt{hPkA_c}\ (T_0 - T_\infty) \tanh mL.$$

8. How is thermal performance of a fin measured ? Define fin efficiency.
9. Two pin fins are identical except that the diameter of one is twice that of other. For which fin will (*a*) fin effectiveness (*b*) fin efficiency be higher ?
10. Under what situations does the fin efficiency become 100% ?
11. What types of boundary conditions are used for various types of fins ?
12. State the various assumptions made in the formation of energy equation for one dimensional heat dissipation from an extended surface.
13. Give a few specific examples of use of fins.
14. What would be the nature of temperature distribution in a fin, if thermal conductivity of fin material is very high ?
15. Show that the fin efficiency for a rectangular fin is given by

$$\eta_{fin} = \frac{\tanh\,[2hL_c^2/kt]^{1/2}}{[2hL_c^2/kt]^{1/2}}$$

where L_c = corrected length = $L + t/2$ or $L + \frac{A_c}{P}$.

16. Show that the total heat transfer rate from a fin wall is given by

$$Q = h\,[A_{total} - (1 - \eta_{fin})\,A_{fin}]\,(T_0 - T_\infty)$$

where A_{total} = total area of fin and unfinned surfaces.

A_{fin} = area of the finned surface.

η_{fin} = fin efficiency.

17. Explain the situation, when addition of fins to a surface is not useful.

PROBLEMS

1. A long rod 6.5 mm in diameter is exposed to an environment at 27°C. The base temperature of the rod is 150°C. The heat transfer coefficient between the rod and environment is 30 W/m^2.K. Calculate the heat loss by the rod. [Ans. 753.5 W/m]
2. One half of a long rod, 25 mm in diameter, was inserted into a furnace, while the other half was projecting into air at 27°C. After steady state had been reached, the temperature at two points 76 mm apart were measured and found to be 126°C and 91°C respectively. The heat transfer coefficient was estimated to be 22.7 W/m^2.K. Calculate thermal conductivity of the rod material. [Ans. 110.2 W/m.K]
3. A long brass rod (k = 104 W/m.K), 25 mm in diameter is heated by inserting its one end into a furnace, while remaining portion is projected into an ambient at 25°C. During steady state, the measurements of temperature at two points 10 cm apart reveal 155°C and 101°C respectively. Calculate the effective heat transfer coefficient. [Ans. 12.6 W/m^2.K]
4. Consider two long, slender rods A and B of the same diameter but different materials. One end of the each rod is attached to a base surface temperature at 100°C, while the surface of the rods are exposed to an ambient air at 20°C. By traversing the length of the each rod with a thermocouple, it was observed that the temperature of rods were equal at the position x_A = 0.15 m and x_B = 0.075 m, where x is measured from the base surface. If the thermal conductivity of the rod A is known to be k_A = 70 W/m.K, determine the value of k_B for rod B. [Ans. 17.5 W/m.K]
5. A very long copper rod (k = 372 W/m.K), 25 mm in diameter has maintained its one end at 100°C. The rod is exposed to a fluid at 40°C with h = 3.5 W/m^2.K. Calculate the heat lost by the rod. [Ans. 13.44 W]
6. A long stainless steel rod (k = 16 W/m.K) has a square cross-section 12.5 cm × 12.5 cm and has one end maintained at 250°C. It is exposed into a fluid at 90°C with h = 40 W/m^2.K. Calculate the heat lost by the rod. [Ans. 357.77 W]
7. A rectangular copper fin has one end maintained at 200°C, while remainder of the fin surface is exposed to convective environment at 25°C with h = 35 W/m^2.K. If the thermal conductivity of the copper is 386 W/m.K, determine the heat lost by fin per unit depth. The length of the fin is 5 cm and thickness is 4 mm. Assume the fin tip to be insulated. [Ans. 612.41 W]
8. An aluminium fin (k = 204 W/m.K), 18 mm thick and 16 cm long has a base temperature of 300°C. The ambient temperature is 20°C, with convection heat transfer coefficient of 30 W/m^2.K. Determine the fin efficiency and heat lost from the fin per unit depth. [Ans. 2681.27 W]

9. A brass rod (k = 100 W/m.K), 100 mm long and 5 mm in diameter extends horizontally from a casting at 200°C. The rod is in an environment at 20°C and convection coefficient of 30 W/m^2.K. What is the temperature of the rod 25 mm, 50 mm and 100 mm from the casting ? *(P.U., Dec. 1994)*

[Ans. 92.1°C, 91.99°C, 91.88°C]

10. An aluminium rod (k = 200 W/m.K), 2.5 cm in diameter, 15 cm long protrudes from a wall, which is maintained at 260°C. The rod is exposed to an environment at 16°C with convection coefficient of 15 W/m^2.K. Calculate the efficiency and heat lost by the rod. [Ans. Q = 78.5 W]

11. A cylindrical rod 2 cm in diameter and 20 cm long protrudes from a heat source at 300°C into air at 40°C. The heat transfer coefficient is 5 W/m^2.K on all exposed surface. Neglecting the radial variation of temperature and heat lost from the tip, find the temperature at the fin tip.
Find the temperature at the fin tip and at the mid-point along the rod made of borosilicate glass (k = 1.09 W/m.K). Also determine fin efficiency.

[Ans. 41.23°C, 52.7°C, 16.5%]

12. A 2 cm diameter glass rod (k = 0.8 W/m.K) is 6 cm long. It has base temperature at 120°C and is exposed to air at 20°C. The temperature at the tip of the rod is measured as 35°C. What is the convection heat transfer coefficient ? How much heat is lost by the rod ? [Ans. 6.32 W/m^2.K, 1.0 W]

13. A straight rectangular fin (k = 55 W/m.K), 1.4 mm thick and 35 mm long is exposed to air at 20°C with h = 50 W/m^2.K. Calculate the maximum possible heat loss for a base temperature of 150°C. What is actual heat loss for this base temperature ?

[Ans. 487.7 W, 309.6 W]

14. The copper fins are installed on an I.C. engine cylinder. The inner and outer radii of fins are 50 mm and 62.5 mm, respectively. The cylinder wall and ambient temperature are 180°C and 36°C, respectively. The thermal conductivity and density of copper are 384 W/m.K and 8800 kg/m^3, respectively. The heat transfer coefficient over the fins surface is 70 W/m^2.K. If the fins are designed to obtain maximum heat transfer rate for a given mass, calculate :
(*a*) rate of heat transfer per fin,
(*b*) saving in mass in kg/fin if aluminium was used in place of copper.
For aluminium take k = 203.5 W/m.K,
ρ = 2670 kg/m^3.

[Ans. (*a*) 55.5 W/fin, (*b*) 42.7% saving in mass]

15. An annular aluminium fin (k = 210 W/m.K) is attached to a circular tube having an outside diameter of 25 mm and a surface temperature of 250°C. The fin is 1 mm thick and 10 mm long. The fin is exposed in an ambient at 25°C with h = 25 W/m^2.K.
(*a*) What is the heat loss from the fin ?
(*b*) If 200 such fins are spaced at 5 mm increments along the tube length, what is the heat loss per metre of the tube length ?

[Ans. (*a*) 12.97 W, (*b*) 2948.73 W]

16. Annular aluminium fins (k = 212 W/m.K) 2 mm thick and 15 mm long are installed on an aluminium tube, 30 mm diameter. If the tube wall is at 100°C and the adjoining fluid is at 25°C with h = 75 W/m^2.K. What is the rate of heat transfer from a fin ?

[Ans. 25 W]

17. Circumferential fin of rectangular cross-section, 37 mm in outer diameter and 3 mm thick is attached to a 25 mm diameter tube. The fin is constructed of mild steel (k = 45 W/m.K). The air circulated over the fin with a heat transfer coefficient of 28.4 W/m^2.K. If the temperature of the base of the fin and air are 260°C, and 38°C, respectively. Calculate the heat transfer rate from the fin. [Ans. 50 W]

18. Circular aluminium fins of constant rectangular profile are attached to an aluminium tube of 50 mm outer diameter and having a surface temperature of 180°C. Fin thickness t = 1 mm, L = 15 mm, k = 200 W/m.K. The fin is exposed to an ambient at 30°C with heat transfer coefficient of 80 W/m^2.K. Find the fin efficiency and heat transfer from each fin. *(Anna University, Dec. 1999)*

[Ans. η_{fin} = 0.93, Q_{fin} = 68.36 W]

$$\left[\textbf{Hint. } L_c^{3/2}\left(\frac{h}{kA_p}\right) = 0.3,\ \frac{r_{2c}}{r_1} \approx 1.6,\ \eta_{fin} = 93\%,\right.$$

$$\left. Q_{fin} = \eta_{fin} \times h \times 2\pi\,(r_{2c}^2 - r_1^2)(T_0 - T_\infty) = 68.36 \text{ W}\right]$$

19. The aluminium fins (k = 206 W/m.K) are installed on an electronic device 1 m wide and 1 m tall. The fins are rectangular in cross-section, 2.5 cm long and 0.25 cm thick. There are 100 fins per metre. The convection heat transfer coefficient is 35 W/m^2.K. Calculate the percentage increase in heat transfer with finned wall in comparison with base wall.

[Ans. 483%]

20. Heat is transferred from water to air through a brass wall (k = 84 W/m.K). The addition of rectangular brass fins 0.08 cm thick and 2.5 cm long, spaced 1.25 cm apart is suggested. Assuming water side heat transfer coefficient is 170 W/m^2.K and that of on air side is 17 W/m^2.K. Compare the heat transfer rate achieved by adding fins to (*a*) water side, (*b*) air side and (*c*) both sides. Neglect temperature drop through the wall. [Ans. (*a*) 6.55%, (*b*) 247%, (*c*) 340%]

21. A 3 mm thick aluminium plate (k = 210 W/m.K) has rectangular fins 1.6 mm × 6 mm on a side spaced 6 mm apart. The finned side is in contact of air at

38°C with h = 28.4 W/m^2.K. On the unfinned side, water flows at 93°C with h = 283.7 W/m^2.K. Calculate :

(*a*) the efficiency of fins,

(*b*) rate of heat transfer per unit area of the wall,

(*c*) comment on the result, if water and air sides are interchanged.

[Ans. (*a*) 99.7%, (*b*) 3598.4 W, (*c*) Heat transfer is reduced by 46.4%, if water and air sides are interchanged]

22. One end of a 30 cm long steel rod (k = 25 W/m.K) is connected to a wall at 204°C. The other end is connected to other wall at 93°C. The air is blown across the rod with h = 17 W/m^2.K. The diameter of the rod is 5 cm and air temperature is 30°C, what is the net rate of heat dissipation to air ?

[Ans. 190.25 W]

23. The both ends of a 0.5 cm diameter copper U-shaped rod (k = 386 W/m.K) are rigidly fixed to a vertical wall. The temperature of the wall is maintained at 90°C. The developed length of the rod is 60 cm and it is exposed in air at 30°C with h = 34 W/m^2.K.

(*a*) Calculate the temperature at the mid-point of the rod

(*b*) Heat transfer rate from the rod.

[Ans. (*a*) 39.6°C, (*b*) 7.53 W]

24. A circular fin of a rectangular profile is used on a 30 cm diameter tube, maintained at 100°C. The outside diameter of the fin is 50 cm and the fin thickness is 1.0 mm. The environment air temperature is 30°C with h = 50 W/m^2.K. Calculate thermal conductivity of the material for fin efficiency of 60%. [Ans. 1.8 W/m.K]

25. The steam in a heating system flows through a tube, 5 cm in outer diameter whose outer surface is maintained at 180°C. The circular aluminium fins (k = 186 W/m.K) of outer diameter 6 cm and of constant thickness of 1 mm are attached to the tube with a spacing 3 mm and thus 250 fins per metre length of the tube. The heat is transferred to surrounding air at 25°C with h = 40 W/m^2.K. Calculate the increase in heat transfer rate from the tube per metre as a result of adding fins.

[Ans. 2750 W]

26. The temperature of air in a reservoir is measured with the aid of a mercury in a glass thermometer placed in a protective steel well filled with oil. The thermometer indicates the temperature at the end of the well as 84°C. The well is 12 cm long, its thickness is 1.5 mm and thermal conductivity of the well material is 55.8 W/m.K. Assume heat transfer coefficient between well and air is 23.5 W/m^2.K. Calculate the error in temperature measurement, if the base of the well is at 40°C. Also calculate the true temperature. [Ans. 16°C, 100°C]

27. An aluminium fin (k = 210 W/m.K), 1.6 mm thick is placed on a circular tube with 2.54 cm OD. The fin is 6.4 mm long. The tube wall is maintained at 150°C, while the ambient is at 20°C with convective coefficient of 25 W/m^2.K. Calculate the heat lost by fin. [Ans. 6.8 W]

28. A thermometer well 22 cm in diameter and 0.5 mm thick is made of steel (k = 27 W/m.K) and it is to be used to measure the temperature of steam flowing through a pipe. Calculate the minimum length of well so that the error is less than 0.5% of the difference between pipe well and the fluid temperature. Take steam temperature as 250°C and h = 98 W/m^2.K. [Ans. L = 1.7 cm]

29. Thin fins of brass (k = 101 W/m.K) are welded longitudinally on a 4 cm brass cylinder, which stands vertically and is surrounded by air at 20°C. The heat transfer coefficient from the metal surface to the air is 20 W/m^2.K. If 20 uniformly spaced fins are used, each 0.8 mm thick and extending 1 cm from the cylinder surface, calculate the heat transfer from the cylinder to the air, when the cylinder surface is maintained at 200°C. [Ans. 1866.22 W]

30. An oil-filled thermometer well made of a steel tube (k = 55.8 W/m.K), 120 mm long and 1.5 mm thick is installed in a tube through which air is flowing. The temperature of the air stream is measured with the help of a thermometer placed in the well. The surface heat transfer coefficient from the air to the well is 23.3 W/m^2.K and the temperature recorded by the thermometer is 88°C. Estimate the measurement error and the percentage error if the temperature at the base of the well is 40°C. [Ans. 17.4°C, 16.5%]

31. A turbine blade 6.25 cm long, 4.5 cm^2 in cross-section, 12 cm in perimeter is made of steel (k = 26.16 W/m.K). The root temperature is 500°C. The blade is exposed to steam at 800°C with convection coefficient of 465 W/m^2.K. Calculate the temperature and rate of heat flow at the root of the blade. [Ans. 243 W]

32. A straight triangular fin of steel (k = 30 W/m.K) is attached to a plane wall maintained at 460°C. The fin thickness is 6.4 mm and it is 25 mm long. It is exposed into a fluid at 90°C with h = 28 W/m^2.K. Calculate the heat loss from the fin. [Ans. 2950 W]

33. A straight rectangular fin 2.0 cm thick and 14 cm long is constructed of steel (k = 45 W/m.K) and placed on a wall at 200°C, exposed to air at 15°C with h = 20 W/m^2.K. Calculate heat lost from the fin per unit depth. [Ans. 845.4 W]

34. A 1 cm diameter steel rod (k = 20 W/m.K) is 20 cm long. Its one end is maintained at 50°C while other at 100°C. It is exposed to convection environment at 20°C with h = 85 W/m^2.K. Calculate the temperature at the centre of the rod. [Ans. 21.8°C]

35. A straight fin (k = 23 W/m.K) with triangular profile has a length of 5 cm and thickness of 4 mm. The fin

is exposed to a fluid at 40°C with h = 20 W/m^2.K. The base of the fin is maintained at 200°C. Calculate the heat loss per unit depth of the fin.

[Ans. 214.16 W]

36. Aluminium fins (k = 200 W/m.K) of rectangular profile are attached on a plane wall with 5 mm spacing (200 fin per metre width). The fins are 1 mm thick, 10 mm long. The wall is maintained at temperature of 200°C and the fins dissipate heat by convection into the ambient air at 40°C with h = 50 W/m^2.K. Determine : (*a*) the fin efficiency, (*b*) the area weighted fin efficiency, (*c*) the heat loss per square metre of the wall. [Ans. (*c*) 37.8 kW/m^2]

37. A 1.6 mm diameter stainless steel rod (k = 22 W/m.K) protrudes from a wall maintained at 80°C. The rod is 12.5 mm long and exposed into a fluid at 25°C with h = 570 W/m^2.K. Calculate the temperature at the tip of the rod. Repeat the calculation with h = 20 W/m^2.K and h = 1200 W/m^2.K. Comment on result. [Ans. 29.12°C, 71.02°C, 25.93°C]

38. Two 30 cm long and 0.4 cm thick cast iron (k = 52 W/m.K) steam pipes of outer diameter 10 cm are connected each other through two 1 cm thick flanges of outer diameter 20 cm. The steam flows inside the tube at an average temperature of 200°C with h = 180 W/m^2.K. The outer of the pipe is exposed to air at 8°C with h = 25 W/m^2.K.
   (*a*) Disregarding the flanges, calculate the average outer surface temperature of pipe.
   (*b*) Using this temperature for the base of the flanges and treating the flanges as fins, calculate the fin efficiency and the rate of heat transfer from the flanges. [Ans. (*a*) 174.53°C, (*b*) 0.93%, 207.4 W]

39. A very long rod, 25 mm in diameter, has one end maintained at 100°C. The surface of the rod is exposed to ambient air at 25°C with convection coefficient of 10 W/m^2.K.
   (*i*) What are the heat losses from the rods, constructed of pure copper (k = 398 W/m.K) and stainless steel (k = 14 W/m.K) ?
   (*ii*) Estimate how long the rods must be to be considered infinite. (*P.U., Nov. 2003*)

[Ans. (*i*) Q_{cu} = 29.37 W, Q_{55} = 5.51 W, (*ii*) L_{cu} = 1.32 m, L_{55} = 0.247 m]

40. Two rods A and B of equal diameter and equal length, but of different materials are used as fins. The both rods are attached to a plain wall maintained at 160°C, while they are exposed to air at 30°C. The end temperature of rod A is 100°C, while that of the rod B is 80°C.
   If the thermal conductivity of rod A is 380 W/m.K, calculate the thermal conductivity of rod B. This fin can be assumed as short with end insulated.

[Ans. 221.94 W/m.K]

41. An electric motor casing has a diameter of 0.36 m and length of 0.4 m. The casing is made from cast steel (k = 60 W/m.K) and number of fins are installed to dissipate 400 W of heat into an ambient air, where the unit surface conductance is 10 W/m^2.K. Each fin is to sum the entire length of the motor casing. Each fin is 8 mm thick and 10 mm long. Calculate the number of fins required to maintain the temperature difference between casing and surrounding air of 30°C. [Ans. 117 Fins]

42. A carbon steel pipe (k = 45 W/m.K), 78 mm in inner diameter and 5.5 mm thick has eight longitudinal fins 1.5 mm thick. Each fin extends 30 mm from the pipe wall. If the wall temperature, ambient temperature and surface heat transfer coefficient are 150°C, 28°C and 75 W/m^2.K, respectively. Calculate the percentage increase in heat transfer rate for the finned tube over the plain tube. [Ans. 104.45%]

43. A copper pipe 100 mm in outer diameter is provided with circular aluminium fins (k = 230 W/m.K) to increase the heat transfer rate. The height of the fin is 80 mm and it is 4 mm thick. The temperature at outer surface of copper pipe is 300°C and the temperature of surrounding air is 38°C. The heat transfer coefficient over the fin surface is 40 W/m^2.K.
   Calculate :
   (*i*) Rate of heat loss from the fin,
   (*ii*) The efficiency of fin,
   (*iii*) The fin effectiveness.

[Ans. (*i*) 776.34 W, (*ii*) 79%, (*iii*) 58.95]

44. Hot oil in a rectangular tank (1 m × 1 m on a side) is exposed to surrounding air at 24°C. The temperature of the tank wall is 110°C. In order to increase the heat dissipation, it is proposed to attach straight rectangular fins to the tank surface. As a result the heat dissipation rate increases by 70% and tank surface temperature drops to 91°C. The fins are 5 mm thick and are spaced 100 mm apart (centre to centre distance). The thermal conductivity of tank and fin material is 230 W/m.K and heat transfer coefficient over fins is 42 W/m^2.K. Heat loss from the fin tip may be neglected. Calculate the minimum height of the fins. [Ans. 68.48 mm]

45. A 1.25 cm diameter 15 cm long iron rod (k = 40 W/m.K) protrudes out from a heat source at 130°C into an ambient at 20°C with convection coefficient of 20 W/m^2.K.
   Determine :
   (*i*) Temperature distribution in the rod,
   (*ii*) Temperature at the free end,
   (*iii*) Heat flow out the source,
   (*iv*) Heat flow rate at the free end.

[Ans. (*ii*) T_L = 51.1°C, (*iii*) 6.54 W, (*iv*) 76.33 mW]

46. Pin fin are provided to increase the heat transfer rate from a hot surface. Which of the following arrangement will give higher heat transfer rate : (*i*) 6 fins of 10 cm length or (*ii*) 12 fins of 5 cm length. For analysis, use fin with insulated tip condition.

Take k_{fin} = 200 W/m°C, h = 20 W/m^2°C, cross-section area of fin = 2 cm^2 perimeter = 4 cm, fin base temp = 230°C, surrounding air temp = 30°C.

(*P.U. May 2013*)

[**Ans.** (*i*) η_{fin} = 48.2%, Q_{fin} = 207 W, (*ii*) η_{fin} = 76.1%, Q_{fin} = 327 W. Shorter fins are effective]

REFERENCES AND SUGGESTED READING

1. Arpaci Vedat S. "*Conduction Heat Transfer*", Addison-Wesley Publishing Company Reading, MA, 1966.
2. Schneider P.J. "*Conduction Heat Transfer*", Addison-Wesley Publishing Company Reading, MA, 1955.
3. Kraus D.A., Aziz A and Welty J., "*Extended Surface Heat Transfer*", Wiley Inc. New York 2001.
4. Serth Robert W, Process "*Heat Transfer-Principles and Applications*", Elsevier Science & Technology Books, 2007.
5. Frank Kreith, Raj M. Manglik, Mark S. Bohn, "*Principles of Heat Transfer*", 7th edition, Cengage Learning, 2011.
6. Incropera Frank. P. And DeWitt David. P., "*Fundamentals of Heat and Mass Transfer*", 5th ed John Wiley & Sons, New York, 2002.
7. Kern Donald Q, and Kraus A.D., "*Extended Surface Heat Transfer*", McGraw Hill, New York, 1972.

Transient Heat Conduction

6

6.1. Approximate Solution—Systems with negligible internal resistance : lumped system analysis—Dimensionless quantities—Thermal time constant and response of thermocouple—The lumped system analysis for mixed boundary conditions—The validity of lumped system analysis. **6.2. Analytical Solution**—Criteria for neglecting internal temperature gradients—Infinite cylinder and sphere with convective boundaries—One term approximation. **6.3. Transient Temperature Charts : Heisler and Gröber Charts**—Transient temperature charts for slab—Transient temperature charts for long cylinder and sphere. **6.4. Transient Heat Conduction in Semi Infinite Solids**—Penetration depth and penetration time. **6.5. Transient Heat Conduction in Multidimensional Systems. 6.6. Summary**—*Review Questions—Problems—References and Suggested Reading.*

When the heat energy is being added or removed to or from a body, its energy content (internal energy) changes, resulting into change in its temperature at each point within the body over the time. During this transient period, the temperature becomes function of time as well as direction in the body. The conduction occurred during this period is called *transient (unsteady state) conduction. Therefore,* in unsteady state

$$T = f(x, t)$$
$$= \text{Function of direction and time}$$

During transient heat conduction, the energy balance on a body yields to

The net rate of heat transfer with the body
= Net rate of internal energy change of the body.

In many engineering applications, the heat transferred is transient. The heat treatment process, like quenching, annealing, normalising etc. are processes of unsteady state heat flow.

The unsteady heat flow is also involved, when the system undergoes a transition from one steady state to another, involving periodic variation in heat flow and temperature, *e.g.*, the periodic heat flow in a building between day and night.

The analysis of heat transfer during unsteady state can be possible by

1. Approximation,
2. Analytical method,
3. Use of transient temperature charts,
4. Product solution,
5. Graphical solution,
6. Numerical technique.

To determine the time dependence of temperature distribution within a solid during transient process, we shall begin by solving the problems that can be simplified by considering the temperature in the solid is only function of time and uniform throughout the system at any instant. In subsequent sections of this chapter, we shall consider the problems of unsteady state, when temperature varies with time as well as it penetrates the interior of the bodies.

6.1. APPROXIMATE SOLUTION

6.1.1. Systems with Negligible Internal Resistance : Lumped System Analysis

If the physical size of the body is very small, the temperature gradient exists in the body is negligible. The small body can be assumed at uniform temperature throughout at any time. The analysis of the unsteady heat transfer with negligible temperature gradients is called the *lumped system analysis.*

Consider a solid of volume V, surface area A_s, thermal conductivity k, density ρ, specific heat C and initially at uniform temperature T_i is suddenly immersed in a well stirred fluid, kept at uniform temperature T_∞. The heat is dissipated by convection into a fluid from its surface, with convection coefficient h.

In absence of any temperature gradient in solid, the energy balance for element is :

The rate of heat flow out the solid through the boundary surface(s)

= The rate of decrease of internal energy of the solid

or $$hA_s(T - T_\infty) = -mC\frac{dT}{dt} \quad ...(6.1)$$

where, $m = \rho V$, mass of the body

and $T = f(t)$, a function of time.

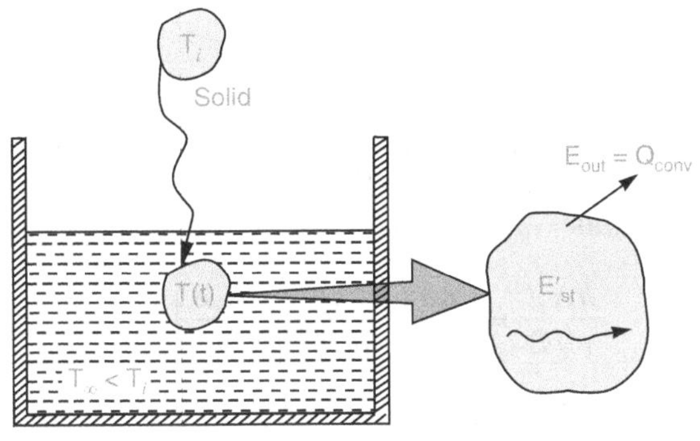

Fig. 6.1. Solid suddenly exposed to convection environment at T_∞

The initial temperature of solid T_i (Fig. 6.1) is greater than ambient fluid temperature, T_∞, the eqn. (6.1) leads to,

$$hA_s(T - T_\infty) = -\rho VC\frac{dT}{dt}$$

Introducing the temperature difference as

$$\theta = T - T_\infty$$

and $$\frac{d\theta}{dt} = \frac{dT}{dt}$$

Thus $$hA_s\theta = -\rho VC\frac{d\theta}{dt}$$

Rearranging, we have

$$\frac{d\theta}{\theta} = -\frac{hA_s}{\rho VC}dt$$

Integrating both sides, we get

$$\ln(\theta) = -\frac{hA_s}{\rho VC}t + P \quad ...(6.2)$$

where P is constant of integration and can be evaluated from initial condition.

At $t = 0$, $T = T_i$ and $\theta_{(t=0)} = \theta_i = T_i - T_\infty$

Applying in eqn. (6.2)

$$\ln(\theta_i) = P + 0$$

or $$P = \ln(\theta_i)$$

Substituting in eqn. (6.2), we get

$$\ln(\theta) = -\frac{hA_s}{\rho VC}t + \ln(\theta_i)$$

or $$\ln\left(\frac{\theta}{\theta_i}\right) = -\frac{hA_s}{\rho VC}t$$

or $$\frac{\theta}{\theta_i} = \exp\left\{-\frac{hA_s t}{\rho VC}\right\}$$

or $$\frac{\theta}{\theta_i} = \frac{T - T_\infty}{T_i - T_\infty} = \exp\left\{-\frac{hA_s t}{\rho VC}\right\} \quad ...(6.3)$$

The eqn. (6.3) is plotted in Fig. 6.2 for different values of $\frac{hA_s}{\rho VC}$ and the observations are :

(1) The eqn. (6.3) can be used to determine the time t required for solid to reach some temperature T.

(2) The temperature of a body approaches the ambient temperature T_∞ exponentially. The temperature of body changes rapidly at the beginning (due to large temperature difference), but slow down later on. The large value of $\frac{hA_s}{\rho VC}$ indicates that the body will approach the ambient temperature in a short time.

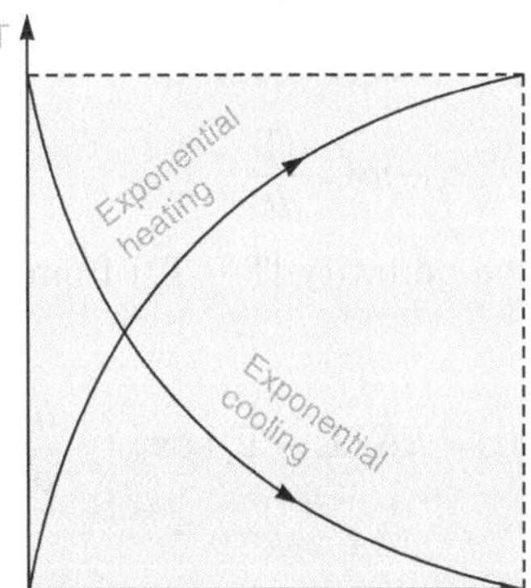

Fig. 6.2. Transient heating and cooling

The lumped system analysis is analogous to the voltage decay that occurs when a capacitor is discharged through a resistor in an electrical R–C circuit. The equivalent circuit is shown in Fig. 6.3. In this network, the capacitor is initially charged to some potential T_i by closing the switch S. When switch is opened, the energy stored in the capacitor is discharged through convection resistance $\frac{1}{hA_s}$. The analogy between thermal system and electrical system is apparent.

where, $R_{th} = \frac{1}{hA_s}$, thermal resistance to convection heat transfer

$C_{th} = \rho VC$, thermal capacitance (stored internal energy).

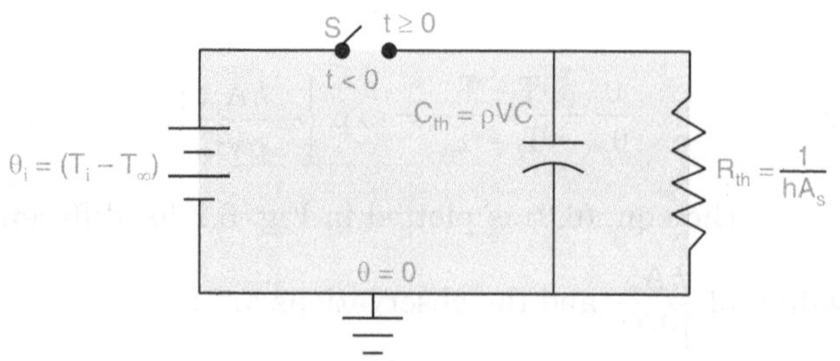

Fig. 6.3. Equivalent thermal (R–C) circuit for lumped capacity method

The instantaneous rate of cooling can be obtained by differentiating eqn. (6.3) with respect to time t,

$$\frac{dT}{dt} = (T_i - T_\infty)\left\{-\frac{hA_s}{\rho VC}\exp\left(-\frac{hA_s t}{\rho VC}\right)\right\} \text{ (°C/s)} \quad ...(6.4)$$

The *instantaneous rate of heat transfer Q(t)* from a solid can be calculated as

$$Q(t) = hA_s[T - T_\infty]$$

$$= -mC\frac{dT}{dt} \quad ...(6.5)$$

Using the quantity $[T - T_\infty]$ from eqn. (6.3), we get

$$Q(t) = hA_s(T_i - T_\infty)\exp\left[-\frac{hA_s t}{\rho VC}\right] \text{ (Watts)} \quad ...(6.6)$$

The *total quantity* of heat transferred during the time t is equal to change in internal energy (ΔU) of the solid. It can be calculated by integrating eqn. (6.6) with respect to time t within limits 0 to t

$$\Delta U = \int_0^t Q(t)\,dt$$

$$= hA_s(T_i - T_\infty)\int_0^t \left\{\exp\left(-\frac{hA_s t}{\rho VC}\right)\right\} dt$$

or

$$\Delta U = hA_s (T_i - T_\infty)\left(-\frac{\rho VC}{hA_s}\right)\left\{\exp\left(-\frac{hA_s t}{\rho VC}\right) - 1\right\}$$

$$\Delta U = -\rho VC(T_i - T_\infty)\left\{\exp\left(-\frac{hA_s t}{\rho VC}\right) - 1\right\} \text{ (Joules)} \quad ...(6.7)$$

Thus the heat transfered during a time period is equal to decrease in internal energy.

6.1.2. Dimensionless Quantities

The exponent quantity $\frac{hA_s}{\rho VC}$ in eqn. (6.3) may be expressed as

$$\frac{hA_s t}{\rho VC} = \left(\frac{h\delta}{k}\right)\left(\frac{kt}{\rho C\delta^2}\right)\left(\frac{A_s\delta}{V}\right)$$

$$= \left(\frac{h\delta}{k}\right)\left(\frac{\alpha t}{\delta^2}\right)\left(\frac{A_s\delta}{V}\right)$$

$$= \text{Bi Fo GF} \quad ...(6.8)$$

where, $Bi = \frac{h\delta}{k}$, Biot number, a dimensionless number.

$Fo = \frac{\alpha t}{\delta^2}$, Fourier number, a dimensionless number.

and $GF = \frac{A_s\delta}{V}$, Geometrical factor, a dimensionless quantity.

The *geometrical factor* GF is considered to be unity for calculation of *characteristic length* δ of the solid as

$$\delta = \frac{V}{A_s} \quad ...(6.9)$$

Then the temperature distribution eqn. (6.3) within the solid can be expressed as

$$\frac{T - T_\infty}{T_i - T_\infty} = \exp\left\{-\frac{ht}{\rho\delta C}\right\} \quad ...(6.10)$$

For certain common body shapes, and their characteristic length δ is shown in Table 6.1.

TABLE 6.1. The characteristic length δ of common geometry exposed to convection environment.

Sr. No.	*Geometry*	*Schematic*	*Volume V*	*Surface Area A_s*	*Characteristic Length δ*	*Equation number*
1	Infinite plate of thickness L exposed on both sides	A, Initially at T_i, O, A, L	AL	2A	$\delta = \frac{AL}{2A} = \frac{L}{2}$	...(6.11)
2	Short cylinder exposed to environment	r_o, L	$\pi r_o^2 L$	$2\pi r_o^2 + 2\pi r_o L$	$\delta = \frac{r_o L}{2(r_o + L)}$	...(6.12)
3	Long cylinder	r_o, L >> D	$\pi r_o^2 L$	$\approx 2\pi r_o L$	$\delta \approx \frac{r_o}{2} = \frac{D}{4}$	...(6.13)
4	Solid sphere	r_o	$\frac{4}{3}\pi r_o^3$	$4\pi r_o^2$	$\delta \approx \frac{r_o}{3} = \frac{D}{6}$	...(6.14)
5	Cube	L, L, L	L^3	$6L^2$	$\delta \approx \frac{L}{6}$	...(6.15)

Biot Number

It is defined as *ratio of internal resistance of the solid to heat flow to convection resistance* at the surfaces.

$$Bi = \frac{\text{Internal resistance to heat flow}}{\text{Convection resistance to heat flow}}$$

$$= \frac{\delta}{kA} \times \frac{hA}{1} = \frac{h\delta}{k} \quad ...(6.16)$$

It can also be interpreted as the ratio of heat transfer coefficient to the internal specific conductance of the solid. The Biot number is required to determine the validity of the lumped heat capacity approach. The lumped system analysis can only be applied when

$$Bi \leq 0.1$$

This criteria indicates that the internal resistance of the solid to heat flow is very small in comparison to convection resistance to heat flow at the surfaces.

Fourier Number

It signifies the degree of penetration of heating or cooling effect through the solid. It is defined as the *ratio of the rate of heat conduction to the rate of the thermal energy storage in the solid*. It is denoted by Fo and expressed as

$$Fo = \frac{kA(\Delta T)/\delta}{\rho VC(\Delta T)/t} = \frac{kAt}{\rho(A\delta)C\delta} = \frac{k}{\rho C}\frac{t}{\delta^2} = \frac{\alpha t}{\delta^2} \quad ...(6.17)$$

6.1.3. Thermal Time Constant and Response of Thermocouple

In eqn. (6.3) the exponent quantity $\left(\frac{hA_s t}{\rho VC}\right)$ is a *dimensionless quantity* and the quantity $\left(\frac{\rho VC}{hA_s}\right)$ has the dimension of time and therefore, it is called the **time constant** of the system. It is denoted by τ, and is given as

$$\tau = \frac{\rho VC}{hA_s} \quad ...(6.18)$$

The temperature difference between a solid and a fluid must decay exponentially to zero as time approaches infinity. The **response of thermocouple** is defined as the *time taken by thermocouple to indicate the source temperature*. At the end of one time constant, the temperature difference between system (T) and source (T_∞) is

$$\frac{T - T_\infty}{T_i - T_\infty} = \exp(-1) = 0.368 \quad ...(6.19)$$

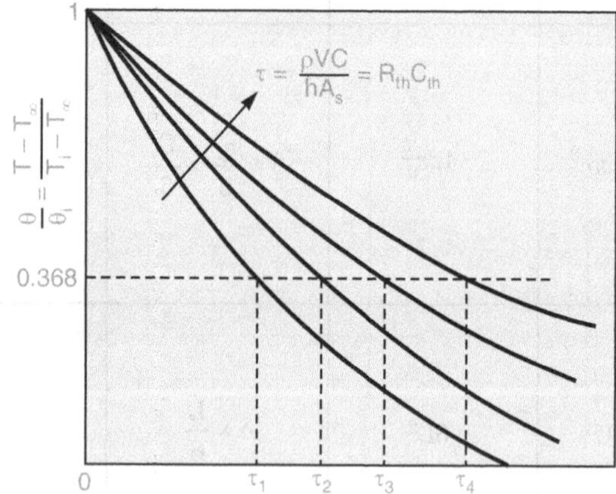

Fig. 6.4. Transient temperature response of lumped systems corresponding to different time constants τ

Thus the temperature difference is reduced by 63.2% (= 1 – 0.368) after one time constant and the time required by thermocouple to indicate the temperature 63.2% of the initial temperature difference is called the *sensitivity of thermocouple.*

For rapid response of thermocouple, the quantity $\left(\frac{hA_s}{\rho VC}\right)$ should be large to make the exponential term least. The low value of the time constant is desirable. It can be achieved for a thermocouple by

1. Decreasing the wire diameter.
2. Using the light metals of low density and low specific heat.
3. Increasing the heat transfer coefficient.

Using time constant, the temperature distribution in the solids can be expressed as

$$\frac{T - T_\infty}{T_i - T_\infty} = \exp\left\{-\frac{hA_s t}{\rho VC}\right\} = \exp\left(-\frac{t}{\tau}\right) \quad ...(6.20)$$

6.1.4. The Lumped System Analysis for Mixed Boundary Conditions

Consider a slab of thickness L, initially at uniform temperature T_i ($t = 0$). For the value of $t > 0$, the surface at $x = 0$ is subjected to constant heat flux q and the surface at $x = L$ dissipates heat by convection into an ambient at T_∞ with a convection coefficient h as shown in Fig. 6.5.

The energy balance for the slab at $t > 0$

$$Aq - hA(T - T_\infty) = \rho VC \frac{dT}{dt} \quad ...(6.21)$$

or
$$Aq - hA(T - T_\infty) = \rho(AL)C \frac{dT}{dt}$$

or
$$q - h(T - T_\infty) = \rho LC \frac{dT}{dt}$$

or
$$\frac{q}{\rho CL} - \frac{h}{\rho CL}(T - T_\infty) = \frac{dT}{dt}$$

Introducing
$$\theta = T - T_\infty$$
$$N = \frac{q}{\rho CL}$$
and
$$M = \frac{h}{\rho CL}$$

Then the above equation is changed to

$$N - M\theta = \frac{d\theta}{dt}$$

or
$$\frac{d\theta}{dt} + M\theta = N$$

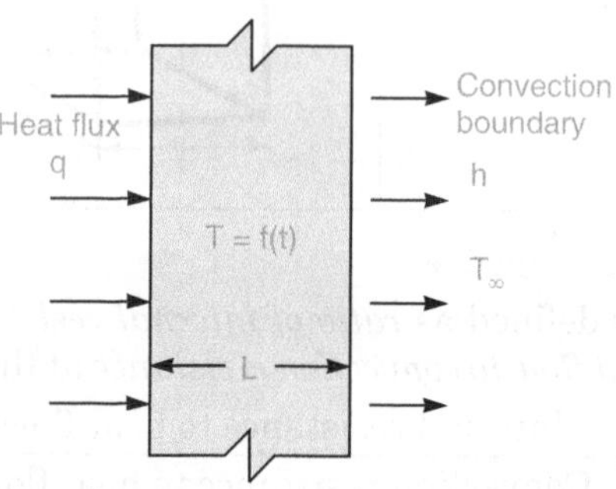

Fig. 6.5. Slab with mixed boundaries

The solution to this equation is the sum of homogeneous and particular solutions as

$$\theta = D \exp(-Mt) + \theta_p \quad ...(6.22)$$

where D is the constant of integration and θ_p is the particular solution. The particular solution is

$$\theta_p = \frac{N}{M}$$

Then $\theta = D \exp(-Mt) + \frac{N}{M}$ ...(6.23)

Using the initial condition

At $t = 0$, $\theta = \theta_i = T_i - T_\infty$

Hence $\theta_i = D + \frac{N}{M}$

It gives $D = \theta_i - \frac{N}{M}$ ...(6.24)

Substituting the value in eqn. (6.23),

$$\theta = \left\{\theta_i - \frac{N}{M}\right\} \exp(-Mt) + \frac{N}{M}$$
$$= \theta_i \exp(-Mt) + \frac{N}{M}\{1 - \exp(-Mt)\}$$

But $\frac{N}{M} = \frac{q}{\rho CL} \times \frac{\rho CL}{h} = \frac{q}{h}$

Hence $$\theta = \theta_i \exp(-Mt) + \frac{q}{h}\{1 - \exp(-Mt)\} \quad ...(6.25)$$

The steady state temperature of the slab can be obtained by setting $t \to \infty$

$$\theta(\infty) = (T - T_\infty) = \frac{q}{h} \quad ...(6.26)$$

Since the exponential term exp(– Mt) becomes zero for $t \to \infty$.

6.1.5. The Validity of Lumped System Analysis

The analysis of transient heat conduction problems becomes very easy by using the lumped heat capacity method due to its simplicity. Hence, it is necessary to specify its limits between which it may be used with reasonable accuracy.

To develop an appropriate criterion, consider steady state heat conduction through a plane wall of area A, thickness L as shown in Fig. 6.6. One surface of the wall is maintained at temperature, T_1 and other is exposed to convection environment at $T_\infty(< T_1)$. Thus the temperature of this surface will be some intermediate temperature say T_2, then energy balance on the wall yields to

$$\frac{kA}{L}(T_1 - T_2) = hA(T_2 - T_\infty)$$

On rearrangement, we get Biot number as defined by eqn. (6.16)

$$Bi = \frac{T_1 - T_2}{T_2 - T_\infty} = \frac{(L/kA)}{(1/hA)} = \frac{hL}{k}$$

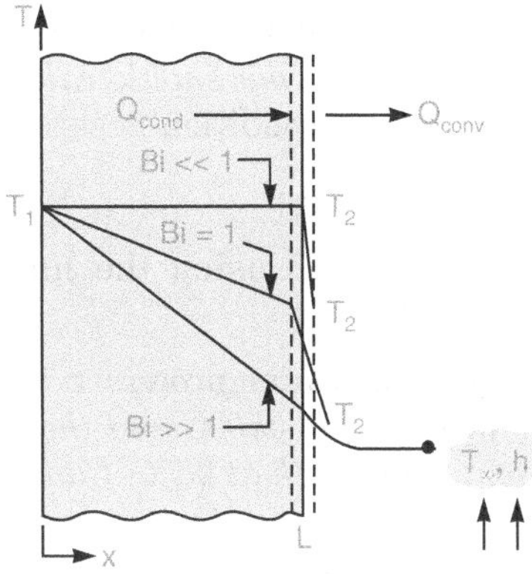

Fig. 6.6. Effect of Biot number on steady-state temperature distribution in a plane wall with surface convection

The Biot number is a measure of the temperature drop in the solid relative to the temperature difference between surface and its ambient. The Fig. 6.6 illustrates, for Bi << 1, the temperature distribution in a solid is uniform at any time during a transient process.

If Bi << 1, the *resistance to conduction with the solid is much less than the resistance to convection across the fluid boundary layer.*

Now consider a plane wall as shown in Fig. 6.7, which is initially at uniform temperature T_i. It is suddenly

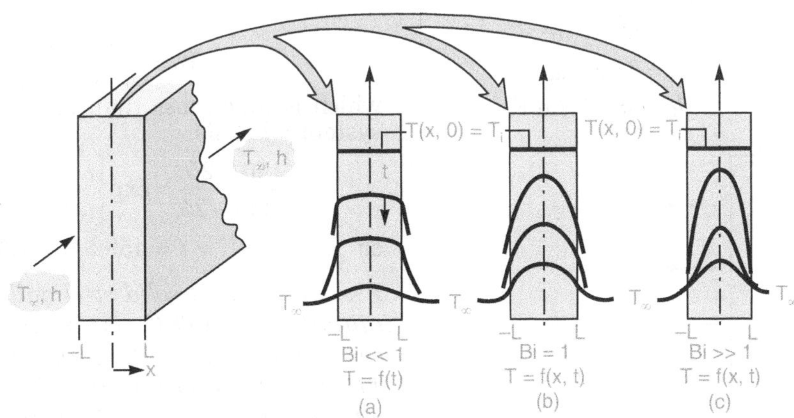

Fig. 6.7. Transient temperature distributions for different Biot numbers in a plane wall symmetrically cooled by convection

exposed for convection cooling in a fluid at T_∞. The temperature variation with position is a strong function of Biot number and three conditions are shown on Fig. 6.7. For Bi << 1, [Fig. 6.7 (*a*)], the temperature gradient in the solid is small and $T(x, t) \approx T(t)$ *i.e.*, the solid temperature remains uniform within the body. For moderate to large value of Biot number, the temperature gradients within the solid are considerable and hence $T = T(x, t)$. For Bi >> 1, the temperature difference across the solid is much larger than that between the surface and the fluid.

We can conclude by emphasizing the importance of Biot number in transient heat conduction. Hence, with each problem, *at very first, one should calculate the Biot number*. If the following condition is satisfied

$$Bi = \frac{h\delta}{k} \le 0.1 \qquad ...(6.27)$$

the error associated with using the lumped system analysis will be small.

Example 6.1. *In a quenching process, a copper plate of 3 mm thick is heated upto 350°C and then suddenly, it is dropped into a water bath at 25°C. Calculate the time required for the plate to reach the temperature of 50°C. The heat transfer coefficient on the surface of the plate is 28 W/m².K. The plate dimensions may be taken as length 40 cm and width 30 cm.*

Also calculate the time required for infinite long plate to cool to 50°C. Other parameters remain same.

Take the properties of copper as

$C = 380$ J/kg.K, $\rho = 8800$ kg/m³,
$k = 385$ W/m.K. (J.N.T.U., May 2004)

Solution

Given : The quenching of a copper plate in water bath.

Size = 40 cm × 30 cm, L = 3 mm,
$T_i = 350°C$, $T_\infty = 25°C$,
$T = 50°C$, $h = 28$ W/m².K,
$C = 380$ J/kg.K, $\rho = 8800$ kg/m³,
$k = 385$ W/m.K.

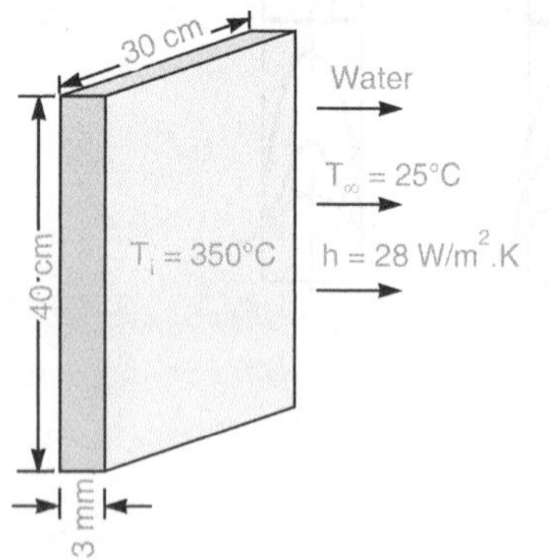

Fig. 6.8. Schematic of plate in example 6.1

To find : Time required to cool the plate to 50°C, if
(*i*) Finite long plate size 40 cm × 30 cm,
(*ii*) Infinite long plate.

Assumptions :
1. The effect of edges of plate for cooling.
2. Internal temperature gradients are negligible.
3. No radiation heat exchange.
4. Constant properties.

Analysis : (*i*) The characteristic length of finite long plate (as shown in Fig. 6.8)

$$\delta = \frac{\text{Volume of plate}}{\text{Exposed area of plate}}$$

$$= \frac{0.4 \times 0.3 \times 0.003}{(2 \times 0.4 + 2 \times 0.3) \times 0.003 + 2 \times 0.3 \times 0.4}$$

$$= 1.474 \times 10^{-3} \text{ m}$$

The Biot number

$$Bi = \frac{h\delta}{k} = \frac{28 \times 1.474 \times 10^{-3}}{385} = 1.072 \times 10^{-4}$$

which is much smaller than 0.1, thus the lumped system analysis can be applied with reasonable accuracy. Using eqn. (6.10) ;

$$\frac{T - T_\infty}{T_i - T_\infty} = \exp\left\{-\frac{ht}{\rho C\delta}\right\}$$

Using numerical values.

$$\frac{50 - 25}{350 - 25} = \exp\left\{-\frac{28t}{8800 \times 380 \times 1.474 \times 10^{-3}}\right\}$$

or
$$t = -\frac{8800 \times 380 \times 1.474 \times 10^{-3}}{28} \times \ln\left(\frac{25}{325}\right)$$

$$= 451.5 \text{ s} = \mathbf{7.52 \text{ min.}} \quad \text{Ans.}$$

(*ii*) Characteristic length of infinite long plate eqn. (6.11)

$$\delta = \frac{L}{2} = 0.0015 \text{ m}$$

$$B_i = \frac{h\delta}{k} = \frac{28 \times 0.0015}{385} = 1.09 \times 10^{-4}$$

which is much less than 0.1, therefore, using lumped system analysis.

$$\frac{50 - 25}{350 - 25} = \exp\left[-\frac{28t}{8800 \times 380 \times 0.0015}\right]$$

or
$$t = 459.5 \text{ s} = \mathbf{7.65 \text{ min.}} \quad \text{Ans.}$$

Example 6.2. *A solid steel ball 5 cm in diameter and initially at 450°C is quenched in a controlled environment at 90°C with convection coefficient of 115 W/m².K. Determine the time taken by centre to reach a temperature of 150°C. Take thermophysical properties as*

$C = 420$ J/kg.K, $\rho = 8000$ kg/m³,
$k = 46$ W/m.K. (P.U., May 2002)

Solution

Given : A solid steel ball quenching with

$T = 150°C$, $T_\infty = 90°C$,
$T_i = 450°C$, $h = 115\ W/m^2.K$,
$C = 420\ J/kg.K$, $\rho = 8000\ kg/m^3$,
$k = 46\ W/m.K$, $D = 5\ cm = 0.05\ m$.

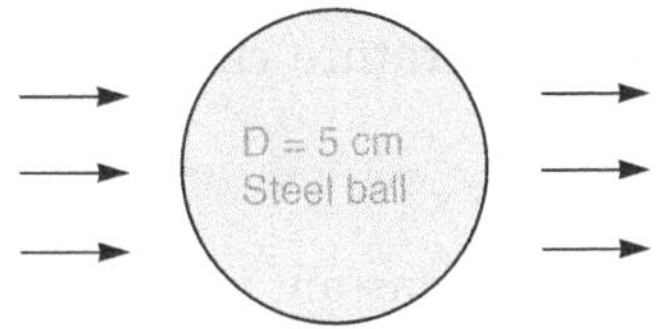

Fig. 6.9. Schematic for example 6.2

To find : Time required by steel ball to reach 150°C.

Assumptions :

1. Internal temperature gradients are negligible.
2. No radiation heat exchange.
3. Constant properties.

Analysis : The characteristic length of the steel ball

$$\delta = \frac{V}{A_s} = \frac{D}{6} = \frac{0.05}{6}\ m = \left(\frac{0.05}{6}\right) m$$

The Biot number

$$Bi = \frac{h\delta}{k} = \frac{(115\ W/m^2.K)}{(46\ W/m.K)} \times \left(\frac{0.05}{6}\ m\right) = 0.0208$$

which is less than 0.1, hence the lumped heat capacity system analysis may be applied.

Using eqn. (6.10) for temperature distribution

$$\frac{T - T_\infty}{T_i - T_\infty} = \exp\left\{-\frac{ht}{\rho\delta C}\right\}$$

Substituting the values

$$\frac{150 - 90}{450 - 90} = \exp\left\{-\frac{115 \times 6t}{8000 \times 0.05 \times 420}\right\}$$

or $\ln (60/360) = -(690/168000)t$

or $t = 440.35\ s =$ **7.34 min.** Ans.

Example 6.3. *A titanium alloy blade of an axial compressor for which k = 25 W/m.K, ρ = 4500 kg/m³ and C = 520 J/kg.K is initially at 60°C. The effective thickness of the blade is 10 mm and it is exposed to gas stream at 600°C, the blade experiences a heat transfer coefficient of 500 W/m².K. Use low Biot number approximation to estimate the temperature of blade after 1, 5, 20 and 100 s.*

(N.M.U., May 2002)

Solution

Given : A titanium alloy blade of compressor with

$k = 25\ W/m.K$, $\rho = 4500\ kg/m^3$,
$C = 520\ J/kg.K$, $h = 500\ W/m^2.K$,
$T_i = 60°C$, $T_\infty = 600°C$,
$L = 10\ mm$, $t = 1, 5, 20$ and $100\ s$.

To find : Temperature attained by compressor blade after 1, 5, 20 and 100 seconds.

Assumptions: 1. Compressor blade as an infinite wall.

2. Negligible internal temperature gradient
3. No. radiation heat exchange.
4. Constant properties.

Analysis : The characteristic length of blade

$$\delta = \frac{L}{2} = \frac{10}{2} = 5\ mm = 5 \times 10^{-3}\ m$$

The Biot number

$$Bi = \frac{h\delta}{k} = \frac{500 \times 5 \times 10^{-3}}{25} = 0.1$$

Hence it is possible to use the low Biot number approximation

$$\frac{T - T_\infty}{T_i - T_\infty} = \exp\left(-\frac{ht}{\rho\delta C}\right)$$

After 1 s

$$\frac{T - 600}{60 - 600} = \exp\left(-\frac{500 \times 1}{4500 \times 5 \times 10^{-3} \times 520}\right)$$

or $T = 600 + (-540) \times \exp(-0.0427)$

$= 600 - 540 \times 0.9581 =$ **82.6°C.** Ans.

similarly the temperature after

t	T
5 s	163.9°C
20 s	370.3°C
100 s	592.5°C. Ans.

Example 6.4. *A long thin glass walled, 0.3 cm diameter, mercury thermometer is placed in a stream of air with convection coefficient of 60 W/m².K for measuring transient temperature of air. Consider cylindrical thermometer bulb consists of mercury only. For which*

k = 8.9 W/m.K and α = 0.016 m²/h

Calculate the time constant and time required for the temperature change to reach half of its initial value.

Solution

Given : A long cylindrical thermometer bulb of mercury with

$D = 0.3\ cm = 0.003\ m$, $h = 60\ W/m^2.K$,
$k = 8.9\ W/m.K$,
$\alpha = 0.016\ m^2/h = 4.444 \times 10^{-6}\ m^2/s$

$$\frac{T - T_\infty}{T_i - T_\infty} = 0.5.$$

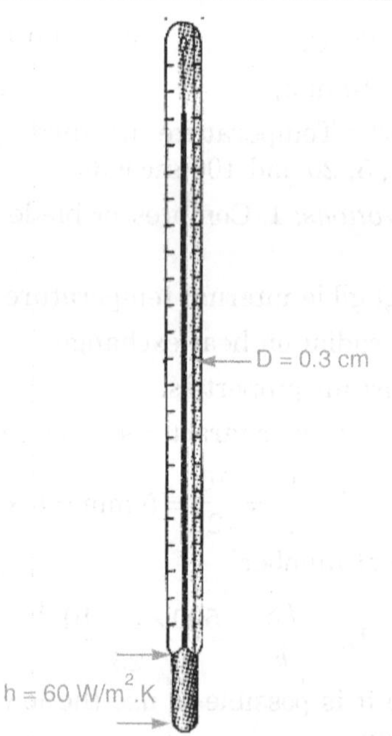

Fig. 6.10. Schematic for example 6.4

To find :

(*i*) Time constant.

(*ii*) Time required for temperature change to reach half of initial temperature change.

Assumptions :

1. Glass thermometer as infinite long thermometer.

2. Internal temperature gradients are negligible.

3. No radiation heat exchange.

4. Constant properties.

Analysis : (*i*) The characteristic length of the cylindrical thermometer

$$\delta = \frac{V}{A_s} = \frac{D}{4} = \frac{0.003}{4} = 7.5 \times 10^{-4} \text{ m}$$

Biot number

$$Bi = \frac{h\delta}{k} = \frac{60 \times 7.5 \times 10^{-4}}{8.9} = 5.056 \times 10^{-3}$$

which is less than 0.1, hence the lumped heat capacity system analysis may be applied.

Using the eqn. (6.18) for time constant

$$\tau = \frac{\rho VC}{hA_s} = \frac{k\delta}{\alpha h}$$

$$= \frac{8.9 \times 7.5 \times 10^{-4}}{4.444 \times 10^{-6} \times 60} = \mathbf{25\ s.} \quad \text{Ans.}$$

$$\therefore \quad \alpha = \frac{k}{\rho C} \quad \text{or} \quad \rho C = \frac{k}{\alpha}.$$

(*ii*) The time required to reach the temperature change to half of its initial value

Using the relation

$$\frac{T - T_\infty}{T_i - T_\infty} = \exp\left\{-\frac{ht}{\rho\delta C}\right\} = \exp\left(-\frac{t}{\tau}\right)$$

Given that

The final temperature change = (1/2) × Initial temperature change

or $\quad T - T_\infty = 0.5 \times (T_i - T_\infty)$

Using the time constant in eqn. (6.20), as

$$0.5 = \exp\left\{-\frac{t}{25}\right\}$$

or $\quad t = -25 \times \ln(0.5) = \mathbf{17.32\ s.}$ Ans.

Example 6.5. *A steel ball of 50 mm diameter and at 900°C temperature is placed in still air at temperature of 30°C. Calculate the initial rate of cooling of ball in °C/min. Take h = 30 W/m^2.K and thermophysical properties of steel as*

$$\rho = 7800 \text{ kg/m}^3, C = 2 \text{ kJ/kg.K}$$

Neglect internal thermal resistance.

(V.T.U., July 2002)

Solution

Given : A steel ball exposed to air

$D = 50$ mm $= 0.05$ m,

$T_i = 900$°C,

$T_\infty = 30$°C,

$h = 30$ W/m^2.K,

$C = 2$ kJ/kg.K $= 2000$ J/kg.K,

$\rho = 7800$ kg/m^3,

$t = 0$ (for initial cooling).

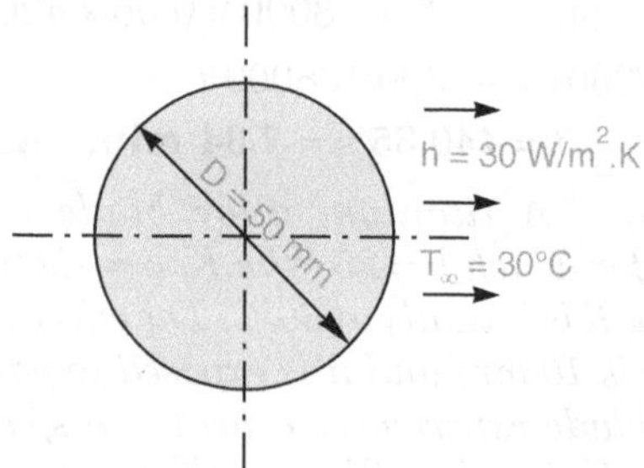

Fig. 6.11. Schematic for example 6.5

To find : The initial rate of cooling of ball in °C/min.

Analysis : The instantaneous cooling rate can be obtained by using eqn. (6.4)

$$\frac{dT}{dt} = (T_i - T_\infty)\left\{-\frac{hA_s}{\rho VC}\exp\left(-\frac{hA_s t}{\rho VC}\right)\right\}$$

For sphere :

$$\frac{A_s}{V} = \frac{1}{\delta} = \frac{6}{D} = \frac{6}{0.05} = 120$$

Using numerical values,

$$\frac{dT}{dt} = (900 - 30)$$

$$\times \left\{ -\frac{30 \times 120}{7800 \times 2000} \times \exp\left(-\frac{30 \times 120 \times 0}{7800 \times 2000} \right) \right\}$$

$$= 870 \times (-2.3077 \times 1) = 0.2°C/s$$

$$= \mathbf{12°C/min.}$$ **Ans.**

Example 6.6. *A person is found dead at 5 p.m. in a room where temperature is 20°C. The temperature of the body is measured to be 25°C when found, and the heat transfer coefficient is estimated to be 8 W/m^2.K. Modelling the human body a 30 cm diameter, 1.70 m long cylinder, calculate actual time of death of the person. Take thermophysical properties of the body :*

k = 6.08 W/m.K, ρ = 900 kg/m^3,

C = 4000 J/kg.K. (N.M.U., Dec. 2002)

Solution

Given : The dead body of a person as a cylinder

T_∞ = 20°C, k = 6.08 W/m.K

T = 25°C, ρ = 900 kg/m^3

h = 8 W/m^2.K, C = 4000 J/kg.K

D = 30 cm = 0.3 m or r_o = 0.15 m

L = 1.70 m.

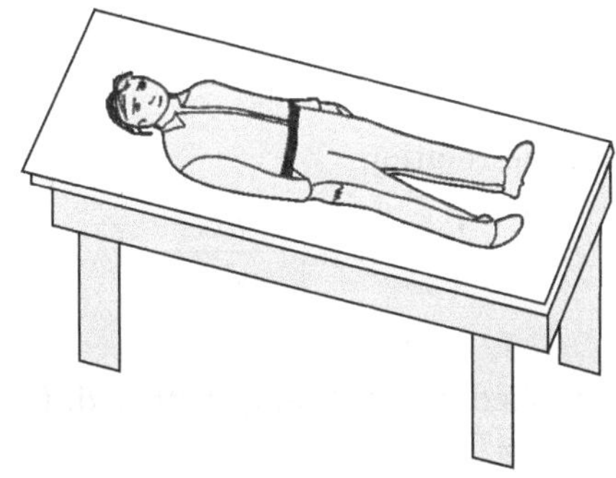

Fig. 6.12. Schematic of a dead body

To find : Actual time of death of the person.

Assumptions :

1. Healthy person, thus body temperature of 37°C at the time of death.

2. Uniform heat transfer coefficient on entire surface of body.

3. No radiation heat transfer.

Analysis : The characteristic length of the body

$$\delta = \frac{V}{A_s} = \frac{\pi r_o^2 L}{2\pi r_o L + 2\pi r_o^2}$$

$$= \frac{\pi \times (0.15 \text{ m})^2 \times (1.7 \text{ m})}{2\pi \times (0.15 \text{ m}) \times (1.7 \text{ m}) + 2\pi \times (0.15 \text{ m})^2}$$

$$= 0.0689 \text{ m}$$

Biot number

$$\text{Bi} = \frac{h\delta}{k} = \frac{(8 \text{ W/m}^2.\text{K}) \times (0.0689 \text{ m})}{(6.08 \text{ W/m.K})} = 0.092$$

The Biot number is less than 0.1, therefore, the lumped system analysis is applicable.

Using eqn. (6.10),

$$\frac{T - T_\infty}{T_i - T_\infty} = \exp\left\{ -\frac{ht}{\rho\delta C} \right\}$$

$$\frac{25 - 20}{37 - 20} = \exp\left[-\frac{8 \times t}{900 \times 0.0689 \times 4000} \right]$$

or $\ln\left(\frac{5}{17}\right) = -3.225 \times 10^{-5} t$

or t = 37,943 s = **10.54 h.** **Ans.**

Example 6.7. *A bearing piece in the form of half of a hollow cylinder of 60 mm ID, 90 mm OD and 100 mm long is to be cooled to –100°C from 30°C using a cryogenic gas at –150°C with a convective heat transfer coefficient of 70 W/m^2.K. Determine the time required. Take properties of bearing material as*

C = 444 J/kg.K, ρ = 8900 kg/m^3,

k = 17.2 W/m.K.

Solution

Given : A piece of bearing as half of hollow cylinder with

D_1 = 60 mm or r_1 = 0.03 m

D_2 = 90 mm or r_2 = 0.045 m

L = 100 mm = 0.1 m, T_i = 30°C

T_∞ = – 150°C, T = – 100°C

C = 444 J/kg.K, ρ = 8900 kg/m^3

k = 17.2 W/m.K, h = 70 W/m^2.K.

To find : Time required to reach – 100°C.

Assumptions :

1. Internal temperature gradients are negligible.

2. No radiation heat exchange.

3. Constant properties.

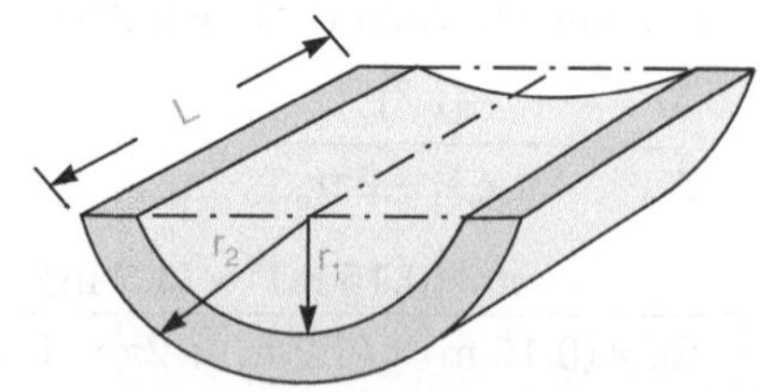

Fig. 6.13. Schematic of half portion of a bearing

Analysis : The characteristic length of the cylinder.

The volume of bearing piece,

$$V = (1/2) \times \pi(r_2^2 - r_1^2)\,L$$
$$= (1/2) \times \pi\{(0.045)^2 - (0.03)^2\} \times 0.1$$
$$= 1.766 \times 10^{-4}\ \text{m}$$

Surface area of the bearing,

$$A_s = (\text{Front + back + lateral + longitudinal}) \text{ area}$$
$$= 2 \times (1/2)\pi\,(r_2^2 - r_1^2) + \pi L(r_1 + r_2) + 2 \times L \times (r_2 - r_1)$$
$$= \pi \times (0.045^2 - 0.03^2) + \pi \times 0.1 \times (0.03 + 0.045) + 2 \times 0.1 \times (0.045 - 0.03)$$
$$= 3.53 \times 10^{-3} + 0.0235 + 3 \times 10^{-3}$$
$$= 0.03003\ \text{m}^2$$

$$\delta = \frac{V}{A_s} = \frac{1.766 \times 10^{-4}}{0.03003}$$
$$= 5.872 \times 10^{-3}\ \text{m}$$

Biot number,

$$\text{Bi} = \frac{h\delta}{k} = \frac{70 \times 5.872 \times 10^{-3}}{17.2} = 0.0239$$

which is less than 0.1, hence the lumped heat capacity system analysis may be applied. Using eqn. (6.10) for temperature distribution

$$\frac{T - T_\infty}{T_i - T_\infty} = \exp\left\{-\frac{ht}{\rho\delta C}\right\}$$

Substituting the values

$$\frac{-100 - (-150)}{30 - (-150)} = \exp\left\{-\frac{70t}{8900 \times 5.872 \times 10^{-3} \times 444}\right\}$$

or t = **424.6 s.** Ans.

Example 6.8. *A thermocouple is used to measure the temperature in a gas stream. The junction may be approximated as a sphere with thermal conductivity of 25 W/m.K, density 9000 kg/m³, and specific heat 0.35 kJ/kg. K. The heat transfer coefficient between the junction and the gas is 250 W/m².K. Calculate the diameter of the junction, if thermocouple should measure 95 per cent of the applied temperature difference in 3 s.*

Solution

Given : A thermocouple junction in form of a sphere with

$$C = 0.35\ \text{kJ/kg.K} = 350\ \text{J/kg.K},$$
$$h = 250\ \text{W/m}^2\text{.K}$$
$$k = 25\ \text{W/m.K},$$
$$\rho = 9000\ \text{kg/m}^3$$
$$T - T_\infty = (1 - 0.95)\,(T_i - T_\infty),$$
$$t = 3\ \text{s}.$$

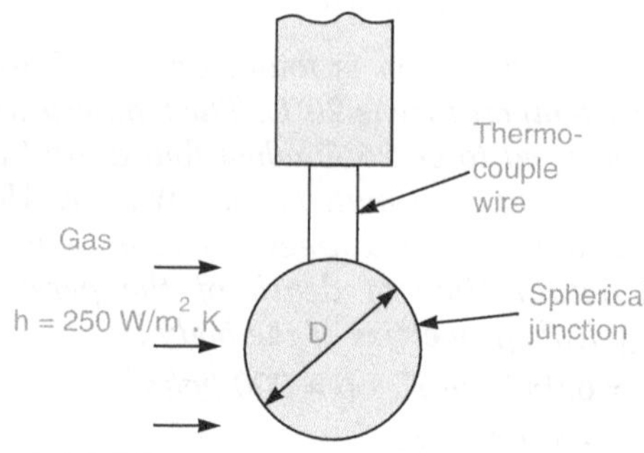

Fig. 6.14. Thermocouple junction for example 6.8

To find : Diameter of the junction to indicate 95% of applied temperature difference.

Assumptions :

1. Internal temperature gradients are negligible.
2. No radiation heat exchange.
3. Constant properties.

Analysis : The characteristic length of the thermocouple junction

$$\delta = \frac{V}{A_s} = \frac{D}{6}$$

Using the relation,

$$\frac{T - T_\infty}{T_i - T_\infty} = \exp\left\{-\frac{ht}{\rho\delta C}\right\}$$

where,

The applied (initial) temperature difference
$= T_i - T_\infty$

Thermocouple measure 95% of applied temperature $(T_i - T_\infty)$, then

The remaining temperature difference
$(T - T_\infty) = 0.05 \times (T_i - T_\infty)$

Hence,

$$\frac{T - T_\infty}{T_i - T_\infty} = 0.05 = \exp\left\{-\frac{ht}{\rho\delta C}\right\}$$
$$= \exp\left\{-\frac{250 \times 3}{9000 \times 350 \times (D/6)}\right\}$$

or $\ln(0.5) = -0.001428/D$

or $D = 4.768 \times 10^{-4}$ m

$= \mathbf{0.4768\ mm.}$ Ans.

Checking the validity of the relation applied above
The Biot number

$$Bi = \frac{h\delta}{k} = \frac{250 \times 4.7687 \times 10^{-4}}{6 \times 25}$$

$$= 7.9 \times 10^{-4}$$

which is less than 0.1, hence the lumped heat capacity system analysis is valid.

Example 6.9. *An aluminium sphere weighing 6 kg and initially at temperature of 350°C is suddenly immersed in a fluid at 30°C with convection coefficient of 60 W/m².K. Estimate the time required to cool the sphere to 100°C. Take thermophysical properties as*

C = 900 J/kg.K, ρ = 2700 kg/m³,
k = 205 W/m.K. (P.U., May 1999)

Solution

Given : An aluminium sphere with

$m = 6$ kg, $C = 900$ J/kg.K,
$\rho = 2700$ kg/m³, $k = 205$ W/m.K,
$T_i = 350$°C, $T_\infty = 30$°C,
$T = 100$°C, $h = 60$ W/m².K.

To find : Time required to reach 100°C.

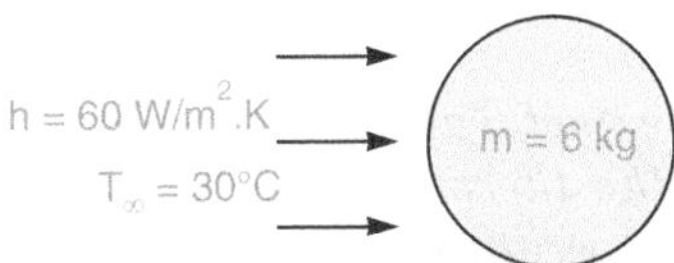

Fig. 6.15. Sphere for example 6.9

Assumptions :

1. Internal temperature gradients are negligible.
2. No radiation heat exchange.
3. Constant properties.

Analysis : The volume of sphere

$$V = \frac{4\pi r_o^3}{3} = \frac{m}{\rho} = \frac{6}{2700}$$

or $r_o = 0.0809$ m

The characteristic length of the sphere

$$\delta = \frac{V}{A_s} = \frac{r_o}{3} = \frac{0.0809\ m}{3}$$

$$= 0.0269\ m$$

Biot number

$$Bi = \frac{h\delta}{k} = \frac{60 \times 0.0269}{205} = 7.89 \times 10^{-3}$$

which is less than 0.1, hence the lumped heat capacity system analysis may be applied.

Using eqn. (6.10) for temperature distribution

$$\frac{T - T_\infty}{T_i - T_\infty} = \exp\left\{-\frac{ht}{\rho\delta C}\right\}$$

Substituting the numerical values

$$\frac{100 - 30}{350 - 30} = \exp\left\{-\frac{60t}{2700 \times 0.0269 \times 900}\right\}$$

or $$\ln\left(\frac{70}{320}\right) = -\frac{60}{65367}t$$

or $t = 1655$ s $= \mathbf{27.6\ min.}$ Ans.

Example 6.10. *A thermocouple junction is in the form of a 4 mm diameter sphere. The properties of the material are C = 420 J/kg.K, ρ = 8000 kg/m³, k = 40 W/m.K, unit surface conductance h = 40 W/m².K. The junction is initially at 40°C is inserted in a stream of hot air at 300°C. Find :*

(i) Time constant.

(ii) Thermocouple is taken out from the hot air after 10 seconds and is kept in still air at 30°C. Assuming the heat transfer coefficient in still air as 10 W/m².K, find the temperature attained by junction 20 seconds after removing it from hot air stream.

(P.U.P., Dec. 2008)

Solution

Given : A thermocouple junction in the form of sphere with

$D = 4$ mm $= 4 \times 10^{-3}$ m, $C = 420$ J/kg.K,
$\rho = 8000$ kg/m³, $k = 40$ W/m.K,
$T_i = 40$°C,

(*a*) $T_{\infty_1} = 300$°C for $t = 10$ s with $h_1 = 40$ W/m².K.
(*b*) $T_{\infty_2} = 30$°C for $t = 20$ s with $h_2 = 10$ W/m².K.

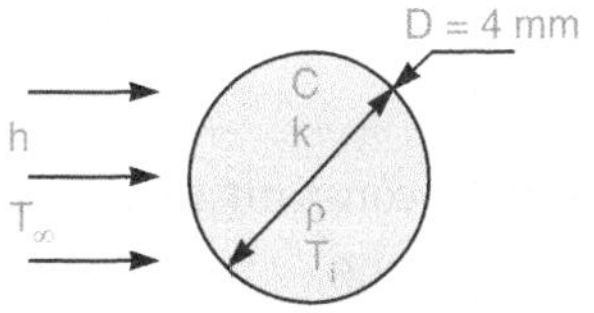

Fig. 6.16 (*a*) Thermocouple junction

To find :

(*i*) Time constant.

(*ii*) Temperature attained by thermocouple after 10 s, when placed in stream of hot air at $T_{\infty_1} = 300$°C.

(*iii*) Temperature attained by thermocouple, taken out from hot air and placed in still air at 30°C for 20 s.

Assumptions :

1. Internal temperature gradients are negligible.
2. No radiation heat exchange.
3. Constant properties.

Analysis : The characteristic length of the sphere

$$\delta = \frac{V}{A_s} = \frac{D}{6} = \frac{0.004}{6}$$

$$= 6.667 \times 10^{-4} \text{ m}$$

The Biot number

$$\text{Bi} = \frac{h_1\delta}{k} = \frac{40 \times 6.667 \times 10^{-4}}{40}$$

$$= 6.667 \times 10^{-4}$$

which is less than 0.1, thus the lumped system analysis is applicable with reasonable accuracy.

(*i*) Using eqn. (6.18) for time constant

$$\tau = \frac{\rho\delta C}{h_1} = \frac{8000 \times 6.667 \times 10^{-4} \times 420}{40}$$

$$= \mathbf{56\ s.} \quad \text{Ans.}$$

(*ii*) When the junction is exposed to hot air stream at 300°C. Temperature attained by junction after 10 s. Using eqn. (6.20)

$$\frac{T_1 - T_{\infty_1}}{T_i - T_{\infty_1}} = \exp\left\{-\frac{h_1 t_1}{\rho\delta C}\right\} = \exp\left(-\frac{t_1}{\tau}\right)$$

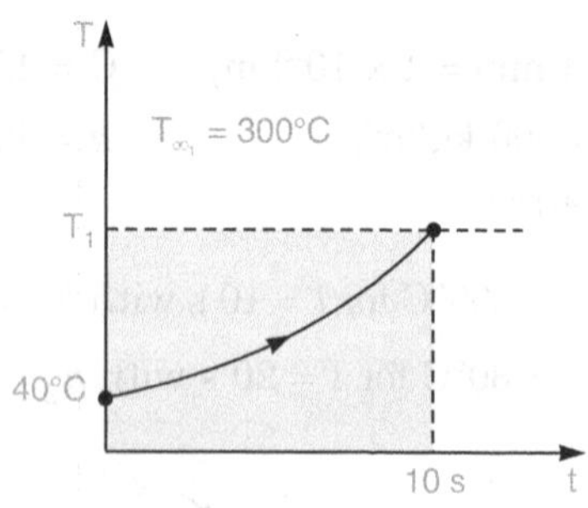

Fig. 6.16 (*b*)

Using the time constant in above equation

$$\frac{T_1 - 300}{40 - 300} = \exp\left\{-\frac{10}{56}\right\} = 0.83646$$

or $$T_1 = 300 - 260 \times 0.83646$$

$$= \mathbf{82.52°C.} \quad \text{Ans.}$$

(*iii*) Now the junction is taken out from hot air stream and placed in stream of still air at 30°C. The temperature after 20 s.

The initial temperature for this case would be 82.52°C. Hence using the relation for temperature distribution, as

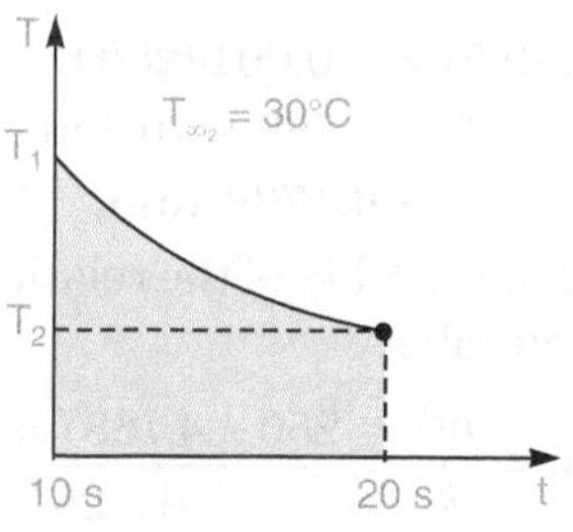

Fig. 6.16 (*c*)

$$\frac{T_2 - T_{\infty_2}}{T_1 - T_{\infty_2}} = \exp\left\{-\frac{h_2 t_2}{\rho C\delta}\right\}$$

$$\frac{T_2 - 30}{82.52 - 30} = \exp\left\{-\frac{10 \times 20}{8000 \times 420 \times 6.667 \times 10^{-4}}\right\}$$

$$= 0.9145$$

or $$T_2 = 30 + 52.52 \times 0.9145 = \mathbf{78°C.} \quad \text{Ans.}$$

Example 6.11. *A thermocouple junction may be approximated as a sphere, is to be used for temperature measurement in a gas stream with convection coefficient of 400 W/m²K. The thermophysical properties of the junction are k = 20 W/m.K, C = 400 J/kg.K, ρ = 8500 kg/m³.*

(a) Determine the diameter needed for the thermocouple junction to have a time constant of 1 s. If the junction is initially at 25°C and placed in a gas stream at 200°C. How long it will take for the junction to reach 199°C ?

(b) If the duct wall temperature is 400°C and the emissivity of the thermocouple bead is 0.9, calculate steady state temperature of the junction. Also calculate the time for junction temperature to increase from an initial condition of 25°C to a temperature that is within 1°C of its steady state value.

Solution

Given : A thermocouple junction in the form of sphere with

(*a*) $T = 199°C$, $T_\infty = 200°C$,
$\tau = 1$ s, $T_i = 25°C = 298$ K.

(*b*) $\varepsilon = 0.9$, $T_{surr} = 400°C = 673$ K.

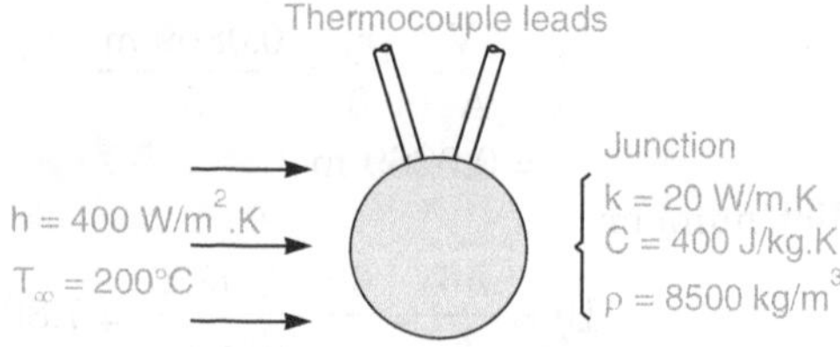

Fig. 6.17. Thermocouple junction.

To find :

Part (a)

(i) Diameter of junction for time constant of 1 s.

(ii) Time required by thermocouple to reach 199°C.

Part (b)

(i) Steady state temperature of the junction.

(ii) Time required for thermocouple junction to reach a temperature that is within 1°C of its steady state value.

Assumptions :

1. Internal temperature gradients are negligible.

2. No radiation heat exchange for part (a) and $\sigma = 5.67 \times 10^{-8}$ W/m².K⁴ for part (b) of the problem.

3. Constant properties.

Analysis : Part (a)

(i) Using eqn. (6.18) for time constant

$$\tau = \frac{\rho VC}{hA_s} = \frac{\rho r_o C}{3h}$$

or
$$r_o = \frac{(1\,s)\times 3\times(400\ W/m^2.K)}{(8500\ kg/m^3)\times(400\ J/kg.K)}$$

$$= 3.529 \times 10^{-4}\ m$$

or Diameter, $D = 2r_o = 7.06 \times 10^{-4}$ m

$= \mathbf{0.706\ mm.}$ Ans.

Checking the validity for the use of lumped system analysis

Biot number,

$$Bi = \frac{h\delta}{k} = \frac{hr_o}{3k} = \frac{400\times 3.529\times 10^{-4}}{3\times 20}$$

$$= 2.35 \times 10^{-3}$$

which is much less than 0.1, hence the lumped heat capacity system analysis is suitable for approximation.

(ii) Using eqn. (6.10) for temperature distribution

$$\frac{T - T_\infty}{T_i - T_\infty} = \exp\left\{-\frac{3ht}{\rho r_o C}\right\}$$

Substituting the numerical values

$$\frac{199 - 200}{25 - 200} = \exp\left\{-\frac{3\times 400t}{8500\times 3.529\times 10^{-4}\times 400}\right\}$$

or
$$\ln\left(\frac{1}{175}\right) = -\frac{3}{2.99965}t$$

It gives $t = \mathbf{5.2\ s.}$ Ans.

Part (b) :

(i) For steady state conditions, the energy balance on the thermocouple junction.

Rate of energy input = Rate of energy dissipation

$$A_s\,\varepsilon\sigma((T_{surr}^4 - T^4) = hA_s(T - T_\infty)$$

or
$$\varepsilon\,\sigma(T_{surr}^4 - T^4) - h(T - T_\infty) = 0$$

Substituting numerical values

$$0.9 \times 5.67 \times 10^{-8} \times (673^4 - T^4) - 400 \times (T - 473) = 0$$

$$10468.5 - 5.103 \times 10^{-8}\,T^4 - 400T + 189200 = 0$$

or
$$5.103 \times 10^{-8}\,T^4 + 400T - 199668.5 = 0$$

Using Newton Raphson's iterative technique for the solution of this non-linear equation.

$$F(T) = 5.103 \times 10^{-8}\,T^4 + 400T - 199668.5 = 0$$

$$F'(T) = 1.5309 \times 10^{-7}\,T^3 + 400$$

Assuming initial guess $T_1 = 485$ K, then

$$T_{i+1} = T_i - \frac{F(T_i)}{F'(T_i)}$$

After two iterations, we get a stable value of

$T = 491.71\ K = \mathbf{218.71°C.}$ Ans.

It is the steady state temperature of thermocouple.

(ii) The temperature to be recorded by thermocouple be

$$T = 218.71 - 1 = 217.71°C$$

$T_\infty = 200°C$, $\quad T_i = 25°C$

Then the energy balance on thermocouple junction

Rate of energy radiated in – Rate of energy convected out

= Rate of change of internal energy

$$E'_{in} - E'_{out} = E'_{st}$$

$$[\varepsilon\sigma(T_{surr}^4 - T^4) - h(T - T_\infty)]\,A_s = \rho VC\,\frac{dT}{dt}$$

$$[0.9 \times 5.67 \times 10^{-8} \times (673^4 - 490.71^4)] - 400 \times (490.71 - 473) \times 4\pi \times r_o^2$$

$$= 8500 \times \frac{4\pi}{3} r_o^2 \times (3.529\times 10^{-4})\times 400\times\frac{dT}{dt}$$

Its solution gives $t = \mathbf{4.9\ s.}$ Ans.

Example 6.12. *An egg with mean diameter of 4 cm is initially at 25°C. It is placed in boiling water for 4 min and found to be consumer's taste. For how long should a similar egg for same consumer be boiled when taken from refrigerator at 2°C. Use lumped system analysis and take thermophysical properties of egg as*

$k = 12$ *W/m.K,* $h = 125$ *W/m².K,*

$C = 2000$ *J/kg.K*

and $\rho = 1250$ *kg/m³.* (P.U., Nov. 1997)

Solution

Given : An egg as sphere

$D = 4 \text{ cm} = 0.04 \text{ m}$

(*i*) $T_i = 25°C$

$t_1 = 4 \text{ min} = 240 \text{ s}$

(*ii*) $T_i = 2°C$

$k = 12 \text{ W/m.K}, \quad h = 125 \text{ W/m}^2\text{.K}$

$C = 2000 \text{ J/kg.K}, \quad \rho = 1250 \text{ kg/m}^3.$

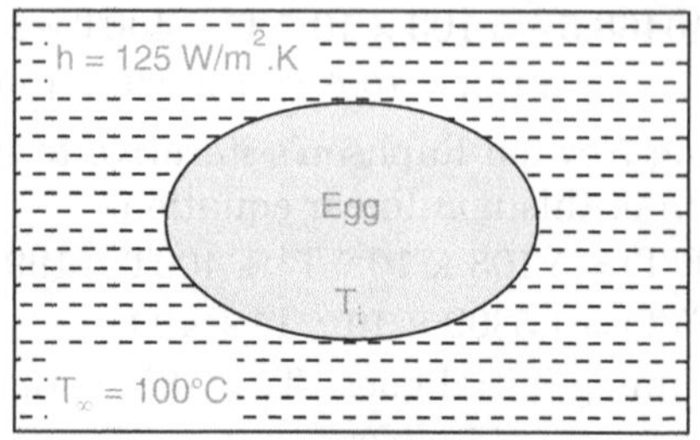

Fig. 6.18. Boiling of an egg

To find : Time required for egg at 2°C to be boiled to consumer's taste.

Assumptions :

(*i*) Egg as a sphere.

(*ii*) Negligible internal temperature gradients.

(*iii*) Constant properties.

(*iv*) Boiling temperature of water as 100°C.

Analysis : The characteristic length of egg

$$\delta = \frac{D}{6} = \frac{0.04}{6} = \frac{0.02}{3} \text{ m}$$

The temperature distribution, using lumped system analysis,

$$\frac{T - T_\infty}{T_i - T_\infty} = \exp\left(-\frac{ht}{\rho\delta C}\right)$$

Temperature of consumer's taste

$$\frac{T - 100}{25 - 100} = \exp\left(-\frac{125 \times 240 \times 3}{1250 \times 0.02 \times 2000}\right)$$

or $$T = 100 - 75 \times 0.1653 = 87.6°C$$

When egg is taken from refrigerator at

$$T_i = 2°C \quad \text{and} \quad T = 87.6°C$$

$$\frac{87.6 - 100}{2 - 100} = \exp\left(-\frac{125 \times 3\,t}{1250 \times 0.02 \times 2000}\right)$$

or $$\ln\left(\frac{12.4}{98}\right) = -\frac{375}{50000}t$$

or $$t = 275.6 \text{ s} = \mathbf{4.6 \text{ min.}}$$ **Ans.**

Example 6.13. *It is proposed to quench the steel balls of bearings, 1 cm in diameter, initially at 400°C is placed in a cold chamber maintained at – 20°C. The steel balls pass through the chamber on a conveyor belt. Optimum bearing production requires that 75% of initial thermal energy content of balls above –15°C be removed. How long the balls should be placed in the chamber ? Take*

$h = 100 \text{ W/m}^2\text{.K}, \quad k = 46 \text{ W/m.K},$

Sp. gravity = 7.8, $\quad C = 420 \text{ J/kg.K.}$

Solution

Given : Steel balls of bearings with

$D = 1 \text{ cm}, \quad \text{Sp. gr.} = 7.8,$

$h = 100 \text{ W/m}^2\text{.K} \quad C = 420 \text{ J/kg.K},$

$T_\infty = -20°C, \quad T_i = 400°C$

$k = 46 \text{ W/m.K}$

Energy to be removed = 75% of initial energy content above – 15°C.

To find : Time required during 75% of initial energy removal.

Assumptions :

1. Internal temperature gradients are negligible.
2. No radiation heat exchange.
3. Constant properties.

Analysis : The radius of steel balls,

$$r_o = D/2 = 0.5 \text{ cm} = 0.005 \text{ m}$$

The density of steel,

$$\rho = \text{Sp. gr.} \times 1000 \text{ kg/m}^3 = 7.8 \times 1000 = 7800 \text{ kg/m}^3$$

The characteristics length of steel balls, $\delta = \frac{r_o}{3}$

Biot number

$$Bi = \frac{h\delta}{k} = \frac{100 \times 0.005}{3 \times 46} = 3.623 \times 10^{-3}$$

which is less than 0.1, hence the lumped heat capacity system analysis can be reasonably used for approximation.

Using eqn. (6.10) for temperature distribution

$$\frac{T - T_\infty}{T_i - T_\infty} = \exp\left\{-\frac{3ht}{\rho r_o C}\right\}$$

Here, Initial energy content of balls above – 15°C $= mC(T_i + 15)$

Energy of balls (above –15°C) is to be removed

$= 75\%$ of initial energy content

$= 0.75 \times mC(T_i + 15)$

The remaining energy (above – 15°C) content of balls,

$$mC(T + 15) = 0.25 \times mC(T_i + 15)$$

Hence $$\frac{(T + 15)}{(T_i + 15)} = 0.25$$

$$= \exp\left\{-\frac{3 \times 100 \times t}{7800 \times 420 \times 0.005}\right\}$$

$$t = \ln(0.25) \times (-54.6)$$

$$= \mathbf{75.7 \text{ s.}}$$ **Ans.**

Example 6.14. *A cylindrical stainless steel ingot ($k = 45$ W/m.K), 15 cm in diameter and 40 cm long passes through a treatment furnace, which is 6 m in length. The temperature of furnace gas is 1300°C. The initial ingot temperature is 100°C. The combined radiant and convective heat transfer coefficient is 100 W/m^2.K. Calculate the maximum speed with which the ingot should pass through the furnace, if it must attain a temperature of 850°C.*

Take $\alpha = 0.46 \times 10^{-5}$ m^2/s. (N.M.U., Nov. 1999)

Solution

Given : A cylindrical stainless ingot with

$D = 15$ cm $= 0.15$ m, $L = 40$ cm $= 0.4$ m,

$h = 100$ W/m^2.K, $k = 45$ W/m.K,

$T_i = 100$°C, $T_\infty = 1300$°C,

$T = 850$°C, $L_{furnace} = 6$ m,

$\alpha = 0.46 \times 10^{-5}$ m^2/s.

To find : The maximum speed of ingot through furnace.

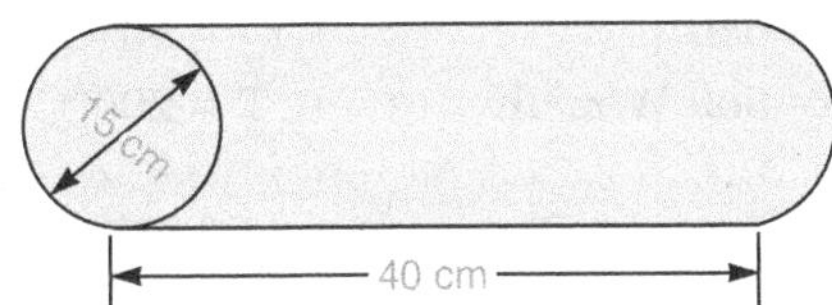

Fig. 6.19. Cylindrical steel ingot for example 6.14

Assumptions :

1. Internal temperature gradients are negligible.
2. Uniform heating throughout the length of furnace.
3. Constant properties.

Analysis : The radius of steel ingot,

$$r_o = \frac{D}{2} = \frac{0.15 \text{ m}}{2} = 0.075 \text{ m}$$

The characteristic length of the cylinder.

$$\delta = \frac{V}{A_s} = \frac{\pi r_o^2 L}{2\pi r_o^2 + 2\pi r_o L}$$

$$= \frac{\pi \times (0.075)^2 \times 0.4}{2\pi \times (0.075)^2 + 2\pi \times 0.075 \times 0.4}$$

$$= 0.0315 \text{ m}$$

$$\text{Biot number Bi} = \frac{h\delta}{k} = \frac{100 \times 0.0315}{45} = 0.070$$

which is less than 0.1, hence the lumped heat capacity system analysis can be reasonably used for approximation.

Using eqn. (6.10) for temperature distribution

$$\frac{T - T_\infty}{T_i - T_\infty} = \exp\left\{-\frac{ht}{\rho\delta C}\right\} = \exp\left\{-\frac{\alpha h t}{k\delta}\right\}$$

$$\frac{850 - 1300}{100 - 1300} = \exp\left\{-\frac{0.46 \times 10^{-5} \times 100t}{45 \times 0.0315}\right\}$$

or $$t = -(3081.52) \times \ln(0.375) = 3022 \text{ s}$$

The velocity of ingot through the furnace

$$u = \frac{L_{furnace}}{\text{time required}} = \frac{6 \text{ m}}{3022 \text{ s}}$$

$= \mathbf{1.985}$ **mm/s.** Ans.

Example 6.15. *A mild steel sphere of 15 mm in diameter initially at 625°C is exposed to a current of air at 25°C with convection coefficient of 120 W/m^2.K. Calculate :*

(*i*) *Time required to cool the sphere to 100°C.*

(*ii*) *Initial rate of cooling in °C/s.*

(*iii*) *Instantaneous heat transfer rate at the end of one minute after the start of cooling.*

(*iv*) *Total energy transferred during first one minute.*

Take properties of mild steel as :

$k = 43$ W/m.K, $\rho = 7850$ kg/m^3,

$C = 474$ J/kg.K, $\alpha = 0.045$ m^2/s.

Solution

Given : The mild steel sphere with ;

$D = 15$ mm, $h = 120$ W/m^2.K,

$k = 43$ W/m.K $T_i = 625$°C,

$T_\infty = 25$°C, $\rho = 7850$ kg/m^3,

$C = 474$ J/kg.K, $T = 100$°C,

$\alpha = 0.045$ m^2/s.

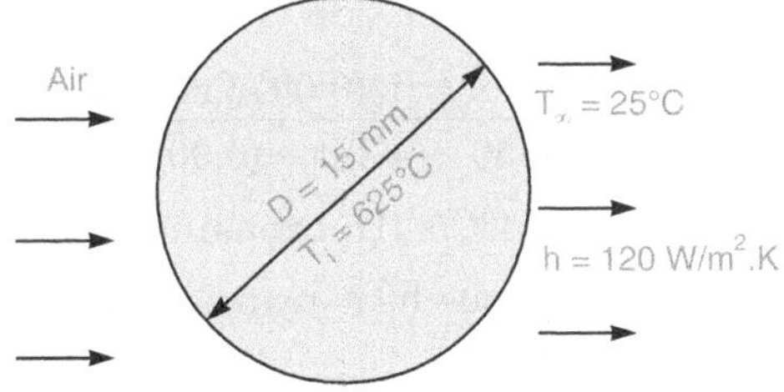

Fig. 6.20. Sphere for example 6.15

To find :

(*i*) The time required to reach the sphere to 100°C.

(*ii*) Initial rate of cooling of the sphere in °C/s

(*iii*) Instantaneous rate of heat transfer at the end of one minute.

(*iv*) Total energy transferred during first one minute.

Assumptions :

1. Internal temperature gradients are negligible.
2. No radiation heat exchange.
3. Constant properties.

Analysis : The radius of mild steel sphere,

$$r_o = \frac{D}{2} = \frac{15}{2} = 7.5 \text{ mm} = 0.0075 \text{ m}$$

The characteristic length of the sphere :

$$\delta = \frac{V}{A_s} = \frac{r_o}{3} = \frac{0.0075}{3} = 0.0025 \text{ m}$$

Biot number

$$\text{Bi} = \frac{h\delta}{k} = \frac{120 \times 0.0025}{43} = 0.00697$$

Which is less than 0.1, hence the lumped heat capacity system analysis can be reasonably used for approximation.

(*i*) Using eqn. (6.10) for temperature distribution

$$\frac{T - T_\infty}{T_i - T_\infty} = \exp\left\{-\frac{ht}{\rho C\delta}\right\}$$

or $$\frac{100 - 25}{625 - 25} = \exp\left\{-\frac{120t}{7850 \times 474 \times 0.0025}\right\}$$

or $t = 161.2 \text{ s} = \mathbf{2.687 \text{ min.}}$ **Ans.**

(*ii*) The initial rate of cooling can be obtained by energy balance as

Rate of decrease of internal energy = Rate of heat convection from the sphere

$$-\rho VC \frac{dT}{dt} = hA_s(T_i - T_\infty)$$

or $$\frac{dT}{dt} = -\frac{120 \times 4\pi \times (0.0075 \text{ m})^2 \times (625 - 25)}{7850 \times (4/3)\,\pi \times (0.0075 \text{ m})^3 \times 474}$$

$$= -\mathbf{7.74°C/s \text{ (Decreasing).}} \quad \textbf{Ans.}$$

(*iii*) Instantaneous heat transfer rate at end of 1 minute :

$$Q(t) = hA_s(T_i - T_\infty) \exp\left\{-\frac{ht}{\rho\delta C}\right\}$$

$$= 120 \times 4\pi \times (0.0075)^2 \times (625 - 25) \times \exp\left\{-\frac{120 \times 60}{7850 \times 0.0025 \times 474}\right\}$$

$$= 50.89 \times 0.461 = \mathbf{23.47 \text{ W.}} \quad \textbf{Ans.}$$

(*iv*) Total heat transferred during first 60 seconds (decrease in internal energy)

$$\Delta U = -\rho VC(T_i - T_\infty)\left\{\exp\left(-\frac{ht}{\rho\delta C}\right) - 1\right\}$$

or $$\Delta U = -7850 \times (4/3)\,\pi \times (0.0075)^3 \times 474 \times (625 - 25) \times \left\{\exp\left(-\frac{120 \times 60}{7850 \times 0.0025 \times 474}\right) - 1\right\}$$

or $\Delta U = \mathbf{2125.8 \text{ W.}}$ **Ans.**

It is the decrease of internal energy of sphere.

Example 6.16. *The steel ball bearing ($k = 50$ W/m.K, $\alpha = 1.3 \times 10^{-5}$ m²/s), 40 mm in diameter are heated to a temperature of 650°C. It is then quenched in an oil bath at 50°C, where the heat transfer coefficient is estimated to be 300 W/m².K.*

Calculate (a) the time required for bearing to reach 200°C, (b) the total amount of heat removed from a bearing during this time, and (c) the instantaneous heat transfer rate from the bearings, when they are first immersed in oil bath and when they reach 200°C.

(P.U.P., Dec. 2009 ; J.N.T.U., May 2000)

Solution: *Given :* Steel ball bearing to be quenched :

$D = 40$ mm	or	$r_o = 0.02$ m
$k = 50$ W/m.K,		$\alpha = 1.3 \times 10^{-5}$ m²/s
$T_i = 650$°C,		$T_\infty = 50$°C
$h = 300$ W/m².K,		$T = 200$°C.

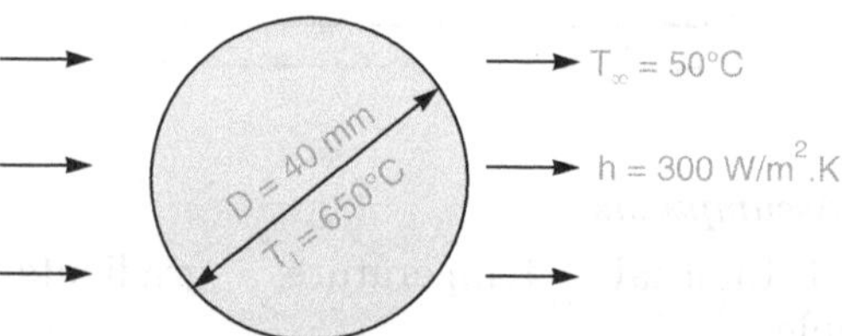

Fig. 6.21. Steel ball bearing for example 6.16

To find :

(*a*) Time required by bearings to reach 200°C.

(*b*) Total heat transferred from a bearing.

(*c*) Instantaneous heat transfer rate from bearings

(*i*) at $t = 0$, and (*ii*) when $T = 200$°C.

Analysis : The characteristic length of bearing (spherical body)

$$\delta = \frac{r_o}{3}$$

$$\text{Bi} = \frac{h\delta}{k} = \frac{300 \times 0.02}{50 \times 3} = 0.04$$

which is less than 0.1, thus the lumped system analysis is applicable.

(*a*) Time required for bearings to reach 200°C, eqn. (6.10)

$$\frac{T - T_\infty}{T_i - T_\infty} = \exp\left\{-\frac{ht}{\rho\delta C}\right\} = \exp\left\{-\frac{3h\alpha t}{r_o k}\right\}$$

or, $\ln\left(\frac{200-50}{650-50}\right) = -\frac{3\times 300\times 1.3\times 10^{-5}t}{0.02\times 50}$

or $t = 118.5 \text{ s} \approx$ **2 min.** Ans.

(*b*) Total heat removed from bearing during a period of 118.5 s can be obtained by eqn. (6.7) ;

$$\Delta U = -\rho VC\,(T_i - T_\infty)\left\{\exp\left(-\frac{h A_s t}{\rho VC}\right) - 1\right\}$$

$$= -\frac{k}{\alpha}\left(\frac{4}{3}\pi r_o^3\right)(T_i - T_\infty)\left\{\exp\left(-\frac{3h\alpha t}{r_o k}\right) - 1\right\}$$

$$= -\frac{50}{1.3\times 10^{-5}} \times \left\{\frac{4}{3}\pi\times(0.02)^3\right\} \times (650 - 50)$$

$$\times\left\{\exp\left(-\frac{3\times 300\times 1.3\times 10^{-5}\times 118.5}{0.02\times 50}\right) - 1\right\}$$

$= 58.0 \times 10^3$ J = **58 kJ.** Ans.

–ve sign indicates decrease of internal energy of bearing.

(*c*) Instantaneous heat transfer rate :

(*i*) At $t = 0$, initial rate of heat transfer

$$Q_{t=0} = hA_s(T_i - T_\infty)$$

$$= 300 \times 4\pi \times (0.02)^2 \times (650 - 50)$$

= **904.7 W.** Ans.

(*ii*) At $t = 118.5$ s or when temperature of bearing reaches to 200°C.

$$Q(t) = hA_s\,(T_{(t)} - T_\infty)$$

$$= 300 \times \{4\pi \times (0.02)^2\} \times (200 - 50)$$

= **226.16 W.** Ans.

Example 6.17. *Two plastic sheets (k = 0.232 W/m.K, C = 1.674 kJ/kg.K, ρ = 1300 kg/m³), each 5 mm thick are to be bonded together with a thin layer of adhesive, which fuses at 140°C. To perform this process, they are pressed between two surfaces at 250°C. Find the time required for which the sheets should be pressed together to complete the process. The initial temperature of sheets is 30°C. Assume perfect contact and neglect resistance of adhesive. Derive the formula used.*

Solution

Given : Two plastic sheets to be bonded ;

$L = 5$ mm, $k = 0.232$ W/m.K,

$T_i = 30$°C, $T_L = 250$°C

$T = 140$°C, $\rho = 1300$ kg/m³,

$C = 1.674$ kJ/kg.K $= 1674$ J/kg.K.

To find : The time required to reach the centre temperature 140°C.

Assumptions :

1. Infinite long plastic sheets.
2. Internal temperature gradients are negligible.
3. No radiation heat exchange.
4. Constant properties.

Analysis : Consider the two plastic sheets as shown in Fig. 6.22.

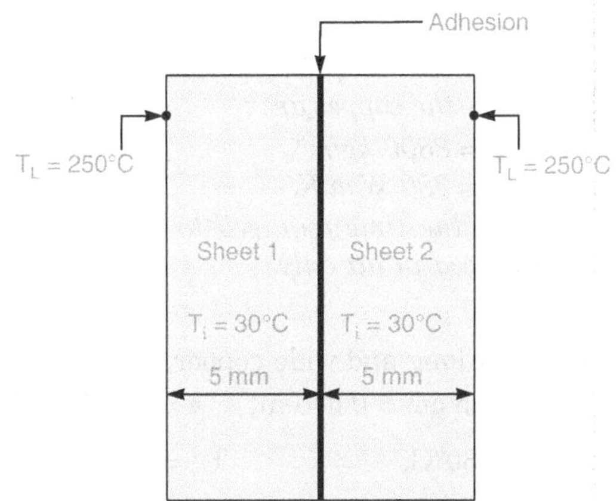

Fig. 6.22. Bonding of two plastic sheets

The thickness of the two sheet

$2L = 5 \text{ mm} + 5 \text{ mm} = 10 \text{ mm} = 0.01$ m.

The energy balance equation for the two sheets together :

Rate of heat inflow

= Rate of internal energy increase

or $$\frac{2kA(T_L - T)}{L} = \rho VC\,\frac{dT}{dt}$$

where T is the temperature, function of time.

Let $\theta = T_L - T$

$$\frac{d\theta}{dt} = -\frac{dT}{dt}$$

Substituting and rearranging,

$$\frac{2kAdt}{\rho VCL} = -\frac{d\theta}{\theta}$$

Integrating both sides within the limits, we get

$$\frac{2kA}{\rho VCL}\int_0^t dt = -\int_{\theta_i}^{\theta_c}\frac{d\theta}{\theta}$$

or $$\frac{2kAt}{\rho VCL} = -\ln\left(\frac{\theta_c}{\theta_i}\right) = -\ln\left[\frac{T_L - T_c}{T_L - T_i}\right]$$

or $$t = \frac{\rho\,(AL)\,C\,L}{2kA} \times \ln\left[\frac{T_L - T_i}{T_L - T_c}\right]$$

$$= \frac{1300\times 1674\times(0.005)^2}{2\times 0.232} \times \ln\left[\frac{250-30}{250-140}\right]$$

$= 81.2 \text{ s} =$ **1.35 min.** Ans.

Example 6.18. *A long and wide copper plate of 4.5 cm thick, at initial temperature of 180°C is held on the water surface so that its one face is in contact with water at 25°C. The other surface is exposed to air side at 25°C. Unit surface conductance on the water and air side are 80 and 8 W/m^2.K, respectively. Neglecting the radiation losses, heat transfer from edges and internal temperature gradients, find the time required to cool the plate to 90°C. The properties of the copper are :*

ρ = *8800 kg/m^3,* *C = 410 J/kg.K*
k = 380 W/m.K.

Also find the time required to cool the plate to 90°C, if it is placed in air only.

Solution

Given : A long and wide copper plate with

L = 4.5 cm = 0.045 m, k = 380 W/m.K,
T_i = 180°C, $T_{\infty_1} = T_{\infty_2}$ = 25°C,
T = 90°C, ρ = 8800 kg/m^3,
C = 410 J/kg.K, h_a = 8 W/m^2.K,
h_w = 80 W/m^2.K.

To find :

(*i*) The time required to cool the plate to 90°C, if one side is in water and other in air.

(*ii*) Time required to cool the plate to 90°C, if it is placed in air only.

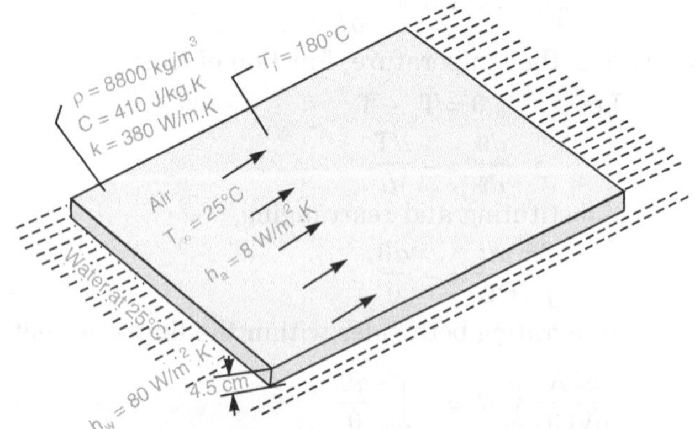

Fig. 6.23. Copper plate exposed to water and air on bottom and top side, respectively.

Assumptions :

1. Infinite long and wide copper plate.
2. Internal temperature gradients are negligible.
3. Constant properties.

Analysis : (*i*) If the one side of the plate is held on water and other side exposed to air. The energy balance for the copper plate :

Rate of heat transfer from its surfaces = Rate of internal energy decrease of the plate.

or $$h_a A_a (T - T_\infty) + h_w A_w (T - T_\infty) = -\rho VC \frac{dT}{dt}$$

where T is the temperature, function of time and

$$A = A_w = A_a$$

Then $$(h_w + h_a)\, A(T - T_\infty) = -\rho VC \frac{dT}{dt}$$

or $$\frac{(h_w + h_a)\, Adt}{\rho VC} = -\frac{dT}{T - T_\infty}$$

Treating h_w, h_a, A, k, ρ, V and C as constants and integrating as

$$\frac{(h_w + h_a)A}{\rho VC}\int_0^t dt = -\int_{(T_i - T_\infty)}^{(T - T_\infty)} \frac{dT}{T - T_\infty}$$

$$\frac{(h_w + h_a)At}{\rho VC} = -\ln\left[\frac{T - T_\infty}{T_i - T_\infty}\right]$$

or $$t = -\frac{\rho VC}{(h_a + h_w)A} \times \ln\left[\frac{T - T_\infty}{T_i - T_\infty}\right]$$

$$= -\frac{8800 \times (0.045\text{ A}) \times 410}{(80 + 8)\text{A}} \times \ln\left[\frac{90 - 25}{180 - 25}\right]$$

= 1603.3 s = **26.72 min. Ans.**

(*ii*) When plate is exposed to air on both sides (A_s = 2A)

$$t = -\frac{\rho VC}{h_a(2A)} \times \ln\left[\frac{T - T_\infty}{T_i - T_\infty}\right]$$

$$= -\frac{8800 \times (0.045\text{ A}) \times 410}{8 \times 2\text{A}} \times \ln\left[\frac{90 - 25}{180 - 25}\right]$$

= 8818.5 s = **2.45 hours. Ans.**

Example 6.19. *A household electric iron has a steel base, weighs 1 kg. The base has an ironing surface of 0.025 m^2 and is heated from the other surface with a 250 W heating element. Initially the iron is at a uniform temperature of 20°C. Suddenly the heating starts, and the iron dissipates heat by convection from ironing surface into an ambient at 20°C with a convection coefficient of 50 W/m^2.K.*

Calculate the temperature of iron 5 minute after the starts of heating. What would be the equilibrium temperature of the iron, if control did not switch off the current ?

The properties of the material are :

ρ = *7840 kg/m^3,* *C = 450 J/kg.K,*
k = 70 W/m.K. (N.M.U., May 2002)

Solution

Given : An electric iron with

k = 70 W/m.K, T_i = 20°C,
T_∞ = 20°C, A = 0.025 m^2,
t = 5 min = 300 s, ρ = 7840 kg/m^3,
C = 450 J/kg.K, h = 50 W/m^2.K,
m = 1 kg, Q = 250 W.

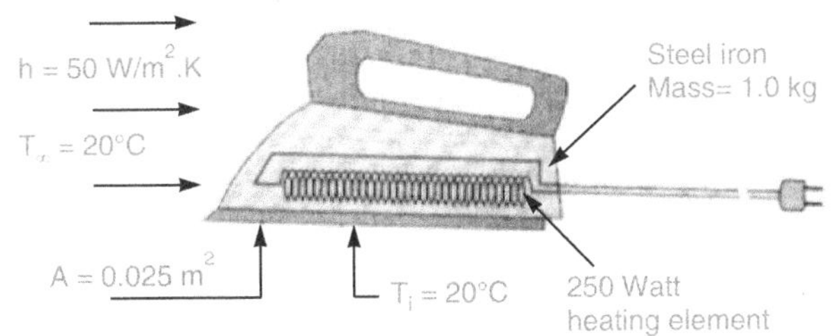

Fig. 6.24. House hold steel iron for example 6.19

To find :

(*i*) The temperature of iron after 5 minute of start of heat supply.

(*ii*) Steady state temperature of iron.

Assumptions :

1. No radiation heat loss.
2. No heat loss from the edges and top face of electric iron.
3. Negligible internal temperature gradients.
4. Constant properties.

Analysis : (*i*) The thickness of the base of the iron can be calculated as

$$L = \frac{m}{A\rho} = \frac{1\,\text{kg}}{(0.025\,\text{m}^2)\times(7840\,\text{kg/m}^3)} = 0.0051\text{ m}$$

Since the iron has convection heat interaction only on one of its surface, hence its characteristic length would be its thickness,

$$\delta = L = 0.0051\text{ m}$$

Biot number $\quad Bi = \frac{h\delta}{k} = \frac{50\times 0.0051}{70} = 0.0036$

which is less than 0.1, hence the lumped heat capacity system analysis can be reasonably used for approximation. Using equation (6.25)

$$\theta = T - T_\infty = \theta_i \exp(-Mt) + \frac{q}{h}\{1-\exp(-Mt)\}$$

where, $\quad \theta_i = T_i - T_\infty = 20 - 20 = 0$

$$Mt = \frac{ht}{\rho CL} = \frac{50\times 300}{7840\times 450\times 0.0051} = 0.834$$

$$q = \frac{Q}{A} = \frac{250}{0.025} = 10{,}000\text{ W/m}^2$$

Hence, $\quad T - 20 = \frac{10{,}000}{50}\times\{1-\exp(-0.834)\}$

or $\quad$ **T = 133°C.** Ans.

(*ii*) The equilibrium temperature becomes for $t\to\infty$

$$T(\infty) = T_\infty + \frac{q}{h} = 20 + \frac{10{,}000}{50} = \textbf{220°C.}\ \text{Ans.}$$

6.2. ANALYTICAL SOLUTION

6.2.1. Criteria for Neglecting Internal Temperature Gradients

Consider an infinite plate of thickness 2L as shown in Fig. 6.25. The plate is initially at uniform temperature T_i at $t = 0$. The plate is suddenly exposed to convection environment at temperature T_∞ and heat transfer coefficient h for all $t > 0$. The governing differential equation for one dimensional time dependent unsteady state heat conduction without heat generation is given by eqn. (2.8).

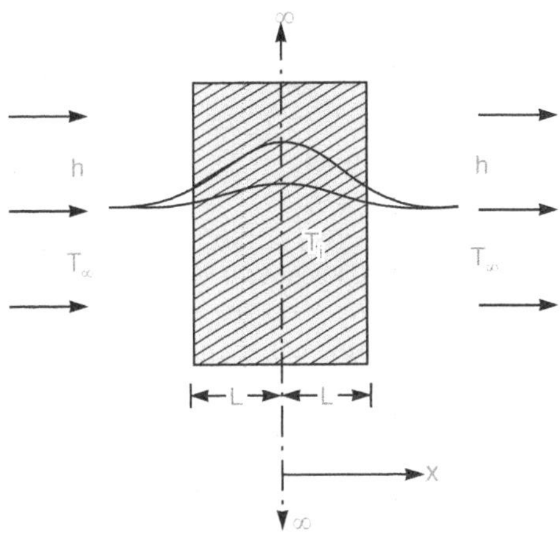

Fig. 6.25. Transient conduction in an infinite plate

$$\frac{\partial}{\partial x}\left\{\frac{\partial T}{\partial x}\right\} = \frac{1}{\alpha}\frac{\partial T}{\partial t}$$

or
$$\alpha\frac{\partial^2 T}{\partial x^2} = \frac{\partial T}{\partial t} \qquad ...(6.28)$$

where, $\quad T = f(x, t)$

and at $t = 0$, $\quad T = T_i$

Introducing the variable

$$\theta(x, t) = \theta = (T - T_\infty)$$

Then $\quad \alpha\frac{\partial^2\theta}{\partial x^2} = \frac{\partial\theta}{\partial t}$

Assuming the product solution as

$$\theta = F(x)\,G(t) \qquad ...(6.29)$$

Substituting in eqn. (6.28), we get

$$\alpha\frac{\partial^2 F(x)}{\partial x^2}G(t) = F(x)\frac{\partial G(t)}{\partial t}$$

or
$$\frac{1}{F(x)}\frac{\partial^2 F(x)}{\partial x^2} = \frac{1}{\alpha G(t)}\frac{\partial G(t)}{\partial t} \qquad ...(6.30)$$

Introducing a separation constant as

$$\frac{1}{F(x)}\frac{\partial^2 F(x)}{\partial x^2} = \frac{1}{\alpha G(t)}\frac{\partial G(t)}{\partial t} = -\lambda^2 \quad ...(6.31)$$

The λ^2 is called separation constant and function G(t) must decay exponentially with time, therefore λ^2 is considered negative.

Each side of eqn. (6.31) is a function of only one variable and each side will be equal to $-\lambda^2$. Taking each equation separately

$$\frac{1}{F(x)}\frac{\partial^2 F(x)}{\partial x^2} = -\lambda^2$$

or
$$\frac{1}{F(x)}\frac{\partial^2 F(x)}{\partial x^2} + \lambda^2 = 0$$

The characteristic equation is in the form of

$$m^2 + \lambda^2 = 0 \quad \text{or} \quad m = \pm \lambda$$

It can be written as

$$\begin{aligned} F(x) &= A_1 e^{-i\lambda x} + A_2 e^{i\lambda x} \\ &= A_1(\cos \lambda x - i \sin \lambda x) + A_2(\cos \lambda x + i \sin \lambda x) \\ &= (A_1 + A_2) \cos \lambda x + (iA_2 - iA_1) \sin \lambda x \\ &= B_1 \cos \lambda x + B_2 \sin \lambda x \quad ...(6.32) \end{aligned}$$

where B_1 and B_2 are new constants.

Again
$$\frac{1}{\alpha}\frac{1}{G(t)}\frac{\partial G(t)}{\partial t} = -\lambda^2$$

or
$$\frac{\partial G(t)}{G(t)} = -\alpha\lambda^2 \partial t$$

Integrating with respect to t, we get

$$\ln[G(t)] = -\alpha\lambda^2 t + A_3$$

or
$$G(t) = A_3 e^{-\alpha\lambda^2 t} \quad ...(6.33)$$

where A_3 is constant of integration.

Substituting these solutions in eqn. (6.29), we have

$$\theta = (B_1 \cos \lambda x + B_2 \sin \lambda x) A_3 e^{-\alpha\lambda^2 t} \quad ...(6.34)$$

Introducing the new constants C_1 and C_2 as

$$C_1 = B_1A_3 \text{ and } C_2 = B_2A_3$$

Then, $\theta = (C_1 \cos \lambda x + C_2 \sin \lambda x)e^{-\alpha\lambda^2 t}$ ...(6.35)

The three constants C_1, C_2 and λ are to be evaluated from initial and boundary conditions

(*i*) At $t = 0$, $\theta = \theta_i = T_i - T_\infty$

(*ii*) At $x = 0$, $\frac{\partial \theta}{\partial x} = 0$

(No heat transfer at mid-plane)

(*iii*) At the surfaces of the wall at $x = L$

$$-k\left(\frac{\partial \theta}{\partial x}\right)_{x=L} = h\,\theta_{x=L} = h\,(T_L - T_\infty)$$

or
$$\left(\frac{\partial \theta}{\partial x}\right)_{x=L} = -\frac{h}{k}\theta_{x=L}$$

Differentiating eqn. (6.35) with respect to x

$$\frac{\partial \theta}{\partial x} = \{-C_1 \sin \lambda x + C_2 \cos \lambda x\}\,\lambda e^{-\alpha\lambda^2 t} \quad ...(6.36)$$

Using boundary condition (*ii*) at $x = 0$

$$\begin{aligned} \left(\frac{\partial \theta}{\partial x}\right)_{x=0} &= \{-C_1 \sin \lambda x + C_2 \cos \lambda x\}_{x=0} \times \lambda e^{-\alpha\lambda^2 t} \\ &= 0 \\ &= \{-C_1 \sin \lambda(0) + C_2 \cos \lambda(0)\} = 0 \end{aligned}$$

It gives $C_2 = 0$

The eqn. (6.35) reduces to

$$\theta = C_1 e^{-\alpha\lambda^2 t} \cos \lambda x \quad ...(6.37)$$

Using boundary condition (*iii*), at $x = L$

$$\left(\frac{\partial \theta}{\partial x}\right)_{x=L} = -C_1 e^{-\alpha\lambda^2 t}\,\lambda \sin \lambda L$$

and
$$\theta_L = C_1 e^{-\alpha\lambda^2 t} \cos \lambda L$$

Therefore,

$$-C_1 e^{-\alpha\lambda^2 t}\,\lambda \sin \lambda L = -\frac{h}{k} C_1 e^{-\alpha\lambda^2 t} \cos \lambda L$$

or
$$\lambda \sin \lambda L = \frac{h}{k} \cos \lambda L$$

or
$$\cot \lambda L = \frac{\lambda k}{h}$$

or
$$\cot \lambda L = \frac{\lambda L}{hL/k} = \frac{\lambda L}{Bi} \quad ...(6.38)$$

Equation (6.38) is a transcedental equation and it has an infinite number of roots. The value of root λ can be obtained by plotting cot λL and λL/Bi against λL as shown in Fig. 6.26. From the intersections of the two functions as many value of λ as $\lambda_1, \lambda_2, \lambda_3$, ... etc. can be determined. The equation cot λL = λL/Bi is satisfied for an infinite succession of values of λL so that for a given λ, the equation defines the value of λ. This succession of values of λ called *eigen values*, will be denoted by λ_n, which depend on Biot number.

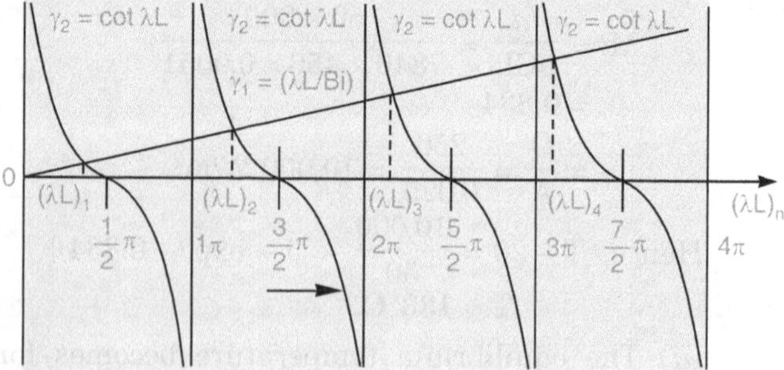

Fig. 6.26. Graphical solution of the transcendental equation $\cot \lambda L = \frac{\lambda L}{Bi}$

The temperature distribution becomes the following series solution.

$$\theta = \sum_{n=1}^{n=\infty} C_n e^{-\alpha\lambda^2 t} \cos \lambda_n x \quad ...(6.39)$$

Using initial condition (*i*), at $t = 0$

$$\theta = \sum_{n=1}^{n=\infty} C_n \cos \lambda_n x \quad ...(6.40)$$

where n is simple integers, 1, 2, 3, ...,

For $n = 1$, at $x = L$ *i.e.*, at outer surface

$$\theta_s = C_1 \cos \lambda_1 L \quad ...(6.41)$$

At centre ($x = 0$)

$$\theta_c = C_1$$

Hence the non dimensional temperature distribution becomes

$$\frac{\theta_s}{\theta_c} = \cos \lambda_1 L \quad ...(6.42)$$

For internal temperature gradients within 5% (negligible), we get

$$\frac{\theta_s}{\theta_c} \geq (1 - 0.05)$$

or $$\frac{\theta_s}{\theta_c} \geq 0.95 \quad ...(6.43)$$

or $\cos \lambda L \geq 0.95$

or $\lambda L \geq 0.3175$ radian

Substituting in eqn. (6.38), we get

$$\cot(0.3175) = \frac{0.3175}{\text{Bi}}$$

or $\text{Bi} \approx 0.1$

Thus when Biot number is less than or equal to 0.1, the internal temperature gradients within the solid can be neglected and the lumped system analysis can be used for unsteady state heat conduction problems.

Further, the constant C_n is determined for each value of λ_n *i.e.*, 1, 2, 3, ... In general

$$C_n = \frac{2\,\theta_i \sin \lambda_n L}{\lambda_n L + \sin \lambda_n L \cos \lambda_n L} \quad ...(6.44)$$

For convenience introducing $\xi_n = \lambda_n L$, where the discrete values of ξ_n are positive roots of the transcendental equation (6.38) in form

$$\xi_n \tan \xi_n = \text{Bi} \quad ...(6.45)$$

The temperature distribution in the slab is finally obtained as

$$\frac{\theta}{\theta_i} = \sum_{n=1}^{n=\infty} e^{-\xi_n^2 (\alpha t/L^2)} \left(\frac{2 \sin \xi_n}{\xi_n + \sin \xi_n \cos \xi_n} \right) \times \cos (\xi_n x/L)$$

where $\theta_i = T_i - T_\infty$;

Using $\text{Fo} = \frac{\alpha t}{L^2}$, Fourier number ; then

$$\frac{\theta}{\theta_i} = \sum_{n=1}^{n=\infty} e^{-\xi_n^2 \text{Fo}} \frac{2 \sin \xi_n \cos (\xi_n x / L)}{\xi_n + \sin \xi_n \cos \xi_n} \quad ...(6.46)$$

At $x = 0$, $\theta = \theta_c$, the temperature distribution at the centre ;

$$\frac{\theta_c}{\theta_i} = \sum_{n=1}^{n=\infty} e^{-\xi_n^2 \text{Fo}} \frac{2 \sin \xi_n}{\xi_n + \sin \xi_n \cos \xi_n} \quad ...(6.47)$$

At $x = L$, $\theta = \theta_L$, the surface temperature distribution

$$\frac{\theta_L}{\theta_i} = \sum_{n=1}^{n=\infty} e^{-\xi_n^2 \text{Fo}} \frac{2 \sin \xi_n \cos \xi_n}{\xi_n + \sin \xi_n \cos \xi_n} \quad ...(6.48)$$

The results using eqns. (6.46), (6.47) and (6.48) have been calculated for different cases and plotted in the form of charts for quick reference by Gröber and Erk, Gurney-Lurie, Heisler and others. Heisler chart for θ_c/θ_i is given in Fig. 6.31(*a*).

Using temperature distribution in eqn. (6.46), the cummulative heat loss from an infinite plate is expressed.

$$\frac{Q}{\rho VC(T_i - T_\infty)} = \frac{Q}{Q_i}$$

$$= \sum_{n=1}^{n=\infty} \frac{2 \sin^2 \xi_n}{\xi_n^2 + \xi_n \sin \xi_n \cos \xi_n} \times (1 - e^{-\xi_n^2 \text{Fo}}) \quad ...(6.49)$$

6.2.2. Infinite Cylinder and Sphere with Convective Boundaries

Similar to the transient temperature distribution and heat flow in an infinite plate, the transient temperature distribution in an infinite cylinder of radius r_o exposed to convection boundary can be obtained. The temperature distribution in an infinite cylinder is given as

$$\frac{\theta}{\theta_i} = \frac{T(r,t) - T_\infty}{T_i - T_\infty}$$

$$= 2 \sum_{n=1}^{\infty} \frac{e^{-\lambda_n^2 \alpha t}}{\lambda_n r_o} \times \frac{J_0(\lambda_n r) J_1(\lambda_n r_o)}{J_0^2(\lambda_n r) + J_1^2(\lambda_n r_o)} \quad ...(6.50)$$

where J_0 and J_1 are zeroth and first order Bessel's function of first kind.

The centre line temperature $J_0(0) = 1$

$$\frac{\theta_c}{\theta_i} = \frac{T(0,t) - T_\infty}{T_i - T_\infty} = 2 \sum_{n=1}^{\infty} \frac{e^{-\lambda_n^2 \alpha t}}{\lambda_n r_o} \times \frac{J_1(\lambda_n r_o)}{1 + J_1^2(\lambda_n r_o)} \quad ...(6.51)$$

Heisler charts for centre line temperature θ_c/θ_i and the position temperature θ/θ_i for a long cylinder are given in Figs. 6.32 (*a*) and 6.32(*b*), respectively. With the use of these charts, the time temperature history at any location in the cylinder can be obtained.

For sphere of radius r_o, the similar relations can be obtained, derived by Schneider. Figs. 6.33 (*a*) and 6.33 (*b*) show Heisler chart for θ_c/θ_i and θ/θ_i to determine the temperature time history at any location in a sphere.

6.2.3. One Term Approximation

The one-dimensional transient heat conduction problems can be solved exactly for any of the three geometries of plane wall, cylinder or sphere. But the solutions involve approximation of an infinite series, which are difficult to deal with. However, the terms in the solutions converge rapidly with increasing time. If Fourier number is greater than 0.2, then the infinite series solution can be reduced to **one term solution** *i.e.*, keeping the first term and ignoring the all other terms in the series. It results into an error less than 2%. Thus it is convenient to express the solution using one term approximation for Fo ≥ 0.2 ; given as

Plane wall :

$$\frac{\theta}{\theta_i} = \frac{T(x,t) - T_\infty}{T_i - T_\infty} = C_1 e^{-\xi_1^2 \text{Fo}} \cos(\xi_1 x/L) \quad ...(6.52)$$

With $C_1 = \dfrac{2\sin\xi_1}{\xi_1 + \sin\xi_1 \cos\xi_1}$

Similarly for

$$\text{Cylinder}: \frac{\theta}{\theta_i} = \frac{T(r,t) - T_\infty}{T_i - T_\infty} = C_1 e^{-\xi_1^2 \text{Fo}} J_0(\xi r/r_o) \quad ...(6.53)$$

$$\text{Sphere}: \frac{\theta}{\theta_i} = \frac{T(r,t) - T_\infty}{T_i - T_\infty} = C_1 e^{-\xi_1^2 \text{Fo}} \frac{\sin(\xi_1 r/r_o)}{\xi_1 (r/r_o)} \quad ...(6.54)$$

where C_1, ξ_1 are functions of Biot number only and their values are presented in Table B-5 of Appendix B, against Biot number for all three geometries. The function J_0 is the zeroth order Bessel function of first kind, its value can be obtained from Table B-6 of Appendix B.

Note: The characteristic length in defining the Biot number must be considered as half thickness L for a plane wall and radius r_o for long cylinder and sphere instead $\dfrac{V}{A}$ as done in lumped heat capacity method.

At the centre of the plane wall, cylinder and sphere

$$x = r = 0, \cos(0) = 1, J(0) = 1 \text{ and } \frac{\sin(x)}{x} = 1$$

The above relations are simplified to

$$\text{Plane wall } (x = 0), \quad \frac{\theta_c}{\theta_i} = \frac{T(0,t) - T_\infty}{T_i - T_\infty} = C_1 e^{-\xi_1^2 \text{Fo}} \quad ...(6.55)$$

Cylinder ($r = 0$),

$$\frac{\theta_c}{\theta_i} = \frac{T(0,t) - T_\infty}{T_i - T_\infty} = C_1 e^{-\xi_1^2 \text{Fo}} \quad ...(6.56)$$

Sphere ($r = 0$),

$$\frac{\theta_c}{\theta_i} = \frac{T(0,t) - T_\infty}{T_i - T_\infty} = C_1 e^{-\xi_1^2 \text{Fo}} \quad ...(6.57)$$

Once the Biot number is known, the above relations can be used to obtain the temperature anywhere in the medium.

The fraction heat transfer can also be determined from the following relations, derived from one term approximations.

$$\text{Plane wall}: \frac{Q}{Q_i} = 1 - \frac{\theta_c}{\theta_i}\frac{\sin\xi_1}{\xi_1} \quad ...(6.58)$$

$$\text{Cylinder}: \frac{Q}{Q_i} = 1 - 2\frac{\theta_c}{\theta_i}\frac{J_1(\xi_1)}{\xi_1} \quad ...(6.59)$$

$$\text{Sphere}: \frac{Q}{Q_i} = 1 - 3\frac{\theta_c}{\theta_i}\frac{\sin\xi_1 - \xi_1\cos\xi_1}{\xi_1^3} \quad ...(6.60)$$

where $Q_i = mC(T_i - T_\infty) = \rho VC(T_i - T_\infty)$ ...(6.61)

Example 6.20. *In a material treatment process, a metallic sphere 10 mm in diameter is initially at 400°C is suddenly subjected to two step cooling process.*

Step 1. *Cooling in stagnant air at 20°C with convective coefficient of 10 W/m².K for a period, until the centre temperature reaches a temperature of 335°C.*

Step 2. *After sphere attains 335°C, it is cooled to 50°C in a well stirred water bath at 20°C, with convective coefficient of 6000 W/m².K.*

The thermophysical properties of material are

ρ = 3000 kg/m³, k = 20 W/m.K,

C = 1000 J/kg.K, α = 6.66 × 10⁻⁶ m²/s.

(i) Calculate the time required for step 1 for cooling process to be completed.

(ii) Calculate the time required during step 2 of the process for centre of sphere to cool from 335°C to 50°C.

Solution

Given : The material treatment of a metallic sphere

$D = 10$ mm

$r_o = 5$ mm $= 0.005$ m

$\rho = 3000$ kg/m³,

$k = 20$ W/m.K

$\alpha = 6.66 \times 10^{-6}$ m²/s,

$C = 1000$ J/kg.K

For step 1. $T_i = 400°C$,

$T_a = 335°C$

$h_a = 10\ W/m^2.K$,

$T_\infty = 20°C$

For step 2. $T_i = 335°C$,

$T_w = 50°C$

$T_\infty = 20°C$,

$h_w = 6000\ W/m^2.K$.

To find :

(*i*) Time required for cooling process in step 1.

(*ii*) Time required for cooling process in step 2.

Assumptions :

1. One-dimensional conduction in radial direction only.
2. No radiation heat exchange in either step of cooling.
3. Constant properties.

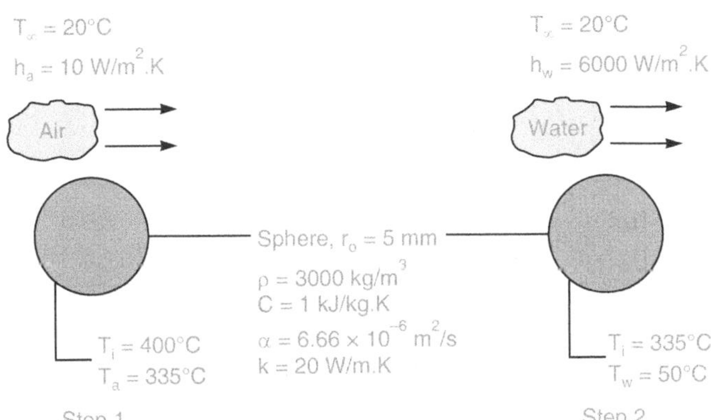

Fig. 6.27. Schematic for example 6.20

Analysis : The characteristic length of the sphere

$$\delta = \frac{r_o}{3} = \frac{0.005}{3}\ \text{m}.$$

Step 1 : Checking the validity of the lumped heat capacity system in air

$$\text{Bi} = \frac{h_1\delta}{k} = \frac{10 \times 0.005}{3 \times 20} = 8.33 \times 10^{-4}$$

It is well within the lumped heat capacity method.

Therefore, using eqn. (6.10)

$$\frac{T_a - T_\infty}{T_i - T_\infty} = \exp\left[-\frac{h_1 t_1}{\rho\delta C}\right]$$

or
$$t_1 = -\frac{\rho r_o C}{3h}\ln\left[\frac{T_a - T_\infty}{T_i - T_\infty}\right]$$

$$= -\frac{3000 \times 0.005 \times 1000}{3 \times 10} \times \ln\left[\frac{335 - 20}{400 - 20}\right]$$

t_1 = **93.8 s.** Ans.

Step 2 : Checking the validity of the lumped heat capacity system in water

$$\text{Bi} = \frac{h_2\delta}{k} = \frac{6000 \times 0.005}{3 \times 20} = 0.5$$

It is greater than 0.1, thus the lumped heat capacity method is not appropriate to use. However, using one term approximation for centre temperature to reach 50°C from 335°C. For this method, using eqn. (6.57) with

$$\text{Bi} = \frac{h_2 r_o}{k} = \frac{6000 \times 0.005}{20} = 1.5$$

From Table B-5

$C_1 = 1.376$,

$\xi_1 = 1.80$ rad.

Thus,
$$\frac{T_w - T_\infty}{T_i - T_\infty} = C_1 e^{-\xi_1^2 \text{Fo}}$$

$$\frac{1}{1.376} \times \left[\frac{50 - 20}{335 - 20}\right] = e^{-(1.8)^2 \text{Fo}}$$

or
$$\text{Fo} = -\frac{1}{3.24} \times \ln\left[\frac{30}{1.376 \times 315}\right] = 0.824$$

and
$$t_2 = \frac{r_o^2 \text{Fo}}{\alpha} = \frac{(0.005)^2 \times 0.824}{6.66 \times 10^{-6}}$$

= **3.09 s.** Ans.

Note that with Fo = 0.824, the use of one term approximation is justified.

Example 6.21. *A rocket engine nozzle is made of high temperature steel 0.64 cm thick, k = 29 W/m.K, $\alpha = 6.39 \times 10^{-6}\ m^2/s$. The flame side surface film coefficient is 8370 $W/m^2.K$. The flame temperature is constant at 2200°C. If the nozzle is initially at uniform temperature of 25°C.*

What should be the duration of combustion in order to limit the operating temperature of steel to 1100°C ?

Solution

Given : Rocket engine nozzle

L = 0.64 cm = 0.0064 m,

$k = 29$ W/m.K

$\alpha = 6.39 \times 10^{-6}\ m^2/s$,

$h = 8370\ W/m^2.K$

$T_\infty = 2200°C$,

$T_i = 25°C$

$T_L = 1100°C$.

To find : Duration of combustion.

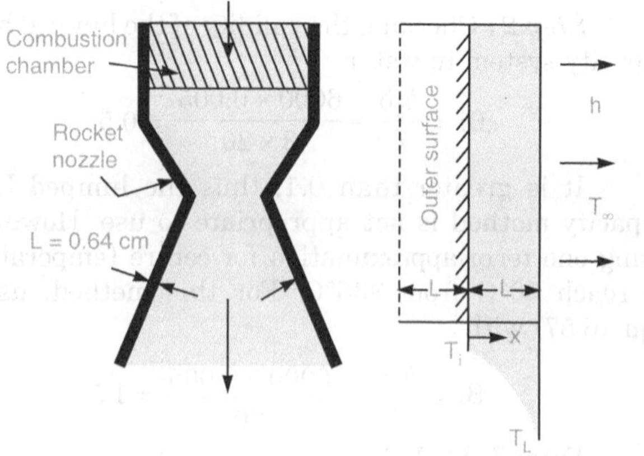

Fig. 6.28. Schematic of a rocket nozzle

Assumptions :

1. Wall thickness of nozzle is very small compared to its diameter, the nozzle wall can be modelled as infinite plane wall of thickness 0.64 cm.

2. The outer surface of the nozzle as insulated, against the heat flow.

3. Uniform heat transfer coefficient.

4. Discarding any radiation heat transfer.

Analysis : The Biot number of wall insulated of thickness L

$$\text{Bi} = \frac{h\text{L}}{k} = \frac{8370 \times 0.0064}{29} = 1.847$$

which is much higher than 0.1, and thus the lumped system analysis cannot be applicable. The Hiesler charts or analytical method can be used for temperature distribution.

Using analytical approach given by eqn. (6.46)

$$\frac{\theta}{\theta_i} = \sum_{n=1}^{n=\infty} e^{-\xi_n^2 \text{Fo}} \cdot \frac{2 \sin \xi_n}{\xi_n + \sin \xi_n \cos \xi_n} \times \cos (\xi_n x/\text{L})$$

At $x = \text{L}, \quad \dfrac{\theta_\text{L}}{\theta_i}$

$$\frac{\text{T}_\text{L} - \text{T}_\infty}{\text{T}_i - \text{T}_\infty} = \sum_{n=1}^{n=\infty} e^{-\xi_n^2 .\text{Fo}} \cdot \frac{2 \sin \xi_n \cos \xi_n}{\xi_n + \sin \xi_n \cos \xi_n}$$

Calculating each term separately

$$\frac{\text{T}_\text{L} - \text{T}_\infty}{\text{T}_i - \text{T}_\infty} = \frac{1100 - 2200}{25 - 2200} = 0.5057$$

$$\frac{1}{\text{Bi}} = \frac{1}{1.847} = 0.5414$$

$$\text{Bi} = \xi_n \tan \xi_n$$

Assuming

$$\xi = \lambda\text{L} = 45° = \frac{\pi}{4}$$

$$\tan \xi = \frac{\text{Bi}}{\xi} = \frac{1.847 \times 4}{\pi} = 2.351$$

$$\xi = 66.96°\text{C}$$

which is greater than assumed value.

Let $\quad \xi = \dfrac{\pi}{3} = 60°$

$$\tan \xi = \frac{3 \times 1.843}{\pi} = 1.7637 \rightarrow \xi = 60.4°\text{C}$$

agrees with assumption.

Taking $n = 1$

$$0.5057 = e^{-\xi_1^2 \text{Fo}} \frac{\sin 2\xi_1}{\xi_1 + \sin \xi_1 \cos \xi_1}$$

$$= e^{-\left(\frac{\pi}{3}\right)^2 \times \text{Fo}} \times \frac{\sin (120°)}{\frac{\pi}{3} + \sin (60°) \cos (60°)}$$

$$= 0.585\, e^{-\left(\frac{\pi}{3}\right)^2 \times \text{Fo}}$$

or $$\ln \left(\frac{0.5057}{0.585}\right) = -\left(\frac{\pi}{3}\right)^2 \times \text{Fo}$$

or $$\text{Fo} = \frac{1.397}{1.0966} = 0.1274$$

or $$t = \frac{\text{Fo L}^2}{\alpha} = \frac{0.1274 \times (0.0064)^2}{6.39 \times 10^{-6}}$$

$$= 0.8175 \text{ s} \simeq 0.825$$

The combustion must complete within **0.82 s.** **Ans.**

Example 6.22. *An egg can be approximated as a sphere, 5 cm in diameter, with thermophysical properties k = 0.6 W/m.K, α = 0.14 × 10⁻⁶ m²/s. The egg is taken from a refrigerator at 2°C and is dropped into boiling water, where the convection heat transfer coefficient is estimated as 1200 W/m².K. Calculate time required to reach the centre temperature of the egg to 75°C.*

Solution

Given : An egg as spherical body :

$\text{D} = 5$ cm,

$r_o = 2.5 \text{ cm} = 0.025$ m

$k = 0.6$ W/m.K

$\alpha = 0.14 \times 10^{-6}$ m²/s

$\text{T}_i = 2°\text{C}$,

$\text{T}_c = 75°\text{C}$,

$h = 1200$ W/m².K.

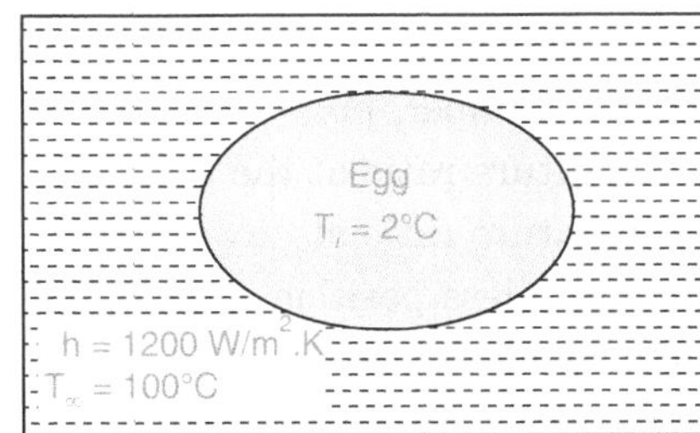

Fig. 6.29. Schematic of egg in boiling water

To find : Time to reach the centre temperature of egg to 75°C.

Assumptions :

1. Boiling water temperature at atmospheric conditions as $T_\infty = 100°C$.

2. Temperature variation in the egg in radial direction only with time.

3. Uniform heat transfer coefficient.

4. Constant properties of egg.

Analysis : Biot number

$$Bi = \frac{hr_o}{k} = \frac{1200 \times 0.025}{0.6} = 50$$

which much greater than 0.1, thus the lumped system analysis is not applicable. Heisler charts or one term solution can be used.

Using one term solution.

For sphere with

$Bi = 50$, $\xi_1 = 3.0788$

and $C_1 = 1.9662$ (From Table B-5)

Substituting these values in eqn. (6.57) and solving for Fo ;

$$\frac{\theta_c}{\theta_i} = \frac{T_c - T_\infty}{T_i - T_\infty} = C_1 e^{-\xi_1^2 Fo}$$

$$\frac{75 - 100}{2 - 100} = 1.9662 \times e^{-(3.0788)^2 Fo}$$

or $$\frac{25}{98} \times \frac{1}{1.9662} = e^{-9.48\ Fo}$$

or $$\ln(0.12974) = -9.48\ Fo$$

It gives $Fo = 0.2154$

Fo is greater than 0.2 and thus the use of one term solution is justified,

$$Fo = \frac{\alpha t}{r_o^2}$$

or $$t = \frac{0.2154 \times (0.025)^2}{0.14 \times 10^{-6}}$$

$$= 961.6\ s = \mathbf{16\ min.}$$ Ans.

It will take 16 min. for centre of egg to reach to 75°C from 2°C.

Example 6.23. *A long cylindrical shaft, 20 cm in diameter is made of steel (k = 14.9 W/m.K), ρ = 7900 kg/m^3, C = 477 J/kg.K and $\alpha = 3.95 \times 10^{-6}$ m^2/s. It comes out of an oven at a uniform temperature of 600°C. The shaft is then allowed to cool slowly in an environment at 200°C with an average heat transfer coefficient of 80 W/m^2.K. Calculate the temeprature at the centre of the shaft, 45 min after the start of cooling process. Also calculate the heat transfered per unit length of the shaft during this period.*

Solution

Given : D = 20 cm

r_o = 10 cm = 0.1 m

k = 14.9 W/m.K, ρ = 7900 kg/m^3

C = 477 W/m.K, $\alpha = 3.95 \times 10^{-6}$ m^2/s

T_i = 600°C, T_∞ = 200°C

h = 80 W/m^2.K, t = 45 min = 2700 s.

To find :

(*i*) Temperature at the centre of the shaft.

(*ii*) Heat transfer from 1 m length of shaft.

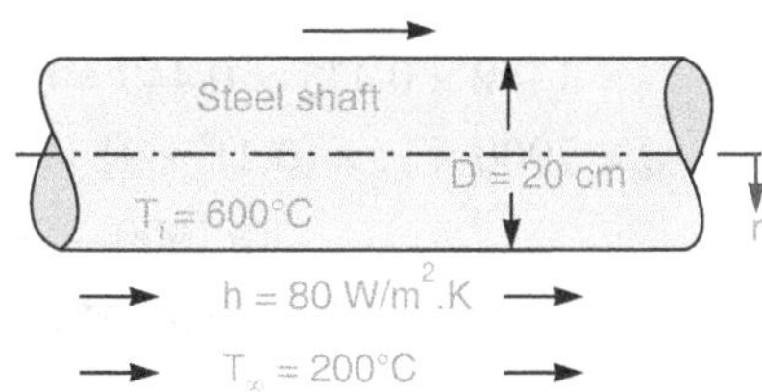

Fig. 6.30. Schematic for shaft exposed to convection ambient

Assumptions :

1. Shaft as an infinite cylinder.

2. Heat conduction in the shaft in radial direction only with time.

3. Uniform heat transfer coefficient.

4. No radiation heat transfer.

Analysis : (*i*) The Biot number for shaft

$$Bi = \frac{hr_o}{k} = \frac{80 \times 0.1}{14.9} = 0.537$$

which is larger than 0.1 and hence the lumped system analysis is not applicable. Either Heisler charts or one term solution may be used.

$$Fo = \frac{\alpha t}{r_o^2} = \frac{3.95 \times 10^{-6} \times 2700}{(0.1)^2} = 1.0665$$

Fo is greater than 0.2, thus using one term solution for cylinder :

At $\quad Bi = 0.537, \quad C_1 = 1.122$

$\quad \xi_1 = 0.970$

Substituting the values in eqn. (6.56)

$$\frac{T_c - T_\infty}{T_i - T_\infty} = C_1 e^{-\xi_1^2 Fo}$$

$$\frac{T_c - 200}{600 - 200} = 1.122 \times e^{-(0.970)^2 \times 1.0665} = 0.411$$

or $\quad T_c = 200 + 400 \times 0.411 =$ **364.5°C.** Ans.

The centre temperature of shaft will reach 364.5°C after 45 min. Ans.

(*ii*) Heat transfer from shaft can be obtained by eqn. (6.59)

$$\frac{Q}{Q_i} = 1 - 2 \times \frac{T_c - T_\infty}{T_i - T_\infty} \frac{J_1(\xi_1)}{\xi_1}$$

where $J_1(\xi_1) = J_1(0.970) = 0.430$

$$\frac{Q}{\rho VC(T_i - T_\infty)} = 1 - 2\left(\frac{T_c - T_\infty}{T_i - T_\infty}\right) \times \frac{0.430}{0.970}$$

$$= 1 - 2 \times 0.411 \times 0.443 = 0.635$$

$$Q = 7900 \times [\pi \times (0.1)^2 \times 1] \times 477 \times (600 - 200) \times 0.635$$

$= 30098500$ J = **30.09 MJ.** Ans.

6.3. TRANSIENT TEMPERATURE CHARTS : HEISLER AND GRÖBER CHARTS

When the internal temperature gradients are large, lumped heat capacity system analysis becomes unsuitable for the analysis of transient heat conduction problems. In such situation the Heisler and Gröber charts are widely used for determination of

1. Centreline temperature.
2. Position temperature.
3. The heat transfer.

To obtain the required value of unknowns, the various dimensionless parameters required are

1. Biot Number, Bi.
2. Fourier Number, Fo.
3. Temperature ratio at the centre.
4. Temperature ratio at any position.
5. Dimensionless position.
6. Dimensionless heat transfer.

For infinite plane wall, long cylinder and sphere, there are three graphs, first one is used to obtain centreline temperature, second one for position temperature and third for determination of heat flow in the geometry.

6.3.1. Transient Temperature Charts for Slab

Consider a slab (*i.e.*, a plane) of thickness 2L, confined to the region $-L \le x \le L$. The slab initially at a temperature T_i, is suddenly exposed to convection environment (for $t > 0$) with a heat transfer coefficient h, on its both boundary surfaces. The heat flows from both surfaces inward. Due to symmetry of problem, only half region $0 \le x \le L$ is considered. The dimensionless parameters for a slab can be expressed as :

1. Biot Number, $Bi = \frac{hL}{k}$
2. Fourier Number, $Fo = \frac{\alpha t}{L^2}$
3. Temperature ratio at the centre,
$$\frac{\theta_c}{\theta_i} = \frac{T_c - T_\infty}{T_i - T_\infty}$$
4. Temperature ratio at any position,
$$\frac{\theta}{\theta_c} = \frac{T(x, t) - T_\infty}{T_c - T_\infty}$$
5. Dimensionless position $= \frac{x}{L}$
6. Dimensionless heat transfer $= \frac{Q}{Q_i}$

where, L = half thickness of a slab, in metres.

x = position in the slab, measured from centre, where temperature is required, m.

T_c = centreline temperature of the slab, °C.

$T(x, t)$ = position temperature in the slab, °C.

Q_i = initial internal energy content in the slab $= \rho(A\ 2L)\ C\ (T_i - T_\infty)$ Joules.

Q = total amount of energy lost by plate during time t.

α = thermal diffusivity of the material, m²/s.

k = thermal conductivity of the material, W/m.K.

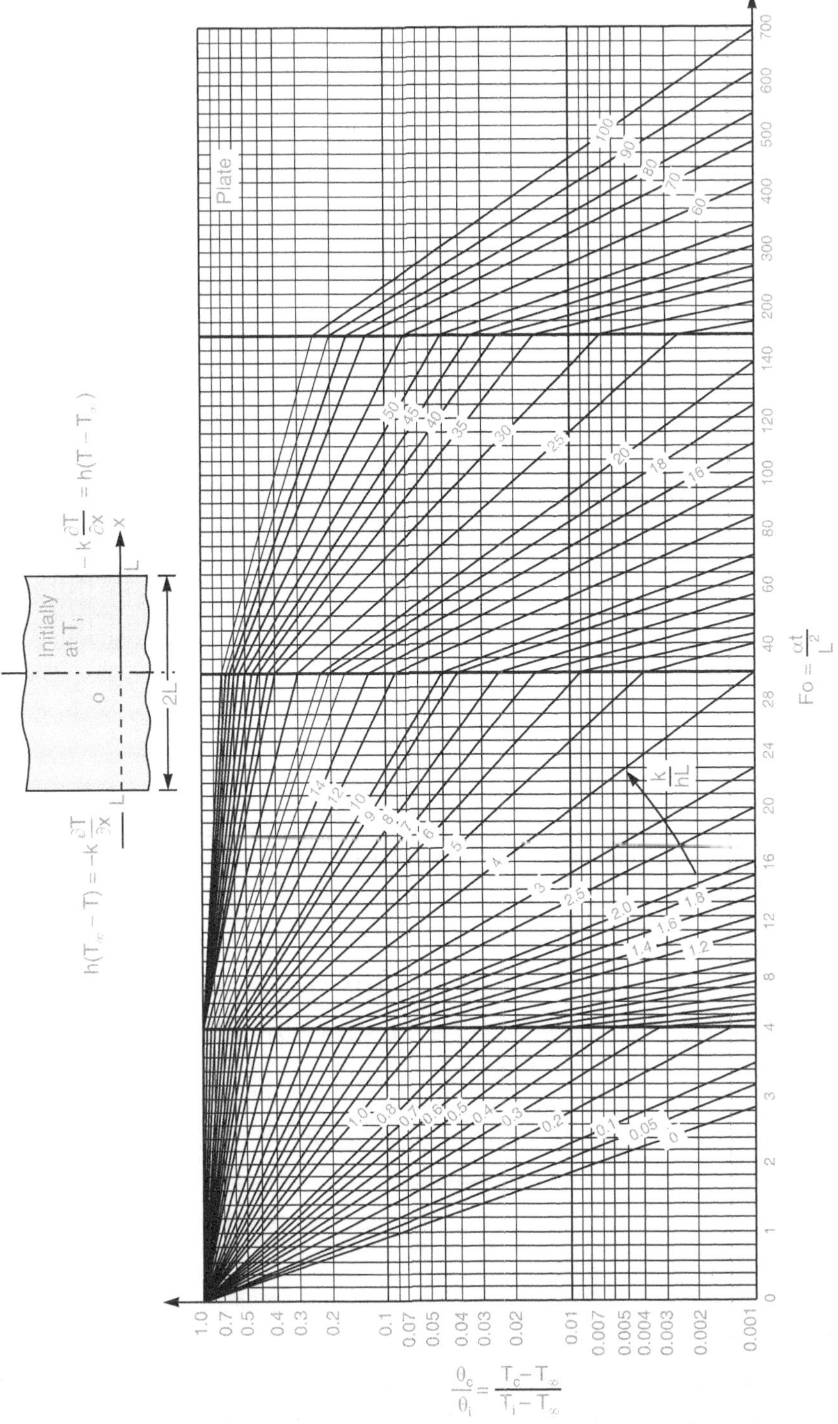

Fig 6.31 (*a*) Centreline temperature for an infinite plate of thickness 2L

The temperature at any position x from the mid-plane can be obtained from position correction temperature chart, Fig. 6.31 (*b*)

$$\frac{\theta}{\theta_i} = \frac{\theta_c}{\theta_i} \times \frac{\theta}{\theta_c} = \frac{T - T_\infty}{T_i - T_\infty}.$$

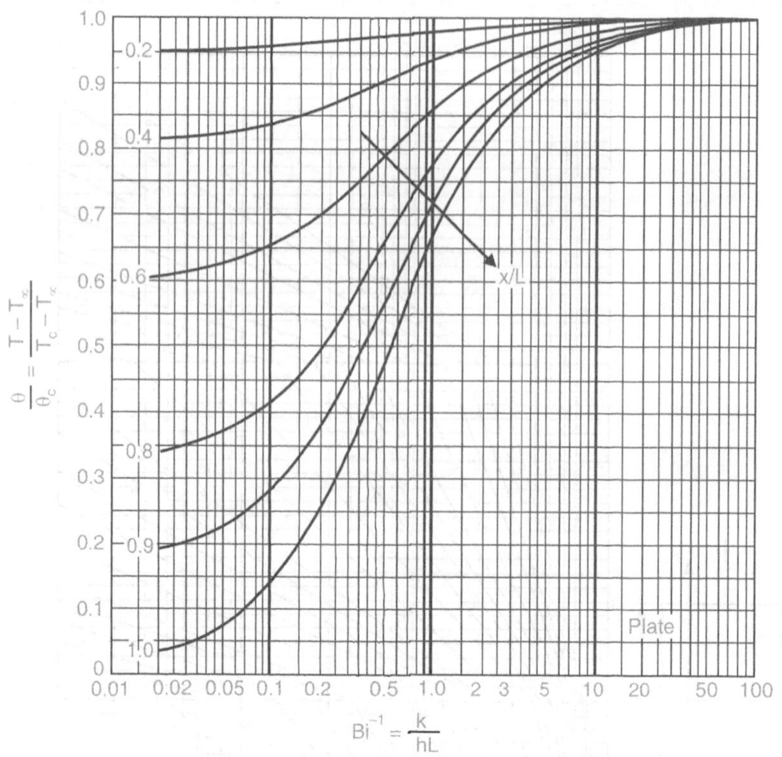

Fig. 6.31 (*b*) Position correction for temperature as a function of centre temperature in an infinite cylinder of radius r_o

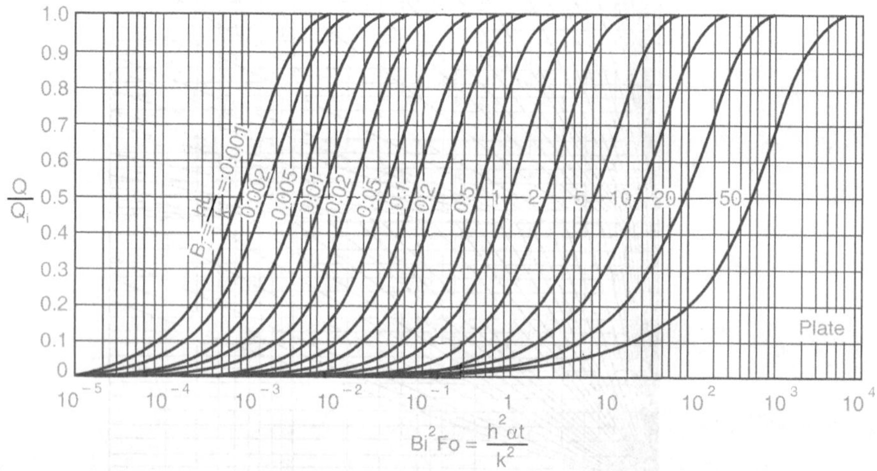

Fig. 6.31 (*c*) Dimensionless heat loss for an infinite plate of thickness 2L

6.3.2. Transient Temperature Charts for Long Cylinder and Sphere

Consider a long cylinder or a sphere of radius r_o, initially at temperature T_i is suddenly subjected to convection environment (for $t > 0$) with heat transfer coefficient h and fluid temperature T_∞. The various dimensionless parameters required for Heisler charts solution are

1. Biot number, $\text{Bi} = \dfrac{hr_o}{k}$

2. Fourier number, $\text{Fo} = \dfrac{\alpha t}{r_o^2}$

3. Temperature ratio at the centre, $\dfrac{\theta_c}{\theta_i} = \dfrac{T_c - T_\infty}{T_i - T_\infty}$

4. Temperature ratio at any position, $\dfrac{\theta}{\theta_c} = \dfrac{T(r, t) - T_\infty}{T_c - T_\infty}$

5. Dimensionless radial position, $\dfrac{r}{r_o}$

6. Dimensionless heat transfer, $\dfrac{Q}{Q_i}$;

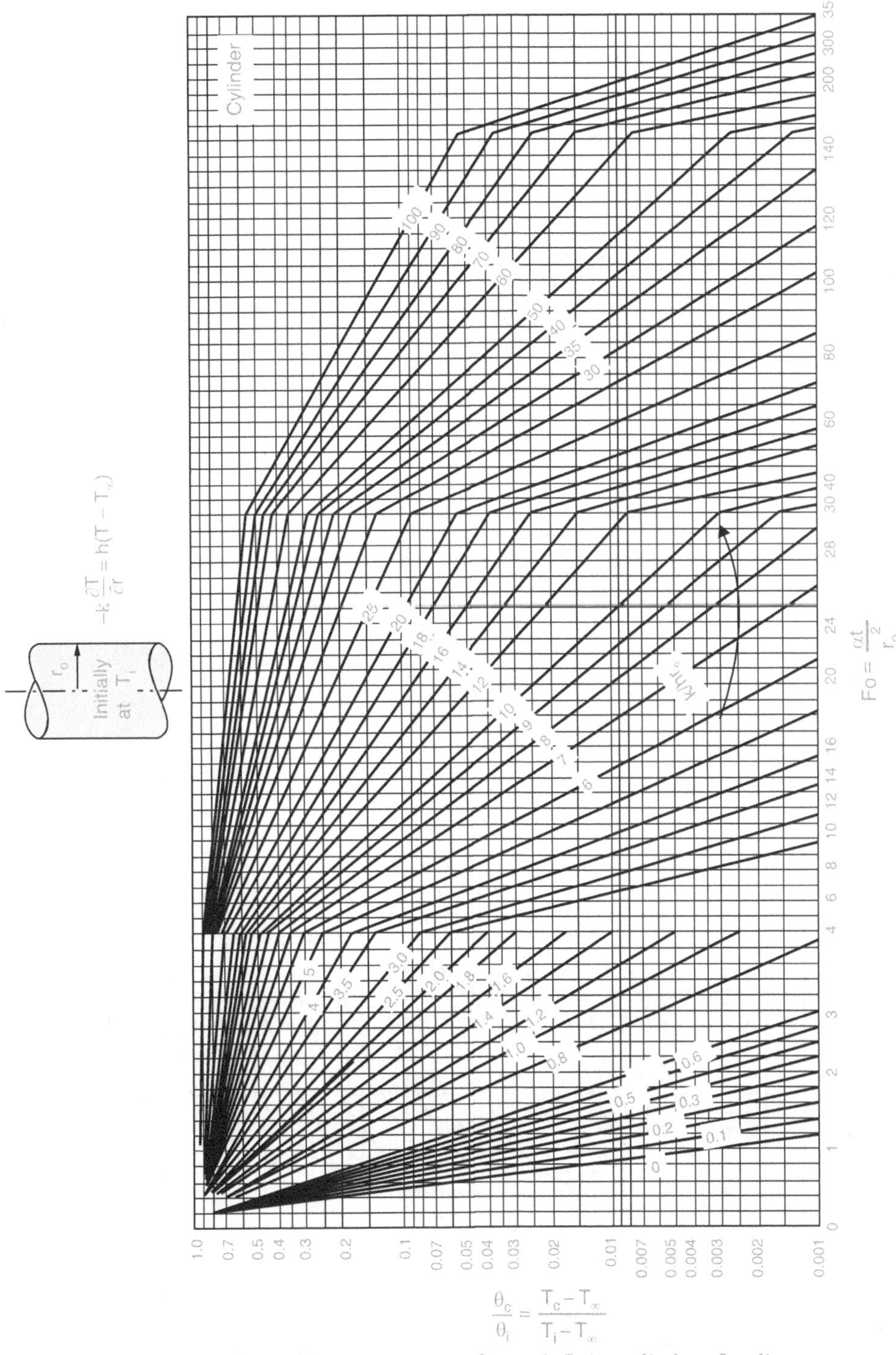

Fig. 6.32. (*a*) Centreline temperature for an infinite cylinder of radius r_o, subjected to convection at its boundary surface

where r_o = radius of cylinder or sphere

r = position radius in cylinder or sphere

T_c = centre temperature, °C

T_i = initial temperature, °C

$T(r, t)$ = position temperature, °C

Q = total amount of heat energy lost by body in time t, Joules

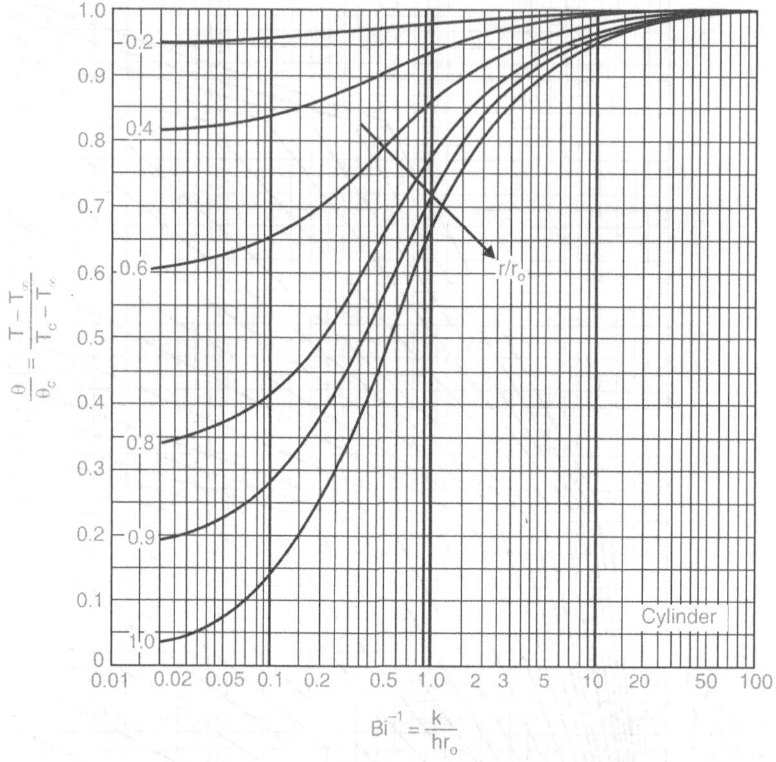

Fig. 6.32. (*b*) Position correction for temperature as a function of centre temperature in an infinite cylinder of radius r_o

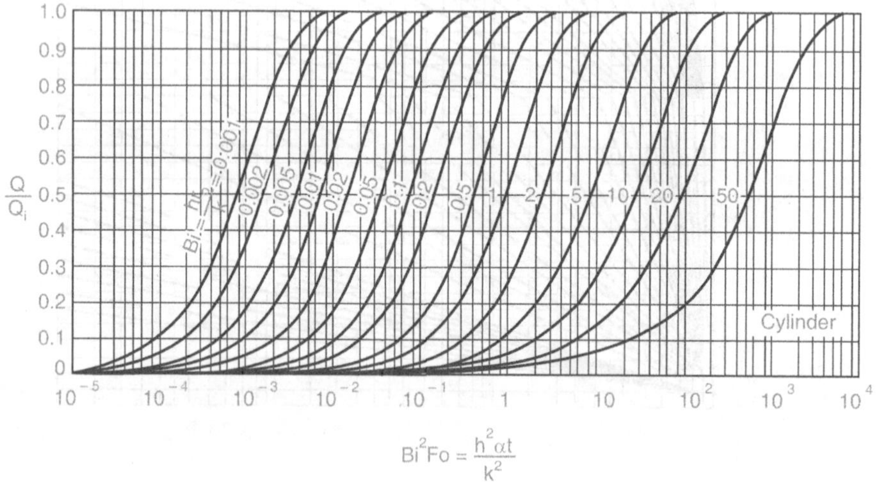

Fig. 6.32. (*c*) Dimensionless heat loss Q/Q_i for an infinite cylinder of radius r_o

Q_i = initial internal energy content of the body = $\rho VC(T_i - T_\infty)$, Joules

t = time, s

α = thermal diffusivity, m^2/s

k = thermal conductivity, W/m.K

h = heat transfer coefficient, $W/m^2.K$.

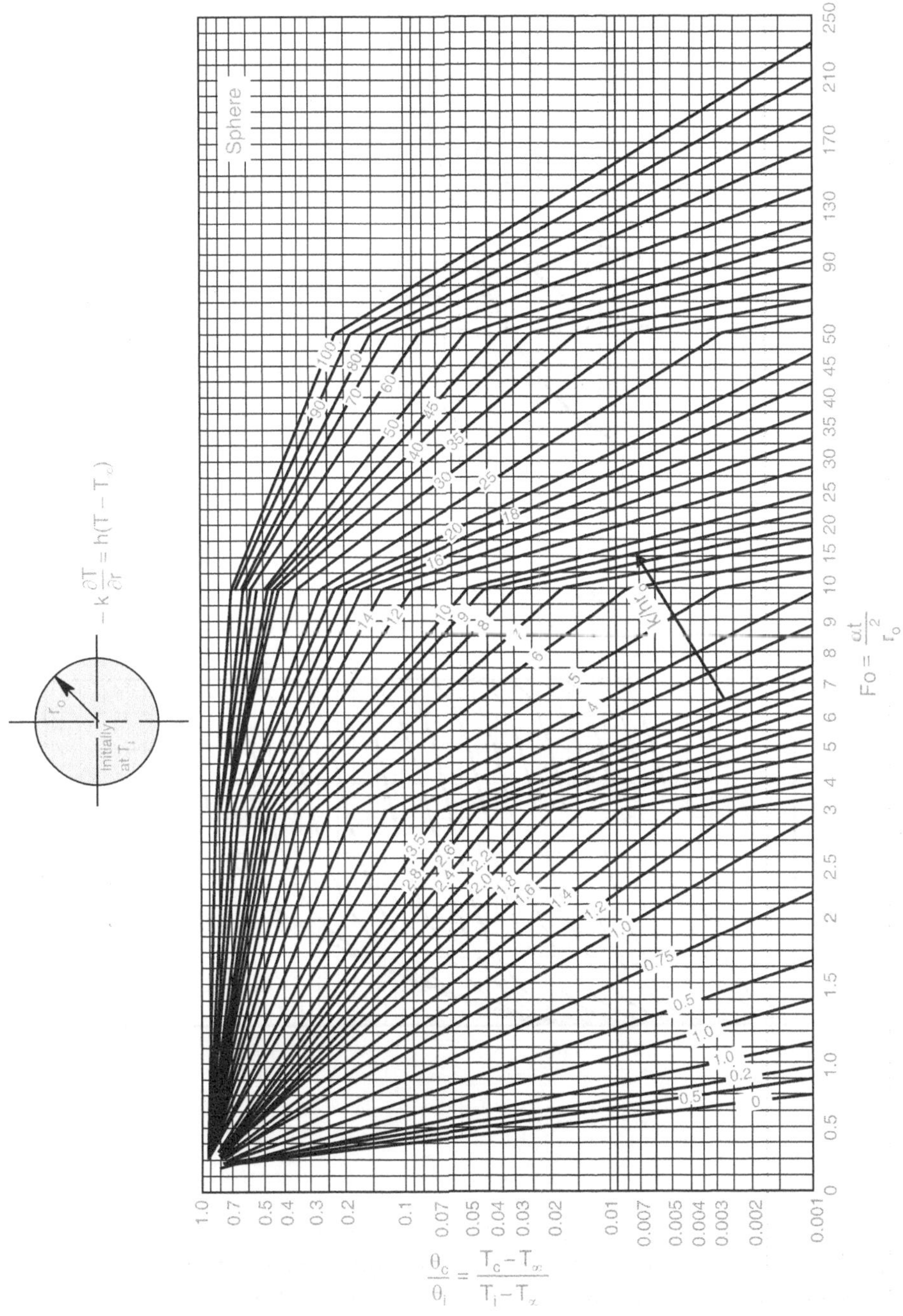

Fig. 6.33 (a) Centre temperature for a sphere of radius r_o, subjected to correction at the boundary surface

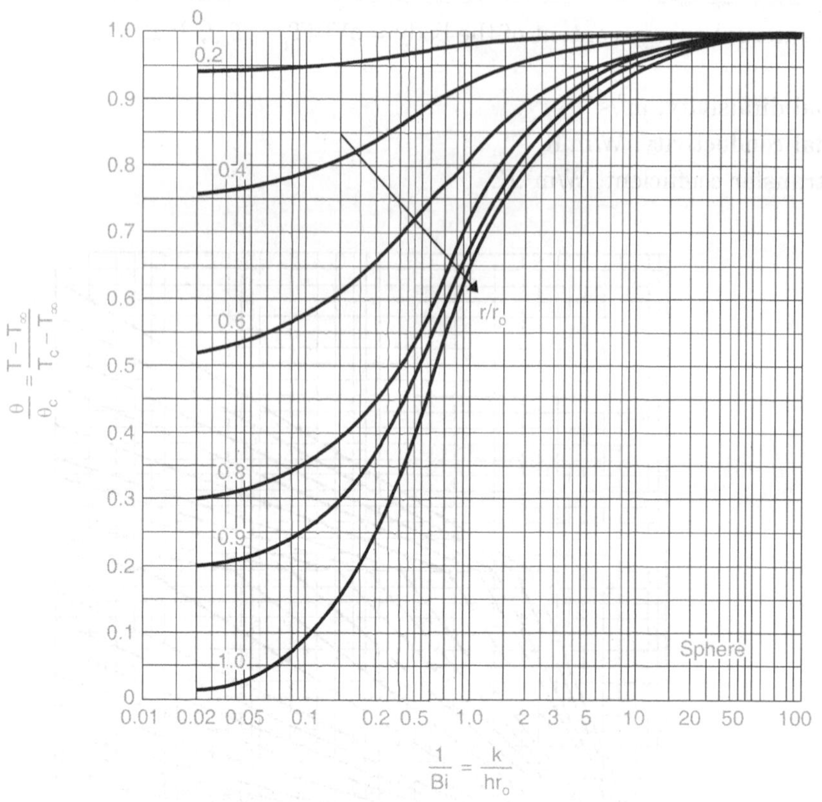

Fig. 6.33 (*b*) Position correction for temperature as a function of centre temperature for a sphere of radius r_o

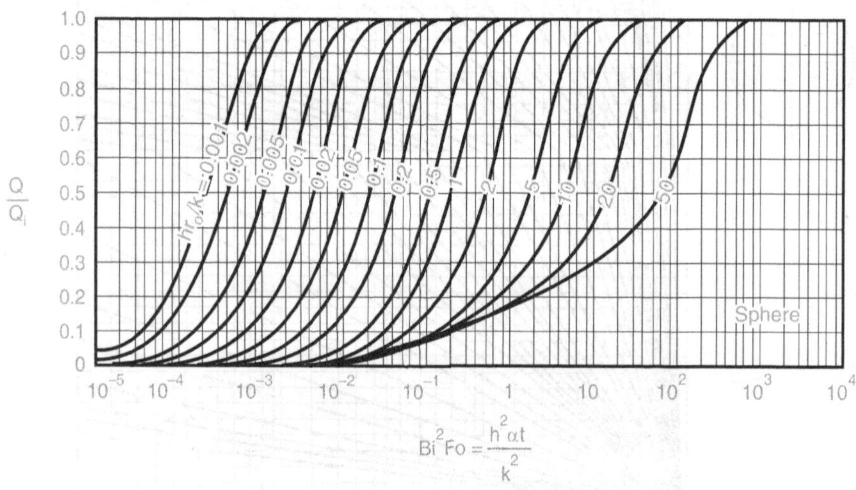

Fig. 6.33 (*c*) Dimensionless heat loss Q/Q_i for a sphere of radius r_o

Example 6.24. *A 50 mm thick iron plate is initially at 225°C. Its both surfaces are suddenly exposed to air at 25°C with convection coefficient of 500 W/m².K.*

(i) Calculate the centre temperature, 2 minute after the start of exposure.

(ii) Calculate the temperature at the depth of 10 mm from the surface, after 2 minute of exposure.

(iii) Calculate the energy removed from the plate per square metre during this period.

Take thermophysical properties of iron plate :

$k = 60$ W/m.K, $\rho = 7850$ kg/m³,

$C = 460$ J/kg, $\alpha = 1.6 \times 10^{-5}$ m²/s.

(Anna Univ., March 2000)

Solution

Given : A hot thick iron plate exposed to air on both surfaces

$2L = 50$ mm or $L = 25$ mm $= 0.025$ m,
$k = 60$ W/m.K, $T_i = 225°C$,
$T_\infty = 25°C$, $t = 2$ min $= 120$ s,
$\rho = 7850$ kg/m^3, $C = 460$ J/kg.K,
$h = 500$ W/m^2.K $\alpha = 1.6 \times 10^{-5}$ m^2/s,

Depth = 10 mm from the surface.

To find :

(*i*) The centreline temperature of the plate, after 2 minute of exposure.

(*ii*) The temperature at the depth of 10 mm from the surface, after 2 minute.

(*iii*) Heat transferred during 2 minute.

Assumptions :

1. The heat transfer area of 1 m^2.
2. Constant properties.

Analysis : (*i*) Consider the plate of thickness 2L, hence considering L as characteristic length

$$\text{Biot number Bi} = \frac{hL}{k} = \frac{500 \times 0.025}{60} = 0.21$$

The Biot number is greater than 0.1, hence the lumped heat system analysis cannot be used. Using the Heisler charts :

$$\frac{1}{\text{Bi}} = \frac{1}{0.21} = 4.8$$

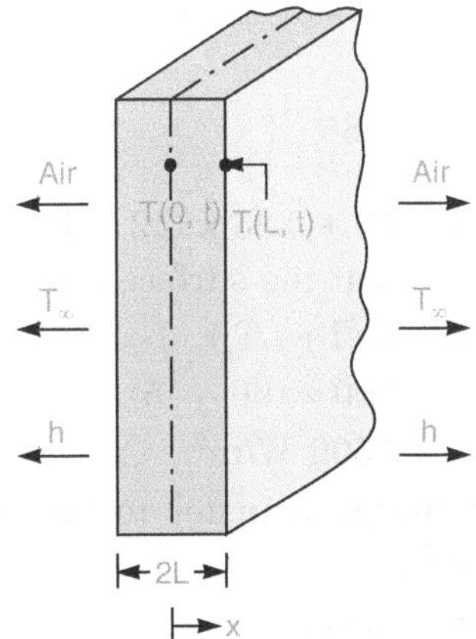

Fig. 6.34. Schematic of thick iron plate

Fourier number

$$\text{Fo} = \frac{\alpha t}{L^2} = \frac{1.6 \times 10^{-5} \times 120}{(0.025)^2} = 3.07$$

From Heisler chart Fig. 6.31 (*a*) for centreline temperature, for $\frac{1}{\text{Bi}} = 4.8$ and Fo = 3.07

$$\frac{\theta_c}{\theta_i} = \frac{T_c - T_\infty}{T_i - T_\infty} = 0.58$$

or $$T_c = 0.58 \times (225 - 25) + 25$$
$$= \mathbf{141°C.} \quad \text{Ans.}$$

(*ii*) Temperature at the depth of 10 mm from the surface,

$$x = L - \text{depth} = 25 \text{ mm} - 10 \text{ mm} = 15 \text{ mm}$$

Hence $$\frac{x}{L} = \frac{15}{25} = 0.6$$

From chart Fig. 6.31 (*b*) for position temperature, for $\frac{1}{\text{Bi}} = 4.8$ and $\frac{x}{L} = 0.6$

Temperature ratio at the location,

$$\frac{\theta}{\theta_c} = \frac{T - T_\infty}{T_c - T_\infty} = 0.95$$

or $$T = 25 + 0.95 \times (141 - 25)$$
$$= \mathbf{135.2°C.} \quad \text{Ans.}$$

(*iii*) Heat loss from the plate during 2 minute exposure ;

$$\text{Bi} = 0.21$$
$$\text{Bi}^2 \text{ Fo} = (0.21)^2 \times 3.07 = 0.135$$

From the Gröber chart Fig. 6.31 (*c*) for heat transfer ratio for plane wall

$$\frac{Q}{Q_i} = 0.45$$

where $$Q_i = \rho VC(T_i - T_\infty) = \rho(A2L)C(T_i - T_\infty)$$
$$= (7850 \text{ kg/m}^3) \times (1 \text{ m}^2 \times 0.05 \text{ m}) \times (460 \text{ J/kg}) \times (225 - 25) \times (\text{K})$$
$$= 35.33 \times 10^6 \text{ J/m}^2 = 35.33 \times 10^3 \text{ kJ/m}^2$$

The heat transferred during 2 minute,

$$Q = 0.45 \times 35.33 \times 10^3 \text{ kJ/m}^2$$
$$= \mathbf{15.9 \times 10^3 \text{ kJ/m}^2.} \quad \text{Ans.}$$

Example 6.25. *Consider a steel pipeline that is 1 m in diameter and has a wall thickness of 40 mm. The pipe is heavily insulated on the outside and before the initiation of flow, the wall of the pipe is at uniform temperature of – 20°C. Suddenly the hot oil at 60°C flows through the pipe creating convective surface condition corresponding h = 500 W/m^2.K at the inner surface of the pipe.*

(i) What is the appropriate Biot and Fourier numbers, 8 minutes after the initiation of flow ?

(ii) At t = 8 minute, what is the temperature of the exterior pipe surface covered by the insulation ?

(iii) What is the heat flux to the pipe from the oil at t = 8 minute ?

(iv) How much energy per metre pipe length has been transferred from the oil to the pipe during the period of 8 minutes ?

The thermophysical properties of the steel :

$k = 63.9$ W/m.K, $\rho = 7823$ kg/m^3 ;
$C = 434$ J/kg, $\alpha = 18.8 \times 10^{-6}$ m^2/s.

Solution

Given : A large steel pipe insulated on its outer surface ;

L = 40 mm = 0.04 m,	k = 63.9 W/m.K,
T_i = – 20°C,	T_∞ = 60°C,
t = 8 min,	ρ = 7823 kg/m^3,
C = 434 J/kg.K,	h = 500 W/m^2.K,
$\alpha = 18.8 \times 10^{-6}$ m^2/s.	

To find :

(*i*) Biot and Fourier numbers after 8 minute of exposure.

(*ii*) The temperature of exterior pipe surface after 8 minute.

(*iii*) Heat flux to the wall at t = 8 minute.

(*iv*) Heat energy transferred to pipe per unit length during 8 minutes period.

Assumptions :

1. Since the pipe diameter is too large as compared to its thickness, therefore, treating it as a plane wall.

2. One surface of the pipe is adiabatic, and hence taking L = 0.04 m.

3. Constant properties.

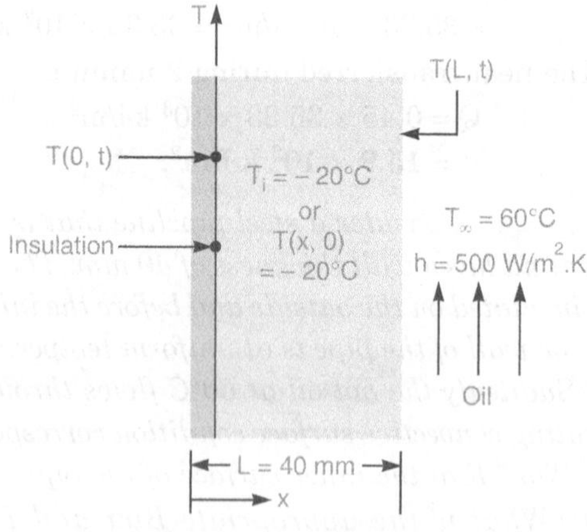

Fig. 6.35. Schematic for example 6.25

Analysis : (*i*) Biot and Fourier numbers after 8 minute of exposure.

Biot number

$$\text{Bi} = \frac{hL}{k} = \frac{500 \times 0.04}{63.9} = \mathbf{0.313.} \quad \textbf{Ans.}$$

Fourier number

$$\text{Fo} = \frac{\alpha t}{L^2} = \frac{18.8 \times 10^{-6} \times (8 \times 60)}{(0.04)^2} = \mathbf{5.64.} \quad \textbf{Ans.}$$

(*ii*) Biot number is greater than 0.1, hence lumped heat system analysis cannot be used.

Using the Heisler charts, Fig. 6.31(*a*)

With $\frac{1}{\text{Bi}} = \frac{1}{0.313} = 3.2$

and Fo = 5.64, the centreline temperature,

$$\frac{T_c - T_\infty}{T_i - T_\infty} = 0.22$$

or $T_c = 0.22 \times (-20 - 60) + 60 = \mathbf{42°C.}$ **Ans.**

(*iii*) Heat flux at the surface requires the determination of temperature at the surface,

Hence, $\frac{x}{L} = 1$

From chart for position temperature, for $\frac{1}{\text{Bi}} = 3.2$

and $\frac{x}{L} = 1.0$ temperature ratio at the location, from Fig. 6.31(*b*)

$$\frac{T - T_\infty}{T_c - T_\infty} = 0.86$$

or $T = 60 + 0.86 \times (42 - 60) = 45°C$

The heat flux at the surface after 8 min

$$q = h(T_\infty - T_s) = 500 \times (60 - 45) = \mathbf{7500\ W/m^2.} \quad \textbf{Ans.}$$

(*iv*) The energy transfer to the pipe wall over 8 minute interval

$$\text{Bi} = 0.313$$

$$\text{Bi}^2\text{Fo} = (0.313)^2 \times 5.64 = 0.55$$

From the Gröber chart Fig. 6.31(*c*) for heat transfer ratio for plane wall

$$\frac{Q}{Q_i} = 0.78$$

where Q_i = initial energy content per unit pipe length

$= \rho VC(T_i - T_\infty) = \rho(\pi DL)C(T_i - T_\infty)$

$= 7823 \times (\pi \times 1 \times 0.04) \times 434 \times \{-20 - (-60)\}$

$= 34.13 \times 10^6$ J/m $= 34.13 \times 10^3$ kJ/m

The heat transferred during 8 minute,

$Q = 0.78 \times 34.13 \times 10^3$ kJ/m

$=$ **26.62×10^3 kJ/m.** Ans.

Example 6.26. *A slab of aluminium 10 cm thick is initially at temperature of 500°C. It is suddenly immersed in a liquid bath at 100°C resulting in a heat transfer coefficient of 1200 W/m².K. Determine the temperature at the centreline and surface 1 min after the immersion. Also calculate the total thermal energy removed per unit area of the slab during this period. The properties of the aluminium for given conditions are:*

$\alpha = 8.4 \times 10^{-5}$ *m/s,* $k = 215$ *W/m.K,*
$\rho = 2700$ *kg/m³,* $C = 0.9$ *kJ/kg.K.*

(Anna Univ., May 2001)

Solution

Given : An aluminium slab as shown in Fig. 6.36,

$2L = 10$ cm, $L = 5$ cm $= 0.05$ m,
$T_i = 500$°C,
$T_\infty = 100$°C,
$h = 1200$ W/m².K,
$\alpha = 8.4 \times 10^{-5}$ m²/s,
$k = 215$ W/m.K,
$C = 0.9$ kJ/kg. K $= 900$ J/kg.K,
$\rho = 2700$ kg/m³,
$t = 1$ min $= 60$ s.

To find :

(*i*) Centreline temperature of slab after 1 min.
(*ii*) Temperature at the surface after 1 min.
(*iii*) Thermal energy removed per unit area of the slab during first one minute.

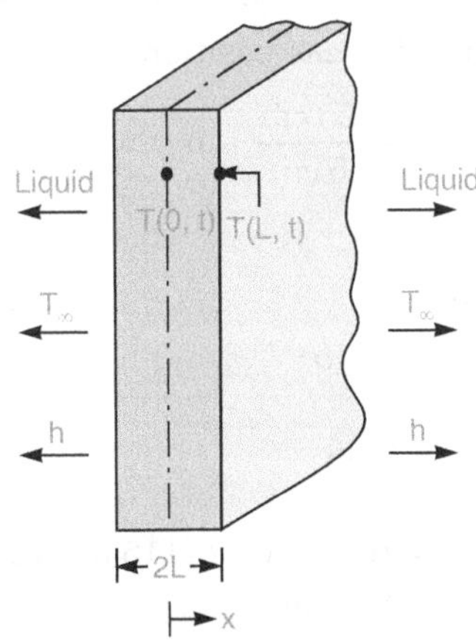

Fig. 6.36. Schematic of aluminium slab of example 6.26

Assumptions :

1. The slab is sufficiently large so it can be treated as an infinite slab.
2. Heat conduction in axial direction only.
3. Uniform heat transfer coefficient on the slab.
4. No radiation heat transfer.

Analysis : (*i*) Biot number for an infinite slab

$$\text{Bi} = \frac{hL}{k} = \frac{1200 \times 0.05}{215} = 0.28$$

It is greater than 0.1, thus lumped system analysis is not applicable.

Fourier number,

$$\text{Fo} = \frac{\alpha t}{L^2} = \frac{8.4 \times 10^{-5} \times 60}{(0.05)^2} = 2.016$$

It is greater than 0.2, thus one term solution as well as Heisler charts solution can be possible. Using Heisler charts for an infinite slab, Fig. 6.31 (*a*)

$$\left.\begin{aligned} \frac{1}{\text{Bi}} &= \frac{1}{0.28} = 3.57 \\ \text{Fo} &= 2.016 \end{aligned}\right\} \quad \frac{\theta_c}{\theta_i} = \frac{T_c - T_\infty}{T_i - T_\infty} = 0.63$$

The centreline temperature of the slab

$T_c = 100 + (500 - 100) \times 0.63$

$=$ **352°C.** Ans.

Alternatively

Using one term solution

At $\text{Bi} = 0.28$, $\xi_1 = 0.504$,
$C_1 = 1.0422$ (From Table B-5)

Using relation

$$\frac{\theta_c}{\theta_i} = \frac{T_c - T_\infty}{T_i - T_\infty} = C_1 e^{-\xi_1^2 \text{Fo}}$$

$= 1.04s22 \times e^{-(0.504)^2 \times 2.016}$

or $T_c = 100 + (500 - 100) \times 0.624$

$=$ **350°C.** Ans.

(*ii*) The surface temperature from Fig. 6.31 (*b*) :

$$\left.\begin{aligned} \text{For surface, } \frac{x}{L} &= 1 \\ \frac{1}{\text{Bi}} &= 3.57 \end{aligned}\right\} \quad \frac{T - T_\infty}{T_c - T_\infty} = 0.87$$

The surface temperature,

$T = 0.87 \times (352 - 100) + 100 =$ **319.24.** Ans.

Alternatively

Surface temperature can also be obtained by using eqn. (6.52)

$$\frac{T(x,t) - T_\infty}{T_i - T_\infty} = C_1 e^{-\xi_1^2 Fo} \cos(\xi_1 x/L)$$

Here $Fo = 2.016$, and $\frac{x}{L} = 1$ (at the surface)

From Table B-5 in Appendix,

For $Bi = 0.28$,

$\xi_1 = 0.505$ rad, $C_1 = 1.0423$

Thus $T(L, t) = 100 + (500 - 100) \times 1.0423$

$$\times e^{-(0.505)^2 \times 2.016} \times \cos(0.505 \times 1)$$

$= \mathbf{318.2.}$ Ans.

(*iii*) Thermal energy removed per unit area of slab during first one minute.

$$\left.\begin{matrix} Bi = 0.28 \\ Bi^2 Fo = 0.158 \end{matrix}\right] \frac{Q}{Q_i} = 0.48 \quad \text{(From Fig. 6.31)}$$

Heat removed

$$Q = 0.48 \times \rho\,(A2L)\,C(T_i - T_\infty)$$

$$= 0.48 \times 2700 \times 1 \times 2 \times 0.05 \times 0.900 \times (500 - 100)$$

$= \mathbf{46.65 \times 10^6\ J/m^2}$. Ans.

Example 6.27. *A long aluminium cylinder 5.0 cm in diameter and initially at 200°C is suddenly exposed to a convection environment at 70°C with heat transfer coefficient of 525 W/m².K. Calculate the temperature at the radius of 1.25 cm 1 minute after the cylinder exposed to the environment.* (J.N.T.U., May 2004)

Solution

Given : A long cylinder

$D = 5.0$ cm, $T_i = 200$°cm,

$T_\infty = 70$°C, $h = 525$ W/m².K,

$r = 1.25$ cm, $t = 1$ min $= 60$ s.

To find : The temperature at the radius of 1.25 cm in the cylinder.

Assumptions :

(*i*) No radiation exchange

(*ii*) The physical properties for aluminium cylinder as

$\rho = 2700$ kg/m³, $C = 900$ J/kg.K, $k = 210$ W/m.K

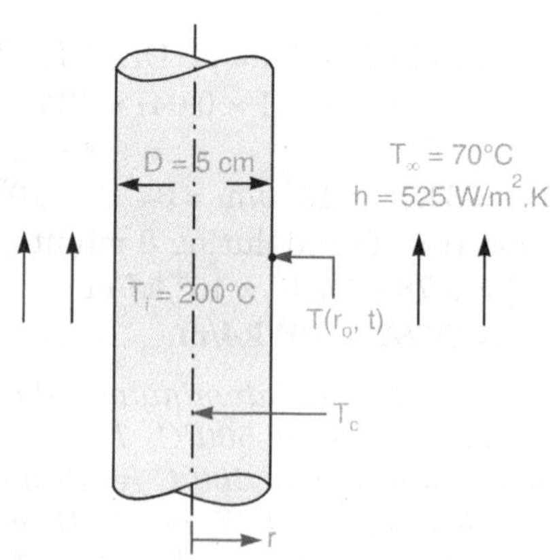

Fig. 6.37. Schematic for example 6.27

Analysis : Since the position temperature is to determine, thus using Heisler charts.

The radius of cylinder

$$r_o = \frac{D}{2} = \frac{0.05\ m}{2} = 0.025\ m$$

Biot number

$$Bi = \frac{hr_o}{k} = \frac{525 \times 0.025}{210} = 0.0625$$

$$\frac{1}{Bi} = 16$$

Fourier number

$$Fo = \frac{\alpha t}{r_o^2} = \left(\frac{k}{\rho C}\right)\frac{t}{r_o^2}$$

$$= \left(\frac{210}{2700 \times 900}\right) \times \frac{60}{(0.025)^2} = 8.29$$

The dimensionless centre temperature from Heisler chart, Fig. 6.32 (*a*)

$$\frac{T_c - T_\infty}{T_i - T_\infty} = 0.35$$

$\therefore \quad T_c = 70 + 0.35 \times (200 - 70) = 115.5$°C

The dimensionless position

$$\frac{r}{r_o} = \frac{1.25\ cm}{2.5\ cm} = 0.5$$

$$\frac{1}{Bi} = 16$$

From Fig. 6.32 (*b*)

$$\frac{T - T_\infty}{T_c - T_\infty} = 0.98$$

$$T = 70 + 0.98 \times (115.5 - 70)$$

$= \mathbf{114.59°C.}$ Ans.

Alternatively

Since Biot number is much less than 0.1, thus this problem can also be solved by using the lumped system analysis, eqn. (6.10)

$$\delta = \frac{r_o}{2} = 0.0125 \text{ m}$$

$$\frac{T - T_\infty}{T_i - T_\infty} = \exp\left[-\frac{ht}{\rho C\delta}\right]$$

$$T = 70 + (200 - 70) \times \exp\left[-\frac{525 \times 60}{2700 \times 900 \times 0.0125}\right]$$

$$= 70 + 130 \times 0.354 = \mathbf{116°C.}$$ **Ans.**

It is temperature in the cylinder with error of 1.2% only.

Example 6.28. *Two long cylinders of 10 cm in diameter, one of copper and other of asbestos are placed in a furnace. The initial temperature of cylinders are 30°C and the temperature in the furnace is 1000°C. Find how much time be kept in furnace to reach its centre temperature 418°C. Also find the temperature at a radius of 4 cm after this time. Assume the following properties :*

Combined convective and radiative heat transfer coefficient = 100 W/m².K.

For copper $k = 350$ *W/m.K,* $\alpha = 114 \times 10^{-7}$ *m²/s*

For asbestos $k = 0.11$ *W/m.K,* $\alpha = 0.28 \times 10^{-7}$ *m²/s.*

(P.U.P., May 1989)

Solution

Given : Two identical cylinders of copper and asbestos with

$D = 10$ cm, or $r_o = 5$ cm $= 0.05$ m,
$h = 100$ W/m².K $T_i = 30°C$,
$T_\infty = 1000°C$, $T_c = 418°C$

For copper $k = 350$ W/m.K, $\alpha = 114 \times 10^{-7}$ m²/s,

For asbestos $k = 0.11$ W/m.K, $\alpha = 0.28 \times 10^{-7}$ m²/s.

To find :

(*i*) The time required to reach for the cylinder centreline temperature 418°C.

(*ii*) Temperature at the radius of 4 cm in each cylinder.

Assumptions :

1. Infinite long cylinders.
2. Constant properties.

Analysis : (A) For copper cylinder :

Biot number

$$\text{Bi} = \frac{hr_o}{k} = \frac{100 \times 0.05}{350} = 0.0142$$

The lumped system analysis or chart solution can be applied. Applying the chart solution, because centreline and position temperature are to be calculated.

Hence $$\frac{1}{\text{Bi}} = \frac{1}{0.0142} = 70$$

(*i*) Temperature ratio at the centre

$$\frac{\theta_c}{\theta_i} = \frac{T_c - T_\infty}{T_i - T_\infty} = \frac{418 - 1000}{30 - 1000} = 0.6$$

From Heisler chart Fig. 6.32 (*a*) for centreline temperature, we get Fourier number,

$$\text{Fo} = 18.8$$

Further, $$\text{Fo} = \frac{\alpha t}{r_o^2}$$

or $$t = \frac{18.8 \times (0.05)^2}{114 \times 10^{-7}} = 4122.85 \text{ s}$$

$$= \mathbf{1.145 \text{ hours.}}$$ **Ans.**

(*ii*) Temperature at the radius of 4 cm

$$\frac{r}{r_o} = \frac{4}{5} = 0.8$$

From chart Fig. 6.32 (*b*), for position temperature of a cylinder, for $\frac{1}{\text{Bi}} = 70$ and $\frac{r}{r_o} = 0.8$

Temperature ratio at the location,

$$\frac{T - T_\infty}{T_c - T_\infty} = 0.985$$

or $$T = 1000 + 0.985 \times (418 - 1000)$$

$$= \mathbf{426.73°C.}$$ **Ans.**

It has very less temperature gradients over 4 cm radius.

(B) For asbestos cylinder

Biot number

$$\text{Bi} = \frac{hr_o}{k} = \frac{100 \times 0.05}{0.11} = 45.45$$

Biot number is too large, hence using chart solution.

Hence, $$\frac{1}{\text{Bi}} = \frac{1}{45.45} = 0.022$$

(*i*) Temperature ratio at the centre

$$\frac{\theta_c}{\theta_i} = \frac{T_c - T_\infty}{T_i - T_\infty} = \frac{418 - 1000}{30 - 1000} = 0.6$$

From Heisler chart Fig. 6.32 (*a*) for centreline temperature, we get fourier number,

$$Fo = 0.21$$

It gives $t = \frac{0.21 \times (0.05)^2}{0.28 \times 10^{-7}} = 18750$ s

$= \textbf{5.2 hours.}$ **Ans.**

(*ii*) Temperature at the radius of 4 cm

$$\frac{r}{r_o} = \frac{4}{5} = 0.8$$

From chart Fig. 6.32 (*b*), for position temperature of cylinder, for $\frac{1}{Bi} = 0.022$ and $\frac{r}{r_o} = 0.8$

Temperature ratio at the location,

$$\frac{T - T_\infty}{T_c - T_\infty} = 0.286$$

or $T = 1000 + 0.286 \times (418 - 1000)$

$= \textbf{833.5°C.}$ **Ans.**

It has large temperature gradients.

Example 6.29. *An iron sphere of diameter 5 cm is initially at a uniform temperature of 225°C. It is suddenly exposed to an ambient at 25°C with convection coefficient of 500 W/m².K.*

(i) Calculate the centre temperature 2 minute after the start of exposure.

(ii) Calculate the temperature at the depth of 1 cm from the surface after 2 minute of exposure.

(iii) Calculate the energy removed from the sphere during this period.

Take thermophysical properties of iron plate :

$k = 60$ W/m.K, $\rho = 7850$ kg/m³,

$C = 460$ J/kg, $\alpha = 1.6 \times 10^{-5}$ m²/s.

Solution

Given : An iron sphere with

D = 5 cm, or $r_o = 2.5$ cm = 0.025 m,

$k = 60$ W/m.K, $T_i = 225$°C,

$T_\infty = 25$°C, $t = 2$ min.,

$\rho = 7850$ kg/m³, C = 460 J/kg.K,

$h = 500$ W/m².K, $\alpha = 1.6 \times 10^{-5}$ m²/s,

depth = 1 cm from the surface.

To find :

(*i*) The centreline temperature of the sphere after 2 minute of exposure.

(*ii*) The temperature at the depth of 1 cm from the surface after 2 minute.

(*iii*) Heat transferred during 2 minutes.

Analysis : Biot number

$$Bi = \frac{hr_o}{k} = \frac{500 \times 0.025}{60} = 0.21$$

Using the Heisler chart :

$$\frac{1}{Bi} = \frac{1}{0.21} = 4.8$$

Fourier number

$$Fo = \frac{\alpha t}{r_o^2} = \frac{1.6 \times 10^{-5} \times 2 \times 60}{(0.025)^2} = 3.07$$

(*i*) From Heisler chart Fig. 6.33 (*a*), for centreline temperature of sphere, at $\frac{1}{Bi} = 4.8$ and Fo = 3.07

$$\frac{T_c - T_\infty}{T_i - T_\infty} = 0.18$$

or $T_c = 0.18 \times (225 - 25) + 25$

$= \textbf{61°C.}$ **Ans.**

(*ii*) Temperature at the depth of 1 cm from the surface,

$$r = r_o - \text{depth} = 25 \text{ mm} - 10 \text{ mm} = 15 \text{ mm}$$

Hence $\frac{r}{r_o} = \frac{15}{25} = 0.6$

From chart for position temperature of sphere Fig. 6.33 (*b*), at $\frac{1}{Bi} = 4.8$ and $\frac{r}{r_o} = 0.6$

Temperature ratio at the location,

$$\frac{T - T_\infty}{T_c - T_\infty} = 0.95$$

or $T = 25 + 0.95 \times (61 - 25)$

$= \textbf{59.2°C.}$ **Ans.**

(*iii*) Heat loss from the sphere during 2 minute exposure

$$Bi = 0.21$$

$$Bi^2 . Fo = (0.21)^2 \times 3.07 = 0.135$$

From the Gröber chart, Fig. 6.33 (*c*) for heat transfer ratio for sphere

$$\frac{Q}{Q_i} = 0.8$$

where, $Q_i = \rho V C (T_i - T_\infty) = \rho \times \frac{4}{3} \pi r_o^3 C (T_i - T_\infty)$

$$= 7850 \times \left\{ \left(\frac{4}{3}\right) \times \pi \times (0.025)^3 \right\} \times 460 \times (225 - 25)$$

$= 47,268$ J $= 47.268$ kJ

The heat transferred during 2 minute,

$Q = 0.8\ Q_i$

$Q = 0.8 \times 47.268$ kJ = **37.814 kJ.** Ans.

6.4. TRANSIENT HEAT CONDUCTION IN SEMI INFINITE SOLIDS

An infinite body extends itself in all direction of space. If such an infinite solid is divided in the middle by a plane, then half is referred as **semi infinite solid.** A semi infinite solid is a body that has a single boundary surface and extends to infinity in one direction as shown in Fig. 6.38. This body is used to estimate the temperature distribution in the part of the body, in which we are interested *i.e.,* region close to surface.

For an example, the earth and the thick slab can be considered as semi infinite body to obtain the temperature variation nearer to its surface.

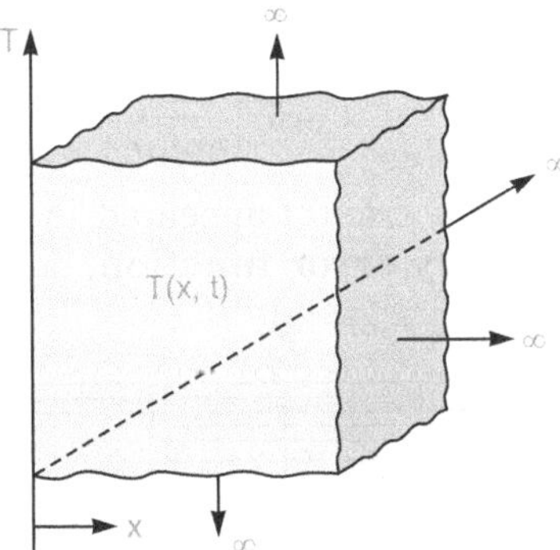

Fig. 6.38. A semi infinite solid with nomenclature

The general criteria for an infinite body to be considered semi infinite subjected to one-dimensional heat conduction is

$$\frac{\delta}{2\sqrt{\alpha t}} \geq 0.5$$

where δ is thickness of the body.

Consider a semi infinite solid, initially at uniform temperature T_i. At time $t > 0$, the surface of the solid is subjected to some boundary condition. The temperature distribution and heat flow at any position x in the solid with time can be obtained by using eqn. (6.28)

$$\frac{\partial^2 T}{\partial x^2} = \frac{1}{\alpha}\frac{\partial T}{\partial t} \qquad 0 \leq x \leq \infty$$

subjected to initial and boundary conditions :

(*i*) *Initial Condition.* At surface $T(x, 0) = T_i$ and at interior $T(\infty, t) = T_i$

(*ii*) *Boundary Conditions.* Three types of conditions may be imposed on the surface.

1. The surface temperature is suddenly changed and maintained at $T = T_s$.

2. The surface is suddenly exposed to constant heat flux q_0.

3. The surface is suddenly exposed to convection environment at $T = T_\infty$ with heat transfer coefficient h.

These three cases are illustrated in Fig. 6.39 and solutions are summarised below.

Case 1. Change in surface temperature

$$T(0, t) = T_s \text{ for } t > 0$$

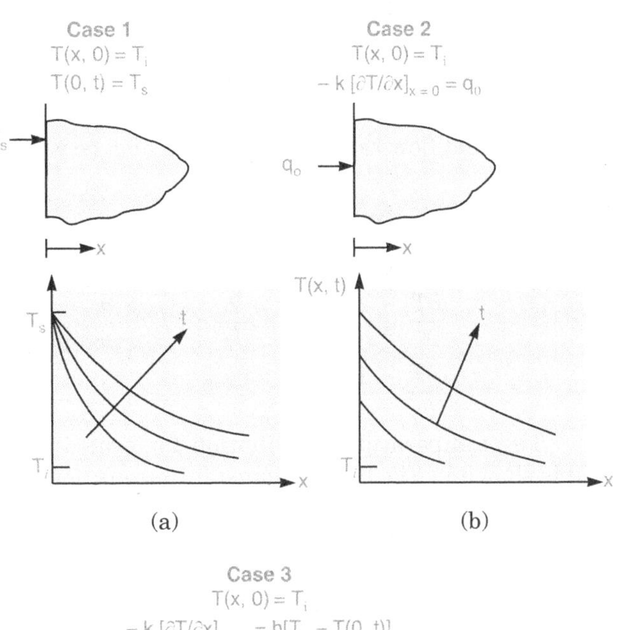

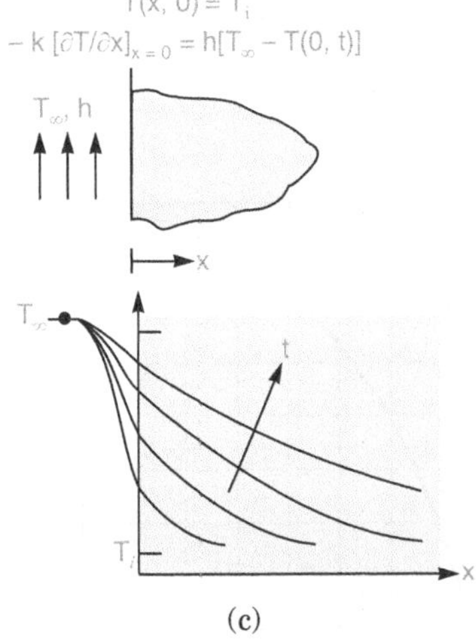

Fig. 6.39. Transient temperature distributions in a semi infinite solid for three surface conditions : (*a*) constant surface temperature, (*b*) constant surface heat flux, and (*c*) surface convection.

Solution to the preceding equation by the Laplace transform technique leads to

$$\frac{T(x, t) - T_s}{T_i - T_s} = \text{erf}\ (\xi) \qquad \text{...(6.62)}$$

where the quantity erf (ξ) is Gauss error function and is defined with a dimensionless dummy variable ξ as

$$\xi = \frac{x}{2\sqrt{\alpha t}}$$

and $$\text{erf}(\xi) = \frac{2}{\sqrt{\pi}} \int_0^{\xi} e^{-\xi^2} d\xi \quad ...(6.63)$$

The numerical values of Gauss error function erf(ξ) are presented as a function of ξ (zeta) in Table B-1 of Appendix B.

Inserting the dummy variable ξ and definition of error function in eqn. (6.62), the expression for temperature distribution becomes :

$$\frac{T(x, t) - T_s}{T_i - T_s} = \frac{2}{\sqrt{\pi}} \int_0^{\xi} e^{-\xi^2} d\xi \quad ...(6.64)$$

The heat flow rate at any position may be worked out as

$$Q = -kA \frac{\partial T}{\partial x}$$

The partial differentiation of eqn. (6.64) yields

$$\frac{\partial T}{\partial x} = (T_i - T_s) \times \frac{2}{\sqrt{\pi}} e^{-x^2/4\alpha t} \frac{\partial}{\partial x}\left(\frac{x}{2\sqrt{\alpha t}}\right)$$

or $$\frac{\partial T}{\partial x} = \frac{T_i - T_s}{\sqrt{\pi \alpha t}} e^{-x^2/4\alpha t} \quad ...(6.65)$$

The temperature distribution for semi infinite solid is shown in Fig. 6.40. The instantaneous heat flow rate can be expressed as

$$Q(t) = \frac{kA(T_s - T_\infty)}{\sqrt{\pi \alpha t}} e^{-x^2/4\alpha t} \quad ...(6.66)$$

The heat flow rate at the surface ($x = 0$)

$$Q = -kA \left.\frac{\partial T}{\partial t}\right|_{x=0} = \frac{kA(T_s - T_i)}{\sqrt{\pi \alpha t}} \quad ...(6.67)$$

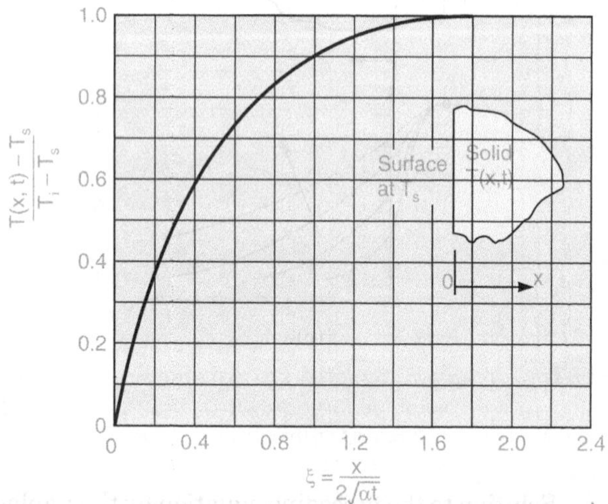

Fig. 6.40. Temperature distribution $T(x, t)$ in a semi infinite solid which is initially at T_i and for $t > 0$ boundary surface at $x = 0$ is maintained at T_s

Case 2. Constant surface heat flux on semi infinite solid

$$q_s = q_{x=0} = q_0$$

$$T(x, t) - T_i = \frac{2q_0 \sqrt{\alpha t/\pi}}{kA} \exp\left(-\frac{x^2}{4\alpha t}\right) - \frac{q_0 x}{kA}\left\{1 - \text{erf}\left(\frac{x}{2\sqrt{\alpha t}}\right)\right\} \quad ...(6.68)$$

Case 3. Convection boundary condition

Heat convected into the surface = Heat conducted into the surface

$$hA(T_\infty - T_{x=0}) = -kA \left.\frac{\partial T}{\partial x}\right|_{x=0}$$

The solution with this boundary condition yields to

$$\frac{T(x, t) - T_i}{T_\infty - T_i} = \text{erfc}\left(\frac{x}{2\sqrt{\alpha t}}\right) - \exp\left\{\frac{hx}{k} + \frac{h^2 \alpha t}{k^2}\right\} \times \text{erfc}\left(\frac{x}{2\sqrt{\alpha t}} + \frac{h\sqrt{\alpha t}}{k}\right) \quad ...(6.69)$$

The quantity erfc (ξ) appeared in eqn. (6.69) is the **complimentary error function,** defined as

$$\text{erfc}(\xi) = 1 - \text{erf}(\xi) \quad ...(6.70)$$

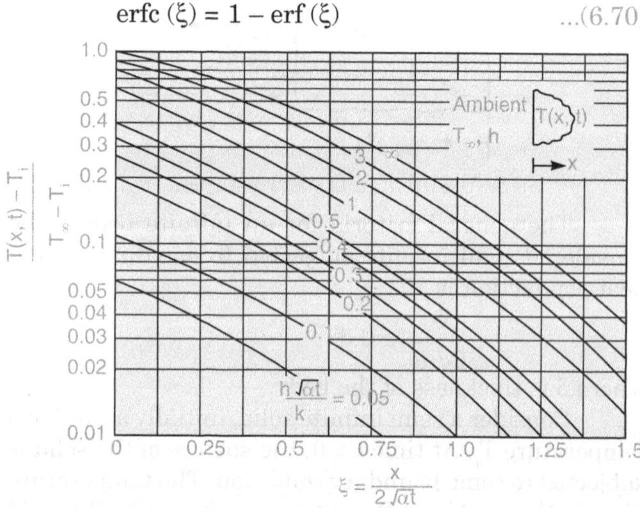

Fig. 6.41. Dimensionless transient temperatures for a semi infinite solid with surface convection to environment T_∞ with h

Despite its simple appearance, the solutions that appear in eqns. (6.62), (6.67), (6.68) and (6.69), the relations cannot be obtained analytically. Therefore, these are evaluated numerically for different values of $\xi\left(= \frac{x}{2\sqrt{\alpha t}}\right)$.

When $h \to \infty$, $T_\infty = T_s$ and eqn. (6.69) reduces to

$$\frac{T(x, t) - T_i}{T_s - T_i} = \text{erfc}\left(\frac{x}{2\sqrt{\alpha t}}\right) = 1 - \text{erf}\left(\frac{x}{2\sqrt{\alpha t}}\right) \quad ...(6.71)$$

equivalent to result obtained in eqn. (6.62). The graphical solution is given in Fig. 6.41 is simply plot of analytical solution given by eqn. (6.69).

6.4.1. Penetration Depth and Penetration Time

The penetration depth is referred to the location, where the temperature changes is within 1% of the applied change in temperature $(T_s - T_i)$ *i.e.*,

$$\frac{T - T_s}{T_i - T_s} = 0.99 = \text{erf}(1.8) \quad ...(6.72)$$

or penetration depth,

$$x = 1.8 \times 2\sqrt{\alpha t} = 3.6\sqrt{\alpha t} \quad ...(6.73)$$

The penetration time at a given depth indicates, the time taken by the surface to get 1% penetration.

i.e.,
$$t = \frac{1}{\alpha}\left(\frac{x}{3.6}\right)^2 \quad ...(6.74)$$

Example 6.30. *The ground at a particular location is covered with snow pack at – 10°C for a continuous period of three months, and the average soil properties at that location are k = 0.4 W/(m.K) and α = 0.15 × 10⁻⁶ m²/s. Assuming an initial uniform temperature of 15°C for the ground, calculate the minimum depth to place the water pipes from the surface to avoid freezing.*

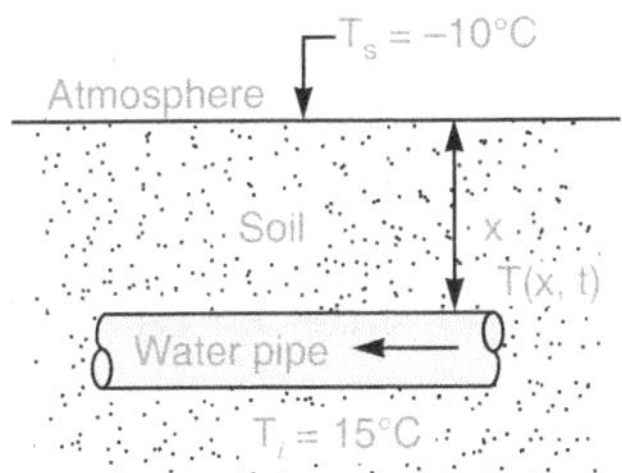

Fig. 6.42. Schematic for example 6.30

Solution

Given : Undergrounded water main (pipe) :

T_s = – 10°C, t = 3 months
k = 0.4 W/m.K, $\alpha = 0.15 \times 10^{-6}$ m²/s
T_i = 15°C, $T(x, t)$ = 0°C

To find : Depth of water pipes in order to avoid freezing.

Assumptions :

1. One-dimensional conduction.
2. Soil is an infinite medium.
3. Uniform and constant properties of soil.
4. Convection heat transfer coefficient $h \to \infty$.
5. The freezing temperature of water as 0°C, which the pipe may attain after three months.

Analysis : For prescribed surface temperature, the temperature distribution in the soil is

$$\frac{T - T_s}{T_i - T_s} = \text{erf}\left(\frac{x}{2\sqrt{\alpha t}}\right)$$

or
$$\frac{0 - (-10)}{15 - (-10)} = 0.4 = \text{erf}\left(\frac{x}{2\sqrt{\alpha t}}\right)$$

From Table B-1. ; for erf (ξ) = 0.4, $\xi \approx 0.37$

Here

$$t = 3 \text{ months} \times 30 \text{ days} \times 24\ h \times 3600 \text{ s} = 7.776 \times 10^6 \text{ s}$$

Then,
$$0.37 = \frac{x}{2\sqrt{\alpha t}} = \frac{x}{2\sqrt{0.15 \times 10^{-6} \times 7.776 \times 10^6}}$$

or x = **0.8 m. Ans.**

The water pipes must be placed 0.8 m below the free surface of earth in order to avoid freezing.

Example 6.31. *A large mass of a material is intially at uniform temperature of 100°C. Its surface is suddenly lowered and maintained at 2°C. The thermal diffusivity of the material is 0.41 m²/h. Calculate the time required for the temperature gradient at the surface to reach 3.5°C/cm.*

Solution

Given : A large mass of material as semi infinite body

T_i = 100°C, T = 2°C,
α = 0.41 m²/h = 1.139×10^{-4} m²/s

$$\frac{\partial T}{\partial t} = 3.5°\text{C/cm}.$$

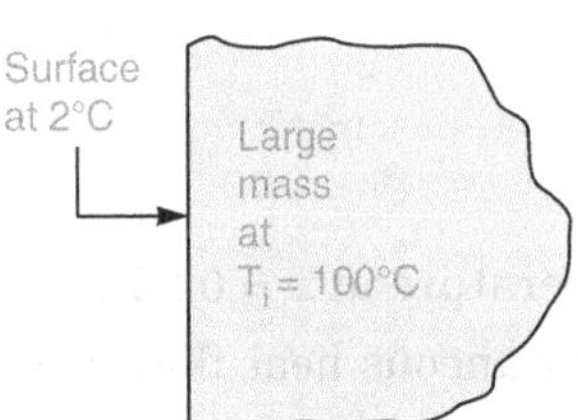

Fig. 6.43. Schematic for example 6.31

To find : Time for surface temperature gradient to 3.5°C/cm.

Analysis : The heat flow rate at the surface is given by eqn. (6.67)

$$Q = -kA\left.\frac{\partial T}{\partial x}\right|_{x=0} = \frac{kA(T_s - T_i)}{\sqrt{\pi \alpha t}}$$

or $$\left.\frac{\partial T}{\partial x}\right|_{x=0} = \frac{T_i - T_s}{\sqrt{\pi \alpha t}}$$

or $$3.5 \times 100 = \frac{100 - 2}{\sqrt{\pi \times 1.1139 \times 10^{-4} \times t}}$$

or $$t = \left(\frac{98}{350}\right)^2 \times \frac{1}{\pi \times 1.139 \times 10^{-4}}$$

$$= \mathbf{219\ s.}$$ **Ans.**

Example 6.32. *A thick steel slab is initially at a uniform temperature of 25°C. When the slab is exposed to hot flue gases, the surface temperature is suddenly changes to 450°C. Calculate the temperature in the plane 250 mm from the slab surface 5 h after the change in surface temperature. Also calculate the heat flow per m² of this plane and total energy flowing the surface during the 5 h period.*

Take $k = 45$ *W/m.K,*

$\rho = 8000$ *kg/m³, and* $C = 480$ *J/kg.K.*

Solution

Given : A thick steel plate as a semi infinite plate

$T_i = 25°C$, $T_s = 450°C$,

$x = 250$ mm $= 0.25$ m

$t = 5\ h = 18000$ s, $A = 1$ m²,

$k = 45$ W/m.K, $\rho = 8000$ kg/m³,

$C = 480$ J/kg.K.

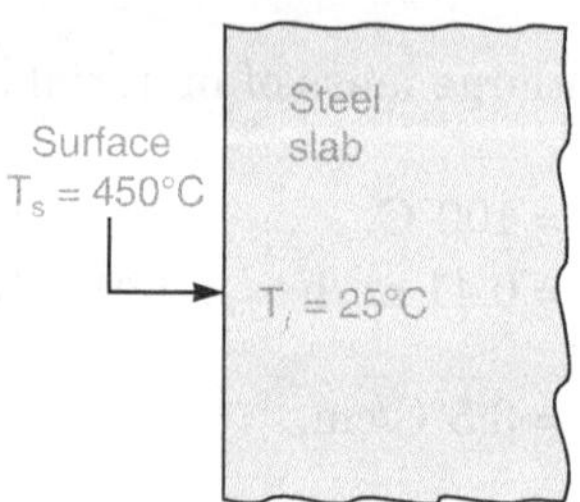

Fig. 6.44. Schematic for example 6.32

To find :

(*i*) Temperature at $x = 0.25$ m

(*ii*) Instantaneous heat flow rate per m².

(*iii*) Total energy flow in 5 *h*.

Analysis : (*i*) Temperature distribution in semi infinite plate :

$$\alpha = \frac{k}{\rho C} = \frac{45}{8000 \times 480}$$
$$= 1.17 \times 10^{-5}\ \text{m}^2/\text{s}$$

$$\frac{T - T_s}{T_i - T_s} = \text{erf}\left(\frac{x}{2\sqrt{\alpha t}}\right)$$

$$= \text{erf}\left(\frac{0.25}{2\sqrt{1.17 \times 10^{-5} \times 18000}}\right)$$

or $$T = 450 + (25 - 450) \times \text{erf}(0.272)$$

From Table B-1, erf (0.272) ≈ 0.3

$$\therefore \quad T = 450 - 425 \times 0.30$$
$$= \mathbf{322.5°C.}$$ **Ans.**

(*ii*) The instantaneous heat flow rate can be obtained by using eqn. (6.66)

$$Q(t) = \frac{kA(T_s - T_i)}{\sqrt{\pi \alpha t}} e^{-\frac{x^2}{4\alpha t}}$$

$$= \frac{45 \times 1 \times (450 - 25) \times e^{(-0.272^2)}}{\sqrt{\pi \times 1.17 \times 10^{-5} \times 18000}}$$

$$= \mathbf{21818.2\ W.}$$ **Ans.**

(*iii*) The total heat flow from the surface during 5 *h*.

$$Q = \frac{kA(T_s - T_i)}{\sqrt{\pi \alpha}} \int_o^t \frac{1}{\sqrt{t}} dt$$

$$= 1.13\, kA(T_s - T_i) \times \sqrt{\frac{t}{\alpha}}$$

$$= 1.13 \times 45 \times 1 \times (450 - 25) \times \sqrt{\frac{18000}{1.17 \times 10^{-5}}}$$

$$= 847 \times 10^6\ \text{J} = \mathbf{847\ MJ.}$$ **Ans.**

6.5. TRANSIENT HEAT CONDUCTION IN MULTIDIMENSIONAL SYSTEMS

The Heisler charts presented earlier may be used to obtain the temperature distribution and heat transfer in one-dimensional transient heat conduction problems associated with large plane of thickness 2L, in long cylinder or in the sphere of radius r_o. When a wall whose height and depth dimensions are not large compared to its thickness or a short cylinder ($L \not> r_o$) is encountered, additional space coordinates are necessary to specify the temperature, the above charts are no longer useful.

But with the use of clever superposition principle called **product solution,** these charts can be used to obtain the solution for two-dimensional transient heat conduction problems such as short cylinder, long rectangular bar, a semi infinite cylinder or plate. The three dimensional problems associated with geometries such as a rectangular prism, semi infinite rectangular bar may also solved by using these charts provided that all the surfaces of the solid is subjected to *same* ambient at T_∞ with *same* heat transfer coefficient h and the body does not involve any heat generation.

The product solution for an infinite rectangular bar, Fig. 6.45 can be formed from two infinite plates of thickness $2L_1$ and $2L_2$, respectively.

$$\left(\frac{T(x, y, t) - T_\infty}{T_i - T_\infty}\right)_{\text{Ractangular bar}} = \left(\frac{T(x, t) - T_\infty}{T_i - T_\infty}\right)_{2L_1,\text{ plate}} \times \left(\frac{T(y, t) - T_\infty}{T_i - T_\infty}\right)_{2L_2,\text{ plate}} \quad ...(6.75)$$

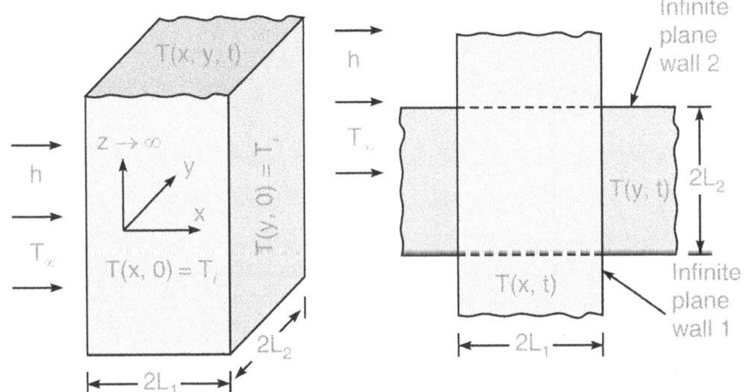

Fig. 6.45. Infinite rectangular bar ($2L_1 \times 2L_2$) is considered intersection of two plane walls of thickness $2L_1$ and $2L_2$ subjected to same convection environment

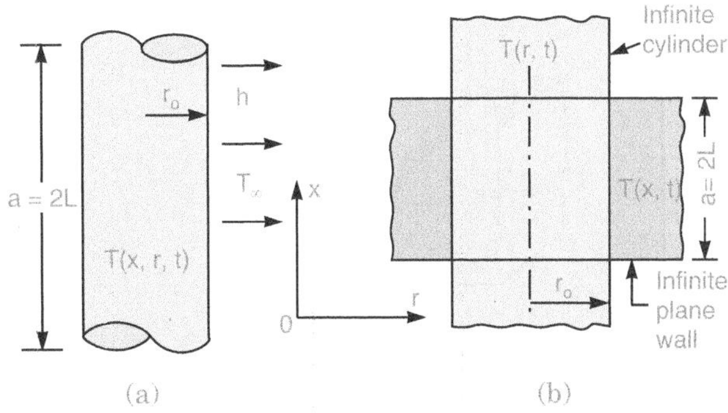

Fig. 6.46. A short cylinder of radius r_o and height (a = 2L) is considered intersection of an infinite plane wall of thickness 2L and infinite cylinder of radius r_o

The product solution for a short cylinder of radius r_o and length (a = 2L), Fig. 6.46 is

$$\frac{T(r, z, t) - T_\infty}{T_i - T_\infty} = \left(\frac{T(x, t) - T_\infty}{T_i - T_\infty}\right)_{\text{plane wall}} \times \left(\frac{T(r, t) - T_\infty}{T_i - T_\infty}\right)_{\text{infinite cylinder}} \quad ...(6.76)$$

The proper form of product solutions for some other geometries given in Table 6.2. It is important to note that x is measured from surface of a semi infinite solid, but from the mid plane of a plane wall and r is measured from centre of cylinder or sphere.

The dimensionless temperature ratio

$$\frac{\theta}{\theta_i} = \frac{T - T_\infty}{T_i - T_\infty}$$

TABLE 6.2. Multidimensional solutions expressed as products of one-dimensional solutions for bodies that are initially at a uniform temperature T_i and exposed to convection on all surfaces to a medium at T_∞

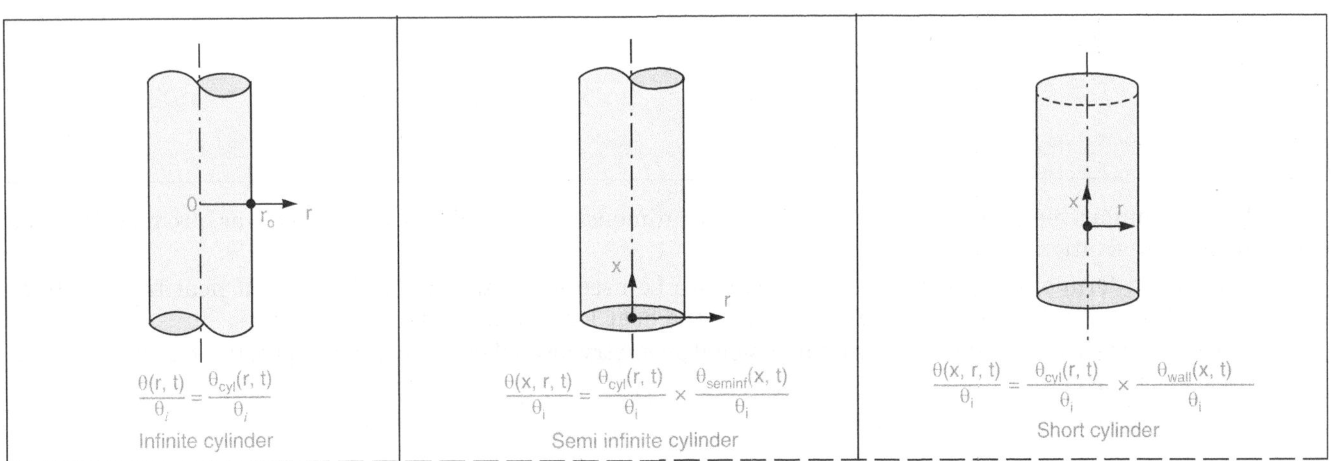

Infinite cylinder	Semi infinite cylinder	Short cylinder
$\frac{\theta(r,t)}{\theta_i} = \frac{\theta_{cyl}(r,t)}{\theta_i}$	$\frac{\theta(x,r,t)}{\theta_i} = \frac{\theta_{cyl}(r,t)}{\theta_i} \times \frac{\theta_{seminf}(x,t)}{\theta_i}$	$\frac{\theta(x,r,t)}{\theta_i} = \frac{\theta_{cyl}(r,t)}{\theta_i} \times \frac{\theta_{wall}(x,t)}{\theta_i}$

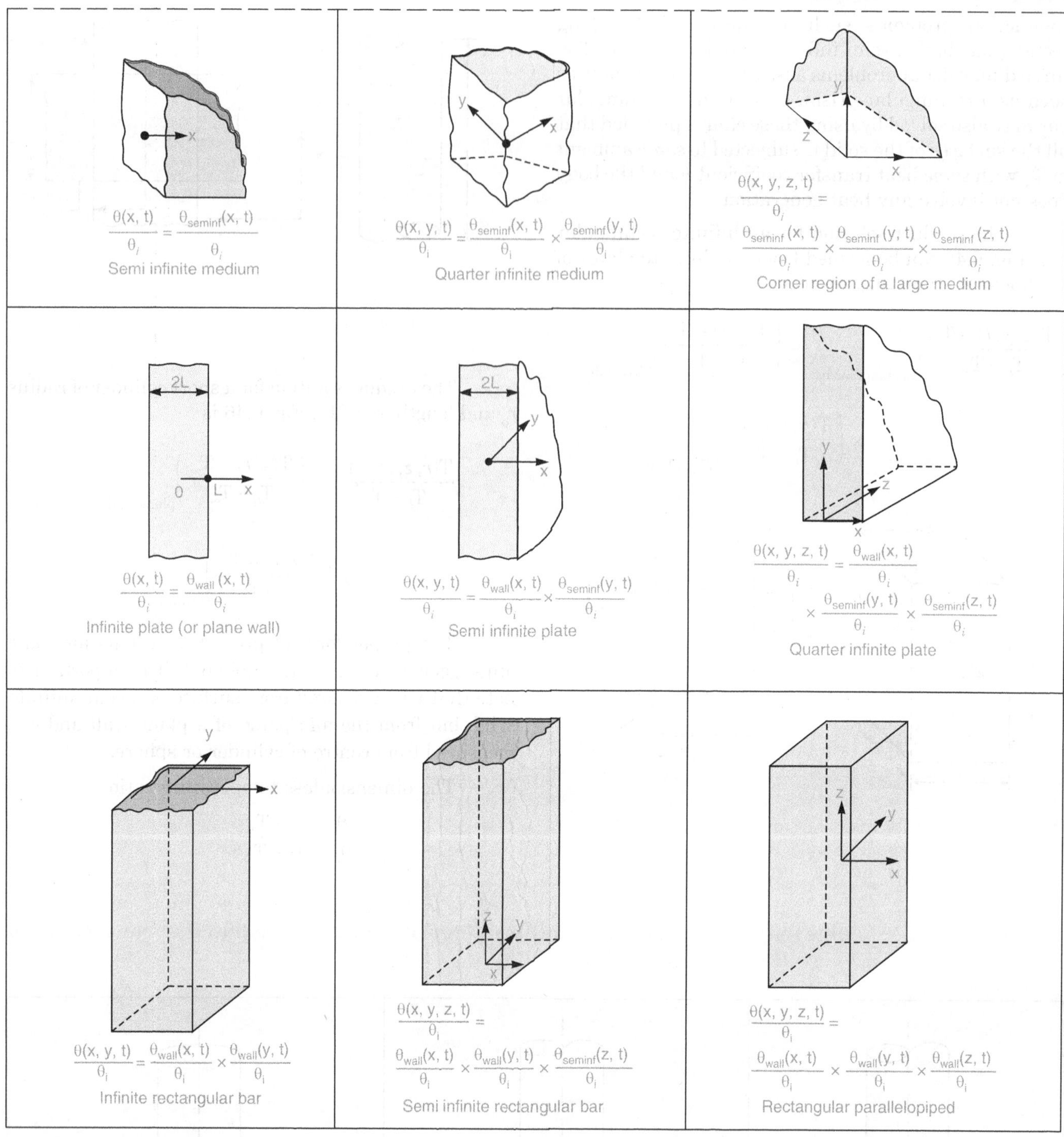

In the similar manner, the solutions for three-dimensional problems is obtained as product of three one-dimensional solutions.

A modified form of the product solutions can also be used to obtain the total transient heat transfer to or from a multidimensional geometry by superimposing the heat loss for one-dimensional bodies.

The transient heat transfer for a two-dimensional geometry formed by the intersection of two one-dimensional geometries 1, 2 is given by

$$\left(\frac{Q}{Q_i}\right)_{2\text{-D, solid}} = \left(\frac{Q}{Q_i}\right)_1 + \left(\frac{Q}{Q_i}\right)_2 \times \left[1 - \left(\frac{Q}{Q_i}\right)_1\right] \quad \ldots(6.77)$$

Transient heat transfer for a three dimensional geometry formed by the intersection of the three one-dimensional geometries is given by

$$\left(\frac{Q}{Q_i}\right)_{3\text{-D, solid}} = \left(\frac{Q}{Q_i}\right)_1 + \left(\frac{Q}{Q_i}\right)_2 \left[1 - \left(\frac{Q}{Q_i}\right)_1\right] + \left(\frac{Q}{Q_i}\right)_3 \times \left[1 - \left(\frac{Q}{Q_i}\right)_1\right] \times \left[1 - \left(\frac{Q}{Q_i}\right)_2\right] \quad \ldots(6.78)$$

Example 6.33. *A semi infinite aluminium cylinder ($k = 237$ W/m.K, $\alpha = 9.71 \times 10^{-5}$ m²/s) 20 cm in diameter is initially at uniform temperature of 200°C. The cylinder is then placed in water at 15°C, with $h = 120$ W/m².K. Calculate the temperature at the centre of the cylinder 15 cm from the end surface 5 minute after the start of cooling.*

Solution

Given : An aluminium cylinder as semi infinite body

$k = 237$ W/m.K

$\alpha = 9.71 \times 10^{-5}$ m²/s

D = 20 cm, $\quad r_o = 0.1$ m

$T_i = 200°C$, $\quad T_\infty = 15°C$

$x = 15$ cm, $\quad h = 120$ W/m².K

$t = 5$ min

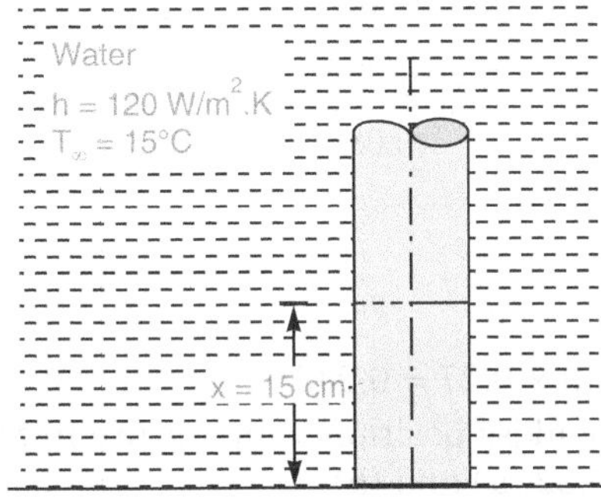

Fig. 6.47

To find : Centre temperature of cylinder after 5 min.

Analysis : A semi infinite cylinder is two-dimensional body, and thus the temperature will vary in both r and x directions within cylinder as well as with time, and its solution is

$$\frac{\theta(x, r, t)}{\theta_i} = \frac{\theta_{\text{cyl}}(r, t)}{\theta_i} \times \frac{\theta_{\text{seminf}}(x, t)}{\theta_i}$$

We will use one term approximate solution for cylinder and analytical solution to semi infinite medium.

For infinite long cylinder :

$$\text{Bi} = \frac{hr_o}{k} = \frac{120 \times 0.1}{237} = 0.05 < 0.1$$

$$\text{Fo} = \frac{\alpha t}{r_o^2} = \frac{9.71 \times 10^{-5} \times (5 \times 60)}{(0.1)^2} = 2.913 > 0.2$$

Thus one term approximation solution is applicable, from Table B-5, at Bi = 0.05 for cylinder

$$C_1 = 1.0124, \quad \xi_1 = 0.3126$$

$$\frac{T_c - T_\infty}{T_i - T_\infty} = C_1 e^{-\xi_1^2 \text{Fo}}$$

$$= 1.0124 \times e^{-(0.3126)^2 \times 2.913} = 0.762$$

The solution for position temperature distribution $\frac{T(x,t) - T_i}{T_\infty - T_i}$ in semi infinite medium is given by eqn. (6.69),

$$\frac{T(x,t) - T_i}{T_\infty - T_i} = \text{erfc}\left[\frac{x}{2\sqrt{\alpha t}}\right] - \exp\left[\frac{hx}{k} + \frac{h^2 \alpha t}{k^2}\right] \times \text{erfc}\left[\frac{x}{2\sqrt{\alpha t}} + \frac{h\sqrt{\alpha t}}{k}\right]$$

Here the quantities

$$\frac{x}{2\sqrt{\alpha t}} = \frac{0.15}{2\sqrt{9.71 \times 10^{-5} \times 300}} = 0.851$$

$$\frac{hx}{k} = \frac{120 \times 0.15}{237} = 0.076$$

$$\frac{h^2 \alpha t}{k^2} = \text{Bi}^2 \text{ Fo} = 0.05^2 \times 2.913 = 0.0073$$

$$\frac{h\sqrt{\alpha t}}{k} = \sqrt{0.0073} = 0.0853$$

$$\frac{T(x,t) - T_i}{T_\infty - T_i} = \text{erfc}\,(0.851) - \exp\,[0.076 + 0.073] \times \text{erfc}\,(0.851 + 0.0853)$$

or
$$\frac{T(x,t) - T_i}{T_\infty - T_i} = 0.2289 - 1.0868 \times 0.187 = 0.0256$$

But position temperature ratio $\frac{T_{(x,t)} - T_\infty}{T_i - T_\infty}$ can be obtained as

$$\frac{T(x,t) - T_\infty}{T_i - T_\infty} = 1 - \frac{T(x,t) - T_i}{(T_\infty - T_i)} = 1 - 0.0256 = 0.9743$$

Therefore, the centre temperature ratio in semi infinite cylinder can be expressed as

$$\frac{T_{(x,\,0,\,t)} - T_\infty}{T_i - T_\infty} = \frac{T(x, t) - T_\infty}{T_i - T_\infty} \times \frac{T_c - T_\infty}{T_i - T_\infty}$$

$$= 0.9743 \times 0.762 = 0.742$$

or $$T_{(x,\,0,\,t)} = 15 + (200 - 15) \times 0.742$$

$$= \mathbf{152.3°C.} \quad \textbf{Ans.}$$

Alternatively

Since Biot number is less than 0.1, thus the internal temperature gradients in the body are negligible and centre and surface temperatures of the cylinder be equal. Using lumped system analysis :

$$\frac{T - T_\infty}{T_i - T_\infty} = e^{-\frac{2ht}{\rho r_o C}} = e^{-\frac{2h\alpha t}{r_o k}}$$

or $$T = 15 + (200 - 15) \times e^{-\frac{2 \times 120 \times 9.71 \times 10^{-5} \times 300}{0.1 \times 237}}$$

$$= \mathbf{152.74°C.} \quad \textbf{Ans.}$$

Example 6.34. *A 10 cm diameter 16 cm long cylinder (k = 0.5 W/m.K and α = 5 × 10⁻⁷ m²/s) is initially at uniform temperature of 20°C. The cylinder is then placed in a furnace where the ambient temperature is 500°C with h = 30 W/m².K. Calculate the minimum and maximum temperature in the cylinder 30 min after it has been placed in the furnace.*

Solution

Given : A short cylinder as two-dimensional body,

$D = 10$ cm or $r_o = 0.05$ m

$2L = 16$ cm or $L = 0.08$ m

$k = 0.5$ W/m.K, $\alpha = 5 \times 10^{-7}$ m²/s

$T_i = 20°C$, $T_\infty = 500°C$

$h = 30$ W/m².K $t = 30$ min $= 1800$ s.

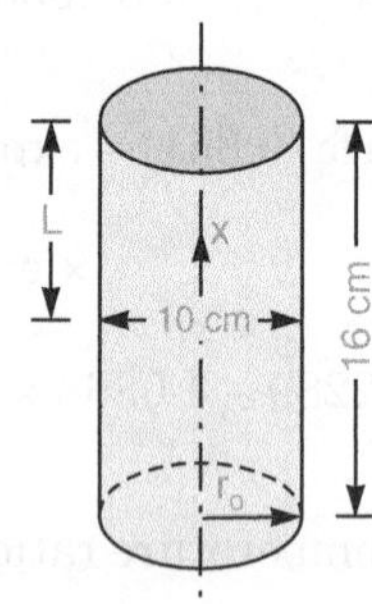

Fig. 6.48. *(a)* Short cylinder

To find :

(i) Minimum temperature in the cylinder.

(ii) Maximum temperature in the cylinder.

Assumptions :

(i) The two-dimensional transient heat conduction.

(ii) Short cylinder is an intersection of infinite cylinder and a plane wall.

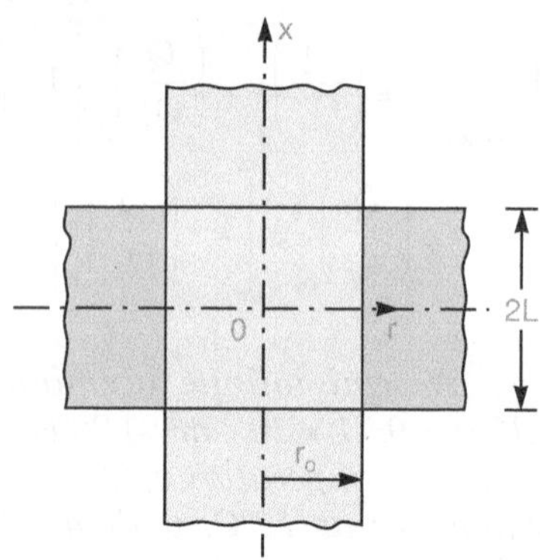

Fig. 6.48. *(b)* Intersection of infinite plane wall and infinite cylinder

Analysis : At any time, the minimum temperature is at the geometric centre of the cylinder *i.e.,* T(0, *t*) or at *x* = 0 and the maximum temperature is at the outer circumference of the cylinder.

T_{min} at $x = 0$, $r = 0$

T_{max} at $x = L$, $r = r_o$

For dimensionless position temperature in the plane wall from Figs. 6.31 *(a)* and *(b)*

$$\frac{1}{Bi} = \frac{k}{hL} = \frac{0.5}{30 \times 0.08} = 0.21$$

$$Fo = \frac{\alpha t}{L^2} = \frac{5 \times 10^{-7} \times 1800}{(0.08)^2} = 0.14$$

$$\left.\right\} \quad \frac{\theta_c}{\theta_i} = 0.90, \quad \frac{x}{L} = 1, \ \frac{\theta}{\theta_c} = 0.27$$

$$\frac{\theta(L, t)}{\theta_i} = \frac{\theta_c}{\theta_i} \times \frac{\theta}{\theta_c}$$

$$= 0.9 \times 0.27 = 0.243$$

For dimensionless position temperature in the cylinder from Figs. 6.32 *(a)* and *(b)*

$$\frac{1}{Bi} = \frac{k}{hr_o} = \frac{0.5}{30 \times 0.05} = 0.33$$

$$Fo = \frac{\alpha t}{r_o^2} = \frac{5 \times 10^{-7} \times 1800}{(0.05)^2} = 0.36$$

$$\left.\right\} \quad \frac{\theta_c}{\theta_i} = 0.47, \quad \frac{r}{r_o} = 1, \ \frac{\theta}{\theta_c} = 0.33$$

$$\frac{\theta(r_o, t)}{\theta_i} = \frac{\theta_c}{\theta_i} \times \frac{\theta}{\theta_c}$$

$$= 0.47 \times 0.33 = 0.155$$

(*i*) Minimum temperature

$$\frac{\theta_{min}}{\theta_i} = \frac{T_{min} - T_\infty}{T_i - T_\infty} = \left(\frac{\theta_c}{\theta_i}\right)_{\text{plane wall}} \times \left(\frac{\theta_c}{\theta_i}\right)_{\text{cylinder}}$$

$$= 0.9 \times 0.47 = 0.423$$

$$T_{min} = 500 + 0.423 \times (20 - 500)$$

$= \mathbf{297°C}$. Ans.

(*ii*) Maximum temperature

$$\frac{\theta_{max}}{\theta_i} = \frac{T_{max} - T_\infty}{T_i - T_\infty} \times \left(\frac{\theta}{\theta_i}\right)_{\text{wall}} = \left(\frac{\theta}{\theta_i}\right)_{\text{cyl}}$$

$$= 0.243 \times 0.155 = 0.0376$$

$$T_{max} = 500 + 0.0376 \times (20 - 500)$$

$= \mathbf{481.92°C}$. Ans.

6.6. SUMMARY

In unsteady state heat conduction, the temperature varies with position as well as time. In the lumped system analysis, the temperature of solid is assumed uniform in the system at any time. The temperature of any solid of mass m, volume V surface area A_s, density ρ, and specific heat C, initially at uniform temperature T_i exposed to convection ambient at T_∞ with h is approximated by the lumped system analysis as

$$\frac{T - T_\infty}{T_i - T_\infty} = \exp\left[-\frac{hA_s t}{\rho VC}\right] \text{ providing that}$$

Biot number,

$$Bi = \frac{h(V/A_s)}{k} \le 0.1$$

The instant heat transfer rate $Q(t)$ between a solid and its ambient at T_∞ with h is expressed as

$$Q(t) = hA_s(T(t) - T_\infty) = \rho VC \frac{\partial T}{\partial t}$$

$$= hA_s(T_i - T_\infty) \exp\left(-\frac{hA_s t}{\rho VC}\right)$$

Total amount of heat transferred between a body and its surroundings at T_∞ is

$$\Delta U = \int_0^t Q(t)\, dt$$

$$= \rho VC(T_i - T_\infty)\left\{\exp\left(-\frac{hA_s t}{\rho VC}\right) - 1\right\}$$

The *Biot number* is expressed as

$$Bi = \frac{\text{Internal resistance to heat flow}}{\text{Convection resistance to heat flow}} = \frac{h\delta}{k}$$

It can also be defined as the ratio of heat transfer coefficient to the internal specific conductance of the solid.

The *Fourier number is* expressed as

$$Fo = \frac{\text{Rate of heat conduction}}{\text{Rate of thermal energy storage}} = \frac{\alpha t}{\delta^2}$$

The *time constant* is defined as $\tau = \dfrac{\rho VC}{hA_s}$

When we attempt a problem of unsteady state heat conduction, the following guidelines should be followed :

Calculate the Biot number for the given solid as $Bi = \dfrac{h\delta}{k}$

With the characteristic length of the solid, $\delta = \dfrac{V}{A_s}$.

When Bi is more than 0.1, then Heisler and Gröber charts are used for the approximation of the solution of the problem. These charts can also be used to obtain the total heat transfer from a body upto time t.

Using one term approximation, the solution of one-dimensional transient heat conduction problems are expressed as

Plane wall :

$$\frac{\theta}{\theta_i} = \frac{T(x,t) - T_\infty}{T_i - T_\infty} = C_1 e^{-\xi_1^2 Fo} \cos(\xi_1 x/L)$$

Cylinder :

$$\frac{\theta}{\theta_i} = \frac{T(r,t) - T_\infty}{T_i - T_\infty} = C_1 e^{-\xi_1^2 Fo} J_0(\xi_1 r/r_o)$$

Sphere :

$$\frac{\theta}{\theta_i} = \frac{T(r,t) - T_\infty}{T_i - T_\infty}$$

$$= C_1 e^{-\xi_1^2 Fo} \frac{\sin(\xi_1 r/r_o)}{(\xi_1 r/r_o)}.$$

where C_1 and ξ_1 are functions of Biot number and their values are listed in Table B-5 of Appendix B.

At the centre of solid, the one term approximation reduces to

$$\frac{\theta}{\theta_i} = \frac{T - T_\infty}{T_i - T_\infty} = e^{-\xi_1^2 Fo}$$

Using one term approximation, the fraction of heat transfer in three geometries are

Plane wall : $\dfrac{Q}{Q_i} = 1 - \dfrac{\theta_c}{\theta_i}\dfrac{\sin\xi_1}{\xi_1}$

Cylinder : $\dfrac{Q}{Q_i} = 1 - 2\dfrac{\theta_c}{\theta_i}\dfrac{J_1(\xi_1)}{\xi_1}$

Sphere : $\dfrac{Q}{Q_i} = 1 - 3\dfrac{\theta_c}{\theta_i}\dfrac{\sin\xi_1 - \xi_1\cos\xi_1}{\xi_1^3}$.

REVIEW QUESTIONS

1. How does transient heat conduction differ from steady state heat conduction ?
2. What is lumped system analysis ? What are the assumptions made in the lumped system analysis and when is it applicable ?
3. Prove that the temperature distribution in a body at time t during a Newtonian heating or cooling is given by

$$\frac{T - T_\infty}{T_i - T_\infty} = e^{-Bi\,Fo}.$$

4. Consider a hot backed chicken piece on a plate. The temperature of the chicken piece is observed to drop by 5°C during first minute. Will the temperature drop during the second minute be less than, equal to or more than 5°C ? Why ? Comment.
5. What is Biot number ? What is its physical significance ? Is the Biot number more likely to larger for highly conducting solids or insulator ones ?
6. What is time constant ? Discuss the response of thermocouple.
7. What is Fourier number ? What is its physical significance ?
8. Discuss the criteria for neglecting internal temperature gradients within a solid during transient heat conduction. Deduce the condition for it.
9. Explain the applications of Heisler and Gröber charts in transient heat conduction.
10. What is the product solution method ? How is it used to determine the transient temperature distribution in a two-dimensional system ?
11. In which situation, one term approximation is suitable to solve unsteady state heat conduction.
12. What do mean by semi infinite body ? What is the general criteria to be considered for a semi infinite body ?

PROBLEMS

1. Steel balls 12 mm in diameter are annealed by heating to 1150 K and then slowly cooling to 400 K in air at 325 K with convection coefficient of 20 W/m².K. Assuming the properties of the steel to be

k = 40 W/m.K, ρ = 7800 kg/m³

and C = 600 J/kg.K

Estimate the time required for the cooling process. [Ans. 18.7 min.]

2. A solid copper sphere (k = 393 W/m.K), 10 mm in diameter, initially at 80°C is placed in an air stream at 30°C. The temperature is dropped to 65°C after 61 seconds. Calculate the value of convection coefficient. Assume properties as ρ = 8925 kg/m³, C = 397 J/kg.K. [Ans. h = 34.53 W/m².K]

3. Glass spheres of radius 2 mm at 600°C are to be cooled in an air stream at 30°C to a temperature of 80°C without any surface crack. Estimate the maximum value of convection coefficient. Also determine the minimum time required for the cooling. Take properties as ρ = 2225 kg/m³, C = 835 J/kg.K and k = 1.4 W/m.K. [Ans. h = 210 W/m².K, t = 14.35 s]

4. Stainless steel ball bearings [ρ = 8085 kg/m³, k = 15.1 W/m.K, C = 480 J/kg.K], 1.2 cm in diameter are taken from an oven at a uniform temperature of 900°C and are exposed to air at 30°C with h = 125 W/m².K, for a short period and then they are dropped into water for quenching. If the temperature of balls does not fall below 850°C prior to quenching, calculate, how long they stand in air before being dropped into water ? [Ans. 3.7 s]

5. The steel balls [k = 54 W/m.K, ρ = 7800 kg/m³, and C = 465 J/kg.K], 8 mm in diameter are annealed by heating them first to 900°C in a furnace and then allowing them to cool slowly to 100°C in ambient air at 30°C with h = 75 W/m².K. Calculate how long the annealing process will take ? If 2500 balls are to be annealed per hour, calculate the rate of heat transfer from the balls to ambient air. [Ans. 162 s, 43.1 MJ/h]

6. Cylindrical pieces of size 30 mm dia and 30 mm height with ρ = 7800 kg/m³, C = 486 J/kg.K and k = 43 W/m.K are to be heat treated. The pieces initially at 35°C are placed in a furnace at 800°C with convection coefficient of 85 W/m².K. Determine the time required to heat the pieces to 650°C. If by mistake the pieces were taken out of the furnace after 300 seconds, determine the shortfall in the requirements. [Ans. 9.08 min, 162°C]

7. It is desired to estimate the batch time for a heat treatment process involved in cooling alloy steel balls of 15 mm diameter from 820°C to 100°C in an oil bath at 40°C with h = 18 W/m².K. The material properties are ρ = 7780 kg/m³, C = 526 J/kg.K and k = 45 W/m.K. Determine the time required. If it is required to be achieved in 10 minutes, determine the value of convection coefficient. [Ans. t = 1457.8 s, h = 43.74 W/m².K]

8. A thermocouple in form of a long cylinder, 2 mm in dia, initially at 30°C is used to measure the temperature of a cold gas at –160°C. The convection coefficient is 60 W/m².K. The material properties are ρ = 8922 kg/m³, C = 410 J/kg.K and k = 22.7 W/m.K. Determine the time it will take to indicate – 150°C. Also calculate the time constant. [Ans. 89.76 s, 30.5 s]

9. A metal plate 10 mm thick at 30°C is suddenly exposed on one face to heat flux of 3000 W/m² and the other side is exposed to convection to a fluid at 30°C with h = 50 W/m².K . Determine the temperature after 10 s. Take properties of the material ρ = 8933 kg/m³, C = 385 J/kg.K and k = 380 W/m.K [Ans. 31.03°C]

10. A thermocouple junction may be approximated as a sphere 2 mm in diameter with k = 30 W/m.K, ρ = 8600 kg/m^3, and C = 400 J/kg.K. The convection coefficient is 280 W/m^2.K. How long will it take for the thermocouple to record 98 per cent of the applied temperature difference ? [Ans. 8 s]

11. A thermocouple is to be used to measure the temperature in a gas stream. The junction may be approximated as sphere having ρ = 8400 kg/m^3, C = 400 J/kg.K and k = 25 W/m.K. The convection coefficient is 560 W/m^2.K. Calculate the diameter of the junction needed to measure the 95 per cent of the applied temperature difference in 3 s. [Ans. 1 mm]

12. An orange of diameter 6 cm initially at a uniform temperature of 30°C. It is placed in a refrigerator in which the air temperature is 2°C. If the convection coefficient is 50 W/m^2.K, determine the time required for the centre of the orange to reach 10°C. Take thermophysical properties of orange as α = 1.4 × 10^{-7} m^2/s, and k = 0.59 W/m.K. [Ans. 45 min.]

13. A chicken piece [α = 1.6 × 10^{-7} m^2/s, and k = 0.5 W/m.K] of diameter 2 cm, initially at a uniform temperature of 7°C, is dropped suddenly in boiling water at 100°C. The heat transfer coefficient is 150 W/m^2.K. The chicken piece is considered cooked when its centre temperature reaches 80°C. How long will it take the centreline temperature to reach 80°C? [Ans. 8 min, 20 s]

14. A 6 cm diameter potato [α = 1.6 × 10^{-7} m^2/s, and k = 0.68 W/m.K], initially at a uniform temperature of 20°C, is suddenly dropped into boiling water at 100°C. The heat transfer coefficient between the water and the potato surface is 6000 W/m^2.K. Determine the time required for the centre temperature of the potato to reach 95°C and energy transferred during this time. [Ans. 33 min., 37.8 kJ]

15. A solid steel ball bearing 25 mm OD, initially at uniform temperature of 600°C is quenched in an oil bath at 40°C. The convective heat transfer coefficient is 1500 W/m^2.K. Determine the centreline temperature and the temperature at 1.25 mm from the surface after the bearing has been in the oil for first half minute. Also, determine the heat lost by spherical ball during the first half minute. [Ans. 900°C, 770°C, and 1390 kJ]

16. A short cylinder 75 mm OD and 10 cm long is at a uniform temperature of 250°C. At the time equal to zero, it is placed in a convection environment with h = 400 W/m^2.K and T_∞ = 40°C. If the material properties are α = 0.046 m^2/h, and k = 37 W/m.K, determine the temperature at the centre of the cylinder after 4 minutes. [Ans. 69°C]

17. A 30 cm × 30 cm slab of copper [k = 370 W/m.K, C = 380 J/kg.K, ρ = 8900 kg/m^3], 5 cm thick is initially at a uniform temperature of 260°C. It is suddenly exposed to a fluid at 35°C with h = 100 W/m^2.K. Calculate the time to reach the centre temperature of the slab to 90°C. [Ans. 14.9 min.]

18. A copper sphere (k = 370 W/m.K, ρ = 8900 kg/m^3, C = 380 J/kg.K), 3 cm in diameter is initially at uniform temperature of 50°C. It is suddenly exposed to an air stream at 10°C with h = 15 W/m^2.K. How long does it take the sphere temperature to drop to 25°C ? [Ans. 18.42 min]

19. An aluminium can (k = 210 W/m.K, ρ = 2700 kg/m^3, C = 900 J/kg.K) having a volume of 350 cm^3 contains beer at 1°C. Using lumped system analysis, calculate the time required to reach the beer temperature to 15°C when place in a room at 22°C with h = 15 W/m^2.K. Assume (k = 0.66 W/m.K, ρ = 1000 kg/m^3, C = 4200 J/kg.K) and surface area of beer can is 650 cm^2. [Ans. 27.6 min.]

20. A solid steel ball (k = 35 W/m.K), 300 mm in diameter is coated with a dieletric material (k = 0.04 W/m.K), 2 mm thick. The coated sphere is initially at a uniform temperature of 500°C and is suddenly quenched in a large oil bath at 100° with h = 3300 W/m^2.K. Calculate the time required for coated steel sphere to reach 140°C.
Take α = 8.72 × 10^{-4} m^2/s, ρ = 8600 kg/m^3,
C = 460 J/kg.K. [Ans. 67.37]
[Hint. Neglect the effect of energy storage in dielectric material, since its ρCV is very small.]

21. A large aluminium plate (k = 210 W/m.K) of thickness 0.15 m, initially at a uniform temperature of 300 K, is placed in a furnace having an ambient temperature of 800 K with h = 500 W/m^2.K.
(*a*) Calculate the time required for the plate mid-plane to reach 700 K.
(*b*) What is the surface temperature of the plate for this condition ?
Take ρ = 2700 kg/m^3, C = 900 J/kg.K,
α = 8.4 × 10^{-5} m^2/s.

22. A copper cylinder 10 cm diameter, 20 cm long is removed from liquid nitrogen bath at −196°C and exposed to air at 25°C with convection coefficient of 20 W/m^2.K. Find the time required by cylinder to attain the temperature of −110°C. Take thermophysical properties as :
C = 380 J/kg.K, ρ = 8800 kg/m^3, k = 360 W/m.K. [Ans. 27.47 min.]

23. The cylindrical steel rods (ρ = 7832 kg/m^3, C = 434 J/kg.K, and k = 63.9 W/m.K), 50 mm in diameter are heat treated by passing them through a furnace, 5 m long in which gases are maintained at 750°C with h = 125 W/m^2.K. The initial temperature of rods is 50°C. Calculate the speed at which the rods must be passed through the furnace in order to achieve 600°C at the centre line. [Ans. 9.55 mm/s]

24. Estimate the time required to cook a hot dog in boiling water. Assume that the hot dog is initially at 6°C and convection heat transfer coefficient is 100 W/m^2.K and the final temperature at the centre line is 80°C. Treat hot dog as a long cylinder of 20 mm diameter with the following properties :
ρ = 880 kg/m^3, C = 3350 J/kg.K, k = 0.52 W/m.K.
[Ans. 7.6 min.]

25. A long pyroceram rod, 20 mm in diameter is initially at a uniform temperature of 627°C and suddenly exposed to a fluid at 27°C with h = 100 W/m^2.K. Calculate the time required to reach the centreline at 327°C. Take thermophysical properties as :
ρ = 2600 kg/m^3, C = 808 J/kg.K,
and k = 3.98 W/m.K. [Ans. 84.36 s]

26. In heat treating to harden steel ball bearings (C = 500 J/kg.K, ρ = 7800 kg/m^3, k = 50 W/m.K) initially at 27°C is desired to increase the surface temperature for a short time without significantly warming the interior of the ball. This type of heating is obtained by sudden immersion in the molten salt bath at 1027°C with h = 5000 W/m^2.K. Calculate the time required to reach the surface temperature of 20 mm diameter ball to 727°C.

27. A sphere, (k = 50 W/m.K, α = 1.5 × 10^{-6} m^2/s), 80 mm in diameter is initially at uniform temperature of 800°C. It is suddenly quenched in an oil bath at 50°C with h = 1000 W/m^2.K. At a certain time, the surface temperature of the sphere is observed to be 150°C. What is the corresponding centre temperature of the sphere ?

28. An aluminium tube, 20 cm long with inner and outer radii as 5 cm and 6 cm, respectively, is quenched from 500°C to 30°C in a large reservoir of water at 10°C. Below 100°C, the heat transfer coefficient is 1500 W/m^2.K and above 100°C, its effective mean value is 500 W/m^2.K. The thermophysical properties of aluminium are
ρ = 2700 kg/m^3, k = 210 W/m.K, C = 900 J/kg.K.
Neglect internal thermal resistance, calculate the quenching time. [Ans. 50.8 s]

29. A 6 mm diameter mild steel rod (k = 54 W/m.K, ρ = 7800 kg/m^3, C = 420 J/kg.K) at 38°C is suddenly immersed in a liquid at 100°C with h = 110 W/m^2.K. Calculate the time required for the rod to get 88°C.
[Ans. 1 min 13 seconds]

30. A 1.4 kg aluminium household iron has 500 W heating element. The surface area is 0.046 m^2. The ambient air temperature is 21°C with h = 11 W/m^2.K. How long after the iron is plugged in will its temperature reach 104°C ?
Take ρ = 2770 kg/m^3, C = 875 J/kg.K, k = 200 W/m.K]
[Ans. 212 s]

31. Consider a household iron of 1000 W heating element whose base plate is made of 5 mm thick aluminium [ρ = 2770 kg/m^3, C = 875 J/kg.K, and α = 7.3 × 10^{-5} m^2/s]. The base plate has surface area of 0.03 m^2. Initially the iron is at a uniform temperature of 22°C, ambient temperature. Assuming heat transfer coefficient at the surface of the base plate to be 12 W/m^2.K and 85 per cent of heat generated the heating element is transfered to the base plate. Calculate the time required for the base plate to reach 140°C. Is it realistic to assume the plate temperature to be uniform at all times ? [Ans. 52 s]

32. During a picnic on a hot summer day all the cold drinks consumed and only available drinks were those at the ambient temperature of 40°C. In an effort to cool a 500 ml drink in a can which is 12.5 cm high and 72 mm in dia, a person grabs the can and start shaking it in the iced water bath at 0°C. The temperature of the drink is assumed to be uniform at all time and the heat transfer coefficient between iced water and aluminium can is 170 W/m^2.K. Calculate the time for the canned drink to cool to 5°C.
Take thermophysical properties of cold drink
k = 0.6 W/m.K, ρ = 1000 kg/m^3, C = 4187 J/kg.K.
[Ans. 31 min.]

33. In order to get some warm milk for a baby, a mother pours the milk into a thin walled metal glass, 6 cm in diameter. The height of the milk in the glass is 7 cm. She then places the glass into a large pan, filled with a hot water at 60°C. The milk is stirred constantly, so that its temperature is uniform throughout. If the heat transfer coefficient between the water and glass is 120 W/m^2.K, calculate the time for milk to warm up from 3°C to 38°C. Can the milk in this case be treated as a lumped system ? Why ?
Take for milk k = 0.56 W/m.K, ρ = 1000 kg/m^3,
C = 4200 J/kg.K. [Ans. 5.83 min.]

34. A spherical stainless steel vessel at 93°C contains 45 kg of water initially at 93°C. If the entire system is suddenly immersed in an iced water, calculate the time required for the water in the vessel to cool to 16°C and the temperature of the walls of the vessel at that time. Assume heat transfer coefficients at inner and outer surfaces are 17 W/m^2.K and 22.7 W/m^2.K, respectively and wall thickness of 25 mm.

35. The temperature of a gas stream is measured by a thermocouple, whose junction can be approximated as a 1 mm diameter sphere. The properties of the junction are k = 35 W/m.K, ρ = 8500 kg/m^3, and C = 320 J/kg.K and the convection heat transfer coefficient between junction and the gas is 210 W/m^2.K. Calculate how long will it take for the

thermocouple to approach the temperature within 1 per cent of the initial temperature difference.

[Ans. 10 s]

36. A 30 cm outer dia 10 m long pipe with a surface temperature of 90°C carries steam. The pipe is buried with its centreline at depth of 1 m. The ground surface is – 6°C and average thermal conductivity of the soil is 0.7 W/m.K. Calculate the heat loss per day and cost of heat loss, if the steam heat is worth ₹ 100 per 10^6 kJ. Also calculate the thickness of 85% magnesia insulation (k = 0.038 W/m.K) necessary to achieve the same insulation as provided by the soil with total heat transfer coefficient of 23 W/m^2.K on the outside of the pipe.

37. In the vulcanization of tires, the carcass is placed into a jig and steam at 150°C is admitted suddenly to both sides as shown in Fig. 6.49. If the tire thickness is 2.5 cm, initial temperature 21°C, h = 150 W/m^2.K, ρ = 1100 kg/m^3, C = 1650 J/kg.K, k = 0.163 W/m.K. Calculate time required for the centre of the rubber to reach 132°C.

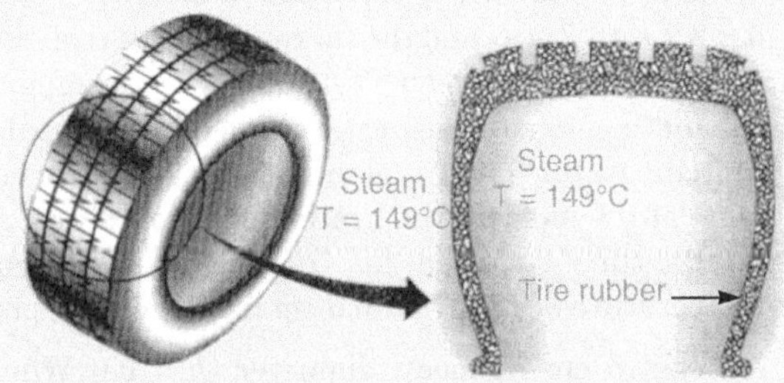

Fig. 6.49. Sechematic for prob. 37

[Ans. 37.12 min.]

38. A person puts a few apples into the freezer at – 15°C to cool them quickly for guests who are about to arrive. Initially, the apples are at a uniform temperature of 20°C, and the heat transfer coefficient on the surfaces is 8 W/(m^2.°C). Treating the apples as 9 cm diameter spheres and taking their properties to be ρ = 840 kg/m^3, C = 3.6 kJ/(kg.°C), k = 0.513 W/(m.°C), and $\alpha = 1.3 \times 10^{-7}$ m^2/s, determine the centre and surface temperatures of the apples in 1 h. Also determine the amount of heat transferred from each apple.

39. (a) An aluminium wire, 1 mm in diameter at 200°C is suddenly exposed to an environment at 30°C with h = 85.5 W/m^2.K. Estimate the time required to cool the wire to 90°C.

(b) If the same wire were to place in air stream (h = 11.65 W/m^2.K). What would be the time required to reach it to 90°C ?

Take thermophysical properties as

C = 900 J/kg.K, ρ = 2700 kg/m^3,

k = 204 W/m.K. [Ans. (a) 7.4 s, (b) 54.3 s]

40. A short brass cylinder [ρ = 8530 kg/m^3, C = 0.389 kJ/(kg.°C), k = 110 W/(m.°C), and $\alpha = 3.39 \times 10^{-5}$ m^2/s] of diameter D = 8 cm and height H = 15 cm is initially at a uniform temperature of

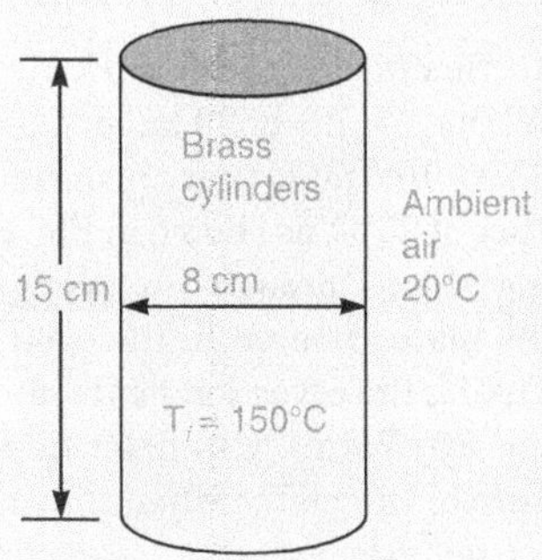

Fig. 6.50. Schematic for prob. 40.

T_i = 150°C. The cylinder is now placed in atmospheric air at 20°C, where heat transfer takes place by convection with a heat transfer coefficient of h = 40 W/(m^2.°C). Calculate (a) the centre temperature of the cylinder, (b) the centre temperature of the top surface of the cylinder, and (c) the total heat transferred from the cylinder 15 min after the start of the cooling.

[Ans. (a) 85.4°C, (b) 85.4°C, (c) 161.37 kJ]

41. A semi infinite aluminium cylinder [k = 237 W/(m.°C), $\alpha = 9.71 \times 10^{-5}$ m^2/s] of diameter D = 15 cm is initially at a uniform temperature of T_i = 150°C. The cylinder is now placed in water at 10°C, where heat transfer takes place by convection with a heat transfer coefficient of h = 140 W/(m^2.°C). Determine the temperature at the centre of the cylinder, 10 cm from the end surface 8 min after the start of the cooling.

42. A hot dog can be considered to be a cylinder 12 cm long and 2 cm in diameter whose properties are ρ = 980 kg/m^3, C = 3.9 kJ/(kg.°C), k = 0.76 W/(m.°C), and $\alpha = 2 \times 10^{-7}$ m^2/s. The hot dog initially at 5°C is dropped into boiling water at 100°C. If the heat transfer coefficient at the surface of the hot dog is estimated to be 600 W/(m^2.°C), determine the centre temperature of the hot dog after 5, 10, and 15 min by treating the hot dog as (a) a finite cylinder and (b) an infinitely long cylinder.

43. A hot dog 12.5 cm long, 2.2 cm in diameter was equipped with two thermocouples, one at the centre and other just under the skin. The initial temperature indicated by both thermocouples was 20°C, which was ambient temperature too. The hot dog was then suddenly dropped into the boiling water at 94°C. After 2 min, the centre and surface temperatures were measured to be 59°C and 88°C, respectively. The thermophysical properties of the hot dog can be taken

as $\rho = 980$ kg/m^3, C = 3900 J/kg.K. Using transient temperature charts, calculate (*a*) thermal diffusivity of the hot dog (*b*) thermal conductivity of hot dog and (*c*) convection heat transfer coefficient.

[Ans. (*a*) 2×10^{-7} m^2/s, (*b*) 0.76 W/m.K, (*c*) 658 W/m^2.K]

44. In a production facility, 3 cm thick large brass plates [k = 110 W/(m.°C), ρ = 8530 kg/m^3, C = 380 J/(kg.°C) and $\alpha = 33.9 \times 10^{-6}$ m^2/s] that are initially at a uniform temperature of 25°C, as shown in Fig. 6.51 are heated by passing them through an oven maintained at 700°C. The plates remain in the oven for a period of 10 min. Taking the covection heat transfer coefficient to be h = 80 W/(m^2.°C), determine the surface temperature of the plates when they come out of the oven.

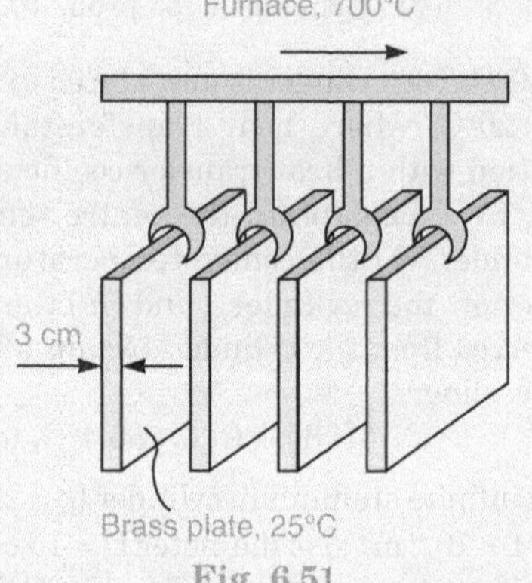

Fig. 6.51

[Ans. 448.5°C]

45. A long 35 cm diameter cylindrical shaft made of stainless steel [k = 14.9 W/(m.°C), ρ = 7900 kg/m^3, C = 477 J/(kg.°C), and $\alpha = 3.95 \times 10^{-6}$ m^2/s] comes out of an oven at a uniform temperature of 400°C, Fig. 6.52. The shaft is then allowed to cool slowly in a chamber at 150°C with an average convection heat transfer coefficient of h = 60 W/(m^2.°C). Determine the temperature at the centre of the shaft 20 min after the start of the cooling process. Also determine the heat transferred per unit length of the shaft during this time period.

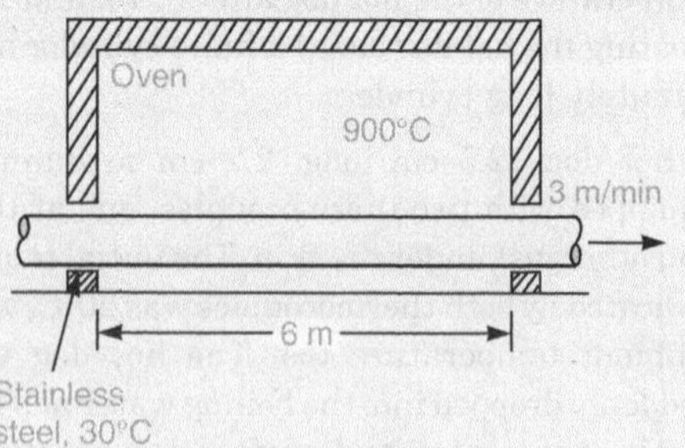

Fig. 6.52

[Ans. 390°C, 15,680 kJ]

46. Long cylindrical stainless steel rods [k = 13.4 W/(m.°C) and $\alpha = 3.48 \times 10^{-6}$ m^2/s] of 10 cm diameter are heat treated by drawing them at a velocity of 3 m/min through a 9 m long oven maintained at 900°C. The heat transfer coefficient in the oven is 90 W/(m^2.C). If the rods enter the oven at 30°C, determine their centreline temperature when they leave.

47. In a heat treatment plant, the balls of bearings 10 mm in diameter are loaded on a conveyor belt. The belt passes through a furnace (inside temperature = 1000°C, h = 200 W/m^2.K) along its length (L = 3 m). If the balls are heated from 30°C to 250°C, such that the temperature gradients should not exceed 5%, find the velocity of the belt required.

Take ρ = 3000 kg/m^3, C = 0.5 kJ/kg.K,

k = 50 W/m.K. [Ans. 0.933 m/s]

48. A 2.5 cm thick sheet of plastic initially at 21°C is placed between two heated steel plates, that are maintained at 138°C. The plastic is heated just long enough for its mid plane temperature to reach 130°C. If thermal conductivity of the plastic is 0.0011 W/m.K, $\alpha = 2.7 \times 10^{-6}$ m^2/s and the thermal contact resistance at the interface between plastic and steel is negligible, calculate, (*a*) the time required for heating (*b*) temperature at the plane 0.6 cm from the steel plate at the moment the heating is discontinued, and (*c*) the time required for the plastic to reach a temperature of 130°C, at 0.6 cm from the steel plate.

49. An egg, 5 cm in mean diameter (k = 0.6 W/m.K, ρ = 1000 kg/m^3, C = 4170 J/kg.K) is initially at a temperature of 4°C. It is dropped in the boiling water at 100°C for 15 min. The heat transfer coefficient from water to egg can be assumed to be 1700 W/m^2.K. What would be the temperature at the centre of egg at the end of cooking period ?

50. A mild steel (k = 54 W/m.K, ρ = 8000 kg/m^3, C = 410 J/kg.K) cylindrical billet, 25 cm in diameter is to be raised to a minimum temperature of 700°C by passing it through a 6 m long furnace. If the furnace gases are at 1600°C with overall heat transfer coefficient of 68 W/m^2.K. Calculate the maximum speed at which a continuous billet entering at 200°C can travel through the furnace.

51. A 6 cm diameter, steel ball is at uniform temperature of 800°C. It is to be hardened by suddenly dropping it into an oil bath at a temperature of 50°C with convection coefficient of 500 W/m^2.K. If the quenching occurs when the ball reaches a temperature of 100°C, determine how long the ball should be kept in the oil bath.

If 100 balls are to be quenched per minute, determine the rate at which the heat must be removed from the oil bath in order to maintain the bath temperature at 50°C.

Take thermophysical properties as : k = 61 W/m.K, ρ = 7850 kg/m^3, C = 460 J/kg.K.

[Ans. 3.26 min, 28.588 MJ/min.]

52. An aluminium cylinder (k = 210 W/m.K) 50 mm in diameter and 10 cm long is initially at uniform temperature of 200°C is plunged into a quenching bath at 10°C. Take h = 530 W/m^2.K, what is the temperature on centreline of the cylinder after one minute? [Ans. 61.33°C]

53. A steel cylinder 20 cm diameter is initially heated to 980°C. It is then quenched in an oil bath at 38°C with convection coefficient of 568 W/m^2.K. Calculate the time required for the cylinder centre to reach a temperature of 260°C. The properties of steel are :

k = 16 W/m.K, ρ = 7816 kg/m^3,
C = 460 J/kg, $\alpha = 4.4 \times 10^{-6}$ m^2/s.

[Ans. 19.7 s]

54. Calculate the total heat transferred from a short brass cylinder 10 cm in diameter and 12 cm long within first fifteen minute of exposure. Take ρ = 8530 kg/m^3, C = 380 J/kg.K, k = 110 W/m.K, T_i = 120°C, T_∞ = 25°C, h = 60 W/m^2.K, $\alpha = 3.39 \times 10^{-5}$ m^2/s.

[Ans. 85.9 kJ]

REFERENCES AND SUGGESTED READING

1. Holman J.P. *"Heat Transfer"*, 8th edition, McGraw Hill Eduction, 2010, New Delhi.
2. Incropera F.P. and DeWitt. D.P., *"Fundamentals of Heat and Mass Transfer"*, 5th edition, John Wiley & Sons, 2002.
3. Cenzel Yunus A., *"Heat Transfer., A Practical Approach"*, 2nd edition, McGraw Hill Education, 2003.
4. Cass law H.S. and Jaeger J.C., *"Conduction of Heat in Solids."*, 2nd edition Oxford University Press. London, 1959.
5. Heisler M.P., *"Temperature Charts for Induction and Constant Temperature Heating."*, ASME Transactions 69, 1947.
6. Gröber H, Erk S., and Grigull U., *"Fundamentals of Heat Transfer"* McGraw Hill New York, 1961.
7. Schneider P.J., *"Conduction Heat Transfer"*, Addision-Wisley, Reading, M.A., 1955.
8. Özisik M.N., *"Heat Transfer—A Basic Approach".*, McGraw Hill, New York, 1985.
9. Suryanarayana N.V. *"Engineering Heat Transfer"*, Penran International Publishing, India, 2008.
10. Kreith Frank and Bohn M.S. *"Principles of Heat Transfer"*, 5th edition, PWS Publishing Company, Boston M.A., 1997.

Principles of Convection

7

The objective of this chapter is to give basic understanding of physics of convection heat transfer and to present them in the form of general equations, which are applied in subsequent chapters for the particular cases.

In the previous chapters, we dealt with heat conduction, which is a mechanism of heat transfer due to random molecular activities through a stationary medium, solid or fluid. The convection heat transfer was restricted to the boundary conditions only and the rate of heat convection at the boundaries was considered constant so far.

The convection heat transfer is of importance to practical problems in industrial application. The flow of a liquid or a gas through a heat exchangers, two phase flow in the boilers and condensers, cooling of electronic chips, heat removal from the condenser of a refrigerator are some common examples of convection heat transfer.

The convection heat transfer is recognised closely related to the fluid flow. Hence understanding of convection should start with basic knowledge of fluid dynamics, momentum transfer, energy transfer, shear stress, pressure drop, friction coefficient and the nature of fluid flow like laminar or turbulent etc.

7.1. MECHANISM OF HEAT CONVECTION

As discussed in chapter one, the heat convection involves two mechanism, simultaneously. One is energy transfer from a hot surface to a adjacent fluid by random molecular motion, it is called *diffusion*. The other one is *advection*, *i.e.*, the transport of energy by bulk movement of the fluid from higher temperature region to lower temperature region. Such motion in presence of temperature gradient will enhance the heat transfer rate. The molecules in aggregate retain their random motion and the fluid motion brings the hotter and colder fluid chunks in contact, thus initiating the high rate of conduction at a large number of sites in the fluid. Therefore, the rate of heat transfer in the convection is due to superposition of energy transfer by random molecular motion (conduction) at the surface as well as the energy transfer by bulk motion of fluid.

7.2. CLASSIFICATION OF CONVECTION

The convection heat transfer is classified as *natural* (or *free*) or *forced convection,* depending on how the fluid motion is initiated. The natural or *free convection* is a process, in which the fluid motion results from heat transfer. When a fluid is heated or cooled, its density changes and the buoyancy effects produce a natural circulation in the affected region, which causes itself the rise of warmer fluid and the fall of colder fluid : Therefore, energy transfers from hotter region to colder region and such process is repeated as long as the temperature difference in the fluid exists.

In the *forced convection*, the fluid is forced to flow over a surface or in a duct by external means such as a pump or a fan. A large number of heat transfer applications utilize forced convection, because the heat transfer rate is much faster than that in free convection.

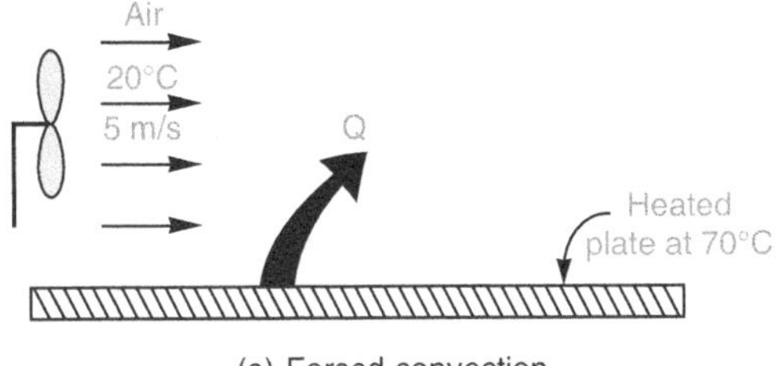

(a) Forced convection

(b) Natural convection

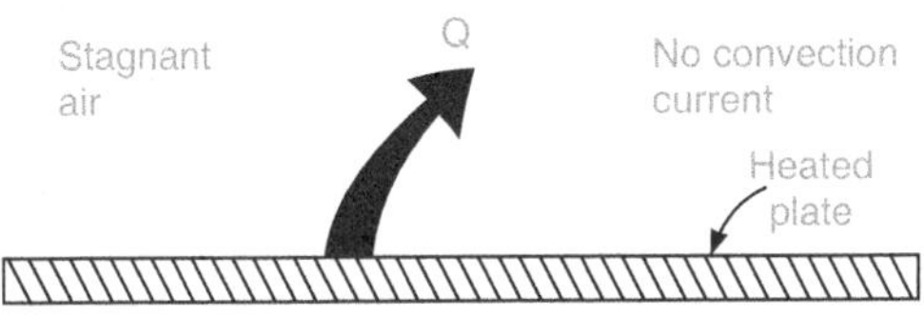

(c) In absence of fluid motion, heat transfer in the fluid is by conduction only

Fig. 7.1. The heat transfer from a hot surface to the surrounding fluid

Further, the convection heat transfer is also classified as *external convection* or *internal convection*. In *external convection*, the fluid surrounds a surface such as flow over a flat or curved surface, while in *internal convection*, the fluid is surrounded by a surface such in a pipe carrying steam or water filled cooling passage in an internal combustion engine. The fluid flows can also be stated as laminar, turbulent or translatory (transition from laminar to turbulent). Forced and natural convection have separate criteria for distinctions of these regims.

Consider the heating of a cold iron block as shown in Fig. 7.2. If there is no significant velocity of hot air surrounds the block, the heat will be transferred from hot air to block by natural convection. If a fan blows air over the block, the heat will be transferred from hot air to cold block by forced convection. If the speed of the air over the block surface increases, the block will be heated up faster. If air is replaced by water, the heat transfer rate by convection will be increased several times.

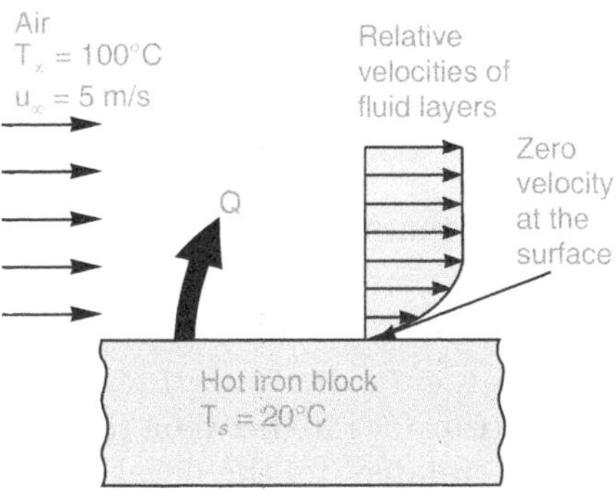

Fig. 7.2. Heating of cold block by forced convection

Experience shows that the convection heat transfer strongly depends on fluid properties, dynamic viscosity μ, thermal conductivity k_f, density ρ, and specific heat C_p, as well as on the fluid velocity. It also depends on geometry and roughness of the solid surface, in addition to type of fluid flow. Thus the convection heat transfer relations are rather complex, because of dependence of convection on so many variables.

7.3. CONVECTION HEAT TRANSFER COEFFICIENT

The rate of heat transfer per unit surface area from a surface to a fluid is proportional to temperature difference and it is expressed as

$$q_{conv} \propto (T_s - T_\infty)$$

$$q_{conv} = h\,(T_s - T_\infty) \quad ...(7.1)$$

where h = constant of proportionality and is called *heat transfer coefficient*,

T_s = temperature of the surface, °C

T_∞ = temperature of free stream fluid, °C.

Based on the interpretation, the convective heat transfer coefficient is expressed as

$$h = \frac{q_{conv}}{(T_s - T_\infty)} \quad ...(7.2)$$

or it is defined as the rate of convection heat transfer per unit surface area per unit temperature difference. It is measured in W/m^2.K or W/m^2. °C.

Consider the flow of a hot fluid at temperature T_∞ over a cold surface at a constant temperature of T_s as shown in Fig. 7.2. It is observed that the fluid layer in contact with the solid surface *sticks* to the surface. It is very thin layer of fluid and has *zero velocity* (*no slip condition*). Therefore, the heat transfer from wall surface to the adjacent, fluid layer is by *pure conduction*, and the conduction heat flux $q(x)$ at the wall surface $y = 0$ is given by

$$q(x) = k_f \left.\frac{\partial T(x, y)}{\partial y}\right|_{y=0} \quad \text{...[7.3(a)]}$$

where $T(x, y)$ is temperature distribution of fluid

$\left.\frac{\partial T}{\partial y}\right|_{y=0}$ = temperature gradient at the surface,

k_f = thermal conductivity of the fluid.

The negative sign is omitted from the eqn. [7.3(*a*)] because the heat flow from fluid to wall, *i.e.*, in negative y direction.

The heat transfer rate between the fluid and the wall surface is related to the local heat transfer coefficient h_x, defined as

$$q(x) = h_x (T_\infty - T_s) \quad \text{...[7.3(b)]}$$

where T_s and T_∞ are the wall surface and free stream fluid temperatures, respectively.

In steady state conditions, the heat flow rate is constant, thus equating eqn. [7.3(*a*)] with eqn. [7.3(*b*)],

$$\underbrace{h_x (T_\infty - T_s)}_{\text{convection}} = \underbrace{k \left.\frac{\partial T(x, y)}{\partial y}\right|_{y=0}}_{\text{conduction}}$$

or
$$h_x = \frac{k \left.\frac{\partial T(x,y)}{\partial y}\right|_{y=0}}{T_\infty - T_s} \quad \text{...(7.4)}$$

where h_x is **local heat transfer coefficient** at a certain position x in flow direction and for given temperature distribution in the flow. It is calculated from eqn. (7.4). It is used to obtain the heat flux at any location in the fluid flow.

The local heat transfer coefficient may vary along the length of flow as a result of changes in the velocity and other parameters in the flow direction. We are usually interested for the heat transfer rate from the entire surface. Which can be obtained by using average heat transfer coefficient over a distance $x = 0$ to $x = L$, determined from

$$h = \frac{1}{L}\int_0^L h_x \, dx \quad \text{...(7.5)}$$

With use of average value of heat transfer coefficient h, the heat transfer rate Q from the fluid to the wall surface is obtained over entire distance $x = 0$ to $x = L$ and width w as

$$Q = h (wL) (T_\infty - T_s) \quad \text{...(7.6)}$$

Example 7.1. *Experimental results for local heat transfer coefficient h_x for flow over a flat plate with an extremely rough surface were found as*

$$h_x = ax^{-0.1}$$

where a is a constant and x is a distance from the leading edge of the plate.

Develop an expression for ratio of average heat transfer coefficient h for a plate of length x to the local heat transfer coefficient h_x at x.

Solution

Given : The variation of local heat transfer coefficient as $h_x = ax^{-0.1}$

To find : The ratio of average heat transfer coefficient to local heat transfer coefficient.

Analysis : The average heat transfer coefficient is given by eqn. (7.5), over a distance 0 to x is

$$h = \frac{1}{x}\int_0^x h_x \, dx$$

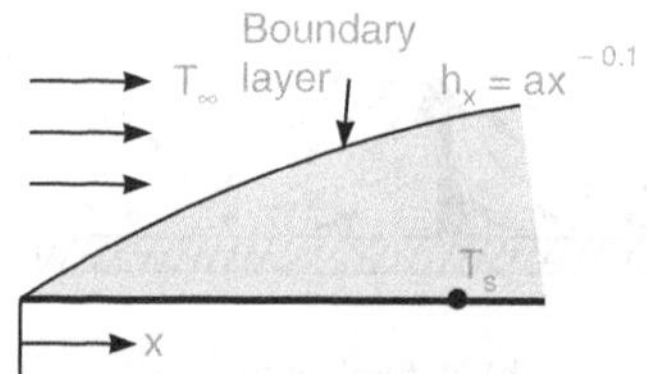

Fig. 7.3. Schematic

Using $h_x = ax^{-0.1}$ and integrating, we get

$$h = \frac{1}{x}\int_0^x ax^{-0.1}\,dx = \frac{a}{x}\int_0^x x^{-0.1}\,dx$$

$$= \frac{a}{x}\left[\frac{x^{0.9}}{0.9}\right] = 1.11\, ax^{-0.1}$$

$$= \mathbf{1.11\, h_x}. \quad \text{Ans.}$$

Example 7.2. *Experimental results for heat transfer over a flat plate with an extremely rough surface were found to be correlated by an expression of the form*

$$Nu_x = 0.04\, Re^{0.9}\, Pr^{1/3}$$

where Nu_x is the local value of Nusselt number at a position x measured from the leading edge of the plate. Derive an expression for ratio of average heat transfer coefficient to local heat transfer coefficient h_x.

Solution

The local Nusselt number for flow over a flat plate is given by

$$Nu_x = 0.04\ Re^{0.9}\ Pr^{1/3}$$

where $Nu_x = \frac{h_x x}{k_f}$,

$$Re_x = \frac{\rho u_\infty x}{\mu} \quad \text{and} \quad Pr = \frac{\mu C_p}{k}$$

The local heat transfer coefficient is expressed as

$$h_x = 0.04 \times \frac{k_f}{x}\left(\frac{\rho u_\infty x}{\mu}\right)^{0.9} Pr^{1/3}$$

$$= 0.04\ k_f \left(\frac{\rho u_\infty}{\mu}\right)^{0.9} x^{-0.1}\ Pr^{1/3} \quad ...(i)$$

The average heat transfer coefficient is obtained as

$$h = \frac{1}{x}\int_0^x 0.04\ k_f \left(\frac{\rho u_\infty}{\mu}\right)^{0.9} x^{-0.1}\ Pr^{1/3}\ dx$$

$$= \frac{0.04\ k_f}{x}\left(\frac{\rho u_\infty}{\mu}\right)^{0.9} Pr^{1/3}\int_0^x x^{-0.1}\ dx$$

$$= \frac{0.04}{x} k_f \left(\frac{\rho u_\infty}{\mu}\right)^{0.9} Pr^{1/3}\left[\frac{x^{0.9}}{0.9}\right]$$

$$= \frac{0.04}{0.9x} k_f \left(\frac{\rho u_\infty x}{\mu}\right)^{0.9} Pr^{1/3} \quad ...(ii)$$

and $\frac{h}{h_x} = \frac{1}{0.9} = \mathbf{1.11.}$ **Ans.**

Example 7.3. *A bare thermocouple is used to measure the temperature of a gas flowing through a hot duct. The heat transfer coefficient between a gas and thermocouple is proportional to $u^{0.8}$, where u is the gas velocity and heat transfer rate by radiation from the walls to the thermocouple is proportional to temperature difference.*

When the gas is flowing at 5 m/s, the thermocouple reads 323 K and when it is flowing at 10 m/s, it reads 313 K. Calculate the appropriate wall temperature at a gas temperature of 298 K.

What temperature will the thermocouple indicate when the gas velocity is 20 m/s ?

Solution

Given : Thermocouple is exposed to gas stream

$$h \propto u^{0.8}$$
$$Q_{rad} \propto \Delta T$$

$u_1 = 5$ m/s, $T_1 = 323$ K

$u_2 = 10$ m/s, $T_2 = 313$ K

$u_3 = 20$ m/s, $T_{\infty_2} = 298$ K.

To find :

(*a*) Wall temperature, when gas temperature is 298 K.

(*b*) Thermocouple reading, when gas velocity is 20 m/s.

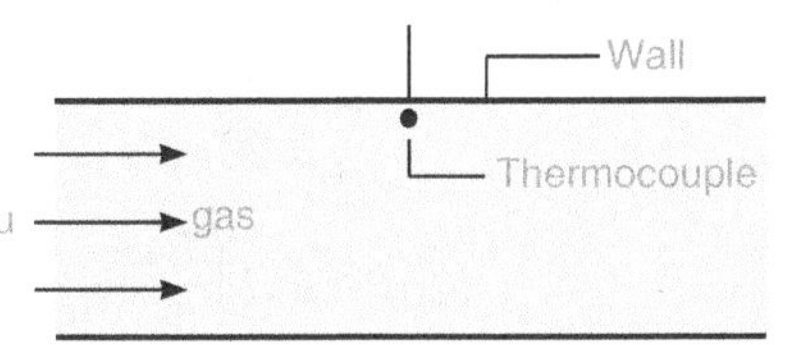

Fig. 7.4. Schematic

Assumptions :

(*i*) Steady state conditions,

(*ii*) Constant properties.

Analysis : The heat transfer coefficient is given by

$$h \propto u^{0.8} \quad \text{or} \quad h = au^{0.8}$$

and the heat transfer rate by radiation

$$Q_{rad} \propto \Delta T \quad \text{or} \quad Q_{rad} = b\ \Delta T$$

where a and b are constants of proportionality under steady state conditions.

Rate of convection heat transfer from thermocouple to gas

= Rate of heat radiation from walls to thermocouple

$$h\ A(T - T_\infty) = Q_{rad}$$

$$a\ u^{0.8}\ A(T - T_\infty) = b(T_w - T)$$

or
$$u^{0.8} = \frac{b}{a}\frac{(T_w - T)}{A(T - T_\infty)} = d\left(\frac{T_w - T}{T - T_\infty}\right)$$

where $d = \frac{b}{aA}$, a new constant.

(*a*) (*i*) At $u_1 = 5$ m/s and $T_1 = 323$ K

$$(5)^{0.8} = d\left(\frac{T_w - 323}{323 - T_\infty}\right)$$

or
$$d = 3.624 \times \left(\frac{323 - T_\infty}{T_w - 323}\right) \quad ...(i)$$

(*ii*) At $u_2 = 10$ m/s, $T_2 = 313$ K

$$(10)^{0.8} = d\left(\frac{T_w - 313}{313 - T_\infty}\right)$$

Using d from eqn. (*i*), we get

$$6.309 = 3.624 \times \left(\frac{323 - T_\infty}{T_w - 323}\right)\times\left(\frac{T_w - 313}{313 - T_\infty}\right)$$

(*iii*) At $T_\infty = 298$ K

$$6.309 = 3.624 \times \left(\frac{323 - 298}{T_w - 323}\right)\times\left(\frac{T_w - 313}{313 - 298}\right)$$

$$1.0446 = \frac{T_w - 313}{T_w - 323}$$

or $1.0446\ T_w - 337.4 = T_w - 313$

or $0.0446\ T_w = 337.41 - 313 = 24$

$T_w = \mathbf{548\ K.}$ **Ans.**

(*b*) At $u_3 = 20$ m/s

$$(20)^{0.8} = d\left(\frac{548 - T}{T - 298}\right)$$

$$10.98 = 3.624 \times \left(\frac{323 - 298}{548 - 323}\right) \times \left(\frac{548 - T}{T - 298}\right)$$

or $10.98\ T - 3272 = 220.66 - 0.40\ T$

$11.37\ T = 3488.7$

or $T = \mathbf{307\ K.}$ Ans.

7.4. CONVECTION BOUNDARY LAYERS

7.4.1. Velocity Boundary Layer

Consider the flow of fluid over a flat plate as shown in Fig. 7.5. The fluid approaches the plate in x direction with a uniform velocity u_∞. The fluid particles in the fluid layer adjacent to the surface get zero velocity. This motionless layer acts to retard the motion of particles in the adjoining fluid layer as a result of friction between the particles of these two adjoining fluid layers at two different velocities. This fluid layer then acts to retard the motion of particles of next fluid layer and so on, until a distance $y = \delta$ from the surface reaches, where these effects become negligible and the fluid velocity u reaches the free stream velocity u_∞. As a result of frictional effects between the fluid layers, the local fluid velocity u will vary from $x = 0, y = 0$ to $y = \delta$.

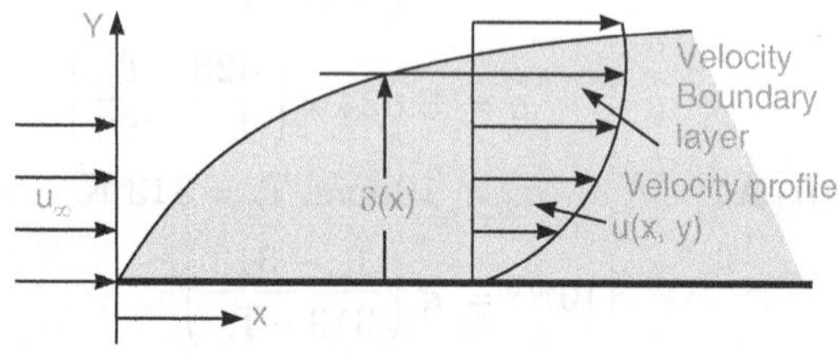

Fig. 7.5. Velocity boundary layer on a flat plate

The region of the flow over the surface bounded by δ *in which the effects of viscous shearing forces caused by fluid viscosity are observed, is called the* **velocity boundary layer or hydrodynamic boundary layer** *or simply the* **boundary layer.** The thickness of boundary layer δ is generally defined as a distance from the surface at which local velocity $u = 0.99$ of free stream velocity u_∞.

The retardation of fluid motion in the boundary layer is due to the shear (viscous) stresses acting in opposite direction. With increasing the distance y from the surface, shear stress decreases, the local velocity u increases until approaches u_∞. With increasing the distance from the leading edge, the effect of viscosity penetrates further into the free stream and boundary layer thickness grows (δ increases with x).

In fluid mechanics, the surface shear stress τ_s in terms of skin friction coefficient C_f is expressed as

$$\tau_s = \frac{C_f}{2}\rho u_\infty^{\,2} \qquad ...(7.7)$$

The surface shear stress may also determined from knowledge of velocity gradient of the fluid at the surface

$$\tau_s = \mu\left[\frac{du}{dy}\right]_{y=0} \qquad ...(7.8)$$

7.4.2. Thermal Boundary Layer

If the fluid flowing on a surface has a different temperature than the surface, the thermal boundary layer is developed in similar manner to velocity boundary layer.

Consider a fluid at temperature T_∞ flows over a surface at a constant temperature T_s. The fluid particles in adjacent layer to the plate get the same temperature that of surface. The particles exchange heat energy with particles in adjoining fluid layer and so on. As a result, the temperature gradients are observed in the fluid layers and a temperature profile is developed in the fluid flow, which ranges from T_s at the surface to fluid temperature T_∞ sufficiently far from the surface in y direction. *The flow region over the surface in which the temperature variation in the direction normal to surface is observed is called* **thermal boundary layer.** The thickness of thermal boundary layer δ_{th} at any location along the length of flow is defined as a distance y from the surface at which the temperature difference $(T - T_s)$ equals 0.99 of $(T_\infty - T_s)$.

or $T_s - T = 0.99(T_s - T_\infty)$ if $T_s > T_\infty$

where T is local temperature in thermal boundary layer, a function of x and y directions.

With increasing the distance from leading edge the effect of heat transfer penetrates further into the free stream and the thermal boundary layer grows as shown in Fig. 7.6 (*a*) and Fig. 7.6 (*b*).

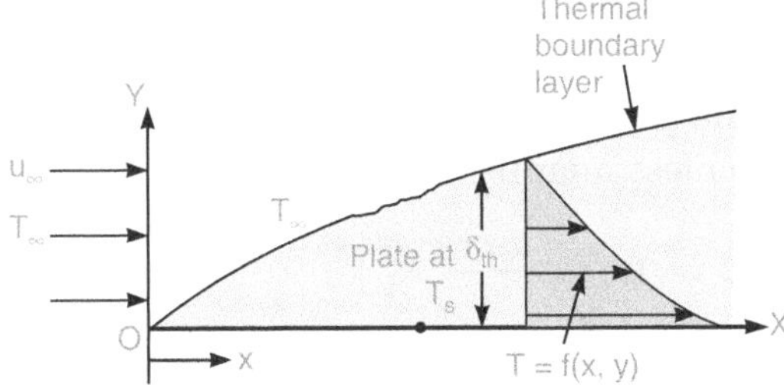

Fig. 7.6. (*a*) Thermal boundary layer for flow of a cold fluid over a hot plate

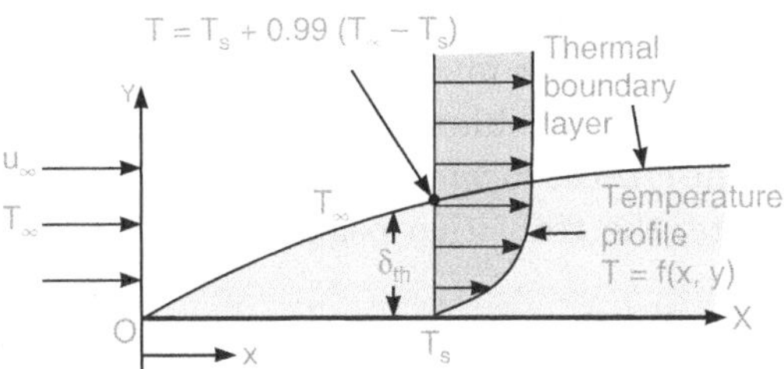

Fig. 7.6. (*b*) Thermal boundary layer for flow of hot fluid on a cold plate

The convection heat transfer rate any where along the surface is directly related to the temperature gradient at that location. Therefore, the shape of the temperature profile in the thermal boundary layer leads to the local convection heat transfer between surface and flowing fluid.

7.4.3. Significance of Boundary Layers

The velocity boundary layer, is of extent $\delta(x)$ and is characterised by the presence of velocity gradients and fluid friction. The thermal boundary layer is of extent $\delta_{th}(x)$ and is characterised by temperature gradient and heat transfer.

For flow over a heated (or cold) surface, both velocity and thermal boundary layers are developed simultaneously. If the effects of fluid viscosity (viscous shear stress) is stronger than thermal effects, then the velocity boundary layer will be thicker than the thermal boundary layer and *vice versa*. The schematic illustrations of relative thickness of $\delta(x)$ and $\delta_{th}(x)$ for liquid metals, gases and oils are shown in Fig. 7.7. For liquid metals, thermal effects are much stronger than viscous effects, and therefore, thermal boundary layer (t.b.l) is much thicker than the velocity boundary layer (v.b.l) Fig. 7.7 (*a*). For gases, the viscous effects are slightly weaker than thermal effects, thereore, thermal boundary layer is a little thicker than velocity boundary layer, Fig. 7.7 (*b*). Similarly for oils, greese etc, the viscous effects are much stronger than thermal effects and thus, the velocity boundary layer is much thicker than the thermal boundary layer Fig. 7.7 (*c*).

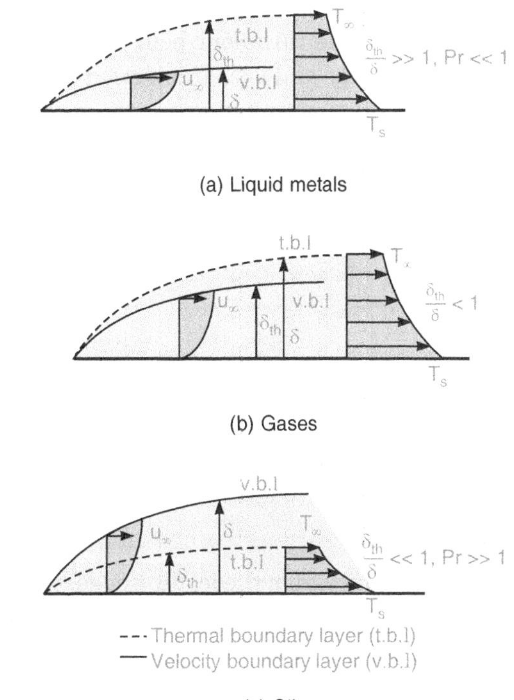

Fig. 7.7. Relative thickness of thermal and velocity boundary layers for different types of fluid

7.5. LAMINAR AND TURBULENT FLOW

The analysis of convection problems requires the knowledge of the type of boundary layer developed, whether it is laminar or turbulent. The type of boundary strongly influences the skin friction and heat transfer coefficient.

The developed boundary layer may consist of laminar boundary, transition region and turbulent boundary layer as shown in Fig. 7.8.

The velocity boundary layer $\delta(x)$ is characterized by the presence of velocity gradients and shear stresses.

The thermal boundary layer $\delta_{th}(x)$ is characterized by temperature gradients and heat transfer.

7.5.1. Laminar Boundary Layer

The velocity boundary layer starts at the leading edge of the plate as a laminar boundary layer, in which the fluid motion is highly ordered and it is possible to identify the stream lines along which particles move. The fluid motion along a stream line is characterized by the velocity components *u* and *v* in both *x* and *y* directions and it influences the momentum and energy transfer through the boundary layer. The velocity profile in laminar boundary layer is approximately parabolic.

7.5.2. Turbulent Boundary Layer

The fluid motion in the turbulent boundary layer has very large disturbances and is characterized by velocity fluctuations. The fluctuations increase the momentum and heat transfer. Due to fluid mixing, the turbulent boundary layer thickness is larger and velocity profiles are flatter with the sharp drop near the surface.

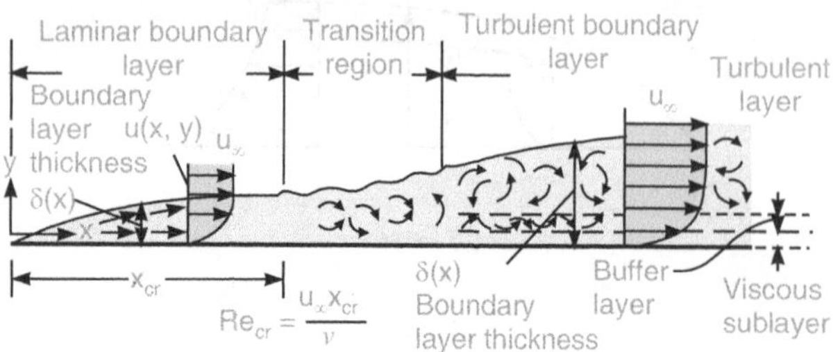

Fig. 7.8. Boundary layer concept for flow along a flat plate

At some distance from the leading edge, the small disturbances in the flow begin to be amplified and the fluid fluctuations begin to develop, it is *transition* from laminar to turbulent boundary layer as shown in Fig. 7.8. The transition to turbulence is attained by significant increase in boundary layer thickness, wall shear stress, and heat transfer coefficient. These effects are shown in Fig. 7.9.

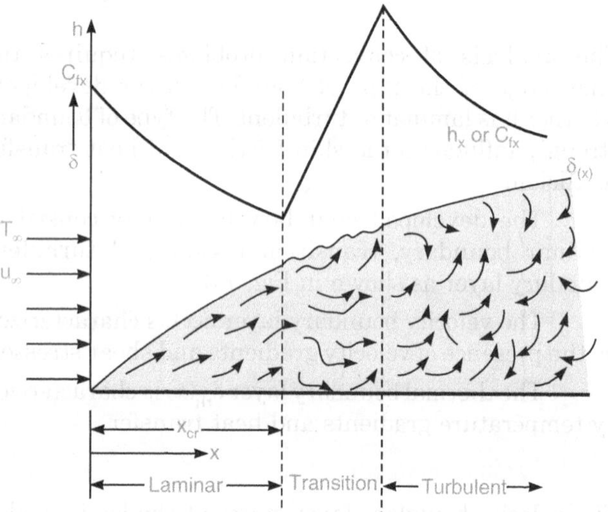

Fig. 7.9. Variation of velocity boundary layer $\delta(x)$, local heat transfer coefficient h_x and local friction coefficient C_{fx} for flow over a flat plate

The **turbulent boundary layer** has three different regions. A *laminar sublayer* is very thin layer next to wall in which flow is laminar. Adjacent to laminar sublayer, there is a *buffer layer* in which small disturbances exist. The buffer layer is followed by the *turbulent layer* with larger turbulences. The momentum and heat transfer mechanisms involve the fluid lumps moving randomly.

The characteristic of fluid flow is governed by dimensionless quantity called the *Reynolds number* as

$$Re_x = \frac{u_\infty x}{\nu} \quad ...(7.9)$$

where, u_∞ = free stream velocity, m/s,

ν = kinematic viscosity, m²/s,

x = distance from the leading edge for flow over a flat plate, m.

The Reynolds number at which the transition from laminar to turbulent boundary layer takes place is called the *critical Reynolds number* and for flow along a flat plate, the transition begins at critical Reynolds number

$$Re_{cr} \approx 5 \times 10^5. \quad ...(7.10)$$

Example 7.4. *Water flows at 20°C at 8 kg/s through the diffuser having 3 cm diameter at the entrance and 7.0 cm diameter at its exit. Calculate the fluid velocity and Reynolds number at the inlet and exit of the diffuser.*

Solution

Given : Flow of water through a diffuser.

D_1 = 3.0 cm = 0.03 m, D_2 = 7.0 cm = 0.07 m

$\dot{m}$ = 8 kg/s.

To find :

(*i*) Velocity of water at inlet and exit of diffuser.

(*ii*) Reynolds number at the inlet and exit of diffuser.

Assumptions :

(*i*) Steady flow conditions.

(*ii*) Constant properties of fluid.

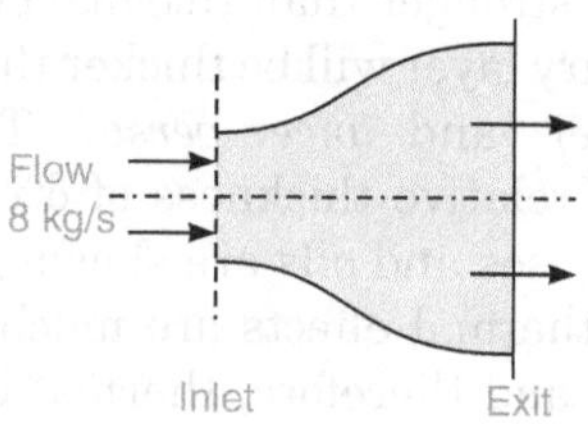

Fig. 7.10. Schematic of diffuser

Analysis : (*i*) The properties of water at 20°C

δ = 1000 kg/m³, μ = 1007.4 × 10⁻⁶ kg/m/s

The flow cross-sectional area at inlet

$$A_1 = \frac{\pi}{4} D_1^2 = \frac{\pi}{4} \times (0.03 \text{ m})^2 = 7.069 \times 10^{-4} \text{ m}^2$$

At exit $A_2 = \frac{\pi}{4} D_2^2 = \frac{\pi}{4} \times (0.07 \text{ m})^2$

$= 3.848 \times 10^{-3} \text{ m}^2$

Using continuity equation, the velocities.

At inlet $u_1 = \frac{\dot{m}}{\rho_1 A_1}$

$$= \frac{8 \text{ kg/s}}{(1000 \text{ kg/m}^3) \times (7.069 \times 10^{-4} \text{m}^2)}$$

$= \mathbf{11.32 \text{ m/s.}}$ **Ans.**

At exit $u_2 = \frac{\dot{m}}{\rho_2 A_2}$

$$= \frac{(8 \text{ kg/s})}{(1000 \text{ kg/m}^3) \times (3.848 \times 10^{-3} \text{ m}^2)}$$

$= \mathbf{2.08 \text{ m/s.}}$ **Ans.**

(*ii*) The respective Reynolds numbers

$$\text{Re}_1 = \frac{\rho u_1 D_1}{\mu} = \frac{1000 \times 11.32 \times 0.03}{1007.4 \times 10^{-6}}$$

$= \mathbf{337105.}$ **Ans.**

and $$\text{Re}_2 = \frac{\rho u_2 D_2}{\mu} = \frac{1000 \times 2.08 \times 0.07}{1007.4 \times 10^{-6}}$$

$= \mathbf{144530.}$ **Ans.**

Example 7.5. *A fan provides air speed upto 50 m/s, is used in low speed wind tunnel with atmospheric air at 27°C. If this wind tunnel is used to study the boundary layer behaviour over a flat plate upto Re = 10⁸. What should be the minimum plate length ? At what distance from the leading edge would transition occur, if critical Reynolds number is* $Re_{cr} = 5 \times 10^5$ *?*

Solution

Given : Flow over a flat plate

$u_\infty = 50$ m/s, $T_\infty = 27$°C

$\text{Re}_x = 10^8$, $\text{Re}_{cr} = 5 \times 10^5$.

To find : (*i*) Minimum length of a flat plate for $\text{Re}_x = 10^8$.

(*ii*) Distance from the leading edge for $\text{Re}_{cr} = 5 \times 10^5$.

Analysis : The properties of air at 27°C from Table A-4

$\rho = 1.16 \text{ kg/m}^3$

$\mu = 184.6 \times 10^{-7}$ kg/m/s

(*i*) The Reynolds number is expressed as

$$\text{Re}_x = \frac{\rho u_\infty x}{\mu} \quad \text{or} \quad x = \frac{\text{Re}_x \, \mu}{\rho u_\infty}$$

For $\text{Re}_x = 10^8$,

$$x = \frac{10^8 \times 184.6 \times 10^{-7}}{1.16 \times 50} = \mathbf{31.82 \text{ m}} \ \textbf{Ans.}$$

The minimum length of the plate for Re = 10^8 is 31.82 m.

(*ii*) For transition to occur at $\text{Re}_{cr} = 5 \times 10^5$

$$x_{cr} = \frac{5 \times 10^5 \times 184.6 \times 10^{-7}}{1.16 \times 50}$$

$= \mathbf{0.159 \text{ m.}}$ **Ans.**

The transition from laminar to turbulent will occur at $x = 0.159$ m.

7.6. MOMENTUM EQUATION FOR LAMINAR BOUNDARY LAYER

Considering two dimensional control volume as shown in Fig. 7.11. The equation of motion for the laminar boundary layer can be obtained by equating force and momentum transfer on the element. The assumptions made in the analysis are

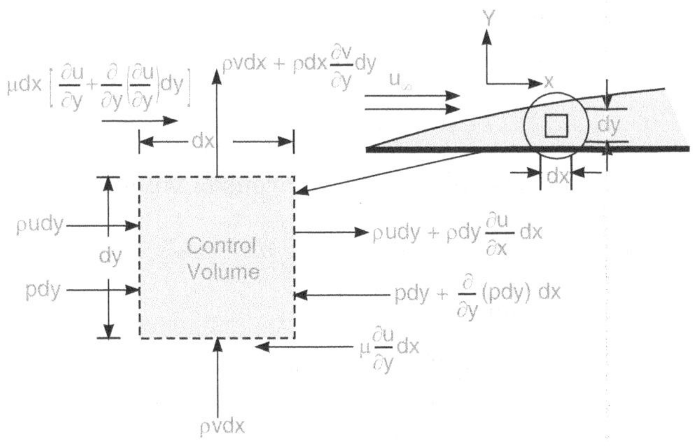

Fig. 7.11. Force and momentum analysis for laminar boundary layer

1. The flow is incompressible and steady ;
2. No pressure variation in perpendicular direction of plate ;
3. Viscosity is constant ;
4. Negligible shear forces in *y* direction ;
5. Unit depth in *z* direction.

According to Newton's second law of motion

$$\Sigma F_x = \frac{d(mv)_x}{dt} \qquad ...(7.11)$$

where ΣF_x = sum of applied forces in x direction, and

$\frac{d(mv)_x}{dt}$ = rate of increase in momentum flux in x direction.

The *momentum flux* in x direction is product of mass flow rate through a particular side of control volume and x directional velocity component at that point.

The rate of mass entering the left face of control volume

$$= \rho u dy$$

The rate of momentum entering the left face of control volume

$$= \rho u dy u = \rho u^2 dy.$$

The rate of mass leaving the right face

$$= \rho u dy + \rho dy \frac{\partial u}{\partial x} dx$$

The rate of momentum leaving the right face

$$= \rho u^2 dy + \rho dy\, dx \frac{\partial}{\partial x}(u^2)$$

$$= \rho u^2 dy + \rho dy dx \left[u \frac{\partial u}{\partial x} + u \frac{\partial u}{\partial x}\right]$$

The rate of mass entering the bottom face = $\rho v dx$

The rate of mass leaving the top face

$$= \rho v dx + \rho dx \frac{\partial v}{\partial y} dy$$

The mass balances on the control volume

$$\rho u dy + \rho v dx = \rho u dy + \rho dy \frac{\partial u}{\partial x} dx + \rho v dx + \rho dx \frac{\partial v}{\partial y} dy$$

Rearranging we get ;

$$\rho \left\{\frac{\partial u}{\partial x} + \frac{\partial v}{\partial y}\right\} dxdy = 0$$

or

$$\left\{\frac{\partial u}{\partial x} + \frac{\partial v}{\partial y}\right\} = 0 \qquad ...(7.12)$$

It is the mass continuity equation for the laminar boundary layer.

The rate of momentum in x direction associated with mass entering the bottom face

$$= \rho v u dx$$

The rate of momentum in x direction leaves the top face

$$= \rho v u dx + \rho dx \frac{\partial}{\partial y}(uv)\, dy$$

$$= \rho u v dx + \rho dx dy \left\{u \frac{dv}{dy} + v \frac{du}{dy}\right\}$$

The net momentum transfer in x direction

$$\frac{d(mv)_x}{dt} = \rho dx dy \left\{u \frac{\partial u}{\partial x} + u \frac{\partial u}{\partial x} + u \frac{\partial v}{\partial y} + v \frac{\partial u}{\partial y}\right\}$$

$$= \rho dx dy \left[u\left(\frac{\partial u}{\partial x} + \frac{\partial v}{\partial y}\right) + \left\{u \frac{du}{dx} + v \frac{du}{dy}\right\}\right]$$

or

$$\frac{d(mv)_x}{dt} = \rho dx dy \left\{u \frac{\partial u}{\partial x} + v \frac{\partial u}{\partial y}\right\} \qquad ...(7.13)$$

Since $u \left\{\frac{\partial u}{\partial x} + \frac{\partial v}{\partial y}\right\} = 0$ (continuity equation)

The forces acting in x direction are viscous and pressure forces.

The pressure force on the left face

$$= p dy$$

The pressure force on the right face

$$= -p dy - \frac{\partial}{\partial x}(p dy)\, dx$$

($\because$ in opposite direction)

The viscous force at bottom face

$$= -\mu \frac{\partial u}{\partial y} dx$$

The viscous force at top face

$$= \mu \frac{\partial u}{\partial y} dx + \mu \frac{\partial}{\partial y}\left\{\frac{\partial u}{\partial y}\right\} dx dy$$

Net forces in x direction

$$\Sigma F_x = -\frac{\partial p}{\partial x} dx dy + \mu \frac{\partial^2 u}{\partial y^2} dx dy \qquad ...(7.14)$$

Substituting eqns. (7.13) and (7.14) in eqn. (7.11), we get

$$\rho \left\{u \frac{\partial u}{\partial x} + v \frac{\partial u}{\partial y}\right\} = \mu \frac{\partial^2 u}{\partial y^2} - \frac{\partial p}{\partial x} \qquad ...(7.15)$$

The eqn. (7.15) is the *momentum equation* for the laminar boundary layer with constant properties. If the pressure changes on two side of control volume is negligible then above equation reduces to

$$u \frac{\partial u}{\partial x} + v \frac{\partial u}{\partial y} = \frac{\mu}{\rho} \frac{\partial^2 u}{\partial y^2} = \nu \frac{\partial^2 u}{\partial y^2} \qquad ...(7.16)$$

7.7. ENERGY EQUATION FOR THE LAMINAR BOUNDARY LAYER

Consider the element control volume as shown in Fig. 7.12

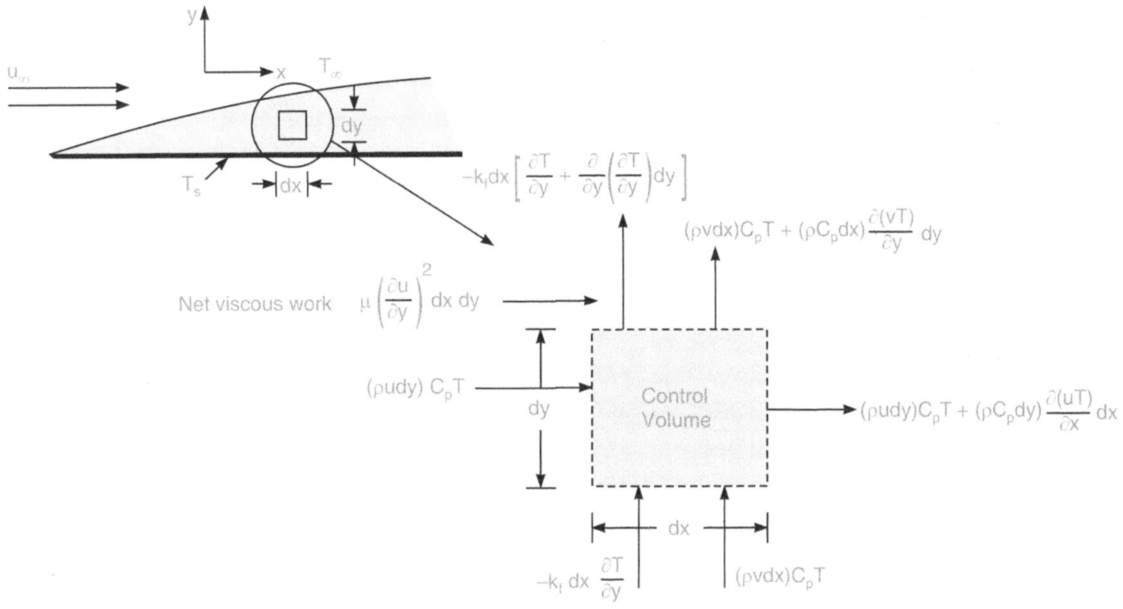

Fig. 7.12. Energy analysis for laminar boundary layer

The assumption made to simplify the analysis :

1. Incompressible steady flow ;
2. Constant properties ;
3. Negligible heat conduction in direction of fluid flow.

The energy balance on the control volume can be expressed as

Energy convected at the left face
+ energy convected at the bottom face
+ heat conducted in the bottom face + net viscous work done on the element
= Energy convected out the right face
+ energy convected out the top face
+ heat conducted out the top face.

Writing the each quantity separately ;

Energy convected in the left face

$$= (\rho u dy) C_p T = \rho C_p (uT) dy$$

Energy convected in the bottom face

$$= (\rho v dx) C_p T = \rho C_p (vT) dx$$

Energy conducted in the bottom face

$$= -k_f dx \frac{\partial T}{\partial y}$$

Energy convected out the right face

$$= \rho C_p (uT)\, dy + \rho C_p\, dy \frac{\partial}{\partial x} (u\, T)\, dx$$

Energy conducted out the top face

$$= \rho C_p (vT)\, dx + \rho C_p\, dx \frac{\partial}{\partial y} (v\, T)\, dy$$

Energy conducted out the top face

$$= -k_f \left[\frac{\partial T}{\partial y} + \frac{\partial^2 T}{\partial y^2}\right] dx\, dy$$

Net energy change rate,

$$E'_{net} = \rho C_p\, dxdy \left\{\frac{\partial}{\partial x}(u\, T) + \frac{\partial}{\partial y}(v\, T)\right\} - k_f \frac{\partial^2 T}{\partial y^2} dx\, dy \quad ...(7.17)$$

The net viscous force

$$F_D = \mu \frac{\partial u}{\partial y} dx$$

Element moves through a distance per unit time

$$= \frac{\partial u}{\partial y} dy$$

Net viscous work done on the element,

$$W_D = \mu \left(\frac{\partial u}{\partial y}\right)^2 dx\, dy$$

Writing the energy balance on the element, we get

$$\rho C_p \left\{ \frac{\partial(uT)}{\partial x} + \frac{\partial(vT)}{\partial y} \right\} = \mu \left(\frac{\partial u}{\partial y} \right)^2 + k_f \frac{\partial^2 T}{\partial y^2}$$

or $$\rho C_p \left[\left(u \frac{\partial T}{\partial x} + v \frac{\partial T}{\partial y} \right) + T \left(\frac{\partial u}{\partial x} + \frac{\partial v}{\partial y} \right) \right]$$

$$= \mu \left\{ \frac{\partial u}{\partial y} \right\}^2 + k_f \frac{\partial^2 T}{\partial y^2} \qquad ...(7.18)$$

Using the continuity eqn. (7.12),

$$\frac{\partial u}{\partial x} + \frac{\partial v}{\partial y} = 0$$

Rearranging eqn. (7.18), we get

$$u \frac{\partial T}{\partial x} + v \frac{\partial T}{\partial y} = \frac{k_f}{\rho C_p} \frac{\partial^2 T}{\partial y^2} + \frac{\mu}{\rho C_p} \left\{ \frac{\partial u}{\partial y} \right\}^2$$

or $$u \frac{\partial T}{\partial x} + v \frac{\partial T}{\partial y} = \alpha \frac{\partial^2 T}{\partial y^2} + \frac{\nu}{C_p} \left\{ \frac{\partial u}{\partial y} \right\}^2 \qquad ...(7.19)$$

For low velocity flow, viscous forces are negligibly small in comparison to conduction term, then

$$u \frac{\partial T}{\partial x} + v \frac{\partial T}{\partial y} = \alpha \frac{\partial^2 T}{\partial y^2} \qquad ...(7.20)$$

Example 7.6. *The velocity profile u(x, y) for a boundary layer flow over a flat plate is given by*

$$\frac{u(x, y)}{u_\infty} = \frac{3}{2} \frac{y}{\delta} - \frac{1}{2} \left[\frac{y}{\delta} \right]^3$$

where the boundary layer thickness δ(x) is the function of x and is given by

$$\delta(x) = \sqrt{\frac{280 \nu x}{13 u_\infty}}$$

(i) Develop an expression for local drag coefficient C_{fx}.

(ii) Develop an expression for average drag coefficient C_f *over a distance x = L from the leading edge of the plate.*

Solution

Given : The velocity profile for the boundary layer as

$$\frac{u}{u_\infty} = \frac{3}{2} \frac{y}{\delta} - \frac{1}{2} \left(\frac{y}{\delta} \right)^3$$

and $$\delta(x) = \sqrt{\frac{280}{13} \frac{\nu x}{u_\infty}}$$

To find :

(*i*) An expression for local friction coefficient C_{fx},

(*ii*) An expression for average friction coefficient C_f.

Analysis : (*i*) The shear stress at the wall is expressed as

$$\tau_s = \mu \left. \frac{\partial u}{\partial y} \right|_{y=0} \qquad ...(i)$$

and shear stress in terms of local friction coefficient as

$$\tau_s = \frac{C_{fx}}{2} \rho u_\infty^2 \qquad ...(ii)$$

Equating two equations for shear stress at the surface, we get

$$C_{fx} = \frac{2\mu}{\rho u_\infty^2} \left. \frac{\partial u}{\partial y} \right|_{y=0}$$

For given velocity profile

$$\left. \frac{\partial u}{\partial y} \right|_{y=0} = u_\infty \left[\frac{3}{2} \times \frac{1}{\delta} - \frac{1}{2} \cdot \frac{3y^2}{\delta^3} \right]_{y=0} = \frac{3u_\infty}{2\delta}$$

Then

$$C_{fx} = \frac{2\mu}{\rho u_\infty^2} \times \frac{3}{2} \frac{u_\infty}{\delta} = \frac{3\nu}{u_\infty \delta}$$

Introducing the expression for δ(*x*), we get,

$$C_{fx} = \frac{3\nu}{u_\infty} \times \sqrt{\frac{13 u_\infty}{280 \nu x}} = 0.646 \sqrt{\frac{\nu}{u_\infty x}}$$

$$= \frac{\mathbf{0.646}}{\sqrt{\mathbf{Re_x}}}. \quad \textbf{Ans.}$$

(*ii*) The average friction coefficient C_f is given by

$$C_f = \frac{1}{L} \int_0^L C_{fx}\, dx = \frac{1}{L} \int_0^L 0.646 \sqrt{\frac{\nu}{u_\infty x}}\, dx$$

$$= \frac{0.646}{L} \sqrt{\frac{\nu}{u_\infty}} \times \int_0^L x^{-1/2}\, dx$$

$$= \frac{0.646}{L} \sqrt{\frac{\nu}{u_\infty}} \left(\frac{x^{-1/2+1}}{-1/2+1} \right)_0^L$$

$$= 2 \times 0.646 \sqrt{\frac{\nu}{u_\infty L}} = \frac{2 \times 0.646}{\sqrt{Re_L}}$$

$$= 2\, \mathbf{C_{fx}} \Big|_{\mathbf{x=L}} \quad \textbf{Ans.}$$

Example 7.7. *The temperature profile in a thermal boundary layer for flow over a flat plate is given by* $\frac{T(x, y) - T_s}{T_\infty - T_s} = \frac{3}{2}\frac{y}{\delta_{th}} - \frac{1}{2}\left(\frac{y}{\delta_{th}}\right)^3$ *and the thickness of thermal boundary layer* δ_{th} *is the function of x and is given by* $\delta_{th}(x) = 4.53 \frac{x}{Re_x^{1/2}\, Pr^{1/3}}$ *where,* $Pr = \frac{\mu C_p}{k_f}$ *and* $Re = \frac{\rho u_\infty x}{\mu}$. *Develop the expressions for local and average heat transfer coefficients.*

Solution

Given : The temperature profile in a thermal boundary layer as

$$\frac{T - T_s}{T_\infty - T_s} = \frac{3}{2}\left(\frac{y}{\delta_{th}}\right) - \frac{1}{2}\left(\frac{y}{\delta_{th}}\right)^3$$

and
$$\delta_{th} = 4.53 \frac{x}{Re_x^{1/2}\, Pr^{1/3}}$$

with
$$Pr = \frac{\mu C_p}{k_f}, \quad \text{and} \quad Re = \frac{\rho u_\infty x}{\mu}$$

To find : The expressions for local and average heat transfer coefficients.

Analysis : The local heat transfer coefficient, h_x is expressed as

$$h_x = \frac{k_f \left.\frac{\partial T}{\partial y}\right|_{y=0}}{T_\infty - T_s}$$

For given temperature profile

$$\left.\frac{\partial T}{\partial y}\right|_{y=0} = (T_\infty - T_s)\left[\frac{3}{2}\frac{1}{\delta_{th}} - \frac{1}{2} \times \frac{3y^2}{\delta_{th}^3}\right]_{y=0}$$

$$= \frac{3(T_\infty - T_s)}{2\delta_{th}}$$

Then
$$h_x = \frac{3k_f(T_\infty - T_s)}{2(T_\infty - T_s)\delta_{th}} = \frac{3}{2}\frac{k_f}{\delta_{th}}$$

Introducing the expression for δ_{th}

$$h_x = \frac{3}{2} \times \frac{k_f \times Re_x^{1/2}\, Pr^{1/3}}{4.53\, x}$$

$$= \mathbf{0.332 \frac{k_f}{x} Re_x^{1/2}\, Pr^{1/3}}. \quad \textbf{Ans.}$$

This expression can be arranged in the dimensionless form as

$$Nu_x = \frac{h_x\, x}{k_f} = 0.332\, Re_x^{1/2}\, Pr^{1/3}$$

where Nu_x is called the local *Nusselt number*.

The average heat transfer coefficient

$$h = \frac{1}{L}\int_0^L h_x\, dx$$

$$= 0.332\, k_f \frac{Pr^{1/3}}{L}\int_0^L \frac{1}{x} \times \sqrt{\frac{\rho u_\infty x}{\mu}}\, dx$$

$$= 0.332\, k_f Pr^{1/3}\sqrt{\frac{\rho u_\infty}{\mu}} \times \frac{1}{L}\int_0^L x^{-1/2}\, dx$$

$$= 0.332\frac{k_f}{L} Pr^{\frac{1}{3}}\sqrt{\frac{\rho u_\infty}{\mu}}\left[\frac{L^{1/2}}{1/2}\right]$$

$$= 2 \times 0.332\frac{k_f}{L}\sqrt{\frac{\rho u_\infty L}{\mu}}\, Pr^{1/3}$$

or
$$h = 2 \times 0.332 \frac{k_f}{L} Re_L^{1/2}\, Pr^{1/3} = 2h_x\Big|_{x=L}$$

Average Nusselt No. Nu = $\mathbf{2Nu_x}$. **Ans.**

7.8. BOUNDARY LAYER SIMILARITIES

The eqn. (7.16) and eqn. (7.20) derived earlier for low speed forced convection flow are of practical importance in many engineering applications. By close examination of above two equations, we find that the two equations are of same form. Each equation is characterised by advection term on the left hand side and a diffusion term on right hand side : These similarity may be extended in a rational manner by non dimensionalizing the governing equations.

The boundary layer equations are normalized by defining dimensionless indenpedent variables of the following form.

$$x^* = \frac{x}{L}, y^* = \frac{y}{L} \qquad ...(7.21)$$

where L is characteristic length for the surface of interest such as length of a flat plate. The dependent dimensionless variables may be defined as

$$u^* = \frac{u}{u_\infty}, v^* = \frac{v}{u_\infty} \quad \text{and} \quad T^* = \frac{T - T_s}{T_\infty - T_s} \qquad ...(7.22)$$

where u_∞ and T_∞ are velocity and temperature of free stream fluid respectively, and T_s is the temperature of the surface. Substituting eqns. (7.21) and (7.22) in eqns. (7.12), (7.16) and (7.20) to obtain the corresponding boundary layer equations in non dimensional form as shown in Table 7.1.

TABLE 7.1. Convection transfer equations and their boundary conditions in non dimensional form

Boundary layer	*Transfer equation*	*Boundary conditions*		*Similarity parameter(s)*
		wall	*free stream*	
Continuity	$\frac{\partial u*}{\partial x*}+\frac{\partial v*}{\partial y*}=0$ ...(7.23)	—	—	—
Velocity	$u*\frac{\partial u*}{\partial x*}+v*\frac{\partial u*}{\partial y*}=\frac{\nu}{u_\infty L}\frac{\partial^2 u*}{\partial y*}$ ...(7.24)	$u*(x*, 0) = 0$ $v*(x*, 0) = 0$	$u*(x*, \infty) = 1$	Re_L
Thermal	$u*\frac{\partial T*}{\partial x*}+v*\frac{\partial T*}{\partial y*}=\frac{\alpha}{u_\infty L}\frac{\partial^2 T*}{\partial y*^2}$ ...(7.25)	$T*(x*, 0) = 0$	$T*(x*, \infty) = 1$	Re_L Pr

From eqns. (7.24) and (7.25), the two similarity parameters may be concluded. These similarity parameters are important, because they permit us to apply solutions from one configuration to another geometrical similar configuration under entirely different conditions. For example, if the Reynolds number is same, the dimensionless velocity distribution for air, water and glycerine etc. flowing over a flat plate will be the same at a given value of $x*$.

Eqn. (7.23) indicates that $v*$ is related to $u*$, $x*$ and $y*$, thus

$$v* = f_1(u*, x*, y*) \qquad ...(7.26)$$

Similarly, from eqn. (7.24), $u*$ can be expressed in the form

$$u* = f_2(x*, y*, Re_L) \qquad ...(7.27)$$

The eqn. (7.24) has a quantity $\frac{\nu}{u_\infty L}$ on its right hand side. This quantity is a dimensionless group and its reciprocal is well known *Reynolds number.*

$$Re_L = \frac{u_\infty L}{\nu} \qquad ...(7.28)$$

From eqn.(7.25) the term $\frac{\alpha}{u_\infty L}$ is also a dimensionless group and it may be expressed as

$$\frac{\alpha}{u_\infty L} = \left(\frac{\nu}{u_\infty L}\right)\times\left(\frac{\alpha}{\nu}\right) = \frac{1}{Re_L}\frac{1}{Pr} \qquad ...(7.29)$$

The ratio of two properties (α/ν) is also a dimensionless property and its reciprocal is referred as **Prandtl number** (Pr)

or
$$Pr = \frac{\nu}{\alpha} \qquad ...(7.30)$$

7.8.1. Friction Coefficient

Shear stress at the surface is given by eqn. (7.8)

$$\tau_s = \mu \left.\frac{\partial u}{\partial y}\right|_{y=0}$$

Substituting u and y from eqns. (7.21) and (7.22), we get

$$\tau_s = \mu \frac{u_\infty}{L}\left.\frac{\partial u*}{\partial y*}\right|_{y*=0} \qquad ...(7.31)$$

Defining local skin friction coefficient using eqn. (7.7)

$$C_{fx} = \frac{\tau_s}{\rho u_\infty^2/2} \qquad ...(7.32)$$

Substituting eqn. (7.31) for τ_s, we get

$$C_{fx} = \frac{2\mu u_\infty}{\rho u_\infty^2 L}\left.\frac{\partial u*}{\partial y*}\right|_{y*=0} = \frac{2}{Re_L}\left.\frac{\partial u*}{\partial y*}\right|_{y*=0} \qquad ...(7.33)$$

From eqn. (7.33) it is also evident that

$$C_{fx} = f_3(x*, Re_L) \qquad ...(7.34)$$

It indicates that for flow over bodies of similar shape the local friction coefficient is function of $x*$ and Re_L and it is independent of fluid or free stream velocity.

7.8.2. Nusselt Number

In convection heat transfer, the local heat transfer coefficient is expressed as

$$h_x = -\frac{k_f\left(\frac{\partial T}{\partial y}\right)_{y=0}}{T_s - T_\infty}$$

In non dimensional form

$$h_x = -k_f \frac{(T_\infty - T_s)}{(T_s - T_\infty)L} \frac{\partial T^*}{\partial y^*}\bigg|_{y^*=0} = \frac{k_f}{L} \frac{\partial T^*}{\partial y^*}\bigg|_{y^*=0} \quad ...(7.35)$$

In appropriate dimensionless form

$$Nu_x = \frac{h_x L}{k_f} = \frac{\partial T^*}{\partial y^*}\bigg|_{y^*=0} = f_4(x^*, Re_L, Pr) \quad ...(7.36)$$

The quantity $h_x L/k_f$ is called the local *Nusselt Number*. It is equal to dimensionless temperature gradient at the surface and thus it provides a measure of convection heat transfer occurring at the surface. The role of local Nusselt number in thermal boundary layer is same as that of local friction coefficient in velocity boundary layer.

The average value of Nusselt number gives average value of heat transfer coefficient h, that is independent of x^*

$$Nu = \frac{hL}{k_f} = f_5(Re_L, Pr) \quad ...(7.37)$$

Example 7.8. *Experimental test on a portion of a turbine blade as shown in Fig 7.13 (a) indicates a heat flux of 95000 W/m².*

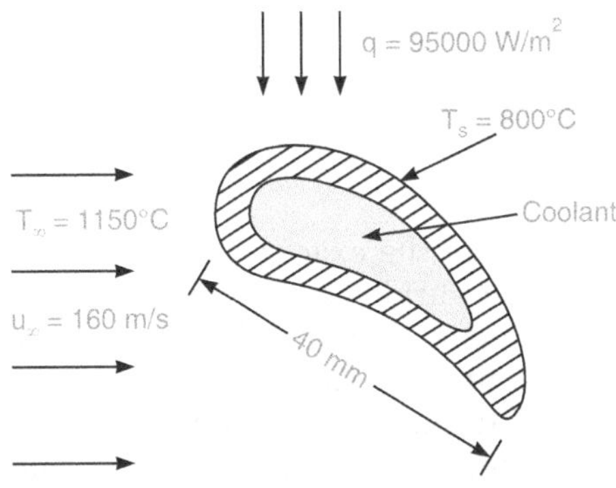

Fig. 7.13. (a) Original condition

The blade is cooled at inside in order to maintain its temperature constant at 800°C.

(i) Determine the heat flux to the blade if its temperature is reduced to 700°C by increasing the coolant flow,

(ii) Calculate the heat flux at same dimensionless location for a similar turbine blade having a chord length of 80 mm when the blade operates in an air flow at T_∞ = 1150°C and u_∞ = 80 m/s with T_s = 800°C.

Solution

Given : Operating conditions of an internally cooled turbine blade as shown in Fig. 7.13 (*a*).

To find :

(*i*) Heat flux to the blade, when surface temperature is lowered to 700°C.

(*ii*) Heat flux to larger identical turbine blade with reduced velocity to 80 m/s.

Schematic :

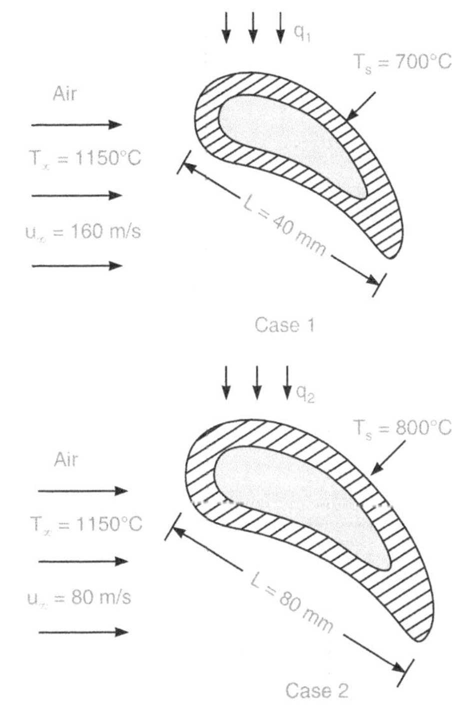

Fig. 7.13 (*b*)

Assumptions :

(*i*) Steady state conditions.

(*ii*) Constant air properties.

Analysis : (*i*) From eqn. (7.36) for given geometry

$$Nu_x = \frac{h_x L}{k_f} = f(x^*, Re_L, Pr)$$

Since there is no change in dimension and environmental conditions, thus x^*, Re_L, Pr will remain same even with change in T_s. The Nusselt number is also unchanged, thus the local heat transfer coefficient h_x also remains same.

For case 1. The heat flux can be given by

$$q_1 = h_{x_1}(T_\infty - T_s),$$

where $h_{x_1} = h = \frac{q_1}{T_\infty - T_s} = \frac{95000}{1150 - 800}$

$= 271.43 \text{ W/m}^2\text{.K}$

and $q_1 = 271.43 \times (1150 - 700)$

$= \mathbf{122143 \text{ W/m}^2\text{.K.}}$ Ans.

(*ii*) For case 2. The blade size is increased to 2L and free stream air velocity is reduced to one half, therefore

$$Re_{L_2} = \frac{1/2\, u_\infty (2L)}{\nu} = Re_L$$

Environment is same, thus Pr remains unchanged therefore, Nu also remains same

$$Nu_2 = Nu$$

or $\frac{h_{x_2} L_2}{k_f} = \frac{h_x L}{k_f}$

or $h_{x_2} = h_x \frac{L}{2L} = 271.43 \times \frac{1}{2}$

$= 135.7 \text{ W/m}^2\text{.K}$

and the heat flux

$$q_2 = h_{x_2} (T_\infty - T_s)$$

$= 135.7 \times (1150 - 800)$

$= \mathbf{47500 \text{ W/m}^2}$. Ans.

7.9. DETERMINATION OF CONVECTION HEAT TRANSFER COEFFICIENT

There are five general methods, that may be used for determination of heat transfer coefficient :

1. Dimensional analysis combined with experimental data.
2. Exact mathematical solution of boundary layer equations.
3. Approximate analysis of boundary layer equations by integral methods.
4. Analogy between heat and momentum transfer.
5. Numerical analysis.

All methods can evaluate the heat transfer coefficient, but no single method can solve all types of problems, because each method has its own limitations that restrict its scope of applications.

7.9.1. Dimensional Analysis

It is a mathematically simple method and has a wide range of applications. Its main limitation is that, the obtained results are incomplete and useless without experimental validation. It does not provide any information about the phenomenon, but facilates the interpretation of variables to obtain them in certain dimensionless groups. The dimensional analysis for forced and natural convection is discussed in next articles.

7.9.2. Exact Mathematical Solutions

This method requires solution of simultaneous equations related with fluid motion and energy transfer in moving fluid. The complete mathematical equations describing fluid flow and heat transfer can only be written for laminar flow. Even for laminar flow, the equations are quite complicated and their solution is very tedious.

Exact solutions are important, because they serve as basis for comparison and a check on approximated solutions.

7.9.3. Approximate Analysis of Boundary Layers

In these methods, the detailed mathematical formulation for flow in the boundary layer is avoided. But simple equations are used to approximate momentum and energy transfer of fluid flow and they are solved by integral method. The method is relatively simple and yielding solutions are well agree with exact solutions within certain range. This technique is used to laminar flow as well as to turbulent flow.

7.9.4. Analogy between Heat and Momentum Transfer

It is very useful tool for analysis of the turbulent flow process. Because our knowledge of turbulent exchange mechanism is quite limited and thus we cannot write equations completely.

7.9.5. Numerical Analysis

It can approximate the exact equations. It requires to express the field variables at descrete points in time and space coordinate. However, the solution can be made sufficiently accurate with proper descretization of problem field. It has one advantage that once the solution procedure is programmed, the solution for different boundary conditions, property variables and so on can easily be handled.

7.10. DIMENSIONAL ANALYSIS

Dimensional analysis differs from the conventional methods of approach in which certain equations are solved for a resulting equation. Instead, it combines several variables affecting a phenomenon in dimensionless group, such as Nusselt number, which facilitates the interpretation and extends its application to experimental data.

The dimensional analysis does not give any information about the nature of phenomenon, hence the success or failure of the method depends on proper selection of affecting variables. It is therefore, important to understand the physics of phenomenon before applying dimensional analysis.

7.10.1. Primary Dimensions and Dimensional Formulae

In SI system of units the primary dimension of length L, time t, temperature T and mass M are used.

The dimensional formula in primary dimensions for a physical quantity is obtained from its definition of physical laws. For an example, the dimensional formula for the length of a rod is (L) by definition, for velocity (distance/time) is Lt^{-1} and so on.

The symbols, units and dimensions of commonly used quantities in heat transfer analysis are listed in Table 7.2.

TABLE 7.2. Important physical quantities with their symbols, units and primary dimensions

Sr. No.	*Quantity*	*Symbol and units*	*Primary dimensions*
1.	Length, diameter	L or D, m	L
2.	Time	t, s	t
3.	Mass	m, kg	M
4.	Temperature	T, °C	T
5.	Area	A, m^2	L^2
6.	Volume	V, m^3	L^3
7.	Density	ρ, kg/m^3	ML^{-3}
8.	Velocity	u, v or u_∞, m/s	Lt^{-1}
9.	Acceleration	a or g, m/s^2	Lt^{-2}
10.	Force	F, N(kg m/s^2)	MLt^{-2}
11.	Pressure	p, N/m^2	$ML^{-1} t^{-2}$
12.	Shear stress	τ, N/m^2	$ML^{-1} t^{-2}$
13.	Heat transfer rate	Q, W(Nm/s)	$ML^2 t^{-3}$
14.	Specific heat	C_p, J/kg. K	$L^2 T^{-1} t^{-3}$
15.	Dynamic viscosity	μ, kg/m.s	$ML^{-1} t^{-1}$
16.	Kinematic viscosity	ν, m^2/s	$L^2 t^{-1}$
17.	Thermal conductivity	k, W/m.K	$ML T^{-1} t^{-3}$
18.	Thermal diffusivity	α, m^2/s	$L^2 t^{-1}$
19.	Coefficient of expansion	β, K^{-1}	T^{-1}
20.	Heat transfer coefficient	h, $W/m^2.K$	$MT^{-1} t^{-3}$
21.	Mass flow rate	$\dot{m}$, kg/s	$M t^{-1}$

7.10.2. Dimensional Homogeneity

A physical equation is the relationship between two or more physical quantities. The principle of dimensional homogeneity states that all equations, describing the behaviour of a physical system must be dimensionally homogeneous *i.e.*, fundamental dimensions of various terms on two sides of the equation are identical. A dimensionally homogeneous equation is independent of fundamental units, if their units are consistent. It states that to a dimensionally homogeneous equation, the quantities of the same units can be added, substracted or equated. Consistency of units demands, if one term of equation is measured in a particular unit say m/s then all term in the equation must be measured in m/s. The dimensional homogeneity is useful in the following ways :

(*i*) It facilates the determination of the dimensions of a physical quantity.

(*ii*) It helps to check the dimensional consistency of an equation.

(*iii*) It facilates the conversion of units from one system to another.

(*iv*) It provides a step towards dimensional analysis.

7.10.3. Rayleigh's Method of Dimensional Analysis

In this method, a functional relationship of the quantities that may influence a dependent variable is expressed in form of an exponential equation.

If y is dependent variable and it depends on variables x_1, x_2, x_3, ..., then the equation in an exponential form

$$y = C\,(x_1^a, x_2^b, x_3^c, ...,) \quad ...(7.38)$$

Where C is dimensional constant, which may be evaluated from physical characteristic of problems or through experimentation, a, b, c,.... are arbitrary exponents and are obtained by comparing the exponents of the primary dimensions on two sides of an equation.

The Rayleigh method does not provide any information regarding the number of dimensionless groups to be obtained as a result of dimensional analysis. Further, this method can only be used for dimensional analysis of a dependent variable which depends upon maximum four independent variables. If number of independent variables exceeds then it becomes tedious work to obtain an expression for the dependent variable. Therefore, this method has become obsolete and is not favoured for use.

Example 7.9. *Find the functional relationship for pressure drop for a fluid flowing through a tube diameter D, fluid density ρ, fluid velocity u and fluid viscosity μ.*

Solution

The pressure drop for a flowing fluid in the functional form.

$$\Delta p = f(D, \rho, u, \mu)$$

Let $\Delta p = C\,[D^p, \rho^q, u^r, \mu^s]$

where C is non dimensional constant.

The dimensional equation of above equation in primary dimensions M, L, and t are

$$ML^{-1} t^{-2} = C\,[L^{p}\,(ML^{-3})^{q}\,(Lt^{-1})^{r}\,(ML^{-1}t^{-1})^{s}]$$

or $$ML^{-1} t^{-2} = C\,[M^{q+s}\,L^{p-3q+r-s}\,.\,t^{-r-s}]$$

For dimensional homogeneity, the exponent of each dimension on both sides of the equation must be identical. Thus

For M : $1 = q + s$

For L : $-1 = p - 3q + r - s$

For t : $-2 = -r - s$

There are four unknowns (p, q, r and s), but three equations. Therefore, it is not possible to find the numerical values of p, q, r and s. However, three unknowns can be expressed in terms of fourth variable. Let choose viscosity with exponent 's' and other exponents are expressed as

$$r = 2 - s$$

$$q = 1 - s \quad \text{and} \quad p = -s,$$

Substituting these values, we get

$$\Delta p = C\,[D^{-s}\,\rho^{1-s}\,u^{2-s}\,\mu^{s}]$$

$$= C\left[\frac{\mu}{\rho u D}\right]^{s} \times \rho u^{2} = C\,\frac{\rho u^{2}}{Re_d^{\,s}}$$

where $Re_D = \dfrac{\rho u D}{\mu}$.

The values of C and s are to be evaluated by experimentation.

Example 7.10. *Find the functional expression for forced convection heat transfer between a fluid flowing through a tube and its wall.*

The heat transfer coefficient in forced convection is influenced by tube diameter (D), fluid velocity (u), and the fluid properties such as density (ρ), dynamic viscosity (μ), thermal conductivity (k_f) and specific heat (C_p).

Solution

The heat transfer coefficient in the functional form

$$h = f(D, u, \rho, \mu, k_f, C_p)$$

Let $$h = C\,[D^{a}\,u^{b}\,\rho^{c}\,\mu^{d}\,k_f^{\,e}\,C_p^{\,f}]$$

where C is non dimensional constant.

Expressing quantities in terms of primary dimensions

$$Mt^{-3}\,T^{-1} = C\,[L^{a}\,(Lt^{-1})^{b}\,(ML^{-3})^{c}\,(ML^{-1}\,t^{-1})^{d}\,(MLt^{-3}\,T^{-1})^{e}\,(L^{2}\,t^{-2}\,T^{-1})^{f}$$

For dimensional homogeneity of equation ;

M : $1 = c + d + e$

L : $0 = a + b - 3c - d + e + 2f$

t : $-3 = -b - d - 3e - 2f$

T : $-1 = -e - f.$

There are four equations and six unknowns, thus complexity starts in such case. Let expressing the exponents in terms of c and f.

$$e = 1 - f, \qquad d = -c + f$$

$$b = c \qquad a = c - 1$$

Therefore, $h = C\,[D^{c-1}\,u^{c}\,\rho^{c}\,\mu^{-c+f}\,k^{1-f}\,C_p^{\,f}]$

Multiplying D/k_f on both sides, we get

$$\frac{hD}{k_f} = C\left[\left(\frac{\rho u D}{\mu}\right)^{c}\left(\frac{\mu C_p}{k_f}\right)^{f}\right]$$

or $$Nu_D = C\,Re_D^{\,c}\,Pr^{f}$$

where $Re_D = \dfrac{\rho u D}{\mu}$, Reynolds numbers,

$Pr = \dfrac{\mu C_p}{k_f}$, Prandtl number

and $Nu_D = \dfrac{hD}{k_f}$, Nusselt number.

Thus in forced convection, the Nusselt number is a function of Reynolds number and Prandtl number.

7.10.4. Buckingham π Theorem

When the large physical quantities are involved in a phenomenon, the Rayleigh method of dimensional analysis becomes tedious. The Buckingham π theorem is an improvement of Rayleigh method. It is simple and more systematic for determining the dimensionless groups.

According to Buckingham π theorem, if n variables are influencing a phenomenon and if they can be expressed in m primary dimensions (M, L, T and t), then the dimensionless independent groups formed will be $n - m$. These dimensionless term are called π terms. The independent dimensionless groups can be arranged as

$$f(\pi_1, \pi_2, \pi_3, \ldots) = 0 \qquad \ldots(7.39)$$

In a particular phenomenon, involving seven variables and if they can be expressed in four primary dimensions, then the number of dimensionless groups formed are

$$n - m = 7 - 4 = 3$$

Hence $$f(\pi_1, \pi_2, \pi_3) = 0 \qquad \ldots(7.40)$$

or $$\pi_1 = \phi(\pi_2, \pi_3) \qquad \ldots(7.41)$$

Each dimensionless π term is formed by m variables, along with one selected variable from remaining $(n - m)$ variables *i.e.*, each π term involves $(m + 1)$ variables. These m variables (equal to primary dimensions) formed a core group, which appear repeatedly in each π term, consequently, called as *repeating variables*. The repeating variables are selected among the n variables in such way that

(*i*) They together must contain all primary dimensions.

(*ii*) They themselves must not form a dimensionless group.

(*iii*) As far as possible, the dependent variable should not be selected as repeating variable.

(*iv*) No repeating variable should have the same dimension.

(*v*) The repeating variables should be chosen in such a way that one variable contains geometrical property, other one contains dynamic (flow) property and third one contains fluid property etc.

Example 7.11. *Reconsider the example 7.10 and develop an expression for pressure drop.*

Solution

Given : The pressure drop for flow through tube as

$$\Delta p = f(D, \rho, u, \mu) \quad \text{or} \quad \phi(\Delta p, D, \rho, u, \mu) = 0$$

There are five (n) variables affecting a phenomenon and they can be expressed in three primary dimensions (M, L, and t). Therefore, the number of dimensionless π groups to be formed are

$$n - m = 5 - 3 = 2 \quad \text{or} \quad \phi(\pi_1, \pi_2) = 0$$

Each π term consists of $m = 3$ common variable, called repeating variables along with one selected variable.

Let $\pi_1 = D^a \rho^b u^c \mu$

$\pi_2 = D^d \rho^e u^f \Delta p$

Expressing the each variable in their primary dimensions for each π term.

Therefore, $\pi_1 = L^a (ML^{-3})^b (Lt^{-1})^c (ML^{-1} t^{-1})$

or $M^0 L^0 t^0 = M^{b+1} L^{a-3b+c-1}, t^{-c-1}$

For dimensional homogeneity the exponent of M, L, t must be equal to zero.

M : $0 = b + 1$

L : $0 = a - 3b + c - 1$

t : $0 = -c - 1$

Solving these simultaneous equations, we get

$$c = -1, b = -1, a = -1$$

or $\pi_1 = D^{-1} \rho^{-1} u^{-1} \mu$

or $$\pi_1 = \frac{\mu}{\rho u D} = \frac{1}{Re_D}$$

Similarly $\pi_2 = L^d (ML^{-3})^e (Lt^{-1})^f (ML^{-2} t^{-2})$

or $M^0 L^0 t^0 = M^{e+1} L^{d-3e+f-2} t^{-f-2}$

For dimensional homogeneity

M : $0 = e + 1$

L : $0 = d - 3e + f - 2$

t : $0 = -f - 2$

On solving, we get

$$e = -1, \quad f = -2, \quad d = 1$$

or $$\pi_2 = D \rho^{-1} u^{-2} \Delta p = \frac{\Delta p D}{\rho u^2}$$

$$\text{Therefore, } \phi\left(\frac{\Delta p D}{\rho u^2}, \frac{\mu}{\rho u D}\right) = 0.$$

7.10.5. Dimensional Analysis for Forced Convection

The forced convection heat transfer phenomenon can be influenced by the variables given in Table 7.3.

TABLE 7.3

Sr. No.	*Parameters*	*Symbol and unit*	*Primary dimensions*
1.	Tube Diameter (Characteristic length)	D, m	L
2.	Fluid density	ρ, kg/m^3	ML^{-3}
3.	Fluid viscosity	μ, kg/m/s	$ML^{-1}t^{-1}$
4.	Fluid velocity	u_∞, m/s	Lt^{-1}
5.	Fluid thermal conductivity	k_f, W/m.K	$MLt^{-3}T^{-1}$
6.	Heat transfer coefficient	h, W/m^2.K	$Mt^{-3}T^{-1}$
7.	Fluid specific heat	C_p, J/kg.K	$L^2t^{-2}T^{-1}$

These seven variables are expressed in four primary dimensions (M, L, T, t), therefore, according to Buckingham π theorem, the independent dimensionless groups formed are :

= No. of variable affecting the phenomenon

− No. of primary dimensions used

= 7 − 4 = 3 (*i.e.*, π_1, π_2, π_3)

Writing these three groups as,

$$\pi_1 = D^a, \rho^b, \mu^c, k_f^d, u_\infty$$

$$\pi_2 = D^e, \rho^f, \mu^g, k_f^h, C_p$$

$$\pi_3 = D^i, \rho^j, \mu^k, k_f^l, h$$

where D, ρ, μ, k_f form a core group (repeating variables) and u_∞, C_p and h are as selected variables.

Since the groups π_1, π_2, π_3 are dimensionless, hence certain exponents are applied on the repeating variable, which are to be determined,

(*i*) Expressing the variable in their primary dimensions for π_1,

$$\pi_1 = L^a . (ML^{-3})^b . (ML^{-1} t^{-1})^c . (MLt^{-3}T^{-1})^d . (Lt^{-1})$$

or $M^0L^0T^0t^0 = M^{b+c+d} . L^{a-3b-c+d+1} . T^{-d} . t^{-c-3d-1}$

Separating the exponents for dimensional homogeneity

M : $0 = b + c + d$

L : $0 = a - 3b - c + d + 1$

T : $0 = -d$

t : $0 = -c - 3d - 1$

Solving these simultaneous equations, we get

$d = 0, \quad c = -1$

$b = 1, \quad a = 1$

Hence the dimensionless group formed is,

$$\pi_1 = \frac{D\rho u_\infty}{\mu} = Re_D \text{ (Reynolds number)} \quad ...(7.42)$$

(*ii*) Expressing the primary dimension for variables of π_2,

$$\pi_2 = L^e . (ML^{-3})^f . (ML^{-1}t^{-1})^g . (MLt^{-3}T^{-1})^h . (L^2t^{-2}T^{-1})$$

or $M^0L^0T^0t^0 = M^{f+g+h} . L^{e-3f-g+h+2} . T^{-h-1} . t^{-g-3h-2}$

Separating the exponents for dimensional homogeneity.

M : $0 = f + g + h$

L : $0 = e - 3f - g + h + 2$

T : $0 = -h - 1$

t : $0 = -g - 3h - 2$

Solving these simultaneous equations, we get

$h = -1, g = 1$

$f = 0, \quad e = 0$

Hence the dimensionless group formed is,

$$\pi_2 = \frac{\mu C_p}{k_f} = Pr \text{ (Prandtl Number)} \quad ...(7.43)$$

(*iii*) Expressing the primary dimension for variables of π_3,

$$\pi_3 = L^i . (ML^{-3})^j . (ML^{-1}t^{-1})^k . (MLt^{-3}T^{-1})^l . (Mt^{-3}T^{-1})$$

or $M^0L^0T^0t^0 = M^{j+k+l+1} . L^{i-3j-k+l} . T^{-l-1} . t^{-k-3l-3}$

Separating the exponents for dimensional homogeneity

M : $0 = j + k + l + 1$

L : $0 = i - 3j - k + l$

T : $0 = -l - 1$

t : $0 = -k - 3l - 3$

Solving these simultaneous equations, we get

$l = -1, k = 0$

$j = 0, \quad i = 1$

Hence the dimensionless group formed is,

$$\pi_3 = \frac{h D}{k_f} = Nu_D \text{ (Nusselt Number)} \quad ...(7.44)$$

Hence for forced convection,

$$Nu_D = f(Re_D, Pr) \quad ...(7.45)$$

7.10.6. Dimensional Analysis for Natural Convection

The relations used for the natural convection are based on practical observations. Hence the dimensional analysis is also useful in natural convection. The parameters which influence the natural convection phenomenon are listed below with their primary dimensions :

TABLE 7.4

Sr. No.	*Parameters*	*Symbol and unit*	*Primary dimensions*
1.	Characteristic length	L_c, m	L
2.	Fluid density	ρ, kg/m^3	M L^{-3}
3.	Fluid viscosity	μ, kg/m.s	M $L^{-1}t^{-1}$
4.	Temperature difference	ΔT, °C	T
5.	Coefficient of volumetric expansion	β, K^{-1}	T^{-1}
6.	Gravitational acceleration	g, m/s^2	L t^{-2}
7.	Fluid thermal conductivity	k_f, W/m.K	$MLt^{-3}T^{-1}$
8.	Heat transfer coefficient	h, W/m^2.K	M $t^{-3}T^{-1}$
9.	Fluid specific heat	C_p, J/kg.K	$L^2t^{-2}T^{-1}$

Out of these nine variables, the product of $g\beta\Delta T$, represents the buoyancy force and considered as a one variable. Thus the variable, affecting the natural convection are now remaining only seven and these can be represented by four primary dimensions (M, L, T, t), therefore, the independent dimensionless groups formed are :

= No. of variables affecting the phenomenon
– No. of primary dimensions used

= 7 – 4 = 3 (*i.e.*, π_1, π_2, π_3)

Writing these three groups as,

$$\pi_1 = L_c^{\ a}, \rho^b, \mu^c, k_f^{\ d}, g\beta\Delta T$$

$$\pi_2 = L_c^{\ e}, \rho^f, \mu^g, k_f^{\ h}, C_p$$

$$\pi_3 = L_c^{\ i}, \rho^j, \mu^k, k_f^{\ l}, h$$

where L_c, ρ, μ, k_f form a core group and $g\beta\Delta T$, C_p and h are as selected variables.

(*i*) Expressing the variable in their primary dimensions for π_1,

$$\pi_1 = L^a.(ML^{-3})^b.(ML^{-1}t^{-1})^c.(MLt^{-3}T^{-1})^d.(Lt^{-2})$$

or $M^0L^0T^0t^0 = M^{b+c+d} . L^{a-3b-c+d+1} . T^{-d} . t^{-c-3d-2}$

Separating the exponents for dimensional homogeneity

M : $0 = b + c + d$

L : $0 = a - 3b - c + d + 1$

T : $0 = -d$

t : $0 = -c - 3d - 2$

Solving these simultaneous equations, we get

$$d = 0, \quad c = -2$$
$$b = 2, \quad a = 3$$

Hence the dimensionless group formed is,

$$\pi_1 = \frac{\rho^2 (g\beta\Delta T)L_c^3}{\mu^2} = \frac{(g\beta\Delta T)L_c^3}{\nu^2}$$
$$= Gr_L \text{ (Grashof Number)} \quad ...(7.46)$$

(*ii*) Expressing the primary dimensions for variables of π_2,

$$\pi_2 = L^e . (ML^{-3})^f . (ML^{-1}t^{-1})^g . (M\,L\,t^{-3}T^{-1})^h . (L^2t^{-2}T^{-1})$$

or $$M^0L^0T^0t^0 = M^{f+g+h} . L^{e-3f-g+h+2} . T^{-h-1} . t^{-g-3h-2}$$

Separating the exponents for dimensional homogeneity

M : $0 = f + g + h$

L : $0 = e - 3f - g + h + 2$

T : $0 = -h - 1$

t : $0 = -g - 3h - 2$

Solving these simultaneous equations, we get

$$h = -1, g = 1$$
$$f = 0, \quad e = 0$$

Hence the dimensionless group formed is,

$$\pi_2 = \frac{\mu C_p}{k_f} = Pr \text{ (Prandtl Number)} \quad ...(7.47)$$

(*iii*) Expressing the primary dimensions for variables of π_3,

$$\pi_3 = L^i. (ML^{-3})^j. (ML^{-1}t^{-1})^k. (MLt^{-3}T^{-1})^l . (Mt^{-3}T^{-1})$$

or $$M^0L^0T^0t^0 = M^{j+k+l+1}\, L^{i-3j-k+l}\, T^{-l-1}\, t^{-k-3l-3}$$

Separating the exponents for dimensional homogeneity

M : $0 = j + k + l + 1$

L : $0 = i - 3j - k + l$

T : $0 = -l - 1$

t : $0 = -k - 3l - 3$

Solving these simultaneous equations, we get

$$l = -1, k = 0$$
$$j = 0, \quad i = 1$$

Hence the dimensionless group formed is,

$$\pi_3 = \frac{hL_c}{k_f} = Nu_L \text{ (Nusselt Number)} \quad ...(7.48)$$

Hence for free convection,

$$Nu_L = f(Gr_L, Pr) \quad ...(7.49)$$

7.11. PHYSICAL SIGNIFICANCE OF THE DIMENSIONLESS PARAMETERS

7.11.1. Reynolds Number

It is the *ratio of inertia forces to viscous forces* in the velocity boundary layer. It is used in forced convection and approximated as :

$$Re = \frac{\text{Inertia forces}}{\text{Viscous forces}} = \frac{\rho u_\infty L_c}{\mu} = \frac{u_\infty L_c}{\nu}$$

where, L_c = characteristic length of flow geometry, m ;
= x, distance from the leading edge in the flow direction for a flat plate;
= D, diameter for flow through or across a cylinder and a sphere;
u_∞ = free stream velocity, m/s ;
ρ = fluid density, kg/m^3 ;
μ = dynamic viscosity of fluid, Ns/m^2 or kg/m/s ;
$\nu = \mu/\rho$ = kinematic viscosity of fluid, m^2/s.

The Reynolds number is a *dimensionless* quantity. It characterises the type of flow, whether it is laminar or turbulent flow.

7.11.2. Critical Reynolds Number Re_{cr}

It is a value of Reynolds number, where *boundary layer changes from laminar to turbulent nature.* It is denoted by Re_{cr}.

For flow over a flat plate, the transition from laminar to turbulent boundary layer occurs roughly when critical Reynolds number is

$$Re_{cr} \geq 5 \times 10^5 \quad ...(7.50)$$

In fluid flow through tubes, the Reynolds number is also used to characterized the fluid flow. The transition from laminar to turbulent boundary layer occurs, when

$$Re_{D,\,cr} = \frac{u_\infty D}{\nu} \geq 2300 \quad ...(7.51)$$

These are generally accepted values of critical Reynolds numbers, which may vary with surface roughness, level of turbulence and the variation of pressure along the flow.

7.11.3. Prandtl Number

It is defined as the *ratio of the momentum diffusivity* ν *to the thermal diffusivity* α or

$$Pr = \frac{\text{Momentum diffusivity}}{\text{Thermal diffusivity}} = \frac{\nu}{\alpha} = \frac{\mu\rho C_p}{\rho k_f} = \frac{\mu C_p}{k_f}$$

It is a dimensionless property, a function of temperature. It provides a measure of relative effectiveness of momentum and energy transfer in the velocity and thermal boundary layers, respectively.

For gases $Pr \cong 1$; *i.e.*, both momentum and heat diffusion through the fluid take place at the same rate.

For liquid metal $Pr << 1$; indicates heat diffuses in the fluid very quickly, and for oils, $Pr >> 1$; indicates heat diffusion is very slow in the fluid relative to momentum.

Consequently, the thermal boundary layer is much thicker for liquid metals, much thinner for oils relative to velocity boundary layer as shown in Fig. 7.7. Further, the thicknesses of two boundary layers can be related as

$$\frac{\delta_{th}(x)}{\delta(x)} = Pr^n \quad \text{where } n \text{ is the exponent} \qquad ...(7.52)$$

7.11.4. Grashof Number

It is defined as the *ratio of the buoyancy forces to the viscous forces* acting in the fluid layer. It is used in free convection and its role is same as that of Reynold number in forced convection. The Grashof number characterises the type of boundary layer developed in natural convection heat transfer. It is denoted by Gr and expressed as

$$Gr = \frac{g\beta\Delta TL_c^{\ 3}}{\nu^2} \qquad ...(7.53)$$

where, g = acceleration due to gravity, m/s^2,

β = coefficient of volumetric expansion $= 1/(T_f + 273)$, K^{-1},

ΔT = temperature difference between surface and fluid, °C or K,

T_f = mean film temperature $= \dfrac{T_s + T_\infty}{2}$ °C,

ν = kinematic viscosity of fluid, m^2/s,

L_c = characteristic length of the body, m

= height, L for vertical plates and cylinders,

= diameter, D for horizontal cylinder and sphere,

$= \dfrac{\text{Surface area}}{\text{Perimeter}} = \dfrac{A_s}{P}$, for other geometries

For free convection, the transition from laminar to turbulent occurs, when $Gr_{cr} \approx 10^9$.

7.11.5. Nusselt Number

It is defined as the *ratio of convection heat flux to conduction heat flux* in the fluid boundary layer or

$$Nu = \frac{\text{Convection heat flux}}{\text{Conduction heat flux}} = \frac{h\Delta T}{k_f \Delta T/L_c} = \frac{hL_c}{k_f}$$

where, ΔT = temperature difference $(T_s - T_\infty)$ between wall surface and fluid, K.

h = heat transfer coefficient; W/m^2.K.

k_f = thermal conductivity of the fluid ; W/m.K.

L_c = characteristic length of fluid flow, m

Based on the interpretation, the value of Nu as unity indicates that there is no convection, the heat transfer is by pure conduction in the boundary layer. Large value of Nu indicates large convection in the fluid.

7.11.6. Stanton Number

It is the *ratio of the heat transfer at the surface to that transported by fluid by its thermal capacity.*

$$St_x = \frac{\text{Heat flux to the fluid}}{\text{Heat transfer capacity of fluid}}$$

$$= \frac{h\,\Delta T}{\rho C_p u_\infty \Delta T} = \frac{h}{\rho C_p u_\infty} \qquad ...(7.54)$$

Mathematically, it is the *ratio of Nusselt number and product of Reynolds number and Prandtl number* and it is also expressed as

$$St_x = \frac{Nu_x}{Re_x\ Pr} \qquad ...(7.55)$$

7.11.7. Peclet Number

It is the *ratio of heat transfer by convection to heat transfer by conduction.* It is denoted by Pe and expressed as

$$Pe = \frac{\text{Heat transfer by convection}}{\text{Heat transfer by conduction}}$$

$$= \frac{mC_p\Delta T}{k_f \Delta T/L} = \frac{\rho VC_p L}{k_f} \qquad ...(7.56)$$

Mathematically, the Peclet numbers is *product of Reynolds number and Prandtl number.*

$$Pe = Re.Pr \qquad ...(7.57)$$

7.11.8. Graetz Number

It is a dimensionless number used in study of stream line fluid flow. It is the ratio of fluid stream thermal capacity of fluid flowing per unit length thermal conductivity of fluid. It is denoted by Gz and expressed as

$$Gz = \frac{\text{Thermal capacity of fluid per unit length}}{\text{Thermal conductivity}}$$

$$= \frac{\dot{m}C_p}{k_f\, x} = \frac{\pi}{4}\ Re\ .\ Pr\ .\ \frac{D}{x} \qquad ...(7.58)$$

where x = hydrodynamic entry length,

D = inside diameter of the tube.

Generally it is associated with thermal entry length of a fully developed flow through tubes.

Example 7.12. *Calculate the approximate Reynolds numbers and state if the flow is laminar or turbulent for the following :*

(i) A 10 m long yatch sailing at 13 km/h in sea water, $\rho = 1000\ kg/m^3$ *and* $\mu = 1.3 \times 10^{-3}$ *kg/ms.*

(ii) A compressor disc of radius 0.3 m rotating at 15000 r.p.m. in air at 5 bar and 400°C and

$$\mu = \frac{1.46 \times 10^{-6}\ T^{3/2}}{(110 + T)}\ kg/ms.$$

(iii) 0.05 kg/s of CO_2 *gas at 400 K flowing in a 20 mm dia. pipe and*

$$\mu = \frac{1.56 \times 10^{-6}\ T^{3/2}}{(233 + T)}\ kg/ms.$$

(N.M.U., May 2002)

Solution

(*i*) *Given :* Yatch sails on sea water

$$L = 10\ m,$$

$$u_\infty = 13\ km/h = \frac{13 \times 10^3}{60 \times 60} = 3.61\ m/s$$

$$\rho = 1000\ kg/m^3,$$

$$\mu = 1.3 \times 10^{-3}\ kg/ms.$$

To find :

(*a*) Reynolds number,

(*b*) Type of flow.

Analysis : (*a*) The Reynolds number can be calculated as

$$Re = \frac{\rho u_\infty L}{\mu} = \frac{1000 \times 3.61 \times 10}{1.3 \times 10^{-3}}$$

$$= \mathbf{2.78 \times 10^7}.\ \text{Ans.}$$

(*b*) The Re > 5×10^5, thus flow is turbulent. Ans.

(*ii*) *Given :* A compressor disc with

$$r_o = 0.3\ m \quad \text{or} \quad D = 0.6\ m$$

$$N = 15000\ rpm,$$

$$p = 5\ bar,\ T = 400°C$$

and

$$\mu = \frac{1.46 \times 10^{-6}\ T^{3/2}}{(110 + T)}$$

To find :

(*a*) Reynolds number,

(*b*) Type of flow.

Assumption : For air R = 0.287 kJ/kg.K.

Analysis : The equivalent linear velocity of compressor disc

$$u_\infty = \frac{\pi DN}{60} = \frac{\pi \times 0.6 \times 1500}{60} = 471.23\ m/s$$

The density of air at 5 bar and 400°C (= 673 K)

$$\rho = \frac{p}{RT} = \frac{5 \times 100\ kPa}{(0.287\ kJ/kg.K) \times (673\ K)}$$

$$= 2.588\ kg/m^3$$

The viscosity at 400°C

$$\mu = \frac{1.46 \times 10^{-6} \times (673)^{3/2}}{(110 + 673)}$$

$$= 3.3 \times 10^{-5}\ kg/ms$$

(*a*) The Reynolds number

$$Re = \frac{\rho u_\infty D}{\mu} = \frac{2.588 \times 471.23 \times 0.6}{3.3 \times 10^{-5}}$$

$$= \mathbf{22.17 \times 10^6}.\ \text{Ans.}$$

(*b*) The Re > 5×10^5, thus flow is turbulent. Ans.

(*iii*) *Given :* CO_2 gas

$$m = 0.05\ kg/s, \quad T = 400\ K,$$

$$D_i = 20\ mm = 20 \times 10^{-3}\ m,$$

and

$$\mu = \frac{1.56 \times 10^{-6}\ T^{3/2}}{(233 + T)}\ kg/ms.$$

To find :

(*a*) Reynolds number,

(*b*) Type of flow.

Analysis : At 400 K, the density of CO_2

$$\rho = 1.3257\ kg/m^3$$

(*a*) The Reynolds number for pipe flow can also be calculated as

$$Re = \frac{4\dot{m}}{\pi D_i \mu}$$

where

$$\mu = \frac{1.56 \times 10^{-6}\ T^{3/2}}{(233 + T)}$$

Then

$$\mu = \frac{1.56 \times 10^{-6} \times (400)^{3/2}}{(233 + 400)}$$

$$= 1.97 \times 10^{-5}\ kg/ms.$$

Then

$$Re = \frac{4 \times 0.05}{\pi \times 20 \times 10^{-3} \times 1.97 \times 10^{-5}}$$

$$= \mathbf{1.61 \times 10^5}.\ \text{Ans.}$$

(*b*) Re ≥ 2300 for tube flow, thus the flow is turbulent. Ans.

Example 7.13. *Calculate the approximate Grashof number and state if the flow is laminar or turbulent for the following :*

(a) A central heating radiator, 0.6 m high with a surface temperature of 75°C in a room at 18°C, (ρ = 1.2 kg/m³, Pr = 0.72, and μ = 1.8 × 10⁻⁵ kg/ms).

(b) A horizontal oil sump with a surface temperature of 40°C, 0.4 m long and 0.2 m wide containing oil at 75°C. Take ρ = 854 kg/m³, Pr = 546, β = 0.7 × 10⁻³ K⁻¹ and μ = 3.56 × 10⁻² kg/ms.

(c) Air at 20°C (ρ = 1.2 kg/m³, Pr = 0.72 and μ = 1.8 × 10⁻⁵ kg/ms) adjacent to a 60 mm dia. horizontal light bulb, with a surface temperature of 90°C. (N.M.U., May 2002)

Solution

The properties are given, thus the Grashof number for any flow situation can be calculated as

$$\text{Gr} = \frac{g\,\beta\Delta T\,L_c^3}{\nu^2} = \frac{g\rho^2\beta\Delta T\,L_c^3}{\mu^2}$$

where L_c = significant length of the body.

(*a*) $L_c = 0.6$ m, $\Delta T = 75 - 18 = 57°C$

$\rho = 1.2$ kg/m³, Pr = 0.72,

$\mu = 1.8 \times 10^{-5}$ kg/ms

Mean film temperature,

$$T_f = \frac{75+18}{2}\ 46.5°C$$

$$\beta = \frac{1}{T_f + 273} = \frac{1}{46.5+273}$$

$$= 3.13 \times 10^{-3}\ K^{-1}$$

The Grashof number

$$\text{Gr}_L = \frac{9.81 \times (1.2)^2 \times 3.13 \times 10^{-3} \times 57 \times (0.6)^3}{(1.8 \times 10^{-5})^2}$$

$$= \mathbf{1.68 \times 10^9}. \quad \textbf{Ans.}$$

The $\text{Gr}_L > 10^9$, the flow is turbulent. **Ans.**

(*b*) A horizontal oil sump

$T_s = 40°C$ $T_\infty = 75°C$,

$L = 0.4$ m, $w = 0.2$ m,

$\rho = 854$ kg/m³, Pr = 546,

$\beta = 0.7 \times 10^{-3}\ K^{-1}$

$\mu = 3.56 \times 10^{-2}$ kg/ms.

For horizontal plate, the characteristic length

Then $$L_c = \frac{A_s}{P} = \frac{L \times w}{2(L+w)}$$

$$= \frac{0.4 \times 0.2}{2 \times (0.4 + 0.2)} = 0.067\ \text{m}$$

(*i*) $$\text{Gr} = \frac{9.81 \times (854)^2 \times 0.7 \times 10^{-3} \times (75-40) \times (0.067)^3}{(3.56 \times 10^{-2})^2}$$

$$= \mathbf{4.1 \times 10^4}. \quad \textbf{Ans.}$$

(*ii*) Gr < 10^9, thus the flow is laminar

(*c*) Air

$T_\infty = 20°C$, $\rho = 1.2$ kg/m³,

Pr = 0.72

$\mu = 1.8 \times 10^{-5}$ kg/ms, $L_c = D = 60$ mm,

$T_s = 90°C$

$$T_f = \frac{90+20}{2} = 55°C,$$

$$\beta = \frac{1}{55+273} = 3.049 \times 10^{-3}\ K^{-1}$$

$$\text{Gr} = \frac{9.81 \times (1.2)^2 \times 3.049 \times 10^{-3} \times (90-20) \times (60 \times 10^{-3})^3}{(1.8 \times 10^{-5})^2}$$

$$= \mathbf{2.0 \times 10^6}, \textbf{ Laminar.} \quad \textbf{Ans.}$$

Example 7.14. *Calculate the Nusselt number in following cases :*

(i) A horizontal electronic component with a surface temperature of 35°C, 5 mm wide and 10 mm long, dissipating 0.1 W heat by free convection from its one side into air at 20°C. Take for air k = 0.026 W/m.K.

(ii) A 1 kW central heating radiator 1.5 m long and 0.6 m high with a surface temperature of 80°C, dissipating heat by radiation and convection into room at 20°C (k = 0.026 W/m.K, assume black body radiation and σ = 5.67 × 10⁻⁸ W/m².K⁴).

(iii) Air at 6°C (k = 0.024 W/m.K) adjacent to a wall 3 m high and 0.15 m thick made of brick with k = 0.3 W/m.K, the inside temperature of the wall is 18°C, the outside wall temperature is 12°C.

Solution

(*i*) *Given :* A horizontal electronic component

$w = 5$ mm, $L = 10$ mm,

$Q = 0.1$ W, $T_s = 35°C$,

$T_\infty = 20°C$, $k_f = 0.026$ W/m.K.

Analysis :

$$Q = h\,A_s\,(T_s - T_\infty)$$

$$0.1 = h \times (10\text{ mm} \times 5\text{ mm} \times 10^{-6}) \times (35 - 20)$$

or $$h = 133.33\text{ W/m}^2\text{.K}$$

The significant length

$$L_c = \frac{A_s}{P} = \frac{10\text{ mm} \times 5\text{ mm} \times 10^{-6}}{2 \times (10\text{ mm} + 5\text{ mm}) \times 10^{-3}}$$

$$= 1.67 \times 10^{-3}\text{ m}$$

The Nusselt number

$$Nu = \frac{hL_c}{k_f} = \frac{133.33 \times 1.67 \times 10^{-3}}{0.026}$$

$= \mathbf{8.54.}$ Ans.

(*ii*) *Given :* Central heating radiator

$Q = 1\text{ kW} = 10^3\text{ W}$,

$w = 1.5\text{ m}$,

$L = L_c = 0.6\text{ m}$,

$T_s = 80°C = 353\text{ K}$,

$T_\infty = 20°C = 293\text{ K}$,

$k_f = 0.026\text{ W/m.K}$,

$\sigma = 5.67 \times 10^{-8}\text{ W/m}^2\text{.K}^4$.

Analysis : The radiation heat transfer for black surface

$$Q_{rad} = \sigma\,A_s\,(T_s^4 - T_\infty^4)$$

$$= (5.67 \times 10^{-8}) \times (1.5 \times 0.6) \times (353^4 - 293^4)$$

$$= 416.2\text{ W}$$

Heat transfer by convection

$$Q_{conv} = Q - Q_{rad} = 1000 - 416.2 = 583.7\text{ W}$$

Then $$Q_{conv} = h\,A_s\,(T_s - T_\infty)$$

$$583.7 = h \times (1.5 \times 0.6) \times (80 - 20)$$

or $$h = 10.80\text{ W/m}^2\text{.K}$$

The Nusselt number

$$Nu = \frac{hL_c}{k_f} = \frac{10.80 \times 0.6}{0.026} = \mathbf{249.4}$$ Ans.

(*iii*) *Given :* Air flow adjacent to a wall

$T_\infty = 6°C$, $k_f = 0.024\text{ W/m.K}$,

$L_c = H = 3\text{ m}$, $L = 0.15\text{ m}$,

$k = 0.3\text{ W/m.K}$, $T_1 = 18°C$,

$T_2 = 12°C$.

Analysis : The heat transfer rate per m² by steady state conduction, through the wall

$$\frac{Q}{A} = \frac{k(T_1 - T_2)}{L} = \frac{0.3 \times (18 - 12)}{0.15}$$

$$= 12\text{ W/m}^2$$

This heat is also transfered by convection, thus

$$\frac{Q}{A} = h(T_s - T_\infty)$$

On inner surface

$$12 = h_1 \times (18 - 6) \text{ or } h_1 = 1\text{ W/m}^2\text{.K}$$

and $$Nu = \frac{h_1 L_c}{k_f} = \frac{1 \times 3}{0.024} = \mathbf{125.}$$ Ans.

On outer wall surface

$$12 = h_2 \times (12 - 6) \text{ or } h_2 = 2\text{ W/m}^2\text{.K}$$

and $$Nu = \frac{h_2 L_c}{k_f} = \frac{2 \times 3}{0.024} = \mathbf{250.}$$ Ans.

7.12. TURBULENT BOUNDARY LAYER HEAT TRANSFER

The flow of fluid in the boundary layer is more often turbulent rather than laminar as shown in Fig. 7.14. In the turbulent flow, the transport mechanism is added by random fluctuation of lumps of fluid. The irregular velocity fluctuations are superimposed upon the motion of main stream and these fluctuations are primarily responsible for transfer of heat and momentum. The rates of momentum and heat transfer in the turbulent flow and associated friction and heat transfer coefficients are many times more than the laminar flow, because of better mixing in which lumps of fluid collide with one another randomly and make multidirectional flow and mix the fluid effectively.

In the turbulent flow, the instantaneous fluid currents are highly torn and fluctuating randomly and it is very difficult to trace the path of an individual fluid element. This behaviour is shown in Fig. 7.15, which plots arbitrary flow property P as a function of time at some location in a turbulent boundary layer. The property P could be a velocity component, or fluid temperature at any instant. The time mean value and fluctuating component may be represented as $\overline{P}$ and P', respectively for steady flow. The instantaneous velocity components u and v can be expressed in the form

$$u = \overline{u} + u' \quad \text{and} \quad v = \overline{v} + v'$$

Similarily instantaneous temperature can be expressed as

$$T = \overline{T} + T' \text{ and so on}$$

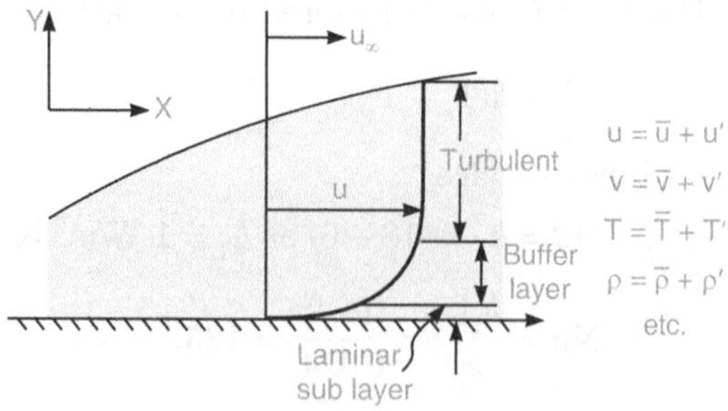

Fig. 7.14. Velocity profile in turbulent boundary layer on a flat plate

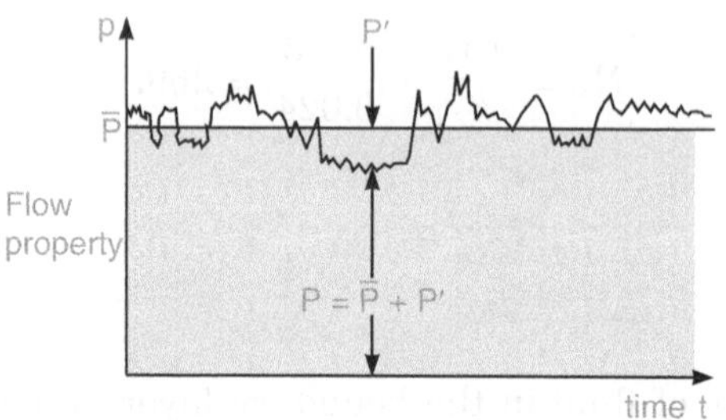

Fig. 7.15. Property variation with time at some points in turbulent boundary layer

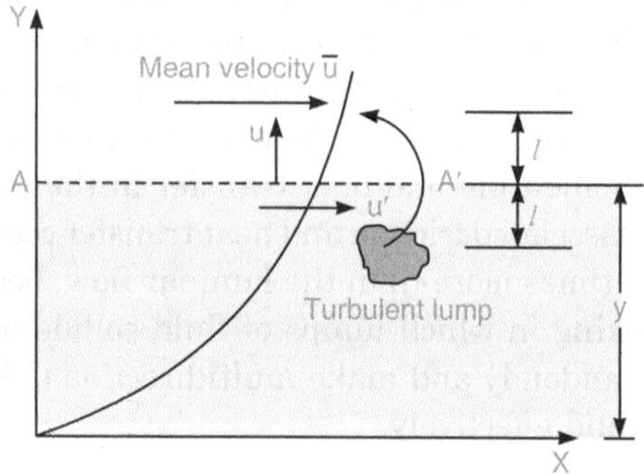

Fig. 7.16. Turbulent shear stress and mixing length

Consider a turbulent lump crosses the plane A – A′ as shown in Fig. 7.16. The fluctuating velocity components continuously transport mass and therefore, momentum across a plane A – A′ normal to y direction.

The instantaneous mass transport per unit area across the plane = $\rho v'$

Instantaneous rate of transfer of x directional momentum per unit area represents shear stress.

$$\tau' = -\rho v' (\overline{u} + u') \qquad ...(7.59)$$

The negative sign is inducted, because, when a turbulent lump moves upward ($v' > 0$), it enters the region of higher $\overline{u}$, it will tend to slow down the fluctuations in u', thus $u' < 0$ and vice-versa so a positive v' is associated with negative u', therefore, the product $u'v'$ is a negative quantity.

Time average of instantaneous rate of x directional momentum transfer per unit area

$$\tau_t = -\frac{1}{t}\int_0^t (\rho v')(\overline{u} + u')\, dt \qquad ...(7.60)$$

It is also called *"apparent turbulent shear stress or Reynolds stress"* and can be rearranged as

$$\tau_t = -\frac{1}{t}\int_0^t \rho v' \overline{u}\, dt - \frac{1}{t}\int_0^t \rho v' u'\, dt$$

$$= -\rho\, \overline{v'}\, \overline{u} - \rho\, \overline{u'v'} \qquad ...(7.61)$$

Since $\overline{u}$ is mean velocity, thus constant and the time averaged $\rho v'$ is zero, therefore

$$\tau_t = -\rho\, \overline{u'v'} \qquad ...(7.62)$$

As stated above $u'v'$ is not zero, but it is negative, thus the turbulent shear stress is positive and analogous to laminar shear stress

$$\tau_l = \mu \frac{d\overline{u}}{dy} = \rho \nu \frac{d\overline{u}}{dy} \qquad ...(7.63)$$

Actually, the laminar shear stress τ_l is true stress whereas the Reynolds stress τ_t is the stress to account for the effects of momentum transfer due to turbulence.

Thus the total shear stress

$$\tau_{\text{total}} = \tau_l + \tau_t = \mu \frac{d\overline{u}}{dy} - \rho\, \overline{u'v'} \qquad ...(7.64)$$

7.12.1 Prandtl Mixing Length Concept

Prandtl postulated that the fluctuations of fluid lumps in turbulent flow on average are analogous to motion of molecules in a gas. The *Prandtl mixing length l* is the distance travelled on an average by the turbulent lumps of fluid in direction perpendicular to mean flow before coming to rest. The Prandtl mixing length l is analogous to the mean free path of molecules in a gas.

Let us imagine a turbulent lump which is located at a distance l above or below plane A – A′ as shown in Fig. 7.16. The fluid lumps move back and forth across the plane and increase turbulent shearing stress effect. At distance $y + l$, the velocity of fluid would be approximately

$$u(y + l) = u(y) + l \frac{d\overline{u}}{dy}$$

and at distance $y - l$

$$u(y - l) = u(y) - l \frac{d\overline{u}}{dy}$$

The Prandtl demonstrated that the turbulent fluctuation u' is proportional to mean of above two quantities, or

$$u' = l\,\frac{d\bar{u}}{dy} \qquad ...(7.65)$$

He also postulated that v' would be of same order of magnitude as u', *i.e.*,

$$v' = l\,\frac{d\bar{u}}{dy}$$

The turbulent shear stress eqn. (7.62)

$$\tau_t = -\rho\,\overline{u'v'} = \rho\, l^2\left(\frac{d\bar{u}}{dy}\right)^2 = \rho\,\varepsilon_M\,\frac{d\bar{u}}{dy} \qquad ...(7.66)$$

where ε_M is *eddy* or *turbulent viscosity,* or

$$\varepsilon_M = l^2\,\frac{d\bar{u}}{dy} \qquad ...(7.67)$$

The *eddy viscosity* ε_M is analogous to kinematic viscosity ν. But ν is a physical property, while ε_M is not, and it depends on dynamics of flow.

Total shearing stress

$$\tau_{total} = \mu\,\frac{d\bar{u}}{dy} + \rho\,\varepsilon_M\,\frac{d\bar{u}}{dy}$$

$$= \rho\nu\,\frac{d\bar{u}}{dy} + \rho\,\varepsilon_M\,\frac{d\bar{u}}{dy}$$

$$= \rho(\nu + \varepsilon_M)\,\frac{d\bar{u}}{dy} \qquad ...(7.68)$$

For turbulent flow

$$\varepsilon_M >> \nu \quad \text{and} \quad \tau_{total} \approx \rho\varepsilon_M\,\frac{d\bar{u}}{dy} \qquad ...(7.69)$$

For laminar flow

$$\varepsilon_M = 0 \quad \text{and} \quad \tau_{total} \approx \rho\nu\,\frac{d\bar{u}}{dy} \qquad ...(7.70)$$

For buffer layer (transition zone),

$$\tau_{total} = \rho(\nu + \varepsilon_M)\,\frac{d\bar{u}}{dy}$$

7.12.2. Turbulent Heat Transfer

The heat transfer in turbulent flow is analogous to momentum transfer. The instantaneous turbulent heat transfer rate per unit area can be expressed as

$$\frac{Q_t}{A} = \frac{1}{t}\int_0^t (\rho v')\,C_p\,(\bar{T} + T')\,dt = \rho C_p\,\overline{v'T'} \qquad ...(7.71)$$

Using Prandtl mixing length concept, the temperature fluctuations T′ can be related with time mean temperature gradient as

$$T' \approx l_T\,\frac{d\bar{T}}{dy} \qquad ...(7.72)$$

when a fluid lump in turbulent flow migrates plane A – A′ by a distance $\pm\, l_T$, the resulting fluctuation is caused by temperature difference between time mean temperature of two planes. The turbulent heat transfer rate per unit area

$$\frac{Q_t}{A} = -\rho C_p\,\overline{v'T'} = -\rho C_p\, l_T^2\,\frac{\partial\bar{u}}{\partial y}\cdot\frac{\partial\bar{T}}{\partial y}$$

$$= -\rho C_p\,\varepsilon_H\,\frac{\partial\bar{T}}{\partial y} \qquad ...(7.73)$$

where ε_H is *eddy* or *turbulent diffusivity* of heat, or

$$\varepsilon_H = l_T^2\,\frac{\partial u}{\partial y} \qquad ...(7.74)$$

The minus sign is due to second law, because $\frac{\partial\bar{T}}{\partial y}$ would be negative.

The total rate of turbulent heat transfer per unit area

$$\left(\frac{Q_t}{A}\right)_{total} = \text{Molecular conduction/area}$$

$$+ \text{ Turbulent heat transfer through eddies/area}$$

$$= -k\,\frac{\partial\bar{T}}{\partial y} - \rho C_p\varepsilon_H\,\frac{\partial\bar{T}}{\partial y}$$

$$= -\rho C_p\,\alpha\,\frac{\partial\bar{T}}{\partial y} - \rho C_p\varepsilon_H\,\frac{\partial\bar{T}}{\partial y}$$

$$= -\rho C_p\,(\alpha + \varepsilon_H)\,\frac{\partial\bar{T}}{\partial y} \qquad ...(7.75)$$

where $\alpha = k/\rho C_p$.

The contribution to the total heat transfer rate by molecular conduction is proportional to α, and turbulent contribution is proportional to ε_H.

For all fluids except liquid metals

$\varepsilon_H >> \alpha$ in turbulent flow and

$$\left(\frac{Q_t}{A}\right)_{total} \approx -\rho C_p\,\varepsilon_H\,\frac{\partial\bar{T}}{\partial y} \qquad ...(7.76)$$

For laminar flow $\varepsilon_H = 0$ and

$$\left(\frac{Q_t}{A}\right)_{total} = -\rho C_p\,\alpha\,\frac{\partial\bar{T}}{\partial y} \qquad ...(7.77)$$

In transition zone

$$\left(\frac{Q_t}{A}\right)_{total} = -\rho C_p\,(\alpha + \varepsilon_H)\,\frac{\partial\bar{T}}{\partial y}$$

The ratio of eddy viscosity to eddy thermal diffusivity is called *turbulent Prandtl number*

$$Pr_t = \frac{\varepsilon_M}{\varepsilon_H} \qquad ...(7.78)$$

This definition is analogous to definition of Prandtl number

$$Pr = \frac{\nu}{\alpha}$$

But the Prandtl number Pr and turbulent Prandtl number are not same. The Prandtl number Pr is a dimensionless physical property of fluid. However, the turbulent Prandtl number Pr is a property of flow field more than a field. Various models have been developed for evaluating of Pr_t. Reynolds model is simplest one, he assumed $Pr_t = 1$ *i.e.*, $\varepsilon_H = \varepsilon_M$.

However, the numerical values of Pr_t may vary between 1 and 2.

For $Pr_t = 1$, the turbulent heat flux eqn. (7.76) and turbulent shear stress eqn. (7.69) can be related as

$$\frac{Q_t}{A\tau_t} = -\frac{\rho C_p \varepsilon_H \dfrac{\partial \overline{T}}{\partial y}}{\rho \varepsilon_M \dfrac{\partial \overline{u}}{\partial y}}$$

or
$$\frac{Q_t}{A} = -\tau_t C_p \frac{\partial \overline{T}}{\partial \overline{u}} \qquad ...(7.79)$$

This relation was first introduced in 1874 by Reynolds and therefore, called *Reynolds analogy* for turbulent flow. This analogy however, does not hold good in viscous sublayer, where the flow is laminar.

7.13. REYNOLDS COLBURN ANALOGY FOR TURBULENT FLOW OVER A FLAT PLATE

To obtain the heat transfer rate for turbulent flow over a flat plate with $Pr_t = 1$, the eqn. (7.79) can be arranged as

$$\frac{Q_s}{A\tau_s C_p} du = -dT \quad \text{or} \quad \frac{q_s}{\tau_s C_p} du = -dT$$

where subscript s indicates that q and τ are taken at surface of the plate. Integrating above equation between $u = 0$, $T = T_s$ and $u = u_\infty$, $T = T_\infty$ yields to

$$\frac{q_s}{\tau_s C_p} u_\infty = T_s - T_\infty \quad \text{or} \quad \frac{q_s}{T_s - T_\infty} = \frac{\tau_s C_p}{u_\infty}$$

Introducing local heat transfer coefficient and friction coefficient as

$$h_x = \frac{q_s}{T_s - T_\infty} \quad \text{and} \quad \tau_s = \frac{C_{fx}}{2} \rho u_\infty^2$$

Then
$$h_x = C_p \frac{C_{fx}}{2} \rho u_\infty \qquad ...(7.80)$$

or
$$\frac{h_x}{\rho u_\infty C_p} = \frac{C_{fx}}{2}$$

or
$$St_x = \frac{C_{fx}}{2} \qquad ...(7.81)$$

where $St_x = \dfrac{Nu_x}{Re_x Pr} = \dfrac{h_x}{\rho C_p u_\infty}$ is called Stanton number.

The eqn. (7.81) is called *Reynolds analogy.* It is satisfactory for gases Pr = 1. Colburn had corrected to fluids having Prandtl number ranging 0.6 to 50 and it is modified to

$$St_x \, Pr^{2/3} = \frac{C_{fx}}{2} \qquad ...(7.82)$$

where subscript x represents the distance from the leading edge. The expression (7.82) is referred as *Reynolds Colburn analogy* for flow over flat pate and $St_x \, Pr^{2/3}$ is called *Colburn's factor.*

For average properties (average heat transfer coefficient and friction coefficient), the above equation is also valid in the form

$$St \, Pr^{2/3} = \frac{C_f}{2} \qquad ...(7.83)$$

valid for all types of flow over a flat plate.

7.14. MEAN FILM TEMPERATURE AND BULK MEAN TEMPERATURE

For external flows such as flow over a flat plate, flow across a cylinder or a sphere, the fluid properties like ρ, C_p, k_f, and μ are generally evaluated at mean film temperature T_f or

$$T_f = \frac{T_s + T_\infty}{2} \qquad ...(7.84)$$

where T_s = surface temperature, °C and

T_∞ = free stream temperature of fluid, °C

For internal flows such as flow through tubes, ducts etc, the fluid properties are evaluated at mean of the bulk inlet and outlet temperature, T_m or

$$T_m = \frac{T_{b,\,in} + T_{b,\,out}}{2} \qquad ...(7.85)$$

where $T_{b,\,in}$ = Bulk mean inlet temperature, °C, and

$T_{b,\,out}$ = Bulk mean outlet temperature, °C.

Sometimes, the correlations may specify some other temperature ; such as for internal flow it may be the mean of fluid temperature T_m and pipe wall surface temperature T_s. If temperature differences (surface to fluid, inlet to outlet) are small enough, then changes in the fluid properties are negligible and the choice of particular temperature becomes unimportant, providing consistancy is maintained.

Example 7.15. *Atmospheric air at 400 K flows with a velocity of 4 m/s along a flat plate, 1 m long, maintained at an uniform temperature of 300 K. The average heat transfer coefficient is estimated to be 7.75 W/m².K. Using Reynolds Colburn analogy, calculate the drag force exerted on the plate per metre width.*

Solution

Given : Flow along a flat plate

$T_\infty = 400$ K, $T_s = 300$ K,

$L = 1$ m, $w = 1$ m,

$h = 7.75$ W/m^2.K, $u_\infty = 4$ m/s.

To find : Drag (shear) force exerted on the plate.

Analysis : Reynolds Colburn analogy for flow over a flat plate is given by

$$\text{St Pr}^{2/3} = \frac{h\,\text{Pr}^{2/3}}{\rho C_p u_\infty} = \frac{C_f}{2}$$

The physical properties of atmospheric air at mean film temperature

$$T_f = \frac{T_s + T_\infty}{2} = \frac{300 + 400}{2} = 350 \text{ K}$$

$\rho = 0.998$ kg/m^3,

$C_p = 1009$ J/kg.K, Pr = 0.697

Then friction coefficient

$$C_f = \frac{2 \times 7.75 \times (0.697)^{2/3}}{0.998 \times 1009 \times 4} = 3.025 \times 10^{-3}$$

The average shear stress

$$\tau = \frac{C_f}{2} \rho u_\infty^2 = \frac{3.025 \times 10^{-3}}{2} \times 0.998 \times (4)^2 = 0.0241 \text{ N/m}^2.$$

The drage (shear) force

$F = wL\tau = 1 \times 1 \times 0.0241$

$=$ **0.0241 N.** Ans.

7.15. SUMMARY

Convection is the mode of heat transfer that involves conduction as well as bulk fluid motion. The rate of convection heat transfer is expressed by Newton's law of cooling as

$$Q = hA(T_s - T_\infty) \text{ (W)}$$

where T_s is surface temperature and T_∞ is free stream fluid temperature.

The convection heat transfer is classified as natural or forced convection. The *natural or free convection* is a process in which fluid motion is set up due to density difference results from heat transfer. While in the *forced convection,* the fluid is forced to flow over a surface or in a duct by external means.

The region of flow in which the effects of viscous shear forces caused by fluid viscosity are observed, is called *velocity boundary layer* (δ). The flow region over the surface in which the temperature variation in the direction normal to surface is observed, is called *thermal boundary layer* (δ_{th}).

The fluid flow over a flat plate starts as a laminar boundary layer, in which the fluid motion is highly ordered and fluid flow can be identified in stream lines. The fluid flow becomes turbulent after some distance from the leading edge, in which large velocity fluctuations and highly disordered motion of the fluid are observed. The intense mixing in turbulent flow enhances both the drag force and heat transfer. The flow regime depends mainly on *Reynolds number*, expressed as

$$\text{Re} = \frac{\text{Inertia forces}}{\text{Viscous forces}} = \frac{u_\infty x}{\nu}$$

where u_∞ is free stream fluid velocity, x is the distance from leading edge and ν is kinematic viscosity.

The Reynolds number for flow through a circular pipe is calculated as

$$\text{Re}_D = \frac{\rho u_m D}{\mu} = \frac{u_m D}{\nu}$$

where u_m = mean fluid velocity, D is inner diameter of tube and ν is kinematic viscosity.

The Reynolds number for non-circular duct is calculated as

$$\text{Re} = \frac{u_m D_h}{\nu}$$

where $D_h = \frac{4A_c}{P}$, hydraulic diameter,

A_c = cross-section area of non-circular tube,

P = wetted perimeter.

The Reynolds number at which the flow turns to be turbulent from laminar flow is called *critical Reynolds number*, Re_{cr} and its value is

$\text{Re}_{cr} = 5 \times 10^5$ for flow over flat plate

$= 2300$ for flow inside tubes.

For flow over a flat plate, the momentum and energy equations are given as

$$u\frac{\partial u}{\partial y} + v\frac{\partial u}{\partial y} = \frac{1}{\nu}\frac{\partial^2 u}{\partial y^2} \quad \text{and} \quad u\frac{\partial T}{\partial x} + v\frac{\partial T}{\partial y} = \frac{1}{\alpha}\frac{\partial^2 T}{\partial y^2}$$

The similarities between velocity and thermal boundary layer indicate that the local skin friction coefficient C_{fx} and Nusselt number Nu_x are function of Reynolds number as

$$C_{fx} = f(x^*, \text{Re}_x)$$

$$\text{Nu}_x = \frac{h_x L}{k_f} = \phi(x^*, \text{Re}_x, \text{Pr})$$

where $x^* = \frac{x}{L}$ and $Pr = \frac{\mu C_p}{k_f}$, Prandtl number.

The local friction coefficient is expressed in terms of local shear stress τ_s as

$$C_{fx} = \frac{\tau_s}{\rho u_\infty^2/2}$$

The dimensional analysis is a method of analysis in which certain variable affecting a phenomenon are combined in dimensionless group such as Nusselt number, which facilates the interpretation and extends its application to experimental data.

The Reynolds Colburn anology for turbulent flow over a flat plate indicates that the heat transfer coefficient and fluid friction are related as

$$\frac{h_x}{\rho u_\infty C_p} Pr^{2/3} = \frac{C_{fx}}{2}$$

Here the quantity

$$\frac{h_x}{\rho u_\infty C_p} = \frac{Nu}{Re_x\ Pr} = St_x \text{ (Stanton number)}$$

$$\therefore \qquad St_x\ Pr^{2/3} = \frac{C_{fx}}{2}$$

TABLE 7.5. Dimensionless groups used in heat transfer

Groups	*Definition*	*Interpretation*
Biot number (Bi)	$\frac{hL_c}{k}$	Ratio of internal thermal resistance of a solid to the boundary layer thermal resistance.
Coefficient of friction (C_f)	$\frac{\tau_s}{\rho u_\infty^2/2}$	Dimensionless surface shear stress.
Colburn *j* factor	$St\ Pr^{2/3}$	Dimensionless heat transfer coefficient.
Fourier number (Fo)	$\frac{\alpha t}{L_c^2}$	Ratio of heat conduction to the rate of thermal energy storage in a solid.
Friction factor (*f*)	$\frac{\Delta p}{(L/D)\rho u_\infty^2/2}$	Dimensionless pressure drop for internal flow.
Grashof number (Gr_L)	$\frac{g\beta(T_s - T_\infty)L_c^3}{\nu^2}$	Ratio of buoyancy to viscous forces of the fluid.
Jacob number (Ja)	$\frac{C_p(T_s - T_{sat})}{h_{fg}}$	Ratio of sensible heat to latent energy absorbed during liquid vapour phase change.
Nusselt number (Nu_L)	$\frac{hL}{k_f}$	Dimensionless temperature gradient at the surface of fluid.
Peclet number (Pe_L)	$Re_L\ Pr$	Dimensionless independent heat transfer parameter.
Prandtl number (Pr)	$\frac{\mu C_p}{k_f} = \frac{\nu}{\alpha}$	Ratio of momentum and thermal diffusivities.
Reynolds number (Re_L)	$\frac{u_\infty L}{\nu}$	Ratio of inertia to viscous forces of a flowing fluid.
Stanton number (St)	$\frac{h}{\rho u_\infty C_p} = \frac{Nu_L}{Re_L\ Pr}$	Modified Nusselt number.

where L_c = characteristic length of the geometry.

REVIEW QUESTIONS

1. Define laminar and turbulent flows. What is Reynolds number ?
2. Explain velocity and thermal boundary layer.
3. Discuss laminar sublayer, buffer layer and turbulent layer in a boundary layer.
4. What is critical Reynolds number? State its approximate values for flow over flat plate and through a circular tube.
5. What do you understand by local and average value of heat transfer coefficient ?
6. Explain local and average value of skin friction coefficient.
7. Show that the Reynolds number for flow through a tube of diameter D can be expressed

$$\text{Re} = \frac{4\dot{m}}{\pi D \mu}.$$

8. Explain the mechanism of convection heat transfer.
9. What are the differences between natural and forced convection ?
10. What is external forced convection ? How does it differ from internal forced convection ?
11. What is physical significance of Prandtl number ?
12. What is physical significance of Reynolds number ? How is it defined for (*a*) flow over a flat plate of length L, (*b*) flow over a cylinder of diameter D, (*c*) flow through a tube of diameter *d*, and flow through a rectangular tube of cross-section $a \times b$?
13. What is physical significance of Nusselt number ? How is it defined for (*a*) flow over a flat plate of length L, (*b*) flow over a cylinder of diameter D, (*c*) flow through a tube of diameter *d*, and flow through a rectangular duct of cross-section $a \times b$?
14. When is heat transfer through a fluid layer by conduction and when is it by convection ? For what case, the rate of heat transfer is higher ?
15. How does the heat transfer coefficient differ from thermal conductivity ?
16. What is no slip condition on a surface ?
17. What property is responsible for development of velocity boundary layer ? What property is for thermal boundary layer ?
18. Consider laminar flow over a flat plate, will the friction coefficient change with position ? How about the heat transfer coefficient ? Explain.
19. In the fully developed region of the flow in a circular tube, will the velocity profile change in the flow direction ?
20. How does surface roughness affect pressure drop and heat transfer in a tube flow ?
21. Derive an expression for momentum transfer equation for flow over a flat plate.
22. Derive an equation for energy transfer for flow over a flat plate.
23. Express the similarities of momentum and energy equations for flow over a flat plate.
24. State the method for evaluation of heat transfer coefficient.
25. State the scope and application of dimensional analysis in heat transfer processes. What are the two methods for obtaining the dimensionless groups ?
26. Show by Rayleigh method of dimensional analysis, that the Nusselt number is function of Reynolds number and Prandtl number.
27. Explain Buckingham π theorem. What are its merits and demerits ? What are repeating variables ? How are they selected ?
28. What do you understand by mean value and fluctuating component of velocity and other properties in turbulent flow ?
29. Explain the Prandtl mixing length concept to describe turbulent flow over a surface.
30. Explain the Reynolds analogy for turbulent flow over a surface.

PROBLEMS

1. Calculate Reynolds numbers and state the type of flow, whether it is laminar or turbulent for the following :

 (*a*) A 15 m long yatch sailing at 15 km/h in sea water ($\rho = 1000$ kg/m^3 and $\mu = 1.3 \times 10^{-3}$ kg/ms).
 [Ans. 48.07×10^6, turbulent]

 (*b*) A compressor disc of radius 0.5 m rotating at 18000 r.p.m in air at 5 bar and 400°C and

 $$\mu = \frac{1.46 \times 10^{-6}\ T^{3/2}}{(110 + T)} \text{ kg/ms.}$$
 [Ans. 14.78×10^7, turbulent]

 (*c*) 0.08 kg/s of CO_2, gas at 400 K flowing in a 40 mm diameter pipe. For viscosity take

 $$\mu = \frac{1.56 \times 10^{-6}\ T^{3/2}}{(233 + T)} \text{ kg/ms.}$$
 [Ans. 1.29×10^5, turbulent]

 (*d*) The roof of a coach 6 m long, travelling at 100 km/h in air ($\rho = 1.2$ kg/m^3 and $\mu = 1.8 \times 10^{-5}$ kg/ms).
 [Ans. 1.11×10^7, turbulent]

2. Calculate Prandtl number $\left(\text{Pr} = \frac{\mu C_p}{k_f}\right)$ for the following :

 (*a*) Water at 20°C : $\mu = 1.002 \times 10^{-3}$ kg/ms, $C_p = 4.183$ kJ/kg.K and $k_f = 0.603$ W/m.K.
 [Ans. 6.95]

 (*b*) Air at 20°C and 1 bar : R = 287 J/kg.K, $\nu = 1.563 \times 10^{-5}$ m^2/s, $C_p = 1005$ J/kg.K and $k_f = 0.02624$ W/m.K. [Ans. 0.719]

 (*c*) Mercury at 20°C ; $\mu = 1520 \times 10^{-6}$ kg/ms, $C_p = 0.139$ kJ/kg.K and $k_f = 0.0081$ W/m.K.
 [Ans. 0.0261]

(*d*) Engine oil at 60°C ; $\mu = 8.36 \times 10^{-2}$ kg/ms, $C_p = 2035$ J/kg.K and $k = 0.141$ W/m.K.

[**Ans.** 1207]

3. Calculate the appropriate Grashof number and state the type of flow for the following :

(*a*) A central heating radiator, 0.8 m high with a surface temperature of 75°C in a room at 18°C ($\nu = 1.5 \times 10^{-5}$ m²/s, Pr = 0.72)

[**Ans.** 3.98×10^9, Turbulent]

(*b*) A horizontal oil sump with a surface temperature of 40°C, 0.5 m long and 0.4 m wide containing oil at 75°C, (Pr = 546, $\beta = 0.7 \times 10^{-3}$ K^{-1} and $\nu = 4.168 \times 10^{-5}$ m²/s) [**Ans.** 18.97×10^4, Laminar]

(*c*) The surface of heating coil 30 mm diameter, having surface temperature of 80°C in water at 20°C ($\rho = 1000$ kg/m³, Pr = 6.95, $\beta = 0.227 \times 10^{-3}$ K^{-1} and $\mu = 1.00 \times 10^{-3}$ kg/ms).

[**Ans.** 3.6×10^6, laminar]

(*d*) Air at 20°C , (Pr = 0.72, and $\nu = 1.5 \times 10^{-5}$ m²/s) adjacent to a 75 mm diameter horizontal light bulb with a surface temperature of 100°C.

[**Ans.** 4.41×10^4, laminar]

4. Calculate appropriate Nusselt number for the following :

A gas flow (Pr = 0.71, $\mu = 4.63 \times 10^{-5}$ kg/ms and $C_p = 1175$ J/kg.K) over a turbine blade of chord length 20 mm, where the average heat transfer coefficient is 1000 W/m².K. [**Ans.** 261]

5. Calculate the distance from the leading edge of a flat plate at which the transition occurs from laminar to turbulent flow for atmospheric air at 27°C with (*a*) 2, (*b*) 10, (*c*) 20 m/s. Assume transition at $Re_{cr} = 5 \times 10^5$.

[**Ans.** (*a*) 4.21 m, (*b*) 0.842 m, (*c*) 0.421 m]

6. Assume transition from laminar to turbulent at $Re_{cr} = 5 \times 10^5$, calculate the distance from the leading edge at which the transition occurs for the flow of each of the following fluids with a velocity of 2 m/s at 40°C (*a*) air at atmospheric pressure, (*b*) hydrogen at atmospheric pressure, (*c*) water, (*d*) ethylene glycol, (*e*) engine oil.

7. The velocity profile $u(x, y)$ for a laminar boundary layer flow along a flat plate is given by

$$\frac{u(x, y)}{u_\infty} = 2\frac{y}{\delta(x)} - 2\left[\frac{y}{\delta(x)}\right]^3 + \left[\frac{y}{\delta(x)}\right]^4$$

where the boundary layer thickness $\delta(x)$ is given by

$$\frac{\delta(x)}{x} = \frac{5.83}{\sqrt{Re_x}}$$

(*a*) Develop an expression for local friction coefficient.

(*b*) Develop an expression for average friction coefficient over a distance x = L form the leading edge of the plate.

(*c*) Calculate the drag force acting on a plate 2 m by 2 m for the flow of air at atmospheric pressure and at 350 K with velocity of 4 m/s.

8. The exact expression for local friction coefficient C_{fx} for laminar flow over a flat plate is given by

$$C_{fx} = \frac{0.664}{\sqrt{Re_x}}.$$

Air at atmospheric pressure and 350 K flows with a velocity of 30 m/s over a flat plate 0.2 m long. Calculate the drag force acting per meter width of the plate.

[**Ans.** 0.222 N]

9. The local heat transfer coefficient h_x for laminar boundary layer flow over a flat plate is given by

$$\frac{xh_x}{k_f} = 0.332\ Re_x^{1/2}\ Pr^{1/3}.$$

Develop an expression for average heat transfer coefficient h over a distance x = L from leading edge of the plate.

10. Engine oil at 40°C ($\mu = 0.21$ kg/(m/s) ; $\rho = 875$ kg/m³) flows inside a 2.5 cm diameter, 50 m long tube with a mean velocity of 1 m/s. Determine the pressure drop for flow through the tube. (*J.N.T.U., May 2004*)

$$\left[\textbf{Hint. } \Delta p = f\frac{L}{D}\frac{\rho u_\infty^2}{2}, f = \frac{64}{Re}\right] \quad [\textbf{Ans. } 537.6 \text{ kPa}]$$

11. For a laminar natural convection from a heated vertical surface, the local convection coefficient may be expressed as $h_x = C\,x^{-1/4}$.

where h_x is heat transfer coefficient at a distance x from leading edge and C is a constant.

Derive an expression for the ratio h/h_x, where h is average heat transfer coefficient between leading edge (x = 0) and x = L location. $\left[\textbf{Ans. } \frac{4}{3}\left[\frac{L}{x}\right]^{-0.25}\right]$

REFERENCES AND SUGGESTED READING

1. Rehsenow W. M, J. P. Harnett and E.N. Ganic, Eds "*Handbook of Heat Transfer*", 2/e, McGraw Hill, New York 1985.

2. Kays W.M. and M.E. Crawford, "*Convective Heat and Mass Transfer*", 2nd ed., McGraw Hill, New York, 1980.

3. Giedt Warren H., "*Investigation of Variation of Point Unit-Heat Transfer Coefficient Around a Cylinder Normal to an Airstream*". Transaction of ASME, vol. 71 pp. 375–301, 1949.
4. Schlichting H., "*Boundary Layer Theory*", 6th ed., McGraw Hill, New York, 1968.
5. Zhukauskas A and A, B, Ambrazyavichyus, Int, J. of "*Heat Mass Transfer*", Vol 3, pp. 305, 1961.
6. Knudsen J.D. and D.L. Katz, "*Fluid Dynamics and Heat Transfer*", McGraw Hill, New York, 1958.
7. McAdams W.M., "*Heat Transmission*", 3rd ed. McGraw Hill, New York, 1954.
8. Jacob M. and G.A. Hawkins, "*Elements of Heat Transfer*", 3rd ed., Wiley, New York, 1957.
9. Krieth Frank and M.S. Bohn, "*Principles of Heat Transfer*", 5th ed., PWS Pub. Company, 1997.
10. Holman J. P., "*Heat Transfer*", 7th ed., McGraw Hill, New York, 1990.
11. Incropera F. P. and D. P. DeWitt, "*Introduction to Heat Transfer*", 2nd ed., John Wiley & Sons, 1990.
12. Bayazitoglu Y. and M. N. Ozisik, "*Elements of Heat Transfer*", McGraw Hill, New York, 1988.
13. Thomas L.C., "*Heat Transfer*", Prentice-Hall, Englewood Cliffs, NJ, 1982.
14. White F.M., "*Heat and Mass Transfer*", Addison Wesley, Reading, MA, 1988.
15. Jacob M., "*Heat Transfer*", Vol I, Wiley, New York, 1949.
16. Suryanarayana N. V., "*Engineering Heat Transfer*" West Pub. Co. New York, 1998.
17. Chapman Alan. J., "*Fundamentals of Heat Transfer*" Macmillan, New York.
18. Christopher Long, "*Essential Heat Transfer*", Addision Wesley Longman, 2001.
19. Giedt Warren H., "*Principles of Engineering Heat Transfer*", Van Nostrand Inc. 2nd ed., 1967.

8 External Flow

When a fluid flows over a body such as plate, cylinder, sphere etc., it is regarded as an *external flow*. In such a flow, the boundary layer develops freely without any constraints imposed by adjacent surfaces. Accordingly, the region of flow, outside the boundary layer in which the velocity and temperature gradients are negligible is called the *free stream region.*

In an external flow forced convection, the relative motion between the fluid and the surface is maintained by external means such as a fan or a pump and not by buoyancy forces due to temperature gradients as in natural convection.

In this chapter, our primary objective is to determine the heat transfer coefficient and coefficient of friction for flow over different geometries such as flat plate, cylinder and sphere, for both laminar and turbulent flow conditions, we will discuss theoretical as well as empirical relation for both quantities.

8.1. LAMINAR FLOW OVER A FLAT PLATE

In the chapter 7, we have discussed that a flow is termed the laminar flow until the critical Reynolds numbers $Re_{cr} \approx 5 \times 10^5$ is reached. Further, the coefficients of friction and heat transfer are related to the velocity and temperature distribution in the flow, respectively. For laminar boundary layer, continuity, momentum, and energy transfer equations with constant properties and zero pressure gradients are:

Continuity : $$\frac{\partial u}{\partial x} + \frac{\partial v}{\partial y} = 0 \quad ...(8.1)$$

Momentum : $$u\frac{\partial u}{\partial x} + v\frac{\partial u}{\partial y} = \nu\frac{\partial^2 u}{\partial y^2} \quad ...(8.2)$$

Energy : $$u\frac{\partial T}{\partial x} + v\frac{\partial T}{\partial y} = \alpha\frac{\partial^2 T}{\partial y^2} \quad ...(8.3)$$

8.1.1. Approximate Analysis of Momentum Equation

Consider two-dimensional steady flow of an incompressible, constant property fluid along a flat plate as shown in Fig. 8.1.

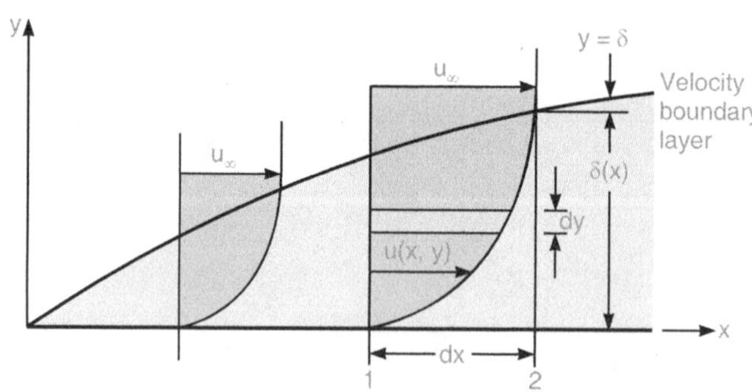

Fig. 8.1. Elemental control volume for integral momentum equation analysis of laminar boundary layer

To obtain momentum equation in the integral form, we must integrate above eqns. (8.1) and (8.2) in the y-direction across the boundary layer. Integrating eqn. (8.1).

$$\int_0^\delta \frac{\partial u}{\partial x}\,dy + \int_0^\delta \frac{\partial v}{\partial y}\,dy = 0$$

The boundary condition, $v = 0$ at $y = 0$

Then $$v(y = \delta) = -\int_0^\delta \frac{\partial u}{\partial x}\,dy \quad ...(8.4)$$

Similarly, integrating eqn. (8.2),

$$\int_0^\delta u\frac{\partial u}{\partial x}dy + \int_0^\delta v\frac{\partial u}{\partial y}dy = \nu\int_0^\delta \frac{\partial^2 u}{\partial y^2}dy$$

$$= \nu\int_0^\delta \frac{\partial}{\partial y}\left(\frac{\partial u}{\partial y}\right)dy$$

Integrating second term on L.H.S. by parts, we get

$$\int_0^\delta u\frac{\partial u}{\partial x}dy + \Big[uv\Big]_0^\delta - \int_0^\delta u\frac{\partial v}{\partial y}dy = \nu\left[\frac{\partial u}{\partial y}\right]_0^\delta$$

Using boundary conditions ;

$$u = u_\infty \quad \text{at} \quad y = \delta$$

$$\frac{\partial u}{\partial y} = 0 \quad \text{at} \quad y = \delta$$

Substituting, we get

$$\int_0^\delta u\frac{\partial u}{\partial x}dy + u_\infty v - \int_0^\delta u\frac{\partial v}{\partial y}dy = -\nu\left.\frac{\partial u}{\partial y}\right|_{y=0} \qquad \text{...(8.5)}$$

The –ve sign is inserted in above equation, because shear force at wall ($y = 0$) acts in opposite direction. Substituting v from eqn. (8.4), we get

$$\int_0^\delta u\frac{\partial u}{\partial x}dy - u_\infty\int_0^\delta \frac{\partial u}{\partial x}dy - \int_0^\delta u\left\{-\frac{\partial}{\partial y}\int_0^\delta \frac{\partial u}{\partial x}dy\right\}dy = -\nu\left.\frac{\partial u}{\partial y}\right|_{y=0}$$

or $$\int_0^\delta u\frac{\partial u}{\partial x}dy - u_\infty\int_0^\delta \frac{\partial u}{\partial x}dy + \int_0^\delta u\frac{\partial u}{\partial x}dy = -\nu\left.\frac{\partial u}{\partial y}\right|_{y=0}$$

or $$u_\infty\int_0^\delta \frac{\partial u}{\partial x}dy - 2\int_0^\delta u\frac{\partial u}{\partial x}dy = \nu\left.\frac{\partial u}{\partial y}\right|_{y=0}$$

It can be written in the form

$$\int_0^\delta \frac{\partial}{\partial x}(u_\infty u - u^2)\,dy = \nu\left.\frac{\partial u}{\partial y}\right|_{y=0}$$

Rearranging, we get

$$\frac{d}{dx}\int_0^\delta (u_\infty - u)\,u\,dy = \nu\left.\frac{\partial u}{\partial y}\right|_{y=0} \qquad \text{...(8.6)}$$

It is known as Von Karman integral equation for momentum transfer in laminar boundary layer.

Assuming velocity distribution in the four term polynomial as

$$u(x) = C_1 + C_2 y + C_3 y^2 + C_4 y^3$$

The constants C_1, C_2, C_3 and C_4 are evaluated with the following boundary conditions :

$$u = 0 \quad \text{at} \quad y = 0 \qquad \text{...}(a)$$

$$u = u_\infty \quad \text{at} \quad y = \delta \qquad \text{...}(b)$$

$$\frac{\partial u}{\partial y} = 0 \quad \text{at} \quad y = \delta \qquad \text{...}(c)$$

and for constant pressure condition at surface

at $\quad y = 0,\; u = v = 0$, therefore $\dfrac{\partial^2 u}{\partial y^2} = 0 \qquad \text{...}(d)$

On solving, we get coefficients as:

$$C_1 = 0,\quad C_2 = \frac{3}{2}\frac{u_\infty}{\delta},\quad C_3 = 0, \text{ and } C_4 = -\frac{u_\infty}{2\delta^3}$$

Therefore, the velocity distribution in the boundary layer becomes

$$\frac{u}{u_\infty} = \frac{3}{2}\left(\frac{y}{\delta}\right) - \frac{1}{2}\left(\frac{y}{\delta}\right)^3 \qquad \text{...(8.7)}$$

Inserting the eqn. (8.7) into eqn. (8.6), we get

$$\frac{d}{dx}\int_0^\delta u_\infty\left[u_\infty - u_\infty\left\{\frac{3}{2}\left(\frac{y}{\delta}\right) - \frac{1}{2}\left(\frac{y}{\delta}\right)^3\right\}\right]\left\{\frac{3}{2}\left(\frac{y}{\delta}\right) - \frac{1}{2}\left(\frac{y}{\delta}\right)^3\right\}dy$$

$$= \nu\frac{\partial}{\partial y}\left\{u_\infty\frac{3}{2}\left(\frac{y}{\delta}\right) - \frac{1}{2}\left(\frac{y}{\delta}\right)^3\right\}_{y=0}$$

or $$\frac{d}{dx}\int_0^\delta u_\infty{}^2\left[\frac{3}{2}\left(\frac{y}{\delta}\right) - \frac{9}{4}\left(\frac{y}{\delta}\right)^2 - \frac{1}{2}\left(\frac{y}{\delta}\right)^3 + \frac{3}{2}\left(\frac{y}{\delta}\right)^4 - \frac{1}{4}\left(\frac{y}{\delta}\right)^6\right]dy$$

$$= \nu u_\infty \frac{3}{2}\frac{1}{\delta} \qquad \text{...(8.8)}$$

On integration it leads to

$$\frac{d}{dx}\left(\frac{39}{280}u_\infty^2\,\delta\right) = \frac{3}{2}\nu\frac{u_\infty}{\delta} \qquad \text{...(8.9)}$$

The free stream velocity u_∞ is constant. The variables may be separated as

$$\delta\,d\delta = \frac{140}{13}\frac{\nu}{u_\infty}dx$$

The δ is function of x only, integrating leads to

$$\frac{\delta^2}{2} = \frac{140}{13}\frac{\nu x}{u_\infty} + C \qquad \text{...(8.10)}$$

where C is constant of integration and it can be evaluated from initial condition,

$$\delta = 0 \quad \text{at } x = 0, \text{ it gives } C = 0$$

Therefore, eqn. (8.10) becomes

$$\delta^2 = \frac{280}{13}\frac{\nu x}{u_\infty} \quad \text{...(8.11)}$$

In dimensionless form

$$\left(\frac{\delta}{x}\right)^2 = \frac{280\,\nu}{13\,u_\infty x}$$

or

$$\frac{\delta}{x} = 4.64\sqrt{\frac{\nu}{u_\infty x}} = \frac{4.64}{\sqrt{\text{Re}_x}} \quad \text{...(8.12)}$$

where $\text{Re}_x = \frac{u_\infty x}{\nu}$, the Reynolds number and $\delta = \delta(x)$, **thickness of velocity boundary layer,** at a distance x from the leading edge of the surface.

The exact solution of the boundary layer equation yields to

$$\frac{\delta}{x} = \frac{5.0}{\sqrt{\text{Re}_x}} \quad \text{...(8.13)}$$

Skin friction coefficient :

To evaluate the *coefficient of friction*, we consider shear stress at the surface,

$$\tau_s = \mu \left.\frac{\partial u}{\partial y}\right|_{y=0}$$

Using eqn. (8.7), we get

$$\tau_s = \frac{3}{2}\frac{\mu u_\infty}{\delta}$$

Substituting δ from eqn. (8.12), we get

$$\tau_s = \frac{3}{2}\frac{\mu u_\infty}{4.64x}\times\sqrt{\frac{u_\infty x}{\nu}} = 0.323\frac{\mu u_\infty}{x}\sqrt{\frac{u_\infty x}{\nu}} \quad \text{...(8.14)}$$

Further, the shear stress can also be expressed in terms of coefficient of friction or skin friction coefficient C_{fx}, as

$$\tau_s = C_{fx}\frac{\rho u_\infty^2}{2} \text{ or } C_{fx} = \frac{2\tau_s}{\rho u_\infty^{\ 2}} \quad \text{...(8.15)}$$

Inserting eqn. (8.14) in eqn. (8.15), we get

$$C_{fx} = 2\times 0.323\frac{\mu u_\infty}{\rho u_\infty^2 x}\sqrt{\frac{u_\infty x}{\nu}}$$

$$= 0.646\sqrt{\frac{\nu}{u_\infty x}} = \frac{0.646}{\sqrt{\text{Re}_x}} \quad \text{...(8.16)}$$

It is the expression for local skin friction coefficient C_{fx}.

The average value of coefficient of friction can be evaluated by integrating eqn. (8.16) over entire plate.

$$C_f = \frac{1}{L}\int_0^L C_{fx}\,dx = \frac{1}{L}\int_0^L \frac{0.646}{\sqrt{\frac{u_\infty x}{\nu}}}\,dx$$

$$= \frac{1}{L}\times\frac{0.646}{\sqrt{u_\infty/\nu}}\int_0^L x^{-1/2}\,dx$$

$$= \frac{1.292}{\sqrt{\frac{u_\infty L}{\nu}}} = \frac{1.292}{\sqrt{\text{Re}_L}} = 2C_{fx}\Big|_{x=L} \quad \text{...(8.17)}$$

where $\text{Re}_L = \frac{u_\infty L}{\nu}$, Reynolds number based on total plate length, L. The average skin friction coefficient or coefficient of friction is often referred as the *drag coefficient.* The *drag force* acting on the plate

$$F = \tau_s A_s = \frac{C_f}{2}\rho u_\infty^2 A_s \quad \text{...(8.18)}$$

Mass flow rate through the boundary layer

The mass flow rate per unit width through the boundary layer at any x position is given by

$$\dot{m}_x = \int_0^\delta \rho u\,dy$$

Inserting eqn. (8.7) for u, we get

$$\dot{m}_x = u_\infty\int_0^\delta \rho\left\{\frac{3}{2}\left(\frac{y}{\delta}\right) - \frac{1}{2}\left(\frac{y}{\delta}\right)^3\right\}dy$$

$$= u_\infty\rho\left[\frac{3}{2}\times\frac{y^2}{2\delta} - \frac{1}{2}\times\frac{1}{4}\frac{y^4}{\delta^3}\right]_0^\delta$$

It gives $$\dot{m}_x = \frac{5}{8}\rho\,\delta\,u_\infty \quad \text{...(8.19)}$$

8.1.2. Approximate Analysis of Energy Equation

The integral energy equation can be derived in similar way as eqn. (8.6). In this case, we consider a control volume for two dimensional steady flow of incompressible fluid as shown in Fig. 8.2.

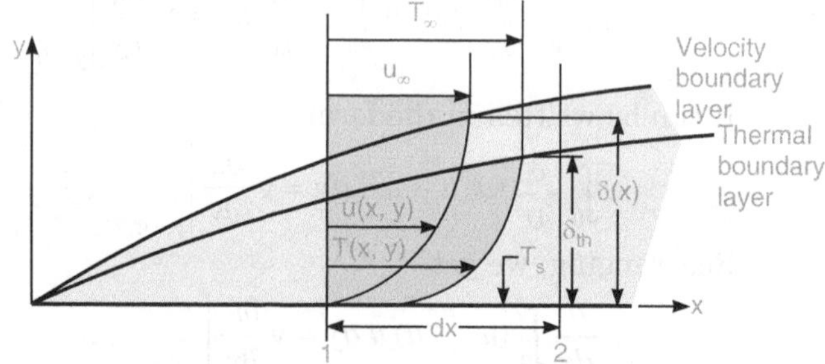

Fig. 8.2. Control volume for integral energy analysis of laminar boundary layer

The energy equation in differential form is given by eqn. (8.3).

$$u\frac{\partial T}{\partial x} + v\frac{\partial T}{\partial y} = \alpha\frac{\partial^2 T}{\partial y^2}$$

For convenience, we introduce a dimensionless temperature $\theta(x, y)$ as:

$$\theta(x, y) = \frac{T(x, y) - T_s}{T_\infty - T_s} \qquad ...(8.20)$$

where $\theta(x,y)$ varies from zero at the wall surface to unity at the edge of thermal boundary layer. Now the energy equation can be written in the form

$$u\frac{\partial \theta}{\partial x} + v\frac{\partial \theta}{\partial y} = \alpha\frac{\partial^2 \theta}{\partial y^2} \qquad ...(8.21)$$

Subjected to boundary conditions

$$\theta = 0 \quad \text{at } y = 0$$
$$\theta = 1 \quad \text{at } y = \delta_{th}$$

Using v from eqn. (8.4), we get resulting energy equation in integral form as

$$\frac{d}{dx}\left[\int_0^{\delta_{th}} u(1-\theta)\, dy\right] = \alpha \left.\frac{\partial \theta}{\partial y}\right|_{y=0} \qquad ...(8.22)$$

Inserting the velocity distribution, eqn. (8.7)

$$\frac{u}{u_\infty} = \frac{3}{2}\left(\frac{y}{\delta}\right) - \frac{1}{2}\left(\frac{y}{\delta}\right)^3$$

Assuming temperature profile as

$$\theta = C_1 + C_2 y + C_3 y^2 + C_4 y^3$$

Subjected to boundary conditions

$$\theta = 0 \quad \text{at } y = 0$$
$$\theta = 1 \quad \text{at } y = \delta_{th}$$
$$\frac{\partial \theta}{\partial y} = 0 \quad \text{at } y = \delta_{th}$$
$$\frac{\partial^2 \theta}{\partial y^2} = 0 \quad \text{at } y = 0$$

Applying, these boundary conditions, we get dimensionless temperature distribution in the form

$$\theta = \frac{3}{2}\left(\frac{y}{\delta_{th}}\right) - \frac{1}{2}\left(\frac{y}{\delta_{th}}\right)^3 \qquad ...(8.23)$$

Introducing velocity and temperature distribution in eqn. (8.22)

$$\frac{d}{dx}\left\{\int_0^{\delta_{th}} u_\infty\left[\frac{3}{2}\left(\frac{y}{\delta}\right) - \frac{1}{2}\left(\frac{y}{\delta}\right)^3\right]\left[1 - \frac{3}{2}\left(\frac{y}{\delta_{th}}\right) + \frac{1}{2}\left(\frac{y}{\delta_{th}}\right)^3\right] dy\right.$$

$$= \alpha \frac{d}{dy}\left[\frac{3}{2}\left(\frac{y}{\delta_{th}}\right) - \frac{1}{2}\left(\frac{y}{\delta_{th}}\right)^3\right]_{y=0}$$

or
$$\frac{d}{dx}\left[\int_0^{\delta_{th}}\left(\frac{3}{2}\frac{y}{\delta} - \frac{9}{4\delta\delta_{th}}y^2 + \frac{3}{4\delta\delta_{th}^3}y^4 - \frac{1}{2\delta^3}y^3 + \frac{3}{4\delta^3\delta_{th}}y^4 - \frac{1}{4\delta^3\delta_{th}^3}y^6\right) dy\right.$$

$$= \frac{3\alpha}{2\delta_{th}\, u_\infty}$$

The integration with respect to y yields to

$$\frac{d}{dx}\left[\frac{3}{4}\frac{\delta_{th}^2}{\delta} - \frac{3}{4}\frac{\delta_{th}^2}{\delta} + \frac{3}{20}\frac{\delta_{th}^2}{\delta} - \frac{1}{8}\frac{\delta_{th}^4}{\delta^3} + \frac{3}{20}\frac{\delta_{th}^4}{\delta^3} - \frac{1}{28}\frac{\delta_{th}^4}{\delta^3}\right] = \frac{3\alpha}{2\delta_{th}\, u_\infty} \qquad ...(8.24)$$

We define new variable as

$$\xi = \frac{\delta_{th}}{\delta} \qquad ...(8.25)$$

Then eqn. (8.24) becomes

$$\frac{d}{dx}\left[\delta\left(\frac{3}{20}\xi^2 - \frac{3}{280}\xi^4\right)\right] = \frac{3}{2}\frac{\alpha}{\xi\delta u_\infty} \qquad ...(8.26)$$

Let we consider the thermal boundary layer is thinner than velocity boundary layer as shown in Fig. 8.2, for Pr > 1 the $\xi < 1$ and term $\frac{3}{280}\xi^4$ becomes least, thus negligible. The eqn. (8.26) simplifies to

$$\xi\delta\frac{d}{dx}(\xi^2\delta) = \frac{10\alpha}{u_\infty}$$

Differentiating w.r.t. x, we get

$$2\xi^2\delta^2\frac{d\xi}{dx} + \xi^3\delta\frac{d\delta}{dx} = \frac{10\alpha}{u_\infty} \qquad ...(8.27)$$

Using $\xi^2\frac{d\xi}{dx} = \frac{1}{3}\frac{d\xi^3}{dx}$, then

$$\frac{2}{3}\delta^2\frac{d\xi^3}{dx} + \xi^3\delta\frac{d\delta}{dx} = \frac{10\alpha}{u_\infty} \qquad ...(8.28)$$

Using velocity boundary layer thickness in the form

$$\delta\, d\delta = \frac{140}{13}\frac{\nu}{u_\infty}dx$$

and
$$\delta^2 = \frac{280}{13}\frac{\nu x}{u_\infty}$$

Then eqn. (8.28) becomes

$$x\frac{d\xi^3}{dx} + \frac{3}{4}\xi^3 = \frac{39\alpha}{56\nu}$$

or
$$\xi^3 + \frac{4}{3}x\frac{d\xi^3}{dx} = \frac{13}{14}\frac{\alpha}{\nu} \qquad ...(8.29)$$

It is linear differential equation of first order in ξ^3 and its solution is

$$\xi^3 = Cx^{-3/4} + \frac{13}{14}\frac{\alpha}{\nu} \qquad ...(8.30)$$

where C is constant of integration and evaluated from boundary conditions

$$\delta_{th} = 0 \quad \text{at } x = 0$$

$\therefore$ $$\xi = 0 \quad \text{at } x = 0$$

Using, we get $C = 0$

Then $$\xi^3 = \frac{13}{14}\frac{\alpha}{\nu} \quad ...(8.31)$$

or $$\xi = \frac{1}{1.026} \Pr^{-1/3}$$

where $\Pr = \frac{\nu}{\alpha}$, Prandtl number and $\xi = \frac{\delta_{th}}{\delta}$

Thus $$\frac{\delta_{th}}{\delta} = \frac{1}{1.026} \Pr^{-1/3} \quad ...(8.32)$$

This relation shows that the ratio of thermal to velocity boundary layer thicknesses for laminar flow along a flat plate is inversely proportional to the cube root of the Prandtl number.

Substituting $\delta(x)$ from eqn. (8.12), we get

$$\delta_{th} = \frac{4.53\,x}{\text{Re}^{1/2}\,\Pr^{1/3}} \quad ...(8.33)$$

Further, the local heat transfer coefficient is defined by

$$h_x = -\frac{k_f\,(\partial T/\partial y)_{y=0}}{T_s - T_\infty} = -\frac{k_f\,(\partial\theta/\partial y)_{y=0}}{T_s - T_\infty}$$

$$= \frac{3}{2}\frac{k_f}{\delta_{th}} = \frac{3}{2}\frac{k_f\,\text{Re}^{1/2}\,\Pr^{1/3}}{4.53x}$$

$$= 0.332\,\frac{k_f}{x}\,\text{Re}^{1/2}\,\Pr^{1/3} \quad ...(8.34)$$

or $$\text{Nu}_x = 0.332\,\text{Re}_x^{1/2}\,\Pr^{1/3} \quad ...(8.35)$$

where $$\text{Nu}_x = \frac{h_x\,x}{k_f} = \text{Local Nusselt number} \quad ...(8.36)$$

Average heat transfer coefficient can be evaluated

$$h = \frac{1}{L}\int_0^L h_x\,dx = 2h_x\big|_{x=L} \quad ...(8.37)$$

Thus $$\text{Nu}_L = 2\,\text{Nu}_x\big|_{x=L}$$

or $$\text{Nu}_L = \frac{hL}{k_f} = 0.664\,\text{Re}_L^{1/2}\,\Pr^{1/3} \quad ...(8.38)$$

where $\text{Re}_L = \frac{u_\infty L}{\nu}$, Reynold number for entire flow length.

The fluid properties should be evaluated at mean film temperature

$$T_f = \frac{T_s + T_\infty}{2} \quad ...(8.39)$$

The eqn. (8.35) is applicable for laminar fluid flowing having Prandtl numbers between 0.6 and 50. It cannot be used for liquid metals with very low Prandtl number and heavy oils or silicons. The *Churchill* and *Ozoe* have suggested the following correlations for laminar flow on an isothermal plate

$$\text{Nu}_x = \frac{0.387\,\text{Re}_x^{1/2}\,\Pr^{1/3}}{\left[1+\left(\frac{0.0468}{\Pr}\right)^{2/3}\right]^{1/4}} \quad ...(8.40)$$

for $\text{Re}_x \Pr > 100$

Constant heat flux boundary condition

In many practical situations, the surface heat flux is constant and the temperature distribution on the plate surface is to be determined. The local Nusselt number for constant heat flux condition on the plate is expressed by

$$\text{Nu}_x = 0.453\,\text{Re}_x^{1/2}\,\Pr^{1/3} \quad ...(8.41)$$

The local Nusselt number can also be expressed in terms of heat flux q and local temperature difference $(T_x - T_\infty)$ as

$$\text{Nu}_x = \frac{q\,x}{k_f\,(T_x - T_\infty)} \quad ...(8.42)$$

or $$T_x - T_\infty = \frac{q\,x}{\text{Nu}_x\,k_f}$$

The average temperature difference over entire plate

$$T_s - T_\infty = \frac{1}{L}\int_0^L (T_x - T_\infty)\,dx = \frac{1}{L}\int_0^L \frac{q\,x}{\text{Nu}_x\,k_f}$$

$$= \frac{1}{L}\int_0^L \frac{q\,x}{0.453x^{1/2}\sqrt{\frac{u_\infty}{\nu}}\,\Pr^{1/3}\,k_f}\,dx$$

$$= \frac{q}{0.453L\sqrt{\frac{u_\infty}{\nu}}\,\Pr^{1/3}\,k_f}\int_0^L x^{1/2}\,dx$$

$$T_s - T_\infty = \frac{qL}{0.6795\,\text{Re}_L^{1/2}\,\Pr^{1/3}\,k_f}$$

or $$q = 0.6795\,\text{Re}_L^{1/2}\,\Pr^{1/3}\,\frac{k_f}{L}\,(T_s - T_\infty) \quad ...(8.43)$$

Comparing eqn. (8.43) with

$$q = h\,(T_s - T_\infty)$$

we get $$h = 0.6795\,\text{Re}_L^{1/2}\,\Pr^{1/3}\,\frac{k_f}{L}$$

$$= 1.5\,\text{Nu}_{x=L}\,\frac{k_f}{L} = 1.5\,h_{x=L} \quad ...(8.44)$$

8.2. REYNOLDS COLBURN ANALOGY : MOMENTUM AND HEAT TRANSFER ANALOGY FOR LAMINAR FLOW OVER FLAT PLATE

If two or more processes are governed by similar dimensionless relations, the processes are said to be analogous. The equations (8.2) and (8.3) for laminar flow over flat plate are of the same form.

The local value of shear stress at the surface may be expressed in terms of skin friction coefficient C_{fx}

$$\tau_x = \left(\frac{C_{fx}}{2}\right)\rho u_\infty^2 \qquad ...(8.45)$$

Further, the shear stress at the surface can also be expressed by equation

$$\tau_x = \mu\left(\frac{\partial u}{\partial y}\right)_{y=0}$$

Using velocity distribution for boundary layer

$$\frac{u}{u_\infty} = \frac{3}{2}\frac{y}{\delta} - \frac{1}{2}\left\{\frac{y}{\delta}\right\}^3$$

we get $$\tau_x = \frac{3\mu u_\infty}{2\delta}$$

Using $$\delta(x) = \frac{4.64x}{\sqrt{\text{Re}_x}}$$

It yields to $$\tau_x = \frac{3}{9.28}\frac{\mu u_\infty}{x} \times \left(\frac{\rho u_\infty x}{\mu}\right)^{1/2} \qquad ...(8.46)$$

Equating equations (8.45) and (8.46) for shear stress at surface, we get

$$\frac{C_{fx}}{2} = 0.323\frac{\mu u_\infty}{x} \times \left(\frac{\rho u_\infty x}{\mu}\right)^{1/2}\frac{1}{\rho u_\infty^2}$$

or $$\frac{C_{fx}}{2} = 0.323\left(\frac{\mu}{\rho u_\infty x}\right)^{1/2} = 0.323\ \text{Re}_x^{-1/2} \qquad ...(8.47)$$

Rewriting eqn. (8.35) as

$$\text{Nu}_x = 0.332\ \text{Pr}^{1/3}\ \text{Re}_x^{1/2}$$

Dividing both sides by Re_x Pr, we get

$$\frac{\text{Nu}_x}{\text{Re}_x\ \text{Pr}} = \frac{h_x}{\rho C_p u_\infty} = 0.332\ \text{Pr}^{-2/3}\ \text{Re}_x^{-1/2}$$

or $$\text{St}_x\ \text{Pr}^{2/3} = 0.332\ \text{Re}_x^{-1/2} \qquad ...(8.48)$$

Comparing eqns. (8.47) and (8.48), we get

$$\text{St}_x\ \text{Pr}^{2/3} \approx \frac{C_{fx}}{2} \qquad ...(8.49)$$

With 3% error in constant. The eqn. (8.49) is called the *Reynolds Colburn analogy*, and it expresses the relation between fluid friction and heat transfer for laminar flow over a flat plate.

For Pr ≅ 1, the eqn. (8.49) reduces to

$$\text{St}_x = \frac{C_{fx}}{2} \qquad ...(8.50)$$

It is called *Reynolds analogy* for laminar flow over a flat plate.

8.3. TURBULENT FLOW OVER A FLAT PLATE

In the turbulent boundary layer, it is very difficult to predict the position of fluid lumps, thus the velocity and temperature profiles can be approximated to give fruitful result. The coefficient of friction and heat transfer coefficient are evaluated from empirical correlations based on experimental data.

1. (*a*) Local coefficient of friction C_{fx}

Based on the boundary layer theory given by *schlichting*, the local skin or friction coefficient within Reynolds number 5×10^5 and 10^7 is related as

$$C_{fx} = 0.0592\ \text{Re}_x^{-1/5}$$

Valid for $5 \times 10^5 < \text{Re}_x < 10^7$ ...(8.51)

At higher Reynolds number, the *Schultz-Grunow* suggested the following correlation

$$C_{fx} = 0.370\ (\ln \text{Re}_x)^{-2.584}$$

Valid for $10^7 < \text{Re}_x < 10^9$ ...(8.52)

(*b*) Average friction coefficient, C_f

The average friction coefficient over the entire plate in the turbulent flow is determined by integrating eqn. (8.51) w.r.t. x,

$$C_f = \frac{1}{L}\int_0^L C_{fx}\,dx \qquad ...(8.53)$$

$$= \frac{1}{L}\int_0^L 0.0592\,x^{-1/5} \times \left(\frac{u_\infty}{\nu}\right)^{-1/5} dx$$

$$= \frac{0.0592}{L} \times \left(\frac{u_\infty}{\nu}\right)^{-1/5}\int_0^L x^{-1/5}\,dx$$

$$= \frac{0.0592}{L} \times \left(\frac{u_\infty}{\nu}\right)^{-1/5}\left[\frac{x^{4/5}}{4/5}\right]_0^L$$

$$= 0.074\left(\frac{u_\infty L}{\nu}\right)^{-1/5}$$

or $$C_f = 0.074\ \text{Re}_L^{-1/5} \qquad ...(8.54)$$

Valid for $5 \times 10^5 < \text{Re}_L < 10^7$

2. The boundary layer thickness in turbulent boundary can be obtained by following correlations.

(*a*) If the boundary layer is completely turbulent, starting from the leading edge, then

$$\frac{\delta}{x} = 0.381\ \text{Re}_x^{-1/5} \qquad ...(8.55)$$

(*b*) If boundary layer is laminar upto $\text{Re}_{cr} = 5 \times 10^5$ and then becomes fully turbulent, for such case, the thickness of boundary layer is given by

$$\frac{\delta}{x} = 0.381\ \text{Re}_x^{-1/5} - 10256\ \text{Re}_x^{-1} \qquad ...(8.56)$$

Valid for $5 \times 10^5 < \text{Re}_x < 10^7$ and $0.6 \le \text{Pr} \le 60$

3. The heat transfer coefficient in turbulent boundary layer can be obtained by using Reynolds Colburn analogy eqn. (8.49) ;

$$\text{St}_x\ \text{Pr}^{2/3} = \frac{C_{fx}}{2}$$

Using C_{fx} from eqn. (8.51), we obtain

$$\text{St}_x\ \text{Pr}^{2/3} = 0.0296\ \text{Re}_x^{-0.2} \qquad ...(8.57)$$

The Local Nusselt number

$$\text{Nu}_x = \text{St}_x\ \text{Re}_x\ \text{Pr}$$

or

$$\text{Nu}_x = 0.0296\ \text{Re}_x^{4/5}\ \text{Pr}^{1/3} \qquad ...(8.58)$$

Valid for $5 \times 10^5 < \text{Re} < 10^7$ and $0.6 < \text{Pr} < 60$

If C_{fx} is used from eqn. (8.52), we get

$$\text{St}_x\ \text{Pr}^{2/3} = 0.185\ \{\ln(\text{Re}_x)\}^{-2.584} \qquad ...(8.59)$$

for $10^7 < \text{Re}_x < 10^9$

The average Nusselt number over the entire plate in turbulent flow is determined by integrating eqn. (8.35), we get

$$\text{Nu} = 0.037\ \text{Re}_L^{4/5}\ \text{Pr}^{1/3} \qquad ...(8.60)$$

Valid for $5 \times 10^5 < \text{Re}_L < 10^7$ and $0.6 < \text{Pr} < 60$

The eqns. (8.54) and (8.60) give average friction and heat transfer coefficients, respectively for the entire plate, when the flow is turbulent over the entire plate.

8.4. COMBINED LAMINAR AND TURBULENT FLOW

In most of the cases, a flat plate is sufficiently long for the flow to become turbulent from laminar as shown in Fig. 8.3. Consider a boundary layer flow along a flat plate such that the flow is laminar over the region $0 \le x \le x_{cr}$ and turbulent over the region $x_{cr} \le x \le L$, Fig. 8.4.

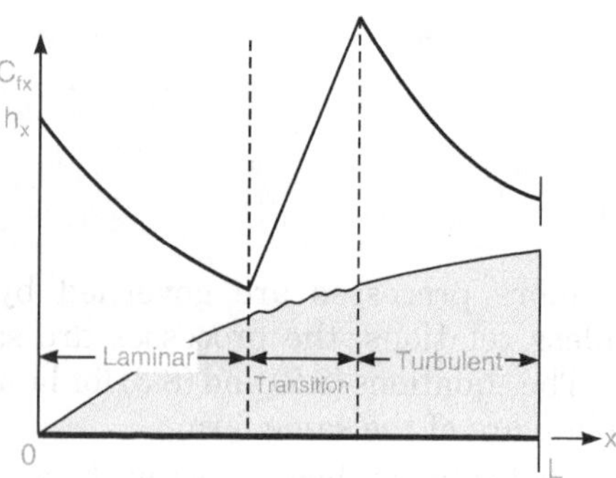

Fig. 8.3. Variation of local friction and local heat transfer coefficients for flow over a flat plate

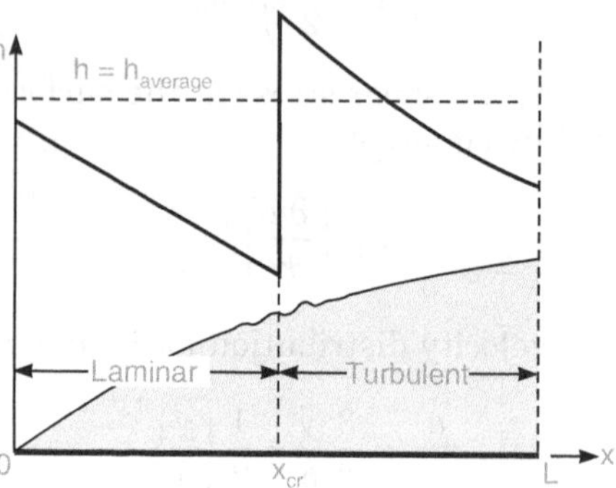

Fig. 8.4. Graphical representation for average heat transfer coefficient for a flat plate with combined laminar and turbulent flow

The friction coefficient for laminar and turbulent regions are

$$C_{fx} = 0.664\ \text{Re}_x^{-1/2} \qquad 0 \le x \le x_{cr}\ \text{(laminar)}$$
$$= 0.0592\ \text{Re}^{-1/5} \qquad x_{cr} \le x \le L\ \text{(turbulent)}$$

The average friction coefficient over the entire plate is obtained as

$$C_f = \frac{1}{L}\left[\int_0^{x_{cr}} C_{fx,\,\text{laminar}}\,dx + \int_{x_{cr}}^{L} C_{fx,\,\text{turbulent}}\,dx\right] \qquad ...(8.61)$$

$$= \frac{1}{L}\left[\int_0^{x_{cr}} 0.664 x^{-1/2}\left(\frac{u_\infty}{\nu}\right)^{-1/2} dx + \int_{x_{cr}}^{L} 0.0592\, x^{-1/5}\left(\frac{u_\infty}{\nu}\right)^{-1/5} dx\right]$$

$$= \frac{1}{L}\left[0.664\left(\frac{u_\infty}{\nu}\right)^{-1/2}\left(\frac{x^{1/2}}{1/2}\right)_0^{x_{cr}} + 0.0592\left(\frac{u_\infty}{\nu}\right)^{-1/5}\left[\frac{x^{4/5}}{4/5}\right]_{x_{cr}}^{L}\right]$$

After simplifying, we get

$$C_f = 0.074\,\mathrm{Re_L}^{-1/5} - \frac{1742}{\mathrm{Re_L}} \quad ...(8.62)$$

Valid for $5 \times 10^5 < \mathrm{Re_L} < 10^7$

In combined boundary conditions, the average convection heat transfer coefficient for entire plate can also be determined by integrating h_x over the laminar region ($0 \le x \le x_{cr}$) and then over turbulent region ($x_{cr} \le x \le L$) as

$$h = \frac{1}{L}\left[\int_0^{x_{cr}} h_{x,\,\mathrm{laminar}}\,dx + \int_{x_{cr}}^{L} h_{x,\,\mathrm{turbulent}}\,dx\right] \quad ...(8.63)$$

Using eqns. (8.35) and (8.58) in laminar and turbulent regions, respectively.

$$h = \frac{k_f}{L}\left[0.332\left(\frac{u_\infty}{\nu}\right)^{1/2}\int_0^{x_{cr}} x^{1/2}\,dx + 0.0296\left(\frac{u_\infty}{\nu}\right)^{4/5}\int_{x_{cr}}^{L} x^{4/5}\,dx\right]\mathrm{Pr}^{1/3}$$

Integrating and rearranging, we get

$$\mathrm{Nu_L} = [0.664\,\mathrm{Re}_{cr}^{1/2} + 0.037(\mathrm{Re}_L^{4/5} - \mathrm{Re}_{cr}^{4/5})]\,\mathrm{Pr}^{1/3} \quad ...(8.64)$$

or $$\mathrm{Nu_L} = (0.037\,\mathrm{Re}_L^{4/5} - A)\,\mathrm{Pr}^{1/3} \quad ...(8.65)$$

where $A = 0.037\,\mathrm{Re}_{cr}^{4/5} - 0.664\,\mathrm{Re}_{cr}^{1/2}$

In typical transition, Reynolds number $\mathrm{Re}_{cr} = 5 \times 10^5$, then the eqn. (8.65) reduces to

$$\mathrm{Nu_L} = \frac{hL}{k_f} = (0.037\,\mathrm{Re_L}^{4/5} - 871)\,\mathrm{Pr}^{1/3} \quad ...(8.66)$$

Valid for $0.6 < \mathrm{Pr} < 60$

$5 \times 10^5 < \mathrm{Re_L} \le 10^8$

$\mathrm{Re}_{cr} = 5 \times 10^5$

When a flat plate is subjected to uniform heat flux instead of uniform temperature, the local Nusselt number in turbulent flow region

$$\mathrm{Nu}_x = 0.0308\,\mathrm{Re}_x^{4/5}\,\mathrm{Pr}^{1/3} \quad ...(8.67)$$

Example 8.1. *The crank case of an automobile is approximated as 0.6 m long, 0.2 m wide, and 0.1 m deep. Assuming that the surface temperature of the crank case is 350 K. Estimate the rate of heat flow from the crank case to atmosphere at 276 K at a road speed of 30 m/s. Assume that the vibration of the engine and chassis induce the transmission from laminar to turbulent flow very near to leading edge that for practical purposes the boundary layer is turbulent over the entire surface. Neglect the radiation and use for the front and rear surfaces, same heat transfer coefficient as for bottom and sides.*

Use relation $Nu_L = 0.036\,Re_L^{0.8}\,Pr^{1/3}$

and $\rho = 1.092\ kg/m^3$,

$\mu = 19.123 \times 10^{-6}\ Ns/m^2$

$k_f = 0.0265\ W/m.K$,

$Pr = 0.71$.

Solution

Given : A crank case of an automobile

$L = 0.6$ m, $w = 0.2$ m, $z = 0.1$ m

$T_s = 350$ K, $T_\infty = 276$ K, $u_\infty = 30$ m/s.

To find: Heat transfer rate.

Analysis: The Reynolds number

$$\mathrm{Re_L} = \frac{\rho u_\infty L}{\mu} = \frac{(1.092\ \mathrm{kg/m^3}) \times (30\ \mathrm{m/s}) \times (0.6\ \mathrm{m})}{(19.123 \times 10^{-6}\ \mathrm{N.s/m^2})} = 1.03 \times 10^6$$

From given relation, the average Nusselt number

$$\mathrm{Nu_L} = 0.036\,\mathrm{Re_L}^{0.8}\,\mathrm{Pr}^{1/3} = 0.036 \times (1.03 \times 10^6)^{0.8} \times (0.71)^{1/3} = 2075$$

Average convective heat transfer coefficient

$$h = \frac{\mathrm{Nu_L}\,k_f}{L} = \frac{2075 \times (0.0265\ \mathrm{W/m{\cdot}K})}{(0.6\ \mathrm{m})} = 91.6\ \mathrm{W/m^2{\cdot}K}$$

Surface area of crank case,

$$A_s = (2 \times 0.6\ \mathrm{m} \times 0.2\ \mathrm{m}) + (2 \times 0.2\ \mathrm{m} \times 0.1\ \mathrm{m}) = 0.28\ \mathrm{m^2}$$

Heat Transfer Rate,

$$Q = hA_s(T_s - T_\infty) = (91.6\ \mathrm{W/m^2.K}) \times (0.28\ \mathrm{m^2}) \times (350 - 276)\ (\mathrm{K}) = \mathbf{1898\ W.}$$ **Ans.**

Example 8.2. *Air at 27°C and 1 atm flows over a flat plate at a speed of 2 m/s. Calculate the boundary layer thickness at distances of 0.2 m and 0.4 m from the leading edge of the plate. Calculate the mass flow rate, which enters the boundary layer between x = 0.2 m and x = 0.4 m. The viscosity of air at 27°C is 1.85 × 10⁻⁵ kg/ms. Assume unit depth in z-direction.*

Solution

Given : Flow over a flat plate

$T_\infty = 27°\mathrm{C}$, $p = 1\ \mathrm{atm} = 101.325\ \mathrm{kN/m^2}$,

$u_\infty = 2$ m/s, $x_1 = 0.2$ m,

$x_2 = 0.4$ m, $\mu = 1.85 \times 10^{-5}$ kg/ms.

To find :

(*i*) Boundary layer at thickness $\delta_{x=0.2}$, and $\delta_{x=0.4}$

(*ii*) Mass flow rate between $\delta_{x=0.2}$, and $\delta_{x=0.4}$.

Assumptions :

1. Steady state conditions.
2. Gas constant R for air as 0.287 kJ/kg·K.
3. Incompressible fluid flow with constant properties.

Analysis : The density of air can be calculated by using equation of state

$$\frac{p}{\rho} = RT_\infty \quad \text{or} \quad \rho = \frac{p}{RT_\infty}$$

or
$$\rho = \frac{101.325 \text{ kN/m}^2}{(0.287 \text{ kJ/kg.K}) \times (27 + 273)(\text{K})} = 1.177 \text{ kg/m}^3$$

The Reynolds number is calculated as

$$Re_x = \frac{\rho u_\infty x}{\mu}$$

At $x = 0.2$ m, $\quad Re_{x_1} = \frac{1.177 \times 2 \times 0.2}{1.85 \times 10^{-5}} = 25448$

At $x = 0.4$ m, $\quad Re_{x_2} = \frac{1.177 \times 2 \times 0.4}{1.85 \times 10^{-5}} = 50897$

(*i*) The boundary layer thickness is calculated by using eqn. (8.13)

$$\delta = \frac{5x}{\sqrt{Re_x}}$$

At $x = 0.2$ m, $\delta_{x=0.2} = \frac{5 \times 0.2 \text{ m}}{\sqrt{25448}} = 6.27 \times 10^{-3}$ m

= **6.027 mm. Ans.**

At $x = 0.4$ m, $\delta_{x=0.4} = \frac{5 \times 0.4 \text{ m}}{\sqrt{50897}} = 8.86 \times 10^{-3}$ m

= **8.86 mm. Ans.**

(*ii*) To calculate the mass flow rate which enters the boundary layer between $x = 0.2$ m and $x = 0.4$ m.

At any x position the mass flow rate in boundary layer can be obtained by using eqn. (8.19)

$$\dot{m} = \frac{5}{8}\rho u_\infty \delta$$

Thus the mass that enters between two boundary layers

$$\Delta \dot{m} = \frac{5}{8}\rho u_\infty \{\delta_{x=0.4} - \delta_{x=0.2}\}$$

$$\Delta \dot{m} = \frac{5}{8} \times (1.177 \text{ kg/m}^3) \times (2 \text{ m/s}) \times [8.86 \times 10^{-3} \text{ m} - 6.27 \times 10^{-3} \text{ m}]$$

= **3.82 × 10⁻³ kg/s. Ans.**

Example 8.3. *Air at 27°C and 1 atm flows over a heated plate with a velocity of 2 m/s. The plate is at uniform temperature of 60°C. Calculate the heat transfer rate from (i) first 0.2 m of the plate, (ii) first 0.4 m of the plate.* (N.M.U., Nov. 2000)

Solution

Given : The flow over a heated flat plate

$T_\infty = 27$°C, $\quad p = 1$ atm,
$u_\infty = 2$ m/s, $\quad T_s = 60$°C,
$x_1 = 0.2$ m, $\quad x_2 = 0.4$ m.

To find :

(*i*) Heat transfer rate from first 0.2 m.
(*ii*) Heat transfer rate from first 0.4 m.

Assumptions :

1. No heat radiation exchange ;
2. The unit depth in z-direction ;
3. Air and surface temperatures are different, taking the properties at mean film temperature.

Properties of air : The mean film temperature

$$T_f = \frac{T_s + T_\infty}{2} = \frac{60 + 27}{2} = 43.5\text{°C}$$

The properties of air at 43.5° C (from Table A-4 of appendix)

$\nu = 17.36 \times 10^{-6}$ m²/s, $\quad k_f = 0.02749$ W/m.K,
Pr = 0.7, $\quad C_p = 1.006$ kJ/kg.K.

Analysis : The Reynolds number at $x = 0.2$ m

$$Re_{x_1} = \frac{u_\infty x_1}{\nu} = \frac{(2 \text{ m/s}) \times (0.2 \text{ m})}{(17.36 \times 10^{-6} \text{ m}^2/\text{s})} = 23041$$

(*i*) The heat transfer rate from first 0.2 m:

The local value of heat transfer coefficient can be calculated as

$$Nu_{x_1} = \frac{h_{x_1} x_1}{k_f} = 0.332 \, Re_{x_1}^{1/2} \, Pr^{1/3}$$

or
$$h_{x_1} = \frac{0.332 \times (0.02749 \text{ W/m.K})}{(0.2 \text{ m})} \times (23041)^{1/2} \times (0.7)^{1/3} = 6.15 \text{ W/m}^2\text{. K}$$

The average value of heat transfer coefficient

$$h_1 = 2\, h_{x_1} = 2 \times 6.15 = 12.3 \text{ W/m}^2\text{. K}$$

The heat transfer rate upto $x = 0.2$ m:

$$Q_1 = h_1 A_s (\Delta T)$$

or
$$\frac{Q_1}{L} = (12.3 \text{ W/m}^2\text{. K}) \times (0.2 \text{ m}) \times (60 - 27)(\text{K})$$

= **81.18 W/m. Ans.**

(*ii*) The heat transfer rate from first 0.4 m:

$$\text{Reynolds no.} \quad Re_{x_2} = \frac{u_\infty x_2}{\nu} = \frac{2 \times 0.4}{17.36 \times 10^{-6}}$$

$$= 46082$$

The local value of heat transfer coefficient

$$Nu_{x_2} = 0.332\ Re_{x_2}^{1/2}\ Pr^{1/3}$$

or

$$h_{x_2} = \frac{0.02749}{0.4} \times 0.332 \times (46082)^{1/2} \times (0.7)^{1/3}$$

$$= 4.35\ W/m^2.\ K$$

Average heat transfer coefficient

$$h_2 = 2h_{x_2} = 2 \times 4.35 = 8.7\ W/m^2.\ K$$

The heat transfer rate

$$\frac{Q_2}{L} = (8.7\ W/m^2.\ K) \times (0.4\ m) \times (60 - 27)(K)$$

$$= \mathbf{114.8\ W/m.}$$ **Ans.**

Example 8.4. *Air at 10°C and at a pressure of 100 kPa is flowing over a plate at a velocity of 3 m/s. If the plate is 30 cm wide and at a temperature of 60°C. Calculate the following quantities at x = 0.3 m.*

(i) Boundary layer thickness,
(ii) Local friction coefficient,
(iii) Local shearing stress,
(iv) Total drag force,
(v) Thermal boundary layer thickness,
(vi) Local convective heat transfer coefficient,
(vii) The heat transfer from the plate.

Solution

Given : Flow over a flat plate

$T_\infty = 10°C$, $u_\infty = 3$ m/s, $T_s = 60°C$,
$w = 30$ cm $= 0.3$ m, $p = 100$ kPa, $x = 0.3$ m.

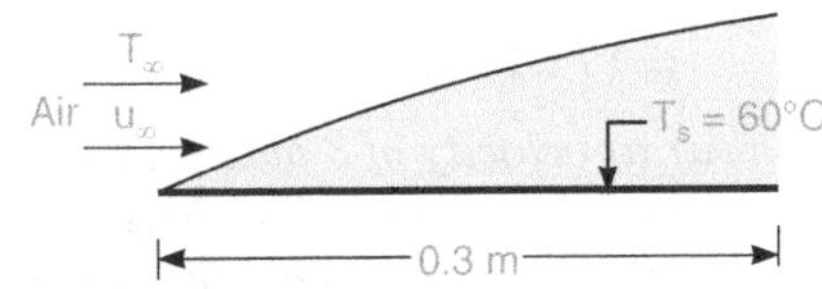

Fig. 8.5. Flow of air over a heated plate

To find :
(*i*) Boundary layer thickness,
(*ii*) Local friction coefficient,
(*iii*) Local shearing stress,
(*iv*) Total drag force,
(*v*) Thermal boundary layer thickness,
(*vi*) Local convective heat transfer coefficient,
(*vii*) The heat transfer from the plate.

Assumptions :
1. No radiation heat exchange.
2. The steady state heat transfer.
3. Air and surface temperatures are different, taking the properties at mean film temperature.

Properties of air : The mean film temperature

$$T_f = \frac{T_s + T_\infty}{2} = \frac{60 + 10}{2} = 35°C$$

The properties of air at 35°C (from Table A-4)

$\rho = 1.1373\ kg/m^3$, $\mu = 19 \times 10^{-6}$ kg/ms,
$k_f = 0.0272$ W/m K, Pr = 0.7,
$C_p = 1.006$ kJ/kg K.

Analysis : The Reynolds number

$$Re_x = \frac{\rho u_\infty x}{\mu}$$

$$= \frac{(1.1373\ kg/m^3) \times (3\ m/s) \times (0.3\ m)}{(19 \times 10^{-6}\ kg/ms)}$$

$$= 53872$$

(*i*) The boundary layer thickness is calculated by eqn. (8.13) :

$$\delta = \frac{5x}{\sqrt{Re_x}} = \frac{5 \times 0.3\ m}{\sqrt{53872}} = 6.46 \times 10^{-3}\ m$$

$$= \mathbf{6.46\ mm.}$$ **Ans.**

(*ii*) The local friction coefficient :

$$C_{fx} = \frac{0.646}{\sqrt{Re_x}} = \frac{0.646}{\sqrt{53872}} = \mathbf{2.783 \times 10^{-3}.}$$ **Ans.**

(*iii*) Local shear stress :

$$\tau_x = \frac{C_{fx}}{2} \rho u_\infty^2 = \frac{2.783 \times 10^{-3}}{2} \times 1.1373 \times (3)^2$$

$$= \mathbf{0.0142\ N/m^2.}$$ **Ans.**

(*iv*) Total drag force :

Drag force,

F = average shear stress × shear area $= \tau_s A_s$

The average shear stress,

$$\tau_s = 2\ \tau_x = 0.0284\ N/m^2$$

and shear area, $A_s = w \times L = 0.3 \times 0.3 = 0.09\ m^2$

Hence $F = 0.0284 \times 0.09$

$$= \mathbf{2.564 \times 10^{-3}\ N.}$$ **Ans.**

(*v*) Thickness of thermal boundary layer :

$$\delta_{th} = \frac{\delta}{1.026} Pr^{-1/3} = \frac{6.46}{1.026} \times (0.7)^{-1/3}$$

$$= \mathbf{7.091\ mm.}$$ **Ans.**

(*vi*) Local heat transfer coefficient :

$$Nu_x = 0.332\ Re_x^{1/2}\ Pr^{1/3}$$

or

$$h_x = \frac{k_f}{x} \times 0.332\ Re_x^{1/2}\ Pr^{1/3}$$

$$= \frac{(0.0272 \text{ W/m.K})}{(0.3 \text{ m})} \times 0.332 \times (53872)^{1/2} \times (0.7)^{1/3}$$

$= \mathbf{6.2\ W/m^2\ K.}$ **Ans.**

(*vii*) Heat transfer rate from the plate :

$$Q = h A_s(\Delta T)$$

where average heat transfer coefficient,

$$h = 2h_x = 2 \times 6.2 = 12.4 \text{ W/m}^2\text{.K}$$

Hence, $Q = (12.4 \text{ W/m}^2\text{.K}) \times (0.09 \text{ m}^2) \times (60 - 10)(\text{K}) = \mathbf{55.8\ W.}$ **Ans.**

Example 8.5. *The air at a temperature of T_∞, flows over a flat plate with a free stream velocity of u_∞. The plate is maintained at a constant temperature of T_s. The velocity u and temperature T of air at any location are given by*

$$\frac{u}{u_\infty} = \sin\frac{\pi y}{2\delta}$$

and

$$\frac{T - T_s}{T_\infty - T_s} = 2\left(\frac{y}{\delta_{th}}\right) - \left(\frac{y}{\delta_{th}}\right)^2$$

where y is the distance measured from the plate along its normal, and δ and δ_{th} are the hydrodynamic and thermal boundary layer thicknesses, respectively. Find the ratio of heat transfer coefficient to shear stress at the plate surface using following data :

$u_\infty = 10 \ m/s$, $\delta/\delta_{th} = Pr^{1/3}$,
$\mu_{(air)} = 2.5 \times 10^{-5} \ kg/ms$, $T_s = 200°C$,
$k_{(air)} = 0.04 \ W/m.K$, $C_{p(air)} = 1000 \ J/kg.K$,
$T_\infty = 50°C$ (N.I.T. Calicut, May 2003)

Solution

Given : Flow of air over a flat plate.

$$\frac{u}{u_\infty} = \sin\frac{\pi y}{2\delta}, \quad \frac{T - T_s}{T_\infty - T_s} = 2\left(\frac{y}{\delta_{th}}\right) - \left(\frac{y}{\delta_{th}}\right)^2$$

$u_\infty = 10$ m/s. $\delta/\delta_{th} = \text{Pr}^{1/3}$
$T_s = 200°C$ $T_\infty = 50°C$
$\mu = 2.5 \times 10^{-5}$ kg/ms, $k_f = 0.04$ W/m.K
$C_p = 1000$ J/kg.K.

To find : Ratio of heat transfer coefficient to shear stress.

Analysis : The shear stress at the wall is given by

$$\tau_s = \mu\left(\frac{du}{dy}\right)_{y=0}$$

Using velocity profile

$$\frac{u}{u_\infty} = \sin\frac{\pi y}{2\delta}$$

or

$$\frac{du}{dy} = u_\infty \frac{d}{dy}\left(\sin\frac{\pi y}{2\delta}\right) = u_\infty\left(\cos\frac{\pi}{2}\frac{y}{\delta}\right)\frac{\pi}{2\delta}$$

and

$$\tau_s = \mu\frac{u_\infty \pi}{2\delta}\left(\cos\frac{\pi}{2}\frac{y}{\delta}\right)_{y=0} = \pi\frac{\mu u_\infty}{2\delta} \quad ...(i)$$

The temperature distribution

$$\frac{T - T_s}{T_\infty - T_s} = 2\left(\frac{y}{\delta_{th}}\right) - \left(\frac{y}{\delta_{th}}\right)^2$$

and

$$\frac{dT}{dy} = (T_\infty - T_s)\frac{d}{dy}\left\{2\left(\frac{y}{\delta_{th}}\right) - \left(\frac{y}{\delta_{th}}\right)^2\right\}$$

$$= (T_\infty - T_s)\left\{2\left(\frac{1}{\delta_{th}}\right) - \frac{2y}{(\delta_{th})^2}\right\}$$

$$\left(\frac{dT}{dy}\right)_{y=0} = (T_\infty - T_s)\frac{2}{\delta_{th}} = \frac{2(T_\infty - T_s)}{\delta_{th}}$$

The heat transfer rate at the plate surface

$$h(T_s - T_\infty) = -k_f \left.\frac{dT}{dy}\right|_{y=0}$$

or

$$h = \frac{-k_f\left[\dfrac{2(T_\infty - T_s)}{\delta_{th}}\right]}{(T_s - T_\infty)} = \frac{2k_f}{\delta_{th}} \quad ...(ii)$$

Dividing eqn. (*ii*) by eqn. (*i*), we get ratio of heat transfer coefficient to shear stress

$$\frac{h}{\tau_s} = \frac{2k_f}{\delta_{th}} \times \frac{2\delta}{\pi\mu u_\infty} = \frac{4k_f}{\pi\mu u_\infty}\left(\frac{\delta}{\delta_{th}}\right) = \frac{4k_f}{\pi\mu u_\infty}\text{Pr}^{1/3} \quad ...(iii)$$

It is the desired ratio. **Ans.**

Now, $$\text{Pr} = \frac{\mu C_p}{k_f} = \frac{2.5 \times 10^{-5} \times 1000}{0.04} = 0.625$$

and

$$\frac{h}{\tau_s} = \frac{4 \times 0.04}{\pi \times 2.5 \times 10^{-5} \times 10} \times (0.625)^{1/3}$$

$= \mathbf{174.18\ m/s\ K.}$ **Ans.**

Example 8.6. *Air at velocity of 3 m/s and at 20°C flows over a flat plate along its length. The length, width and thickness of the plate are 100 cm, 50 cm, and 2 cm, respectively. The top surface of the plate is maintained at 100°C. Calculate the heat lost by the plate and temperature of bottom surface of the plate for the steady state conditions. The thermal conductivity of the plate may be taken as 23 W/m.K.* (P.U., Nov. 1999)

Solution

Given : Flow over a flat plate

$T_\infty = 20°C$, $u_\infty = 3$ m/s,
$T_s = 100°C$, $k = 23$ W/m.K

$w = 0.5$ m, $z = 0.02$ m,
$L = 100$ cm $= 1$ m.

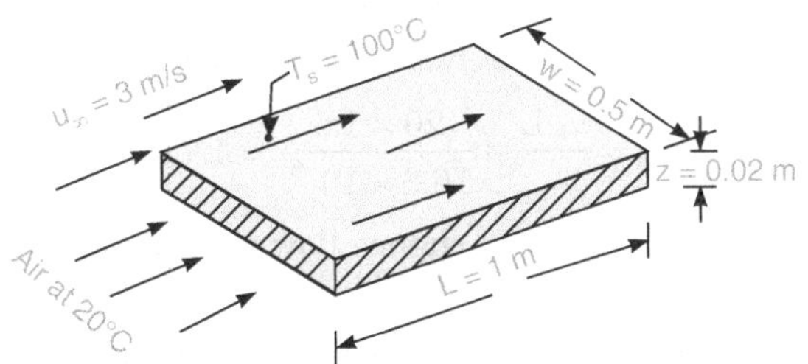

Fig. 8.6. Schematic for example 8.6

To find :

(*i*) Heat transfer rate from the plate.

(*ii*) Temperature of bottom surface of the plate.

Assumptions :

1. No radiation heat exchange.
2. Steady state heat transfer conditions.
3. Air and surface temperatures are different, taking the properties at mean film temperature.

Properties of fluid : The film temperature

$$T_f = \frac{T_s + T_\infty}{2} = \frac{100 + 20}{2} = 60°C$$

The properties of air at 60°C from Table A-4 :

$\rho = 1.06$ kg/m^3, $\nu = 18.97 \times 10^{-6}$ m^2/s,
$k_f = 0.02894$ W/m K, Pr = 0.696,
$C_p = 1.005$ kJ/kg.K.

Analysis : The Reynolds number

$$Re_L = \frac{u_\infty L}{\nu} = \frac{3 \times 1}{18.97 \times 10^{-6}} = 1.58 \times 10^5$$

The Re_L is less than 5×10^5, hence the flow is laminar.

Average heat transfer coefficient :

$$Nu_L = 0.664\, Re_L^{1/2}\, Pr^{1/3}$$

or
$$h = \frac{k_f}{L} \times 0.664\, Re_L^{1/2}\, Pr^{1/3}$$

$$h = \frac{0.02894}{1} \times 0.664 \times (1.58 \times 10^5)^{1/2} \times (0.696)^{1/3}$$

$$= 6.77 \text{ W/m}^2\text{.K}$$

(*i*) The average heat transfer rate from the plate:

$$Q = hA_s\,(\Delta T) = 6.77 \times (1 \times 0.5) \times (100 - 20)$$
$$= \mathbf{270.7\ W/m.}\quad \textbf{Ans.}$$

(*ii*) The temperature of the bottom surface of the plate:

Making the energy balance for the plate;

$$Q = \frac{kA(T_b - T_s)}{z}$$

or
$$T_b = T_s + \frac{Q \times z}{kA} = 100 + \frac{270.7 \times 0.02}{23 \times (1 \times 0.5)}$$
$$= \mathbf{100.47°C.}\quad \textbf{Ans.}$$

Example 8.7. *A flat plate 1 m wide and 1.5 m long is maintained at 90°C in air with free stream temperature of 10°C flowing along 1.5 m side of the plate. Determine the velocity of the air required to have a rate of energy dissipation as 3.75 kW.*

Use correlations

$Nu_L = 0.664\, Re^{1/2}\, Pr^{1/3}$ *for laminar flow;*

$Nu_L = [0.036\, Re^{0.8} - 836]\, Pr^{1/3}$ *for turbulent flow.*

Take properties of air:

$\rho = 1.0877$ *kg/m*^{3}, $\mu = 2.029 \times 10^{-5}$ *kg/ms,*
$k_f = 0.028$ *W/m K,* $Pr = 0.703$,
$C_p = 1.007$ *kJ/kg K.* (P.U., May 1995)

Solution

Given : Flow along a flat plate:

$L = 1.5$ m, $w = 1$ m,
$T_\infty = 10°C$ $T_s = 90°C$,
$Q = 3.75$ kW.

To find : The velocity of air.

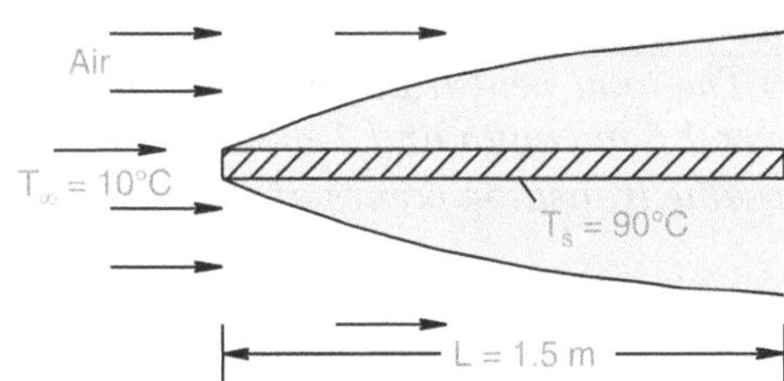

Fig. 8.7. Schematic for example 8.7

Assumptions :

1. No radiation heat exchange.
2. Steady state heat transfer conditions.
3. Air flow on both sides of the plate.

Analysis : The heat transfer rate by convection is given by :

$$Q = hA_s(\Delta T)$$

For two sides of the plate

$$3.75 \times 10^3 = 2h \times (1.5 \times 1) \times (90 - 10)$$
$$(\because\ A_s = 2 \text{ sides} \times 1.5 \times 1 \text{ m}^2)$$

or
$$h = 15.625 \text{ W/m}^2\text{.K.}$$

The Nusselt number,

$$Nu_L = \frac{hL}{k_f} = \frac{15.625 \times 1.5}{0.028} = 837.05$$

Assuming the laminar flow along the plate ;

$$\text{Nu}_L = 0.664\ \text{Re}^{1/2}\ \text{Pr}^{1/3}$$

or $$837.05 = 0.664\ \text{Re}^{1/2} \times (0.703)^{1/3}$$

or $$\text{Re}_L = 2.01 \times 10^6$$

The Reynolds number Re_L is greater than critical Reynolds number 5×10^5, hence assumption made is wrong. The fluid flow is turbulent, using the relation ;

$$\text{Nu}_L = [0.036\ \text{Re}^{0.8} - 836]\ \text{Pr}^{1/3}$$

Using the values ;

$$837.05 = [0.036\ \text{Re}^{0.8} - 836] \times (0.703)^{1/3}$$

or $$\text{Re}_L^{0.8} = 49371.8$$

or $$\text{Re}_L = 7.36 \times 10^5$$

Assumption made is correct. The velocity of air :

$$\text{Re}_L = \frac{\rho u_\infty L}{\mu}$$

$$u_\infty = \frac{\mu\,\text{Re}}{\rho L} = \frac{2.029 \times 10^{-5} \times 7.36 \times 10^5}{1.0877 \times 1.5}$$

$$= \mathbf{9.15\ m/s.}\ \text{Ans.}$$

Example 8.8. *Atmospheric air at 275 K and free stream velocity of 20 m/s flows over a 1.5 m long flat plate maintained at a uniform temperature of 325 K, calculate:*

(a) The average heat transfer coefficient over the region of laminar boundary layer ;

(b) The average heat transfer coefficient over the entire length of 1.5 m ;

(c) The total heat transfer rate from the plate to the air over 1.5 m length and 1 m wide.

Assume transition occurs at $Re_{cr} = 2 \times 10^5$.

(Mumbai University, May 2003)

Solution

Given : $T_\infty = 275$ K, $u_\infty = 20$ m/s, $L = 1.5$ m, $w = 1$ m, $T_s = 325$ K, $\text{Re}_{cr} = 2 \times 10^5$

To find :

(a) The average h over the region of laminar boundary layer ;

(b) The average h over the entire length of 1.5 m ;

(c) The total heat transfer rate from the plate to the air over 1.5 m length and 1 m wide.

Assumptions :

1. No heat exchange by thermal radiation and heat conduction.

2. The steady state heat transfer.

3. Air and surface temperatures are different, taking the properties at mean film temperature.

Properties of air : The film temperature

$$T_f = \frac{T_s + T_\infty}{2} = \frac{275 + 325}{2} = 300\ \text{K}$$

The properties of air at 300 K (from Table A-4)

$k_f = 0.026$ W/m.K, $\text{Pr} = 0.708$,

$\nu = 16.8 \times 10^{-6}$ m²/s, $\mu = 1.98 \times 10^{-5}$ kg/ms.

Analysis : The Reynolds number

$$\text{Re}_L = \frac{u_\infty L}{\nu} = \frac{20 \times 1.5}{16.8 \times 10^{-6}} = 1.785 \times 10^6$$

or $\text{Re}_L > 2 \times 10^5$, hence flow is turbulent at $x = 1.5$ m

The critical length of flow for laminar boundary layer can be calculated by using critical Reynolds number.

$$\text{Re}_{cr} = \frac{u_\infty x_{cr}}{\nu}$$

or $$x_{cr} = \frac{2 \times 10^5 \times 16.8 \times 10^{-6}}{20} = 0.168\ \text{m}$$

(*a*) The average heat transfer coefficient for the laminar boundary layer :

$$\text{Nu}_x = \frac{h\,x_{cr}}{k_f} = 0.664\ \text{Re}_{cr}^{1/2}\ \text{Pr}^{1/3}$$

$$h = 0.664\ \frac{k_f}{x_{cr}}\ \text{Re}_{cr}^{1/2}\ \text{Pr}^{1/3}$$

$$= \frac{0.664 \times 0.026}{0.168} \times (2 \times 10^5)^{1/2} \times (0.708)^{1/3}$$

$$= \mathbf{41.0\ W/m^2.K.}\ \text{Ans.}$$

(*b*) The average heat transfer coefficient over entire plate :

Since the flow is turbulent at $x = 1.5$ m and Reynolds number

$$\text{Re} = 1.785 \times 10^6$$

Using eqn. (8.66) for average heat transfer coefficient

$$\text{Nu}_L = (0.037\ \text{Re}_L^{0.8} - 871)\text{Pr}^{1/3}$$

$$h = \frac{0.026}{1.5} \times [0.037 \times (1.785 \times 10^6)^{0.8} - 871] \times (0.708)^{1/3}$$

$$= \mathbf{43.8\ W/m^2.K.}\ \text{Ans.}$$

(*c*) The total heat transfer rate

$$Q = h\,A_s\,(\Delta T) = (43.8\ \text{W/m}^2\cdot\text{K}) \times (1.5\ \text{m} \times 1\ \text{m}) \times (325 - 275)(\text{K})$$

$$= \mathbf{3290\ W.}\ \text{Ans.}$$

Example 8.9. *The local atmospheric pressure at Mahableshwar hill station in Maharashtra (1610 m from sea level) is 83.4 kPa. Air at this pressure and 20°C flows with a velocity of 8 m/s over a 1.5 m × 6 m flat plate whose temperature is 134°C. Determine the rate of heat transfer from the plate, if the air flows parallel to (a) 6 m long side, and (b) the 1.5 m side.*

Solution

Given :

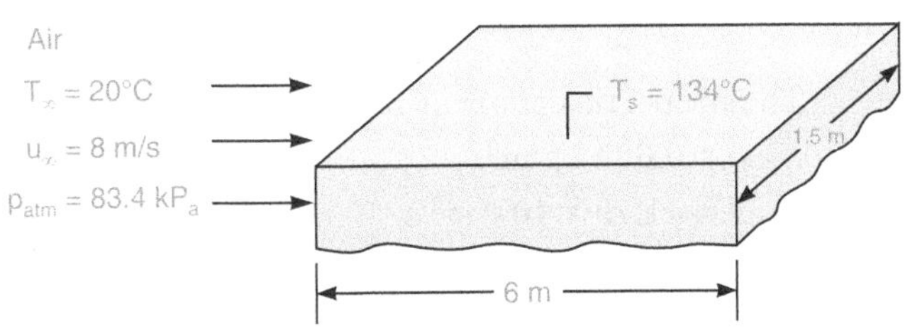

Fig. 8.8. Flow over a flat plate

To find : Rate of heat transfer from the plate, if (*a*) L = 6 m, and (*b*) L = 1.5 m.

Analysis : The film temperature

$$T_f = \frac{T_\infty + T_s}{2} = \frac{20 + 134}{2} = 77°C = 350 \text{ K}$$

The density of air

$$\rho = \frac{p}{RT} = \frac{83.4 \text{ kPa}}{(0.287 \text{ kJ/kg.K}) \times (20 + 273)(\text{K})}$$

$$= 0.991 \text{ kg/m}^3$$

The density is only the function of pressure and other properties are independent of pressure. Thus from Table A-4 ;

$$k_f = 0.030 \text{ W/m.K}, \quad \mu = 2.075 \times 10^{-5} \text{ kg/ms}$$
$$\text{Pr} = 0.697$$

(*a*) When air flows parallel to 6 m side, the Reynolds number

$$\text{Re}_L = \frac{\rho u_\infty L}{\mu} = \frac{0.991 \times 8 \times 6}{2.075 \times 10^{-5}} = 2292434$$

which is greater than 5 × 10^5, thus there would be a combined laminar and turbulent flow. The average Nusselt number

$$\text{Nu} = (0.037 \text{ Re}_L^{0.8} - 871) \text{ Pr}^{1/3}$$
$$= [0.037 \times (2292434)^{0.8} - 871] \times (0.697)^{1/3} = 3248$$

$$\text{Then } h = \frac{\text{Nu } k_f}{L} = \frac{3248 \times 0.030}{6} = 16.25 \text{ W/m}^2\text{.K}$$

The heat transfer rate from the plate

$$Q = h\,(wL)\,(T_s - T_\infty)$$
$$= 16.25 \times (1.5 \times 6) \times (134 - 20)$$
$$= \mathbf{16{,}672 \text{ W.}} \quad \text{Ans.}$$

(*b*) When air flows is parallel to 1.5 m side of plate

$$\text{Re}_L = \frac{0.991 \times 8 \times 1.5}{2.075 \times 10^{-5}} = 5.73 \times 10^5$$

which is again slightly greagter than 5 × 10^5, thus using

$$\text{Nu} = [0.037 (\text{Re}_L)^{0.8} - 871] \text{ Pr}^{1/3} = 553.5$$

and $$h = \frac{553.5 \times 0.030}{1.5} = 11.07 \text{ W/m}^2\text{.K}$$

The heat transfer rate

$$Q = 11.07 \times (1.5 \times 6) \times (134 - 20)$$
$$= \mathbf{11{,}358 \text{ W.}} \quad \text{Ans.}$$

Example 8.10. *An air stream at 0°C is flowing along a heated plate at 90°C at a speed of 75 m/s. The plate is 45 cm long and 60 cm wide. Assuming the transition of boundary layer takes plate at Re_{cr} = 5 × 10^5. Calculate the average value of friction coefficient and heat transfer coefficient for full length of the plate. Also calculate the heat dissipation from the plate.*

(Anna Univ., March 2000)

Solution

Given : Air flows along a heated plate

$$L = 45 \text{ cm} = 0.45 \text{ m}, \quad T_\infty = 0°C$$
$$w = 60 \text{ cm} = 0.6 \text{ m}, \quad T_s = 90°C$$
$$\text{Re}_{cr} = 5 \times 10^5, \quad u_\infty = 75 \text{ m/s}$$

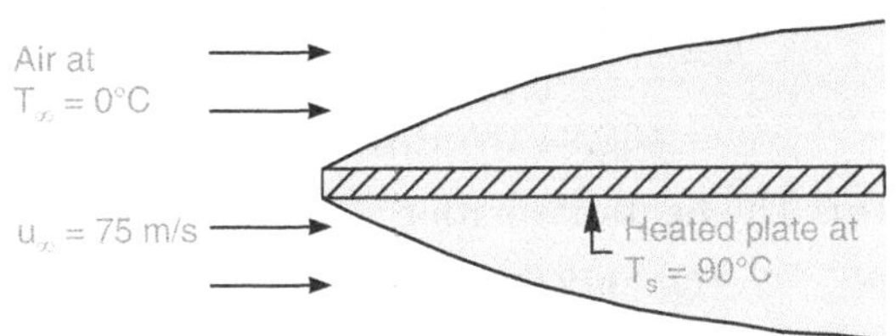

Fig. 8.9. Schematic for example 8.10

To find :

(*i*) Average value of friction coefficient,

(*ii*) Average heat transfer coefficient,

(*iii*) Heat dissipation from the plate.

Assumptions :

(*i*) Steady state conditions.

(*ii*) Due to symmetry, the analysis for friction coefficient and heat transfer rate on one side of plate only.

(*iii*) Constant properties.

Analysis : The film temperature of fluid

$$T_f = \frac{T_s + T_\infty}{2} = \frac{90 + 0}{12} = 45°C \approx 318 \text{ K}$$

The properties of air from Table A-4 :

$$\rho = 1.113 \text{ kg/m}^3, \quad C_p = 1.007 \text{ kJ/kg.K},$$
$$\mu = 1.928 \times 10^{-5} \text{ kg/ms}$$
$$k_f = 0.0276 \text{ W/m.K}, \quad \text{Pr} = 0.693$$

(*i*) The Reynolds number for fluid flow

$$\text{Re}_L = \frac{\rho u_\infty L}{\mu} = \frac{1.113 \times 75 \times 0.45}{1.928 \times 10^{-5}} = 1.95 \times 10^6$$

which is greater than 5×10^5, thus the flow becomes turbulent at $x = 0.45$ m. For combined region of laminar (upto Re = 5×10^5) and turbulent (Re > 5×10^5), the average friction coefficient is determined by eqn. (8.62)

$$C_f = 0.074\, Re_L^{-1/5} - \frac{1742}{Re_L}$$

$$= 0.074 \times (1.95 \times 10^6)^{-1/5} - \frac{1742}{1.95 \times 10^6}$$

$$= \mathbf{3.19 \times 10^{-3}}. \quad \text{Ans.}$$

(*ii*) The average Nusselt number is given by eqn. (8.66)

$$Nu = \frac{hL}{k_f} = (0.037\, Re_L^{4/5} - 871)\, Pr^{1/3}$$

$$= \{0.037 \times (1.95 \times 10^6)^{4/5} - 871\} \times (0.693)^{1/3} = 2754.4$$

and the heat transfer coefficient

$$h = \frac{2754.4 \times 0.0276}{0.45}$$

$$= \mathbf{168.94\ W/m^2.K}. \quad \text{Ans.}$$

(*iii*) The heat dissipation rate from the plate

$$Q = h(2A)(T_s - T_\infty) = h\,(2wL)\,(T_s - T_\infty)$$

$$= 168.94 \times (2 \times 0.6 \times 0.45) \times (90 - 0)$$

$$= \mathbf{8210.34\ W}. \quad \text{Ans.}$$

Example 8.11. *Hot air at 470°C flows over a flat plate 40 cm × 20 cm and 3 mm thick at a velocity of 2 m/s along the 40 cm side. The initial plate temperature is 30°C. The specific heat of the plate is 450 J/kg.K and density of the plate material is 8000 kg/m³. Calculate the initial temperature rise of the plate in °C/min, if the plate receives heat due to convection and radiation from both sides.*

Assume emissivity of the plate is 0.85.

Solution

Given : Hot air flows over a flat plate

$T_i = 30°C = 303$ K, $\quad T_\infty = 470°C = 743$ K.

To find : The initial temperature rise of the plate in °C/min.

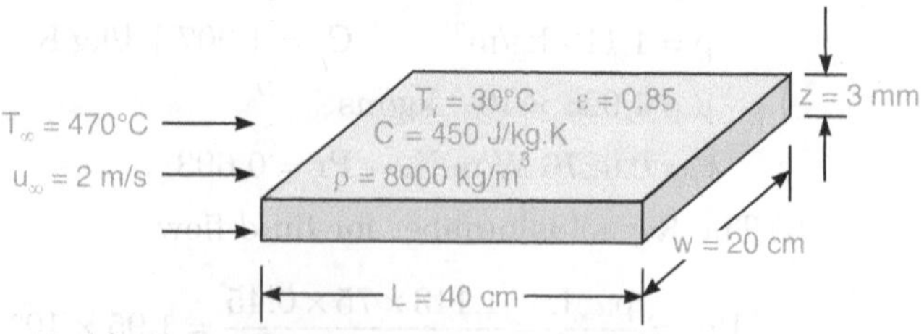

Fig. 8.10. Schematic for example 8.11

Analysis : The film temperature of air over the plate

$$T_f = \frac{T_i + T_\infty}{2} = \frac{30 + 470}{2} = 250°C$$

The properties of air at 250°C from Table A-4

$\rho = 0.674$ kg/m³, $\quad C_p = 1035$ J/kg.K,

$\nu = 41.79 \times 10^{-6}$ m²/s, $\quad k_f = 0.0427$ W/m.K,

Pr = 0.68

The Reynolds number

$$Re_L = \frac{u_\infty L}{\nu} = \frac{2 \times 0.4}{41.79 \times 10^{-6}} = 19143.33$$

which is less than 5×10^5, thus the flow is laminar.

For average Nusselt number, using

$$Nu = \frac{hL}{k_f} = 0.664\, Re_L^{1/2}\, Pr^{1/3}$$

or $$h = 0.664 \times \frac{0.0427}{0.4} \times (19143.33)^{1/2} \times (0.68)^{1/3}$$

$$= 8.62\ W/m^2.K$$

Making the energy balance on the plate

Rate of internal energy gain = Rate of heat transfer to plate by convection and radiation

or $$mC\frac{dT}{dt} = hA_s\,(T_\infty - T_i) + \sigma\varepsilon\, A(T_\infty^4 - T_i^4)$$

or $$\rho(Lwz)\, C\frac{dT}{dt} = h(2wL)\,(T_\infty - T_i) + \sigma\varepsilon\,(2wL)\,(T_\infty^4 - T_i^4)$$

or $$8000 \times (0.4 \times 0.2 \times 0.003) \times 450 \times \frac{dT}{dt}$$

$$= 8.62 \times (2 \times 0.4 \times 0.2) \times (470 - 30) + 5.667 \times 10^{-8} \times 0.85 \times (2 \times 0.4 \times 0.2) \times (743^4 - 303^4)$$

or $$864\frac{dT}{dt} = 606.85 + 2283.84 = 2890.7$$

or $$\frac{dT}{dt} = \frac{2890.7}{864} = 3.345°C/sec$$

$$= \mathbf{199.76°C/min}. \quad \text{Ans.}$$

Example 8.12. *In a glass making process, a plate of glass 0.5 m × 2 m and 3 mm in thickness is cooled by blowing hot air with velocity 1 m/s in direction parallel to plate, such that the rate of cooling is slow. The initial glass plate temperature is 425°C and hot air temperature is 200°C. Estimate :*

(*i*) *Initial rate of cooling in °C/min.*

(*ii*) *Time required for cooling from 425°C to 375°C.*

Assume properties of glass as

$\rho = 2500$ kg/m³, $\quad C = 0.76$ kJ/kg.K

and properties of air may be taken from following table:

T °C	$\nu \times 10^6$ m^2/s	k_f $(W/m.K)$	Pr	ρ kg/m^3
200	34.85	0.039	0.68	0.746
300	48.33	0.046	0.67	0.615
400	63.09	0.051	0.66	0.524

Assume that air flow takes place on both sides of plate.

Solution

Given : Hot air is flowing across a glass plate

$T_\infty = 200°C$ $\quad T_i = 425°C$

$z = 3$ mm $\quad L = 2$ m

$w = 0.5$ m $\quad u_\infty = 1$ m/s

$T = 375°C$.

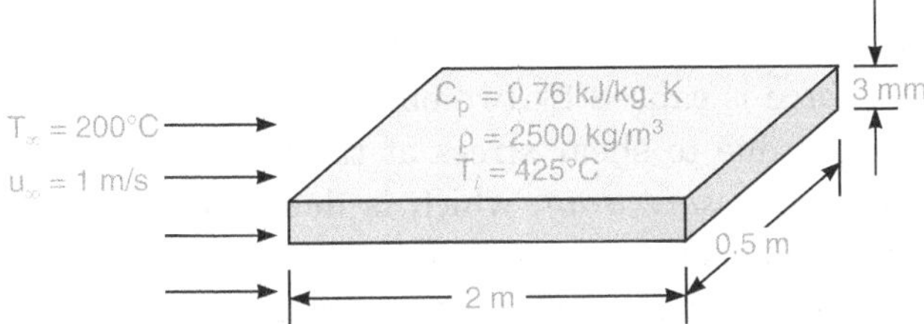

Fig. 8.11. Schematic for example 8.12

To find :

(*i*) Initial rate of cooling,

(*ii*) Time required for cooling of glass plate from 425°C to 375°C.

Analysis : The film temperature of air

$$T_f = \frac{T_s + T_\infty}{2} = \frac{400 + 200}{2} = 300°C$$

where mean surface temperature of plate

$$T_s = \frac{425 + 375}{2} = 400°C$$

Thus the air properties should be used at 300°C.

$\nu = 48.33 \times 10^{-6}$ m²/s, $\quad k_f = 0.046$ W/m.K,

$Pr = 0.67,$ $\quad \rho_f = 0.615$ kg/m³

The Reynolds number

$$Re_L = \frac{u_\infty L}{\nu} = \frac{1 \times 2}{48.33 \times 10^{-6}} = 41382$$

which is less than $Re_{cr} = 5 \times 10^5$, thus the flow is laminar. Thus for average Nusselt number, using correlation

$$Nu_L = \frac{hL}{k_f} = 0.664\, Re_L^{1/2}\, Pr^{1/3}$$

$$= 0.664 \times (41382)^{1/2} \times (0.67)^{1/3} = 118.2$$

The average heat transfer coefficient

$$h = \frac{Nu_L k_f}{L} = \frac{118.2 \times 0.046}{2}$$

$$= 2.72 \text{ W/m}^2\text{.K.}$$

(*i*) The initial rate of cooling :

The energy balance on glass plate yields :

Rate of decrease of internal energy of plate

= Rate of heat convection from both sides of plate.

$$-mC\frac{dT}{dt} = h(2A)(T_i - T_\infty)$$

or $$-\rho VC\frac{dT}{dt} = h(2wL)(T_i - T_\infty)$$

or $$-\rho(wL\,z)C\frac{dT}{dt} = h(2wL)(T_i - T_\infty)$$

$$-2500 \times (0.5 \times 2 \times 3 \times 10^{-3}) \times (0.76 \times 10^3)\frac{dT}{dt}$$

$$= 2.72 \times (2 \times 0.5 \times 2) \times (425 - 200)$$

or $$\frac{dT}{dt} = -\frac{1224}{5700} = -0.214 \text{ °C/s}$$

$$= \mathbf{-12.88\ °C/min}$$

The temperature decreases at the rate of 12.88°C per minute initially. **Ans.**

(*ii*) Time required to cool the glass plate from 425°C to 375°C :

Using lumped system analysis

$$\frac{T - T_\infty}{T_i - T_\infty} = \exp\left(-\frac{h A_s t}{\rho VC}\right)$$

$$\frac{375 - 200}{425 - 200} = \exp\left(-\frac{2.72 \times (2 \times 2 \times 0.5)\,t}{2500 \times (2 \times 0.5 \times 3 \times 10^{-3}) \times 0.76 \times 10^3}\right)$$

or $$0.778 = \exp\left[-\frac{5.44\,t}{5700}\right]$$

or $$t = -\frac{5700}{5.44} \times \ln(0.778) = 263.33 \text{ s}$$

$$= \mathbf{4.39\ min.}$$ **Ans.**

8.5. FLOW ACROSS CYLINDERS AND SPHERES

Another external flow involves fluid flow across circular cylinders and spheres. The characteristic length for a circular cylinder or sphere is taken to be the *external diameter* D. Thus the Reynolds number is defined as:

$$Re_D = \frac{\rho u_\infty D}{\mu} = \frac{u_\infty D}{\nu} \quad ...(8.68)$$

where u_∞ is the uniform velocity of fluid as it approaches the cylinder or sphere. The critical Reynolds number for flow across a circular cylinder or sphere is

$Re_{cr} = 2 \times 10^5$.

The cross flow over a cylinder or sphere involves complex flow pattern as shown in Fig. 8.12.

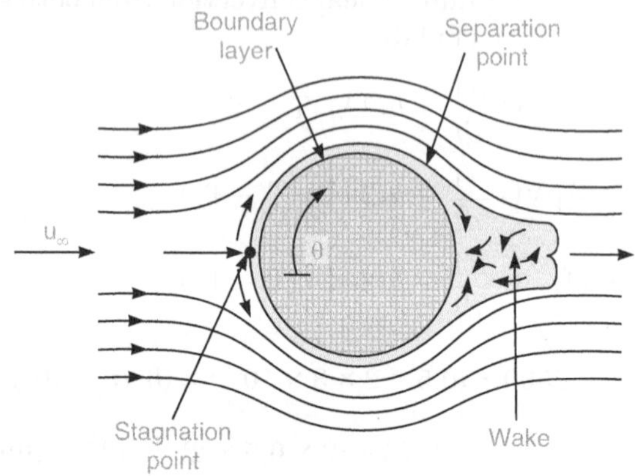

Fig. 8.12. Typical flow pattern in cross flow over a circular cylinder

As free stream fluid approaches the cylinder, it is brought to rest at the forward stagnation point with an increase in fluid pressure. Thus the fluid branches out and encircle cylinder forming a boundary layer, that wraps around the cylinder. The pressure decreases in the flow direction, while fluid velocity increases.

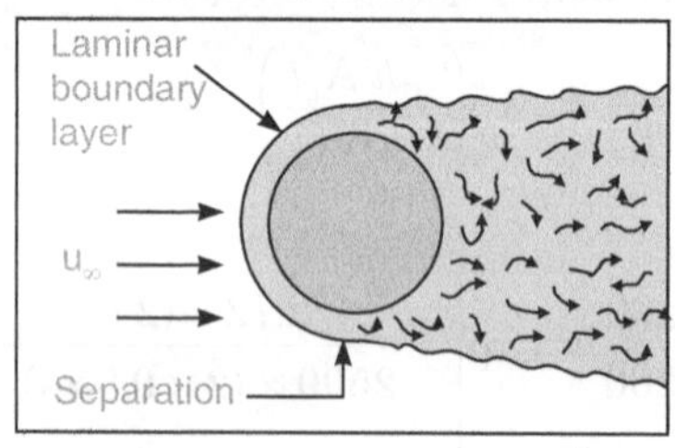

(a) Laminar flow (Re < 2 × 10^5)

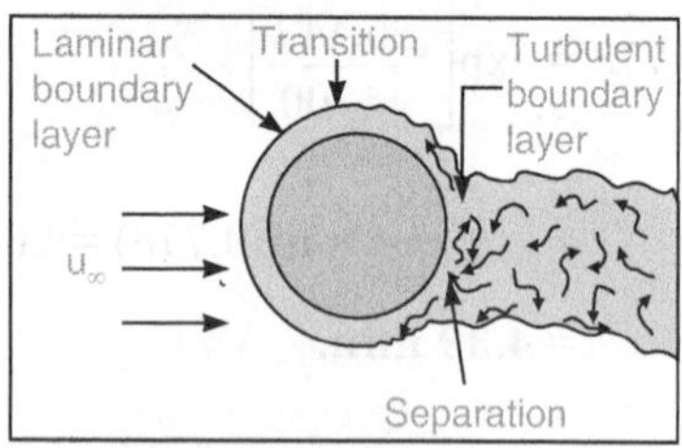

(b) Turbulence occurs (Re > 2 × 10^5)

Fig. 8.13. Flow pattern for cross flow over a cylinder for various Reynolds number

At very low free stream (Re < 2 × 10^5), the fluid completely wraps around the cylinder as shown in Fig. 8.13 (*a*), the fluid flows the curvature of the cylinder. At higher velocities (Re > 2 × 10^5), the fluid wraps the cylinder on frontal side and it is attached to the surface of cylinder as it approaches the top of the cylinder. As boundary layer detaches from the surface, forming a wake behind the cylinder as shown in Fig. 8.13 (*b*). This point is called *separation point.*

Flow in wake region is characterised by random vortex formation and pressure is much lower than the stagnation point pressure.

The flow separation occurs at about θ = 80°, when the boundary layer is laminar and at about θ = 140°, when it is turbulent.

8.5.1. Drag Coefficient

The flow across a circular cylinder or sphere strongly influences the drag force F_D acting on the body. This drag force is caused by two effects : the *friction drag*, which is due to shear stress at the boundary surface and the *pressure drag*, which is due to the pressure differential between front and rear side of the body when wake is formed in the rear. The variation of average drag coefficient C_D for cross flow over a single circular cylinder and a sphere with Reynolds number are presented in Fig. 8.14. The large decrease in C_D for Re > 2 × 10^5 is caused by transition to turbulent flow, which moves the separation point further on the rear of the body reducing the size of wake and thus magnitude of pressure drag. The drag coefficient C_D may be defined as

$$C_D = \frac{F_D}{A_f \dfrac{\rho u_\infty^2}{2}} \quad ...(8.69)$$

where A_f = cylinder frontal area, normal to direction of flow.

A_f = LD for cylinder of length L

$= \frac{\pi}{4} D^2$ for a sphere.

The drag coefficient plays very important role in design of high speed vehicles like racing cars and aeroplanes. The cars are manufactured with low inclined walls and glasses to reduce the drag coefficient. Aeroplanes are designed in the shape of birds and submarines in shape of fish in order to minimise drag coefficient, thus fuel consumption.

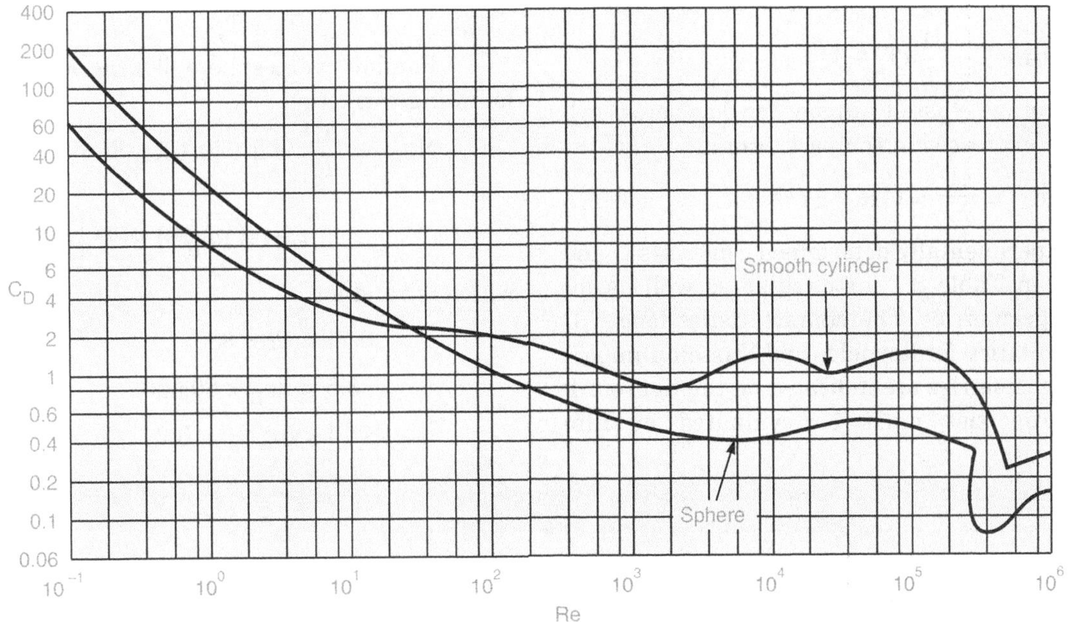

Fig. 8.14. Average drag coefficient for cross flow over a smooth circular cylinder and a smooth sphere

8.5.2. Heat Transfer Coefficient

Flow across cylinders and spheres involves flow separation, which make the analysis complicated. Therefore, such flow must be studied experimentally. The complicated flow pattern across a cylinder discussed earlier influences the heat transfer. The experimental results of variation of the local Nusselt number Nu_θ around the circumference of cylinder, subjected to cross flow of air is shown in Fig. 8.15. Here, for all the cases, the Nusselt number Nu_θ is relatively high at the stagnation point ($\theta = 0°$), but it decreases with increase in θ due to thickening of boundary layer. The Nusselt number becomes minimum at the separation point (between 80° to 100°) and then it further increases with increase in θ due to intense mixing of fluid in the separated flow region. When transition from laminar to turbulent takes place, there is observed a sharp rise in Nusselt number, and once again due to increase in thickness of turbulent boundary layer, the Nusselt number decreases.

The engineers are always interested in average value of heat transfer coefficient over the entire surface. Several relations are available in the literature. The *Churchill* and *Bernstein* suggested the following empirical relation for average Nusselt number, when cylinder is in cross flow

$$Nu_{cyl} = \frac{hD}{k_f}$$

$$= 0.3 + \frac{0.62\, Re_D^{1/2}\, Pr^{1/3}}{[1 + (0.4/Pr)^{2/3}]^{1/4}} \left[1 + \left(\frac{Re_D}{28{,}200}\right)^{5/8}\right]^{4/5} \quad ...(8.70)$$

Valid for $Re_D\,Pr > 0.2$.

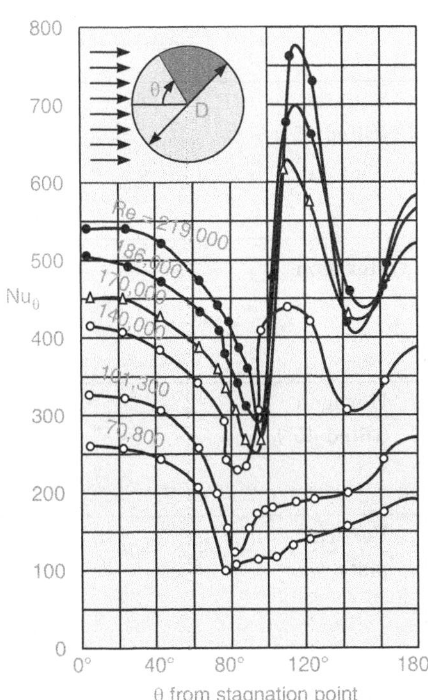

Fig. 8.15. Variation of the local heat transfer coefficient along the circumference of a circular cylinder in cross flow of air

All the fluid properties must be evaluated at the film temperature $T_f = \frac{1}{2}(T_s + T_\infty)$.

The average Nusselt number for flow across cylinders can be expressed in compact form as

$$Nu_{cyl} = \frac{hD}{k_f} = C\, Re^m\, Pr^{1/3} \quad ...(8.71)$$

The experimentally determined constants C and m are given in Table 8.1 for circular as well as for non-circular geometries. The characteristics length D for use in calculation for Reynolds and Nusselt numbers for different geometries are indicated on the figure. All the fluid properties must be evaluated at film temperature

$$T_f = \frac{1}{2}(T_\infty + T_s).$$

For flow over a sphere, *Whitaker* recommands the following correction.

$$Nu_{sph} = \frac{hD}{k_f} = 2 + [0.4\, Re_D^{1/2} + 0.06\, Re_D^{2/3}]\, Pr^{0.4} \left(\frac{\mu_\infty}{\mu_s}\right)^{1/4} \quad ...(8.72)$$

which is valid for

$$0.71 < Pr < 380$$

$$3.5 < Re_D < 80{,}000$$

$$1 < (\mu_\infty/\mu_s) < 3.2$$

TABLE 8.1. Empirical correlations for the average Nusselt number for forced convection over circular and non-circular cylinders in cross-flow

Cross-section of the cylinder	*Fluid*	*Range of Re*	*Nusselt number*
Circle	Gas or liquid	0.4 – 4 4 – 40 40 – 4000 4000 – 40,000 40,000 – 400,000	$Nu = 0.989\, Re^{0.330}\, Pr^{1/3}$ $Nu = 0.911\, Re^{0.385}\, Pr^{1/3}$ $Nu = 0.683\, Re^{0.466}\, Pr^{1/3}$ $Nu = 0.193\, Re^{0.618}\, Pr^{1/3}$ $Nu = 0.027\, Re^{0.805}\, Pr^{1/3}$
Square	Gas	5000 – 100,000	$Nu = 0.102\, Re^{0.675}\, Pr^{1/3}$
Square (tilted 45°)	Gas	5000 – 100,000	$Nu = 0.246\, Re^{0.588}\, Pr^{1/3}$
Hexagon	Gas	5000 – 100,000	$Nu = 0.153\, Re^{0.638}\, Pr^{1/3}$
Hexagon (tilted 45°)	Gas	5000 – 19,500 19,500 – 100,000	$Nu = 0.160\, Re^{0.638}\, Pr^{1/3}$ $Nu = 0.0385\, Re^{0.782}\, Pr^{1/3}$
Vertical plate	Gas	4000 – 15,000	$Nu = 0.228\, Re^{0.731}\, Pr^{1/3}$
Ellipse	Gas	2500 – 15,000	$Nu = 0.248\, Re^{0.612}\, Pr^{1/3}$

A special case of convection heat transfer from sphere when liquid droplets freely fall on the sphere, the *Ranz* and *Marshall* suggested

$$Nu_{sph} = 2 + 0.6\, Re_D^{1/2}\, Pr^{1/3} \quad ...(8.73)$$

Example 8.13. *A long 10 cm diameter steam pipe is exposed to atmospheric air at 4°C. The outer surface of the pipe is at 110°C and air is flowing across the pipe at the velocity of 8 m/s. Determine the rate of heat loss from the pipe per unit of its length.*

Solution

Given : A long pipe exposed to air in cross flow.

To find : The heat transfer rate from pipe per unit length.

Analysis : The properties of air at 1-atm pressure and the film temperature of $T_f = \frac{1}{2}(T_\infty + T_s) = \frac{1}{2}(4 + 110)$ = 57°C = 330 K are:

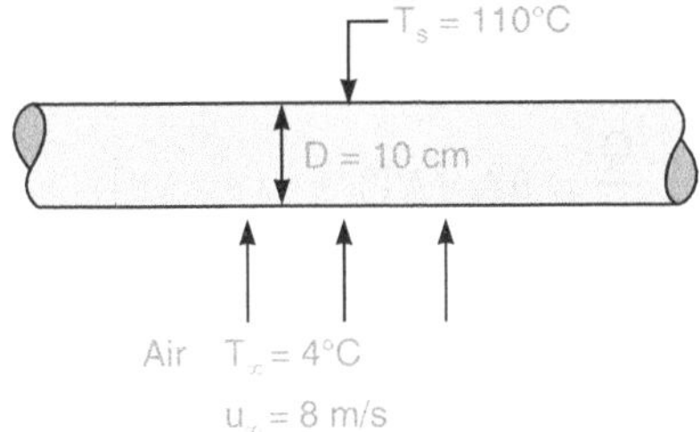

Fig. 8.16. Schematic for example 8.13

$$k_f = 0.0283 \text{ W/(m.K)}$$
$$\nu = 1.86 \times 10^{-5} \text{ m}^2\text{/s}$$
$$Pr = 0.708$$

The Reynolds number of the flow is

$$Re_D = \frac{u_\infty D}{\nu} = \frac{(8 \text{ m/s})(0.1 \text{ m})}{1.86 \times 10^{-5} \text{ m}^2\text{/s}} = 43{,}011$$

The Nusselt number for flow across a cylinder can be calculated as

$$Nu = 0.3 + \frac{0.62\, Re_D^{1/2}\, Pr^{1/3}}{[1 + (0.4/Pr)^{2/3}]^{1/4}} \times \left[1 + \left(\frac{Re_D}{28{,}200}\right)^{5/8}\right]^{4/5}$$

$$= 0.3 + \frac{0.62 \times (43011)^{1/2} \times (0.708)^{1/3}}{[1 + (0.4/0.708)^{2/3}]^{1/4}} \times \left[1 + \left(\frac{43011}{28{,}200}\right)^{5/8}\right]^{4/5} = 196.34$$

and $$h = \frac{Nu\, k_f}{D} = \frac{196.34 \times 0.0283}{0.1}$$
$$= 55.56 \text{ W/m}^2\text{.K}$$

The heat dissipation rate

$$Q = h\, (\pi DL)\, (T_s - T_\infty)$$
$$= 55.56 \times (\pi \times 0.1 \times 1) \times (110 - 4)$$
$$= \mathbf{1850.35\ W.}$$ **Ans.**

Example 8.14. *A metallic bar of 25 mm diameter is cooled by air at 30°C, cross-flowing past the bar with a velocity of 2.5 m/s. If the surface temperature of the bar is not to exceed 85°C and resistivity of the metal is 0.015 × 10⁻⁶ Ohm per metre. Calculate (i) the heat transfer coefficient from the surface to air, and (ii) the permissible current intensity for the bus bar.*

(P.U.P., May 2002)

Solution

Given : Flow across a metallic bar

$D = 25$ mm $= 0.025$ m, $T_\infty = 30°C$

$u_\infty = 2.5$ m/s, $T_s = 85°C$

$\rho = 0.015 \times 10^{-6}$ ohm/m

To find :

(*i*) The heat transfer coefficient,

(*ii*) Permissible current intensity for the bus bar.

Analysis : The properties of air at

$$T_f = \frac{1}{2}(T_s + T_\infty) = \frac{1}{2}(85 + 30)$$
$$= 57.5°C = 330.5 \text{ K}$$
$$\nu = 18.65 \times 10^{-6} \text{ m}^2\text{/s},$$
$$k_f = 0.0288 \text{ W/m.K}, \quad Pr = 0.696$$

(*i*) Heat transfer coefficient :

The Reynolds number

$$Re_D = \frac{u_\infty D}{\nu} = \frac{2.5 \times 0.025}{18.65 \times 10^{-6}} = 3351.2$$

The average Nusselt number for flow across a cylinder can be given by

$$Nu_D = C Re_D^m\, Pr^{1/3}$$

where $m = 0.466$ and $C = 0.683$ for Re = 3351.2 from Table 8.1, thus

$$Nu_D = 0.683 \times (3351.2)^{0.466} \times (0.696)^{1/3} = 26.6$$

The heat transfer coefficient

$$h = \frac{Nu_D\, k_f}{D} = \frac{26.6 \times 0.0288}{0.025}$$
$$= \mathbf{30.63\ W/m^2.K.}$$ **Ans.**

(*ii*) Permissible current intensity :

Heat dissipation rate for 1 metre of bar

$$Q = hA_s\, (\Delta T) = h\, (\pi DL)\, (\Delta T)$$
$$= 30.63 \times (\pi \times 0.025 \times 1) \times (85 - 30)$$
$$= 132.33 \text{ W}$$

Heat generation rate in bus bar for 1 metre of bar.

$$= I^2 R_e = I^2 \left(\frac{\rho L}{A_c}\right) = I^2 \left[\frac{\rho L}{(\pi/4) D^2}\right]$$

$$= I^2 \left[\frac{0.015 \times 10^{-6} \times 1}{(\pi/4) \times (0.025)^2}\right] = 30.557 \times 10^{-6} I^2$$

In steady state conditions

Heat dissipation rate = Heat generation rate

$132.33 = 30.557 \times 10^{-6} I^2$

or $I^2 = 4330489$ or $I =$ **2081 A.** Ans.

Example 8.15. *Air at 27°C is flowing across a tube with a velocity of 25 m/s. The tube could be either a square of 5 cm side or a circular cylinder of 5 cm dia. Compare the rate of heat transfer in each case, if the tube surface is at 127°C.*

Use the correlation : $Nu = C\, Re^n\, Pr^{1/3}$

where, $C = 0.027$, $n = 0.805$ *for cylinder*

$C = 0.102$, $n = 0.675$ *for square tube.*

Solution

Given : Flow across the square and circular tube

$D = 5$ cm, $T_s = 127°C$,

$T_\infty = 27°C$, $u_\infty = 25$ m/s.

To find : Heat transfer rate in each case.

Assumptions :

1. No radiation heat exchange.
2. The steady state heat transfer conditions.
3. Air and surface temperatures are different, taking the properties at mean film temperature.

Properties of air : The mean film temperature

$$T_f = \frac{T_s + T_\infty}{2} = \frac{27 + 127}{2} = 77°C$$

The properties of air at 77°C = 350 K

$\rho = 0.955$ kg/m³, $k_f = 0.03$ W/m.K,

$\nu = 20.92 \times 10^{-6}$ m²/s. Pr = 0.7,

$C_p = 1.009$ kJ/kg. K.

Analysis : (*i*) The cylindrical tube :

The Reynolds number,

$$Re = \frac{u_\infty D}{\nu} = \frac{25 \times 0.05}{20.92 \times 10^{-6}} = 5.975 \times 10^4$$

The Nusselt number,

$$Nu = \frac{hD}{k_f} = 0.027\, Re^{0.805}\, Pr^{1/3}$$

or
$$h = \frac{0.03}{0.05} \times 0.027 \times (5.975 \times 10^4)^{0.805} \times (0.7)^{1/3}$$

$$= 100.6 \text{ W/m}^2\text{.K}$$

The rate of heat transfer from cylinder surface per metre length

$$\frac{Q}{L} = (\pi D)\, h\, (\Delta T)$$

$$= (\pi \times 0.05 \text{ m}) \times (100.6 \text{ W/m}^2\text{. K}) \times (127 - 27)(\text{K})$$

$=$ **1581 W/m.** Ans.

(*ii*) Square tube : With vertical height of 0.05 m

$$Re = \frac{u_\infty L}{\nu} = 5.975 \times 10^4$$

$$Nu = 0.102\, Re^{0.675}\, Pr^{1/3}$$

or
$$h = \frac{0.102 \times 0.03 \times (5.975 \times 10^4)^{0.675} \times (0.7)^{1/3}}{0.05}$$

$$= 91.02 \text{ W/m}^2\text{. K}$$

The rate of heat transfer from cylinder surface per metre length

$$\frac{Q}{L} = hA_s\, (\Delta T)$$

$$= (91.02 \text{ W/m}^2\text{. K}) \times (4 \times 0.05 \text{ m}) \times (127 - 27)(\text{K})$$

$=$ **1820 W/m.** Ans.

The heat transfer rate from square tube is higher than that from circular tube.

Example 8.16. *Air stream at 27°C moving at 0.3 m/s across 100 W incandescent bulb, glowing at 127°C. If the bulb is approximated by a 60 mm diameter sphere, estimate the heat transfer rate and the percentage of power lost due to convection.*

Use correlation $Nu = 0.37\, Re_D^{0.6}$.

(N.M.U., Dec. 2002)

Solution

Given : Flow over a electric bulb :

$T_\infty = 27°C$, $u_\infty = 0.3$ m/s, P = 100 W,

$T_s = 127°C$, D = 60 mm = 0.06 m.

To find :

(*i*) The heat transfer rate.

(*ii*) Percentage of power lost due to convection.

Properties of fluid : The film temperature

$$T_f = \frac{T_s + T_\infty}{2} = \frac{127 + 27}{2} = 77°C = 350 \text{ K}$$

The properties of air at 77°C are

$\nu = 2.09 \times 10^{-5}$ m²/s,

$k_f = 0.03$ W/m.K.

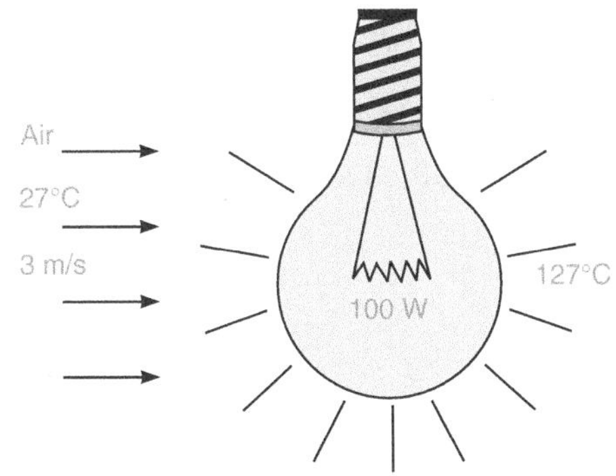

Fig. 8.17. Schematic of an incandescent bulb

Assumptions :

1. Spherical shape of the electric bulb.
2. No radiation heat exchange.
3. The steady state heat transfer conditions.

Analysis : The Reynolds number

$$Re_D = \frac{u_\infty D}{\nu} = \frac{0.3 \times 0.06}{2.09 \times 10^{-5}} = 865.3$$

$$Nu_D = 0.37\ Re_D^{0.6} = 0.37 \times (865.3)^{0.6} = 21.4$$

The heat transfer coefficient

$$h = \frac{k_f}{D} Nu_D = \frac{0.03}{0.06} \times 21.4 = 10.7 \text{ W/m}^2\text{. K}$$

(*i*) The heat transfer rate:

$$Q = hA_s(T_s - T_\infty) = h(\pi D^2)(T_s - T_\infty)$$
$$= (10.7 \text{ W/m}^2\text{. K}) \times \pi \times (0.06 \text{ m})^2 \times (127 - 27)(\text{K})$$

= **12.10 W. Ans.**

(*ii*) The percentage of heat lost by forced convection:

$$= \frac{Q}{P} \times 100 = \frac{12.10}{100} \times 100 = \mathbf{12.1\%}.$$ **Ans.**

Example 8.17. *A decorative plastic film on a copper sphere, 10 mm in diameter is cured in an oven at 75°C. The sphere is suddenly removed from the oven and exposed to an air stream at 1 atm. and 23°C, flowing at 10 m/s. Estimate how long, it will take to cool the sphere to 35°C.*

Take properties of copper as :

$\rho = 8933$ *kg/m*^{3}, $\quad k = 399$ *W/m.K,*

$C = 387$ *J/kg.K.*

and properties of air at 296 K as :

$\nu = 15.36 \times 10^{-6}$ *m*^{2}*/s,* $k_f = 0.0258$ *W/m.K,*

$Pr = 0.709,$ $\quad \mu = 181.6 \times 10^{-7}$ *kg/ms*

and for air at 328 K, $\mu = 197.8 \times 10^{-7}$ *kg/ms.*

(N.M.U., May 1998)

Solution

Given : A decorative plastic film on a copper sphere to be cured ;

$D = 10$ mm, $\quad u_\infty = 10$ m/s,

$T_i = 75$°C, $\quad T_\infty = 23$°C

$T = 35$°C, $\quad p = 1$ atm.

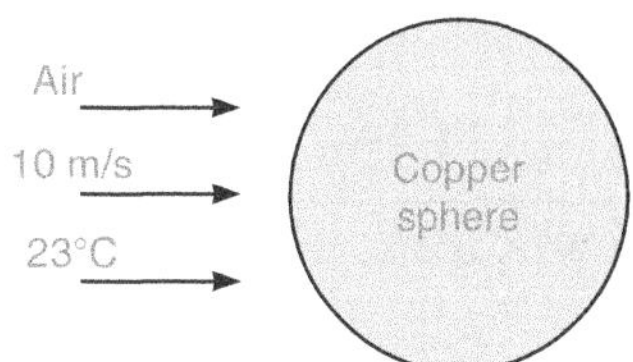

Fig. 8.18. Schematic for example 8.17

To find : The time required to cool the sphere to 35°C.

Assumptions :

1. No radiation heat exchange.
2. The steady state heat transfer conditions.
3. Internal temperature gradients within the sphere are negligible.

Analysis : The time required to cool the sphere can be calculated by

$$\frac{\theta}{\theta_i} = \frac{T - T_\infty}{T_i - T_\infty} = \exp\left\{-\frac{ht}{\rho\delta C}\right\}$$

where, δ = characteristic length, and it is calculated for sphere as ;

$$\delta = \frac{V}{A_s} = \frac{D}{6} = \frac{0.01\text{m}}{6} = 0.001667 \text{ m}$$

and h is the heat transfer coefficient, can be calculated by using eqn. (8.72) :

$$Nu_D = 2 + (0.4\ Re_D^{1/2} + 0.06\ Re_D^{2/3})\ Pr^{0.4}\ [\mu_\infty/\mu_s]^{1/4}$$

where, $Re_D = \frac{u_\infty D}{\nu} = \frac{(10 \text{ m/s}) \times (0.01 \text{ m})}{15.36 \times 10^{-6} \text{ m}^2\text{/s}} = 6510$

Hence the Nusselt number ;

$$Nu_D = 2 + \{0.4 \times (6510)^{1/2} + 0.06 \times (6510)^{2/3}\} \times (0.709)^{0.4} \times \left[\frac{181.6 \times 10^{-7} \text{ kg/ms}}{197.8 \times 10^{-7} \text{ kg/ms}}\right]^{1/4}$$

$$= 47.4$$

and the heat transfer coefficients ;

$$h = \text{Nu}\,\frac{k_f}{\text{D}} = 47.4 \times \frac{0.0258 \text{ W/m.K}}{0.01 \text{ m}}$$

$$= 122 \text{ W/m}^2\text{.K}$$

Now using the values for calculation of time required for cooling ;

$$\frac{35-23}{75-23} = \exp\left\{-\frac{122t}{8933 \times 0.001667 \times 387}\right\}$$

or $\ln(12/52) = -(122/5762)t$

or $t = \mathbf{69.13}$ **sec.** Ans.

8.6. SUMMARY

TABLE 8.2. Summary of convection heat transfer correlations for external flow

Correlation	*Geometry*	*Conditions and properties at*
$\delta = 5x/\text{Re}_x^{-1/2}$	Flat plate	Laminar, T_f
$C_{fx} = 0.646\ \text{Re}_x^{-1/2}$	Flat plate	Laminar, local, T_f
$\text{Nu}_x = 0.332\ \text{Re}_x^{1/2}\ \text{Pr}^{1/3}$	Flat plate	Laminar, local, T_f, $\text{Pr} \gtrsim 0.6$
$\delta_{th} \approx \delta\ \text{Pr}^{-1/3}$	Flat plate	Laminar, T_f
$C_f = 1.292\ \text{Re}_L^{-1/2}$	Flat plate	Laminar, average, T_f
$\text{Nu}_L = 0.664\ \text{Re}_L^{1/2}\ \text{Pr}^{1/3}$	Flat plate	Laminar, average, T_f, $\text{Pr} \gtrsim 0.6$
$\text{Nu}_x = 0.565\ \text{Pe}_x^{1/2}$	Flat plate	Laminar, local, T_f, $\text{Pr} \lesssim 0.05$
$C_{fx} = 0.0592\ \text{Re}_x^{-1/5}$	Flat plate	Turbulent, local, T_f, $\text{Re}_x \lesssim 10^8$
$\delta = 0.37x\ \text{Re}_x^{-1/5}$	Flat plate	Turbulent, local, T_f, $\text{Re}_x \lesssim 10^8$
$\text{Nu}_x = 0.0296\ \text{Re}_x^{4/5}\ \text{Pr}^{1/3}$	Flat plate	Turbulent, local, T_f, $\text{Re}_x \lesssim 10^8$, $0.6 \lesssim \text{Pr} \lesssim 60$
$C_f = 0.074\ \text{Re}_L^{-1/5} - 1742\ \text{Re}_L^{-1}$	Flat plate	Mixed, average, T_f, $\text{Re}_{cr} = 5 \times 10^5$, $\text{Re}_L \lesssim 10^8$
$\text{Nu}_L = (0.037\ \text{Re}_L^{4/5} - 871)\ \text{Pr}^{1/3}$	Flat plate	Mixed, average, T_f, $\text{Re}_{cr} = 5 \times 10^5$, $\text{Re}_L \lesssim 10^8$, $0.6 < \text{Pr} < 60$
$\text{Nu}_D = C\text{Re}_D^m\text{Pr}^{1/3}$ (and m from Table 8.1)	Cylinder	Average, T_f, $0.4 < \text{Re}_D < 4 \times 10^5$, $\text{Pr} \gtrsim 0.7$
$\text{Nu}_D = 0.3 + \{0.62\ \text{Re}_D^{1/2}\ \text{Pr}^{1/3}/[1 + (0.4/\text{Pr})^{2/3}]^{1/4}\}[1 + (\text{Re}_D/28{,}200)^{5/8}]^{4/5}$	Cylinder	Average, T_f, $\text{Re}_D\ \text{Pr} > 0.2$
$\text{Nu}_D = 2 + (0.4\ \text{Re}_D^{1/2} + 0.06\ \text{Re}_D^{2/3})\text{Pr}^{0.4} \cdot (\mu_\infty/\mu_s)^{1/4}$	Sphere	Average, T_∞, $3.5 < \text{Re}_D < 7.6 \times 10^4$, $0.71 < \text{Pr} < 380$, $1.0 < (\mu_\infty/\mu_s) < 3.2$
$\text{Nu}_D = 2 + 0.6\ \text{Re}_D^{1/2}\text{Pr}^{1/3}\ [25(x/D)^{-0.7}]$	Falling drop	Average, T_∞
$\text{Nu}_D = C\text{Re}^m_{D,\ max}\text{Pr}^{0.36}(\text{Pr}_\infty/\text{Pr}_s)^{1/4}$	Tube bank	Average, T_∞, $1000 < \text{Re}_D < 2 \times 10^6$, $0.7 < \text{Pr} < 500$

REVIEW QUESTIONS

1. Where is the heat flux be higher for laminar forced convection from a horizontal flat plate at leading edge or trailing edge ?
2. How does the flow in thermal boundary layer over a flat plate differ from the flow outside the thermal boundary layer ?
3. How are average friction and heat transfer coefficients determined in flow over a flat plate ?
4. When a fluid flows over a cylinder, why does the drag coefficient suddenly drop, when the flow becomes turbulent ?
5. Why is the flow separation in flow over cylinders delayed in turbulent flow ?
6. Why are racing cars and airplanes designed aerodynamic ?
7. Which car will consume less fuel; one with sharp corners and other one is contoured in shape of an ellipse ?
8. Explain the Reynolds Colburn analogy for laminar flow over a plate.
9. Explain Reynolds Chilton analogy for turbulent flow over a flat plate.
10. Explain the heat transfer phenomenon, when a fluid flows across a bluff body.

PROBLEMS

1. Air at atmospheric pressure and 400 K flows over a flat plate with a velocity of 5 m/s. The transition from laminar to turbulent occurs at a Reynold number of 5×10^5; determine the distance from the leading edge of the plate at which transition takes place.
[Ans. 2.59 m]
2. Atmospheric air at 27°C flows along a flat plate with a velocity of 8 m/s. The critical Reynolds number at which transition from laminar to turbulent takes place is 5×10^5.
(*a*) Determine the distance from the leading edge of the plate at which the transition occurs ;
(*b*) Determine the local coefficient of friction at the location where transition occurs ; and
(*c*) Determine the average drag coefficient over the distance where the flow is laminar.
[Ans. (*a*) 1.05 m, (*b*) 9.1358×10^{-4}, (*c*) 0.072 N]
3. Air at atmospheric pressure and 54°C flows with a velocity over a 1 m long flat plate maintained at 200°C, calculate the average shear stress and heat transfer coefficient over 1 m length of the plate. Determine the rate of heat transfer between the plate and air per metre width of the plate.
[Ans. 1790 W/m]
4. Air at 24°C flows along a 4 m long flat plate with a velocity of 5 m/s. The plate is maintained at 130°C. Calculate the heat transfer coefficient over the entire length of the plate and the heat transfer rate per metre with of the plate. [Ans. 9.73 W/m^2. K, 4120 W/m]
5. Air flows along a thin plate with a velocity of 2.5 m/s. The plate is 1 m long and 1 m wide. Estimate the boundary layer thickness at the trailing edge of the plate and the force necessary to hold the plate in the stream of air. The air has a viscosity of 0.86×10^{-5} kg/ms and a density of 1.12 kg/m^3.
[Ans. 8.1 mm, 0.0158 N]
6. Air flows along a thin flat plate 1 m wide and 1.5 m long at a velocity of 1 m/s. The free stream temperature is 4°C. Calculate the amount of heat that must be supplied to plate in order to maintain it at a uniform temperature of 50°C. [Ans. 441.6 W]
7. Atmospheric air at 300 K and free stream velocity of 30 m/s flows across a single tube. Water at 60°C enters a tube of 25 mm diameter at mean fluid velocity of 2 m/s. Calculate the exit temperature of water, if the tube is 3 m long and wall temperature is constant at 100°C. [Ans. 77°C]
8. Forced air at 30°C flows over a square flat plate maintained at 110°C. The drag force experienced by the plate is 12 N. Using the Reynolds Colburn analogy, to calculate the heat transfer coefficient and heat loss from the plate surface. Assume flow is turbulent. [Ans. 80.22 W/m^2.K, 10285.84 W]
9. A refrigerated truck carrying the food stuff is speeding on a highway at 95 km/h in a desert area, where the ambient air temperature is 50°C. The body of the truck may be considered as a rectangular box, 10 m long, 4 m wide and 3 m high. Consider the boundary layer on four walls to be turbulent and the heat is transferred from these four surfaces. If the wall surfaces of the truck are maintained at 10°C. Calculate the following :
(*a*) Heat loss from the four vertical surfaces ;
(*b*) Power required to overcome the resistance acting on surfaces. [Ans. (*a*) 320.08 W, (*b*) 3.88 kW]
10. The atmospheric air at 30°C flows past a flat plate with a sharp leading edge. The velocity of the air is 4 m/s. The plate is heated uniformly throughout its entire length and is maintained at a surface temperature of 50°C. Assuming that the transition occurs at a critical Reynolds number of 5×10^5, find the distance from the leading edge at which the flow is in the boundary layer changes from laminar to turbulent. At this location, calculate (*a*) thickness of hydrodynamic and thermal boundary layers ; (*b*) total drag force per unit width of the plate; (*c*) heat transfer rate ; (*d*) mass enters the layer.
[Hint. Consider surface area of both sides of the plate] (*N.M.U., Nov. 1999*)
[Ans. (*a*) 13.9 mm, 15.258 mm ; (*b*) 6.98 N ; (*c*) 445.2 W ; (*d*) 140.76 kg/h]

11. Air at 27°C and 1 bar flows over a flat plate at a speed of 2 m/s,

(*i*) Calculate the boundary layer thickness at 400 mm from the leading edge of the plate. Find the mass flow rate per unit width of the plate. Take $\mu = 19.8 \times 10^{-6}$ kg/ms at 27°C.

(*ii*) The plate is maintained at 60°C, calculate the heat transfer rate per hour. Use following properties of air

$\nu = 17.36 \times 10^{-6}$ m²/s,

$k_f = 0.0275$ W/m.K

$C_p = 1006$ J/kg.K,

R = 287 J/kg.K, Pr = 0.7.

[Ans. (*i*) $\delta = 8.57$ mm, $\dot{m} = 0.01242$ kg/s, (*ii*) Q = 416.7 kJ/h]

12. Air at 20°C and at atmospheric pressure flows at a velocity of 4.5 m/s past a flat plate with a sharp leading edge. The entire plate surface is maintained at a temperature of 60°C. Assuming that the transition occurs at Re = 5×10^5, find the distance from the leading edge, at which the flow in the boundary layer changes from laminar to turbulent, at this location calculate :

(*i*) Thickness of hydrodynamic and thermal boundary layers,

(*ii*) Local and average heat transfer coefficients,

(*iii*) Heat transfer rate from both sides for unit width of the plate,

(*iv*) The skin friction coefficient.

[Ans. $x_{cr} = 1.88$ m, (*i*) 12.34 mm, 13.55 mm, (*ii*) 3.05 W/m².K, 6.1 W/m².K, (*iii*) 917.44 W, (*iv*) 9.136×10^{-4}]

13. Air at 20°C and at a pressure of 1 bar is flowing over a flat plate at a velocity of 3 m/s. If the plate is 280 mm wide and at 56°C. Calculate the following quantities at x = 280 mm

(*i*) Boundary layer thickness.

(*ii*) Local friction coefficient,

(*iii*) Average friction coefficient,

(*iv*) Shearing stress due to friction,

(*v*) Thickness of thermal boundary layer.

(*vi*) Local heat transfer coefficient,

(*vii*) Average heat transfer coefficient,

(*viii*) Rate of heat transfer by convection,

(*ix*) Drag force on the plate, and

(*x*) Total mass flow rate through the boundary.

[Ans. (*i*) 6.26 mm, (*ii*) 0.00296, (*iii*) 0.00594, (*iv*) 0.0152 N/m², (*v*) 7.05 mm, (*vi*) 6.43 W/m².K, (*vii*) 12.86 W/m².K, (*viii*) 36.3 W, (*ix*) 0.0012 N (*x*) 0.01335 kg/s]

14. Engine oil at 100°C and at a velocity of 0.1 m/s flows past a flat plate maintained at 20°C. Determine:

(*i*) The velocity and thermal boundary layer thicknesses at the trailing edge,

(*ii*) The local heat flux and surface shear stress at the trailing edge,

(*iii*) The total drag force and heat transfer per unit width of the plate.

15. Air at 25°C and atmospheric pressure flows at a velocity of 25 m/s over both surfaces of a 1 m long flat plate, maintained at 125°C. Calculate the rate of heat transfer per unit width from the plate for the value of critical Reynolds number corresponding to 10^5, 5×10^5 and 10^6.

16. A circular pipe 25 mm outside diameter is exposed to an air stream at 27°C and 1 atm. The air moves in cross flow over the pipe at 15 m/s, while the outer surface of the pipe is maintained at 100°C. What is the drag force exerted on the pipe per unit length? What is the rate of heat transfer from the pipe per unit length ?

17. Atmospheric air at 27°C is flowing at a velocity of 15 m/s. What is the rate of heat transfer per unit length from the following surfaces, each at 77°C, when the air is in cross flow over the surface ?

(*i*) A circular cylinder 10 mm in diameter,

(*ii*) A square cylinder 10 mm on a side,

(*iii*) A vertical plate 10 mm high.

18. An uninsulated steam pipe is used to transport steam from one building to another. The pipe is 0.5 m in diameter has a surface temperature of 150°C and is exposed to atmospheric air at 30°C. The air moves in cross-flow over the pipe with a velocity of 5 m/s. What is the heat loss per unit length of the pipe ?

19. Water at 20°C flows over a 20 mm diameter sphere with a velocity of 5 m/s. The surface of the sphere is at 60°C. What is the drag force on the sphere ? What is the heat transfer rate from the sphere ?

20. Atmospheric air at 27°C and at a velocity of 0.5 m/s flows over a 40 W bulb, whose surface is at 140°C. The bulb may be approximated as a sphere of 50 mm diameter. What is the heat loss by convection to air?

21. A 25 mm diameter sphere is to be maintained at 50°C in either an air stream or a water stream, both at 20°C, and 2 m/s, velocity. Compare the rate of heat transfer and drag force for two fluids.

22. Steam at 1 atm and 100°C is flowing across a 5 cm outer diameter tube at a velocity of 6 m/s. Estimate Nusselt number, heat transfer coefficient and heat transfer rate per metre length of the pipe, if the pipe is at 200°C.

23. An electric transmission line of 1.2 cm diameter carries a current of 200 A and has a resistance of 3×10^{-4} ohm per metre length. If air at 16°C and 33 km/h is flowing across it, calculate surface temperature of wire.

24. A long hexagonal copper extrusion is removed from a oven at 400°C and exposed to air at 50°C, flowing across it at 10 m/s. The surface of the copper has an emissivity of 0.9 due to oxidation. The rod is 3 cm across opposite flat sides and has an cross-sectional

area of 7.79 cm^2 and perimeter of 10.4 cm. Calculate the time required for the centre of the copper to cool to 100°C.

25. A stainless steel pin fin 5 cm long, 6 mm outer diameter extends from a flat plate into an air stream flowing at 175 m/s. Estimate (*i*) average heat transfer coefficient, (*ii*) temperature at the end of fin, (*iii*) the rate of heat flow from the fin.

Take plate temperature as 650°C and air steam temperature as 30°C.

26. During a cold winter day air at 15.27 m/s is blowing parallel to 4 m high and 10 m long wall of a house. If outside air is at 5°C and wall is maintained at 12°C, calculate the rate of heat loss from the wall by convection. What would be the heat dissipation rate, if air velocity is doubled ? [**Ans.** 9212 W, 16,408 W]

27. The top surface of a container truck moving at 70 km/h is 2.8 m wide and 8 m long. The top surface is absorbing net solar radiation at the rate of 200 W/m^2, while it is exposed to ambient air at 30°C. Assuming the roof of the truck is perfectly insulated and the radiation heat exchange with the surroundings to be negligible, calculate the equilibrium temperature of the top surface of the container.

28. A stainless steel ball (ρ = 8055 kg/m^3, C = 480 J/kg.K) of diameter 15 cm is removed from the oven at a uniform temperature of 350°C. The ball is then exposed to atmospheric air at 30°C with velocity of 6 m/s. The surface temperature of the ball eventually drops to 250°C. Determine the average heat transfer coefficient during the cooling process and calculate the time for cooling process.

29. A person extends his uncovered arms into an air stream at 6°C and 30 km/h in order to feel cooling. Initially the skin temperature of the arm is at 37°C. Consider the arm as 60 cm long and 7.5 cm diameter cylinder, calculate the rate of heat dissipation from an arm.

30. A long aluminium wire of 3 mm diameter is extruded at a temperature of 350°C. The air at 35°C flow across the wire at 6 m/s, velocity. Calculate the rate of heat transfer from a wire to air per metre length, when it is first exposed to air.

31. Consider a person who is trying to keep his body cool in a hot summer day by turning a fan on and exposing his entire body to an air stream at 30°C. The fan is blowing air at a velocity of 2 m/s. If the person is doing some light work and generating sensible heat at a rate of 100 W, calculate the average temperature of outer skin of the person. The average human body can be treated as a 30 cm diameter cylinder with an exposed surface area of 1.7 m^2. Neglect heat loss by radiation. Also calculate the heat dissipation rate from the body if air velocity is doubled.

[**Ans.** 33.8°C, 32.2 °C]

32. A 4 m long, 1.5 kW electric resistance wire is made of 0.25 cm diameter stainless steel (k = 15.1 W/m.K). The resistance wire is exposed to an air stream at 30°C, with velocity of 7 m/s. Calculate the surface temperature of the wire.

33. Air at 20°C and atmospheric pressure is flowing over a flat plate at a velocity of 3 m/s. If the plate is 30 cm wide and at a temperature of 60°C, calculate at x = 0.3 m;

(*i*) Thickness of velocity and thermal boundary layers,

(*ii*) Local and average friction coefficients,

(*iii*) Local and average heat transfer coefficients,

(*iv*) Total drag force on the plate,

(*v*) Heat transfer rate.

Take the following properties of air at 313 K

ρ = 1.18 kg/m^3, $\nu = 17 \times 10^{-6}$ m^2/s

k_f = 0.0272 W/m.K, C_p = 1.007 kJ/kg.K

Pr = 0.705. (*V.T.U., Karnataka, July 2002*)

[**Ans.** (*i*) 6.52 mm, 7.14 mm
(*ii*) 2.80×10^{-3}, 5.61×10^{-3}
(*iii*) 6.16 W/m^2.K, 12.32 W/m^2.K
(*iv*) 2.68×10^{-3} N
(*v*) 44.46 W]

34. The air at atmospheric pressure and at 40°C flows with a velocity 8 m/s along a flat plate, 3 m long, which is maintained at a uniform temperature of 100°C. Calculate the local heat transfer coefficient at the end of the plate and average heat transfer coefficient over entire length of the plate. Assume $Re_{cr} = 2 \times 10^5$. [**Ans.** 18.93 W/m^2.K, 20.62 W/m^2.K]

35. Assuming a man as a cylinder of 40 cm diameter and 1.72 m high with a surface temperature of 37°C. Calculate the heat lost from its body, while standing in wind flowing at 20 km per hour at 17°C.

Use the relation : $Nu_D = 0.027\, Re_D^{0.805}\, Pr^{1/3}$.

[**Ans.** 947.43 W]

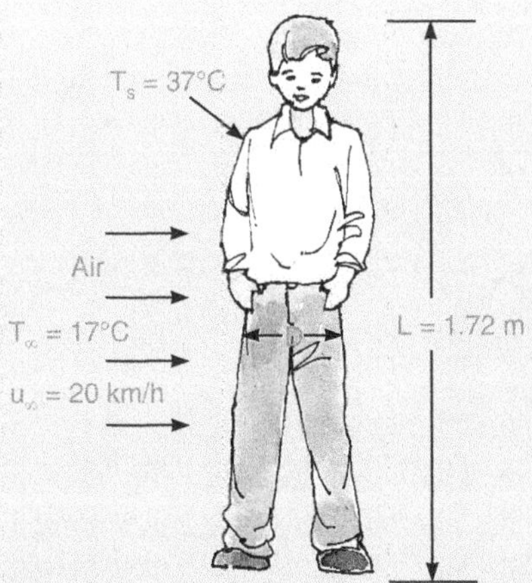

Fig. 8.19. Schematic for problem 35

REFERENCES AND SUGGESTED READING

1. Rehsenow, W.M., J.P. Harnett and E.N. Genic, eds. *"Handbook of Heat Transfer"*, 2/e, McGraw Hill, New York, 1985.
2. Kays, W.M. and M.E. Crawford, *"Convective Heat and Mass transfer"*, 2nd ed. McGraw Hill, New York, 1980.
3. Giedt Warren H., *"Investigation of Variation of Point Unit-Heat Transfer Coefficient Around a Cylinder Normal to an Airstream." "Transaction of ASME"*, Vol. 71, pp. 375–381, 1949.
4. Christopher, Long, *"Essential Heat Transfer"*, Addison-Wesley, Longman, 2001.
5. Zhukauskas, A. and A. B. Ambrazyavichyus, *"Int. J. of Heat and Mass Transfer"*, Vol 3. pp. 305, 1961.
6. Giedt, Warren H., *"Principles of Engineering Heat Transfer"*, Van Nostrand Inc., 2nd ed., 1967.
7. Knudsen, J.D. and D.L. Katz, *"Fluid Dynamics and Heat Transfer"*, McGraw Hill, New York, 1958.
8. McAdams, W.M., *"Heat Transmission"*, 3rd ed. McGraw Hill, New York, 1954.
9. Jacob, M. and G.A. Hawkins, *"Elements of Heat Transfer"* 3rd ed. Wiley, New York, 1957.
10. Krieth Frank and M.S. Bohn, *"Principles of Heat Transfer"*, 5th ed., PWS Pub. Company, 1997.
11. Holman, J.P., *"Heat Transfer"*, 7th ed. McGraw Hill, New York, 1990.
12. Incropera, F.P., and D.P. DeWitt, *"Introduction to Heat Transfer"*, 2/e, John Wiley and Sons, 1990.
13. Ozisik, M.N., *"Heat Transfer—A Basic Approach"*, McGraw Hill, New York, 1985.
14. Bayazitoglu, Yand M.N. Ozisik, *"Elements of Heat Transfer"*, McGraw Hill, New York, 1988.
15. Thomas, L.C., *"Heat Transfer"*, Prentice-Hall, Englewood Cliffs, N.J., 1982.
16. White, F.M., *"Heat and Mass Transfer"*, Addison-Wesley, Reading, MA, 1988.
17. Jacob, F. M., *"Heat Transfer"*, Vol. 1, Wiley, New York, 1949.
18. Suryanarayana, N.V., *"Engineering Heat Transfer"*, West Pub. Co., New York, 1998.
19. Chapman, Alan. J., *"Fundamentals of Heat Transfer"*, Macmillan, New York.

Internal Flow

9

9.1. **Flow Inside Ducts.** **9.2. Hydrodynamic Considerations**—Mean velocity u_m—Hydrodynamic entry length—Velocity profile in fully developed region—Friction factor—Pressure drop and friction factor in fully developed flow. **9.3. Thermal Considerations**—The mean temperature or bulk temperature. **9.4. The Heat Transfer in Fully Developed Flow.** **9.5. General Thermal Analysis**—Constant surface heat flux—Constant surface temperature. **9.6. Heat Transfer in Laminar Tube Flow.** **9.7. Flow Inside a Non-circular Duct.** **9.8. Thermally Developing, Hydrodynamically Developed Laminar Flow.** **9.9. Heat Transfer in Turbulent Flow Inside a Circular Tube**—Analogy between heat and momentum transfer in turbulent flow through tube—Correlation for turbulent flow. **9.10. Heat Transfer to Liquid Metal Flow in Tube.** **9.11. Summary**—*Review Questions—Problems—References and Suggested Reading.*

9.1. FLOW INSIDE DUCTS

The flow of fluid through the tubes and ducts for transporting cooling and heating fluids, etc., is of engineering importance. Most heat exchangers involve the heating or cooling of fluids flowing in the tubes. The fluid in such applications is forced to flow by a fan or pump through a tube that is sufficiently long to accomplish desired heating or cooling. Pressure drop and heat flux are associated with forced flow through the tubes and friction factor and heat transfer coefficient are used to determine the pumping power and length of tube.

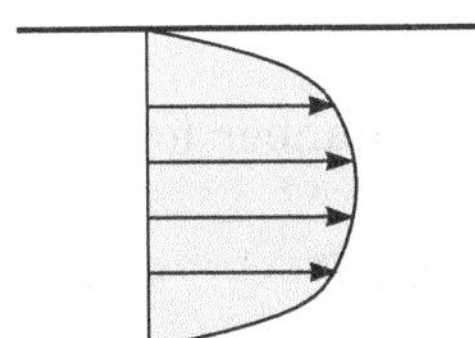

Fig. 9.1. Flow through duct

There is a fundamental difference between external and internal flows. In the external flow, we have studied so far, the fluid had a free surface, thus its boundary layer growth is not restricted by any confining surface. However, in internal flow, such as in tubes, the fluid is completely confined by inner surfaces of the tube. Thus there is a limit to velocity and thermal boundary layer thicknesses, the radius of tube.

There are large changes in the value of heat transfer coefficient in the region, where, the boundary layer thickness increases, but smaller changes in the region, where the boundary layer has reached its maximum value.

9.2. HYDRODYNAMIC CONSIDERATIONS

When a fluid enters a tube, with a velocity, the boundary layer develops along the surface of the tube. The growth of boundary layer at the entrance of a larger diameter tube is much similar to that is for flow along a flat surface. However, the flow velocity cannot be same due to presence of boundary layer on opposite wall, the development of boundary layer occurs at the expense of shrinking the flow region and concludes the boundary layer merger at the centre line of the tube, where the velocity profile becomes independent of flow length, that

$$\frac{\partial u}{\partial x} = 0$$

and flow is called *hydrodynamically developed flow.*

To illustrate the concept of fully developed region, developing region and hydrodynamic entry length, consider the flow of an incompressible fluid through a tube as shown in Fig. 9.2.

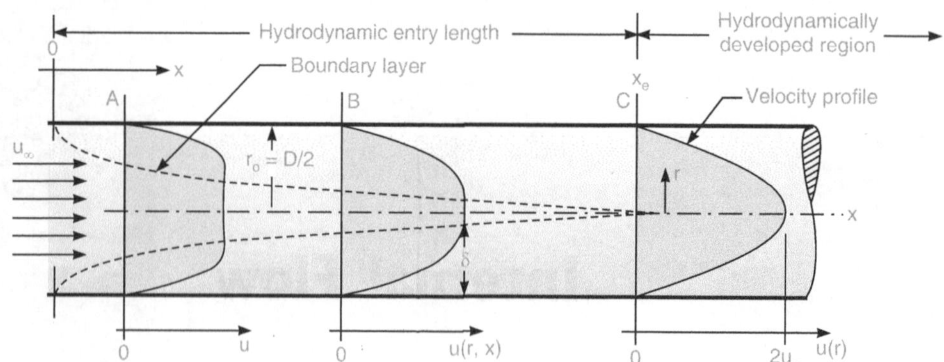

Fig. 9.2. The development of a laminar velocity profile in a pipe

At the entrance to tube at section A, the fluid velocity is uniform as u_∞. As fluid proceeds in the tube, the viscous forces at the wall retard the motion of particles in the fluid layer near to the wall. The boundary layer begins to develop along the flow length and thus the velocity profile changes continuously in the direction of flow. For instance at section B, the velocity profile indicates zero velocity at the surface and some value u at a distance δ from the surface of the tube. Here u becomes greater than u_∞ due to shrinkage in flow area. Further, for down stream flow, the boundary layer thickness δ increases and becomes equal to radius r_o of the tube at section C. From section C onward the velocity profile remains unchanged. This velocity profile is called the **fully developed velocity profile.** The region of fully developed velocity profile is known as the **hydrodynamically developed region.** The region from the tube inlet to the point at which the boundary layer merges at the centre line is called the **hydrodynamic entrance region** or the **hydrodynamically developing region,** and the length of this region is called the **hydrodynamic entry length.** Beyond this length, the viscous effects are extended over the entire cross-section and the *velocity profile* becomes *parabolic* for the *laminar flow*.

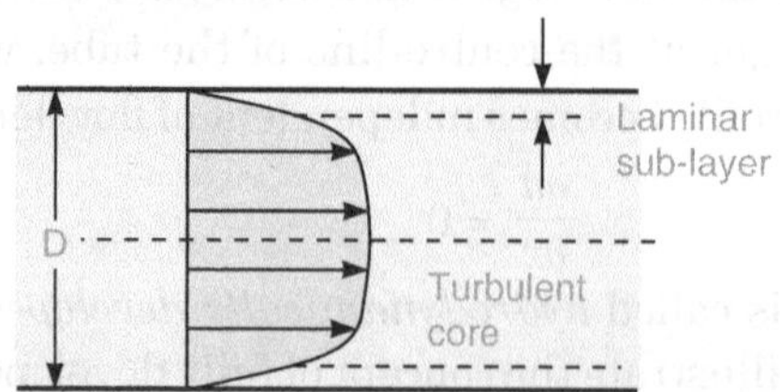

Fig. 9.3. Turbulent flow through a tube

When the fluid flow through the tube becomes *turbulent* a somewhat blunter profile is observed as shown in Fig. 9.3.

9.2.1. Mean Velocity u_m

In above illustration, throughout the length of the tube the *mass flow rate* $\dot{m}$ (kg/s) is assumed to be constant, and the *mean velocity*, u_m (m/s) of fluid is also constant

$$\dot{m} = \rho u_m A_c = \rho u_m \left(\frac{\pi}{4} D^2\right)$$

and
$$u_m = \frac{4\dot{m}}{\pi \rho D^2} \qquad \text{...(9.1)}$$

where ρ = density of fluid, kg/m³,

A_c = cross-section area of tube $\left(= \frac{\pi}{4} D^2\right)$, m²

D = inner diameter of the tube, m.

If the velocity profile, $u(r, x)$ at any location is known, then the mass flow rate

$$\dot{m} = \int_{A_c} \rho\, u(r, x)\, dA_c \qquad \text{...(9.2)}$$

For an incompressible fluid flow, the mean velocity, u_m

$$u_m = \frac{\int_{A_c} u(r, x)\, dA_c}{A_c} = \frac{2}{r_o^2} \int_0^{r_o} u(r, x)\, r\, dr \qquad \text{...(9.3)}$$

The Reynolds number for flow through circular tube

$$Re_D = \frac{\rho u_m D}{\mu} \qquad \text{...(9.4)}$$

where μ = dynamic viscosity of fluid, kg/ms

D = tube diameter, m

For flow through a circular tube, using u_m from eqn. (9.1), the Reynolds number is given by

$$Re_D = \frac{4\dot{m}}{\pi D \mu} \qquad \text{...(9.5)}$$

The Reynolds number again provides convenient criteria for distinguishing the flow regime in the tube. A range of Reynolds number for transition may be observed, depending on surface roughness of tube and velocity fluctuations in the flow. Generally, the accepted critical Reynolds number is 2300, at which the transition from laminar to turbulent begins. Therefore,

$Re_D < 2300$, Laminar flow,

$2300 < Re_D < 4000$, Transition to turbulence,

$Re_D > 4000$, Turbulent flow.

9.2.2. Hydrodynamic Entry Length

For laminar flow, the hydrodynamic entry length x_e is given by

$$\left(\frac{x_e}{D}\right)_{lam} = 0.05\, Re_D \qquad ...(9.6)$$

For turbulent flow, the hydrodynamic entry length is independent of Reynolds number and is expressed as

$$(x_e)_{turb} = 10D \qquad ...(9.7)$$

9.2.3. Velocity Profile in Fully Developed Region

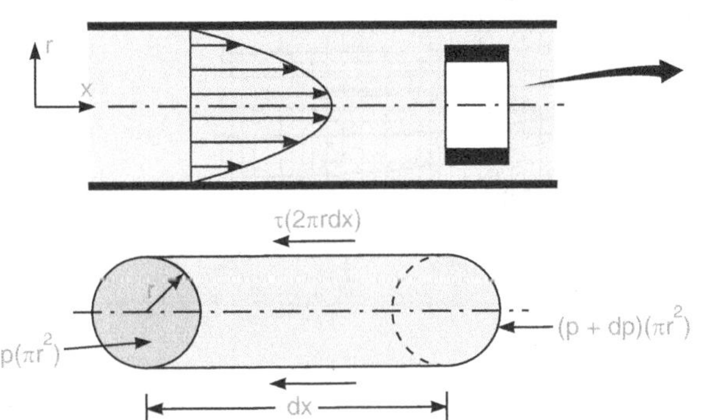

Fig. 9.4. Force balance on the fluid element for laminar, fully developed flow in a circular tube

Consider the flow of an incompressible fluid inside a tube of radius r_o, as shown in Fig. 9.4. It is assumed that velocity at the centre of the tube is u_0 and pressure is uniform at any section. For annular differential element, the pressure forces are balanced by viscous shear forces, so that

$$-\pi r^2 dp = 2\pi r\tau\, dx \quad \text{or} \quad \frac{dp}{dx} = -\frac{2\tau}{r} \qquad ...(9.8)$$

Substituting from Newton's law of viscosity

$$\tau = -\mu \frac{du}{dr} \qquad ...(9.9)$$

The eqn. (9.8) becomes

$$\frac{dp}{dx} = \frac{2\mu}{r}\frac{du}{dr} \quad \text{or} \quad du = \frac{r}{2\mu}\frac{dp}{dx}\, dr$$

Integrating w.r.t. r to obtain

$$u(r) = \frac{r^2}{4\mu}\frac{dp}{dx} + C$$

where C is constant of integration and is evaluated from boundary condition at the tube surface, *i.e.*,

$$u = 0 \quad \text{at} \quad r = r_o$$

we get

$$C = -\frac{r_o^2}{4\mu}\frac{dp}{dx}$$

Thus

$$u(r) = \frac{r^2 - r_o^2}{4\mu}\frac{dp}{dx} \qquad ...(9.10)$$

The velocity at the centre of the tube ($r = 0$) is given by

$$u_0 = -\frac{r_o^2}{4\mu}\frac{dp}{dx} \qquad ...(9.11)$$

and the velocity distribution is given by

$$\frac{u(r)}{u_0} = 1 - \frac{r^2}{r_o^2} \qquad ...(9.12)$$

which is the parabolic distribution for laminar flow inside a tube.

The mean velocity of the flow can be obtained by substituting eqn. (9.12) in eqn. (9.3)

$$u_m = \frac{2u_0}{r_o^2}\int_0^{r_o}\left\{1 - \left(\frac{r}{r_o}\right)^2\right\} r\, dr$$

$$= \frac{2u_0}{r_o^2} \times \left[\frac{r_o^2}{2} - \frac{r_o^4}{4r_o^2}\right] = \frac{u_0}{2}$$

Substituting u_0 from eqn. (9.11), we get

$$u_m = -\frac{r_o^2}{8\mu}\frac{dp}{dx} \qquad ...(9.13)$$

Using this result in eqn. (9.10), the velocity profile becomes

and

$$\frac{u(r)}{u_m} = 2\left[1 - \left(\frac{r}{r_o}\right)^2\right] \qquad ...(9.14)$$

9.2.4. Friction Factor

The shear stress at the wall is normally expressed as

$$\tau_s = -\mu \left.\frac{du}{dr}\right|_{r=r_o}$$

Using eqn. (9.14), we get

$$\tau_s = -\mu 2u_m \left[-\frac{2r}{r_o^2}\right]_{r=r_o} = \frac{4\mu u_m}{r_o} = \frac{8\mu u_m}{D} \qquad ...(9.15)$$

Further the practical definition of shear stress at surface

$$\tau_s = \frac{C_f}{2}\rho u_m^2 \qquad ...(9.16)$$

where C_f is the friction coefficient, equating eqn. (9.15) with eqn. (9.16) and solving for C_f, we get

$$C_f = \frac{2}{\rho u_m^2} \times \frac{8\mu u_m}{D} = \frac{16\,\mu}{\rho u_m D} = \frac{16}{Re_D} \qquad ...(9.17)$$

The *friction factor*, f is a parameter of practical interest, used to calculate the pressure drop of fluid flow in the tube and it is related to the friction coefficient for fully developed laminar flow as

$$f = 4C_f = \frac{64}{Re_D} \qquad ...(9.18)$$

Note that the friction factor, f is associated with pressure drop in fluid flow through the ducts whereas the friction coefficient, C_f is associated with drag force on the surfaces.

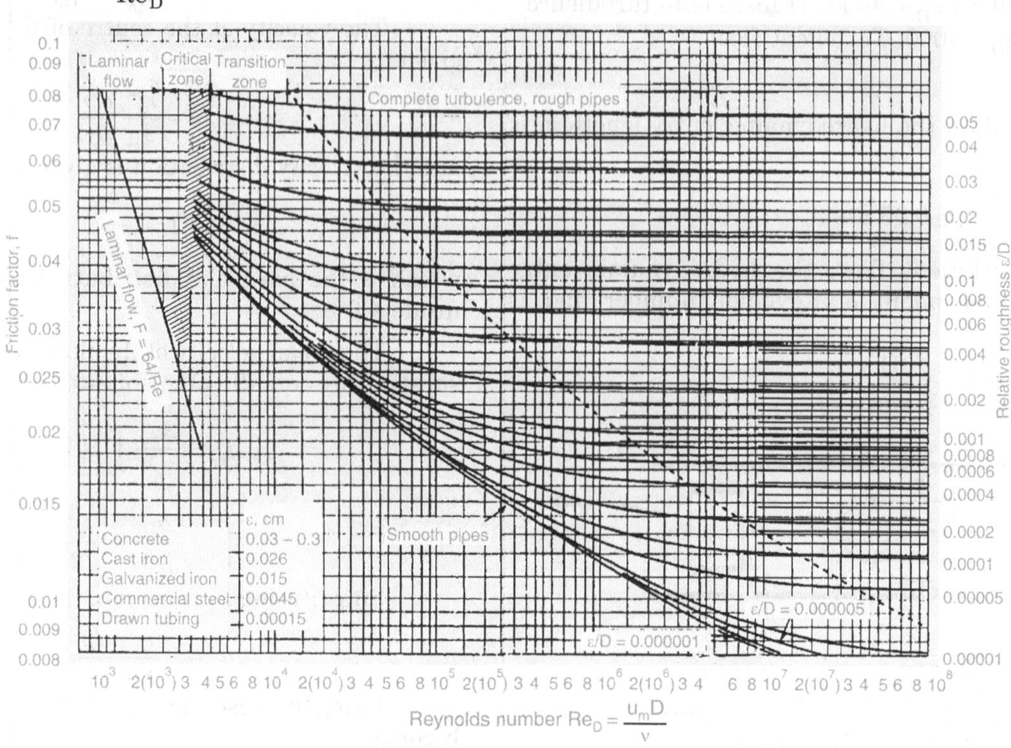

Fig. 9.5. Friction factor for fully developed flow in circular tubes (The Moody chart)

For fully developed turbulent flow ($Re_D > 4000$), the analysis is much more complicated and we must rely on the experimental results. The friction factors for a wide range of Reynolds number are presented in Moody diagram, Fig. 9.5. For a smooth tube, the friction factor for fully developed turbulent flow can also be determined from

$$f = 0.316\, Re_D^{-1/4} \quad 2300 \le Re_D \le 2 \times 10^4 \qquad ...(9.19)$$

$$= 0.184\, Re_D^{-1/5} \quad Re_D \ge 2 \times 10^4 \qquad ...(9.20)$$

9.2.5. Pressure Drop and Friction Factor in Fully Developed Flow

The *pressure drop* sustains an internal flow and is a quantity of practical interest. It is used to calculate the *pumping power* of a fan or pump. The pressure drop during flow in a tube of length L is expressed as

$$\Delta p = f \frac{L}{D} \frac{\rho u_m^2}{2} \ (\text{N/m}^2) \qquad ...(9.21)$$

where f is the *friction factor* and the pumping power to overcome the pressure drop, Δp is

$$\dot{W}_{pump} = \dot{V} \Delta p = \frac{\dot{m}}{\rho} \Delta p \qquad ...(9.22)$$

where $\dot{V} = u_m A_c = \frac{\dot{m}}{\rho}$ is volume flow rate of fluid through the tube.

9.3. THERMAL CONSIDERATIONS

When a fluid at a uniform temperature T_i enters a circular tube that is maintained at some different temperature T_s ($T_i > T_s$) at section A. At a short distance, in down stream at section B, the fluid particles adjacent to tube is cooled by tube surface and attains the tube temperature T_s. The fluid temperature varies from T_s at the tube surface to T_i at a small distance δ_{th} from the tube surface. It will initiate the convection in fluid and development of the *thermal boundary layer*. The boundary layer thickness, δ_{th} increases in the direction of flow, until at some location C, where it reaches the tube centre and thus fills the entire tube as shown in Fig. 9.6(a). Upto section C, the centre line, temperature remains constant at T_i but beyond this section, the centre line temperature changes in the direction of flow. But the dimensionless temperature profile does not change in the x direction

$$\frac{\partial}{\partial x}\left(\frac{T_s - T}{T_s - T_m}\right) = 0 \qquad ...(9.23)$$

where T_s is tube surface temperature, T is local fluid temperature and T_m is mean temperature of fluid over the cross-section of the tube.

Fig. 9.6 (*b*) shows the thermal boundary layer development for a cold fluid flowing through a heated tube.

The region of unchanged temperature profile is called the **thermally developed region** and the region of flow over which the thermal boundary develops, is called the **thermally developing region** or **thermal entrance region,** and length of this region is called the **thermal entry length** x_{eth}. In the region A to C, thermal entry region, the velocity profile is fully developed and temperature profile is developing. Beyond section C, the flow is both hydrodynamically and thermally developed and thus this region is called the **fully developed region.**

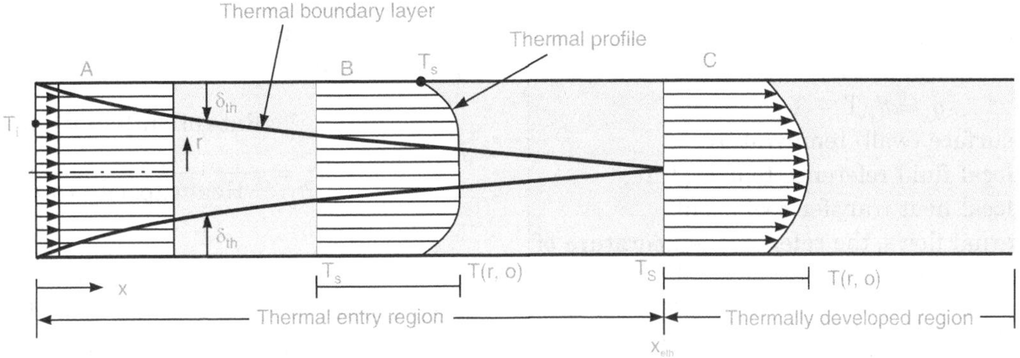

(a) The development of the thermal boundary layer in a cold tube. (The fluid is hotter than tube surface)

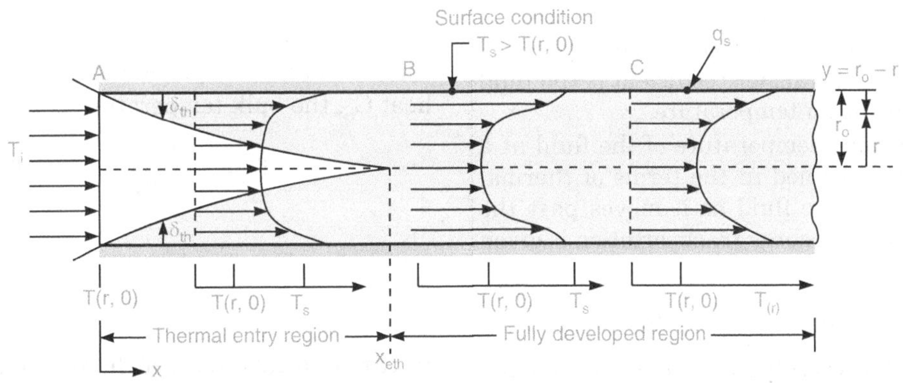

(b) Thermal boundary layer development in a heated circular tube

Fig. 9.6. Developing and fully developed thermal boundary layer in the tube

For laminar flow the thermal entry length may be expressed as

$$\left(\frac{x_{eth}}{D}\right)_{lam} \approx 0.05\, Re_D Pr \qquad ...(9.24)$$

Comparing eqn. (9.24) with eqn. (9.6), it is evident that, for Pr > 1, the hydrodynamic boundary layer develops more rapidly than the thermal boundary layer and the hydrodynamic entry length is shorter than the thermal entry length as shown in Fig. 9.7. The inverse is also true for Pr < 1. In contrast for turbulent flow conditions, the thermal entry length is independent of Prandtl number and is approximated as

$$\left(\frac{x_{eth}}{D}\right)_{turb} = 10 \qquad ...(9.25)$$

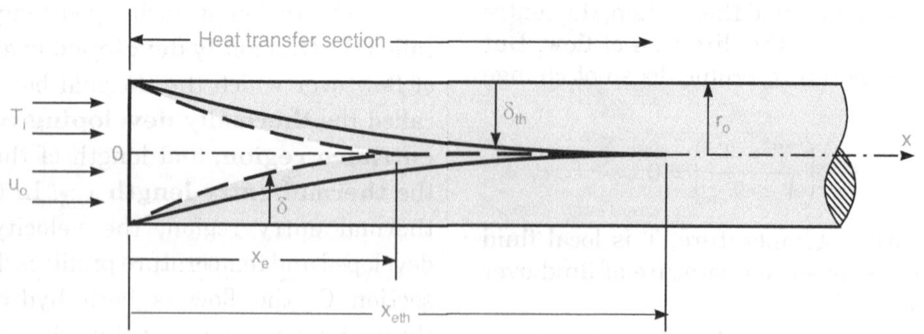

Fig. 9.7. Development of hydrodynamic and thermal boundary layers for Pr > 1

9.3.1. The Mean Temperature or Bulk Temperature

The convective heat transfer rate at any location in the internal flow is determined as

$$q_x = h_x(T_s - T_{ref}) \quad ...(9.26)$$

where T_s = surface (wall) temperature,
T_{ref} = local fluid reference temperature,
h_x = local heat transfer coefficient.

For external flows, the reference temperature of the fluid is the free stream temperature (T_∞), which is constant. But in internal flows, the temperature of fluid varies in the direction of flow due to continuous heat transfer.

The fluid temperature varies not only in direction of flow, but normal to direction of flow. This variation depends on thermal boundary conditions imposed, type of fluid flow and entry length effects.

Therefore, the accepted reference temperature of fluid for computing heat transfer coefficient is the fluid bulk temperature or mean temperature.

The mean or bulk temperature of the fluid at a given cross-section is defined in the terms of thermal energy transported by the fluid as it moves past the cross-section. The rate of energy transportation is given by

$$E'_{th} = \int_{A_c} \rho u(r)\, C_p\, T(r, x)\, dA_c \quad ...(9.27)$$

where $u(r)$ is axial flow velocity and A_c is cross-sectional area perpendicular to the flow direction.

If a mean or bulk temperature T_m is defined, then the energy transfer in terms of mean or bulk enthalpy

$$E'_{th} = \dot{m}\, h_m = \dot{m}\, C_p T_m \quad ...(9.28)$$

Equating eqn. (9.27) and eqn. (9.28) to obtain

$$T_m = \frac{\int_{A_c} \rho\, u C_p\, T\, dA_c}{\dot{m}\, C_p}$$

$$= \frac{\int_{A_c} \rho\, u C_p\, T\, dA_c}{(\rho\, u_m A_c) C_p} \quad ...(9.29)$$

In other words, the *mean temperature* can be defined as the ratio of flow rate of enthalpy to the heat capacity rate or

$$T_m = \frac{\text{Rate of enthalpy flow through a cross-section}}{\text{Heat capacity rate through a cross-section}}$$

This definition of bulk temperature is general and can be applied to either laminar or turbulent flow.

For a circular pipe $dA_c = 2\pi r dr$, the eqn. (9.29) becomes

$$T_m = \frac{\int_0^{r_o} \rho\, u C_p\, T\, (2\pi r)\, dr}{\rho\, u_m (\pi r_o^2) C_p} \quad ...(9.30)$$

For an incompressible fluid with constant specific heat C_p, the bulk temperature is given by

$$T_m = \frac{2}{u_m r_o^2} \int_0^{r_o} u(r, x)\, T(r, x)\, r\, dr \quad ...(9.31)$$

9.4. HEAT TRANSFER IN FULLY DEVELOPED FLOW

In case of a tube flow, if there is a difference between the tube wall temperature and fluid temperature, the heat transfer takes place. The temperature difference produces a temperature profile in the direction of fluid flow as shown in Fig. 9.8. Here, the tube surface is hotter than the fluid, the fluid temperature varies from the value at the surface to that at the centre line.

The heat flux from the surface to the fluid is given by

$$q_s = k_f \left.\frac{\partial T(r, x)}{\partial r}\right|_{wall} \quad ...(9.32)$$

where k_f the thermal conductivity of the fluid and $T(r, x)$ = local fluid temperature. Further, the heat flux can also be expressed in terms of the local heat transfer coefficient h_x and fluid, mean temperature, as

$$q_x = h_x(T_s - T_m) \qquad ...(9.33)$$

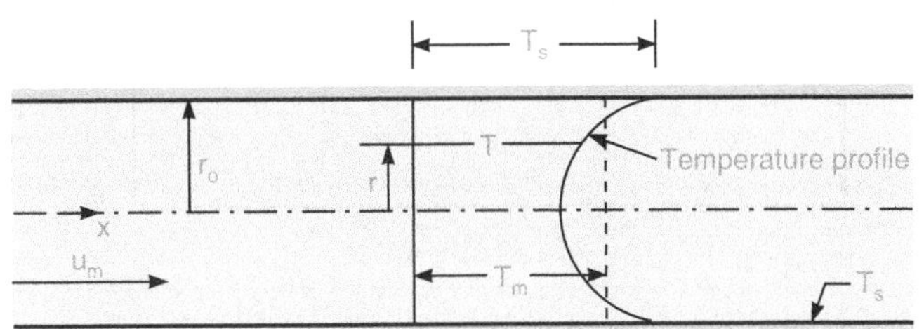

Fig. 9.8. Temperature distribution in fully developed tube flow ($T_s > T_m$)

where T_s = local tube wall temperature,

T_m = local bulk mean fluid temperature.

Equating eqns. (9.32) and (9.33), we get the local heat transfer coefficient as

$$h_x = \frac{k_f}{(T_s - T_m)} \times \left.\frac{\partial T(r, x)}{\partial r}\right|_{wall} \qquad ...(9.34)$$

In thermally developed region, the dimensionless temperature $\theta(r, x)$ is defined as

$$\theta(r, x) = \frac{T_s - T(r, x)}{T_s - T_m} \qquad ...(9.35)$$

The eqn. (9.34) can be expressed in terms of dimensionless temperature θ as

$$h_x = -k_f \left.\frac{\partial \theta(r, x)}{\partial r}\right|_{wall} \qquad ...(9.36)$$

In the fully developed region, the dimensionless temperature θ is independent of x and eqn. (9.36) reduces to

$$h = -k_f \left.\frac{\partial \theta(r)}{\partial r}\right|_{wall} \qquad ...(9.37)$$

9.5. GENERAL THERMAL ANALYSIS

Fig. 9.9 and Fig. 9.10 show the development of thermal boundary layer for two cases, in which the wall surface is *hotter* than the fluid. As flow proceeds in the tube, a thermal boundary layer develops at the surface, in which the fluid temperature T varies due to heat transfer at the surface, but the central core retains the free stream temperature T_i. As the thermal boundary layer fills the tube at thermal entry length x_{eth}, the temperature profile becomes fully developed and it does not change further in flow direction.

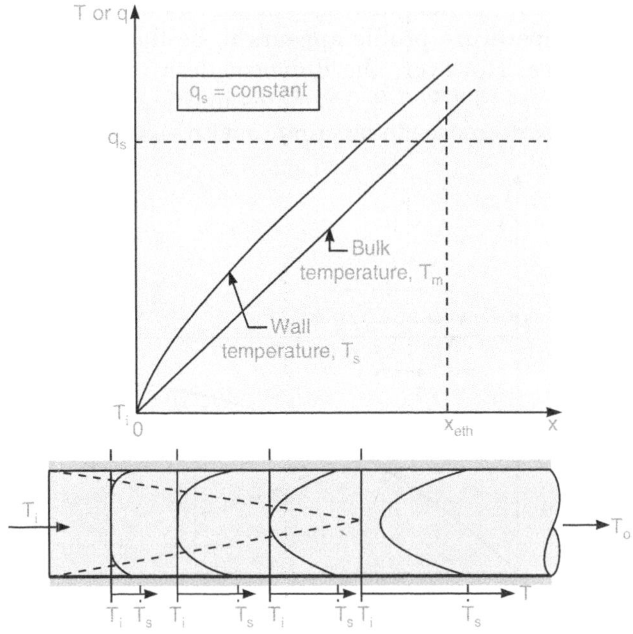

Fig. 9.9. Developing and fully developed temperature profile in tube flow for constant wall heat flux

9.5.1. Constant Surface Heat Flux

Fig. 9.9 depicts the case, in which the *heat flux at the surface* (q_s = Q/A_s) remains *constant*. As the heat is transferred to the fluid, the bulk temperature increases, but the difference between surface temperature T_s and bulk mean temperature T_m remains constant. Thus both T_s and T_m increase in flow direction, while the temperature profile $(T_s - T)/(T_s - T_m)$ remains unchanged.

For constant heat flux at the wall, the heat transfer rate can be expressed as

$$Q = q_s A_s = \dot{m} C_p (T_o - T_i) \qquad ...(9.38)$$

where A_s is the surface area of tube and T_i and T_o are mean fluid temperatures at the inlet and exit of the tube, respectively. The mean fluid temperature at the exit of the tube

$$T_o = T_i + \frac{q_s A_s}{\dot{m} C_p} \qquad ...(9.39)$$

Note that the bulk mean temperature T_m increases linearly in the flow direction, since the surface area increases and it becomes T_o at the tube exit. The properties are evaluated at mean bulk temperature, given by

$$T_m = \frac{T_i + T_o}{2} \qquad ...(9.40)$$

9.5.2. Constant Surface Temperature

Fig. 9.10 depicts the case, in which the *tube wall temperature* T_s remains *constant*. The bulk or mean temperature of fluid increases in the flow direction and the temperature profile appears to be flatten, as T_m increases. However, the dimensionless temperature profile $(T_s - T)/(T_s - T_m)$ remains unchanged, but the heat flux decreases in accordance with eqn. (9.33).

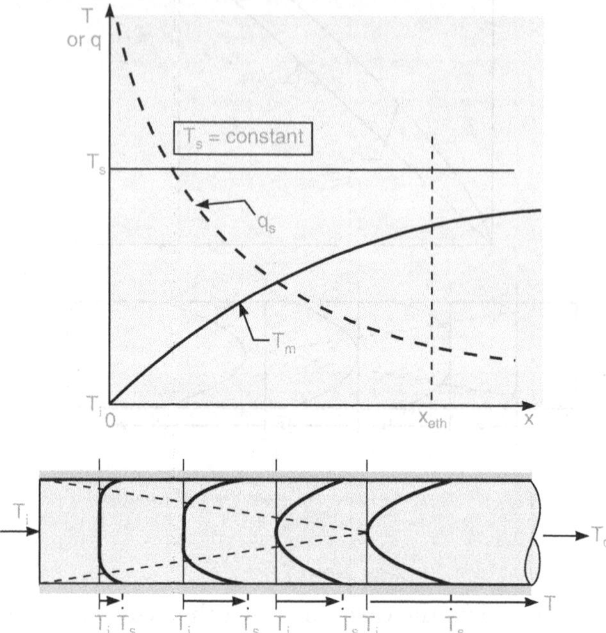

Fig. 9.10. Developing and developed temperature profile in tube flow for constant wall temperature

The rate of heat transfer to or from the fluid flowing in the tube can be determined as

$$Q = hA_s \Delta T_{av} = hA_s(T_s - T_m)_{av} \quad ...(9.41)$$

where, h is an average heat transfer coefficient, A_s is the heat transfer surface area (= πDL for a circular tube of length L) and ΔT_{av} is some appropriate average temperature difference between surface and fluid.

Since the fluid temperature varies almost linearly along the tube, when tube surface temperature is kept constant. Therefore, the average appropriate temperature difference between fluid and surface needs a better evaluation of ΔT_{av}.

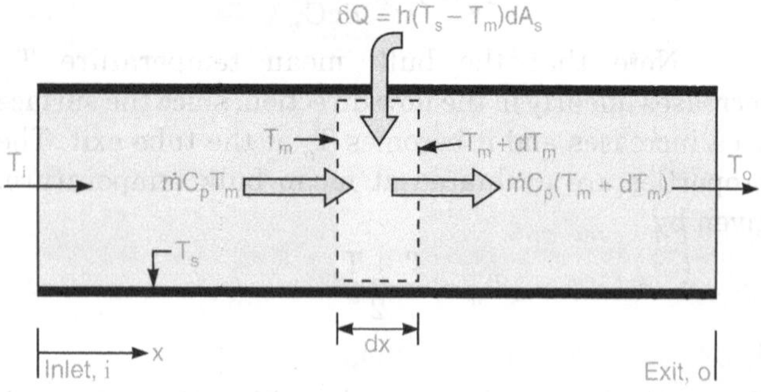

Fig. 9.11. (*a*) Energy interactions for a differential control volume in a tube flow

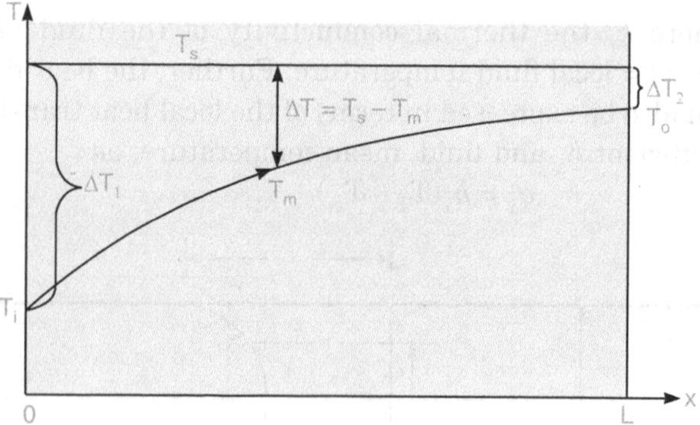

Fig. 9.11. (*b*) Variation of bulk mean fluid temperature along the tube for the constant surface temperature

Consider the heating of the fluid in a tube of constant cross-section, whose inner wall surface is maintained at constant temperature T_s. The bulk or mean fluid temperature T_m increases in the flow direction as a result of heat transfer as shown in Fig. 9.11.

The energy balance on a differential control volume gives.

Increase in enthalpy of fluid = Heat convected to fluid from the surface.

$$\dot{m} C_p \, dT_m = h(T_s - T_m)\, dA_s \quad ...(9.42)$$

Substituting $dA_s = Pdx$, where P is the perimeter of tube. Rearranging as

$$\frac{dT_m}{T_s - T_m} = \frac{hP}{\dot{m}C_p}dx \quad ...(9.43)$$

Using $\Delta T = T_s - T_m$ and $dT_m = -d(\Delta T)$, then

$$\frac{d(\Delta T)}{\Delta T} = -\frac{hP}{\dot{m}C_p}dx$$

Integrating from inlet to exit conditions, treating h, C_p, $\dot{m}$, P as constant quantities,

$$\int_{\Delta T_1}^{\Delta T_2} \frac{d(\Delta T)}{\Delta T} = -\frac{hP}{\dot{m}C_p}\int_0^L dx$$

we get $$\ln\left(\frac{\Delta T_2}{\Delta T_1}\right) = -\frac{hP}{\dot{m}C_p}L$$

where $\Delta T_1 = T_s - T_i$ and $\Delta T_2 = T_s - T_o$, then

$$\ln\left(\frac{T_s - T_o}{T_s - T_i}\right) = -\frac{hPL}{\dot{m}C_p} \quad ...(9.44)$$

or $$\frac{T_s - T_o}{T_s - T_i} = \exp\left(-\frac{hPL}{\dot{m}C_p}\right) \quad ...(9.45)$$

The quantity $\frac{hPL}{\dot{m}C_p}$ is a dimensionless parameter.

It is called the number of transfer units, denoted by NTU. It is the measure of the effectiveness of the heat transfer systems.

The mean fluid temperature at the exit can be determined as

$$T_o = T_s - (T_s - T_i) \exp\left(-\frac{hPL}{\dot{m}C_p}\right) \quad ...(9.46)$$

This relation can also be used to determine mean fluid temperature at any x by replacing L by x as

$$T_x = T_s - (T_s - T_i) \exp\left(-\frac{hPx}{\dot{m}C_p}\right) \quad ...(9.47)$$

The eqns. (9.46) and (9.47) reveal that the mean temperature of fluid varies exponentially in the direction of flow.

Further, solving the eqn. (9.44) for $\dot{m}C_p$ as

$$\dot{m}C_p = \frac{hPL}{\ln\left(\frac{T_s - T_i}{T_s - T_o}\right)} \quad ...(9.48)$$

The heat transfer rate to the fluid can also be given by

$$Q = \dot{m}C_p (T_o - T_i)$$

which can be arranged as

$$Q = \dot{m}C_p[(T_s - T_i) - (T_s - T_o)]$$

Substituting eqn. (9.48) for $\dot{m}C_p$, we get

$$Q = hPL \frac{(T_s - T_i) - (T_s - T_o)}{\ln\left(\frac{T_s - T_i}{T_s - T_o}\right)} \quad ...(9.49)$$

or $$Q = hPL\,(\Delta T)_{lm} = hA_s\,(\Delta T)_{lm} \quad ...(9.50)$$

where ΔT_{lm} is the log mean temperature difference and is given by

$$\Delta T_{lm} = \frac{\Delta T_1 - \Delta T_2}{\ln\left(\frac{\Delta T_1}{\Delta T_2}\right)} \quad ...(9.51)$$

where $\Delta T_1 = T_s - T_1$ and $\Delta T_2 = T_s - T_2$ are the temperature difference between surface and fluid at the inlet and exit of the tube, respectively.

Example 9.1. *Water at 20°C flows through a small tube, 1 mm in diameter at a uniform speed of 0.2 m/s. The flow is fully developed at a point beyond which a constant heat flux of 6000 W/m² is imposed. How much farther down the tube will the water reach 74°C as its hottest point?*

Solution

Given : Water flows through a tube of constant wall heat flux.

$T_i = 20°C$, $D = 1$ mm,

$u_m = 0.2$ m/s, $q = 6000$ W/m²,

$T_o = 74°C$.

To find : The distance of tube at which water is heated to a temperature of 74°C.

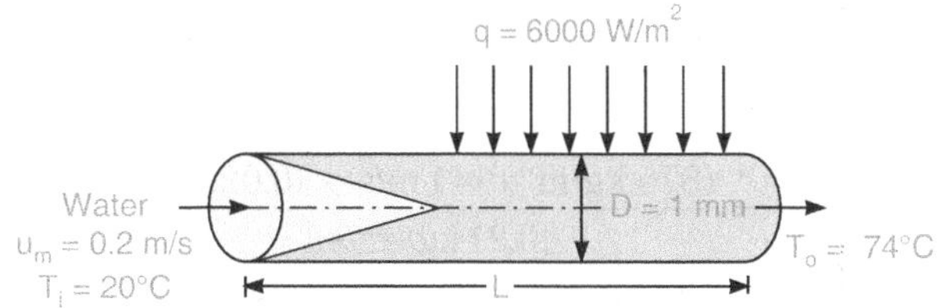

Fig. 9.12. Schematic for example 9.1

Analysis : The mean fluid temperature

$$T_m = \frac{T_i + T_o}{2} = \frac{20 + 74}{2}$$
$$= 47°C\ (320\ K)$$

The physical properties of water at 320 K from Table A-7

$\rho = 989$ kg/m³, $C_p = 4180$ J/kg.K,

$\mu = 577 \times 10^{-6}$ kg/ms, $k_f = 0.640$ W/m.K,

Pr = 3.77.

Mass flow rate through tubes

$$\dot{m} = \rho u_m A_c = \rho u_m \left(\frac{\pi}{4} D^2\right)$$

$$= 989 \times 0.2 \times \left(\frac{\pi}{4}\right) \times (0.001)^2$$

$$= 1.55 \times 10^{-4}\ \text{kg/s}$$

Surface area $A_s = \pi DL = \pi \times 0.001$ L

The length of tube for exit temperature $T_o = 74°C$ can be determined by using eqn. (9.39)

$$T_o = T_i + \frac{q_s A_s}{\dot{m}C_p}$$

or $$74 = 20 + \frac{6000 \times \pi \times 0.001}{1.55 \times 10^{-4} \times 4180} L$$

or **L = 1.86 m. Ans.**

Example 9.2. *Engine oil at 40°C (μ = 0.21 kg/(ms) ; ρ = 875 kg/m³) flows inside 2.5 cm diameter, 50 m long tube with a mean velocity of 1 m/s. Determine the pressure drop for flow through the tube.*

(J.N.T.U., May 2004)

Solution

Given : Flow of engine oil through a tube.

$T_m = 40°C$, $\mu = 0.21$ kg/(ms),

$\rho = 875$ kg/m³, $D = 2.5$ cm,

$L = 50$ m, $u_m = 1$ m/s.

To find : The pressure drop for fluid flow through the tube.

Assumptions :

1. Steady state conditions,
2. Constant properties.

Analysis : The Reynolds number for the fluid flow

$$\text{Re}_D = \frac{\rho u_m D}{\mu}$$

$$= \frac{(875\ \text{kg/m}^3) \times (1\ \text{m/s}) \times (0.025\ \text{m})}{[0.21\ \text{kg/(m/s)}]}$$

$$= 104.16$$

Thus the flow is laminar through the tube.

Using eqn. (9.18) for fully developed laminar flow

$$f = \frac{64}{\text{Re}_D} = \frac{64}{104.16} = 0.6144$$

The pressure drop during the fluid flow can be obtained by eqn. (9.21)

$$\Delta p = f \frac{L}{D} \frac{\rho u_m^2}{2}$$

$$= 0.6144 \times \frac{50}{0.025} \times \frac{875 \times 1}{2}$$

$$= 537600\ \text{N/m}^2 = \mathbf{5.376\ bar.}$$ Ans.

Example 9.3. *Explain, why the Nusselt number remains constant for fully developed laminar tube flow.*

Solution

For a fully developed flow, the temperature profile is given by

$$\frac{\partial}{\partial x}\left(\frac{T_s - T}{T_s - T_m}\right) = 0$$

It indicates that the dimensionless temperature gradient at the surface is constant along the flow. Hence for fluids with constant properties, the heat transfer coefficient and Nusselt numbers are constant in fully developed tube flow.

Example 9.4. *A system for heating of water from an inlet temperature of 20°C to an outlet temperature of 60°C involves passing the water through a thick walled tube of inner and outer diameters of 20 and 40 mm. The outer surface of the tube is well insulated, and electrical heating within the wall provides a uniform heat generation at the rate of 10^6 W/m³.*

(i) What is the length of the tube to achieve the desired outlet temperature, if water mass flow rate is 0.1 kg/s ?

(ii) What is the local heat transfer coefficient at the outlet, if the inner surface temperature of the tube at the outlet is 70°C ?

Solution

Given :

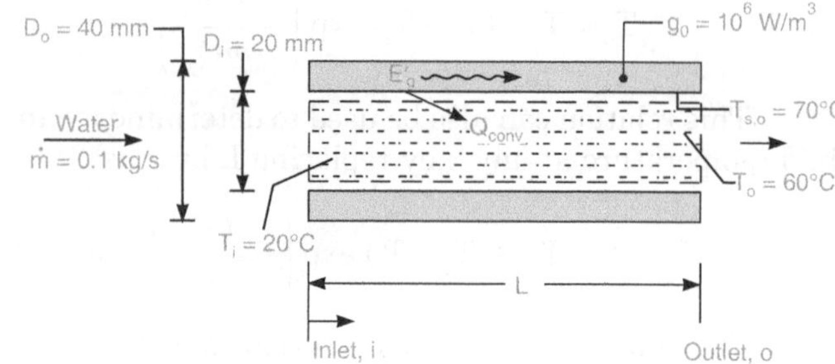

Fig. 9.13. Schematic for example 9.4

To find :

(*i*) Length of the tube to achieve the water outlet temperature 60°C,

(*ii*) Local convection heat transfer coefficient at the outlet.

Assumptions:

1. Steady state conditions,
2. Uniform heat generation over entire length of the tube,
3. Constant properties of the fluid,
4. No heat transfer to surroundings,
5. Specific heat of water as 4180 J/kg.K.

Analysis : (*i*) Making the energy balance for the heating system ;

Heat generation rate = Enthalpy rise rate of water

$$g_0 V = \dot{m} C_p (T_o - T_i)$$

or $$g_0 \frac{\pi}{4} (D_o^2 - D_i^2) L = \dot{m} C_p (T_o - T_i)$$

or $$10^6 \times \left(\frac{\pi}{4}\right) \times (0.04^2 - 0.02^2) L = 0.1 \times 4180 \times (60 - 20)$$

or $$L = \mathbf{17.7\ m.}$$ Ans.

(*ii*) Local convection heat transfer coefficient at the tube exit :

Making energy balance at the tube exit ;

Heat generation rate = Heat convection rate

$$g_0 V = h_o A_s (T_{s,o} - T_o)$$

or $$g_0 \times \left(\frac{\pi}{4}\right) (D_o^2 - D_i^2) L = h_o (\pi D_i L)(T_{s,o} - T_o)$$

or $$10^6 \times \left(\frac{\pi}{4}\right) \times (0.04^2 - 0.02^2) L$$

$$= h_o \times (\pi \times 0.02 \times L) \times (70 - 60)$$

or $$h_o = \mathbf{1500\ W/m^2.K.}$$ Ans.

9.6. HEAT TRANSFER IN LAMINAR TUBE FLOW

Consider the heat transfer process for a laminar flow system inside a tube, with a uniform heat flux at the tube surface.

Consider an annular fluid element as shown in Fig. 9.14.

The heat conducted into the annular fluid element radially

$$q_r = -k_f(2\pi r dx)\frac{\partial T}{\partial r} \quad \text{...(i)}$$

The heat conducted out the annular fluid element

$$q_{r+dr} = q_r + \frac{\partial}{\partial r}(q_r)\,dr + \text{......}$$

$$\approx -k_f(2\pi r dx)\frac{\partial T}{\partial r} + \frac{\partial}{\partial r}\left(-2\pi r dx\, k_f \frac{\partial T}{\partial r}dr\right) \quad \text{...(ii)}$$

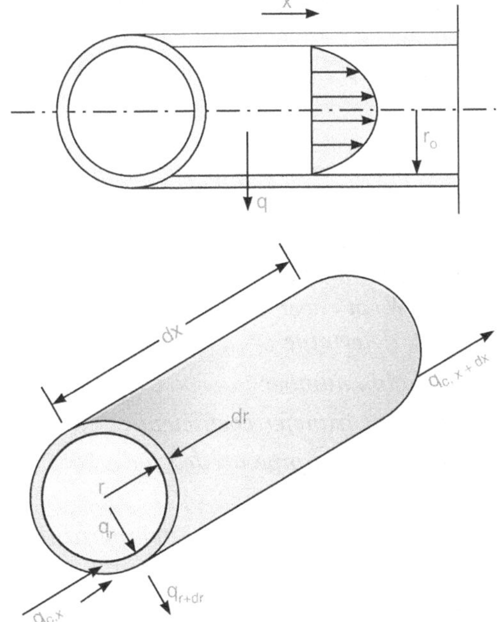

Fig. 9.14. Control volume for energy analysis in the tube flow

The net heat conducted into the element is

$$q_r - q_{r+dr} = 2\pi k_f \frac{\partial}{\partial r}\left(r\frac{\partial T}{\partial r}\right)drdx \quad \text{...(iii)}$$

The heat convected into the annular element axially

$$q_{c,x} = \rho(2\pi r dr)\,uC_pT \quad \text{...(iv)}$$

The heat convected out the annular fluid element at $x + dx$

$$q_{c,\,x+dx} = q_{c,x} + \frac{\partial}{\partial x}(q_{c,x})\,dx + \text{......}$$

$$\approx \rho(2\pi r dr)\,uC_pT + \rho(2\pi r dr)\,uC_p\frac{\partial T}{\partial x}dx \quad \text{...(v)}$$

The net heat convected out the annular element is

$$q_{c,\,x+dx} - q_{c,x} = 2\pi r\rho uC_p\frac{\partial T}{\partial x}dx\,dr \quad \text{...(vi)}$$

The energy balance on the fluid element is

$$2\pi r\rho uC_p\frac{\partial T}{\partial x}drdx = 2\pi k_f\frac{\partial}{\partial r}\left(r\frac{\partial T}{\partial r}\right)drdx$$

or

$$ur\frac{\rho C_p}{k_f}\frac{\partial T}{\partial x} = \frac{\partial}{\partial r}\left(r\frac{\partial T}{\partial r}\right)$$

or

$$\frac{\partial}{\partial r}\left(r\frac{\partial T}{\partial r}\right) = \frac{ur}{\alpha}\frac{\partial T}{\partial x} \quad \text{...(9.52)}$$

where $\alpha = \dfrac{k_f}{\rho C_p}$, thermal diffusivity of fluid, m²/s,

k_f = thermal conductivity of the fluid, W/m.K,

ρ = density of fluid, kg/m³,

C_p = specific heat of fluid J/kg.K,

$u = u(r)$, local velocity of fluid,

r = radial coordinate,

$T = T(r, x)$, function of radial and flow directions.

For constant wall heat flux, the bulk fluid temperature increases linearly, and

$$\left.\frac{\partial T}{\partial x}\right|_{wall} = \text{constant}$$

Inserting the velocity distribution $u(r)$ from eqn. (9.12) into eqn. (9.52), we have

$$\frac{\partial}{\partial r}\left(r\frac{\partial T}{\partial r}\right) = u_0\left(1 - \frac{r^2}{r_o^2}\right)r \times \frac{1}{\alpha}\frac{\partial T}{\partial x}$$

Integrating w.r.t. r,

$$r\frac{\partial T}{\partial r} = \frac{u_0}{\alpha}\left(\frac{r^2}{2} - \frac{r^4}{4r_o^2}\right)\frac{\partial T}{\partial x} + C_1 \quad \text{...(9.53)}$$

and second integration leads to

$$T = \frac{u_0}{\alpha}\left(\frac{r^2}{4} - \frac{r^4}{16r_o^2}\right)\frac{\partial T}{\partial x} + C_1\ln r + C_2 \quad \text{...(9.54)}$$

where C_1 and C_2 are constants of integrations and are evaluated from boundary conditions.

The boundary conditions are

At $r = 0$; $\dfrac{\partial T}{\partial r} = 0$

and $\left.k_f\dfrac{\partial T}{\partial r}\right|_{r=r_o} = q_s$ = constant.

Using first boundary condition at $r = 0$, it gives

$$C_1 = 0$$

The second boundary condition is satisfied, if the axial temperature gradient $\frac{\partial T}{\partial x}$ is constant, let the temperature at the centre of the tube be T_c, then

$$T = T_c \quad \text{at} \quad r = 0$$

It leads to $C_2 = T_c$

Then the temperature distribution in the fluid element becomes

$$T - T_c = \frac{u_0 r_o^2}{4\alpha} \frac{\partial T}{\partial x} \left[\left(\frac{r}{r_o} \right)^2 - \frac{1}{4} \left(\frac{r}{r_o} \right)^4 \right] \qquad ...(9.55)$$

The heat transfer coefficient is given by eqn. (9.34)

$$h = \frac{k_f}{T_s - T_m} \left. \frac{\partial T}{\partial r} \right|_{r=r_o}$$

where T_s = surface temperature

T_m = bulk fluid temperature.

The temperature gradient at the wall is given by

$$\left. \frac{\partial T}{\partial r} \right|_{r=r_o} = \frac{u_0}{\alpha} \frac{\partial T}{\partial x} \left[\frac{r}{2} - \frac{r^3}{4r_o^2} \right]_{r=r_o} = \frac{u_0 r_o}{4\alpha} \frac{\partial T}{\partial x} \qquad ...(9.56)$$

The bulk fluid temperature T_m, can be obtained by using eqn. (9.55) for temperature distribution in eqn. (9.31). For constant heat flux at the wall, we get

$$T_m = T_c + \frac{7}{96} \frac{u_0 r_o^2}{\alpha} \frac{\partial T}{\partial x} \qquad ....(9.57)$$

and wall temperature,

$$T_s = T_c + \frac{3}{16} \frac{u_0 r_o^2}{\alpha} \frac{\partial T}{\partial x} \qquad ...(9.58)$$

Using eqns. (9.56), (9.57) and (9.58) in eqn. (9.34), we have

$$h = k_f \frac{\frac{u_0 r_o}{4\alpha} \frac{\partial T}{\partial x}}{\left(\frac{7}{96} - \frac{3}{16} \right) \frac{u_0 r_o^2}{\alpha} \frac{\partial T}{\partial x}}$$

$$= \frac{96}{44} \frac{k_f}{r_o} = \frac{24}{11} \frac{k_f}{r_o}$$

or

$$h = \frac{48}{11} \frac{k_f}{D} \qquad ...(9.59)$$

and the Nusselt number for constant heat flux condition is given by

$$Nu_D = \frac{hD}{k_f} = \frac{48}{11} = 4.36, \qquad ...(9.60)$$

It remains constant for fully developed laminar tube flow, that is subjected to uniform heat flux at the wall.

If eqn. (9.54) for temperature distribution is solved for constant wall temperature, *i.e.*,

$$T = T_s \quad \text{at} \quad r = r_o$$

and

$$\frac{\partial T}{\partial r} = 0 \quad \text{at} \quad r = 0$$

Then the corresponding analysis leads to a Nusselt number as

$$Nu_D = 3.66 \quad \text{for } T_s = \text{constant} \qquad ...(9.61)$$

It is also constant for fully developed laminar tube flow, that is subjected to constant wall temperature. A general relation for average Nusselt number for hydrodynamically and/or thermally developing laminar flow in circular tube is suggested by *Sieder* and *Tate* as

$$Nu_D = 1.86 \left(Re_D \, Pr \frac{D}{L} \right)^{1/3} \left(\frac{\mu}{\mu_s} \right)^{0.14}$$

for $\quad Pr > 0.5 \qquad ...(9.62)$

In eqns. (9.60), (9.61) and (9.62), the fluid properties may be evaluated at mean fluid temperature, $\left(\frac{T_i + T_o}{2} \right)$ except μ_s, which is evaluated at surface temperature T_s.

Example 9.5. *Water entering at 10°C is heated to 40°C in the tube of 0.02 m ID at a mass flow rate of 0.01 kg/s. The outside of the tube is covered with an insulated electric heating element that produces a uniform heat flux of 15000 W/m² over the surface. Neglecting any entrance effect, determine ;*

(*a*) *Reynolds number ;*

(*b*) *The heat transfer coefficient ;*

(*c*) *The length of pipe needed for a 30°C increase in average temperature ;*

(*d*) *The inner tube surface temperature at the outlet ;*

(*e*) *The friction factor ;*

(*f*) *The pressure drop in the pipe ;*

(*g*) *The pumping power required, if the pump is 50% efficient.*

Solution

Given : Flow through pipe ;

$D_i = 0.02$ m, $\dot{m} = 0.01$ kg/s,

$T_o = 40°C$, $T_i = 10°C$,

$q = 15000$ W/m², $\eta_{pump} = 0.5$.

To find :

(*a*) Reynolds number,

(*b*) The heat transfer coefficient,

(c) The length of pipe needed for a 30°C increase in average temperature,

(d) The inner tube surface temperature at the outlet,

(e) The friction factor,

(f) The pressure drop in the pipe,

(g) The pumping power required, if the pump is 50% efficient.

Assumptions :

1. No heat exchange by thermal radiation and heat conduction,
2. The steady state heat transfer conditions,
3. Entrance and exit temperatures of water are different, taking the properties at mean film temperature.

Properties : The properties of water at its mean temperature

$$T_m = \frac{40+10}{2} = 25°C$$

The properties of water at 25°C (from Table A-7)

$$\rho = 997 \text{ kg/m}^3, \quad k_f = 0.608 \text{ W/m.K},$$

$$C_p = 4180 \text{ J/kgK}, \quad \mu = 910 \times 10^{-6} \text{ Ns/m}^2.$$

Analysis : (a) The Reynolds number:

$$\text{Re} = \frac{u_m D_i}{\nu} = \frac{4\dot{m}}{\pi D_i \mu}$$

$$= \frac{4 \times (0.01 \text{ kg/s})}{\pi \times (0.02 \text{ m}) \times (910 \times 10^{-6} \text{ Ns/m}^2)}$$

$$= \mathbf{700.} \quad \text{Ans.}$$

It indicates the flow is laminar.

(b) Since uniform flux is on the pipe surface, hence using eqn. (9.60) ;

$$\text{Nu}_D = 4.36$$

And heat transfer coefficient

$$h = \frac{\text{Nu}_D\, k_f}{D_i} = \frac{4.36 \times (0.608 \text{ W/m.K})}{(0.02 \text{ m})}$$

$$= \mathbf{132.5 \text{ W/m}^2\text{.K.}} \quad \text{Ans.}$$

(c) Length of the pipe needed for 30°C temperature rise :

$$Q = q\,A_s = q(\pi D_i L) = \dot{m}\, C_p(T_o - T_i)$$

or
$$L = \frac{(0.01 \text{ kg/s}) \times (4180 \text{ J/kg.K}) \times (30 \text{ K})}{(15000 \text{ W/m}^2) \times (\pi \times 0.02 \text{ m})}$$

$$= \mathbf{1.33 \text{ m.}} \quad \text{Ans.}$$

(d) Inner tube surface temperature at the outlet :

$$q = h(T_{s,o} - T_o)$$

$$T_{s,o} = \frac{q}{h} + T_o = \frac{15000 \text{ W/m}^2}{132.5 \text{ W/m}^2\text{. K}} + 40$$

$$= \mathbf{153.6°C.} \quad \text{Ans.}$$

(e) Friction factor :

$$f = \frac{64}{\text{Re}_D} = \frac{64}{700} = \mathbf{0.0914.} \quad \text{Ans.}$$

(f) Pressure drop in pipe :

$$\Delta p = f\frac{L}{D_i}\frac{\rho u_m^{\,2}}{2}$$

where,
$$u_m = \frac{4\dot{m}}{\pi D_i^{\,2}\rho} \text{ since } \dot{m} = \rho u_m A_c$$

$$= \frac{(4 \times 0.01 \text{ kg/s})}{\pi \times (0.02 \text{ m})^2 \times (997 \text{ kg/m}^3)}$$

$$= 0.032 \text{ m/s}$$

Hence
$$\Delta p = 0.0914 \times \frac{(1.33 \text{ m})}{0.02} \times \frac{(997 \text{ kg/m}^3) \times (0.032 \text{ m/s}^2)}{2}$$

$$= \mathbf{3.1 \text{ N/m}^2.} \quad \text{Ans.}$$

(g) The pumping power :

$$= \frac{\dot{m}}{\rho}\Delta p = \frac{(0.01 \text{ kg/s})}{(997 \text{ kg/m}^3)} \times (3.1 \text{ N/m}^2)$$

$$= 3.11 \times 10^{-5} \text{ W}$$

Actual power required

$$= \frac{3.11 \times 10^{-5} \text{ W}}{0.5} = \mathbf{6.22 \times 10^{-5} \text{ W.}} \quad \text{Ans.}$$

Example 9.6. *A water heater is fabricated by a resistance wire wound uniformly over a 10 mm diameter and 4 m long tube. The resistance element maintains a uniform heat flux of 1000 W/m². The mass flow rate of water is 12 kg/h, and its inlet temperature is 10°C. Estimate the surface temperature of the tube at exit.*

Solution

Given : A water heater

Fluid : water $\quad T_i = 10°C$

$\dot{m}$ = 12 kg/h, $\quad D = 10$ mm

$q = 1000 \text{ W/m}^2$, $\quad L = 4$ m.

To find : The exit surface temperature, $T_{s,o}$.

Assumptions :

1. Steady state conditions,
2. Constant properties.

Analysis : With a known heat flux, the local surface temperature at the exit of the tube can be obtained by

$$q = h_o(T_{s,o} - T_o) \quad \text{...}(i)$$

where h_o is the local heat transfer coefficient at exit. The exit bulk temperature is found from

$$Q = q\,(\pi\, DL) = \dot{m}\, C_p\,(T_o - T_i) \quad \text{...}(ii)$$

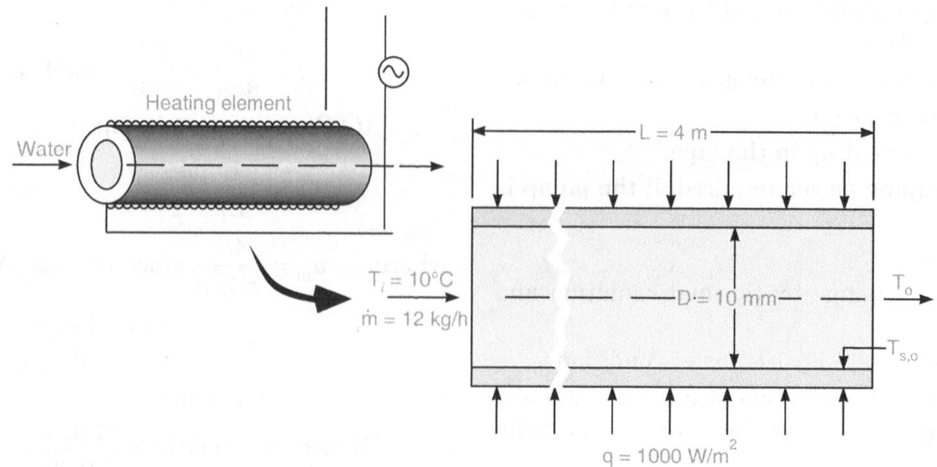

Fig. 9.15. Water at 10°C enters a 10 mm I.D. tube subjected to a uniform heat flux of 1000 W/m²

For obtaining fluid properties assume,

$$T_o = 20°C \quad \text{and} \quad T_m = 15°C$$
$$C_p = 4182 \text{ at } 15°C$$

Using eqn. (9.39) for determination of exit temperature

$$T_o = T_i + \frac{qA_s}{\dot{m}C_p}$$

with numerical values

$$T_o = 10 + \frac{1000 \times \pi \times 0.01 \times 4}{(12/3600) \times 4182}$$
$$= 10 + 9 = 19°C$$

It is very close to assumed value. With this exit temperature

$$T_m = 14.5°C \text{ and } C_p = 4187 \text{ J/kg. K.}$$

It also gives T_o = 19°C, the other properties of fluid at 14.5°C

$$\rho = 999 \text{ kg/m}^3, \qquad k_f = 0.595 \text{ W/m.K},$$
$$\mu = 1167 \times 10^{-6} \text{ kg/ms}, \quad Pr = 8.31.$$

$$Re_D = \frac{\rho u_m D}{\mu} = \frac{4\dot{m}}{\pi D \mu}$$
$$= \frac{4 \times (12/3600)}{\pi \times 0.01 \times 1167 \times 10^{-6}} = 363$$

which is less than 2300, thus the flow is laminar. The thermal entry length

$$x_{eth} = 0.05 \, Re_D Pr.D$$
$$= 0.05 \times 363 \times 8.31 \times 0.01 = 1.5 \text{ m}$$

Hence at exit of 4 m long tube, the flow is fully developed.

For constant wall heat flux, eqn. (9.60) gives

$$Nu = 4.36$$

and local heat transfer coefficient at exit

$$h_o = \frac{Nu_D \, k_f}{D} = \frac{4.36 \times 0.595}{0.01} = 259.42$$

and temperature of surface at the exit of tube

$$T_{s,o} = T_o + \frac{q}{h_o} = 19 + \frac{1000}{259.42}$$
$$= \mathbf{22.85°C.} \quad \textbf{Ans.}$$

Example 9.7. *The oil at 20°C flows at an average velocity of 2 m/s through a pipeline, 30 cm diameter. A 200 m long section of the pipeline passes through icy water of a lake at 0°C. The measurements reveal that the surface temperature of the pipe is very near to 0°C. Neglecting the thermal resistance of the pipe material, determine,*

(*i*) *Temperature of oil when pipe leaves the lake,*

(*ii*) *The rate of heat transfer from the oil, and*

(*iii*) *The pumping power required to overcome the pressure losses and to maintain the flow of oil in the pipe.*

Solution

Given : T_i = 20°C $\qquad u_m$ = 2 m/s

D = 30 cm = 0.3 m $\qquad$ L = 200 m

T_s = 0°C.

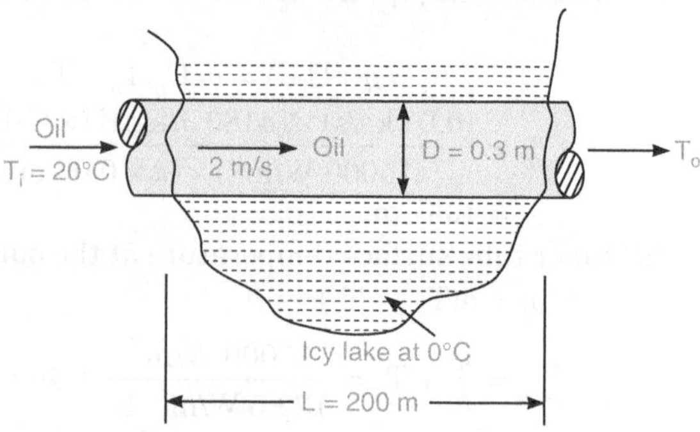

Fig. 9.16. Schematic of pipeline passing icy lake

To find :

(*i*) The temperature of oil leaves the lake.

(*ii*) Heat transfer rate to lake.

(*iii*) Pumping power to overcome the frictional losses.

Assumptions :

(*i*) Steady state conditions.

(*ii*) The fully developed flow.

(*iii*) Constant properties.

Analysis : The temperature of oil leaving the icy lake is not known, hence, the mean temperature of oil cannot be evaluated, we take properties of oil at 20°C from Table A-7

$\rho = 888$ kg/m³, $k_f = 0.145$ W/m.K,

$\mu = 0.8$ kg/m/s, $\nu = 901 \times 10^{-6}$ m²/s,

$C_p = 1880$ J/kgK, Pr = 10400,

$\mu_s = 3.85$ kg/m/s (at 0°C).

(*i*) The Reynolds number

$$\text{Re}_D = \frac{u_m D}{\nu} = \frac{2 \times 0.3}{901 \times 10^{-6}} = 666$$

which is less than 2300, thus flow is laminar. The entry length of tube in this case

$$x_{eth} = 0.05\,\text{Re}_D\,\text{Pr}.\,D = 0.05 \times 666 \times 10400 \times 0.3 = 103885 \text{ m}$$

which is much greater than the total length of pipe. We assume thermally developed flow and use eqn. (9.62)

$$\text{Nu}_D = 1.86\left(\text{Re}_D\,\text{Pr}\frac{D}{L}\right)^{1/3}\left(\frac{\mu}{\mu_s}\right)^{0.14}$$

$$= 1.86 \times \left(666 \times 10400 \times \frac{0.3}{200}\right)^{1/3} \times \left(\frac{0.8}{3.85}\right)^{0.14}$$

$$= 32.47$$

and

$$h = \frac{\text{Nu}_D\,k_f}{D} = \frac{32.47 \times 0.145}{0.3} = 15.70 \text{ W/m}^2\text{.K}$$

$$\dot{m} = \rho\,u_m\,A_c = \rho\,u_m\,\frac{\pi}{4}D^2$$

$$= 888 \times 2 \times \frac{\pi}{4} \times (0.3)^2 = 125.5 \text{ kg/s.}$$

The temperature of oil leaving icy lake can be obtained by eqn. (9.46)

$$T_o = T_s - (T_s - T_i)\exp\left(-\frac{h\,PL}{\dot{m}\,C_p}\right)$$

$$= 0 - (0 - 20)\exp\left(-\frac{15.70 \times \pi \times 0.3 \times 200}{125.5 \times 1880}\right)$$

$$= 19.75°\text{C}$$

which is very close to assumed value, and now

$$T_m = \frac{20 + 19.75}{2} = 19.875°\text{C}$$

The physical properties of oil hardly change for temperature difference of 0.125°C. We consider it as a temperature of oil leaving the lake. **Ans.**

(*ii*) The rate of heat transfer from oil can be calculated as

$$Q = \dot{m}\,C_p\,(T_i - T_o)$$

$$= 125.5 \times 1880 \times (20 - 19.75)$$

$$= \mathbf{58985\ W.}$$ **Ans.**

(*iii*) For laminar, hydrodynamically developed flow

$$f = \frac{64}{\text{Re}_D} = \frac{64}{666} = 0.0961$$

Fluid pressure drop in pipe

$$\Delta p = f\frac{L}{D}\frac{\rho u_m^2}{2}$$

$$\Delta p = 0.0961 \times \frac{200}{0.3} \times \frac{888 \times (2)^2}{2}$$

$$= 113782 \text{ Pa} = 113.78 \text{ kPa}$$

The pumping power required

$$\dot{W}_{pump} = \frac{\dot{m}}{\rho}\Delta p$$

$$\dot{W}_{pump} = \frac{125.5}{888} \times 113.78 = \mathbf{16.08\ kW.}$$ **Ans.**

9.7. FLOW INSIDE A NON-CIRCULAR DUCT

The friction factor, heat transfer coefficient and other quantities have been discussed so far for fully developed tube flow. Many engineering applications involve flow of fluid inside ducts of non-circular cross-section. All the expressions and charts (like Moody chart) are equally applicable to ducts of non-circular cross-section, if the tube diameter D is replaced by the **hydraulic diameter** D_h as

$$D_h = \frac{4A_c}{P} \quad \text{...(9.63)}$$

where A_c = cross-sectional area of flow, and

P = wetted perimeter

D_h = D for a circular tube,

since $A_c = (\pi/4)\,D^2$ and

$P = \pi D = 2w$ (= 2 width) for flow between two parallel plates

$= \dfrac{2ab}{(a+b)}$, for a rectangular duct of sides a and b.

Then the Reynolds number and Nusselt number for non-circular ducts are defined as

$$\text{Re} = \frac{u_m D_h}{\nu} \quad \text{...(9.64)}$$

$$\text{Nu} = \frac{h D_h}{k_f} \quad \text{...(9.65)}$$

For non-circular ducts, the turbulent flow occurs for Re > 2300. With non-circular ducts, the heat transfer coefficient varies around the perimeter and approaches zero near the sharp corners. Therefore, for certain situations, the approximation for non-circular ducts, by using hydraulic diameter concept is not proper.

The fully developed laminar flow equations for friction factor, heat transfer coefficient, etc., can readily be solved for any shape of cross-section. *Shah* and *London* discussed an outstanding number of such solution to almost every possible duct shape. A partial list of their solutions is presented in Table 9.1.

TABLE 9.1. Nusselt number and friction factor for fully developed laminar flow in tubes of various cross-sections

Cross-section of tube	*a/b or θ°*	*Nusselt number* $T_s = const.$	$q_s = const.$	*Friction factor f*	*Remark*
Circle	—	3.66	4.36	$\frac{64}{\text{Re}}$	$D_h = \frac{4A_c}{P}$
Hexagon	—	3.35	4.00	$\frac{60.2}{\text{Re}}$	$\text{Re} = \frac{u_m D_h}{\nu}$
Square	—	2.98	3.61	$\frac{56.92}{\text{Re}}$	$\text{Nu} = \frac{hD_h}{k_f}$
Rectangle	*a/b* 1	2.98	3.61	56.92/Re	
	2	3.39	4.12	62.20/Re	
	3	3.96	4.79	68.36/Re	
	4	4.44	5.33	72.92/Re	
	6	5.14	6.05	78.80/Re	
	8	5.60	6.49	82.32/Re	
	∞	7.54	8.24	96.00/Re	
Ellipse	*a/b* 1	3.66	4.36	64.00/Re	
	2	3.74	4.56	67.28/Re	
	4	3.79	4.88	72.96/Re	
	8	3.72	5.09	76.60/Re	
	16	3.65	5.18	78.16/Re	
Triangle	θ 10°	1.61	2.45	50.80/Re	
	30°	2.26	2.91	52.28/Re	
	60°	2.47	3.11	53.32/Re	
	90°	2.34	2.98	52.60/Re	
	120°	2.00	2.68	50.96/Re	

Example 9.8. *Engine oil at 60°C flows at 0.5 kg/s in a duct with constant surface temperature of 20°C. Assuming fully developed flow, calculate (i) heat flux at entry (ii) pressure drop per metre length for 3 cm diameter tube and for a 3 × 1 rectangular duct of equal wall area.*

Solution

Given : Flow of oil in a duct at constant surface temperature

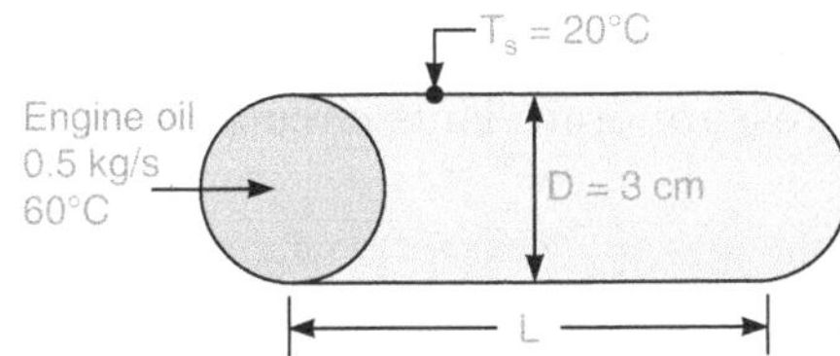

Fig. 9.17. (*a*) Flow through circular tube

T_i = 60°C, $\dot{m}$ = 0.5 kg/s,

T_s = 20°C.

To find :

(*a*) For tube (D = 0.03 m), flow

(*i*) heat flux, and (*ii*) pressure drop per metre length,

(*b*) For rectangular duct (3 × 1),

(*i*) heat flux, and (*ii*) pressure drop per metre length.

Assumptions :

1. Steady state conditions,

2. Constant properties.

Analysis : The exit temperature of fluid is not known. The flow is fully developed and we consider that the fluid exit temperature will be close to

$$T_s = 20°C,$$

Thus $T_m = \frac{60+20}{2} = 40°C$

The properties of engine oil, at 40°C from Table A-5

ρ = 876 kg/m^3, k_f = 0.144 W/m.K,

ν = 0.24 × 10^{-3} m^2/s. C_p = 1.96 kJ/kg.K.

(*a*) For flow through tube of D = 0.03 m

$$D_h = D = 0.03 \text{ m}$$

The velocity of fluid

$$u_m = \frac{\dot{m}}{\rho A_c} = \frac{\dot{m}}{\rho \frac{\pi}{4} D^2} = \frac{0.5}{876 \times \frac{\pi}{4} \times (0.03)^2}$$

$$= 0.8074 \text{ m/s}$$

The Reynolds number of tube flow

$$Re_D = \frac{u_m D}{\nu} = \frac{0.8074 \times 0.03}{0.24 \times 10^{-3}} = 101$$

Re < 2300. The flow is definitely laminar.

(*i*) With assumption of fully developed flow in tube of constant surface temperature. From Table 9.1

$$Nu = 3.66$$

The heat transfer coefficient

$$h = \frac{Nu\, k_f}{D} = \frac{3.66 \times 0.144}{0.03}$$

$$= 17.57 \text{ W/m}^2\text{.K.}$$

The heat flux at entry of tube

$$q = h(T_s - T_i) = 17.57 \times (20 - 60)$$

$$= \mathbf{-\ 702.7\ W/m^2}. \quad \text{Ans.}$$

(*ii*) The friction factor

$$f = \frac{64}{Re} = \frac{64}{101} = 0.633$$

The pressure drop per metre length, eqn. (9.21)

$$\frac{\Delta p}{L} = \frac{f}{D} \cdot \frac{\rho u_m^2}{2} = \frac{0.633}{0.03} \times \frac{876 \times (0.8074)^2}{2}$$

$$= \mathbf{6031\ Pa/m.} \quad \text{Ans.}$$

(*b*) Rectangular duct (3 × 1) with same wall area

For duct of side *a*

$$A_s = PL = 2(a + 3a)\, L = \pi\, DL$$

or $$8a = \pi \times 0.03 \times 1$$

or $$a = 0.0118 \text{ m} = 1.18 \text{ mm}$$

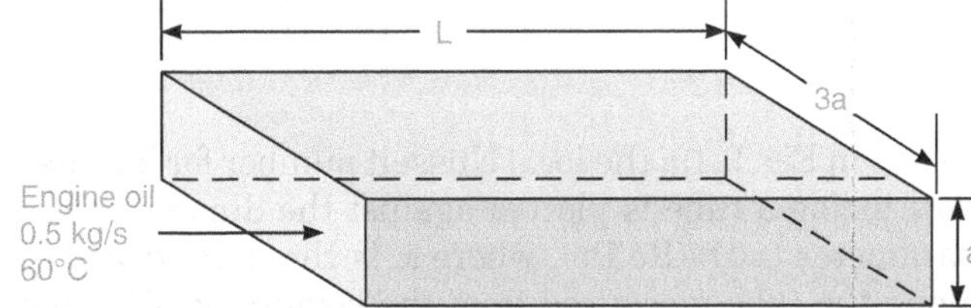

Fig. 9.17. (*b*) Flow through rectangular (3 × 1) duct

Its hydraulic diameter,

$$D_h = \frac{4A_c}{P} = \frac{4(a \times 3a)}{2(a+3a)} = \frac{12a}{8}$$
$$= 1.5 \times 0.0118 = 0.0177 \text{ m}$$

The fluid velocity

$$u_m = \frac{\dot{m}}{\rho A_c} = \frac{\dot{m}}{\rho(3a^2)} = \frac{0.5}{876 \times 3 \times (0.0118)^2}$$
$$= 1.367 \text{ m/s}$$

The corresponding Reynolds number

$$Re_{D_h} = \frac{u_m D_h}{\nu} = \frac{1.367 \times 0.0177}{0.24 \times 10^{-3}} = 100.77$$

The flow is again laminar, for fully developed flow through constant temperature duct. From Table 9.1 for $\frac{a}{b} = 3$.

$$Nu = 3.96 \quad \text{and} \quad f = \frac{68.36}{Re_{D_h}}$$

Thus $$h = \frac{Nu\, k_f}{D_h} = \frac{3.96 \times 0.144}{0.0177} = 32.2 \text{ W/m}^2.\text{K}$$

(*i*) The heat flux at the entry of duct

$$q = h(T_s - T_i) = 32.2 \times (20 - 60)$$
$$= -\mathbf{1288.7\ W/m^2}. \quad \textbf{Ans.}$$

(*ii*) The friction factor

$$f = \frac{68.36}{100.77} = 0.678$$

The pressure drop per metre

$$\frac{\Delta p}{L} = \frac{f}{D_h}\frac{\rho u_m^2}{2} = \frac{0.678}{0.0177} \times \frac{876 \times (1.367)^2}{2}$$
$$= \mathbf{31352\ Pa/m}. \quad \textbf{Ans.}$$

Comparison : The rectangular duct has more heat flux and pressure drop. For the tube

$$\frac{q}{(\Delta p/L)} = \frac{702.7}{6031} = 0.1164 \text{ W/Pa.m}$$

For rectangular duct

$$\frac{q}{\Delta p/L} = \frac{1288.7}{31352} = 0.0411 \text{ W/Pa.m}$$

or $$\frac{0.1164 - 0.411}{0.1164} \times 100 = 65\%$$

The rectangular duct requires 65% more power for same pressure drop. Thus circular duct is effective.

9.8. THERMALLY DEVELOPING, HYDRODYNAMICALLY DEVELOPED LAMINAR FLOW

Consider a fluid flow inside a duct as shown in Fig. 9.18. There is an isothermal section, in which velocity boundary layer develops. Then fluid enters into heat transfer zone, where thermal boundary layer begins to develop. The region of thermal entry length x_{eth} is called the *hydrodynamically developed, thermally developing region.* It is the representation of physical situation for fluids such as oils, that have a large Prandtl number, for which the hydrodynamic entry length is smaller than the thermal entry length. A classic solution for laminar forced convection inside a circular tube, in thermally developing region for constant wall temperature and constant wall heat flux was given by *Graetz* and the results are shown in Fig. 9.19.

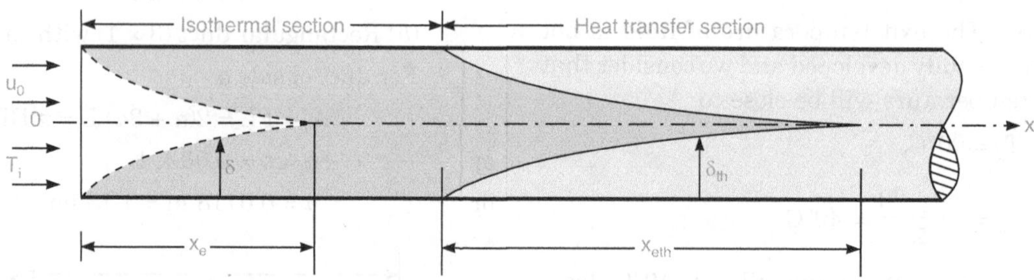

Fig. 9.18. Hydrodynamically developed, thermal developing region concept

In Fig. 9.19, the local Nusselt number for laminar flow inside a tube is plotted against the dimensionless parameter $(x/D)/(Re.Pr)$, where x is the axial distance along the tube measured from the beginning of heated section. The inverse of this dimensionless parameter is called the Graetz number, Gz :

$$(Gz)^{-1} = \frac{x/D}{Re.\,Pr} = \frac{x/D}{Pe} \qquad ...(9.66)$$

where Pe = the Peclet number, a product of Reynolds and Prandtl numbers.

D = inside diameter of the tube

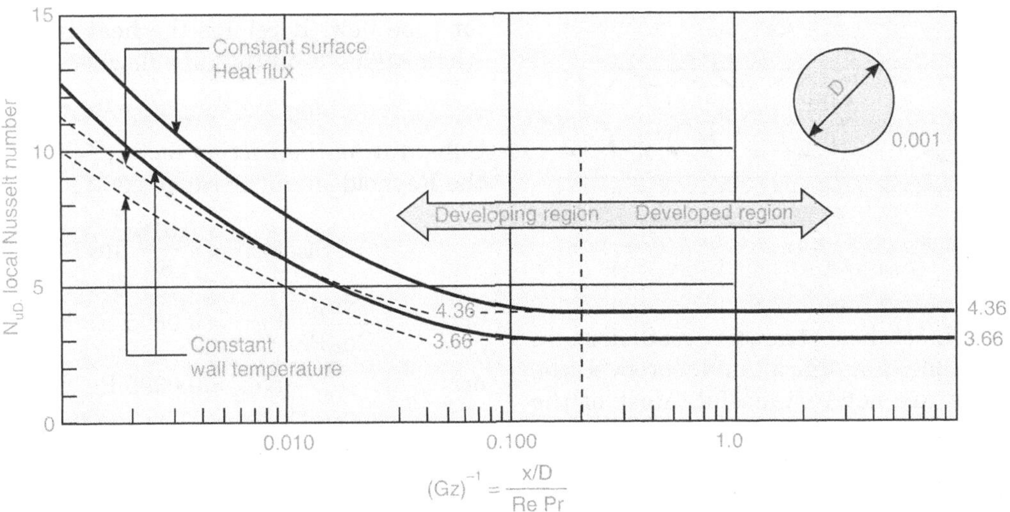

Fig. 9.19. Local Nusselt number for thermally developing, hydrodynamically developed laminar flow inside a circular tube

For an isothermal wall, *Hausen* suggested an empirical relation for Nusselt number in thermally developing region.

$$\text{Local} \quad Nu_D = 3.66 + \frac{1 + 0.0018\, Gz^{1/3}}{(0.04 + Gz^{-2/3})^2} \qquad ...(9.67)$$

$$\text{Mean} \quad Nu = 3.66 + \frac{0.668\, Gz}{1 + 0.04(Gz)^{2/3}} \qquad ...(9.68)$$

where $Gz = \frac{Re\ Pr}{L/D}$,

L = distance from the inlet ...(9.69)

The eqns. (9.67) and (9.68) are valid for Gz < 100 and all properties are evaluated at fluid bulk mean temperature. As length L increases, Nu approaches the asymptotic value 3.66.

A rather simple empirical correlation has been suggested by *Sieder* and *Tate* to predict the mean Nusselt number in the thermally developing region for laminar flows, for constant wall temperature,

$$Nu_D = 1.86\, (Gz)^{1/3} \left(\frac{\mu}{\mu_s}\right)^{0.14} \qquad ...(9.70)$$

It is recommended for

$$0.48 < Pr < 16{,}700$$

$$0.0044 < \frac{\mu}{\mu_s} < 9.75$$

$$(Gz)^{1/2} \left(\frac{\mu}{\mu_s}\right)^{0.14} > 2$$

$$T_s = \text{constant.}$$

All the fluid properties are evaluated at fluid bulk mean temperature except μ_s, which is evaluated at T_s.

9.9. HEAT TRANSFER IN TURBULENT FLOW INSIDE A CIRCULAR TUBE

The fully developed laminar flow has limited applications to engineering problems. However, the turbulent flow is more commonly used in practice because of the high heat transfer coefficient associated with it.

A qualitative illustration for turbulent behaviour can be understood by schematic shown in Fig. 9.20. At Reynolds number above 4000, the flow inside a tube becomes fully turbulent, except for a very thin layer of fluid adjacent to the wall, called the *viscous (laminar) sublayer*. The turbulent eddies are damped in the viscous sublayer as a result of viscous forces, and therefore, the heat transfer through this layer is mainly by conduction. The flow beyond this layer is turbulent in central core. The turbulent eddies sweep the edges of layer and carry along them the fluid at layer temperature. The eddies mix the hotter and colder fluids effectively and the heat is transferred rapidly between viscous sublayer and turbulent bulk of the fluid. It is evident that the thermal resistance of viscous layer controls the rate of heat transfer, because most of the temperature drop takes place across it, while turbulent portion of flow field offers a little resistance to heat flow. Thus the heat transfer in the turbulent flow is composite of heat transfer in viscous sublayer and turbulent core and it increases with increase in Reynolds number.

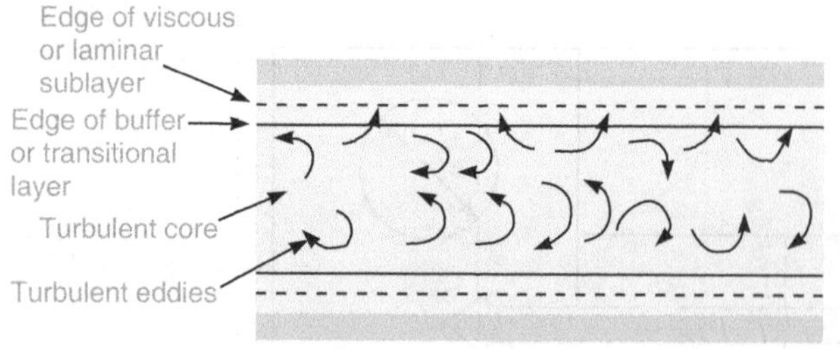

Fig. 9.20. Flow structure for a fluid in turbulent flow through a pipe

The analysis of heat transfer coefficient in turbulent flow is somewhat difficult and moreover, the theoretical results are not very useful. Most of the correlations for the friction factor and heat transfer coefficient are based on experimental studies. An analogy between heat and momentum transfer is discussed below.

9.9.1. Analogy between Heat and Momentum Transfer in Turbulent Flow through Tube

The analogy between heat and momentum transfer for turbulent flow inside a circular tube is very similar to that for laminar flow over a flat plate and given by eqn. (9.71).

$$\mathrm{St}_x \, \mathrm{Pr}^{2/3} = \frac{\mathrm{C}_{fx}}{2} \qquad \ldots(9.71)$$

The definition of skin coefficient C_f can be obtained from eqn.(9.16)

$$\frac{\mathrm{C}_f}{2} = \frac{\tau_s}{\rho\, u_m^{\,2}}$$

and the definition of friction factor f is given by eqn. (9.18)

$$f = 4\mathrm{C}_f$$

$$\therefore \qquad \frac{\mathrm{C}_f}{2} = \frac{f}{8} = \frac{\tau_s}{\rho u_m^{\,2}} \qquad \ldots(9.72)$$

Substituting C_f from eqn. (9.72) into eqn. (9.71), it takes the form for flow in tube as

$$\mathrm{St_D}\, \mathrm{Pr}^{2/3} = \frac{f}{8} \qquad \ldots(9.73)$$

It is called the **Chilton Colburn analogy** for *turbulent* flow inside a smooth tube. The dimensionless quantity $\mathrm{St_D}$ is the Stanton number for tube flow, given as

$$\mathrm{St_D} = \frac{\mathrm{Nu_D}}{\mathrm{Re_D\, Pr}} = \frac{h}{\rho\, u_m \mathrm{C}_p} \qquad \ldots(9.74)$$

For gases with Prandtl number very close to unity, (Pr = 1), the eqn. (9.73) reduces to

$$\mathrm{St_D} = \frac{f}{8} \qquad \ldots(9.75)$$

The eqn. (9.75) is called the **Reynolds analogy** for tube flow. It relates the heat transfer rate to the frictional losses in the tube flow for gases with $\mathrm{Pr} \approx 1$.

9.9.2. Correlation for Turbulent Flow

Substituting the friction factor, f from eqn. (9.19), for the Reynolds number range $2300 \le \mathrm{Re_D} \le 1 \times 10^4$

$$\mathrm{St_D}\, \mathrm{Pr}^{2/3} = \frac{0.316}{8}\, \mathrm{Re}^{-1/4}$$

or
$$\frac{\mathrm{Nu_D}}{\mathrm{Re_D Pr}}\, \mathrm{Pr}^{2/3} = 0.0395\, \mathrm{Re}^{-1/4}$$

or
$$\mathrm{Nu_D} = 0.0395\, \mathrm{Re_D}^{3/4}\, \mathrm{Pr}^{1/3} \qquad \ldots(9.76)$$

A similar equation exists by substituting f from eqn. (9.20) for the $\mathrm{Re} > 2 \times 10^4$

$$\mathrm{Nu_D} = 0.023\, \mathrm{Re_D}^{0.8}\, \mathrm{Pr}^{1/3} \qquad \ldots(9.77)$$

which is known as the **Colburn equation.** The accuracy of this equation is improved by Dittus and Boelter by modifying it as

$$\mathrm{Nu_D} = 0.023\, \mathrm{Re}^{0.8}\, \mathrm{Pr}^{n} \qquad \ldots(9.78)$$

where n = 0.4 for heating of fluid ($\mathrm{T}_s > \mathrm{T}_m$)

= 0.3 for cooling of fluid ($\mathrm{T}_s < \mathrm{T}_m$).

The eqn. (9.78) is known as the **Dittus-Boelter equation,** and it is preferred to Colburn equation. Its validity for

$$0.7 \le \mathrm{Pr} \le 160;\ \mathrm{Re_D} \ge 10{,}000 \quad \text{and} \quad \frac{\mathrm{L}}{\mathrm{D}} \ge 10$$

This equation should be used only for small to moderate temperature difference ($\mathrm{T}_s - \mathrm{T}_m$), with all properties are evaluated at mean temperature of fluid *i.e.,*

$$\mathrm{T}_m = \frac{\mathrm{T}_i + \mathrm{T}_o}{2}$$

In situation of large variation in the properties the *Sieder-Tate equation* should be used

$$\mathrm{Nu} = 0.027\, \mathrm{Re}^{0.8}\, \mathrm{Pr}^{1/3} \left(\frac{\mu}{\mu_s}\right)^{0.14} \qquad \ldots(9.79)$$

Its validity is $0.7 \le \mathrm{Pr} \le 16{,}700$

$$\mathrm{Re_D} \ge 10{,}000$$

$$\frac{\mathrm{L}}{\mathrm{D}} \ge 10$$

All the fluid properties are evaluated at mean fluid temperature except μ_s, which is evaluated at wall temperature T_s.

A correlation similar to eqn. (9.78), but restricted to gases was proposed by *Kays* and *London* for long ducts

$$\mathrm{Nu_D} = \mathrm{C}\, \mathrm{Re_D^{0.8}}\, \mathrm{Pr}^{0.3} \left(\frac{\mathrm{T}_m}{\mathrm{T}_s}\right)^{n} \qquad \ldots(9.80)$$

where all the fluid properties are evaluated at bulk or mean fluid temperature T_m. The constant C and exponent n are

$$C = \begin{cases} 0.020 & \text{for uniform surface temperature,} \\ 0.021 & \text{for uniform heat flux} \end{cases}$$

$$n = \begin{cases} 0.575 & \text{for heating} \\ 0.150 & \text{for cooling} \end{cases}$$

The eqns. (9.76), (9.77), (9.78) and (9.79) have significant uncertainties as high as 20%. For internal turbulent flows, the correlations suggested by *Gnielinsky* have much better accuracy with uncertainties upto 6% only. The correlations are

$$Nu_D = 0.0214\,(Re^{0.8} - 100)\,Pr^{0.4}\left[1 + \left(\frac{D}{L}\right)^{2/3}\right] \quad ...(9.81)$$

All the fluid properties at T_m are valid for

$0.5 < Pr < 1.5, \quad 2300 < Re_D < 10^6,$

and $0 < \dfrac{D}{L} < 1.$

and

$$Nu_D = 0.012\,(Re_D^{0.87} - 280)\,Pr^{0.4}\left[1 + \left(\frac{D}{L}\right)^{2/3}\right] \quad ...(9.82)$$

The fluid properties at T_m are valid for

$1.5 < Pr < 500, \quad 2300 < Re_D < 10^6$

and $0 < D/L < 1$

The *Gnielinsky* eqns. (9.81) and (9.82) can be expressed in composite form and can be used for rough surface tubes.

$$Nu_D = \frac{(f/8)\,(Re_D - 1000)\,Pr}{1 + 12.7\,(f/8)^{1/2}\,(Pr^{2/3} - 1)}\left[1 + \left(\frac{D}{L}\right)^{2/3}\right] \quad ...(9.83)$$

valid for $0 < D/L < 1$, $0.6 < Pr < 2000$, and $Re_D > 2300$, where the friction factor f is obtained either by using the Moody chart (Fig. 9.5) or the *Colebrook* formula to evaluate the friction factor,

$$f^{-1/2} = 1.74 - 2\log_{10}\left\{\left(\frac{\varepsilon}{r_o}\right) + \frac{1.87}{Re_D\sqrt{f}}\right\} \quad ...(9.84)$$

where ε = average surface asperty height.

In order to improve the accuracy by accounting the variation of fluid properties due to temperature change, *Gnielinsky* suggested that Nu_D should be multiplied by

$$\left(\frac{T_m}{T_s}\right)^{0.45} \text{ for gases and } \left(\frac{Pr}{Pr_s}\right)^{0.11} \text{ for liquids.}$$

The absolute temperature T_s and Pr_s should be evaluated at surface temperature T_s.

More complex correlations have been proposed by Petukhov and Popov, and Sleicher and Rouse. Their results are presented in Table 9.2.

TABLE 9.2. Heat transfer correlations for liquids and gases in incompressible fluid flow through tubes

Name of equation	*Formula*[a]	*Conditions for use*
Dittus-Boelter	$Nu_D = 0.023\,Re_D^{0.8}\,Pr^n$ $n \begin{cases} = 0.4 \text{ for heating} \\ = 0.3 \text{ for cooling} \end{cases}$	$0.7 < Pr < 160$ $6000 < Re_D < 10^7$
Sieder-Tate	$Nu_D = 0.027\,Re_D^{0.8}\,Pr^{0.3}\left(\dfrac{\mu}{\mu_s}\right)^{0.14}$	$6000 < Re_D < 10^7$ $0.7 < Pr < 10^4$
Petukhov-Popov	$Nu_D = \dfrac{(f/8)\,Re_D\,Pr}{K_1 + K_2\,(f/8)^{1/2}\,(Pr^{2/3} - 1)}$ where $f = (1.82\log_{10} Re_D - 1.64)^{-2}$ $K_1 = 1 + 3.4f$ $K_2 = 11.7 + \dfrac{1.8}{Pr^{1/3}}$	$0.5 < Pr < 2000$ $10^4 < Re_D < 5 \times 10^6$
Sleicher-Rouse	$Nu_D = 5 + 0.015\,Re_D^{a}\,Pr_s^{b}$ where $a = 0.88 - \dfrac{0.24}{4 + Pr_s}$ $b = 1/3 + 0.5e^{-0.6\,Pr_s}$	$0.1 < Pr < 10^5$ $10^4 < Re_D < 10^6$

"All properties are evaluated at the bulk fluid temperature T_m except properties with subscript, they are evaluated at surface temperature T_s."

Fig. 9.21 shows a comparison of these equations with experimental data at Pr = 0.6 for water at 26.7°C.

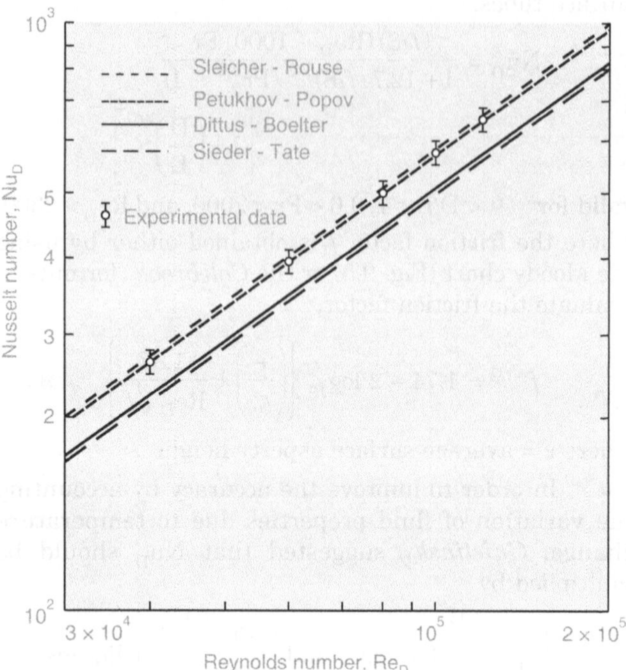

Fig. 9.21. Comparison of predicted and measured Nusselt number for turbulent flow of water in a tube (26.7°C ; Pr = 6.0)

Example 9.9. *Steam is generated on the surface of tubes (surrounded by water) with pressurized water flowing inside the tubes of a heat exchanger. At a particular section, the velocity of the water in the tubes is 3 m/s. The inside diameter of the tubes is 25 mm and the tube surfaces are at 250°C. Find the convective heat transfer coefficient by different correlations at a section where the bulk temperature of the pressurized water is 280°C. This section is 2.5 m from the entrance of the water to the tube.*

Solution

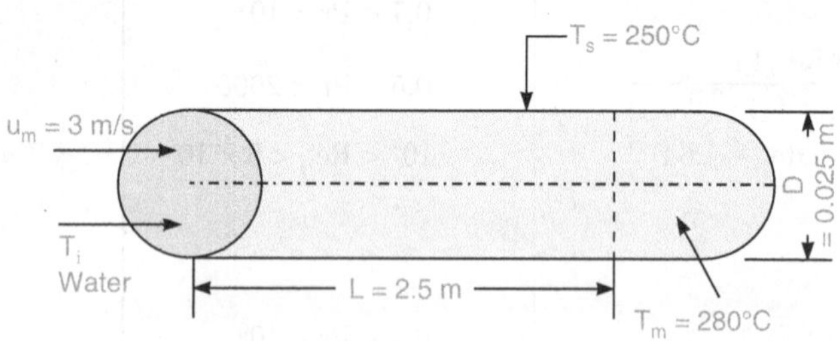

Fig. 9.22. Schematic for example 9.9

Given : For given section

u_m = 3 m/s, T_m = 280°C, T_s = 250°C,

D = 25 mm, L = 2.5 m.

To find : The heat transfer coefficient by using different correlations.

Assumptions :

(*i*) Steady state conditions,

(*ii*) Conduction and radiation effects are negligible,

(*iii*) Constant Properties.

Analysis : The properties of water at T_m = 280°C from Table A-7 ;

$\rho = 756$ kg/m³, $C_p = 5.24$ kJ/kg.K,

$\mu = 97 \times 10^{-6}$ kg/ms, $k_f = 0.580$ W/m.K,

Pr = 0.87, $\mu_s = 106 \times 10^{-6}$ kg/ms (at 250°C).

Reynolds number

$$Re_D = \frac{\rho u_m D}{\mu} = \frac{756 \times 3 \times 0.025}{97 \times 10^{-6}} = 584536$$

Re > 2300, thus the flow is turbulent.

(*i*) Using Colburn eqn. (9.77)

$$Nu_D = 0.023 Re_D^{0.8} \, Pr^{1/3}$$

$$= 0.023 \times (584536)^{0.8} \times (0.87)^{1/3} = 901$$

Average heat transfer coefficient

$$h = \frac{Nu_D k_f}{D} = \frac{901 \times 0.580}{0.025}$$

$$= \mathbf{20917\ W/m^2.K.} \quad \text{Ans.}$$

(*ii*) Using Dittus-Boelter eqn. (9.78), with $n = 0.3$ for cooling ;

$$Nu_D = 0.023 \, Re_D^{0.8} \, Pr^{0.3} = 906$$

and h = **21015 W/m².K.** Ans.

(*iii*) Using Sieder-Tate eqn. (9.79)

$$Nu_D = 0.027 \, Re_D^{0.8} \, Pr^{1/3} \left(\frac{\mu}{\mu_s}\right)^{0.14}$$

$$= 0.027 \times (584536)^{0.8} \times (0.87)^{1/3} \times \left(\frac{97 \times 10^{-6}}{106 \times 10^{-6}}\right)^{0.14} = 1045$$

and h = **24252 W/m².K.** Ans.

(*iv*) Using Gnielinsky eqn. (9.81) for 0 < Pr < 1.5

$$Nu_D = 0.0214 \, (Re_D^{0.8} - 100) \, Pr^{0.4} \left[1 + \left(\frac{D}{L}\right)^{2/3}\right]$$

$$= 0.0214 \times [(584536)^{0.8} - 100) \times (0.87)]^{0.4} \times \left[1 + \left(\frac{0.025}{2.5}\right)^{2/3}\right] = 867.6$$

and h = **20128 W/m².K.** Ans.

The Colburn, Dittus-Boelter and Gnielinsky equations give almost similar result, except, the result obtained by Sieder-Tate equation which is over predicted.

Example 9.10. *Determine the Nusselt number for water flowing at an average velocity of 3 m/s in an annulus formed between a 25 mm -OD tube and a 38 mm in-ID tube. The water enters at 100°C and is being cooled. The temperature of the inner wall is 50°C, and the outer wall of the annulus is insulated. Neglect entrance effects and compare the results obtained from all four equations in Table 9.2. The properties of water are given below :*

T (°C)	*μ, kg/ms*	*k_f, W/m.K*	*ρ, kg/m³*	*C_p, J/kg.K*	*Pr*
50	555.1 × 10⁻⁶	0.647	988.1	4178	3.55
75	376.6 × 10⁻⁶	0.671	974.9	4190	2.23
100	277.5 × 10⁻⁶	0.682	958.4	4211	1.71

Solution

Given : Flow of water through an annulus space between two pipes as shown in Fig. 9.23

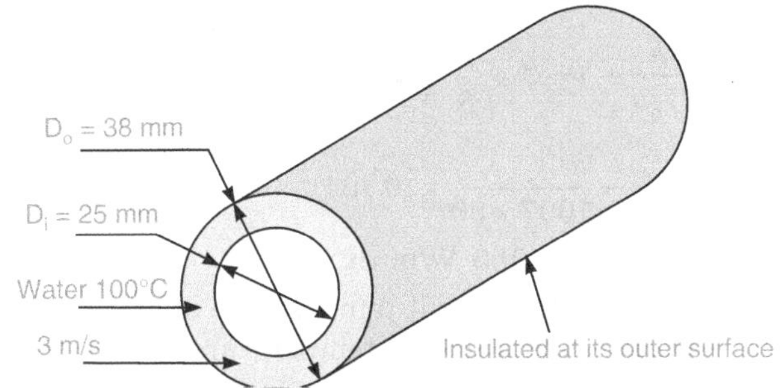

Fig. 9.23. Water flow through annulus space

$T_i = 100°C$, $u_m = 3$ m/s,

$T_s = 50°C$, $D_i = 25$ mm $= 0.025$ m,

$D_o = 38$ mm $= 0.038$ m.

To find : The Nusselt number for water, by using

(*i*) Dittus-Boelter equation,

(*ii*) Sieder-Tate equation,

(*iii*) Petukhov-Popov equation, and

(*iv*) Sleicher-Rouse equation.

Analysis : The hydraulic diameter for annulus space

$$D_h = \frac{4A_c}{P} = \frac{4 \times (\pi/4)(D_o^2 - D_i^2)}{\pi(D_o + D_i)} = D_o - D_i$$

$$= 38 \text{ mm} - 25 \text{ mm} = 13 \text{ mm} = 0.013 \text{ m}.$$

Reynolds number based on D_h and bulk temperature properties at 100°C

$$Re = \frac{\rho u_m D_h}{\mu} = \frac{958.4 \times 3 \times 0.013}{277.5 \times 10^{-6}} = 134694$$

The Prandtl number

$$Pr = \frac{\mu C_p}{k_f} = \frac{277.5 \times 10^{-6} \times 4211}{0.682} = 1.71$$

(*i*) The Nusselt number by using Dittus-Boelter equation for cooling ($n = 0.3$)

$$Nu = 0.023\, Re^{0.8}\, Pr^{0.3}$$

$$= 0.023 \times (134694)^{0.8} \times (1.71)^{0.3}$$

$$= \mathbf{342.85.} \quad \text{Ans.}$$

(*ii*) The Nusselt number by using Sieder-Tate equation

$$Nu = 0.027\, Re^{0.8}\, Pr^{0.3} \left(\frac{\mu}{\mu_s}\right)^{0.14}$$

$$= 0.027 \times (134694)^{0.8} \times (1.71)^{0.3}$$

$$\times \left(\frac{275.5 \times 10^{-6}}{555.1 \times 10^{-6}}\right)^{0.14} = \mathbf{365.} \quad \text{Ans.}$$

(*iii*) The Nusselt number by using Petukhov-Popov equation

$$f = (1.82 \log_{10} Re - 1.64)^{-2}$$

$$= [1.82 \log_{10} \times (134694) - 1.64]^{-2} = 0.0168$$

$$K_1 = 1 + 3.4 f = 1 + 3.4 \times 0.0168 = 1.0573$$

$$K_2 = 11.7 + \frac{1.8}{Pr^{0.33}} = 11.7 + \frac{1.8}{(1.71)^{0.33}} = 13.2$$

$$Nu = \frac{(f/8) Re\, Pr}{K_1 + K_2 (f/8)^{1/2} (Pr^{0.67} - 1)}$$

$$= \frac{(0.0168/8) \times 134694 \times 1.71}{1.0573 + 13.2 \times (0.0168/8)^{1/2} \times [(1.71)^{0.67} - 1]}$$

$$= \mathbf{370.} \quad \text{Ans.}$$

(*iv*) The Nusselt number by using Sleicher-Rouse equation

$$\text{Nu} = 5 + 0.015\ \text{Re}^a\ \text{Pr}_s^{\ b}$$

$$a = 0.88 - \frac{0.24}{4 + \text{Pr}_s} = 0.88 - \frac{0.24}{4 + 3.55} = 0.848$$

$$b = \frac{1}{3} + 0.5\ e^{-0.6\ \text{Pr}_s} = \frac{1}{3} + \frac{0.5}{e^{0.6 \times 3.55}} = 0.392$$

$$\text{Nu} = 5 + 0.015 \times (134694)^{0.848} \times (3.55)^{0.392}$$

$$= \mathbf{556.4}. \quad \text{Ans.}$$

It is over prediction.

Example 9.11. *Air flows with 30 m/s velocity through a tube of 2 cm dia. and 1 m length. The air inlet temperature is 20°C and its pressure is 101.3 kPa. The pressure loss in the tube is 80 mm of water column. How much heat is transferred from the wall to the air, when the wall is kept at 95°C. Use Chilton Colburn analogy.*

Solution

Given : Flow of air through a tube.

To find : The heat transfer rate from tube to air.

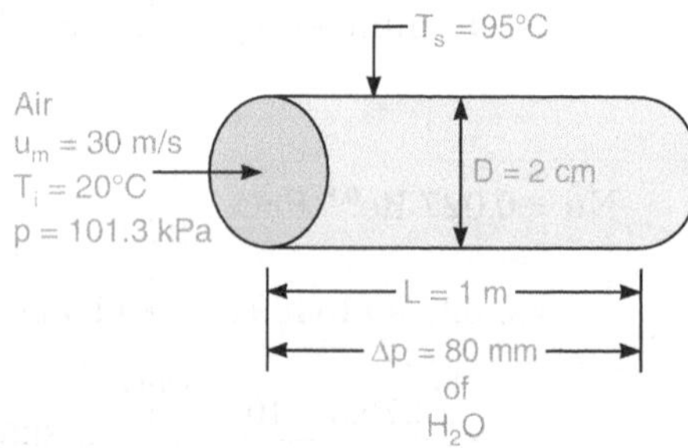

Fig. 9.24. Schematic for example 9.11

Assumptions :

(*i*) Steady state conditions.

(*ii*) Fully developed flow.

(*iii*) Density of water as 1000 kg/m³.

Analysis : The pressure drop in N/m² (Pa)

$$\Delta p = (\rho g h)_{H_2O}$$

$$= 1000 \times 9.81 \times \frac{80}{1000} = 784.8 \text{ Pa}$$

The exit temperature is unknown, the mean fluid temperature cannot be obtained, and thus appropriate physical properties of air.

Assuming exit temperature of air to be 66°C

$$T_m = \frac{66 + 20}{2} = 43°\text{C} = 316 \text{ K}$$

The physical properties of air (from Table A-7)

$\rho = 1.112$ kg/m³, $C_p = 1007$ J/kg.K,

$\mu = 191.68 \times 10^{-7}$ Ns/m²

$k_f = 27.41 \times 10^{-3}$ W/m.K, Pr = 0.704

The friction factor is obtained by eqn. (9.21)

$$\Delta p = f \frac{L}{D} \frac{\rho u_m^{\ 2}}{2}$$

or

$$f = \frac{2\Delta p D}{\rho u_m^{\ 2} L} = \frac{2 \times 784.8 \times 0.02}{1.112 \times (30)^2 \times 1} = 0.0313$$

The Reynolds number

$$\text{Re}_D = \frac{\rho u_m D}{\mu} = \frac{1.112 \times 30 \times 0.02}{191.68 \times 10^{-7}} = 34808$$

which is greater than 2300, thus the flow is turbulent, using Chilton Colburn analogy

$$\frac{h}{\rho C_p u_m} \text{Pr}^{2/3} = \frac{f}{8} \left\{ \text{where} \quad st_D = \frac{h}{\rho C_p u_\infty} \right\}$$

$$\frac{h}{1.112 \times 1007 \times 30} \times (0.704)^{2/3} = \frac{0.0313}{8}$$

It gives $h = 166$ W/m².K.

Now calculating exit temperature of air by using eqn. (9.46) for constant wall temperature

$$T_o = T_s - (T_s - T_i) \exp\left(-\frac{hPL}{\dot{m} C_p}\right)$$

where

$$\dot{m} = \rho u_m \left(\frac{\pi}{4}\right) D^2$$

$$= 1.112 \times 30 \times \left(\frac{\pi}{4}\right) \times (0.02)^2 = 0.010 \text{ kg/s}$$

Then
$$T_o = 95 - (95 - 20) \exp\left(-\frac{166 \times (\pi \times 0.02) \times 1}{0.010 \times 1007}\right)$$

$$= 68.38°\text{C}$$

which is very close to assumed value of 66°C. Thus keeping it 68.38°C as exit temperature of air for any further calculations.

Heat transfer rate by pipe = Heat gain by air

$$Q = \dot{m} C_p (T_o - T_i)$$

$$= 0.010 \times 1007 \times (68.38 - 20)$$

$$= \mathbf{487\ W}. \quad \text{Ans.}$$

Example 9.12. *Water at 50°C enters 1.5 cm diameter and 3 m long tube with a velocity of 1.5 m/s. The tube wall is maintained at 100°C. Calculate the heat transfer coefficient and total amount of heat transferred if the water exit temperature is 70°C.*

Solution

Given : Flow through tube with.

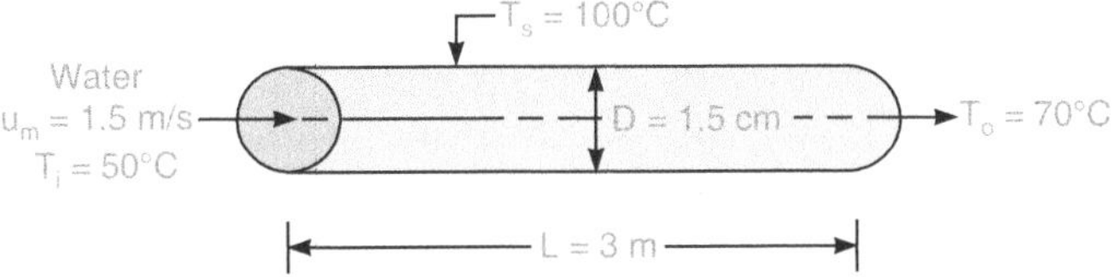

Fig. 9.25. Schematic for example 9.12

To find :

(*i*) The heat transfer coefficient.

(*ii*) Heat transfer rate.

Properties : The mean temperature

$$T_m = \frac{50 + 70}{2} = 60°C.$$

The properties of water at 60°C (from Table A-7)

$\rho = 990$ kg/m³, $\nu = 0.517 \times 10^{-6}$ m²/s,
$C_p = 4184$ J/kg.K, $k_f = 0.65$ W/m.K,
Pr = 3.15.

Analysis : The Reynolds number

$$Re_D = \frac{u_m D}{\nu} = \frac{(1.5 \text{ m/s}) \times (0.015 \text{ m})}{(0.517 \times 10^{-6} \text{ m}^2/\text{s})} = 43520$$

$Re_D > 2300$, hence flow is turbulent.

(*i*) Using Dittus Boelter equation, for heating of water

$$Nu_D = 0.023 Re_D^{0.8} Pr^{0.4}$$
$$= 0.023 \times (43520)^{0.8} \times (3.15)^{0.4} = 187$$

$$h = \frac{Nu_D \, k_f}{D} = \frac{187 \times 0.65}{0.015}$$

$= \mathbf{8106}$ **W/m².K. Ans.**

$(\Delta T)_{lm}$ log mean temperature difference is calculated as

$$(\Delta T)_{lm} = \frac{\Delta T_i - \Delta T_o}{\ln\left(\frac{\Delta T_i}{\Delta T_o}\right)}$$

$$= \frac{(100-50)-(100-70)}{\ln\left(\frac{100-50}{100-70}\right)} = 39.15°C$$

(*ii*) The rate of heat transfer:

$$Q = h(\pi DL)(\Delta T)_{lm}$$
$$= 8106 \times (\pi \times 0.015 \times 3) \times 39.15$$

$= 44867$ W $\approx$ **44.86 kW. Ans.**

Example 9.13. *Water at 20°C enters a 2 cm diameter tube with a velocity of 1.5 m/s. The tube is maintained at 100°C. Find the tube length required to heat the water to a temperature of 60°C.* (Anna Univ, March 2001)

Solution

Given : Flow through tube.

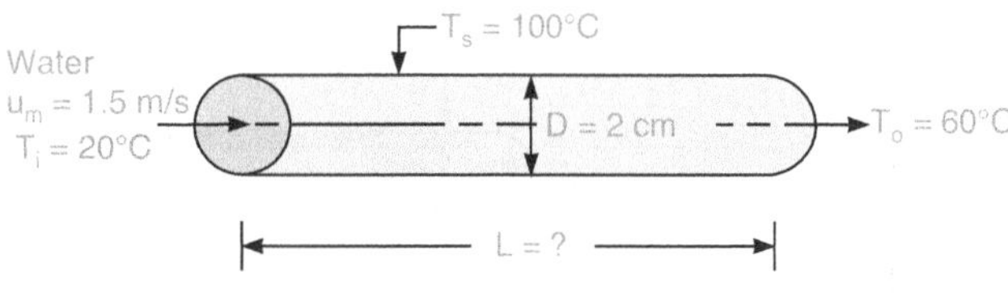

Fig. 9.26. Schematic for tube flow

To find : Length of tube to exit temperature at 60°C.

Properties : The mean water temperature

$$T_m = \frac{T_i + T_o}{2} = \frac{20 + 60}{2} = 40°C$$

The properties of water at 40°C from Table A-7

Pr = 4.31, $\rho = 992.2$ kg/m³,
$C_p = 4174$ J/kg.K $k_f = 0.634$ W/m.K,
$\nu = 0.659 \times 10^{-6}$ m²/s.

Analysis : The Reynolds number of fluid flow

$$Re_D = \frac{u_m D}{\nu} = \frac{(1.5 \text{ m/s}) \times (0.02 \text{ m})}{(0.659 \times 10^{-6} \text{ m}^2/\text{s})} = 45523.5$$

$Re_D > 2300$, flow is turbulent and Dittus Boelter equation can be used to obtain Nu_D.

$$Nu_D = 0.023 \, Re_D^{0.8} \, Pr^{0.4}$$
$$= 0.023 \times (45523.5)^{0.8} \times (4.31)^{0.4} = 219.84$$

And the heat transfer coefficient.

$$h = Nu_D \times \frac{k_f}{D} = 219.84 \times \frac{0.634}{0.02}$$
$$= 6969 \text{ W/m}^2.\text{K}$$

The mass flow rate;

$$\dot{m} = \rho u_m A_c = \rho u_m (\pi/4).\, D^2$$
$$= 992.2 \times 1.5 \times (\pi/4) \times (0.02 \text{ m})^2$$
$$= 0.467 \text{ kg/s}$$

Using the relation (9.48) for constant tube wall temperature in the form

$$L = \frac{\dot{m} C_p}{\pi D h} \times \ln\left(\frac{T_s - T_i}{T_s - T_o}\right)$$

$$L = \frac{0.467 \times 4174}{(\pi \times 0.02) \times 6969} \times \ln\left(\frac{100-20}{100-60}\right)$$

= **3.08 m. Ans.**

Example 9.14. *A water heater consists of a 25 mm diameter tube inside a second coaxial tube. Water at 10°C enters the inner tube at 0.8 kg/s. Condensing steam in the annulus maintains the temperature of the inner tube at 90°C.*

(i) Determine the exit temperature and the heat transfer rate, if the tube is 10 m long.

(ii) What should be the length of the tube for the exit temperature of the water to be 70°C ?

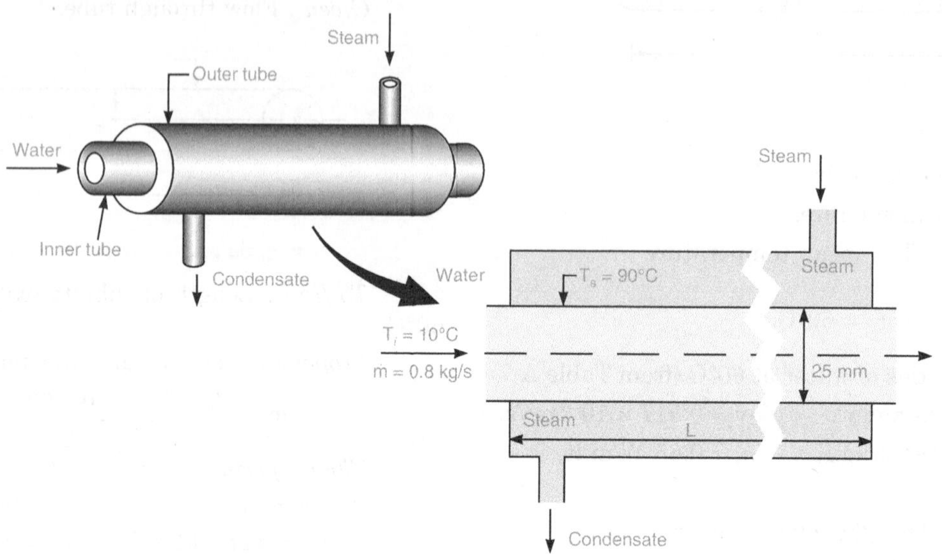

Fig. 9.27. Schematic of coaxial tube for example 9.14

Solution

Given : Fluid flow through a water heater

$T_i = 10°C$

$\dot{m} = 0.8$ kg/s, $\quad T_s = 90°C$

$D = 25$ mm $= 0.025$ m.

To find :

(*i*) Exit temperature, T_o and heat transfer rate, if L = 10 m.

(*ii*) Length of the tube for exit temperature $T_o = 70°C$.

Assumptions :

1. Steady state conditions,
2. Heat transfer begins at inlet to the tube,
3. Entrance length is small compared with the total length of the tube, and the correlations for fully developed conditions are applicable,
4. The convective heat transfer coefficient, determined at the mean of the bulk temperatures at inlet and exit, is uniform.

Analysis : (*i*) Since the *exit temperature of water* is unknown, it is difficult to predict the mean temperature of fluid, and thus calculation of heat transfer coefficient.

Assuming water exit temperature to be 70°C, then

$$T_m = \frac{10+70}{2} = 40°C,$$

and properties of water from Table A-7

$\rho = 992.2$ kg/m^3, $\quad C_p = 4175$ J/kg.K,

Pr = 4.19, $\quad \mu = 633.7 \times 10^{-6}$ kg/ms,

$k_f = 0.631$ W/m.K.

The Reynolds number for flow inside a tube

$$Re_D = \frac{\rho u_m D}{\mu} = \frac{4\dot{m}}{\pi D \mu}$$

$$= \frac{4 \times 0.8}{\pi \times 0.025 \times 633.7 \times 10^{-6}} = 64295$$

The $Re_D > 2300$, the flow is turbulent, thus using Dittus-Boelter equation for heating of water ($n = 0.4$)

$$Nu_D = 0.023\ Re^{0.8}\ Pr^{0.4}$$

$$= 0.023 \times (64295)^{0.8} \times (4.19)^{0.4} = 286.5$$

and
$$h = \frac{Nu_D\ k_f}{D} = \frac{286.5 \times 0.631}{0.025}$$

$$= 7232 \text{ W/m}^2\text{.K}$$

For constant surface temperature, the exit temperature of water can be determined from eqn. (9.46)

$$T_o = T_s - (T_s - T_i)\exp\left(-\frac{hPL}{\dot{m}C_p}\right)$$

$$= 90 - (90 - 10)$$

$$\times \exp\left(-\frac{7232 \times \pi \times 0.025 \times 10}{0.8 \times 4175}\right) = 75.4°C$$

which is very close to assumed value, an improvement can be made by assuming $T_o = 75.4°C$, and $T_m = 42.7°C$, we get

$$h = 7406 \quad \text{and} \quad C_p = 4174 \text{ J/kg.K}$$

T_o = **76°C.** Ans.

The heat transfer rate

$$Q = \dot{m}C_p(T_o - T_i) = 0.8 \times 4174 \times (76 - 10)$$

= **220387.2 W.** Ans.

(*ii*) Length of heat exchanger for $T_o = 70°C$

The heat transfer rate to water

$$Q = \dot{m}C_p(T_o - T_i) = 0.8 \times 4175 \times (70 - 10)$$
$$= 200400 \text{ W}$$

The heat transfer coefficient as determined above for assumed exit temperature of 70°C.

$$h = 7232 \text{ W/m}^2\text{.K.}$$
$$\Delta T_1 = 90 - 10 = 80°C$$
$$\Delta T_2 = 90 - 70 = 20°C$$

$$\Delta T_{lm} = \frac{\Delta T_1 - \Delta T_2}{\ln\left(\frac{\Delta T_1}{\Delta T_2}\right)} = \frac{80 - 20}{\ln\left(\frac{80}{20}\right)} = 43.2°C$$

Then the length of tube can be determined by eqn. (9.50)

$$Q = hA_s\,\Delta T_{lm} = h(\pi DL)\,\Delta T_{lm}$$

or
$$L = \frac{200400}{7232 \times (\pi \times 0.025) \times 43.2}$$

= **8.15 m.** Ans.

Example 9.15. *Air at 200 kPa and 200°C is heated as it flows through a tube with a diameter of 25 mm at a velocity of 10 m/s. Calculate the heat transfer rate per unit length of the tube, if a constant heat flux condition is maintained at the wall and the wall temperature is 20°C above the air temperature, all along the length of the tube. How much would the bulk temperature increase over 3 m length of the tube ?*

Solution

Given : Uniform heating of the tube:

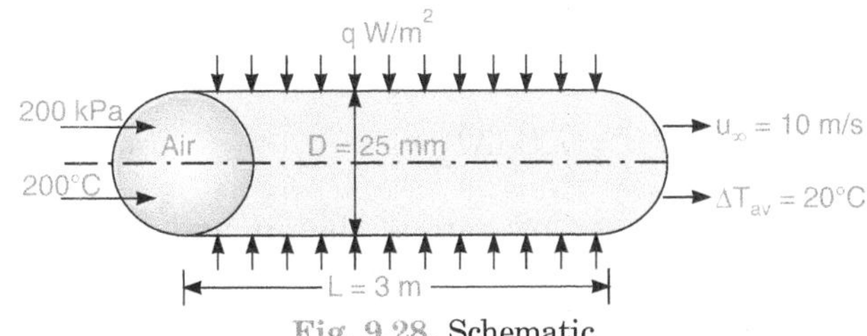

Fig. 9.28. Schematic

To find :

(*i*) Heat transfer rate per unit length of the tube.

(*ii*) Bulk temperature rise over 3 m length of the tube.

Assumptions :

1. Steady state heat transfer conditions.
2. Fully developed flow through a tube.
3. Conduction and radiation effects are negligible.

Properties of air : The properties of air at temperature of 200°C (from Table A-4)

$$C_p = 1.025 \text{ kJ/kg.K}, \qquad k_f = 0.0386 \text{ W/m.K},$$
$$\mu = 2.57 \times 10^{-5} \text{ kg/ms}, \quad Pr = 0.681.$$

Analysis : The density of air at 200 kPa and 200°C is

$$\rho = \frac{p}{RT} = \frac{200 \text{ kPa}}{(0.287 \text{ kJ/kg.K}) \times (473 \text{ K})}$$
$$= 1.473 \text{ kg/m}^3$$

The Reynolds number of flow is

$$Re_D = \frac{\rho u_m D}{\mu} = \frac{1.473 \times 10 \times 0.025}{2.57 \times 10^{-5}} = 14332$$

Since $Re_D > 2300$, hence the flow is turbulent. Using Dittus Boelter equation ;

$$Nu_D = \frac{hD}{k_f} = 0.023\, Re_D^{0.8}\, Pr^{0.4}$$
$$= 0.023\,(14332)^{0.8} \times (0.681)^{0.4} = 41.69$$

or
$$h = \frac{k_f}{D} Nu_D = \frac{0.0386}{0.025} \times 41.69$$
$$= 64.37 \text{ W/m}^2\text{.K.}$$

(*i*) The heat transfer rate per metre length :

$$\frac{Q}{L} = h(\pi D)\Delta T_{av}$$
$$= 64.37 \times (\pi \times 0.025) \times (20)$$

= **101.1 W/m.** Ans.

(*ii*) Bulk temperature rise :

Making the energy balance over 3 m length of the tube ;

Heat supply rate = Enthalpy rise rate of the fluid

$$\frac{Q}{L} \times L = \dot{m}\, C_p (\Delta T_m)$$

where $\dot{m}$ is mass flow rate of the air and it can be calculated by Continuity equation,

$$\dot{m} = \rho u_m A_c = \rho u_m \frac{\pi D^2}{4}$$

$$= (1.473) \times (10) \times \frac{\pi \times (0.025)^2}{4}$$

$$= 7.329 \times 10^{-3} \text{ kg/s}$$

Using mass flow rate in the energy balance

$$101.1 \times 3 = 7.329 \times 10^{-3} \times (1025) \times (\Delta T_m)$$

or $\Delta T_m = \mathbf{40.37°C.}$ **Ans.**

Example 9.16. *Water at 20°C is to be heated by passing it through the tube. Surface of the tube is maintained at 90°C. The diameter of the tube is 4 cm, while its length is 9.0 m. Find the mass flow rate so that the exit temperature of the water will be 60°C.*

The properties of water are :

$\rho = 995 \text{ kg/m}^3$, $C_p = 4.174 \text{ kJ/kg.K}$,

$k_f = 0.64 \text{ W/m.K}$, $\nu = 0.62 \times 10^{-6} \text{ m}^2\text{/s}$,

$\beta = 4.25 \times 10^{-3} \text{ K}^{-1}$.

Use relation $Nu = 0.023\, Re^{0.8}\, Pr^{0.4}$.

(N.M.U., May 1998)

Solution

Given : The flow through tube as shown in Fig. 9.29.

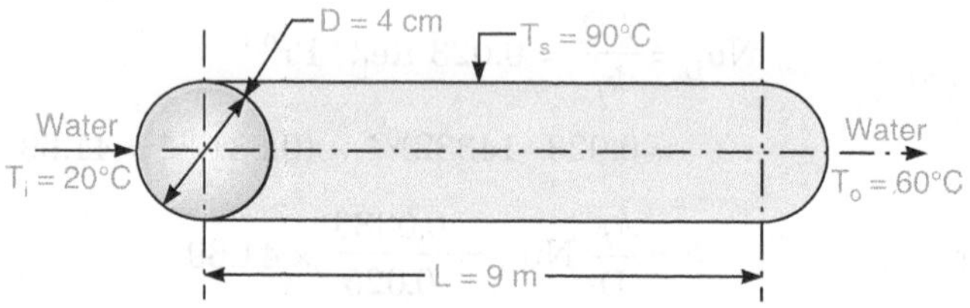

Fig. 9.29. Schematic

To find : The mass flow rate of water.

Assumptions :

1. Steady state heat transfer conditions,
2. Fully developed turbulent flow,
3. Conduction and radiation effects are negligible.

Analysis : The Reynolds number of fluid flow is

$$Re = \frac{u_m D}{\nu} = \frac{4\dot{m}}{\pi D\mu} = \frac{4\dot{m}}{\pi D\nu\rho}$$

$$= \frac{4\dot{m}}{\pi \times 0.04 \times 0.62 \times 10^{-6} \times 995}$$

$$= 51598.3\, \dot{m}$$

And $$Pr = \frac{\mu C_p}{k_f} = \frac{\nu \rho C_p}{k_f}$$

$$= \frac{0.62 \times 10^{-6} \times 995 \times 4174}{0.64}$$

$$= 4.02$$

Using relation,

$$Nu = \frac{hD}{k_f} = 0.023\, Re^{0.8}\, Pr^{0.4}$$

or $$h = 0.023 \times (51598.3\, \dot{m})^{0.8} (4.02)^{0.4} \times \frac{0.64}{0.04}$$

or $$h = 0.023 \times 5890 \times 1.7451\, (\dot{m})^{0.8} \times 16$$

$$= 3781.3\, (\dot{m})^{0.8} \qquad \ldots(i)$$

Further using the eqn. (9.48) in the form

$$h = \frac{\dot{m}\, C_p}{\pi D L} \times \ln\left[\frac{T_s - T_i}{T_s - T_o}\right]$$

or $$h = \frac{\dot{m} \times 4174}{\pi \times 0.04 \times 9} \times \ln\left[\frac{90 - 20}{90 - 60}\right]$$

or $$h = 3127\, \dot{m} \qquad \ldots(ii)$$

Equating eqns. (*i*) and (*ii*) ;

$$3127\, \dot{m} = 3781.3\, (\dot{m})^{0.8}$$

or $$(\dot{m})^{0.2} = \frac{3781.3}{3127} = 1.2096$$

or $\dot{m} = \mathbf{2.59\ kg/s.}$ **Ans.**

Checking the validity of assumed turbulent flow.

$$Re_D = \frac{4\dot{m}}{\pi D \rho \nu} = \frac{4 \times 2.59}{\pi \times 0.04 \times 995 \times 0.62 \times 10^{-6}}$$

$$= 1.32 \times 10^5$$

The flow is turbulent, and above calculations are valid.

Example 9.17. *A rectangular tube, 30 mm × 50 mm carries water at a rate of 2 kg/s. Determine the length of tube required to heat water from 30°C to 50°C, if the wall temperature is maintained at 90°C.*

Use following properties of water at 40°C

$\rho = 992.2 \text{ kg/m}^3$, $k_f = 0.634 \text{ W/m.K}$

$C_p = 4.174 \text{ J/kg.K}$ $\mu = 6.531 \times 10^{-4} \text{ kg/ms}$.

(P.U., Nov. 2001)

Solution

Given : A rectangular tube wall at constant temperature.

$\dot{m}$ = 2 kg/s, T_s = 90°C,

T_i = 30°C, T_o = 50°C,

A_c = 30 mm × 50 mm.

To find : Length of the tube.

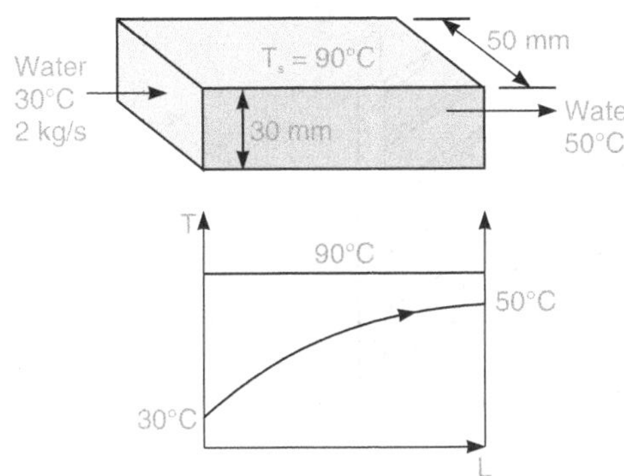

Fig. 9.30. Schematic for example 9.17

Analysis : The cross-section area of tube

$$A_c = 0.03 \text{ m} \times 0.05 \text{ m} = 0.0015 \text{ m}^2$$

Perimeter, $P = 2 \times (0.03 + 0.05) = 0.16$ m

Surface area,

$$A_s = PL = 0.16 \text{ L}$$

Hydraulic diameter of the rectangular tube

$$D_h = \frac{4A_c}{P} = \frac{4 \times 0.0015}{0.16} = 0.0375 \text{ m}$$

The length of the tube can be determined by using eqn. (9.50)

$$Q = hA_s\Delta T_{lm} \quad ...(i)$$

where

$$Q = \dot{m} C_p(T_o - T_i)$$

$$= 2 \times 4.174 \times (50 - 30)$$

$$= 166.96 \text{ kW} = 166960 \text{ W}$$

$$\Delta T_1 = 90 - 30 = 60°C,$$

$$\Delta T_2 = 90 - 50 = 40°C$$

$$\Delta T_{lm} = \frac{\Delta T_1 - \Delta T_2}{\ln\left(\dfrac{\Delta T_1}{\Delta T_2}\right)} = \frac{60 - 40}{\ln\left(\dfrac{60}{40}\right)} = 49.32°C$$

Further, the h is not available and is to calculate. The velocity of water through rectangular tube:

$$\dot{m} = \rho A_c u_m$$

$$u_m = \frac{2}{992.2 \times 0.0015} = 1.343 \text{ m/s} \quad ...(ii)$$

Reynolds number

$$Re = \frac{\rho u_m D_h}{\mu}$$

$$= \frac{992.2 \times 1.343 \times 0.0375}{6.531 \times 10^{-4}} = 76557$$

Re > 2300, thus flow is turbulent,

Prandtl number

$$Pr = \frac{\mu C_p}{k_f} = \frac{6.531 \times 10^{-4} \times 4174}{0.634} = 4.3$$

Using Dittus Boelter Equation

$$Nu = \frac{hD_h}{k_f} = 0.023 \text{ Re}^{0.8} \text{ Pr}^{0.4}$$

or

$$h = \frac{0.634}{0.0375} \times 0.023 \times (76557)^{0.8} \times (4.3)^{0.4}$$

$$= 5628 \text{ W/m}^2\text{.K.}$$

Using numerical values in eqn. (*i*)

$$166960 = 5628 \times 0.16 \text{ L} \times 49.32$$

or L = **3.76 m.** **Ans.**

Example 9.18. *A square channel of side 15 mm and length 2.0 m is a heat transfer problem carries water at a velocity of 6 m/s. The mean temperature of water along the length of channel is found to be 30°C, while the inner channel surface temperature is 70°C. Calculate the heat transfer coefficient from channel wall to the water and heat transfer rate from channel to water. Use correlation*

$$Nu = 0.021 \, Re^{0.8} \, Pr^{0.43} \left(\frac{Pr}{Pr_s}\right)^{0.25}$$

The thermophysical properties are evaluated at mean bulk temperature except Pr_s, which is evaluated at channel surface temperature. Take equivalent diameter as characteristic length of channel. The properties of water are :

$\rho = 995.7 \text{ kg/m}^3$, $k_f = 0.6175$ W/m.K,

$\nu = 0.805 \times 10^{-6} \text{ m}^2/\text{s}$, $Pr = 5.42$ and

$Pr_s = 2.55$ at 70°C. (P.U., May 2000)

Solution

Given : Flow of water through a square channel.

To find :

(*i*) The heat transfer coefficient between channel wall to flowing water.

(*ii*) Heat transfer rate from channel to water.

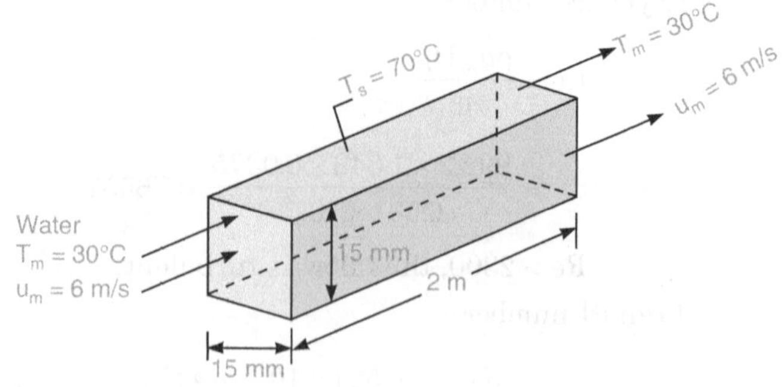

Fig. 9.31. Schematic of flow through square duct

Assumptions :

(*i*) Steady state conditions,

(*ii*) Constant properties,

(*iii*) Smooth surface of channel.

(*iv*) No radiation heat transfer.

Analysis : (*i*) For square channel, the hydraulic diameter

$$D_h = \frac{4A_c}{P} = \frac{4 \times 0.015 \times 0.015}{2 \times (0.015 + 0.015)} = 0.015 \text{ m}$$

The Reynolds number

$$Re = \frac{u_m D_h}{\nu} = \frac{6 \times 0.015}{0.805 \times 10^{-6}} = 111801.2$$

Using given correlation

$$Nu = 0.021 \times (111801.2)^{0.8} \times (5.42)^{0.43} \times \left(\frac{5.42}{2.55}\right)^{0.25} = 573.4$$

The heat transfer coefficient

$$h = \frac{Nu\, k_f}{D_h} = \frac{573.4 \times 0.6175}{0.015} = \mathbf{23605\ W/m^2.K.}$$

(*ii*) Heat transfer rate

$$Q = h\,(PL)\,(T_s - T_\infty)$$
$$= 23605 \times (4 \times 0.015 \times 2) \times (70 - 30)$$
$$= \mathbf{113305\ W.}$$ **Ans.**

Example 9.19. *In the manufacturing of modern computers, one of the limiting factors is the rise in the temperature of the chips due to internal energy generation in the chips. The reliability of the chips decreases rapidly when the temperature goes above a certain value, typically between 85°C and 100°C. Some high capacity components may require high thermal dissipation rate, which may reach 50 W/cm² and, in some cases, even 200 W/cm².*

One of the methods to cool a computer is to mount the circuit boards on a metal plate. An array of such metal plates is attached to two end-plates, which are cooled by a liquid.

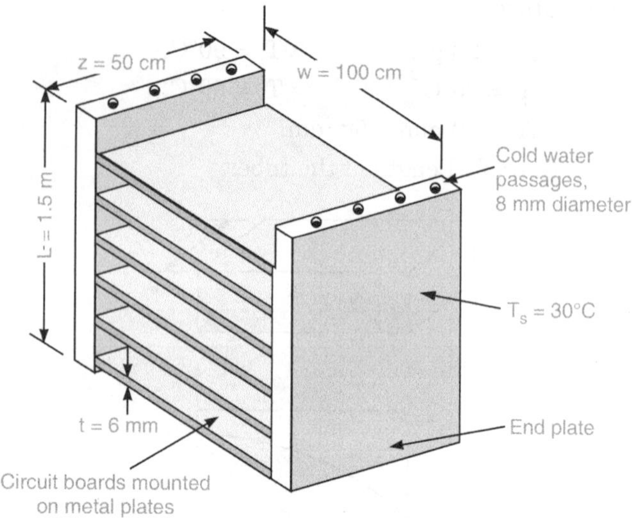

Fig. 9.32. Cold water is circulated through passages in the end plates of a computer to cool the circuit boards mounted on metal plates attached to the end plate

Consider a plate 100 cm wide, 50 cm deep, and 6 mm thick on which a circuit board is mounted (Fig. 9.32). Several such plates are attached to two heavy end plates through each of which four, 8 mm diameter passages are drilled. Cold water (with an additive to suppress the freezing point) flows through the passages and cools the end plates, which, in turn, cools the plates on which the circuit boards are mounted as shown in Fig. 9.32. Water enters each passage at 0°C at 0.15 kg/s. The end plates, which are 1.5 m high, are at 30°C. Estimate the heat transfer rate from the end plates to the cooling water.

Solution

Given : Cooling arrangement for computer chips

$T_i = 0°C$, $T_s = 30°C$, $t = 6$ mm

$\dot{m} = 0.15$ kg/s, $L = 1.5$ m, $w = 100$ cm

$D = 8$ mm, $N_p = 8$ (passages), $z = 50$ cm

To find : Heat transfer rate to the cooling water.

Assumptions :

(*i*) Steady state conditions.

(*ii*) Constant Properties.

(*iii*) Fully developed flow of water in the tubes.

Analysis : The exit temperature of water is unknown and thus mean temperature of water cannot be calculated for obtaining physical properties.

Let us assume $T_o = 14°C$

$$T_m = \frac{0 + 14}{2} = 7°C = 280 \text{ K}$$

The thermophysical properties of water at 280 K from Table A-7

$\rho = 1000$ kg/m^3, $C_p = 4198$ J/kg.K,

$\mu = 1422 \times 10^{-6}$ kg/ms,

$k_f = 0.582$ W/m.K, Pr = 10.26

The Reynolds number for tube flow

$$Re_D = \frac{4\dot{m}}{\pi D \mu}$$

$$= \frac{4 \times 0.15}{\pi \times 8 \times 10^{-3} \times 1422 \times 10^{-6}} = 16788$$

It is greater than 2300, thus flow is turbulent, and Nusselt number for heating of water by Dittus-Boelter equation

$$Nu_D = 0.023\ Re_D^{0.8}\ Pr^{0.4}$$

$$= 0.023 \times (16788)^{0.8} \times (10.26)^{0.4} = 140$$

The heat transfer coefficient

$$h = \frac{Nu_D\ k_f}{D} = \frac{140 \times 0.582}{8 \times 10^{-3}}$$

$$= 10152 \text{ W/m}^2.\text{K}$$

Before calculation of heat transfer rate, checking exit temperature of water by using eqn. (9.46)

$$T_o = T_s - (T_s - T_i) \exp\left(-\frac{hPL}{\dot{m}\, C_p}\right)$$

$$= 30 - (30 - 0) \times \exp\left(-\frac{10152 \times (\pi \times 8 \times 10^{-3}) \times 1.5}{0.15 \times 4198}\right)$$

$$= 13.86°\text{C}$$

which is very close to assumed value of 14°C, thus taking 14°C as exit temperature of water, the heat transfer rate

$$Q = N_p\ \dot{m}\, C_p (T_o - T_i)$$

$$= 8 \times 0.15 \times 4198 \times (14 - 0)$$

= **70526 W.** Ans.

Example 9.20. *A water heater uses 15 mm diameter copper pipe with a mean roughness height of 0.075 mm, which is heated electrically to a constant surface temperature of 95°C. The water enters the pipe at 0.1 kg/s and at a temperature of 10°C. What is the length of the pipe required to achieve the water exit temperature of 75°C ? Compare the result with length required for a perfectly smooth pipe.*

Solution

Given : Flow of water through a rough tube

$\varepsilon = 0.075$ mm, D = 15 mm = 0.015 m,

$T_s = 95°$C, $\dot{m} = 0.1$ kg/s,

$T_i = 10°$C, $T_o = 75°$C.

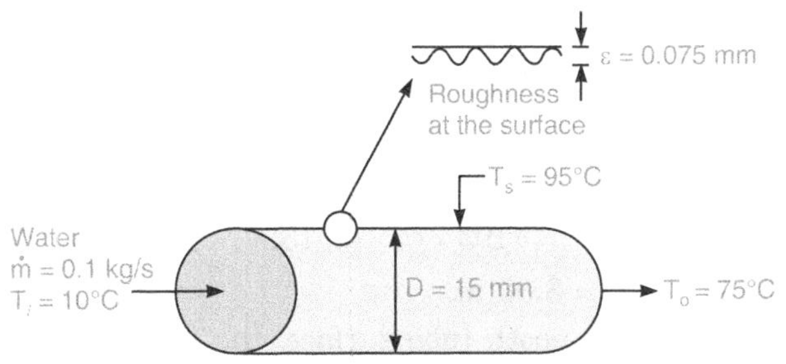

Fig. 9.33. Schematic

To find : Length of tube for,

(*i*) Rough tube surface,

(*ii*) Smooth tube surface.

Assumptions :

(*i*) Steady state conditions,

(*ii*) Constant properties at T_m,

(*iii*) Fully developed flow,

(*iv*) Heating starts as fluid enters the tube.

Analysis. The mean temperature of fluid

$$T_m = \frac{T_i + T_o}{2} = \frac{10 + 75}{2} = 42.5°\text{C} \approx 315 \text{ K}$$

The properties of water, at 315 K from Table A-7

$\rho = 991$ kg/m^3, $C_p = 4179$ J/kg.K,

$\mu = 631 \times 10^{-6}$ kg/ms, $k_f = 0.634$ W/m.K,

Pr = 4.16.

The Reynolds number

$$Re_D = \frac{4\dot{m}}{\pi D \mu} = \frac{4 \times 0.1}{\pi \times 0.015 \times 631 \times 10^{-6}} = 13452$$

It is greater than 2300, thus the flow is turbulent, from Moody diagram, Fig. 9.5.

$$\left.\begin{array}{l} \frac{\varepsilon}{D} = \frac{0.075}{15} = 0.005 \\ Re_D = 13452 \end{array}\right\} f = 0.036$$

(*i*) For rough tube surface,

Using *Gnielinsky* eqn. (9.83) to predict the Nusselt number

$$Nu_D = \frac{(f/8)\{Re_D - 1000\}\, Pr}{1 + [12.7\sqrt{(f/8)}\,(Pr^{2/3} - 1)]}$$

$$= \frac{(0.036/8)\{13452 - 1000\} \times 4.16}{1 + [12.7 \times \sqrt{(0.036/8)} \times \{(4.16)^{2/3} - 1\}]}$$

$$= 99.12$$

The heat transfer coefficient

$$h = \frac{\text{Nu}_D\, k_f}{D} = \frac{99.12 \times 0.634}{0.015}$$

$$= 4193 \text{ W/m}^2.\text{K}$$

The length of the tube can be determined by using eqn. (9.48) in the form

or $$L = -\frac{\dot{m}\, C_p}{hP} \ln\left(\frac{T_s - T_o}{T_s - T_i}\right)$$

$$= -\frac{0.1 \times 4179}{4193 \times (\pi \times 0.015)} \times \ln\left(\frac{95-75}{95-10}\right)$$

= **3.06 m.** Ans.

(*ii*) For smooth tube surface, the Dittus-Boelter equation can be used to predict Nu (n = 0.4 for heating)

$$\text{Nu}_D = 0.023\, \text{Re}_D^{0.8}\, \text{Pr}^{0.4}$$

$$= 0.023 \times (13452)^{0.8} \times (4.16)^{0.4} = 81.73$$

$$h = \frac{81.73 \times 0.634}{0.015} = 3454.56 \text{ W/m}^2.\text{K}$$

and $$L = -\frac{0.1 \times 4179}{3454.56 \times \pi \times 0.015} \times \ln\left(\frac{95-75}{95-10}\right)$$

= **3.71 m.** Ans.

Example 9.21. *1000 kg/h of cream cheese at 15°C is pumped through 1.5 m length of 8 cm inner diameter tube, which is maintained at 95°C. Estimate the temperature of cheese leaving the heated section.*

The thermophysical properties of cheese are

$$\rho = 1150 \text{ kg/m}^3, \quad C_p = 2750 \text{ J/kg.K},$$

$$\mu = 22.5 \text{ kg/ms}, \quad k_f = 0.42 \text{ W/m.K.}$$

Use the following correlation for laminar flow inside the tube.

$$Nu = 3.65 + \frac{0.067\left(\frac{D}{L} Re_D\, Pr\right)}{1 + 0.04\left[\frac{D}{L} Re_D\, Pr\right]^{1/3}}.$$

(P.U., May 2001)

Solution

Given : Flow of cream cheese through a heated pipe

$\dot{m}$ = 1000 kg/h, L = 1.5 m,

T_i = 15°C D = 8 cm = 0.08 m,

T_s = 95°C (Constant).

Properties of cheese and correlation

To find : The temperature of cheese leaving the pipe.

Assumptions :

1. Fully developed flow.
2. Steady state conditions.
3. Constant properties.

Analysis : The Reynolds number

$$\text{Re}_D = \frac{4\,\dot{m}}{\pi D \mu} = \frac{4 \times (1000/3600)}{\pi \times 0.08 \times 22.5} = 0.196$$

Prandtl number

$$\text{Pr} = \frac{\mu C_p}{k_f} = \frac{22.5 \times 2750}{0.42} = 147321.42$$

The quantity

$$\frac{D}{L}\, \text{Re}_D\, \text{Pr} = \frac{0.08}{1.5} \times 0.196 \times 147321.42 = 1540$$

Using given correlation

$$\text{Nu} = 3.65 + \frac{0.067 \times 1540}{1 + 0.04 \times (1540)^{1/3}} = 74.28$$

The average heat transfer coefficient

$$h = \frac{\text{Nu}\, k_f}{D} = \frac{74.28 \times 0.42}{0.08} = 390 \text{ W/m}^2.\text{K}$$

Using eqn. (9.46) for exit temperature of cheese

$$T_o = T_s - (T_s - T_i) \exp\left(-\frac{h\, A_s}{\dot{m}\, C_p}\right)$$

$$= 95 - (95 - 15) \times \exp\left[-\frac{390 \times \pi \times 0.08 \times 1.5}{(1000/3600) \times 2750}\right]$$

= **29.0°C.** Ans.

Example 9.22. *Hot air flows with a mass flow rate of 0.05 kg/s through an uninsulated sheet metal duct of diameter 0.15 m, which is located in a large room. The hot air enters at 103°C and, after a distance of 5 m, cools to 77°C. The heat transfer coefficient between the duct outer surface and ambient air at 0°C is 6 W/m².K.*

(*i*) *Calculate the heat loss from the duct over its length,*

(*ii*) *Determine the heat flux and the duct surface temperature at x = L.*

Solution

Given : Hot air flowing in a duct :

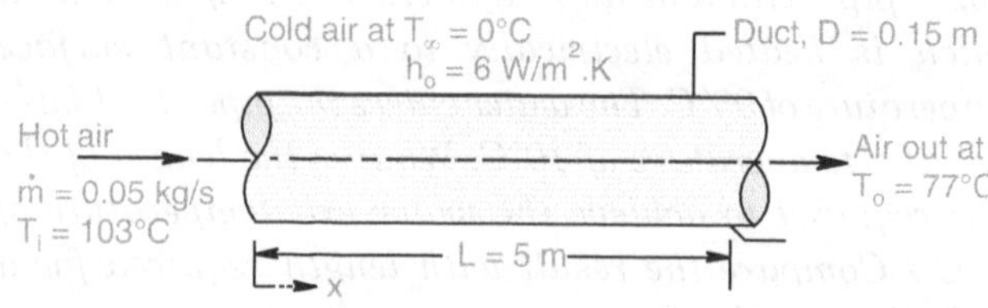

Fig. 9.34. (*a*) Schematic

To find :

(*i*) Heat loss from the duct over the length of 5 m,

(*ii*) Heat flux and surface temperature at x = L.

Assumptions :

1. Steady state conditions.
2. Negligible duct wall thermal resistance.
3. Constant properties of the fluid.
4. Negligible kinetic and potential energy changes.

Analysis : Properties of air at mean temperature

$$T_m = \frac{T_i + T_o}{2} = \frac{103 + 77}{2} = 90°C = 363 \text{ K}$$

The specific heat, C_p = 1010 J/kg.K

Properties at outlet temperature of 77°C are

k_f = 0.030 W/m.K, $\mu = 208 \times 10^{-7}$ kg/ms,

Pr = 0.70.

(*i*) The heat lost by air over the entire duct ;

$$Q = \dot{m} C_p (T_i - T_o)$$
$$= 0.05 \times 1010 \times (103 - 77)$$
$$= \textbf{1313 W.} \quad \text{Ans.}$$

(*ii*) The heat flux at x = L can be calculated from the resistance network.

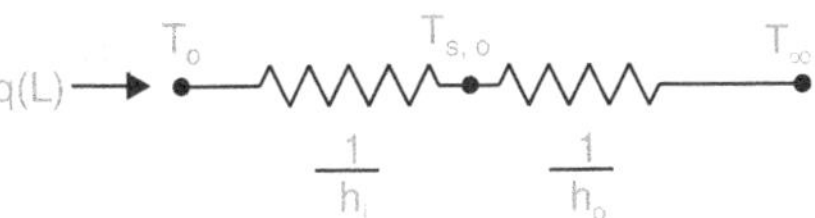

Fig. 9.34 (*b*) Resistance network

For inside heat transfer coefficient h_i in forced convection

$$Re_D = \frac{4\dot{m}}{\pi D \mu} = \frac{4 \times 0.05}{\pi \times 0.15 \times 208 \times 10^{-7}} = 20404$$

Using Dittus-Boelter equation for air cooling (n = 0.3)

$$Nu_D = 0.023\, Re_D^{0.8} \,.\, Pr^{0.3}$$
$$= 0.023 \times (20404)^{0.8} \times (0.70)^{0.3} = 57.94$$

$$h_i = Nu_D \times \frac{k_f}{D} = 57.94 \times \frac{0.030}{0.15}$$
$$= 11.58 \text{ W/m}^2\text{.K}$$

$$\text{Heat flux } q(L) = \frac{T_o - T_\infty}{\frac{1}{h_i} + \frac{1}{h_o}} = \frac{77 - 0}{\frac{1}{11.58} + \frac{1}{6}}$$
$$= \textbf{304.4 W/m}^2\textbf{.} \quad \text{Ans.}$$

With the help of resistance network

$$q(L) = \frac{T_o - T_{s,o}}{\frac{1}{h_i}}$$

or

$$T_{s,o} = T_o - \frac{q(L)}{h_i} = 77 - \frac{304.4}{11.58}$$
$$= \textbf{50.71°C.} \quad \text{Ans.}$$

9.10. HEAT TRANSFER TO LIQUID METAL FLOW IN TUBE

If the liquid metal flows through the tube, then the above relations are not applicable. Since liquid metals have very low Prandtl number, thus very small thermal entry length.

For fully developed turbulent flow (L/D ≥ 10) for metals in a smooth circular tube with uniform surface heat flux, *Skupinski et al* recommended

$$Nu_D = 4.82 + 0.0185\, (Re_D\, Pr)^{0.287}$$
$$q_s = \text{Constant} \qquad \ldots(9.85)$$

Validity $3.6 \times 10^3 < Re_D < 9.05 \times 10^5$

$100 < Re_D \,Pr < 10000$

For constant surface temperature condition of turbulent metal flow in tube, *Seban* and *Shimazaki* recommended the following relation for Re_D Pr > 100

$$Nu_D = 5.0 + 0.025\, (Re_D\, Pr)^{0.8}$$
$$T_s = \text{Constant.} \qquad \ldots(9.86)$$

Example 9.23. *The liquid metal flows at a rate of 270 kg/min through a 5 cm diameter stainless steel tube. It enters at 415°C and is heated to 440°C, when it passes through the tube. The tube wall temperature is kept 20°C higher than the liquid bulk temperature and a constant heat flux is maintained along the tube. Calculate the length of the tube required to effect the transfer. Use*

For constant wall temperature

$$Nu = 5.0 + 0.025\, Pe^{0.8}$$

For constant heat flux

$$Nu = 4.82 + 0.0185\, Pe^{0.827}$$

Use the following properties

$\mu = 1.34 \times 10^{-3}$ *kg/ms,* C_p = *149 J/kg.K,*

Pr = 0.013, k_f = *15.6 W/m.K.*

(P.U., Dec. 1997)

Solution

Given : Liquid metal flow through a circular tube.

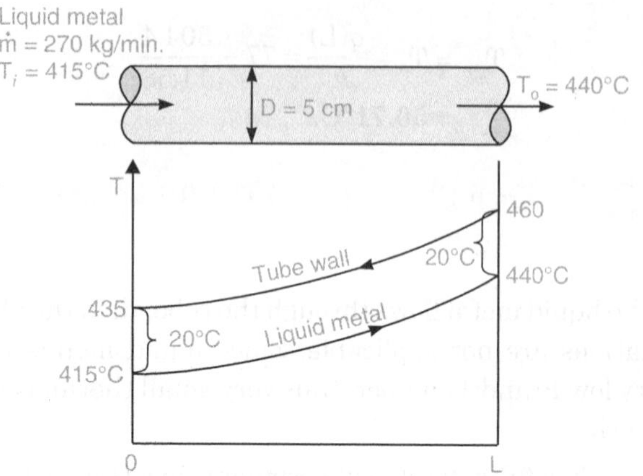

Fig. 9.35. Schematic for liquid metal heating

To find : Length of the tube.

Assumptions.

(*i*) Steady state conditions,

(*ii*) Fully developed flow,

(*iii*) Constant properties.

Analysis : The heat transfer rate

$$Q = \dot{m}\,C_p(T_o - T_i)$$

$$= \frac{270}{60} \times 149 \times (440 - 415)$$

$$= 16{,}762.5 \text{ W}$$

The Reynolds number

$$Re_D = \frac{4\dot{m}}{\pi D\mu} = \frac{4 \times (270/60)}{\pi \times 0.05 \times 1.34 \times 10^{-3}} = 85{,}516$$

The Peclet number

$$Pe_D = Re_D\,Pr = 85516 \times 0.013 = 1111.7$$

The constant temperature difference between surface and fluid leads to constant wall heat flux, thus using

$$Nu_D = 4.82 + 0.0185 \times (1111.7)^{0.827} = 10.93$$

The heat transfer coefficient

$$h = \frac{Nu_D\,k_f}{D} = \frac{10.93 \times 15.6}{0.05} = 3410 \text{ W/m}^2\text{.K}$$

The heat transfer rate for $\Delta T = 20°C$, can be expressed as

$$Q = hA_s\,\Delta T = h(\pi DL)\Delta T$$

or $$L = \frac{16762.5}{3410 \times \pi \times 0.05 \times 20} = \mathbf{1.56\ m.}$$ **Ans.**

Example 9.24. *The liquid sodium flows with a mean velocity of 3 m/s inside a smooth tube of 25 mm dia. and is heated by the tube wall maintained at a uniform temperature of 120°C.*

Determine the heat transfer coefficient at a location, where bulk mean fluid temperature is 93°C and the flow is fully developed.

Solution

Given : Liquid sodium flows through a smooth tube maintained at constant temperature

u_m = 3 m/s, D = 25 mm,

T_s = 120°C, T_m = 93°C.

To find : Heat transfer coefficient.

Analysis : The properties of liquid sodium at 93°C (366 K) from Table A-8

ρ = 929.1 kg/m³, k_f = 86.2 W/m.K,

$\mu = 0.698 \times 10^{-3}$ kg/ms, Pr = 0.011.

The Reynolds number

$$Re_D = \frac{\rho u_m D}{\mu} = \frac{929.1 \times 3 \times 0.025}{0.698 \times 10^{-3}} = 99{,}832$$

and Peclet number

$$Pe_D = Re_D\,Pr = 99832 \times 0.011 = 1098.15$$

For constant wall temperature, using eqn. (9.86)

$$Nu_D = 5.0 + 0.025\,(Re_D\,Pr)^{0.8}$$

$$= 5.0 + 0.025 \times (1098.15)^{0.8} = 11.77$$

The heat transfer coefficient

$$h = \frac{Nu_D\,k_f}{D} = \frac{11.77 \times 86.2}{0.025}$$

$$= \mathbf{40576.5\ W/m^2.K.}$$ **Ans.**

9.11. SUMMARY

The fluid flow through tubes or ducts for transporting, cooling, heating, etc., is of engineering importance. In internal flows, the fluid is completely confined by inner surfaces of the tube, thus the velocity and thermal boundary layers merge at the centre of the tube, after certain distance from the entrance in the direction of flow. For hydrodynamically developed flow, the velocity profile becomes independent of x, *i.e.*,

$$\frac{\partial u}{\partial x} = 0$$

Similarly, for thermally developed boundary layer, the temperature profile becomes independent of x, *i.e.*,

$$\frac{\partial}{\partial x}\left(\frac{T_s - T}{T_s - T_m}\right) = 0$$

The hydrodynamic thermal entry lengths, in which respective profiles develop are given by

$$x_{e,\,lam} \approx 0.05\ Re_D\ D$$

$$x_{eth,\,lam} \approx 0.05\ Re_D\ Pr\ D$$

For turbulent boundary layer

$$x_{e,\,turb} = x_{eth,\,turb} \approx 10\ D$$

The mean velocity, u_m is the average velocity of the fluid. The mean temperature T_m at a cross-section is the average temperature at that cross-section. The mean velocity u_m is considered constant, but mean temperature, T_m changes in the direction of flow, unless the fluid gets tube temperature. The heat transfer rate to a fluid during steady flow in tube can be expressed as

$$Q = \dot{m}C_p(T_o - T_i)\ (kW)$$

For *constant surface heat flux,* the energy balance on the tube is

$$Q = q_s A_s = \dot{m}\,C_p(T_o - T_i)$$

where A_s = surface area of the tube. The exit temperature of the fluid can be calculated as

$$T_o = T_i + \frac{q_s A_s}{\dot{m}C_p}$$

For constant surface temperature, the rate of heat transfer is expressed as

$$Q = h\ A_s\ \Delta T_{lm}$$

where $$\Delta T_{lm} = \frac{\Delta T_1 - \Delta T_2}{\ln\left(\frac{\Delta T_1}{\Delta T_2}\right)}$$

$$\Delta T_1 = T_s - T_i \text{ and } \Delta T_2 = T_s - T_o$$

and the exit temperature of the fluid is given by

$$T_o = T_s - (T_s - T_i)\exp\left(-\frac{hPL}{\dot{m}C_p}\right)$$

For fully developed laminar flow, the friction factor f is expressed as

$$f = 4C_f = \frac{64}{Re}$$

The pressure drop during the flow in the tube is expressed as

$$\Delta p = f\frac{L}{D}\cdot\frac{\rho u_m^{\,2}}{2}\ (N/m^2)$$

The pumping power required to overcome the pressure drop is given by

$$\dot{W}_{pump} = \frac{\dot{m}}{\rho} \times \Delta p\ (W)$$

For fully developed laminar flow, the Nusselt number and friction factor can be obtained from Table 9.1. A general relation for average Nusselt number for hydrodynamically and/or thermally developing laminar flow in a circular tube is

$$Nu_D = 1.86\left(\frac{Re_D\,Pr\,D}{L}\right)^{1/3}\left(\frac{\mu}{\mu_s}\right)^{0.14} \quad (Pr > 0.5)$$

For turbulent flow, in smooth circular tube, the most commonly used relations are

$$f = 0.184Re_D^{-0.2}$$

$$Nu_D = 0.023Re_D^{0.8}\ Pr^n \begin{pmatrix} 0.7 \le Pr \le 160 \\ Re > 10000 \end{pmatrix}$$

where n = 0.3 for cooling, and n = 0.4 for heating of fluid.

For rough surface tube, the Nusselt number can be determined by *Gnielinsky* equation

$$Nu_D = \frac{(f/8)(Re_D - 1000)\,Pr}{1 + 12.7\sqrt{(f/8)}\,(Pr^{2/3} - 1)}\left[1 + \left(\frac{D}{L}\right)^{2/3}\right]$$

for $\frac{D}{L} \le 1 \quad 0.6 < Pr < 2000,\ Re_D \ge 2300$

The fluid properties should be evaluated at bulk mean fluid temperature

$$T_m = \frac{T_i + T_o}{2}.$$

TABLE 9.3. Summary of convection heat transfer correlations for flow in tube

Correlations	*Conditions*	*Remark*
$f = \dfrac{64}{Re_D}$	Laminar fully developed	Friction factor
$Nu_D = 4.36$	Laminar fully developed	Constant heat flux, $Pr \geq 0.6$
$Nu_D = 3.36$	Laminar fully developed	Constant T_s, $Pr \geq 0.6$
$Nu_D = 1.86 + \left[\dfrac{Re_D\, Pr}{L/D}\right]^{1/3} \left[\dfrac{\mu}{\mu_s}\right]^{0.14}$	Laminar combined entry length	Constant T_s
$f = 0.316\, Re_D^{-1/4}$	Turbulent fully developed	$Re_D \leq 2 \times 10^4$
$f = 0.184\, Re_D^{-1/5}$	Turbulent fully developed	$Re_D \geq 2 \times 10^4$
$Nu_D = 0.023\, Re_D^{0.8}\, Pr^{1/3}$	Turbulent fully developed	$0.7 \leq Pr \leq 160$, $Re_D > 10000$, $L/D > 10$
$Nu_D = 0.023\, Re_D^{0.8}\, Pr^{n}$	Turbulent fully developed	$n = 0.4$ for heating, $n = 0.3$ for cooling $0.7 < Pr < 160$, $Re_D \geq 10000$, $L/D \geq 10$
$Nu_D = 0.027\, Re_D^{0.8}\, Pr^{1/3} \left(\dfrac{\mu}{\mu_s}\right)^{0.14}$	Turbulent fully developed	$0.7 < Pr < 16700$, $Re \geq 100,00$, $L/D \geq 10$
$Nu_D = 0.0214\, (Re_D^{0.8} - 100) \times Pr^{0.4} \left[1 + \left(\dfrac{D}{L}\right)^{2/3}\right]$	Turbulent fully developed	$0.5 < Pr < 1.5$ $2300 < Re_D < 10^6$, $D/L < 1$
$Nu_D = 0.012\, [Re_D^{0.87} - 280] \times Pr^{0.4} \left[1 + \left(\dfrac{D}{L}\right)^{2/3}\right]$	Turbulent fully developed	$1.5 < Pr < 500$ $2300 < Re_D < 10^6$, $\dfrac{D}{L} < 1$
$Nu = \dfrac{(f/8)(Re_D - 1000)\, Pr}{1 + 12.7\sqrt{(f/8)}\,(Pr^{2/3} - 1)} \times \left[1 + \left(\dfrac{D}{L}\right)^{2/3}\right]$	Rough surface, turbulent fully developed	$0.6 < Pr < 2000$ $Re_D > 2300$, $\dfrac{D}{L} < 1$
$Nu_D = 4.82 + 0.0185\, (Re_D\, Pr)^{0.827}$	Liquid metals, turbulent fully developed	Constant heat flux
$Nu_D = 5.0 + 0.025\, (Re_D Pr)^{0.8}$	Liquid metals, turbulent fully developed	Constant temperature T_s

REVIEW QUESTIONS

1. Prove that the Reynolds number for flow in a circular tube of diameter D can be expressed as

$$Re_D = \frac{4\dot{m}}{\pi D \mu}$$

where $\dot{m}$ = mass flow rate of fluid, μ is viscosity of fluid.

2. What do you mean by hydrodynamically developed flow in a circular tube ? Explain.
3. What do you understand by thermally developed flow in a circular tube ? Explain.
4. Explain velocity developing region and hydrodynamic entry length for flow in a circular tube.
5. Explain thermally developing region and thermal entry length.
6. What is the difference between friction factor and coefficient of friction ? Show that for laminar flow in a tube

$$f = 4C_f \quad \text{and} \quad f = \frac{64}{Re_D}.$$

7. What do you mean by mean velocity u_m and mean temperature T_m ?
8. How is the friction factor related to pressure drop and pumping power ?
9. What does the log mean temperature difference represent ?
10. How does surface roughness affect the pressure drop and heat transfer rate in the tube, if fluid flow is turbulent ?
11. Why are heat transfer rates higher in turbulent flow, inside a tube ? Explain.
12. What is the Stanton number ? What is the Chilton Colburn analogy ?
13. Derive an expression for Colburn equation with the help of Chilton Colburn analogy.
14. Explain the dimensional analysis for forced flow through a circular tube.
15. In a constant surface temperature tube, the fluid enters at temperature T_i and leaves the tube at temperature T_o. Prove that

$$\frac{T_s - T_o}{T_s - T_i} = \exp\left(-\frac{hA_s}{\dot{m}C_p}\right).$$

16. To measure the mass flow rate of a fluid in a laminar flow through a circular tube, a hot wire type velocity anemometer is placed in the centre of the tube. Assume that the measuring station is far from the entrance of the pipe, the velocity distribution is parabolic

$$\frac{u_{(r)}}{u_{max}} = 1 - \left(\frac{2r}{D}\right)^2$$

where u_{max} is centre line velocity of the fluid, r is the radial distance from the centre of tube and D is pipe diameter :

(*i*) Derive an expression for average fluid velocity at the cross-section in terms of u_{max} and D,

(*ii*) Obtain an expression for mass flow rate.

17. Derive an expression for the heat transfer coefficient for the turbulent flow through a long tube in the moderate temperature range.
18. How does the fluid flow inside the duct differ from fluid flow over the bodies ?

PROBLEMS

1. Water at 20°C with a flow rate of 0.01 kg/s enters a 2 cm diameter tube, which is maintained at 100°C. Assume hydrodynamically developed flow, determine the tube length required to heat the water to 70°C. [Ans. 5 m]
2. Water at an inlet temperature of 50°C with a flow rate of 0.01 kg/s flows through a duct 2 cm by 2 cm in a cross-section, which is maintained at a uniform temperature of 100°C. Assume hydrodynamically developed flow, determine the length of the duct needed to heat the water to 70°C. [Ans. 1.06 m]
3. Water at a mean temperature of 60°C flows inside a 2.5 cm ID, 10 m long tube with a velocity of 6 m/s. The tube wall is maintained at a uniform temperatures of 100°C by condensing steam. Determine the heat transfer rate to water. Assume an inlet temperature of 30°C. [Ans. 670 W]
4. Air at atmospheric pressure and 27°C enters a 12 m long, 1.5 m ID tube with a mass flow rate of 0.1 kg/s. The tube surface is maintained at a uniform temperature of 80°C. Calculate the average heat transfer coefficient and the rate of heat transfer to air. [Ans. 5.3 kW]
5. Water flows at a rate of 0.01 kg/s through an equilateral triangular duct with sides of 2 cm whose walls are kept at uniform temperature of 100°C. Assume that the flow is hydrodynamically and thermally developed. Determine the duct length required to heat the water from 20°C to 70°C. [Ans. 5 m]
6. Water at a velocity of 15 m/s flows through a straight tube of 50 mm diameter. The tube surface is maintained at a uniform temperature of 60°C and the flowing water is heated from an inlet temperature of 20°C to an outlet temperature of 40°C. Find the heat transfer coefficient from the tube surface to the water, the heat transferred and the tube length. [Ans. 29715 W/m^2.K, 2447.6 kW, 17.5 m]
7. Water at a rate of 0.5 kg/s is passed through a smooth 25 mm inner diameter tube, 15 m long. The inlet water temperature is 10°C and tube wall is at a constant temperature of 40°C. What is the exit water temperature ?

Average properties of water are

$C_p = 4180$ J/kg.K, $\mu = 0.8 \times 10^{-3}$ Pas,

$k_f = 0.57$ W/m.K. *(AMIE, Summer, 1998)*

[**Ans.** 37.28°C]

8. A pipeline heater, 3 m long, 40 mm diameter, is used to heat a supply of carbon dioxide from 250 K to 500 K. The walls of heater are kept at constant temperature of 600 K.

Determine the mass flow rate through the pipeline.

For carbon dioxide at 375 K and 1 bar, take

$\rho = 1.41$ kg/m^3, $k_f = 0.023$ W/m.K,

Pr = 0.737, $\mu = 1.8 \times 10^{-5}$ kg/ms.

9. Calculate the Nusselt number and convection heat transfer coefficient by three different correlations for water at a bulk temperature of 32°C, flowing at a velocity of 1.5 m/s through a 2.54 cm ID duct with a wall temperature of 43°C. Compare the results.

10. A fully developed flow of air at 27°C moves at 2 m/s in a 1 cm ID pipe. An electric resistance heater surrounds the last 20 cm of the pipe and supplies a constant heat flux to bring the air out at $T_m = 40$°C. What power input is needed to do this ? What will be the wall temperature at the exit ?

[**Ans.** 378 W/m^2, 68.1°C]

11. A surface condenser consists of 200 thin walled circular tubes, each 22.5 mm in diameter and 5 m long, arranged in parallel, through which water flows. If the mass flow rate of water through the tube bank is 160 kg/s and its inlet and exit temperatures are 21°C and 29°C, respectively, calculate the average heat transfer coefficient associated with flow of water.

[**Ans.** 7563 W/m^2.K]

12. The square channel of side 20 mm and length 2.5 m, carries water at a velocity of 45 m/s. The mean temperature of water along the length of the channel is found to be 30°C, while the inner surface of the channel is at 70°C.

Calculate the heat transfer coefficient from the channel wall to water. Use correlation :

$$Nu = 0.021\, Re^{0.8}\, Pr^{0.43} \left[\frac{Pr}{Pr_s}\right]^{0.25}$$

The thermophysical properties of water at 30°C

$\rho = 995.7$ kg/m^3, $k_f = 0.6175$ W/m.K, $\nu = 0.805 \times 10^{-6}$ m^2/s, Pr = 5.42 and Pr_s at 70°C = 2.55.

[**Ans.** 1770 W/m^2.K] *(P.U. May 1997)*

13. Gasolene at a mean bulk temperature of 27°C flows inside a circular tube, 19 mm ID. The average bulk velocity is 0.061 m/s. The tube surface temperature is kept constant at 38°C. Assume fully developed flow and determine the average heat transfer coefficient over 1.5 m length of the tube. Are the temperature and velocity profiles developing or developed in 1.5 m length of the tube. Take following fluid properties

$\mu_s = 5.223 \times 10^{-4}$ kg/ms at 38°C,

and at bulk temperature

$\mu = 5.892 \times 10^{-4}$ kg/ms, $k_f = 0.1591$ W/m^2.K,

$\rho = 874.6$ kg/m^3, $C_p = 1757$ J/kg.K,

Pr = 6.5 ; use correlation

$$Nu = 1.86 \left[Re\, Pr \,.\, \frac{D}{L}\right]^{1/3} . \left(\frac{\mu}{\mu_s}\right)^{0.14} .$$ *(M.U., 1997)*

[**Ans.** 81.2 W/m^2.K, 1.63 m, 10.62 m]

14. Air at 30°C enters a rectangular duct 1 m long and 4 mm by 16 mm in cross-section at a rate of 0.0004 kg/s. If the uniform heat flux of 500 W/m^2 is imposed on both the long sides of the duct, calculate
(*a*) the air outlet temperature, (*b*) average duct surface temperature, and (*c*) pressure drop.

15. Water enters a double pipe heat exchanger at 60°C. The water flows through inner copper tube, 2.54 cm ID at a velocity of 2 cm/s. Steam flows in the annulus and condenses on the outside of the copper tube at a temperature of 80°C. Calculate the outlet temperature of water, if the heat exchanger is 3 m long.

[**Ans.** 68.4°C]

16. An electronic device is cooled by water flowing through the small holes drilled in the casing. The temperature of the device casing is constant at 80°C. The holes are 0.3 m long and 2.54 mm in diameter. If water enters at a temperature of 60°C and flows at a velocity of 0.2 m/s, calculate the outlet temperature of water. [**Ans.** 72°C]

17. Atmospheric air is heated in a long annulus 25 mm ID, 38 cm OD, by steam condenses at 149°C, on the inner surface. If the velocity of the air is 6 m/s and its bulk temperature is 38°C, calculate the heat transfer coefficient.

18. Water at an average temperature of 27°C flowing through a smooth 50 mm ID tube at a velocity of 1 m/s. If the temperature of inner surface of the tube is 50°C, determine (*a*) heat transfer coefficient, (*b*) the rate of heat flow per metre length of tube, (*c*) bulk temperature rise per metre, and (*d*) the pressure drop per metre.

19. The intake manifold of an automobile engine can be approximated as a 4 cm ID tube, 30 cm long. The air at 20°C enters the manifold at a flow rate of 0.01 kg/s. The manifold is heavy aluminium casting and is at a uniform temperature of 40°C. Determine the temperature of air at the end of the manifold.

20. Engine oil at a rate of 0.02 kg/s flows through a 3 mm diameter tube 30 m long. The oil has an inlet

temperature of 60°C, while the tube wall temperature is maintained at 100°C by steam condensing on its outer surface.

(*a*) Estimate the average heat transfer coefficient, for internal flow of the oil,

(*b*) Determine the outlet temperature of the oil.

21. Engine oil flows through a 25 mm diameter, 10 m long tube a rate of 0.5 kg/s. The oil enters the tube at 25°C, while the tube surface is maintained at 100°C. Determine the total heat transfer to the oil and oil outlet temperature.

22. A thick walled stainless steel pipe of inside and outer radii as 20 mm and 40 mm, respectively is heated electrically to provide a uniform heat generation rate of 10^6 W/m^2. The outer surface of the pipe is insulated. The water flows through the pipe at a rate of 0.1 kg/s.

(*a*) If water inlet temperature is 20°C and desired outlet temperature is 40°C, what is the required pipe length ?

(*b*) What are the location and value of the maximum pipe temperature ?

23. Water flows at 0.25 kg/s through a thin walled, 40 mm diameter tube, that is 4 m long. The water enters at 30°C and is heated by hot gases moving in cross flow over the tube with a velocity of 100 m/s and free stream temperature of 225°C. Estimate the outlet temperature of the water. The gas properties may be approximated to be those of atmospheric air.

24. Consider an air solar collector that is 1 m wide and 5 m long and has a constant spacing of 3 cm between the glass cover and collector plate. Air enters the collector at 30°C at a rate of 0.15 m^3/s through the 1 m wide edge and flows along 5 m long passage way. If the passage temperature of glass cover and collector plate are 20°C and 60°C, respectively, determine, (*a*) net rate of heat transfer to the air in the collector and (*b*) the temperature rise of the air as it flows through the collector.

25. The water in a house is to be heated from 15°C to 90°C by a parabolic solar collector, flowing at a rate of 2 kg/s. The water flows through a 3 cm diameter thin aluminium tube whose outer surface is black anodised in order to maximise its solar absorption ability.

If the solar energy is transferred to water at a rate of 300 W per metre length of the tube, determine the required length of the parabolic collector to meet the hot water requirement of this house. Also estimate the surface temperature of the tube at the exit.

26. Air enters a 7 m long section of a rectangular duct of cross-section 15 cm × 20 cm at 50°C at an average velocity of 7 m/s. If the walls of the duct are maintained at 10°C, determine (*a*) outlet temperature of the air, (*b*) the rate of heat transfer from the air, and (*c*) the fan power needed to overcome the pressure losses in this section of the duct.

[**Ans.** (*a*) 32.8°C, (*b*) 3674 W, (*c*) 4.22 W]

27. Hot air at 60°C leaving a furnace of a house enters a 12 m long section of a sheet metal duct of a square cross-section 20 cm × 20 cm at an average velocity of 4 m/s. The thermal resistance of the duct is negligible and the outer surface of the duct, whose emissivity is 0.3 is exposed to cold air at 10°C in the basement with convection heat transfer coefficient of 10 W/m^2.K. Taking the walls of the basement to be at 10°C also, determine (*a*) temperature at which the hot air will leave the basement, and (*b*) the rate of heat loss from the hot air in the duct to the basement.

28. A device that recovers heat from the high temperature combustion gases involves passing the combustion gas between parallel plates each of which is maintained at 350 K by water flow on opposite surface. The plate separation is 40 mm and the gas flow is fully developed. The gas may be assumed to have the properties of atmospheric air and its mean temperature and velocity are 1000 K and 60 m/s, respectively.

(*a*) What is the heat flux at the plate surface ?

(*b*) If a third plate, 20 mm thick is suspended midway between the original plates, what is the surface heat flux for the horizontal plates? Assume the temperature and flow rate of the gas to be unchanged and radiation effects are negligible.

29. Consider a thin walled metallic tube, 1 m long and 3 mm ID. The water enters the tube at 97°C with a mass flow rate of 0.015 kg/s.

(*a*) What is the outlet temperature of water, if the tube surface is maintained at 27°C ?

(*b*) If 0.5 mm thick layer of insulation (k = 0.05 W/m.K) is applied to the tube and its outer surface is maintained at 27°C, what is the outlet temperature of water ?

(*c*) If the outer surface of the insulation is no longer maintained at 27°C, but allowed to exchange the heat by free convection with ambient air at 27°C, what is the outlet temperature of water ? The free convection heat transfer coefficient is 5 W/m^2.K.

30. In a particular solar collector, energy collected by placing a tube at the focal line of parabolic collector

and passing fluid through the tube. The arrangement resulting in a uniform heat flux of 2000 W/m^2 along the axis of the tube of diameter 60 mm. Determine :

(*i*) Length of the tube required to heat the water from 20°C to 80°C which flows at the rate of 0.01 kg/s.

(*ii*) Surface temperature at the outlet of tube.

The properties of water are :

$\mu = 352 \times 10^{-6}$ Ns/m^2, $C_p = 4187$ J/kg.K,

$k_f = 0.67$ W/m.K, Pr = 2.2.

[Ans. (*i*) 6.66 m, (*ii*) 121.08°C]

31. Calculate average heat transfer coefficient and friction factor for flow of *n*-butyl alcohol at a mean temperature of 20°C through a 0.1 m × 0.1 m square duct, 5 m long with walls at 27°C, if the average velocity is 0.03 m/s. Use physical properties of *n*-butyl alcohol at 293 K. [Ans. 4.98 W/m^2.K, 0.069]

32. The water flows through a tube, 25 mm dia, and 6 m long with a velocity of 2.40 m/s. It is observed that the pressure loss due to frictional losses is 1.22 m of water. Determine the heat transfer coefficient by using Chilton Colburn analogy.
Compare the result with the Reynolds analogy.
Take $\rho = 998$ kg/m^3, $C_p = 4187$ J/kg.K, Pr = 6.62.
[Ans. 6156 W/m^2.K, and 21687 W/m^2.K]

33. Air at atmospheric pressure and 120°C enters 40 mm diameter, and 2 m long tube with a velocity of 10 m/s. A 1 kW electric heater is wound on the outer surface of the tube. Find (*i*) mass flow rate of air, (*ii*) exit temperature of air, and the wall temperature at the outlet. Assume that the heat absorption rate by air is uniform throughout the length of the tube.
Take R = 0.287 kJ/kg.K and $C_p = 1.005$ kJ/kg.K for air. [Ans. (*i*) 0.011 kg/s, (*ii*) 208°C and 319°C]

34. Liquid sodium at 180°C with a mass flow rate of 3 kg/s enters a 2.5 cm ID tube whose wall is maintained at a uniform temperature of 240°C. Calculate the tube length required to heat the liquid sodium to 230°C.
Use the properties of liquid sodium as :
$C_p = 1339$ J/kg.K, $k = 80.81$ W/m.K,
$\rho = 907.5$ kg/m^3, $\nu = 0.501 \times 10^{-6}$ m^2/s,
Pr = 0.0075. [Ans. 1.56 m]

35. In a long annulus (3.5 cm ID and 5 cm OD), the water is heated by maintaining the outer surface of inner tube at 60°C. The water enters at 20°C and leaves at 34°C. While its flow rate is 2 m/s. Estimate the heat transfer coefficient. [Ans. 8212 W/m^2.K]

36. Hot air at atmospheric pressure and 80°C enters an 8 m long uninsulated square duct of cross-section 0.2 m × 0.2 m, that passes through the roof of a house at a rate of 0.15 m^3/s. The duct is observed to be nearly isothermal at 60°C. Determine the exit temperature of air and rate of heat loss from the duct to roof.
[Ans. 71.2°C, 1340 W]

REFERENCES AND SUGGESTED READING

1. Kays W.M. and H.C. Perkins, in "*Handbook of Heat Transfer*", by Rehsenow W.M., J.P. Harnett and E.N. Ganic, eds, 3/*e*, McGraw-Hill, New York, 1985.
2. Kays W.M. and M.E. Crawford, "*Convective Heat and Mass Transfer*", 2nd ed. McGraw Hill, New York, 1980.
3. Knudsen J.D. and D.L. Katz, "*Fluid Dynamics and Heat Transfer*", McGraw Hill, New York, 1958.
4. Shah R.K. and A.L. London, "*Laminar Flow: Forced Convection in Ducts*", Academic press, New York, 1978.
5. Petukhov B.S., "*Heat Transfer and Friction in Turbulent Pipe Flow with Variable Physical Properties*", Academic Press, New York, 1970.
6. Shah R.K., "*Thermal Entry Length Solution for a Circular Tube and Parallel Plates, Proceeding of National Heat and Mass Transfer*", IIT Mumbai, Vol. I, 1975, McAdams W.M."*Heat Transmission*", 3rd ed. McGraw Hill, New York, 1954.
7. Jacob M. and G. A. Hawkins, "*Elements of Heat Transfer*", 3rd Ed. Wiley, New York, 1957.
8. Krieth Frank and M.S. Bohn, "*Principles of Heat Transfer*", 5th ed., PWS Pub. Company, 1997.
9. Holman J.P., "*Heat Transfer*", 7th ed. McGraw Hill, New York, 1990.
10. Incropera F.P. and D.P. Dewitt, "*Introduction to Heat Transfer*" Z/e, John Wiley and Sons, 1990.
11. Bayazitoglu Y and M.N. Ozisik "*Elements of Heat Transfer*", McGraw Hill, New York, 1988.
12. Thomas L.C., "*Heat Transfer*", Prentice-Hall, Englewood Cliffs, NJ, 1982.
13. White F.M., "*Heat and Mass Transfer*", Addison Wesley, Reading, MA, 1988.
14. Jacob M., "*Heat Transfer*", Vol. I Wiley, New York, 1949.
15. Suryanarayana N.V., "*Engineering Heat Transfer*" West Pub. Co. New York, 1998.
16. Cenzel Yunus A, "*Introduction to Thermodynamics and Heat Transfer*", McGraw Hill, New York, 1997.

10 Natural Convection

10.1. PHYSICAL MECHANISM

In natural convection, the fluid motion is due to buoyancy forces within the fluid. The buoyancy forces are developed due to density variation in the fluid caused by temperature difference between the fluid and adjacent surface. The larger the temperature difference in adjacent fluid, the larger the buoyancy force and stronger natural convection currents and higher the heat transfer rate. Whenever a heated object for an example a hot egg, is exposed to atmospheric air, the air adjacent to the hot egg gets heated and becomes lighter (less dense) and thus rises up as shown in Fig. 10.1. This motion leads to the formation of the boundary layer on the surface of the egg and the heat is transferred from the warmer boundary layer to outer atmospheric air by *natural convection.* The velocity of air is zero at the boundary surface and it is significant outside the boundary layer.

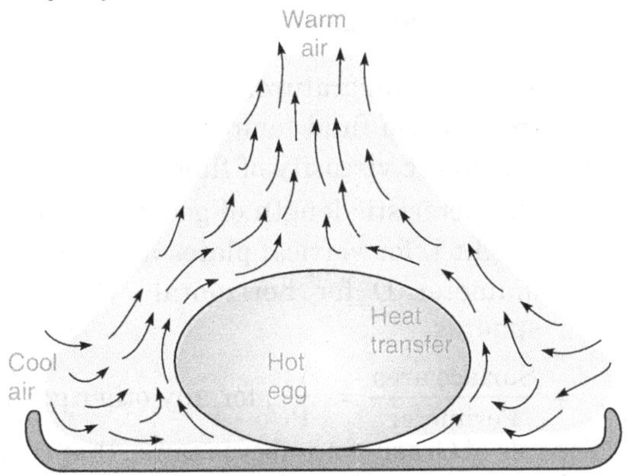

Fig. 10.1. The cooling of a boiled egg in a cooler environment by natural convection

The motion that results from continuous replacement of heated air in the vicinity of hot body by the adjacent cooler air is called a **natural or free convection current** and the resulting heat transfer is called **natural convection heat transfer.**

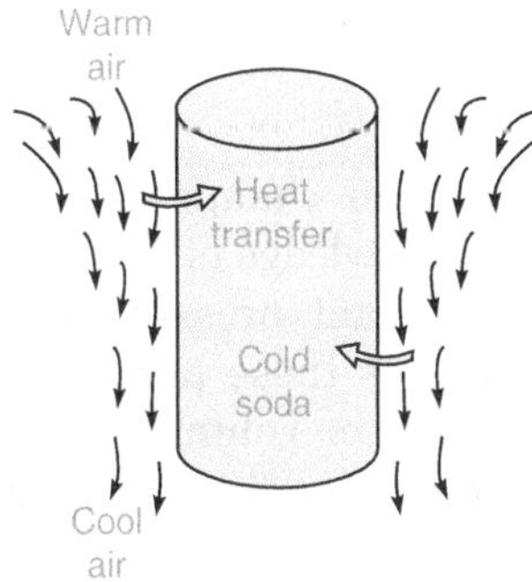

Fig. 10.2. The warming up of a cold drink in a warmer environment by natural convection

The natural convection is an effective way to heat the cold surfaces in the warmer environment and to cool the hot surfaces in the colder environment as shown in Fig. 10.2. Here the direction of fluid motion is reversed.

There are many situations, where the heat is transferred by free convection to the surrounding air. Heat transfer from a heater to heat a room, heat transfer from pipes, transmission line, condenser coil of a refrigerator, electric transformer, electric motors and electronic equipments are some typical examples of natural convection heat transfer.

In free convection, the fluid motions setup by buoyancy forces are much smaller than those associated with forced convection, therefore, the heat transfer rate in natural convection is also smaller.

10.2. DEFINITIONS

10.2.1. Buoyancy Force

In gravitational field, there is a net force that pushes the light fluid upward from the heavier fluid. The upward force exerted by a fluid on a body that is completely or partially immersed in it, is called the **buoyancy force.** The magnitude of buoyancy force is equal to the weight of fluid displaced by the body

$$F_{buoyancy} = \rho_{fluid}\, g\, V_{body} \qquad ...(10.1)$$

where ρ_{fluid} is the average density of fluid, g is the acceleration due to gravity and V_{body} is the volume portion of the body immersed in the fluid. In absence of other effects, the net vertical force acting on the body is the difference between the weight of the body and buoyancy force,

i.e.,

$$F_{net} = mg - F_{buoyancy}$$

$$= \rho_{body}\, g\, V_{body} - \rho_{fluid}\, g\, V_{body}$$

$$= (\rho_{body} - \rho_{fluid})\, g\, V_{body} \qquad ...(10.2)$$

The net force is proportional to the density difference between the fluid and body immersed in it. Thus a body immersed in a fluid experiences a weight loss equal to weight of the fluid it displaces. It is known as *Archimedes principle.*

10.2.2. Volumetric Expansion Coefficient

The density of a fluid changes as its temperature changes. Thus the knowledge of density variation with temperature at constant pressure is essential. The property which relates these properties is called the *coefficient of thermal* (or volumetric) *expansion*. It is defined as the ratio of fractional change in volume to change in temperature at constant pressure, and it is denoted by β and is expressed as

$$\beta = -\frac{1}{v}\left(\frac{\partial v}{\partial T}\right)_p$$

using $v = \dfrac{1}{\rho}$

Then $$\beta = \rho\left[\frac{\partial(1/\rho)}{\partial T}\right]_p = \rho\left[-\frac{1}{\rho^2}\frac{\partial \rho}{\partial T}\right]_p$$

$$\beta = -\frac{1}{\rho}\left(\frac{\partial \rho}{\partial T}\right)_p \qquad ...(10.3)$$

For an ideal gas;

$$\rho = \frac{p}{RT}, \text{ and } \left(\frac{\partial \rho}{\partial T}\right)_p = -\frac{p}{RT^2}$$

Therefore, $$\beta = \frac{1}{T}\,(K^{-1}) \qquad ...(10.4)$$

where T is the absolute temperature.

10.2.3. Grashof Number

The flow regime in natural convection is characterised by a dimensionless number called the *Grashof number,* which is defined as the ratio of buoyancy force to viscous force acting on the fluid. It is denoted as Gr and is given by

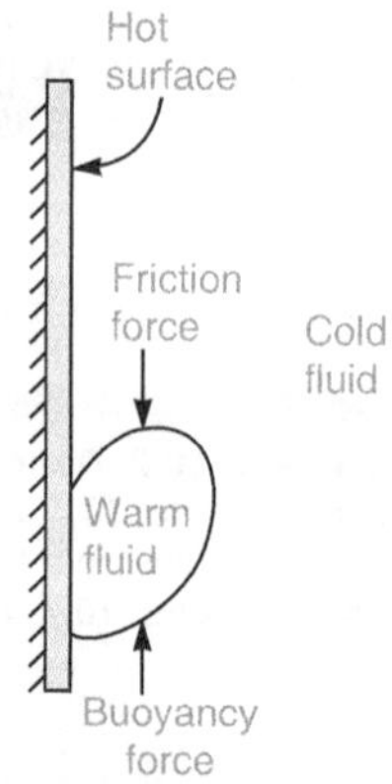

Fig. 10.3. The Grashof number measures the relative magnitude of buoyancy force and friction force acting on the fluid

$$Gr = \frac{\text{Buoyancy force}}{\text{Viscous force}} = \frac{g\Delta\rho V}{\rho\nu^2} = \frac{g\beta\Delta T\, V}{\nu^2}$$

where $\beta\Delta T = \dfrac{\Delta\rho}{\rho}$; the fraction of volume change of fluid corresponding to temperature change ΔT at constant pressure.

Using $V = L_c^3$ and $\Delta T = T_s - T_\infty$ then it is formally expressed as

$$Gr = \frac{g\beta\,(T_s - T_\infty)\,L_c^3}{\nu^2} \qquad ...(10.5)$$

where g = acceleration due to gravity, m/s²
β = coefficient of volumetric expansion, K^{-1}
$= \dfrac{1}{T_f}$ for ideal gases
T_s = surface temperature, °C,
T_∞ = free stream fluid temperature, °C,
ν = kinematic viscosity of fluid, m²/s,
L_c = characteristic length of geometry, m
= height L for vertical plates and cylinders
= diameter D for horizontal cylinders and spheres,
$= \dfrac{\text{Surface area}}{\text{Perimeter}} = \dfrac{A_s}{P}$, for any other geometry

The role of Grashof number is same, that is played by Reynolds number in forced convection. The Grashof number provides the criteria to distinguish the type of flow: laminar or turbulent in natural convection. The *critical Grashof number* for flow over plates is considered to be 10^9. Therefore, the flow on a vertical plate becomes turbulent, if Grashof number exceeds 10^9. Fig. 10.4 shows an interferometer produced map of interference fringes of constant temperature lines over a hot plate in air. The smooth and parallel lines in (*a*) indicate the laminar flow, whereas, the eddies and irregularities in, (*b*) indicate the turbulent flow.

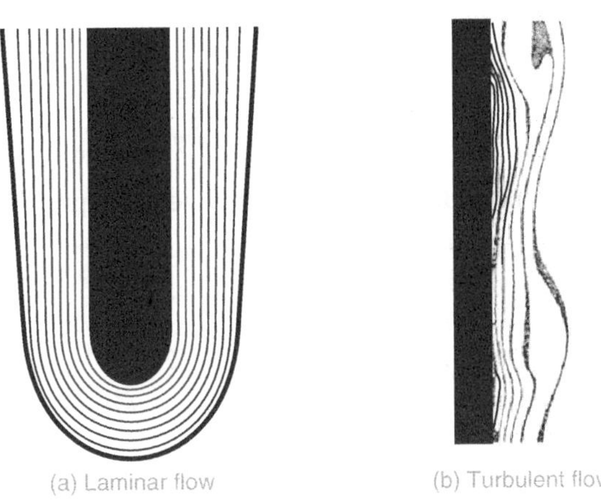

Fig. 10.4. Isotherms in natural convection over a hot plate in air

10.3. NATURAL CONVECTION OVER A VERTICAL PLATE

The heat transfer rate in natural convection depends on geometry of the surface as well as on its orientation. It also depends on temperature variation on the surface and thermophysical properties of the fluid. Fig. 10.5 shows development of the velocity boundary layer for natural convection on a vertical plate.

Consider a heated vertical plate at temperature T_s, exposed to a stagnant fluid at temperature T_∞ ($T_s > T_\infty$) as shown in Fig. 10.5. The fluid adjacent to the plate is heated and its density decreases. The buoyancy force, therefore, induces a free convection boundary layer, in which heated fluid rises up, leaving the space for fluid from the cold region. This boundary layer grows in the flow direction. The fluid velocity is zero at the surface of the plate ($y = 0$) because of no slip condition at the surface. It is also zero at outer edge of boundary layer ($y = \delta$), because fluid is stationary beyond the boundary layer. The fluid velocity increases with distance from the surface, reaches a maximum value and again gradually decreases to zero at the outer edge of boundary layer. The temperature of the fluid equals the plate temperature at the surface and gradually decreases to temperature of surrounding fluid at a distance at the outer edge of the boundary layer.

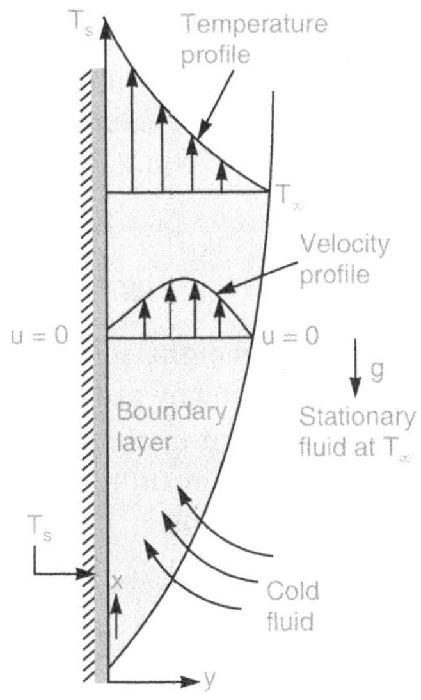

Fig. 10.5. Typical velocity and temperature profile for natural convection flow over a hot vertical plate at T_s, exposed to fluid at T_∞

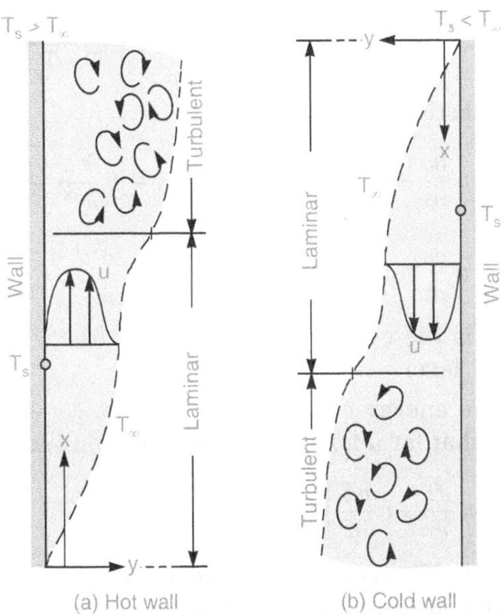

Fig. 10.6. Free convection on vertical plates

The boundary layer developed initially is laminar, but after certain distance from the leading edge, depending on the fluid properties and temperature difference the turbulent eddies are formed and transition to turbulent layer begins and further, it becomes fully turbulent as shown in Fig. 10.6.

Fig. 10.6 (*a*) shows the natural convection boundary layers on a heated vertical wall, while Fig. 10.6 (*b*) shows convection currents, and boundary layers on a cold vertical wall.

In order to develop the governing equation, we choose x coordinate along the vertical wall and y coordinate perpendicular to the wall. The new force is to be considered as weight of the element fluid (gravitational force), then the momentum eqn. (7.15) derived earlier becomes ;

$$\rho\left(u\frac{\partial u}{\partial x}+v\frac{\partial u}{\partial y}\right)=-\frac{\partial p}{\partial x}-\rho g+\mu\frac{\partial^2 u}{\partial y^2} \quad ...(10.6)$$

where $-\rho g$ represents weight force exerted on the element per unit area in downward direction.

The pressure gradient at the edge of the vertical boundary layer in x direction ($u \to 0$ and $\rho \to \rho_\infty$) are due to change in density. Thus ;

$$\frac{\partial p}{\partial x}=-\rho_\infty g \quad ...(10.7)$$

Substituting eqn. (10.7) in eqn. (10.6), we get ;

$$\rho\left(u\frac{\partial u}{\partial x}+v\frac{\partial u}{\partial y}\right)=(\rho_\infty-\rho)\,g+\mu\frac{\partial^2 u}{\partial y^2} \quad ...(10.8)$$

The density difference $(\rho_\infty - \rho)$ can be expressed in term of coefficient of volumetric expansion as

$$\beta=-\frac{1}{\rho}\left(\frac{\partial \rho}{\partial T}\right)_p=\frac{\rho_\infty-\rho}{\rho\,(T-T_\infty)}$$

Substituting, we get

$$\rho\left(u\frac{\partial u}{\partial x}+v\frac{\partial u}{\partial y}\right)=\beta\rho g(T-T_\infty)+\mu\frac{\partial^2 u}{\partial y^2}$$

or
$$\left(u\frac{\partial u}{\partial x}+v\frac{\partial u}{\partial y}\right)=\beta g(T-T_\infty)+\nu\frac{\partial^2 u}{\partial y^2} \quad ...(10.9)$$

It is the equation of motion for free convection boundary layer.

The energy equation for the free convection is same as that for a forced convection system, eqn. (7.20)

$$\left(u\frac{\partial T}{\partial x}+v\frac{\partial T}{\partial y}\right)=\alpha\frac{\partial^2 T}{\partial y^2} \quad ...(10.10)$$

Von Korman integral technique can also be applied to natural convection from a vertical surface with the following assumptions :

1. The density variation is within boundary layer only. The flow is laminar and steady.
2. The buoyancy effects are confined to boundary layer region only and velocity v in y direction is almost negligible.
3. The analysis is made for Pr = 1 *i.e.,* $\delta_{th} = \delta$.
4. Since the magnitude of the velocity is small, thus the viscous dissipation is negligible at any y.

After incorporating above assumptions in eqn. (10.9) and (10.10), the integral momentum and energy equations for control volume shown in Fig. 10.7 become

$$\frac{d}{dx}\int_0^\delta u^2\,dy=-\nu\left(\frac{\partial u}{\partial y}\right)_{y=0}+\beta g\int_0^\delta (T-T_\infty)\,dy \quad ...(10.11)$$

and
$$\frac{d}{dx}\int_0^\delta u(T-T_\infty)\,dy=-\alpha\left[\frac{\partial (T-T_\infty)}{\partial y}\right]_{y=0} \quad ...(10.12)$$

The boundary conditions for velocity profile are

$u = 0$ at $y = 0$
$u = 0$ at $y = \delta$
$\frac{\partial u}{\partial y} = 0$ at $y = \delta$.

And the boundary conditions for temperature profile are

$T = T_s$ at $y = 0$
$T = T_\infty$ at $y = \delta$
$\frac{\partial (T-T_\infty)}{\partial y} = 0$ at $y = \delta$
$\because \quad q = 0$

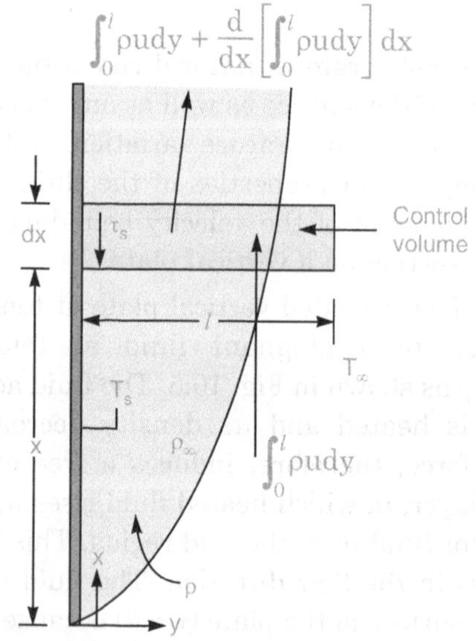

Fig. 10.7. Control volume in boundary layer for natural convection flow over a heated plate at T_s, exposed to fluid at T_∞

The temperature porfile is approximated by quadratic equation

$$T = C_1 + C_2 y + C_3 y^2$$

and the velocity profile is assumed to be cubical parabola

$$u = u_1(a_1 + a_2 y + a_3 y^2 + a_4 y^3)$$

where $u_1 = u(x)$, reference velocity

Then the velocity and temperature profiles are given by

$$\frac{u}{u_0} = \frac{y}{\delta}\left(1-\frac{y}{\delta}\right)^2 \quad ...(10.13)$$

where $$u_0 = u_1 \frac{g\,\beta\,(T_s - T_\infty)\delta^2}{4\nu} \quad ...(10.14)$$

and $$\frac{T - T_\infty}{T_s - T_\infty} = \left(1-\frac{y}{\delta}\right)^2 \quad ...(10.15)$$

Using the velocity profile, the location of maximum velocity is given by

$$\frac{du}{dy} = u_0 \frac{d}{dy}\left\{\frac{y}{\delta}\left(1-\frac{y}{\delta}\right)^2\right\}$$

$$= 0 = \frac{1}{\delta} - \frac{4y}{\delta^2} + \frac{3y^2}{\delta^3} = 0$$

or $3y^2 - 4\delta y + \delta^2 = 0$

Its solution $$y = \frac{4\delta \pm \sqrt{16\delta^2 - 4\times 3\delta^2}}{2\times 3} = \frac{4\delta \pm 2\delta}{6}$$

$$y = \delta \text{ or } \frac{1}{3}\delta \quad ...(10.16)$$

Since $u = 0$ at $y = \delta$, therefore, the velocity u will be maximum at $y = \delta/3$

$$u_{max} = \frac{u_0}{3}\left(1-\frac{1}{3}\right)^2 = \frac{4}{27}u_0 \quad ...(10.17)$$

On solution, the individual terms in momentum integral equation becomes

$$\int_0^\delta u^2 dy = \int_0^\delta u_0^2 \frac{y^2}{\delta^2}\left(1-\frac{y}{\delta}\right)^4 dy = \frac{u_0^2\,\delta}{105} \quad ...(10.18)$$

And for energy equation

$$\int_0^\delta u(T - T_\infty)\,dy = \int_0^\delta u_0 \frac{y}{\delta}\left(1-\frac{y}{\delta}\right)^2 (T_s - T_\infty)\left(1-\frac{y}{\delta}\right)^2 dy$$

$$= \frac{u_0 (T_s - T_\infty)\,\delta}{30} \quad ...(10.19)$$

Substituting in momentum eqn. (10.11)

$$\frac{d}{dx}\left(\frac{u_0^2\delta}{105}\right) = -\nu\frac{u_0}{\delta} + \frac{g\,\beta\,(T_s - T_\infty)\,\delta}{3} \quad ...(10.20)$$

and the energy eqn. (10.12),

$$\frac{d}{dx}\left[\frac{u_0\,(T_s - T_\infty)\,\delta}{30}\right] = 2\alpha\frac{(T_s - T_\infty)}{\delta} \quad ...(10.21)$$

where u_0 and δ are function of x.

Let $u_0 = C_1 x^m$ and $\delta = C_2 x^n$

Solution gives the value of constants C_1 and C_2 as

$$C_1 = 5.17\,\nu\left[\frac{20}{21}+\frac{\nu}{\alpha}\right]^{-1/2}\left\{\frac{g\,\beta(T_s - T_\infty)}{\nu^2}\right\}^{1/2} \quad ...(10.22)$$

and $$C_2 = 3.93\left(\frac{20}{21}+\frac{\nu}{\alpha}\right)^{1/4}\left\{\frac{g\,\beta(T_s - T_\infty)}{\nu^2}\right\}^{-1/4}\left(\frac{\nu}{\alpha}\right)^{-1/2}$$

Using constant C_2, the thickness of velocity boundary layer in natural convection over a vertical surface is given by

$$\frac{\delta}{x} = 3.93(0.952 + Pr)^{1/4}\,\frac{1}{Gr_x^{1/4}\,.\,Pr^{1/2}} \quad ...(10.23)$$

The heat flux,

$$q = -k\left(\frac{\partial T}{\partial y}\right)_{y=0} = \frac{2k(T_s - T_\infty)}{\delta} = h(T_s - T_\infty) \quad ...(10.24)$$

or $$h = \frac{2k}{\delta}$$

The Nusselt number

$$Nu_x = \frac{h_x x}{k_f} = \frac{2x}{\delta}$$

$$= \frac{2}{3.93\,(0.952 + Pr)^{1/4}\,Pr^{-1/2}\,.\,Gr_x^{-1/4}}$$

$$Nu_x = 0.508\,Pr^{1/2}\,(0.952 + Pr)^{-1/4}\,.\,Gr_x^{1/4} \quad ...(10.25)$$

The eqn. (10.23) yields to

$$\delta \propto x^{1/4}$$

Thus as x increases, the boundary layer thickness δ increases, and eqn. (10.25) results into

$$h_x \propto x^{-1/4}$$

As x increases, the local heat transfer coefficient decreases. The average heat transfer coefficient

$$h = \frac{1}{L}\int_0^L h_x\,dx = \frac{1}{L}\int_0^L C\,x^{-1/4}\,dx = \frac{4}{3}\,[C\,x^{3/4}]_{x=L}$$

$$= \frac{4}{3}h\Big|_{x=L} \text{ or } \frac{4}{3}h_L \quad ...(10.26)$$

The average Nusselt number for plate of height L is given by

$$Nu_L = 0.677\,Pr^{1/2}\,(0.952 + Pr)^{-1/4}\,Gr_L^{1/4} \quad ...(10.27)$$

For air, Pr ≈ 0.72, the eqns. (10.25) and (10.27) reduce to

$$Nu_x = 0.378\ Gr_x^{1/4} \quad ...(10.28)$$

$$Nu_L = 0.504\ Gr_L^{1/4} \quad ...(10.29)$$

Some more relations extracted from above equations are the mean fluid velocity

$$u_{mean} = \frac{27}{48}\ u_{max} \quad ...(10.30)$$

The local velocity of fluid

$$u_x = 5.17\ \nu \left(\frac{20}{21} + Pr\right)^{-1/2} \left(\frac{g\ \beta(T_s - T_\infty)}{\nu^2}\right)^{1/2} \quad ...(10.31)$$

The mass flow rate of fluid in boundary layer at any location

$$\dot{m}_x = \frac{\rho\ \delta\ u_x}{12} \quad ...(10.32)$$

The total mass flow rate through the boundary

$$\dot{m} = 1.7\ \rho\nu \left[\frac{Gr_L}{(Pr)^2\ (0.592 + Pr)}\right]^{1/4} \quad ...(10.33)$$

All the fluid properties are evaluated at the film temperature

$$T_f = \frac{T_s + T_\infty}{2} \quad ...(10.34)$$

10.4. EMPIRICAL CORRELATIONS FOR EXTERNAL FREE CONVECTION FLOW

Some analytical solutions for natural convection can only be obtained for simple geometries under some simplified assumptions. Therefore, the correlations are developed with the help of experimental data. Here some of the recommended empirical correlations for determining natural convection heat transfer coefficient on certain geometries are presented. The correlations are in the form of

$$Nu = C(GrPr)^n = C\ Ra^n \quad ...(10.35)$$

where Ra is the **Rayleigh number**, a product of Grashof and Prandtl numbers ;

$$Ra = GrPr = \frac{g\ \beta(T_s - T_\infty)\ L_c^{\ 3}}{\nu^2}.Pr \quad ...(10.36)$$

The values of constants C and n depend on the geometry of the surface and flow regime. The value of n is generally 1/4 for laminar flow and 1/3 for turbulent flow. All the properties are evaluated at the film temperature $T_f = (T_s + T_\infty)/2$.

10.4.1. Vertical Plate

1. **Uniform wall temperature.** The most useful correlation for a vertical plate maintained at *uniform temperature* T_s and exposed to a fluid at T_∞ in natural convection is proposed by *Churchill* and *Chu* as

$$Nu = \left[0.825 + \frac{0.387\ Ra_L^{\ 1/6}}{\{1 + (0.492 / Pr)^{9/16}\}^{8/27}}\right]^2 \quad ...(10.37)$$

It may be used for entire range of Ra_L.

Some simplified empirical correlations for laminar and turbulent natural convection are given below :

Laminar free convection :

$$Nu_L = 0.59 Ra_L^{\ 1/4} \quad \text{for} \quad 10^4 < Ra_L < 10^9 \quad ...(10.38)$$

Turbulent free convection

$$Nu_L = 0.13 Ra_L^{\ 1/3} \quad \text{for} \quad 10^9 < Ra_L < 10^{13} \quad ...(10.39)$$

2. **Uniform surface heat flux.** For free convection over a vertical plate subjected to *uniform heat flux* q_s at the wall surface has been studied and empirical correlations have been proposed for average Nusselt number in laminar and turbulent regims ;

$$Nu = 0.75\ (Gr_L^*\ .\ Pr)^{1/5} \quad ...(10.40)$$

for $10^5 < Gr^*.Pr < 10^{11}$ (laminar)

$$Nu = 0.645\ (Gr_L^*\ .\ Pr)^{0.22} \quad ...(10.41)$$

for $2 \times 10^{13} < Gr^*.Pr < 10^{16}$ (turbulent)

where the *modified Grashof number* Gr^* is given by

$$Gr^* = \frac{g\beta q_s L_c^4}{k\nu^2} \quad ...(10.42)$$

10.4.2. Horizontal Surfaces

The natural convection currents associated with heated horizontal surfaces are different from those, that occurred on a vertical surface. The buoyancy force acts normal to the surface and flow field depends on heating configuration.

1. **Uniform surface temperature.** Consider a horizontal heated surface at a uniform temperature T_s, exposed to an ambient at T_∞ ($T_\infty < T_s$), as shown in Fig. 10.8.

When heated horizontal surface is exposed to surrounding air, the heated lighter fluid adjacent to the surface tends to rise from the surface, but its motion is supressed by the heavier, cooler fluid above it. After a small disturbance, the natural convection currents are setup, then the heated fluid rises up and cooler fluid moves down to occupy the space vacated by the heated fluid as shown in Fig. 10.8 (*a*). On the contrast, if the heated surface is facing down, the transfer rate is reduced in comparison with hot surface facing up. The flow pattern of fluid on a cold horizontal surface facing up is similar to that of the heated surface facing down, and the flow with horizontal, cold surface facing down is also similar to that of a heated surface facing up as shown in Fig. 10.8 (*b*).

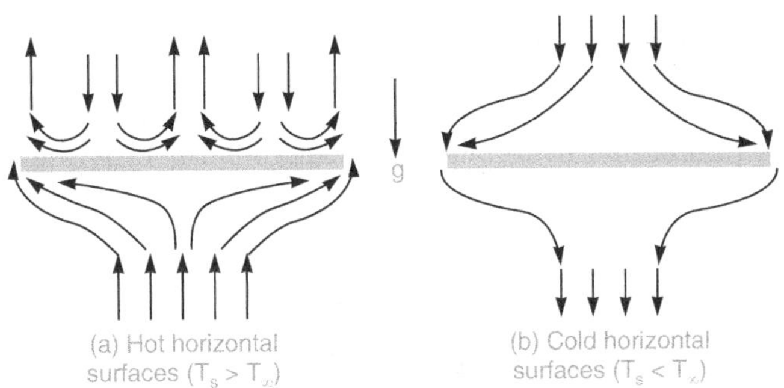

Fig. 10.8. Free convection from horizontal surfaces

The average Nusselt number for different configuration of horizontal plate is expressed in a form of

$$Nu = C(Gr\ Pr)^n \quad ...(10.43)$$

where $$Nu = \frac{h\,L_c}{k_f} \text{ and } Gr = \frac{g\,\beta\,\Delta T\,L_c^3}{\nu^2} \quad ...(10.44)$$

L_c = Characteristic length of the surface

$$= \frac{A_s}{P} = \frac{\text{Surface area of the plate}}{\text{Perimeter of the plate}} \quad ...(10.45)$$

The constant C and exponent n are listed in Table 10.1. The physical properties of the fluid are evaluated at film temperature T_f defined by eqn. (10.34).

TABLE 10.1. Constant C and exponent n used in eqn. (10.43) for natural convection on a horizontal surface at uniform temperature

Orientation of plate	L_c	*Range of Ra*	*C*	*n*	*Flow regime*
1. Hot horizontal surface facing up or cold surface facing down	$\frac{A_s}{P}$	$1 < Ra_L < 200$	0.96	1/6	Laminar
		$200 < Ra_L < 10^4$	0.59	1/4	Laminar
		$10^4 < Ra_L < 10^7$	0.54	1/4	Laminar
		$10^7 < Ra_L < 10^{11}$	0.15	1/3	Turbulent
2. Hot horizontal surface facing down or cold surface facing up	$\frac{A_s}{P}$	10^5 to 10^{11}	0.27	1/4	Laminar

2. **Uniform surface heat flux.** *Fujii* and *Imura* studied the average Nusselt number for natural convection on a horizontal surfaces subjected to uniform heat flux q_s and exposed to an ambient at T_∞. The following correlations are proposed for the cases in which heated surface facing up and facing down.

Horizontal surface with the heated surface facing up

$$Nu = 0.13\,(Gr_L\,Pr)^{1/3}$$

for $$Ra_L < 2 \times 10^8 \quad ...(10.46)$$

$$Nu = 0.16\,(Gr_L\,Pr)^{1/3}$$

for $$5 \times 10^8 < Ra_L < 10^{11} \quad ...(10.47)$$

For downward facing heated surface or upward facing cooled surface

$$Nu = 0.58\,(Gr_L\,Pr)^{1/5}$$

for $$10^6 < Ra_L < 10^{11} \quad ...(10.48)$$

The physical properties of fluid are evaluated at mean temperature defined as

$$T_m = T_s - 0.25\,(T_s - T_\infty) \quad ...(10.49)$$

The coefficient of volumetric expansion β is evaluated at $T_\infty + 0.25\,(T_s - T_\infty)$.

10.4.3. Inclined Plates

The heat transfer coefficient for a downward facing heated or upward facing cooled inclined plate at uniform temperature can be predicted from the correlation given for vertical plate by replacing gravitational term g by $g\cos\theta$, where θ is the angle of inclination of the surface with vertical as defined in Fig. 10.9.

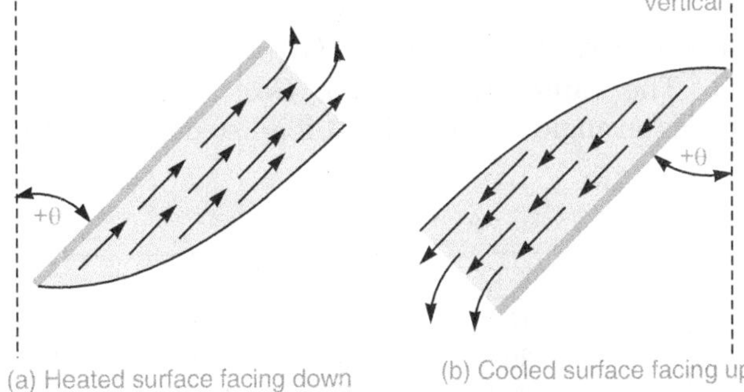

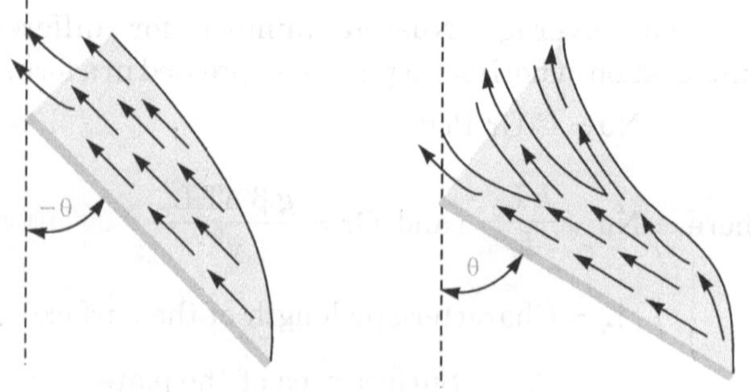

(c) Upward facing heated surface, $Gr_L < Gr_c$ (d) Upward facing heated surface, $Gr_L > Gr_c$

Fig. 10.9. Natural convection from surfaces in various orientation

The characteristic length of the inclined plate is the length along the plate and

$$Gr_L = \frac{g \cos\theta\, \beta\, \Delta T\, L^3}{\nu^2} \quad ...(10.50)$$

$$Ra_L = Gr_L Pr, \quad Nu = \frac{hL}{k_f} \quad ...(10.51)$$

When $\theta > 88°$, the plate is slightly inclined with the horizontal, and heated surface is facing downward, then the correlations for horizontal plates may be used.

The orientation of heated surface facing up or down affects the Nusselt number. The natural convection from upward facing heated surface is more complex than with downward facing heated plate. On upward facing heated surface, for small value of Gr_L (formed with $g\cos\theta$), the fluid motion parallel to plate is similar to downward facing heated plate as shown in Fig. 10.9 (*c*). But when the value of Gr_L exceeds a critical value Gr_c, the boundary layer detaches itself from the heated surface due to strong buoyancy force perpendicular to the plate. The value of Gr_cPr depends on θ and tabulated in Table 10.2. Further, the correlations used for upward facing heated surface may also be applied to downward facing cooled surface and for such orientation (for angle between – 15 and – 75°), a suitable correlation for average Nusselt number is recommended

$$Nu = 0.145\,[(Gr_L Pr)^{1/3} - (Gr_c Pr)^{1/3}] + 0.56\,(Gr_c Pr \cos\theta)^{1/4} \quad ...(10.52)$$

valid for $Gr_L\, Pr < 10^{11}$, $Gr_L > Gr_c$

The values of *transition Grashof number* Gr_c depends on angle of inclination θ, it is listed in Table 10.2.

TABLE 10.2. Transition Grashof number Gr_c used in eqn. (10.52)

θ, *degree*	Gr_c
– 15	5×10^9
– 30	2×10^9
– 60	10^8
– 75	10^6

Here, for inclined surfaces, the fluid properties are evaluated at mean temperature defined by eqn. (10.49) as

$$T_f = T_s - 0.25\,(T_s - T_\infty)$$

and β is evaluated at $T_\infty + 0.25(T_s - T_\infty)$.

10.4.4. Free Convection on a Long Cylinders

1. **Horizontal cylinder at uniform temperature.** The natural convection flow over the surface of a horizontal cylinder shown in Fig. 10.10 is similar to that occurs over a vertical wall, only the difference

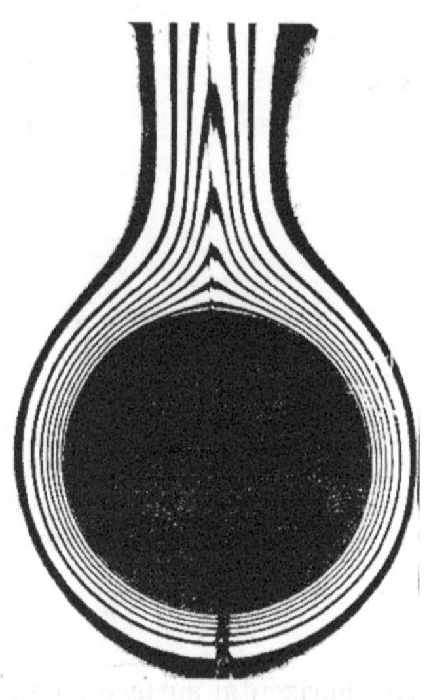

Fig. 10.10. Natural convection around a horizontal cylinder

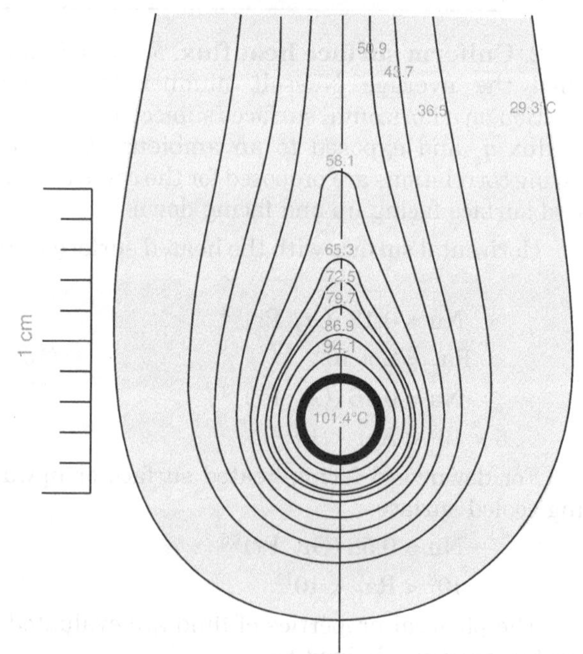

Fig. 10.11. Measured isotherms around a cylinder in air when $Gr_D \approx 585$ in natural convection

being that the surface of the cylinder is curved. Thus the Nusselt number and Grashof number are calculated by using diameter D of cylinder as a characteristic length. For a wide range of Rayleigh number $10^{-3} < Ra < 10^{13}$, the *Churchill* and *Chu* proposed the following correlation for average Nusselt number for natural convection over a cylinder at uniform temperature at T_s, exposed to ambient at T_∞;

$$Nu = \left[0.6 + \frac{0.387\, Ra_D^{1/6}}{\{1 + (0.559/Pr)^{9/16}\}^{8/27}}\right]^2 \quad \ldots(10.53)$$

where $$Ra_D = Gr_D\, Pr = \frac{g\beta(T_s - T_\infty)\, D^3}{\nu^2}\, Pr \quad \ldots(10.54)$$

Evaluate β at T_∞ and all other properties at $T_f = (T_s + T_\infty)/2$.

A simple correlation for natural convection from a horizontal isothermal cylinder is proposed by *Morgan* in the form

$$Nu = \frac{hD}{k_f} = C\, Ra_D^{m} \quad \ldots(10.55)$$

where the value of constant C and exponent m are function of Ra_D and are given in Table 10.3. All properties are evaluated at film temperature T_f.

TABLE 10.3. Constant C and exponent *m* used in eqn. (10.55)

Ra_D	C	m
10^{-10}–10^{-2}	0.675	0.058
10^{-2}–10^{2}	1.02	0.148
10^{2}–10^{4}	0.85	0.188
10^{4}–10^{8}	0.53	0.25
10^{8}–10^{12}	0.13	0.333

2. **Vertical cylinder at uniform temperature.** The average Nusselt number for natural convection on a vertical cylinder is very similar to that for a vertical plate, if the curvature effects are negligible. The correlation for vertical plate may be used for vertical cylinder given by eqns. (10.37), (10.38) or (10.39).

10.4.5. Free Convection on a Spheres

The natural convection around the spheres is very similar to that for horizontal cylinders. A simple correlation for calculation of average Nusselt number for natural convection on a single sphere at uniform temperature is given by *Yuge* as

$$Nu_D = \frac{hD}{k_f} = 2 + 0.43\, Ra_D^{1/4} \quad \ldots(10.56)$$

for $1 < Ra_D < 10^5$ and $Pr \approx 1$ and all properties at film temperature T_f, and characteristic length as diameter D of sphere.

For a wide range of Rayleigh number *Churchill* recommended

$$Nu_D = 2 + \frac{0.589\, Ra_D^{1/4}}{[1 + (0.469/Pr)^{9/16}]^{4/9}} \quad \ldots(10.57)$$

for $Ra_D < 10^{11}$, $Pr \geq 0.7$.

All the properties at T_f except β at T_∞.

The summary of correlations for average Nusselt number in natural convection over various geometries and orientations are presented in Table 10.4.

TABLE 10.4. Summary of empirical correlations for the average Nusselt number for natural convection over surfaces

Geometry	*Characteristic length L_c*	*Range of Ra*	*Nu*
Vertical plate (figure labels: T_s, L)	L	10^4–0^9	$Nu = 0.59\, Ra^{1/4}$
		10^9–10^{13}	$Nu = 0.13\, Ra^{1/3}$
		Entire range	$Nu = \left\{0.825 + \frac{0.387\, Ra^{1/6}}{[1 + (0.492/Pr)^{9/16}]^{8/27}}\right\}^2$ (complex but more accurate)

Geometry	Characteristic length	Range of Ra	Nu
Inclined plate	L		Use vertical plate equations as a first degree of approximation Replace g by $g\cos\theta$ for $\text{Ra} < 10^9$
Horizontal plate (Surface area A_s and perimeter P) (*a*) Upper surface of a hot plate (or lower surface of a cold plate) (*b*) Lower surface of a hot plate (or upper surface of a cold plate)	$\frac{A_s}{P}$	10^4–10^7 10^7–10^{11} 10^5–10^{11}	$\text{Nu} = 0.54\ \text{Ra}^{1/4}$ $\text{Nu} = 0.15\ \text{Ra}^{1/3}$ $\text{Nu} = 0.27\ \text{Ra}^{1/4}$
Vertical cylinder	L		A vertical cylinder can be treated as a vertical plate when $D \geq \frac{35L}{\text{Gr}^{1/4}}$
Horizontal cylinder	D	10^5–10^{12}	$\text{Nu} = \left\{0.6 + \frac{0.387\ \text{Ra}^{1/6}}{[1 + (0.559/\text{Pr})^{9/16}]^{8/27}}\right\}^2$
Sphere	$\frac{1}{2}\pi D$	$\text{Ra} \leq 10^{11}$ $(\text{Pr} \geq 0.7)$	$\text{Nu} = 2 + \frac{0.589\ \text{Ra}^{1/4}}{[1 + (0.469/\text{Pr})^{9/16}]^{4/9}}$

Example 10.1. *Vertical door of a hot oven is 0.5 m high and is maintained at 200°C. It is exposed to atmospheric air at 20°C. Find (a) local heat transfer coefficient half way up the door; (b) average heat transfer coefficient for entire door ; (c) thickness of free convection boundary layer at the top of the door.*

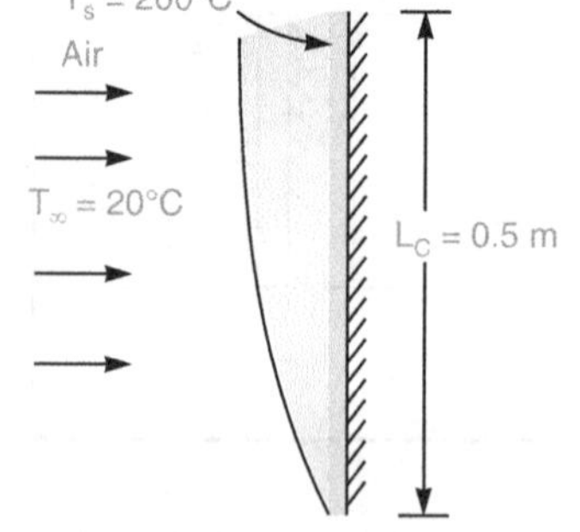

Fig. 10.12. Schematic of vertical door

Solution

Given : A vertical door of an oven

$L = L_c = 0.5$ m $\qquad T_s = 200°C$

$T_\infty = 20°C.$

To find :

(*a*) Local heat transfer coefficient half way of door *i.e.*, $x = 0.25$ m.

(*b*) Average heat transfer coefficient.

(*c*) Thickness of free convection boundary layer at the top of door *i.e.*, $x = L_c = 0.5$ m.

Assumptions :

(*i*) Heat convection from one side of the door only.

(*ii*) Negligible radiation heat transfer.

(*iii*) Constant properties and steady state conditions.

Analysis : The film temperature

$$T_f = \frac{T_s + T_\infty}{2} = \frac{200 + 20}{2} = 110°C$$

The properties of atmospheric air at 110°C from Table A-4

$\rho = 0.922$ kg/m^3, $C_p = 1000$ J/kg.K,

$\mu = 2.24 \times 10^{-5}$ kg/ms $\nu = 2.429 \times 10^{-5}$ m^2/s,

$k_f = 0.0332$ W/m.K, Pr = 0.687

$$\beta = \frac{1}{383} \text{ K}^{-1}$$

(*a*) Local heat transfer coefficient at $x = 0.25$ m

The Grashof number at $x = 0.25$ m

$$Gr_x = \frac{g\beta(T_s - T_\infty)L_c^3}{\nu^2}$$

$$= \frac{9.81 \times \frac{1}{383} \times (200-20) \times (0.25)^3}{(2.429 \times 10^{-5})^2} = 12.2 \times 10^7$$

The Rayleigh number

$$Ra_x = Gr_x \, Pr$$
$$= 12.2 \times 10^7 \times 0.687 = 8.388 \times 10^7$$

The boundary layer is laminar ($Ra_x \leq 10^9$), thus using eqn. (10.25) for calculation of local Nusselt number

$$Nu_x = 0.508 \, Pr^{1/2} (0.952 + Pr)^{-1/4} \cdot Gr_x^{1/4}$$
$$= 0.508 \times (0.687)^{1/2} \times (0.952 + 0.687)^{-1/4} \times (12.2 \times 10^7)^{1/4} = 39.11$$

The local heat transfer coefficient

$$h_x = \frac{Nu_x \, k_f}{x} = \frac{39.11 \times 0.0332}{0.25}$$

$= \mathbf{5.2 \; W/m^2.K.}$ Ans.

(*b*) Average heat transfer coefficient

The Grashof number at $x = L_c$

$$Gr_L = \frac{9.81 \times \frac{1}{383} \times (200-20) \times (0.5)^3}{(2.429 \times 10^{-5})^2}$$
$$= 9.76 \times 10^8$$

The Rayleigh number

$$Ra_L = Gr_L \, Pr = 9.76 \times 10^8 \times 0.687 = 6.71 \times 10^8$$

Again the flow is laminar, using the eqn. (10.38) for determination of average Nusselt number

$$Nu = 0.59 \, Ra_L^{1/4} = 0.59 \times (6.71 \times 10^8)^{1/4}$$
$$= 94.96$$

The average heat transfer coefficient h

$$h = \frac{Nu \, k_f}{L_c} = \frac{94.96 \times 0.0332}{0.5}$$

$= \mathbf{6.30 \; W/m^2.K.}$ Ans.

(*c*) Thickness of free convection boundary layer at $x = 0.5$ m using eqn. (10.23)

$$\frac{\delta}{x} = 3.93(0.952 + Pr)^{1/4} \frac{1}{Gr_x^{1/4} \, . \, Pr^{1/2}}$$

or $$\frac{\delta}{0.5} = 3.93 \times (0.952 + 0.687)^{1/4}$$

$$\times \frac{1}{(9.76 \times 10^8)^{1/4} \times (0.687)^{1/2}} = 0.0303$$

or $\delta = 0.01517$ m = **15.17 mm.** Ans.

Example 10.2. *Derive a relationship between Grashof number and Reynolds number, assuming that the heat transfer coefficients over vertical plate for pure forced and natural convection are equal in laminar flow.*

Solution

Given : For laminar forced convection :

$$Nu = 0.664 \, Re^{1/2} \, Pr^{1/3} \quad ...(i)$$

For laminar natural convection, eqn. (10.27)

$$Nu = 0.677 \, Pr^{1/2} (0.952 + Pr)^{-1/4} \, Gr_L^{1/4} \quad ...(ii)$$

where Nu = average Nusselt number and is expressed as

$$Nu = \frac{hL}{k_f}$$

For equal heat transfer coefficients in natural and forced convection; equating eqns. (*i*) and (*ii*)

$$0.664 \, Re^{1/2} \, Pr^{1/3} = 0.667 \frac{Gr^{1/4} \, Pr^{1/2}}{(0.952 + Pr)^{1/4}}$$

or $$Gr \approx Re^2 \frac{(0.952 + Pr)}{Pr^{2/3}}$$

It is the required relationship between Grashof number and Reynolds number.

Example 10.3. *A vertical plate 15 cm high and 10 cm wide is maintained at 140°C. Calculate the maximum heat dissipation rate from the both sides of the plate in an ambient of at 20°C. The radiation heat transfer* *coefficient is 9.0 W/m^2.K. For air at 80°C, take ν = 21.09 × 10^{-6} m^2/s, Pr = 0.692, k_f = 0.03 W/m.K.*

Solution

Given : A vertical plate is exposed to air on its both sides

$$L = 15 \text{ cm} = 0.15 \text{ m}, \quad w = 10 \text{ cm} = 0.1 \text{ m}$$
$$T_s = 140°\text{C}, \quad T_\infty = 20°\text{C},$$
$$h_r = 9.0 \text{ W/m}^2\text{.K} \quad \nu = 21.09 \times 10^{-6} \text{ m}^2\text{/s}$$
$$\text{Pr} = 0.692, \quad k_f = 0.03 \text{ W/m. K}$$

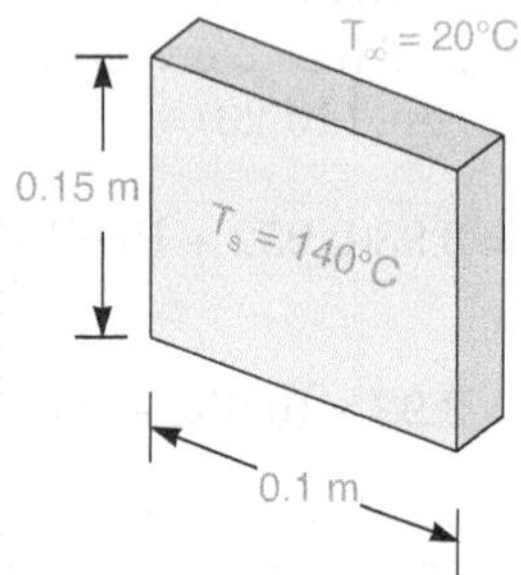

Fig. 10.13

To find : Maximum heat dissipation rate from both sides of a vertical plate.

Analysis : For a vertical plate, the characteristic length

$$L_c = \text{Height of plate} = 0.15 \text{ m}$$

The film temperature

$$T_f = \frac{T_s + T_\infty}{2} = \frac{140 + 20}{2} = 80°\text{C} = 353 \text{ K}$$

The coefficient of volumetric expansion

$$\beta = \frac{1}{T_f} = \frac{1}{353} \text{ K}^{-1}$$

The Grashof number at $L_c = L$

$$Gr_L = \frac{g\,\beta\,(\Delta T)\,L_c^3}{\nu^2}$$
$$= 9.81 \times \frac{1}{353} \times \frac{(140 - 20) \times (0.15)^3}{(21.09 \times 10^{-6})^2}$$
$$= 2.53 \times 10^7$$

The Rayleigh number

$$Ra_L = Gr_L Pr = 2.53 \times 10^7 \times 0.692$$
$$= 1.751 \times 10^7$$

The Rayleigh number is less than 10^9, thus the flow is laminar using eqn. (10.38)

$$Nu = 0.59\, Ra_L^{1/4}$$
$$= 0.59 \times (1.751 \times 10^7)^{1/4} = 38.167$$

The average convective heat transfer coefficient

$$h_c = Nu \frac{k_f}{L_c} = 38.167 \times \frac{0.030}{0.15}$$
$$= 7.63 \text{ W/m}^2\text{.K.}$$

The convective heat dissipation rate from two sides of the plate

$$Q_{conv} = h_c\,(2wL)\,(T_s - T_\infty)$$
$$= 7.63 \times (2 \times 0.1 \times 0.15) \times (140 - 20)$$
$$= 27.48 \text{ W}$$

The radiative heat dissipation rate from two sides of the plate

$$Q_{rad} = h_r\,(2\,wL)\,(T_s - T_\infty)$$
$$= 9.0 \times (2 \times 0.1 \times 0.15) \times (140 - 20)$$
$$= 32.4 \text{ W}$$

Total heat dissipation rate from the plate

$$= Q_{conv} + Q_{rad} = 27.48 + 32.4$$
$$= \mathbf{59.88 \text{ W.}} \quad \text{Ans.}$$

Example 10.4. *Water at the rate of 0.8 kg/s at 90°C flows through a steel tube having 25 mm ID and 30 mm OD. The outside surface temperature of the pipe is 84°C and temperature of surrounding air is 20°C. The room pressure is 1 atm and pipe is 15 m long. How much heat is lost by free convection in the room ?*

You may use correlation

$$Nu = 0.53\,(Gr\,Pr)^{0.25} \text{ for } 10^4 < Gr\,Pr < 10^9$$
$$= 0.10\,(Gr\,Pr)^{1/3} \text{ for } 10^9 < Gr\,Pr < 10^{12}$$

Take properties of air as

ρ = 1.0877 kg/m^3, C_p = 1.0073 kJ/kg.K
μ = 1.9606 × 10^{-5} kg/ms,
k_f = 0.02813 W/m.K. (P.U., Dec. 2010)

Solution

Given : A hot pipe passes through a room as shown in Fig. 10.14.

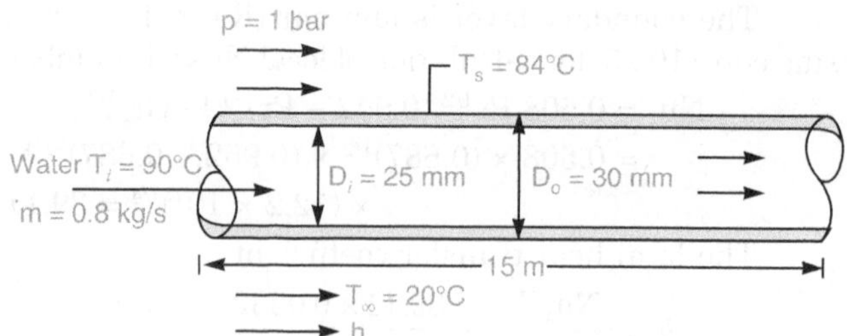

Fig. 10.14. Schematic for a pipe passing a room

To find : Heat dissipation by natural convection to room.

Analysis : The film temperature

$$T_f = \frac{T_s + T_\infty}{2} = \frac{84 + 20}{2} = 52°\text{C}$$

$$\beta = \frac{1}{T_f + 273} = \frac{1}{52 + 273} = \frac{1}{325} \text{ K}^{-1}$$

The characteristic length

$L_c = D_o = 30 \text{ mm} = 0.03 \text{ m}$

The Grashof number

$$Gr_L = \frac{g\,\beta\,(T_s - T_\infty)\,L_c^3}{\nu^2}$$

$$= \frac{\rho^2 g\,\beta(T_s - T_\infty)\,L_c^3}{\mu^2}$$

$$= \frac{(1.0877)^2 \times 9.81 \times \frac{1}{325} \times (84 - 20) \times (0.03)^3}{(1.9606 \times 10^{-5})^2}$$

$= 1.60 \times 10^5$

The Prandtl number

$$Pr = \frac{\mu\,C_p}{k_f} = \frac{1.9606 \times 10^{-5} \times 1007.3}{0.02813} = 0.702$$

The Rayleigh number

$Ra_L = Gr\,Pr = 1.60 \times 10^5 \times 0.702 = 1.12 \times 10^5$

Thus using given correlation

$Nu = 0.53\,(Gr\,Pr)^{0.25}$

$= 0.53 \times (1.12 \times 10^5)^{0.25} = 9.70$

The average natural heat transfer coefficient

$$h = \frac{Nu\,k_f}{L_c} = \frac{9.70 \times 0.02813}{0.03} = 9.1 \text{ W/m}^2.\text{K}$$

The heat dissipation rate by natural convection

$Q = h\,A_s\,(T_s - T_\infty) = h\,(\pi\,D_o\,L)\,(T_s - T_\infty)$

$= 9.1 \times (\pi \times 0.03 \times 15) \times (84 - 20)$

$=$ **823.0 W.** Ans.

Example 10.5. *Consider a rectangular plate 0.2 m × 0.4 m is maintained at a uniform temperature of 80°C. It is placed in atmospheric air at 24°C. Compare the heat transfer rates from the plate for the cases when the vertical height is (a) 0.2 m and (b) 0.4 m.*

Solution

Given : A rectangular plate of size 0.2 m × 0.4 m;

$L_1 = 0.2$ m, $\quad L_2 = 0.4$ m

$T_s = 80°C$, $\quad T_\infty = 24°C$.

To find : Comparison of heat transfer rates when the vertical height is

(*a*) 0.2 m and (*b*) 0.4 m.

Properties of fluid : The mean film temperature;

$$T_f = \frac{T_s + T_\infty}{2} = \frac{80 + 24}{2} = 52°C = 325 \text{ K}$$

The properties of air at 325 K from Table A-4 ;

$\nu = 1.822 \times 10^{-5}$ m²/s, $\quad Pr = 0.703$

$k_{air} = 0.02814$ W/m.K,

$$\beta = \frac{1}{325} = 3.077 \times 10^{-3} \text{ K}^{-1}$$

Analysis : The Grashof number with characteristic length L_c of plate :

$$Gr = \frac{g\,\beta\,\Delta T\,L_c^3}{\nu^2}$$

$$= \frac{(9.81) \times (3.077 \times 10^{-3}) \times (80 - 24) \times L_c^3}{(1.822 \times 10^{-5})^2}$$

$= 5.092 \times 10^9\,L_c^3$

$Ra = Gr\,Pr = (5.092 \times 10^9\,L_c^3) \times (0.703)$

$= 3.579 \times 10^9\,L_c^3$

(*a*) When 0.2 m side is vertical : $L_c = L_1 = 0.2$ m

$Ra_{L_1} = 3.579 \times 10^9 \times (0.2)^3 = 28.637 \times 10^6$

Thus the flow is laminar, and using eqn. (10.38)

$Nu_{L_1} = 0.59(Ra_{L_1})^{1/4}$

$= 0.59 \times (28.637 \times 10^6)^{1/4} = 43.176$

The average heat transfer coefficient :

$$h_1 = Nu_{L_1}\frac{k_{air}}{L_1} = 43.176 \times \frac{0.02814}{0.2}$$

$= 6.075$ W/m².K

The heat transfer rate :

$Q_1 = h_1\,A_s\,(T_s - T_\infty)$

$= 6.075 \times (0.2 \times 0.4) \times (80 - 24)$

$=$ **27.216 W.** Ans.

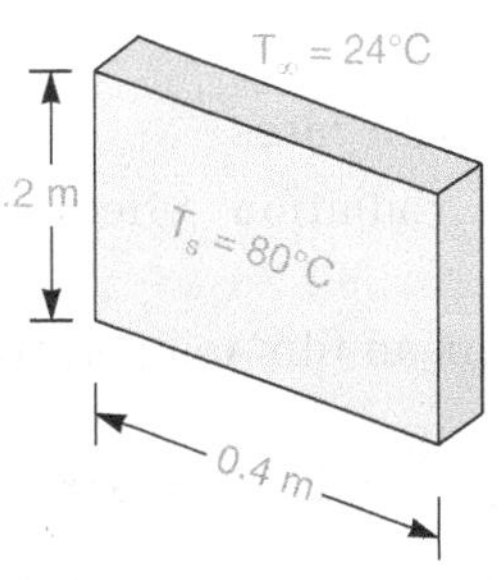

(a) Rectangular plate with 0.2 m side vertical

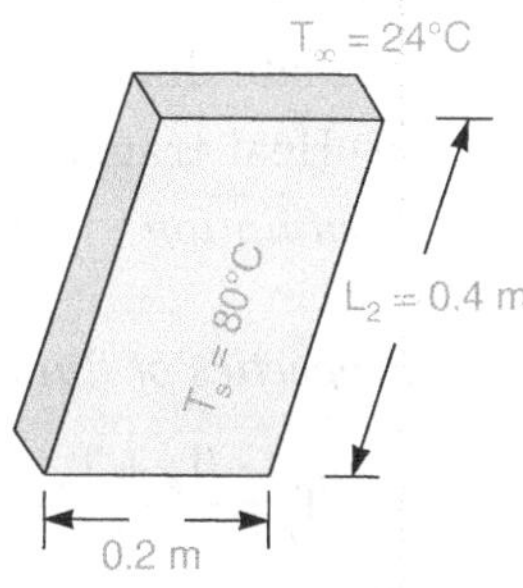

(b) Rectangular plate with 0.4 m side vertical

Fig. 10.15. Schematic for example 10.5

(*b*) For the different vertical orientation of the plate of $L_c = 0.4$ m. The relevant Rayleigh number is

$Ra_{L_2} = 3.579 \times 10^9 \times (0.4)^3 = 229.0 \times 10^6$

The boundary layer is laminar, hence using eqn. (10.38)

$$Nu_{L_2} = 0.59(Ra_{L_2})^{1/4}$$
$$= 0.59 \times (229.0 \times 10^6)^{1/4} = 72.58$$

and
$$h_2 = Nu_{L_2}\frac{k_{air}}{L_2} = 72.58 \times \frac{0.02814}{0.4}$$
$$= 5.10 \text{ W/m}^2\text{.K}$$

The heat transfer rate :
$$Q_2 = h_2 A_s (T_s - T_\infty)$$
$$= 5.10 \times (0.2 \times 0.4) \times (80 - 24)$$
$$= \mathbf{22.848\ W.}$$ Ans.

The percentage decrease in heat transfer :

$$\frac{(Q_1 - Q_2)}{Q_1} \times 100 = \frac{(27.216 - 22.848)}{27.216} \times 100 = 16\%$$

Heat transfer is 16% higher when the vertical side is 0.2 m instead of 0.4 m side. Ans.

Example 10.6. *A hot plate of 15 cm² area maintained at temperature of 200°C is exposed to still air at 30°C. When smaller side of the plate is held vertical, the convective heat transfer rate is 14% higher than that when bigger side of the plate, held vertical. Determine the dimensions of the plate. Neglecting the internal temperature gradients of the plate. Also calculate the heat transfer rate in both cases. Use relation*

$$Nu_L = 0.59\ (GrPr)^{1/4}$$

Use air properties

Temperature °C	*ρ, kg/m³*	*C_p, kJ/kg.K*	*μ, kg/ms*	*k_f, W/m.K*
30	*1.165*	*1.005*	*18.6 × 10⁻⁶*	*0.0267*
115	*0.910*	*1.009*	*22.65 × 10⁻⁶*	*0.0331*
200	*0.746*	*1.026*	*26.0 × 10⁻⁶*	*0.0393*

(P.U., Dec. 2006)

Solution

Given : A hot plate exposed to air ;

$A = 15\ cm^2 = 15 \times 10^{-4}\ m^2$, $T_s = 200°C$,

$T_\infty = 30°C$, $Q_s = 1.14 \times Q_b$

where, Q_s = heat transfer rate when small side is vertical

Q_b = heat transfer rate when bigger side is vertical.

To find :

(*i*) Plate dimensions.

(*ii*) Heat transfer rate in both cases.

Assumption : Surface radiation effect are negligible.

Properties of fluid : The mean film temperature;

$$T_f = \frac{T_s + T_\infty}{2} = \frac{200 + 30}{2} = 115°C = 388\ K$$

The properties of air at 115°C from given table

$\rho = 0.910\ kg/m^3$, $k_f = 0.0331$ W/m.K,

$C_p = 1.009$ kJ/kg.K = 1009 J/kg.K,

$\mu = 22.65 \times 10^{-6}$ kg/ms,

$$\beta = \frac{1}{T_f} = \frac{1}{338}K^{-1}$$

Analysis : (*i*) Considering the smaller side of the plate is (L_s) cm, then bigger side $15/L_s$ cm.

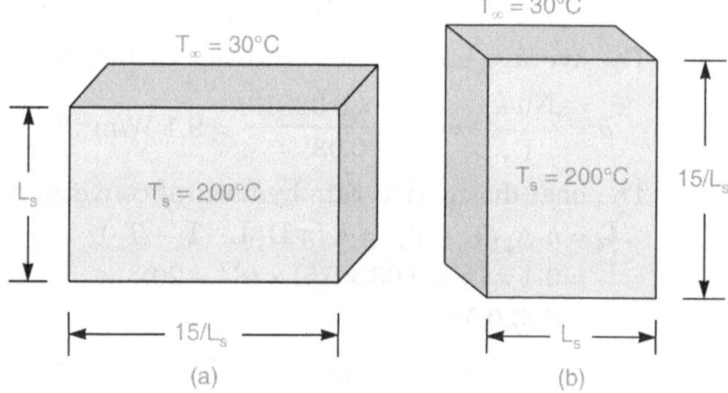

Fig. 10.16

The Grashof number for smaller side (L_s) vertical,

$$Gr = \frac{g\rho^2\beta\Delta T L_s^3}{\mu^2}$$

$$= \frac{(9.81)\times(0.910)^2\times\left(\frac{1}{388}\right)\times(200-30)\times L_s^3}{(22.65\times10^{-6})^2}$$

$$= 6.937 \times 10^9\ L_s^3$$

Prandtl number:

$$Pr = \frac{\mu C_p}{k_f} = \frac{22.65\times10^{-6}\times1009}{0.0331} = 0.69$$

$$Ra = Gr\ Pr$$
$$= (6.937 \times 10^9\ L_s^3) \times 0.69 = 4.789 \times 10^9\ L_s^3$$

We also have, $Q_s = 1.14 \times Q_b$

$$h_s A(\Delta T) = 1.14\, h_b A(\Delta T)$$

or $$h_s = 1.14\, h_b$$

Thus using the given relation

$$\text{Nu} = \frac{h_s L_s}{k_f} = 0.59 \times (4.789 \times 10^9 L_s^3)^{1/4} \quad ...(i)$$

or $$h_s = 0.59 \frac{k_f}{L_s} \times (4.789 \times 10^9 L_s^3)^{1/4} \quad ...(ii)$$

Similarly, for bigger side $(15/L_s)$ cm vertical,

$$h_b = 0.59 \frac{k_f L_s}{15} \times \left[4.789 \times 10^9 \left\{\frac{15}{L_s}\right\}^3\right]^{1/4} \quad ...(iii)$$

Using the values of h_s and h_b in equation (*i*), we get

$$\frac{1}{L_s}(L_s^3)^{1/4} = 1.14 \frac{L_s}{15} \times \left\{\frac{15}{L_s}\right\}^{3/4}$$

Solving we get smaller side, $L_s = 3$ cm

Hence bigger side $L_b = \frac{15}{3}$ = **5 cm.** Ans.

(*ii*) The average heat transfer coefficient with smaller side vertical :

$$h_s - 0.59 \times \frac{0.0331}{0.03} \times [4.789 \times 10^9 \times (0.03)^3]^{1/4}$$
$$= 12.344 \text{ W/m}^2.\text{K}$$

The heat transfer rate :

$$Q_s = h_s A (T_s - T_\infty)$$
$$= 12.344 \times 15 \times 10^{-4} \times (200 - 30)$$
$$= \mathbf{3.147 \text{ W.}} \text{ Ans.}$$

The heat transfer rate with bigger side vertical :

$$Q_s = 1.14.\ Q_b$$

or $$Q_b = \frac{Q_s}{1.14} = \mathbf{2.76 \text{ W.}} \text{ Ans.}$$

Example 10.7. *A 2.5 kW plate heater of size 10 cm × 20 cm is held vertical with 20 cm side in a water bath at 40°C. Assuming the properties of water remains constant and the heat transfer takes place by convection only, find the steady state temperature attained by the heater.*

Use relation $Nu = 0.13(GrPr)^{1/3}$

The properties of water are

Temp. °C	C_p, J/kg.K	k_f, W/m.K	ν, m²/s	Pr	β, K⁻¹
60	4179	0.659	0.478×10^{-6}	2.98	5.11×10^{-4}
70	4187	0.668	0.415×10^{-6}	2.55	5.7×10^{-4}
80	4195	0.675	0.365×10^{-6}	2.21	6.32×10^{-4}

(P.U., Nov. 2008)

Solution

Given: 2.5 kW plate heater of size 10 cm × 20 cm ;

w = 10 cm = 0.1 m, L_c = 20 cm = 0.2 m

Q = 2.5 kW = 2500 W, T_∞ = 40°C.

To find : Surface temperature of the heater plate.

Assumptions :

1. Steady state conditions.
2. No radiation heat transfer.
3. Heat transfer from one side of the plate and other side as insulated.

Analysis : The heat transfer rate can be given by

$$Q = hA_s(\Delta T) \quad ...(i)$$

and given relation, $$\text{Nu} = \frac{hL_c}{k_f} = 0.13(\text{GrPr})^{1/3}$$

or $$h = 0.13 \frac{k_f}{L_c} (\text{GrPr})^{1/3} \quad ...(ii)$$

Trial 1. Assuming the surface temperature as 100°C

The properties at (100 + 40)/2 = 70°C can be used

The Grashof number, $$\text{Gr} = \frac{g\beta\Delta T L_c^3}{\nu^2}$$

$$= \frac{(9.81) \times 5.7 \times 10^{-4} \times (100-40) \times (0.2)^3}{(0.415 \times 10^{-6})^2}$$

$$= 1.5584 \times 10^{10}$$

Substituting in eqn. (*ii*),

$$h = 0.13 \times \frac{0.668}{0.2} \times (1.5584 \times 10^{10} \times 2.55)^{1/3}$$
$$= 1480 \text{ W/m}^2.\text{K}$$

The heat transfer rate with this value of convection coefficient

$$Q = 1480 \times 10 \times 20 \times 10^{-4} \times (100 - 40) = 1776 \text{ W}$$

which is less than the heater rating of 2500 W, hence our assumption was wrong.

Trial 2. Assuming heater surface temperature as 120°C

Then $$T_f = \frac{120 + 40}{2} = 80°\text{C}$$

The properties of water from given table ;

k_f = 0.675 W/m.K, ν = 0.365×10^{-6} m²/s,

Pr = 2.21, β = 6.32×10^{-4} K⁻¹

Grashof number,

$$Gr = \frac{9.81 \times 6.32 \times 10^{-4} \times (120 - 40) \times (0.2)^3}{(0.365 \times 10^{-6})^2}$$

$$= 2.98 \times 10^{10}$$

Substituting in eqn. (*ii*)

$$h = 0.13 \times \frac{0.675}{0.2} \times (2.98 \times 10^{10} \times 2.21)^{1/3}$$

$$= 1771.5 \text{ W/m}^2\text{.K}$$

The heat transfer rate

$$Q = 1771.5 \times (10 \times 20 \times 10^{-4}) \times (120 - 40)$$

$$= 2834.4 \text{ W}$$

Which is higher than the heater rating, thus this assumption was also wrong.

Trial 3. Assuming heater surface temperature as 114°C

$$\text{Then } T_f = \frac{114 + 40}{2} = 77°C$$

The properties of water at 77°C by interpolation

$k_f = 0.673$ W/m.K, $\nu = 0.38 \times 10^{-6}$ m²/s

Pr = 2.312 $\beta = 6.134 \times 10^{-4}$ K⁻¹

Grashof number

$$Gr = \frac{9.81 \times 6.134 \times 10^{-4} \times (114 - 40) \times (0.2)^3}{(0.38 \times 10^{-6})^2}$$

$$= 2.467 \times 10^{10}$$

The heat transfer coefficient from eqn. (*ii*)

$$h = 0.13 \times \frac{0.673}{0.2} \times (2.467 \times 10^{10} \times 2.312)^{1/3}$$

$$= 1684 \text{ W/m}^2\text{.K}$$

The heat transfer rate with this value of heat transfer coefficient

$$Q = 1684 \times (10 \times 20 \times 10^{-4}) \times (114 - 40)$$

$$= 2495 \text{ W}$$

Which is very nearer to the value of heater rating, thus keeping the heater surface temperature as **114°C.** Ans.

Example 10.8. *Consider an electrical heated square plate (60 cm × 60 cm) with one of its surface thermally insulated and the other surface dissipating heat by free convection into atmospheric air at 30°C. The heat flux over the surface of the plate is uniform and results in a mean temperature of 50°C. The plate is inclined at an angle of 50° from vertical. Determine the heat loss from the plate for the following cases: (a) Heated surface facing up; (b) Heated surface facing down.*

Solution

Given : An electrical heated plate insulated on one of its side;

$L = 60$ cm $= 0.60$ m, $w = 60$ cm $= 0.60$ m

$T_s = 50°C$, $T_\infty = 30°C$, $\theta = -50°$.

To find : The heat transfer rate when

(*i*) heated surface facing up,

(*ii*) heated surface facing down.

Properties of fluid : The mean temperature

$$T_f = T_s - 0.25\,(T_s - T_\infty)$$

$$= 50 - 0.25 \times (50 - 30) = 45°C = 318 \text{ K}$$

The physical properties of air at 318 K from Table A-4 ;

$\nu = 1.751 \times 10^{-5}$ m²/s, Pr = 0.704

$k_{air} = 0.0276$ W/m.K,

$$\beta = \frac{1}{T_\infty + 0.25\,(T_s - T_\infty)}$$

$$= \frac{1}{30 + 0.25 \times (50 - 30) + 273} = \frac{1}{308} \text{K}^{-1}.$$

Analysis : Characteristic length $L_c = L = 0.6$ m

Grashof number,

$$Gr = \frac{g\,\beta\,\Delta T\,L_c^3}{\nu^2}$$

$$= \frac{(9.81) \times \left(\frac{1}{308}\right) \times (50 - 30) \times (0.6)^3}{(1.751 \times 10^{-5})^2}$$

$$= 4.487 \times 10^8$$

(*i*) For hot surface facing up and inclined at – 50°. From Table 10.2, we have

$$Gr_c = 3.33 \times 10^8$$

Using the relation,

$$Nu = 0.145\,[(Gr\,Pr)^{1/3} - (Gr_c\,Pr)^{1/3}] + 0.56(Gr_c\,Pr\cos\theta)^{1/4}$$

$$= 0.145 \times [(4.487 \times 10^8 \times 0.704)^{1/3} - (3.33 \times 10^8 \times 0.704)^{1/3}] + 0.56 \times \{3.33 \times 10^8 \times 0.704 \times \cos(-50°)\}^{1/4} = 71.39$$

[**Note :** It can also be obtained by replacing g by $g\cos\theta$ in eqn. (10.38)]

Therefore, the value of average heat transfer coefficient,

$$h = Nu.\frac{k_f}{L_c} = 71.39 \times \frac{0.0276}{0.6} = 3.284 \text{ W/m}^2\text{.K}.$$

The heat transfer rate from the plate,

$$Q = h\,A_s(T_s - T_\infty)$$

$$= 3.284 \times (0.6 \times 0.6) \times (50 - 30)$$

$$= \mathbf{23.64\ W.}$$ Ans.

(*ii*) For the hot surface facing down, ($\theta = +50°$) the relation is

$$\text{Nu} = 0.56\,(\text{Gr Pr} \cos\theta)^{1/4}$$
$$= 0.56 \times [4.487 \times 10^8 \times 0.704 \times \cos(50°)]^{1/4}$$
$$= 66.85$$

and $$h = \text{Nu}\frac{k_f}{L_c} = 34.7 \times \frac{0.0276}{0.6} = 3.075 \text{ W/m}^2\text{.K}$$

The heat transfer rate from the plate,

$$Q = h\,A_s(T_s - T_\infty) = 3.075 \times (0.6 \times 0.6) \times (50 - 30)$$
$$= \mathbf{22.14\ W.}$$ **Ans.**

Example 10.9. *The size of CPU of a personal computer is 40 cm wide, 50 cm deep, and 10 cm high. Its top surface is dissipating 25 W to its surrounding air at 20°C. Calculate the temperature of the top surface.*

Solution

Given : $T_\infty = 20°C$ $\quad Q = 25$ W $\quad L = 40$ cm

$w = 50$ cm $\quad H = 10$ cm.

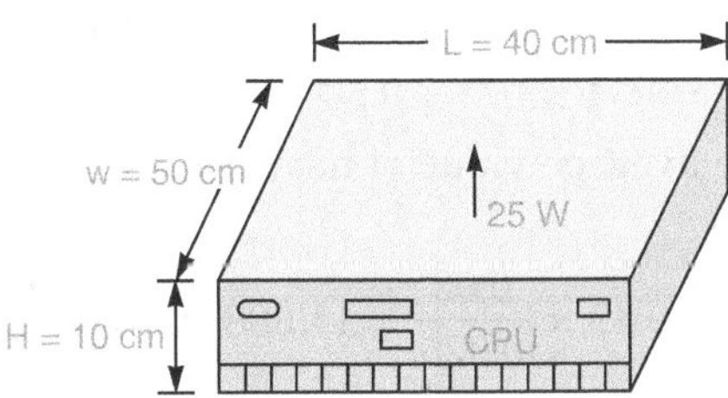

Fig. 10.17. Top surface of a computer to dissipate 25 W

To find : T_s, the top surface temperature.

Assumptions :

(*i*) Steady state conditions,

(*ii*) Uniform surface temperature,

(*iii*) The conduction and radiation heat transfer from top and sides are negligible

(*iv*) No monitor above the CPU.

(*v*) Air as an ideal gas.

Analysis : The convection heat transfer rate is given by

$$Q = h\,A_s\,(T_s - T_\infty) \qquad \ldots(i)$$

The surface temperature T_s is unknown and it is required to evaluate the properties of air for determination of heat transfer coefficient h.

We assume T_s as 40°C, then film temperature

$$T_f = \frac{T_s + T_\infty}{2} = \frac{40 + 20}{2} = 30°C$$

The physical properties of air at 30°C from Table A-4 ;

$\rho = 1.165$ kg/m³, $\quad C_p = 1005$ J/kg.K

$\mu = 1.865 \times 10^{-5}$ kg/ms. $\quad \nu = 16.0 \times 10^{-6}$ m²/s,

$k_f = 0.0264$ W/m.K $\quad$ Pr = 0.72,

$$\beta = \frac{1}{303}\ \text{K}^{-1}$$

The characteristic length of the geometry

$$L_c = \frac{A_s}{P} = \frac{0.4 \times 0.5}{2 \times (0.4 + 0.5)} = 0.111 \text{ m}$$

Grashof number

$$\text{Gr} = \frac{g\,\beta\,\Delta T\,L_c^3}{\nu^2}$$
$$= \frac{9.81 \times \frac{1}{303} \times (40 - 20) \times (0.111)^3}{(16.0 \times 10^{-6})^2}$$
$$= 3.46 \times 10^6$$

Rayleigh number

$$\text{Ra} = \text{Gr Pr} = 3.46 \times 10^6 \times 0.72 = 2.49 \times 10^6$$

Thus for hot horizontal plate facing up from Table 10.1

$$C = 0.54, \quad n = 1/4$$
$$\text{Nu} = 0.54(\text{Gr Pr})^{1/4} = 0.54 \times (\text{Ra})^{1/4}$$
$$= 0.54 \times (2.49 \times 10^6)^{1/4} = 21.45$$

The heat transfer coefficient

$$h = \frac{\text{Nu}\,k_f}{L_c} = \frac{21.45 \times 0.0264}{0.111} = 5.10 \text{ W/m}^2\text{.K.}$$

Using values in eqn. (*i*)

$$25 = 5.10 \times (0.4 \times 0.5) \times (T_s - 20)$$

we get $T_s = 44.5°C$

which is greater than assumed value, thus repeating calculation with 44°C

$$T_f = \frac{44 + 20}{2} = 32°C \text{ (305 K)}$$

The properties of air from Table A-4 ;

$\rho = 1.157$ kg/m³, $\quad C_p = 1005$ J/kg.K.

$\mu = 1.885 \times 10^{-5}$ kg/ms, $\quad \nu = 16.192 \times 10^{-6}$ m²/s

$k_f = 0.028$ W/m.K, $\quad$ Pr = 0.7

$$\text{Ra}_L = \frac{g\,\beta\,(T_s - T_\infty)\,L_c^3}{\nu^2}\,\text{Pr}$$
$$= \frac{9.81 \times \frac{1}{305} \times (44 - 20) \times (0.111)^3}{(16.192 \times 10^{-6})^2} \times 0.7$$
$$= 2.82 \times 10^6$$
$$\text{Nu} = 0.54(\text{Ra}_L)^{1/4} = 22.12$$
$$h = \frac{\text{Nu}\,k_f}{L_c} = \frac{22.12 \times 0.028}{0.111} = 5.58 \text{ W/m}^2\text{.K}$$

The heat transfer rate

$$Q = h\,A_s\,(T_s - T_\infty)$$

or $$25 = 5.58 \times (0.4 \times 0.5) \times (T_s - 20)$$

or $$T_s = 42.4°C$$

which is very close to the assumed value of 44°C, thus keeping the temperature of top surface as **42.4°C.** Ans.

Example 10.10. *A block 10 cm × 10 cm × 10 cm in size is suspended in still air at 10°C with one of its surface in horizontal position. All surfaces of the block are maintained at 150°C. Determine the total heat transfer rate from the block.* (N.M.U., Dec. 2002)

Solution

Given : $L = 10\text{ cm} = 0.1\text{ m}$,

$w = 0.1\text{ m}$, $z = 0.1\text{ m}$,

$T_s = 150°C$, and $T_\infty = 10°C$.

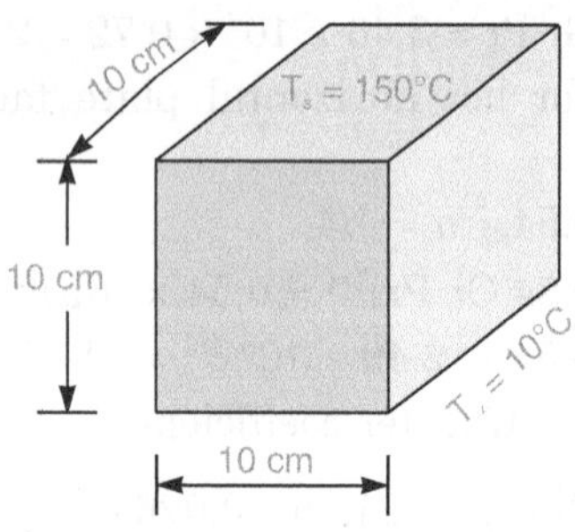

Fig. 10.18. Schematic of cubical block

To find :

Heat transfer rate from the cubical block.

Properties of fluid : The film temperature

$$T_f = \frac{T_s + T_\infty}{2} = \frac{(150 + 10)}{2} = 80°C = 353\text{ K}$$

The properties of air from Table A-4 ;

$\nu = 2.107 \times 10^{-5}\text{ m}^2/\text{s}$, $Pr = 0.697$

$k_{air} = 0.03\text{ W/m.K}$,

$$\beta = \frac{1}{353} = 2.832 \times 10^{-3}\text{ K}^{-1}.$$

Analysis : The Grashof number,

$$Gr = \frac{g\,\beta\,\Delta T\,L_c^3}{\nu^2}$$

$$= \frac{(9.81) \times (2.832 \times 10^{-3}) \times (150 - 10) \times L_c^3}{(2.107 \times 10^{-5})^2}$$

$$= 8.763 \times 10^9\,L_c^3$$

The Rayleigh number,

$$Ra = Gr\,Pr = (8.763 \times 10^9\,L_c^3) \times (0.697)$$

$$= 6.108 \times 10^9\,L_c^3$$

For four vertical surface of the cube, $L_c = L = 0.1$ m

$$Ra_L = 6.108 \times 10^9 \times (0.1)^3 = 6.11 \times 10^6$$

Using the relation from Table 10.4

$$Nu_L = 0.59 \times (Ra_L)^{1/4} = 0.59 \times (6.11 \times 10^6)^{1/4}$$

$$= 29.33$$

The average value of heat transfer coefficient on vertical surfaces

$$h_v = Nu_L\,\frac{k_{air}}{L_c} = 29.33 \times \frac{0.03}{0.1} = 8.8\text{ W/m}^2\text{.K}$$

The heat transfer rate from 4 vertical faces :

$$Q_v = h_v\,A_s\,(T_s - T_\infty)$$

$$= (8.8\text{ W/m}^2\text{.K}) \times (4 \times 0.1\text{ m} \times 0.1\text{ m}) \times (150 - 10)(\text{K}) = 49.28\text{ W}.$$

For top surface of the cube, the characteristic length is

$$L_c = \frac{A_s}{P} = \frac{0.1 \times 0.1}{2 \times (0.1 + 0.1)} = 0.025\text{ m}$$

$$Ra = 6.108 \times 10^9 \times (0.025)^3 = 95437.5$$

Using the relation from Table 10.4

$$Nu = 0.54(Ra)^{1/4} = 0.54 \times (95437.5)^{1/4} = 9.49$$

The average value of heat transfer coefficient on top surfaces

$$h_T = 9.49 \times \frac{0.03}{0.025} = 11.388\text{ W/m}^2\text{.K}$$

The heat transfer rate from the top surface :

$$Q_T = h_T\,A_s\,(T_s - T_\infty)$$

$$= 11.388 \times (0.1 \times 0.1) \times (150 - 10) = 15.94\text{ W}.$$

For bottom surface of the cube, the characteristic length is

$$L_c = \frac{A_s}{P} = 0.025\text{ m}$$

$$Ra = 6.108 \times 10^9 \times (0.025)^3 = 95437.5$$

Using the relation from Table 10.4

$$Nu = 0.27(Ra)^{1/4} = 0.27 \times (95437.5)^{1/4} = 4.75$$

The average value of heat transfer coefficient

$$h_B = 4.75 \times \frac{0.03}{0.025} = 5.7\text{ W/m}^2\text{.K}$$

The heat transfer rate from the plate,

$$Q_B = h_B\,A_s\,(T_s - T_\infty)$$

$$= 5.7 \times (0.1 \times 0.1) \times (150 - 10) = 7.97\text{ W}.$$

Total heat transfer rate from the block

$$Q_v + Q_T + Q_B = 49.28 + 15.94 + 7.97$$

$$= \mathbf{73.2\text{ W}.}$$ Ans.

Example 10.11. *A circular disc heater 0.2 m in diameter is exposed to ambient air at 25°C. One surface of the disc is insulated and other surface is maintained at 130°C. Calculate the amount of heat transferred from the disc when it is (i) horizontal with hot surface facing up, (ii) horizontal with hot surface facing down, and (iii) vertical.*

Solution

Given : A circular disc in different configuration exposed to air :

$$D = 0.2 \text{ m}, \quad T_s = 130°C, \quad T_\infty = 25°C.$$

To find : The heat transfer rate from the disc when ;

(*i*) horizontal with hot surface facing down,

(*ii*) horizontal with hot surface facing up and,

(*iii*) vertical.

Properties of fluid : The mean film temperature

$$T_f = \frac{T_s + T_\infty}{2} = \frac{130 + 25}{2} = 77.5°C = 350.5 \text{ K}$$

The physical properties of air :

$$\nu = 2.08 \times 10^{-5} \text{ m}^2/\text{s}, \qquad Pr = 0.697$$

$$k_{air} = 0.03 \text{ W/m.K}, \qquad \beta = \frac{1}{T_f} = \frac{1}{350.5} \text{ K}^{-1}$$

Analysis : The Grashof number with characteristic length L_c :

$$Gr = \frac{g\beta \Delta T L_c^3}{\nu^2}$$

$$= \frac{(9.81) \times \left(\frac{1}{350.5}\right) \times (130 - 25) \times L_c^3}{(2.08 \times 10^{-5})^2}$$

$$= 6.79 \times 10^9 L_c^3$$

The Rayleigh number,

$$Ra = Gr\,Pr = (6.79 \times 10^9 L_c^3) \times (0.697)$$

$$= 4.734 \times 10^9 L_c^3.$$

(*i*) For horizontal disc facing down :

$$L_c = \frac{A_s}{P} = \frac{(\pi/4)D^2}{\pi D} = \frac{D}{4} = 0.05 \text{ m}$$

Hence $Ra = 4.734 \times 10^9 \times (0.05)^3 = 591.81 \times 10^3$

Thus the flow is laminar, and for horizontal disc facing down, the correlation from Table 10.2

$$Nu = 0.54(Ra)^{1/4}$$

$$= 0.54 \times (591.81 \times 10^3)^{1/4} = 14.977$$

The average heat transfer coefficient of air,

$$h_1 = \frac{k_{air}}{L_c} Nu = \frac{0.03}{0.05} \times 14.977 = 8.98 \text{ W/m}^2.\text{K}$$

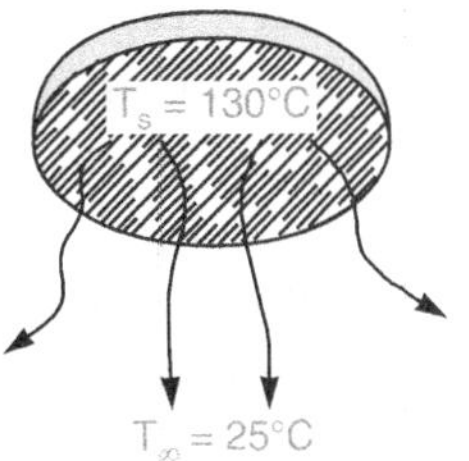

Fig. 10.19 (*i*) Horizontal disc facing down

The heat transfer rate from the disc ;

$$Q_1 = h_1 A_s (T_s - T_\infty)$$

$$= 8.98 \times (\pi/4) \times (0.2)^2 \times (130 - 25)$$

$$= \mathbf{29.64 \text{ W}}.$$ **Ans.**

(*ii*) For horizontal disc facing up :

The significant length remains same.

Hence $Ra = 4.734 \times 10^9 \times (0.05)^3 = 591.81 \times 10^3$

Thus the flow is laminar, and for horizontal disc facing up the correlation from Table 10.2

$$Nu = 0.27(Ra)^{1/4}$$

$$= 0.27 \times (591.81 \times 10^3)^{1/4} = 7.488$$

The average heat transfer coefficient,

$$h_2 = \frac{k_{air}}{L_c} Nu = \frac{0.03}{0.05} \times 7.488 = 4.493 \text{ W/m}^2.\text{K}$$

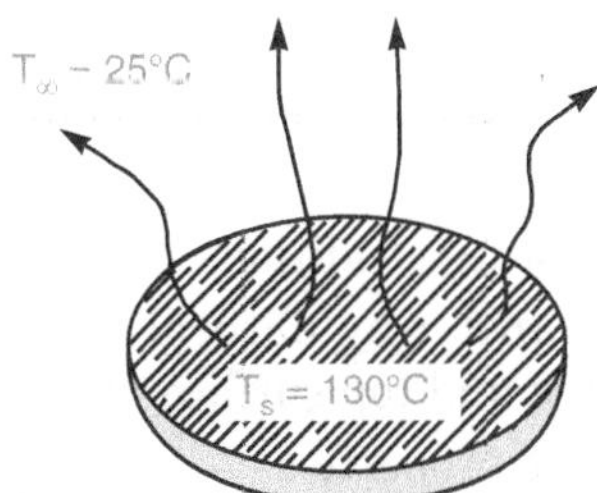

Fig. 10.19 (*ii*) Horizontal disc facing up

The heat transfer rate from the disc ;

$$Q_2 = h_2 A_s (T_s - T_\infty)$$

$$= 4.493 \times (\pi/4) \times (0.2)^2 \times (130 - 25)$$

$$= \mathbf{14.82 \text{ W}}.$$ **Ans.**

(*iii*) For vertical disc :

$$L_c = D = 0.2 \text{ m}$$

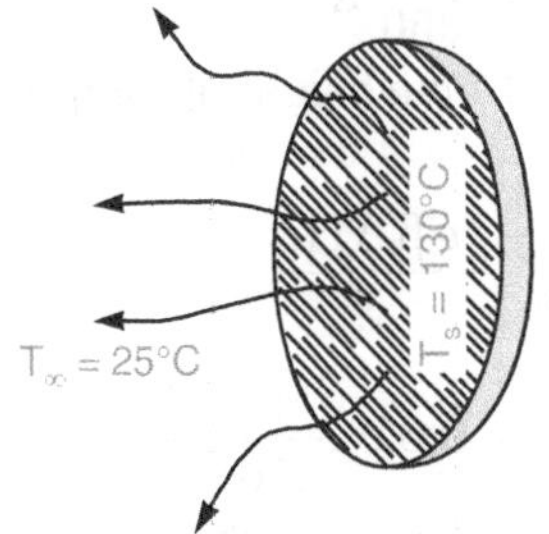

Fig. 10.19 (*iii*) Vertical disc

Hence $Ra_D = 4.734 \times 10^9 \times (0.2)^3 = 37.872 \times 10^6$

Thus the flow is laminar, and for vertical disc, from Table 10.4;

$$Nu_D = 0.59\,(Ra_D)^{1/4}$$
$$= 0.59 \times (37.872 \times 10^6)^{1/4} = 46.28$$

The average heat transfer coefficient,

$$h_3 = \frac{k_{air}}{L_c} Nu_D = \frac{0.03}{0.2} \times 46.28 = 6.94 \text{ W/m}^2\text{.K}$$

The heat transfer rate from the disc;

$$Q_3 = h_3 A_s (T_s - T_\infty)$$
$$= 6.94 \times (\pi/4) \times (0.2)^2 \times (130 - 25)$$
$$= \mathbf{22.9\ W.}\ \text{Ans.}$$

Example 10.12. *A computer chip, square in horizontal position, produces heat, while functioning. It was found that there are two cooling solutions : (i) air, and (ii) water. Calculate, which is the better, when chip temperature is 127°C and exposed in air at 27°C. The chip protrudes from the base. The chip is 1 cm high and 5 cm × 5 cm in size.*

Solution

Given : A computer chip with

$L_c = L = 1$ cm $= 0.01$ m

$w = 5$ cm $= 0.05$ m, $z = 0.05$ m

$T_s = 127$°C, $T_\infty = 27$°C

To find : Better solution of cooling

Assumptions :

1. Steady state conditions.

2. Heat transfer by natural convection from all four vertical sides and top surface of chip.

3. Constant properties.

Analysis :

Mean film temperature

$$T_f = \frac{T_s + T_\infty}{2} = 77°\text{C or } 350 \text{ K}$$

(*i*) Properties of air

$\nu = 20.92 \times 10^{-6}$ m²/s,

$k_f = 0.03$ W/m.K, Pr = 0.7

$$\beta = \frac{1}{350} \text{ K}^{-1}$$

For vertical 1 cm height of computer chip, Grashof number

$$Gr_L = \frac{g\beta\,\Delta T L_c^3}{\nu^2}$$
$$= 9.81 \times \frac{1}{350} \times \frac{(127 - 27) \times (0.01)^3}{(20.92 \times 10^{-6})^2}$$
$$= 6.404 \times 10^3$$

Rayleigh number

$$Ra_L = Gr_L Pr$$
$$= 6.404 \times 10^3 \times 0.7 = 4.483 \times 10^3$$

Flow is laminar.

Nusselt number

$$Nu_L = 0.59(Ra_L)^{1/4}$$
$$= 0.59 \times (4.483 \times 10^3)^{1/4} = 4.82$$

Average heat transfer coefficient

$$h = Nu_L \times \frac{k_f}{L_c} = 4.82 \times \frac{0.03}{0.01}$$
$$= 14.48 \text{ W/m}^2\text{.K}$$

For top surface (5 cm × 5 cm) of chip

$$L_c = \frac{A_s}{p} = \frac{0.05 \times 0.05}{(4 \times 0.05)} = 0.0125 \text{ m}$$

Then $Ra_t = 4.483 \times 10^3 \times \left(\frac{0.0125}{0.01}\right)^3$

$$= 8.755 \times 10^3$$

For hot surface facing up from Table 10.1

$$Nu_t = 0.54\,Ra^{1/4}$$
$$= 0.54 \times (8.755 \times 10^3)^{1/4} = 5.223$$

$$h_t = 5.223 \times \frac{0.03}{0.0125} = 12.536 \text{ W/m}^2\text{.K.}$$

Heat convection rate to air from sides and top surface of chip

$$Q = h_L \times 4 \text{ side area} \times \Delta T + h_t \times \text{top surface area} \times \Delta T$$
$$= 14.48 \times (4 \times 0.05 \times 0.01) \times 100 + 12.536 \times (0.05 \times 0.05) \times 100$$
$$= 2.896 + 3.134 = \mathbf{6.03\ W.}\ \text{Ans.}$$

(*ii*) Properties of water at 350 K.

$\mu_f = 343 \times 10^{-6}$ kg/ms, $\rho_f = 973.9$ kg/m³

$k_f = 0.668$ W/m.K, $Pr_f = 2.29$,

$\beta = 624.2 \times 10^{-6}$ K⁻¹

$$\nu = \frac{\mu_f}{\rho_f} = \frac{343 \times 10^{-6}}{973.9} = 3.522 \times 10^{-7}$$

For sides $L_c = 0.01$ m

$$Gr_L = \frac{9.81 \times 624.2 \times 10^{-6} \times (127 - 27) \times (0.01)^3}{(3.522 \times 10^{-7})^2}$$
$$= 4.934 \times 10^6$$

$Ra_L = 11.30 \times 10^6$ (Laminar flow)

$$Nu_L = 0.59(Ra_L)^{1/4}$$
$$= 0.59 \times (11.30 \times 10^6)^{1/4} = 34.20$$

$$h_L = 34.20 \times \frac{0.668}{0.01} = 2285 \text{ W/m}^2\text{.K}$$

For top surface $L_c = 0.0125$ m

$$Ra_t = 11.30 \times 10^6 \times \left(\frac{0.0125}{0.01}\right)^3$$
$$= 22.07 \times 10^6$$
$$Nu_t = 0.54\,(Ra_t)^{1/4}$$
$$= 0.54 \times (22.07 \times 10^6)^{1/4} = 37.012$$
$$h_t = 37.012 \times \frac{0.668}{0.0125} = 1977.94 \text{ W/m}^2\text{.K}$$

Heat transfer to water by free convection

$$Q = 2285 \times (4 \times 0.05 \times 0.01) \times 100 + 1977.94 \times (0.05 \times 0.05) \times 100$$
$$= 457 + 494.5 = \mathbf{951.5\ W.}$$ Ans.

Hence water is better coolant. Ans.

Example 10.13. *A hot plate 1 m × 0.5 m at 180°C is kept in still air at 20°C. Find :*

(i) The heat transfer coefficient.

(ii) Initial rate of cooling of the plate in °C/min.

(iii) Time required to cool the plate from 180°C to 80°C, if the heat transfer is due to convection only.

Mass of the plate is 20 kg and specific heat is 400 J/kg.K. Assume that the 0.5 m sides is vertical.

Solution

Given : $L = 0.5$ m, $w = 1$ m, $T_s = 180°C$, $T_\infty = 20°C$, $m = 20$ kg, $C = 400$ J/kg K.

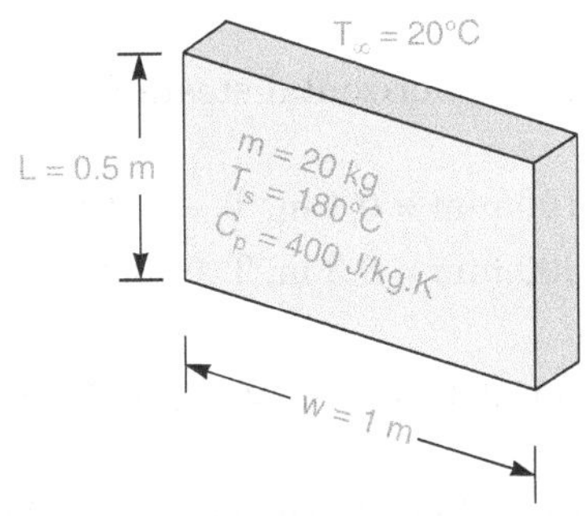

Fig. 10.20

To find :

(*i*) Heat transfer coefficient,

(*ii*) Initial rate of cooling of plate in °C/min,

(*iii*) Time required to cool the plate to 80°C.

Properties of fluid : The mean film temperature

$$T_f = \frac{T_s + T_\infty}{2} = \frac{180 + 20}{2} = 100°C = 373 \text{ K}$$

The properties of air,

$$\beta = \frac{1}{373} = 2.68 \times 10^{-3} \text{ K}^{-1},$$
$$\nu = 23.18 \times 10^{-6} \text{ m}^2/\text{s}$$
$$k_f = 0.0321 \text{ W/m.K}, \qquad Pr = 0.688.$$

Assumptions :

1 Radiation heat transfer is negligible.

2. Heat transfer from both sides of plate.

3. Transient heat conduction.

4. Constant properties.

Analysis : (*i*) The characteristic length

$$L_c = L = 0.5 \text{ m}$$

The Grashof number,

$$Gr_L = \frac{g\,\beta\,\Delta T\,L_c^3}{\nu^2}$$
$$= \frac{(9.81) \times (2.68 \times 10^{-3}) \times (180 - 20) \times (0.5)^3}{(23.18 \times 10^{-6})^2}$$
$$= 978.95 \times 10^6$$

Rayleigh number,

$$Ra_L = Gr_L.Pr = (978.95 \times 10^6 \times 0.688)$$
$$= 673.52 \times 10^6$$

The boundary layer is laminar, hence using the relation,

$$Nu_L = 0.59\,(Ra_L)^{1/4} = 0.59 \times (673.52 \times 10^6)^{1/4} = 95$$

The average value of heat transfer coefficient

$$h = Nu_L \frac{k_f}{L_c} = 95 \times \frac{0.0321}{0.5}$$
$$= \mathbf{6.1\ W/m^2.K.}$$ Ans.

(*ii*) The initial rate of cooling can be obtained by energy balance as

Rate of decrease of internal energy = Rate of heat convection from the plate

or $$-m\,C\,\frac{dT}{dt} = h\,A_s\,(T_s - T_\infty)$$

or $$\frac{dT}{dt} = \frac{6.1 \times (2 \times 1 \times 0.5\ \text{m}^2) \times (180 - 20)}{20 \times 400}$$
$$= -\,0.122°C/s$$
$$= \mathbf{-\ 7.322°C/min.}$$ Ans.

(*iii*) The time taken by plate to cool to 80°C :

$$\frac{T - T_\infty}{T_i - T_\infty} = \exp\left[-\frac{h\,A_s\,t}{\rho\,VC}\right] = \exp\left[-\frac{h\,A_s\,t}{m\,C}\right]$$

or $$t = -\frac{m\,C}{h\,A_s} \ln\left(\frac{T - T_\infty}{T_i - T_\infty}\right)$$
$$= -\frac{20 \times 400}{6.1 \times 2 \times 1 \times 0.5} \times \ln\left[\frac{80 - 20}{180 - 20}\right]$$
$$= 1286 \text{ s} = \mathbf{21.43\ min.}$$ Ans.

Example 10.14. *Estimate the heat transfer rate from a 100 W incandescent bulb at 140°C to an ambient at 24°C. Approximate the bulb as 60 cm diameter sphere. Calculate the percentage of power lost by natural convection.*

Use following correlation and air properties ;

$$Nu = 0.60\ (GrPr)^{1/4}$$

The properties of air at 82°C are

$\nu = 21.46 \times 10^{-6}\ m^2/s$,

$k_f = 30.38 \times 10^{-3}\ W/m.K$,

$Pr = 0.699$. (M.U., May 2002)

Solution

Given : The heat convection rate from a 100 W bulb (sphere)

$D = 60\ mm = 0.06\ m$, $T_s = 140°C$

$T_\infty = 24°C$, $Q_{gen} = 100\ W$.

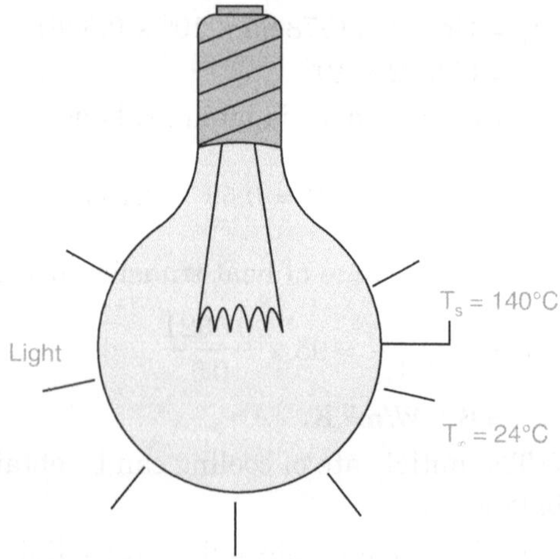

Fig. 10.21. Schematic of an incandescent bulb

To find : Percentage power lost by natural convection.

Assumptions :

(*i*) Negligible radiation heat transfer.

(*ii*) Steady state condition and constant properties.

Analysis. The film temperature

$$T_f = \frac{T_s + T_\infty}{2} = \frac{140 + 24}{2} = 82°C = 355\ K$$

$$\beta = \frac{1}{T_f} = \frac{1}{355}\ K^{-1}$$

Characteristic length, $L_c = D = 0.06$ m

$$Gr = \frac{g\beta(T_s - T_\infty)L_c^3}{\nu^2}$$

$$= 9.81 \times \frac{1}{355} \times \frac{(140-24)\times(0.06)^3}{(21.46\times10^{-6})^2}$$

$$= 1.503 \times 10^6$$

The average Nusselt number

$$Nu = 0.60\ (Gr\ Pr)^{1/4}$$

$$= 0.60 \times (1.503 \times 10^6 \times 0.699)^{1/4}$$

$$= 19.21\ W/m^2.K$$

The average heat transfer coefficient

$$h = \frac{Nu\ k_f}{L_c} = \frac{19.21\times30.38\times10^{-3}}{0.06}$$

$$= 9.73\ W/m^2.K.$$

The heat dissipation rate by natural convection

$$Q_{conv} = h\ (\pi D^2)\ (T_s - T_\infty)$$

$$= 9.73 \times [\pi \times (0.06)^2] \times (140 - 24)$$

$$= 12.76\ W$$

Percentage of power lost by natural convection

$$= \frac{Q_{conv}}{Q_{gen}} \times 100 = \frac{12.76}{100} \times 100$$

$= \mathbf{12.76\%}$. **Ans.**

Example 10.15. *Two horizontal steam pipes having diameters 100 mm and 300 mm are so laid in a boiler house that the mutual heat transfer may be neglected. The surface temperature of each of the steam pipe is 480°C. If these pipes are exposed in an ambient at 30°C. Calculate the ratio of heat transfer coefficients and heat losses per metre length of the pipes.*

Solution

Given : Two horizontal steam pipes exposed in a boiler house.

$D_1 = 100\ mm = 0.1\ m$, $T_s = 480°C$

$D_2 = 300\ mm = 0.3\ m$, $T_\infty = 30°C$.

To find :

(*i*) Ratio of heat transfer coefficients over two pipes.

(*ii*) Ratio of heat losses per metre length of two steam pipes.

Analysis : In natural convection heat transfer, the Nusselt number is expressed as

$$Nu = C(Gr\ Pr)^{1/n}$$

$$= C\left[\frac{g\ \beta \Delta T\ D^3}{\nu^2}\ Pr\right]^{1/4}$$

$$Nu \propto D^{3/4}$$

or $$\frac{hD}{k_f} \propto D^{3/4} \quad \text{or} \quad h \propto \frac{1}{D^{1/4}}$$

(*i*) The ratio of heat transfer coefficients

$$\frac{h_1}{h_2} = \left(\frac{D_2}{D_1}\right)^{1/4} = \left(\frac{300}{100}\right)^{1/4}$$

$= \mathbf{1.316.}$ Ans.

(*ii*) Similarly $Q = h(\pi DL)\,\Delta T$

Ratio $\dfrac{Q_1}{Q_2} = \dfrac{h_1 D_1}{h_2 D_2} = 1.316 \times \dfrac{100}{300}$

$= \mathbf{0.438.}$ Ans.

10.5. SIMPLIFIED EQUATIONS FOR AIR

At atmospheric pressure and moderate temperature, range, some simplified expressions given in Table 10.5, can be used for natural convection on isothermal surfaces exposed to air. The use of these relations can be extended to CO, CO_2, O_2, N_2 and the flue gases for the temperature ranges from 20°C to 800°C. For more precise approximation, the expressions presented in Table 10.4 must be used.

TABLE 10.5. Simplified relations for free convection to air at atmospheric pressure and moderate temperature

Sr. No.	*Geometry*	*Characteristic Length,* L_c	*Type of Flow*	*Range of Gr Pr*	*Correlation h =*
1.	Vertical Planes and Vertical Cylinders	Height, L	Laminar Turbulent	$10^4 \le Ra \le 10^8$ $10^8 \le Ra \le 10^{12}$	$1.42(\Delta T/L)^{1/4}$ $1.31(\Delta T)^{1/3}$
2.	Horizontal Cylinder	Diameter, D	Laminar Turbulent	$10^4 \le Ra \le 10^8$ $10^8 \le Ra \le 10^{12}$	$1.32(\Delta T/D)^{1/4}$ $1.24(\Delta T)^{1/3}$
3.	Horizontal Plates	$\frac{A_s}{P}$			
	(*i*) Heated surface facing down or cold surface facing up		Laminar Turbulent	$10^4 \le Ra \le 10^7$ $10^7 \le Ra \le 10^{11}$	$1.32(\Delta T/L_c)^{1/4}$ $1.52(\Delta T)^{1/3}$
	(*ii*) Heated surface facing up or cold surface facing down		Laminar	$10^5 \le Ra \le 10^{10}$	$0.59(\Delta T/L_c)^{1/4}$

Example 10.16. *1 cm O.D. horizontal copper tube carries liquid freon at – 30°C. If 2 m length of this tube must pass uninsulated through the still air at 40°C, determine the heat leakage when outside tube surface emissivity is 0.8. Use the following properties and correlations for determination of convection coefficient ;*

Air properties : $\beta = 3.597 \times 10^{-3}\ K^{-1}$, $Pr = 0.69$, $\nu = 1.66 \times 10^{-5}\ m^2/s$, $k_f = 0.028\ W/m.K$

Correlation for free convection ;

$h = 1.32(\Delta T/D)^{1/4}$ for $10^3 < Ra < 10^9$

$h = 1.24\ (\Delta T/D)^{1/3}$ for $10^9 < Ra < 10^{12}$.

(J.N.T.U., Nov. 2003)

Solution

Given : Horizontal copper tube carries liquid Freon :

$D = 1\ cm = 0.01\ m$, $L = 2\ m$,

$T_\infty = 40°C = 313\ K$, $\varepsilon = 0.8$,

$T_s = -30°C = 243\ K$,

To find : The heat loss from the tube.

Assumptions :

(*i*) Negligible convection resistance at the inner side of tube.

(*ii*) Constant properties.

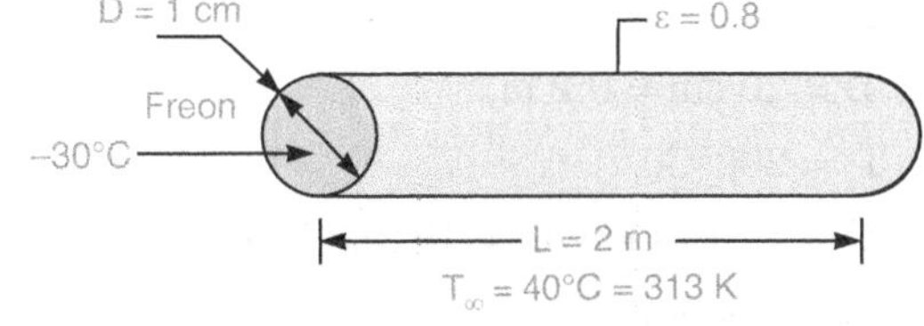

Fig. 10.22

Analysis : The Grashof number with characteristic length $L_c = D$

$$Gr = \frac{g\beta\Delta T\, D^3}{\nu^2}$$

$$= \frac{(9.81) \times (3.597 \times 10^{-3}) \times (40 + 30) \times (0.01)^3}{(1.66 \times 10^{-5})^2}$$

$= 8963.78$

The Rayleigh number,

$$Ra = Gr\ Pr = (8963.78 \times 0.69) = 6185$$

The Ra lies between 10^3 and 10^9, hence using the relation,

$$h = 1.32 \left(\frac{\Delta T}{D}\right)^{1/4} = 1.32 \times \left(\frac{40+30}{0.01}\right)^{1/4}$$

$$= 12.074 \text{ W/m}^2.\text{K}$$

The heat loss rate from the horizontal pipe by convection ;

$$Q_c = h(\pi DL)(T_s - T_\infty)$$
$$= 12.074 \times (\pi \times 0.01 \times 2) \times (40 + 30)$$
$$= 53.1 \text{ W}$$

The heat loss rate from horizontal pipe by radiation ;

$$Q_r = \sigma\, \varepsilon\, A_s(T_s^4 - T_\infty^4)$$
$$= 5.67 \times 10^{-8} \times (0.8) \times (\pi \times 0.01 \times 2) \times (313^4 - 243^4)$$
$$= 17.41 \text{ W}$$

The total heat loss rate from the pipe

$Q_c + Q_r = 53.1 + 17.41 =$ **70.5 W. Ans.**

Example 10.17. *A pipe carrying steam runs in a large room and is exposed to air at a temperature of 30°C. The pipe surface temperature is 200°C. The pipe diameter is 20 cm. If total heat loss rate from the pipe per metre length is 1.9193 kW/m, determine the pipe surface emissivity. Use correlation*

$$Nu = 0.53\ (Gr\ Pr)^{1/4}$$

and properties of air at 115°C

$k_f = 0.03306$ W/m².K, $\nu = 24.93 \times 10^{-6}$ m²/s

$Pr = 0.687$. (P.U., May 2001)

Solution

Given : A hot pipe is exposed in a large room.

$D = 20$ cm $= 0.2$ m, $L = 1$ m

$T_s = 200°C$, $T_\infty = 30°C$

$Q = 1.9193$ kW/m $= 1919.3$ W/m

$k_f = 0.03306$ W/m.K, $\nu = 24.93 \times 10^{-6}$ m²/s

$Pr = 0.687$

and relation for Nu.

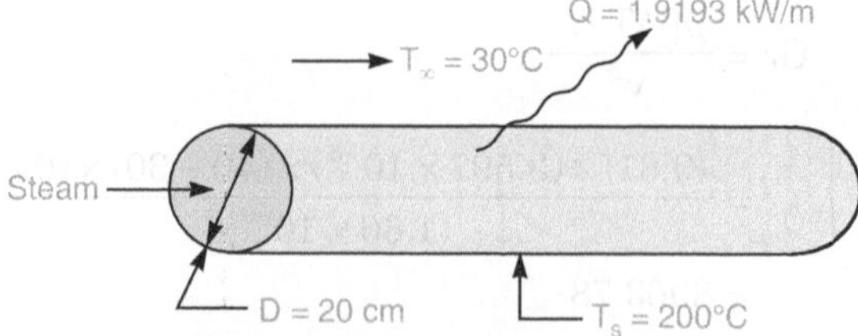

Fig. 10.23. Steam pipe in a room

To find :

(*i*) Natural convection heat transfer rate, then

(*ii*) Emissivity of the pipe surface.

Assumptions :

1. Steady state conditions.
2. Stefan Boltzmann constant as 5.67×10^{-8} W/m².K⁴.
3. Room walls are at 30°C.
4. Constant properties.

Analysis : The film temperature

$$T_f = \frac{T_s + T_\infty}{2} = \frac{200+30}{2} = 115°C$$

$$\beta = \frac{1}{T_f + 273} = \frac{1}{115+273} = \frac{1}{388} \text{ K}^{-1}$$

The characteristic length

$$L_c = D = 0.2 \text{ m}$$

The Grashof number

$$Gr = \frac{g\beta(T_s - T_\infty)L_c^3}{\nu^2}$$

$$= 9.81 \times \frac{1}{388} \times \frac{(200-30)\times(0.2)^3}{(24.93\times 10^{-6})^2}$$

$$= 5.53 \times 10^7$$

The Nusselt number

$$Nu = 0.53(Gr\ Pr)^{1/4}$$
$$= 0.53 \times (5.53 \times 10^7 \times 0.687)^{1/4} = 41.61$$

The heat transfer coefficient

$$h = \frac{Nu\, k_f}{L_c} = \frac{41.61 \times 0.03306}{0.2} = 6.88 \text{ W/m}^2.\text{K}$$

The heat dissipation rate by free convection,

$$Q_{conv} = h\ (\pi DL)\ (T_s - T_\infty)$$
$$= 6.88 \times (\pi \times 0.2 \times 1) \times (200 - 30)$$
$$= 734.77 \text{ W/m}$$

The heat dissipation rate by thermal radiation

$$Q_{rad} = Q - Q_{conv}$$
$$= 1919.3 - 734.77 = 1184.53 \text{ W/m}$$

The radiation heat transfer rate is expressed as

$$Q_{rad} = \sigma\, \varepsilon\, A_s(T_s^4 - T_\infty^4)$$

where T is in K, and

$$\therefore \quad 1184.53 = 5.67 \times 10^{-8} \times \varepsilon\ (\pi \times 0.2 \times 1) \times [(200 + 273)^4 - (30 + 273)^4]$$

$$= 1482.94\ \varepsilon$$

or $\varepsilon =$ **0.798. Ans.**

Example 10.18. *Beer cans (diameter 65 mm, length 150 mm) are to be cooled from an initial temperature of 20°C by placing them in a bottle cooler with an ambient air temperature of 1°C. Compare the initial cooling rates, when the cans are laid horizontally, to when they are laid vertically.* (N.M.U., Nov. 1997)

Solution

Given : Beer cans are to be cooled as :

$D = 65 \text{ mm} = 0.065 \text{ m}, \quad L = 150 \text{ mm} = 0.15 \text{ m}$

$T_s = 20°C, \quad T_\infty = 1°C.$

To find :

(*i*) Heat transfer rate from horizontal cans.

(*ii*) Heat transfer rate from vertical cans.

(*iii*) Comparison of heat transfer rate from cans in above two orientations.

Properties of fluid : The film temperature

$$T_f = \frac{T_s + T_\infty}{2} = \frac{20+1}{2} = 10.5°C = 283.5 \text{ K}$$

The properties of air are :

$\nu = 15.55 \times 10^{-6} \text{ m}^2/\text{s}, \quad \alpha = 0.19 \times 10^{-4} \text{ m}^2/s$

$\rho = 1.25 \text{ kg/m}^3$ and $k_f = 0.024$ W/m.K

$\beta = 1/283.5 = 3.527 \times 10^{-3} \text{ K}^{-1}.$

Analysis : The Grashof number with characteristic length L_c ;

$$Gr = \frac{g\beta\Delta T L_c^3}{\nu^2}$$

$$= \frac{(9.81)\times(3.527\times10^{-3})\times(20-1)\times L_c^3}{(15.55\times10^{-6})^2}$$

$$= 2.719 \times 10^9 L_c^3$$

The Prandtl number,

$$Pr = \frac{\nu}{\alpha} = \frac{15.55\times10^{-6}}{0.19\times10^{-4}} = 0.818$$

The Rayleigh number,

$$Ra = Gr\,Pr = 2.719 \times 10^9 L_c^3 \times 0.818$$

$$= 2.225 \times 10^9 L_c^3$$

(*i*) For horizontally laid cylinders :

$L_c = D = 0.065$ m

$Ra_D = 2.225 \times 10^9 \times (0.065)^3 = 611.0 \times 10^3$

Thus the flow is laminar, and relation from Table 10.4

$$Nu_D = 0.53(Ra_D)^{1/4}$$

$$= 0.53 \times (611.0 \times 10^3)^{1/4} = 14.82$$

The heat transfer coefficient,

$$h_1 = Nu_D \frac{k_f}{D} = 14.82 \times \frac{0.024}{0.065} = 5.47 \text{ W/m}^2\text{.K.}$$

The cooling rate,

$$Q_1 = h_1 A_s (\Delta T)$$

$$= 5.47 \times (\pi \times 0.065 \times 0.15) \times (20-1)$$

$= \mathbf{3.18\ W.}$ **Ans.**

(*ii*) For the vertical orientation, the cylindrical cans can be approximated as vertical wall of L = 0.15 m. The relevant Rayleigh number

$$Ra_L = 2.225 \times 10^9 \times (0.15)^3 = 7.51 \times 10^6$$

The boundary layer is laminar, hence using relation

$$Nu_L = 0.59(Ra_L)^{1/4}$$

$$= 0.59 \times (7.51 \times 10^6)^{1/4} = 30.88$$

and $$h_2 = Nu_L \frac{k_f}{L} = 30.88 \times \frac{0.024}{0.15} = 4.94 \text{ W/m}^2\text{.K.}$$

The cooling rate,

$$Q_2 = h_2 A_s (\Delta T)$$

$$= 4.94 \times (\pi \times 0.065 \times 0.15) \times (20-1)$$

$= \mathbf{2.87\ W.}$ **Ans.**

(*iii*) The percentage change in cooling rate, when cans are laid horizontally

$$\frac{Q_1 - Q_2}{Q_1} = \frac{3.18 - 2.87}{3.18} \times 100$$

$= \mathbf{9.6\%\ higher.}$ **Ans.**

Example 10.19. *A pipe 8 cm diameter is covered with 3 cm thick layer of insulation, which has surface emissivity of 0.9. The surface temperature of the insulation is 80°C and the pipe is placed in air at 20°C. Considering heat loss by radiation and natural convection, Calculate,*

(*i*) *Heat loss from 5 m length of pipe,*

(*ii*) *The overall heat transfer coefficient,*

(*iii*) *Heat transfer coefficient due to radiation.*

The properties of air are

T°C	*ρ kg/m³*	*C_p kJ/kg.K*	*μ × 10⁶ Ns/m²*	*k_f W/m.K*
20	1.205	1.005	18.1	0.0259
30	1.1625	1.005	18.6	0.02673
50	1.092	1.007	19.57	0.02781
80	1.00	1.009	21.1	0.0305
90	0.972	1.009	21.5	0.0313

The following correlations may be used :

$Nu = 0.53(Gr\,Pr)^{1/4}$ for $10^4 < Gr\,Pr < 10^7$

$= 0.15\,(Gr\,Pr)^{1/3}$ for $10^7 < Gr\,Pr < 10^9$

$= 0.22\,(Re)^{0.6}$ for $10^3 < Re < 10^5$.

(P.U., May 2002)

Solution

Given : An insulated pipe is exposed to air :

$D_1 = 8$ cm, $L = 5$ m,

$D_2 = 8$ cm + 2 × 3 cm = 14 cm = 0.14 m

$\varepsilon = 0.9$, $T_s = 80°C = 353$ K,

$T_\infty = 20°C = 293$ K.

To find :

(*i*) The heat dissipation rate by natural convection and thermal radiation for 5 m long insulated surface of pipe.

(*ii*) Overall heat transfer coefficient, and

(*iii*) Radiation heat transfer coefficient.

Analysis : The film temperature

$$T_f = \frac{T_s + T_\infty}{2} = \frac{80 + 20}{2} = 50°C$$

The properties of air at 50°C from Table A-4 are

$\rho = 1.092$ kg/m^3, $C_p = 1.007$ kJ/kg.K,

$\mu = 19.57 \times 10^{-6}$ Ns/m^2, $k_f = 0.02781$ W/m.K.

$$\beta = \frac{1}{T_f + 273} = \frac{1}{50 + 273} = \frac{1}{323} \text{ K}^{-1}$$

(*i*) The characteristic length of the geometry

$L_c = D_2 = 0.14$ m

The Grashof number

$$Gr = \frac{\rho^2 g \beta (T_s - T_\infty) L_c^3}{\mu^2}$$

$$= \frac{(1.092)^2 \times 9.81 \times \frac{1}{323} \times (80 - 20) \times (0.14)^3}{(19.57 \times 10^{-6})^2}$$

$= 15.569 \times 10^6$

The Prandtl number

$$Pr = \frac{\mu C_p}{k_f} = \frac{19.57 \times 10^{-6} \times 1007}{0.02781} = 0.708$$

Rayleigh number

$Ra = Gr\,Pr = 15.569 \times 10^6 \times 0.708 = 11.03 \times 10^6$

which is greater than 10^7, thus using

$Nu = 0.15(Gr\,Pr)^{1/3}$

$= 0.15 \times (15.59 \times 10^6 \times 0.708)^{1/3} = 33.4$

The heat transfer coefficient

$$h = \frac{Nu\,k_f}{L_c} = \frac{33.4 \times 0.02781}{0.14} = 6.63 \text{ W/m}^2\text{.K.}$$

The heat dissipation rate by natural convection.

$Q_{conv} = hA_s (T_s - T_\infty)$

$= 6.63 \times (\pi \times 0.14 \times 5) \times (80 - 20)$

$= 875.23$ W.

The heat dissipation rate by radiation

$Q_{rad} = \sigma \varepsilon A_s (T_s^4 - T_\infty^4)$

where T is in K and $\sigma = 5.67 \times 10^{-8}$ W/m^2.K^4

$Q_{rad} = 5.67 \times 10^{-8} \times 0.9 \times (\pi \times 0.14 \times 5)$

$(353^4 - 293^4) = 915.42$ W

The total heat dissipation rate by natural convection and radiation

$Q = Q_{conv} + Q_{rad} = 875.23 + 915.42$

$=$ **1790.66 W. Ans.**

(*ii*) The overall heat transfer coefficient

$Q = U A_s (\Delta T)$

$1790.66 = U \times (\pi \times 0.14 \times 5) \times (80 - 20)$

or $U =$ **13.57 W/m^2.K. Ans.**

(*iii*) Radiation heat transfer coefficient.

$Q_{rad} = h_r A_s (T_s - T_\infty)$

$915.42 = h_r (\pi \times 0.14 \times 5) \times (80 - 20)$

or $h_r =$ **6.94 W/m^2.K. Ans.**

Example 10.20. *A two stroke motor cycle petrol engine cylinder consists of 15 annular fins. If outside and inside diameters of each fin are 200 mm and 100 mm, respectively. The average fin surface temperature is 475°C and they are exposed in air at 25°C. Calculate the heat transfer rate from the fins for the following conditions :*

(*i*) *When motorcycle is at rest.*

(*ii*) *When motorcycle is running at a speed of 60 km/h.*

The fin may be idealised as a single horizontal flat plate of the same area.

Solution

Given : Fins as horizontal flat plate

$N_{fin} = 15$, $D_o = 200$ mm, $D_i = 100$ mm

$T_s = 475°C$, $T_\infty = 25°C$.

To find : Heat dissipation rate from fins in

(*i*) natural convection, and

(*ii*) forced convection.

Analysis : The film temperature

$$T_f = \frac{T_s + T_\infty}{2} = \frac{475 + 25}{2} = 250°C = 523 \text{ K.}$$

The thermophysical properties of air at 250°C

$k_f = 0.0427$ W/m.K, $\nu = 40.61 \times 10^{-6}$ m^2/s

$Pr = 0.677$, $\beta = \frac{1}{T_f} = \frac{1}{523}$ K^{-1}.

Case I : Motorcycle at rest :

The characteristic length for horizontal fin

$$L_c = \frac{A_s}{P} = 2 \times \frac{\pi}{4} \frac{(D_o^2 - D_i^2)}{\pi (D_o - D_i)}$$

$$= \frac{D_o + D_i}{2} = \frac{0.2 + 0.1}{2} = 0.15 \text{ m}$$

The Grashof number

$$Gr = \frac{g \beta \Delta T L_c^3}{\nu^2}$$

$$= 9.81 \times \frac{1}{523} \times \frac{(475 - 25) \times (0.15)^3}{(40.61 \times 10^{-6})^2}$$

$$= 17.27 \times 10^6$$

The Rayleigh number

$$Ra = Gr\, Pr = 17.27 \times 10^6 \times 0.677$$

$$= 11.694 \times 10^6$$

which is less than 10^9, thus the flow is laminar. For horizontal surface : $10^4 < Ra < 10^7$, from Table 10.1

$$Nu = 0.54\, Ra^{1/4}$$

$$= 0.54 \times (11.694 \times 10^6)^{1/4} = 31.58$$

The heat transfer coefficient

$$h = Nu \frac{k_f}{L_c} = 31.58 \times \frac{0.0427}{0.15} = 9.0 \text{ W/m}^2\text{.K}$$

The heat dissipation rate from both sides of fin

$$Q = h \left[2 \times \frac{\pi}{4} (D_o^2 - D_i^2) \right] N_{fin} (T_s - T_\infty)$$

$$= 9.0 \times \left[2 \times \frac{\pi}{4} \times (0.2^2 - 0.1^2) \right] \times 15 \times (475 - 25)$$

$$= \mathbf{2859.36 \text{ W}}. \quad \text{Ans.}$$

Case II : When motorcycle is running at a speed of 60 km/h

$$u_m = \frac{60 \times 10^3}{3600} = 16.67 \text{ m/s}$$

The hydraulic diameter

$$D_h = \frac{4A_c}{P} = \frac{4 \times (\pi/4)(D_o^2 - D_i^2)}{\pi (D_o + D_i)}$$

$$= D_o - D_i = 0.2 - 0.1 = 0.1 \text{ m}$$

The Reynolds number

$$Re = \frac{u_m D_h}{\nu} = \frac{16.67 \times 0.1}{40.61 \times 10^{-6}} = 41.04 \times 10^3$$

For $Re > 4 \times 10^4$

$$Nu = 0.027\, Re^{0.805}\, Pr^{1/3}$$

$$= 0.027 \times (41.04 \times 10^3)^{0.805} \times (0.677)^{1/3}$$

$$= 122.02$$

$$h = Nu \frac{k_f}{D_h} = 122.02 \times \frac{0.0427}{0.1}$$

$$= 52.1 \text{ W/m}^2\text{.K}$$

The heat dissipation rate from fins surface

$$Q = h\, [2 \times (\pi/4)(D_o^2 - D_i^2)]\, N_{fin} (T_s - T_\infty)$$

$$= 52.1 \times \left[\frac{2 \times \pi}{4} \times (0.2^2 - 0.1^2) \right] \times 15 \times (475 - 25)$$

$$= 16.572 \times 10^3 \text{ W} = \mathbf{16.57 \text{ kW}}. \quad \text{Ans.}$$

Example 10.21. *In a wind tunnel, 15°C air at 5 m/s flows over a flat plate, 1 m × 0.8 m in size. The plate temperature is 35°C. One of the side of the plate is arranged parallel to the flow direction, such that the heat transfer is lesser, estimate :*

(*i*) *Rate of heat transfer from the one side of plate.*

(*ii*) *Initial rate of cooling per hour of the plate, if mass of the plate is 5 kg and specific heat is 875 J/kg.K.*

(*iii*) *If the flow is turned off, compute the heat flow rate from the upper surface of the plate in still air at 15°C.*

(*iv*) *What is the percentage change in heat flow rate ?*

Use the following thermophysical properties of air and correlations

$\rho = 1.1707$ kg/m³, $C_{pf} = 1007$ J/kg.K,

$\nu = 15.712 \times 10^{-6}$ m²/s $k_f = 0.02614$ W/m.K

$Pr = 0.7075$

$Nu = 0.664\, Re_L^{1/2}\, Pr^{1/3}$ *for forced convection*

$= 0.27\, (Gr_L\, Pr)^{1/4}$ *for natural convection.*

(P.U., Dec. 2001)

Solution

Given : Flow over a flat plate

$u_\infty = 5$ m/s, $T_\infty = 15$°C, $L = 1$ m

$w = 0.8$ m $T_s = 35$°C $m = 5$ kg,

$C_p = 875$ J/kg.K.

and fluid properties.

To find :

(*i*) Rate of forced convection heat transfer from one side of the plate.

(*ii*) Initial rate of cooling of plate.

(*iii*) Heat flow rate for natural convection condition.

(*iv*) Percentage change in heat flow.

Assumptions :

(*i*) For lesser heat transfer rate in forced convection, the side with longer length to be consider as flow length.

(*ii*) Steady state conditions.

(*iii*) Constant properties.

(*iv*) No radiation heat transfer.

Analysis :

(*i*) The Reynolds number

$$Re_L = \frac{u_\infty L}{\nu} = \frac{5 \times 1}{15.712 \times 10^{-6}} = 318228.1$$

which is less than $Re_{cr} = 5 \times 10^5$, the flow is laminar, using correlation for average Nusselt number

$$Nu = 0.664\ Re_L^{1/2}\ Pr^{1/3}$$
$$= 0.664 \times (318228.1)^{1/2} \times (0.7075)^{1/3}$$
$$= 333.77$$

The average heat transfer coefficient

$$h = Nu\frac{k_f}{L} = 333.77 \times \frac{0.02614}{1}$$
$$= 8.725\ W/m^2.K.$$

The rate of heat transfer from one side of plate

$$Q_{forced} = h(wL)(T_s - T_\infty)$$
$$= 8.725 \times (1 \times 0.8) \times (35 - 15)$$
$$= \mathbf{139.6\ W.}\quad \text{Ans.}$$

(*ii*) Initial rate of cooling

$$Q_{forced} = mC_p\frac{dT}{dt}$$

$$139.6 = 5 \times 875 \times \frac{dT}{dt}$$

or $$\frac{dT}{dt} = 0.0319°C/s \simeq \mathbf{114.86°C/h.}\quad \text{Ans.}$$

(*iii*) Heat flow rate in natural convection from heated surface facing up.

The Grashof number

$$Gr_L = \frac{g\beta\Delta T L_c^3}{\nu^2}$$

where $$\beta = \frac{1}{T_f} \quad\text{and}\quad T_f = \frac{35+15}{2}$$
$$= 25°C = 298\ K$$
$$\Delta T = 35 - 15 = 20°C$$
$$L_c = \frac{A_s}{P} = \frac{0.8 \times 1}{2 \times (1+0.8)} = 0.222\ m$$

$$Gr_L = \frac{9.81 \times \frac{1}{298} \times 20 \times (0.222)^3}{(15.712 \times 10^{-6})^2} = 2.926 \times 10^7$$

$$Ra_L = Gr\ Pr = 2.926 \times 10^7 \times 0.7075$$
$$= 20.70 \times 10^6$$

Using given relation

$$Nu = \frac{hL_c}{k_f} = 0.27(Gr\ Pr)^{1/4}$$
$$= 0.27 \times (20.70 \times 10^6)^{1/4} = 18.21$$

and heat transfer coefficient

$$h = Nu\frac{k_f}{L_c} = 18.21 \times \frac{0.02614}{0.222}$$
$$= 2.144\ W/m^2.K.$$

The rate of heat convection from the plate

$$Q_{natural} = h(wL)(T_s - T_\infty)$$
$$= 2.144 \times (1 \times 0.8) \times (35 - 15)$$
$$= \mathbf{34.31\ W.}\quad \text{Ans.}$$

(*iv*) Percentage change in heat flow

$$= \frac{Q_{forced} - Q_{natural}}{Q_{forced}} \times 100$$

$$= \frac{139.6 - 34.31}{139.6} \times 100 = \mathbf{75.42\%.}\quad \text{Ans.}$$

Example 10.22. *A 12 cm-wide and 18 cm-high vertical hot surface in 25°C air is to be cooled by a heat sink with equally spaced fins of rectangular profile. The fins are 1 mm thick, 18 cm long in the vertical direction, and have a height of 2.4 cm from the base. Determine the optimum fin spacing, and the rate of heat transfer by natural convection from the heat sink, if the base temperature is 80°C.*

Use following relation for fin spacing and heat transfer coefficient

$$S_{opt} = 2.714\frac{L}{Ra^{1/4}}$$

Convection coefficient $h = 1.31\dfrac{k_f}{S_{opt}}$.

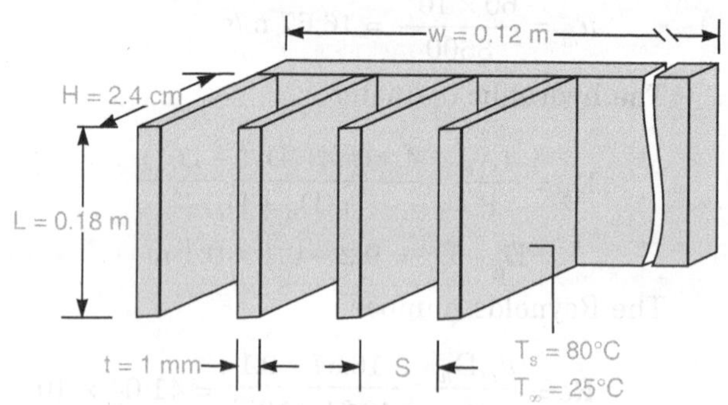

Fig. 10.24. Schematic for example 10.22

Solution

Given : A vertical hot surface with rectangular fins.

$L_c = 0.18$ m, $T_s = 80°C$, $H = 2.4$ cm,
$T_\infty = 25°C$, $w = 0.12$ m, $t = 1$ mm.

To find :

(*i*) Optimum fin spacing, and

(*ii*) Heat transfer rate from heat sink.

Assumptions :

(*i*) The fin thickness is very small as compared to fin spacing.

(*ii*) Fins are as vertical plate.

(*iii*) Constant properties and steady state conditions.

Analysis : The film temperature

$$T_f = \frac{T_s + T_\infty}{2} = \frac{80 + 25}{2} = 52.5°C = 325.5 \text{ K}$$

At this temperature, the properties of air

$k_f = 0.0279$ W/m.K

$\nu = 1.82 \times 10^{-5}$ m²/s

Pr = 0.709

$$\beta = \frac{1}{T_f} = \frac{1}{325.5 \text{ K}} = 0.003072 \text{ K}^{-1}$$

(*i*) The characteristic length for vertical fin

$L_c = L = 0.18$ m

The Grashof number

$$Gr_L = \frac{g\,\beta\,(\Delta T)\,L_c^3}{\nu^2}$$

$$= \frac{9.81 \times 0.003072 \times (80 - 25) \times (0.18)^3}{(1.82 \times 10^{-5})^2}$$

$$= 29.18 \times 10^6$$

The Rayleigh number

$$Ra_L = Gr_L Pr = 29.18 \times 10^6 \times 0.709$$

$$= 2.609 \times 10^7$$

The optimum fin spacing

$$S_{opt} = 2.714 \frac{L}{Ra^{1/4}} = 2.714 \times \frac{0.18}{(2.609 \times 10^7)^{1/4}}$$

$= 0.00724$ m = **7.24 mm.** Ans.

(*ii*) The number of fins

$$N_{fin} = \frac{\text{Width of plate}}{\text{Fin spacing + Fin thickness}} = \frac{w}{S + t}$$

$$= \frac{0.12}{0.00724 + 1 \times 10^{-3}} = 15 \text{ fins}$$

The heat transfer coefficient

$$h = 1.31 \frac{k_f}{S_{opt}} = 1.31 \times \frac{0.0279}{0.00724}$$

$= 5.04$ W/m².K.

Then heat transfer rate by natural convection

Q = Heat transfer from two sides of fins + Heat transfer from unfinned surface

$$Q = h\,(2N_{fin}\,LH)\,(T_s - T_\infty) + h\,(wL - Lt\,N_f)\,(T_s - T_\infty)$$

$$= [5.04 \times (2 \times 15 \times 0.18 \times 0.024) \times (80 - 25) + 5.04 \times (0.12 \times 0.18 - 0.18) \times 1 \times 10^{-3} \times 15] \times (80 - 25)$$

= 35.925 W + 5.24 W = **41.16 W.** Ans.

10.6. NATURAL CONVECTION IN ENCLOSED SPACES

The heat transfer through enclosures is of practical interest. The typical examples are natural convection in wall cavity, between window glazing and flat plate solar collectors. The heat transfer in enclosed spaces is complicated, due to movement of fluid in the enclosure. In a vertical enclosure, the fluid adjacent to hotter surface rises and fluid adjacent to cooler surface falls setting of rotationary motion within the enclosure, that increases the heat transfer rate through the enclosure. The typical flow pattern in vertical rectangular cavity is shown in Fig. 10.25.

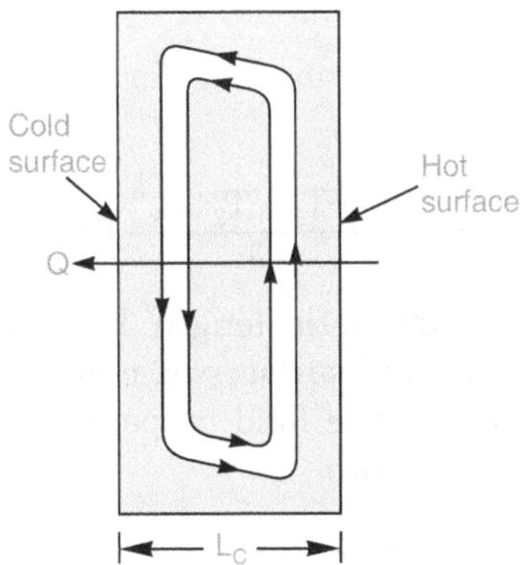

Fig. 10.25. Convection currents in vertical rectangular cavity

The flow pattern of fluid in a horizontal enclosure depends on the position of hotter surface. When the hotter surface in a horizontal rectangular enclosure is at the top, the convection current does not develop in the cavity, since the lighter fluid is always at the top of heavier fluid. Thus the heat is transferred in such a situation by pure conduction. But when the hotter surface is at the bottom of enclosure, then the fluid adjacent to surface is heated and becomes lighter, thus rises up and comes in contact of cooler surface at top, where it cools down. If $Ra_L < 1708$, the heat transfer is essentially by pure conduction. For $Ra_L > 1708$, the buoyancy force overcomes the fluid resistance and convection current starts in the cavity. The two situations of horizontal cavity are shown in Fig. 10.26.

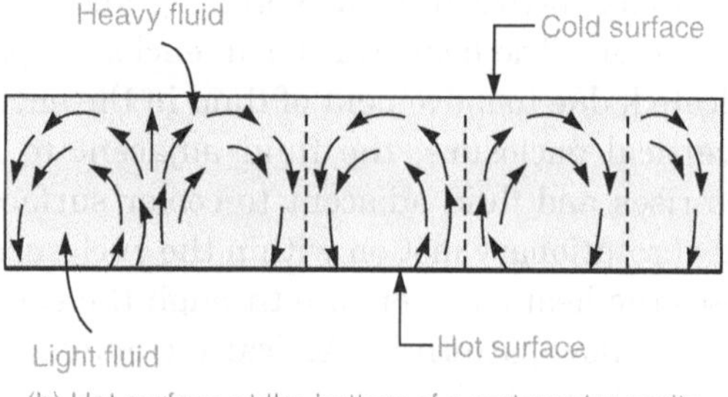

Fig. 10.26. Convection current in a horizontal enclosure

The Rayleigh number for an enclosure is calculated as

$$Ra = \frac{g\,\beta\,(T_1 - T_2)\,L_c^{\,3}}{\nu^2} Pr \qquad ...(10.58)$$

where the characteristic length L_c is the distance between hot and cold surfaces at temperature T_1 and T_2, respectively. All the fluid properties are evaluated at average temperature

$$T_f = \frac{1}{2}(T_1 + T_2)$$

The empirical correlations for Nusselt number for various enclosures are presented in Table 10.6.

The heat transfer rate

$$Q = hA_s\,(T_1 - T_2) = k_f\,Nu\,A_s\left(\frac{T_1 - T_2}{L_c}\right) \qquad ...(10.59)$$

$$\text{where} \quad A_s = \begin{cases} Hw & \text{Rectangular cavity} \\ \dfrac{\pi L(D_2 - D_1)}{\ln\left(\dfrac{D_2}{D_1}\right)} & \text{Concentric cylinders} \\ \pi D_1 D_2 & \text{Concentric spheres} \end{cases} \qquad ...(10.60)$$

where, w = width of rectangular cavity

H = height of cavity

L = length of cylinder

D_1 = inner diameter

D_2 = outer diameter.

Fig. 10.27. Isotherms in natural convection between concentric cylinders

Inclined cavity. For an **inclined rectangular cavity,** the complex correlation are available in the literature for accurate value of Nusselt number. But in Table 10.6, the Nusselt number for inclined rectangular cavities heated from the below and inclined upto 20° is determined from correlations for vertical rectangular cavity with the replacement of g by $g\cos\theta$ in Rayleigh number (Ra) relation given by eqn. (10.58).

TABLE 10.6. Empirical correlations for the average Nusselt number for natural convection in enclosures (the characteristic length L_c is as indicated on the respective diagram)

Geometry	Fluid	H/L_c	Range of Pr	Range of Ra	Nusselt number = $\frac{hL_c}{k_f}$
Vertical rectangular enclosure (or vertical cylindrical enclosure)	Gas or liquid	—	—	Ra < 2000	Nu = 1
		11–42	0.5–2	2×10^3–2×10^5	$\text{Nu} = 0.197\ \text{Ra}^{1/4} \times \left(\frac{H}{L_c}\right)^{-1/9}$
	Gas	11–42	0.5–2	2×10^5–10^7	$\text{Nu} = 0.073\ \text{Ra}^{1/3} \times \left(\frac{H}{L_c}\right)^{-1/9}$
	Liquid	10–40	1–20,000	10^4–10^7	$\text{Nu} = 0.42\ \text{Pr}^{0.012}\ \text{Ra}^{1/4} \times \left(\frac{H}{L_c}\right)^{-0.3}$
		1–40	1–20	10^6–10^9	$\text{Nu} = 0.046\ \text{Ra}^{1/3}$
Inclined rectangular enclosure					Use the correlations for vertical enclosures as a first-degree approximation for $\theta \le 20°$ by replacing g in the Ra relation by $g \cos\theta$
Horizontal rectangular enclosure (hot surface at the top)	Gas or liquid	—	—	—	Nu = 1
Horizontal rectangular enclosure (hot surface at the bottom)	Gas or liquid	—	—	Ra < 1708	Nu = 1
	Gas	—	0.5–2	1.7×10^3–7×10^3	$\text{Nu} = 0.059\ \text{Ra}^{0.4}$
		—	0.5–2	7×10^3–3.2×10^5	$\text{Nu} = 0.212\ \text{Ra}^{1/4}$
		—	0.5–2	$\text{Ra} > 3.2 \times 10^5$	$\text{Nu} = 0.061\ \text{Ra}^{1/3}$
	Liquid	—	1–5000	1.7×10^3–6×10^3	$\text{Nu} = 0.012\ \text{Ra}^{0.6}$
		—	1–5000	6×10^3–3.7×10^4	$\text{Nu} = 0.375\ \text{Ra}^{0.2}$
		—	1–20	3.7×10^4–10^8	$\text{Nu} = 0.13\ \text{Ra}^{0.3}$
		—	1–20	$\text{Ra} > 10^8$	$\text{Nu} = 0.057\ \text{Ra}^{1/3}$
Concentric horizontal cylinders	Gas or liquid	—	1–5000	6.3×10^3–10^6	$\text{Nu} = 0.11\ \text{Ra}^{0.29}$
			1–5000	10^6–10^8	$\text{Nu} = 0.40\ \text{Ra}^{0.20}$
Concentric spheres	Gas or liquid	—	0.7–4000	10^2–10^9	$\text{Nu} = 0.228\ \text{Ra}^{0.226}$

Effective thermal conductivity. We know that the steady state heat conduction rate, Q, in a stationary fluid layer is given by

$$Q = \frac{k_f A(T_1 - T_2)}{L_c} \quad ...(10.61)$$

where k_f = thermal conductivity of the fluid,
A = area normal to heat transfer,
L_c = thickness of the fluid layer,
T_1, T_2 = temperature on two sides of the fluid layer.

The comparison of eqn. (10.61) with eqn. (10.59), indicates that convection heat transfer in an enclosure or cavity is identical to heat conduction across the fluid layer, if thermal conductivity of the fluid k_f is replaced by k_fNu as a result of convection current. Therefore, the quantity k_f Nu is called the **effective thermal conductivity** of the cavity. That is

$$k_{eff} = k_f \, Nu \quad ...(10.62)$$

when Nu = 1, then

$$k_{eff} = k_f$$

It indicates pure conduction in the fluid layer. The heat transfer rate by natural convection between two long, horizontal concentric cylinders at constant temperatures T_1 and T_2, respectively is expressed as

$$Q = \frac{2\pi L \, k_{eff} \, (T_1 - T_2)}{\ln\left(\frac{D_2}{D_1}\right)} \quad ...(10.63)$$

and for two concentric spheres.

$$Q = \frac{\pi D_1 D_2 \, k_{eff}}{L_c}(T_1 - T_2) \quad ...(10.64)$$

where $L_c = \frac{1}{2}(D_2 - D_1)$.

Rotating disc. The rotating disc provides a good example of fluid flow that changes from pure natural convection when disc is at the rest to mixed and forced convection, when disc is rotating. For a disc at uniform surface temperature and exposed to air (Pr = 0.72) the correlation is suggested in the form

$$Nu = 0.47(Re_\omega^2 + Gr)^{1/4} \quad ...(10.65)$$

where

$$Gr = \frac{g\beta \Delta T L_c^3}{\nu^2}$$

$$Re_\omega = \frac{\omega r_o^2}{\nu} \quad ...(10.66)$$

$$Nu = \frac{h r_o}{k_f}$$

ω = angular velocity of disc = $\frac{2\pi N}{60}$
N = rotation per minute (r.p.m.)
L_c = characteristic length of disc
= r_o for horizontal disc
= $\frac{1}{2}\pi r_o$ for vertical disc

where, r_o = radius of the disc.

Example 10.23. *A vertical 0.8 m high, 2 m wide, double pane window consists of two sheets of glass separated by 2 cm air gap at atmospheric pressure. If the glass surface temperatures across the air gap are measured to be 12°C and 2°C, determine the rate of heat transfer through the window.*

Solution

Given : A vertical rectangular enclosure.

H = 0.8 m, w = 2 m
L_c = 2 cm = 0.02 m, T_1 = 12°C
T_2 = 2°C.

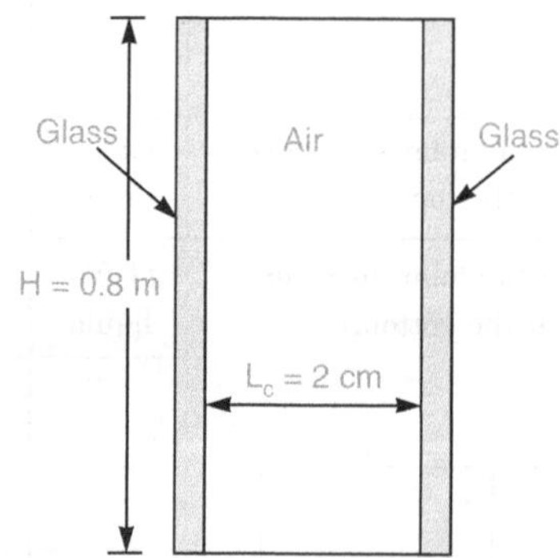

Fig. 10.28. Schematic of double pane glass window

To find : Rate of heat transfer through the window.

Analysis : The average temperature of two surfaces.

$$T_f = \frac{T_1 + T_2}{2} = \frac{12+2}{2} = 7°C = 280 \text{ K}$$

The properties of air at 7°C from Table A-4

k_f = 0.0246 W/m.K
$\nu = 1.40 \times 10^{-5}$ m²/s
Pr = 0.717

$$\beta = \frac{1}{T_f + 273} = \frac{1}{280} = 0.00357 \text{ K}^{-1}$$

The characteristic length L_c = 2 cm = 0.02 m

The Grashof number

$$Gr = \frac{g\,\beta\,\Delta T\, L_c^3}{\nu^2}$$

$$= \frac{9.81 \times 0.00357 \times (12-2) \times (0.02)^3}{(1.40 \times 10^{-5})^2}$$

$$= 14.29 \times 10^3$$

The Rayleigh number

$$Ra = Gr\,Pr = 14.29 \times 10^3 \times 0.717$$

$$= 10.25 \times 10^3$$

Then from Table 10.6, the Nusselt number is given by

$$Nu = 0.197\, Ra^{1/4} \left(\frac{H}{L_c}\right)^{-1/9}$$

$$= 0.197 \times (10.25 \times 10^3)^{1/4} \times \left(\frac{0.8}{0.02}\right)^{-1/9}$$

$$= 1.315$$

The heat transfer coefficient

$$h = Nu\,\frac{k_f}{L_c} = 1.315 \times \frac{0.0246}{0.02}$$

$$= 1.62\ \text{W/m}^2.\text{K}$$

The convection heat transfer rate

$$Q = hwH(T_1 - T_2)$$

$$= 1.62 \times (0.8 \times 2) \times (12 \quad 2)$$

$$= \mathbf{25.9\ W.}$$ **Ans.**

Example 10.24. *A 10 cm diameter sphere is maintained at 120°C. It is enclosed in a 12 cm diameter concentric spherical surface maintained at 100°C. The space between two spheres is filled with air at 200 kPa. Calculate the convective heat transfer rate from inner sphere.*

Solution

Given : Two concentric sphere and air in the annular gap.

D_1 = 10 cm, D_2 = 12 cm

T_1 = 120°C, T_2 = 100°C

p = 200 kPa

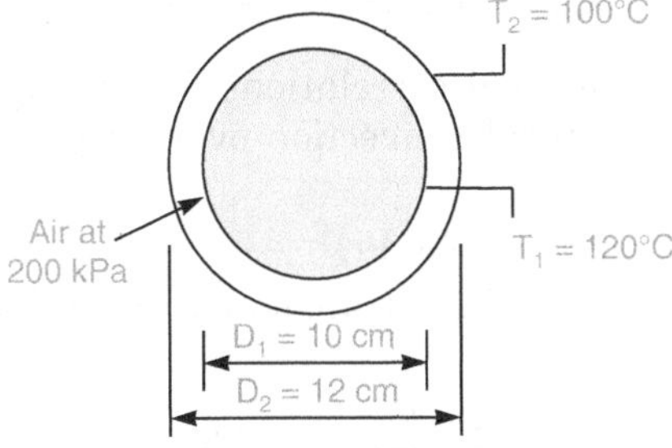

Fig. 10.29. Schematic of two isothermal concentric spheres

To find : Convective heat transfer rate.

Analysis : The film temperature.

$$T_f = \frac{T_1 + T_2}{2} = \frac{120 + 100}{2} = 110°C = 383\ K$$

$$\rho = \frac{p}{RT} = \frac{200\ \text{kPa}}{(0.287\ \text{kJ/kg.K}) \times (383\ \text{K})}$$

$$= 1.819\ \text{kg/m}^3$$

and other properties at 110°C are

$k_f = 0.0319$ W/m.K, $\mu = 2.22 \times 10^{-5}$ kg/ms.

$Pr = 0.703$, $\beta = \frac{1}{T_f + 273} = \frac{1}{383}\ K^{-1}$

The characteristic length,

$$L_c = \frac{1}{2}(D_2 - D_1) = \frac{1}{2} \times (0.12 - 0.1) = 0.01\ m$$

The Grashof number

$$Gr = \frac{\rho^2 g\,\beta\,\Delta T\, L_c^{\,3}}{\mu^2}$$

$$= (1.819)^2 \times 9.81 \times \frac{1}{383} \times \frac{(120-100) \times (0.01)^3}{(2.22 \times 10^{-5})^2} = 3441$$

The Rayleigh number

$$Ra = Gr\,Pr = 3441 \times 0.703 = 2419$$

Using equation from Table 10.6,

$$Nu = 0.228\, Ra^{0.226}$$

$$= 0.228 \times (2419)^{0.226} = 1.326$$

$$h = Nu\,\frac{k_f}{L_c} = 1.326 \times \frac{0.0319}{0.01} = 4.23\ \text{W/m}^2.\text{K}$$

The convective heat transfer rate from inner sphere

$$Q = h(\pi D_1^{\,2})\,(T_1 - T_2)$$

$$= 4.23 \times \pi \times (0.1)^2 \times (120 - 100)$$

$$= \mathbf{2.65\ W.}$$ **Ans.**

Example 10.25. *A flat plate solar collector has 8 cm high and 1 m wide and 1.6 m depth is tilted at 40° to the horizontal. The inner wall is at 70°C and the outer wall at 10°C and the enclosure is filled with air at 1 atm. Estimate the heat loss.*

Solution

Given : An inclined flat plate solar collector with air as working fluid

L_c = 8 cm = 0.08 m, w = 1 m

H = 1.6 m, θ = 40° (with horizontal)

T_1 = 70°C, T_2 = 10°C.

To find : Heat loss from the solar flat plate collector.

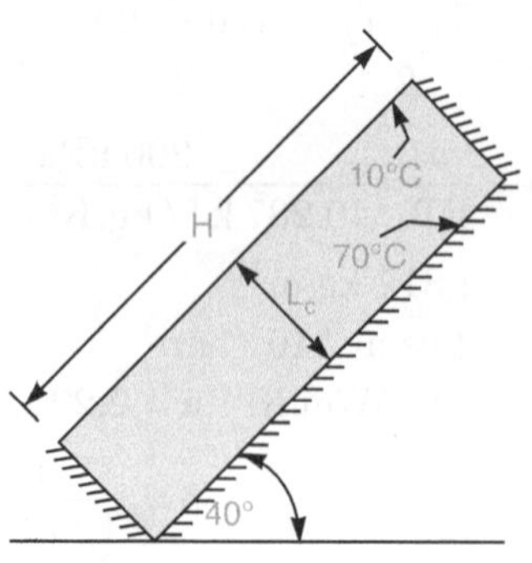

Fig. 10.30

Assumptions :

(*i*) The bottom plate and sides of solar collector are well insulated.

(*ii*) Heat transfer by natural convection only.

(*iii*) Steady state conditions.

(*iv*) Constant properties.

Analysis : The film temperature

$$T_f = \frac{70 + 10}{2} = 40°C$$

The properties of air at 40°C,

$\rho = 1.128$ kg/m^3, $C_p = 1005$ J/kg.K

$\mu = 1.91 \times 10^{-5}$ kg/ms, $\nu = 16.96 \times 10^{-6}$ m^2/s

$k_f = 0.0276$ W/m.K Pr = 0.699, $\beta = \frac{1}{313}$ K^{-1}

The Rayleigh number

$$Ra_L = \frac{g\,\beta\,(T_1 - T_2)\,L_c^{\,3}}{\nu^2}\,Pr$$

$$= \frac{9.81 \times \frac{1}{313} \times (70-10) \times (0.08)^3}{(16.96 \times 10^{-6})^2} \times 0.699$$

$$= 2.34 \times 10^6$$

$$Ra_\theta = Ra_L \sin(40°) = 1.504 \times 10^6$$

The average Nusselt number (Table 10.6)

$$Nu = 0.073\, Ra_\theta^{1/3} \left(\frac{H}{L_c}\right)^{-1/9}$$

$$= 0.073 \times (1.504 \times 10^6)^{1/3} \times \left(\frac{1.6}{0.08}\right)^{-1/9} = 5.97$$

The average heat transfer coefficient

$$h = Nu\,\frac{k_f}{L_c} = 5.97 \times \frac{0.0276}{0.08} = 2.05$$

The heat dissipation rate

$$Q = h(wH)\,(T_1 - T_2)$$
$$= 2.05 \times (1 \times 1.6) \times (70 - 10)$$
$$= \mathbf{197.7\ W.}\quad \textbf{Ans.}$$

10.7. SUMMARY

In natural convection heat transfer, the fluid motion is induced by buoyancy effects, developed due to density variation in the fluid. The fluid velocity associated with natural convection is usually much lower, therefore, the heat transfer rate is also much lower than in forced convection.

The buoyancy force is upward force exerted by a fluid on a body that is immersed in it. Its magnitude is equal to weight of fluid displaced by the body. The coefficient of volumetric expansion β of fluid represents the variation of density of fluid with temperature at constant pressure.

where T_f is absolute film temperature = $\frac{T_s + T_\infty}{2}$.

The flow regime in natural convection is characterised by a dimensionless number, called the Grashof number, which represents the ratio of buoyancy force to viscous force acting on the fluid and is expressed as

$$Gr = \frac{g\,\beta\,(T_s - T_\infty)\,L_c^3}{\nu^2}$$

where g = gravitational acceleration, m/s^2

β = coefficient of volumetric expansion, for an ideal gas, $\beta = \frac{1}{T_f}$ K^{-1}

$T_s - T_\infty$ = temperature difference between surface and its ambient, °C or K

L_c = characteristic length of the geometry, m

ν = kinematic viscosity, m^2/s.

The Rayleigh number is also a dimensionless number given as

$$Ra = Gr\,Pr = \frac{g\,\beta\,(T_s - T_\infty)\,L_c^{\,3}}{\nu^2}\,Pr$$

The empirical correlations for average Nusselt number for natural convection over surfaces given in the form

$$Nu = C(Gr\,Pr)^n$$

The average heat transfer coefficient is obtained as

$$h = Nu\,\frac{k_f}{L_c}\ \text{(W/m}^2\text{.K)}$$

where k_f = thermal conductivity of fluid, W/m.K

The convection heat transfer rate between a surface and its surrounding is expressed as

$$Q = hA_s(T_s - T_\infty)$$

where A_s = the heat transfer surface area, m^2.

For various enclosures, the simple correlations to obtain average Nusselt number are presented in Table 10.6. The heat transfer through an enclosure is given by

$$Q = hA_s(T_1 - T_2) = k_f \, Nu \, A_s \frac{(T_1 - T_2)}{L_c}$$

where $$A_s = \begin{cases} H w & \text{Rectangular cavity} \\ \dfrac{\pi L(D_2 - D_1)}{\ln(D_2/D_1)} & \text{Concentric cylinders} \\ \pi D_1 D_2 & \text{Concentric spheres.} \end{cases}$$

The quantity $k_f Nu$ is called effective thermal conductivity of the enclosure.

REVIEW QUESTIONS

1. What is the natural convection ? How does it differ from the forced convection ? What force causes natural convection currents ?
2. Show that the coefficient of volumetric expansion for an ideal gas is

   $\beta = \dfrac{1}{T}$, where T is absolute temperature of gas.
3. What is Rayleigh number ?
4. Why the heat transfer coefficient for natural convection is much less than that for forced convection ?
5. How is the velocity field developed for natural flow of fluid over a vertical plate when its surface is maintained at temperature (*i*) higher, and (*ii*) lower than its surroundings ?
6. Show that for a laminar flow of air with Pr = 0.72, the local and average value of Nusselt numbers are given by

   $Nu_x = 0.378 \, Gr_x^{1/4}$ and $Nu = 0.504 \, Gr_L^{1/4}$.
7. What is the modified Grashof number ? Where does it use ?
8. Explain the heat transfer mechanism in a vertical rectangular cavity consisting of two isothermal parallel planes.
9. Why does heat transfer rate decrease drastically if double pane window with an air gas is used instead of a single wheel window ?
10. How does the effective thermal conductivity of an enclosure define ?
11. Beginning with the natural convection correlation of the form

    $$Nu = \frac{hL}{k_f} = C \, Ra_L^n$$

    Show that for air at atmospheric pressure and a film temperature of 400 K, the average heat transfer coefficient for a vertical plate can be expressed as

    $$h = 1.40 \left(\frac{\Delta T}{L}\right)^{1/4} \qquad 10^4 < Ra_L < 10^9$$

    $$= 0.98 \, (\Delta T)^{1/3} \qquad 10^9 < Ra_L < 10^{13}.$$

PROBLEMS

1. A vertical plate 4 m high and 1 m wide is maintained at 60°C in an ambient of still air at 10°C. Determine the value of heat transfer coefficient. [Ans. 4.82 W/m^2.K]
2. Water is heated in a tank using horizontal pipes, 50 mm diameter with wall temperature of 60°C maintained by condensing steam on the inside of the tubes. The water in the tank is at 20°C. Calculate the value of natural convection coefficient, if the water is stagnant. [Ans. 795 W/m^2.K]
3. Consider a object of characteristic length of 0.01 m and a situation for which the temperature difference is 30°C. Evaluate the thermophysical properties at the given conditions and determine the Rayleigh number for the following fluids : (*i*) air at 1 atm and 400 K, (*ii*) helium at 1 atm and 400 K, and (*iii*) water at 310 K. [Ans. (*i*) 615.3, (*ii*) 12, (*iii*) 1.658 × 10^6]
4. Estimate the coefficient of free convection on a wire, 2 mm in diameter, immersed in water at 20°C, if the wire surface is maintained at 300°C. [Ans. 3366 W/m^2.K]
5. A flat square electrical heater of 0.5 m × 0.5 m is placed vertically in still air at 20°C. The heat generated is 1200 W/m^2. Determine the value of natural convection coefficient and average temperature of the plate. [Ans. 33.02 W/m^2.K, 56.5°C]
6. A plate heater 0.4 m × 0.4 m, using electrical elements, has a constant heat flux of 1.3 kW/m^2. It is placed in a room air at 20°C with hot side facing up. Determine the value of heat transfer coefficient and average temperature of the plate. [Ans. 9.13 W/m^2.K, 151.4°C]

7. A vertical pipe of 10 cm diameter and 3 m long, at a surface temperature of 100°C, is in a room where the air is at 20°C. What is the rate of heat loss per unit length of the pipe ? [Ans. 119.7 W/m]
8. A circular disk of 0.2 m diameter with a constant heat generation rate of 1.2 kW/m^2, is kept in ambient air at 20°C, with its heated surface facing downward and the plate is inclined at 15 degree to the horizontal. Determine the value of convection coefficient.
9. The heat transfer rate per unit length due to free convection from a horizontal tube is 200 W/m, when its surface is maintained at 70°C in the ambient air at 20°C. Estimate the heat transfer rate per unit length, when the tube surface is maintained at 145°C. Neglect the heat transfer rate by radiation and any influence of temperature on thermophysical properties of air. [Ans. 625 W/m]
10. Air flows through a long 0.3 m square duct maintains the outer duct surface temperature at 10°C. If the duct is uninsulated and exposed to air at 35°C, what is the heat gain per unit length of the duct ? [Ans. 109.25 W]
11. The warm air in a heating system is circulated through a sheet metal duct of size 60 cm × 40 cm × 7 cm. The duct carries the warm air at 60°C and the air surrounding the duct is at 15°C. Determine the heat loss from the duct surface to the surroundings. [Ans. 2677.5 W]
12. Beer in cans 160 mm long and 75 mm in diameter is initially at 30°C and is to be cooled in a refrigerator to 2°C. In the interest of maximizing the cooling rate, should the cans be laid horizontally or vertically in the compartment ? As a first approximation, neglect the heat transfer from ends. [Ans. 5.57 W horizontally]
13. A horizontal tube of 125 mm diameter with an outer surface temperature of 240°C is located in a large room with an air temperature of 20°C. Estimate the heat transfer rate per unit length of the tube due to free convection. [Ans. 705 W]
14. A horizontal uninsulated steam pipe passes through a large room, whose walls and ambient air are at 30°C. The pipe of the 150 mm diameter has an emissivity of 0.85 and an outer surface temperature of 170°C. Calculate the heat loss per unit length from the pipe. [Ans. 1153.2 W]
15. A sphere of 15 mm diameter contains an embedded electrical heater. Calculate the power required to maintain its surface temperature at 94°C, when the sphere is exposed to an ambient at 20°C for (*a*) air at atmospheric pressure, (*b*) water. [Ans. (*a*) 0.66 W]
16. A flat horizontal plate 0.5 m × 3.5 cm is exposed to atmospheric air at 6°C. The plate receives a net radiant energy flux from the sun of 750 W/m^2. The surface emissivity of the plate is 0.85. There is convection heat transfer from both upper and lower surface of the plate. What average temperature, will be attained by the plate ? [Ans. 61°C]
17. A 50 mm × 50 mm plate is maintained at 50°C and inclined at 60° with the horizontal. Calculate the heat loss from both sides of the plate to water at 20°C. [Ans. 1.18 W]
18. A 1 m × 1 m plate is maintained at 150°C and inclined at 45° with the horizontal. Calculate the heat loss from both sides of the plate to air at 20°C. [Ans. 913.3 W]
19. A thin, 16 cm diameter horizontal plate is maintained at 130°C in a large body of water at 70°C. The plate convects heat from both its top and bottom surfaces. Determine the rate of heat input into the plate necessary to maintain the temperature of 130°C. [Ans. 3.41 kW]
20. Solar energy at a rate of 280 W/m^2, is incident on a roof inclined at an angle of 40° with the horizontal. Assume that the back of the roof is insulated and the surface of the roof behaves as a blackbody. Determine the equilibrium temperature of the roof, if the ambient air is at 0°C and length of the roof is 3 m. [Ans. 94°C]
21. The dimension of the brick made in a factory are 7.5 cm (height) by 22.5 cm by 15 cm. The temperature of the brick leaving the kiln is 350°C and the brick is exposed to still air at 35°C. Calculate the instantaneous rate of cooling as the brick leaves the kiln. [Ans. 445.3 W]
22. A 25 mm OD electrical transmission line carries 100 A and having a resistance of 400 × 10^{-5} ohms per metre length is situated horizontal in the atmosphere. Neglect the radiation losses, determine the temperature of the surface of the cable, if the ambient temperature is (*a*) 26°C, (*b*) –26°C. [Ans. (*a*) 83°C]
23. A vertical plate 10 cm high and 5 cm wide is cooled by natural convection. The rate of heat transfer is 5.55 W and air temperature is 38°C. Calculate the maximum temperature of the plate. Assume uniform heat flux. [Ans. 175°C]
24. Calculate the rate of convection heat loss from top and bottom of flat 1 m^2 horizontal restaurant grill heated to 227°C in an ambient of 27°C. [Ans. 2268 W]

25. One surface of a panel 0.915 m × 0.915 m is insulated and other surface is kept at a uniform temperature of 65.5°C. Calculate the mean heat transfer coefficient due to free convection between heated surface of panel and the atmospheric air at 10°C, when

(*a*) Heated surface is vertical,

(*b*) Panel is horizontal with heated surface facing up,

(*c*) Panel is horizontal with hot surface facing down.

26. A thin vertical plate 3 m high and 1 m wide is thermally insulated on one side and exposed to solar radiation flux q_s = 750 W/m^2 on otherside. The exposed surface has an absorptivity of 0.8 for solar radiation. Assuming that the energy absorbed by the plate is dissipated by free convection into an ambient at 300 K, calculate the surface temperature of the panel.

The physical properties of ambient air are

$\nu = 2.076 \times 10^{-5}$ m^2/s, $k_f = 0.03$ W/m.K,

Pr = 0.697

Use correlation

$Nu_x = 0.568\,(Gr_x\,Pr)^{0.22}$ [Ans. 87.7°C]

27. The heat transfer rate due to free convection from a vertical surface, 1 m high and 0.6 m wide to stagnant air, which is 20°C colder than the surface is known. What is the ratio of heat transfer rate for the situation of vertical surface 0.6 m high and 1 m wide, when stagnant air is 20°C warmer than the surface ? Neglect any radiation effect and influence of temperature on the properties of air.

28. An aluminium alloy (k = 190 W/m.K) plate, heated to a uniform temperature of 227°C, is allowed to cool, while vertically suspended in a room where the ambient air and surroundings are at 27°C. The plate is 0.3 m square and 15 mm thick and has an emissivity of 0.25.

(*a*) Develop an expression for the rate of change of plate temperature, assuming the temperature to be uniform at any time.

(*b*) Determine the initial rate of cooling (°C/s), when the temperature is 227°C.

29. An electrical heater in the form of horizontal disc of 400 mm diameter is used to heat the bottom of a tank filled with engine oil at 5°C. Calculate the power required to maintain the heater surface temperature at 70°C.

30. A 1.2 m high and 2 m wide glass window, whose thickness is 6 mm and has thermal conductivity of 0.78 W/m.K and emissivity ε = 0.9. The room and the walls which face the window are maintained at 22°C, and the average temperature of the inner surface of the window is measured to be 4°C. If the temperature of the outdoor is – 6°C, calculate

(*a*) the convection heat transfer coefficient on the inner surface of the window

(*b*) the rate of total heat transfer through the window

(*c*) the combined natural convection and radiation heat transfer coefficient on the outer surface of the window.

31. A vertical 1.2 m high and 2 m wide double pane window consists of two sheets of glass separated by 2.5 cm air gap at atmospheric pressure. If the glass surface temperatures across the air gap are measured to be 18°C and 5°C, calculate the rate of heat transfer through the window by

(*a*) natural convection and

(*b*) radiation. Also calculate the effective thermal conductivity of the air space of this double paned window, which also accounts the radiation effect. The effective emissivity for two glass plates may be taken as 0.82.

[Ans. (*a*) 49.6 W, (*b*) 134 W, 0.147 W/m°C]

32. A large box containing ethylene glycol is heated by an electric heating element that consists of a square horizontal plate 15 cm × 15 cm in size. The glycol is at 0°C. Find the power input to the heating element required to maintain its top and bottom surface at 40°C.

[Hint. $L_c = \dfrac{A_s}{P}$, $Nu_T = 0.54\,Ra^{1/4}$, $Nu_H = 0.27\,Ra^{1/4}$,

$Q = (h_T + h_B)\,A(\Delta T)$] [Ans. 253 W]

33. A vertical plate 0.5 m high and 1 m wide is maintained at uniform temperature of 124°C. It is exposed to ambient air at 30°C. Calculate the heat transfer rate from the plate. [Ans. 252.774 W]

34. Estimate the electrical power required to maintain a vertical plate resistance heater at 130°C in an ambient air at 20°C. The plate is 15 cm high and 10 cm wide. Compare result with 450 cm high plate. The radiation heat transfer coefficient is 8.5 W/m^2.K for specified surface temperature.

[Ans. 52.87 W, 1346 W]

35. A helicopter plateform for a hospital is 12 m square and covered with a non-slip coating with a solar absorptivity of $\alpha_s = 1$ and thermal emissivity of ε = 1. Estimate the surface temperature, when it is subjected to incident solar flux of 1 kW/m^2 and surrounding air temperature is 20°C. [Ans. 87.16°C]

36. Two vertical plates each 80 mm high and at 85°C are placed in a tank of water at 15°C. Calculate the minimum spacing which will prevent interference of free convection boundary layer. [Ans. 3.22 mm]

37. An ornament space heater is in the form of a 60 cm diameter sphere, which is freely suspended in a large room. The surface of the sphere is maintained at 100°C and the room air is at 20°C. Calculate the convective heat transfer rate. [Ans. 371.5 W]

38. A copper heating coil is used to heat a large cylinder of water. The coil may be considered to be a horizontal cylinder, 1.5 m long and an outer diameter of 0.025 m. It has uniform surface temperature of 80°C. Estimate the heat transfer to water at 10°C. [Ans. 70.65 W]

39. Air flow through a long rectangular heating duct of width and height 0.75 m and 0.3 m, respectively, maintains the outer duct surface temperature at 45°C. If the duct is uninsulated and exposed to air 15°C, what is the heat loss from the duct surface per metre length ? [Ans. 142 W/m]

40. Compare the rate of heat loss from a human body with a typical energy intake from consumption of food (5440 kJ/day). Consider the body as a vertical cylinder 30 cm in diameter and 175 cm high in still air. Assume the skin temperature to be 37°C. Consider the emissivity at the skin surface as 0.4 and neglect, sweating and effect of clothing. The ambient air is taken at 13ºC. [Ans. 233.18 W]

Fig. 10.31. Schematic of a human body

41. A horizontal 40 W fluorescent tube which is 3.8 cm in diameter and 120 cm long stands in still air at 1 atm and 20°C. If the surface temperature is 40°C and radiation is neglected, what percentage of power is being dissipated by convection ? [Ans. 46.5%]

42. Calculate the cooling capacity by natural convection in air of a heat sink, having four rectangular thin fins of size 20 mm × 25 mm. The fins may be assumed to have a constant surface temperature of 60°C in ambient air at 20°C. Take the fin efficiency as 60%. [Ans. 0.875 W]

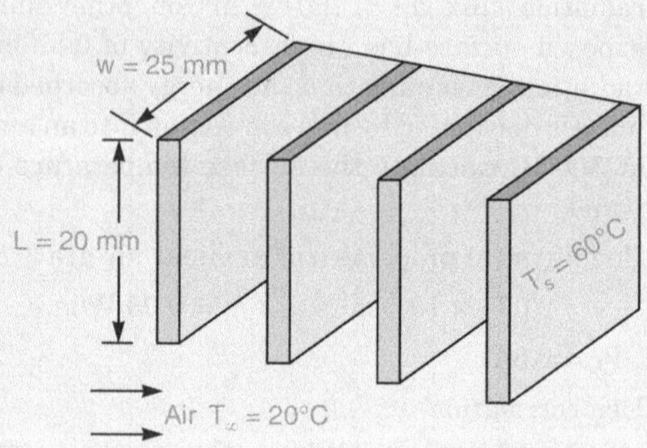

Fig. 10.32

43. Determine the coefficient of heat transfer by free convection convection heat transfer rate per mere length and maximum current density for a nichrom wire 0.5 mm in diameter. The surface of the wire is maintained at 300ºC. The wire is exposed to still air at 20°C and resistance per metre length of the wire is 6 Ω/m.

Use relation : $Nu = 1.18(Gr\ Pr)^{1/8}$

Use properties of air at 160°C

$k_f = 0.0361$ W/m.K,

$\nu = 30.35 \times 10^{-6}$ m²/s,

Pr = 0.687

(Anna. Univ., May 2003)

[Ans. 79.77 W/m².K, 35 W/m and 2.41 A]

REFERENCES AND SUGGESTED READING

1. Jaluria Y., *"Natural Convection Heat and Mass Transfer"*, Pergamon Press, New York, 1980.
2. Sehlichting H., *"Boundary Layer Theory"*, McGraw Hill, New York, 1968.
3. Gebhart B., *"Heat Transfer"*, 2nd ed, McGraw Hill, New York, 1970.
4. Churchill S.W. and H.H.S. Chu, *"Correlating Equations for Laminar and Turbulent Free Convection from a Vertical Plate"*, Int. J. of Heat and Mass Transfer vol. 18, 1975.

5. McAdams W.H., "*Heat Transmission*", 3rd ed, McGraw Hill, New York, 1954.
6. Fujii T. and H. Imura, "*Natural Convection Heat Transfer from a Plate with Arbitrary Inclination*", Int. J. of Heat and Mass Transfer vol. 15, 1972.
7. Moran W.R. and J.R. Lloyd,"*Natural Convection Mass Transfer Adjacent to Vertical and Downward Facing Surfaces*", J. Heat Transfer, vol. 94C, 1974.
8. Özisik M.N., "*Heat Transfer—A Basic Approach*", McGraw Hill, 1985.
9. Suryanarayana N.V., "*Engineering Heat Transfer*", West Pub. Co. New York, 1999.
10. Bayazitoglu Y and M.N. Özisik, "*Elements of Heat Transfer*", McGraw Hill, New York, 1988.
11. M.M. Rathore, "*Thermal Engineering*", McGraw Hill Education, 2010.
12. Adrian Bejan, "*Convective Heat Transfer*", 3rd ed. John Wiley & Sons, 2004.
13. Welty J.R, Wicks C.E., Wilson R.E. and Rorrer G.L., "*Fundamentals of Momentum, Heat and Mass Transfer*", 5th ed. John Wiley & Sons, 2007.
14. Raithby G.D. and Hollands K.G. Terry. "*Convective Heat Transfer*", CRC Press, 1999.

Condensation and Boiling

11

The condensers and boilers are widely used heat transfer equipments in the industries. The condensation and boiling involve convection processes associated with change of phase of fluid. Because there is a phase change during the process, the fluid transfers the latent heat only at its saturation temperature.

The determination of heat transfer coefficient is more complex than that of the convection without phase change. All the variables associated with convection also play role along with those associated with phase change. The variables affecting the phase phenomenon are the surface tension, latent heat of phase change, density difference, buoyancy forces, etc.

11.1. CONDENSATION

The *condensation* is a phase change process from vapour to liquid. It occurs when the vapour strikes a surface which is at temperature below the vapour saturation temperature, the vapour releases its latent heat and immediately converts into liquid phase.

The condensation may occur in two possible ways depending on the condition of the surface : filmwise condensation, and dropwise condensation, as shown in Fig. 11.1.

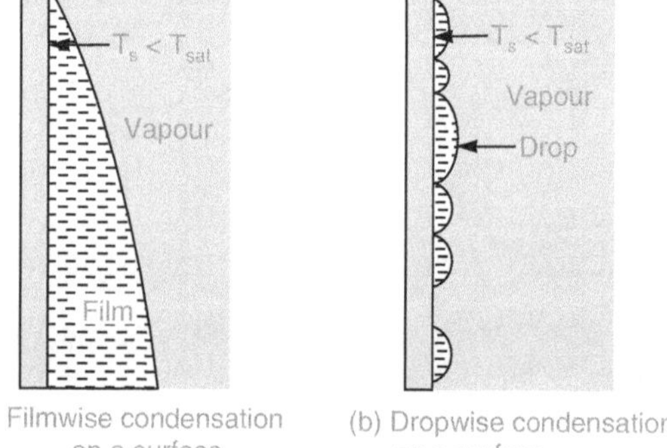

Fig. 11.1. Modes of condensation

11.1.1. Filmwise Condensation

If the condensation takes place continuously over a surface cooled by some process and the condensate film covers entire condensing surface and falls down under the action of gravity, the situation is called *filmwise condensation.* The presence of condensate layer acts as a resistance to heat transfer between vapour and surface. This resistance increases with condensate thickness, which increases in the flow direction, hence it is desirable to use short vertical surfaces of horizontal cylinders in situations involving film condensation.

11.1.2. Dropwise Condensation

If traces of oil are present during the condensation on highly polished surface, the film of condensate is broken into droplets and the condensation is called *dropwise condensation*. The droplets grow, collapse, and increase in size and they run off the surface under the action of gravity and carry other droplets in their path. Therefore, with dropwise condensation, the condensing surface almost exposed to condensing vapour, hence, the heat transfer rate with dropwise condensation is several times higher than that with filmwise condensation.

Film and dropwise condensation of steam on a vertical copper surface are shown in Fig. 11.2. A thin coating of cupric oleate was applied to left portion of the surface to promote drop wise condensation. A thin thermocouple probe extends in the photograph.

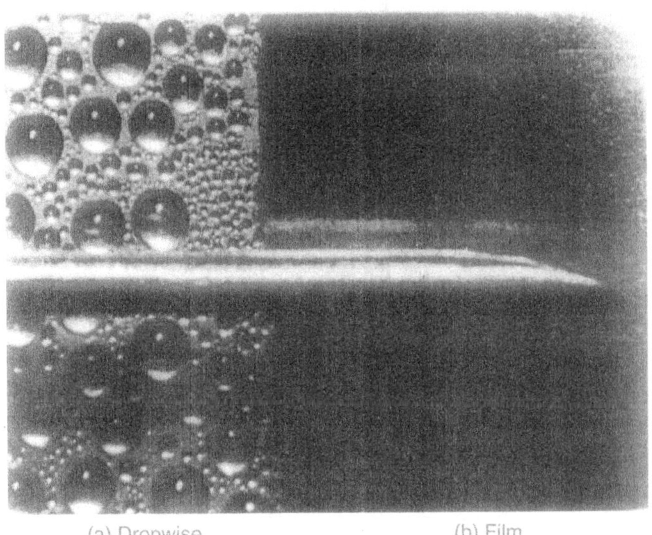

(a) Dropwise (b) Film

Fig. 11.2. Condensation on a vertical surface

Although, it is desirable to have dropwise condensation, but it is very difficult to maintain this type of condensation. The heat transfer coefficient for film wise condensation is comparatively small, still the condenser design calculations are based on the assumption of film condensation.

11.2. LAMINAR FILM CONDENSATION ON A VERTICAL PLATE

11.2.1. Nusselt Analysis

Consider a cold vertical plate at a surface temperature, T_s is exposed to saturated vapours at temperature T_{sat} ($T_s < T_{sat}$). The condensate film starts at the top of the plate and flows downward under the influence of gravity and its thickness $\delta(x)$ and condensate mass rate $\dot{m}(x)$ increase with increasing the distance x, along the surface. The velocity and temperature profiles are shown in Fig. 11.3.

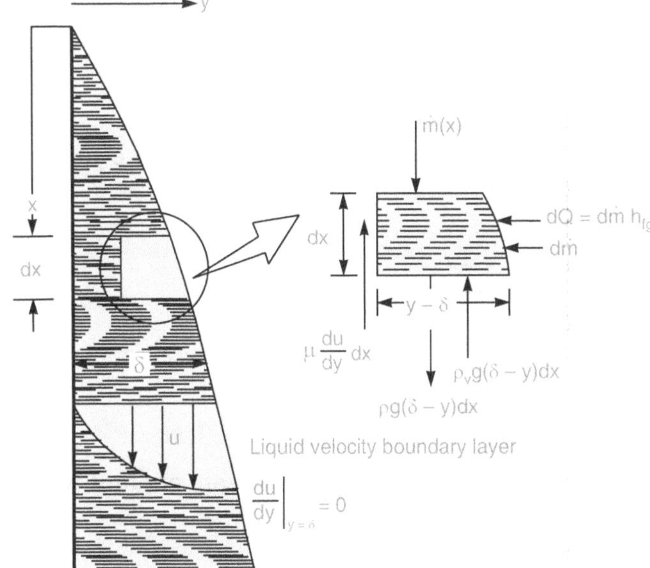

Fig. 11.3. Film condensation on a vertical plate

The condensation problem was first analysed by Nusselt under the following assumptions :

1. The plate is maintained at constant temperature T_s, that is less than the saturation temperature of vapour.
2. The vapour has low viscosity and it does not exert viscous shear force at liquid, vapour interface (*i.e.*, at $y = \delta$).
3. The downward flow of condensate under the action of gravity is laminar.
4. Fluid properties are constant.
5. Heat transfer across the condensate layer is by pure conduction, hence the liquid temperature distribution is linear.
6. Unit depth in z direction.

The weight of the fluid element of thickness dx between y and δ is balanced by viscous shear force at y and buoyancy force due to displaced vapour. Thus

$$\rho g(\delta - y)\, dx = \rho_v g(\delta - y)\, dx + \mu \frac{du}{dy}\, dx \quad \text{...(11.1)}$$

where,
ρ = density of condensate (liquid) ;
ρ_v = density of vapour ;
μ = viscosity of the condensate (liquid) ;
δ = thickness of boundary layer at any x ;
u = local velocity of condensate.

or

$$(\rho - \rho_v)\, g\, (\delta - y) = \mu \frac{du}{dy}$$

Integrating both sides, we get

$$(\rho - \rho_v) g \left(\delta y - \frac{y^2}{2} \right) = \mu u + C$$

where C is constant of integration and can be evaluated with boundary condition at $y = 0$, $u = 0$, we get

$$C = 0$$

$$(\rho - \rho_v) g \left(\delta y - \frac{y^2}{2} \right) = \mu u$$

Hence $$u = \frac{(\rho - \rho_v) g \left(\delta y - \frac{y^2}{2} \right)}{\mu} \quad ...(11.2)$$

The mass flow rate of condensate per unit width position of the film

$$\dot{m} = \int_0^\delta \rho u \, dy$$

or $$\dot{m} = \int_0^\delta \rho \left[\frac{(\rho - \rho_v) g \left(\delta y - \frac{y^2}{2} \right)}{\mu} \right] dy$$

It gives $$\dot{m} = \frac{\rho(\rho - \rho_v) g}{\mu} \left[\frac{\delta^3}{2} - \frac{\delta^3}{6} \right] = \frac{\rho(\rho - \rho_v) g \delta^3}{3\mu} \quad ...(11.3)$$

Since the liquid temperature distribution is linear, the elemental heat flux at the wall

$$q_s = k_f \left[\frac{dT}{dy} \right]_{y=0} = k_f \frac{(T_{sat} - T_s)}{\delta}$$

The negative sign is omitted, since heat flows in opposite y direction. The heat transfer rate :

$$Q = \frac{k_f dx \, (T_{sat} - T_s)}{\delta} \quad ...(11.4)$$

The amount of condensate added between x and $x + dx$, can be obtained by differentiating eqn. (11.3) with respect to x ;

$$d\dot{m} = \frac{d}{dx} \left\{ \frac{\rho (\rho - \rho_v) g \delta^3}{3\mu} \right\} dx$$

$$= \frac{d}{d\delta} \left\{ \frac{\rho (\rho - \rho_v) g \delta^3}{3\mu} \right\} \frac{d\delta}{dx} dx$$

$$= \frac{\rho (\rho - \rho_v) g \delta^2 \, d\delta}{\mu}$$

For the element,

The heat transfer rate Q at the wall = Heat transfer rate during condensation ($= d\dot{m} h_{fg}$)

$$\frac{k_f dx \, (T_{sat} - T_s)}{\delta} = \frac{\rho (\rho - \rho_v) g \delta^2 \, d\delta}{\mu} h_{fg}$$

or $$\delta^3 \, d\delta = \frac{k_f (T_{sat} - T_s)}{\rho (\rho - \rho_v) g} \frac{\mu}{h_{fg}} dx$$

Integrating both sides as

$$\int_0^\delta \delta^3 \, d\delta = \int_0^x \frac{k_f (T_{sat} - T_s)}{\rho (\rho - \rho_u) g} \frac{\mu}{h_{fg}} dx$$

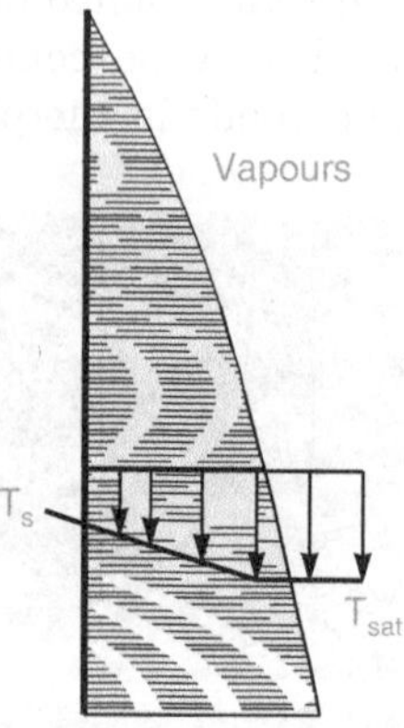

Fig. 11.4. Thermal boundary layer in laminar condensation

Treating k_f, h_{fg}, ρ, ρ_v, T_{sat}, T_s and μ as constant quantities, then integration leads to

$$\frac{\delta^4}{4} = \frac{k_f (T_{sat} - T_s) \mu x}{\rho (\rho - \rho_v) g h_{fg}}$$

or $$\delta = \left[\frac{4 k_f (T_{sat} - T_s) \mu x}{\rho (\rho - \rho_v) g h_{fg}} \right]^{1/4} \quad ...(11.5)$$

It gives $\delta \propto x^{1/4}$ ...(11.6)

The surface heat transfer rate per unit width of the plate by convection can be expressed as :

$$Q = h_x \, dx \, (T_{sat} - T_s) \quad ...(11.7)$$

Equating it with eqn. (11.4), we get

$$h_x = \frac{k_f}{\delta}$$

Substituting δ, we get

$$h_x = \left\{ \frac{\rho (\rho - \rho_v) g h_{fg} k_f^3}{4 \mu x (T_{sat} - T_s)} \right\}^{1/4} \quad ...(11.8)$$

It gives $h_x \propto x^{-1/4}$ ...(11.9)

The average heat transfer coefficient on entire plate

$$h = \frac{1}{L} \int_0^L h_x \, dx = \frac{4}{3} h_{x=L}$$

It gives,

$$h = 0.943\left[\frac{\rho(\rho-\rho_v)\, g\, h_{fg}\, k_f^{\,3}}{\mu\, L\,(T_{sat}-T_s)}\right]^{1/4} \quad ...(11.10)$$

where, μ = viscosity of liquid, Ns/m² or kg/ms

k_f = thermal conductivity of liquid, W/m.K

ρ = density of liquid, kg/m³

ρ_v = density of vapour, kg/m³

L = height of the plate, m

h_{fg} = latent heat of vapour; J/kg

T_s = plate surface temperature °C

T_{sat} = saturation temperature of vapour °C

g = acceleration due to gravity, m/s²

The physical properties of liquid film should be evaluated at an effective film temperature $T_f = \frac{T_s + T_{sat}}{2}$

The average Nusselt number can be expressed as

$$Nu = \frac{hL}{k_f} = 0.943\left[\frac{\rho(\rho-\rho_v) g h_{fg} L^3}{\mu k_f (T_{sat}-T_s)}\right]^{1/4} \quad ...(11.11)$$

The laminar condensation eqn. (11.10), presented above match experimental data very well as long as the film remains smooth and well behaved. But for low value of Reynolds number less than 1800, experiments has shown that the measured heat transfer coefficient is 20% higher than the theoretical value. Therefore, *McAdams* recommendation :

$$h = 1.13\left[\frac{\rho(\rho-\rho_v) g h_{fg} k_f^3}{\mu L(T_{sat}-T_s)}\right]^{1/4} \quad \text{for Re} < 1800 \quad ...(11.12)$$

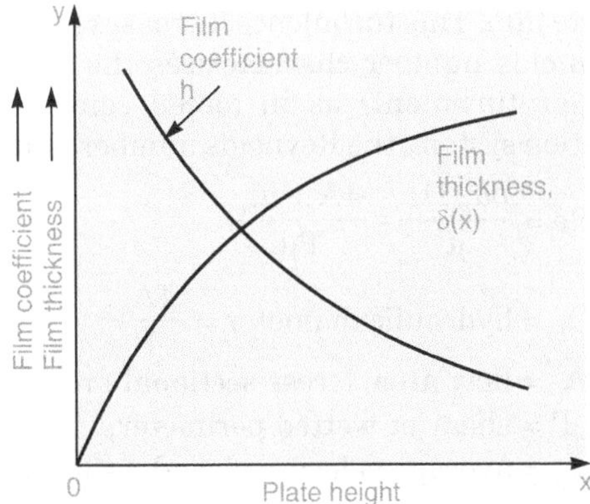

Fig 11.5. Film thickness and film coefficient for condensation on a vertical plate

Generally $\rho_v << \rho$, therefore, ρ_v is neglected from above expression, and

$$h = 1.13\left[\frac{\rho^2 g h_{fg} k_f^{\,3}}{\mu L(T_{sat}-T_s)}\right]^{1/4} \quad ...(11.13)$$

which can be arranged in the form

$$h\left(\frac{\mu^2}{k_f^{\,3}\rho^2 g}\right)^{1/3} = 1.76\ Re^{-1/3} \text{ for Re} > 1800 \quad ...(11.14)$$

Fig. 11.5 shows the variation of film thickness and film coefficient for film wise condensation on a vertical plate.

Condensation on inclined plates

If a surface is inclined at angle θ with the horizontal as shown in Fig. 11.6, the average convection coefficient is given by

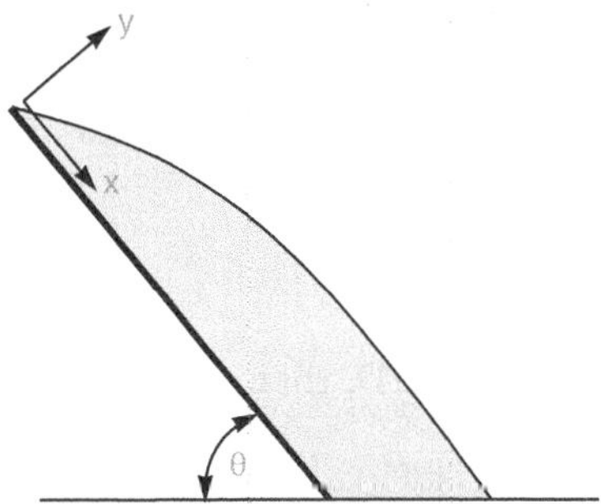

Fig. 11.6. Filmwise condensation on an inclined surface

$$h = 1.13\left[\frac{\rho(\rho-\rho_v) h_{fg} k_f^{\,3}}{\mu L(T_{sat}-T_s)}\, g \sin\theta\right]^{1/4} \quad ...(11.15)$$

If $\rho_v << \rho$ then

$$h = 1.13\left[\frac{\rho^2 h_{fg} k_f^{\,3}}{\mu L\,(T_{sat}-T_s)}\, g \sin\theta\right]^{1/4} \quad ...(11.16)$$

11.3. CONDENSATION ON A SINGLE HORIZONTAL TUBE

The analysis of condensation on outside surface of a horizontal tube as shown in Fig. 11.7 (*b*) is more complicated than that on a vertical surface. Nusselt analysis for laminar filmwise condensation on the surface of a horizontal tube of outside diameter D gives the average heat transfer coefficient as

$$h = 0.725\left[\frac{\rho(\rho-\rho_v)\, g\, h_{fg} k_f^{\,3}}{\mu D\,(T_{sat}-T_s)}\right]^{1/4} \quad ...(11.17)$$

For $\rho_v << \rho$, the eqn.(11.17) reduces to

$$h = 0.725 \left[\frac{\rho^2 g\, h_{fg} k_f{}^3}{\mu D\,(T_{sat} - T_s)} \right]^{1/4} \quad ...(11.18)$$

Fig. 11.7. Film condensation on (*a*) a sphere, (*b*) a single horizontal tube, (*c*) a vertical tier of horizontal tubes with a continuous condensate sheet

Comparing eqns.(11.10) and (11.17) for filmwise condensation on vertical tube of length L and a horizontal tube of diameter D, yields to

$$\frac{h_{vert}}{h_{horz}} = 1.30 \left(\frac{D}{L} \right)^{1/4} \quad ...(11.19)$$

If $\quad L = 2.87D$, then

$$h_{vert} = h_{horz}$$

The laminar condensation on outside surface of a sphere of diameter D is expressed as

$$h = 0.815 \left[\frac{\rho(\rho - \rho_v)\, g\, h_{fg} k_f^3}{\mu\, D\,(T_{sat} - T_s)} \right]^{1/4} \quad ...(11.20)$$

11.3.1. Condensation on Horizontal Tube Banks

If the condensation occurs on N number of horizontal tubes so arranged that the condensate from one tube flows directly on to the next tube below it as shown in Fig. 11.7 (*c*), the average heat transfer coefficient for such system can be estimated by replacing the tube diameter D in eqn. (11.17) by ND, we get

$$h_{N\ tubes} = 0.725 \left[\frac{\rho\,(\rho - \rho_v) g h_{fg} k_f^3}{\mu ND\,(T_{sat} - T_s)} \right]^{1/4} \quad ...(11.21)$$

This equation generally gives a conservative result, because condensate does not fall in smooth sheets from one tube row to another, but it drips from tube to tube and adds some turbulence and disturbances in the phenomenon.

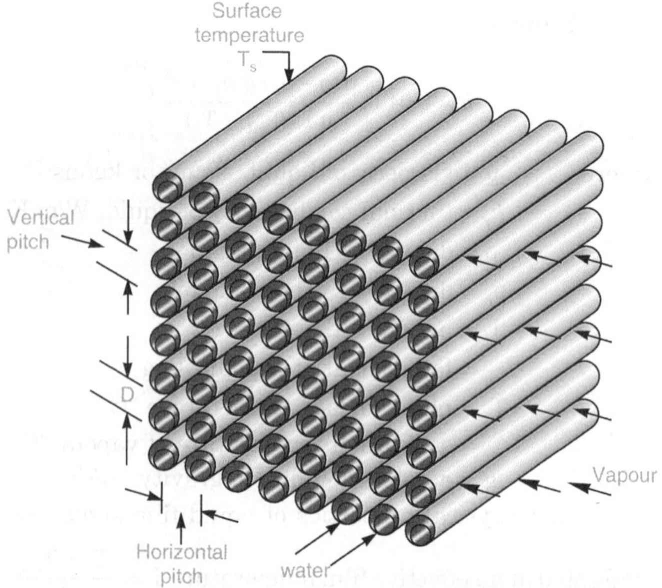

Fig. 11.8. Condensation on horizontal tube bank

For Re > 1800, then analysis yields to

$$h = 0.725\,[1 + 0.2(N - 1)Ja] \left[\frac{\rho\,(\rho - \rho_v)\, g h_{fg} k_f{}^3}{\mu ND\,(T_{sat} - T_s)} \right]^{1/4} \quad ...(11.22)$$

where $\quad Ja = \dfrac{C_p (T_{sat} - T_s)}{h_{fg}}$ is the **Jacob number,** a ratio of sensible heat to evaporation heat.

C_p = specific heat of liquid

h_{fg} = enthalpy of evaporation.

11.3.2. Calculation of Reynolds Number

When a plate, on which condensation takes place is sufficiently large or there is sufficient amount of condensate flow, the turbulence may appear in the condensate film. This turbulence increases heat transfer rate. Reynolds number characterises the type of flow (laminar or turbulent) as in forced convection. For condensation system, the Reynolds number is defined as

$$Re = \frac{\rho u_m D_h}{\mu} = \frac{4A_c \rho\, u_m}{P\mu} \quad ...(11.23)$$

where $\quad D_h$ = hydraulic diameter = $\dfrac{4A_c}{P}$

A_c = flow area, (cross-sectional area),

P = shear or wetted perimeter,

u_m = average velocity of condensate flow.

The mass flow rate of condensation at any section is given by

$$\dot{m} = \rho A_c u_m$$

Substituting in eqn. (11.23), we get

$$\therefore \quad \text{Re} = \frac{4\dot{m}}{P\,\mu} \qquad ...(11.24)$$

The mass flow rate may be related to total heat transfer rate during condensation as

$$Q = \dot{m}\,h_{fg} = hA_s(T_{sat} - T_s) \qquad ...(11.25)$$

where A_s = total heat transfer area, thus

$$\dot{m} = \frac{Q}{h_{fg}} = \frac{hA_s(T_{sat} - T_s)}{h_{fg}} \qquad ...(11.26)$$

and

$$\text{Re} = \frac{4hA_s(T_{sat} - T_s)}{h_{fg}\,P\mu} \qquad ...(11.27)$$

where the wetted perimeter defined earlier as

$$P = \begin{cases} \pi D = \text{for vertical tube of outside diameter D,} \\ w = \text{for vertical or inclined plate of width } w, \\ 2L = \text{for horizontal tubes, each of length L and arranged in vertical tiers.} \end{cases}$$

For a plate of height L and width w

$$A_s = wL \quad \text{and} \quad P = w$$

Then

$$\text{Re} = \frac{4hL\,(T_{sat} - T_s)}{h_{fg}\,\mu} \qquad ...(11.28)$$

11.4. TURBULENT FILMWISE CONDENSATION

If the Reynolds number exceeds 1800, during filmwise condensation, then the empirical relation proposed by *Kirkbridge* can be used

$$h\left[\frac{\mu^2}{\rho^2 g\,k_f^{\,3}}\right]^{1/3} = 0.0077\,(\text{Re})^{0.4} \qquad ...(11.29)$$

Fig. 11.9 shows a plot of eqns. (11.14) and (11.29) as a function of Reynolds number for condensate flow in laminar and turbulent regimes, respectively.

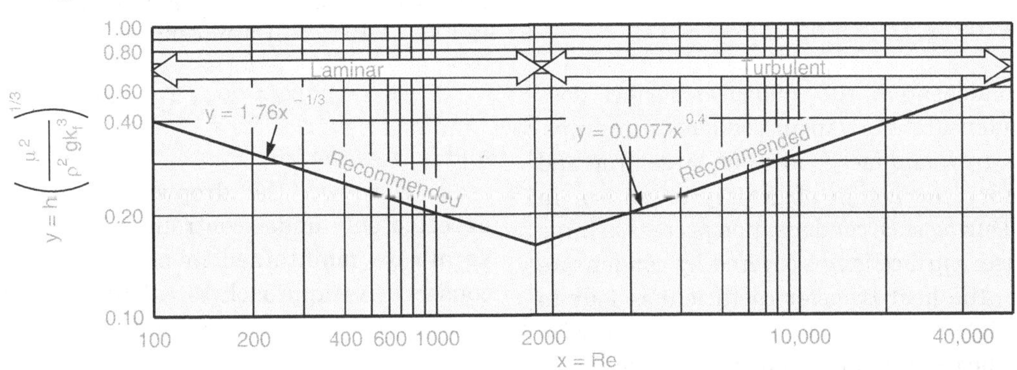

Fig. 11.9. Average heat transfer coefficient for filmwise condensation on a vertical surface for laminar and turbulent flows

11.5. CONDENSATE NUMBER

The Reynolds number signifies the condensation behaviour : laminar or turbulent. It is convenient to express the heat transfer coefficient directly in terms of Reynolds number Re. Reconsider eqn. (11.15) in the form

$$h = C\left[\frac{\rho(\rho - \rho_v)\,g\sin\theta\,h_{fg}\,k_f^3}{\mu L\,(T_{sat} - T_s)}\right]^{1/4} \qquad ...(11.30)$$

where constant C is evaluated for vertical plate or a horizontal cylinder. From eqn. (11.25) ;

$$T_{sat} - T_s = \frac{\dot{m}\,h_{fg}}{hA_s} \qquad ...(11.31)$$

where A_s = heat transfer surface area. Substituting eqn. (11.31) in eqn. (11.30), we get

$$h^{3/4} = C\left[\frac{\rho(\rho - \rho_v)\,g\sin\theta\,k_f^3}{\mu\,\dot{m}}\,\frac{A_s}{L}\right]^{1/4} \qquad ...(11.32)$$

It may be arranged as

$$h^{3/4} = C\left[\frac{\rho(\rho - \rho_v)\,g\,k_f^3}{\mu^2} \times \frac{\mu P}{4\dot{m}} \times 4\sin\theta\left(\frac{A_s}{PL}\right)\right]^{1/4}$$

Using $\frac{\mu P}{4\dot{m}} = \text{Re}^{-1}$, then

$$h = C^{4/3}\left[\frac{\rho(\rho - \rho_v)\,g\,k_f^3}{\mu^2} \times \text{Re}^{-1} \times 4\sin\theta\left(\frac{A_s}{PL}\right)\right]^{1/3} \qquad ...(11.33)$$

We may now define a new dimensionless group, called *condensation number* Co, as

$$\text{Co} = h\left[\frac{\mu^2}{\rho(\rho - \rho_v)\, g\, k_f^{\,3}}\right]^{1/3} \qquad ...(11.34)$$

Then the expression (11.33), takes the form

$$\text{Co} = \text{C}^{4/3}\left[4\sin\theta\left(\frac{\text{A}_s}{\text{PL}}\right)\right]^{1/3} \text{Re}^{-1/3} \qquad ...(11.35)$$

For a vertical plate, C = 0.943

$$\frac{\text{A}_s}{\text{PL}} = 1.0$$

$$\text{Co} = 1.47\ \text{Re}^{-1/3} \qquad ...(11.36)$$

For a horizontal cylinder, C = 0.725, and

$$\frac{\text{A}_s}{\text{PL}} = \pi \quad \text{and} \quad \text{Co} = 1.514\ \text{Re}^{-1/3} \qquad ...(11.37)$$

For turbulent flow, the empirical relation proposed by *Kirkbridge,* can be rearranged as

$$\text{Co} = 0.0077\ (\text{Re})^{0.4} \quad \text{for Re} > 1800.$$

11.6. DROPWISE CONDENSATION

In dropwise condensation, the condenser surface does not wet by condensate, the vapour condenses in drops. These droplets grow and merge into one large drop and falls down under the action of gravity as shown in Fig. 11.10 (*a*). During such condensation process, a large part of condenser surface is not covered by condensate film, therefore, the heat transfer coefficient is four to eight times higher than that results in filmwise condensation. The ratio of condensate mass flux for dropwise condensation, $\dot{m}_\text{D}$ on outside surface of tube of diameter D to that for film condensation, $\dot{m}_f$ can be related as

$$\frac{\dot{m}_\text{D}}{\dot{m}_f} = \left(\frac{\rho^2\,\text{D}^2\,g}{24\,\mu\,\dot{m}_f}\right)^{1/9} \qquad ...(11.38)$$

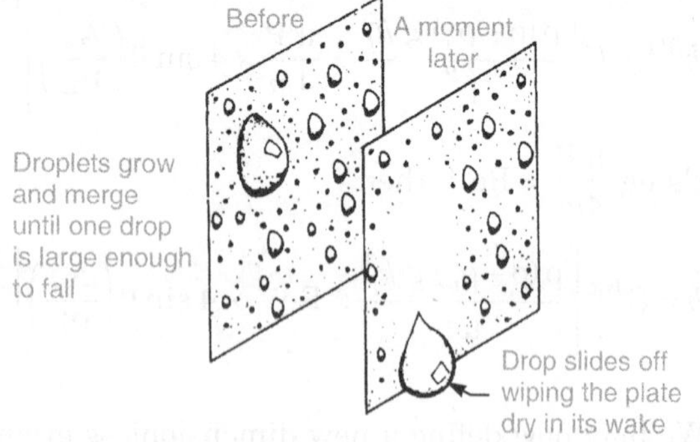

(a) The process of liquid removal condensation during dropwise condensation

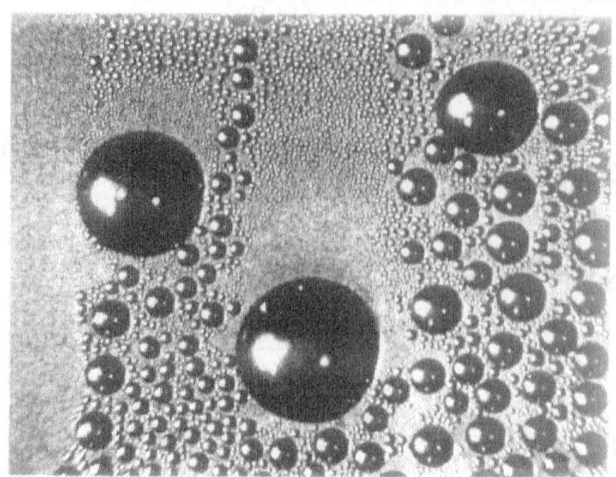
(b) Typical photograph of dropwise condensation under ideal conditions

Fig. 11.10. Dropwise condensation

Many of the surface fluid systems are studied for dropwise condensation. Most of them are for steam condensation on well promoted copper surfaces as shown in Fig. 11.10 (*b*). At atmospheric pressure, the data can be expressed in the form

$$h = 51{,}104 + 2044\ \text{T}_s \quad \text{for } 22°\text{C} < \text{T}_s < 100°\text{C} \qquad ...(11.39)$$

and $$h = 255{,}510 \quad \text{for } 100°\text{C} < \text{T}_s$$

However, the dropwise condensation can be expected only under controlled conditions, that cannot be always maintained in practice. Therefore, it is a conservative approach to calculate the heat transfer coefficient for filmwise condensation.

11.7. FILM CONDENSATION INSIDE HORIZONTAL TUBES

The condensers used for refrigeration and air-conditioning system usually involve vapour condensation inside horizontal or vertical tubes. Unfortunately, the phenomena of condensation inside the tubes is very complex and strongly depends on velocity of vapour flowing through the tube. If the velocity is small, the vapour condensation occurs in the manner as shown in Fig. 11.11(*a*) for a horizontal tube, where the condensate flows from upper portion of the tube to the bottom and then in longitudinal direction with vapour. For low vapour velocities :

$$\text{Re}_{v,\,i} = \left(\frac{\rho_v u_{m,\,v}\text{D}}{\mu_v}\right) < 3500$$

where i refers to tube inlet and v pertains to vapour, the *Chato* proposed and presented expression in the form of

$$h = 0.555\left[\frac{\rho(\rho-\rho_v)\,g\,k_f^3\,h'_{fg}}{\mu\,D\,(T_{sat}-T_s)}\right]^{1/4} \quad \text{...(11.40)}$$

where $h'_{fg} = h_{fg} + \frac{3}{8}C_p(T_{sat} - T_s)$ ...(11.41)

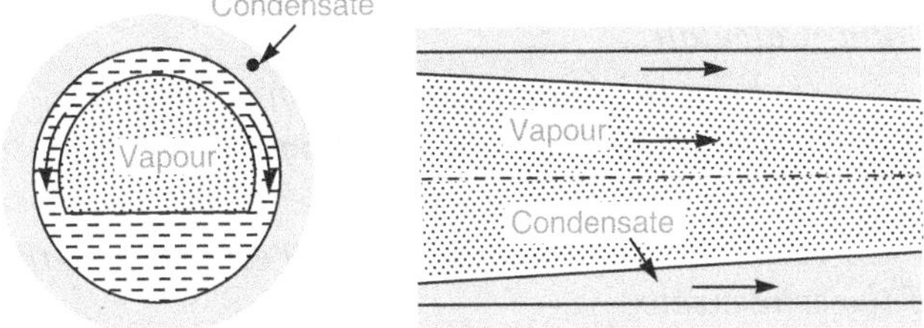

(a) Cross-section of condensate flow for low vapour velocities (b) Longitudinal section of condensate flow for large vapour velocities

Fig. 11.11. Film condensation in a horizontal tube

At higher vapour velocities, the two phase flow regime becomes annular as shown in Fig. 11.11 (*b*). The vapour flows in the core of tube, while the condensate occupies annular space and its thickness grows in flow direction. For higher flow rates, the approximate empirical expression is given by *Akers, Deans,* and *Crosser* as

$$\frac{hD}{k_f} = 0.026\,(\mathrm{Re}_m)^{0.8}\,\mathrm{Pr}_f^{1/3} \quad \text{...(11.42)}$$

where $R_m = \frac{D}{\mu}\left[G_f + G_v\left(\frac{\rho}{\rho_v}\right)^{1/2}\right]$ ...(11.43)

and $G = \rho u_m$ (mass velocity)

The mass velocity for liquid G_f and vapour G_v can be approximated, when

$$\mathrm{Re}_v = \frac{DG_v}{\mu_v} > 20{,}000$$

$$\mathrm{Re}_f = \frac{DG_f}{\mu} > 5000.$$

Example 11.1. *The outer surface of a vertical tube 80 mm in outer diameter and 1 m long is exposed to saturated steam at atmospheric pressure. The tube surface is maintained at 50°C by flow of water through the tube. What is the rate of heat transfer to coolant and what is the rate of condensation of steam ?*

Solution

Given : Condensation of saturated steam on a vertical condenser :

$T_{sat} = 100°C$, $L = 1$ m, $T_s = 50°C$,
$D = 80$ mm $= 0.08$ m.

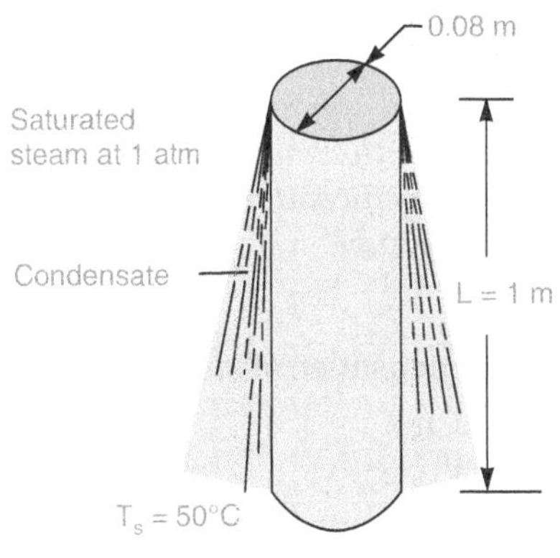

Fig. 11.12

To find :

(*i*) Heat transfer rate on tube surfaces.

(*ii*) Steam condensation rate.

Assumptions :

1. Laminar film condensation.
2. Negligible convection and radiation heat transfer.

Properties of saturated vapour and water

$h_{fg} = 2257$ kJ/kg, $\rho = 975$ kg/m^3,
$\rho_v = 0.596$ kg/m^3, $k_f = 0.668$ W/m.K,
$\mu = 375 \times 10^{-6}$ kg/ms.

Analysis : The heat transfer coefficient for laminar filmwise condensation on a vertical tube surface can be calculated by eqn. (11.12).

$$h = 1.13\left\{\frac{g\,\rho\,(\rho-\rho_v)\,h_{fg}\,k_f^3}{\mu\,L\,(T_{sat}-T_s)}\right\}^{1/4}$$

$$h = 1.13 \times \left\{\frac{(9.81)\times(975)\times(975-0.596)\times(2257\times10^3)\times(0.668)^3}{375\times10^{-6}\times 1\times(100-50)}\right\}^{1/4}$$

$= 4832.21$ W/m^2.K

(*i*) The heat transfer rate :

$Q = hA_s(T_{sat} - T_s) = h(\pi DL)(T_{sat} - T_s)$
$= (4832.21 \text{ W/m}^2\text{.K}) \times (\pi \times 0.08 \text{ m} \times 1 \text{ m}) \times (100 - 50)(\text{K})$
$= 60{,}723$ W $=$ **60.723 kW. Ans.**

(*ii*) The condensation rate :

$$\dot{m} = \frac{Q}{h_{fg}} = \frac{60.723 \text{ kW}}{2257 \text{ kJ/kg}} = \mathbf{0.0269\ kg/s.} \quad \textbf{Ans.}$$

Example 11.2. *Dry saturated steam at a pressure of 2.45 bar condenses on the surface of a vertical tube of height 1 m. The tube surface temperature is kept at 117°C. Estimate the thickness of condensate film and local heat transfer coefficient at a distance of 0.2 m from the upper end of the tube.* (V.T.U., May 2009)

Solution

Given : Condensation of steam over a vertical tube

$L = 1$ m, $x = 0.2$ m

$T_s = 117°C$, $p = 2.45$ bar.

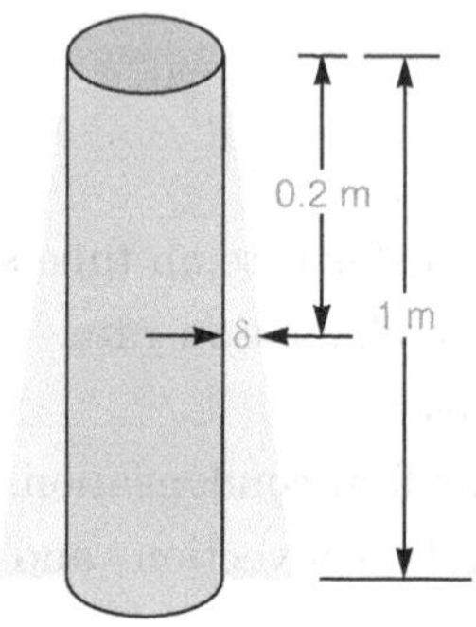

Fig. 11.13. Schematic of condensation over vertical surface

To find :

(*i*) Thickness of condensate at $x = 0.2$ m

(*ii*) Local heat transfer coefficient at $x = 0.2$ m.

Properties of steam at $p = 2.45$ bar

$$T_{sat} = 126.7°C$$
$$h_{fg} = 2183.5 \times 10^3 \text{ J/kg}$$
$$T_f = \frac{126.7 + 117}{2} = 121.85°C$$
$$\rho = 943 \text{ kg/m}^3$$
$$\mu = 237 \times 10^{-6} \text{ kg/ms},$$
$$k_f = 0.686 \text{ W/m.K.}$$

Analysis : (*i*) The thickness of condensate at $x = 0.2$ m

$$\delta(x) = \left[\frac{4k_f (T_{sat} - T_s)\,\mu\, x}{\rho(\rho - \rho_v)\, g\, h_{fg}}\right]^{1/4}$$

Neglecting ρ_v from expression, then

$$\delta(x) = \left[\frac{4 \times 0.686 \times (126.7 - 117) \times 237 \times 10^{-6} \times 0.2}{(943)^2 \times 9.81 \times 2183.5 \times 10^3}\right]$$

$= 9.02 \times 10^{-5}$ m $=$ **0.09 mm.** Ans.

(*ii*) Local heat transfer coefficient

$$h_x = \left[\frac{\rho(\rho - \rho_v)\, g\, h_{fg}\, k_f^{\,3}}{4\mu\, x\,(T_{sat} - T_s)}\right]^{1/4}$$

$$= \left[\frac{(943)^2 \times 9.81 \times 2183.5 \times 10^3 \times (0.686)^3}{4 \times 237 \times 10^{-6} \times 0.2 \times (126.7 - 117)}\right]^{1/4}$$

$=$ **7604 W/m².K.** Ans.

Example 11.3. *A vertical plate 350 mm high and 420 mm wide, at 40°C, is exposed to saturated steam at 1 atm. Calculate:*

(*i*) *Film thickness at the bottom of the plate,*

(*ii*) *Maximum velocity at the bottom of the plate,*

(*iii*) *Total heat flux to the plate.*

Assume vapour density is small compared to that of the condensate.

Solution

Given : Condensation on a vertical plate

$L = 350$ mm, $w = 420$ mm

$T_s = 40°C$, $p = 1$ atm ($T_{sat} = 100°C$).

To find :

(*i*) Film thickness at bottom of the plate,

(*ii*) Maximum velocity at the bottom of the plate,

(*iii*) Total heat flux to plate.

Properties of steam and condensate from Table A-7 ;

At $T_{sat} = 100°C$, $h_{fg} = 2257 \times 10^3$ J/kg

At $T_f = \dfrac{100 + 40}{2} = 70°C$

$\mu = 406 \times 10^{-6}$ kg/ms, $k_f = 0.668$ W/m.K

$\rho = 977.8$ kg/m³

Analysis : (*i*) Film thickness δ at the bottom of the plate

$$x = 350 \text{ mm} = 0.35 \text{ m}$$

$$\delta(x) = \left[\frac{4k_f\, \mu(T_{sat} - T_s)\, x}{g\,(\rho - \rho_v)\,\rho\, h_{fg}}\right]^{1/4}$$

Neglecting ρ_v,

$$\delta(x) = \left[\frac{4 \times 0.668 \times 406 \times 10^{-6} \times (100 - 40) \times 0.35}{9.81 \times (977.8)^2 \times 2257 \times 10^3}\right]^{1/4}$$

$= 1.82 \times 10^{-4}$ m $=$ **0.182 mm.** Ans.

(*ii*) The maximum velocity at the bottom of the plate, u_{max} :

$$u = \frac{(\rho - \rho_v)g\left\{\delta y - \dfrac{y^2}{2}\right\}}{\mu}$$

At $y = \delta$, $u = u_{max}$, neglecting ρ_v,

$$u_{max} = \frac{\rho g \delta^2}{2\mu} = \frac{977.8 \times 9.81 \times (1.82 \times 10^{-4})^2}{2 \times 406 \times 10^{-6}}$$

$= \mathbf{0.391\ m/s.}$ **Ans.**

(*iii*) Total heat flux to the plate, q

The average heat transfer coefficient

$$h = 1.13 \left[\frac{\rho(\rho - \rho_v) g h_{fg} k_f^3}{\mu L (T_{sat} - T_s)} \right]^{1/4}$$

Neglecting

$$\rho_v \; h = 1.13 \times \left[\frac{(977.8)^2 \times 9.81 \times 2357 \times 10^3 \times (0.668)^3}{406 \times 10^{-6} \times 0.35 \times (100 - 40)} \right]^{1/4}$$

$= 5958$ W/m^2.K

The heat flux

$q = h(T_{sat} - T_s) = 5958 \times (100 - 40)$

$= 357485$ W/m^2 = **357.485 kW/m^2.** **Ans.**

Example 11.4. *A horizontal tube 50 mm in diameter with a surface temperature of 34°C is exposed to steam at 0.2 bar. Estimate the condensation rate and heat transfer rate per unit length of the tube.*

Solution

Given : Condensation on horizontal tube :

D = 50 mm = 0.05 m, $T_s = 34$°C,

L = 1 m $p = 0.2$ bar = 20 kPa.

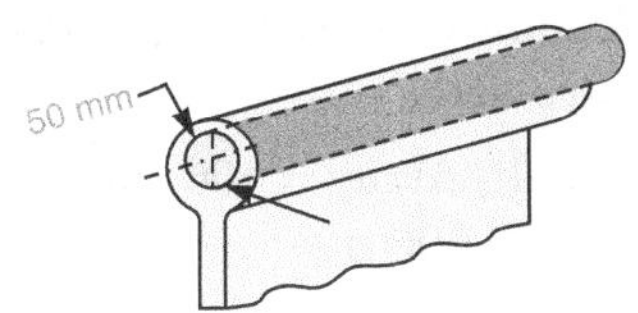

Fig. 11.14. Schematic for example 11.4

To find :

(*i*) Condensation rate, and (*ii*) Heat transfer rate.

Properties of steam at $p = 0.2$ bar = 20 kPa :

$T_{sat} = 60$°C, $h_{fg} = 2358 \times 10^3$ J/kg

$$T_f = \frac{60 + 34}{2} = 47\text{°C}$$

At 47°C (320 K)

$\rho = 989.1$ kg/m^3, $\mu = 577 \times 10^{-6}$ kg/ms

$k_f = 0.640$ W/m.K.

Analysis : The heat transfer coefficient for condensation on horizontal tube

$$h = 0.725 \left[\frac{\rho(\rho - \rho_v) g h_{fg} k_f^3}{\mu D (T_{sat} - T_s)} \right]^{1/4}$$

Neglecting ρ_v,

$$h = 0.725 \left[\frac{(989.1)^2 \times 9.81 \times 2358 \times 10^3 \times (0.640)^3}{577 \times 10^{-6} \times 0.05 \times (60 - 34)} \right]^{1/4}$$

$= 6837$ W/m^2.K

(*i*) Condensation rate of steam

$$\dot{m} = \frac{Q}{h_{fg}} = \frac{8888.1}{2358 \times 10^3}$$

$= 3.77 \times 10^{-3}$ kg/s = **13.56 kg/h.** **Ans.**

(*ii*) The heat transfer rate

$Q = h A_s (T_{sat} - T_s)$

$= 6837 \times 0.05 \times 1 \times (60 - 34)$

= **8888.1 W.** **Ans.**

Example 11.5. *Compare the values of heat transfer coefficient for laminar film wise condensation, when a pipe 6.25 cm in diameter and 1 m long is used as condenser ; when it (i) vertical, (ii) horizontal.*

(P.U., May 2002)

Solution

For vertical tube condensation

$$h_v = 1.13 \left\{ \frac{\rho(\rho - \rho_v) g h_{fg} k_f^3}{\mu L (T_{sat} - T_s)} \right\}^{1/4} \quad ...(i)$$

For horizontal tube condensation

$$h_h = 0.725 \left\{ \frac{\rho(\rho - \rho_v) g h_{fg} k_f^3}{\mu D (T_{sat} - T_s)} \right\}^{1/4} \quad ...(ii)$$

Dividing (*ii*) by (*i*), $\dfrac{h_h}{h_v} = \dfrac{0.725}{1.13} \left(\dfrac{L}{D} \right)^{0.25}$

Using the given values ;

$$\frac{h_h}{h_v} = \frac{0.725}{1.13} \times \left(\frac{1}{0.0625} \right)^{0.25} = 1.283$$

The horizontal position provided 28.3% more condensation rate. **Ans.**

Example 11.6. *A steam condenser consist of 16 tubes arranged in 4 × 4 array. The tubes are 25 mm in diameter and 1.2 m long. Water flows through the tube at 65°C while steam condenses at 75°C over the tube surface. Find the rate of condensation, if (a) tubes are horizontal ; (b) tubes are vertical.*

Take latent heat of steam as 2300 kJ/kg and properties of water at 70°C

$\rho = 977.8$ *kg/m^3*, $C_p = 4.187$ *kJ/kg.K*

$k_f = 0.668$ *W/mK*, $\nu = 0.415 \times 10^{-6}$ *m^2/s*

$\beta = 5.7 \times 10^{-3}$ *K^{-1}*,

(P.U., May 2008)

Solution

Given : A vertical condenser with

$T_{sat} = 75°C$, $L = 1.2$ m,

$T_s = 65°C$, $D = 0.025$ m.

$N^2 = 16$ in 4×4 array and properties of fluid.

$h_{fg} = 2300 \times 10^3$ J/kg

To find : The rate of condensation, if

(*a*) The tubes are horizontal.

(*b*) Tubes are vertical.

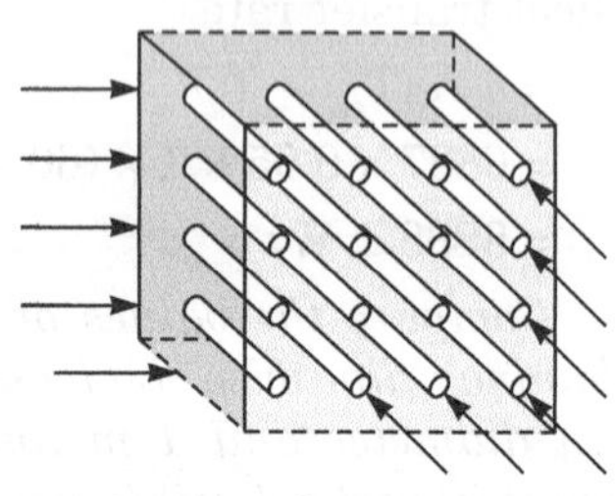

Fig. 11.15

Assumptions :

1. Laminar film wise condensation.
2. Negligible film thermal resistance.
3. Vapour density ρ_v negligible.

Analysis : (*a*) For horizontal tube bank, the eqn. (11.21) may be used in the form

$$h = 0.725 \left\{ \frac{\rho^2 g\, h_{fg} k_f^{\,3}}{\mu N D (T_s - T_{sat})} \right\}^{1/4} \qquad (\because \quad \mu = \rho v)$$

using $\mu = \rho v$ in above equation;

$$h = 0.725 \times \left\{ \frac{(977.8)^2 \times (9.81) \times (2300 \times 10^3) \times (0.668)^3}{(0.415 \times 10^{-6} \times 977.8) \times 4 \times 0.025 \times (75 - 65)} \right\}^{1/4}$$

$= 8134.5$ W/m^2.K

The surface area of tubes :

$A_s = \pi D L N^2 = \pi \times (0.025 \text{ m}) \times (1.2 \text{ m}) \times 16$ tubes

$= 1.508$ m^2

The heat transfer rate :

$Q = hA_s(\Delta T) = (8134.5 \text{ W/m}^2\text{.K}) \times (1.508 \text{ m}^2) \times (75 - 65)(\text{K})$

$=$ **122,669.1 W. Ans.**

The condensation rate :

$$\dot{m} = \frac{Q}{h_{fg}} = \frac{122{,}669.1 \text{ W}}{2300 \times 10^3 \text{ J/kg}} = 0.053 \text{ kg/s}$$

$=$ **192 kg/h. Ans.**

(*b*) For condensation on vertical tubes;

$$h = 1.13 \left\{ \frac{\rho^2 g\, h_{fg}\, k_f^{\,3}}{\mu L (T_s - T_{sat})} \right\}^{1/4}$$

$$h = 1.13 \times \left\{ \frac{(977.8)^2 \times 9.81 \times (2300 \times 10^3) \times (0.668)^3}{(0.415 \times 10^{-6} \times 977.8) \times 1.2 \times (75 - 65)} \right\}^{1/4}$$

$= 6811.85$ W/m^2.K

The surface area

$A_s = \pi D L N^2 = 1.508$ m^2

The heat transfer rate :

$Q = hA_s\,(\Delta T) = 6811.85 \times 1.508 \times (75 - 65)$

$=$ **102,722.2 W. Ans.**

The condensation rate :

$$\dot{m} = \frac{Q}{h_{fg}} = \frac{102{,}722.2 \text{ W}}{2300 \times 10^3 \text{ kJ/kg}} = 0.446 \text{ kg/s}$$

$=$ **160.78 kg/h. Ans.**

Example 11.7. *A condenser is designed to condense 500 kg/h of dry and saturated steam at 0.1 bar. A square array of 400 tubes, 6 mm in diameter is used. The tube surface is maintained at 24°C by flowing water. Calculate the heat transfer coefficient and length of the each tube.*

Solution

Given : A square array for condensation of steam on tubes in square array

$\dot{m} = 500$ kg/h, Array = 400 tubes ; $p = 0.1$ bar

$D = 6$ mm $= 0.006$ m ; $T_s = 24°C$.

To find :

(*i*) Heat transfer coefficient.

(*ii*) Length of each tube.

Assumptions :

1. Laminar film wise condensation.
2. Effect of thermal resistance of film thickness is negligible.
3. Negligible vapour density.

Properties of fluid : At pressure of 0.1 bar ;

$h_{fg} = 2393$ kJ/kg, $T_{sat} = 45.74°C$

$$T_f = \frac{45.74 + 24}{2} \approx 35°C$$

The properties of saturated water at 35°C from Table A-7.

$\rho = 993.95$ kg/m^3, $k_f = 0.625$ W/m.K

$\mu = 728.15 \times 10^{-6}$ kg/ms.

Analysis : Number of horizontal tubes in a column :

$N = \sqrt{400} = 20$ tubes in a row column.

(*i*) The heat transfer coefficient for horizontal tube bank

$$h = 0.725 \left\{ \frac{\rho^2 g\, h_{fg}\, k_f^{\,3}}{ND\mu\,(T_{sat} - T_s)} \right\}^{1/4}$$

$$h = 0.725 \times \left\{ \frac{(993.95)^2 \times (9.81) \times (2393 \times 10^3) \times (0.625)^3}{20 \times (0.006) \times (728.15 \times 10^{-6}) \times (45.74 - 24)} \right\}^{1/4}$$

$= \mathbf{5357\ W/m^2.K.}$ Ans.

(*ii*) The length of the tubes :

The heat transfer rate during condensation is given by

$$Q = \dot{m} h_{fg} = \left\{ \frac{500}{3600} \text{ kg/s} \right\} \times (2393 \times 10^3)$$

$= 332{,}361.11$ W

The heat transfer rate can also be given by

$$Q = hA_s (T_{sat} - T_s) = h(\pi N^2 DL)(T_{sat} - T_s)$$

$$332361.11 = (5357 \text{ W/m}^2.\text{K}) \times (\pi \times 400 \times 0.006 \text{ m} \times L) \times (45.74 - 24)(\text{K})$$

or $$L = \frac{332{,}361.11}{878{,}089} = \mathbf{0.376\ m.}$$ Ans.

Example 11.8. *The dry and saturated steam at 42°C is condensed over a 60 cm square vertical plate maintained at 28°C. Calculate the following :*

(*i*) *Film thickness, local heat transfer coefficient at 30 cm from the top edge.*

(*ii*) *Average heat transfer coefficient and total heat transfer rate.*

(*iii*) *What would be the heat transfer coefficient, if plate is inclined at 30 degree with the horizontal ?*

Solution

Given :

1. For vertical plate :

$T_{sat} = 42°C$, $L = w = 60$ cm $= 0.6$ m

$T_s = 28°C$, $x = 0.3$ m.

2. Inclined plate at 30 degree with horizontal.

To find :

(*i*) (A) Film thickness δ, (B) heat transfer coefficient h_x at 30 cm from the top edge.

(*ii*) Average heat transfer coefficient, h and total heat transfer rate.

(*iii*) The average heat transfer coefficient, if plate is inclined at 30 degree with the horizontal.

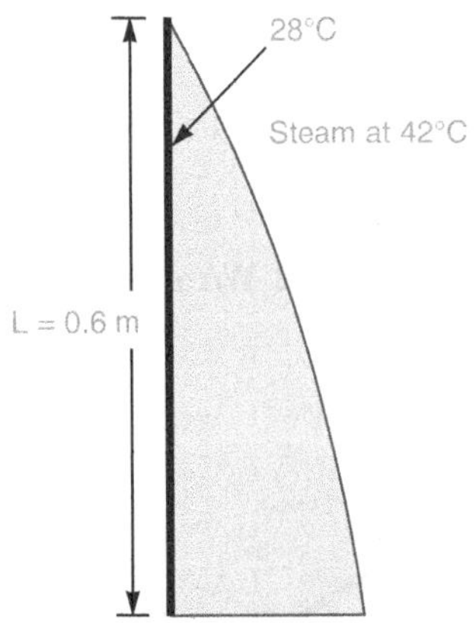

Fig. 11.16 (*a*) Vertical plate

Assumptions :

1. Laminar film wise condensation ;
2. Negligible film resistance ;
3. Constant properties.

Properties of fluid : The properties of steam at 42°C ;

$h_{fg} = 2402$ kJ/kg, $\rho_v = 0.0561$ kg/m³

$$T_f = \frac{42 + 28}{2} = 35°C$$

The properties of saturated fluid at 35°C from Table A-7;

$\rho = 994$ kg/m³, $k_f = 0.625$ W/m.K

$\mu = 728.15 \times 10^{-6}$ kg/m³.

Analysis : (*i*) (A) The thickness of condensate film can be calculated by using eqn. (11.5) ;

$$\delta = \left\{ \frac{4k_f (T_{sat} - T_s)\,\mu\, x}{\rho(\rho - \rho_v) g h_{fg}} \right\}^{1/4}$$

$$= \left\{ \frac{4 \times 0.625 \times (42 - 28) \times 728.15 \times 10^{-6} \times 0.3}{994 \times (994 - 0.0561) \times 9.81 \times 2402 \times 10^3} \right\}^{1/4}$$

$= 1.346 \times 10^{-4}$ m

or $\delta = \mathbf{0.134\ mm.}$ Ans.

(B) The local value of heat transfer coefficient

$$h_x = \frac{k_f}{\delta} = \frac{0.625}{1.346 \times 10^{-4}}$$

$= \mathbf{4642.74\ W/m^2.K.}$ Ans.

(*ii*) Average value of heat transfer coefficient :

$$h = \frac{4}{3} h_{x=L} = \frac{4}{3} \left\{\frac{L}{x}\right\}^{1/4} h_x$$

$$= \frac{4}{3} \times \left\{\frac{0.6}{0.3}\right\}^{1/4} \times 4642.74$$

$$= \mathbf{5521.2\ W/m^2.K.}\ \ \text{Ans.}$$

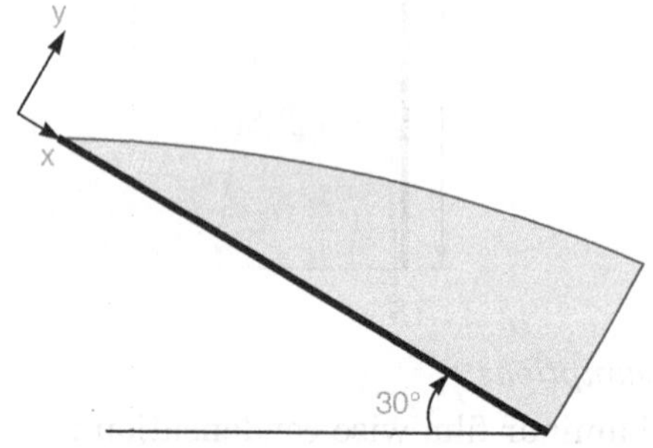

Fig. 11.16 (*b*) Inclined plate

The heat transfer rate :

$$Q = hA_s(T_{sat} - T_s)$$

$$= 5521.2 \times (0.6 \times 0.6) \times (42 - 28)$$

$$= 27.82 \times 10^3\ W = \mathbf{27.82\ kW.}\ \ \text{Ans.}$$

(*iii*) If the plate is inclined at 30 degree with horizontal then g is replaced by $g \sin \theta$, hence

$$h_{inclined} = h_{vertical} \times (\sin \theta)^{1/4}$$

$$= 5521.2 \times (\sin 30)^{1/4}$$

$$= \mathbf{4642.75\ W/m^2.\ K.}\ \ \text{Ans.}$$

Example 11.9. *A vertical tube 40 mm diameter and 1 m long is used for condensing dry steam at atmospheric pressure. The tube surface temperature is 60°C.*

(*i*) *Determine the mass of condensate.*

(*ii*) *If the tube is held in horizontal position, will there be any change in mass of condensate ? If yes, calculate the value and change.*

Use the properties of the fluid as

$k_f = 0.675\ W/mK$, $\rho = 971.8\ kg/m^3$.

$h_{fg} = 2257\ kJ/kg$, $\mu = 355 \times 10^{-6}\ kg/m.s$

(P.U., May. 2008)

Solution

Given : Condensation on a vertical tube :

D = 40 mm = 0.04 m, L = 1 m

p = 1 atm, T_{sat} = 100°C

T_s = 60°C.

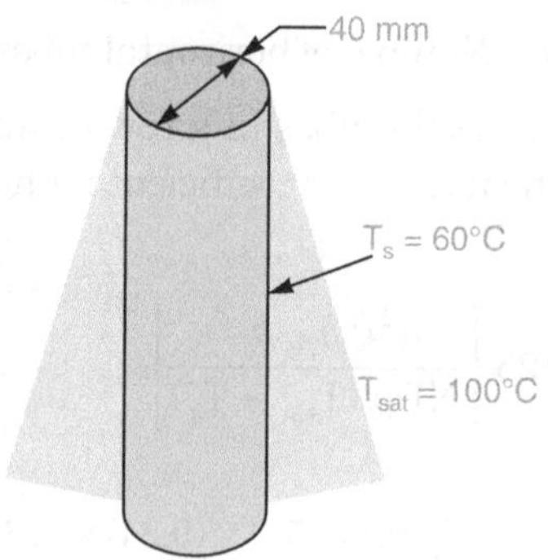

Fig. 11.17 Schematic for example 11.9

To find : Mass of condensate in

(*i*) Vertical tube

(*ii*) Horizontal tube

(*iii*) If mass changes in horizontal position tube, then percentage change in condensation rate.

Assumptions :

(*i*) Laminar condensation over the tube surface.

(*ii*) Negligible convection and radiation.

Analysis : (*i*) For laminar condensation, the average heat transfer coefficient on vertical tube is given by

$$h = 1.13 \left[\frac{\rho(\rho - \rho_v)\, g h_{fg} k_f^3}{\mu L (T_{sat} - T_s)}\right]^{1/4}$$

Neglecting ρ_v ;

$$= 1.13 \times \left[\frac{(971.8)^2 \times 9.81 \times 2257 \times 10^3 \times (0.675)^3}{355 \times 10^{-6} \times 1 \times (100 - 60)}\right]^{1/4}$$

$$= 1.13 \times 4613.1 = 5212\ W/m^2.K.$$

The heat transfer rate

$$Q = hA_s(T_{sat} - T_s) = 5212 \times (0.04 \times 1) \times (100 - 60) = 8340.5\ W$$

The condensation rate

$$\dot{m} = \frac{Q}{h_{fg}} = \frac{8340.5}{2257 \times 10^3} = 3.69 \times 10^{-3}\ kg/s$$

$$= \mathbf{13.3\ kg/h.}\ \ \text{Ans.}$$

(*ii*) If tube orientation changes to horizontal, then the heat transfer coefficient will also change. It can be related by eqn. (11.19)

$$\frac{h_{vert}}{h_{horz}} = 1.30 \left[\frac{D}{L}\right]^{1/4}$$

or $$\frac{5212}{h_{horz}} = 1.30 \times \left[\frac{0.04}{1}\right]^{1/4} = 0.581$$

or $$h_{horz} = 8965 \text{ W/m}^2\text{.K.}$$

Heat transfer rate,

$$Q = 8965 \times (0.04 \times 1) \times (100 - 60)$$
$$= 14,344 \text{ W}$$

and condensation rate

$$\dot{m} = \frac{14,344}{2257 \times 10^3} = 6.355 \text{ kg/s}$$
$$= \mathbf{22.88 \text{ kg/h.}} \quad \text{Ans.}$$

(*iii*) Percentage change in condensation rate

$$= \frac{22.88 - 13.3}{13.3} \times 100 = \mathbf{72\%.} \quad \text{Ans.}$$

The horizontal orientation *i.e.,* short length of condensation results into better condensation of steam.

11.8. BOILING

The phase change from liquid to vapour at liquid vapour interface is called the boiling. Actually boiling is a convective heat transfer process that involves phase change at a constant temperature from liquid to vapour. During the boiling, the heat added to liquid for its phase change is the *latent heat*. The boiling is sometimes termed as *evaporation* when the vapour pressure is less than the saturation pressure of liquid at the liquid vapour interface and then liquid itself extracts its latent heat for phase change from its surroundings. For examples, the drying of moisture from the wet clothes exposed to air, cooling of our body by evaporation of sweat etc.

Sometimes, the boiling process is referred as *vaporisation,* when the latent heat of liquid is supplied for its phase change such as process of steam generation in the boiler.

The boiling takes place, when the temperature T_s, of solid surface in contact of a liquid exceeds the saturation temperature T_{sat}, of the liquid. The heat flux during boiling depends on temperature excess ΔT_e ($= T_s - T_{sat}$) and is given by

$$q = h(T_s - T_{sat}) = h\Delta T_e \quad \text{...(11.44)}$$

As temperature excess ΔT_e increases, the liquid boiling rate also increases.

11.8.1. Boiling Modes

The boiling is mode of heat transfer that involves the formation of vapour bubbles at the solid-liquid interface. The growth of vapour and its dynamics depend on temperature excess, nature of surface and thermophysical properties of liquid, such as its surface tension, latent heat etc. Thus the boiling may occur under various conditions. The various modes of boiling are

(*i*) pool boiling, (*ii*) forced convection boiling, (*iii*) sub-cooled or local boiling, and (*iv*) saturated boiling.

(*i*) **Pool boiling.** It refers to a situation in which the heated surface is submerged below the free surface of a stagnant liquid and its motion near the surface is due to free convection only and the mixing is induced by bubble growth and its detachment.

(*ii*) **Forced convection boiling.** It refers to a situation of boiling, in which the fluid motion is induced artificially by an external means as well as by natural convection and the bubble induced mixing. This type of boiling occurs in high pressure water tube boilers involving forced convection.

(*iii*) **Sub-cooled or local boiling.** It refers to a situation of boiling, in which the temperature of the liquid is below its saturation temperature and the bubbles formed at the heated surface are condensed in the liquid as they leave the surface.

(*iv*) **Saturated boiling.** It refers to a situation of boiling, in which the temperature of the liquid is equal to its saturation temperature. The bubbles formed at the heated surface are pushed through the liquid by buoyancy effects to escape from free surface of the liquid.

11.9. POOL BOILING REGIMES

When boiling occurs as the pool boiling or forced convection boiling, there are six definite regimes of boiling associated with progressive increasing heat fluxes. These different regimes are shown in Fig. 11.18 (*a*), where the heat flux *q* from an electrically heated platinum wire submerged in the water is plotted against temperature excess, ΔT_e.

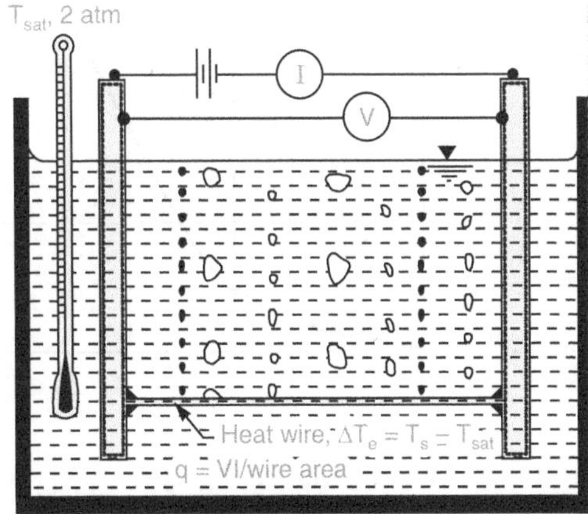

Fig. 11.18. (*a*) Nukiyama's power-controlled heating apparatus for demonstrating the boiling curve

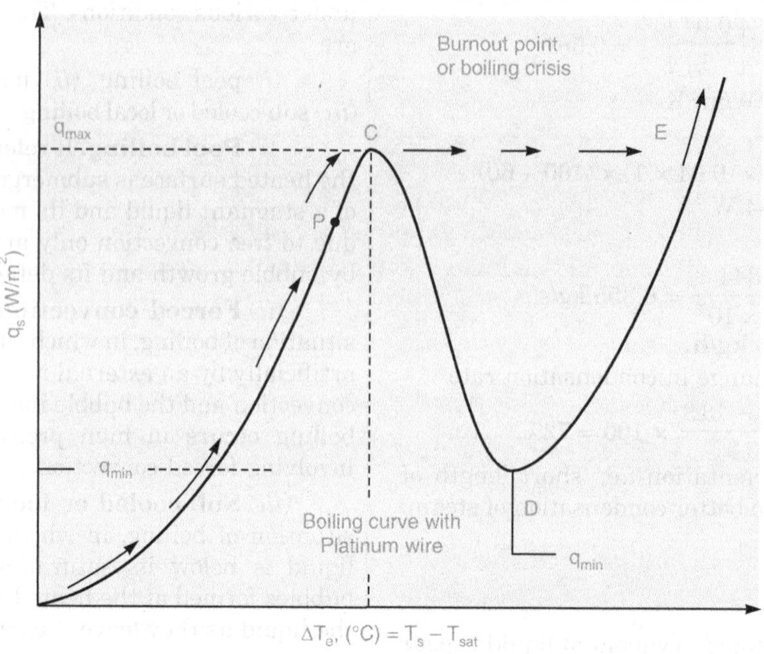

Fig. 11.18. *(b)* Nukiyama's boiling curve for saturated water

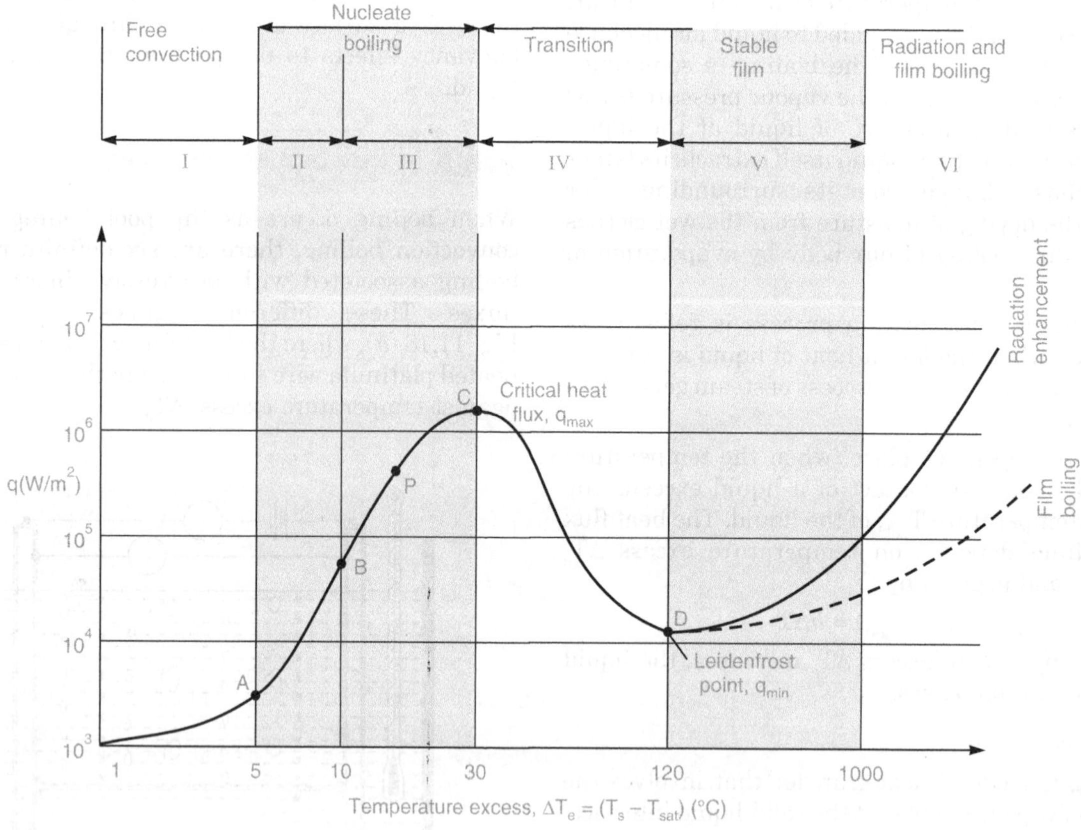

Fig. 11.19. Typical boiling curve for saturated water at 1 atmosphere

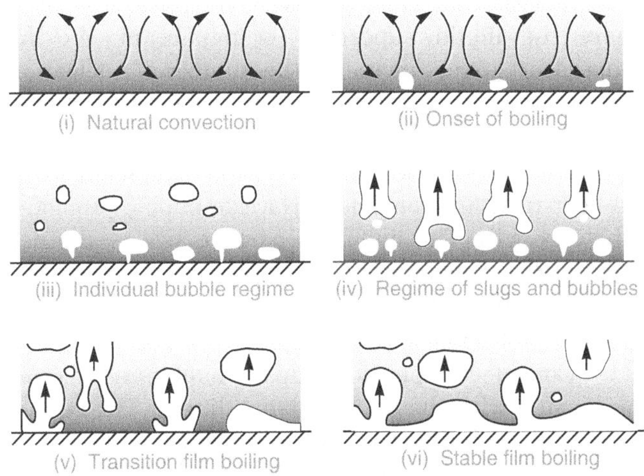

Fig. 11.20. Schematic representation of each boiling regime

Free convection boiling. In the region I, the temperature excess $\Delta T_e \le \Delta T_{e,A} \cong 5°C$, the free convection currents, shown in Fig. 11.20(*i*) are responsible for fluid motion and heat transfer. The liquid near the heated surface is slightly superheated and evaporates when it rises the surface. The heat transfer in this region can be calculated from relations for free convection.

Nucleate boiling. The nucleate boiling as shown in Fig. 11.20(*ii*) exists when $\Delta T_{e,A} \le \Delta T_e \le \Delta T_{e,C} \cong 30°C$. In this range, the two different flow regimes may be distinguished. In the region II, the bubbles begin to form at some nucleation sites on the surface of the wire and are collapsed in the liquid after detaching from the surface as shown in Fig. 11.20(*iii*). This region II is referred as *unstable or isolated nucleate boiling (A – B).*

As the temperature excess is increased further in the region III, more nucleation sites become active and the bubbles form more rapidly and rise to free surface of the liquid and form slugs of vapour as shown in Fig. 11.20(*iv*). It is *stable nucleate boiling (B – C)*. The interference between the densely populated bubbles retard the motion of the liquid near the surface. The point P corresponds to an *inflection point* in the boiling curve at which the heat transfer coefficient h reaches maximum value.

After this point P, the heat transfer coefficient h begins to decrease with increasing ΔT_e causing imbalance in the heat transfer rate and heat generation rate, thus the ΔT_e increases at a faster rate. In this region, the increase in ΔT_e is balanced by reduction in the heat transfer coefficient h. The heat flux q reaches to its maximum value at point C, which is called the **critical heat flux** and in water at atmospheric pressure it is generally more than 1 MW/m^2.

In practical situations, the nucleate boiling nearer to point P is desirable, because large heat is transferred with small temperature difference. In the nucleate boiling regime, the heat flux increases rapidly with increasing temperature difference until the peak heat flux is reached.

Transition boiling. It is also called *unstable film boiling* or *partial film boiling (C – D)*. In this region IV, $\Delta T_{e,C} \le \Delta T_e \le \Delta T_{e,D} \cong 120°C$. Bubble formation is so high that the vapour film form on the fraction of surface as shown in Fig. 11.20(*v*). At any point on the surface, the film and nucleate boiling may appear alternatively but the growth of film increases with increase in ΔT_e. This film prevents the inflow of fresh liquid taking their place. Further this film also offers strong thermal resistance in path of heat flow, hence the heat transfer rate decreases.

Film boiling. It appears in region V, when $\Delta T_e \ge 120°C$, the surface is completely covered by vapour blanket, Fig. 11.20(*vi*). The heat flux becomes minimum at point D. The heat transfer from the surface to the liquid occurs by conduction through the stable vapour film. As the surface temperature increases, the heat flux slowly increases due to increase in h. Then the heating surface becomes glowing bright due to large temperature excess and the thermal radiation from the surface becomes gradually dominant and heat flux again increases with increasing ΔT_e.

In this film boiling region V and VI, the heating surface is separated from the liquid by a vapour layer, across which heat must be transferred by conduction only. Since the thermal conductivity of vapour is very low, its layer offers strong thermal resistance, thus large temperature differences are needed for transfer of heat energy in this region.

11.9.1. Critical Heat Flux

The peak of curve at point C is the locus of *maximum heat flux,* where the excess temperature is raised to about 35°C. After this point, any increase in temperature excess, causing the heat flux to decrease. At this point, the temperature of the surface exceeds the melting point of solid and thus the melting of wire may occur. For this reason the locus (Point C) is often termed as *burnout point or boiling crisis.* The heat flux (maximum heat flux) at this point is called *critical heat flux,* q_{max}. The corresponding temperature excess is referred as the *critical temperature excess* $\Delta T_{e,\,cr}$. We are always interested to operate the heating surface of an evaporator or a boiler close to this peak to get maximum heat transfer from minimum area with smaller temperature difference.

11.9.2. Leidenfrost Point

It is point D in the film boiling region, where the vapour completely covers the surface and the heat flux reaches to its *minimum value.* This point is referred as the *Leidenfrost point.* The J.G. Leidenfrost pointed out that the water droplets on a very hot metal surface, dance about and boil away very slowly and thus being insulated by a thin blanket of vapour.

11.10. MECHANISM OF NUCLEATE BOILING

In the boiling heat transfer, the nucleate boiling has its own importance. The nucleate boiling induces good fluid mixing near the heated surface, resulting into increase in heat transfer coefficient and heat flux. The nucleate boiling involves two separate processes–the formation of bubbles, and subsequent growth and motion of these bubbles as shown in Fig. 11.21(*a*).

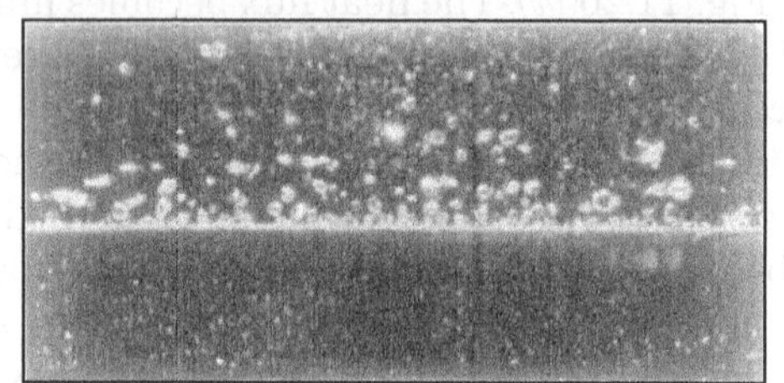

(a) Formation of bubbles, their growth and departure

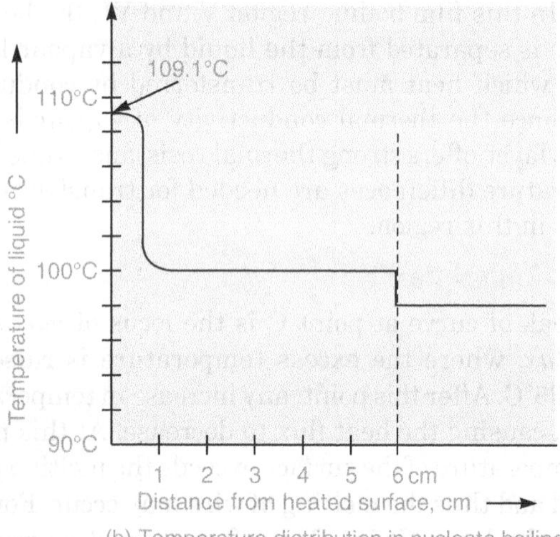

(b) Temperature distribution in nucleate boiling

Fig. 11.21. Nucleate boiling of water at 1 atm

The formation of bubbles is called the *nucleation.* As the surface temperature exceeds few degree the temperature of liquid, the nucleation starts at some favoured spots, called *nucleate sites* on the heated surface. The liquid layer immediately adjacent to heated surface becomes superheated as shown in Fig. 11.21(*b*).

The rate of bubble formation is large at the spots, where the liquid superheat is maximum. As the temperature excess increases the number of nucleation sites increase resulting into increase in the rate of bubble formation.

Once the nucleation process is completed, then the heat transfer from the surface to bubble promotes bubble growth. The bubble departure size is another important parameter that affects the heat transfer in nucleate boiling. The mechanisms of bubble growth and its collapse are discussed below.

Consider a spherical vapour bubble in thermal equilibrium with the liquid surrounding it. The bubble of radius r is split into two halves as shown in Fig. 11.22. The pressure force on the bubble must be balanced by the surface tension at the interface. Thus

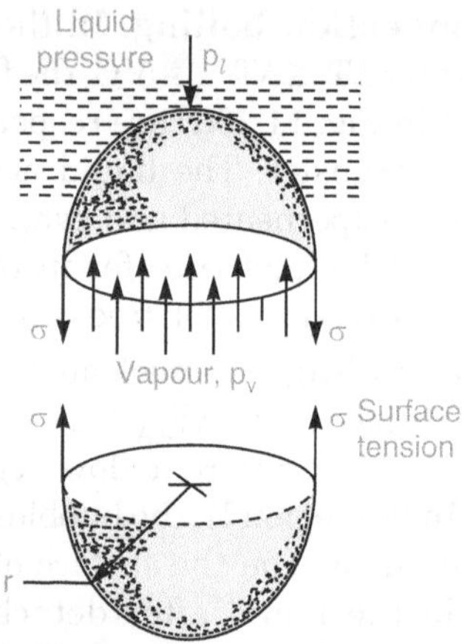

Fig. 11.22. Force balance on a spherical bubble

$$\pi r^2(p_v - p_l) = 2\pi r\sigma \quad \text{or} \quad p_v - p_l = \frac{2\sigma}{r} \qquad ...(11.45)$$

where p_v = vapour pressure inside the bubble,

p_l = liquid pressure over the surface of bubble, and

σ = surface tension at vapour liquid interface.

The liquid gets superheated at the heated surface and the vapour bubbles formed are also superheated. For superheated vapour Claussius-Clapeyron equation with the ideal gas approximation, relates T and v along the saturation line as

$$\frac{dp}{dT} = \frac{h_{fg}\rho_v}{T_{sat}} = \frac{h_{fg}\,p_v}{R_v T_{sat}^{\;2}} \qquad ...(11.46)$$

where R_v = gas constant for vapour,

h_{fg} = latent heat of vaporisation,

T_{sat} = saturation temperature of liquid.

The eqn. (11.46) can be rearranged as

$$\frac{dp}{dT} = \frac{p_v - p_l}{T_v - T_{sat}} = \frac{p_v\, h_{fg}}{R_v {T_{sat}}^2} \quad ...(11.47)$$

where T_v = vapour temperature inside the bubble,

Combining eqn. (11.45) with eqn. (11.47), we get

$$T_v - T_{sat} = \frac{2\sigma}{r}\left[\frac{R_v}{p_v}\frac{T_{sat}^2}{h_{fg}}\right] \quad ...(11.48)$$

If $(T_{liq} - T_{sat}) > (T_v - T_{sat})$, the bubble of radius r will grow, otherwise it will collapse.

11.10.1. Critical Diameter of a Bubble

The size of bubble formed on a heated surface depends on the followings:

(*i*) Surface tension at liquid-vapour interface, σ_{lv}.

(*ii*) Surface tension at solid-liquid interface, σ_{sl}.

(*iii*) Surface tension at vapour-solid interface, σ_{vs}.

(*iv*) The angle formed by bubbles with the solid surface, β

(*v*) Buoyancy force, $g(\rho_l - \rho_v)$.

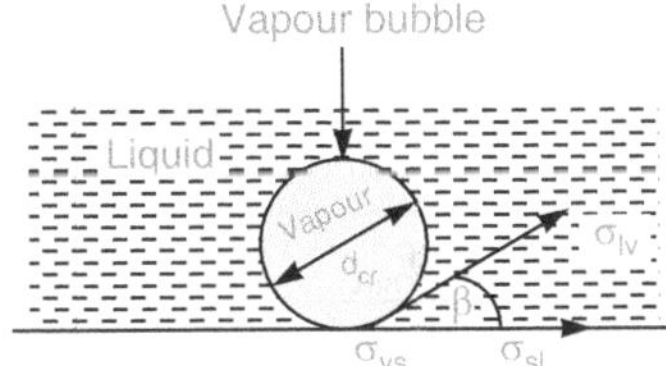

Fig. 11.23. Critical radius of bubble

If d_{cr} is critical diameter of bubble, then

$$d_{cr} = f\{\beta, \sigma_{lv}, \sigma_{sl}, \sigma_{vs}, g(\rho_l - \rho_v)\}$$

The dimensional analysis gives us

$$d_{cr} = C\beta\left(\frac{\sigma_{lv}}{\sigma_{sl}}\right)\sqrt{\frac{\sigma_{lv}}{g(\rho - \rho_v)}} \quad ...(11.49)$$

where C is a constant and its value is evaluated experimentally. For water bubbles, *Fritz* suggested

C = 0.0148.

11.11 POOL BOILING CORRELATIONS

The factors which affects the pool boiling process are

(*i*) Specific heat of saturated liquid, C_p, J/kg. K

(*ii*) Latent heat of vaporisation, h_{fg}, J/kg

(*iii*) Temperature excess between surface and saturated liquid $\Delta T_e = (T_s - T_{sat})$, (K)

(*iv*) Prandtl number of saturated liquid, Pr

(*v*) Viscosity of liquid, μ, kg/ms

(*vi*) Surface tension at liquid-vapour interface, σ, N/m

(*vii*) Density of saturated liquid, ρ, kg/m^3

(*viii*) Density of saturated vapour, ρ_v, kg/m^3

(*ix*) Acceleration due to gravity, g, m/s^2 and

(*x*) Length of heater surface, L, m.

Therefore, heat flux during the boiling process is

$$q = f(\Delta T_e, L, (\rho - \rho_v)\,g, h_{fg}, \sigma, k_f, C_p, \mu) \quad ...(11.50)$$

The dimensional analysis gives

$$\frac{C_p \Delta T_e}{h_{fg}} = f\left[\frac{q}{\mu h_{fg}}, \sqrt{\frac{\sigma}{g(\rho - \rho_v)}}\,L, \frac{\mu C_p}{k_f}\right] \quad ...(11.51)$$

where, $\frac{g(\rho - \rho_v)}{\sigma} L^2$ = Bond number,

$\frac{C_p \Delta T_e}{h_{fg}}$ = Jakob number,

$\frac{\mu C_p}{k_f}$ = Prandtl number.

The *Bond number* represents the ratio of buoyancy forces to surface tension force. The *Jacob number* represents the ratio of sensible heat to the latent heat and the quantity $\frac{q}{\mu h_{fg}}$ is an unnamed dimensionless quantity. The combination of these dimensionless quantities makes *Nusselt number*.

$$\frac{qL}{\mu h_{fg}} \times \frac{\mu C_p}{k_f} \times \frac{h_{fg}}{C_p \Delta T_e} = \frac{q}{\Delta T_e} \times \frac{L}{k_f} = \frac{hL}{k_f}$$

= Nusselt number.

During the pool boiling in natural convection region A-B in Fig. 11.19, the correlations for natural convection are applicable.

11.11.1. Correlation for Nucleate Boiling

In the nucleate boiling region B-C of Fig. 11.19, the analysis requires prediction of nucleation sites and the rate at which the bubbles orginate from each side. An alternate form of correlation obtained by manipulating different dimensionless parameters of eqn. (11.51) had developed by the *Rehsenow,* most successful relation which is widely used in nucleate boiling regimes.

$$\frac{C_p \Delta T_e}{h_{fg} \Pr^n} = C_{sf} \left[\frac{q}{\mu h_{fg}} \sqrt{\frac{\sigma}{g(\rho - \rho_v)}} \right]^{1/3} \quad ...(11.52)$$

where C_{sf} is an empirical constant, depends on nature of heating surface-fluid combination and to be determined from experimental data.

It gives

$$q = \mu h_{fg} \left\{ \left[\frac{g(\rho - \rho_v)}{\sigma} \right]^{1/2} \times \left(\frac{C_p \Delta T_e}{h_{fg} C_{sf} \Pr^n} \right)^3 \right\} \quad ...(11.53)$$

In eqn. (11.53) the exponent n and constant C_{sf} are provisionally adjusted. The value of n for water is approximately 1.0 and for other surface-fluid combination the value of constant C_{sf} are given in Table 11.1.

TABLE 11.1. Value of constant C_{sf} for surface-fluid combinations

Liquid-surface combination	C_{sf}	*Exponent n*
Water-scored copper	0.0068	1.0
Water-copper	0.0130	1.0
Water-emery polished copper	0.0128	1.0
Water-brass	0.0060	1.0
Water-emery polished paraffin treated copper	0.0470	1.0
Water-teflon coated steel	0.0058	1.0
Water-ground and polished stainless steel	0.0080	1.0
Water-chemically etched stainless steel	0.0133	1.0
Water-mechanically polished stainless steel	0.0132	1.0
Water-platinum	0.0130	1.0
Water-nickel	0.0060	1.0
n-pentane-lapped copper	0.0049	1.7
n-pentane-emery rubbed copper	0.0074	1.7
n-pentane-emery polished copper	0.0154	1.7
n-pentane-emery polished nickel	0.0127	1.7
n-pentane-chromium	0.0150	1.7
Isopropyl alcohol-copper	0.0025	1.7
Benzene-chromium	0.0100	1.7
Ethyl alcohol-chromium	0.0027	1.7
Carbon tetrachloride-copper	0.0130	1.7
Carbon tetrachloride-emery polished copper	0.0070	1.7

11.11.2. Correlation for Critical Heat Flux

The point C represents the maximum heat flux on the pool boiling curve in Fig. 11.19, it is referred as the *critical point.* The *Zuber* had obtained the following relation through hydraulic stability. It may be used for the determination of the peak heat flux.

$$q_{max} = \left(\frac{\pi}{24} \right) \rho_v^{1/2} h_{fg} [\sigma g (\rho - \rho_v)]^{1/4} \times \left[\frac{\rho + \rho_v}{\rho} \right]^{1/2} \quad ...(11.54)$$

where, q_{max} = critical heat flux, W/m².

σ = surface tension on liquid-vapour interface, N/m.

ρ = density of saturated liquid, kg/m³.

ρ_v = density of saturated vapour, kg/m³.

h_{fg} = latent heat of evaporation, J/kg.

The Zuber constant ($\pi/24$) = 0.131 is replaced by an experimental value of 0.149 and approximating the last term $\frac{\rho + \rho_v}{\rho} \approx 1$, the relation takes the form

$$q_{max} = 0.149 \rho_v^{1/2} h_{fg} [\sigma g (\rho - \rho_v)]^{1/4} \quad ...(11.55)$$

The properties of vapour should be taken at mean temperature as

$$T_f = \frac{T_s + T_{sat}}{2}.$$

11.11.3. Pool Film Boiling

At the excess temperature greater than the *Leidenfrost point* D, a continuous film blankets the heated surface and there is no direct contact between liquid and surface. Similar to the film vwise condensation on vertical surfaces, the following expression for *stable film* boiling on the outside of horizontal cylinder and sphere in absence of any radiation may be used.

$$Nu = \frac{hD}{k_v} = C \left\{ \frac{\rho_v (\rho - \rho_v) g h'_{fg} D^3}{k_v \mu_v \Delta T_e} \right\}^{1/4} \quad ...(11.56)$$

where, Suffix v indicates properties of vapour,

ρ = density of saturated liquid,

h'_{fg} = corrected latent heat of evaporation,

D = diameter of the cylinder or sphere,

$\Delta T_e = T_s - T_{sat}$, temperature excess.

The value of constant

C = 0.62 for cylinders,

and C = 0.67 for spheres.

The liquid properties are evaluated at T_{sat} and vapour properties should be taken at mean temperature as

$$T_f = \frac{T_s + T_{sat}}{2}$$

Thus for stable film boiling on the horizontal cylinders, the average heat transfer coefficient is given by

$$h = 0.62 \left\{ \frac{\rho_v (\rho - \rho_v) g h'_{fg} k_v^3}{D \mu_v \Delta T_e} \right\}^{1/4} \quad ...(11.57)$$

The corrected latent heat of vaporisation is calculated as

$$h'_{fg} = h_{fg} + 0.4 C_{pv} \Delta T_e \quad ...(11.58)$$

The eqn. (11.56) can be rearranged as

$$Nu = \frac{hD}{k_v} = C(Ra')^{1/4} \quad ...(11.59)$$

where Ra′ represents the modified Rayleigh number and

$$Ra' = \frac{\rho_v (\rho - \rho_v) g h'_{fg} D^3}{\mu_v k_v \Delta T_e} \quad ...(11.60)$$

For the film boiling from the large (infinite) horizontal surfaces

$$Nu = \frac{hL^*}{k_v} = 0.425 \left\{ \frac{\rho_v (\rho - \rho_v) g h'_{fg} L^{*3}}{k_v \mu_v \Delta T_e} \right\}^{1/4} \quad ...(11.61)$$

where, $$L^* = \left[\frac{\sigma}{g(\rho - \rho_v)} \right]^{1/2} \quad ...(11.62)$$

At the elevated temperatures (T_s > 300°C), the radiation heat transfer across the vapour film becomes significant. The radiation heat transfer causes the film thickness to increase. The thicker vapour film results into reduced convective heat transfer. On account for the interaction between radiative and convective heat transfer, *Bromley* suggested

$$q_{total} = q_{conv} + 0.75\, q_{rad}$$

or $$h_{total} = h_{conv} + 0.75\, h_{rad} \quad ...(11.63)$$

where h_{conv} is computed from eqn. (11. 57). The radiation heat transfer coefficient h_{rad} is evaluated as

$$h_{rad} = \frac{\varepsilon \sigma (T_s^4 - T_{sat}^4)}{T_s - T_{sat}}$$

where ε is the emissivity of the solid surface and σ is the Stefan Boltzmann constant (= 5.67×10^{-8} W/m².K⁴).

11.11.4. Minimum Heat Flux

Fig. 11.24 shows actual boiling curve in nucleate transition and film transition boiling region. With start of transition boiling, the liquid-solid contact decreases. The transition boiling regime is of little interest and it is not used in engineering applications and thus no sufficient theory has been developed for this regime. However, the upper portion of the regime is still important, because it corresponds to formation of a stable vapour blanket or film and to minimum heat flux condition. If the heat flux drops below this minimum value, the film will collapse and causing the surface to cool and nucleate boiling to be re-established. The expression for the minimum heat flux in the film boiling regime, derived by *Zuber*

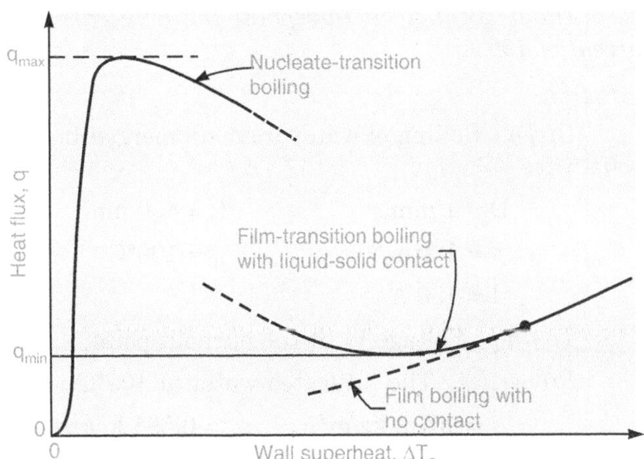

Fig. 11.24. The transition boiling regime

$$q_{min} = C \rho_v h'_{fg} \left[\frac{\sigma(\rho - \rho_v) g}{(\rho + \rho_v)^2} \right]^{1/4} \quad ...(11.64)$$

where $h'_{fg} = h_{fg} + 0.4\, C_{pv}\, \Delta T_e$

For infinite horizontal surfaces, the constant C = 0.09 and correlation for minimum heat flux for infinite horizontal surfaces in the film boiling regime becomes

$$q_{min} = 0.09\, \rho_v\, h'_{fg} \left[\frac{\sigma(\rho - \rho_v) g}{(\rho + \rho_v)^2} \right]^{1/4} \quad ...(11.65)$$

Combining eqn. (11. 61) with eqn. (11.65) and solving for ΔT_e corresponding to q_{min}, we get

$$\Delta T_{min} = 0.127 \frac{\rho_v h'_{fg}}{k_v} \left[\frac{\mu_v (\rho - \rho_v) g L^{*3}}{(\rho + \rho_v)^2} \right]^{1/3} \quad ...(11.66)$$

where $$L^* = \left[\frac{\sigma}{g(\rho - \rho_v)}\right]^{1/2}$$

For horizontal wire of radius r_o, *Licnhard* and *Wong* suggested

$$q_{min} = 0.0464\, \rho_v h'_{fg} \left[\frac{\sigma(\rho - \rho_v)\, g}{(\rho + \rho_v)^2}\right]^{1/4} \times \left[\frac{18}{R'^2\,(2R'^2 + 1)}\right] \quad ...(11.67)$$

where $$R' = r_o \left[\frac{(\rho - \rho_v)\, g}{\sigma}\right]^{1/2} \quad ...(11.68)$$

Example 11.10. *A nickel wire of 1 mm diameter and 400 mm long, carrying current is submerged in a water bath open to atmospheric air. Calculate the voltage at the burnout point, if at this point the wire carries a current of 190 A.*

Solution

Given : Boiling of water from submerged heated nickel wire

D = 1 mm, L = 400 mm

p = 1 atm, T_{sat} = 100°C

I = 190 A

To find : Voltage V_b at the burnout point.

Properties : The saturated water at 100°C

$\rho = 957.9$ kg/m³, $\rho_v = 0.595$ kg/m³

$h_{fg} = 2257 \times 10^3$ J/kg, $\sigma = 58.9 \times 10^{-3}$ N/m.

Analysis : At burnout point, the critical heat flux is given by Zuber, eqn. (11.55)

$$q_{max} = 0.149\, h_{fg}\, \rho_v^{1/2}\, [\sigma g (\rho - \rho_v)]^{1/4}$$
$$= 0.149 \times (2257 \times 10^3) \times (0.595)^{1/2} \times [58.9 \times 10^{-3} \times 9.81 \times (957.9 - 0.595)]^{1/4}$$
$$= 12{,}58{,}012.5 \text{ W/m}^2$$

The electrical energy input to the wire

$$Q = qA_s = V_b I \Rightarrow 12{,}58{,}012.5 \times (\pi \times 0.001 \times 0.4) = V_b \times 190$$

V_b = **8.32 V.** Ans.

Example 11.11. *An electric wire of 1.5 mm diameter and 20 cm long is laid horizontally and submerged in water at atmospheric pressure. The current flowing through the wire is 40 A, while voltage drop is 16 V. Calculate the heat flux, heat transfer coefficient and excess temperature.*

Use correlation : $h = 1.54\,(Q/A)^{3/4} = 5.58(\Delta T_e)^3$.

(P.U., May 1998)

Solution

Given : An electric wire :

D = 1.5 mm = 0.0015 m, L = 20 cm = 0.2 m

I = 40 A, ΔV = 16 V.

To find :

(*i*) The heat flux ;

(*ii*) Heat transfer coefficient ; and

(*iii*) The temperature excess.

Analysis : The surface area of the wire, $A_s = \pi DL$
$= \pi \times (0.0015 \text{ m}) \times (0.2 \text{ m})$
$= 9.42 \times 10^{-4}$ m²

Electrical power input to wire,

$$Q = \Delta V.I = 16 \text{ V} \times 40 \text{ A} = 640 \text{ W}$$

(*i*) The heat flux,

$$q = \frac{Q}{A} = \frac{640 \text{ W}}{9.42 \times 10^{-4} \text{ m}^2}$$

= **6.79 × 10⁵ W/m².** Ans.

(*ii*) The heat transfer coefficient can be calculated by given correlation

$$h = 1.54\,(Q/A)^{3/4} = 1.54 \times (6.79 \times 10^5)^{3/4}$$

= **36,429.42 W/m².K.** Ans.

(*iii*) The temperature excess can be calculated as

$$h = 5.58(\Delta T_e)^3$$

$$\Delta T_e = \left[\frac{36{,}429.42}{5.58}\right]^{1/3} = \mathbf{18.68°C.}$$ Ans.

Example 11.12. *Calculate nucleate boiling heat transfer coefficient for water boiling on the tube, whose wall temperature is maintained at 20°C, above saturation temperature. Assume water to be at 20 bar.*

Also calculate the heat transfer coefficient, when,

(*a*) *Temperature is reduced by 10°C at 20 bar.*

(*b*) *Pressure is reduced by 10 bar at 20°C temperature difference.*

Comment on the result.

Use correlation : $h_A = 5.56\,(\Delta T)^3$ *W/m².K*

and $$h_p = h_A \left\{\frac{p}{p_0}\right\}^{0.4} \text{ W/m}^2\text{.K}$$

Suffix, A corresponds to atmospheric pressure ;
p corresponds to fluid pressure

Assume atmospheric pressure to be 100 kPa.

(P.U., May 2002)

Solution

Given : ΔT = 20°C, p_1 = 20 bar = 2000 kPa

(*a*) ΔT = 10°C, p_1 = 20 bar,

(*b*) ΔT = 20°C, p_1 = 10 bar = 1000 kPa

Correlations as above.

To find : Nucleate heat transfer coefficient in given conditions and

(*a*) When temperature excess is 10°C at 2000 kPa.

(*b*) When pressure is reduced by 1000 kPa at 20°C.

Analysis : When the temperature excess is 20°C at atmospheric pressure ;

The air heat transfer coefficient at 20°C :

$$h_A = 5.56\ (\Delta T)^3\ \text{W/m}^2\text{.K}$$

$$h_A = 5.56 \times (20)^3\ \text{W/m}^2\text{.K}$$

$$= \mathbf{44{,}480\ W/m^2.K.}\quad \text{Ans.}$$

and at 20 bar, $h_p = h_A \left\{\dfrac{p}{p_0}\right\}^{0.4}$

$$= 44{,}480\ \text{W/m}^2\text{.K} \times \left\{\frac{20}{1}\right\}^{0.4}$$

$$= \mathbf{1{,}47{,}427\ W/m^2.K.}\quad \text{Ans.}$$

(*a*) When temperature excess is reduced to 10°C at 20 bar.

$$\Delta T = 10°\text{C}$$

The heat transfer coefficient of air at 10°C ;

$$h_A = 5.56 \times (10)^3\ \text{W/m}^2\text{.K}$$

$$= \mathbf{5560\ W/m^2.K.}\quad \text{Ans.}$$

The fluid heat transfer coefficient at 20 bar and 10°C :

$$h_p = h_A \left\{\frac{p}{p_0}\right\}^{0.4}$$

$$= 5560\ \text{W/m}^2\text{.K} \times \left\{\frac{20}{1}\right\}^{0.4}$$

$$= \mathbf{18{,}428.36\ W/m^2.K.}\quad \text{Ans.}$$

(*b*) When pressure is reduced by 10 bar at 20°C

Now the new pressure, $p_1 = 10$ bar,

and temperature difference, $\Delta T = 20°C$

Hence, at 20°C the heat transfer coefficient h_A remains unchanged ;

or $h_A = 5.56 \times (20)^3\ \text{W/m}^2\text{.K}$

$$= \mathbf{44{,}480\ W/m^2.K.}\quad \text{Ans.}$$

The heat transfer coefficient of fluid, h_p;

$$h_p = h_A \left\{\frac{p}{p_0}\right\}^{0.4}$$

$$= 44{,}480\ \text{W/m}^2\text{.K} \times \left\{\frac{10}{1}\right\}^{0.4}$$

$$= \mathbf{1{,}11{,}728.7\ W/m^2.K.}\quad \text{Ans.}$$

Example 11.13. *Water is boiled at a rate of 30 kg/h in a copper pan, 30 cm in diameter, at atmospheric pressure. Estimate the temperature of bottom surface of the pan, assuming nucleate boiling conditions. Also determine peak heat flux.* (V.T.U., April 1999)

Solution

Given : The evaporation of water in a copper pan.

$\dot{m}_e = 30$ kg/h, D = 30 cm = 0.3 m

$p = 1$ atm, $T_{sat} = 100°C$.

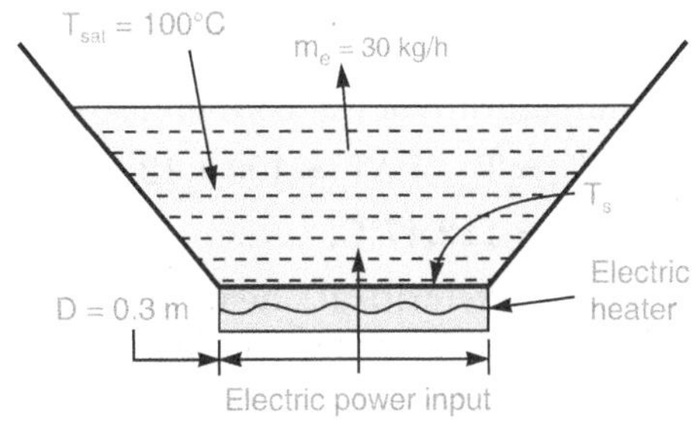

Fig. 11.25. Schematic of water heating in a pan

To find :

(*i*) The temperature of bottom surface of the pan.

(*ii*) Critical heat flux.

Assumptions :

(*i*) Steady state conditions.

(*ii*) Pan bottom surface is polished copper.

(*iii*) Pan bottom surface at uniform temperature.

Properties : Saturated water at 100°C, (from Table A-7)

$\rho = 957.9$ kg/m³, $\rho_v = 0.595$ kg/m³,

$C_p = 4217$ J/kg.K, $h_{fg} = 2257 \times 10^3$ J/kg.K,

$\mu = 279 \times 10^{-6}$ kg/ms, Pr = 1.76

$\sigma = 58.9 \times 10^{-3}$ N/m.

Analysis : (*i*) The heat transfer rate is

$$Q = \dot{m}_e h_{fg}$$

$$= (30\ \text{kg/h}) \times (2257 \times 10^3\ \text{J/kg})$$

$$= 6771 \times 10^4\ \text{J/h} = 18{,}808.3\ \text{W}$$

The heat flux is

$$q = \frac{Q}{A} = \frac{Q}{(\pi/4)\,D^2} = \frac{18{,}808.3}{(\pi/4) \times (0.3)^2}$$

$$= 266{,}083.5\ \text{W/m}^2$$

During nucleate boiling, the heat flux is given by eqn. (11.53)

$$q = \mu\, h_{fg} \left[\frac{g(\rho - \rho_v)}{\sigma}\right]^{1/2} \times \left[\frac{C_p \Delta T_e}{C_{sf}\, h_{fg}\, \text{Pr}^n}\right]^3$$

Using numerical values and taking C_{sf} = 0.0130 for water-polished copper interface, and n = 1, from Table 11.1,

$$266{,}083.5 = 279 \times 10^{-6} \times (2257 \times 10^3) \times \left[\frac{9.81 \times (957.9 - 0.595)}{58.9 \times 10^{-3}}\right]^{1/2} \times \left[\frac{4217 \times \Delta T_e}{0.0130 \times 2257 \times 10^3 \times 1.76}\right]^3$$

or $266{,}083.5 = 629.703 \times 399.30 \times 5.445 \times 10^{-4}\, \Delta T_e^{\,3}$

or $\Delta T_e^{\,3} = 1943.27$ or $\Delta T_e = 12.47°C$

The temperature of the bottom heated surface

$$T_s = T_{sat} + \Delta T_e = 100 + 12.47$$

$= $ **112.47°C. Ans.**

(*ii*) The peak (critical) heat flux for nucleate boiling is determined by using relation

$$q_{max} = 0.149\, \rho_v^{1/2}\, h_{fg}\, [\sigma g(\rho - \rho_v)]^{1/4}$$

$$= 0.149 \times (0.595)^{1/2} \times 2257 \times 10^3 \times [58.9 \times 10^{-3} \times 9.81 \times (957.9 - 0.595)]^{1/4}$$

$$= 259403.78 \times 4.85 = 1258012.5 \text{ W/m}^2$$

$= $ **1.258 MW/m². Ans.**

Example 11.14. *A metal clad heating element 8 mm in diameter and emissivity ε = 0.95 is horizontally submerged in a water bath. The surface temperature of the metal is maintained at 250°C under steady state conditions. Estimate the power dissipation per unit length of the heater. Assume the water is exposed to atmospheric pressure and is at a uniform temperature.*

Solution

Given : The boiling from submerged horizontal metal clad (cylinder)

D = 8 mm = 0.008 m, ε = 0.95

T_s = 250°C = 523 K, p = 1 atm

T_{sat} = 100°C = 373 K, L = 1 m.

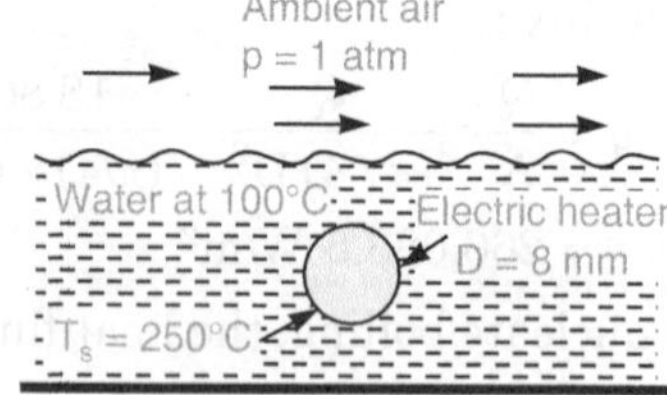

Fig. 11.26. Schematic

To find : Power dissipation per unit length for the cylinder.

Properties : Saturated water liquid at 100°C (from Table A-7)

ρ = 957.9 kg/m³, h_{fg} = 2257 × 10³ J/kg

Saturated water vapour at

$$T_f \approx \frac{250 + 100}{2} = 175°C = 448 \text{ K}$$

ρ_v = 4.808 kg/m³

C_{pv} = 2.56 kJ/kg.K,

k_v = 0.0331 W/m.K

μ_v = 14.85 × 10⁻⁶ Ns/m².

Analysis : The excess temperature

$$\Delta T_e = T_s - T_{sat} = 250 - 100 = 150°C$$

According to pool boiling curve, at ΔT_e = 150°C, heat transfer is due to both convection and radiation and the heat transfer rate is given by

$$Q = hA\Delta T_e = h(\pi DL)\, \Delta T_e$$

The heat transfer coefficient h is combined heat transfer coefficient for convection and radiation effects and it is calculated as

$$h = h_{conv} + (0.75)\, h_{rad}$$

The convection coefficient by eqn. (11.57) in combination with eqn. (11.58)

$$h_{conv} = 0.62 \left[\frac{k_v^3\, \rho_v (\rho - \rho_v)\, g\, (h_{fg} + 0.4\, C_{p_v}\, \Delta T_e)}{\mu_v\, D\, \Delta T_e}\right]^{1/4}$$

$$= 0.62 \times \left[\frac{(0.0331)^3 \times 4.808 \times (957.9 - 4.808) \times 9.81 \times (2257 \times 10^3 + 0.4 \times 2.56 \times 10^3 \times 150)}{14.85 \times 10^{-6} \times 0.008 \times 150}\right]^{1/4}$$

$= 425$ W/m².K

The radiation coefficient

$$h_{rad} = \frac{\varepsilon\sigma (T_s^{\,4} - T_{sat}^4)}{T_s - T_{sat}}$$

$$= \frac{0.95 \times 5.67 \times 10^{-8} \times (523^4 - 373^4)}{523 - 373}$$

$= 19.91$ W/m².K

Then $h = h_{conv} + 0.75\, h_{rad}$

$= 425 + (0.75) \times 19.91 = 440$ W/m^2.K

Hence the heat transfer rate is

$$Q = 440 \times (\pi \times 0.008 \times 1) \times 150$$

$$= \mathbf{1658.76\ W/m.}$$ **Ans.**

Example 11.15. *The heater for a steam boiler to produce saturated steam at 170°C is made of an electrical heating element inside a 15 mm OD mechanically polished stainless steel tube. The power input to the heater is 5 kW. If the surface temperature of heater not to exceed 175°C, find the length of the heater.*

Solution

Given : Boiling of water on polished stainless heater

$T_{sat} = 170°C$, $D = 15$ mm

$P = Q = 5$ kW, $T_s = 175°C$.

To find : Length of the heating element.

Assumptions :

(*i*) Steady state conditions,

(*ii*) Pool boiling.

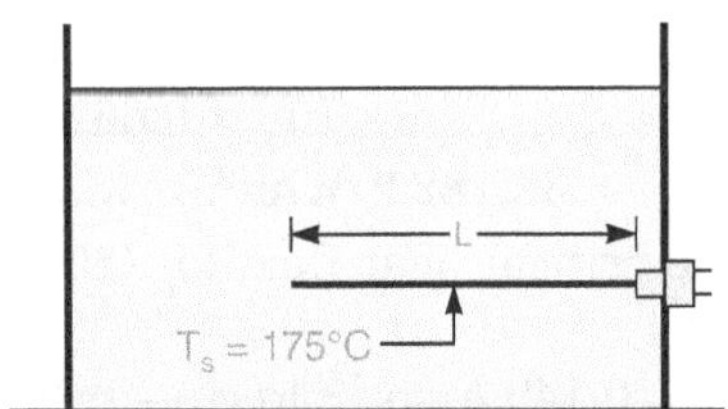

Fig. 11.27

Properties : Properties of saturated water and steam at 170°C from Table A-7

$\rho = 897.3$ kg/m^3, $C_p = 4340$ J/kg.K

$k_f = 0.681$ W/m.K $\mu = 157.6 \times 10^{-6}$ Ns/m^2

Pr = 1.0, $h_{fg} = 2049.5 \times 10^3$ J/kg

$\rho_v = 4.119$ kg/m^3, $\sigma = 0.04406$ N/m

For mechanically polished stainless steel $C_{sf} = 0.0132$ and $n = 1$.

Analysis : The temperature excess is

$$\Delta T_e = T_s - T_{sat} = 175 - 170 = 5°C$$

The heat flux during nucleate boiling

$$q = \mu h_{fg} \left\{\frac{g(\rho - \rho_v)}{\sigma}\right\}^{1/2} \times \left[\frac{C_p\, \Delta T_e}{C_{sf} h_{fg}\, Pr^n}\right]^3$$

$$= 157.6 \times 10^{-6} \times (2049.5 \times 10^3) \times \left[\frac{9.81 \times (897.3 - 4.119)}{0.04406}\right]^{1/2} \times \left[\frac{4340 \times 5}{0.0132 \times 2049.5 \times 10^3 \times 1}\right]^3$$

$$= 74{,}325 \text{ W/m}^2$$

The heat transfer rate is given by

$$Q = q\, \pi D L$$

or $5000 = 74325 \times \pi \times 0.015\, L$

or $L = \mathbf{1.43\ m.}$ **Ans.**

Example 11.16. *The long 3 cm diameter carbon steel cylindrical rods ($\varepsilon = 0.66$) at 300°C are rapidly cooled by immersing them (one at a time) horizontally in a water bath at atmospheric pressure. Determine*

(*a*) *The minimum heat flux, in the film boiling region and the temperature at which it occurs.*

(*b*) *The heat flux, when the surface temperature of the cylinder is 300°C.*

(*c*) *Maximum heat flux.*

Solution

Given : Water : $p = 1$ atm, $D = 3$ cm

$T_{sat} = 100°C$, $\varepsilon = 0.66$

$T_s = 300°C$.

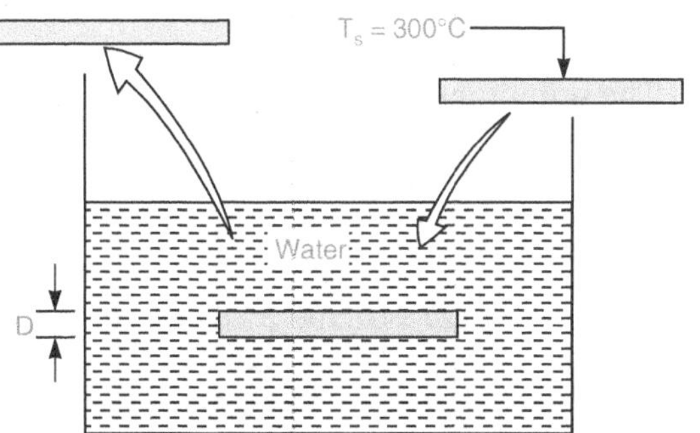

Fig. 11.28. Cooling of horizontal cylinders

To find :

(*i*) q_{min} and T_s during film boiling,

(*ii*) q when $T_s = 300°C$, and

(*iii*) Maximum heat flux.

Analysis : (*i*) At the time of minimum heat flux, the surface temperature of horizontal rods is not known. We assume $T_s = 180°C$

and $\Delta T = T_s - T_{sat} = 180 - 100 = 80°C$

$$T_f = \frac{180 + 100}{2} = 140°C$$

The properties of vapour at 140°C from Table A-7

$\rho_v = 0.5296$ kg/m³, $C_{pv} = 2243$ J/kg.K

$k_v = 0.0258$ W/m.K, $\mu_v = 13.54 \times 10^{-6}$ N.s/m²

$Pr_v = 1.177$

At 100°C, for water

$\sigma = 0.06106$ N/m, $\rho = 958.3$ kg/m³

$h_{fg} = 2257 \times 10^3$ J/kg

The corrected latent heat

$$h'_{fg} = h_{fg} + 0.4\, C_{pv} (T_s - T_{sat})$$
$$= 2257 \times 10^3 + 0.4 \times 2243 \times (180 - 100)$$
$$= 2.329 \times 10^6 \text{ J/kg}$$

The minimum heat flux by using eqn. (11.65)

$$q_{min} = 0.09\, \rho_v\, h'_{fg} \left[\frac{g\sigma(\rho - \rho_v)}{(\rho + \rho_v)^2}\right]^{1/4}$$
$$= 0.09 \times 0.5296 \times 2.329 \times 10^6 \times \left[\frac{9.81 \times 0.06106 \times (958.3 - 0.5296)}{(958.3 + 0.5296)^2}\right]^{1/4}$$

$= \mathbf{17{,}545\ W/m^2}$. **Ans.**

The corrected length

$$L^* = \left[\frac{\sigma}{g(\rho - \rho_v)}\right]^{1/2} = \left[\frac{0.06106}{9.81 \times (958.3 - 0.5296)}\right]^{1/2}$$
$$= 0.00255 \text{ m}$$

The minimum temperature, by eqn. (11.66)

$$\Delta T_{min} = \frac{0.127\, \rho_v h'_{fg}}{k_v} \left[\frac{g(\rho - \rho_v)\mu_v L^{*3}}{(\rho + \rho_v)^2}\right]^{1/3}$$
$$= \frac{0.127 \times 0.5296 \times 2.329 \times 10^6}{0.0258} \times \left[\frac{9.81 \times (958.3 - 0.5296) \times 13.54 \times 10^{-6} \times 0.00255^3}{(958.3 + 0.5296)^2}\right]^{1/3}$$
$$= 80.1°C$$

and $T_{s,\,min} = T_{sat} + \Delta T_{min} = 100 + 80.1$

$= \mathbf{180.1°C}$. **Ans.**

which approaching the assumed value, thus no more computations are required.

(*ii*) At 300°C film boiling, the properties of vapour at $\frac{300 + 100}{2} = 200°C$

$\rho_v = 0.4156$ kg/m³, $C_{pv} = 2837$ J/kg.K

$k_v = 0.0317$ W/m.K, $\mu_v = 15.65 \times 10^{-6}$ N.s/m²

The corrected latent heat

$$h'_{fg} = h_{fg} + 0.4\, C_{pv}\, \Delta T_e$$
$$= 2257 + 0.4 \times 2837 \times (300 - 100)$$
$$= 2.484 \times 10^6 \text{ J/kg}$$

$$Ra' = \frac{\rho_v (\rho - \rho_v)\, g\, h'_{fg}\, D^3}{\mu_v\, k_v\, \Delta T_e}$$
$$= \frac{0.4156 \times (958.3 - 0.4156) \times 9.81 \times 2.484 \times 10^6 \times 0.03^3}{15.65 \times 10^{-6} \times 0.0317 \times (300 - 100)}$$
$$= 2.639 \times 10^9$$

For horizontal cylinders

$$Nu_D = \frac{hD}{k_v} = 0.62\, [Ra']^{1/4}$$
$$= 0.62 \times [2.639 \times 10^9]^{1/4} = 140.5$$

and $$h = \frac{140.5 \times 0.0317}{0.03} = 148.5 \text{ W/m}^2\text{.K}$$

Total heat flux

$$q = q_{conv} + 0.75\, q_{rad} = h\, \Delta T_e + 0.75\, \sigma\, \varepsilon\, (T_s^4 - T_{sat}^4)$$
$$= 148.5 \times (300 - 100) + 0.75 \times 5.67 \times 10^{-8} \times 0.66 \times (573^4 - 373^4)$$

$= \mathbf{32{,}182.2\ W/m^2}$. **Ans.**

(*iii*) The critical heat flux, [ρ_v (100°C, 100 kPa) = 0.5977 kg/m³]

$$q_{max} = 0.149\, h_{fg}\, \rho_v^{1/2}\, [\sigma g(\rho - \rho_v)]^{1/4}$$
$$= 0.149 \times 2257 \times 10^3 \times \sqrt{0.5977} \times [0.06106 \times 9.81 \times (958.3 - 0.5977)]^{1/4}$$

$= \mathbf{1.27 \times 10^6\ W/m^2}$. **Ans.**

11.12 FORCED CONVECTION BOILING

In the pool boiling, the fluid flow is governed by the buoyancy driven motion of bubbles originated from heated surface. In contrast, for the forced convection boiling, the fluid is forced to flow on the heated surface. The forced flow boiling is of two types—*external forced convection boiling* and *internal forced convection boiling*. The internal forced convection boiling is also called two phase flow and is characterised by rapid changes from liquid to vapour in the flow direction.

Two phase flow. The internal forced convection boiling is associated with bubble formation at the inner surface of a heated tube, through which the liquid is forced to flow. The growth of bubbles and its separation from the heated surface are strongly influenced by flow velocity and hydrodynamic effects. The process depends on a large number of variables and the complexity of two phase flow pattern, thus, is very complicated.

Consider the flow development in a heated vertical tube as shown in Fig. 11.29 (*a*). The fluid at a temperature below its boiling enters the tube, in which it is heated so that the progressive vaporisation occurs. The heat transfer coefficient at the tube inlet can be evaluated from the correlation of forced convection. As fluid proceeds in the tube, its temperature in surface vacinity increases to saturation temperature and the bubbles begin to form at nucleation sites and they enter into main stream of the liquid. This regime is known as the *bubbly flow regime.* There is a sharp increase in the heat transfer coefficient due to start of nucleate boiling as shown in Fig. 11.29 (*b*). As the vapour volume fraction increases, the individual bubbles coalesce (merge) to form plugs or slugs of vapour. This is known as *slug flow regime.* The fluid velocity in the slug flow regime increases and the bubble formation produces oscillations within the tube, thus the heat transfer rate and heat transfer coefficient increase.

As the fluid flows further along in the tube, the volume of bubbles increases continuously and the *annular flow regime* appears, in which the wall of the tube is covered by a thin film of liquid and the heat is transferred through this liquid film. The vapour is flowing in the centre of the tube, at a higher velocity and there may be a number of nucleation sites at the wall and the vapour is generated by vaporisation at the liquid vapour interface inside the tube. In addition, there may be a significant liquid dispersion through the vapour core as droplets and dry spots appear on the inner surface until the surface is completely dry. It is the transition to the *mist flow regime.*

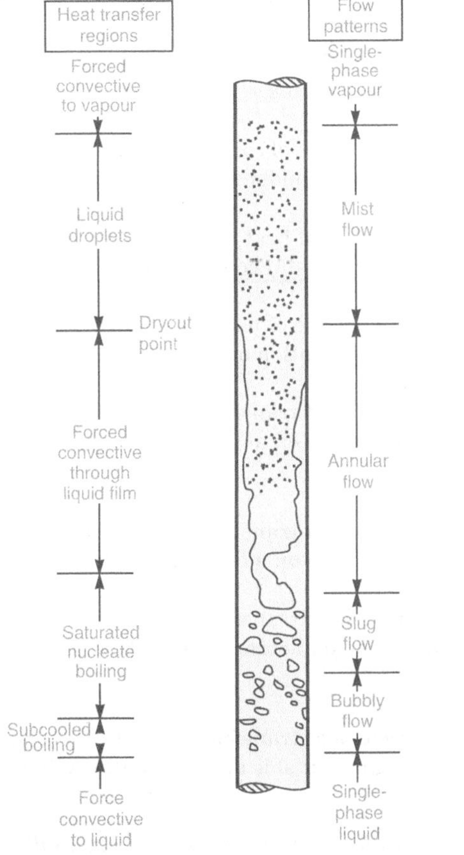

(a) The development of a two-phase flow in a vertical tube with a uniform wall heat flux

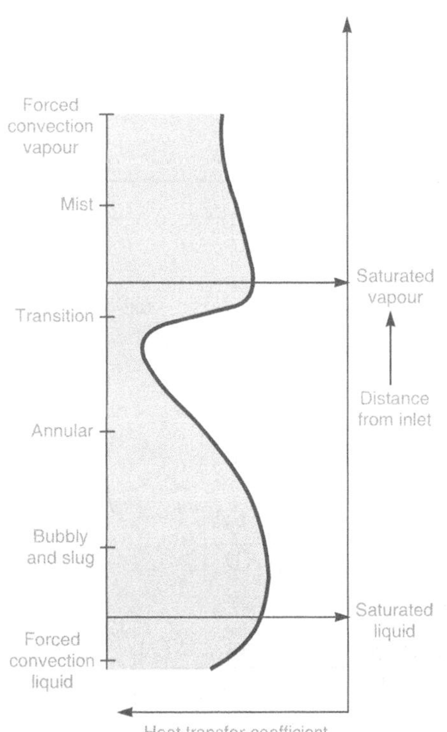

(b) Heat transfer coefficient versus type of flow

Fig. 11.29. Forced convection boiling inside a tube

The heat transfer coefficient experiences a sharp decrease in the mist flow regime, because the wall is covered by relatively low conductivity vapour and tube wall is no longer wetted by liquid. The vapour is then superheated by forced convection from the surface. The point in the system where the maximum heat flux appears is known as critical point and corresponding heat flux is called critical heat flux.

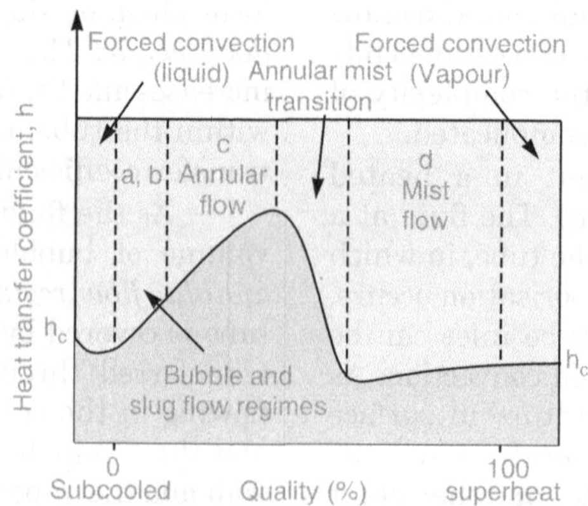

Fig. 11.29 (c) Heat transfer coefficient versus quality and type of flow regime

11.13. SUMMARY

TABLE 11.2. Summary of the relations

Correlation	*Remark*
$\dot{m} = \dfrac{\rho(\rho-\rho_v)\, g\, \delta^3}{3\mu}$	The mass flow rate of the condensate on a vertical plate for laminar condensation.
$\delta = \left[\dfrac{4k\,(T_{sat}-T_s)\,\mu\, x}{\rho(\rho-\rho_v)\, g\, h_{fg}}\right]^{1/4}$	The thickness of condensate film on a vertical plate for laminar condensation.
$h_x = \left\{\dfrac{\rho(\rho-\rho_v) g h_{fg} k_f^3}{4\mu\, x\,(T_{sat}-T_s)}\right\}^{1/4}$	Local value of heat transfer coefficient on a vertical plate for laminar condensation.
$h = 1.13\left[\dfrac{\rho(\rho-\rho_v) g h_{fg} k_f^3}{\mu\, L\,(T_{sat}-T_s)}\right]^{1/4}$	Average value of heat transfer coefficient for laminar film condensation on a vertical surface.
$h = 1.13\left[\dfrac{\rho(\rho-\rho_v)\, h_{fg}\, k_f^3}{\mu\, L\,(T_{sat}-T_s)}\, g \sin\theta\right]^{1/4}$	Average heat transfer coefficient for laminar film condensation on an inclined surface at angle θ.
$h = 0.725\left[\dfrac{g\,\rho(\rho-\rho_v)\, h_{fg}\, k_f^3}{\mu\,(T_{sat}-T_s)\, D}\right]^{1/4}$	Average heat transfer coefficient for laminar film condensation on a horizontal cylinder.
$h = 0.725\left[\dfrac{g\,\rho(\rho-\rho_v)\, h_{fg}\, k_f^3}{\mu\,(T_{sat}-T_s)\, ND}\right]^{1/4}$	Average heat transfer coefficient for laminar condensation on horizontal tube banks.
$h_o = 0.62 \times \left\{\dfrac{\rho_v(\rho-\rho_v)\, g\, h_{fg}\, k_v^3}{\mu_v\, \Delta T_e\, D}\left(1+\dfrac{0.4\, C_{pv}\,\Delta T_e}{h_{fg}}\right)\right\}^{1/4}$	Average heat transfer coefficient for stable film boiling on out side of a horizontal cylinder.
$\dfrac{C_p\,\Delta T_e}{h_{fg}\, Pr^n} = C_{sf}\left[\dfrac{q}{\mu\, h_{fg}}\sqrt{\dfrac{\rho}{g(\rho-\rho_v)}}\right]^{0.33}$	Heat flux in nucleate pool boiling, coefficient C_{sf} is is given in Table 11.1, $n = 1$ for water, $n = 1.7$ for other liquids.
$q_{max} = 0.149\, h_{fg}\, \rho_v^{1/2}\,[\sigma g(\rho-\rho_v)]^{1/4}$	Maximum heat flux in pool boiling.
$q_{min} = 0.09\, \rho_v\, h_{fg}\left[\dfrac{\sigma g(\rho-\rho_v)}{(\rho+\rho_v)^2}\right]^{1/4}$	Minimum heat flux for large horizontal surfaces.

REVIEW QUESTIONS

1. Explain, why are heat transfer rates high for phase change process ?
2. Discuss the following dimensionless number used in condensation.
   (*i*) Jacob number (*ii*) Condensation number.
3. Explain bond number with its physical significance.
4. Discuss modes of condensation. Why is dropwise condensation preferred ? What are the practical difficulties in retaining dropwise condensation on a surface ?
5. Discuss the conditions under which the dropwise condensation can take place. Why the rate of heat transfer in dropwise condensation is many time that of film condensation ?
6. State the assumption made in deriving Nusselt's equation for film condensation.
7. Using Nusselt's theory of laminar film condensation show that $\delta \propto x^{1/4}$ for a flat vertical surface. Where x is the distance from the leading edge of the film and δ is the film thickness.
8. Explain the mechanism of laminar film condensation on a vertical plate.
9. Derive an expression for Nusselt number for laminar film condensation on a vertical surface.
10. Prove that the local heat transfer coefficient during filmwise condensation is

$$h_x = \left[\frac{k_f^3 \rho^2 g\, h_{fg}}{4\mu\, x\,(T_{sat} - T_s)}\right]^{1/4}.$$

11. Discuss (*a*) filmwise and dropwise condensation, (*b*) pool boiling phenomenon.
12. Distinguish between
    (*a*) Subcooled and saturated boiling.
    (*b*) Nucleate and film boiling.
13. Discuss the various regimes of pool boiling.
14. Explain critical heat flux and Leidenfrost point during pool boiling process.
15. Why radiation heat transfer plays dominant role during film boiling process ?
16. What are the effects of superheat of vapour on the condensation parameters ?
17. The steam condenses on a vertical plane wall, derive expression for the followings:
    (*i*) average heat transfer coefficient,
    (*ii*) film thickness,
    (*iii*) rate of condensation, and
    (*iv*) rate of heat flow.
18. Assuming laminar film condensation, calculate the ratio of the heat transfer on a vertical tube to that for a horizontal tube of same diameter D and length L.
19. What is an excess temperature ?
20. Distinguish the pool boiling from forced convection boiling.
21. Explain the film boiling, why is it avoided in practice ? What is the boiling crisis ?

PROBLEMS

1. Saturated steam at 50°C and 12.35 kPa condenses on outside surface of a condenser tube 2.5 cm OD, 2 m long vertical. The tube surface is maintained at 30°C by flow of cooling water. Assuming filmwise condensation calculate :
   (*a*) Average heat transfer coefficient,
   (*b*) Rate of condensation.
   [Ans. (*a*) 4350 W/m^2.K, (*b*) 5.685 × 10^{-3} kg/s]
2. Saturated steam at 80°C (p = 47.39 kPa) condenses on outer surface of 1.2 m long, 0.1 m diameter vertical tube maintained at room temperature of 40°C.
   (*a*) Calculate average heat transfer coefficient,
   (*b*) Rate of condensation,
   (*c*) Condensate thickness at the bottom of the tube.
   [Ans. (*a*) 4600 W/m^2.K, (*b*) 0.03 kg/s, (*c*) 0.189 mm]
3. A chilled water pipe of 6.25 cm OD with outside surface at 5°C passes through an area where steam is saturated at 35°C. Determine the condensation rate per metre length. [Ans. 45.76 kg/h]
4. Ammonia vapour at 35°C is to be condensed on outside of horizontal tubes of 2.5 cm OD with a surface temperature of 25°C. A square array of 10 × 10 tubes of 1.2 m long is used. Determine the rate of condensation of ammonia.
   Use properties :
   ρ = 596 kg/m^3, k_f = 0.5071 W/m.K
   μ = 2.086 × 10^{-4} kg/ms, h_{fg} = 1123.46 × 10^3 J/kg,
   ρ_v = 1.042 kg/m^3. [Ans. 1224.2 kg/h]
5. R—12 is condensed at the rate of 10000 kg/h at 35°C using water at 25°C. The condenser uses a square array of 25 × 25 tubes of 12 mm OD. Calculate the length of the tube bundle. Use properties of refrigerant R—12 as
   ρ = 1295 kg/m^3, k_f = 0.0709 W/m.K
   μ = 2.5123 × 10^{-4} kg/ms, h_{fg} = 133.22 × 10^3 J/kg
   ρ_v = 48.08 kg/m^3. [Ans. 2.17 kg/h]
6. A vertical plate 0.3 m wide and 11.2 m high is maintained at 70°C and exposed to saturated steam at 100 kPa. Calculate the heat transfer rate and mass flow rate of the condensate.
7. A condenser is to design to condense 1.3 kg/s of steam at atmospheric pressure. A square array of 12.5 mm OD tubes is to be used with the outside tube walls maintained at 93°C. The spacing of the tube is 19 mm between centres and their length is 3 times the square dimension. How many tubes are required for the condenser and what are the outside dimensions of square array ?

8. Saturated Freon-12 vapour condense on outside of a bank of 25 horizontal tubes having OD of 1 cm arranged in 5 × 5 square array. Calculate the rate of condensation per metre length of the array, if its surface temperature is maintained at 40°C. Use properties of Freon-12 as

$\rho = 1218$ kg/m^3, $k_f = 0.0686$ W/m.K
$\mu = 0.0248$ kg/ms, $h_{fg} = 128.12 \times 10^3$ J/kg.

[Ans. 68.81 kg/h]

9. Steam condenser at 0.08132 bar is arranged to condense over a 60 cm square vertical surface. The surface temperature is maintained at 28°C. Calculate the following :
   (*a*) Film thickness, local heat transfer coefficient and mean flow velocity of the condensate at 30 cm from the top of the plate.
   (*b*) Average heat transfer coefficient and total heat transfer from the entire plate.
   (*c*) Total steam condensation rate.
   (*d*) What would be the heat transfer coefficient, if the plate is inclined at 30 degree with the horizontal plane ?

[Ans. (*a*) 0.04256 mm, 14692.2 W/m^2. K, 0.808 × 10^{-2} m/s; (*b*)16477.2 W/m^2.K, 100662.1 W; (*c*) 0.0419 kg/s ; (*d*) 16626.74 W/m^2.K]

10. The saturated steam at 0.1 bar condenses with convection coefficient of 6800 W/m^2.K on the outside of a vertical brass tube (k = 110 W/m.K) having inner and outer diameters of 16.5 mm and 19 mm, respectively. The convection coefficient for water flowing inside the tube is 5200 W/m^2.K.
Estimate the steam condensation rate per unit length of the tube, when the mean water temperature is 30°C. [Ans. 3.72 kg/h]

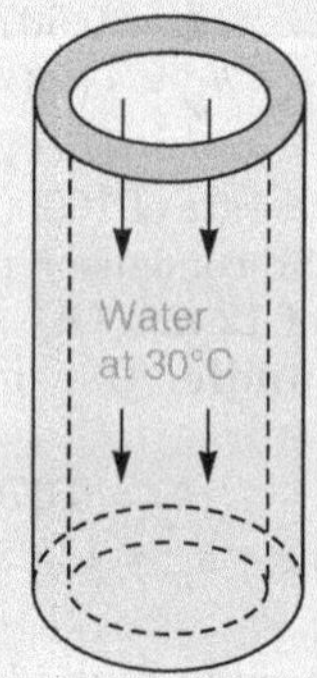

Fig. 11.30. Condensation on a vertical brass tube

11. Steam at 110°C condenses on the outside of a tube having 80 mm outside diameter. The tube wall temperature is 96°C. Calculate the ratio of heat transfer coefficient for condensing of steam when tube is horizontal position and when tube is in vertical position. The properties of condensate are

$\rho = 960$ kg/m^3, $h_{fg} = 2255$ kJ/kg
$\mu = 0.29 \times 10^{-3}$ kg/ms, $\rho_v = 0.6$ kg/m^3
$k_f = 0.68$ W/m.K. *(N.M.U., Nov. 2000)*

12. Estimate the heat transfer coefficient for the tube, when it is (*i*) in horizontal position, and (*ii*) in vertical position. The tube is 1.7 m long and has 1.25 cm outer diameter. It is used to condense steam at 0.4 bar. The water flows through the tube maintaining the wall surface at 54°C.

13. Saturated steam at 90°C and 70 kPa is condensed on outer surface of a 1.5 m long, 2.5 m diameter vertical tube maintained at uniform temperature of 70°C. Assuming film wise condensation, calculate the heat transfer rate on the tube surface.
Use properties of water at 80°C

$h_{fg} = 2309$ kJ/kg, $\rho = 974$ kg/m^3
$k_f = 0.668$ W/mK, $\mu = 0.355 \times 10^{-3}$ kg/ms.

(P.U., May 2003)

[Ans. 1318.5 kW]

14. A tube 13 mm in outer diameter and 1.5 m long is used to condense the steam at 40 kPa (T_{sat} = 76°C). Calculate the heat transfer coefficient for this tube in (*a*) horizontal position, and (*b*) vertical position. Take average tube wall temperature as 52°C.

[Ans. (*a*) 10,328 W/m^2.K; (*b*) 5330 W/m^2.K]

15. A vertical plate 500 mm high and 200 mm wide is used to condense saturated steam at 1 atm. At what surface temperature must the plate be maintained to achieve a condensation rate of 24.4 kg/h ? [Ans. 80°C]

16. A vertical cooling fin approximates a flat plate 40 cm in height is exposed to saturated steam at 100°C (h_{fg} = 2257 kJ/kg). The fin is maintained at a temperature of 90°C. Calculate :
   (*i*) Thickness of film at bottom of the fin,
   (*ii*) Average heat transfer coefficient, and
   (*iii*) Heat transfer rate, after incorporating McAdam's correction,

The relevant properties are

$\rho = 965.3$ kg/m^3, $k_f = 0.68$ W/m.K,
$\mu = 3.153 \times 10^{-4}$ kg/ms.

(A.M.I.E., Summer, 1999)

[Ans. (*i*) 0.1135 mm; (*ii*) 7984.4 W/m^2.K; (*iii*) 38.32 kW]

17. A horizontal copper electric heater 30 cm long and 1 cm in diameter is submerged below the free layer of water at 100°C.
   (*i*) Calculate the maximum power input to heater for nucleate boiling, and
   (*ii*) Heat flux corresponds to a ΔT_e = 10°C.

[Ans. (*i*) 9520 W, (*ii*) 1.513 × 10^5 W/m^2]

18. A pool of saturated water at 130°C boils off a horizontal brass plate at a temperature of 140°C. Assume the nucleate boiling, calculate the heat flux. Take for water

$\rho_v = 1.50$ kg/m^3 [Ans. 2.735 × 10^6 W/m^2]

19. A vessel with a flat bottom 0.1 m^2 in area is used for boiling water at atmospheric pressure. Calculate the temperature, at which the vessel must be maintained for the boiling rate of 80 kg/h. Assume nucleate boiling and take ρ_v = 0.6 kg/m^3 and C_{sf} = 0.01.

[Ans. 111.7°C]

20. (*a*) Calculate the heat flux and heat transfer coefficient associated with the pool boiling of water at 100°C and 1 bar, with an excess temperature of 10°C, in a stainless steal container with a ground and polished surface. Take σ = 0.058 N/m, h_{fg} = 2257 kJ/kg.

(*b*) Compare this with maximum heat flux and

(*c*) repeat the calculation for mechanically polished surface.

[Ans. (*a*) 1.03 × 10^6 W/m^2, 103 × 10^3 W/m^2.K;
(*b*) 1.25 × 10^6 W/m^2, 14.7 × 10^3 W/m^2.K]

21. The surface temperature of the horizontal surface in problem 13 is increased to 300°C. Calculate the heat transfer coefficient for teflon coated stainless steel surface (ε = 0.9). [Ans. 8.18 × 10^4 W/m^2.K]

22. A design for refrigerator, where the condensate is pumped through tubes and condenses inside them, has thermal load of 400 Watts. Calculate the length of 5 mm diameter tube required for two different refrigerants. In each case, the saturation temperature of the refrigerant is 30°C and tube wall temperature is 26°C.

(*a*) Freon—12

ρ = 1295 kg/m^3, C_p = 984 J/kg.K
μ = 2.47 × 10^{-4} kg/ms, k_v = 0.071 W/m.K
h_{fg} = 135 × 10^3 J/kg. [Ans. 3.2 m]

(*b*) Ammonia :

ρ = 602 kg/m^3, C_p = 4900 J/kg.K
μ = 1.31 × 10^{-4} kg/ms k_v = 0.51 W/m.K
h_{fg} = 1.146 × 10^6 J/kg. [Ans. 0.53 m]

23. A 10 cm diameter ground and polished stainless steel tube (ε = 0.05) is maintained at a surface temperature of 300°C, while the boiling water at atmospheric pressure. Identify the regime of pool boiling and calculate the heat transfer coefficient and heat flux.

[Ans. Film boiling, 461 W/m^2.K, 92.2 × 10^3 W/m^2]

24. The bottom of a copper pan, 300 mm in diameter is maintained at 120°C by an electric heater. Calculate the power required to boil water in this pan. What is the evaporation rate ? Estimate the critical heat flux.

[Ans. 77.43 kW, 123.5 kg/h, 1.258 MW/m^2]

25. An aluminium cylinder, 2 cm diameter and 15 cm long is heated to a temperature of 500°C and immersed horizontally in a liquid nitrogen bath at −196°C. Neglect the heat transfer from end faces, calculate the initial heat transfer rate. Take emissivity of the aluminium surface to be 0.4. Take density and latent heat of liquid nitrogen as 800 kg/m^3 and 201 kJ/kg, respectively. [Ans. 636 W]

REFERENCES AND SUGGESTED READING

1. R.G., Goldsten et.al., "*Heat Transfer*"—A review of 2003 litrature Int. Journal of Heat and Mass Transfer, vol. 49, 2006.
2. M.Kemal Atesman, "*Every day Heat Transfer Problems*", ASME, 2009.
3. William S. Janna, "*Engineering Heat Transfer*", 2/e, CRC Press, 2000.
4. Incropera F.P. and Dewitt-D.P., "*Fundamentals of Engineering Heat and Mass Transfer*", 7/e John Wiley & Sons, 2007.

12 Thermal Radiation: Properties and Processes

12.1. **Theories of Radiation**—Maxwell's theory—Max Planck's theory. **12.2. Spectrum of Electromagnetic Radiation. 12.3. Black body Radiation. 12.4. Spectral and Total Emissive Power. 12.5. Surface Absorption, Reflection and Transmission. 12.6. Black body Radiation Laws**—Black body spectral emissive power—Wien's displacement law—Stefan Boltzmann law—Radiation function and band emission. **12.7. Emissivity**—Hemispherical and total emissivity—Spectral emissivity—Directional emissivity—Kirchhoff's law—Gray and diffuse surfaces : Gray Lambert body approximation. **12.8. Radiation From a Surface**—Solid angle—Spectral intensity of radiation ($I_{b\lambda}$)—Radiation intensity (I_b). **12.9. Radiosity. 12.10. Solar Radiation**—Solar radiation on the earth—Atmospheric emission—Green house effect—Selective surfaces. **12.11. Summary**—*Review Questions—Problems—References and Suggested Reading.*

Thermal radiation or *radiation heat transfer* is a distinct separate mechanism from conduction and convection for transfer of heat energy. It refers to the heat energy emitted by the bodies because of their temperatures. All bodies at a temperature above absolute zero temperature emit energy by a process of electromagnetic radiation. The intensity of such radiation depends upon the temperature and nature of the surface. The energy transfer by radiation does not require any medium between hot and cold surfaces. The energy transfer by radiation is the fastest (at the speed of light) and it does not suffer any attenuation even in the vacuum. In fact, the heat transfer through an evacuated space can occur only by radiation. When a person sits infront of a fire, he gets most of the heat energy by radiation as shown in Fig. 12.1. Further, it is also interesting that the radiation heat transfer can also occur between two bodies separated by a medium that is colder than the both bodies. For an example, the energy emitted by sun reaches the earth surface after travelling through space and extremely cold air layers at high altitudes.

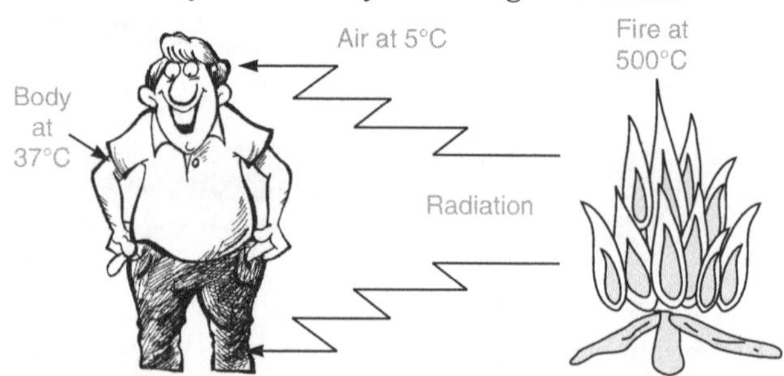

Fig. 12.1. Radiation heat transfer between two bodies separated by a colder medium

12.1. THEORIES OF RADIATION

The actual mechanism of radiation propagation is not fully understood, but two theories; *Maxwell theory* and *Max Planck's theory* are in use. Both concepts are used in study of thermal radiation.

12.1.1. Maxwell's Theory

According to *Maxwell electromagnetic theory,* the energy is transferred from a hot body to cold body in the form of *electromagnetic waves.* The *electromagnetic waves* possess the energy emitted by a body as a result of the change in electronic configuration of atoms or molecules. These electromagnetic waves transport energy like other waves and these electromagnetic waves travel with the speed of light.

The electromagnetic waves are characterised by their frequency ν and wavelength λ, in a medium as:

$$\lambda = \frac{c}{\nu} \qquad ...(12.1)$$

where c is the speed of light in the medium. In vacuum, $c = c_0 = 2.998 \times 10^8$ m/s. This concept is useful in studies for the prediction of the radiation properties of the surfaces and materials.

12.1.2. Max Planck's Theory

According to *Max Planck's concept*, the propagation of thermal radiation takes place in form of discrete quanta called photons, each quantum having an energy of

$$E = h\nu = \frac{hc}{\lambda} \qquad ...(12.2)$$

Where h is Planck's constant = 6.6256×10^{-34} J-s, ν is frequency of photons, and c is a constant. It is also revealed that the energy of the photons is inversely proportional to its wavelength. Therefore, the shorter wavelength radiation possesses the larger photon energy. This theory is used to predict the magnitude of emitted energy by a body at a given temperature under ideal conditions.

12.2. SPECTRUM OF ELECTROMAGNETIC RADIATION

The radiation energy in form of electromagnetic waves, is emitted at all wave length from $\lambda = 0$ to $\lambda = \infty$. A portion of radiation spectrum is shown in Fig. 12.2. The bulk of thermal energy emitted by a body lies in wavelength between $\lambda \approx 0.1$ and $\lambda \approx 100$ μm. For this reason, this portion of the spectrum is generally referred as *thermal radiation.* The sun emits thermal radiation at an effective surface temperature of 5760 K and bulk of this energy lies between $\lambda = 0.1$ to $\lambda = 3$ μm, therefore, this spectrum is referred as the *solar radiation.* The radiation energy emitted by the sun is in wavelength between $\lambda = 0.4$ to $\lambda = 0.76$ μm, is visible to human eye, therefore, this spectrum is referred as the *visible radiation* (light). Almost half of the solar radiation is light, falls in the visible range and remaining being ultraviolet and infrared.

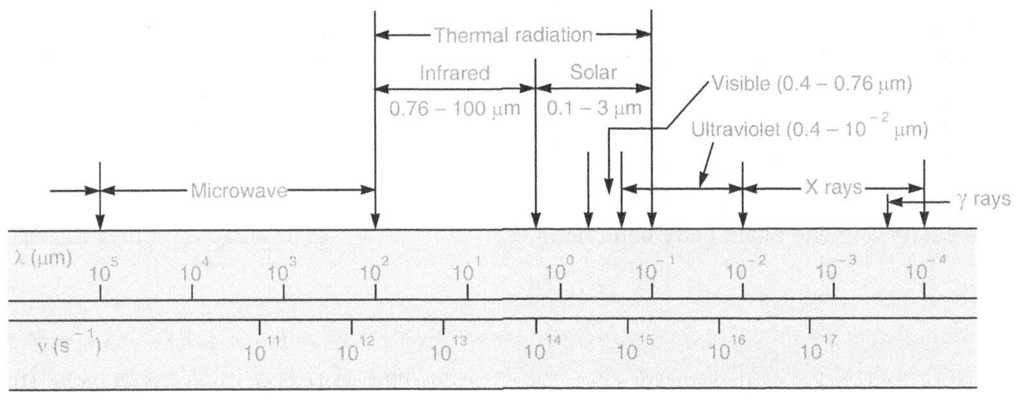

Fig. 12.2. Typical spectrum of electromagnetic radiation

The visible spectrum (light) consists of narrow bands of colour from violet (0.40–0.44 μm) to red (0.63–0.76 μm) as shown in Table 12.1 below :

TABLE 12.1. Wavelength ranges of different colours in light

Colour	*Wavelength band*
Violet	0.40–0.44 μm
Blue	0.44–0.49 μm
Green	0.49–0.54 μm
Yellow	0.54–0.60 μm
Orange	0.60–0.63 μm
Red	0.63–0.76 μm

The colour of a surface depends on its ability to reflect the radiation in certain wavelength. For an example, a surface that reflects a fraction of radiation in the wavelength ranges from 0.63 μm to 0.76 μm, while absorbing rest of the visible radiation appears *red* to human eye. Similarly, if a surface reflects all the radiation in the visible range (0.4 μm to 0.76 μm) appears *white* and a surface that absorbs all the light incident on it, appears black.

The radiation emitted by bodies at atmospheric temperature falls into **infrared** region, ($0.76 \le \lambda \le 100$ μm). The bodies emit radiation in the visible range usually at temperature above 800 K. The **ultraviolet** radiation is low wavelength ($0.1 \le \lambda \le 0.4$ μm) radiation in the thermal radiation range. These rays are harmful, since they can kill micro-organisms and cause serious damage to human, and other living beings. The solar radiation contains about 12% ultraviolet rays and it would be extremely harmful, if it will reach to earth's surface. Fortunately, the ozone (O_3) layer in the atmosphere absorbs most of the ultraviolet solar radiation. Ultraviolet rays remaining in the sunlight is still sufficient to cause sun burn, skin cancer etc.

12.3. BLACK BODY RADIATION

A *black body* is defined as a body which is a perfect emitter and absorber of radiation. It has an ideal surface with the following properties :

1. A black body absorbs all incident radiation from all directions at all wavelengths.

2. At a specified temperature and wavelength, nobody can emit energy more than a black body.
3. Although the radiation emitted by a blackbody depends upon wavelength and temperature, but it is independent of direction.
4. A black body neither reflects nor transmits any amount of incident radiation.

In actual, no surface has all the properties mentioned above for a blackbody. The term black is different from visual observations. A surface is coated with lampblack appears to be black-in colour to the human eye in visible range of spectrum, but turns out to be black for thermal radiation in certain range of wavelength. On the other hand, the ice and snow appear quite bright to human eye but almost black to thermal radiation at all wavelength from $\lambda = 0$ to $\lambda = \infty$.

Consider a radiation beam enters the cavity of an enclosure as shown in Fig. 12.3. It experiences many reflections within the enclosure and almost entire beam is absorbed by the cavity and the black body behaviour is experienced.

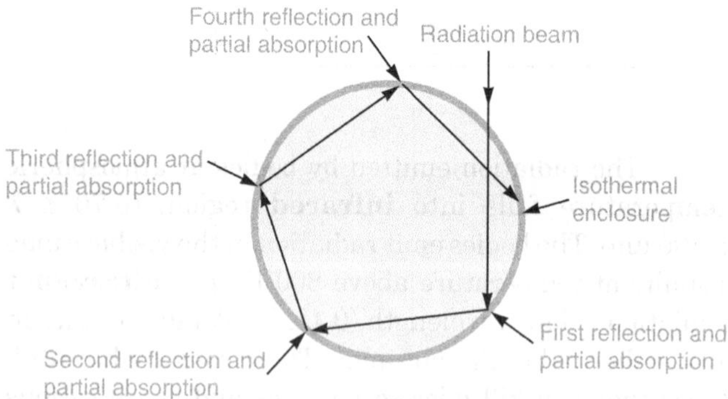

Fig. 12.3. The concept of a blackbody radiation

12.4. SPECTRAL AND TOTAL EMISSIVE POWER

All surfaces at a temperature above absolute zero temperature emit energy in all directions over a wide range of wavelength. At a given temperature, the total amount of heat energy emitted by a surface in all direction over entire wavelength per unit area, per unit time is called the *emissive power*. The emissive power depends on characteristics and temperature of the surface. It is designated as E and measured in W/m^2, and

$$E = f(\varepsilon, \lambda, T) \quad ...(12.3)$$

where ε = emissivity, a surface characteristics,

λ = wavelength of radiation, μm,

T = absolute temperature of the surface, K.

Spectral or monochromatic emissive power. The amount of radiation energy emitted from a surface at a given temperature also varies with the wavelength. The **spectral** or **monochromatic emissive power** is defined as amount of radiation energy emitted by a surface at an absolute temperature T, per unit time, per unit surface area and per unit wavelength $d\lambda$ about the wavelength λ. It is designated as E_λ and measured in W/m^2. μm.

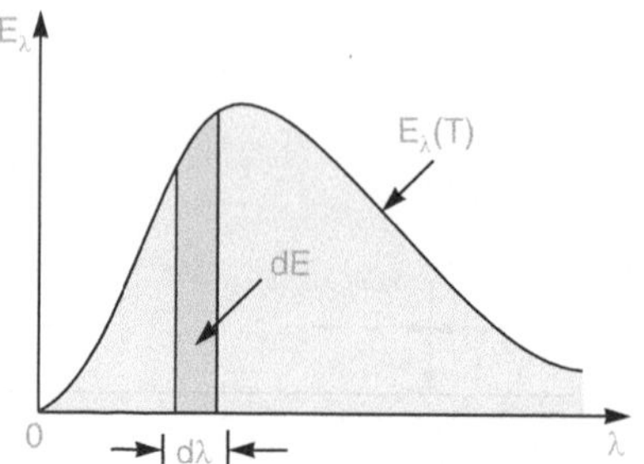

Fig. 12.4. Area under the curve for a given temperature represents the total emissive power of the surface

The radiation is emitted over the wavelength ranges from $\lambda = 0$ to $\lambda = \infty$. The emissive power of surface over the wavelength ranges from λ to $\lambda + d\lambda$ is given by

$$dE = E_\lambda d\lambda$$

where dE = emissive power in given wave band between λ to $\lambda + d\lambda$, (W/m^2)

E_λ = monochromatic emissive power, (W/m^2.μm).

The total emissive power of a surface over entire spectrum of wavelength is

$$E = \int_0^\infty E_\lambda d\lambda \quad ...(12.4)$$

12.5. SURFACE ABSORPTION, REFLECTION AND TRANSMISSION

The **irradiation** is the total radiation energy incident per unit area per unit time over entire wave length from all directions. It is denoted by G and measured in W/m^2.

For most of the surfaces, when the radiation incidents on a body, part of it is absorbed, part of it is reflected and remaining part is transmitted as shown in Fig. 12.5.

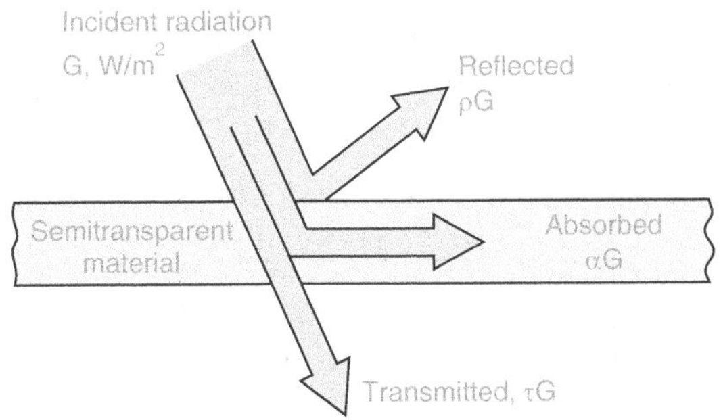

Fig. 12.5. Reflection, Absorption and transmission of incident radiation

Absorptivity. The *total* or *average* or *hemispherical absorptivity* α is defined as fraction of radiation energy incident on the surface from all directions, over entire wavelength spectrum, that is absorbed by the surface. Mathematically

$$\alpha = \frac{G_\alpha}{G} \qquad ...(12.5)$$

where, G_α = energy absorbed by the surface, W/m^2, and

G = irradiation, W/m^2.

A blackbody absorbs all incident radiation, hence its absorptivity is considered unity. But real surfaces do not absorb all energy incident on it.

Reflectivity. When radiation is incident on a non black surface, a fraction is always reflected by the surface. The *reflectivity* of a surface is defined as the fraction of radiation energy incident on a surface from all directions over entire wavelengths, that is reflected. It is designated as ρ and is expressed as :

$$\rho = \frac{G_\rho}{G} \qquad ...(12.6)$$

where G_ρ is energy reflected by the surface. If the surface is perfectly smooth and the angle θ of incident and reflected rays is equal, then the reflection is called the **specular** (or *mirror like*) **reflection** as shown in Fig. 12.6 (*a*).

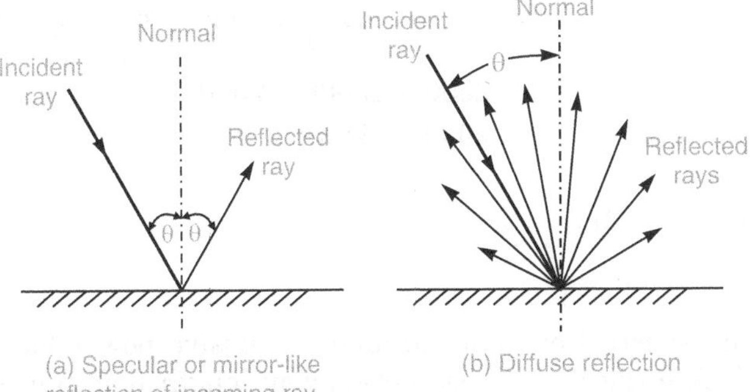

(a) Specular or mirror-like reflection of incoming ray.

(b) Diffuse reflection

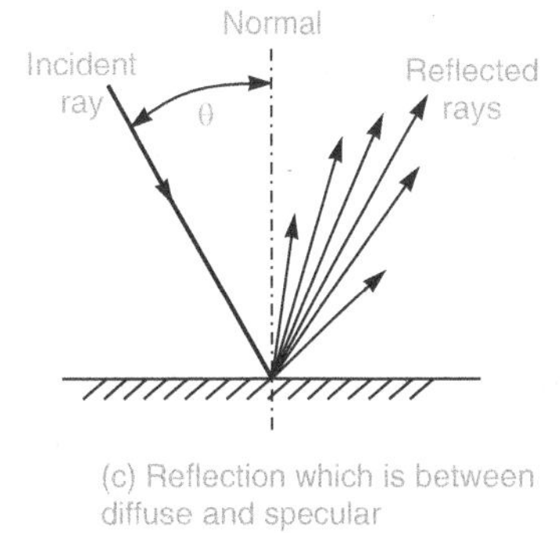

(c) Reflection which is between diffuse and specular (a real surface).

Fig. 12.6. Types of reflections from a surface

If the surface has some roughness, the incident radiation is scattered in all directions after reflection, such reflection is called the **diffuse reflection** as shown in Fig. 12.6 (*b*).

The reflection from real surfaces is neither specular nor diffuse but combination of diffuse and specular behaviour as shown in Fig. 12.6 (*c*).

Transmissivity. When radiation is incident on a semi-transparent surface, a part is reflected, a part is absorbed and remaining is transmitted. Hence the *transmissivity*, τ is the fraction of incident energy transmitted through the surface. Mathematically

$$\tau = \frac{G_\tau}{G} \qquad ...(12.7)$$

where G_τ is energy transmitted by the surface.

With above considerations, for a surface, the sum of absorbed, reflected and transmitted radiation energy is equal to the radiation energy incident on the surface;

$$G_\alpha + G_\rho + G_\tau = G \qquad ...(12.8)$$

Dividing each term in above relation by G, we get

$$\alpha + \rho + \tau = 1 \qquad ...(12.9)$$

Monochromatic irradiation. The irradiation G defined above is *total hemispherical* property. Thus α, ρ, and τ are average properties of a surface for all directions and all wavelengths. However, for a specific wavelength or direction, the irradiation is referred as *monochromatic irradiation.* It is also called *spectral irradiation* and is defined as the radiant heat flux incident on a surface per unit wavelength about a wavelength λ from all directions. It is denoted by G_λ and measured in W/m^2.μm. Mathematically it is expressed as :

$$G_\lambda = \frac{dG}{d\lambda} \qquad ...(12.10)$$

Total hemispherical irradiation may be evaluated as:

$$G = \int_0^\infty G_\lambda \, d\lambda \quad \text{...(12.11)}$$

The spectral absorptivity, reflectivity and transmissivity of a surface are defined in a similar manner as

1. *Spectral absorptivity* α_λ is the fraction of monochromatic irradiation absorbed.

2. *Spectral reflectivity* ρ_λ is the fraction of monochromatic irradiation reflected.

3. *Spectral transmissivity* τ_λ is the fraction of monochromatic energy transmitted.

$$\alpha_\lambda = \frac{G_{\lambda,\alpha}}{G_\lambda} \quad \rho_\lambda = \frac{G_{\lambda,\rho}}{G_\lambda} \quad \alpha_\lambda = \frac{G_{\lambda,\tau}}{G_\lambda} \quad \text{...(12.12)}$$

where $G_{\lambda,\alpha}$, $G_{\lambda,\rho}$ and $G_{\lambda,\tau}$ are the absorbed, reflected and transmitted portion of spectral irradiation G_λ. Similar to average properties, the sum of spectral properties is always equal to unity as :

$$\alpha_\lambda + \rho_\lambda + \tau_\lambda = 1 \quad \text{...(12.13)}$$

The average absorptivity, reflectivity and transmissivity of a surface can also be expressed in terms of their spectral counterparts as :

$$\alpha = \frac{\int_0^\infty \alpha_\lambda G_\lambda d\lambda}{\int_0^\infty G_\lambda d\lambda}, \qquad \rho = \frac{\int_0^\infty \rho_\lambda G_\lambda d\lambda}{\int_0^\infty G_\lambda d\lambda}$$

$$\tau = \frac{\int_0^\infty \tau_\lambda G_\lambda d\lambda}{\int_0^\infty G_\lambda d\lambda} \quad \text{...(12.14)}$$

Opaque body. For an opaque surface, there is no transmission thus the reflectivity and absorptivity are :

$$\alpha + \rho = 1 \quad \text{...(12.15)}$$

and $$\alpha_\lambda + \rho_\lambda = 1 \quad \text{...(12.16)}$$

White body. A body is called white body which reflects almost all radiation incident upon it and does not absorb or transmit any part of it. For white body :

$$\alpha = 0 \,; \quad \tau = 0 \quad \text{...(12.17)}$$

thus $\rho \equiv 1$

Black body. A black body neither reflects nor transmits any part of the incident radiation but it absorbs all of it, *i.e.,*

$$\rho = 0 \,; \tau = 0 \quad \text{and} \quad \alpha = 1 \quad \text{...(12.18)}$$

Example 12.1. *The spectral distribution of surface irradiation is shown in Fig. 12.7.*

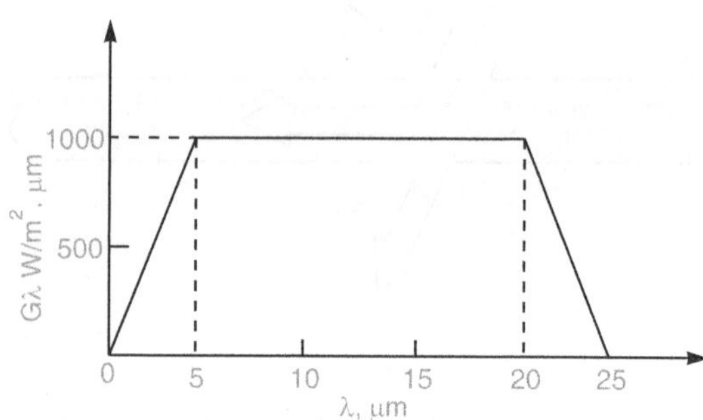

Fig. 12.7. Spectral distribution of irradiation

What is the total irradiation ?

Solution

Given : Spectral distribution of the surface irradiation.

$$G_{\lambda_1} = 0, \quad G_{\lambda_2} = 1000 \text{ W/m}^2.\mu\text{m}, \quad G_{\lambda_3} = 0.$$

To find : Total irradiation on the surface.

Analysis : Total irradiation on the surface may be obtained by using eqn. (12.11).

$$G = \int_0^\infty G_\lambda d\lambda$$

Evaluating the integral into parts

$$G = \int_0^{5\,\mu m} G_\lambda d\lambda + \int_{5\,\mu m}^{20\,\mu m} G_\lambda \, d\lambda + \int_{20\,\mu m}^{25\,\mu m} G_\lambda \, d\lambda + \int_{25\,\mu m}^{\infty} G_\lambda \, d\lambda$$

Each integral represents area under the curve, first and third integral has trapezoid and its area

$$= \frac{1}{2} \times \text{height} \times \text{width}.$$

Thus, $$G = \frac{1}{2} \times 1000 \times (5 - 0) + 1000 \times (20 - 5) + \frac{1}{2} \times 1000 \times (25 - 20) + 0$$

$$= 2500 + 15000 + 2500$$

$$= \mathbf{20{,}000 \text{ W/m}^2}. \quad \text{Ans.}$$

12.6. BLACK BODY RADIATION LAWS

12.6.1. Black body Spectral Emissive Power

The spectral or monochromatic emissive power for a black surface is highest at every wavelength at any given temperature and it is given by *Max Planck's distribution law,* based on quantum theory, as :

$$E_{b\lambda}(T) = \frac{C_1}{\lambda^5\{\exp[C_2/(\lambda T)] - 1\}} \quad ...(12.19)$$

where C_1 and C_2 are constant ;

$C_1 = 2\pi hc_0^2 = 3.742 \times 10^8$ W.µm⁴/m², and

$C_2 = hc_0/k_B = 1.438 \times 10^4$ µm.K ;

$k_B = 1.3805 \times 10^{-23}$, *Boltzmann constant.*

T = absolute temperature, K

λ = wavelength, µm ;

$E_{b\lambda}(T)$ = spectral black body emissive power at absolute temperature T, W/m².µm.

This relation is valid for black body spectral emission in a vacuum or a gas. For other mediums, the constant C_1 should be replaced by C_1/n^2, where n is the index of refraction of the medium.

The Fig. 12.8 is a plot of black body spectral emissive power $E_{b\lambda}(T)$ against the wavelength λ of radiation for some selected temperatures.

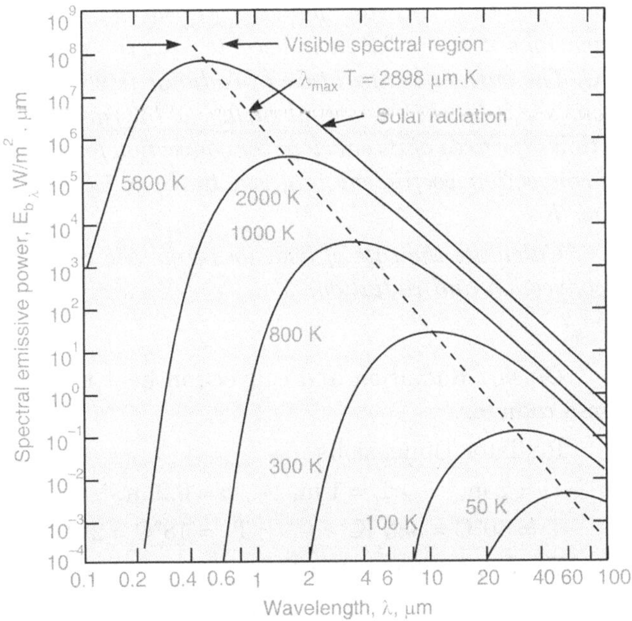

Fig. 12.8. Spectral black body emissive power

The plot indicates the following facts.

1. The spectral emissive power of a blackbody is continuous function of wavelength. At any given temperature, it increases with wavelength, reaches a maximum value and then decreases with increasing wavelength.

2. At any given wavelength, the emissive power increases with increase in absolute temperature.

3. Each curve has a peak value of emissive power.

4. As temperature increases the peaks shift toward smaller wavelengths.

5. The peak of the solar radiation (a black body radiation) at T ≈ 5800 K reaches in visible range of the spectrum. Therefore, the sun is in tune with our eye. On the other hand, the surfaces at temperatures T ≤ 800 K emit almost entirely in the infrared region and thus the radiation is not visible to eye.

12.6.2. Wien's Displacement Law

Fig. 12.8 shows spectral blackbody emissive power distribution over a certain range of wavelength. It is observed that, for a given temperature, there is a definite peak, at a particular wavelength. The relationship between the wavelength λ_{max} and absolute temperature T at which $E_{b\lambda}$ reaches a maximum value is given by the **Wien's displacement law.** "It can be derived from *Planck's distribution law* by applying the condition of maxima, *i.e.,* differentiating $E_{b\lambda}$ with respect to λ and setting it to zero ;

$$\frac{dE_{b\lambda}(T)}{d\lambda} = \frac{d}{d\lambda}\left[\frac{C_1\lambda^{-5}}{\exp[C_2/(\lambda T)] - 1}\right] = 0$$

$$-\frac{5C_1\lambda^{-6}}{\exp\left(\frac{C_2}{\lambda T}\right) - 1} - \frac{C_1\lambda^{-5}\exp\left(\frac{C_2}{\lambda T}\right)}{\left[\exp\left(\frac{C_2}{\lambda T}\right) - 1\right]^2}\left(-\frac{C_2}{\lambda^2 T}\right) = 0$$

Simplifying and rearranging, we get

$$\frac{\exp\left(\frac{C_2}{\lambda T}\right)}{\exp\left(\frac{C_2}{\lambda T}\right) - 1} = \frac{5\lambda T}{C_2}$$

Using $x = \frac{C_2}{\lambda T}$, we get

$$\frac{e^x}{e^x - 1} = \frac{5}{x} \quad \text{or} \quad x = 5(1 - e^{-x})$$

It is a transcendental equation and its solution by trial and error method converges to

$$x = \frac{C_2}{\lambda_{max} T} = 4.9651$$

Substituting constant $C_2 = 1.438 \times 10^4$ µm, we get

$$\lambda_{max} T = 2897.6 \text{ µm.K.} \quad ...(12.20)$$

where λ_{max} represents wavelength corresponding to maximum spectral black body emissive power at a given temperature T. The eqn. (12.20) is valid for entire spectrum of wavelength for black body.

Substituting λ_{max} as 2897.6/T in eqn. (12.19) to obtain **maximum spectral emissive power** at temperature T,

$$E_{b\lambda}(T) = \frac{C_1 T^5}{(2897.6)^5 \times \left[\exp\left(\frac{C_2}{2897.6}\right) - 1\right]}$$

Using C_1 and C_2 ;

$$E_{b\lambda}(T) = \frac{3.742 \times 10^8\ T^5}{(2897.6)^5 \left[\exp\left(\frac{1.438 \times 10^4}{2897.6}\right) - 1\right]}$$

$$E_{b\lambda\ max} = 12.87 \times 10^{-10}\ T^5\ (W/m^2.\mu m) \quad ...(12.21)$$

12.6.3. Stefan Boltzmann Law

The total emissive power of a blackbody E_b may be obtained by integrating Planck's distribution eqn. (12.19) over entire wavelength as $E_b = \int_0^{\infty} E_{b\lambda}\, d\lambda$

or
$$E_b = \int_0^{\infty} \frac{C_1}{\lambda^5 [\exp(C_2/\lambda T) - 1]} d\lambda \quad ...(12.22)$$

$$= T^4 \int_0^{\infty} \frac{C_1 T d\lambda}{(\lambda T)^5 [\exp(C_2/\lambda T) - 1]}$$

Put $\lambda T = x$, $T d\lambda = dx$, then

$$E_b = C_1 T^4 \int_0^{\infty} \frac{dx}{x^5 [\exp(C_2/x) - 1]}$$

Using $C_2/x = y$, and $dx = -C_2 \cdot \frac{dy}{y^2}$, then

$$E_b = -\frac{C_1 T^4}{C_2^4} \int_0^{\infty} \frac{y^3 dy}{e^y - 1}$$

Its integration yields to :

$$E_b = \frac{C_1}{C_2^4} T^4 \times 6 \sum_{n=1}^{\infty} \frac{1}{n^4}$$

We have $\sum_{n=1}^{\infty} \frac{1}{n^4} = \frac{\pi^4}{90}$

which gives $E_b = \frac{C_1}{C_2^4} T^4 \frac{6\pi^4}{90}$

$$E_b = \sigma T^4 \quad ...(12.23)$$

where
$$\sigma = \frac{C_1}{C_2^4} \times \frac{\pi^4}{15} = \frac{3.742 \times 10^8}{(1.438 \times 10^4)^4} \times \frac{\pi^4}{15}$$

$$= 5.672 \times 10^{-8}\ W/m^2.K^4 \quad ...(12.24)$$

The constant σ is called the *Stefan Boltzmann constant.*

Example 12.2. *One side of metallic plate is insulated, while the other side absorbs a radiant heat flux of 900 W/m². The convective heat transfer coefficient between the plate and ambient air is 10 W/m².K. The surface emissivity of the plate is 0.8. The surrounding and ambient air are at 27°C. Determine the temperature of the plate under steady state conditions.*

Solution

Given : $q_r = 900$ W/m², $h_c = 10$ W/m².K
$\varepsilon = 0.8$, $T_\infty = 27°C = 300$ K.

To find : The temperature of the plate.

Assumptions :

1. Steady state conditions.
2. One side of the plate is adiabatic.
3. Constant properties.

Analysis : The radiant heat flux absorbed by the plate, will be dissipated by convection and radiation.

Thus, $q_r = h_c(T_s - T_\infty) + \varepsilon\,\sigma\,(T_s^4 - T_\infty^4)$

Using the numerical values,

$$900 = 10 \times (T_s - 300) + 0.8 \times 5.67 \times 10^{-8} \times (T_s^4 - 300^4)$$

It gives, T_s = **354.8 K. Ans.**

Example 12.3. *A hot water radiator of overall dimensions 2 × 1 × 0.2 m is used to heat the room at 18°C. The surface temperature of radiator is 60°C and its surface is black. The actual surface of the radiator is 2.5 times the area of its envelope for convection for which the convection coefficient is given by $h_c = 1.3(\Delta T)^{1/3}$ W/m².K.*

Calculate the rate of heat loss from the radiator by convection and radiation.

Solution

Given : Radiation and convection heat transfer from a radiator.

Radiator Dimensions :

H = 2 m, L = 1 m, w = 0.2 m

$T_s = 60°C = 333$ K, $T_\infty = 18°C = 291$ K

For convection $A_s = 2.5\ A_{rad}$

$h_c = 1.3(\Delta T)^{1/3}$.

To find : Rate of heat transfer by convection and radiation.

Assumptions :

(*i*) The radiator as a blackbody.

(*ii*) Steady state conditions.

(*iii*) Uniform heat transfer coefficient.

Analysis : The area of the radiator ;

$$A_{rad} = 2 \times \{2 \times 1 + 2 \times 0.2 + 1 \times 0.2\} = 5.2\ m^2$$

Radiation heat transfer rate ;

$$Q_{rad} = A_{rad}\,\sigma\,(T_s^4 - T_\infty^4)$$
$$= 5.2 \times 5.67 \times 10^{-8} \times (333^4 - 291^4)$$
$$= 1511.2 \text{ W}$$

The convection heat transfer coefficient,

$$h_c = 1.3(\Delta T)^{1/3}$$
$$= 1.3 \times (60 - 18)^{1/3}$$
$$= 4.51 \text{ W/m}^2\text{.K}$$

The convection heat transfer area,

$$A_s = 2.5 \times A_{rad} = 2.5 \times 5.2 = 13 \text{ m}^2$$

The convection heat transfer rate ;

$$Q_{conv} = h_c\,A_s(T_s - T_\infty)$$
$$= 4.51 \times 13 \times (60 - 18) = 2462.24 \text{ W}$$

The total heat transfer rate from the radiator,

$$Q = Q_{rad} + Q_{conv}$$
$$= 1511.2 + 2462.24$$
$$= \mathbf{3973.67\ W.}\quad \text{Ans.}$$

Example 12.4. *Calculate the following quantities for an industrial furnace (black body) emitting radiation at 2650°C.*

(i) Spectral emissive power at $\lambda = 1.2$ μm,

(ii) Wavelength at which the emissive power is maximum,

(iii) Maximum spectral emissive power,

(iv) Total emissive power,

(v) Total emissive power of the furnace, if it is treated as gray and diffuse body with an emissivity of 0.9.

Solution

Given : An industrial furnace as black body radiating at

$$T = 2650°\text{C} = 2923 \text{ K}.$$

To find :

(*i*) $E_{b\lambda}$ at $\lambda = 1.2$ μm

(*ii*) λ_{max}

(*iii*) $E_{b\lambda\,max}$

(*iv*) E_b (Total emissive power)

(*v*) $\varepsilon\,E_b$.

Analysis : (*i*) The spectral emissive power at $\lambda = 1.2$ μm :

The Planck's distribution law, eqn. (12.19)

$$E_{b\lambda} = \frac{C_1}{\lambda^5\,[\exp(C_2/\lambda T) - 1]}$$

where $C_1 = 3.742 \times 10^8$ W.μm^4/m^2

$C_2 = 1.438 \times 10^4$ μm.K.

Substituting the values,

$$E_{b\lambda} = \frac{3.742 \times 10^8}{(1.2)^5 \times \left[\exp\left(\frac{1.438 \times 10^4}{1.2 \times 2923}\right) - 1\right]}$$
$$= \mathbf{2.53 \times 10^6\ W/m^2.\mu m.}\quad \text{Ans.}$$

(*ii*) The wavelength at which the emissive power is maximum :

Using Wien's displacement law, eqn.(12.20)

$$\lambda_{max}\,T = 2897.6 \text{ μm.K}$$
$$\lambda_{max} = \frac{2897.6}{2923} = \mathbf{0.9913\ \mu m.}\quad \text{Ans.}$$

(*iii*) Maximum spectral emissive power, by eqn.(12.21)

$$E_{b\lambda\,max} = 12.87 \times 10^{-10}\,T^5 \text{ W/m}^2\text{.μm}$$
$$= 12.87 \times 10^{-10} \times (2923)^5$$
$$= \mathbf{2.746 \times 10^8\ W/m^2.\mu m.}\quad \text{Ans.}$$

(*iv*) Total emissive power

$$E_b = \int_0^\infty E_{b\lambda}\,d\lambda = \sigma\,T^4$$
$$= 5.67 \times 10^{-8} \times (2923)^4$$
$$= \mathbf{4.139 \times 10^6\ W/m^2.}\quad \text{Ans.}$$

(*v*) Total emissive power with $\varepsilon = 0.9$

$$E = \varepsilon E_b = \varepsilon\,\sigma\,T^4$$
$$= 0.9 \times 4.139 \times 10^6$$
$$= \mathbf{3.725 \times 10^6\ W/m^2.}\quad \text{Ans.}$$

Example 12.5. *The average solar radiation flux on the earth's atmosphere is 1353 W/m^2 and it is known as solar constant. Calculate the temperature of sun (a black body), 1.392×10^6 km in diameter, when it has mean distance of 1.496×10^8 km from the earth's atmosphere.*

Solution

Given : Average solar constant for determination of temperature of sun

$$(Q/A)_{sun} = 1353 \text{ W/m}^2$$
$$D = 1.392 \times 10^9 \text{ m}$$
$$s = 1.496 \times 10^{11} \text{ m}$$

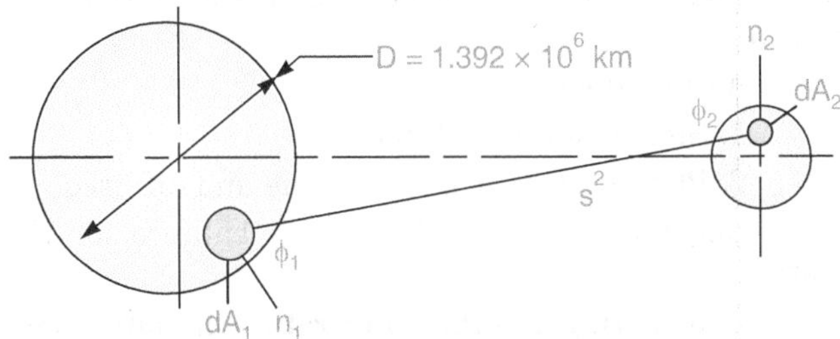

Fig. 12.9

To find : The temperature of sun.

Assumptions :

(*i*) The negligible emissive power of earth in comparison of sun.

(*ii*) Due to large distance, all rays of sun falling on the earth's surface.

(*iii*) Sun has spherical surface.

Analysis : The energy radiated by sun (black body)

$$Q_{sun} = A_{sun}\, \sigma\, T_{sun}^4 = \pi (D_{sun})^2 \sigma T_{sun}^4$$

$$= \pi \times (1.392 \times 10^9 \text{ m})^2 \times 5.67 \times 10^{-8}\, T_{sun}^4$$

$$= 3.45 \times 10^{11}\, T_{sun}^4 \quad ...(1)$$

The sun is considered as source at a distance $s = 1.496 \times 10^{11}$ m from earth's surface.

Mean area, which receives solar radiation

$$A = 4\pi s^2 = 4\pi \times (1.496 \times 10^{11})^2$$

$$= 2.812 \times 10^{23} \text{ m}^2$$

The solar flux incidence on the earth is :

$$q = \frac{Q_{sun}}{A} \quad \text{or} \quad 1353 = \frac{3.45 \times 10^{11}\, T_{sun}^4}{2.812 \times 10^{23}}$$

or $T_{sun}^4 = 1.102 \times 10^{15}$

or T_{sun} = **5762.2 K.** Ans.

Example 12.6. *Calculate the equilibrium temperature for a plate, exposed to a solar flux of 700 W/m² and convection environment at 25°C, with convection coefficient of 10 W/m².K. If the plate is coated with*

(*a*) *White paint :* $\alpha_{sun} = 0.12$; $\alpha_{plate} = 0.9$.

(*b*) *Flat black paint :* $\alpha_{sun} = 0.96$; $\alpha_{plate} = 0.95$.

(N.M.U., Nov. 2000)

Solution

Given : A plate exposed to solar flux and convection environment ;

Solar flux = 700 W/m²,

T_∞ = 25°C = 298 K, h = 10 W/m².K,

(*a*) $\alpha_{sun} = 0.12$; $\alpha_{plate} = 0.9$

(*b*) $\alpha_{sun} = 0.96$; $\alpha_{plate} = 0.95$.

To find : The equilibrium temperature in above two cases.

Assumptions :

1. Steady state conditions.
2. Plate surface is gray, opaque and diffused.

Analysis : Making the energy balance for the plate :

Considering T is the temperature of plate, then

Solar flux on the plate

= Convection flux + Radiation flux (Emissive power)

$$\alpha_{sun} \frac{Q_{sun}}{A} = h(T - T_\infty) + \varepsilon\, \sigma\, (T^4 - T_\infty^4) \quad ...(1)$$

(*a*) When the plate is coated with white paint :

$\alpha_{sun} = 0.12$; $\alpha_{plate} = 0.9$ or $\varepsilon_{plate} = 0.9$ (By Kirchhoff's law)

$$0.12 \times 700 = 10 \times (T - 298) + 0.9 \times 5.67 \times 10^{-8} \times (T^4 - 298^4)$$

$$84 = 10\,T - 2980 + 5.103 \times 10^{-8}\, T^4 - 402.43$$

or $$T^4 + 195.963 \times 10^6\, T - 67.929 \times 10^9 = 0$$

It is a non linear equation and its numerical (Newton Raphson method) solution gives ;

T = **303.40 K** or **30.4°C.** Ans.

(*b*) When plate is coated with black paint :

$\alpha_{sun} = 0.96$; $\alpha_{plate} = 0.95$

Using these values in eqn. (1) ;

$$0.96 \times 700 = 10 \times (T - 298) + 0.95 \times 5.67 \times 10^{-8} \times (T^4 - 298^4)$$

$$672 = 10\,T - 2980 + 5.3865 \times 10^{-8}\, T^4 - 424.787$$

or $$T^4 + 185.65 \times 10^6\, T - 75.6852 \times 10^9 = 0$$

It is a non linear equation and its numerical (Newton Raphson method) solution gives ;

T = **337.65 K** or **64.65°C.** Ans.

12.6.4. Radiation Function and Band Emission

Eqn. (12.23) gives the total amount of radiant energy emitted by a blackbody at temperature T over wavelength $\lambda = 0$ to $\lambda = \infty$. There are often situations, (Fig. 12.10) when it is necessary to evaluate the energy over certain wavelength band, like 0 to λ or λ_1 to λ_2.

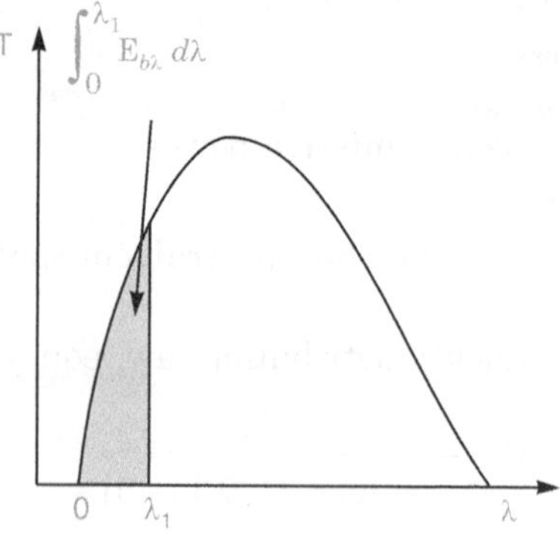

Fig. 12.10. Radiation emission from a black body in spectral band of 0 to λ_1

The radiation energy emitted by a blackbody per unit area, over a wavelength band from $\lambda = 0$ to λ is determined as :

$$E_{b,0-\lambda_1} = \int_0^{\lambda_1} E_{b\lambda}\, d\lambda \ (\text{W/m}^2) \qquad \text{...(12.25)}$$

The eqn. (12.25) is evaluated numerically by using eqn. (12.19). But the integration does not have simple closed form solution and therefore, performing integration is not practical solution. Therefore, a dimensionless quantity $f_{0-\lambda}$ called the **black body radiation function** is used, which is defined as :

$$f_{0-\lambda} = \frac{\int_0^{\lambda} E_{b\lambda}\, d\lambda}{\sigma T^4} \qquad \text{...(12.26)}$$

The function $f_{0-\lambda}$ represents fraction of radiation energy emitted from a black body at temperature T in the wavelength band from $\lambda = 0$ to λ. A table of computed black body radiation function $f_{0-\lambda}$ as a function of λT is given in Table 12.2.

TABLE 12.2. Black body radiation functions

λT (μm K)	$f_{0-\lambda}$	λT (μm K)	$f_{0-\lambda}$	λT (μm K)	$f_{0-\lambda}$
100	0.000000	4800	0.607589	9500	0.903124
200	0.000000	4900	0.620937	9600	0.905490
300	0.000000	5000	0.633777	9700	0.907782
400	0.000000	5100	0.646127	9800	0.910002
500	0.000000	5200	0.658001	9900	0.912153
600	0.000000	5300	0.669417	10000	0.914238
700	0.000002	5400	0.680392	11000	0.931929
800	0.000016	5500	0.690940	12000	0.945138
900	0.000087	5600	0.701079	13000	0.955179
1000	0.000321	5700	0.710824	14000	0.962938
1100	0.000911	5800	0.720192	15000	0.969021
1200	0.002134	5900	0.729196	16000	0.973855
1300	0.004317	6000	0.737852	17000	0.977741
1400	0.007791	6100	0.746173	18000	0.980901
1500	0.012850	6200	0.754174	19000	0.983494
1600	0.019720	6300	0.761869	20000	0.985643
1700	0.028535	6400	0.769268	21000	0.987437
1800	0.039344	6500	0.776386	22000	0.988947
1900	0.052111	6600	0.783234	23000	0.990227
2000	0.066733	6700	0.789823	24000	0.991319
2100	0.083058	6800	0.796164	25000	0.992256
2200	0.100895	6900	0.802268	26000	0.993064
2300	0.120037	7000	0.808144	27000	0.993765
2400	0.140266	7100	0.813803	28000	0.994376
2500	0.161366	7200	0.819253	29000	0.994911
2600	0.183132	7300	0.824504	30000	0.995381
2700	0.205370	7400	0.829563	35000	0.997044
2800	0.227904	7500	0.834439	40000	0.998008
2900	0.250577	7600	0.839139	45000	0.998605
3000	0.273248	7700	0.843671	50000	0.998994
3100	0.295797	7800	0.848042	55000	0.999259
3200	0.318120	7900	0.852258	60000	0.999444
3300	0.340130	8000	0.856325	65000	0.999579
3400	0.361755	8100	0.860251	70000	0.999678
3500	0.382937	8200	0.864040	75000	0.999754
3600	0.403628	8300	0.867699	80000	0.999812
3700	0.423794	8400	0.871232	85000	0.999857
3800	0.443406	8500	0.874645	90000	0.999893
3900	0.462446	8600	0.877943	95000	0.999922
4000	0.480902	8700	0.881129	100000	0.999946
4100	0.498767	8800	0.884210		
4200	0.516040	8900	0.887188		
4300	0.532723	9000	0.890068		
4400	0.548823	9100	0.892853		
4500	0.564348	9200	0.895548		
4600	0.579309	9300	0.898156		
4700	0.593718	9400	0.900680		

The fraction of radiation energy emitted by a black body at a temperature T over a finite wavelength band from $\lambda = \lambda_1$ to $\lambda = \lambda_2$ (Fig. 12.11) is evaluated as :

$$E_{b_{\lambda_1-\lambda_2}} = \int_{\lambda_1}^{\lambda_2} E_{b\lambda}\, d\lambda$$

$$= \int_0^{\lambda_2} E_{b\lambda}\, d\lambda - \int_0^{\lambda_1} E_{b\lambda}\, d\lambda \qquad \text{...(12.27)}$$

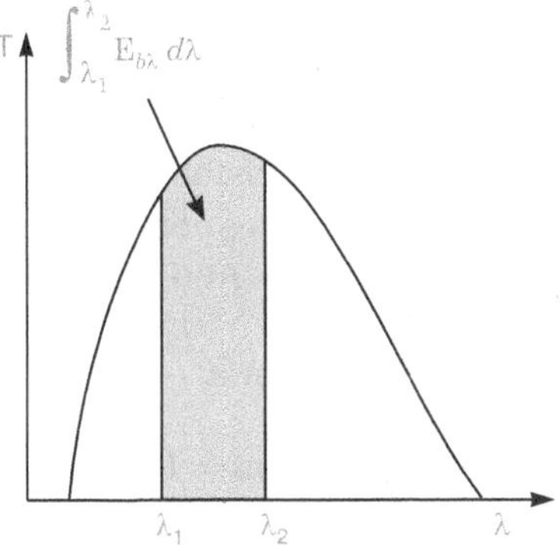

Fig. 12.11. Black body radiation in wavelength band $\lambda = \lambda_1$ to $\lambda = \lambda_2$

In terms of black body radiation function

$$f_{\lambda_1-\lambda_2} = \frac{\int_0^{\lambda_2} E_{b\lambda}\, d\lambda - \int_0^{\lambda_1} E_{b\lambda}\, dy}{\sigma T^4}$$

$$= \frac{\int_0^{\lambda_2} E_{b\lambda}\, d\lambda}{\sigma T^4} - \frac{\int_0^{\lambda_1} E_{b\lambda}\, d\lambda}{\sigma T^4}$$

or $$f_{\lambda_1-\lambda_2} = f_{0-\lambda_2} - f_{0-\lambda_1} \qquad \text{...(12.28)}$$

where $f_{0-\lambda_1}$ and $f_{0-\lambda_2}$ are the black body radiation functions corresponding to $\lambda_1 T$ and λ_2 T, respectively.

Example 12.7. *The temperature of a filament of an incandescent light bulb (a black body) is maintained at 2500 K. Calculate the fraction of radiant energy emitted by the filament in the visible spectrum. Also calculate the wavelength at which the emission from the filament reaches a maximum value.*

Solution

Given : Radiation from a filament of an incandescent light bulb in visible range.

$$\left.\begin{aligned} T &= 2500 \text{ K} \\ \lambda_1 &= 0.4 \ \mu\text{m} \\ \lambda_2 &= 0.76 \ \mu\text{m} \end{aligned}\right\} \text{ For visible range.}$$

To find :

(*i*) Fraction of radiation energy emitted in visible range.

(*ii*) Wavelength corresponding to maximum emissive power.

Analysis : (*i*) The black body radiation function corresponds to $\lambda_1 T$ and $\lambda_2 T$ are :

$$\lambda_1 T = (0.4 \ \mu\text{m}) \times (2500 \text{ K})$$
$$= 1000 \ \mu\text{m.K} \longrightarrow f_{0-\lambda_1} = 0.000321$$
$$\lambda_2 T = (0.76 \ \mu\text{m}) \times (2500 \text{ K})$$
$$= 1900 \ \mu\text{m K} \longrightarrow f_{0-\lambda_2} = 0.052111$$
$$f_{\lambda_1-\lambda_2} = f_{0-\lambda_2} - f_{0-\lambda_1}$$
$$= 0.052111 - 0.000321 = 0.05179$$

It indicates that only 5.18% of the radiation energy emitted falls in the visible range. **Ans.**

(*ii*) Wavelength corresponding to maximum emissive power is obtained by using Wien's displacement law

$$\lambda_{max} T = 2897.8 \ \mu\text{m.K}$$

or $$\lambda_{max} = \frac{2897.8}{2500} \ \mu\text{m} = \mathbf{1.16 \ \mu m.} \quad \textbf{Ans.}$$

Example 12.8. *Solar radiation has approximately same spectral distribution as an ideal radiating body at temperature of 5800 K. Determine the amount of solar radiation, which is in the visible range of 0.4 μm to 0.7 μm, use following data :*

Range	*Black body radiation function*
$0 \le \lambda \le 0.4$	$f_{0-0.4} = 0.1245$
$0 \le \lambda \le 0.7$	$f_{0-0.7} = 0.4914.$

(Pune Univ., Dec. 1999)

Solution

Given : Solar radiation at T_s = 5800 K, and blackbody radiation functions with wavelengths.

To find : Amount of solar radiation in visible spectrum.

Analysis : The total radiation emitted from the sun in visible spectrum

$$\int_{0.4\,\mu m}^{0.7\,\mu m} E_{b\lambda} d\lambda = \sigma T^4 [f_{0-0.7} - f_{0-0.4}]$$
$$= 5.67 \times 10^{-8} \times (5800)^4 \times [0.4914 - 0.1245]$$
$$= \mathbf{23.54 \times 10^6 \ W/m^2.} \quad \textbf{Ans.}$$

Example 12.9. *Determine (a) the wavelength at which the spectral emissive power of a tungsten filament at 1400 K is maximum, (b) the spectral emissive power at that wavelength, and (c) the spectral emissive power at 5 μm.*

Solution

Given : For a radiating surface

$$T_s = 1400 \text{ K}.$$

To find :

(*a*) λ_{max} corresponds to peak emissive power,

(*b*) Peak spectral emissive power corresponding to λ_{max}

(*c*) Spectral emissive power at λ = 5 μm.

Assumption : Black body radiation and σ = 5.67 × 10^{-8} W/m². K⁴.

Analysis : (*a*) The wavelength corresponds to maximum emissive power.

$$\lambda_{max} T = 2897.6 \ \mu\text{m.K}$$

or $$\lambda_{max} = \frac{2897.6}{1400} = \mathbf{2.07 \ \mu m.} \quad \textbf{Ans.}$$

(*b*) Spectral emissive power at $\lambda = \lambda_{max}$ can be obtained from eqn. (12.21) ;

$$E_{b\lambda,\,max} = 12.87 \times 10^{-10} T^5$$
$$= 12.87 \times 10^{-10} \times (1400)^5$$
$$= \mathbf{69.23 \times 10^3 \ W/m^2.\mu m.} \quad \textbf{Ans.}$$

(*c*) Monochromatic emissive power at 5 μm.

$$\lambda T = 5 \times 1400$$
$$= 7000 \ \mu\text{m.K} \longrightarrow f_{0-\lambda} = 0.808144$$
$$E_{b\lambda} = \sigma T^4 . f_{0-\lambda}$$
$$= 5.67 \times 10^{-8} \times (1400)^4 \times 0.808144$$
$$= \mathbf{1.76 \times 10^5 \ W/m^2.\mu m.} \quad \textbf{Ans.}$$

Example 12.10. *A window glass 0.3 cm thick has a monochromatic transmissivity of 0.9 in the range of 0.3 μm to 2.5 μm and nearly zero elsewhere. Estimate the total transmissivity of the window for (a) near black solar radiation at 5800 K, and (b) black room radiation at 300 K.*

Solution

Given : Transmission through a glass window

$$\tau_\lambda = 0.9, \quad \lambda_1 = 0.3 \ \mu\text{m}, \quad \lambda_2 = 2.5 \ \mu\text{m}$$

(*a*) T = 5800 K, (*b*) T = 300 K.

To find: Transmissivity in range of $\lambda_1 = 0.3$ μm to $\lambda_2 = 2.5$ μm for (*i*) T = 5800 K, and (*ii*) T = 300 K.

Assumptions :

(*i*) Black body behaviour,

(*ii*) Stefan Boltzmann constant,

$$\sigma = 5.67 \times 10^{-8} \text{ W/m}^2.\text{ K}^4.$$

Analysis : The transmissivity of a surface is defined as :

$$\tau = \frac{\text{Energy transmitted through body}}{\text{Energy incident on the body}}$$

For a black body

$$\tau = \frac{\int_{\lambda_1}^{\lambda_2} \tau_\lambda E_{b\lambda} d\lambda}{\sigma T^4} = 0.9 \times [f_{0-\lambda_2} - f_{0-\lambda_1}]$$

(*i*) At 5800 K

$$\lambda_1 T = 0.3 \times 5800$$

$$= 1740 \text{ μm.K} \longrightarrow f_{0-\lambda_1} = 0.03285$$

$$\lambda_2 T = 2.5 \times 5800$$

$$= 14500 \text{ μm.K} \longrightarrow f_{0-\lambda_2} = 0.96597$$

$\therefore \quad \tau = 0.9 \times [0.96597 - 0.03285] = \mathbf{0.834.}$ Ans.

(*ii*) At T = 300 K

$$\lambda_1 T = 0.3 \times 300 = 90 \text{ μm.K} \longrightarrow f_{0-\lambda_1} = 0.000$$

$$\lambda_2 T = 2.5 \times 300$$

$$= 750 \text{ μm.K} \longrightarrow f_{0-\lambda_2} = 0.000005$$

$\therefore \quad \tau = 0.9 \times (0.000005 - 0)$

$\approx \mathbf{4.5 \times 10^{-6} \approx 0.0.}$ Ans.

Example 12.11. *The aluminium paint is used to cover the surface of a body that is maintained at 27°C. In one installation this body is irradiated by the sun, in another by a source at 527°C. Calculate the effective absorptance of the surface for both conditions, assuming the sun is a black body at 5800 K.*

Take $\alpha = 0.4$ *for* $0 \le \lambda < 3$ μm.

$\alpha = 0.8$ *for* $3\mu \le \lambda < \infty$.

Solution

Given : Irradiation on a aluminium painted cover surface.

$T_s = 27°\text{C} = 300$ K, $\quad T_{sun} = 5800$ K

$T_{source} = 527°\text{C} = 800$ K

$\alpha_{\lambda_1} = 0.4 \quad$ for $\quad 0 \le \lambda < 3$ μm

$\alpha_{\lambda_2} = 0.8 \quad$ for $\quad 3 \text{ μm} \le \lambda < \infty$.

To find : The effective absorptivity of the surface for

(*i*) Solar radiation, and

(*ii*) Source radiation.

Anslysis : (*i*) For solar radiation

$$\lambda_1 T = 3 \text{ μm} \times 5800 \text{ K}$$

$$= 17400 \text{ μm.K} \longrightarrow f_{0-\lambda_1} = 0.98$$

and $\quad f_{\lambda_1} - \infty = 1 - f_{0-\lambda_1} = 1 - 0.98 = 0.02$

Thus 98% solar radiation falls below 3 μm and the remaining 2% between λ = 3 μm to λ = ∞.

The effective absorptivity of the surface

$$\alpha_{solar} = \frac{\int_0^{3\text{ μm}} \alpha_{\lambda_1} G_\lambda d\lambda + \int_{3\text{ μm}}^{\infty} \alpha_{\lambda_2} G_\lambda d\lambda}{\int_0^\infty G_\lambda d\lambda}$$

For black surface (sun)

$$\alpha_{solar} = \frac{\alpha_{\lambda_1}\int_0^{3\mu} E_{b\lambda} d\lambda}{\sigma T^4} + \frac{\alpha_{\lambda_2}\int_{3\mu m}^{\infty} E_{b\lambda} d\lambda}{\sigma T^4}$$

$$= \alpha_{\lambda_1} f_{0-\lambda_1} + \alpha_{\lambda_2} f_{\lambda_1 - \infty}$$

$= 0.4 \times 0.98 + 0.8 \times 0.02 = \mathbf{0.408.}$ Ans.

(*ii*) Source condition, $T_{source} = 527°\text{C} = 800$ K

$$\lambda_1 T = 3 \text{ μm} \times 800 \text{ K}$$

$$= 2400 \text{ μm.K} \longrightarrow f_{0-\lambda_1} = 0.140266$$

and $\quad f_{\lambda_1-\infty} = 1 - f_{0-\lambda_1} = 1 - 0.140266 = 0.859734$

Absorptivity of the surface for source

$$\alpha_{source} = \alpha_{\lambda_1} f_{0-\lambda_1} + \alpha_{\lambda_2} f_{\lambda_1-\infty}$$

$$= 0.4 \times 0.140266 + 0.8 \times 0.859734$$

$= \mathbf{0.7439.}$ Ans.

12.7 EMISSIVITY

The radiation emitted by a real surface at temperature T is always less than that of blackbody. Therefore, the blackbody emission is considered as reference. The **emissivity** is defined as the *ratio of the radiation energy emitted by a surface to that emitted by a black body at the same temperature.* It is a dimensionless quantity, a property of a radiating surface to measure of how closely a surface approximates a black surface for which ε = 1. It is designated as ε and varies between 0 and 1.

The emissivity of a real surfaces is not constant. It varies with temperature of the surface, as well as wavelength and direction of the emission. Therefore, different emissivities may be defined for a surface, depending upon the effect considered.

12.7.1. Hemispherical and Total Emissivity

The emissivity of a surface that is averaged over all directions is called the *hemispherical emissivity* and the emissivity averaged over all wavelengths is called the *total emissivity*. Thus the *total hemispherical emissivity* ε(T) of a surface is defined as ratio of the radiation heat flux emitted over all wavelengths into a hemispherical space (all directions) to that which would have been emitted by a blackbody at same temperature. Mathematically

$$\varepsilon(T) = \frac{E(T)}{\int_0^{\infty} E_{b\lambda}(T)d\lambda} = \frac{E(T)}{E_b(T)} = \frac{E(T)}{\sigma T^4} \quad ...(12.29)$$

For a given value of emissivity, the emissive power of a real surface at a temperature T is determined by

$$E(T) = \varepsilon\, E_b(T) = \varepsilon\, \sigma\, T^4 \quad ...(12.30)$$

The total, normal emissivities for some selected materials are shown in Figs. 12.12 and 12.13. and listed in Table 12.3. The following observations are listed below :

1. The emissivity of the metallic surfaces is very small having the values as low as 0.02 for highly polished gold and silver.
2. The presence of oxide layers may improve the emissivity of metallic surfaces.
3. The non conductors have the large value of emissivity, generally exceeding 0.6.
4. The emissivity of conducting materials increase with increase in temperature, but emissivity of non-conducting materials may either increase or decrease with increasing temperature.

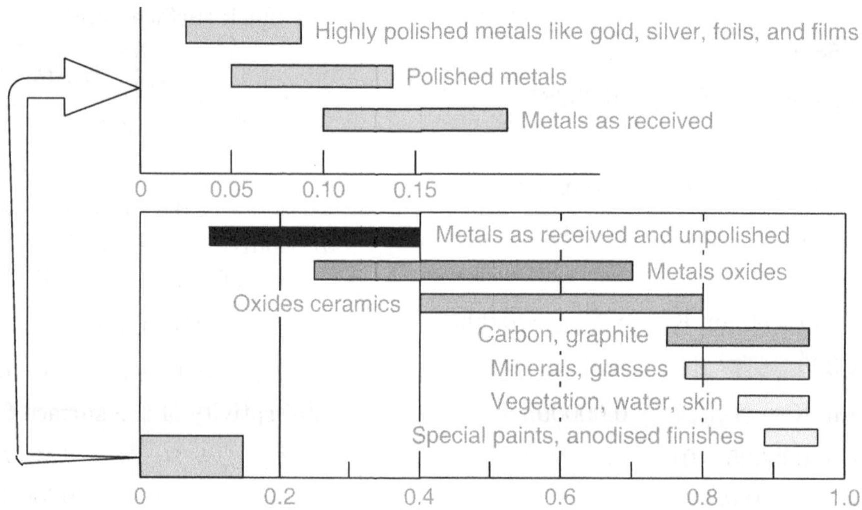

Fig. 12.12. Total, normal emissivity of selected materials

The value of emissivity strongly depends on the nature of the surface, which can be influenced by the method of fabrication, thermal cycling and chemical reaction with the environment.

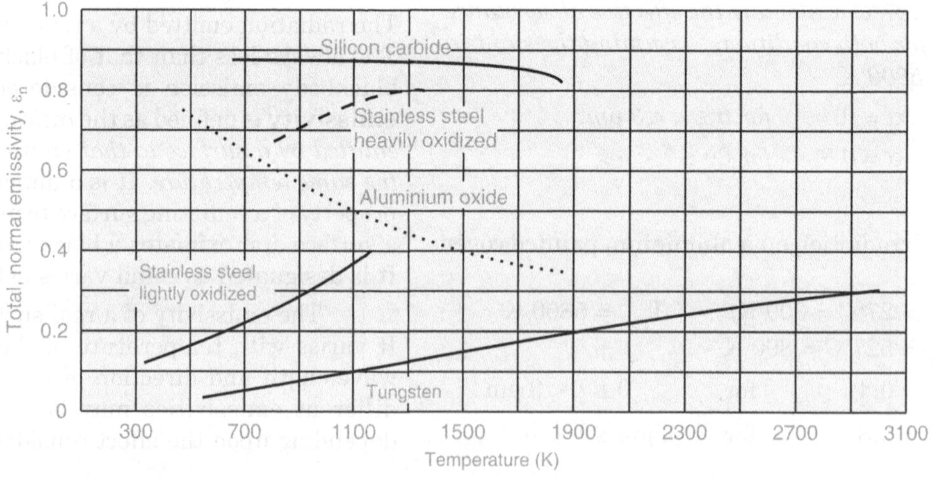

Fig. 12.13. Temperature dependence of the total, normal emissivity ε_n of selected materials

12.7.2. Spectral Emissivity

The emissivity of a surface at a specified wavelength is called the *spectral emissivity*. It is the ratio of spectral emissive power of a real surface to that of black surface at the same temperature. It is denoted by ε_λ and expressed as :

$$\varepsilon_\lambda(T) = \frac{E_\lambda(T)}{E_{b\lambda}(T)} \quad \text{or} \quad E_\lambda(T) = \varepsilon_\lambda\, E_{b\lambda}(T) \qquad ...(12.31)$$

Average value or hemispherical emissivity :

$$\varepsilon = \frac{\int_0^\infty \varepsilon_\lambda E_{b\lambda}(T)\, d\lambda}{\int_0^\infty E_{b\lambda}\, d\lambda} = \frac{\int_0^\infty \varepsilon_\lambda E_{b\lambda}(T)\, d\lambda}{E_b} = \frac{\int_0^\infty \varepsilon_\lambda E_{b\lambda}(T)\, d\lambda}{\sigma T^4} \qquad ...(12.32)$$

The spectral emissivity from a real surface depends upon temperature and wavelength. The spectral distribution of the emissivity against wavelength is shown in Fig. 12.14.

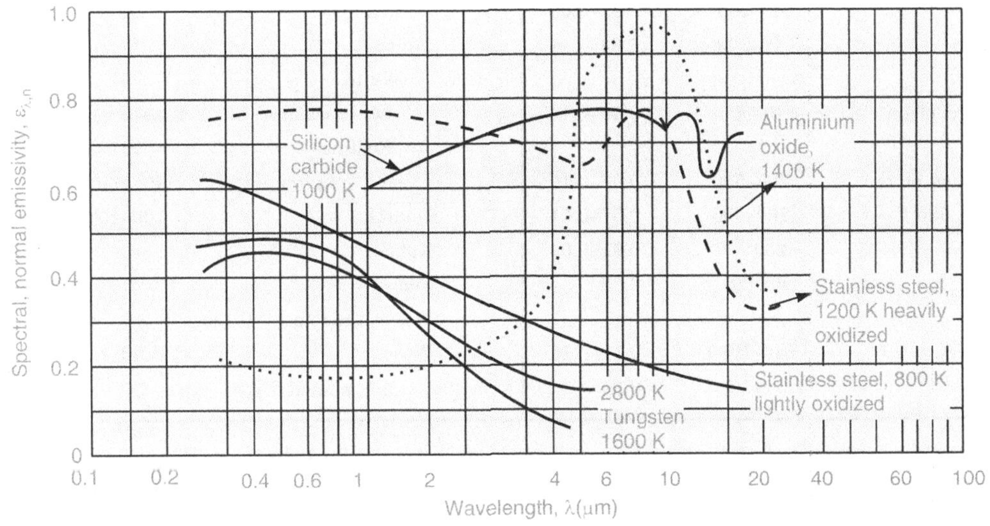

Fig. 12.14. Spectral dependence of the spectral, normal emissivity $\varepsilon_{\lambda,n}$ of selected materials

TABLE 12.3. Total emissivity for a variety of surfaces

Metals			*Nonmetals*		
Surface	*Temp. (°C)*	ε	*Surface*	*Temp. (°C)*	ε
Aluminium			Asbestos	40	0.93–0.97
Polished, 98% pure	200–600	0.04–0.06	Brick		
Commercial sheet	90	0.09	Red, rough	40	0.93
Heavily oxidized	90–540	0.20–0.33	Silica	980	0.80–0.85
Brass			Fireclay	980	0.75
Highly polished	260	0.03	Ordinary refractory	1090	0.59
Dull plate	40–260	0.22	Magnesite refractory	980	0.38
Oxidized	40–260	0.46–0.56	White refractory	1090	0.29
Copper			Carbon		
Highly polished electrolytic	90	0.02	Filament	1040–1430	0.53
Slightly polished to dull	40	0.12–0.15	Lampsoot	40	0.95
Black oxidized	40	0.76	Concrete, rough	40	0.94
Gold : Pure, polished	90–600	0.02–0.035	Glass		
Iron and steel			Smooth	40	0.94

Mild steel, polished	150–480	0.14–0.32
Steel, polished	40–260	0.07–0.10
Sheet steel, rolled	40	0.66
Sheet steel, strong rough oxide	40	0.80
Cast iron, oxidized	40–260	0.57–0.66
Iron, rusted	40	0.61–0.85
Wrought iron, smooth	40	0.35
Wrought iron, dull oxidized	20–360	0.94
Stainless, polished	40	0.07–0.17
Lead		
Polished	40–260	0.05–0.08
Oxidized	40–200	0.63
Mercury : pure, clean	40–90	0.10–0.12
Platinum		
Pure, polished plate	200–590	0.05–0.10
Oxidized at 590°C	260–590	0.07–0.11
Silver	200	0.01–0.04
Tin	40–90	0.05
Tungsten		
Filament	540–1090	0.11–0.16
Filament	2760	0.39
Gypsum	40	0.80–0.90
Ice	0	0.97–0.98
Limestone	40–260	0.92–0.95
Marble	40	0.93–0.95
Mica	40	0.75
Paints	40	
Black glass	40	0.90
White paint	40	0.89–0.97
Various oil paints	40	0.92–0.96
Red lead	90	0.93
Paper		
White	40	0.95–0.98
Other colours	40	0.92–0.94
Roofing	40	0.91
Plaster, rough lime	40–260	0.92
Quartz	100–1000	0.89–0.93
Rubber	40	0.86–0.94
Snow	10–20	0.82
Water, thickness ≥ 0.1 mm	40	0.96
Wood	40	0.80–0.90
Oak, planed	20	0.90

12.7.3. Directional Emissivity

The emissivity in a specified direction is called the *directional emissivity*. It is denoted by ε_θ, where θ is the angle between the direction of radiation and normal of the surface. The variation of emissivity with direction for conductors and non conductors is shown in Fig. 12.15. It is observed that the directional emissivity ε_θ is almost constant for about $\theta < 40°$ for conductors such as metals, and for $\theta < 70°$ for non conductors such as plastics. The directional emissivity ε_θ of a surface in normal direction $(\theta = 0)$ represents the hemispherical emissivity of the surface.

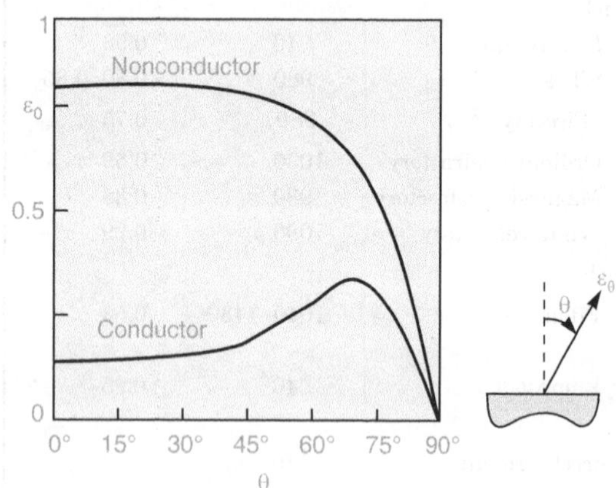

Fig. 12.15. Variations of emissivity with direction for electrical conductors and nonconductors

12.7.4. Kirchhoff's Law

It states that *at thermal equilibrium, the ratio of the total emissive power to the total absorptivity is constant for all bodies.*

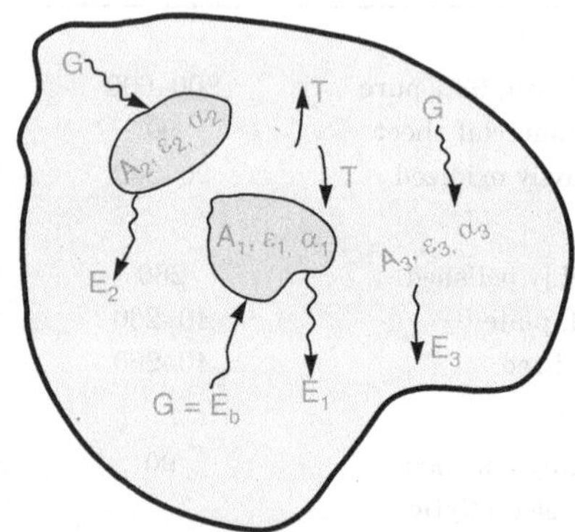

Fig. 12.16. Radiation exchange by small bodies in an isothermal enclosure

Consider a small body 1 of surface area A_1, emissivity ε_1 and absorptivity α_1 at temperature T within a large radiating (black) enclosure at the same temperature as shown in Fig. 12.16. The energy incident on the small body is E_b (= σT^4) and of this energy, the radiation absorbed by the small body per unit surface area at temperature T is :

$$G_{abs} = \alpha_1 E_b = \alpha_1 \sigma T^4 \quad ...(12.33)$$

The radiation emitted by the small body is :

$$E_1 = \varepsilon_1 \sigma T^4 \quad ...(12.34)$$

Thus at thermal equilibrium, the radiation emitted by the body must be equal to the radiation absorbed by it, or

$$E_1 = \alpha_1 E_b$$

Similarly, if small body 1 is replaced by a small body 2 with absorptivity α_2 in the enclosure, then energy incident on the body is again E_b, energy absorbed is $\alpha_2 E_b$ and at thermal equilibrium, the energy emitted E_2 is :

$$E_2 = \alpha_2 E_b$$

Similarly, for other bodies, it can be shown that at thermal equilibrium, the energy emitted by a surface must be equal to energy absorbed by the surface. Therefore, we may write

$$\frac{E_1}{\alpha_1} = \frac{E_2}{\alpha_2} = \frac{E_3}{\alpha_3} = E_b \quad ...(12.35)$$

or

$$\frac{E_1}{E_b} = \alpha_1 \quad \text{or} \quad \varepsilon_1 = \alpha_1 \quad ...(12.36)$$

Thus the total emissivity of a surface at temperature T is always equal to its total absorptivity for radiation coming from a blackbody at the same temperature. This relation simplifies the radiation analysis and is developed by Gaustav Kirchhoff in 1860, and thus known as *Kirchhoff's law.*

Hence it is also true at a specific wavelength that the spectral emissivity is equal to spectral absorptivity at thermal equilibrium for all bodies. This law is applicable when the radiation properties are independent of wavelength *i.e.,* for graybodies or when incident and emitted radiation have same spectral distribution.

12.7.5. Gray and Diffuse Surfaces : Gray Lambert Body Approximation

The radiation properties of a real surface such as absorptivity, emissivity etc. depend on wavelength and direction of radiation. Most emissivity data are either averaged over direction or over wavelength. But for some real surfaces, the spectral emissivity ε_λ is a complex function of wavelength λ and its calculation for all wavelengths becomes very tedious. Therefore, the gray and diffuse approximations are commonly used in radiation calculations. A surface is called to be *diffuse surface,* if its properties are *independent of directions,* and a *gray surface,* if its properties are *independent of wavelength.*

Consider the spectral emissivity of a real surface is represented by a wiggly line as shown in Fig. 12.17. It differs from Planck's distribution and it has large variation with wavelength and consists of several peaks and valleys. Further, the spectral emissivity also varies with temperature. Thus, its evaluation to average emissivity becomes very complicated. A useful simplification is to replace the wiggly line by an average emissivity line as shown by dashed line in Fig. 12.17. It is the **gray body approximation.** The effect of approximation on emissive power is shown in Fig. 12.18. A gray surface should emit same radiation as the real surface. That is :

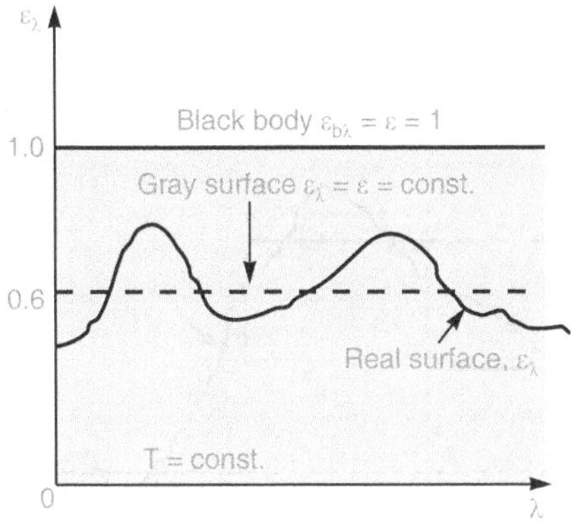

Fig. 12.17. Comparison of the emissivity of a real surface, a gray surface and a black body at the same temperature

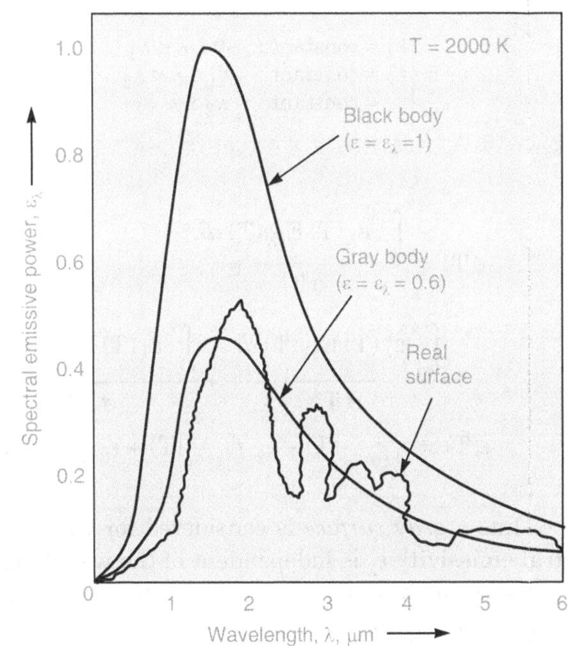

Fig. 12.18. Comparison of hemispherical spectral emission for black, gray and real surfaces

$$\varepsilon(T)\, \sigma T^4 = \int_0^\infty \varepsilon_\lambda(T)\, E_{b\lambda}(T)\, d\lambda \quad ...(12.37)$$

Then the *average emissivity* for a gray surface can be expressed as :

$$\varepsilon(T) = \frac{\int_0^\infty \varepsilon_\lambda(T)\, E_{b\lambda}(T)\, d\lambda}{\sigma T^4} \quad ...(12.38)$$

If the variation of spectral emissivity is quite large, but it is constant over certain wavelength as shown in Fig. 12.19, then ε(T) is expressed as step function and eqn. (12.38) is integrated by dividing the spectrum into a number of wavelength bands and assuming the emissivity ε(T) remains constant over each wavelength band. Then the average emissivity can be determined as :

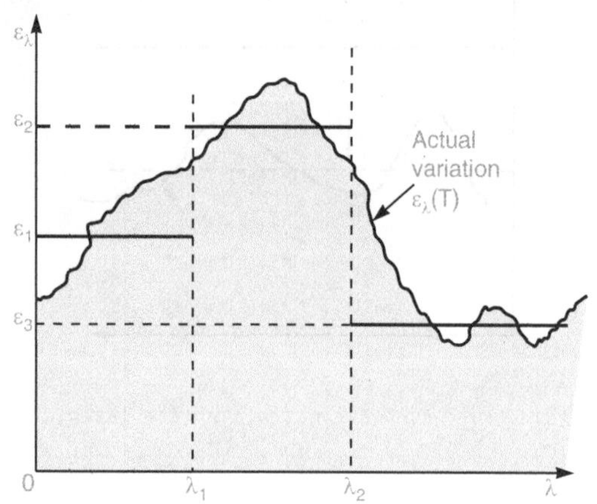

$$\varepsilon_\lambda = \begin{cases} \varepsilon_1 = \text{constant}; & 0 \le \lambda < \lambda_1 \\ \varepsilon_2 = \text{constant}; & \lambda_1 \le \lambda < \lambda_2 \\ \varepsilon_3 = \text{constant}; & \lambda_2 \le \lambda < \infty \end{cases}$$

Fig. 12.19. Approximation of actual variation of spectral emissivity with wavelength by a step function

$$\varepsilon(T) = \frac{\int_0^{\lambda_1} \varepsilon_\lambda(T)\, E_{b\lambda}(T)\, d\lambda}{\sigma T^4} + \frac{\int_{\lambda_1}^{\lambda_2} \varepsilon_\lambda(T)\, E_{b\lambda}(T)\, d\lambda}{\sigma T^4} + \frac{\int_{\lambda_2}^{\infty} \varepsilon_\lambda(T)\, E_{b\lambda}(T)\, d\lambda}{\sigma T^4}$$

or $$\varepsilon(T) = \varepsilon_1 f_{0-\lambda_1}(T) + \varepsilon_2 f_{\lambda_1-\lambda_2}(T) + \varepsilon_3 f_{\lambda_2-\infty}(T) \quad ...(12.39)$$

Thus a *gray surface* is considered for which the spectral emissivity ε_λ is independent of the wavelength and the spectral emissivity is equal to total emissivity ε

$$\varepsilon(T) = \varepsilon_\lambda(T) \quad ...(12.40)$$

Further, for a gray surface, the absorptivity, reflectivity and transmissivity are also independent of wavelength. For such a surface ;

$$\varepsilon(T) = \alpha(T) \quad ...(12.41)$$

Example 12.12. *In an isothermal enclosure at uniform temperature, two small surfaces A and B are placed. The irradiation to the surface by the enclosure is 6200 W/m². The absorption rates by the surfaces A, and B are 5500 W/m² and 620 W/m². When steady state is established, calculate the following :*

(i) What are the heat fluxes to each surfaces ? What are their temperatures ?

(ii) Absorptivity of both surfaces,

(iii) Emissive power of each surface, and

(iv) Emissivity of each surface.

Solution

Given : Two small body A, and B are placed in an enclosure.

$$G = 6200 \text{ W/m}^2$$
$$G_{\alpha A} = 5500 \text{ W/m}^2$$
$$G_{\alpha B} = 620 \text{ W/m}^2.$$

To find :

(*i*) Heat flux to each surface and surface temperatures.

(*ii*) α_A and α_B

(*iii*) E_A and E_B

(*iv*) ε_A and ε_B.

Analysis : (*i*) At steady state, the net heat flux is zero. The enclosure is considered to be a black body and isothermal, thus $T_A = T_B = T$

Irradiation to the surface = Emission from the surface

$$6200 = \sigma T^4 = 5.67 \times 10^{-8} (T)^4$$

or $$T = \mathbf{575\ K.} \quad \text{Ans.}$$

$$\mathbf{q_A = 0, \quad q_B = 0.} \quad \text{Ans.}$$

(*ii*) Absorptivity of surface, A,

$$\alpha_A = \frac{G_{\alpha A}}{G} = \frac{5500}{6200} = \mathbf{0.887.} \quad \text{Ans.}$$

Absorptivity of surface B,

$$\alpha_B = \frac{G_{\alpha B}}{G} = \frac{620}{6200} = \mathbf{0.1.} \quad \text{Ans.}$$

(*iii*) Emissive power of each surface

According to Kirchhoff's law, at thermal equilibrium

Energy absorbed = Energy emitted

$$\therefore \quad E_A = \alpha_A G = \mathbf{5500\ W/m^2.} \quad \text{Ans.}$$
$$E_B = \alpha_B G = \mathbf{620\ W/m^2.} \quad \text{Ans.}$$

(*iv*) Emissivity of each surface

According to Kirchhoff's law

$$\varepsilon = \alpha$$

$$\therefore \quad \varepsilon_A = \alpha_A = \mathbf{0.887.} \quad \text{Ans.}$$
$$\varepsilon_B = \alpha_B = \mathbf{0.1.} \quad \text{Ans.}$$

Example 12.13. *A solar collector surface has an absorptivity of 0.85 for wavelength $0 \le \lambda < 3$ μm and a value of 0.15 for wavelength $\lambda > 3$ μm. (i) Calculate the energy absorbed from the source at 5000 K, (ii) If the flux is 800 W/m², calculate the energy radiated by the body, if its temperature was 350 K.*

Solution

Given : A solar collector

(*i*) $\alpha_{\lambda_1} = \alpha_1 = 0.85$ for $0 \le \lambda \le 3$ μm

$\alpha_{\lambda_2} = \alpha_2 = 0.15$ for $\lambda > 3$ μm

$T_s = 5000$ K.

(*ii*) $T_s = 350$ K, $G = q = 800$ W/m²

To find :

(*i*) Energy absorbed by solar collector surface.

(*ii*) Emissive power, if surface at 350 K.

Assumptions :

(*i*) Diffuse, gray surfaces.

(*ii*) Kirchhoff's law holds good.

Analysis : (*i*) At temperature 5000 K, from Table 12.2.

$$\lambda_1 T = 3 \text{ μm} \times 5000$$
$$= 15000 \text{ μm.K} \longrightarrow f_{0-\lambda_1} = 0.969$$

The average absorptivity is

$$\alpha = \alpha_1 f_{0-\lambda_1} + \alpha_2(1 - f_{0-\lambda_1})$$
$$= 0.85 \times 0.969 + 0.15 \times (1 - 0.969)$$
$$= 0.8283$$

For flux of 800 W/m².

The energy absorbed

$$= \alpha G = 0.8283 \times 800$$

$= \mathbf{662.64\ W/m^2}$. **Ans.**

(*ii*) When surface temperature is 350 K

$$\lambda_1 T = 3 \text{ μm} \times 350 \text{ K}$$
$$= 1050 \text{ μm} \longrightarrow f_{0-\lambda_1} = 0.00057$$

From Kirchhoff's law

$$\alpha_1 = \varepsilon_1 \text{ and } \alpha_2 = \varepsilon_2$$

Energy radiated

$$= [\varepsilon_1 f_{0-\lambda_1} + \varepsilon_2(1 - f_{0-\lambda_1})]\, \sigma T^4$$
$$= [0.85 \times 0.00057 + 0.15 \times (1 - 0.00057)] \times 5.67 \times 10^{-8} \times (350)^4$$

$= \mathbf{128\ W/m^2}$. **Ans.**

12.8. RADIATION FROM A SURFACE

The radiation emitted from a surface propagates in all possible direction. Similarly, the radiation incident on a surface may come from different directions. In both cases, effect of radiation on the surface depends on the directional distribution. Such directional effect may be treated by introducing the concept of *radiation intensity*.

12.8.1. Solid Angle

A *solid angle* is defined as a ratio of area of spherical surface enclosed by a conical surface with the vertex of the cone at the centre of the sphere to the square of radius of the sphere. Consider the emission from the differential area dA_1 towards the normal area dA_n as shown in Fig. 12.20. The differential area dA_n, through which the emission passes, subtends an angle called the *solid angle* when viewed from dA_1. Mathematically it is expressed as :

$$d\omega = \frac{dA_n}{r^2} \quad \text{...(12.42)}$$

where A_n = normal area

r = radius of sphere

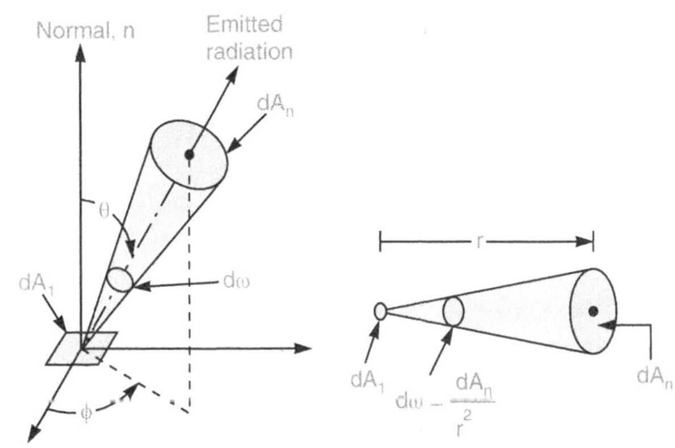

(a) Emission from a differential area dA_1 into a solid angle subtended by dA_n (b) Definition of solid angle

Fig. 12.20

The solid angle is measured in *steradian* (sr). The solid angle subtended by hemisphere from its centre is

$$\frac{2\pi r^2}{r^2} = 2\pi$$

and by full sphere is $4\pi \left(\text{as } \dfrac{4\pi r^2}{r^2}\right)$.

12.8.2. Spectral Intensity of Radiation ($I_{b\lambda}$)

It is the radiant energy emitted by a black body at a temperature T, streaming through a unit area normal to direction of propagation per unit wavelength about a wavelength per unit solid angle about the direction of propagation of beam. It is denoted by $I_{b\lambda}$ and can be expressed as :

$$I_{b\lambda} = \frac{\text{Energy emitted}}{(\text{Projected area}) \times (\text{Wavelength}) \times (\text{Solid angle})} \left(\frac{\text{W}}{\text{m}^2.\text{μm.sr}}\right) \quad \text{...(12.43)}$$

12.8.3. Radiation Intensity (I_b)

It is the radiation emitted by a blackbody at temperature T over all wavelength per unit projected area per unit solid angle. Mathematically,

$$I_b = \int_0^{\infty} I_{b\lambda}\, d\lambda \text{ [W/(m}^2\text{.sr)]} \qquad ...(12.44)$$

Consider the emission from an elemental area dA_1 located at the centre of the hemisphere as shown in Fig. 12.21. Let this emission is absorbed by elemental area dA_n, a portion of hemisphere. The area dA_n subtends an angle $d\theta$ from the perpendicular line joining dA_1.

The solid angle subtended by dA_n

$$d\omega = \frac{dA_n}{r^2}$$

where the normal area, $dA_n = r^2 \sin\theta\, d\theta\, d\phi$

$$\therefore \qquad d\omega = \sin\theta\, d\theta\, d\phi \qquad ...(12.45)$$

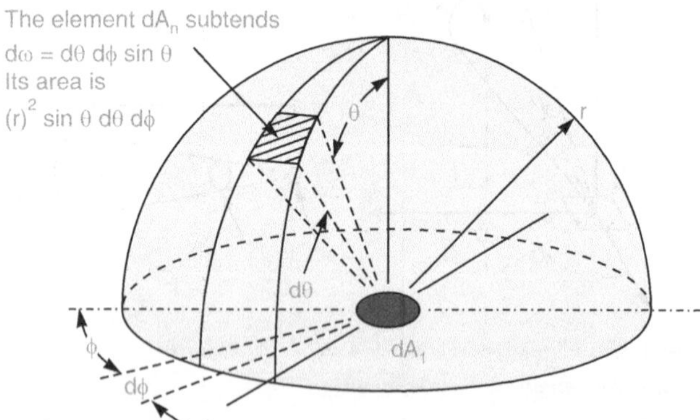

(a) Radiation intensity through a unit sphere

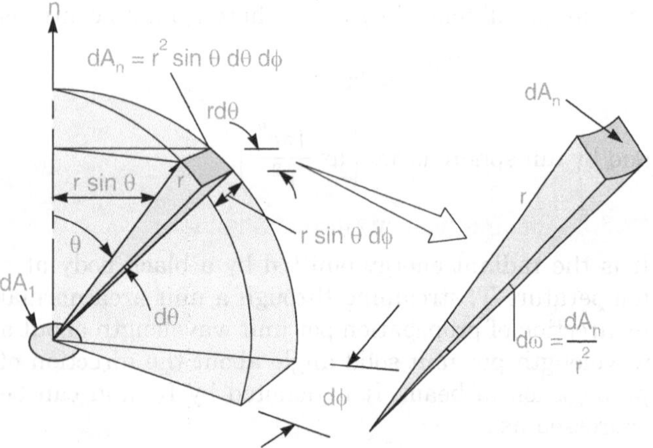

(b) Solid angle subtended by dA_n at a point on dA_1

Fig. 12.21

The projected area on a plane normal to the line joining dA_1 and $dA_n = dA_1 \cos\theta$

The spectral intensity of radiation according to its definition eqn. (12.43).

$$I_{b\lambda} = \frac{(dQ/d\lambda)}{dA_1 \cos\theta \dfrac{dA_n}{r^2}} \qquad ...(12.46)$$

where $\dfrac{dQ}{d\lambda} = dQ_\lambda$ is the rate of spectral radiation energy emitted by the surface dA_1 and passes through $dA_n = r^2 \sin\theta\, d\theta\, d\phi$.

Rearranging eqn. (12.46), we get

$$dQ_\lambda = I_{b\lambda}\, dA_1 \cos\theta\, (\sin\theta\, d\theta\, d\phi) \text{ (W/}\mu\text{m)} \qquad ...(12.47)$$

The spectral emissive power associated with dA_1 is

$$dE_\lambda = \frac{dQ_\lambda}{dA_1} = I_{b\lambda} \sin\theta \cos\theta\, d\theta\, d\phi \qquad ...(12.48)$$

If spectral and directional distribution of $I_{b\lambda}$ is known, then the spectral emissive power associated with the emission into a hemisphere shown in Fig. 12.21 (*b*) is given by

$$E_{b\lambda} = \int_{\phi=0}^{\phi=2\pi} \int_{\theta=0}^{\theta=\pi/2} I_{b\lambda} \sin\theta \cos\theta\, d\theta\, d\phi \text{ (W/m}^2.\mu\text{m)} \qquad ...(12.49)$$

The total hemispherical emissive power E_b(W/m^2) is given by

$$E_b = \int_{\lambda=0}^{\lambda=\infty} E_{b\lambda}\, d\lambda$$

Then

$$E_b = \int_{\lambda=0}^{\lambda=\infty} \int_{\phi=0}^{\phi=2\pi} \int_{\theta=0}^{\theta=\pi/2} I_{b\lambda}(\lambda, \phi, \theta) \sin\theta \cos\theta\, d\theta\, d\phi\, d\lambda \qquad ...(12.50)$$

For a diffuse surface, the intensity of radiation is independent of direction for such a surface

$$I_{b\lambda}(\lambda, \phi, \theta) = I_{b\lambda}(\lambda),$$

Then eqn. (12.50) is rearranged as

$$E_b = \int_{\theta=0}^{\theta=\pi/2} \int_{\theta=0}^{\theta=2\pi} \sin\theta \cos\theta\, d\theta\, d\phi \times \int_{\lambda=0}^{\lambda=\infty} I_{b\lambda}(\lambda) d\lambda \qquad ...(12.51)$$

The total intensity of blackbody radiation is defined as

$$I_b = \int_{\lambda=0}^{\lambda=\infty} I_{b\lambda}(\lambda) d\lambda$$

and integration of remaining part of eqn. (12.51), yields to

$$\int_{\theta=0}^{\theta=\pi/2} \int_{\phi=0}^{\phi=2\pi} \sin\theta \cos\theta\, d\theta d\phi$$

$$= \pi \int_{\theta=0}^{\theta=\pi/2} 2 \sin\theta \cos\theta\, d\theta = \pi$$

using in eqn. (12.51), we get

$$E_b = \pi I_b \qquad ...(12.52)$$

Thus the total emissive power of a black body is equal to π times the intensity of radiation (I_b).

12.8.4. Lambert Cosine Law

It states that the total emission from a surface in any direction is directly proportional to the cosine of the angle of the emission.

The angle of emission θ is the angle subtended by the normal to the radiating surface and the direction vector of emission of the receiving surface.

If E_n is total emissive power in normal direction, then the emissive power is :

$$E = E_n \cos\theta \qquad ...(12.53)$$

It is true only for diffuse radiating surfaces. For the surfaces, obey Lambert cosine law, the intensity of radiation in any direction is same, *i.e.*,

$$I_n = I = \text{const.} \qquad ...(12.54)$$

Example 12.14. *A surface emits as a blackbody at 1500 K. Calculate the rate of emission per unit area, if radiation corresponds to 0° ≤ θ ≤ 60° and wavelength interval 2 μm ≤ λ ≤ 4 μm.*

Solution

Given : Black body emission

$T_s = 1500$ K

$\lambda_1 = 2$ μm, $\qquad \lambda_2 = 4$ μm

$\phi_1 = 0°$ and $\qquad \phi_2 = 60° = \dfrac{\pi}{3}$

To find : Rate of emission per unit area.

Analysis : The emission from the black surface may be obtained by using eqn. (12.50) within limits $\lambda_1 = 2$ μm to $\lambda_2 = 4$ μm, θ = 0° to 2π, and φ = 0 to π/3.

$$\Delta E_b = \int_{2\,\mu m}^{4\,\mu m}\int_0^{2\pi}\int_0^{\pi/3} I_{b\lambda}\cos\theta\sin\theta\, d\theta\, d\phi\, d\lambda$$

For diffuse black body

$$\Delta E_b = \int_{2\,\mu m}^{4\,\mu m} I_{b\lambda}\, d\lambda \int_0^{2\pi}\int_0^{\pi/3} \cos\theta\sin\theta\, d\theta\, d\phi$$

$$= \int_{2\,\mu m}^{4\,\mu m} I_{b\lambda}\, d\lambda \left[2\pi\frac{\sin^2\theta}{2}\right]_0^{\pi/3}$$

$$= 0.75\int_{2\,\mu m}^{4\,\mu m} \pi I_{b\lambda}\, d\lambda$$

We know $E_{b\lambda} = \pi I_{b\lambda}$. Thus

$$\Delta E_b = 0.75\int_{2\,\mu m}^{4\,\mu m} E_{b\lambda}\, d\lambda$$

$$= 0.75\, E_b \int_{2\,\mu m}^{4\,\mu m} \frac{E_{b\lambda}\, d\lambda}{E_b}$$

$$= 0.75\, E_b\,(f_{0-4} - f_{0-2})$$

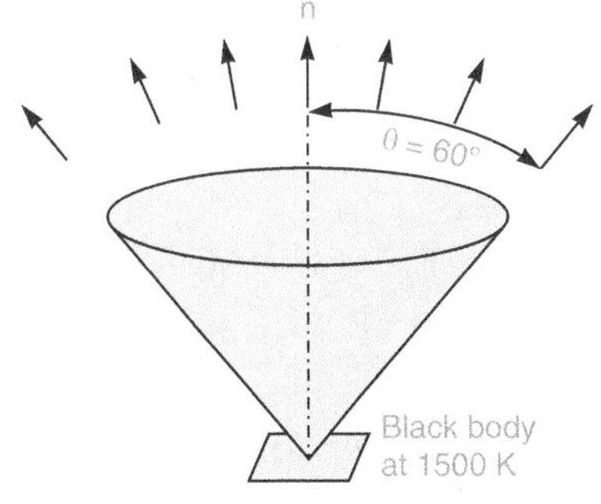

Fig. 12.22. Schematic

From Table 12.2.

$$\lambda_1 T = 2\ \mu m \times 1500\ K$$
$$= 3000\ \mu m\, K \longrightarrow f_{0-2} = 0.273$$
$$\lambda_2 T = 4\ \mu m \times 1500\ K$$
$$= 6000\ \mu m\, K \longrightarrow f_{0-4} = 0.738$$
$$\Delta E_b = 0.75 \times 5.67 \times 10^{-8} \times (1500)^4 \times (0.738 - 0.273)$$
$$= \mathbf{100.16 \times 10^3\ W/m^2}. \quad \text{Ans.}$$

12.9. RADIOSITY

Radiosity, J, is the total radiant energy leaving a surface per unit area per unit time. The total radiant energy leaving a surface consists of the emitted energy and the reflected part of the incident energy shown in Fig. 12.23. Thus, radiosity is defined by the relation.

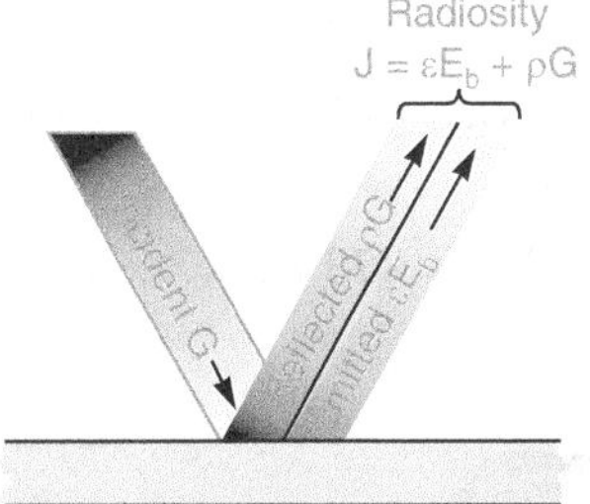

Fig. 12.23. Definition of radiosity of a surface (specular reflection not implied)

$$J = \varepsilon\, E_b + \rho G \qquad ...(12.55)$$

where $\quad$ J = radiosity of the surface (W/m²),

ε = emissivity of the surface,

E_b = black body emissive power at the temperature of the surface (W/m²),

ρ = reflectivity of the surface,

G = incident radiant flux (W/m²),

For a gray, diffuse opaque surface ($\tau = 0$),

$$\alpha + \rho = 1 \quad \text{or} \quad \rho = 1 - \alpha$$

and at thermal equilibrium

$$\alpha = \varepsilon$$

Thus $\quad J = \varepsilon E_b + (1 - \varepsilon)G$

or
$$G = \frac{J - \varepsilon E_b}{1 - \varepsilon} \qquad \text{...(12.56)}$$

The total rate of energy leaving the surface = AJ

The total rate of energy incident on the surface = AG

Thus the net radiant energy leaving the surface

$$Q = AJ - AG = A(J - G)$$

$$= A\left[J - \frac{J - \varepsilon E_b}{1-\varepsilon}\right] = \frac{A\varepsilon(E_b - J)}{1 - \varepsilon} \qquad \text{...(12.57)}$$

This equation is not valid for black surface for which $\varepsilon = \alpha = 1$; $\rho = 0$

For a black surface, $J = E_b$

$$Q = A(E_b - G) \qquad \text{...(12.58)}$$

Example 12.15. *A gray, diffuse opaque surface ($\alpha = 0.8$) is at 100°C and receives an irradiation 1000 W/m². If the surface area is 0.1 m². Calculate*

(i) Radiosity of the surface, and

(ii) Net radiative heat transfer rate from the surface.

(iii) Calculate above quantities, if surface is black.

Solution

Given : A gray, diffuse opaque surface

$\alpha = 0.8$, $\qquad G = 1000$ W/m²

$T_s = 100°C = 373$ K, $\qquad A_s = 0.1$ m².

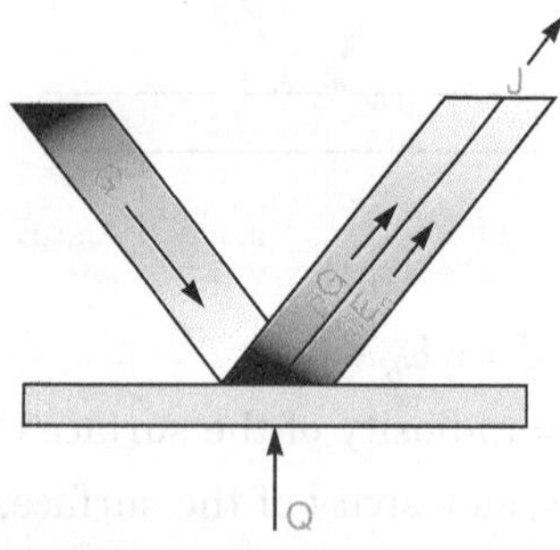

Fig. 12.24. Surface energy balance with radiant energy

To find :

(*i*) Radiosity J,

(*ii*) Net heat transfer rate,

(*iii*) For black surface J and Q_{net}.

Analysis : (*i*) The radiosity of the surface

$$J = \varepsilon E_b + \rho G$$

For a gray, diffuse and opaque surface, $\alpha = \varepsilon$

$$\tau = 0, \quad \alpha + \rho = 1$$

or $\quad \rho = 1 - \alpha = 1 - 0.8 = 0.2$

$\therefore \quad J = \alpha \sigma T^4 + \rho G$

$$= 0.8 \times 5.67 \times 10^{-8} \times (373)^4 + 0.2 \times 1000$$

$$= \mathbf{1078 \ W/m^2}. \quad \text{Ans.}$$

(*ii*) The net heat transfer rate using eqn. (12.58)

$$Q_{net} = A(J - G)$$
$$= 0.1 \times (1078 - 1000)$$
$$= \mathbf{7.8 \ W/m^2}. \quad \textbf{Ans.}$$

Alternatively using eqn. (12.57)

$$Q_{net} = \frac{A\varepsilon(E_b - J)}{1 - \varepsilon}$$

$$= 0.1 \times 0.8 \times \frac{[5.67 \times 10^{-8} \times (373)^4 - 1078]}{1 - 0.8}$$

$$= \mathbf{7.81 \ W/m^2}. \quad \text{Ans.}$$

(*iii*) For black surface $\varepsilon = \alpha = 1$

(*a*) Radiosity $J = E_b = \sigma T^4$

$$= 5.67 \times 10^{-8} \times (373)^4$$
$$= \mathbf{1097.5 \ W/m^2}. \quad \text{Ans.}$$

(*b*) The heat transfer rate, eqn. (12.58)

$$Q = A(E_b - G)$$
$$= 0.1 \times (1097.5 - 1000)$$
$$= \mathbf{9.753 \ W/m^2}. \quad \text{Ans.}$$

Example 12.16. *A plane , gray, diffuse and, opaque surface (absorptivity = 0.7) with a surface area of 0.5 m², is maintained at 500°C and receives radiant energy at a rate of 10,000 W/m². Determine per unit time*

(i) The energy absorbed.

(ii) The radiant energy emitted.

(iii) The total energy leaving the surface per unit area.

(iv) The radiant energy emitted by the surface in the wave band 0.2 μm to 4 μm.

(v) The net radiative heat transfer from the surface.

Solution

Given : Plane, gray, diffuse, opaque surface

$$\alpha = 0.7$$

A_s(surface area) = 0.5 m²

$$G = 10000 \text{ W/m}^2$$

$$T_s = 500°C = 773 \text{ K}$$

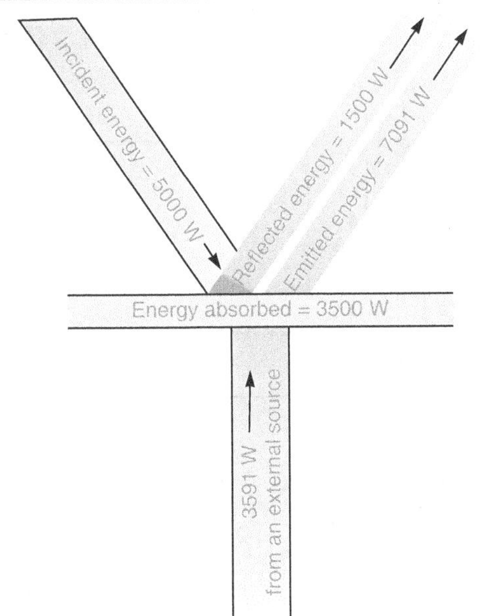

Fig. 12.25. Schematic for energy transactions on the surface

To find :

(*i*) Rate of energy absorbed,

(*ii*) Emitted radiant energy,

(*iii*) Total radiant energy leaving the surface per unit area

(*iv*) Emitted radiant energy in the wave band 0.2 μm–4 μm

(*v*) Q(net radiative heat transfer rate from the surface).

Analysis : (*i*) Rate of energy absorbed = αAG

$= 0.7 \times 0.5 \times 10000 =$ **3500 W. Ans.**

(*ii*) Rate of radiant energy emitted = $A\,\varepsilon\,\sigma\,T^4$

(For a gray surface, $\varepsilon = \alpha = 0.7$)

$= 0.5 \times 0.7 \times 5.67 \times 10^{-8} \times (773)^4$

= **7091 W. Ans.**

(*iii*) Total energy flux leaving the surface is the sum of the emitted energy flux and the reflected energy flux

$$J = \varepsilon E_b + \rho G$$

where $\rho = 1 - \alpha = 1 - 0.7 = 0.3$

∴ Radiosity,

$$J = 0.7 \times 5.67 \times 10^{-8} \times (773)^4 + 0.3 \times 10000$$

= **17182 W/m². Ans.**

(*iv*) Rate of radiant energy emitted in the wave band 0.2 μm to 4 μm is given by

$$E_{\lambda_1-\lambda_2} = \varepsilon[f_{0-\lambda_2} - f_{0-\lambda_1}]E_b \text{ W/m}^2$$

At λ_1 and λ_2, T = 773, from Table 12.2

$\lambda_1 T = 0.2 \times 773$

$= 154.6\ \mu\text{m.K} \longrightarrow f_{0-\lambda_1} = 0.0$

$\lambda_2 T = 4.0 \times 773$

$= 3092\ \mu\text{m.K} \longrightarrow f_{0-\lambda_2} = 0.294$

$E_{\lambda_1-\lambda_2} = 0.7 \times [0.294 - 0] \times 5.67 \times 10^{-8} \times (773)^4$

$= 4166.2\ \text{W/m}^2.$

Rate of radiant energy on surface area $A_s = 0.5$ m² is :

$= AE_{\lambda_1-\lambda_2} = 0.5 \times 4166.2$

= **2083.13 W. Ans.**

(*v*) From an energy balance on the surface

Net radiative heat transfer rate from the surface

= Total energy leaving the surface – Energy reaching the surface

$= 0.5 \times 17182 - 0.5 \times 10000 =$ **3591 W. Ans.**

12.10. SOLAR RADIATION

The sun is our primary source of energy. The energy coming out the sun is called *solar energy* and it reaches to earth in the form of electromagnetic waves. The sun is considered as a *nuclear reactor,* where the heat being generated due to continuous fusion reaction of hydrogen atoms to form helium. The sun experiences very large temperature in its core region, but its temperature drops to approximately 5800 K in its outer region, due to continuous dissipation of energy by radiation.

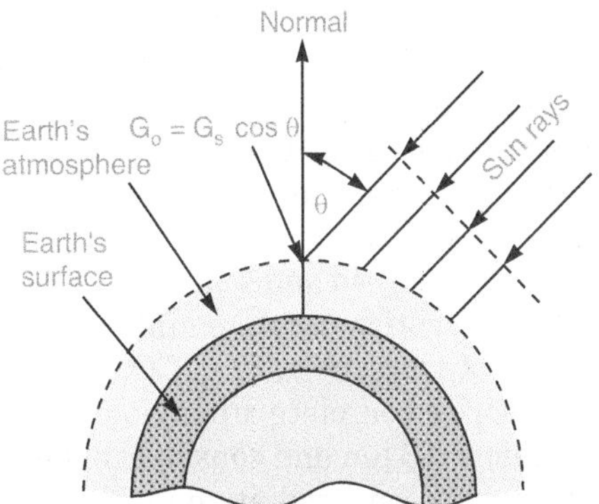

Fig. 12.26. Solar radiation reaching the earth's atmosphere, solar constant G_s, and extraterrestrial solar irradiation G_0

The sun is nearly spherical body of diameter of 1.392×10^6 km and a mass of 2×10^{30} kg. It is located at a mean distance of 1.496×10^8 km from the earth. The earth has its mean diameter of 1.27×10^4 km and its surface gets only small fraction of sun's energy, because, the sun subtends only an angle of 32 minute at the earth's surface. The intensity of solar radiation reaches outside the earth's atmosphere is almost constant. *The solar constant G_s is the rate at which the solar radiation* *flux is received on a surface normal to the sun rays just outside the earth's atmosphere, when the earth is its mean distance from the sun.* The radiation coming from the sun is equivalent to blackbody radiation. Using the *Stefan Boltzmann law,* the solar constant can be calculated as :

$$G_s = \left[\frac{r_{sun}}{r_{orbit}}\right]^2 \sigma T_s^4 = 1353 \text{ W/m}^2 \quad ...(12.59)$$

where, r_{sun} = radius of sun = 6.9598×10^8 m.

r_{orbit} = distance between sun and earth = 1.496×10^{11} m.

$\sigma = 5.67 \times 10^{-8}$ W/m²-K⁴, Stefan Boltzmann constant.

T_s = effective temperature of sun = 5762 K

Due to very small eccentricity in the earth, the distance between the earth and sun varies throughout the year. Therefore, solar constant also varies from its maximum value of 1399 W/m² on December 21 to a minimum of 1310 W/m² on June 21 and on any day of year, it can be calculated as :

$$G_s = 1367\left[1 + 0.033\cos\frac{360n}{365}\right] \quad ...(12.60)$$

where n is the day of year.

The *extraterrestrial solar irradiation* G_0 incident normal to the outer surface of the earth's atmosphere is calculated as :

$$G_0 = G_s \cos\theta \quad (\text{W/m}^2) \quad ...(12.61)$$

The spectral distribution of solar radiation on the earth's atmosphere and physical significance of G_s and G_0 are illustrated in Fig. 12.27. The solar radiation travels in atmosphere about 30 km outside the earth's surface. As solar radiation passes through this atmosphere, it is absorbed and scattered by atmospheric material. The absorption occurs mainly due to presence of ozone, water vapours, CO_2, NO_2, CO, O_2 and CH_4 etc. The *ozone* absorbs complete ultraviolet radiation at wavelength below 0.3 μm and considerably in the range of 0.3 to 0.4 μm and some radiation in the visible range. Thus the *ozone layer* in the upper regions of atmosphere guards biological systems on the earth from the harmful ultraviolet radiation. In turn the ozone layer must be protected from the destructive chemicals commonly used as refrigerants, cleaning agents and propellants in the aerosol cans. The carbon dioxide and water vapour absorb mainly *longer wavelength radiation* (infrared radiation).

As a results of these absorption, the solar energy reaching the earth surface is weakened considerably. Its absorption is large as 950 W/m² on a clear day and much less on cloudy or smoggy days.

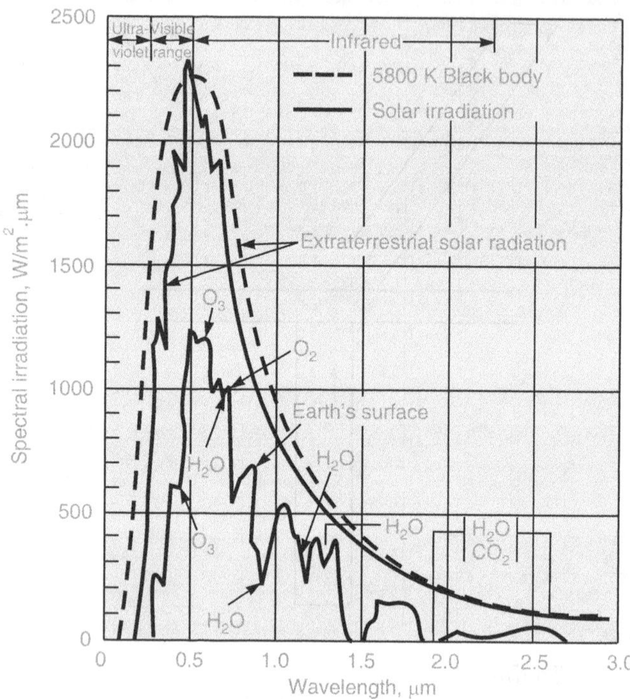

Fig. 12.27. Spectral distribution of solar radiation

The solar energy reaching the earth's surface is also weakened by *scattering* or *reflection* when it passes through the atmosphere. The *scattering* or *reflection* occurs due to all gaseous molecules as well as particulate matter in the atmosphere. The radiation at wavelength corresponding to violet and blue colours is scattered most. These scattered radiation is redistributed in all directions and gives *bluish colour sky.* The same phenomenon is responsible for *red sunrise* and *sunset.* Early in the morning and late in evening, the sun rays pass through a larger thickness of atmosphere. Therefore, the violet and blue colours of light experience a large number of scattering and thus do not reach the earth's surface. While the colours correspond to longer wavelength such as yellow, orange and red reach the earth's atmosphere and making *sunset* and *sunrise* as reddish. Similarly, the *red traffic light* can be seen from longer distance as compared to green light.

12.10.1. Solar Radiation on the Earth

The Fig. 12.28 shows the distribution of solar energy on the earth's surface. Actually 45% of sun's energy reaches the earth's surface. The lower left hand figure shows how this sun energy is in turn returned to atmosphere and space.

The solar energy reaches the earth surface is sum of direct and scattered (diffused) components. The part of solar radiation that reaches the earth's surface without being scattered or absorbed is called the **direct solar radiation** G_D. The scattered part (**diffuse solar**

radiation G_d) reaches the earth surface uniformly from all directions and it varies from about 10% of total on clear day to almost 100% on totally cloudy day. Then the total solar radiation that reaches the horizontal earth's surface per unit area is shown in Fig. 12.29 and is expressed as :

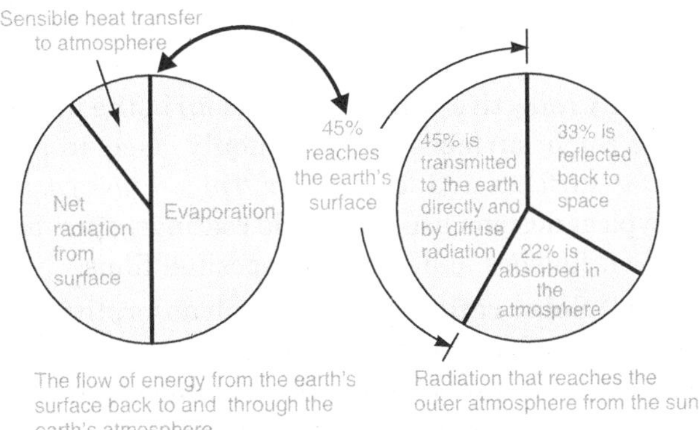

Fig. 12.28. The approximate distribution of sun's energy to the earth's surface

$$G_{solar} = G_D \cos\theta + G_d \qquad ...(12.62)$$

where θ is the angle of incidence of direct solar radiation.

Further, the quantity of solar energy reaching a surface on the earth depends on its orientation angle in relation to sun, the hour of the day, the day of year and latitude of the place and atmospheric conditions. In the early morning and late afternoon, the solar radiation reaches the earth surface follows an oblique, longer path through atmosphere thus reduces the intensity of radiation.

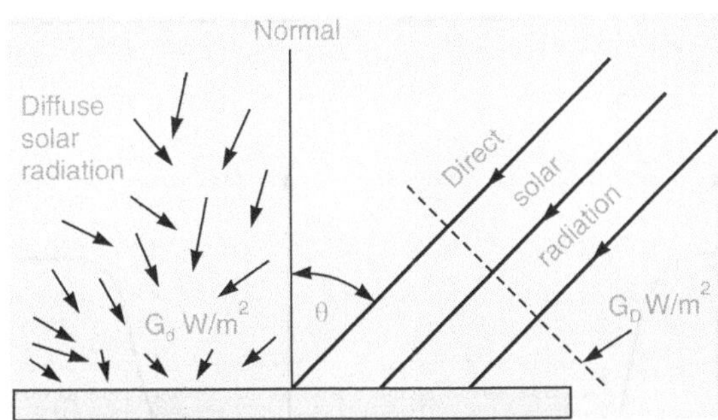

Fig. 12.29. Direct and diffuse radiation incident on horizontal earth's surface

12.10.2. Atmospheric Emission

Since the solar radiation passes through the atmosphere undergoes through absorption and scattering, thus the atmospheric emission is due to temperature of gas constituents like CO_2 and H_2O molecules and suspended particle in atmospheric. The emission of atmosphere to the earth surface G_{sky} can be obtained as:

$$G_{sky} = \sigma\, T_{sky}^4 \text{ W/m}^2\text{.K} \qquad ...(12.63)$$

where T_{sky} = effective atmospheric temperature. It ranges from 230 K for cold, clear sky conditions to a high of 285 K for warm, cloudy-sky conditions.

12.10.3. Green House Effect

The surface of the earth warms up during the day as a result of absorption of solar radiation and cools down during night by radiating its energy back into deep atmosphere as infrared radiation. The sun emits most of its energy in shorter wavelength (λ = 0.3 – 3 μm), while earth's surface and other objects are relatively at much lower temperature, and thus emit radiation in longer wavelength infrared regions (λ > 5 μm). The constituents of atmosphere, carbon dioxide, water vapour, methane, and pollutants are practically transparent for *shorter wavelength radiation,* while they are opaque for *longer wavelength radiation.* Therefore, they easily pass on the *short wavelength radiation* that the earth receives from the sun, but they absorb the *longer wavelength radiation* remitted by the surface of the earth, when it cools. This absorbed energy is once again reradiated to the earth surface Fig. 12.30. In this process an additional amount of heat energy is retained within the lower atmosphere, thus causing a slight temperature increase of the earth. Such warming up and remission of radiation is called the **green house effect** and these effects eventually resulting in *global warming.*

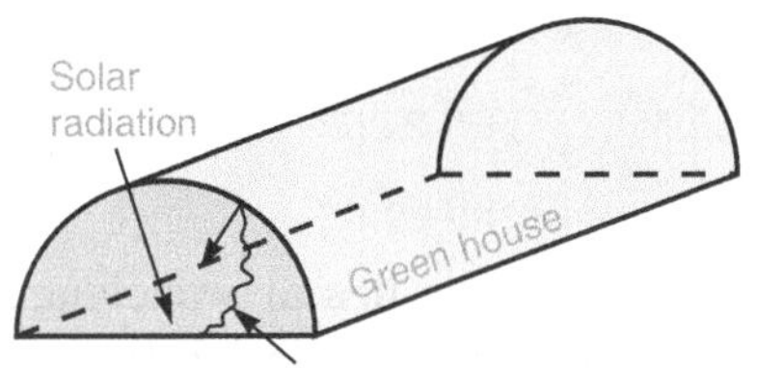

Fig. 12.30. Green house traps energy by allowing solar radiation to enter but not allowing infrared radiation to leave

The coastal areas are humid and therefore, there is no drastic change between day and night temperature, because the *humidity* acts as a *barrier in the path of infrared radiation* coming from the earth's surface. A similar trend is also followed for large pollutant places, where CO_2 affects the natural cooling process. In clean areas with clear sky such as *deserts,* there is *large variation in day and night temperature,* because absence of such barriers for infrared radiation. The causes of green house effect are :

1. Excessive deforestation,
2. Industrialisation and burning of fuel.

The average global temperature has been risen up by about 0.3°C to 0.6°C over last 100 years due to green house effect (Fig. 12.31) as a result of concentration of CO_2, methane, CFCs etc. in the atmosphere and the rate of rise is increasing alarmingly due to increasing atmospheric pollution. In this directions, some corrective measures are taken seriously. The production of CFCs has been banned worldwide. Pollution norms are set to automobiles, and industries. However, the efforts put for normal climatic conditions are not sufficient and need to be more effective enhance, because a small increase in global temperature may lead to series of catastrophic consequences.

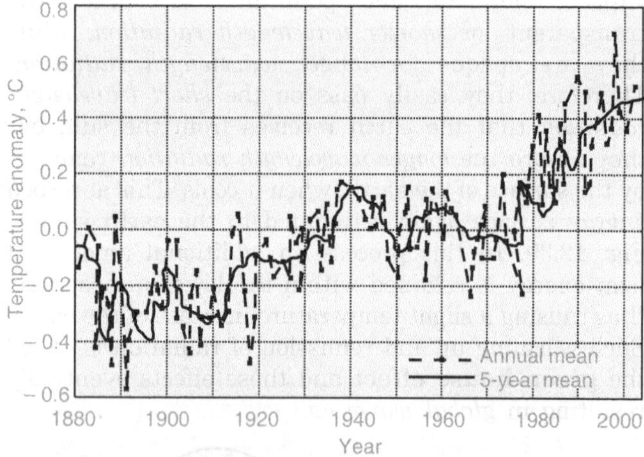

Fig. 12.31. Global surface temperature over last 120 years

12.10.4. Selective Surfaces

Most of the sun's energy lies at wavelength near the visible region of electromagnetic spectrum and most of the radiation from the objects and earth's surface is coming on much longer infrared wavelength. Thus engineers are often interested to use a selective radiating surface to promote specific spectral or directional characteristics. The Table 12.4 shows some infrared emittance and solar absorptance for several materials. There are four basic types of selective surfaces.

(*i*) Flat absorber, (*ii*) flat reflector, (*iii*) solar absorber, and (*iv*) solar reflector.

(*i*) **Flat absorber.** It has high absorptivity over entire spectrum of wavelength. It is intended to capture all incident radiation. The black paints, enamels and black or dark oxides or black anodised coating are some flat absorber.

(*ii*) **Flat reflector.** It has low absorptivity and transmissivity over entire spectrum of wavelength, such a surface reflects most of irradiation and is obtained by bare or polished metals and foils, electroplated or vacuum deposited metal coatings, aluminium films and metallic paints.

(*iii*) **Solar absorber.** It is a truly selective surface, and it absorbs low wavelength solar irradiation but highly reflective (with low emission) in the infrared range. Such surfaces are commonly used in *solar collector* surfaces and for *satellite temperature control.* The typical designs use thin oxide coatings, sprayed or baked on finishes and vacuum deposited films.

(*iv*) **Solar reflector.** It has high absorptivity and emission in infrared range, but reflects most of the low wavelength solar irradiation. It is obtained by white or light coloured paints, enamels, or ceramic coatings or anodised aluminium. Such surfaces cool the exterior structure such as building, a fuel tanks or cryogenic containers by reflecting solar irradiation and emitting low temperature heat from within.

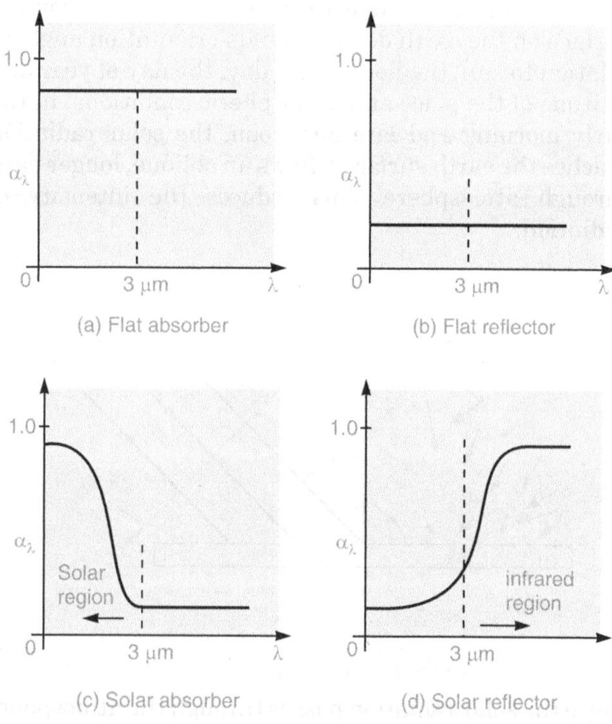

Fig. 12.32. Selective radiating surfaces

The surface shown in Fig. 12.32 are assumed to be opaque. Transparent or semitransparent material can also be selective with somewhat different designs.

TABLE 12.4. Solar absorbance and infrared emittance for several surfaces near 300 K

Surface	α_{solar}	ε
Aluminium, pure	0.09	0.1
Carbon black in acrylic binder	0.94	0.83
Copper, polished	0.3	0.04
Selective solar absorbers		
Black Cr on Ni plate	0.95	0.09
CuO on Cu (Ebanol C)	0.90	0.16
Nickel black on steel	0.81	0.17
Sputtered cermet on steel	0.96	0.16
Selective solar reflectors		
Magnesium oxide	0.14	0.7
Snow	0.2–0.35	0.82
White paint		
Acrylic	0.26	0.90
Zinc oxide	0.12–0.18	0.93

Example 12.17. *A flat plate solar collector with no cover has a selective absorber surface with $\varepsilon = 0.1$ and $\alpha_s = 0.95$. At a particular time of the day, the absorber surface temperature T_s is 120°C, when the solar irradiation is 750 W/m², the effective sky temperature is – 10°C and ambient air temperature T_∞ is 30°C. Assume the natural convection is given by*

$$q = 0.22(T_s - T_\infty)^{4/3} \text{ W/m}^2\text{. K.}$$

Calculate the useful heat removal rate (W/m²) from the collector for these conditions. What is the corresponding efficiency of the collector ?

Solution

Given: A flat plate solar collector with its operating conditions.

$\varepsilon = 0.1$, $\alpha_s = 0.95$

$T_s = 120°C = 393$ K, $T_\infty = 30°C$

$T_{sky} = -10°C = 263$ K, $G_s = 750$ W/m²

and relation for calculation of h.

Fig. 12.33. Schematic

To find :

(*i*) Useful heat removal rate per unit area, W/m².

(*ii*) Efficiency of the collector.

Assumptions :

(*i*) Steady state conditions.

(*ii*) Bottom of the collector is well insulated.

(*iii*) Diffuse absorber surface.

(*iv*) Sky radiation is in approximately same spectral region that of surface emission *i.e.,*

$$\varepsilon = \alpha_{sky} = 0.1.$$

Analysis : (*i*) The rate of incoming energy on the absorber surface/m²

$$\frac{E'_{in}}{A} = \alpha_s G_s + \alpha_{sky} G_{sky}$$
$$= 0.95 \times 750 + 0.1 \times 5.67 \times 10^{-8} \times (263)^4$$
$$= 739.62 \text{ W/m}^2$$

The energy outgoing the absorber surface per unit area

$$\frac{E'_{out}}{A} = q_{conv} + q_{rad} + q_{use}$$
$$= h(T_s - T_\infty) + E + q_{use}$$
$$= 0.22 \times (T_s - T_\infty)^{4/3} + \varepsilon \sigma T_s^4 + q_{use}$$
$$= 0.22 \times (120 - 30)^{4/3} + 0.1 \times 5.67 \times 10^{-8} \times (393 \text{ K})^4 + q_{use}$$
$$= 88.73 + 135.25 + q_{use} = 223.98 + q_{use}$$

In steady state conditions,

$$\frac{E'_{in}}{A} = \frac{E'_{out}}{A}$$

or $739.62 = 223.98 + q_{use}$

or $q_{use} =$ **515.65 W/m².** **Ans.**

(*ii*) The collector efficiency is defined as the fraction solar radiation extracted as useful energy.

$$\eta = \frac{q_{use}}{G_s} = \frac{515.65}{750} = 0.687$$

$\simeq$ **68.7%.** **Ans.**

Example 12.18. *The white paint on a roof as a selective solar absorber ($\alpha_s = 0.26$). Consider now a bare roof under a sunlight sky. The solar radiation on the plane of the roof is 600 W/m², the air temperature is 35°C, and a light breeze produces a convective heat transfer coefficient of h = 8 W/m².K. The sky temperature is 18°C. Find the temperature of the roof, if it is painted with either white acrylic paint or a non-selective black paint having $\varepsilon = 0.9$.*

Solution

Given : A roof surface with operating condition

$G_s = 600$ W/m², $T_\infty = 35°C = 308$ K

$h = 8$ W/m².K, $\alpha_s = 0.26$

$T_{sky} = 18°C = 291$ K, $\varepsilon = 0.9$

for non-selective black or white acrylic paint.

To find : The surface temperature of roof with

(*i*) White acrylic paint, and

(*ii*) Non-selective black paint.

Assumptions :

(*i*) Steady state conditions.

(*ii*) No heat transfer to interior of roof.

(*iii*) For non-selective black paint $\alpha_s = \varepsilon$.

Analysis : In steady state condition, the energy balance yields to

Incident (solar + sky radiation) energy on the roof surface = Heat loss by (convection + radiation) from the roof surface

For 1 m² surface area

$$\alpha_s G_s + \varepsilon \sigma T_{sky}^4 = h(T_s - T_\infty) + \varepsilon \sigma T_s^4$$

(*i*) For white acrylic paint $\alpha_s = 0.26$, $\varepsilon = 0.9$

$$0.26 \times 600 + 0.9 \times 5.67 \times 10^{-8} \times (291)^4$$
$$= 8 \times (T_s - 308) + 0.9 \times 5.67 \times 10^{-8} \times T_s^4$$

or $\quad 5.103 \times 10^{-8} T_s^4 + 8T_s - 2986 = 0$

or $\quad T_s^4 + 156.77 \times 10^6 T_s - 5.851 \times 10^{10} = 0$

It is a transcendental equation and its solution by iterative technique converge to

$$T_s = 312 \text{ K} = \mathbf{39°C.} \quad \text{Ans.}$$

(*ii*) For non-selective black paint, $\alpha_s = \varepsilon = 0.9$ and then above equation leads to

$$T_s^4 + 156.77 \times 10^6 T_s - 6.6038 \times 10^{10} = 0$$

Its solution converges to

$$T_s = 338\text{K} = \mathbf{65°C.} \quad \text{Ans.}$$

Example 12.19. *An artificial spherical satellite flies around the earth. Calculate the temperature of the satellite surface, assuming that there is no heat sources and surface temperature is uniform all over the surface. The solar radiation reflected from the earth and radiation emitted from the earth should also be ignored.*

(i) If $\alpha_s = 0.2$ and $\varepsilon = 0.1$;

(ii) If surface of the satellite is gray ;

(iii) Find the ratio α_s/ε, when the temperature of the satellite surface becomes 30°C.

The incident solar radiation is 1500 W/m².

Solution

Given : A satellite around the earth ;

$G_s = 1500$ W/m²,

To find : Satellite surface temperature if

(*i*) $\alpha_s = 0.2$ and $\varepsilon = 0.1$

(*ii*) Surface of the satellite is gray, and

(*iii*) α_s/ε ratio for $T_s = 30°C$.

Analysis : For steady state conditions, the energy balance

Heat gain by incident radiation = Heat lost by emission

$$\alpha_s A_{proj} G = \varepsilon A_s \sigma T_s^4$$

where A_{proj} = projected area of satellite for solar irradiation = πr^2

A_s = surface area of satellite for emission = $4\pi r^2$

Therefore,

(*i*) $0.2 \times \pi r^2 \times 1500 = 0.1 \times 4\pi r^2 \times 5.67 \times 10^{-8} T_s^4$

or $\quad T_s^4 = 1.3227 \times 10^{10}$

or $\quad T_s = 339.13 \text{ K} = \mathbf{66.13°C.}$ Ans.

(*ii*) For gray surface $\alpha = \varepsilon = 0.2$

$$0.2 \times \pi r^2 \times 1500 = 0.2 \times 4\pi r^2 \times 5.67 \times 10^{-8} T_s^4$$

or $\quad T_s^4 = 6.6137 \times 10^9$

or $\quad T_s = 285.17$ K

$= \mathbf{12.17°C.}$ Ans.

(*iii*) For given temperature

$$T_s = 30°C = 303 \text{ K}$$

$$\alpha_s \times \pi r^2 \times 1500 = \varepsilon \times 4\pi r^2 \times 5.67 \times 10^{-8} \times (303)^4$$

or $\quad \alpha_s/\varepsilon = \mathbf{1.27.}$ Ans.

Example 12.20. *In the process of estimating the emission from the sun, it may be treated as blackbody with the surface temperature of 5800 K at a mean distance of 15 × 10¹⁰ m from the earth. The diameter of the sun is 1.4 × 10⁹ m and that of the earth is 12.8 × 10⁶ m.*

Estimate the following :

(i) The total energy emitted by the sun.

(ii) The emission received per m² just outside the earth's surface.

(iii) The total energy received by the earth, if no radiation is blocked by the earth's atmosphere.

(iv) The energy received by a 1.5 m × 1.5 m, solar collector, whose normal is inclined at 45° to the sun. The energy loss through the atmosphere is 40% and the diffused radiation is 20% of the direct radiation.

(N.M.U., Nov. 1999)

Solution

Given : Average solar constant for determination of temperature of sun

$T_{sun} = 5800$ K, $D_{sun} = 1.4 \times 10^9$ m

$r_{orbit} = 15 \times 10^{10}$ m, $D_{earth} = 12.8 \times 10^6$ m

$A = 1.5 \times 1.5 = 2.25$ m^2, $\phi_2 = 45°$,

Energy loss through the atmosphere = 40%

Diffuse radiation = 20% of direct radiation.

To find :

(*i*) Total energy emitted by the sun.

(*ii*) Emission received by the earth per m^2, just outside the earth surface.

(*iii*) Total energy received by the earth, if no radiation is blocked by the earth's atmosphere.

(*iv*) Energy received by solar collector.

Assumptions :

(*i*) The negligible emissive power of earth in comparison of sun.

(*ii*) No convection and conduction effects.

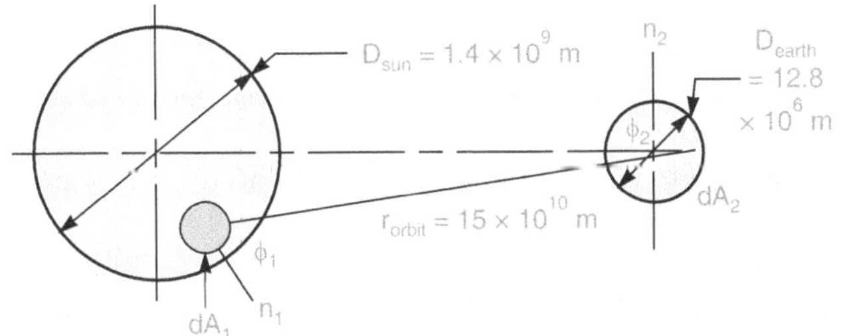

Fig. 12.34

Analysis : (*i*) Total energy emitted by the sun :

Surface area of the sun $= 4\pi r_{sun}^2 = \pi D_{sun}^2$

$= \pi \times (1.4 \times 10^9 \text{ m})^2 = 6.1575 \times 10^{18}$ m^2

The energy emitted by the sun,

$$Q = \sigma A T_{sun}^4 = 5.67 \times 10^{-8} \times 6.1575 \times 10^{18} \times (5800)^4$$

$= \mathbf{3.95 \times 10^{26}}$ **W.** Ans.

(*ii*) Emission received by 1 m^2, just outside the earth's atmosphere :

The sun emits radiation in all direction over a distance of 15×10^{10} m from the earth. The orbit area, just out side the earth's atmosphere :

$A_{orbit} = 4\pi (r_{orbit})^2$

$= 4\pi \times (15 \times 10^{10} \text{ m})^2 = 2.8274 \times 10^{23}$ m^2

The radiation received per m^2 outside the earth's surface

$$G_s = \frac{Q}{A_{orbit}} = \frac{3.95 \times 10^{26}}{2.8274 \times 10^{23}}$$

$=$ **1397 W/m^2.** Ans.

(*iii*) The earth may be assumed as a spherical body and energy received by the earth will be proportional to the projected area *i.e.*, area of the earth ;

Energy received by the earth

$= G_s \times A_{earth}$

$= 1397 \times (\pi/4) \times (12.8 \times 10^6 \text{ m})^2$

$= \mathbf{1.797 \times 10^{17}}$ **W.** Ans.

(*iv*) Energy received by a solar collector :

The direct energy blocked by atmosphere = 40%

Hence, the direct energy reaching the earth surface = 60% = $0.6\, G_s$

$= 0.6 \times 1397 = 838.2$ W/m^2

The diffuse radiation

$= 0.2 \times 838.2 = 167.64$ W/m^2

The total radiation energy reaching the collector

$=$ Projected area of collector $(\cos \phi_2) \times (838.2 + 197.64)$ W/m^2

$= 1.5 \times 1.5 \cos (45°) \times (1005.84)$

$=$ **1600.28 W.** Ans.

12.11. SUMMARY

The radiation refers to the energy emitted in form of electromagnetic waves by the bodies because of their temperature.The radiation energy emitted in wavelength between $\lambda \approx 0.1$ and $\lambda \approx 100$ μm is *referred thermal radiation.* The sun emits thermal radiation at an effective surface temperature of 5760 K and bulk of this energy lies between $\lambda = 0.1$ to $\lambda = 3$ μm, therefore this spectrum is referred as the *solar radiation.* The radiation emitted by the sun is in wavelength between $\lambda = 0.4$ to $\lambda = 0.76$ μm, is visible to human eye, therefore, this spectrum is referred as the *visible radiation* (light).

The glossary of radiation terms and their definition are given in Table 12.5.

TABLE 12.5. Glossary of the radiation terms

Terms	*Definition*
Absorption	The process of converting the radiation intercepted by the matter to internal thermal energy.
Absorptivity, α	Fraction of the incident radiation absorbed by the surface.
Black body	An ideal body which absorbs all incident radiation and emits maximum energy.
Black body radiation function	Fraction of radiation energy emitted by a black body at temperature T in wavelength band $\lambda = 0$ to λ.
Diffuse	A surface, whose properties are independent of directions.
Directional	The property pertains to a particular direction, denoted by θ.
Emission	The process of radiation production by the surface at a finite temperature.
Emissive power, E	The rate of radiant energy emitted by a surface in all direction per unit area of the surface. It is measured in W/m^2.
Emissivity, ε	Ratio of the emissive power of a surface to the emissive power of the blackbody at the same temperature.
Gray surface	A surface for which the spectral absorptivity and emissivity are independent of the wavelength over the spectral region of the surface irradiation and emission.
Green house effect	The warming up process due to remission between earth's surface and atmosphere.
Hemispherical	The quantity pertains to all directions above the surface.
Irradiation	The rate at which the radiation is incident on a surface from all direction per unit area of the surface, G(W/m^2).
Kirchhoff 's law	Relation between emission and absorption properties of a surface at thermal equilibrium.
Planck's law	It is associated with spectral distribution of emission from a black body.
Radiation intensity	The rate of radiation energy propagation in a particular direction, per unit area normal to the direction, per unit solid angle about the direction, I (W/m^2.sr).
Radiosity	Rate at which energy leaving the surface due to emission and reflection in all directions per unit area per unit time, J (W/m^2).
Reflection	The process of redirection of radiation energy incident on a surface.
Reflectivity, ρ	The fraction of incident radiation energy reflected by the surface.
Semitransparent	It is a medium in which radiation absorption is the volumetric process.
Solar energy	It is energy coming out the sun.
Solar constant	Rate at which the solar radiation flux is received on a surface normal to sun's rays just outside the earth's atmosphere, G_s(W/m^2).
Solid angle, ω	Ratio of area of spherical surface enclosed by a conical surface with vertex of the cone at the centre of sphere to square of radius of sphere. It is measured in straradian (sr).
Spectral	It refers to a single wavelength (monochromatic) radiation. The quantity is denoted by subscript λ.
Spectral distribution	It refers to properties variation with wavelength.
Specular	It refers to the surface for which the angle of reflected radiation is equal to the angle of incident radiation.
Stefan Boltzmann law	The emissive power of the blackbody is directly proportional to fourth power of the absolute temperature ; $E_b = \sigma T^4$, where $\sigma = 5.67 \times 10^{-8}$ W/m^2.K^4, is Stefan Boltzmann constant, and T is an absolute temperature in K.
Thermal radiation	It is the electromagnetic energy emitted by a matter at a finite temperature in the spectral region from approximately from 0.1 to 100 μm.
Total solar radiation	Sum of direct and diffuse solar radiation.
Transmission	It is process of the thermal radiation passing through the matter.
Transmissivity	It is the fraction of radiation energy transmitted by the matter.
Wien's displacement law	Relation between wavelength λ_{max} and absolute temperature T at which $E_{b\lambda}$ reaches a maximum ; λ_{max} T = 2897.6 μm.K.

REVIEW QUESTIONS

1. What is an electromagnetic wave ? How does it differs from a sound wave ?
2. What are the ranges of wavelengths of electromagnetic waves covering ultraviolet, visible, infrared and thermal radiation ?
3. What is the speed of energy propagation between two bodies when the space between them is evacuated ?
4. What do you mean by ultraviolet, visible and infrared radiation ?
5. What is a black body ? What are its properties ? Why does a cavity with a small hole behave as a black body ?
6. Why are microwave oven suitable for cooking ?
7. What are the total and spectral emissive power of a black body ?
8. What do you mean by spectral, terms used in thermal radiation ?
9. State Planck's distribution law and list down its features.
10. What is Wien's displacement law ? Derive an expression for its relation. What is a diffuse body ?
11. State and explain Stefan Boltzmann law. Derive an expression for total emissive power of a black body.
12. What is radiation intensity ? How do you distinguish between spectral emissive power and spectral radiation intensity ?
13. What do you mean by spectral, total emissivity ?
14. Discuss the effect of temperature on emissivity of surfaces.
15. What is directional emissivity ?
16. Explain Kirchhoff 's law.
17. What do you mean by gray body approximation ?
18. What is a blackbody radiation function ? Why is it important ?
19. Define the terms irradiation and radiosity. Establish a relationship between them.
20. Define solid angle.
21. Define total emissive power and intensity of radiation and show that $E = \pi I$.
22. State and explain Lambert cosine law.
23. Define absorptivity, reflectivity and transmissivity.
24. What do you mean by opaque body and white body ?
25. What is solar constant ? How is it used to determine effective temperature of sun ? How would the value of solar constant change, if the distance between the earth and sun doubled ?
26. Explain why sky is blue in the day and sunset is yellow orange ?
27. What is an effective sky temperature ?
28. What is a green house effect ? Discuss its importance.
29. Why the ozone layer in the atmosphere must be protected ?
30. How does solar radiation attenuate as it passes through the atmosphere ?

PROBLEMS

1. Calculate the heat flux emitted due to thermal radiation from a black surface at 6000°C. At what wavelength is the monochromatic emissive power maximum and what is the maximum value ?

   [Ans. 87, 798 kW/μ^2 ; 0.462 μm ; 1.25×10^{14} W/m^2]

2. A 10 cm diameter peephole in the side of a furnace acts as a blackbody. The furnace interior is at 500°C, and the surrounding temperature is 25°C, what is the net radiant heat loss from the furnace through the peephole ? [Ans. 155.2 W]

3. Estimate the rate at which the sun emits the radiant energy. What fraction of this energy is absorbed by the earth and in what amount ? If effective temperature of the sun is 560 K and surface of the sun is treated black. The diameter of the sun is 1.39×10^6 km. The diameter of the earth is 1.27×10^4 km and the distance between sun and earth is 1.5×10^8 km.

   [Ans. 3.81×10^{26} W ; 4.48×10^{-10} ; 1.71×10^{17} W]

4. A cubical body of 20 cm side at 1200 K is suspended in air. Assuming body closely approximates a blackbody, Calculate (*a*) rate of radiation energy emitted from the body, in W, (*b*) spectral black body emissive power at a wavelength of 4 μm.

   [Ans. (*a*) 28.21×10^3 W, (*b*) 19.23×10^3 W/m^2.μm]

5. The sun can be treated as a blackbody at an effective surface temperature of 5762 K. Determine the fraction of radiation energy emitted by sun that falls in

   (*a*) ultraviolet range (λ = 0.01 mm–0.4 mm), and

   (*b*) visible range (λ = 0.4 μm–0.76 μm). Also calculate the wavelength at which the emission of radiation from the sun reaches a maximum value.

   [Ans. (*a*) 0.121, (*b*) 0.425, 0.503 μm]

6. A balck body at 727°C emits radiation. Calculate the wavelength at which the radiation from the body becomes maximum. [Ans. 2.897 μm]

7. The sun is located at 150×10^6 km from the earth and it has radius of 0.7×10^6 km and its temperature of 6000 K. Calculate the mean temperature of earth. Assume that the rate of radiative heat transfer from

the sun to the earth is equal to rate of radiant transfer from earth to outer space at 0°C. Consider earth and sun as black body. [Ans. 289.82 K]

8. A gray body at 2500 K has total emissive power of 2.0×10^6 W/m^2. What is maximum spectral emissive power and at what wavelength does it occur ?

[Ans. 1.145×10^7 W/m^2, 1.16 μm]

9. Radiant energy with an intensity of 700 W/m^2 strikes a flat plate normally. The absorptivity is twice the transmissivity and 2.9 times its reflectivity. Determine the rate of absorption, transmission and reflection of energy in W/m^2.

[Ans. 379.5 W/m^2, 189.7 W/m^2, 130.8 W/m^2]

10. A black body is at 1000°C ; calculate

(*a*) The wavelength at which the body has the maximum spectral emissive power, and the corresponding emissive power.

(*b*) The total emissive power of the black body.

(*c*) The fraction of total radiant energy emission between the wavelength 2.0 and 4.5 μm.

(*d*) The percentage reduction in the emissive power when the temperature of the body falls down to 900°C.

(*e*) Hemispherical emissive power.

[Ans. (*a*) 2.276 μm, (*b*) 1.489×10^5 W/m^2, (*c*) 0.5422, (*d*) 28%, (*e*) 1.489×10^5 W/m^2]

11. A black body has a total emissive power of 1000 W/m^2, calculate

(*a*) Its surface temperature

(*b*) The wavelength above which (*i*) 50% (*ii*) 75% radiant energy occurs, and

(*c*) The wavelength at which the body has maximum spectral emissive power.

[Ans. (*a*) 364.4 K, (*b*) (*i*) 9.45 μm, (*ii*) 11.24 μm, (*c*) 7.95 μm]

12. The spectral emissivity function of an opaque surface at 1000 K is approximated by

$$\varepsilon_\lambda = \begin{cases} \varepsilon_1 = 0.4 & 0 \le \lambda < 2\ \mu m \\ \varepsilon_2 = 0.7 & 2\ \mu m \le \lambda < 6\ \mu m \\ \varepsilon_3 = 0.3 & 6\ \mu m \le \lambda < \infty \end{cases}$$

Calculate average emissivity and the emissive power of the surface. [Ans. 0.575, 32.6 kW/m^2]

13. The spectral emissivity for a gray surface at 1000 K is approximated as

$$\varepsilon_\lambda = \begin{cases} \varepsilon_1 = 0.3 & 0 \le \lambda < 1.5\ \mu m \\ \varepsilon_2 = 0.5 & 1.5 \le \lambda < 4\ \mu m \\ \varepsilon_3 = 0.4 & 4\ \mu m \le \lambda < \infty \end{cases}$$

Calculate

(*i*) Total hemispherical emissivity,

(*ii*) Total hemispherical emissive power,

(*iii*) The average wavelength at which the body has maximum emissive power.

[Ans. (*i*) 0.445, (*ii*) 2.523 W/m^2, (*iii*) 2.898 μm]

14. A pipe carrying steam runs in a large room and exposed to air at 30°C. The pipe surface temperature is 200°C. Diameter of the pipe is 20 cm. If the total heat loss per metre length of the pipe is 1.9193 kW/m, determine the emissivity of the pipe surface.

Given that

$$Nu_D = 0.53(Gr_D\,Pr)^{1/4}$$

and air properties at 115°C are

$k_f = 0.03306$ W/m.K, $\nu = 24.93 \times 10^{-6}$ m^2/s, Pr = 0.687. [Ans. 0.8]

15. The emissivity of a tungsten filament can be approximated to be 0.5 for radiation at wavelength less than, 1 μm and 0.15 for radiation at greater than 1 μm. Calculate the average emissivity of the filament at (*a*) 1500 K, and (*b*) 3000 K. Also determine the absorptivity and reflectivity of the filament at both temperature.

16. Solar irradiation of 1100 W/m^2 is incident on a large flat horizontal metal roof on a day when air flowing over the roof causing a heat transfer coefficient of 25 W/m^2.K. The outside air temperature is 27°C, the metal surface absorptivity for solar radiation is 0.6, and the metal surface emissivity is 0.2. If roof is well insulated from below, calculate the roof temperature under steady state condition. [Ans. 48.5°C]

17. Consider an opaque, horizontal flat plate, that is well insulated on its backside. The irradiation on the plate is 2500 W/m^2, of which 500 W/m^2 is reflected. The plate is at 227°C and has an emissive power of 1200 W/m^2. Air at 127°C flow over the plate with $h = 15$ Wm^2.K. Calculate emissivity, absorptivity, reflectivity and radiosity of the plate. What is the net heat transfer rate per unit area ?

[Ans. $\varepsilon = 0.338$, $\alpha = 0.80$, $\rho = 0.2$, $J = 1700$ W/m^2, $q = -700$ W/m^2]

18. A 10 cm diameter spherical ball emits radiation at a rate of 30 W when its surface temperature is 400 K. Calculate the average emissivity of the ball at this temperature. [Ans. 0.658]

19. A surface at 365 K has an absorptivity of $\alpha_s = 0.85$ for solar radiation and an emissivity 0.5 at room temperature. The direct and diffuse components of solar radiation are 350 and 400 W/m^2, respectively. The direct radiation makes and angle of 30° with normal of the surface. Taking the effective sky temperature of 280 K, calculate the net rate of radiation to the surface at that time.

20. Solar radiation is incident on the outer surface of a spaceship at a rate of 1320 W/m^2. The surface has an absorptivity of 0.1 for solar radiation and an emissivity of 0.8 at room temperature. The outer surface radiates heat into a space at 0 K. If there is no heat transfer to the spaceship, calculate the equilibrium temperature of the surface.

[Ans. 232.3 K]

21. A central heating radiator has a surface temperature of 70°C and heats a room maintained at 20°C. Calculate the contribution of convection and radiation to heat transfer from the radiator.

Use following correlation for determination of natural convection coefficient.

$Nu_L = 0.118(Gr_L\, Pr)^{1/3}$

The properties of fluid in the room are

$\rho = 1.2$ kg/m^3, $\mu = 1.8 \times 10^{-5}$ kg/ms

$k_f = 0.026$ W/m.K, $Pr = 0.71$

[Ans. 634 W/m^2]

22. The spectral emissivity as a function of an opaque surface at 800 K is approximated by step function and is given below :

$\varepsilon_1 = 0.3$ for $0 \le \lambda \le 3\ \mu m$

$\varepsilon_\lambda = \varepsilon_2 = 0.8$ for $3\ \mu m \le \lambda \le 7\ \mu m$

$\varepsilon_3 = 0.1$ for $7\ \mu m \le \lambda \le \infty$.

Calculate the average emissivity of the surface and the emissive power.

[Ans. 0.5206, 12.09 × 10^3 W/m^2]

REFERENCES AND SUGGESTED READING

1. J.A. Duffy and W.A. Backman, *"Solar Energy Thermal Process"* Wiley, New York, 1974.
2. Rehsenow W.M, J.P. Harnett and E.N. Ganic, eds. *"Handbook of Heat Transfer"*, 2/e, McGraw Hill, NewYork, 1985.
3. F. Krieth and M.S. Bohn, *"Principles of Heat Transfer"*, 5th ed. PWS Pub. Company, 1997.
4. J.P. Holman, *"Heat Transfer"*, 7th ed. McGraw Hill, New York, 1990.
5. F.P. Incropera and D.P. DeWitt, *"Introduction to Heat Transfer"*, 2/e, John Wiley & Sons, 1990.
6. M.N. Ozisik, *"Heat Transfer–A Basic Approach"*, McGraw Hill, New York, 1985.
7. Y. Bayazitoglu and M.N. Ozisik, *"Elements of Heat Transfer"*, McGraw Hill, New York, 1988.
8. Suryanarayana N.V., *"Engineering Heat Transfer"*, West Pub. Co. New York, 1998.
9. F.M. White, *"Heat and Mass Transfer"*, Addison-Wesley, Reading, MA, 1998.
10. M. Jacob, *"Heat Transfer"*, Vol. 1, Wiley, New York, 1949.
11. L.C. Thomas, *"Heat Transfer"*, Prentice-Hall, Englewood Cliffs, NJ, 1982.
12. Chapman Alan. J., *"Fundamentals of Heat Transfer"*, Macmillan, New York.
13. Christopher Long, *"Essential Heat Transfer"*, Addison Wesley Longman, 2001.
14. McAdams W.M., *"Heat Transmission"*, 3rd ed. McGraw Hill, New York, 1954.
15. Jacob M. and G.A. Hawkins, *"Elements of Heat Transfer"*, 3rd ed. Wiley, New York, 1957.
16. John H. Lienhard IV and J.H. Lienhard, *"A Heat Transfer Textbook"*, 3/e, Phlogiston Press, 2003.
17. Robert Siegel and John R. Howell, *"Thermal Radiation Heat Transfer"*, 3/e, Hemisphere Pub. Corporation, 1992.

Radiation Exchange between Surfaces

13

In the previous chapter, our discussion was restricted to radiation properties, physical relation, and radiation processes that occur at a single surface. In this chapter, we will consider the radiation heat exchange between two or more surfaces. Such type of radiation exchange depends on the surface geometries, surface orientation as well as their temperatures and radiative properties. We will consider that the surfaces are separated by non-participating medium. Such medium neither emits, absorbs nor scatters any amount of radiation energy. Moreover, such medium has no effect on radiative heat transfer between bodies.

Initially, the geometrical features of radiating surfaces is considered by introducing a new parameter, called the *view factor*. Then radiation heat transfer between blackbodies and thereafter between diffuse, gray surfaces are discussed.

13.1 RADIATION VIEW FACTOR

It is also known as *shape factor, angle factor, geometric factor and configuration factor*. It is purely a function geometry of two surfaces, the orientation of one surface with respect to other and the space between the two radiating surfaces. It is independent of surface properties and temperature.

The view factor of surface i with respect to surface j is denoted by F_{i-j} and is defined as the ratio of the radiation energy leaving surface i and that reaches the surface j directly, to the total energy leaving the surface i.

View factor,

$$F_{i-j} = \frac{\text{Radiation energy leaving surface } i \text{ and that reaches the surface } j \text{ directly}}{\text{Total energy leaving the surfaces } i}$$

Therefore, the view factor F_{1-2} represents the fraction of radiation energy leaving surface 1 and that strikes the surface 2 and view factor F_{2-1} represents the fraction of radiation energy leaving surface 2 and that strikes the surface 1 directly. For a special case

F_{i-i} = the fraction of radiation energy leaving surface i that strikes itself directly.

13.1.1. View Factor Integral

Consider the arbitrarily oriented surfaces A_1 and A_2 of Fig. 13.1. The elemental areas on each surface dA_1 and dA_2 are joined by a line of length s, which form the angles β_1 and β_2, with the surface normals n_1 and n_2, respectively. The value of s, β_1 and β_2 may vary with position of the elemental area on A_1 and A_2.

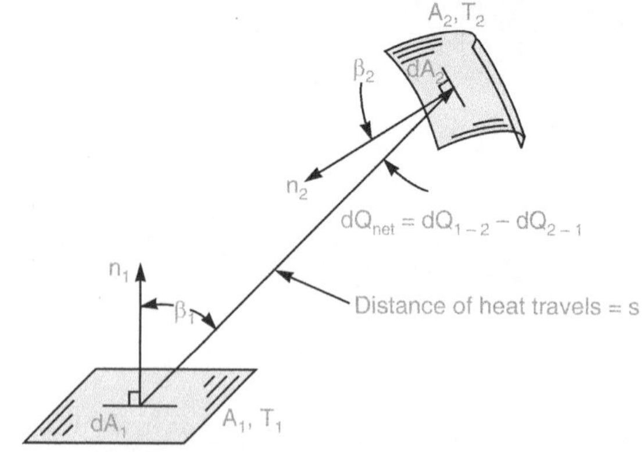

Fig. 13.1. Radiation heat exchange between two elemental surfaces of area dA_1 and dA_2

The projection of dA_1 on the line joining the centre is

$$dA_1 \cos \beta_1$$

From the definition of radiation intensity, the rate at which radiation energy leaving dA_1 and strikes the dA_2 may be expressed as

$$dQ_{1-2} = I_{b_1} dA_1 \cos \beta_1 \, d\omega_{1-2}$$

where I_{b_1} is the *intensity of radiation* leaving surface 1 and $d\omega_{1-2}$ is the *solid angle* subtended on dA_2, when viewed from dA_1, *i.e.*,

$$d\omega_{1-2} = \frac{dA_2 \cos \beta_2}{s^2}$$

It follows

$$dQ_{1-2} = I_{b_1} \cos \beta_1 \cos \beta_2 \frac{dA_1 \, dA_2}{s^2}$$

The intensity of radiation I_b may be expressed in terms of emissive power E_b. Eqn. (12.52), as

$$I_b = \frac{E_b}{\pi}$$

Then $dQ_{1-2} = E_{b_1} \cos \beta_1 \cos \beta_2 \dfrac{dA_1 \, dA_2}{\pi s^2}$

The total rate of radiation energy, which leaves the surface 1 and is intercepted by surface 2, may then be obtained by integrating the above equation over the two surfaces, *i.e.*,

$$Q_{1-2} = \frac{E_{b_1}}{\pi} \int_{A_1} \int_{A_2} \frac{\cos \beta_1 \cos \beta_2 \, dA_1 \, dA_2}{s^2}$$

From the definition of view factor

$$F_{1-2} = \frac{Q_{1-2}}{A_1 E_{b_1}}$$

It follows

$$F_{1-2} = \frac{1}{A_1} \int_{A_1} \int_{A_2} \frac{\cos \beta_1 \cos \beta_2}{\pi s^2} dA_1 \, dA_2 \quad \text{...(13.1)}$$

Similarly, the view factor F_{2-1} is defined as the fraction of radiation energy that leaves A_2 and strikes A_1. By same token, we get

$$F_{2-1} = \frac{1}{A_2} \int_{A_1} \int_{A_2} \frac{\cos \beta_1 \cos \beta_2}{\pi s^2} dA_1 \, dA_2 \quad \text{...(13.2)}$$

where s = distance between two elemental surfaces dA_1 and dA_2.

β_1 = angle between normal to surface A_1 and the line joining dA_1 and dA_2.

β_2 = angle between normal to surface A_2 and line joining dA_2 and dA_1.

The value of view factor ranges between zero and one.

Either eqn. (13.1) or eqn. (13.2) may be used to determine the view factor associated with any two *black surfaces.* The net radiation heat exchange between two blackbodies then may be expressed as

$$Q_{net} = Q_{1-2} - Q_{2-1}$$

or $$Q_{net} = A_1F_{1-2} [E_{b_1} - E_{b_2}] - A_2F_{2-1} [E_{b_1} - E_{b_2}] \quad \text{...(13.3)}$$

The concept of view factor F_{1-2} may also be extended for calculation of radiation heat exchange among diffuse, nonblack bodies ; which is only the correction to the Stefan Boltzmann law, that we use frequently for blackbodies. Thus the net heat exchange between two gray, diffuse surfaces

$$Q_{net} = A_1F_{1-2} (J_1 - J_2) - A_2F_{2-1} (J_1 - J_2) \quad \text{...(13.4)}$$

where J_1 and J_2 are radiosities for surface 1 and surface 2, respectively.

13.1.2. The View Factor Relations

1. Some evident results. For flat and convex surfaces, the view factor F_{1-1} or F_{i-i} with respect to itself is always zero, *i.e.*, the energy once leaving the surface never returns to same surface.

$$F_{1-1} = F_{i-i} = 0$$

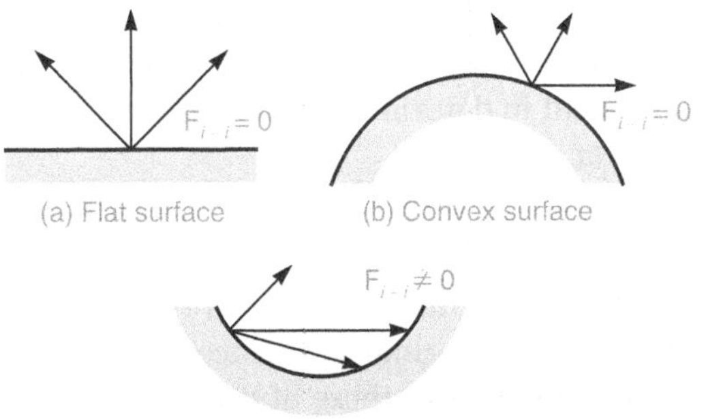

$F_{i-i} = 0$

(a) Flat surface

$F_{i-i} = 0$

(b) Convex surface

$F_{i-i} \neq 0$

(c) Concave surface

Fig. 13.2. The view factor from surface to itself is zero for *flat* and *convex* surfaces, and non-zero for *concave* surface

For the concave surface, the shape factor with respect to itself is never zero *i.e.*, radiation energy coming out a portion of the surface will be absorbed by other portion or

$$F_{1-1} \neq 0$$

Fig. 13.3 shows three elementary situations in which the value of F_{1-2} is evident by using just the definition:

F_{1-2} = fraction of field of view of (1) occupied by (2).

When the surfaces are isothermal and diffuse, this corresponds to

F_{1-2} = fraction of energy leaving (1) that reaches (2)

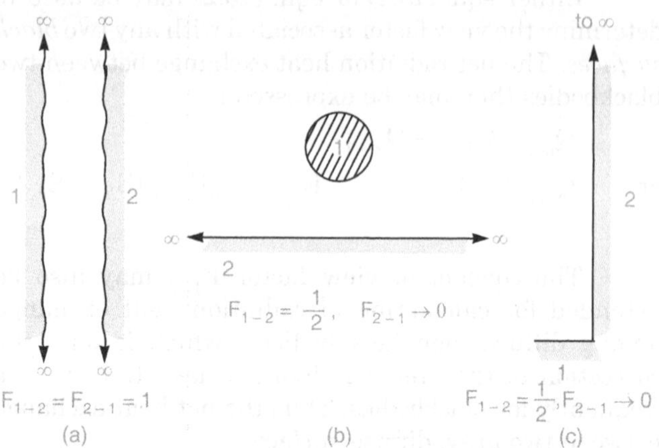

Fig. 13.3. Some configurations for which the value of the view factor is immediately apparent

A second apparent result in Fig. 13.3 (*b*) and (*c*) in regard to the view factor is that half the energy leaving surface-1 reaches the surface-2. While two infinite surfaces facing each other in Fig. 13.3 (*a*) will have $F_{1-2} = F_{2-1} = 1$ since all the energy leaving one surface will be intercepted by other one.

2. Reciprocity rule. Comparing eqns. (13.1) and (13.2) yields to

$$A_1F_{1-2} = A_2F_{2-1} \quad ...(13.5)$$

This expression is termed as *reciprocity rule* and it is very useful in determining one view factor from the knowledge of other.

3. The summation rule. If an enclosure consisting of N surfaces, interacting radiatively as shown in Fig. 13.4. In such enclosures, the opening is treated as an imaginary surface with radiation properties equivalent to those of opening.

The conservation energy principle requires that the entire radiation leaving a surface *i* must be absorbed by all surfaces of the enclosure. Therefore, the sum of view factors from surface *i* of an enclosure with respect to all surfaces of the enclosure including to itself must be unity. This is known as the *summation rule* for an enclosure, and is expressed as

$$\sum_{i=1}^{N} F_{i-j} = 1 \quad ...(13.6)$$

where N is the number of surfaces of an enclosure. The summation rule can be applied to each surface of an enclosure by varying *i* from 1 to N. Therefore, the summation rule applied to each of N surfaces and gives N equations for determination of N^2 view factors. These equations are

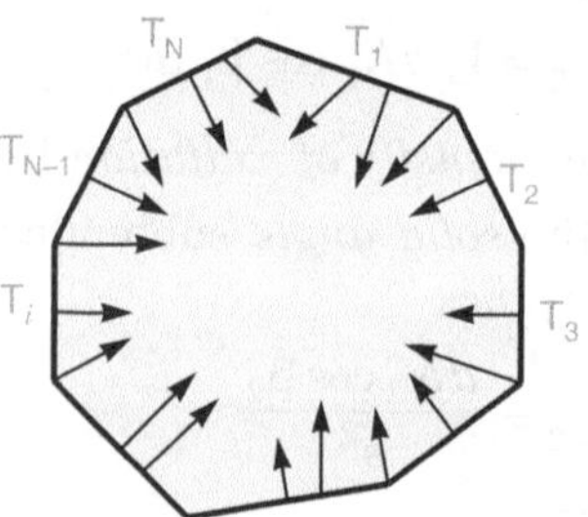

Fig. 13.4. Radiation exchange in an enclosure of N surfaces

$$F_{1-1} + F_{1-2} + F_{1-3} + \ldots\ldots + F_{1-N} = 1$$
$$F_{2-1} + F_{2-2} + F_{2-3} + \ldots\ldots + F_{2-N} = 1$$
$$F_{3-1} + F_{3-2} + F_{3-3} + \ldots\ldots + F_{3-N} = 1$$
$$\ldots\ldots\ldots\ldots\ldots\ldots$$
$$\ldots\ldots\ldots\ldots\ldots\ldots$$
$$F_{N-1} + F_{N-2} + F_{N-3} + \ldots\ldots + F_{N-N} = 1$$

The $\frac{1}{2}N(N-1)$ view factors may be obtained from $\frac{1}{2}N(N-1)$ *reciprocity* relations. Accordingly, the total number of view factors are to be evaluated directly for N surfaces enclosure are

$$N^2 - [N + \tfrac{1}{2}N(N-1)] = \tfrac{1}{2}N(N-1) \quad ...(13.7)$$

The remaining view factors may be obtained from summation rule.

4. Superposition or additive rule. Sometimes the view factors for given geometries are not available with standard tables and graphs. In such cases the given geometry is approximated as sum or difference of some geometries with known view factors and then to apply the *superposition rule.*

The view factor F_{1-2} between the two surfaces A_1 and A_2 is equal to the sum of the view factors F_{1-3} and F_{1-4}, if two areas A_3 and A_4 together make up the area A_2 as shown in Fig. 13.5. It is expressed as *additive relation*.

$$F_{1-2} = F_{1-3} + F_{1-4} \quad ...(13.8)$$

and if both sides are multiplied by A_1, then

$$A_1F_{1-2} = A_1F_{1-3} + A_1F_{1-4}$$

And by reciprocity rule;

$$A_2F_{2-1} = A_3F_{3-1} + A_4F_{4-1}$$

or

$$F_{2-1} = \frac{A_3F_{3-1} + A_4F_{4-1}}{A_2} \quad ...(13.9)$$

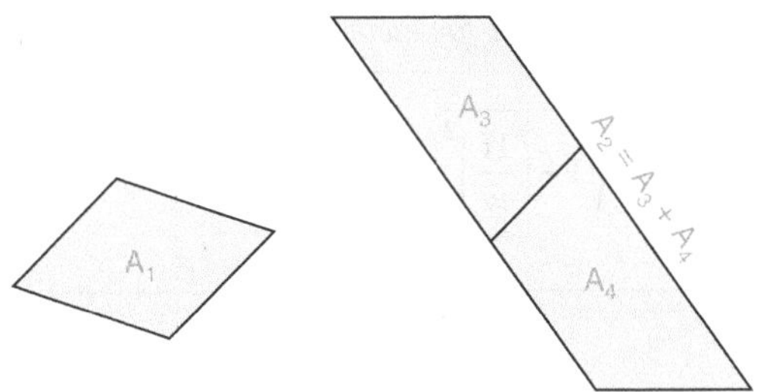

Fig. 13.5. Additive relation between shape factors

Suppose area A_2 is divided into N parts,

$$A_2 = A_3 + A_4 + ... + A_N$$

then eqn. (13.9) becomes

$$F_{2-1} = \frac{A_3F_{3-1} + A_4F_{4-1} + ... + A_NF_{N-1}}{A_2} \quad ...(13.10)$$

5. Symmetry rule. If geometry involves some symmetry then the determination of view factor is rather simplified. The presence of symmetry can be judged by inspection, keeping the definition of view factor in mind. The identical surfaces 2 and 3 that are oriented in an identical manner with respect to surface 1 shown in Fig. 13.6, will intercept identical amount of radiation leaving the surface 1. That is $F_{1-2} = F_{1-3}$

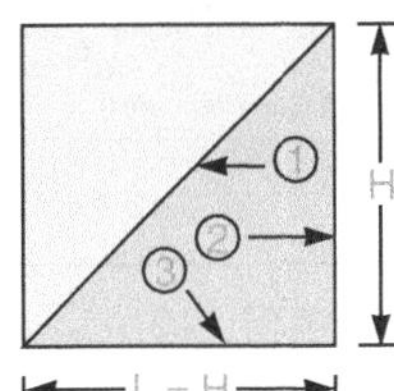

Fig. 13.6. Radiation exchange with identical surfaces

Therefore, the *symmetry* rule states that two (or more) surfaces that possesses symmetry about a third surface will have identical view factors with respect to third surface.

For, illustration of determination of view factor, consider a simple two surface enclosure involving the spherical surfaces as shown in Fig. 13.7. Then enclosure is characterised by N^2 view factors (F_{1-1}, F_{1-2}, F_{2-1} and F_{2-2}). Here $N(N - 1)/2 = 1$ view factor need to be determined directly. It is also evident from inspection. Since,

$$F_{1-1} = 0 \qquad \text{(convex surface)}$$

$$F_{1-2} = 1$$

Since entire radiation leaving the inner sphere surface 1 will be intercepted by inner surface of outer sphere 2.

$$\therefore \qquad F_{2-1} \neq 1$$

Because entire radiation leaving outer surface, 2 does not fall on inner surface 1, but a fraction of energy also strikes the surface 2. Therefore, $F_{2-2} \neq 0$

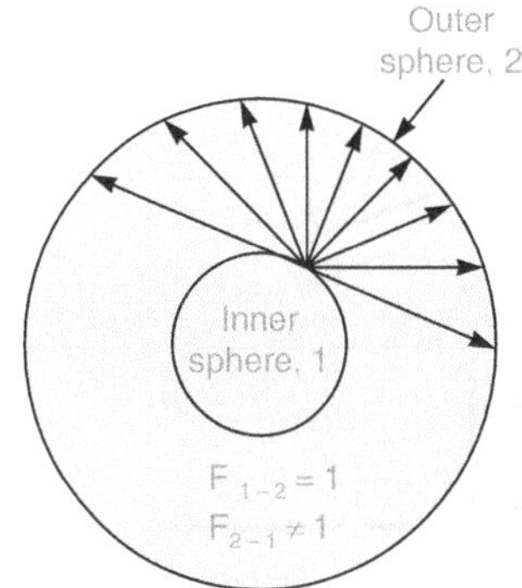

Fig. 13.7. View factor for the enclosure formed by two concentric spheres

From summation rule ;

$$F_{1-1} + F_{1-2} = 1$$

$$F_{2-1} + F_{2-2} = 1.$$

By reciprocity relation

$$A_1F_{1-2} = A_2 F_{2-1}$$

using $F_{1-2} = 1$; which leads to

$$F_{2-1} = \frac{A_1}{A_2} \quad \text{and} \quad F_{2-2} = 1 - \frac{A_1}{A_2}$$

For complicated geometries, the view factor must be determined by solving the double integral of eqn. (13.1). However, this approach is not practical, to evaluate the complicated integration for each geometries. The solutions have been obtained for different surface arrangements and are available in analytical, graphical and tabular forms. The view factors for some selected geometries are presented in Table 13.1 and Table 13.2 in *analytical form* and in Figs. 13.8 to 13.12 in *graphical form*.

The view factors in Table 13.1 for geometries that are infinite long in direction perpendicular to the plane of paper and therefore, are *two-dimensional*. The view factors in Table 13.2 and Figs. 13.8 to 13.11 are for *three-dimensional* geometries.

TABLE 13.1. View factors for a variety of two-dimensional configurations (infinite in extent normal to the paper)

	Configuration	*Analytical equation for view factor*
1.	(parallel plates 1 and 2, width w, separation H)	$F_{1-2} = F_{2-1} = \sqrt{1+\left(\frac{H}{w}\right)^2} - \left(\frac{H}{w}\right)$
2.	(plates 1 and 2 of width w at angle α)	$F_{1-2} = F_{2-1} = 1 - \sin(\alpha/2)$
3.	(perpendicular plates: 2 of height H, 1 of width w)	$F_{1-2} = \frac{1}{2}\left[1 + \frac{H}{w} - \sqrt{1+\left(\frac{H}{w}\right)^2}\right]$
4.	(triangle with sides 1, 2, 3)	$F_{1-2} = \frac{(A_1 + A_2 - A_3)}{2A_1}$
5.	(plate 1 of length b and cylinder 2 of radius r; distances a, c)	$F_{1-2} = \frac{r}{b-a}\left[\tan^{-1}\left(\frac{b}{c}\right) - \tan^{-1}\left(\frac{a}{c}\right)\right]$
6	(cylinders 1 and 2 of diameter D, spacing s)	Let $X = 1 + \frac{s}{D}$. Then $F_{1-2} = F_{2-1} = \frac{1}{\pi}\left[\sqrt{X^2 - 1} + \sin^{-1}\left(\frac{1}{X}\right) - X\right]$
7.	(concentric cylinders: inner 1 of radius r_1, outer 2 of radius r_2)	$F_{1-2} = 1,\ F_{2-1} = \frac{r_1}{r_2}$, and $F_{2-2} = 1 - F_{2-1} = 1 - \left(\frac{r_1}{r_2}\right)$

TABLE 13.2. View factors for some three-dimensional configurations

	Configuration	*Analytical equation for view factor*
1.	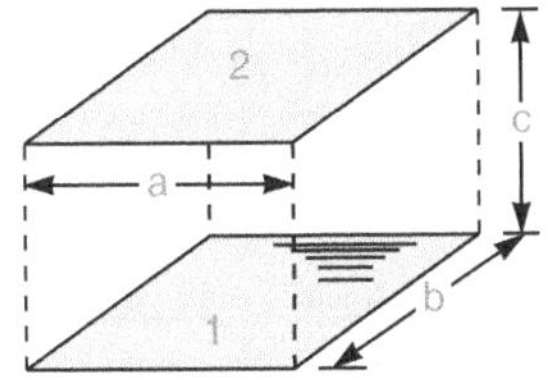 	Let X = a/c and Y = b/c. Then $F_{1-2} = \frac{2}{\pi XY}\left\{\ln\left[\frac{(1+X^2)(1+Y^2)}{1+X^2+Y^2}\right]^{1/2} - X\tan^{-1}X - Y\tan^{-1}Y + X\sqrt{1+Y^2}\,\tan^{-1}\frac{X}{\sqrt{1+Y^2}} + Y\sqrt{1+X^2}\,\tan^{-1}\frac{Y}{\sqrt{1+X^2}}\right\}$
2.	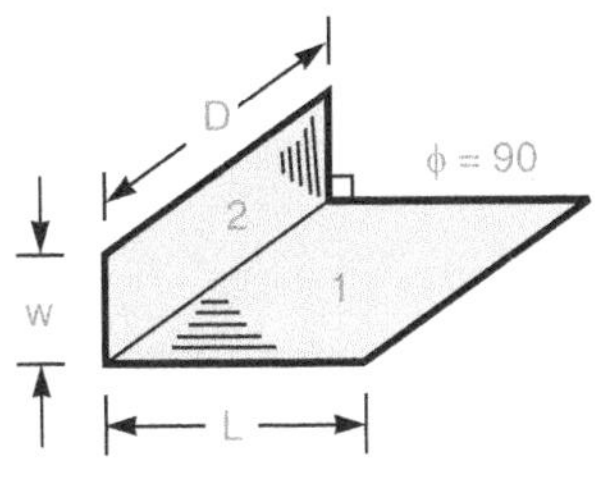 	Let $H = \frac{w}{D}$ and $w = \frac{L}{D}$. Then $F_{1-2} = \frac{1}{\pi w}\left\{w\tan^{-1}\frac{1}{w} - \sqrt{H^2+w^2}\,\tan^{-1}(H^2+w^2)^{-1/2} + H\tan^{-1}\frac{1}{H} + \frac{1}{4}\ln\left\{\left[\frac{(1+w^2)(1+H^2)}{1+w^2+H^2}\right] \times \left[\frac{w^2(1+w^2+H^2)}{(1+w^2)(w^2+H^2)}\right]^{W^2}\left[\frac{H^2(1+H^2+w^2)}{(1+H^2)(H^2+w^2)}\right]^{H^2}\right\}\right\}$
3.	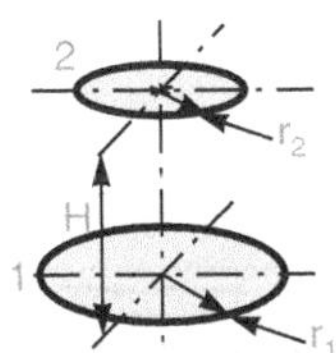 	Let $R_1 = r_1/H$, $R_2 = r_2/H$, and $X = 1 + \left(1+R_2^{\,2}\right)/R_1^{\,2}$. Then $F_{1-2} = \frac{1}{2}\left[X - \sqrt{X^2 - 4(R_2/R_1)^2}\right]$
4.	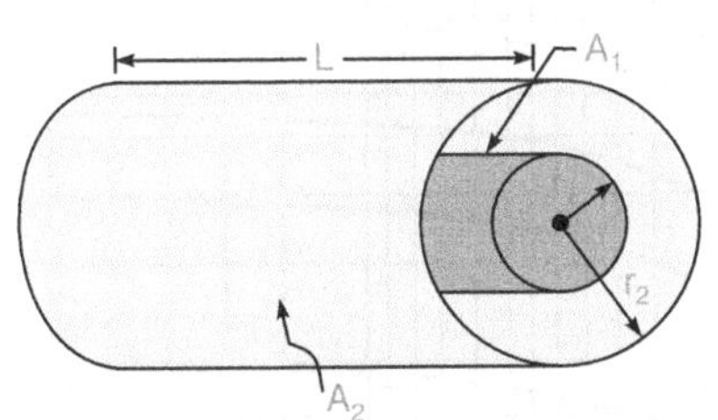 	Let $R = \frac{r_2}{r_1}$, $M = \frac{L}{r_1}$, $A = M^2 + R^2 - 1$, $B = M^2 - R^2 + 1$ and $X = \frac{4(R^2-1) + (M^2/R^2)(R^2-2)}{M^2 + 4(R^2-1)}$ $F_{1-2} = \frac{1}{R} - \frac{1}{\pi R}\left\{\cos^{-1}\left(\frac{B}{A}\right) - \frac{1}{2M}\left[\sqrt{(A+2)^2 - 4R^2}\,\cos^{-1}\left(\frac{B}{RA}\right) + B\sin^{-1}\left(\frac{1}{R}\right) - \frac{\pi A}{2}\right]\right\}$ $F_{2-1} = 1 - \frac{1}{R} + \frac{2}{\pi R}\tan^{-1}\left(\frac{2\sqrt{R^2-1}}{M}\right) - \frac{M}{2\pi R}\left[\frac{\sqrt{4R^2+M^2}}{M}\sin^{-1} - \sin^{-1}\left(\frac{R^2-2}{R^2}\right) + \frac{\pi}{2}\left(\frac{\sqrt{4R^2+M^2}}{M} - 1\right)\right]$

5.	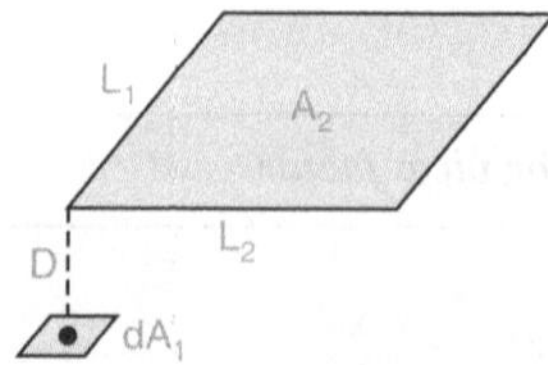 Differential surface parallel to a finite rectangular surface	$$F_{1-2} = \frac{1}{2\pi}\left(\frac{X}{\sqrt{1+X^2}}\tan^{-1}\frac{Y}{\sqrt{1+X^2}} + \frac{Y}{\sqrt{1+Y^2}}\tan^{-1}\frac{X}{\sqrt{1+Y^2}}\right)$$ where $X = L_1/D$ and $Y = L_2/D$ are in radians.
6.	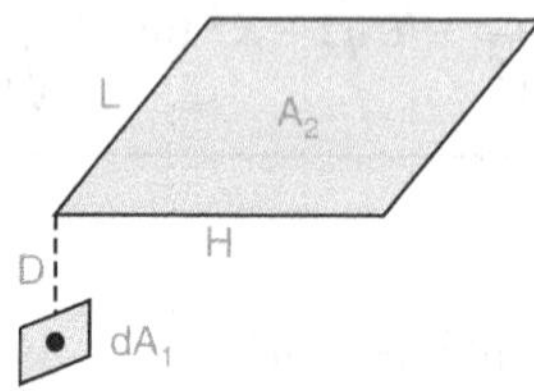 Differential surface perpendicular to a rectangular finite surface	$$F_{1-2} = \frac{1}{2}\left[\tan^{-1}\frac{1}{X} - \frac{1}{\sqrt{1-(Y/X)^2}}\tan^{-1}\frac{1}{\sqrt{X^2+Y^2}}\right]$$ where $X = D/L$ and $Y = H/L$ are in radians.
7.	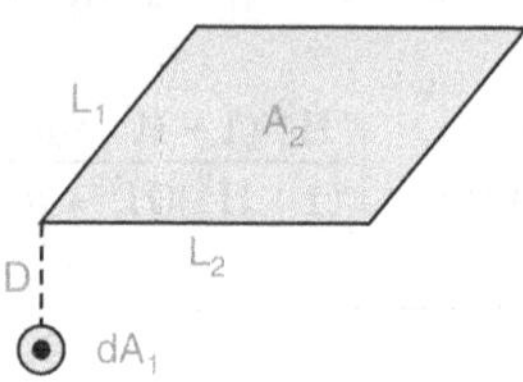 A differential spherical surface and a finite rectangular surface	$$F_{1-2} = \frac{1}{4\pi}\sin^{-1}\frac{XY}{\sqrt{1+X^2+Y^2+X^2Y^2}}$$ where $X = L_1/D$ and $Y = L_2/D$ are in radians.

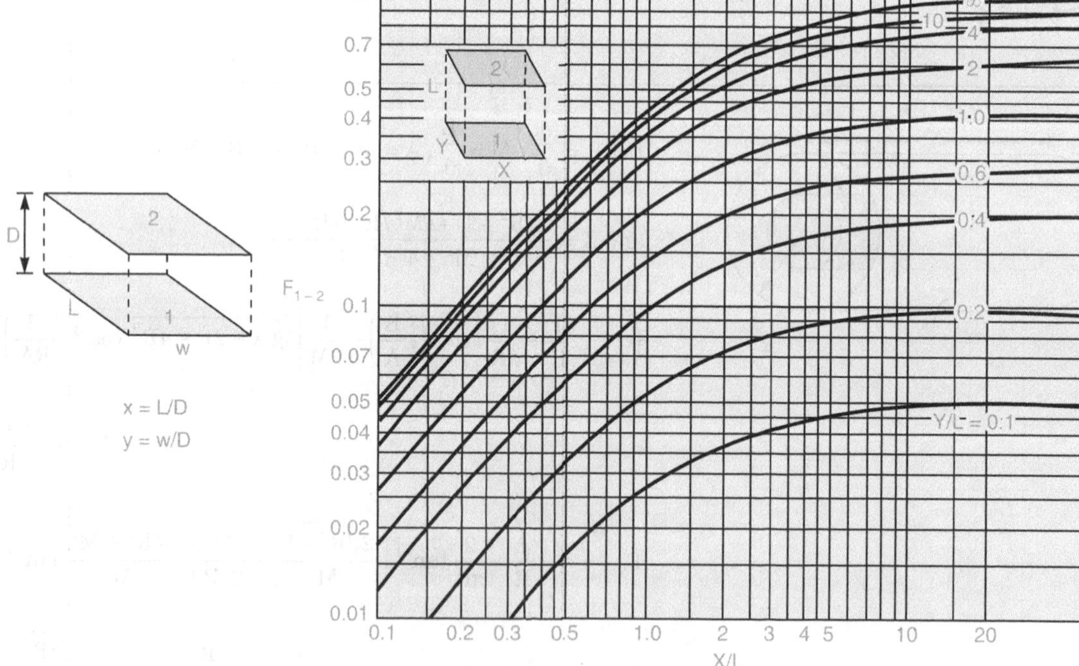

Fig. 13.8. View factors for two identical, rectangular surfaces directly facing each other

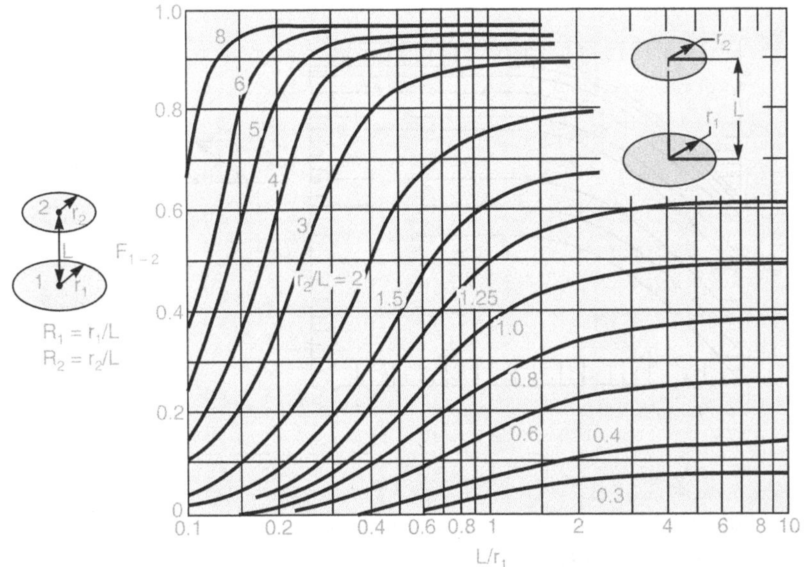

Fig. 13.9. View factor for coaxial parallel circular disks

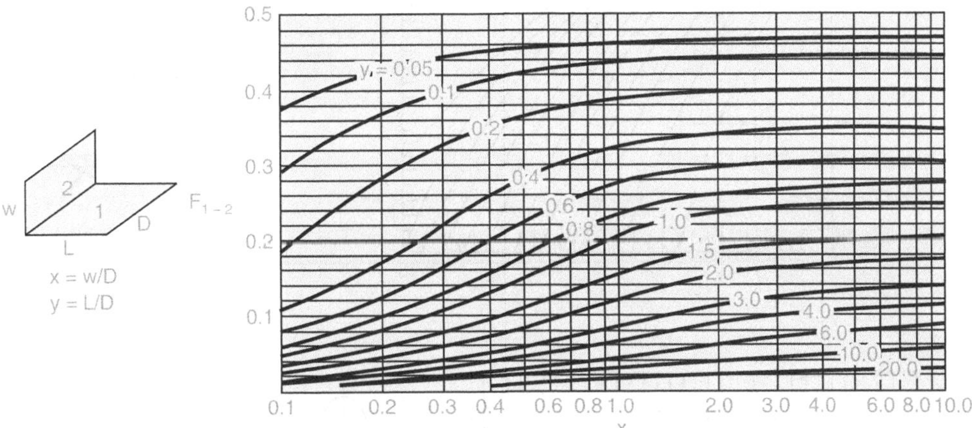

Fig. 13.10. View factors for two rectangular surfaces sharing a common side

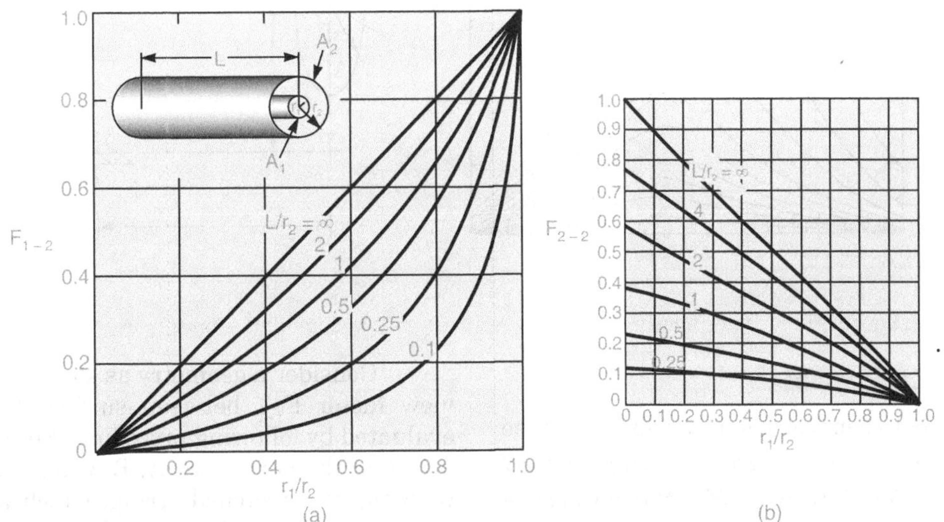

Fig. 13.11. View factors for two concentric cylinders of finite length : (*a*) outer cylinder to inner cylinder ; (*b*) outer cylinder to itself

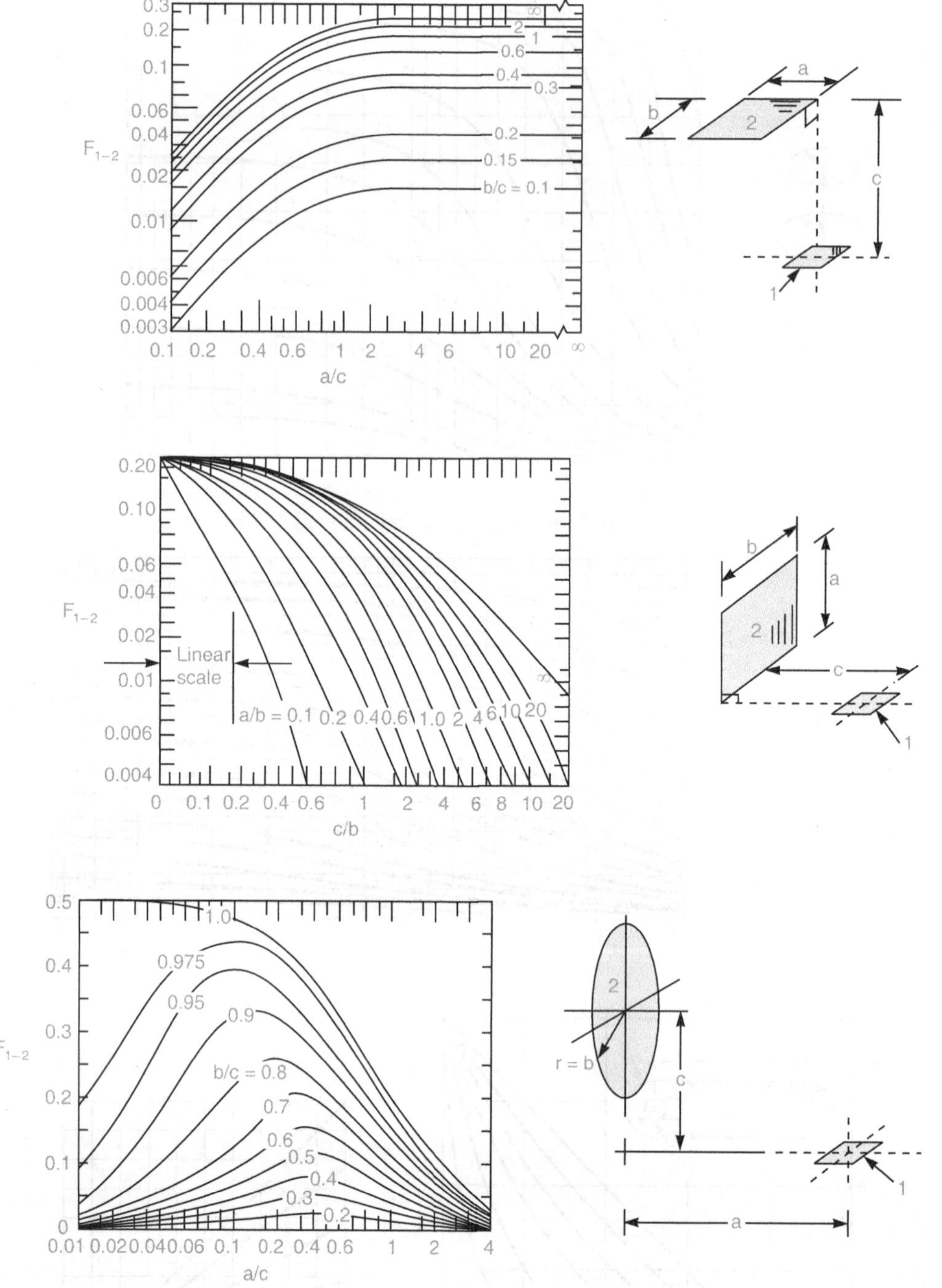

Fig. 13.12. The view factor for three very small surfaces "looking at" the large surfaces ($A_1 << A_2$)

13.1.3. The Cross String Method

The view factor between very long surfaces can be determined by a very simple *crossed string method* developed by *H.C. Hottel* in the 1950s, the surfaces of the geometry do not need to be flat ; they may be convex, concave, or any irregular shape.

Consider a geometry as shown in Fig. 13.13. The view factor F_{1-2} between surface 1 and 2 can be evaluated by following procedure. First identify the end points of the surface as A, B, C and D. Connect them with tightly stretched string, which are indicated by dashed line. Hottel has proved that the view factor F_{1-2} can be expressed in terms of lengths of these stretched strings which are straight lines as

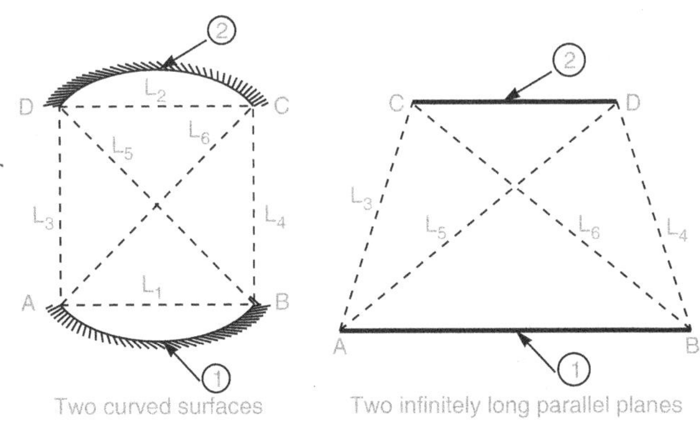

Fig. 13.13

$$F_{1-2} = \frac{(L_5 + L_6) - (L_3 + L_4)}{2L_1} \quad ...(13.11)$$

where $L_5 + L_6$ is sum of lengths of crossed strings and $L_3 + L_4$ is sum of lengths of uncrossed strings attached to the end points. Therefore, Hottel cross string method can be expressed as

$$F_{i-j} = \frac{\Sigma\,(\text{crossed strings}) - \Sigma\,(\text{uncrossed strings})}{2 \times (\text{string on surface } i)}.$$

Example 13.1. *Calculate the view factor F_{1-2} between a small area dA_1 and a parallel circular disc A_2. The elemental area dA_1 is located at the axis of the disc A_2, at a distance L.*

Solution

The elemental area of the disc $dA_2 = 2\pi r dr$

Here $\beta_1 = \beta_2$ (due to symmetry)

Using the relation

$F_{A_1 - A_2}$ or F_{1-2}

$$= \frac{1}{A_1}\int_{A_1}\int_{A_2} \frac{\cos\beta_1 \cos\beta_2 \, dA_1 \, dA_2}{\pi s^2}$$

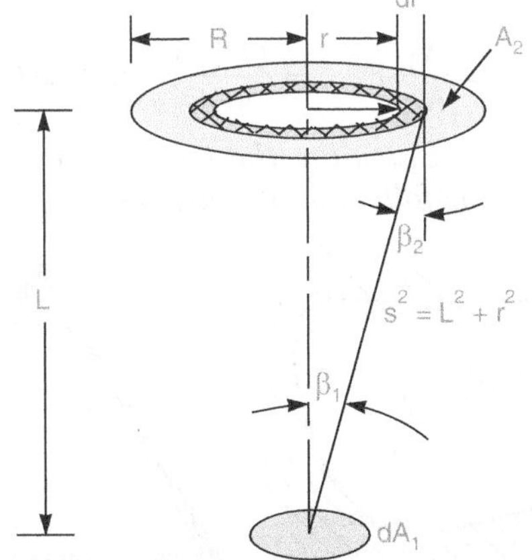

Fig. 13.14

Since the area A_1 is small, hence assuming it as constant.

$$F_{1-2} = \int_{A_2} \frac{\cos^2\beta\,(2\pi r)\,dr}{\pi s^2}$$

where $s^2 = r^2 + L^2$ and $\cos^2\beta = \dfrac{L^2}{L^2 + r^2}$

and limits for A_2 (or r) are from $r = 0$ to $r = R$

Substituting in above relation,

$$F_{1-2} = \int_0^R \left\{\frac{L^2}{L^2 + r^2}\right\} \frac{2rdr}{(L^2 + r^2)}$$

Put $r^2 + L^2 = t$; $2rdr = dt$

and limits : $t = 0 + L^2$ to $R^2 + L^2$

Then the integration :

$$F_{1-2} = L^2 \int_{L^2}^{R^2 + L^2} \frac{dt}{t^2} = 1 - \frac{L^2}{L^2 + R^2}$$

$$= \frac{R^2}{L^2 + R^2} = \textbf{sin}\,\boldsymbol{\beta}. \quad \textbf{Ans.}$$

Example 13.2. *Determine the view factor between a small area dA_1 and a rectangular surface of dimension a and b, where the rectangular surface is in horizontal plane and the small area dA_1 lies in the vertical plane and below one corner of the rectangle at a height H.*

(P.U., May 2000)

Solution

The view factor between two surfaces is given by

$$F_{1-2} = \frac{1}{A_1}\int_{A_1}\int_{A_2} \frac{\cos\beta_1 \cos\beta_2 \, dA_1 \, dA_2}{\pi s^2}$$

For element $dA_1 = A_1$;

$$F_{1-2} = \frac{1}{\pi}\int_{A_1}\int_{A_2} \frac{\cos\beta_1 \cos\beta_2 \, dA_2}{s^2}$$

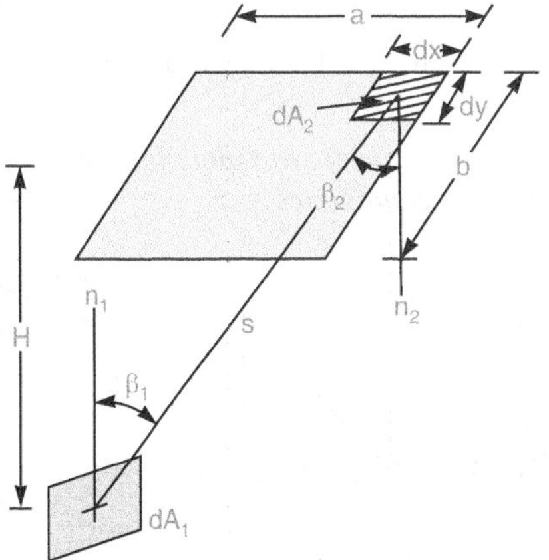

Fig. 13.15

For given geometry

$$\cos \beta_1 = \frac{H}{s}, \cos \beta_2 = \frac{b}{s}$$

and $s = \sqrt{H^2 + a^2}$, $dA_2 = dx\,dy$

Then,

$$F_{1-2} = \frac{1}{\pi} \int_0^a \int_0^b \frac{Hb}{(H^2 + a^2)^2} dx\,dy$$

$$= \frac{Hab}{\pi(H^2 + a^2)^2} \int_0^b dy = \frac{\mathbf{Hab^2}}{\pi\,(\mathbf{H^2 + a^2})^2}. \quad \text{Ans.}$$

Example 13.3. *A flat surface, 1 is completely enclosed by a second surface, 2 Fig. 13.16. Determine the view factors F_{1-2}, F_{2-1} and F_{2-2}.*

Solution

As no part of surface 1 "sees" itself, all the energy leaving surface 1 reaches surface 2, and from the definition of view factor,

$$F_{1-2} = 1$$

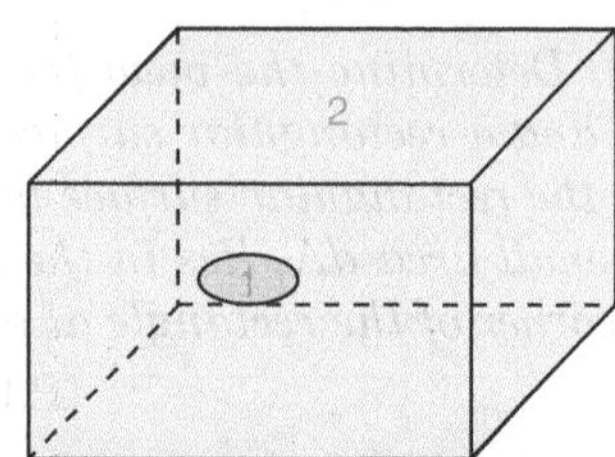

Fig. 13.16. A flat surface, 1, completely enclosed by a second surface, 2.

To determine F_{2-1}, from eqn. (13.5)

$$A_2 F_{2-1} = A_1 F_{1-2}$$

$$F_{2-1} = \frac{A_1 F_{1-2}}{A_2} = \frac{A_1}{A_2}$$

From eqn. (13.6)

$$F_{2-1} + F_{2-2} = 1$$

and $F_{2-2} = 1 - F_{2-1} = \mathbf{1 - \frac{A_1}{A_2}}$. Ans.

Example 13.4. *Calculate the view factor F_{1-2} and F_{2-1} for the following geometries :*

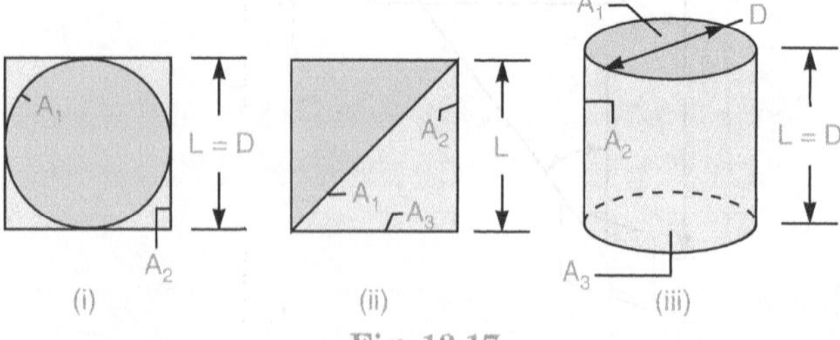

Fig. 13.17

(i) Sphere of diameter D inside a cubical box of length D.

(ii) Diagonal partition within a long square duct.

(iii) End and side of a circular tube of equal length and diameter.

Solution

Given : Surface geometries.

To find : The view factors F_{1-1} and F_{2-1}.

Analysis : (*i*) Sphere within a cube :

By inspection

$$F_{1-2} = 1$$

By reciprocity rule,

$$F_{2-1} = \frac{A_1}{A_2} F_{1-2} = \frac{\pi D^2}{6L^2} \times 1 = \frac{\pi}{\mathbf{6}}. \quad \text{Ans.}$$

(*ii*) Partition within a square duct :

From summation rule,

$$F_{1-1} + F_{1-2} + F_{1-3} = 1$$

Where for flat surface,

$$F_{1-1} = 0$$

By symmetry,

$$F_{1-2} = F_{1-3}$$

Hence,

$$F_{1-2} = 0.5$$

By reciprocal theorem,

$$F_{2-1} = \frac{A_1}{A_2} F_{1-2} = \frac{L\sqrt{2}}{L} \times 0.5 = \mathbf{0.71}. \quad \text{Ans.}$$

(*iii*) Circular tube :

From Fig. 13.9 with $r_3/L = 0.5$

and $L/r_1 = 2$, $F_{1-3} = 0.17$

From summation rule,

$$F_{1-1} + F_{1-2} + F_{1-3} = 1$$

Where for flat surface,

$$F_{1-1} = 0$$

$$F_{1-2} = 1 - F_{1-3} = 1 - 0.17 = 0.83$$

By reciprocity rule,

$$F_{2-1} = \frac{A_1}{A_2} F_{1-2} = \frac{\pi D^2/4}{\pi DL} \times 0.83 = \mathbf{0.21}. \quad \text{Ans.}$$

Example 13.5. *Calculate the view factor F_{1-1}, F_{1-2} and F_{2-1} for the following geometries :*

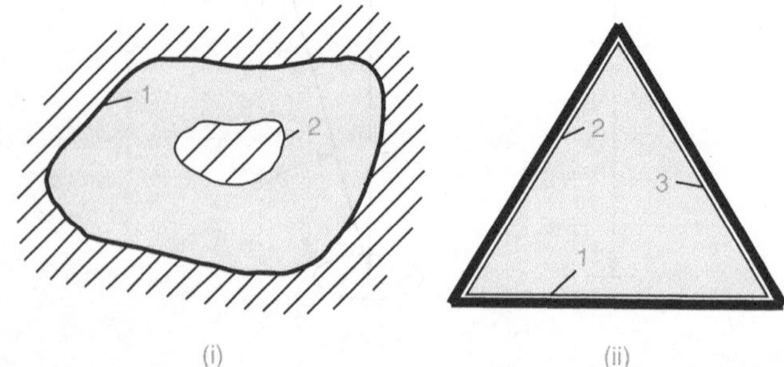

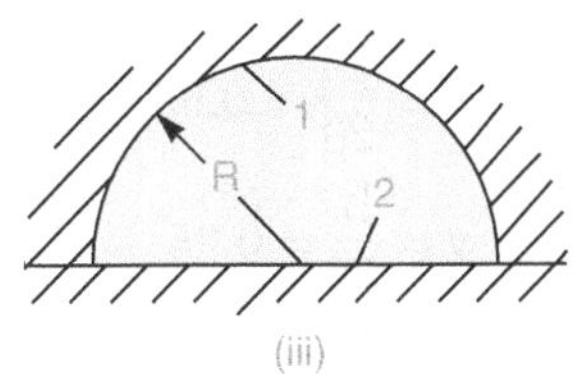

Fig. 13.18

(i) A black body inside a black enclosure.

(ii) A tube whose section is equilateral triangle.

(iii) Radiation exchange between a hemisphere and a plane surface.

Solution

Given : Surface geometries.

To find : The view factor.

Analysis : (i) A black body inside a black enclosure :

By inspection, $F_{2-1} = 1$

By summation rule, $F_{1-1} + F_{1-2} = 1$

By reciprocity rule, $A_2F_{2-1} = A_1F_{1-2}$

$$F_{1-2} = \frac{A_2}{A_1}$$

or $$F_{1-1} = \mathbf{1 - \frac{A_2}{A_1}}. \text{ Ans.}$$

(ii) A tube of equilateral triangle :

From summation rule, $F_{1-1} + F_{1-2} + F_{1-3} = 1$

Where for flat surface, $F_{1-1} = 0, \quad F_{2-2} = 0,$

$F_{3-3} = 0$

and $F_{2-1} + F_{2-3} = 1$

or $F_{2-3} = 1 - F_{2-1}$

By symmetry, $F_{1-2} = F_{1-3}$

Hence, $F_{1-2} = 0.5$

or $$F_{2-1} = \frac{A_1}{A_2} F_{1-2} = 0.5$$

Hence, $F_{2-3} = 1 - 0.5$

$= \mathbf{0.5.}$ Ans.

(iii) The radiation heat exchange between a hemisphere surface 1 and a flat surface 2.

From summation rule, $F_{1-1} + F_{1-2} = 1$

and $F_{2-1} + F_{2-2} = 1$

But for flat surface, $F_{2-2} = 0,$

Hence, $F_{2-1} = 1$

By reciprocity rule, $A_2F_{2-1} = A_1F_{1-2}$

To flat surface hemisphere,

$$F_{1-2} = \frac{A_2F_{2-1}}{A_1} = \frac{\pi R^2}{2\pi R^2} = 0.5$$

$$F_{1-1} = 1 - F_{1-2} = 1 - 0.5 = \mathbf{0.5.} \text{ Ans.}$$

Example 13.6. *Calculate the shape factor of a conical hole.*

Solution

Analysis : Considering an imaginary flat plane covers the hole, hence by summation rule

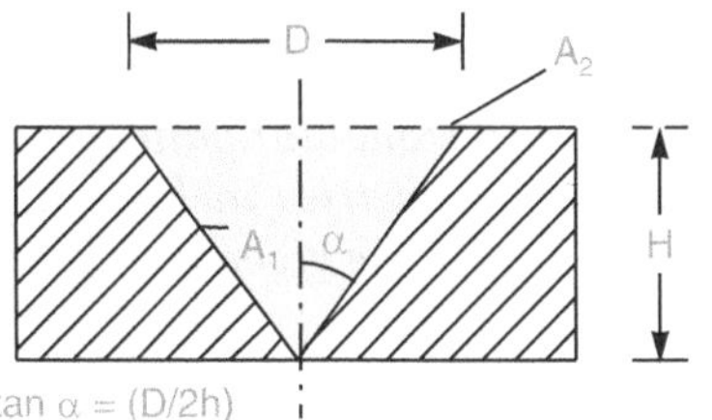

Fig. 13.19

$F_{1-1} + F_{1-2} = 1$ and $F_{2-1} + F_{2-2} = 1$

But $F_{2-2} = 0$, hence $F_{2-1} = 1.$

By reciprocity relation,

$A_1F_{1-2} = A_2F_{2-1}$ or $F_{1-1} = 1 - F_{1-2}$

$$\text{Therefore, } F_{1-1} = 1 - \frac{A_2}{A_1} F_{2-1} = 1 - \frac{A_2}{A_1}$$

$$= 1 - \frac{(\pi/4)D^2}{\pi D L/2} = 1 - \frac{D}{2L}$$

(where L = H/cos α)

or $$F_{1-1} = 1 - \frac{\pi H^2 \tan^2 \alpha}{\dfrac{\pi H^2 \sin \alpha}{\cos^2 \alpha}}$$

$= \mathbf{1 - \sin \alpha.}$ Ans.

Example 13.7. *Calculate the shape factor for cylindrical cavity as shown in Fig. 13.20 with respect to itself.*

Solution

Assuming cylindrical surface 1 and plane surface 2

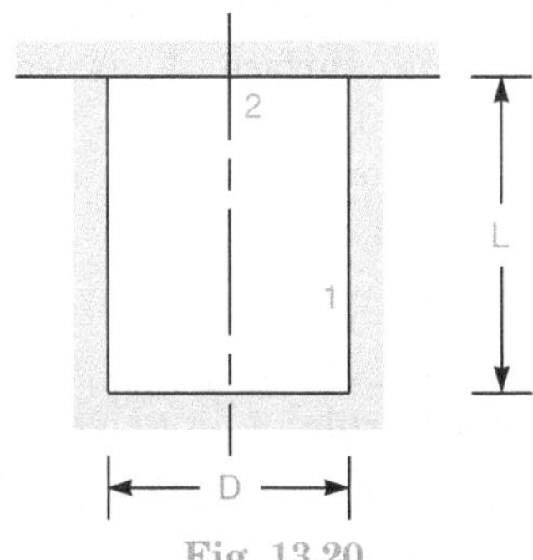

Fig. 13.20

From summation rule, $F_{1-1} + F_{1-2} = 1$

or $F_{1-1} = 1 - F_{1-2}$

and $F_{2-1} + F_{2-2} = 1$

But for flat surface, $F_{2-2} = 0$

From surface 2 to surface 1, $F_{2-1} = 1$

By reciprocity rule, $A_2F_{2-1} = A_1F_{1-2}$

Flat surface of hemisphere,

$$F_{1-2} = \frac{A_2}{A_1} F_{2-1} = \frac{A_2}{A_1}$$

$$F_{1-1} = 1 - F_{1-2} = 1 - \frac{A_2}{A_1} = 1 - \frac{(\pi/4)D^2}{(\pi/4)D^2 + \pi DL}$$

$$F_{1-1} = 1 - \frac{D}{D+4L} = \mathbf{\frac{4L}{D+4L}}. \quad \text{Ans.}$$

Example 13.8. *Two rectangular plates are at right angle in following shown geometries in Fig. 13.21. Calculate the view factor F_{1-2} for the given configurations.*

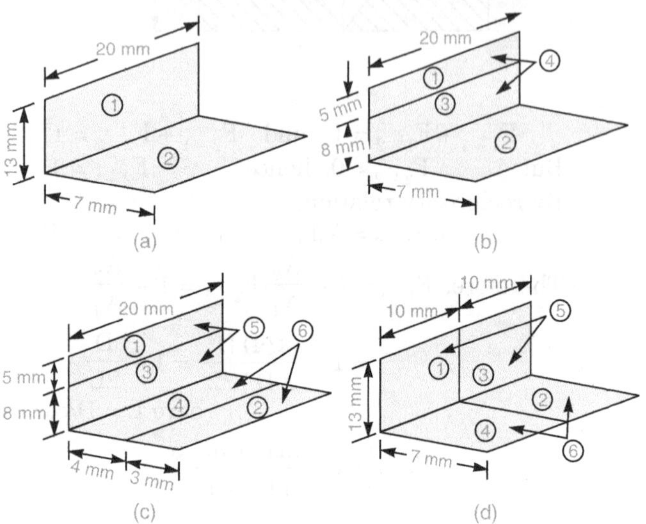

Fig. 13.21

Solution

(*a*)
$$\left.\begin{array}{r} L = 7 \text{ mm} \\ w = 13 \text{ mm} \\ D = 20 \text{ mm} \end{array}\right\} \quad \begin{array}{l} x = \frac{w}{D} = \frac{13}{20} = 0.65 \\ y = \frac{L}{D} = \frac{7}{20} = 0.35 \end{array}$$

From Fig. 13.10, view factor, F_{1-2} = **0.34.** Ans.

(*b*) Assuming surface 4 as combination of surface 1 and surface 3 ;

$$A_4 = A_1 + A_3$$

By additive rule

$$F_{2-4} = F_{2-1} + F_{2-3}$$

or
$$F_{2-1} = F_{2-4} - F_{2-3}$$

By reciprocity rule

$$A_1F_{1-2} = A_2 F_{2-1} = A_2 (F_{2-4} - F_{2-3})$$

or
$$F_{1-2} = \frac{A_2}{A_1}(F_{2-4} - F_{2-3})$$

For surface 1 and surface 2 :

$$A_1 = 5 \text{ mm} \times 20 \text{ mm} = 100 \text{ mm}^2$$
$$A_2 = 7 \text{ mm} \times 20 \text{ mm} = 140 \text{ mm}^2$$

For F_{2-3}
$$\left\{\begin{array}{l} w = 8 \text{ mm}, \quad L = 7 \text{ mm}, \quad D = 20 \text{ mm} \\ x = \frac{w}{D} = \frac{8}{20} = 0.4, \quad y = \frac{L}{D} = \frac{7}{20} = 0.35 \end{array}\right.$$

From Fig. 13.10 ; $F_{2-3} = 0.3$

For $F_{2-4} : x = \frac{13}{20} = 0.65,$

$$y = \frac{L}{D} = 0.35 \rightarrow F_{2-4} = 0.35$$

Then $F_{1-2} = \frac{140}{100} \times (0.35 - 0.3) = \mathbf{0.07.}$ Ans.

(*c*) Assuming surface 5 and surface 6 as

$$A_5 = A_1 + A_3 \quad \text{and} \quad A_6 = A_2 + A_4$$

By additive rule

$$F_{1-6} = F_{1-2} + F_{1-4}$$
$$F_{1-2} = F_{1-6} - F_{1-4}$$

By reciprocity rule

$$A_1F_{1-6} = A_6 F_{6-1} \longrightarrow F_{1-6} = \frac{A_6}{A_1} F_{6-1}$$

$$A_1 F_{1-4} = A_4 F_{4-1} \longrightarrow F_{1-4} = \frac{A_4}{A_1} F_{4-1}$$

From additive (difference) rule

$$F_{6-5} = F_{6-3} + F_{6-1} \longrightarrow F_{6-1} = F_{6-5} - F_{6-3}$$
$$F_{4-5} = F_{4-1} + F_{4-3} \longrightarrow F_{4-1} = F_{4-5} - F_{4-3}$$

Therefore,

$$F_{1-6} = \frac{A_6}{A_1}(F_{6-5} - F_{6-3})$$
$$= \frac{7 \times 20}{5 \times 20}(F_{6-5} - F_{6-3})$$
$$= 1.4\,(F_{6-5} - F_{6-3})$$

$$F_{1-4} = \frac{A_4}{A_1}(F_{4-5} - F_{4-3})$$
$$= \frac{4 \times 20}{5 \times 20}(F_{4-5} - F_{4-3})$$
$$= 0.8\,(F_{4-5} - F_{4-3})$$

From Fig. 13.10 ;

For $F_{6-5} ; x = \frac{13}{20} = 0.65 ;$

$$y = \frac{7}{20} = 0.35 \longrightarrow F_{6-5} = 0.35$$

For $F_{6-3} ; x = \frac{8}{20} = 0.4 ;$

$$y = \frac{7}{20} = 0.35 \longrightarrow F_{6-3} = 0.3$$

For $F_{4-5} ; x = \frac{13}{20} = 0.65 ;$

$$y = \frac{4}{20} = 0.2 \longrightarrow F_{4-5} = 0.367$$

For $F_{4-3} ; x = \frac{8}{20} = 0.4 ;$

$$y = \frac{4}{20} = 0.2 \longrightarrow F_{4-3} = 0.358$$

Then $F_{1-6} = 1.4(0.35 - 0.3) = 0.07$

$$F_{1-4} = 0.8(0.367 - 0.358) = 0.0072$$

and $$F_{1-2} = 0.07 - 0.0072 = \mathbf{0.063.} \quad \text{Ans.}$$

(*d*) Assuming surface 5 and 6 as

$$A_5 = A_1 + A_3$$
$$A_6 = A_2 + A_4$$

By additive rule

$$F_{1-6} = F_{1-2} + F_{1-4} \longrightarrow F_{1-2} = F_{1-6} - F_{1-4}$$

By reciprocal rule

$$A_1F_{1-6} = A_6F_{6-1} \longrightarrow F_{1-6} = \frac{A_6}{A_1} F_{6-1}$$

$$\Rightarrow F_{1-2} = \frac{A_6}{A_1} F_{6-1} - F_{1-4}$$

Again by additive rule

$$F_{6-5} = F_{6-1} + F_{6-3} \longrightarrow F_{6-1} = F_{6-5} - F_{6-3}$$

$$\Rightarrow F_{1-2} = \frac{A_6}{A_1} (F_{6-5} - F_{6-3}) - F_{1-4}$$

Reciprocity rule

$$A_3 F_{3-6} = A_6 F_{6-3} \longrightarrow F_{6-3} = \frac{A_3}{A_6} F_{3-6}$$

$$\Rightarrow F_{1-2} = \frac{A_6}{A_1}\left[F_{6-5} - \frac{A_3}{A_6} F_{3-6}\right] - F_{1-4}$$

Additive rule

$$F_{3-6} = F_{3-2} + F_{3-4} \longrightarrow F_{1-2}$$

$$= \frac{A_6}{A_1}\left[F_{6-5} - \frac{A_3}{A_6}(F_{3-2} + F_{3-4})\right] - F_{1-4}$$

By symmetry rule

$$F_{3-4} = F_{2-1} = \frac{A_1}{A_2} . F_{1-2} \quad \text{and} \quad F_{3-2} = F_{1-4}$$

Therefore,

$$F_{1-2} = \frac{A_6}{A_1}\left[F_{6-5} - \frac{A_3}{A_6}\left(F_{3-2} + \frac{A_1}{A_2}.F_{1-2}\right)\right] - F_{1-4}$$

or $$F_{1-2}\left(1 - \frac{A_3}{A_2}\right) = \frac{A_6}{A_1} F_{6-5} - \frac{A_3}{A_1} F_{1-4} - F_{1-4}$$

$$= \frac{A_6}{A_1} F_{6-5} - F_{1-4}\left(1 + \frac{A_3}{A_1}\right)$$

From Fig. 13.10, the view factor F_{6-5}, F_{3-2} and F_{1-4} can be obtained

For F_{6-5}, $x = \frac{13}{20} = 0.65$,

$$y = \frac{7}{20} = 0.35 \longrightarrow F_{6-5} = 0.35$$

For F_{1-4}, $x = \frac{13}{10} = 1.3$,

$$y = \frac{7}{10} = 0.7 \longrightarrow F_{3-2} = 0.24$$

Now,

$$F_{1-2}\left(1 - \frac{13 \times 10}{7 \times 10}\right) = \left(\frac{7 \times 20}{13 \times 10}\right) \times 0.35 - 0.24\left(1 + \frac{13 \times 10}{13 \times 10}\right)$$

or $$F_{1-2} = \frac{-0.103}{-0.857} = \mathbf{0.120.} \quad \text{Ans.}$$

Example 13.9. *Fig. 13.22 depicts window in the wall of a room. Find the view factor F_{B-A}, where A is the window and B is the floor.*

Solution

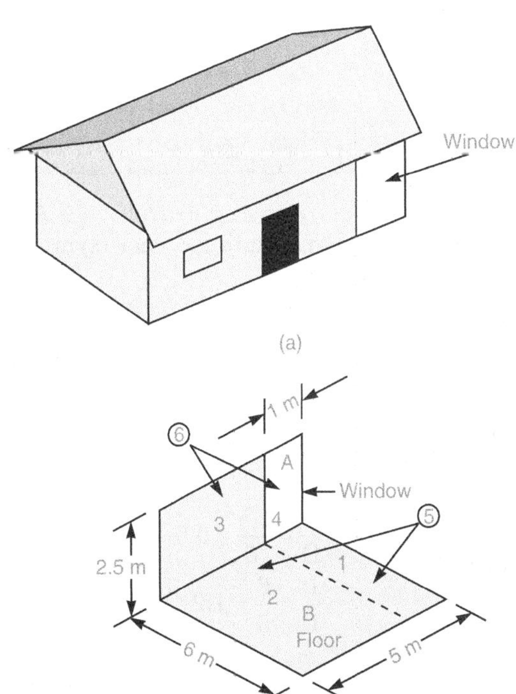

Fig. 13.22. Schematic for example 13.9

Given : The geometry as shown above.

$L = 6$ m, $w = 2.5$ m

$D = 5$ m, $D_1 = 4$ m

$D_2 = 1$ m.

To find : View factor $F_{B-A} = F_{5-4}$

Assumption : Surface 5 as $A_5 = A_1 + A_2$

Surface 6 as $A_6 = A_3 + A_4$.

Analysis : For convenience, divide the floor into two parts, 1 and 2 as shown in Fig. 13.22 (*b*). The window is represented by 4 and the remaining part of the wall by 3 and we need F_{5-4} ; using reciprocity and additive rule simultaneously.

$$A_5 F_{5-4} = A_4 F_{4-5} = A_4 [F_{4-1} + F_{4-2}]$$

or
$$F_{5-4} = \frac{A_4}{A_5}\left[\frac{A_1}{A_4} F_{1-4} + \frac{A_2}{A_4} F_{2-4}\right]$$

$$= \frac{A_1}{A_5} F_{1-4} + \frac{A_2}{A_5} F_{2-4} \quad ...(i)$$

We can obtain F_{1-4} from Fig. 13.10, now we need F_{2-4};

$$F_{5-6} = F_{5-3} + F_{5-4}$$

$$= \frac{A_3}{A_5} F_{3-5} + \frac{A_4}{A_5} F_{4-5}$$

$$= \frac{A_3}{A_5} [F_{3-1} + F_{3-2}] + \frac{A_4}{A_5} [F_{4-1} + F_{4-2}]$$

or
$$F_{5-6} = \frac{1}{A_5} [A_1 F_{1-3} + A_2 F_{2-3} + A_1 F_{1-4} + A_2 F_{2-4}]$$

From law of corresponding corner (symmetry)

$$A_2 F_{2-4} = A_1 F_{1-3}$$

$$\therefore\ A_1 F_{1-3} = A_2 F_{2-4}$$

$$= \frac{A_5 F_{5-6} - A_2 F_{2-3} - A_1 F_{1-4}}{2} \quad ...(ii)$$

From Fig. 13.10.

For F_{5-6} : $\left.\begin{aligned} x &= \frac{w}{D} = \frac{2.5}{5} = 0.5 \\ y &= \frac{L}{D} = \frac{6}{5} = 1.2 \end{aligned}\right\}$ $\longrightarrow F_{5-6} = 0.14$

For F_{2-3} : $\left.\begin{aligned} x &= \frac{w}{D_1} = \frac{2.5}{4} = 0.625 \\ y &= \frac{L}{D_1} = \frac{6}{4} = 1.5 \end{aligned}\right\}$ $\longrightarrow F_{2-3} = 0.12$

For F_{1-4} : $\left.\begin{aligned} x &= \frac{w}{D_2} = \frac{2.5}{1} = 2.5 \\ y &= \frac{L}{D_2} = \frac{6}{1} = 6 \end{aligned}\right\}$ $\longrightarrow F_{1-4} = 0.06$

Now from eqn. (*ii*)

$$A_2 F_{2-4} = \frac{(6 \times 5) \times 0.14 - (6 \times 4) \times 0.12 - (6 \times 1) \times 0.06}{2} = 0.48$$

Now from eqn. (*i*)

$$F_{B-A} = F_{5-4} = \left(\frac{6 \times 1}{6 \times 5}\right) \times 0.06 + \frac{0.48}{6 \times 5}$$

$= \mathbf{0.028.}$ **Ans.**

Example 13.10. *Calculate all view factors for conical geometry shown in Fig. 13.23.*

Solution

Given : A conical geometry

$r_1 = 30$ mm, $r_2 = 15$ mm, H = 30 mm.

To find : The view factors

$F_{1-1}, F_{1-2}, F_{1-3}, F_{2-1}, F_{2-2}, F_{2-3}, F_{3-1}, F_{3-2}, F_{3-3}$.

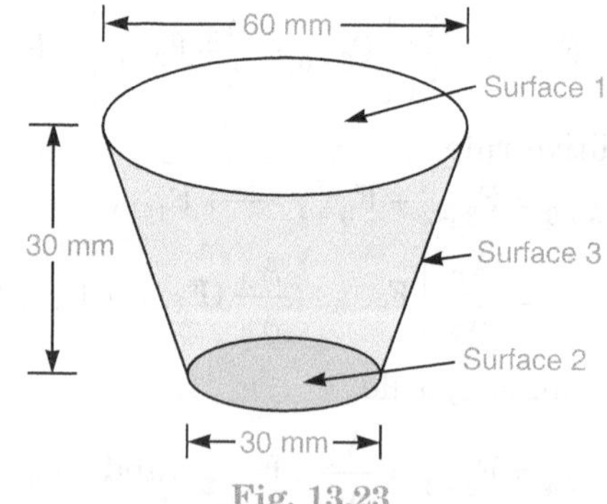

Fig. 13.23

Analysis : For given geometry

$$A_1 = \pi r_1^2 = \pi \times (30 \text{ mm})^2 = 900\ \pi \text{ mm}^2$$

$$A_2 = \pi r_2^2 = \pi \times (15 \text{ mm})^2 = 225\ \pi \text{ mm}^2$$

$$A_3 = \pi (r_1 + r_2) \sqrt{(r_1 - r_2)^2 + H^2}$$

$$= \pi (30 + 15) \times \sqrt{(30 - 15)^2 + 30^2}$$

$$= 1509.34\ \pi \text{ mm}^2.$$

For surface 1. (Top surface), For two parallel discs from Table 13.2

$$R_1 = \frac{r_1}{H} = \frac{30}{30} = 1, \quad R_2 = \frac{r_2}{H} = \frac{15}{30} = 0.5$$

$$X = 1 + (1 + R_2^2)/R_1^2 = 2.25$$

$$F_{1-2} = \frac{1}{2}\left[X - \sqrt{X^2 - 4(R_2/R_1)^2}\right]$$

$$= \frac{1}{2}\left[2.25 - \sqrt{2.25^2 - 4\left(\frac{0.5}{1}\right)^2}\right]$$

$= \mathbf{0.117.}$ **Ans.**

Alternatively from Fig. 13.9

$R_1 = 1, \quad R_2 = 0.5 \longrightarrow F_{1-2} = 0.117$

Now, by summation rule

$$F_{1-1} + F_{1-2} + F_{1-3} = 1.0$$

Surface 1 is flat, thus $F_{1-1} = 0$

$\therefore \quad F_{1-3} = 1 - F_{1-2} = \mathbf{0.883.}$ **Ans.**

For surface 2: From reciprocity rule

$$A_1F_{1-2} = A_2F_{2-1}$$

or
$$F_{2-1} = \frac{A_1}{A_2}F_{1-2} = \frac{900\pi}{225\pi} \times 0.117$$

$= \mathbf{0.468.}$ **Ans.**

By summation rule

$$F_{2-1} + F_{2-2} + F_{2-3} = 1.0$$

Surface 2 is flat ; $F_{2-2} = 0$

$\therefore \quad F_{2-3} = 1 - F_{2-1}$

$= 1 - 0.468 = \mathbf{0.532.}$ **Ans.**

For surface 3. By reciprocity rule

$$A_1F_{1-3} = A_3F_{3-1}$$

or
$$F_{3-1} = \frac{A_1}{A_3}F_{1-3} = \frac{900\pi}{1509.34\pi} \times 0.883$$

$= \mathbf{0.526.}$ **Ans.**

and
$$A_2F_{2-3} = A_3F_{3-2}$$

$$F_{3-2} = \frac{A_2}{A_3}F_{2-3} = \frac{225\pi}{1509.34\pi} \times 0.532$$

$= \mathbf{0.0793.}$ **Ans.**

By summation rule

$$F_{3-3} = 1 - F_{3-1} - F_{3-2}$$

$= 1 - 0.526 - 0.0793 = \mathbf{0.388.}$ **Ans.**

Example 13.11. *Find the shape factor between two areas 1 and 2 which are in the form of circular ring, coaxial and are in two parallel planes at a distance 10 cm. Area 1 has inner radius of 5 cm and outer radius of 10 cm. Area 2 has inner radius of 8 cm and outer radius of 20 cm. Use following formula for calculating the shape factor between two circular areas located coaxially in two parallel planes is given by*

$$F_{1-2} = \frac{1}{2B^2}[X - \sqrt{X^2 - 4B^2C^2}]$$

where, $B = \frac{R_1}{H}$, $C = \frac{R_2}{H}$ *and* $X = (1 + B^2 + C^2)$

where, R_1 *and* R_2 *are the radii of the circular planes and H is the distance between them.* (P.U., Nov. 1992)

Solution

Given : Two circular parallel surfaces located coaxially ;

$R_3 = 5$ cm, $\quad R_4 = 10$ cm, $\quad H = 10$ cm

$R_5 = 8$ cm, $\quad R_6 = 20$ cm.

To find : The shape factor between the two parallel surfaces.

Assumptions : $A_4 = A_1 + A_3$

$A_6 = A_5 + A_2$

Analysis : Using the property relations for shape factor

$$F_{3-6} = F_{3-5} + F_{3-2}$$

or
$$F_{3-2} = F_{3-6} - F_{3-5}$$

$$F_{2-4} = F_{2-3} + F_{2-1}$$

or
$$F_{2-1} = F_{2-4} - F_{2-3}$$

$$F_{4-6} = F_{4-5} + F_{4-2}$$

Applying the reciprocal theorem

$$A_1F_{1-2} = A_2F_{2-1}$$
$$= A_2(F_{2-4} - F_{2-3})$$

But $\quad A_1 = A_4 - A_3$

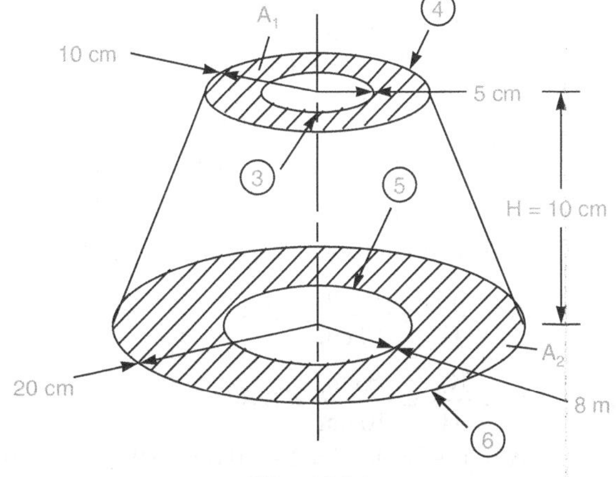

Fig. 13.24

Hence, the radiation from surface 1 *i.e.,* surface 3 and 4 will deliver energy to surface 4 and 5 ;

$$A_1F_{1-2} = A_4F_{4-2} - A_3F_{3-2}$$

or
$$A_1F_{1-2} = A_4(F_{4-6} - F_{4-5}) - A_3(F_{3-6} - F_{3-5}) \quad \text{...(A)}$$

where, the surface area,

$$A_3 = \pi(R_3^2) = \pi(0.05^2) = 0.00785 \text{ m}^2$$

the surface area, $A_4 = \pi(R_4^2) = \pi(0.1^2) = 0.0314 \text{ m}^2$
the surface area of ring 1,

$$A_1 = (A_4 - A_3) = 0.0236 \text{ m}^2$$

Using given relation for calculation of F_{4-6} :

$$B = \frac{R_4}{H} = \frac{10 \text{ cm}}{10 \text{ cm}} = 1$$

and $$C = \frac{R_6}{H} = \frac{20 \text{ cm}}{10 \text{ cm}} = 2$$

$$X = 1 + B^2 + C^2 = 1 + (1)^2 + (2)^2 = 6$$

$$F_{4-6} = \frac{1}{2(1)^2}\left[6 - \sqrt{(6)^2 - 4 \times (1)^2 \times (2)^2}\right]$$
$$= 0.764$$

Using given relation for calculation of F_{4-5} :

$$B = \frac{R_4}{H} = \frac{10 \text{ cm}}{10 \text{ cm}} = 1$$

and $$C = \frac{R_5}{H} = \frac{8 \text{ cm}}{10 \text{ cm}} = 0.8$$

$$X = 1 + B^2 + C^2 = 1 + (1)^2 + (0.8)^2 = 2.64$$

$$F_{4-5} = \frac{1}{2(1)^2}\left[2.64 - \sqrt{(2.64)^2 - 4 \times (1)^2 \times (0.8)^2}\right] = 0.27$$

Using given relation for calculation of F_{3-6} :

$$B = \frac{R_3}{H} = \frac{5 \text{ cm}}{10 \text{ cm}} = 0.5$$

and $$C = \frac{R_6}{H} = \frac{20 \text{ cm}}{10 \text{ cm}} = 2$$

$$X = 1 + B^2 + C^2 = 1 + (0.5)^2 + (2)^2 = 5.25$$

$$F_{3-6} = \frac{1}{2(0.5)^2}\left[5.25 - \sqrt{(5.25)^2 - 4 \times (0.5)^2 \times (2)^2}\right]$$
$$= 0.7917$$

Using given relation for calculation of F_{3-5}:

$$B = \frac{R_3}{H} = \frac{5 \text{ cm}}{10 \text{ cm}} = 0.5$$

and $$C = \frac{R_5}{H} = \frac{8 \text{ cm}}{10 \text{ cm}} = 0.8$$

$$X = 1 + B^2 + C^2 = 1 + (0.5)^2 + (0.8)^2 = 1.89$$

$$F_{3-5} = \frac{1}{2(0.5)^2}\left[1.89 - \sqrt{(1.89)^2 - 4 \times (0.5)^2 \times (0.8)^2}\right]$$
$$= 0.3553$$

Substituting the values in eqn. (A) ;

$$0.0236 \times F_{1-2} = 0.0314 \times (0.764 - 0.27) - 0.00785 \times (0.7917 - 0.3553)$$

or $F_{1-2} = \mathbf{0.5121.}$ **Ans.**

Example 13.12. *Two concentric cylinders have inner and outer radii 5 cm and 10 cm, respectively and length 20 cm. Calculate the all possible view factors.*

Solution

Given : Two concentric cylinders as shown below :

To find : All possible view factors.

Analysis : For parallel cylinders from Fig. 13.25.

For surface 2. $\frac{L}{r_2} = \frac{20}{10} = 2$

and $$\frac{r_1}{r_2} = \frac{5}{10} = 0.5,$$

$$F_{2-1} = 0.43$$

and $$F_{2-2} = 0.34$$

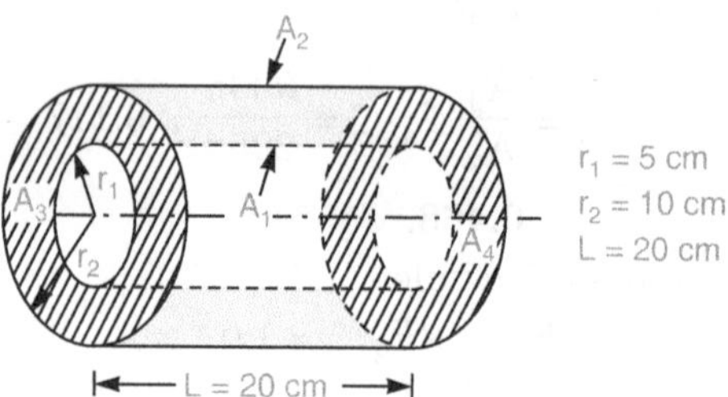

Fig. 13.25

For surface 1. By reciprocity

$$A_1F_{1-2} = A_2 F_{2-1}$$

$$F_{1-2} = \frac{A_2}{A_1} F_{2-1} = \frac{2\pi r_2 L}{2\pi r_1 L} . F_{2-1}$$

$$= \left(\frac{10}{5}\right) \times 0.43 = 0.86.$$

By summation rule

$$F_{2-1} + F_{2-2} + F_{2-3} + F_{2-4} = 1$$

By symmetry $F_{2-3} = F_{2-4}$

Then, $$F_{2-3} = F_{2-4} = \frac{1}{2}(1 - 0.43 - 0.34)$$
$$= 0.115$$

By summation rule

$$F_{1-1} + F_{1-2} + F_{1-3} + F_{1-4} = 1$$

Convex surface 1, thus $F_{1-1} = 0$

By symmetry $F_{1-3} = F_{1-4}$

$\therefore$ $0.86 + 2F_{1-3} = 1$

or $$F_{1-3} = \frac{1}{2}[1 - 0.86] = 0.07$$

For surface 3. By reciprocity :

$$A_3 F_{3-1} = A_1F_{1-3}$$

$$F_{3-1} = \frac{A_1}{A_3} . F_{1-3} = \frac{2\pi r_1 L}{\pi (r_2^2 - r_1^2)} \times F_{1-3}$$

$$= \frac{2 \times 5 \times 20}{(10^2 - 5^2)} \times 0.07 = 0.187$$

$$A_3 F_{3-2} = A_2 F_{2-3}$$

$$F_{3-2} = \frac{A_2}{A_3} F_{2-3} = \frac{2\pi r_2 L}{\pi (r_2^2 - r_1^2)} F_{2-3}$$

$$= \frac{2 \times 10 \times 20}{(10^2 - 5^2)} \times 0.115 = 0.613$$

For surface 4. Due to symmetry

$$F_{3-j} = F_{4-j}, j = 1, 2, 3, 4.$$

By summation rule

$$F_{3-1} + F_{3-2} + F_{3-3} + F_{3-4} = 1$$

Flat surface 3, thus

$$F_{3-3} = 0$$

$$\therefore \quad 0.187 + 0.613 + 0 + F_{3-4} = 1$$

or $$F_{3-4} = 0.2$$

View factor matrix

$$\begin{vmatrix} F_{1-1} & F_{1-2} & F_{1-3} & F_{1-4} \\ F_{2-1} & F_{2-2} & F_{2-3} & F_{2-4} \\ F_{3-1} & F_{3-2} & F_{3-3} & F_{3-4} \\ F_{4-1} & F_{4-2} & F_{4-3} & F_{4-4} \end{vmatrix} = \begin{vmatrix} \mathbf{0} & \mathbf{0.86} & \mathbf{0.07} & \mathbf{0.07} \\ \mathbf{0.43} & \mathbf{0.34} & \mathbf{0.115} & \mathbf{0.115} \\ \mathbf{0.187} & \mathbf{0.613} & \mathbf{0} & \mathbf{0.2} \\ \mathbf{0.187} & \mathbf{0.613} & \mathbf{0} & \mathbf{0.2} \end{vmatrix}.$$ **Ans.**

13.2. BLACK BODY RADIATION EXCHANGE

The radiation may leave a surface due to reflection and emission and on reaching on the second surface, there may be reflection as well as absorption. For a black body radiation, there is no reflection ($\rho = 0$). Hence energy leaves a surface as a result of emission, while it absorbs all incident energy.

Considering the radiation exchange between two black bodies having surface area A_1 and A_2.

The rate of energy leaves the surface 1 and reaches the surface 2

$$Q_{1-2} = A_1 F_{1-2} E_{b1} = A_1 F_{1-2} \sigma T_1^4 \quad \text{...[13.11 (a)]}$$

Similarly, the energy leaves the surface 2 and reaches the surface 1

$$Q_{2-1} = A_2 F_{2-1} E_{b2} = A_2 F_{2-1} \sigma T_2^4 \quad \text{...[13.11(b)]}$$

The net radiation exchange between the two surfaces

$$Q_{net} = A_1 F_{1-2} \sigma (T_1^4 - T_2^4) = A_2 F_{2-1} \sigma (T_1^4 - T_2^4) \quad \text{...(13.12)}$$

Since by reciprocity relation $A_1 F_{1-2} = A_2 F_{2-1}$.

Consider an enclosure consists of N black surfaces maintained at specified temperatures. The radiation heat transfer from any surface i of the enclosure is sum of radiation heat transfers to each surface of the enclosure from surface i.

$$Q_i = \sum_{j=1}^{N} Q_{i-j} = \sum_{j=1}^{N} A_i F_{i-j} (T_i^4 - T_j^4) \quad \text{...(13.13)}$$

Example 13.13. *Two black discs of diameter 50 cm each are placed parallel to each other concentrically at a distance of 1 m. The discs are maintained at 727°C and 227°C, respectively. Calculate the heat transfer between the discs per hour, when no other surface is present except the discs.*

Solution

Given : Two parallel discs of 50 cm diameter and 1 m apart.

$$r_1 = r_2 = 25 \text{ cm} = 0.25 \text{ m}$$
$$L = 1 \text{ m}$$
$$T_1 = 727°C = 1000 \text{ K}$$
$$T_2 = 227°C = 500 \text{ K}.$$

To find : Heat transfer rate when no other surface is present between the discs.

Analysis : When no other surface is present except two parallel discs. Fig. 13.26. The view factor between the two parallel plates, from Fig. 13.9

$$\frac{L}{r_1} = \frac{1}{0.25} = 4.0$$

and $$\frac{r_2}{L} = \frac{0.25}{1} = 0.25 ; \longrightarrow F_{1-2} = 0.06$$

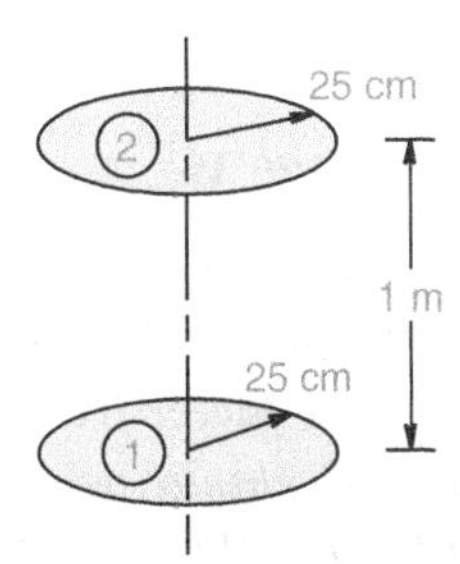

Fig. 13.26

The heat transfer rate,

$$Q = F_{1-2} A \sigma (T_1^4 - T_2^4)$$
$$= 0.06 \times (\pi/4) \times (0.5)^2 \times 5.67 \times 10^{-8} \times (1000^4 - 500^4)$$
$$= 626.24 \text{ W} = 2.254 \times 10^6 \text{ J/h}$$
$$= \mathbf{2254.46 \text{ kJ/h}}.$$ **Ans.**

Example 13.14. *A 5 cm diameter sphere at 600°C is placed near an infinite wall at 100°C. Both surface are black. Calculate the net radiant heat transfer between the two bodies.*

Solution

Given : A hot sphere near a wall

$T_1 = 600°C = 873 K, \quad T_2 = 100°C = 373 K$

$D = 5 cm = 0.05 m$

To find : Net radiant heat transfer.

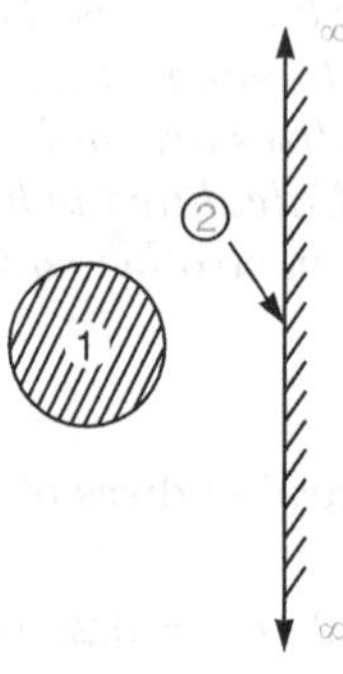

Fig. 13.27

Analysis : The net radiation heat exchange

$$Q_{1-2} = A_1 F_{1-2} E_{b1} - A_2 F_{2-1} E_{b2}$$
$$= A_1 F_{1-2} (E_{b1} - E_{b2})$$
$$= A_1 F_{1-2} \sigma (T_1^4 - T_2^4)$$

Since, the sphere is half covered by an infinite wall thus half of the radiation emitted by sphere will fall on infinite wall *i.e.,*

$$F_{1-2} = 0.5$$

and $A_1 = \pi D^2 = \pi \times 0.05^2$

$$= 7.853 \times 10^{-3} m^2$$

Then $Q_{1-2} = 7.853 \times 10^{-3} \times 0.5 \times 5.67 \times 10^{-8} \times (873^4 - 373^4)$

$$= \mathbf{125\ W.}$$ Ans.

Example 13.15. *A heater as shown in Fig. 13.28 radiates heat partially to conical shield that surrounds it.*

(i) Determine view factor from shield to heater.

(ii) If heater and shield are black and are at temperatures 1200°C and 100°C, respectively, what is the net heat transfer rate from heater to shield ?

Solution

Given : A heater at the base of a conical shield as shown in Fig 13.28.

$r_1 = 0.05$ m, $r_2 = 0.1$ m

$H = 0.2$ m, $T_1 = 1200°C = 1473$ K

$T_2 = 100°C = 373$ K

Both surfaces are black.

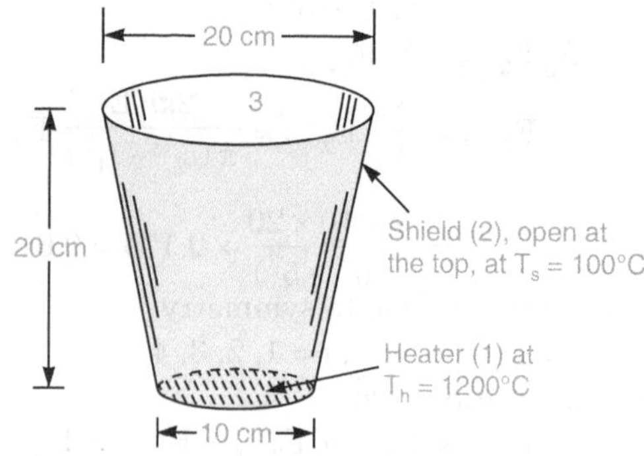

Fig. 13.28. Heat transfer from a disc heater to its radiation shield

To find :

(*i*) View factor F_{1-2}

(*ii*) Radiation heat transfer from heater to shield.

Analysis : (*i*) The view factor, by reciprocity

$$A_1 F_{1-2} = A_2 F_{2-1}$$

$$F_{2-1} = \frac{A_1}{A_2} F_{1-2}$$

and $F_{1-1} + F_{1-2} + F_{1-3} = 1$

or $F_{1-2} = 1 - F_{1-3} \quad \because \quad F_{1-1} = 0$

For two parallel discs 1 and 3, from Fig. 13.9 ;

$$\frac{L}{r_1} = \frac{0.2}{0.05} = 4.0$$

and $\frac{r_2}{L} = \frac{0.1}{0.2} = 0.5 \longrightarrow F_{1-3} = 0.196$

Then,

$$F_{1-2} = 1 - 0.196 = 0.804$$

Area $A_1 = \pi r_1^2 = \pi \times (0.05)^2 = 0.00785 m^2$

Area $A_2 = \pi(r_1 + r_2) \sqrt{(r_2 - r_1)^2 + H^2}$

$$= \pi(0.05 + 0.1) \times \sqrt{(0.1 - 0.05)^2 + (0.2)^2}$$
$$= 0.09715 m^2$$

Then

$$F_{2-1} = \frac{0.00785}{0.09715} \times (0.804) = \mathbf{0.065.}$$ Ans.

(*ii*) Net radiation heat transfer from heater to shield

$$Q_{net} = A_1 F_{1-2} \sigma (T_1^4 - T_2^4)$$
$$= 0.00785 \times 0.804 \times 5.67 \times 10^{-8} \times (1473^4 - 373^4) = \mathbf{1678\ W.}$$ Ans.

Example 13.16. *An outlet shoe store with a display window in the front is shown, with dimensions, in Fig. 13.29. The store is to be heated by making the floor a black radiant heating panel at 45°C. The glass window acts as a black plane at 10°C and the other walls and the ceiling act as black planes at 25°C. Find the net heat given up by the floor. What difference results, if the ceiling height is raised to 4.5 m, the other dimensions remaining unchanged ?*

Solution

Given : The display window of a shoe store. We designate floor as surface 1, window as surface 2, and other walls and ceiling as surface 3, then data

$$T_1 = 45 + 273 = 318 \text{ K}$$
$$A_1 = (6 + 2 + 2) \times 12 = 120 \text{ m}^2$$
$$T_2 = 10 + 273 = 283 \text{ K}$$
$$A_2 = 6 \times 3 = 18 \text{ m}^2$$
$$T_3 = 25 + 273 = 298 \text{ K}$$

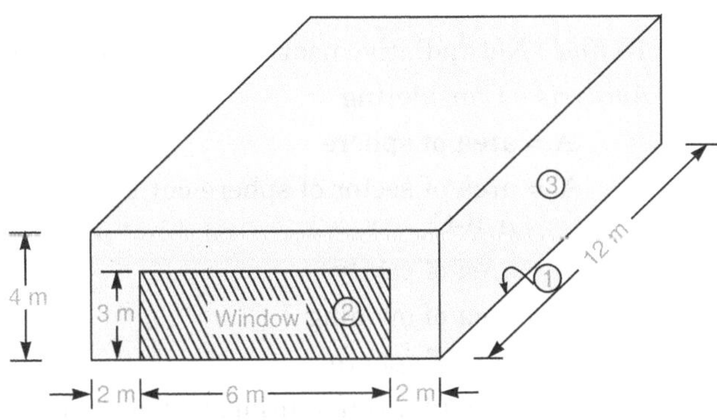

Fig. 13.29. Geometry for example 13.16

To find :

(*i*) Net heat given up by floor,

(*ii*) Difference in heat radiation by floor, if ceiling height is increased to 4.5 m.

Analysis : (*i*) Since all surfaces are black, thus various emissive powers are

$$E_{b1} = \sigma T_1^4 = 5.67 \times 10^{-8} \times (318)^4 = 579.8 \text{ W/m}^2$$
$$E_{b2} = \sigma T_2^4 = 5.67 \times 10^{-8} \times (283)^4 = 363.7 \text{ W/m}^2$$
$$E_{b3} = \sigma T_3^4 = 5.67 \times 10^{-8} \times (298)^4 = 447.14 \text{ W/m}^2.$$

The view factors : The geometrical arrangement for floor and window is very similar to example 13.9.

$$F_{1-1} = 0 \quad \text{and} \quad F_{1-2} = 0.058$$

For an enclosure of floor, window, wall and ceiling

$$F_{1-1} + F_{1-2} + F_{1-3} = 1$$
$$\longrightarrow F_{1-3} = 1 - F_{1-2} = 1 - 0.058 = 0.942$$

For complete enclosure

Heat radiated by floor = Heat absorbed by window and ceiling

Thus,

$$Q_1 = A_1 F_{1-2} (E_{b1} - E_{b2}) + A_1 F_{1-3} (E_{b1} - E_{b3})$$
$$= 120 \times 0.058 \times (579.8 - 363.7) + 120 \times 0.942 \times (579.8 - 447.14)$$
$$= \mathbf{16500 \text{ W}}. \quad \text{Ans.}$$

(*ii*) When ceiling height in increased to 4.5 m, neither view factor F_{1-2} nor F_{1-3} will change, thus the heat transfer rate from floor to window and ceiling, walls will remain same.

13.3. RADIATION FROM CAVITIES

Reconsider examples 13.6 and 13.7 solved for radiation view factor of conical and cylindrical cavities, respectively. The radiative heat transfer from such cavities can be estimated by using view factor F_{1-2} obtained.

Now, we go for formulation of some standard relation for such cavities as shown in Fig. 13.30.

Let us consider cavity as surface 1 and opening as surface 2. For any of following geometries, all the energy emitted from cavity (surface A_1) does not coming out the opening. But a part of radiation emitted from the surface falls on its other part, of which a portion is absorbed and remainder is reflected back.

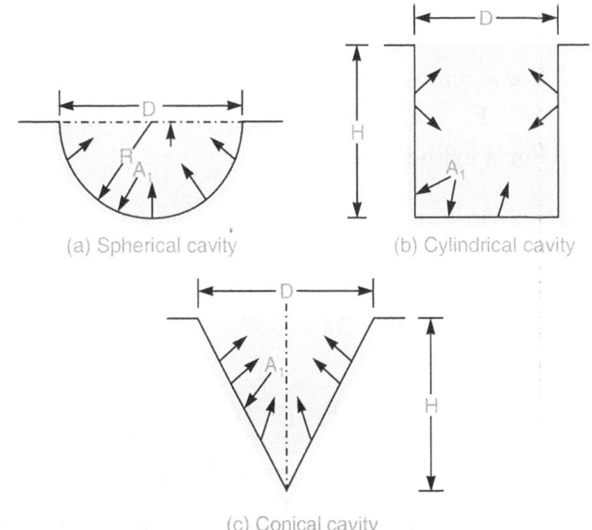

Fig. 13.30. Radiations from cavities

Rate of emission from a cavity surface = $A_1 \sigma \varepsilon_1 T^4$; of this energy, a part falling on A_1 and absorbed by it $= A_1 \sigma \varepsilon_1 T_1^4 (\alpha_1 F_{1-1}) = A_1 \sigma \varepsilon_1 T^4 \varepsilon_1 F_{1-1}$

where F_{1-1} is view factor of the cavity surface with respect to itself and $\alpha_1 = \varepsilon_1$ from Kirchhoff's law.

Now, the reflected radiation from the surface $A_1 = (1 - \varepsilon_1) A_1 \sigma \varepsilon_1 T_1^4 F_{1-1}$ of this reflected radiation, a part again falls on cavity surface and absorbed = $(1-\varepsilon_1) A_1 \sigma \varepsilon_1 T_1^4 F_{1-1}^2 \varepsilon_1$

Then again reflected

$$= (1-\varepsilon_1) A_1 \sigma \varepsilon_1 T_1^4 F_{1-1}^2 (1-\varepsilon_1)$$

Again absorbed

$$= (1-\varepsilon_1) A_1 \sigma \varepsilon_1 T_1^4 F_{1-1}^3 (1-\varepsilon_1) \varepsilon_1$$

Again reflected

$$= (1-\varepsilon_1) A_1 \sigma \varepsilon_1 T_1^4 F_{1-1}^3 (1-\varepsilon_1)^2 \text{ and so on.}$$

Net rate of radiation coming out the surface Q

= Total emission rate – Total absorption rate

$$= A_1 \sigma \varepsilon_1 T_1^4 - A_1 \sigma \varepsilon_1 T_1^4 [\varepsilon_1 F_{1-1} + (1-\varepsilon_1) \varepsilon_1 F_{1-1}^2 + (1-\varepsilon_1)^2 \varepsilon_1 F_{1-1}^3 + ...]$$

$$= A_1 \sigma \varepsilon_1 T_1^4 [1 - \varepsilon_1 F_{1-1} - (1-\varepsilon_1) \varepsilon_1 F_{1-1}^2 - (1-\varepsilon_1)^2 \varepsilon_1 F_{1-1}^3 - ......]$$

$$= A_1 \sigma \varepsilon_1 T_1^4 [1 - \varepsilon_1 F_{1-1} \{1 + (1-\varepsilon_1) F_{1-1}\} + (1-\varepsilon_1)^2 F_{1-1}^2 + ...]$$

It is an infinite series, and it can be arranged as

$$= A_1 \sigma \varepsilon_1 T_1^4 \left[1 - \frac{\varepsilon_1 F_{1-1}}{1-(1-\varepsilon_1) F_{1-1}}\right]$$

$$= A_1 \sigma \varepsilon_1 T_1^4 \left[\frac{1-F_{1-1}}{1-F_{1-1}(1-\varepsilon_1)}\right] \quad ...(13.14)$$

For a conical cavity

$F_{1-1} = 1 - \sin \alpha$ [see example 13.6]

For a cylindrical cavity

$F_{1-1} = \dfrac{4L}{D + 4L}$ [see example 13.7]

For hemispherical cavity

$$\mathbf{F_{1-1} = 1 - \frac{A_2}{A_1}}$$

$$\left[\because \mathbf{F_{2-1} = 1, F_{1-2} = \frac{A_2}{A_1}, \quad F_{1-1} = 1 - F_{1-2}}\right]. \quad \textbf{Ans.}$$

Example 13.17. *A spherical cavity of radius 10 cm is made in a large flat metal plate. The cavity has circular opening to the atmosphere. The radius of circular opening is 5 cm. The surface of the cavity is at 227°C. There is no heat conduction through the metal plate and there is no convection in the cavity. If the emissivity of the cavity surface is 0.8. Calculate the net radiative heat loss to the surroundings. Assume atmospheric temperature as 27°C.*

Solution

Given : A spherical cavity in a metal plate.

$$T_s = 227°C = 500 \text{ K}$$
$$T_\infty = 27°C = 300 \text{ K}$$
$$\varepsilon = 0.8, R = 10 \text{ cm}$$

opening radius $r_1 = 5$ cm.

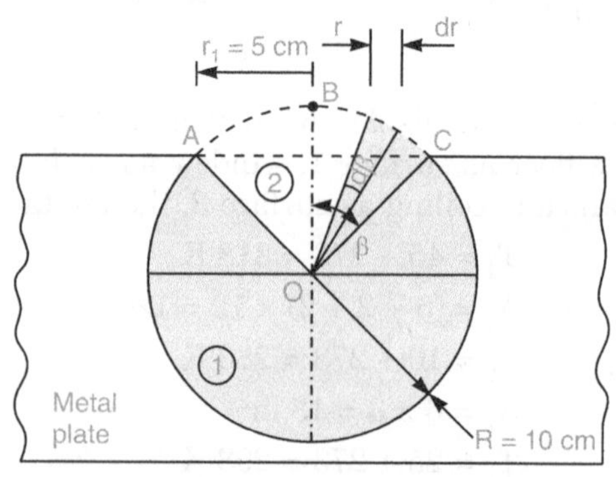

Fig. 13.31

To find : Net radiative heat loss to surroundings.

Analysis : Considering

A = area of sphere

A_s = area of sector of sphere cut by opening A-B-C.

A_1 = area of cavity

A_2 = area of opening

Area of sector of sphere :

Consider a circular differential strip of thickness *dr* at radius *r* from vertical centre of sphere. It substends an angle of $d\beta$ at the centre.

Here, $r = R \sin \beta$

$dr = R\, d\beta$ (since chord is very small)

Area of element

$$dA = 2\pi r\, dr = 2\pi R \sin \beta\, R\, d\beta \quad ...(i)$$
$$= 2\pi R^2 \sin \beta\, d\beta$$

Semi (half) angle substended by sector A-B-C

$$\angle BOC = \sin^{-1}\left(\frac{BC}{OC}\right) = \sin^{-1}\left(\frac{5}{10}\right) = 30°$$

Integrating to obtain the area of sector of sphere

$$A_s = \int_0^{30} dA = \int_0^{30} 2\pi R^2 \sin \beta\, d\beta$$

$$= 2\pi R^2 \left[-\cos \beta\right]_0^{30}$$

$= 2\pi R^2 [1 - 0.8660]$

$= 0.268\ \pi R^2\ m^2 = 0.00842\ m^2$

Area of cavity

A_1 = Area of sphere – Area of sector of sphere

$= 4\pi R^2 - A_s = 4\pi R^2 - 0.268\ \pi R^2$

$= 3.732\ \pi R^2 = 0.1172\ m^2$

Area of opening,

$A_2 = \pi r^2 = \pi \times (0.05)^2 = 0.00785\ m^2$

If we mark cavity opening as surface 2 and sphere inner surface as cavity 1 then

$F_{1-1} + F_{1-2} = 1$

$F_{2-1} + F_{2-2} = 1 \longrightarrow F_{2-2} = 0$

since cavity opening is flat.

$F_{2-1} = 1$

By reciprocity,

$$F_{1-2} = \frac{A_2}{A_1} F_{2-1} = \frac{0.00785}{0.1172} \times 1 = 0.067$$

and $F_{1-1} = 1 - 0.067 = 0.933.$

The rate of radiant heat loss through cavity

$$Q = A_1 \sigma \varepsilon_1 (T_s^4 - T_\infty^4) \times \frac{1 - F_{1-1}}{1 - F_{1-1}(1-\varepsilon_1)}$$

$$= 0.1172 \times 5.67 \times 10^{-8} \times 0.8$$

$$\times (500^4 - 300^4) \times \frac{1 - 0.933}{1 - 0.933 \times (1 - 0.8)}$$

$= \mathbf{23.863\ W.}$ Ans.

13.4. RADIATION HEAT EXCHANGE BETWEEN DIFFUSE, GRAY SURFACES

In the preceding section, the heat exchange between blackbodies has been discussed. The analysis was simpler, because blackbodies do not reflect any amount of incident radiation. But in practical applications, most of enclosures have nonblack surfaces, which allow multiple surface reflections. Therefore, the analysis of radiation heat transfer for such enclosures becomes very complicated. The analysis of radiation exchange in such enclosures may be simplified by making certain assumptions. These assumptions are

1. Each surface of enclosure is *opaque, diffuse and gray.*
2. Each surface of the enclosure is *isothermal.*
3. Each surface of the enclosure is characterised by *uniform radiosity* and *irradiation.*

13.4.1. The Net Radiation Exchange by a Surface

Consider a surface with the following properties during radiation exchange :

ε = emissivity of the surface.

E_b = emissive power of the black surface, W/m^2.

ρ = reflectivity of the surface.

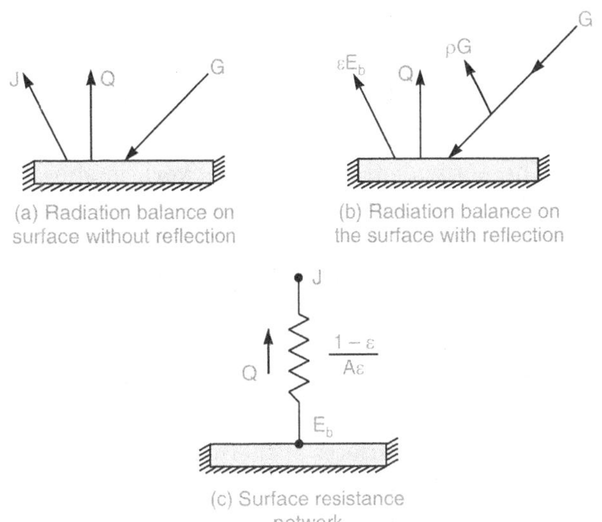

Fig. 13.32. Electrical analogy of radiation exchange with a surface

During radiation exchange, the energy leaves a surface by emission and reflection, while the surface gains energy by absorbing radiation incident (irradiation) on it. Therefore, the rate of net energy leaving a surface Q is the difference between the radiosity J and irradiation G :

$$Q = A(J - G) \quad ...(13.15)$$

where, G = irradiation on the surface, W/m^2.

J = radiosity of the surface as sum of the energy emitted and energy reflected per unit area, W/m^2. That is

$$J = \varepsilon E_b + \rho G$$

For an opaque, diffuse and gray surface ;

$\rho = 1 - \alpha = 1 - \varepsilon$, then

$$J = \varepsilon E_b + (1 - \varepsilon)G \quad ...(13.16)$$

The irradiation G in terms of radiosity J can be expressed as

$$G = \frac{J - \varepsilon E_b}{1 - \varepsilon} \quad ...(13.17)$$

Substituting in eqn. (13.15)

$$Q = A\left(J - \frac{J - \varepsilon E_b}{1 - \varepsilon}\right)$$

$$Q = \frac{A\varepsilon (E_b - J)}{1 - \varepsilon} \quad ...(13.18)$$

In electrical analogy to Ohm's law, this equation can be arranged as

$$Q = \frac{E_b - J}{R} \qquad \text{...(13.19)}$$

where $$R = \frac{1-\varepsilon}{A\,\varepsilon} \qquad \text{...(13.20)}$$

is the **surface resistance** to the radiation and $E_b - J$ corresponds to the potential difference of the surface, and Q corresponds to the heat current in electrical analogy.

The equivalent radiation network is shown in Fig. 13.32(*c*). The direction of radiation heat transfer depends on relative magnitude of radiosity J and blackbody emissive power, E_b. It will flow from surface, if $E_b > J$ and to the surface, if $J > E_b$. Hence, a negative value for Q will indicate the heat flow to the surface.

The surface resistance for a *blackbody* is *zero.* For a black surface ; $\varepsilon = 1$ and thus $J = E_b$. Therefore, the net *heat flow* for *black* surfaces at thermal equilibrium, will be *zero*.

13.4.2. Radiation Exchange between Two Gray Surfaces

Consider the radiation exchange between the two *opaque, diffuse* and *gray* surfaces with areas A_1 and A_2 maintained at uniform temperatures T_1 and T_2 as shown in Fig. 13.33(*a*)

The fraction of energy leaves the surface 1 and reaches the surface 2 = $A_1J_1F_{1-2}$

The fraction of energy leaves the surface 2 and reaches the surface 1 = $A_2J_2F_{2-1}$

The net radiation energy exchange between the two surfaces ;

$$Q_{1-2} = A_1F_{1-2}\,J_1 - A_2F_{2-1}\,J_2$$

But by reciprocity relation,

$$A_1F_{1-2} = A_2F_{2-1}$$

Hence, $$Q_{1-2} = A_1F_{1-2}(J_1 - J_2) \qquad \text{...(13.21)}$$

Rearranging ;

$$Q_{1-2} = \frac{J_1 - J_2}{R_{1-2}} \qquad \text{...(13.22)}$$

where $$R_{1-2} = \frac{1}{A_1F_{1-2}} \qquad \text{...(13.23)}$$

is the **space resistance** to radiation. Again $J_1 - J_2$ corresponds to space or radiosity potential difference, W/m² and net heat flow between surfaces corresponds to current flow in the radiation circuit as shown in Fig. 13.33 (*b*)

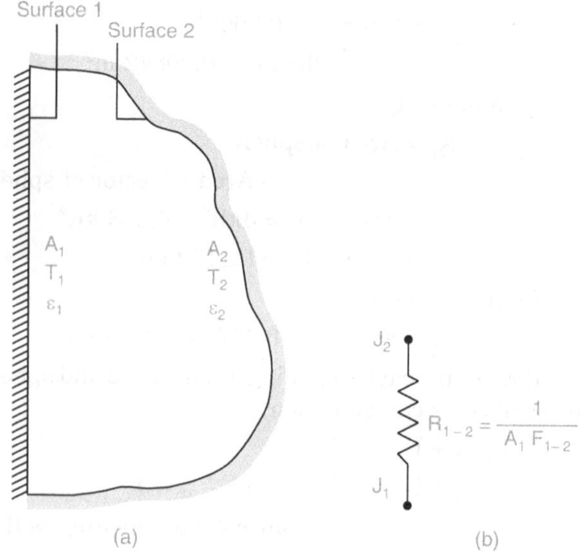

Fig. 13.33 (*a*) Enclosure consists of two surfaces, which are opaque, diffuse and gray (*b*) Equivalent radiation network for space resistance

Now combining eqns. (13.19) and (13.22) for two surfaces 1 and 2, we get

$$\frac{E_{b1} - J_1}{R_1} = \frac{J_1 - J_2}{R_{1-2}} = \frac{J_2 - E_{b2}}{R_2} \qquad \text{...(13.24)}$$

and the thermal network between two surfaces in steady state conditions takes the shape as shown in Fig. 13.33 (*c*)

Q_{12} → E_{b1} — $R_1 = \frac{1-\varepsilon_1}{\varepsilon_1 A_1}$ — J_1 (1) — $R_{12} = \frac{1}{A_1F_{1-2}}$ — J_2 (2) — $R_2 = \frac{1-\varepsilon_2}{\varepsilon_2 A_2}$ — E_{b2} ← Q_{21}

Fig. 13.33 (*c*) Thermal network for radiation exchange between two diffuse, gray surfaces

The steady state radiation heat transfer rate between two surfaces

$$Q \text{ (Heat current)} = \frac{\text{Overall potential difference}}{\text{Total thermal resistance}}$$

$$Q = \frac{E_{b1} - E_{b2}}{R_1 + R_{1-2} + R_2}$$

or $$Q = \frac{\sigma\,(T_1^{\,4} - T_2^{\,4})}{\dfrac{1-\varepsilon_1}{\varepsilon_1 A_1} + \dfrac{1}{A_1F_{1-2}} + \dfrac{1-\varepsilon_2}{\varepsilon_2 A_2}} \qquad \text{...(13.25)}$$

The radiation network is solved on each node by *network method.* That is net heat flow to any node (surface) must be zero. For node (surface) 1

$$\frac{E_{b1} - J_1}{R_1} - \frac{J_2 - J_1}{R_{1-2}} = 0 \quad ...(13.26)$$

and for node (surface) 2

$$\frac{J_1 - J_2}{R_{2-1}} - \frac{E_{b2} - J_2}{R_2} = 0 \quad ...(13.27)$$

13.4.3. Radiation Heat Exchange between Two Parallel Infinite Planes

For two *infinite long parallel plates*, area $A_1 = A_2$; the radiation view factor F_{1-2} is also unity, since all amount of energy leaving one plate reaches the other. The network remains same as shown in Fig. 13.33(*c*) and heat flow per unit area can be expressed as

$$\frac{Q}{A} = \frac{\sigma(T_1^4 - T_2^4)}{\frac{1}{\varepsilon_1} + \frac{1}{\varepsilon_2} - 1}$$

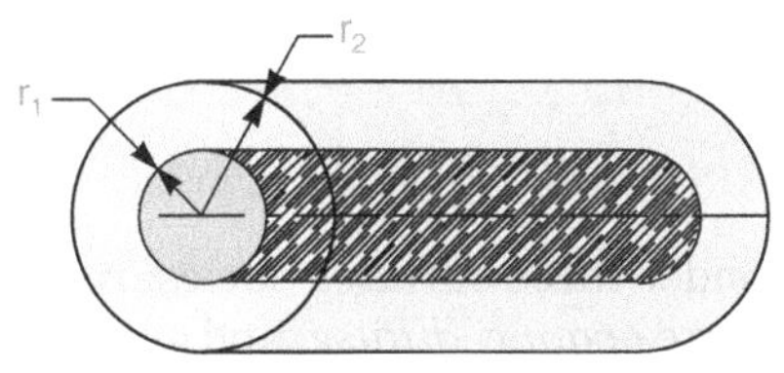

Fig. 13.34. Two concentric cylinders

When two infinite long concentric cylinders ($F_{1-2} = 1$) as shown in Fig. 13.34 exchange radiation heat energy, eqn. (13.25) can be rewritten as

$$Q = \frac{A_1 \sigma (T_1^4 - T_2^2)}{\frac{1}{\varepsilon_1} + \frac{A_1(1-\varepsilon_2)}{\varepsilon_2 A_2}}$$

This relation can also be applied to two concentric spheres.

Table 13.3 shows the area, view factor and radiation energy exchange relations between two opaque, diffuse gray surface enclosures.

TABLE 13.3. Radiation heat exchange between opaque, diffuse, gray two surface enclosures

Enclosure	*View factor*	*Radiation heat exchange*
Small object in a large cavity (A_1, T_1, ε_1; A_2, T_2, ε_2)	$\frac{A_1}{A_2} \approx 0$ $F_{1-2} = 1$	$\dot{Q}_{12} = A_1 \sigma \varepsilon_1 (T_1^4 - T_2^4)$ ...(13.28)
Infinitely large parallel plates (A_1, T_1, ε_1; A_2, T_2, ε_2)	$A_1 = A_2 = A$ $F_{1-2} = 1$	$\dot{Q}_{12} = \frac{A\sigma(T_1^4 - T_2^4)}{\frac{1}{\varepsilon_1} + \frac{1}{\varepsilon_2} - 1}$ ...(13.29)
Infinitely long concentric cylinders (r_1, r_2)	$\frac{A_1}{A_2} = \frac{r_1}{r_2}$ $F_{1-2} = 1$	$\dot{Q}_{12} = \frac{A_1 \sigma (T_1^4 - T_2^4)}{\frac{1}{\varepsilon_1} + \frac{1-\varepsilon_2}{\varepsilon_2}\left(\frac{r_1}{r_2}\right)}$ ...(13.30)
Concentric spheres (r_1, r_2)	$\frac{A_1}{A_2} = \left(\frac{r_1}{r_2}\right)^2$ $F_{1-2} = 1$	$\dot{Q}_{12} = \frac{A_1 \sigma (T_1^4 - T_2^4)}{\frac{1}{\varepsilon_1} + \frac{1-\varepsilon_2}{\varepsilon_2}\left(\frac{r_1}{r_2}\right)^2}$ ...(13.31)

13.5. THE RADIATION EXCHANGE BETWEEN THREE SURFACE ENCLOSURES

We now consider three surfaces enclosure for which the two surfaces are *opaque, diffuse,* and *gray,* and the third one is an *adiabatic*, since its back side is well insulated, and in absence of any convection effects, the net radiation exchange for such a surface is *zero*. But its presence, influences the heat transfer process in the enclosure. Because it absorbs and radiates energy back to two surfaces. Such a surface is called a **reradiating surface.**

For an example, the burning fuel bed (surface 1) in a boiler exchanges radiation heat transfer with boiler tubes containing water (surface 2). These two surfaces are enclosed by refractory brick walls (surface 3) of very low thermal conductivity, thus it conducts negligible heat energy. Under steady state conditions, in absence of convection effects, the refractory brick walls may be approximated as an *adiabatic* surface, and therefore, the net radiation heat transfer for this surface becomes zero. But this surface influences the heat transfer by absorbing and reradiating energy as shown in Fig. 13.35(*a*).

The radiation heat exchange for such an enclosure is given by

$$Q_{1-2} = A_1F_{1-2}(J_1 - J_2) + A_1F_{1-3}(J_1 - J_3) \quad ...(13.32)$$

For reradiating surface net heat exchange is zero :

$$Q_3 = 0 = A_3F_{3-1}(J_3 - J_1) + A_3F_{3-2}(J_3 - J_2)$$

By reciprocity rule ;

$$A_3F_{3-1} = A_1F_{1-3} \quad \text{and} \quad A_3F_{3-2} = A_2F_{2-3}$$

Therefore,

$$A_1F_{1-3}(J_1 - J_3) - A_2F_{2-3}(J_3 - J_2) = 0 \quad ...(13.33)$$

Hence,

$$\frac{J_1 - J_3}{\dfrac{1}{A_1F_{1-3}}} = \frac{J_3 - J_2}{\dfrac{1}{A_2F_{2-3}}} \quad ...(13.34)$$

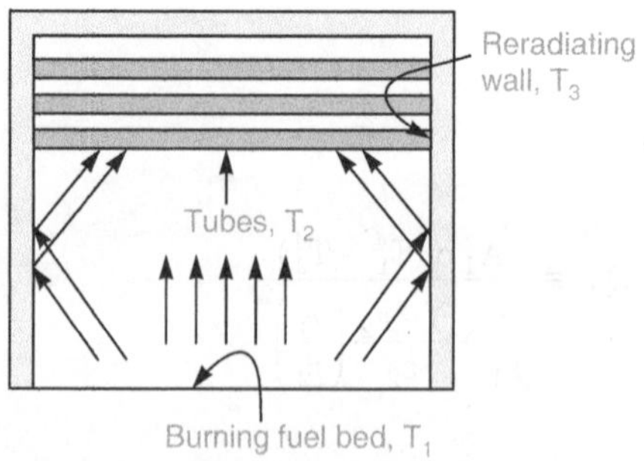

(a) A fuel bed, water tubes and refractory walls make an enclosure in a boiler

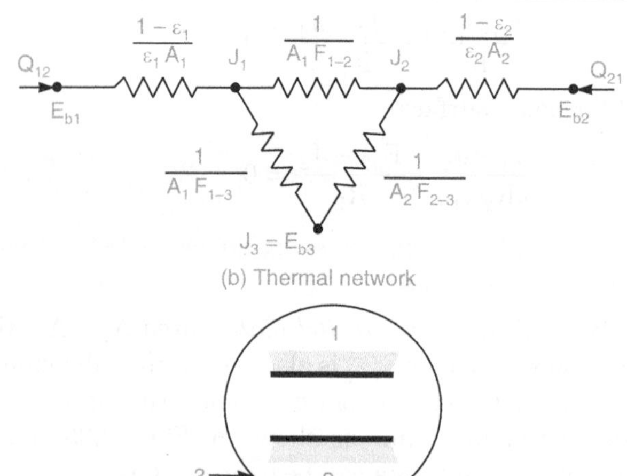

(b) Thermal network

1
3
2

(c) A typical three-body configuration

Fig. 13.35. Three surface enclosure with one surface reradiating

An equivalent thermal network for three surface enclosure with a reradiating surface is shown in Fig. 13.35(*b*). It is simple series parallel arrangement and the total resistances can be expressed as :

The resistances $1/A_1F_{1-3}$ and $1/A_2F_{2-3}$ for the reradiating surface are in series, therefore, its total resistance is

$$R_{s_1} = \frac{1}{A_1F_{1-3}} + \frac{1}{A_2F_{2-3}} \quad ...(13.35)$$

Further, the resistance R_{s_1} is in parallel with resistance $1/A_1F_{1-2}$, therefore, its equivalent resistance

$$\frac{1}{R_{eq}} = \frac{1}{\dfrac{1}{(A_1F_{1-2})}} + \frac{1}{R_{s_1}}$$

$$= \frac{1}{\dfrac{1}{A_1F_{1-2}}} + \frac{1}{\dfrac{1}{A_1F_{1-3}} + \dfrac{1}{A_2F_{2-3}}}$$

$$= A_1F_{1-2} + \left[\frac{1}{A_1F_{1-3}} + \frac{1}{A_2F_{2-3}}\right]^{-1}$$

or

$$R_{eq} = \frac{1}{A_1F_{1-2} + \left[\dfrac{1}{A_1F_{1-3}} + \dfrac{1}{A_2F_{2-3}}\right]^{-1}} \quad ...(13.36)$$

The total radiation resistance of the network

$$\Sigma R_{th} = \frac{1-\varepsilon_1}{\varepsilon_1 A_1} + \frac{1}{A_1F_{1-2} + \left[\dfrac{1}{A_1F_{1-3}} + \dfrac{1}{A_2F_{2-3}}\right]^{-1}} + \frac{1-\varepsilon_2}{\varepsilon_2 A_2} \quad ...(13.37)$$

The net radiation heat exchange

$$Q_{12} = -Q_{21} = \frac{E_{b1} - E_{b2}}{\Sigma R_{th}} = \frac{\sigma (T_1^4 - T_2^4)}{\Sigma R_{th}} \quad ...(13.38)$$

Since $Q_{12} = -Q_{21}$

or for a *reradiating* surface, net heat transfer is zero, therefore, its temperature can be calculated as

$$J_3 = E_{b3} = \sigma T_3^4 \quad ...(13.39)$$

Example 13.18. *A spherical liquid oxygen tank 0.3 m in diameter is enclosed concentrically in a spherical container of 0.4 m diameter and the space in between is evacuated. The tank surface is at – 183°C and has an emissivity of 0.2. The container surface is at 15°C and has an emissivity of 0.25. Determine the net radiant heat transfer rate and rate of evaporation of liquid oxygen if its latent heat is 220 kJ/kg.*

Solution

Given : A spherical oxygen tank with $\varepsilon_1 = 0.2$, $\varepsilon_2 = 0.25$, $h_{fg} = 220$ kJ/kg.

To find : Net radiation heat transfer and rate of evaporation of oxygen.

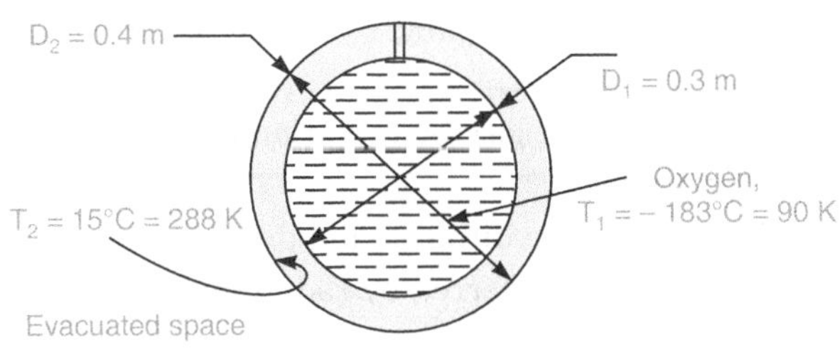

Fig. 13.36

Assumptions :

1. Surfaces are opaque, diffuse and gray.

2. Space between two concentric spheres is evacuated.

3. No conduction and convection heat transfer.

Analysis : The net radiation heat exchange between two concentric can be expressed as ;

$$Q = \frac{A_1 \sigma (T_1^4 - T_2^4)}{\frac{1}{\varepsilon_1} + \left(\frac{1}{\varepsilon_2} - 1\right)\frac{A_1}{A_2}}$$

where, $A_1 = \pi D_1^2 = \pi \times (0.3)^2 = 0.2827 \text{ m}^2$

and $\frac{A_1}{A_2} = \left(\frac{D_1}{D_2}\right)^2 = \left(\frac{0.3}{0.4}\right)^2 = 0.5625$

$$Q = \frac{0.2827 \times 5.67 \times 10^{-8} \times (90^4 - 288^4)}{\frac{1}{0.2} + \left(\frac{1}{0.25} - 1\right) \times 0.5625}$$

$$= -16.33 \text{ W}$$

The rate of evaporation of oxygen,

$$\dot{m} = \frac{Q}{h_{fg}} = \frac{16.33}{220 \times 10^3}$$

$$= 7.423 \times 10^{-5} \text{ kg/s}$$

$$= \mathbf{0.267 \text{ kg/h}}. \quad \text{Ans.}$$

Example 13.19. *Two parallel, infinite gray surfaces are maintained at temperature of 127°C and 227°C respectively. If the temperature of the hot surface is increased to 327°C. By what factor is the net radiation exchange per unit area increased ? Assume the emissivities of colder and hotter surfaces to be 0.9 and 0.7, respectively.* (N.M.U., May 1997)

Solution

Given : Two parallel infinite surfaces with

$T_3 = 327°C = 600$ K, $\varepsilon_1 = 0.9$, $\varepsilon_2 = 0.7$.

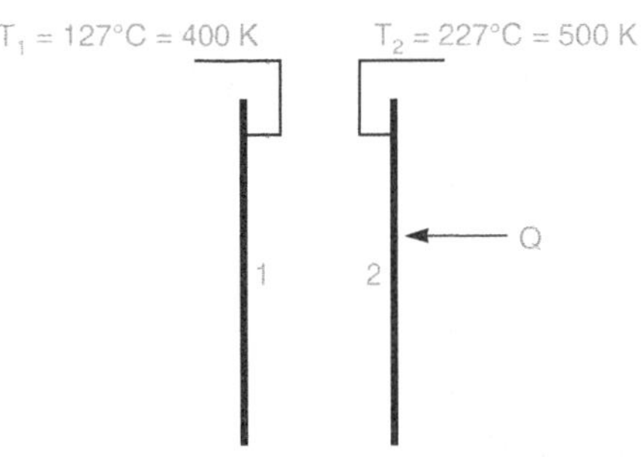

Fig. 13.37. Two parallel infinite gray surfaces

To find : Net radiation heat transfer.

Assumptions :

1. Surfaces are diffused and gray.

2. Heat is transferred by radiation only.

Analysis : The net radiation heat exchange between two parallel plates can be expressed as ;

$$\frac{Q}{A} = \frac{\sigma (T_1^4 - T_2^4)}{\frac{1}{\varepsilon_1} + \left(\frac{1}{\varepsilon_2} - 1\right)}$$

$$\frac{Q_1}{A} = \frac{5.67 \times 10^{-8} \times (500^4 - 400^4)}{\frac{1}{0.9} + \left(\frac{1}{0.7} - 1\right)}$$

$$= 1359 \text{ W/m}^2$$

When the hot plate temperature is raised to 600 K, then

$$\frac{Q_2}{A} = \frac{5.67 \times 10^{-8} \times (600^4 - 400^4)}{\frac{1}{0.9} + \left(\frac{1}{0.7} - 1\right)}$$

$$= 3830 \text{ W/m}^2$$

Therefore, $\frac{Q_2}{Q_1} = \frac{3830}{1359} = \mathbf{2.82}$. Ans.

Example 13.20. *A cubical room 4 m by 4 m by 4 m is heated through the ceiling by maintaining it at uniform temperature of 350 K, while walls and the floor are at 300 K. Assuming that the all surfaces have an emissivity of 0.8, determine the rate of heat loss from ceiling by radiation.*

Solution

Given : A cubical room with sides of 4 m each

$L = H = w = 4$ m, $T_1 = 350$ K

$T_2 = 300$ K, $\varepsilon_1 = \varepsilon_2 = 0.8$

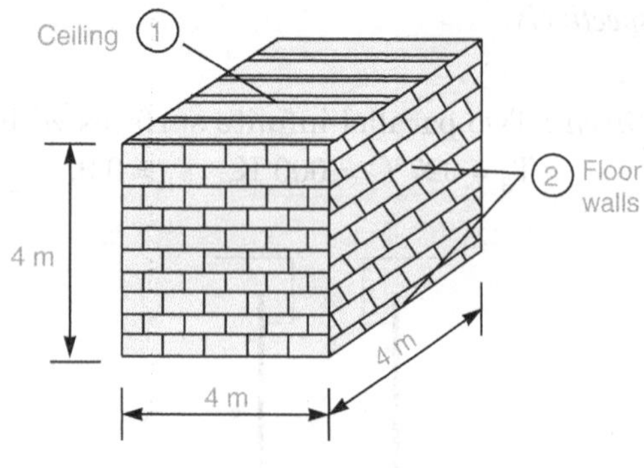

Fig. 13.38

To find : Heat loss from ceiling to room.

Assumptions :

1. Surfaces are diffused and gray.
2. Heat is transferred by radiation only.

Analysis : Considering ceiling as surface 1 and other surfaces of room as surface 2.

The surface area,

$$A_1 = (4 \text{ m} \times 4 \text{ m}) = 16 \text{ m}^2$$

The surface area,

$$A_2 = 5 \text{ surface} \times (4 \text{ m} \times 4 \text{ m}) = 80 \text{ m}^2$$

The net radiation heat exchange between two surfaces can be calculated by electrical analogy ;

The calculation of resistances

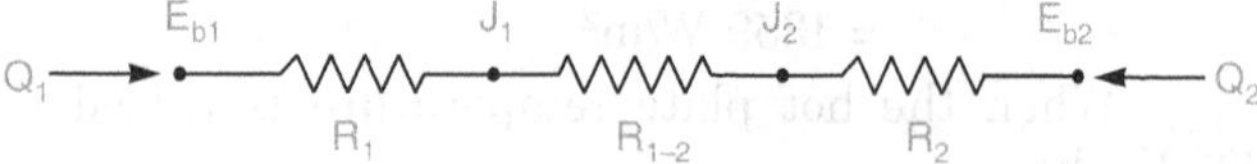

Fig. 13.38 (*a*) Radiation network

$$R_1 = \frac{1-\varepsilon_1}{A_1\,\varepsilon_1} = \frac{1-0.8}{16\times 0.8} = 0.0156 \text{ m}^{-2}$$

$$R_2 = \frac{1-\varepsilon_2}{A_2\,\varepsilon_2} = \frac{1-0.8}{80\times 0.8} = 0.003125 \text{ m}^{-2}$$

Since, all energy leaving the ceiling will be absorbed by room walls and floor,

Hence, $F_{1-2} = 1$

and $$R_{1-2} = \frac{1}{A_1 F_{1-2}} = \frac{1}{16\times 1} = 0.0625 \text{ m}^{-2}$$

$$Q = \frac{\sigma\,(T_1^4 - T_2^4)}{R_1 + R_{1-2} + R_2}$$

$$Q = \frac{5.67\times 10^{-8}\times (350^4 - 300^4)}{0.0156 + 0.0625 + 0.003125}$$

$$= \mathbf{4815.6\ W.}$$ **Ans.**

Example 13.21. *A cubical room 4 m by 4 m by 4 m is heated through the floor by maintaining it at uniform temperature of 350 K, while side walls are well insulated. The heat loss takes place through the ceiling at 300 K. Assuming that the all surfaces have an emissivity of 0.8, determine the rate of heat loss by radiation through the ceiling.*

Solution

Given : A cubical room with sides of 4 m each

$L = H = w = 4$ m, $T_1 = 350$ K

$T_2 = 300$ K, $\varepsilon_1 = \varepsilon_2 = 0.8.$

To find : Heat loss by radiation through ceiling.

Assumptions :

1. Surfaces are diffused and gray.
2. Heat is transferred by radiation only.

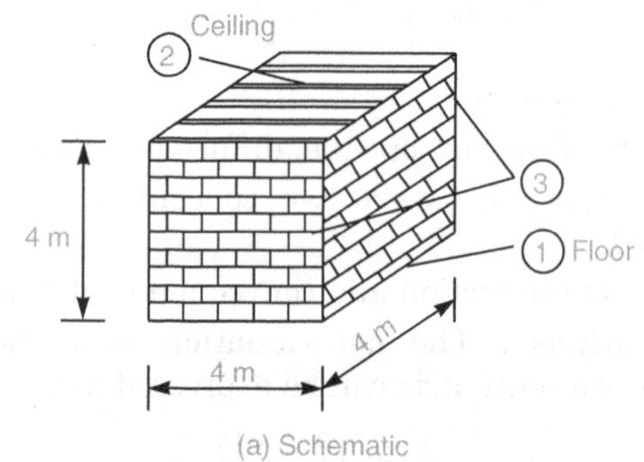

(a) Schematic

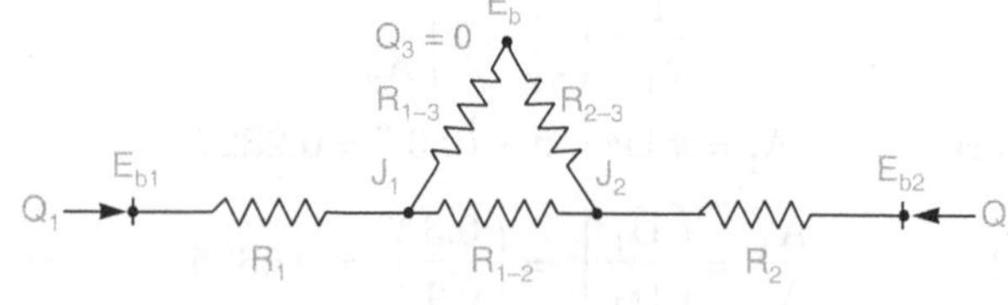

(b) Radiation network

Fig. 13.39

Analysis : Considering floor as surface 1, ceiling as surface 2 and walls of room as surface 3.

The surface area of floor ;

$A_1 = (4 \text{ m} \times 4 \text{ m}) = 16 \text{ m}^2$

The surface area of ceiling ;

$A_2 = (4 \text{ m} \times 4 \text{ m}) = 16 \text{ m}^2$

The surface area of walls ;

$A_3 = 4 \text{ walls} \times (4 \text{ m} \times 4 \text{ m}) = 64 \text{ m}^2$

The net radiation heat exchange between two surfaces can be calculated by electrical analogy ;

The calculation of resistances

$$R_2 = R_1 = \frac{1-\varepsilon_1}{A_1 \varepsilon_1} = \frac{1-0.8}{16 \times 0.8}$$
$$= 0.0156 \text{ m}^{-2}$$

Since, all energy leaving the floor will not reach the ceiling and hence the view factor F_{1-2} and F_{1-3} are to be determined from Fig. 13.8 ;

$$F_{1-2} = 0.2$$

By summation rule,

$$F_{1-1} + F_{1-2} + F_{1-3} = 1$$

But $\quad F_{1-1} = 0$

Hence $\quad F_{1-3} = 1 - F_{1-2} = 1 - 0.2 = 0.8$

and
$$R_{1-2} = \frac{1}{A_1 F_{1-2}} = \frac{1}{16 \times 0.2}$$
$$= \frac{1}{3.2} \text{ m}^{-2}$$

and
$$R_{2-3} = R_{1-3} = \frac{1}{A_1 F_{1-3}}$$
$$= \frac{1}{16 \times 0.8} = \frac{1}{12.8} \text{ m}^{-2}$$

The total radiation resistance between E_{b1} and E_{b2} is

$$\Sigma R_{th} = R_1 + \frac{1}{1/R_{1-2} + 1/\{R_{1-3} + R_{2-3}\}} + R_2$$
$$= 0.0156 + \frac{1}{3.2 + 1/\{(1/12.8) + (1/12.8)\}} + 0.0156 = 0.1353 \text{ m}^{-2}$$

$$Q = \frac{\sigma (T_1^4 - T_2^4)}{\Sigma R_{th}}$$

$$Q = \frac{5.67 \times 10^{-8} \times (350^4 - 300^4)}{0.1353}$$
$$= \mathbf{2892.77 \ W.} \text{ Ans.}$$

Example 13.22. *Two parallel discs 50 cm in diameter are spaced 40 cm apart with one disc located directly above the other disc. One disc is maintained at 500°C and other at 227°C. The emissivities of the discs are 0.2 and 0.4, respectively. The curved cylindrical surface approximates a black body and is maintained at a temperature of 67°C. Determine the rate of heat loss by radiation from the inside surfaces of each disc, and explain how these surfaces can be maintained at specified temperatures.*

Solution

Given : Two parallel discs spaced at 40 cm apart, located in a large room with

$D_1 = 50 \text{ cm} = 0.5 \text{ m}$,	$D_2 = 0.5 \text{ m}$
$\varepsilon_1 = 0.2$,	$T_1 = 500°C = 773 \text{ K}$
$\varepsilon_2 = 0.4$,	$T_2 = 227°C = 500 \text{ K}$
$L = 40 \text{ cm} = 0.4 \text{ m}$,	$T_3 = 67°C = 340 \text{ K}$

To find : The heat loss by radiation from each disc to room wall.

Assumptions :

1. Steady state conditions.
2. Diffuse and gray surfaces.
3. All surfaces are opaque.
4. Room has reradiating surfaces.

Analysis : The thermal network is shown in Fig. 13.40 (*b*) ;

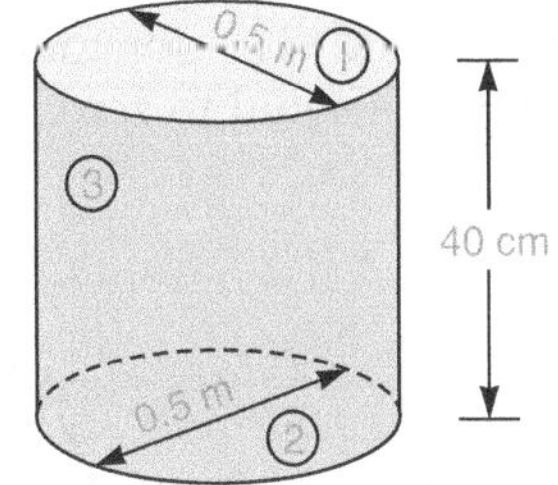

(a) Schematic for example 13.22

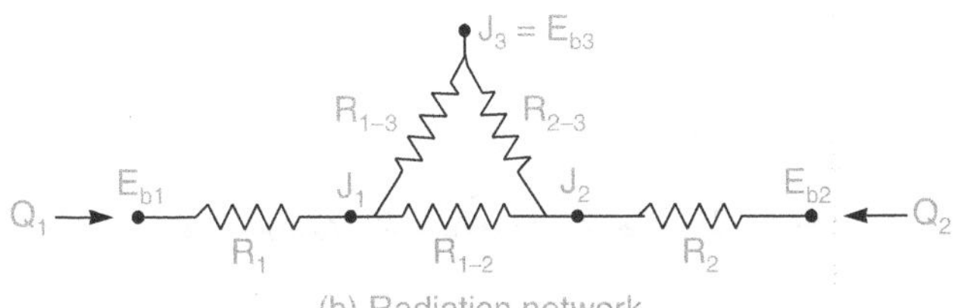

(b) Radiation network

Fig. 13.40

Area of each disc,

$$A_1 = A_2 = (\pi/4)\, D_1^2$$
$$= (\pi/4) \times (0.5 \text{ m})^2 = 0.1963 \text{ m}^2$$

The shape factor from Fig. 13.9

$$L/r_1 = 40/25 = 1.6$$

and $\quad r_2/L = 25/40 = 0.625$

Hence,

$$F_{1-2} = F_{2-1} = 0.24$$

By summation rule,

$$F_{1-3} = 1 - F_{1-2} = 1 - 0.24 = 0.76$$
$$F_{2-3} = 1 - F_{2-1} = 1 - 0.24 = 0.76$$

The various resistances ;

$$R_1 = \frac{1-\varepsilon_1}{\varepsilon_1 A_1} = \frac{1-0.2}{0.2 \times 0.1963} = 20.37 \text{ m}^{-2}$$

$$R_2 = \frac{1-\varepsilon_2}{\varepsilon_2 A_2} = \frac{1-0.4}{0.4 \times 0.1963} = 7.64 \text{ m}^{-2}$$

$$R_{1-2} = \frac{1}{A_1 F_{1-2}} = \frac{1}{0.1963 \times 0.24} = 21.22 \text{ m}^{-2}$$

$$R_{1-3} = \frac{1}{A_1 F_{1-3}} = \frac{1}{0.1963 \times 0.76} = 6.703 \text{ m}^{-2}$$

$$R_{2-3} = R_{1-3}$$

The emissive powers,

$$E_{b1} = \sigma T_1^4 = 5.67 \times 10^{-8} \times (773)^4 = 20244.22 \text{ W/m}^2$$

$$E_{b2} = \sigma T_2^4 = 5.67 \times 10^{-8} \times (500)^4 = 3543.75 \text{ W/m}^2$$

$$J_3 = E_{b3} = \sigma T_3^4 = 5.67 \times 10^{-8} \times (340)^4 = 757.7 \text{ W/m}^2$$

Applying Kirchhoff's law of electrical current at each node of Fig. 13.40 (*b*) *i.e.*, summation of incoming currents to each node is equal to zero.

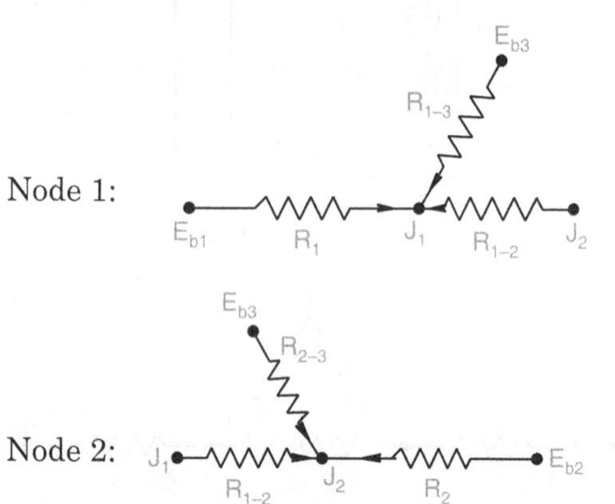

Fig. 13.40 (*c*)

Node 1, for J_1 :

$$\frac{E_{b1} - J_1}{R_1} + \frac{J_2 - J_1}{R_{1-2}} + \frac{E_{b3} - J_1}{R_{1-3}} = 0$$

$$\frac{20244.22 - J_1}{20.37} + \frac{J_2 - J_1}{21.22} + \frac{757.7 - J_1}{6.703} = 0$$

Node 2, for J_2 :

$$\frac{E_{b2} - J_2}{R_2} + \frac{J_1 - J_2}{R_{1-2}} + \frac{E_{b3} - J_2}{R_{2-3}} = 0$$

$$\frac{3543.75 - J_2}{7.64} + \frac{J_1 - J_2}{21.22} + \frac{757.7 - J_2}{6.703} = 0$$

Solution, leads to simultaneous equations as

$$-0.2454 J_1 + 0.0471 J_2 + 1106.82 = 0$$
$$0.0471 J_1 - 0.3272 J_2 + 576.88 = 0$$

The solution to these equations are

$$J_1 = 4986.426 \text{ W/m}^2, \; J_2 = 2480.87 \text{ W/m}^2$$

(*i*) The heat loss from hot disc at 500°C

$$Q_1 = \frac{E_{b1} - J_1}{R_1} = \frac{20244.22 - 4986.426}{20.37}$$

= **749 W. Ans.**

(*ii*) Heat loss from disc at 227°C

$$Q_2 = \frac{E_{b2} - J_2}{R_2} = \frac{3543.75 - 2480.87}{7.64}$$

= **139.81 W. Ans.**

(*iii*) The net rate of radiation heat transfer at the side curved surface is net radiation flow to node 3, *i.e.*,

$$Q_3 + \frac{J_1 - J_3}{R_{1-3}} + \frac{J_2 - J_3}{R_{2-3}} = 0$$

or
$$Q_3 = \frac{J_3 - J_1}{R_{1-3}} + \frac{J_3 - J_2}{R_{2-3}}$$

$$= \frac{757.7 - 4986.426}{6.703} + \frac{757.7 - 2480.87}{6.703}$$

= **–888 W. Ans.**

For the surfaces to be at specified temperatures, the algebraic sum of Q_1, Q_2 and Q_3 must be zero.

$$Q_1 + Q_2 + Q_3 = 749 + 139.81 - 888 \approx 0$$

Example 13.23. *Two very long strips 1 m wide and 2.40 m apart face each other, as shown in Fig. 13.41 (a) Find $Q_{net\ 1-2}$ (W/m) if the surroundings are black and at 250 K. (b) Find $Q_{net\ 1-2}$ (W/ m), if they are connected by an insulated diffuse reflector between the edges on both sides. Also evaluate the temperature of the reflector in part (b).*

Solution

Given : Two very long strips parallel to each other

$w = 1$ m, $T_1 = 400$ K, $L = 2.4$ m
$T_2 = 300$ K, $\varepsilon_1 = 0.3$, $\varepsilon_2 = 0.5$

To find :

(*i*) $Q_{net\ 1-2}$ if open

(*ii*) $Q_{net\ 1-2}$, if connected by an insulated diffuse reflector

(*iii*) Temperature of reflector.

Assumptions :

1. Each surface has uniform radiosity and the enclosure can be treated as a three surface enclosure.

2. Diffuse, gray and opaque surfaces.

3. Medium separating the surfaces does not participate in radiation.

4. Negligible convection inside the enclosure.

Analysis : (i) When two infinite long parallel plates are spaced at a distance L and are open. The view factor from Fig. 13.8

$$\left.\begin{aligned}\frac{L_2}{D} &= \frac{w}{D} = \frac{1}{2.4} = 0.416\\ \frac{L_1}{D} &= \frac{\infty}{0} = \infty\end{aligned}\right\} F_{1-2} = 0.2 = F_{2-1}$$

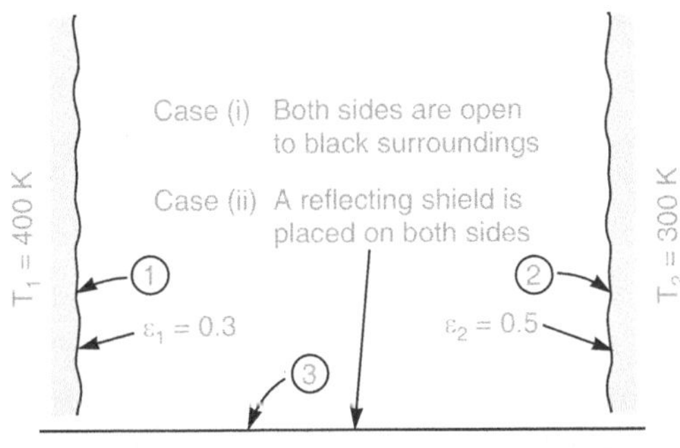

Fig. 13.41

In addition, surface 3 may represent reflector or surroundings, then for an enclosure

$$F_{1-1} + F_{1-2} + F_{1-3} = 1$$

$$F_{1-3} = 1 - F_{1-2} = 1 - 0.2 = 0.8 \qquad [F_{1-1} = 0]$$

By symmetry $F_{2-3} = F_{1-3} = 0.8$

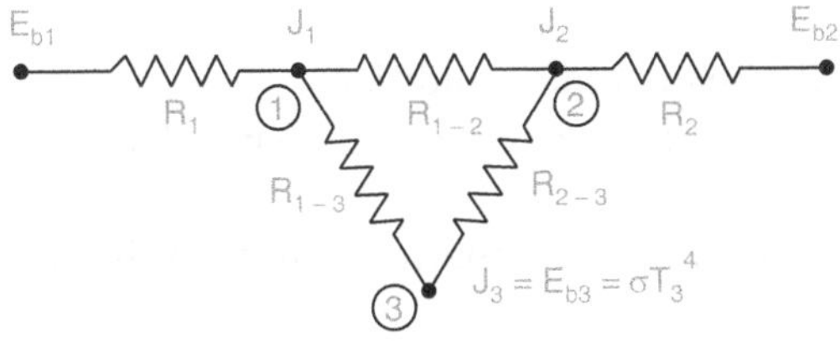

Fig. 13.41 *(a)* Radiation network for case *(i)*

The nodal equations for nodes 1 and 2 are :

$$\text{Node 1}: \frac{E_{b1} - J_1}{R_1} + \frac{J_2 - J_1}{R_{1-2}} + \frac{E_{b3} - J_1}{R_{1-3}} = 0$$

$$\text{Node 2}: \frac{J_1 - J_2}{R_{1-2}} + \frac{E_{b2} - J_2}{R_2} + \frac{E_{b3} - J_2}{R_{2-3}} = 0$$

where for 1 m^2 area

$$R_1 = \frac{1-\varepsilon_1}{\varepsilon_1} = \frac{1-0.3}{0.3} = 2.333 \text{ m}^{-2}$$

$$R_{1-2} = \frac{1}{F_{1-2}} = \frac{1}{0.2} = 5 \text{ m}^{-2}$$

$$R_2 = \frac{1-\varepsilon_2}{\varepsilon_2} = \frac{1-0.5}{0.5} = 1 \text{ m}^{-2}$$

$$R_{1-3} = \frac{1}{F_{1-3}} = \frac{1}{0.8} = 1.25 \text{ m}^{-2}$$

$$R_{2-3} = \frac{1}{F_{2-3}} = \frac{1}{0.8} = 1.25 \text{ m}^{-2}$$

$$E_{b1} = \sigma T_1^4 = (5.67 \times 10^{-8}) \times (400)^4 = 1451.5 \text{ W/m}^2$$

$$E_{b2} = \sigma T_2^4 = (5.67 \times 10^{-8}) \times (300)^4 = 459.3 \text{ W/m}^2$$

$$E_{b3} = \sigma T_\infty^4 = (5.67 \times 10^{-8}) \times (250)^4 = 221.5 \text{ W/m}^2$$

Thus,

$$\frac{1451.5 - J_1}{2.333} + \frac{J_2 - J_1}{5} + \frac{221.5 - J_1}{1.25} = 0$$

$$\frac{J_1 - J_2}{5} + \frac{459.3 - J_2}{1} + \frac{221.5 - J_2}{1.25} = 0$$

Simplification leads to

$$J_1 - 0.14\, J_2 - 0.56 \times 221.5 = 435.6$$

$$-J_1 + 10.0\, J_2 - 4.0 \times 221.5 = 2296.5$$

or

$$J_1 - 0.14\, J_2 = 559.6$$

$$-J_1 + 10.0 J_2 = 3182.5$$

Its solution leads to

$$J_1 = 612.1 \text{ W/m}^2, \quad J_2 = 379.5 \text{ W/m}^2$$

Thus the net heat flow from plane 1 to plane 2 is

$$Q_{net\,1-2} = \frac{J_1 - J_2}{R_{1-2}} = \frac{612.1 - 379.5}{5}$$

$$= \mathbf{46.52 \ W/m.} \quad \text{Ans.}$$

(ii) When two parallel strips are connected by an insulated diffuse reflector (a radiation shield) between the edges of both sides.

Then $Q_3 = 0$

and nodel equations

$$\text{Node 1}: \frac{E_{b1} - J_1}{R_1} + \frac{J_2 - J_1}{R_{1-2}} + \frac{J_3 - J_1}{R_{1-3}} = 0$$

$$\text{Node 2}: \frac{J_1 - J_2}{R_{1-2}} + \frac{E_{b2} - J_2}{R_{3-2}} + \frac{J_3 - J_2}{R_{2-3}} = 0$$

Node 3 : $$\frac{J_1 - J_3}{R_{1-3}} + \frac{J_2 - J_3}{R_{2-3}} = 0$$

The resistances R_1, R_2, R_{1-2}, R_{1-3} and R_{2-3} remain unchanged, because the surroundings is replaced by radiation shield,

The solution of these simultaneous equation leads to

$J_1 = 987.7\ W/m^2$, $J_2 = 657.4\ W/m^2$

$J_3 = 822.6\ W/m^2$

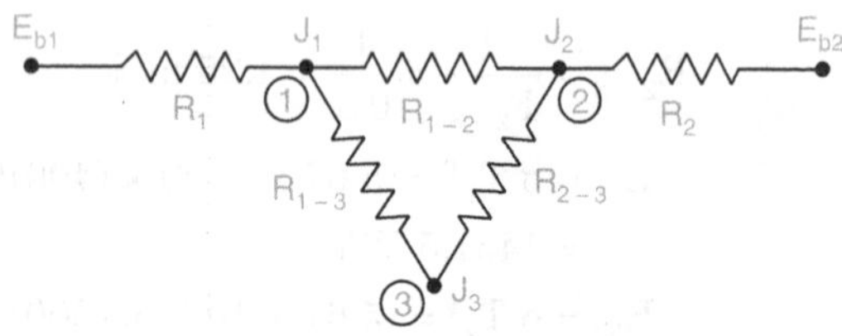

Fig. 13.41 *(b)* Radiation network for case *(ii)*

In this case, the surface 3 is adiabatic, thus all the heat leaving the surface 1 will be transferred to surface 2.

$$Q_{net\ 1-2} = \frac{J_1 - J_2}{R_{1-2}} + \frac{J_1 - J_3}{R_{1-3}}$$

$$= \frac{987.7 - 657.4}{5} + \frac{987.7 - 822.6}{1.25}$$

$$= \mathbf{198\ W/m.}$$ **Ans.**

Since the surface 3 is reradiating, thus we can also obtain $Q_{net\ 1-2}$ by relation

$$Q_{net\ 1-2} = \frac{A\,\sigma\,(T_1^4 - T_2^4)}{\dfrac{1-\varepsilon_1}{\varepsilon_1} + \dfrac{1}{F_{1-2} + \left[\dfrac{1}{R_{1-3}} + \dfrac{1}{R_{2-3}}\right]^{-1}} + \dfrac{1-\varepsilon_2}{\varepsilon_2}}$$

$$= \frac{1 \times 5.67 \times 10^{-8} \times [400^4 - 300^4]}{\dfrac{1-0.3}{0.3} + \dfrac{1}{0.2 + \left[\dfrac{1}{1.25} + \dfrac{1}{1.25}\right]^{-1}} + \dfrac{1-0.5}{0.5}}$$

$$= \mathbf{198.4\ W/m.}$$ **Ans.**

(iii) Temperature of reflector or reradiating surface

$$Q_3 = 0 \quad \text{or} \quad J_3 = E_{b3}$$

Thus, $5.67 \times 10^{-8}\ T_3^4 = 822.6$

$$T_3 = \mathbf{347\ K.}$$ **Ans.**

Example 13.24. *A heater of 1 m diameter is covered by a hemisphere of 4 m diameter. The surface of hemisphere is maintained at 400 K. The emissivity of the surface is 0.8. The heater surface is maintained at 1000 K. The remaining base area is open to surroundings at 300 K. The surroundings may be considered black. The emissivity of heater surface is also 0.8. Determine the heat exchange from heater to the hemisphere and to the surroundings.*

Solution

Given : A heater (1) is covered by a hemisphere (2)

$T_1 = 1000$ K, $\varepsilon_1 = 0.8$, $D_1 = 1$ m

$T_2 = 400$ K, $\varepsilon_2 = 0.8$, $D_2 = 4$ m

$T_3 = 300$ K, $\varepsilon_3 = 1$.

To find :

(i) Heat exchange with hemisphere, and

(ii) Heat exchange with surroundings.

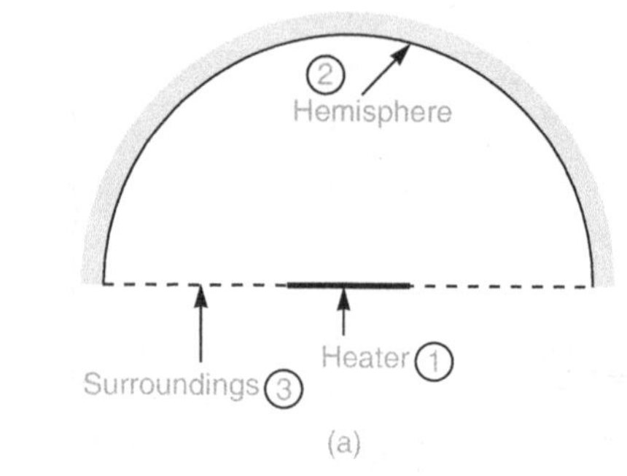

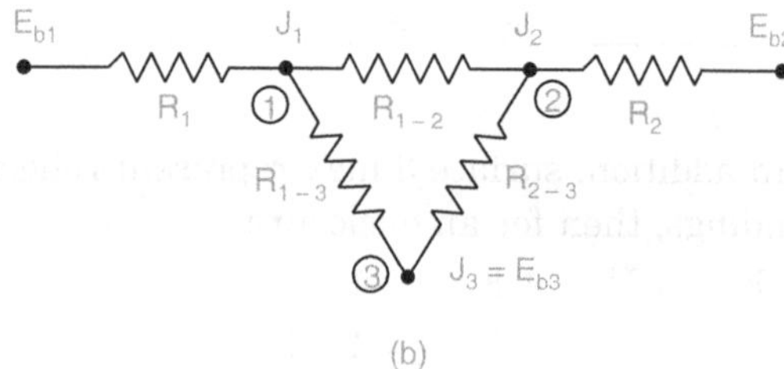

Fig. 13.42. Schematic and radiation network

Assumptions :

1. Heater and hemispherical surfaces are opaque, diffuse and gray.

2. Steady state conditions.

3. Assuming circular disc heater, facing towards hemisphere.

Analysis : The thermal network is shown in Fig. 13.42 *(b)* :

The areas are :

Area of circular heater,

$$A_1 = \frac{\pi}{4} D_1^2 = \frac{\pi}{4} \times (1)^2 = 0.7853\ m^2$$

Area of hemisphere,

$$A_2 = \frac{\pi}{2} D_2^2 = \frac{\pi}{2} \times (4)^2 = 25.132\ m^2$$

Area of opening of hemisphere,

$$A_3 = \frac{\pi}{4}(D_2^2 - D_1^2) = \frac{\pi}{4}(4^2 - 1^2) = 11.78 \text{ m}^2$$

View factors. If circular heater faces hemisphere, then

$$F_{1-2} = 1, F_{1-1} = 0, F_{1-3} = 0$$

$$F_{2-1} + F_{2-2} + F_{2-3} = 1$$

$$F_{2-1} = \frac{A_1}{A_2} F_{1-2} = \frac{0.7853}{25.132} \times 1 = 0.0312$$

There will be negligible radiation from surroundings to heater surface, thus

$$F_{3-1} = F_{3-3} = 0$$

Thus $\quad F_{3-2} = 1$

$$F_{2-3} = \frac{A_3}{A_2} F_{3-2} = \frac{11.78}{25.132} \times 1 = 0.468$$

Applying Network theorem (Kirchhoff 's law) at each node

$$\text{Node 1: } \frac{E_{b1} - J_1}{R_1} + \frac{J_2 - J_1}{R_{1-2}} + \frac{E_{b3} - J_1}{R_{1-3}} = 0 \quad ...(i)$$

$$\text{Node 2: } \frac{E_{b2} - J_2}{R_2} + \frac{J_1 - J_2}{R_{1-2}} + \frac{E_{b3} - J_2}{R_{2-3}} = 0 \quad ...(ii)$$

where $E_{b1} = \sigma T_1^4 = 5.67 \times 10^{-8} \times (1000)^4$
$= 56700 \text{ W/m}^2$

$E_{b2} = \sigma T_2^4 = 5.67 \times 10^{-8} \times (400)^4$
$= 1451.51 \text{ W/m}^2$

$E_{b3} = \sigma T_3^4 = 5.67 \times 10^{-8} \times (300)^4$
$= 459.27 \text{ W/m}^2$

$$R_1 = \frac{1 - \varepsilon_1}{\varepsilon_1 A_1} = \frac{1 - 0.8}{0.8 \times 0.7853} = 0.3183 \text{ m}^{-2}$$

$$R_{1-2} = \frac{1}{A_1 F_{1-2}} = \frac{1}{0.7853 \times 1} = 1.273 \text{ m}^{-2}$$

$$R_2 = \frac{1 - \varepsilon_2}{\varepsilon_2 A_2} = \frac{1 - 0.8}{0.8 \times 25.132} = 0.0099 \text{ m}^{-2}$$

$$R_{1-3} = \frac{1}{A_1 F_{1-3}} = \frac{1}{0.7853 \times 0} = \infty$$

$$R_{2-3} = \frac{1}{A_2 F_{2-3}} = \frac{1}{25.132 \times 0.468} = 0.085 \text{ m}^{-2}$$

Using value in eqns. (*i*) and (*ii*)

$$\frac{56700 - J_1}{0.3183} + \frac{J_2 - J_1}{1.273} + \frac{459.27 - J_1}{\infty} = 0$$

$$\frac{1451.51 - J_2}{0.0099} + \frac{J_1 - J_2}{1.273} + \frac{459.27 - J_2}{0.085} = 0$$

Simplification leads to

$$226764.37 - 4.99 J_1 + J_2 = 0$$

$$163.45 - J_2 + 0.7884 = 0$$

Its solution leads to

$$J_1 = 5.39 \times 10^4 \text{ W/m}^2, J_2 = 42556 \text{ W/m}^2$$

(*i*) Now net rate of radiation from heater to hemisphere

$$\frac{E_{b1} - J_1}{R_1} = \frac{56700 - 5.39 \times 10^4}{0.3183}$$

$= \mathbf{8796.7 \text{ W}}$. **Ans.**

(*ii*) Net rate of heat radiation from heater to surroundings

$$\frac{J_2 - E_{b3}}{R_{2-3}} = \frac{42556 - 459.27}{0.085}$$

$= \mathbf{4.95 \times 10^5 \text{ W}}$. **Ans.**

Comment. Heater does not radiate heat directly ($F_{13} = 0$), but the heat transfer to hemisphere is transferred to surroundings.

Example 13.25. *A short cylinder enclosure is formed with three surfaces, a circular plane surface 1 of radius 20 cm maintained at 2000 K and having emissivity of 0.8, another circular plane surface 2 of same size as surface 1 having emissivity of 0.5 and maintained at 500 K. The surface 1 and 2 are parallel to each other and the distance between them is 5 cm. The third surface is reradiating, which forms an enclosure. Draw an equivalent circuit and compute all resistances. Also find, (i) temperature attained by reradiating surface, and (ii) net heat transfer rate between surface 1 and 2 due to radiation.*

Use the following expression for finding the shape factor between two circular discs, coaxial and parallel areas :

$$F_{1-2} = \frac{1}{2B^2}\left[X - \sqrt{X^2 - 4B^2C^2}\right]$$

where, $B = \frac{r_1}{H}$, $C = \frac{r_2}{H}$ *and* $X = (1 + B^2 + C^2)$

where, r_1 and r_2 are the radii of the circular planes and H is the distance between them. (P.U., Nov. 1992)

Solution

Given : A short circular cylinder consists of a two parallel plane and a reradiating lateral surface with

$r_1 = r_2 = 20 \text{ cm} = 0.2 \text{ m}$, $\quad H = 5 \text{ cm} = 0.05 \text{ m}$

$T_1 = 2000 \text{ K}$, $\quad T_2 = 500 \text{ K}$

$\varepsilon_1 = 0.8$, $\quad \varepsilon_2 = 0.5$.

To find :

(*i*) To draw an equivalent electric circuit.

(*ii*) Net heat transfer rate between two parallel surfaces.

(*iii*) Temperature, T_3 attained by reradiating surface.

Assumptions :

1. Steady state conditions.
2. All surfaces are opaque, diffuse and gray.
3. Convection effects are negligible.

Analysis : (*i*) The schematic and thermal network are shown below :

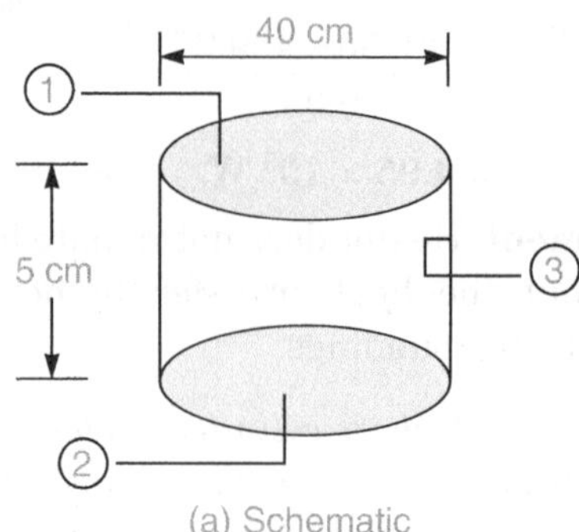

(a) Schematic

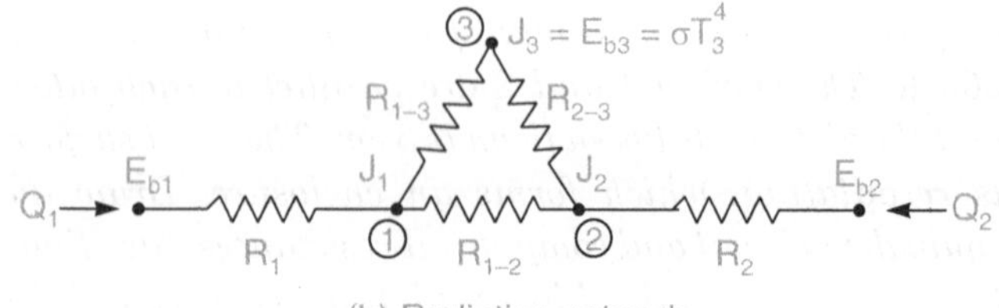

(b) Radiation network

Fig. 13.43

(*ii*) The shape factor :

$$B = C = \frac{r_1}{H} = \frac{20 \text{ cm}}{5 \text{ cm}} = 4$$

and $$X = 1 + (4)^2 + (4)^2 = 33$$

Using given relation :

$$F_{1-2} = \frac{1}{2B^2}\,[X - \sqrt{X^2 - 4\,B^2C^2}\,]$$

$$F_{1-2} = \frac{1}{2 \times (4)^2}\,[33 - \sqrt{(33)^4 - 4 \times (4)^2 \times (4)^2}\,]$$

$$= 0.78$$

From symmetry,

$$F_{1-2} = F_{2-1} = 0.78$$

and by summation rule,

$$F_{1-2} + F_{1-3} = 1 \quad \text{or} \quad F_{1-3} = 1 - 0.78 = 0.22$$

By symmetry,

$$F_{2-3} = F_{1-3} = 0.22$$

The area of surfaces :

$$A_1 = A_2 = \pi\, r_1^2 = (\pi) \times (0.2)^2 = 0.12566$$

The various resistances :

$$R_1 = \frac{1-\varepsilon_1}{\varepsilon_1 A_1} = \frac{1-0.8}{0.8 \times 0.12566} = 1.989 \text{ m}^{-2}$$

$$R_2 = \frac{1-\varepsilon_2}{\varepsilon_2 A_2} = \frac{1-0.5}{0.5 \times 0.12566} = 7.957 \text{ m}^{-2}$$

$$R_{1-2} = \frac{1}{A_1 F_{1-2}} = \frac{1}{0.12566 \times 0.78} = 10.2 \text{ m}^{-2}$$

$$R_{1-3} = \frac{1}{A_1 F_{1-3}} = \frac{1}{0.12566 \times 0.22} = 36.17 \text{ m}^{-2}$$

$$R_{2-3} = R_{1-3}$$

The emissive powers,

$$E_{b1} = \sigma\, T_1^4 = 5.67 \times 10^{-8} \times (2000)^4 = 907200 \text{ W/m}^2$$

$$E_{b2} = \sigma\, T_2^4 = 5.67 \times 10^{-8} \times (500)^4 = 3543.75 \text{ W/m}^2$$

The resistance R_{1-3} and R_{2-3} are in series, thus total resistance,

$$R_{s1} = 36.17 + 36.17 = 72.34 \text{ m}^{-2}$$

This total series resistance R_{s1} and R_{1-2} are in parallel to each other, thus

$$\frac{1}{R_{eq}} = \frac{1}{R_{s1}} + \frac{1}{R_{1-2}}$$

$$= \frac{1}{72.34} + \frac{1}{10.2} = 0.11186$$

$$R_{eq} = 8.94 \text{ m}^{-2}$$

The heat transfer rate between two surfaces,

$$Q_{1-2} = \frac{E_{b1} - E_{b2}}{R_1 + R_{eq} + R_2}$$

$$= \frac{907200 - 3543.75}{1.989 + 8.94 + 7.957}$$

$$= 47849 \text{ W} = \mathbf{47.849 \text{ kW}}.$$ **Ans.**

(*iii*) The temperature of reradiating surface may be calculated by obtaining J_3.

$$Q_{1-2} = \frac{E_{b1} - J_1}{R_1}$$

or $$J_1 = (907200 \text{ W/m}^2) - (47849 \text{ W}) \times (1.989 \text{ m}^{-2}) = 812028.34 \text{ W/m}^2$$

$$Q_{2-1} = -Q_{1-2} = \frac{E_{b2} - J_2}{R_2}$$

or $J_2 = (47849 \text{ W}) \times (7.957 \text{ m}^{-2}) + (3543.75 \text{ W/m}^2)$

$= 384278.24 \text{ W/m}^2$

For reradiating surface

$$\frac{J_1 - J_3}{R_{1-3}} = \frac{J_3 - J_2}{R_{2-3}}$$

$$\frac{812028.34 - J_3}{36.17} = \frac{J_3 - 384278.24}{36.17}$$

or $J_3 = 598153.3 \text{ W/m}^2$

$= 5.67 \times 10^{-8} \times T_3^4$

or $T_3 =$ **1802.2 K.** Ans.

13.6. RADIATION HEAT TRANSFER IN THREE SURFACE ENCLOSURES

Now we consider an enclosure consisting of three opaque, diffuse and gray surfaces as shown in Fig. 13.44

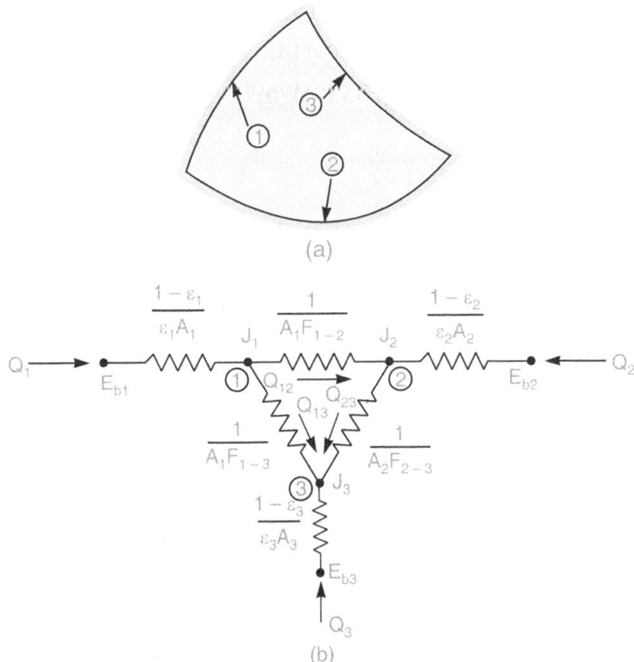

Fig. 13.44. Three surface enclosure and equivalent radiation network

Surface 1, 2 and 3 have surface area A_1, A_2 and A_3, emissivities ε_1, ε_2 and ε_3, and uniform temperatures T_1, T_2 and T_3, respectively. The triangular circuit for three bodies radiation network is not so easy to analysis as in line circuit for two surface radiation problem. The basic approach is to apply energy conservation on each radiosity node in the circuit. The three equations for determination of three unknowns J_1, J_2 and J_3 are obtained and net heat transfer at each node is set to zero. That is

At node J_1: $$\frac{E_{b1} - J_1}{\dfrac{1-\varepsilon_1}{\varepsilon_1 A}} + \frac{J_2 - J_1}{\dfrac{1}{A_1F_{1-2}}} + \frac{J_3 - J_1}{\dfrac{1}{A_1F_{1-3}}} = 0 \quad \text{...[13.40 (a)]}$$

At node J_2: $$\frac{J_1 - J_2}{\dfrac{1}{A_1F_{1-2}}} + \frac{J_3 - J_2}{\dfrac{1}{A_2F_{2-3}}} + \frac{E_{b2} - J_2}{\dfrac{1-\varepsilon_2}{\varepsilon_2 A_2}} = 0 \quad \text{...[13.40 (b)]}$$

At node J_3: $$\frac{J_1 - J_3}{\dfrac{1}{A_1F_{1-3}}} + \frac{J_2 - J_3}{\dfrac{1}{A_2F_{2-3}}} + \frac{E_{b3} - J_3}{\dfrac{1-\varepsilon_3}{\varepsilon_3 A_3}} = 0 \quad \text{...[13.40 (c)]}$$

These three equations are solved simultaneously for determination of J_1, J_2 and J_3. Then net rate of heat transfer at each surface can be determined from eqn. (13.26).

The solution of set of equations can be simplified, if one or more surfaces are special in some way. For an example, for a black or reradiating surface $J = E_b = \sigma T^4$ and $Q = 0$ for such a surface at thermal equilibrium. If net rate of radiation heat transfer is specified at a surface instead of the temperature, then the term $\dfrac{(E_b - J)}{\dfrac{1-\varepsilon}{\varepsilon A}}$ should be replaced by Q.

Example 13.26. *A long square duct has its three surfaces 1, 2, and 3 maintained at uniform temperatures of 400 K, 500 K and 600 K, respectively, their respective emissivities are 0.9, 1.0 and 0.1. The surface 4 is subjected to a uniform heat flux of 5000 W/m² and emissivity of 0.8. Determine the net radiative heat fluxes from surfaces 1, 2, and 3 and temperature of surface 4. Assume all surfaces are gray and diffuse.*

Solution

Given : A long square duct

$T_1 = 400$ K	$\varepsilon_1 = 0.9$
$T_2 = 500$ K	$\varepsilon_2 = 1.0$
$T_3 = 600$ K	$\varepsilon_3 = 0.1$
$q_4 = 5000 \text{ W/m}^2$	$\varepsilon_4 = 0.8$.

To find :

(*i*) Radiation heat flux q_1,

(*ii*) Radiation heat flux q_2,

(*iii*) Radiation heat flux q_3,

(*iv*) Temperature of surface 4.

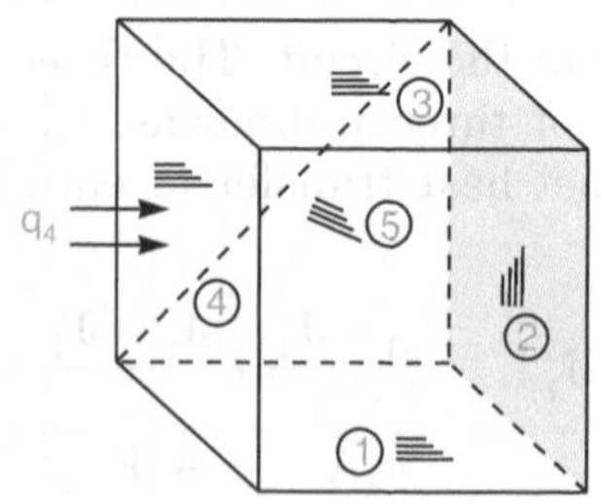

Fig. 13.45. Schematic for example 13.26

Analysis : The energy balance

For surface 1:

$$\frac{E_{b1}-J_1}{\frac{1-\varepsilon_1}{\varepsilon_1 A_1}}+\frac{J_2-J_1}{\frac{1}{A_1F_{1-2}}}+\frac{J_3-J_1}{\frac{1}{A_1F_{1-3}}}+\frac{J_4-J_1}{\frac{1}{A_1F_{1-4}}}=0 \quad ...(i)$$

For surface 2:

$$\frac{E_{b2}-J_2}{\frac{1-\varepsilon_2}{\varepsilon_2 A_2}}+\frac{J_1-J_2}{\frac{1}{A_2F_{2-1}}}+\frac{J_3-J_2}{\frac{1}{A_2F_{2-3}}}+\frac{J_4-J_2}{\frac{1}{A_2F_{2-4}}}=0 \quad ...(ii)$$

For surface 3:

$$\frac{E_{b3}-J_3}{\frac{1-\varepsilon_3}{\varepsilon_3 A_3}}+\frac{J_1-J_3}{\frac{1}{A_3F_{3-1}}}+\frac{J_2-J_3}{\frac{1}{A_3F_{3-2}}}+\frac{J_4-J_3}{\frac{1}{A_3F_{3-4}}}=0 \quad ...(iii)$$

For surface 4:

$$q_4+\frac{J_1-J_4}{\frac{1}{A_4F_{4-1}}}+\frac{J_2-J_4}{\frac{1}{A_4F_{4-2}}}+\frac{J_3-J_4}{\frac{1}{A_4F_{4-3}}}=0 \quad ...(iv)$$

where,

$E_{b1} = \sigma T_1^4 = 5.67 \times 10^{-8} \times (400)^4 = 1451.52$ W/m^2

$E_{b2} = \sigma T_2^4 = 5.67 \times 10^{-8} \times (500)^4 = 3543.75$ W/m^2

$E_{b3} = \sigma T_3^4 = 5.67 \times 10^{-8} \times (600)^4 = 7348.32$ W/m^2

The view factors

We assume diagonally imaginary surface 5 as shown in Fig. 13.45

$$F_{5-1} = F_{5-2} = F_{5-4} = F_{5-3} = 0.5$$

$$F_{1-5} = \frac{A_5}{A_1} F_{5-1} = \sqrt{2} \times 0.5 = \frac{1}{\sqrt{2}} = 0.7071$$

$$F_{1-1} + F_{1-2} + F_{1-5} = 1$$

$$\longrightarrow F_{1-2} = 1 - F_{1-5} = 0.2928$$

$$F_{1-4} = F_{1-2} = 0.2928$$

$$F_{2-1} = \frac{A_1}{A_2} F_{1-2} = 0.2928, \quad F_{2-3} = 0.2928$$

(By symmetry)

$$F_{2-4} = F_{1-3} = 1 - F_{1-2} - F_{1-4}$$

$$= 1 - 0.2928 - 0.2928 = 0.4142$$

Areas : $A_1 = A_2 = A_3 = A_4 = A$

Solution of above simultaneous equations gives

$J_1 = 1843.1$ W/m^2

$J_3 = 4555.2$ W/m^2

$J_4 = 8342.1$ W/m^2

$J_2 = E_{b2} = 3543.75$ W/m^2

(since black $\varepsilon_2 = 1$)

The heat fluxes :

$$q_1 = \frac{E_{b1}-J_1}{\frac{1-\varepsilon_1}{\varepsilon_1}} = \frac{\varepsilon_1}{1-\varepsilon_1}(E_{b1}-J_1)$$

$$= \frac{0.9}{1-0.9} \times (1451.2 - 1843.1)$$

$$= \mathbf{-3527.1\ W/m^2}. \text{ Ans.}$$

Since surface 2 is black, its surface resistance will be zero, the emissive power equals the heat flux.

$$q_2 = F_{2-1}(E_{b2}-J_1) + F_{2-3}(E_{b2}-J_3) + F_{2-4}(E_{b2}-J_4)$$

$$= 0.2928 \times (3543.75 - 1843.1) + 0.2928 \times (3543.75 - 4555.2) + 0.4142 \times [13543.75 - 8342.1]$$

$$= \mathbf{-1785.6\ W/m^2}. \text{ Ans.}$$

$$q_3 = \frac{\varepsilon_3}{1-\varepsilon_3}(E_{b3}-J_3)$$

$$= \frac{0.1}{1-0.1} \times (7348.32 - 4555.2)$$

$$= 310.34 \text{ W/m}^2$$

$$q_4 = \frac{\varepsilon_4}{1-\varepsilon_4}(E_{b4}-J_4)$$

or

$$E_{b4} = J_4 + \frac{1-\varepsilon_4}{\varepsilon_4} q_4$$

$$= 8342.1 + \frac{1-0.8}{0.8} \times 5000 = 9592.1$$

Further $E_{b4} = \sigma T_4^4$

or

$$T_4^4 = \frac{9592.1}{5.67 \times 10^{-8}} \longrightarrow T_4$$

$$= \mathbf{641.3\ K}. \text{ Ans.}$$

Example 13.27. *Two sides of a long triangular duct, as shown in Fig. 13.46 (a), are made of stainless steel ($\varepsilon = 0.5$) and are maintained at 500°C. The third side is of copper ($\varepsilon = 0.15$) and has a uniform temperature of 100°C. Calculate the rate of heat transfer to the copper base per metre of length of the duct.*

Solution

Given : A triangular duct as shown in Fig. 13.46 (*a*)

$T_1 = 100°C = 373$ K, $\varepsilon_1 = 0.15$

$L_1 = 0.5$ m, $T_2 = 500°C = 773$ K

$\varepsilon_2 = 0.5$, $L_2 = 0.3$ m

$T_3 = 500°C = 773$ K, $\varepsilon_3 = 0.5$

$L_3 = 0.4$ m.

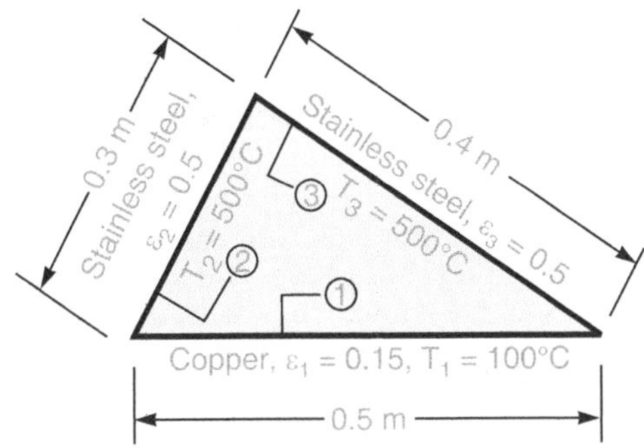

Fig. 13.46. (*a*) Illustration for example 13.27

To find: Net rate of heat transfer to copper surface 1 per metre length of duct

Assumptions :

(*i*) The duct surfaces are opaque, diffuse and gray.

(*ii*) Negligible convection from surfaces.

(*iii*) Steady state conditions.

Analysis: The view factors for the enclosure surfaces.

From Table 13.1 for triangular duct

$$F_{1-2} = \frac{A_1 + A_2 - A_3}{2\,A_1} = \frac{0.5 + 0.3 - 0.4}{2 \times 0.5} = 0.4$$

By reciprocity

$$F_{2-1} = \frac{A_1}{A_2} F_{1-2} = \frac{0.5}{0.3} \times 0.4 = 0.67$$

By summation

$$F_{1-1} + F_{1-2} + F_{1-3} = 1 \longrightarrow F_{1-3} = 1 - F_{1-2} = 0.6$$

$$F_{3-1} = \frac{A_1}{A_3} F_{1-3} = \frac{0.5}{0.4} \times 0.6 = 0.75$$

$$F_{3-1} + F_{3-2} + F_{3-3} = 1$$

$$F_{3-2} = 1 - F_{3-1} = 1 - 0.75 = 0.25$$

$$F_{2-3} = \frac{A_3}{A_2} F_{3-2} = \frac{0.4}{0.3} \times 0.25 = 0.333$$

The radiation network is shown in Fig. 13.46 (*b*):

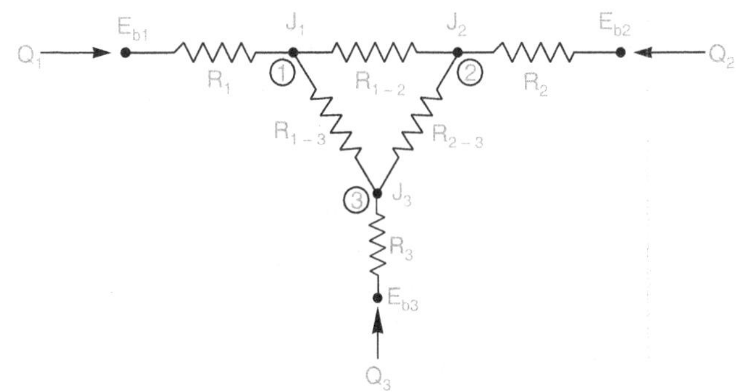

Fig. 13.46 (*b*) Radiation network

The nodel equations at three nodes are

$$\text{Node 1 :} \quad \frac{E_{b1} - J_1}{R_1} + \frac{J_2 - J_1}{R_{1-2}} + \frac{J_3 - J_1}{R_{1-3}} = 0$$

$$\text{Node 2 :} \quad \frac{E_{b2} - J_2}{R_2} + \frac{J_1 - J_2}{R_{1-2}} + \frac{J_3 - J_2}{R_{2-3}} = 0$$

$$\text{Node 3 :} \quad \frac{E_{b3} - J_3}{R_3} + \frac{J_1 - J_3}{R_{1-3}} + \frac{J_2 - J_3}{R_{2-3}} = 0$$

where J_1, J_2 and J_3 are unknowns and for per metre depth of the duct

$$R_1 = \frac{1 - \varepsilon_1}{L_1 \varepsilon_1} = \frac{1 - 0.15}{0.5 \times 0.15} = 11.33 \text{ m}^{-1}$$

$$R_{1-2} = \frac{1}{L_1 F_{12}} = \frac{1}{0.5 \times 0.4} = 5 \text{ m}^{-1}$$

$$R_2 = \frac{1 - \varepsilon_2}{L_2 \varepsilon_2} = \frac{1 - 0.5}{0.3 \times 0.5} = 3.33 \text{ m}^{-1}$$

$$R_{1-3} = \frac{1}{L_1 F_{13}} = \frac{1}{0.5 \times 0.6} = 3.33 \text{ m}^{-1}$$

$$R_{2-3} = \frac{1}{L_2 F_{23}} = \frac{1}{0.3 \times 0.333} = 10.0 \text{ m}^{-1}$$

$$R_3 = \frac{1 - \varepsilon_3}{L_3 \varepsilon_3} = \frac{1 - 0.5}{0.4 \times 0.5} = 2.5 \text{ m}^{-1}$$

$$E_{b1} = \sigma T_1^4 = (5.67 \times 10^{-8}) \times (373)^4 = 1097.5 \text{ W/m}^2$$

$$E_{b2} = \sigma T_2^4 = (5.67 \times 10^{-8}) \times (773)^4 = 20244.2 \text{ W/m}^2$$

$$E_{b3} = \sigma T_3^4 = (5.67 \times 10^{-8}) \times (773)^4 = 20244.2 \text{ W/m}^2$$

Thus,

$$\frac{1097.5 - J_1}{11.33} + \frac{J_2 - J_1}{5.0} + \frac{J_3 - J_1}{3.33} = 0$$

$$\frac{20244.2 - J_2}{3.333} + \frac{J_1 - J_2}{5.0} + \frac{J_3 - J_2}{10.0} = 0$$

$$\frac{20244.2 - J_3}{2.5} + \frac{J_1 - J_3}{3.33} + \frac{J_2 - J_3}{10.0} = 0$$

Simplification leads to three simultaneous equations

$$6.67\,J_1 - 2.268\,J_2 - 3.395\,J_3 = 1097.5$$
$$-2.0\,J_1 + 56\,J_2 - J_3 = 60732.65$$
$$-3.0\,J_1 - J_2 + 8.0\,J_3 = 80976.8$$

Solving these equations

$$J_1 = 15767.3\ \text{W/m}^2$$
$$J_2 = 18434.4\ \text{W/m}^2$$
$$J_3 = 18340\ \text{W/m}^2$$

$$q_{net-1} = \frac{E_{b1} - J_1}{R_1} = \frac{1097.5 - 15767.3}{11.33}$$

$$= -\mathbf{1295.1\ W/m.}\quad \text{Ans.}$$

Heat energy received by copper plate is 1295.1 W/m.

13.7. RADIATION SHIELDS

The radiation heat transfer rate between the two surfaces can be reduced significantly by placing a thin sheet of high reflectivity (very low emissivity) material, between them Fig. 13.47. Such highly reflective thin sheets are called the **radiation shields.** The radiation shields increase the thermal resistance in path of radiation heat transfer and hence reduce the heat flow rate. Multilayer radiation shields constructed of about 20 sheets per cm thickness separated by an evacuated space are commonly used in cryogenic and space applications.

Consider a radiation shield is placed between two large plates as shown in Fig. 13.48 (*a*). Let the emissivities of the radiation shield facing plate 1 and 2 be $\varepsilon_{1,3}$ and $\varepsilon_{3,2}$, respectively. Since, the radiation shield does not deliver or remove any amount of heat from the system. Thus the heat transfer rate between plate 1 and radiation shield must be equal to the heat transfer rate between the shield and plate 2.

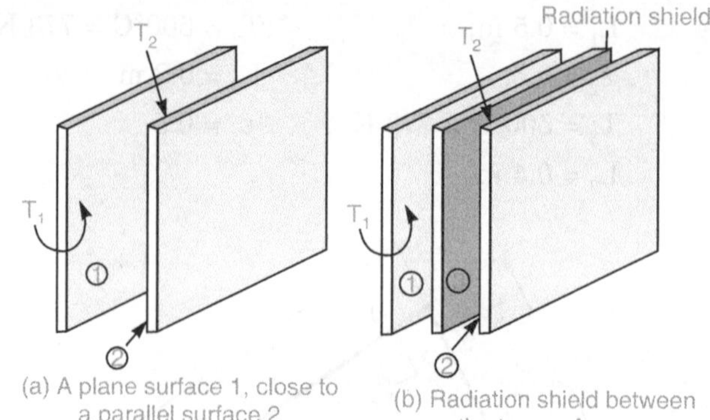

Fig. 13.47. Radiation shield between two parallel surfaces to reduce radiative heat transfer

$$\left\{\frac{Q}{A}\right\}_{1-3} = \left\{\frac{Q}{A}\right\}_{3-2}$$

or
$$\frac{\sigma(T_1^4 - T_3^4)}{\dfrac{1}{\varepsilon_1} + \dfrac{1}{\varepsilon_{3,1}} - 1} = \frac{\sigma(T_3^4 - T_2^4)}{\dfrac{1}{\varepsilon_{3,2}} + \dfrac{1}{\varepsilon_2} - 1} \quad \text{...(13.41)}$$

The temperature of radiation shield T_3 is unknown. For instance, if all emissivities are equal :

or
$$\varepsilon_1 = \varepsilon_2 = \varepsilon_{3,1} = \varepsilon_{3,2}$$

Then
$$T_3^4 = \frac{(T_1^4 - T_2^4)}{2} \quad \text{...(13.42)}$$

The heat transfer rate between the plates with a shield becomes

$$Q_{1-2,\text{ one shield}} = \frac{(1/2)\,\sigma A(T_1^4 - T_2^4)}{\dfrac{1}{\varepsilon_1} + \dfrac{1}{\varepsilon_2} - 1} \quad \text{...(13.43)}$$

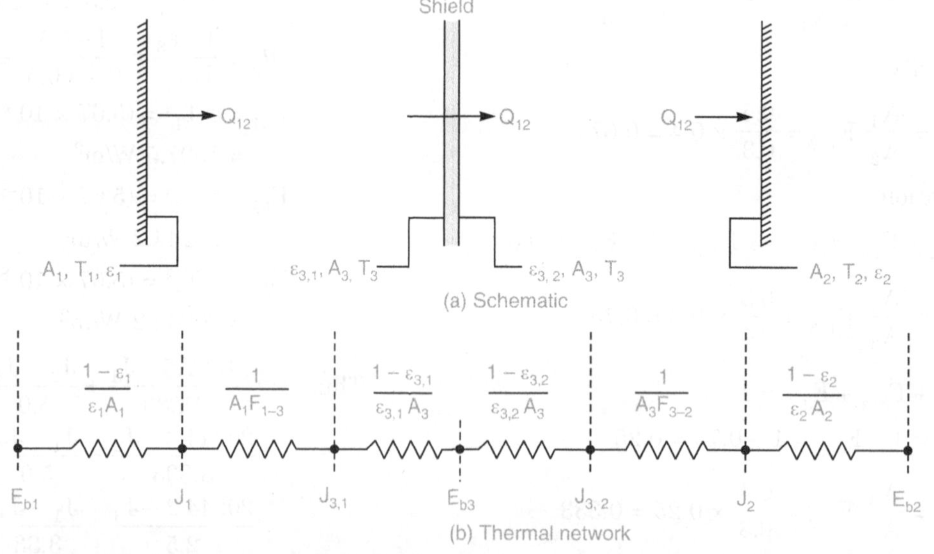

Fig. 13.48. Radiation shield placed between two large parallel plates

Thus by inserting a shield, the radiation heat flow rate becomes just half of that would be experienced without radiation shield. The radiation network with one radiation shield is shown in Fig. 13.48 (*b*). All the resistances are in series, and thus the radiation heat transfer rate is

$$Q_{1-2,\text{ one shield}} = \frac{E_{b1} - E_{b2}}{\dfrac{1-\varepsilon_1}{\varepsilon_1 A_1} + \dfrac{1}{A_1 F_{1-3}} + \dfrac{1-\varepsilon_{3,1}}{\varepsilon_{3,1} A_3} + \dfrac{1-\varepsilon_{3,2}}{\varepsilon_{3,2} A_3} + \dfrac{1}{A_3 F_{3-2}} + \dfrac{1-\varepsilon_2}{\varepsilon_2 A_2}} \quad ...(13.44)$$

For large parallel plates, $A_1 = A_2 = A_3 = A$ and $F_{1-3} = F_{3-2} = 1$, then eqn. (13.44) simplifies to

$$Q_{1-2,\text{ one shield}} = \frac{\sigma A(T_1^4 - T_2^4)}{\left\{\dfrac{1}{\varepsilon_1} + \dfrac{1}{\varepsilon_2} - 1\right\} + \left\{\dfrac{1}{\varepsilon_{3,1}} + \dfrac{1}{\varepsilon_{3,2}} - 1\right\}} \quad ...(13.45)$$

where the terms in the second set of parentheses in the denominator represent the additional resistance to radiation heat transfer introduced by shield. If N radiation shields are inserted between two large parallel plates, then eqn. (13.45) becomes

$$Q_{1-2,\text{ N, shield}} = \frac{\sigma A(T_1^4 - T_2^4)}{\left\{\dfrac{1}{\varepsilon_1} + \dfrac{1}{\varepsilon_2} - 1\right\} + \left\{\dfrac{1}{\varepsilon_{3,1}} + \dfrac{1}{\varepsilon_{3,2}} - 1\right\} + \ldots\ldots + \left\{\dfrac{1}{\varepsilon_{N,1}} + \dfrac{1}{\varepsilon_{N,2}} - 1\right\}}$$

If the emissivities of all the surfaces are equal then above expression reduces to

$$Q_{1-2,\text{ N, shield}} = \frac{\sigma A(T_1^4 - T_2^4)}{(N+1)\left\{\dfrac{2}{\varepsilon} - 1\right\}} \quad ...(13.46)$$

A radiation shield may have same emissivity on its two faces or the emissivity of one face of shield may differ from that associated with opposite side. If the emissivities associated with two sides of shield are very low, the heat flow can be reduced drastically.

Example 13.28. *Two large parallel plates at temperature 1000 K and 600 K have emissivity of 0.5 and 0.8 respectively. A radiation shield having emissivity 0.1 on one side and 0.05 on the other side is placed between the plates. Calculate the heat transfer rate by radiation per square metre with and without radiation shield.*

Solution

Given : Two parallel infinite surfaces with and without radiation shield

$T_1 = 1000$ K, $T_2 = 600$ K, $\varepsilon_1 = 0.5$

$\varepsilon_2 = 0.8$, $\varepsilon_{3,1} = 0.1$, $\varepsilon_{3,2} = 0.05$.

To find :

(*i*) The radiation heat transfer rate without radiation shield.

(*ii*) The radiation heat transfer rate with radiation shield.

Assumptions:

1. Surfaces are diffused and gray.
2. Heat is transferred by radiation only.
3. The conduction resistance of radiation shield is negligible.

Analysis: (*i*) The net radiation heat exchange between two parallel plates without radiation shield can be expressed as :

$$\frac{Q_1}{A} = \frac{\sigma(T_1^4 - T_2^4)}{\dfrac{1}{\varepsilon_1} + \left(\dfrac{1}{\varepsilon_2} - 1\right)}$$

$$\frac{Q_1}{A} = \frac{5.67 \times 10^{-8} \times (1000^4 - 600^4)}{\dfrac{1}{0.5} + \left(\dfrac{1}{0.8} - 1\right)}$$

$= \mathbf{21934\ W/m^2}$. **Ans.**

(*ii*) When a radiation shield is placed between the parallel plates, then the radiation heat transfer can be calculated as $Q_2 = \dfrac{E_{b1} - E_{b2}}{\Sigma R_{th}}$

For 1 m^2 area of the plates

where $\Sigma R_{th} = \dfrac{1}{\varepsilon_1} + \dfrac{1}{\varepsilon_2} - 1 + \dfrac{1}{\varepsilon_{3,1}} + \dfrac{1}{\varepsilon_{3,2}} - 1$

$$\Sigma R_{th} = \frac{1}{0.5} + \frac{1}{0.8} - 1 + \frac{1}{0.1} + \frac{1}{0.05} - 1$$

$$\Sigma R_{th} = 2 + 1.25 - 1 + 10 + 20 - 1 = 31.25\ m^{-2}$$

The heat transfer gain with radiation shield

$$\frac{Q_2}{A} = \frac{5.67 \times 10^{-8} \times (1000^4 - 600^4)}{31.251}$$

$= \mathbf{1579.25\ W/m^2}$. **Ans.**

The presence of radiation shield reduces the heat transfer rate

$$\frac{21934 - 1579.25}{21934} = 92.8\%\ \text{(reduction)}.$$

Example 13.29. *Two large parallel planes with emissivity 0.6 are at 900 K and 300 K. A radiation shield with one side polished and having emissivity of 0.05, while the emissivity of other side is 0.4 is proposed to be used. Which side of the shield to face the hotter plane, if the temperature of shield is to be kept minimum ? Justify your answer.* (P.U., May 2001)

Solution

Given : A radiation shield between two large parallel planes.

$\varepsilon_1 = \varepsilon_2 = 0.6,$ $T_1 = 900$ K

$T_2 = 300$ K, $\varepsilon_3 = \varepsilon_{polished} = 0.05$

$\varepsilon_4 = \varepsilon_{unpolished} = 0.4$

T_3 = Temperature of shield.

To find : Which side of the shield should face the hotter plane?

Assumptions :

1. Each surface has a uniform radiosity ; the configuration can be considered as an enclosure with two surfaces.

2. The surfaces are gray, diffuse and opaque.

3. The medium between the surfaces does not participate in radiation.

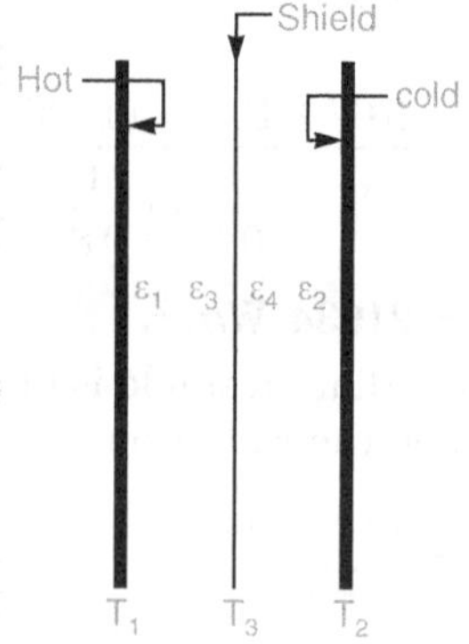

Fig. 13.49. Schematic for example 13.29

Analysis : Arrangement 1 : Let polished surface of the shield face hotter plane, the energy balance on the radiation shield :

$$\left(\frac{Q}{A}\right)_{1-3} = \left(\frac{Q}{A}\right)_{3-2}$$

$$\Rightarrow \quad \frac{\sigma(T_1^4 - T_3^4)}{\dfrac{1}{\varepsilon_1} + \dfrac{1}{\varepsilon_3} - 1} = \frac{\sigma(T_3^4 - T_2^4)}{\dfrac{1}{\varepsilon_4} + \dfrac{1}{\varepsilon_2} - 1}$$

or
$$\frac{(900\text{ K})^4 - T_3^4}{\dfrac{1}{0.6} + \dfrac{1}{0.05} - 1} = \frac{T_3^4 - (300\text{ K})^4}{\dfrac{1}{0.4} + \dfrac{1}{0.6} - 1}$$

$$\frac{6.561 \times 10^{11} - T_3^4}{20.67} = \frac{T_3^4 - 8.1 \times 10^9}{3.167}$$

$$\Rightarrow \quad 6.561 \times 10^{11} - T_3^4 = 6.526\, T_3^4 - 5.286 \times 10^{10}$$

$$\Rightarrow \quad 7.526\, T_3^4 = 7.0896 \times 10^{11}$$

$$\Rightarrow \quad T_3 = 554 \text{ K}$$

Arrangement 2 : Let the polished side of shield face colder side : Energy balance on shield.

$$\frac{\sigma(T_1^4 - T_3^4)}{\dfrac{1}{\varepsilon_1} + \dfrac{1}{\varepsilon_4} - 1} = \frac{\sigma(T_3^4 - T_2^4)}{\dfrac{1}{\varepsilon_3} + \dfrac{1}{\varepsilon_2} - 1}$$

$$\Rightarrow \quad \frac{(900)^4 - T_3^4}{\dfrac{1}{0.6} + \dfrac{1}{0.4} - 1} = \frac{T_3^4 - (300)^4}{\dfrac{1}{0.05} + \dfrac{1}{0.6} - 1}$$

$$\Rightarrow \quad \frac{6.561 \times 10^{11} - T_3^4}{3.167} = \frac{T_3^4 - 8.1 \times 10^9}{20.67}$$

or $\quad 4.2486 \times 10^{12} - 6.526\, T_3^4 = T_3^4 - 8.1 \times 10^9$

or $\quad 7.526\, T_3^4 = 4.2567 \times 10^{12}$

or $\quad T_3 = 867.2$ K

which is greater than the temperature of shield, when it faces the hotter plane. Therefore, the radiation shield will be effective, when its polished side will face the hotter plane.

Example 13.30. *A cryogenic fluid flows through a long tube of 20 mm diameter, the outer surface of which is diffuse and gray (ε_1 = 0.02) at 77 K. This tube is concentric with a larger tube of 50 mm diameter, the inner surface of which is diffuse and gray (ε_2 = 0.05) and at 300 K. The space between the surfaces is evacuated. Calculate the heat gain by cryogenic fluid per unit length of tubes. If a thin radiation shield of 35 mm diameter (ε_3 = 0.02) both sides is inserted midway between the inner and outer surfaces, calculate the percentage change in heat gain per unit length of the tube.* (P.U., Dec. 2001 ; N.M.U., May 1998)

Solution

Given : Concentric tube arrangement with diffuse and gray surfaces.

To find :

(*i*) Heat gain by cryogenic fluid passing through the inner tube.

(*ii*) Percentage change in heat gain with radiation shield inserted midway between inner and outer pipe.

Assumptions :

1. Space between tubes is evacuated.

2. No conduction and convection involve.

3. Infinite long concentric tubes.

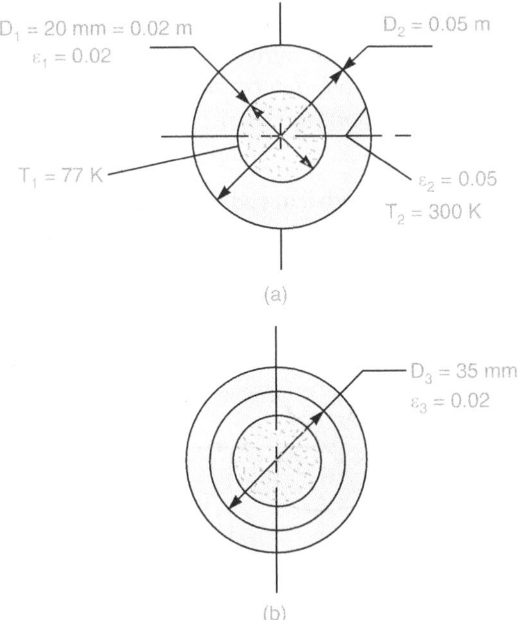

Fig. 13.50. Schematic

Analysis : For long concentric cylinders

$$F_{1-2} = F_{1-3} = F_{3-2} = 1$$

(*i*) The radiation exchange without radiation shield can be calculated as

$$Q = \frac{A_1\sigma(T_1^4 - T_2^4)}{\frac{1}{\varepsilon_1} + \left(\frac{1}{\varepsilon_2} - 1\right)\frac{A_1}{A_2}}$$

where $\frac{A_1}{A_2} = \left(\frac{D_1}{D_2}\right) = \left(\frac{0.02}{0.05}\right) = 0.4$

$$Q_1 = \frac{\pi \times 0.02 \times 1 \times 5.67 \times 10^{-8} \times (77^4 - 300^4)}{\frac{1}{0.02} + \left(\frac{1}{0.05} - 1\right) \times 0.4}$$

$= -$ **50 W/m.** Ans.

(*ii*) When radiation shield is placed in midway between two tubes,

$$Q_2 = \frac{E_{b1} - E_{b2}}{\Sigma R_{th}}$$

where
$$\Sigma R_{th} = \frac{1-\varepsilon_1}{\varepsilon_1 A_1} + \frac{1}{A_1 F_{1-3}} + \frac{2(1-\varepsilon_3)}{\varepsilon_3 A_3} + \frac{1}{A_3 F_{3-2}} + \frac{1-\varepsilon_2}{\varepsilon_2 A_2}$$

For 1 m length of tube

$$\Sigma R_{th} = \frac{1-0.02}{0.02 \times \pi \times 0.02} + \frac{1}{\pi \times 0.02 \times 1} + \frac{2 \times (1-0.02)}{0.02 \times \pi \times 0.035} + \frac{1}{\pi \times 0.035 \times 1} + \frac{1-0.05}{0.05 \times \pi \times 0.05}$$

$$\Sigma R_{th} = 779.9 + 15.9 + 891.3 + 121.0 = 1817 \text{ m}^{-2}$$

The heat gain with radiation shield

$$Q_2 = \frac{\pi \times 0.02 \times 1 \times 5.67 \times 10^{-8} \times (77^4 - 300^4)}{1817}$$

$= -$ 25 W/m

The percentage reduction in heat gain

$$= \frac{(Q_1 - Q_2)}{Q_1} = \frac{(-50 + 25)}{-50} = \mathbf{50\%}. \text{ Ans.}$$

Example 13.31. *A pipe carrying steam having an outside diameter of 20 cm runs in a large room and is exposed to air at a temperature of 30°C. The pipe surface temperature is 400°C. Calculate the heat loss to the surroundings per metre length of pipe due to thermal radiation. The emissivity of the pipe surface is 0.8.*

(i) What would be the loss of heat due to radiation, if the pipe is enclosed in a 50 cm diameter brick conduit of emissivity of 0.9 ?

(ii) What would be the radiation heat transfer from the pipe, if it is enclosed within a square conduit of 0.5 m side of emissivity of 0.9 ? (Anna Univ., May 2001)

Solution

Given : A pipe carrying steam with

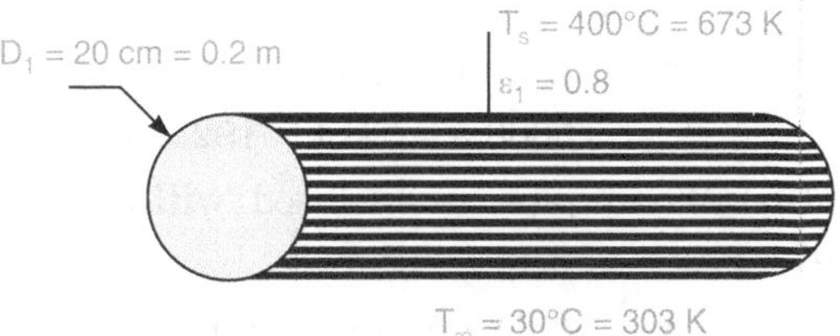

Fig. 13.51

1. Brick conduit,

D_2 = 50 cm = 0.5 m, $\varepsilon_2 = 0.9$

2. Square conduit of side

w = 0.5 m, $\varepsilon_3 = 0.9$.

To find :

(*i*) Net radiation heat transfer from pipe surface.

(*ii*) The radiation heat exchange when pipe is enclosed within a 50 cm diameter, brick conduit.

(*iii*) The radiation heat exchange, when pipe is enclosed within a square conduit.

Assumptions :

1. Surfaces are opaque, diffuse and gray.
2. Space between two concentric pipes is evacuated.
3. No conduction and convection heat transfer.

Analysis : (*i*) The net radiation heat exchange from pipe surface to room can be expressed as :

$$\frac{Q}{L} = \varepsilon\,(\pi D_1)\,\sigma\,(T_s^4 - T_\infty^4)$$

$$= 0.8 \times \pi \times 0.2 \times 5.67 \times 10^{-8} \times (673^4 - 303^4)$$

$$= \mathbf{5606.5\ W/m.}$$ **Ans.**

(*ii*) The radiation heat exchange between pipe and a conduit can be calculated as :

$$Q = \frac{A_1 \sigma (T_s^4 - T_\infty^4)}{\frac{1}{\varepsilon_1} + \left(\frac{1}{\varepsilon_2} - 1\right)\frac{A_1}{A_2}}$$

When pipe is enclosed within brick conduit :

$$\frac{A_1}{A_2} = \left(\frac{D_1}{D_2}\right) = \left(\frac{0.2}{0.5}\right) = 0.4$$

$$\frac{Q}{L} = \frac{\pi \times 0.2 \times 5.67 \times 10^{-8} \times (673^4 - 303^4)}{\frac{1}{0.8} + \left(\frac{1}{0.9} - 1\right) \times 0.4}$$

$$= 5414\ W/m$$

The reduction in heat radiation

$$= 5606.5 - 5414 = \mathbf{192.5\ W/m.}$$ **Ans.**

(*iii*) When pipe is enclosed within a square conduit :

$$\frac{A_1}{A_2} = \left(\frac{\pi D_1 L}{4wL}\right) = \left(\frac{\pi \times 0.2}{4 \times 0.5}\right) = 0.314$$

$$\frac{Q}{L} = \frac{\pi \times 0.2 \times 5.67 \times 10^{-8} \times (673^4 - 303^4)}{\frac{1}{0.8} + \left(\frac{1}{0.9} - 1\right) \times 0.314}$$

$$= 5454.2\ W/m$$

The reduction in heat radiation

$$= 5606.5 - 5454.2 = \mathbf{152.3\ W/m.}$$ **Ans.**

Example 13.32. *A cylindrical rod (ε = 0.7) of 50 mm diameter is maintained at 1000°C by an electric resistance heating and is kept in a room, the walls (ε = 0.6) of which are at 15°C. Determine the energy which must be supplied per metre length of the rod. If an insulated half circular reflector of 0.45 m diameter. is placed around the rod, determine the energy supplied to the rod per metre length.*

Solution

Given: A cylindrical rod (1) exposed in a room (2).

$\varepsilon_1 = 0.7$, $D_1 = 50$ mm

$T_1 = 1000°C = 1273$ K, $\varepsilon_2 = 0.6$

$T_2 = 15°C = 288$ K

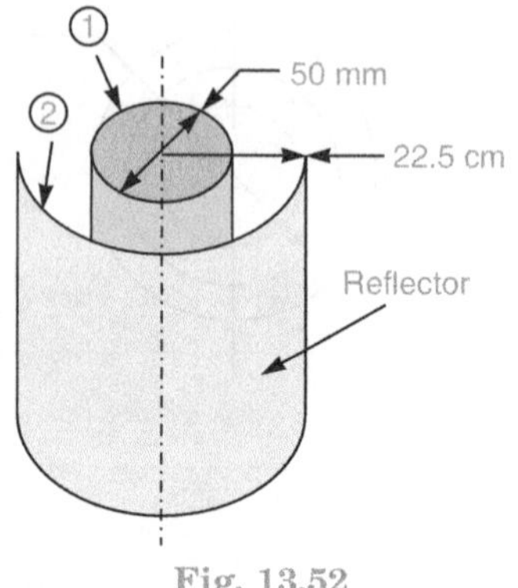

Fig. 13.52

If reflector is placed

$D_2 = 0.45$ mm.

To find : Energy supplied to rod per metre, if

(*i*) Rod is exposed to room, with negligible convection effects, and

(*ii*) Rod is covered by half circular reflector.

Fig. 13.52(*a*). Thermal network

Assumptions :

1. Rod and room surfaces are opaque, diffuse and gray.
2. Steady state conditions.
3. Room air does not participate in radiation.

Analysis: (*i*) When cylindrical rod is exposed to room

$$Q = \frac{\sigma (T_1^4 - T_2^4)}{R_1 + R_{1-2} + R_2}$$

where R_1 = surface resistance of rod

$$= \frac{1 - \varepsilon_1}{\varepsilon_1 A_1} = \frac{1 - 0.7}{0.7 \times \pi \times 0.05 \times 1} = 2.728\ m^{-2}$$

$$R_{1-2} = \frac{1}{A_1 F_{1-2}} = \frac{1}{\pi \times 0.05 \times 1} = 6.366\ m^{-2}$$

($F_{1-2} = 1.0$. All heat transfers to room)

$R_2 \approx 0$, negligible heat transmission to rod from room.

$$\text{Then } Q_1 = \frac{5.67 \times 10^{-8} \times (1273^4 - 288^4)}{2.728 + 6.366 + 0}$$

$= \mathbf{16{,}330\ W.}$ **Ans.**

(*ii*) When half circular reflector is placed around the rod.

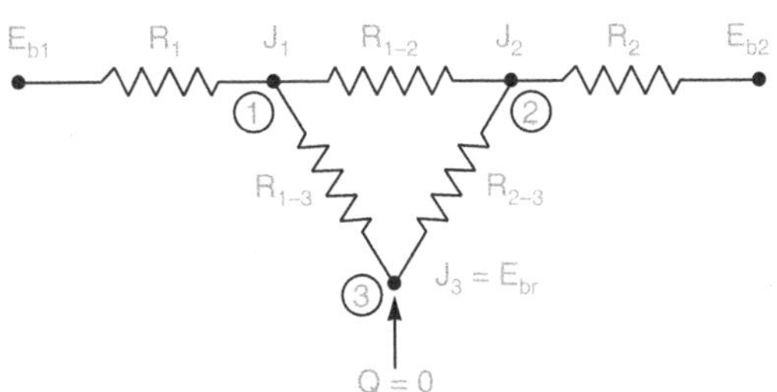

Fig. 13.52(*b*). Thermal network

$R_1 = 2.728 \quad R_2 = 0.0$

Since reflector is half circular so half of the energy radiated by rod falls on it

$$R_{1-3} = \frac{1}{A_1F_{1-3}} = \frac{1}{\pi \times (0.05) \times (0.5)} = 12.732\ m^{-2}$$

Resistance to radiation from half reflector to room

$$R_{2-3} = \frac{1}{A_2F_{2-3}} = \frac{1}{\frac{\pi}{2}(D_2 - D_1)L\,F_{2-3}}$$

$$= \frac{1}{\frac{\pi}{2} \times (0.45 - 0.05) \times 1 \times 1.0} = 1.591\ m^{-2}$$

Half of the radiation from the rod reaches the room directly

$$R_{1-2} = \frac{1}{A_1F_{1-2}} = \frac{1}{\pi \times 0.05 \times 0.5} = 12.732\ m^{-2}$$

Total of series resistances R_{1-3} and R_{2-3}

$R_{s_1} = 12.732 + 1.591 = 14.323\ m^{-2}$

The resistance R_{s_1} acts parallel to resistance R_{1-2}, thus

$$\frac{1}{R_{eq}} = \frac{1}{R_{1-2}} + \frac{1}{R_{s_1}} = \frac{1}{12.732} + \frac{1}{14.323} = 0.148$$

or $\quad R_{eq} = 6.74\ m^{-2}$

Total thermal resistances

$$\Sigma R_{th} = R_1 + R_{eq} + R_2$$

$$= 2.73 + 6.74 + 0 = 9.47\ m^{-2}$$

The radiation heat transfer rate

$$Q = \frac{\sigma(T_1^4 - T_2^4)}{\Sigma R_{th}}$$

$$= \frac{5.67 \times 10^{-8} \times (1273^4 - 288^4)}{9.47}$$

$= \mathbf{15{,}681\ W.}$ **Ans.**

13.8. TEMPERATURE MEASUREMENT OF A GAS BY THERMOCOUPLE: COMBINED CONVECTIVE AND RADIATION HEAT TRANSFER

The temperature of the flowing fluid through a duct or a pipe is measured by thermocouple as shown in Fig. 13.53. The thermocouple bead is placed in direct contact of gas, the heat is convected from gas to thermocouple sensor tries to gain steady state. Like use of thermometer, the temperature measured by thermocouple is less than the true gas temperature, because a part of heat gain by thermocouple sensor is emitted to wall at low temperature. In absence of conduction, the energy balance on the thermocouple bead yields to

$$h_c(T_\infty - T_c) = \sigma\varepsilon_c(T_c^4 - T_w^4)$$

or

$$T_\infty = T_c + \frac{\sigma\,\varepsilon_c(T_c^4 - T_w^4)}{h_c} \text{ (K)} \quad ...(13.47)$$

where, h_c = convective heat transfer coefficient, W/m².K

T_c = temperature recorded by thermocouple, K

T_w = wall temperature, K

T_∞ = gas temperature, K

$\sigma = 5.67 \times 10^{-8}$ W/m².K⁴, the Stefan Boltzmann constant

ε_c = emissivity of thermocouple sensor.

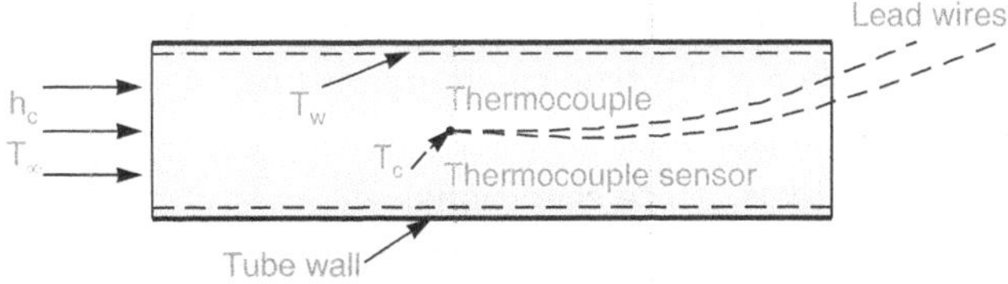

Fig. 13.53. Thermocouple in a gas stream

The last term in eqn. (13.47) is due to the *radiation effect* and represents the radiation correction. When the convection coefficient is small, then radiation correction term becomes most important, if $T_\infty >> T_w$. The large error in temperature measurement can be reduced significantly by using

1. The low emissivity thermocouple junction. The special coating of low emissivity metal like aluminium, zinc, chromium etc. can be used.

2. Placing the sensor in a radiation shield without interfering the fluid flow.

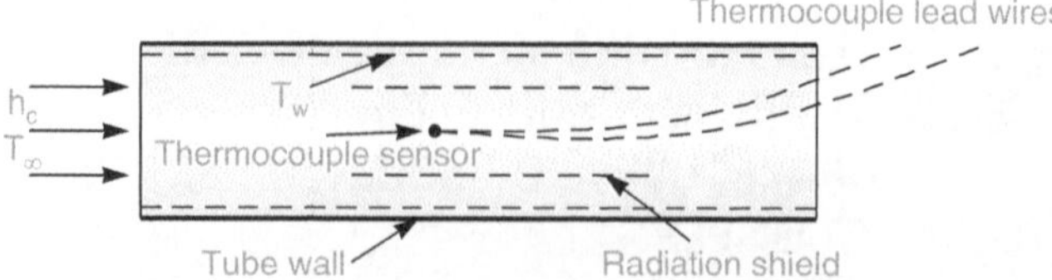

Fig. 13.54. Thermocouple with radiation shield

If thermocouple sensor is surrounded by a radiation shield as shown in Fig. 13.54. The radiation shield receives heat by convection on its two sides and it reradiates. Thus the energy balance for the radiation shield (a small body in compare to enclosure) is

$$2h_c(T_\infty - T_s) = \sigma\, \varepsilon_s(T_w^4 - T_s^4) \quad ...(13.48)$$

where, ε_s = emissivity of the radiation shield.

T_s = temperature of the radiation shield.

The energy balance on the thermocouple bead now yields

$$h_c(T_\infty - T_c) = \sigma\, \varepsilon_c(T_s^4 - T_c^4) \quad ...(13.49)$$

where T_c = temperature recorded by thermocouple.

Example 13.33. *A thermocouple is used to measure the temperature of a hot gas flowing in a tube maintained at 100°C. The thermocouple indicates a temperature of 500°C. If the emissivity of thermocouple junction is 0.5 and the convective heat transfer coefficient is 250 W/m².K, determine the actual temperature of the gas.*

Solution

Given :

T_w = 100°C = 373 K, T_c = 500°C = 773 K

ε_c = 0.5, h_c = 250 W/m².K.

To find : The true gas temperature.

Assumptions :

1. Steady state conditions.
2. Junction surface is gray and diffuse.
3. Constant properties.

Analysis : The energy balance on thermocouple bead is

$$h_c(T_\infty - T_c) = \sigma\, \varepsilon_c(T_c^4 - T_w^4)$$

Using the numerical values,

$$250 \times (T_\infty - 500) = 5.67 \times 10^{-8} \times 0.5 \times (773^4 - 373^4)$$

or $T_\infty = 500 + 38 =$ **538°C. Ans.**

Example 13.34. *A thermocouple is used to measure the temperature of gas flowing through a duct, records 280°C. If the emissivity of the junction is 0.4 and convection coefficient is 150 W/m².K. Find the true gas temperature. The duct wall temperature is 140°C. What should be the emissivity of the junction in order to reduce the error by 30% ?*

Solution

Given : Measurement of temperature by a thermocouple

(*i*) T_c = 280°C = 553 K, ε_c = 0.4

h_c = 150 W/m².K, T_w = 140°C = 413 K

(*ii*) Error to be reduced by 30%.

To find :

(*i*) True temperature of the gas, and

(*ii*) Emissivity of the junction in order to reduce the error by 30%.

Analysis: (*i*) The true gas temperature is determined by eqn. (13.47)

$$T_\infty = T_c + \frac{\sigma\varepsilon_c\,(T_c^4 - T_w^4)}{h_c}$$

$$= 553 + \frac{5.67 \times 10^{-8} \times 0.4 \times (553^4 - 413^4)}{150}$$

$$= 562.74 \text{ K}$$

The true temperature of gas is **289.74°C** (562.74 K). **Ans.**

(*ii*) The error in temperature measurement

$$= 289.74 - 280 = 9.74°C$$

Error to be reduced by 30%, thus the remaining error is 70% of 9.74°C

$$= 0.7 \times 9.74 = 6.818°C$$

Thus T_c = 286.818°C = 559.818 K

T_∞ = 562.74 K, T_w = 413 K

Then, $$562.74 = 559.81 + \frac{5.67 \times 10^{-8}\, \varepsilon_c\,(559.818^4 - 413^4)}{150}$$

or $\varepsilon_c =$ **0.11. Ans.**

Example 13.35. *A thermocouple (ε = 0.6) is used to measure the temperature of exhaust gas in a large duct. The temperature of the duct wall is at 20°C and temperature measured by thermocouple is 500°C. Calculate the true temperature of the gas, if the convection coefficient between gas and thermocouple bead is 200 W/m².K.*

To measure the temperature of the gas more correctly, it is enveloped by a thin radiation shield (ε = 0.3). Estimate the error between the thermocouple temperature and gas temperature with the shielded thermocouple arrangement.

Solution

Given : Thermocouple without radiation shield :

$T_w = 20°C = 293$ K, $T_c = 500°C = 773$ K

$\varepsilon_c = 0.6$, $h_c = 200$ W/m^2.K

Thermocouple with radiation shield :

$T_w = 20°C = 293$ K, $T_c = 500°C = 773$ K

$\varepsilon_s = 0.3$, $h_c = 200$ W/m^2.K.

To find : The true gas temperature.

Assumptions :

1. Steady state conditions.
2. Junction surface is gray and diffuse.
3. Constant properties.

Analysis : The energy balance on thermocouple sensor without radiation shield is

$$h_c(T_\infty - T_c) = \sigma\, \varepsilon_c(T_c^4 - T_w^4)$$

Using the numerical values,

$$200 \times (T_\infty - 773) = 5.67 \times 10^{-8} \times 0.6 \times (773^4 - 293^4)$$

or $T_\infty = 773 + 56.67 = \mathbf{829.67\ K.}$ **Ans.**

The energy balance on thermocouple sensor with radiation shield is

$$2h_c(T_\infty - T_s) = \sigma\, \varepsilon_s(T_w^4 - T_s^4)$$

$$2 \times 200 \times (829.67 - T_s) = 5.67 \times 10^{-8} \times 0.3 \times (293^4 - T_s^4)$$

or $331856 - 400\, T_s = 125.36 - 1.701 \times 10^{-8}\, T_s^4$

or $T_s^4 - 23515579070\, T_s + 1.95 \times 10^{13} = 0$

or $T_s = 829.3$ K

Using the shield temperature to calculate the temperature measured by thermocouple.

$$h_c(T_\infty - T_c) = \sigma\, \varepsilon_c(T_s^4 - T_c^4)$$

$$200 \times (829.67 - T_c) = 5.67 \times 10^{-8} \times 0.6 \times (829.3^4 - T_c^4)$$

It yields $T_c = 827$ K

The error between thermocouple temperature and gas temperature is of only 2°C. **Ans.**

13.9. SUMMARY

The *view factor* from surface 1 to surface 2 is designated as F_{1-2} and is defined as fraction of radiation leaving the surface 1 and that strikes surface 2, directly. The view factor F_{1-1} represents the fraction of radiation leaving surface 1 and that strikes itself directly. $F_{1-1} = 0$ for flat or convex surfaces while $F_{1-1} \neq 0$ for concave surfaces. For view factor the *reciprocity rule* is expressed as

$$A_1F_{1-2} = A_2F_{2-1}$$

For an *enclosure*, the sum of view factors from surface *i* to all surfaces of enclosure including itself must be equal to unity *i.e.*, $\sum_{i=1}^{N} F_{i-j} = 1$

or $$F_{i-1} + F_{i-2} + F_{i-3} + \ldots\ldots + F_{i-N} = 1$$

It is known as summation rule. The superposition or additive rule states that the view factor F_{1-2} is equal to the sum of view factors from surface 1 to the parts of surface 2. The symmetry rule states that if two or more surfaces are symmetric about the surface 1, then

$$F_{1-2} = F_{1-3} = \ldots\ldots$$

The rate of radiation heat transfer between *two black surfaces* is expressed as

$$Q_{1-2} = A_1\, F_{1-2}\, \sigma\, (T_1^4 - T_2^4)\ \text{(W)}$$

The net radiation heat transfer rate from a surface *i* of a *black enclosure* is sum of radiation heat transfers from surface *i* to each of the surface of enclosure :

$$Q_i = \sum_{j=1}^{N} Q_{i-j} = \sum_{j=1}^{N} A_i\, F_{i-j}\, \sigma\, (T_i^4 - T_j^4)$$

The radiation heat transfer from a *cavity* surface 1 is given by

$$Q_1 = A_1\, \sigma\, \varepsilon_1\, T_1^4 \left[\frac{1 - F_{1-1}}{1 - F_{1-1}(1 - \varepsilon_1)}\right]\ \text{(W)}$$

The net rate of radiation heat transfer from a *surface* is expressed as

$$Q = \frac{E_b - J}{R}\ \text{(W)}$$

where $R = \dfrac{1 - \varepsilon}{\varepsilon A}$ is the *surface resistance,* to radiation.

The net rate of radiation heat transfer from surface 1 to surface 2 can be expressed as

$$Q = \frac{J_1 - J_2}{R_{1-2}}\ \text{(W)}$$

where $R_{1-2} = \dfrac{1}{A_1F_{1-2}}$ is the *space resistance* to radiation.

The network method is used to solve the radiation

network. The radiation heat transfer between any two opaque, gray, and diffused surfaces is given by

$$Q_{1-2} = \frac{\sigma (T_1^4 - T_2^4)}{\frac{1-\varepsilon_1}{\varepsilon_1 A_1} + \frac{1}{A_1 F_{1-2}} + \frac{1-\varepsilon_2}{\varepsilon_2 A_2}} \text{ (W)}$$

The radiation heat transfer rate between two surfaces can be reduced drastically by placing thin, high reflectivity (low emissivity) material sheets between these two surfaces, called *radiation shield.* The radiation heat transfer rate between two large parallel planes separated by N radiation shield is

$$Q_{1-2,\text{ N shield}} = \frac{A\,\sigma (T_1^4 - T_2^4)}{\left(\frac{1}{\varepsilon_1} + \frac{1}{\varepsilon_2} - 1\right) + \left(\frac{1}{\varepsilon_{3,1}} + \frac{1}{\varepsilon_{3,2}} - 1\right) + \ldots\ldots + \left(\frac{1}{\varepsilon_{N,1}} + \frac{1}{\varepsilon_{N,2}} - 1\right)}.$$

The radiation effect in temperature measurement can be properly accounted by the relation

$$T_\infty = T_c + \frac{\varepsilon_c\, \sigma(T_c^4 - T_w^4)}{h} \text{ (K)}$$

where T_∞ is actual temperature of fluid, T_c is temperature measured by thermocouple, and T_w is the temperature of surrounding walls.

REVIEW QUESTIONS

1. What does the view factor represent ? When is the view factor from a surface to itself not zero ?
2. Explain the various relations for determination of view factor.
3. Show that

$$F_{1-2} = \frac{1}{A_1} \int_{A_1} \int_{A_2} \frac{\cos \beta_1 \cos \beta_2}{\pi s^2}\, dA_1\, dA_2$$

4. What is the cross string method ? For what kind of geometries is the crossed string method applicable ?
5. Consider an enclosure consisting of 12 surfaces. How many view factors does the geometry involve ? How many these view factors can be determined by application of reciprocity and summation rules ? [Ans. 144,78]
6. Prove that the view factor for two surfaces 1 and 2 connected by a reradiating surface is given by

$$\overline{F}_{1-2} = \frac{A_2 - A_1 F_{1-2}^2}{A_1 + A_2 - 2AF_{1-2}}$$

where $\overline{F}_{1-2}$ is modified view factor between surface 1 and 2.

7. Consider a hemispherical furnace with a flat circular base of diameter D. Determine the view factor from the dome of this furnace to its base. [Ans. 0.5]
8. Determine the view factors from very long grooves as shown in Fig. 13.55 to surroundings without using any view factor tables or charts.

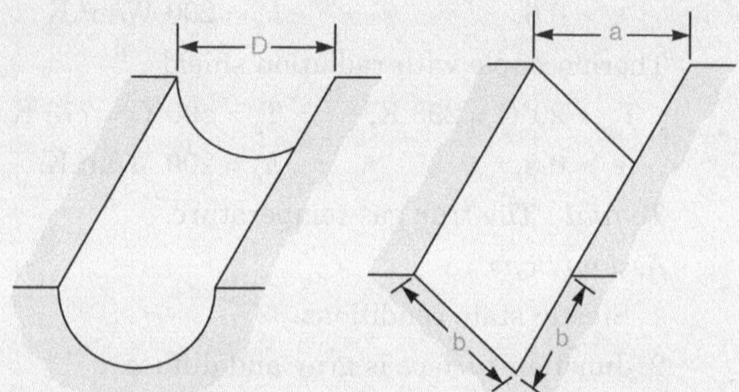

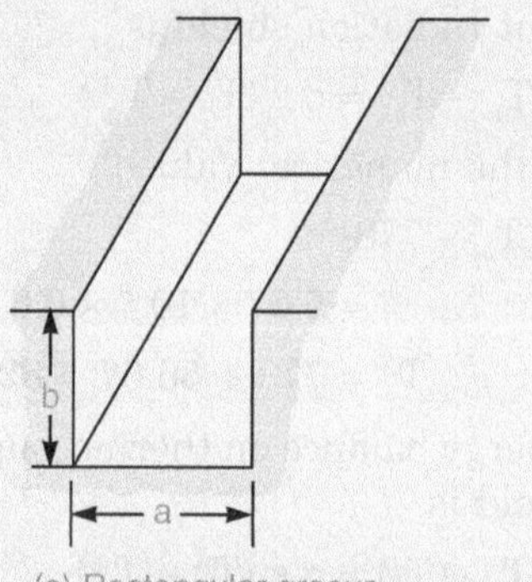

Fig. 13.55

9. What are the radiation surface and space resistances ? How are they expressed ? For what kind of surface, is radiation surface resistance zero ?
10. What is a reradiating surface ? What simplifications does a reradiating surface offer in the radiation analysis ?
11. What is a radiation shield ? Where is it used ?
12. What is the radiation effect ? How does it influence the temperature measurements ?
13. What benefit can be derived from a radiation shield and a reradiating surface ?
14. Derive an expression for a rate of radiation exchange, when a radiation shield is inserted between two large parallel plates.
15. What is the difference between radiation from gases and radiation from solid surfaces ?
16. Prove that the net radiation from a conical cavity of diameter D, height H and lateral length L, semi vertex angle α and surface A_1 is given by

$$Q = A_1 \sigma\, \varepsilon_1 T_1^4 \frac{1 - F_{1-1}}{1 - F_{1-1}(1 - \varepsilon_1)}.$$

PROBLEMS

1. Calculate the view factor between two opposite sides of a hollow cube as shown in Fig. 13.56 if view factor between two adjacent sides of it is 0.2. *(P.U., 2000)*

[Ans. $F_{1-3} = 0.2$]

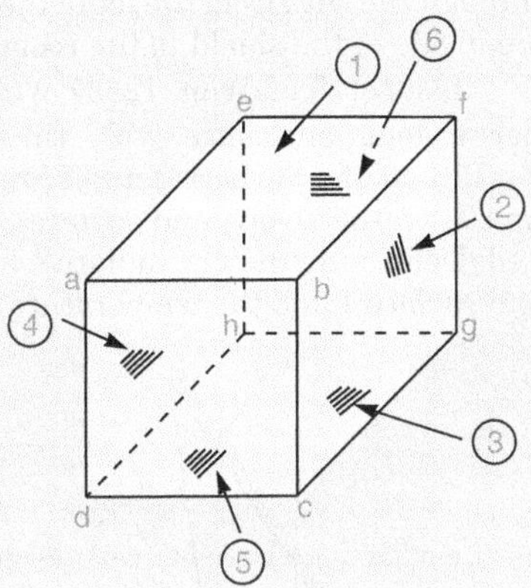

Fig. 13.56

2. Consider a triangular duct of length L as shown in Fig. 13.57. For given dimensions, prove that $F_{2-3} = 0.75$ where, $ab = ac = x$ and $bc = x/2$.

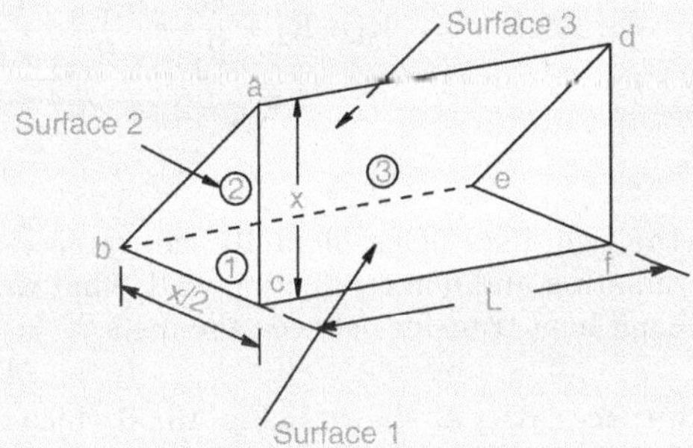

Fig. 13.57

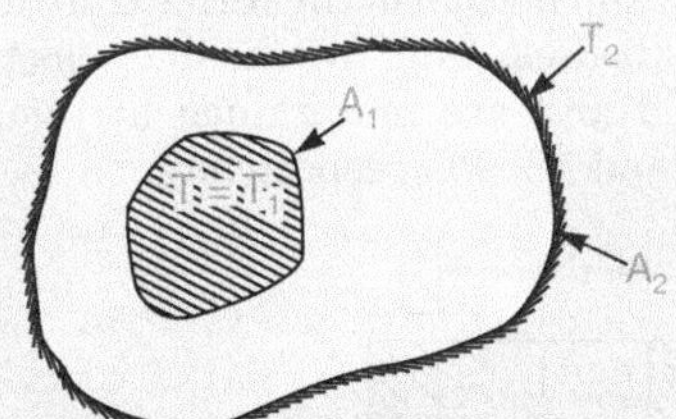

Fig. 13.58. Heat transfer between an enclosed body and the body surrounding it

3. Derive F_{1-2} for a small gray object (1) in a large isothermal environment (2), for the geometry shown in Fig. 13.58.

4. Find F_{1-2} for the configuration of two offset squares of area A, as shown in Fig. 13.59.

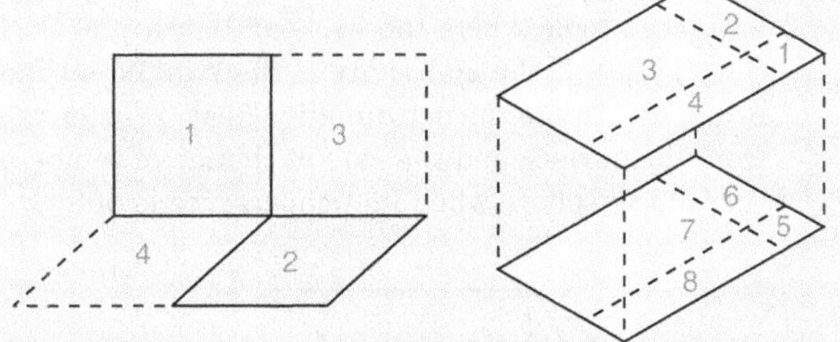

Fig. 13.59

Fig. 13.60. A room with the floor represented by 5, 6, 7, 8 and the skylight by 1

5. Fig. 13.60 shows the floor and roof of a house. The roof has a skylight in one corner, shown as 1. The house has radiant heating with heated floors. To determine the radiant energy that reaches the skylight from the floor, the view factor of the floor to the skylight is needed. Determine the view factor.

6. What is the radiation shape factor between the inside surface of one hemisphere and that of another, both form a spherical enclosure ? What is F_{1-1} ?

[Ans. 0.730]

7. Calculate the view factor F_{1-2} for the following geometry shown in Fig. 13.61. [Ans. $F_{1-2} = 0.057$]

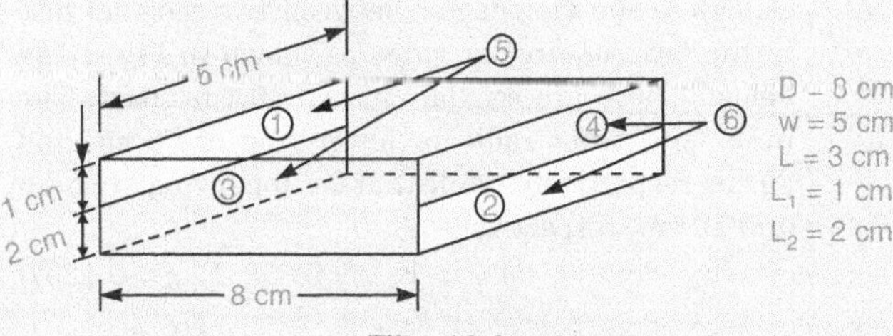

Fig. 13.61

8. Consider an enclosure formed by closing one end of cylinder as show in Fig. 13.62 (diameter = D, Height = L) by a flat surface and other end by a hemispherical dome. Determine the view factors of all the surfaces of the enclosure, if height is the twice the diameter.

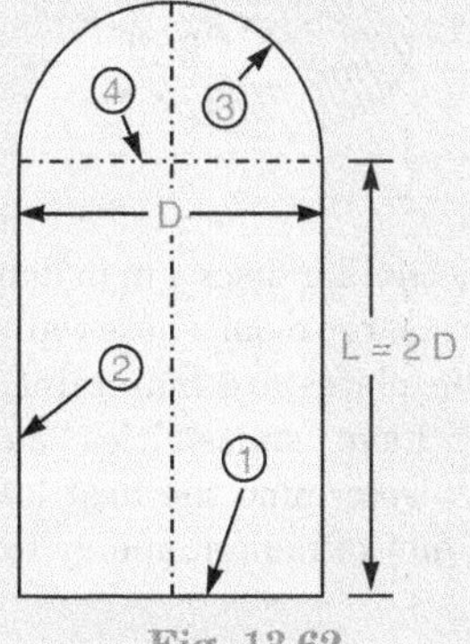

Fig. 13.62

9. A 3 cm diameter painted copper sphere is suspended in a large room where the walls and air are at 20°C (Fig. 13.63). If the emissivity of the painted surface is 0.9, determine the net radiative heat transfer rate from the surface and the rate of change of temperature of the sphere when its temperature is 600°C:

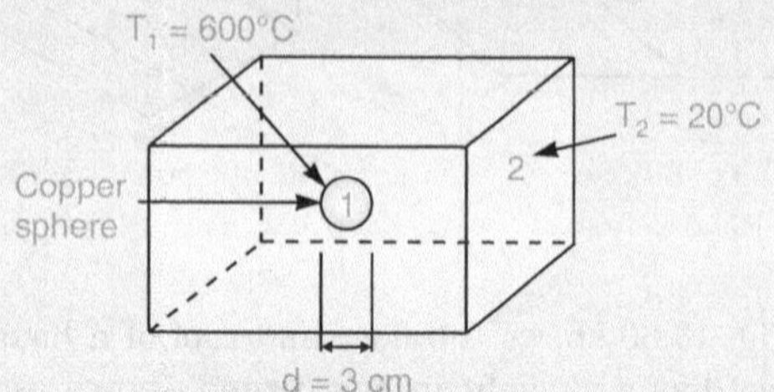

Fig. 13.63. A copper sphere 1, is freely suspended in a large room. The walls are represented as surface 2

(*a*) Neglecting convection,

(*b*) Including convection.

[**Ans.** q_1 = 82.8 W, (*a*) – 1.82°C/s; (*b*) – 2.41°C/s]

10. A 10 cm diameter peephole in the side of a furnace acts as a blackbody. The furnace interior is at 500°C, and the surrounding temperature is 25°C, what is the net radiant heat loss from the furnace through the peephole ? [**Ans.** 155.2 W]

11. Calculate the view factor between two parallel disc in the form of circular rings as shown in Fig. 13.64 below. These are coaxial, spaced 10 cm apart. The inner and outer radii for lower ring are 8 cm and 20 cm, respectively, while that for upper ring are 5 cm and 10 cm, respectively.

[**Ans.** F_{3-2} = 0.563, F_{2-3} = 0.1257]

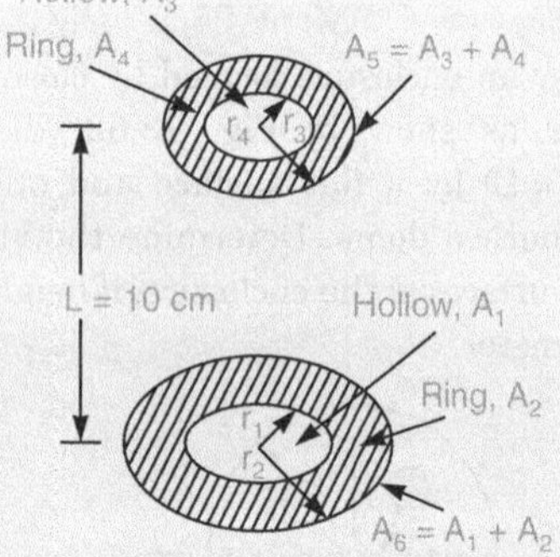

Fig. 13.64. Schematic of parallel, coaxial circular rings

12. Two parallel circular discs 4 m in diameter 1 m apart are placed in a large room whose walls are maintained at 25°C. The plates are maintained at 500°C and 300°C and have emissivities as 0.5 and 0.6, respectively. Determine the heat loss by each plate and the net flow of radiant energy to the walls of the room.

[**Ans.** Q_1 = 109 kW, Q_2 = – 32 kW, Q_{net} = 77 kW]

13. A jet of liquid metal at 2000°C pours from a crucible. It is 3 mm in diameter. A long cylindrical radiation shield, 5 cm diameter, surrounds the jet through an angle of 330°, but there is a 30° slit in it. The jet and the shield radiate as black bodies. The slit in a room is at 30°C, and the shield has a temperature of 700°C. Calculate the net heat transfer : from the jet to the room through the slit ; from the jet to the shield ; and from the inside of the shield to the room.

[**Ans.** 1188.2 W/m, 12637 W/m, 619 W/m]

14. Consider a long enclosure with three radiation surfaces (Fig. 13.65). Surfaces 1 and 2 are maintained at T_1 and T_2, respectively, and surface 3 is perfectly insulated. Determine the net radiative heat transfer rate from surface 1 and the temperature of surface 3.

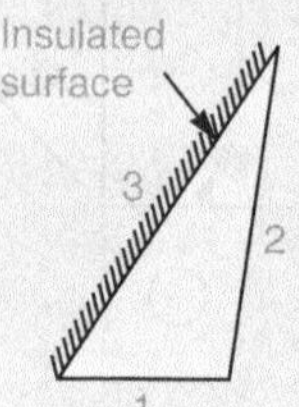

Fig. 13.65

$$\left[\textbf{Ans. } q_1 = -q_2 = \frac{E_{b1} - E_{b2}}{R_1 + R_2 + \dfrac{R_{1-2}(R_{1-3} + R_{2-3})}{R_{1-2} + R_{1-3} + R_{2-3}}} \quad E_{b3} = J_3 = \sigma T_3^4 \right]$$

15. If the two discs of problem 10 are connected by a reradiating and non conducting wall, what would be the net heat transfer between the discs ?

[**Ans.** 74.8 kW]

16. A furnace cavity as shown in Fig. 13.66 which in form of a hollow cylinder of 75 cm in diameter and 150 cm long, is open at one end to the surroundings at 27°C. The side and bottom of the furnace are approximated as blackbodies, and are heated electrically, well insulated and are maintained at temperatures of 1500°C and 1800°C respectively.

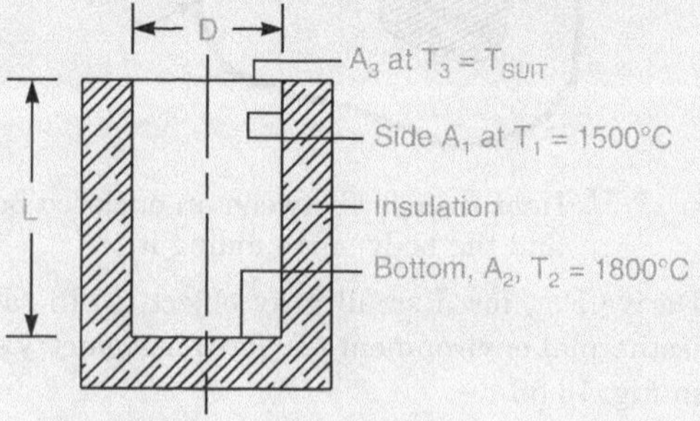

Fig. 13.66

How much power is required to maintain the furnace conditions ? [**Ans.** 354.6 kW]

17. Fig. 13.67 shows a cavity having surface temperature of 900°C and emissivity as 0.6. Find the rate of emission from the cavity to the surroundings.
(P.U., May 1994) [**Ans.** 200.61 W]

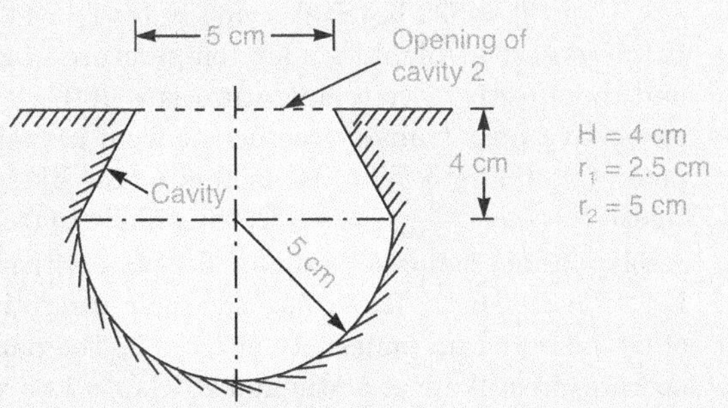

Fig. 13.67

18. Consider three large infinite parallel plates. The plate 1 is maintained at 1500 K and the plate 3 is maintained at 100 K. Take $\varepsilon_1 = 0.8$, $\varepsilon_2 = 0.5$ and $\varepsilon_3 = 0.8$. The plate 2 is placed between plates 1 and 3 and receive no heat from external sources. What would be the temperature of the plate ? [**Ans.** 987 K]

19. Two parallel plates 1 m × 1 m, insulated on their back sides and separated by 4 m, may be approximated as blackbodies at 500 K and 750 K. These plates are located in large room, whose walls are maintained at 300 K. Determine the net radiation heat transfer from each plate and net radiative heat transfer to the room.

20. Two parallel rectangular surfaces of dimension 1 m × 2 m, are opposite to each other at a distance of 4 m. The surfaces are black and at temperatures of 100°C and 200°C, respectively. Calculate the net rate of heat exchange by radiation between the two surfaces. [**Ans.** 122 W]

21. A very long electrical conductor 10 mm in diameter is concentric with a cooled cylindrical tube 50 mm in diameter whose surface is diffused, gray with an emissivity of 0.9 and temperature of 27°C. The electrical conductor has diffused, gray surface with an emissivity of 0.6 and is dissipating 6.0 W per metre length. Assuming that the space between the two surfaces is evacuated, calculate the surface temperature of the conductor.

22. Two concentric spheres of diameter 0.8 m and 1.2 m are separated by an air space and have surface temperatures of 400 K and 300 K, respectively.
   (*a*) If the surfaces are black, what is the net rate of radiation exchange between the spheres ?
   (*b*) What is the net rate of radiation exchange between the surfaces, if they are diffuse, gray with $\varepsilon_1 = 0.5$ and $\varepsilon_2 = 0.05$?
   [**Ans.** (*a*) 1995 W, (*b*) 191 W]

23. A furnace can be approximated as an equilateral triangular duct, which is sufficiently long thus the end effects are negligible. The hot wall is maintained at 900 K and has an emissivity of 0.8. The cold wall is at 400 K and has an emissivity of 0.8. The third wall is reradiating zone for which net heat transfer is zero. Calculate the net radiation heat flux leaving the hot wall. [**Ans.** 19.5 kW/m^2]

24. Two square plates, each 1 m by 1 m, are parallel and directly opposite to each other at a distance of 1 m. The hot plate is at 800 K and has an emissivity of 0.8. The colder plate is at 600 K and has an emissivity of 0.8. The radiation heat exchange takes place between the plates as well as with a large ambient at 300 K through the opening between the plates. Calculate the net heat exchange by radiation at each plate and to the ambient.
[**Ans.** 17.212 kW, 2.55 kW, – 19.769 kW]

25. A spherical tank with diameter 40 cm filled with a cryogenic fluid at 100 K is placed inside a container of diameter of 60 cm and is maintained at 300 K. The emissivity of the inner and outer tanks are 0.15 and 0.2, respectively. A spherical radiation shield of diameter 50 cm with an emissivity of 0.05 on both surfaces is placed between the spheres.
Calculate the rate of heat loss from the system by radiation. Then find the rate of evaporation of the cryogenic liquid for h_{fg} = 210 kJ/kg.
[**Ans.** – 6.83 W, 3.25 × 10^{-5} kg/s]

26. Calculate the heat transfer rate per unit area by the radiation between the surfaces of two long concentric cylinders having radii 5 cm and 10 cm, respectively. The inner surface is maintained at 400 K and outer surface at 300 K. The emissivity for both surfaces is 0.5. [**Ans.** 124.7 W/m]

27. In a boiler, the heat is radiated from burning fuel (surface 1) to the side walls (surface 3) and boiler tubes (surface 2) at the top.
   (*a*) Assuming blackbody behaviour and assuming side wall is perfectly insulated ; derive an expression for the temperature of side wall T_3, as a function temperature of fuel bed T_1 and the boiler tubes T_2 and corresponding areas A_1 and A_2.
   (*b*) If $A_1 = A_2 = 0.25$ m^2, $F_{1-2} = F_{2-3} = F_{1-3} = 0.5$ and T_1 = 1700°C, T_2 = 300°C, what is the net radiative heat transfer to the boiler tubes ?

Ans. (*a*) $$T_3 = \left\{ \frac{A_1 F_{1-3} T_1^4 + A_2 F_{2-3} T_2^4}{A_1 F_{1-3} + A_2 F_{2-3}} \right\}^{1/4}$$
(*b*) – 160 kW

28. The inside dimensions of a hollow cylinder are 6 cm diameter and 12 cm long. The cylindrical surface 2, is perfectly insulated. End surface 1 is maintained at 300°C, and end surface 3 at 100°C. Determine the total radiative heat transfer rate from surface 1 to surface 3, if each surface has an emissivity of 0.8.
[Ans. $q_1 = -q_3 = 5.93$ W]

29. Consider a cylindrical furnace, whose radius is 1 m and equal to its height. The base and top surface of the furnace have emissivities 0.4, and 0.8, respectively, and are maintained at uniform temperatures of 700 K and 500 K. The curved cylindrical surface approximates a blackbody and is maintained at a temperature of 400 K. Calculate the net rate of radiation heat transfer at each surface during steady state conditions.
[Ans. $Q_1 = 27588$ W, $Q_2 = -2131$ W, $Q_3 = -25455$ W]

30. Reconsider the configuration in problem 28, where $\varepsilon_1 = \varepsilon_3 = 0.8$, $T_1 = 300°C$, and $T_3 = 100°C$.
(a) Determine q_1, q_2, and q_3 if $T_2 = 500°C$ and $\varepsilon_2 = 0.8$.
(b) Determine q_1, q_2, and q_3 if $T_2 = 500°C$ and $\varepsilon_2 = 1$.
(c) Determine q_1, q_3, and T_2 if $q_2 = 30$ W and $\varepsilon_2 = 1$.
[Ans. (a) $q_1 = -28.39$ W, $q_2 = 68.66$ W, $q_3 = -40.24$ W (b) $q_1 = -30.03$ W, $q_3 = -41.89$ W, $q_2 = 71.93$ W (c) $q_1 = -9.1$ W, $T_2 = 657$ K, $q_3 = -20.9$ W]

31. Reconsider above problem 28. Determine q_1 and T_2 if all the surfaces are black.
[Ans. $q_1 = 7.5$ W, $T_2 = 502.3$ W]

32. High temperature gas flows through a pipe of outer radius $r_2 = 30$ mm. To reduce the thermal radiation emitted to an electrical control panel mounted nearby, a semi circular radiation shield of radius $r_1 = 100$ mm is placed concentrically around the pipe. The thermal radiation emitted from the pipe is radiated to both the shield and the surroundings which are at 310 K. The radiation view factor to the shield to itself is $F_{1-1} = 0.3345$.
(a) Determine the remaining view factors.
(b) Consider a radiation balance on the system. The pipe temperature is 900 K and the shield temperature is T_1; the emissivity of inner surface of the shield is $\varepsilon_1 = 0.8$ and the emissivity of the outer surface of the pipe is $\varepsilon_2 = 0.5$. Assuming the surroundings as blackbody at 310 K, prove that for surface 1 and 2 :
$$J_1 = 4.861 \times 10^{-8} T_1^4 + 0.0321 J_2 + 57.9 \text{ W/m}^2$$
$$J_2 = 18731 + 0.25 J_1 \text{ W/m}^2.$$
(c) Neglect any heat loss by convection, estimate the surface temperature of the shield, when the emissivity of the outer surface of the radiation shield is $\varepsilon_0 = 0.1$.
(d) Recalculate part (c) with $\varepsilon_0 = 0.8$ and $\varepsilon_1 = 0.1$.
(e) Recalculate part (c) with convection loss with $h = 10$ W/m².K.
[Ans. (a) $F_{2-1} = F_{2-3} = 0.5$, $F_{1-2} = 0.15$, $F_{1-3} = 0.5155$, $F_{3-2} = 0.15$, $F_{3-3} = F_{1-1} = 0.3345$, $F_{3-1} = 0.5155$ (c) $T_1 = 522$ K, (d) $T_1 = 465$ K, (e) $T_1 = 442$ K]

33. In the desert at night, the air temperature is 20°C and the effective sky temperature is – 40°C. If the convective heat transfer coefficient from a shallow pool of water is 5 W/m².K. Is this water likely to freeze ? [Ans. $T_{water} = -5°C$, thus likely to freeze]

34. A solar pannel comprises a square flat plate with sides L = 0.7 m in length. Its surface has solar absorptivity of $\alpha_s = 0.9$ and an emissivity of $\varepsilon = 0.2$. The pannel operates in still air at 1 atm and 25°C ($\mu = 1.85 \times 10$ kg/ms and $k = 0.026$ W/m.K) and in still surroundings air, the convective heat loss from the surface is given by
$$Nu_L = 0.5\, Gr_L\, \eta_4$$
At a particular time of day, the solar irradiation is 500 W/m³ and the effective sky temperature is – 3°C. If the efficiency of the pannel is 50%, calculate the surface temperature of the cover plate.
[Hint. $q_{rad,\,in} + q_{rad,\,sky} - q_{rad,\,out} - q_{conv} = q_{useful}$]
[Ans. $T_s = 330$ K ≈ 57°C]

35. Consider two concentric cylinders, 50 mm long. The diameter of inner and outer cylinders are 5 mm and 20 mm, respectively. The surface temperature of inner cylinder is 1800 K, while the ambient temperature is $T_\infty = 300$ K. The emissivity of outer surface of inner cylinder is $\varepsilon_1 = 0.22$, and for the inner surface of outer cylinder is $\varepsilon_2 = 0.5$ and for outer surface of outer cylinder is $\varepsilon_0 = 0.17$.
(a) Using network theorem, prove that the radiosity J_1 is
$$J_1 = 130996 + 0.702 J_2$$
and $$J_2 = \frac{\sigma T_2^4 + 0.225 J_1 + 72.56}{1.383}.$$
(b) Neglect convection and consider a heat balance on the outer surface of the outer cylinder. Show that
$$J_2 = \sigma(\varepsilon_0 + 1) T_2^4 - \sigma \varepsilon_0 T_\infty^4$$
(c) Calculate the surface temperature of outer cylinder. [Ans. 1050 K]

36. A helicopter platform (16 m × 16 m) is subjected to an incident radiative heat flux from a nearby gas flame of 6 kW/m². The absorptivity α, of the platform surface to thermal radiation at wavelength of 0.7 μm and average emissivity ε at this wavelength is 0.6.
By making energy balance on the surface of platform, show that for an ambient temperature of 300 K, the surface temperature of the platform T_s is given by
$$4475.6 - 3.402 \times 10^{-8} T_s^4 - 1.551 (T_s - 300)^{4/3} = 0.$$
Calculate surface temperature of the platform

Take for air
$\mu = 1.846 \times 10^{-5}$ kg/m.s, $k = 0.02624$ W/m.K,
$\rho = 1.177$ kg/m^2, and
Pr = 0.707, and use following correlation for calculation of h

$$Nu_L = 0.13\,(Gr_L\,Pr)^{1/3}, \text{ where } L = \frac{A_s}{P}.$$ [Ans. 518 K]

37. Two parallel discs of 1 m diameter are situated 2 m apart in the surroundings at a temperature of 20°C. The one side of one disc has an emissivity of 0.5 and is maintained at 500°C and other side is insulated. The other disc is open to radiation on both sides. Determine the equilibrium temperature of the second disc and heat flow rate from the first disc.
[Ans. 62°C, 7.8 kW]

38. Two parallel directly opposed, coaxial, annular discs are maintained at temperatures 1000 K and 300 K, respectively. The discs have the following dimensions and emissivities :

Disc 1 (1000 K) ID = 10 cm,
OD = 20 cm $\varepsilon_1 = 0.8$
Disc 2 (300 K) ID = 12 cm,
OD = 25 cm $\varepsilon_2 = 0.7$.

If these discs are 8 cm apart, what is the net rate of exchange of radiation between the discs ? Assume that the discs are placed in a radiation free environment.
[Ans. 2286 W/m^2]

39. The temperature of the top and bottom surfaces of a frustum of a cone shown in Fig. 13.68 are maintained at 600 K and 1200 K, respectively, while the side curved surface is insulated. If all the surfaces are gray and diffuse, determine net radiative exchange between the top and bottom surfaces *i.e.*, A_3 and A_1.

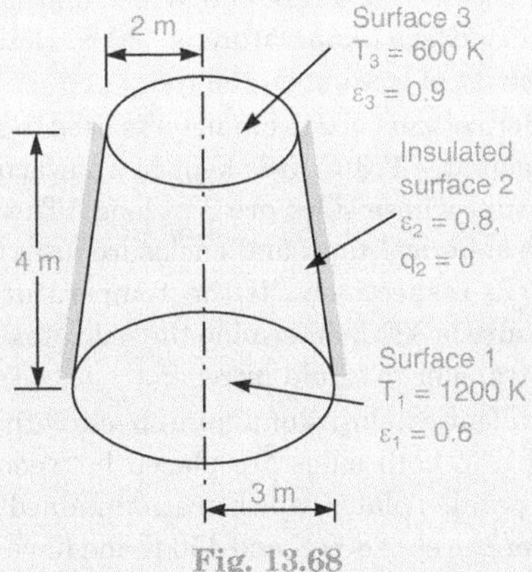

Fig. 13.68

40. The ceiling of many furnaces is in the shape of an arch. The arch is constructed with suspended firebricks (Fig. 13.69). The bricks deteriorate due to cracking and disintegration, which leads to significantly higher temperature of the surface of the arch exposed to the surroundings. A reduction in the heat transfer rate from the surface of the arch to the surroundings reduces the fuel requirements of the furnace and also results in a more comfortable working environment. In one such furnace it is found that the average surface temperature of the top surface is 200°C. The emissivity of the surface is 0.9. To reduce the radiative heat transfer rate from the surface, it is proposed to add a radiation shield of steel (ε = 0.85) close to the top surface. The temperature of the surroundings is 35°C. Determine, per unit surface area,

(*a*) The radiative heat flux without and with the radiation shield.
(*b*) The temperature of the shield.

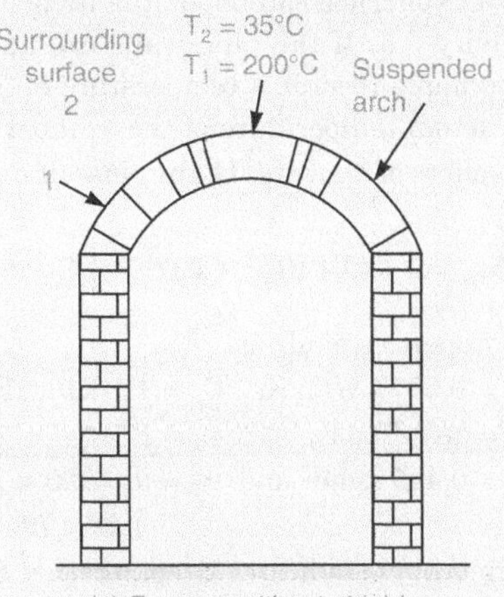

(a) Furnace without shield

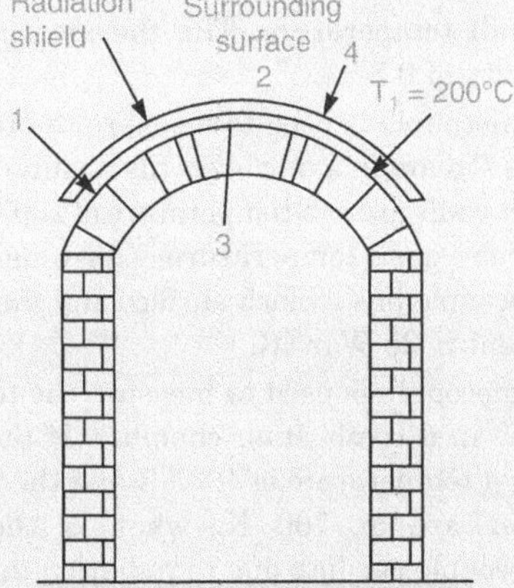

(b) Furnace with a radiation shield to reduce the radiant heat transfer rate from the top surface

Fig. 13.69. Furnace with suspended arch

41. The room in the accompanying Fig. 13.70 is 610 cm by 610 cm wide and 275 cm high. The floor is at 37.7°C, the walls are at 15.5°C, and the ceiling is at 4.4°C. All surfaces are assumed black. Estimate the

net radiation heat transfer (*a*) from floor to walls and (*b*) from floor to ceiling.

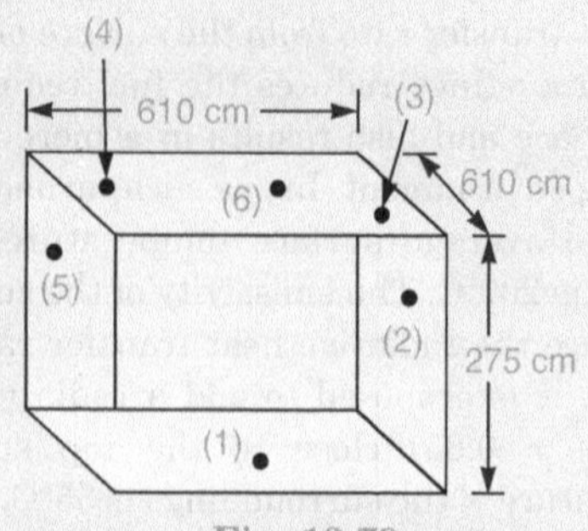

Fig. 13.70

42. Exhaust gas from a diesel engine flows at a rate of 0.066 m³/s along an exhaust duct with a diameter of 38 mm. The temperature of the gas is measured by using a spherical probe of diameter 3 mm and emissivity 0.5. If the pipe walls are at 500 K and thermocouple records a temperature of 504°C, what is the actual temperature of the exhaust gas ?

For a sphere of diameter D, the Nusselt number given

$$Nu_D = 2 + (0.4\, Re_D^{1/2} + 0.06\, Re_D^{2/3})\, Pr^{0.4} \left(\frac{\mu_\infty}{\mu_s}\right)^{0.25}$$

Take properties of air as

$k_f = 0.0577$ W/m.K, $C_p = 1.1$ kJ/kg.K,

R = 287 J/kg.K,

$\rho = 0.435$ kg/m³ and $\mu_\infty = \mu_s = 3.6 \times 10^{-5}$ kg/ms.

[Ans. 794 K ≈ 521°C]

43. A thermocouple indicates a temperature 800°C when placed in a duct where a hot gas at 870°C is flowing. The convective coefficient is 60 W/m².K, estimate the duct wall temperature. The thermocouple has an emissivity of 0.5. [Ans. 768.67°C]

44. A thermocouple is used to measure the temperature of a gas stream in a duct, records a value of 250°C. If the duct walls are at a temperature of 200°C, calculate the true gas temperature. Assume that the thermocouple has a black surface and the convection coefficient is 25 W/m².K. [Ans. 306°C]

45. A thermocouple is used to measure the temperature of flame in a combustion chamber. If thermocouple records a temperature of 1033 K and the walls of the chamber are at 700 K, what is the error in thermocouple reading due to radiation to the walls ? Assume all surfaces are black and convection coefficient is 568 W/m².K. [Ans. 89.7°C]

46. Three thin sheets of polished aluminium are placed parallel to each other, so that the distance between them is very small to the size of the sheets. If one of the outer sheet is at 280°C and other outer sheet is at 60°C, calculate the temperature of intermediate sheet and the net rate of heat flow by radiation.

47. A 1 m × 1 m square solar collector is placed on the roof of a house. The collector receives a solar radiation flux of 800 W/m². Assuming that the surroundings acts as a blackbody at an effective sky temperature of 30°C, calculate the equilibrium temperature of the collector.

(*a*) Assuming collector surface is black and the conduction and convection are negligible

(*b*) Assuming collector is horizontal and heat is lost by natural convection only.

48. A long triangular duct has width of each side is 1 m. The base surface has an emissivity of 0.7 and is maintained at uniform temperature of 600 K. The heated left side surface is black at 1000 K. The right side surface is well insulated. Determine the rate at which the energy must be supplied to the heated side externally per unit length of the duct in order to maintain these conditions. [Ans. 28009 W]

49. Consider a hemispherical furnace of diameter 5 m with a flat base. The dome of the furnace is black and the base has an emissivity of 0.7. The base and dome of the furnace are maintained at uniform temperatures of 400 K and 1000 K, respectively. Determine the net rate of radiation heat transfer from the dome to the base surface during steady state operation. [Ans. 759 W]

50. An experiment is conducted to determine the emissivity of a material. A long cylindrical rod of diameter 0.01 m is coated with this material and is placed in an evacuated long cylindrical enclosure of diameter 0.1 m and emissivity $\varepsilon_2 = 0.95$, which is cooled externally and is maintained at 200 K at all times. The rod is heated by passing an electric current through it. When steady operating conditions are reached, it is observed that the rod is dissipating electric power at a rate of 8 W per unit of its length and its surface temperature is 500 K. Determine the emissivity of coating on the rod.

51. A spherical tank of 2 m diameter is used to store liquid nitrogen at – 173°C. It is kept in an evacuated cubic enclosure whose sides are 3 m long. The emissivity of the spherical tank and enclosure are $\varepsilon_1 = 0.1$ and $\varepsilon_2 = 0.8$, respectively. If the temperature of cubic enclosure is 33°C, determine the net rate of radiation heat transfer to liquid nitrogen. [Ans. 228 W]

52. Fire indentical thin aluminium sheets with emissivity $\varepsilon = 0.1$ on both sides are placed between two very large parallel plates, which are maintained at uniform temperatures of 800 K and 450 K and have emissivity $\varepsilon_1 = \varepsilon_2 = 0.1$, respectively. Determine the net rate of radiation heat transfer between the two plates per unit surface area of the plates, and compare the result to that without the shield.

53. A thermocouple shielded by an aluminium foil ($\varepsilon = 0.1$) is used to measure the temperature of the hot gases flowing in a duct whose walls are maintained at 380 K. The thermocouple records a temperature of 530 K. Assuming emissivity of thermocouple junction as 0.8 and convection coefficient of 120 W/m^2.K. Determine the actual temperature of gas. What would be the thermocouple reading, if no radiation shield was used ?

54. A paint baking oven consists of a long triangular duct in which a heated surface is maintained at 1200 K and another surface is insulated. Painted panels, which are maintained at 500 K, occupy the third surface. The triangle of width of 1 m on a side, and heated and insulated surfaces have an emissivity of 0.8. The emissivity of the panels is 0.4. During steady state operation, at what rate must energy be supplied to the heated side per unit length of the duct to maintained its temperature at 1200 K ? What is the temperature of insulated surface ?

(N.M.U., Nov. 1994)

[Ans. 36.98 kW/m, 1102.2 K]

55. Two parallel plates have emissivity of 0.8 and 0.5. A radiation shield having same emissivity on both sides is placed between them. Calculate the emissivity of the shield in order to reduce the radiation losses from the system to one-tenth of that of without shield.

[Ans. 0.094]

56. A physics experiment uses liquid nitrogen as a coolant. Saturated liquid nitrogen at 80 K flows through 6.35 mm O.D. stainless steel line ($\varepsilon_1 = 0.2$) inside a vacuum chamber. The chamber walls are at $T_c = 230$ K and are at some distance from the line. Determine the heat gain of the line per unit length. If a second stainless steel tube, 12.7 mm in diameter, is placed around the line to act as radiation shield, to what rate is the heat gain reduced ? Find the temperature of the shield. [Ans. 0.328 W/m, 213 K]

REFERENCES AND SUGGESTED READING

1. Modest M. Michael, "*Radiative Heat Transfer*", 2/e, Academic Press (Elsevier Science), 2003.
2. Oppenheim A.K., "*Radiation Analysis by Network Method*", Transactions of ASME, Journal of Heat Transfer, vol. 78, pp. 725–735, 1956.
3. Özisik M.N., "*Radiative Heat Transfer and Interactions with Conduction and Convection*", John Wiley & Sons, New York, 1973.
4. Sparrow E.M., "*Application to Variation Methods to Radiation Heat Transfer Calculations*", ASME Journal of Heat Transfer, vol. 82. pp. 375–380, 1960.
5. Jacob M., "*Heat Transfer*", vol. 2, John Wiley & Sons, New York, 1957.
6. Sparrow E.M. and R.D. Cess, "*Radiation Heat Transfer*", Hemisphere, New York, 1978.
7. Siegel R. and J.R. Howell,"*Radiation Heat Transfer*" 3rd, ed. Hemisphere, New York, 1992.
8. Howell J.R., "*A Catalog of Radiation Configuration Factors*", McGraw Hill, New York, 1982.
9. Wen Jei Yang, H. Taniguchi, K. Kudo, "*Radiative Heat Transfer*", by the Monte Carlo, Method Academic Press, 1995.

14 Heat Exchangers

A device used for exchange of heat between the two fluids that are at different temperatures, is called the *heat exchanger*. The heat exchangers are commonly used in wide range of applications, for example, in a car as radiator, where hot water from the engine is cooled by atmospheric air. In a refrigerator, the hot refrigerant from the compressor is cooled by convection into atmosphere by passing it through finned tubes. In a steam condenser, the latent heat of condensation is removed by circulating water through the tubes. The heat exchangers are also used in space heating and air-conditioning, waste heat recovery and chemical processing. Therefore, the different types of heat exchangers are needed for different applications.

The heat transfer in a heat exchanger usually involves convection on each side of fluids and conduction through the wall separating the two fluids. Thus for analysis of a heat exchanger, it is very convenient to work with an *overall heat transfer coefficient* U, that accounts for the contribution of all these effects on heat transfer. The rate of heat transfer between two fluids at any location in a heat exchanger depends on the magnitude of temperature difference at that location. This temperature difference varies along the length of heat exchanger. Therefore, it is also convenient to work with *logarithmic mean temperature difference LMTD,* which is an equivalent temperature difference between two fluids for entire length of heat exchanger.

In this chapter, the classification of heat exchangers, determination of overall heat transfer coefficient, the analysis of heat exchangers by logarithmic mean temperature difference and *effectiveness—NTU* methods for sizing and rating of heat exchangers are discussed.

14.1. CLASSIFICATION OF HEAT EXCHANGER

Heat exchangers are designed in so many sizes, types, configurations and flow arrangements and used for so many purposes. These are classified according to heat transfer process, flow arrangement and type of construction.

(A) **According to Heat Transfer Process :** These are :

(*i*) **Direct contact type heat exchanger.** In this type of heat exchanger, the two immiscible fluids at different temperatures are come in direct contact. For the heat exchange between two fluids, one fluid is sprayed through the other. Cooling towers, jet condensers, desuperheaters, open feed water heaters and scrubbers are the best examples of such heat exchangers. A direct contact type heat exchanger (cooling tower) is shown in Fig. 14.1.

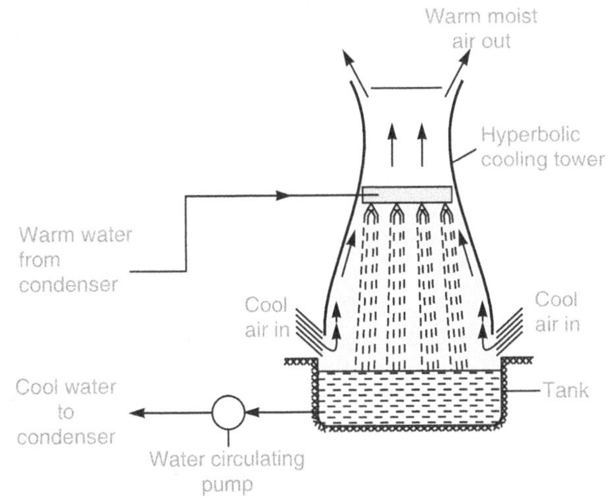

Fig. 14.1. Direct contact type heat exchanger (cooling tower)

(*ii*) **Transfer type heat exchangers or recuperators.** In this type of heat exchanger, the cold and hot fluids flow simultaneously through the device and the heat is transferred through the wall separating them. These types of heat exchangers are most commonly used in almost all fields of engineering.

(*iii*) **Regenerators or storage type heat exchangers.** In these type of heat exchangers, the heat energy is transferred directly from one fluid to another across a solid surface, which is alternatively in contact with both fluids. When hot fluid flows in an interval of time, it gives its heat to the surface, which stores it in the form of an increase in its internal energy. This stored energy is transferred to cold fluid as it flows over the surface in next interval of time. Thus the same surface is subjected to periodic heating and cooling. In many applications, a rotating disc type matrix is used, the continuous flow of both the hot and cold fluids are maintained. These are preheaters for steam power plants, blast furnaces, oxygen producers, etc. A stationary and rotating matrix shown in Fig. 14.2, are examples of storage type of heat exchangers.

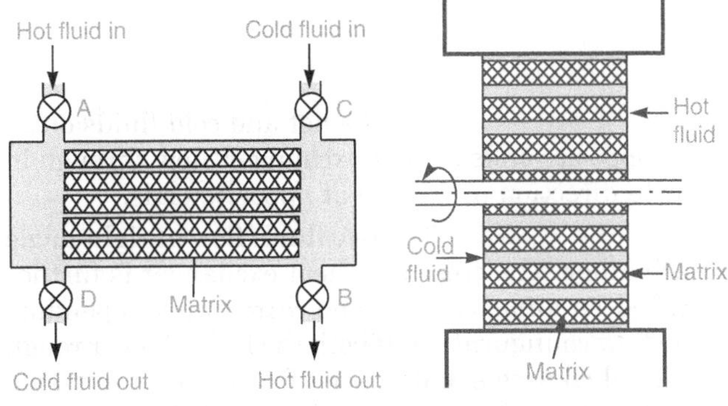

(a) Single matrix regenerator

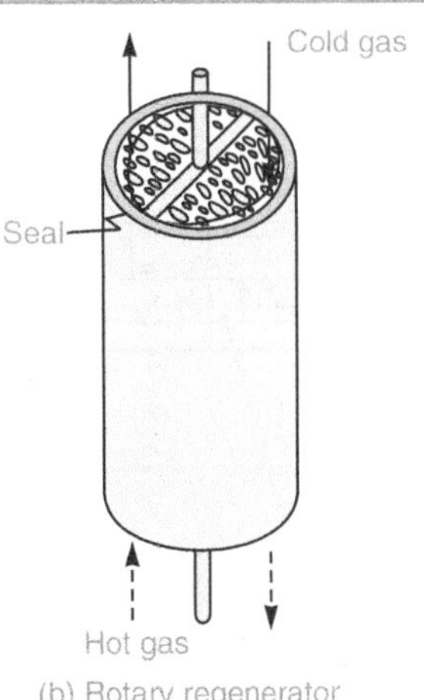

(b) Rotary regenerator

Fig. 14.2. Storage type heat exchangers

The storage type of heat exchangers are more compact than the transfer type of heat exchangers with more surface area per unit volume. However, some mixing of hot and cold fluids is always there.

(B) **According to Constructional Features :** These are :

(*i*) **Tubular heat exchangers.** These are also called *tube in tube* or *concentric tube* or *double pipe heat exchanger* as shown in Fig. 14.3. These are widely used in many sizes and different flow arrangements and type.

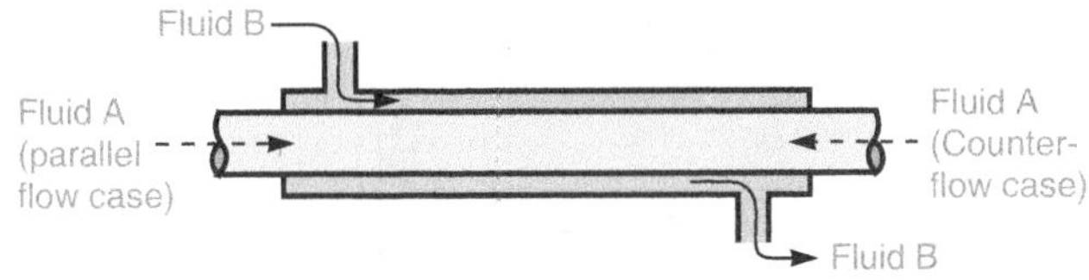

Fig. 14.3. Tubular heat exchanger

(*ii*) **Shell and tube type heat exchangers.** These are also called *surface condensers* and are most commonly used for heating, cooling, condensation or evaporation applications. It consists of a shell and a large number of parallel tubes housing in it. The heat transfer takes place as one fluid flows through the tubes and other fluid flows outside the tubes through the shell. The baffles are commonly used on the shell to create turbulence and to keep the uniform spacing between the tubes and thus to enhance the heat transfer rate. They are having large surface area in small volume. A typical shell and tube type heat exchanger is shown in Fig. 14.4.

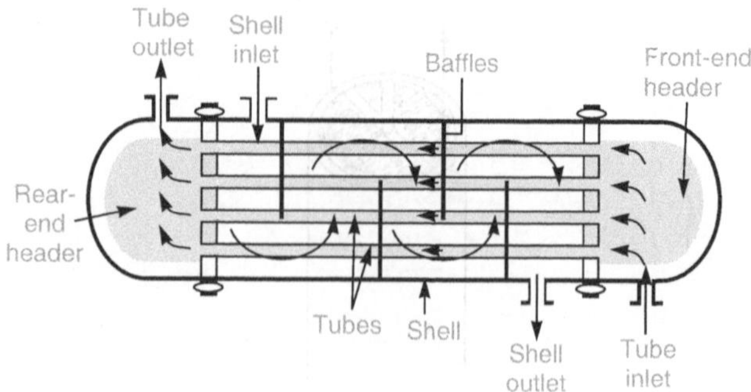

Fig. 14.4. Shell and tube type heat exchanger : One shell and one tube pass

The shell and tube type heat exchangers are further classified according to number of shell and tube passes involved. A heat exchanger with all tubes make one U turn in a shell is called *one shell pass and two tube pass* heat exchanger. Similarly, a heat exchanger that involves two passes in the shell and four passes in the tubes is called a *two shell pass and four tube pass* heat exchanger as shown in Fig. 14.5.

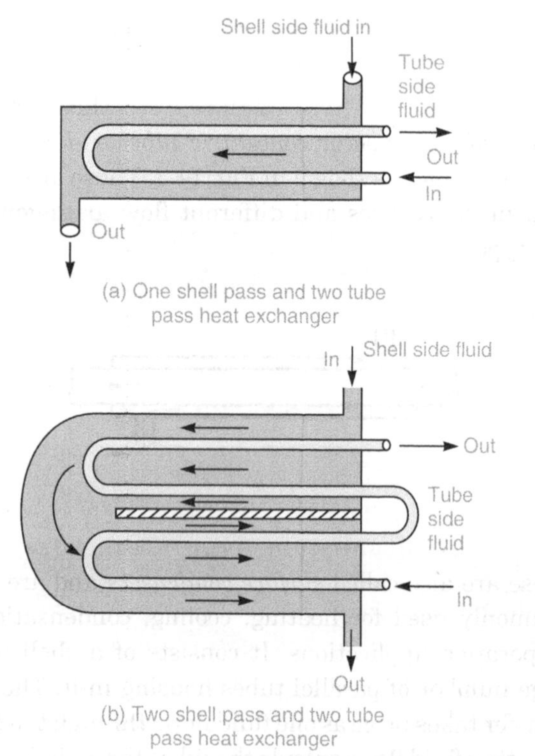

Fig. 14.5. Multipass flow arrangement in shell and tube type heat exchangers

(*iii*) **Finned tube type.** When a high operating pressure or an enhanced heat transfer rate is required, the extended surfaces are used on one side of the heat exchanger. These heat exchangers are used for liquid to gas heat exchange. Fins are always added on gas side. The finned tubes are used in gas turbines, automobiles, aeroplanes, heat pumps, refrigeration, electronics, cryogenics, air-conditioning systems etc. The radiator of an automobile is an example of such heat exchanger.

(*iv*) **Compact heat exchanger.** These are special class of heat exchangers in which the heat transfer surface area per unit volume is very large. The ratio of heat transfer surface area to the volume is called *area density*. A heat exchanger with an area density greater than 700 m^2/m^3 is called *compact heat exchanger*. The compact heat exchangers are usually cross flow, in which the two fluids usually flow perpendicular to each other. These heat exchangers have dense arrays of finned tubes or plates, where at least one of the fluid used is gas. For example, automobile radiators have an area density in order of 1100 m^2/m^3.

(C) **According to Flow Arrangement :** These are :

(*i*) **Parallel flow.** The hot and cold fluids enter at same end of the heat exchanger, flow through in same direction and leave at other end. It is also called the *concurrent heat exchanger*.

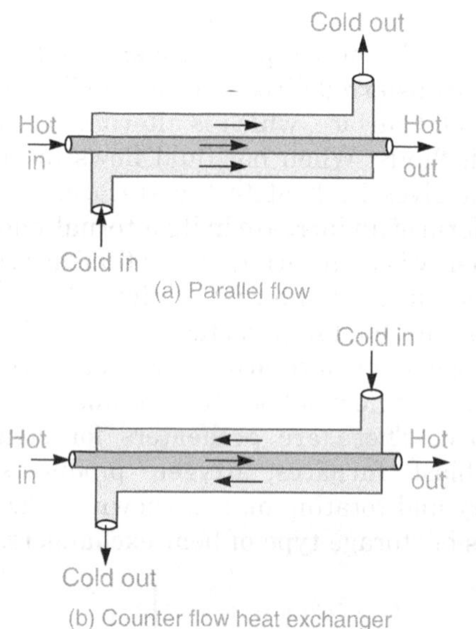

Fig. 14.6. Concentric tube heat exchanger

(*ii*) **Counter flow.** The hot and cold fluids enter at the opposite ends of heat exchanger, flow through in opposite direction and leave at opposite ends.

(*iii*) **Cross flow.** The two fluids flow at right angle to each other. The cross flow heat exchanger is further classified as *unmixed flow* and *mixed flow* depending on the flow configuration. If both the fluids flow through individual channels and are not free to move in transverse direction, the arrangement is called *unmixed* as shown in Fig. 14.7 (*a*). If any fluid flows on the surface

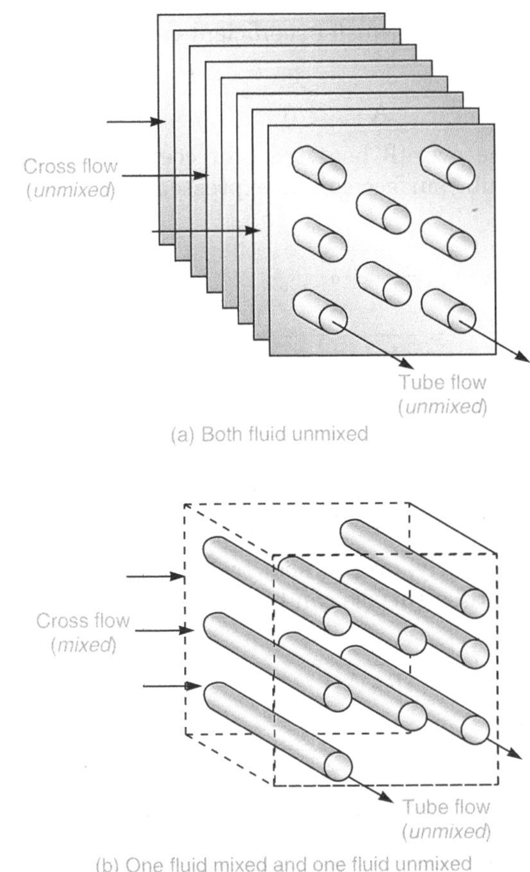

Fig. 14.7. Different flow configurations in cross-flow heat exchangers

and free to move in transverse direction, then this fluid stream is said to be *mixed* as shown in Fig. 14.7 (*b*).

14.2. TEMPERATURE DISTRIBUTION

The heat transfer from the hot to the cold fluid causes a change in temperature of one or both fluids flowing through the heat exchanger. In some common cases, the temperature is plotted against the distance from the cold fluid inlet in Fig. 14.8.

From the temperature distributions for various types of heat exchangers, the common observations are:

(*i*) In all other cases, the temperature difference between hot and cold fluids varies with the position along the path of flow.

(*ii*) In counter flow arrangement, the temperature rise in the cold fluid is almost equal to temperature drop in the hot fluid, if both fluids have equal heat capacity rate as shown in Fig. 14.8 (*e*).

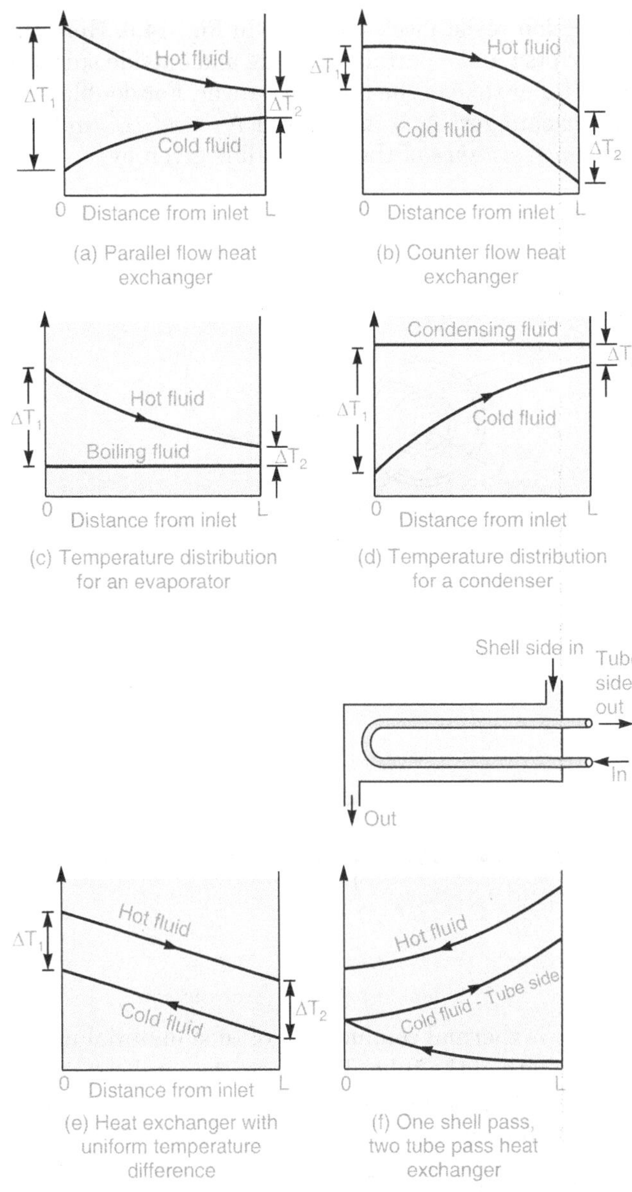

Fig. 14.8. Temperature distribution for different types of heat exchangers

14.3. OVERALL HEAT TRANSFER COEFFICIENT

A tubular heat exchanger usually involves two flowing fluids separated by a solid wall. The heat is first transferred from hot fluid to wall by convection, through the wall by conduction, and from the wall to cold fluid by convection again.

The total thermal resistances associated with this heat transfer process involves two convection and one conduction resistances as shown in Fig. 14.9. Here the subscripts *i* and *o* pertain to inside and outside surfaces of the inner tube of the heat exchanger. For double pipe heat exchanger, $A_i = \pi d_i L$, and $A_o = \pi d_o L$, and the *thermal resistance* of the tube wall is given by

$$R_{wall} = \frac{\ln(d_o/d_i)}{2\pi L k} \quad ...(14.1)$$

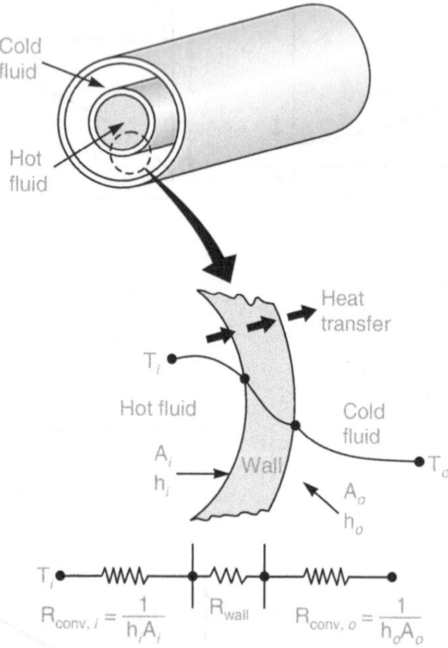

Fig. 14.9. Thermal resistance network for heat transfer in a double-pipe heat exchanger

where *k* is thermal conductivity of tube material and L is the length of the tube. Therefore, the total resistance for a double pipe heat exchanger with clean surfaces can be expressed as :

$$\Sigma R_{th} = R_{total} = R_{conv,i} + R_{wall} + R_{conv,o}$$

$$\Sigma R_{th} = \frac{1}{h_i A_i} + \frac{\ln(d_o/d_i)}{2\pi L k} + \frac{1}{h_o A_o} \quad ...(14.2)$$

where A_i is the area of the *inner surface* of the tube wall that separates the two fluids and A_o is area of its *outer surface.*

In the analysis of heat exchanger, it is convenient to combine all the thermal resistances in the path of heat flow from hot fluid to cold fluid as discussed in chapter 3 and the heat transfer rate is expressed as

$$Q = \frac{\Delta T}{R_{th}} = UA\Delta T = U_i A_i \Delta T = U_o A_o \Delta T \quad ...(14.3)$$

where U is **overall heat transfer coefficient.** It is measured in W/m^2.K. After cancelling ΔT from eqn. (14.3), the overall heat transfer coefficient can be expressed as

$$\Sigma R_{th} = \frac{1}{UA} = \frac{1}{U_i A_i} = \frac{1}{U_o A_o} \quad ...(14.4)$$

The overall heat transfer coefficient based on outside tube surface can be expressed as :

$$U_o = \frac{1}{\Sigma R_{th} A_o}$$

$$= \frac{1}{\dfrac{A_o}{A_i h_i} + \dfrac{A_o}{2\pi L k}\ln\left(\dfrac{d_o}{d_i}\right) + \dfrac{1}{h_o}}$$

$$= \frac{1}{\dfrac{d_o}{d_i h_i} + \dfrac{d_o}{2k}\ln\left(\dfrac{d_o}{d_i}\right) + \dfrac{1}{h_o}} \quad ...(14.5)$$

Similarly, overall heat transfer coefficient based on inside tube surface can be expressed as :

$$U_i = \frac{1}{\Sigma R_{th} A_i}$$

$$= \frac{1}{\dfrac{1}{h_i} + \dfrac{A_i}{2\pi L k}\ln\left(\dfrac{d_o}{d_i}\right) + \dfrac{A_i}{A_o h_o}}$$

$$= \frac{1}{\dfrac{1}{h_i} + \dfrac{d_i}{2k}\ln\left(\dfrac{d_o}{d_i}\right) + \dfrac{d_i}{d_o h_o}} \quad ...(14.6)$$

When wall thickness of the tube is small and its thermal conductivity is high, the thermal resistance of the tube material can be neglected ($R_{wall} \approx 0$) and then inner and outer surfaces of the tube are identical ($A_i \approx A_o$). The eqn. (14.6) for overall heat transfer coefficient simplifies to

$$U = \frac{1}{\dfrac{1}{h_i} + \dfrac{1}{h_o}} \quad ...(14.7)$$

The overall heat transfer coefficient U is dominated by the smaller heat transfer coefficient, since the inverse of the small value is large. If one of the convection coefficient is much smaller than the other, the value of overall heat transfer coefficient is almost equal to smaller convection coefficient *i.e.*, $U \approx h_{smaller}$.

The representative values of the overall heat transfer coefficient U are given in Table 14.1. The values of overall heat transfer coefficient is very small (U = 10 W/m^2.K) for gas to gas heat exchanger and it is very high about 10,000 W/m^2.K for heat exchanger during phase change.

TABLE 14.1. Representative values of overall heat transfer coefficient

Fluid combination	*U (W/m².K)*
Water to water	850–1700
Water to oil	100–350
Feedwater heaters	1000–8500
Steam to light fuel oil	200–400
Steam to heavy fuel oil	50–200
Freon condensers (water cooled)	300–1000
Steam condenser	1000–6000
Gas to gas	10–40
Finned tube heat exchanger (water in tube and air in cross flow)	25–50

14.4 FOULING FACTOR

After certain period of operation of a heat exchanger, its surfaces are often subjected to deposition of fluid impurities, rust formation and other reactions between the fluids and surfaces. The subsequent deposition of film or scale on the surface offers an additional resistance to heat transfer between the fluids. This effect is approximated by introducing one more thermal resistance called as *fouling factor* R_f. Its value depends on the operating temperature, fluid velocity and length of service of heat exchanger.

The most common type of fouling is deposition of solid particles, suspended in the flowing fluids. The scale of solid deposits can be cleaned off by scratching and chemical treatment of surfaces.

The *corrosion* and other *chemical fouling* are common in chemical process industries. Such type of fouling can be eliminated by coating metal pipes by glass layer or by use of plastic pipes. However, it may affect the heat transfer rate. The *biological fouling* due to growth of algae in warm fluid and deposition of ash particles in the flue gases are some other forms of fouling.

The fouling factor can be obtained experimentally by determining the overall heat transfer coefficient for both clean and dirty conditions in the heat exchanger. The fouling factor is thus defined as :

$$R_f = \frac{1}{U_{dirty}} - \frac{1}{U_{clean}}$$

If the surface is fouled by deposit formation on both sides, then the total thermal resistance in the path of heat flow between hot and cold fluids is expressed as:

$$\Sigma R_{th} = \frac{1}{A_i h_i} + \frac{R_{f,i}}{A_i} + \frac{\ln(d_o/d_i)}{2\pi L k} + \frac{R_{f,o}}{A_o} + \frac{1}{A_o h_o} \quad ...(14.8)$$

where $R_{f,i}$ and $R_{f,o}$ are the fouling factors at the inside and outside surfaces of the heat exchanger tube.

The overall heat transfer coefficients ;

$$U_o = \frac{1}{\dfrac{d_o}{d_i h_i} + \dfrac{d_o}{d_i} R_{f,i} + \dfrac{d_o}{2k}\ln\left(\dfrac{d_o}{d_i}\right) + R_{f,o} + \dfrac{1}{h_o}} \quad ...(14.9)$$

and
$$U_i = \frac{1}{\dfrac{1}{h_i} + R_{f,i} + \dfrac{d_i}{2k}\ln\left(\dfrac{d_o}{d_i}\right) + \dfrac{d_i}{d_o} R_{f,o} + \dfrac{d_i}{d_o h_o}} \quad ...(14.10)$$

The eqns. (14.9) and (14.10) represent overall heat transfer coefficients, include all possible resistances in the path of heat flow. The values of overall heat transfer coefficient may vary according to applications.

Based on the experience of manufacturers and users, the Tubular Equipment Manufacturer Association (TEMA) prepared the tables of fouling factors as a guide in heat transfer calculations and are available in handbook. The representative values of fouling factors are given in Table 14.2.

TABLE 14.2. Unit fouling resistance (or fouling factor) R_f

Type of fluid	*R_f (m².K/W)*
Distilled water, sea water, river water and treated boiler feed water	
Below 50°C	0.0001
Above 50°C	0.0002
Fuel oil	0.0009
Organic liquids	0.0002
Refrigerating liquid	0.0002
Refrigerant vapour	0.0004
Steam (oil free)	0.0001
Air	0.0004
Vegetables oil	0.0005

Example 14.1. *Calculate the overall heat transfer coefficient based on outer surface of a steel pipe (k = 54 W/m.K) with inner and outer diameters as 25 mm and 35 mm respectively. The inside and outside heat transfer coefficients are 1200 W/m².K and 2000 W/m².K respectively.*

Solution

Given : A fluid flow through a steel pipe :

k = 54 W/m.K

d_i = 25 mm = 0.025 m,

h_i = 1200 W/m^2.K

d_o = 35 mm = 0.035 m,

h_o = 2000 W/m^2.K.

To find : Overall heat transfer coefficient based on outer surface of the tube.

Analysis : The overall heat transfer coefficient U_o based on outer tube surface is calculated by using eqn. (14.5)

$$U_o = \frac{1}{\frac{d_o}{d_i h_i} + \frac{d_o}{2k}\ln\left(\frac{d_o}{d_i}\right) + \frac{1}{h_o}}$$

$$= \frac{1}{\frac{0.035 \text{ m}}{(0.025 \text{ m}) \times (1200 \text{ W/m}^2.\text{K})} + \frac{0.035 \text{ m}}{2 \times (54 \text{ W/m.K})} \times \ln\left(\frac{0.035 \text{ m}}{0.025 \text{ m}}\right) + \frac{1}{(2000 \text{ W/m}^2.\text{K})}}$$

$$= \frac{1}{1.1667 \times 10^{-3} + 1.090 \times 10^{-4} + 5 \times 10^{-4}}$$

$$= \frac{1}{1.776 \times 10^{-3}} = \textbf{563 W/m}^2\textbf{.K.} \quad \text{Ans.}$$

Example 14.2. *Water at 25°C and 1.5 m/s enters a long brass (k = 110 W/m.K) condenser tube with inner diameter of 1.34 cm and outer diameter of 1.58 cm. The heat transfer coefficient for condensation at outer surface of the tube is 12000 W/m^2.K. Calculate overall heat transfer coefficient based on outer surface of the tube.*

Solution

Given : Water flow through a tube of a condenser

$T_{c,i}$ = 25°C

u_m = 1.5 m/s

d_i = 1.34 cm = 0.0134 m

d_o = 1.58 cm = 0.0158 m

h_o = 12000 W/m^2.K

k = 110 W/m.K.

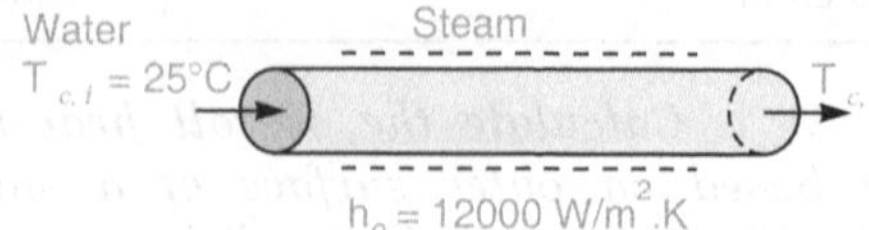

Fig. 14.10. Schematic for example 14.2

To find : Overall heat transfer coefficient U_o based on outer surface of the tube.

Assumption :

(*i*) Clean inner and outer surfaces of the tube

(*ii*) Steady state conditions.

Analysis : The properties of water at 25°C

ρ = 996 kg/m^3,

μ = 8.6 × 10^{-4} kg/ms,

k_f = 0.614 W/m.K, Pr = 5.85

For flow through tube,

Reynolds number

$$\text{Re} = \frac{\rho u_m d_i}{\mu} = \frac{996 \times 1.5 \times 0.0134}{8.6 \times 10^{-4}} = 23279$$

The Re > 2300, thus flow is turbulent and we may use Dittus-Boelter equation.

$$\text{Nu} = 0.023 \text{ Re}^{0.8} \text{ Pr}^{0.4}$$
$$= 0.023 \times (23279)^{0.8} \times (5.85)^{0.4}$$
$$= 145.26$$

$$h_i = \frac{\text{Nu } k_f}{d_i} = \frac{145.26 \times 0.614}{0.0134}$$
$$= 6656 \text{ W/m}^2.\text{K}.$$

The overall heat transfer coefficient based on outer surface of the tube can be calculated by using eqn. (14.5)

$$U_o = \frac{1}{\frac{d_o}{d_i h_i} + \frac{d_o}{2k}\ln\left(\frac{d_o}{d_i}\right) + \frac{1}{h_o}}$$

$$= \frac{1}{\frac{0.0158}{0.0134 \times 6656} + \frac{0.0158}{2 \times 110} \times \ln\left(\frac{0.0158}{0.0134}\right) + \frac{1}{12000}}$$

$$= \frac{1}{2.723 \times 10^{-4}} = \textbf{3672 W/m}^2\textbf{.K.} \quad \text{Ans.}$$

Example 14.3. *A double pipe heat exchanger is constructed of a stainless steel (k = 15.1 W/m.K) inner tube of inner diameter of 1.5 cm and outer diameter of 1.9 cm. It is concentric to an outer tube of diameter 3.2 cm.*

The inside and outside heat transfer coefficients are 800 and 1200 W/m^2.K, respectively. Due to continuous operation of the heat exchanger, the inner and outer surfaces of tube are fouled and respective fouling factors are 0.0004 m^2.K/W and 0.0001 m^2. K/W. Calculate:

(i) Thermal resistance of heat exchanger per unit length, and

(ii) Overall heat transfer coefficient based on inner and outer surface areas of the tube.

Solution

Given : A double pipe heat exchanger ;

Inner tube :

$k = 15.1$ W/m.K

$d_i = 1.5$ cm $d_o = 1.9$ cm

$h_i = 800$ W/m^2.K

$R_{f,i} = 0.0004$ m^2.K/W

$h_o = 1200$ W/m^2.K

$R_{f,o} = 0.0001$ m^2.K/W

$D_i = 3.2$ cm.

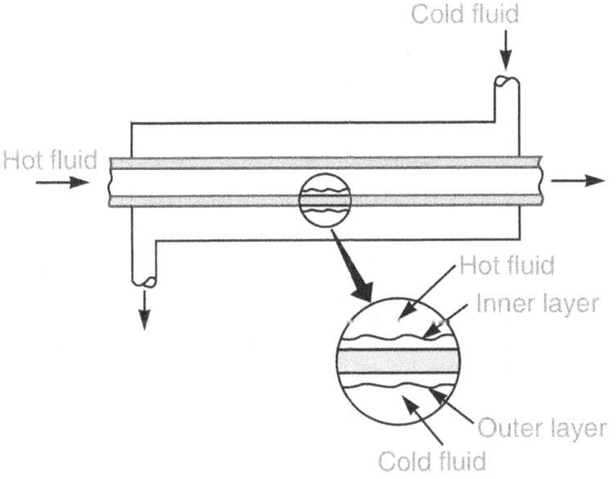

Fig. 14.11

To find :

(*i*) Thermal resistance of heat exchanger per unit length,

(*ii*) Overall heat transfer coefficient based on inner and outer surface areas of the tube.

Analysis : (*i*) The total thermal resistance of double pipe heat exchanger with fouling on both side surfaces consists of

(*a*) Convection resistances on both sides of tube

(*b*) Fouling resistances from impurities deposits on both side surfaces of steel tube.

(*c*) Conduction resistance of tube wall material.

Thus

$$\Sigma R_{th} = \frac{1}{h_i A_i} + \frac{R_{f,i}}{A_i} + \frac{\ln(d_o/d_i)}{2\pi L k} + \frac{R_{f,o}}{A_o} + \frac{1}{h_o A_o}$$

where $A_i = \pi d_i L = \pi \times (0.015 \text{ m}) \times (1 \text{ m}) = 0.0471 \text{ m}^2$

$A_o = \pi d_o L = \pi \times (0.019 \text{ m}) \times (1 \text{ m}) = 0.0597 \text{ m}^2$

Using, the total thermal resistance

$$\Sigma R_{th} = \frac{1}{800 \times 0.0471} + \frac{0.0004}{0.0471} + \frac{\ln\left(\frac{0.019}{0.015}\right)}{2\pi \times 1 \times 15.1} + \frac{0.0001}{0.0597} + \frac{1}{1200 \times 0.0597}$$

$$= 0.02654 + 0.00849 + 0.00249 + 0.00168 + 0.01396$$

$$= \mathbf{0.0532 \text{ k/W}}. \quad \text{Ans.}$$

(*ii*) Overall heat transfer coefficient :

We have

$$\Sigma R_{th} = \frac{1}{U_i A_i} = \frac{1}{U_o A_o}$$

$$U_i = \frac{1}{\Sigma R_{th} A_i} = \frac{1}{0.0532 \times 0.0471}$$

$$= \mathbf{399 \text{ W/m}^2\text{.K}}. \quad \text{Ans.}$$

and

$$U_o = \frac{1}{\Sigma R_{th} A_o} = \frac{1}{0.0532 \times 0.0597}$$

$$= \mathbf{314.85 \text{ W/m}^2\text{.K}}. \quad \text{Ans.}$$

14.5. HEAT EXCHANGER ANALYSIS

Heat exchangers usually operate for long periods of time without change in their operating conditions. Therefore, they can be modelled as *steady flow* devices with the following assumptions:

1. The mass flow rate of each fluid remains constant.
2. The fluid properties like temperature, velocity etc. at inlet and outlet also remain unchanged.
3. The changes in kinetic energy, potential energy, specific heat and conduction along the length are negligible.
4. The outer surface of outer tube of heat exchanger is well insulated and it does not allow any heat transfer to the surroundings, therefore, heat transfer occurs between hot and cold fluids only.
5. No fluid is undergoing phase change within heat exchanger.

The first law of thermodynamics states that the rate of heat transfer from the hot fluid be equal to the rate of heat transfer to the cold fluid.

The rate of heat transfer by hot fluid

$$Q = \dot{m}_h C_{p,h} (T_{h,i} - T_{h,o}) \quad ...(14.11)$$

The rate of heat transfer to cold fluid

$$Q = \dot{m}_c C_{p,c} (T_{c,o} - T_{c,i}) \quad ...(14.12)$$

where $\dot{m}$ = mass flow rate, kg/s

C_p = specific heat of the fluid, J/kg.K

T = temperature, °C

Suffix h = for hot fluid,

c = cold fluid,

i = for inlet condition,

o = for outlet condition.

Usually, heat exchangers are analysed either for their size using log mean temperature difference method or for their rating using effectiveness—NTU method.

14.6. LOG MEAN TEMPERATURE DIFFERENCE METHOD

As discussed in chapter 9, in tubular heat exchanger, the temperature difference between the hot and cold fluids varies as fluids proceed in the heat exchanger, therefore, it is convenient to determine log mean temperature difference ΔT_{lm}.

The total heat transfer rate between the hot and cold fluids can also be calculated by using overall heat transfer coefficient and surface area as :

$$Q = UA\Delta T_{lm} \quad ...(14.13)$$

where U = overall heat transfer coefficient,

A = surface area for heat exchange,

ΔT_{lm} = log mean temperature difference.

14.6.1. Parallel Flow Heat Exchanger

Consider a parallel flow double pipe heat exchanger as shown in Fig. 14.12. The temperature difference ΔT between hot and cold fluids is large at the inlet of heat exchanger and it decreases exponentially towards the outlet. The temperature of hot fluid decreases and that of cold fluid increases along the length of heat exchanger, thus the temperature of cold fluid can never exceed that of hot fluid in any case.

Applying the energy balance to differential element between hot and cold fluids.

The rate of heat transfer δQ from hot fluids

$$\delta Q = -\dot{m}_h C_{p,h}\, dT_h \quad ...(14.14)$$

and for cold fluid

$$\delta Q = \dot{m}_c C_{p,c}\, dT_c \quad ...(14.15)$$

The temperature difference of hot fluid is a negative, while temperature difference of cold fluid is positive as shown in Fig. 14.12. Let for differential surface area dA, the temperature difference ΔT between hot and cold fluid is expressed as

$$\Delta T = T_h - T_c \quad ...(14.16)$$

In differential form :

$$d(\Delta T) = dT_h - dT_c \quad ...(14.17)$$

Solving the eqn. (14.14) and eqn. (14.15) for dT_h and dT_c as

$$dT_h = -\frac{\delta Q}{\dot{m}_h C_{p,h}} \quad ...(14.18)$$

$$dT_c = \frac{\delta Q}{\dot{m}_c C_{p,c}} \quad ...(14.19)$$

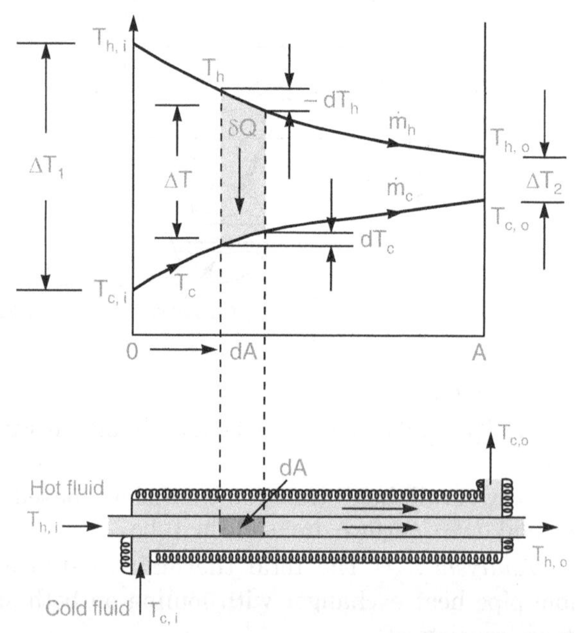

Fig. 14.12. Temperature distribution for a parallel flow double pipe heat exchanger

Substituting the values of dT_h and dT_c from eqns. (14.18) and (14.19) in eqn. (14.17), we get

$$d(\Delta T) = -\delta Q\left[\frac{1}{\dot{m}_h C_{p,h}} + \frac{1}{\dot{m}_c C_{p,c}}\right] \quad ...(14.20)$$

The heat transfer rate across the differential surface area dA of heat exchanger can also be expressed as

$$\delta Q = U\Delta T dA \quad ...(14.21)$$

using δQ in eqn. (14.20), we get

$$d(\Delta T) = -U\Delta T dA \left[\frac{1}{\dot{m}_h C_{p,h}} + \frac{1}{\dot{m}_c C_{p,c}}\right] \quad ...(14.22)$$

Rearranging,

$$\frac{d(\Delta T)}{\Delta T} = -U dAB \quad ...(14.23)$$

where, $B = \frac{1}{\dot{m}_h C_{p,h}} + \frac{1}{\dot{m}_c C_{p,c}}$

Integrating eqn. (14.23) from inlet to outlet conditions of the heat exchanger.

$$\int_{\Delta T_1}^{\Delta T_2} \frac{d(\Delta T)}{\Delta T} = -UB \int_0^A dA$$

or,
$$\ln\left[\frac{\Delta T_2}{\Delta T_1}\right] = -UBA$$

Substituting B, we get

$$\ln\left[\frac{\Delta T_2}{\Delta T_1}\right] = -UA\left\{\frac{1}{\dot{m}_h C_{p,h}} + \frac{1}{\dot{m}_c C_{p,c}}\right\} \quad ...(14.24)$$

Substituting for $\dot{m}_h C_{p,\ h}$ and $\dot{m}_c C_{p,\ c}$ from eqns. (14.11) and (14.12), respectively, we get

$$\ln\left[\frac{\Delta T_2}{\Delta T_1}\right] = -UA\left\{\frac{T_{h,i} - T_{h,o}}{Q} + \frac{T_{c,o} - T_{c,i}}{Q}\right\}$$

or
$$Q = \frac{UA}{\ln\left[\frac{\Delta T_1}{\Delta T_2}\right]}\{(T_{h,i} - T_{c,i}) - (T_{h,o} - T_{c,o})\}$$

or
$$Q = UA\frac{\Delta T_1 - \Delta T_2}{\ln\left[\frac{\Delta T_1}{\Delta T_2}\right]} \quad ...(14.25)$$

Comparing above expression with eqn. (14.13), we get

$$\Delta T_{lm} = \frac{\Delta T_1 - \Delta T_2}{\ln\left[\frac{\Delta T_1}{\Delta T_2}\right]} = \frac{\Delta T_2 - \Delta T_1}{\ln\left[\frac{\Delta T_2}{\Delta T_1}\right]} \quad ...(14.26)$$

The ΔT_{lm} is called the *log mean temperature difference,* which is suitable form of temperature difference for tubular heat exchanger. Where ΔT_1 and ΔT_2 represent the temperature difference between hot and cold fluids at two ends (inlet and outlet) of a heat exchanger. It makes no difference which end of heat exchanger is designed as inlet or outlet. In this book, ΔT_1 and ΔT_2 are designated as temperature difference at left and right ends, respectively.

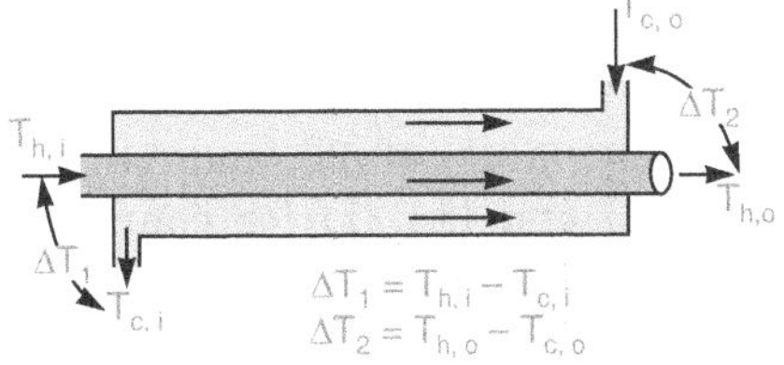

(a) Parallel flow heat exchanger

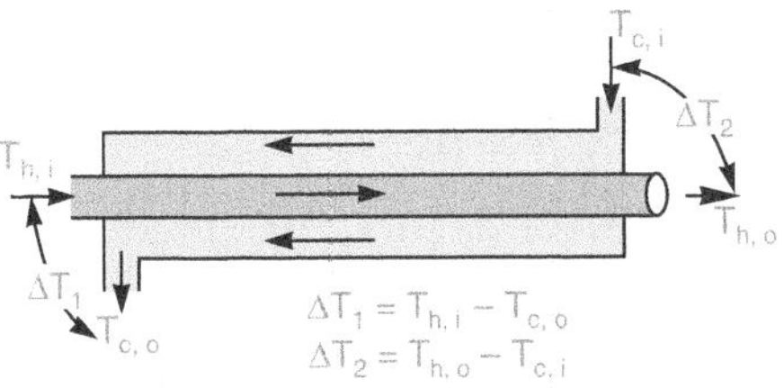

(b) Counter flow heat exchanger

Fig. 14.13. Definition of ΔT_1 and ΔT_2 for parallel and counter flow arrangements

Using eqn. (14.26), the heat transfer rate in double pipe heat exchanger can be expressed as

$$Q = UA\Delta T_{lm} = U_o A_o \Delta T_{lm} = U_i A_i \Delta T_{lm} \quad ...(14.27)$$

14.6.2. Counter Flow Heat Exchanger

In counter flow arrangement, the hot and cold fluids enter the heat exchanger from opposite ends, two fluids flow in opposite direction. Therefore, the outlet temperature of cold fluid may exceed the outlet temperature of hot fluid. The temperature distribution for counter flow heat exchanger is shown in Fig. 14.14.

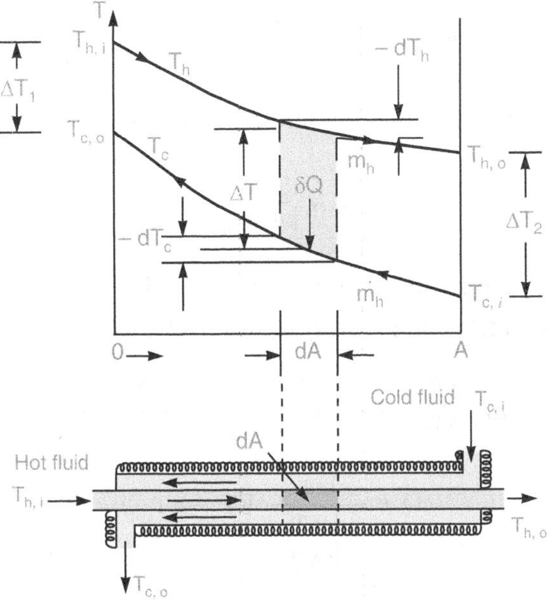

Fig. 14.14. Temperature distribution for a counter flow heat exchanger

Here, $\Delta T_1 = T_{h,i} - T_{c,o}$

$\Delta T_2 = T_{h,o} - T_{c,i}$

Using these values of ΔT_1 and ΔT_2, we may use the eqns. (14.26) and (14.27) for counter flow heat exchanger. Its derivation can also be taken as ;

Applying the energy balance to differential elements in hot and cold fluids.

The rate of heat transfer δQ from hot fluid

$$\delta Q = -\dot{m}_h C_{p,h} dT_h \quad \text{...(14.28)}$$

and for cold fluid,

$$\delta Q = -\dot{m}_c C_{p,c} dT_c \quad \text{...(14.29)}$$

The temperature difference ΔT between hot and cold fluids within the differential area dA can be expressed as :

$$\Delta T = T_h - T_c$$

In differential form :

$$d(\Delta T) = dT_h - dT_c$$

Substituting the values of dT_h and dT_c from eqns. (14.28) and (14.29), respectively, we get

$$d(\Delta T) = -\delta Q \left\{ \frac{1}{\dot{m}_h C_{p,h}} - \frac{1}{\dot{m}_c C_{p,c}} \right\}$$

Substituting $\delta Q = U(\Delta T)dA$ for differential element, we get

$$d(\Delta T) = -U(\Delta T)\, dA \left[\frac{1}{\dot{m}_h C_{p,h}} - \frac{1}{\dot{m}_c C_{p,c}} \right]$$

Rearranging and integrating, we get

$$\ln\left[\frac{\Delta T_2}{\Delta T_1}\right] = -UA \left\{ \frac{1}{\dot{m}_h C_{p,h}} - \frac{1}{\dot{m}_c C_{p,c}} \right\} \quad \text{...(14.30)}$$

Substituting for $\dot{m}_c C_{p,c}$ and $\dot{m}_h C_{p,h}$ from eqns. (14.11) and (14.12), respectively, we get

$$\ln\left[\frac{\Delta T_2}{\Delta T_1}\right] = -UA \left\{ \frac{T_{h,i} - T_{h,o}}{Q} - \frac{T_{c,o} - T_{c,i}}{Q} \right\}$$

or

$$Q = \frac{UA}{\ln\left[\dfrac{\Delta T_1}{\Delta T_2}\right]} \{(T_{h,i} - T_{c,o}) - (T_{h,o} - T_{c,i})\} \quad \text{...(14.31)}$$

or

$$Q = U_i A_i \Delta T_{lm} = U_o A_o \Delta T_{lm}$$

The ΔT_{lm} is called log mean temperature difference for counter flow heat exchanger and is expressed as

$$\Delta T_{lm,\ \text{counter}} = \frac{(T_{h,i} - T_{c,o}) - (T_{h,o} - T_{c,i})}{\ln\left[\dfrac{T_{h,i} - T_{c,o}}{T_{h,o} - T_{c,i}}\right]} \quad \text{...(14.32)}$$

14.6.3. Condenser

Temperature distribution for a condenser is shown in Fig. 14.8 (*d*). The temperature of the condensing fluid (hot fluid) remains constant. Hence the temperature differences

$$\Delta T_1 = T_h - T_{c,i} \quad \text{...(14.33)}$$

$$\Delta T_2 = T_h - T_{c,o} \quad \text{...(14.34)}$$

can be used with eqn. (14.26) or eqn. (14.32).

14.6.4. Evaporator

Temperature distribution for an evaporator is shown in Fig. 14.8(*c*), the temperature of cold fluid (evaporating fluid) remains constant, hence, the log mean temperature difference given by eqn. (14.26) or eqn. (14.32) can be used with

$$\Delta T_1 = T_{h,i} - T_c \quad \text{...(14.35)}$$

$$\Delta T_2 = T_{h,o} - T_c \quad \text{...(14.36)}$$

14.7. MULTIPASS AND CROSS FLOW HEAT EXCHANGERS

Although the temperature distribution and flow conditions are complicated in multipass and cross flow heat exchangers. In such cases, the appropriate mean temperature difference can be obtained by introducing a correction factor F to the log mean temperature difference (LMTD) for counter flow arrangement with same hot and cold fluid temperatures. In such case

$$\Delta T_{lm} = F\, \Delta T_{lm,\ \text{counter flow}} \quad \text{...(14.37)}$$

The correction factor F depends on geometry of heat exchanger and inlet and outlet temperatures of hot and cold fluids. The value of correction factor F varies from unity to zero. The correction factor F = 1, for counter flow heat exchangers and less than unity for cross flow and multipass arrangement for shell and tube type heat exchangers. It represents the degree of departure of true mean temperature difference from the LMTD for a counter flow heat exchanger. For some commonly used heat exchangers, the correction factor F can be obtained from plot shown in Fig. 14.15. In these figures, the abscissa is a dimensionless ratio P, which represents the *thermal effectiveness* of the tube side fluid. It is expressed as

$$P = \frac{t_2 - t_1}{T_1 - t_1} \quad ...(14.38)$$

where T represents to shell side temperature, t represents the tube side temperature and subscripts 1 and 2 for inlet and outlet conditions, respectively.

The parameter R on the curve is equal to *heat capacity ratio* and is expressed as

$$R = \frac{(\dot{m}C_p)_{\text{tube side}}}{(\dot{m}C_p)_{\text{shell side}}} = \frac{T_1 - T_2}{t_2 - t_1} \quad ...(14.39)$$

If one fluid remains at constant temperature as in case of condensation and evaporation processes, then either value P or R will be zero. In this case, the correction factor, F is always unity and simple LMTD can be used.

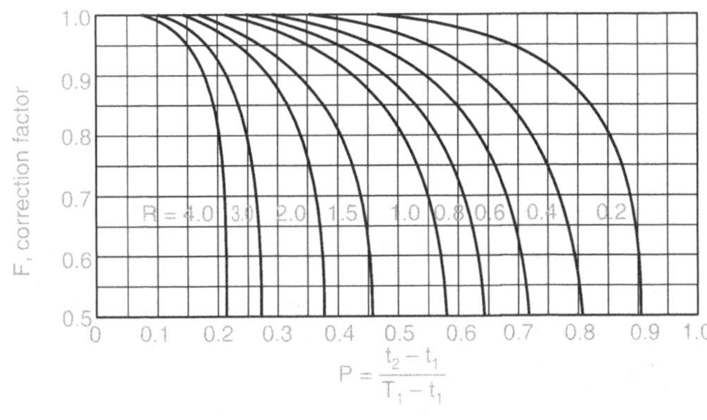

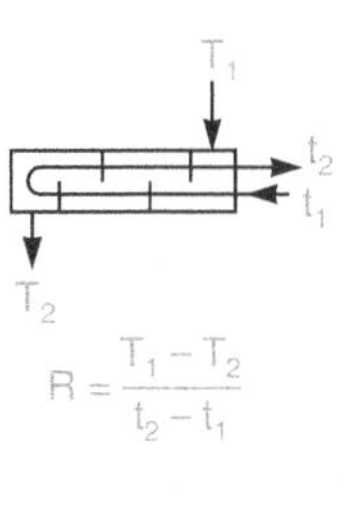

(a) One shell pass and two tube pass or multiple of two tube pass

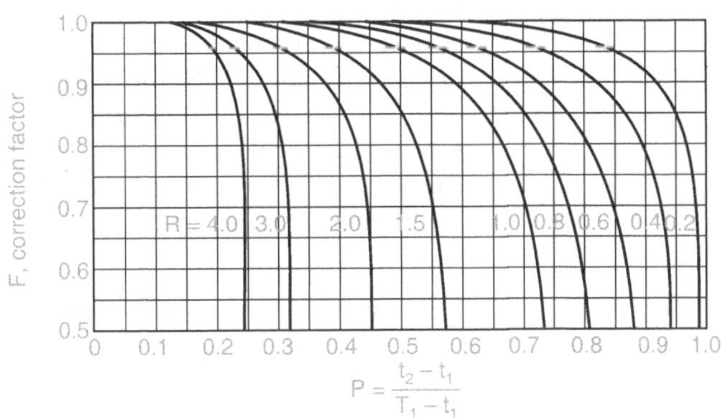

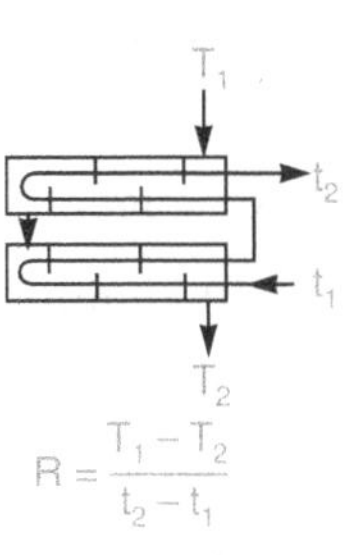

(b) Two shell pass and four tubes pass or multiple of four tube pass

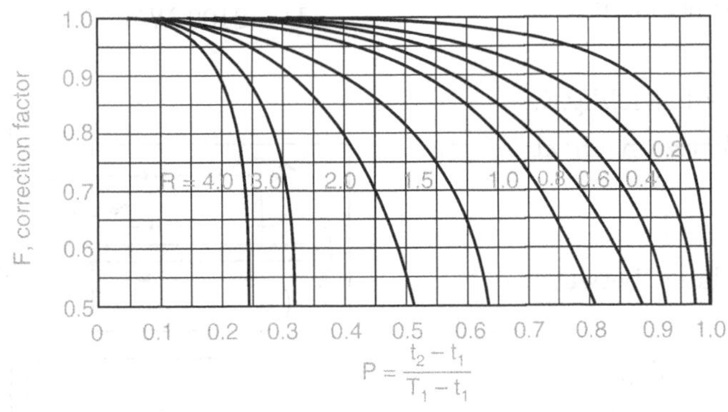

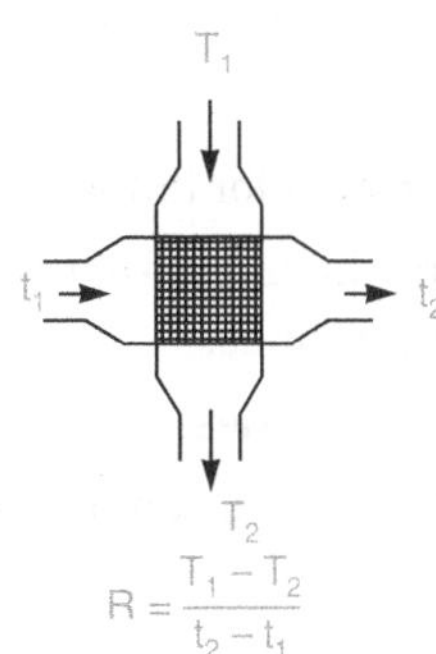

(c) Cross flow, both fluid unmixed

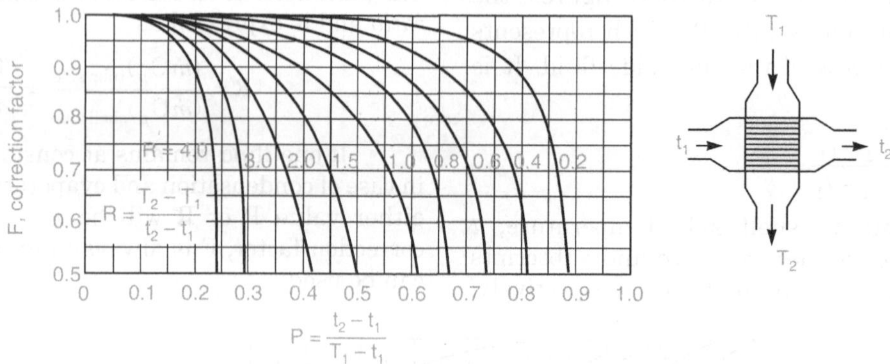

(d) Single-pass cross-flow exchanger, one fluid mixed, the other unmixed

Fig. 14.15. Correction factor F

Example 14.4. *A heat exchanger is to be designed to transfer heat from hot water to a certain cold liquid. The given conditions are : hot water flow rate 5 kg/min, cold liquid flow rate 10 kg/min, hot water inlet temperature = 85°C, cold liquid inlet temperatures = 27°C, cold liquid outlet temperature is 55°C and specific heat of cold liquid = 2.93 kJ / kg.K.*

The engineer is assigned to do the work to design a parallel flow configuration, but he is getting difficulty in doing the calculations. Can you explain why ?

Solution

Given : A heat exchanger

$$\dot{m}_c = 10 \text{ kg/min},$$

$$\dot{m}_h = 5 \text{ kg/min}$$

$$C_{p,c} = 2.93 \text{ kJ/kg.K},$$

$$T_{h,i} = 85°C$$

$$T_{c,i} = 27°C$$

$$T_{c,o} = 55°C.$$

To find : Find the type of flow arrangement.

Assumptions :

1. The specific heat for hot water as 4.18 kJ/kg.K.

2. Negligible heat transfer between heat exchanger and surroundings.

3. Constant properties.

Analysis : Making the energy balance between the two fluids

$$\dot{m}_h C_{p,h}(T_{h,i} - T_{h,o}) = \dot{m}_c C_{p,c}(T_{c,o} - T_{c,i})$$

$$5 \text{ (kg/min)} \times 4.18 \text{ (kJ/kg.K)} \times (85 - T_{h,o})(°C)$$

$$= 10 \text{ (kg/min)} \times 2.93 \text{ (kJ/kg.K)} \times (55 - 27)(°C)$$

or $$T_{h,o} = 45.74°C$$

Since $$T_{c,o} > T_{h,o}$$

The cold fluid exit temperature is greater than the hot water exit temperature. The parallel flow arrangement is impossible with such data. ***Hence counter flow arrangement would be suitable.***

Example 14.5. *Steam in a condenser of a steam power plant is to be condensed at a temperature of 30°C with cooling water from a nearby lake, which enters the tubes of condenser at 14°C and leaves at 22°C. The surface area of the tubes is 45 m² and overall heat transfer coefficient is 2100 W/m².K. Calculate the mass flow rate of cooling water needed and rate of steam condensation in the condenser.*

Solution

Given : A steam condenser

$$T_h = 30°C$$

$$T_{c,i} = 14°C$$

$$T_{c,o} = 22°C$$

$$A = 45 \text{ m}^2$$

$$U = 2100 \text{ W/m}^2\text{.K}.$$

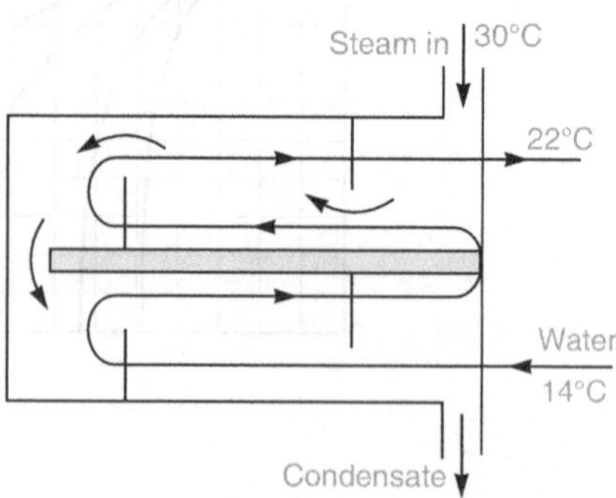

Fig. 14.16. Schematic of a condenser

To find :

(*i*) Mass flow rate of water into condenser

(*ii*) Rate of steam condensation in the condenser.

Assumptions :

(*i*) Condenser is well insulated, so that the heat loss from steam is transferred to water only.

(*ii*) Treating condenser as counter flow heat exchanger.

Analysis : The properties of water at $\frac{22+14}{2}$ = 18°C, from Table A-7 of Appendix A.

$$C_p = 4.18 \text{ kJ/kg.K}$$

and latent heat of vaporisation at 30°C

$$h_{fg} = 2430.5 \text{ kJ/kg}$$

For counter flow heat exchanger, the temperature difference

$$\Delta T_1 = T_{h,i} - T_{c,o} = 30 - 22 = 8°C$$

$$\Delta T_2 = T_{h,o} - T_{c,i} = 30 - 14 = 16°C$$

and log mean temperature difference

$$\Delta T_{lm} = \frac{\Delta T_1 - \Delta T_2}{\ln\left(\frac{\Delta T_1}{\Delta T_2}\right)} = \frac{8-16}{\ln\left(\frac{8}{16}\right)} = 11.5°C$$

The heat transfer rate in the condenser is

$$Q = UA\,\Delta T_{lm} = 2100 \times 45 \times 11.5$$

$$= 1{,}086{,}750 \text{ W} = 1086.75 \text{ kW}$$

This heat will be absorbed by cooling water, thus

$$Q = \dot{m}_c C_{p,c}(T_{c,o} - T_{c,i})$$

or $$1086.75 = \dot{m}_c \times 4.18 \times (22 - 14)$$

or $$\dot{m}_c = \mathbf{32.5 \text{ kg/s.}} \quad \text{Ans.}$$

and steam condensation rate is

$$\dot{m}_{steam} = \frac{Q}{h_{fg}} = \frac{1086.75}{2430.5} = \mathbf{0.45 \text{ kg/s.}} \quad \text{Ans.}$$

Example 14.6. *A heat exchanger is required to cool 55,000 kg/h of alcohol from 66°C to 40°C using 40,000 kg/h of water entering at 5°C. Calculate (i) exit temperature of water, (ii) heat transfer rate, (iii) surface area required for (a) parallel flow type, (b) counter flow type of heat exchanger.*

Take overall heat transfer coefficient

$$U = 580 \text{ W/m}^2\text{.K}$$

$$C_{p\,(alcohol)} = 3760 \text{ J/kg.K}$$

$$C_{p\,(water)} = 4180 \text{ J/kg.K.}$$ (V.T.U., July 2002)

Solution

Given: A double pipe heat exchanger

Alcohol, $\dot{m}_h = 55000$ kg/h

Water, $\dot{m}_c = 40000$ kg/h

$T_{h,i} = 66°C$

$C_{p,c} = 4180$ J/kg.K

$T_{h,o} = 40°C$

$T_{c,i} = 5°C$

$C_{p,h} = 3760$ J/kg.K

$U = 580$ W/m².K.

To find :

(*i*) Exit temperature $T_{c,o}$ of water,

(*ii*) Heat transfer rate Q,

(*iii*) Surface area of heat exchanger if it is (*a*) parallel type, (*b*) counter flow type.

Analysis : (*i*) Making energy balance on the heat exchanger

$$\dot{m}_h C_{p,h}(T_{h,i} - T_{h,o}) = \dot{m}_c C_{p,c}(T_{c,o} - T_{c,i})$$

$$55000 \times 3760 \times (66 - 40) = 40{,}000 \times 4180 \times (T_{c,o} - 5)$$

or $$T_{c,o} = \mathbf{37.15°C.} \quad \text{Ans.}$$

(*ii*) Heat transfer rate

$$Q = \dot{m}_h C_{p,h}(T_{h,i} - T_{h,o})$$

$$= 55000 \times 3760 \times (66 - 40)$$

$$= 537.68 \times 10^7 \text{ J/h}$$

$$= \mathbf{1493.55 \text{ kW.}} \quad \text{Ans.}$$

(*iii*) Surface area of heat exchanger,

(*a*) Parallel flow type

$$\Delta T_1 = T_{h,i} - T_{c,i} = 66 - 5 = 61°C$$

$$\Delta T_2 = T_{h,o} - T_{c,o} = 40 - 37.15 = 2.85°C$$

Log mean temperature difference

$$\Delta T_{lm,\,parallel} = \frac{\Delta T_1 - \Delta T_2}{\ln\left(\frac{\Delta T_1}{\Delta T_2}\right)}$$

$$= \frac{61 - 2.85}{\ln\left(\frac{61}{2.85}\right)} = 18.98°C$$

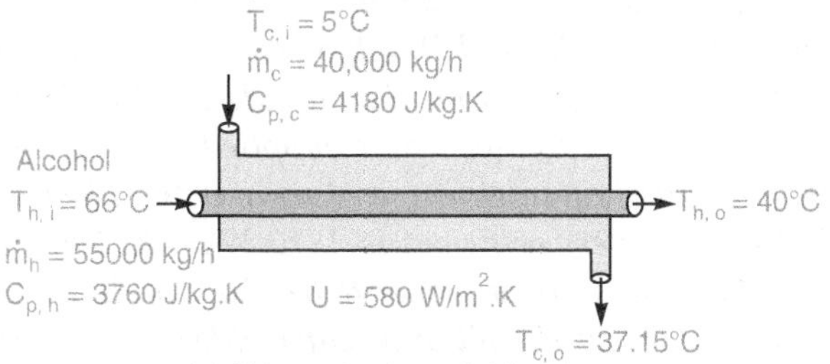

(a) Schematic of parallel flow heat exchanger for example 14.6

Heat transfer area,

$$A = \frac{Q}{U\,\Delta T_{lm,\,\text{parallel}}}$$

$$= \frac{1493.55 \times 10^3}{580 \times 18.98}$$

$$= \mathbf{135.66\ m^2}.\ \text{Ans.}$$

(*b*) Counter flow type

$$\Delta T_1 = T_{h,\,i} - T_{c,\,o} = 66 - 37.15 = 28.85°C$$

$$\Delta T_2 = T_{h,\,o} - T_{c,\,i} = 40 - 5 = 35°C$$

$$\Delta T_{lm,\,\text{counter}} = \frac{\Delta T_1 - \Delta T_2}{\ln\left(\dfrac{\Delta T_1}{\Delta T_2}\right)}$$

$$= \frac{28.85 - 35}{\ln\left(\dfrac{28.85}{35}\right)} = 31.82°C$$

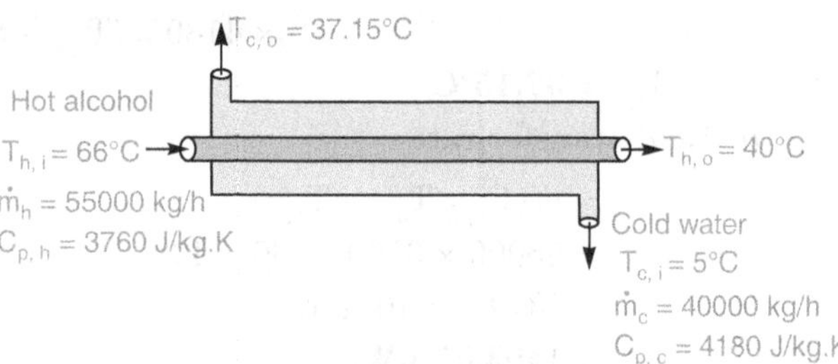

(b) Schematic of counter flow heat exchanger for example 14.6

Fig. 14.17

Surface area,

$$A = \frac{Q}{U\,\Delta T_{lm,\,\text{counter}}}$$

$$= \frac{1493.55 \times 10^3}{580 \times 31.82}$$

$$= \mathbf{80.91\ m^2}.\ \text{Ans.}$$

Example 14.7. *A simple heat exchanger consisting of two concentric flow passages is used for heating 1110 kg/h of oil (sp. heat = 2.1 kJ/kg.K) from a temperature of 27°C to 49°C. The oil flows through the inner pipe made of copper (O.D. = 2.86 cm, I.D. = 2.54 cm) and the surface heat transfer coefficient on the oil side is 635 W/m².K. The oil is heated by hot water supplied at the rate of 390 kg/h and at an inlet temperature of 93°C. The water side heat transfer coefficient is 1270 W/m².K. Take the thermal conductivity of copper to be 350 W/m.K and the fouling factors on the oil and water sides to be 0.0001 and 0.0004 m².K/W. What is the length of the heat exchanger for : (i) parallel flow, and (ii) counter flow ?*

Solution

Given : Double heat exchanger

$\dot{m}_c$ = 1110 kg/h,

$\dot{m}_h$ = 390 kg/h,

$C_{p,\,c}$ = 2.1 kJ/kg. K = 2100 J/kg.K,

$T_{h,\,i}$ = 93°C,

$T_{c,\,i}$ = 27°C,

$T_{c,\,o}$ = 49°C,

d_i = 2.54 cm = 0.0254 m,

d_o = 2.86 cm = 0.0286 m,

h_i = 635 W/m².K,

h_o = 1270 W/m².K,

k = 350 W/m.K,

$R_{f,\,i}$ = 0.0001 m².K/W,

$R_{f,\,o}$ = 0.0004 m².K/W.

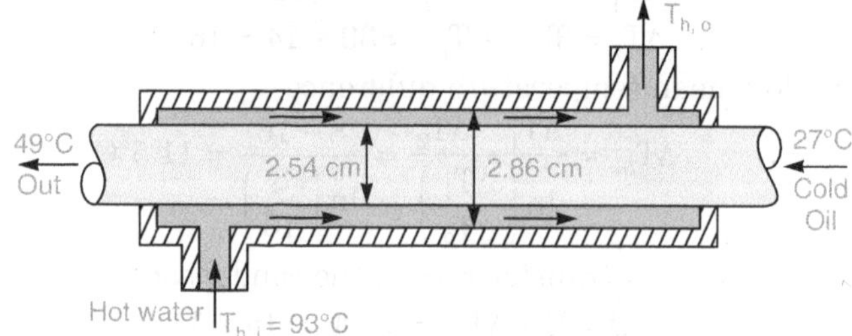

Fig. 14.18. Schematic of double pipe counter flow heat exchanger for example 14.7

To find : Length of heat exchanger for

(*i*) Parallel flow arrangement,

(*ii*) Counter flow arrangement.

Assumptions :

1. The specific heat of hot fluid water as 4180 J/kg.K.

2. Negligible heat transfer between heat exchanger and surroundings.

3. Constant properties.

Analysis : The exit temperature of water : The energy balance between the two fluids ;

$$\dot{m}_h C_{p,\,h}(T_{h,\,i} - T_{h,\,o}) = \dot{m}_c C_{p,\,c}(T_{c,\,o} - T_{c,\,i})$$

$$390 \times 4180 \times (93 - T_{h,\,o}) = 1110 \times 2100 \times (49 - 27)$$

or $$T_{h,\,o} = 93 - 31.45 = 61.55°C$$

The heat transfer rate,

$$Q = \dot{m}_c C_{p,\,c}(T_{c,\,o} - T_{c,\,i})$$

$$Q = (1100/3600)(\text{kg/s}) \times (2100\ \text{J/kg.K}) \times (49 - 27)(\text{K})$$

$$= 14245\ \text{W}$$

The overall heat transfer coefficient, U_o

$$U_o = \frac{1}{\dfrac{d_o}{d_i h_i} + \dfrac{d_o}{d_i} R_{f,i} + \dfrac{d_o \ln (r_o/r_i)}{2k} + R_{f,o} + \dfrac{1}{h_o}}$$

$$= \frac{1}{\dfrac{0.0286}{0.0254} \times \left\{ \dfrac{1}{635} + 0.0001 \right\} + \dfrac{0.0286}{2 \times 350} \times \ln \left(\dfrac{0.0286}{0.0254} \right) + 0.0004 + \dfrac{1}{1270}}$$

$= 325$ W/m^2.K

$$\text{LMTD, } \Delta T_{lm} = \frac{\Delta T_1 - \Delta T_2}{\ln \left[\dfrac{\Delta T_1}{\Delta T_2} \right]}$$

(i) For parallel flow arrangement :

$$\Delta T_1 = (T_{h,i} - T_{c,i}) = 93 - 27 = 66°C$$

and $$\Delta T_2 = (T_{h,o} - T_{c,o}) = 61.55 - 49 = 12.55$$

$$\Delta T_{lm} = \frac{66 - 12.55}{\ln \left(\dfrac{66}{12.55} \right)} = 32.2°C$$

The length of heat exchange can be calculated as

$$Q = U_o A_o \Delta T_{lm} = U_o (\pi d_o L)\, \Delta T_{lm}$$

or $$14245 = 325 \times (\pi \times 0.0286\ L) \times 32.2$$

or $$L = 15.15 \text{ m.}$$ **Ans.**

(ii) For counter flow arrangement :

$$\Delta T_1 = T_{h,o} - T_{c,i} = 61.55 - 27 = 34.55°C$$

$$\Delta T_2 = T_{h,i} - T_{c,o} = 93 - 49 = 44°C$$

$$\text{LMTD} \Delta T_{lm} = \frac{34.55 - 44}{\ln \left[\dfrac{34.55}{44} \right]} = 39.07°C$$

The length of heat exchanger

$$Q = U_o A_o \Delta T_{lm} = U_o (\pi d_o L)\, \Delta T_{lm}$$

or $$14245 = 325 \times (\pi \times 0.0286\ L) \times 39.07$$

or $$L = 12.48 \text{ m.}$$ **Ans.**

The counter flow heat exchanger will require less space than parallel flow due to more effective temperature difference.

Example 14.8. *In a balanced counter flow heat exchanger, where $\dot{m}_c C_{p,c} = \dot{m}_h C_{p,h}$. Show that $\Delta T_1 = \Delta T_2 = \Delta T$, at any section, and the temperature profile of two fluids are parallel and linear. Also prove that $\Delta T_{lm} = \Delta T_1 = \Delta T_2$.*

Solution

Analysis : The temperature distribution for balance counter flow heat exchanger is shown in Fig. 14.19.

For differential element of heat exchanger

$$\delta Q = -\dot{m}_h C_{p,h} dT_h$$
$$= -\dot{m}_c C_{p,c} dT_c$$
$$= U dA \Delta T$$

Since $\dot{m}_c C_{p,c} = \dot{m}_h C_{p,h}$

$\therefore$ $$dT_h = dT_c$$

or $$d(T_h - T_c) = 0$$

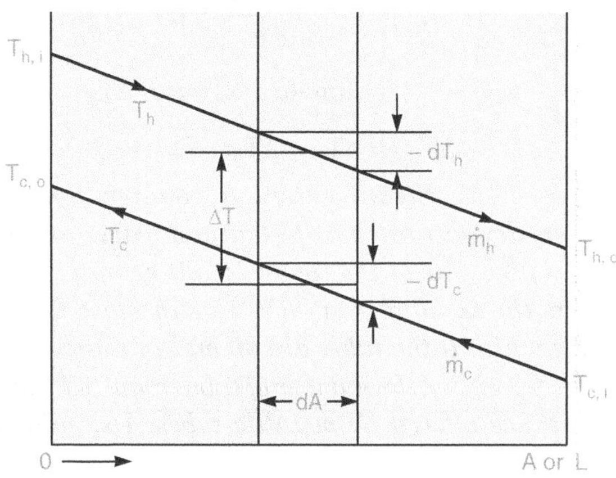

Fig. 14.19. Schematic of counter flow heat exchanger subjected to uniform temperature difference

On integration, we get

$$T_h - T_c = \text{constant}$$

or $$\Delta T = \Delta T_1 = \Delta T_2$$
$$= \text{constant at any section.}$$

Proved.

Further, from energy balance

$$-\dot{m}_h C_{p,h} dT_h = -\dot{m}_c C_{p,c} dT_c$$
$$= U dA \Delta T$$

or $$\frac{dT_h}{dA} = -\frac{U \Delta T}{\dot{m}_h C_{p,h}} = \text{constant}$$

and $$\frac{dT_c}{dA} = -\frac{U \Delta T}{\dot{m}_c C_{p,c}} = \text{constant}$$

Since temperature difference is uniform and slope $\dfrac{dT}{dA}$ is constant for both lines thus these lines are parallel and linear.

The LMTD is given by

$$\Delta T_{lm} = \frac{\Delta T_1 - \Delta T_2}{\ln\left(\frac{\Delta T_1}{\Delta T_2}\right)}$$

Let $\Delta T_1 = a$ and $\Delta T_1 - \Delta T_2 = x$

$\therefore$ $\Delta T_2 = a - x$

Then $\Delta T_{lm} = \dfrac{x}{\ln [a/(a-x)]}$

$$\lim_{x \to 0} (\Delta T_{lm}) = \lim_{x \to 0} \left[\frac{x}{\ln \{a/(a-x)\}}\right]$$

$$= \lim_{x \to 0} \frac{1}{\frac{a-x}{a}(-a)(-1)\frac{1}{(a-x)^2}}$$

$$= \lim_{x \to 0} (a-x) = a = \Delta T_1$$

$\therefore$ $\Delta T_{lm} = \Delta T_1 = \Delta T_2$. **Proved.**

Example 14.9. *Steam enters a counter flow heat exchanger, dry saturated at 10 bar and leaves at 350°C. The mass flow rate of the steam is 720 kg/min. The hot gas enters the exchanger at 650°C with mass flow rate of 1320 kg/min. If the tubes are 30 mm in diameter and 3 m long, determine the number of tubes required. Neglect the resistance offered by metallic tubes. Use following data:*

For steam $T_{sat} = 180°C$ *(at 10 bar),*

$C_{p,s} = 2.71$ *kJ/kg.K,*

$h_i = 600$ *W/m²K*

For gas $C_{p,g} = 1$ *kJ/kg.K,*

$h_o = 250$ *W/m².K.* (P.U., May 2006)

Solution

Given : A counter flow heat exchanger ;

For steam ; $\dot{m}_c = 720$ kg/min = 12 kg/s

$T_{c,i} = 180°C$

$T_{c,o} = 350°C$

$C_{p,c} = 2.71$ kJ/kg.K,

$h_i = 600$ W/m².K

For gas; $\dot{m}_h = 1320$ kg/min = 22 kg/s

$C_{p,h} = 1$ kJ/kg.K

$T_{h,i} = 650°C$,

$h_o = 250$ W/m².K

$d = 30$ mm = 0.03 m,

L = 3 m.

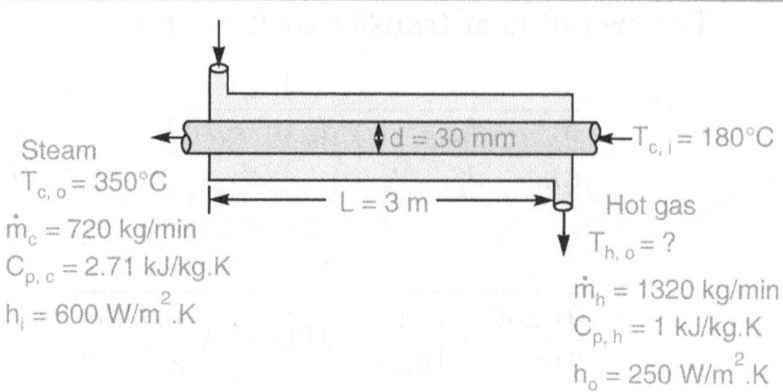

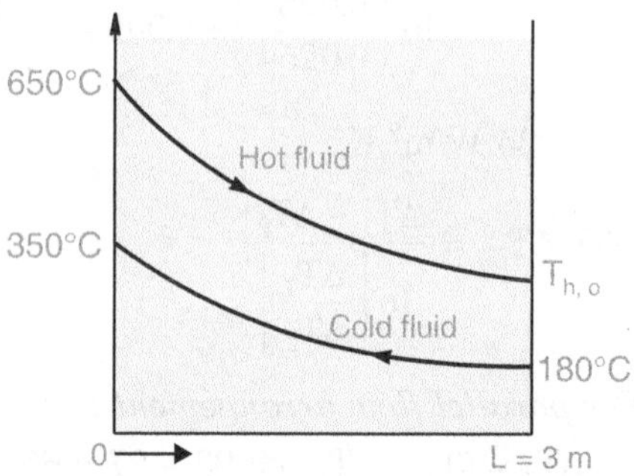

Fig. 14.20. Schematic and temperature profile for counter flow heat exchanger for example 14.9

To find : Number of tubes required.

Analysis : The exit temperature of hot gas can be determined by energy balance on the heat exchanger.

Heat lost rate by gas = Heat gain rate by steam

$$\dot{m}_h C_{p,h} (T_{h,i} - T_{h,o}) = \dot{m}_c C_{p,c}(T_{c,o} - T_{c,i})$$

$$22 \times 1 \times (650 - T_{h,o}) = 12 \times 2.71 \times (350 - 180)$$

$$T_{h,o} = 398.7°C$$

Heat transfer rate :

$$Q = \dot{m}_h C_{p,h}(T_{h,i} - T_{h,o})$$

$$= 22 \times 1 \times (650 - 398.7)$$

$$= 5528.6 \text{ kW}$$

Log mean temperature difference (LMTD)

$$\Delta T_1 = T_{h,i} - T_{c,o} = 650 - 350$$

$$= 300°C$$

$$\Delta T_2 = T_{h,o} - T_{c,i} = 398.7 - 180$$

$$= 218.7°C$$

$$\Delta T_{lm} = \frac{\Delta T_1 - \Delta T_2}{\ln\left(\frac{\Delta T_1}{\Delta T_2}\right)}$$

$$= \frac{300 - 218.7}{\ln\left(\frac{300}{218.7}\right)} = 257.21°C$$

Overall heat transfer coefficient U ; without tube wall resistance

$$U = \frac{1}{\frac{1}{h_i} + \frac{1}{h_o}} = \frac{1}{\frac{1}{600} + \frac{1}{250}}$$
$$= 176.47 \text{ W/m}^2\text{.K}$$

Further, the heat transfer rate in terms of U and ΔT_{lm} is expressed as

$$Q = UA\Delta T_{lm}$$

or
$$A = \frac{Q}{U\Delta T_{lm}}$$
$$= \frac{5528.6 \times 10^3}{176.47 \times 257.21} = 121.8 \text{ m}^2$$

Further, $A = \pi d LN$

$$N = \frac{121.8}{\pi \times 0.03 \times 3} = \textbf{431 tubes.} \quad \text{Ans.}$$

Example 14.10. *A heat exchanger contains 4500 tubes, each having a diameter of 2.54 cm, through which 11.3 kg/s of air to be cooled from 538°C to 148°C. Water passes in counterflow over the tubes, rising in temperature from 38°C to 85°C. Determine the tube length required for this duty, if the water side resistance to heat flow is negligible. The physical properties of air at average temperature are as :*

$$\rho = 1.009 \ kg/m^3,$$
$$\mu = 2.075 \times 10^{-5} \ kg/ms$$
$$k = 3.003 \times 10^{-5} \ kW/m.K,$$
$$C_p = 1.0082 \ kJ/kg.K.$$

For turbulent flow inside tubes, use relation

$$Nu = 0.023 \ Re^{0.8} Pr^{0.4}.$$

(N.M.U., May 1997)

Solution

Given : A counter flow heat exchanger

$T_{h,i} = 538°C,$

$T_{c,i} = 38°C$

$T_{h,o} = 148°C,$

$T_{c,o} = 85°C$

$C_{p,h} = 1.0082$ kJ/kg.K,

$\rho_{air} = 1.009$ kJ/m^3

$\dot{m}_h = 11.3$ kg/s,

$d = 2.54$ cm $= 0.0254$ m

$N = 4500,$

$\mu_{air} = 2.075 \times 10^{-5}$ kg/m.s

$k_f = 0.03003$ W/m.K, $\quad h_o = 0$

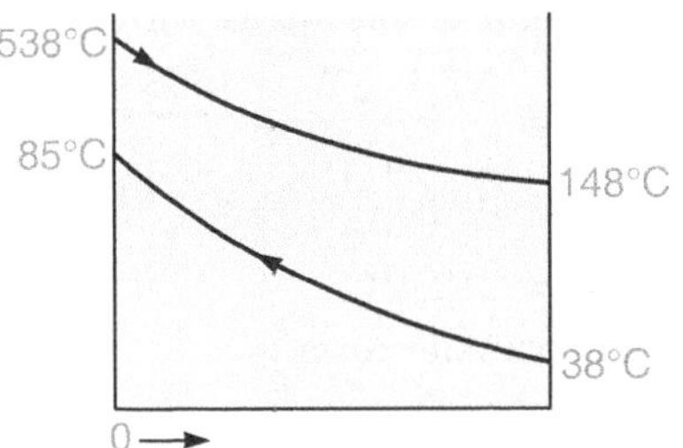

Fig. 14.21. Temperature distribution for example 14.10.

To find : Length of the heat exchanger.

Assumptions :

(*i*) No heat loss to the surroundings.

(*ii*) No scaling on heat transfer surfaces.

(*iii*) The specific heat of the water as 4180 J/kg.K.

(*iv*) Assuming one pass of tubes.

(*v*) Tube material resistance is negligible.

Analysis : The heat transfer rate through the heat exchanger can be calculated as :

$$Q = \dot{m}_h C_{p,h}(T_{h,i} - T_{h,o})$$
$$= (11.3 \text{ kg/s}) \times (1.0082 \text{ kJ/kg.K})$$
$$\times (538 - 148)(\text{K}) = 4443.13 \text{ kW}$$

The calculation of heat transfer coefficient on air side :

The mass flow rate of air per tube :

$$\dot{m}_{\text{tube}} = \frac{11.3}{4500} = 2.51 \times 10^{-3} \text{ kg/s}$$

$$Re = \frac{4\dot{m}_{\text{tube}}}{\pi \, d \, \mu}$$
$$= \frac{4 \times (2.51 \times 10^{-3} \text{ kg/s})}{\pi \times (0.0254 \text{ m}) \times (2.075 \times 10^{-5} \text{ kg/ms})}$$
$$= 6066$$

and
$$Pr = \frac{\mu \, C_{p,h}}{k_f} = \frac{2.075 \times 10^{-5} \times 1.0082 \times 10^3}{0.03003}$$
$$= 0.6966$$

The Nusselt number

$$Nu = \frac{h_i \, d}{k_f} = 0.023 \times (Re)^{0.8} \times (Pr)^{0.4}$$

or
$$h_i = \frac{0.03003}{0.0254} \times 0.023 \times (6066)^{0.8} \times (0.6966)^{0.4}$$
$$= 25.92 \text{ W/m}^2\text{.K} = 0.02592 \text{ kW/m}^2\text{.K}$$

The overall heat transfer coefficient; if water side and tube material resistances are neglected.

$$U = h_i = 0.02592 \text{ kW/m}^2\text{.K}$$

The log mean temperature difference ;

$$\Delta T_{lm} = \frac{\Delta T_1 - \Delta T_2}{\ln\left(\frac{\Delta T_1}{\Delta T_2}\right)} = \frac{(538-85)-(148-38)}{\ln\left[\frac{(538-85)}{(148-38)}\right]}$$

$$= 242.4°C$$

The heat transfer area ;

$$A = \frac{Q}{U\,\Delta T_{lm}}$$

or $$A = \frac{4443.13}{0.02592 \times 242.4} = 707.16 \text{ m}^2$$

The length of each pipe ;

$$A = \pi d L\, pN$$

(where p is number of tube passes)

or $$L = \frac{707.16}{\pi \times 0.0254 \times 1 \times 4500} = \mathbf{1.97\ m.}$$ **Ans.**

Example 14.11. *A pipe (k = 59 W/m.K) with an inner diameter of 3.75 cm and wall thickness of 0.318 cm is externally heated by steam at a temperature of 180°C. The water flows through the pipe with a velocity of 1.22 m/s. Calculate the length of pipe required to heat water from 30°C to 90°C. Assume the heat transfer coefficient on the steam side to be 11.3 kW/m².K.*

Solution

Given : Water heating by steam through a steel pipe.

$k = 59$ W/m.K, $d_i = 3.75$ cm

$d_o = d_i + 2t = 4.068$ cm, $T_h = 180°C$

$T_{c,i} = 30°C$, $T_{c,o} = 90°C$

$u_m = 1.22$ m/s, $h_o = 11.3$ kW/m².K

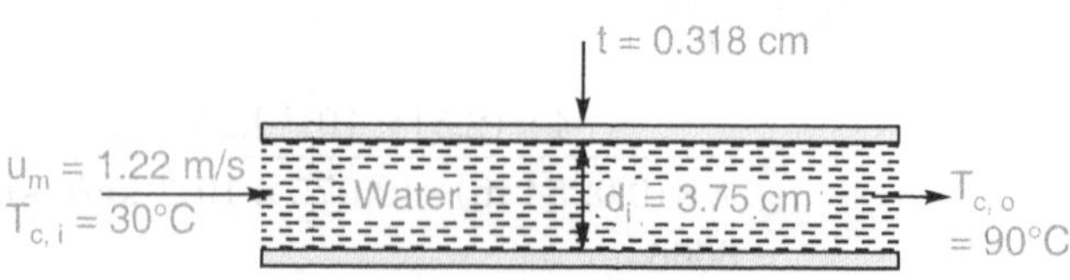

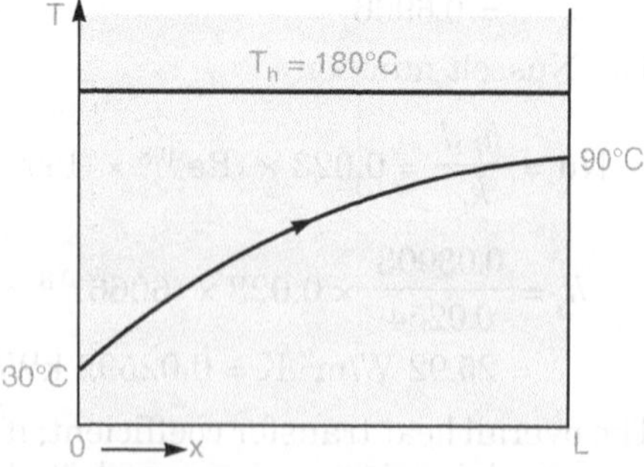

Fig. 14.22. Schematic and temperature distribution for example 14.11

To find : Length of heat exchanger.

Assumptions :

(*i*) Steam losses its total latent heat at temperature of 180°C.

(*ii*) Clean and fouling free surfaces.

Analysis : The properties of water at mean temperature

$$T_m = \frac{30+90}{2} = 60°C$$

$\rho = 982$ kg/m³,

$C_p = 4.186$ kJ/kg.K,

$\mu = 528 \times 10^{-6}$ kg/ms

Pr = 3.42,

$k_f = 0.645$ W/m.K

The mass flow rate of water

$$\dot{m}_c = \rho A_c u_m = \rho\,[(\pi/4)\,d_i^2]\,u_m$$

$$= 982 \times \left(\frac{\pi}{4}\right) \times (0.0375)^2 \times 1.22$$

$$= 1.323 \text{ kg/s}$$

The heat transfer rate

$$Q = \dot{m}_c C_{p,c}(T_{c,o} - T_{c,i})$$

$$= 1.323 \times 4.186 \times (90-30)$$

$$= 332.333 \text{ kW}$$

The temperature difference

$$\Delta T_1 = T_h - T_{c,i} = 180 - 30 = 150°C$$

$$\Delta T_2 = T_h - T_{c,o} = 180 - 90 = 90°C$$

$$\Delta T_{lm} = \frac{\Delta T_1 - \Delta T_2}{\ln\left(\frac{\Delta T_1}{\Delta T_2}\right)} = \frac{150-90}{\ln\left(\frac{150}{90}\right)} = 117.5°C$$

The heat transfer coefficient h_i on water side :

$$\text{Re} = \frac{\rho u_m d_i}{\mu} = \frac{982 \times 1.22 \times 0.0375}{528 \times 10^{-6}}$$

$$= 85088$$

Re > 2300, thus flow is turbulent and fully developed.

$$\text{Nu} = 0.023\ \text{Re}^{0.8}\ \text{Pr}^{0.4}$$

$$= 0.023 \times (85088)^{0.8} \times (3.42)^{0.4}$$

$$= 330.54$$

and $$h_i = \frac{\text{Nu}\,k_f}{d_i} = \frac{330.54 \times 0.645}{0.0375}$$

$$= 5685.4 \text{ W/m}^2\text{.K.}$$

The overall heat transfer coefficient based on inner surface

$$U_i = \frac{1}{\frac{1}{h_i} + \frac{d_i}{2k}\ln\left(\frac{d_o}{d_i}\right) + \frac{d_i}{d_o h_o}}$$

$$= \frac{1}{\frac{1}{5685.4} + \frac{0.0375}{2\times 59}\times\ln\left(\frac{0.04068}{0.0375}\right) + \frac{0.0375}{0.04068\times 11.3\times 10^3}}$$

$$= 3529.4 \text{ W/m}^2\text{.K}$$

Further, the heat transfer rate is expressed as :

$$Q = U_i A_i \Delta T_{lm}$$

and $$A_i = \frac{Q}{U_i\,\Delta T_{lm}} = \frac{332.333\times 10^3}{3529.4\times 117.5} = 0.801 \text{ m}^2$$

$$L = \frac{A_i}{\pi d_i} = \frac{0.801}{\pi \times 0.0375} = \mathbf{6.8\ m.}\quad \text{Ans.}$$

Example 14.12. *A one tonne window air conditioner removes 3.5 kW heat from a room and in process, it rejects 4.2 kW heat in an air cooled condenser. The ambient temperature is 30°C, whereas, the refrigerant condenser at 45°C. For the condenser, the product of overall heat transfer coefficient and corresponding area is 350 W/K. Calculate the temperature rise of air as it flows over the condenser tubes.* (M.U., Winter 1998)

Solution

Given : A condenser of (cross flow heat exchanger) of a window air conditioner.

$UA = 350$ W/K

$Q = 4.2$ kW

$T_{h,i} = T_{h,o} = T_h = 45°C$

$T_{c,i} = 30°C.$

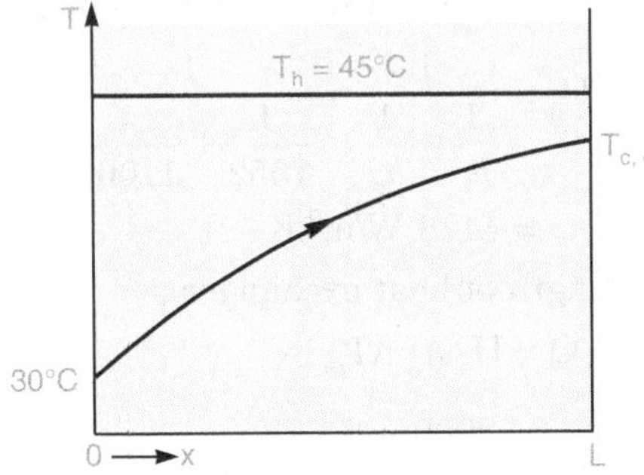

Fig. 14.23. Temperature distribution in condenser of window air conditioner

To find : Temperature of air leaving the condenser.

Analysis : The heat transfer rate can be expressed in terms of overall heat transfer coefficient as

$$Q = UA(\Delta T)_{lm} \quad ...(i)$$

where for condenser

$$\Delta T_{lm} = \frac{(T_h - T_{c,i}) - (T_h - T_{c,o})}{\ln\left(\frac{T_h - T_{c,i}}{T_h - T_{c,o}}\right)}$$

$$= \frac{(45-30)-(45-T_{c,o})}{\ln\left(\frac{45-30}{45-T_{c,o}}\right)} = \frac{T_{c,o}-30}{\ln\left(\frac{15}{45-T_{c,o}}\right)}$$

Using in eqn. (*i*)

$$4.2\times 10^3 \text{ W} = (350 \text{ W/K}) \times \frac{T_{c,o}-30}{\ln\left(\frac{15}{45-T_{c,o}}\right)}$$

or $$\frac{T_{c,o}-30}{\ln(15)-\ln(45-T_{c,o})} = 12$$

or $$T_{c,o} - 30 = 32.5 - 12\ \ln(45 - T_{c,o})$$

or $$\ln(45 - T_{c,o}) + 0.0833\ T_{c,o} - 5.2 = 0$$

It is a trancedental equation and its numerical solution gives

$$T_{c,o} = 34.8°C$$

Temperature rise of air = 34.8 – 30 = **4.8°C.** Ans.

Example 14.13. *Assume the condenser of a large power plant to be a shell and tube type heat exchanger consisting of a single shell and 30,000 tubes, each executing two passes. The tubes are of thin wall constructed with 25 mm in diameter, and steam condenses on their outer surface with an associated convection coefficient of 11,000 W/m².K. The heat transfer rate that must be effected by exchanger is 2000 MW and this is accomplished by cooling water through the tubes at the rate of 30,000 kg/s. The water enters at 20°C, while the steam condenses at 50°C. What is the temperature of cooling water coming out the condenser ? What is the required tube length per pass ?* (N.M.U., Nov. 1996)

Solution

Given : A shell and tube type condenser of a power plant ;

No. of shell = 1,

No. of tubes,	$N = 30{,}000$
No. of passes,	$p = 2$,
$d = 25$ mm,	$Q = 2000$ MW
Cold fluid	Hot fluid
$\dot{m}_c = 30{,}000$ kg/s,	$h_o = 11{,}000$ W/m².K
$T_{c,i} = 20°C$,	$T_h = 50°C$.

To find :

(*i*) Outlet temperature of the cooling water.

(*ii*) Tube length per pass.

Schematic :

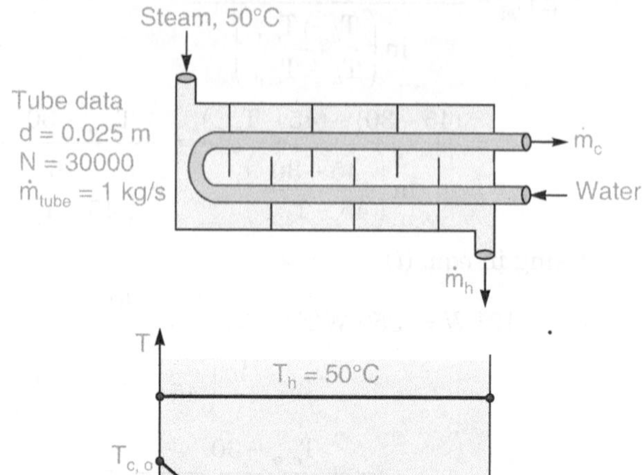

Fig. 14.24. One shell pass two tube pass heat exchanger for example 14.13

Assumptions :

(*i*) Assuming no heat loss to the surroundings.

(*ii*) Negligible thermal resistance of tube material and fouling effects.

(*iii*) Constant properties.

(*iv*) Fully developed flow through tubes.

(*v*) Specific heat of water as 4180 J/kg.K.

Properties : Properties of water at $T_m \approx 300$ K

$$C_p = 4180 \text{ J/kg.K},$$
$$\mu = 855 \times 10^{-6} \text{ Ns/m}^2,$$
$$k_f = 0.613 \text{ W/m.K},$$
$$Pr = 5.83.$$

Analysis : (*i*) The heat transfer rate through heat exchanger

$$Q = \dot{m}_c C_{p,c} (T_{c,o} - T_{c,i})$$

$$2000 \times 10^6 \text{ W} = (30{,}000 \text{ kg/s}) \times (4180 \text{ J/kg.K}) \times (T_{c,o} - 20)(\text{K or °C})$$

or $T_{c,o}$ = **36.0°C.** **Ans.**

(*ii*) The LMTD for counter flow arrangement

$$\Delta T_1 = T_{h,i} - T_{c,o} = 50 - 36 = 14°C$$
$$\Delta T_2 = T_{h,o} - T_{c,i} = 50 - 20 = 30°C$$

$$\Delta T_{lm,\text{ counter}} = \frac{\Delta T_1 - \Delta T_2}{\ln\left(\dfrac{\Delta T_1}{\Delta T_2}\right)} = \frac{14 - 30}{\ln\left(\dfrac{14}{30}\right)} = 21°C$$

The correction factor F can be obtained from Fig. 14.15(*a*), where

$$P = \frac{t_2 - t_1}{T_1 - t_1} = \frac{36 - 20}{50 - 20} = 0.466$$

and
$$R = \frac{T_1 - T_2}{t_2 - t_1} = \frac{50 - 50}{36 - 20} = 0$$

Hence the correction factor F = 1, and the corrected LMTD

$$\Delta T_{lm} = F\,\Delta T_{lm,\text{ counter}} = 1 \times 21 = 21°C$$

Since the heat transfer coefficient on waterside is not given, thus its calculation is required to determine overall heat transfer coefficient.

The mass flow rate per tube,

$$\dot{m}_{tube} = \frac{30{,}000 \text{ (kg/s)}}{30{,}000 \text{ tubes}} = 1 \text{ kg/s per tube}$$

Reynolds number

$$Re = \frac{4\dot{m}_{tube}}{\pi d \mu} = \frac{4 \times 1}{\pi \times 0.025 \times 855 \times 10^{-6}}$$
$$= 59{,}567$$

Using relation,

$$Nu = 0.023 \, Re^{0.8} \, Pr^{0.4}$$

or
$$h_i = 0.023 \times \frac{k_f}{d} \times Re^{0.8} \, Pr^{0.4}$$

$$= \frac{0.023 \times 0.613}{0.025} \times (59{,}567)^{0.8} \times (5.83)^{0.4}$$
$$= 7552 \text{ W/m}^2\text{.K}$$

The overall heat transfer coefficient ;

$$U_o = \frac{1}{\dfrac{1}{h_i} + \dfrac{1}{h_o}} = \frac{1}{\dfrac{1}{7552} + \dfrac{1}{11000}}$$
$$= 4478 \text{ W/m}^2\text{.K}$$

The length of heat exchanger ;

$$Q = U_o A_o \Delta T_{lm}$$

or
$$A_o = \frac{2000 \times 10^6}{4478 \times 21} = 21{,}268 \text{ m}^2$$

The length of each pipe ;

$$A_o = \pi d L p N$$

$$L = \frac{21,268}{\pi \times 0.025 \times 2 \times 30,000}$$

$$= \mathbf{4.51\ m.}\quad \text{Ans.}$$

The length of each tube per pass is 4.51 m.

Example 14.14. *A condenser is employed in a steam power plant to handle 35,000 kg/h of dry and saturated steam at 50°C. The cooling water enters the condenser at 15°C and leaves at 25°C. The tubes are 22.5 mm inside diameter and 25 mm outside diameter. The water flows through the tubes at an average velocity of 2 m/s. The heat transfer coefficient on steam side is 5000 W/m²·K. Calculate :*

(i) The mass flow rate of water.

(ii) Heat transfer surface area.

(iii) Number of tubes required for water flow.

(iv) Number of tube passes in condenser if the length of each tube per pass should not be more than 2.5 m.

Ignore thermal resistance of wall material and assume that the condensate coming out the condenser is saturated water i.e., steam loose only its latent heat.

Use relation ; $Nu = 0.023\ Re^{0.8}\ Pr^{0.4}$ for determination of waterside heat transfer coefficient.

Take latent heat of steam as 2374 kJ/kg

Following properties of water at mean bulk temperature of 20°C may be used :

$\rho = 998.8\ kg/m^3$,

$C_p = 4180\ J/kg.K$,

$\nu = 1.0006 \times 10^{-6}\ m^2/s$,

$k_f = 0.59859\ W/m.K$.

(P.U., June 1999)

Solution

Given : A shell and tube type steam condenser of a steam power plant ;

No. of shell = 1

Hot fluid	*Cold fluid*
$\dot{m}_h = 35,000$ kg/h,	$u_m = 2$ m/s
$h_o = 5,000$ W/m²·K,	$T_{c,i} = 15°C$
$T_h = 50°C$,	$T_{c,o} = 25°C$,
$d_o = 25$ mm,	$d_i = 22.5$ mm,
	$h_{fg} = 2374$ kJ/kg.

and properties of water and relation.

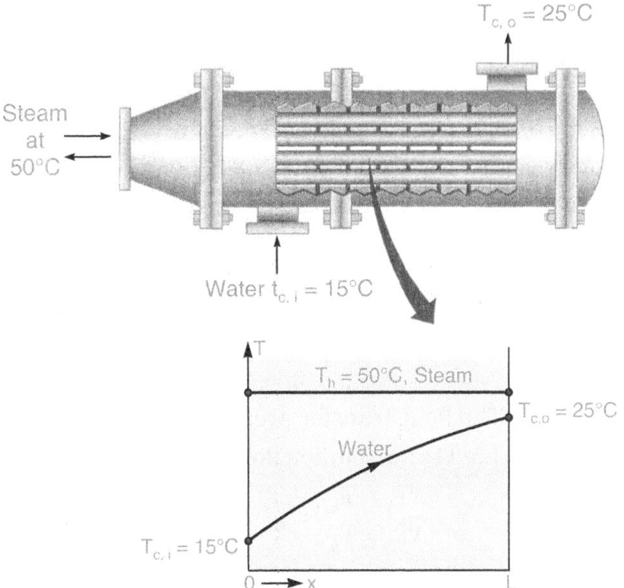

Fig. 14.25. Schematic and temperature distribution for steam condenser for example 14.14.

To find :

(*i*) Mass flow rate of water.

(*ii*) Number of tubes required for given water flow rate.

(*iii*) Heat transfer surface area.

(*iv*) Number of tube passes for tube length per pass within 2.5 m.

Assumptions :

(*i*) Assuming no heat loss to the surroundings.

(*ii*) Fully developed flow through tubes.

(*iii*) Constant properties.

(*iv*) Negligible resistances of tube material and scale formation.

Analysis : The heat transfer rate through heat exchanger

$$Q = \dot{m}_h h_{fg}$$

$$= (35,000/3600)(\text{kg/s}) \times (2374\ \text{kJ/kg})$$

$$= 23.08 \times 10^3\ \text{kW}$$

(*i*) The mass flow rate of water :

$$Q = \dot{m}_c C_{p,c} (T_{c,o} - T_{c,i})$$

$$23.08 \times 10^3\ \text{kW} = \dot{m}_c \times (4.180\ \text{kJ/kg.K}) \times (25 - 15)(\text{K or °C})$$

or $\dot{m}_c = \mathbf{552\ kg/s.}$ Ans.

(*ii*) The mass flow rate of water through each tube;

$$\dot{m}_{\text{tube}} = \rho\, A_c\, u_m = \rho\left(\frac{\pi}{4}d_i^2\right) u_m$$

$$= (998.8 \text{ kg/m}^2) \times \left\{\frac{\pi}{4}\times(0.0225\text{ m})^2\right\}$$

$$\times (2 \text{ m/s}) = 0.7942 \text{ kg/s}$$

Hence, the number of tubes required for mass flow rate of 552 kg/s are

$$N = \frac{\dot{m}_c}{\dot{m}_{\text{tube}}} = \frac{552}{0.7942} = \mathbf{695\ tubes.}\quad \text{Ans.}$$

(*iii*) The heat transfer area :

The LMTD for counter flow arrangement

$$\Delta T_1 = T_{h,o} - T_{c,i} = 50 - 15 = 35°C$$

$$\Delta T_2 = T_{h,i} - T_{c,o} = 50 - 25 = 25°C$$

$$\Delta T_{lm,\text{ counter}} = \frac{\Delta T_1 - \Delta T_2}{\ln\left(\frac{\Delta T_1}{\Delta T_2}\right)} = \frac{35-25}{\ln\left(\frac{35}{25}\right)} = 29.72°C$$

The correction factor F for one fluid condensing through condenser is unity.

The corrected LMTD

$$\Delta T_{lm} = F\,\Delta T_{lm,\text{ counter}}$$

$$= 1 \times 29.72 = 29.72°C.$$

Since the heat transfer coefficient on waterside is not given, thus, its calculation is required to determine overall heat transfer coefficient.

Reynolds number

$$Re = \frac{4\dot{m}_{\text{tube}}}{\pi\, d_i\, \rho\, \nu}$$

$$= \frac{4\times 0.7942}{\pi\times 0.0225\times 998.8\times 1.0006\times 10^{-6}}$$

$$= 44970$$

Prandtl number,

$$Pr = \frac{\mu C_p}{k_f} = \frac{\rho \nu C_p}{k_f}$$

$$= \frac{998.8\times 1.0006\times 10^{-6}\times 4180}{0.59859} = 6.97$$

Using relation

$$Nu = \frac{h_i d_i}{k_f} = 0.023\ Re^{0.8}\ Pr^{0.4}$$

or

$$h_i = 0.023 \times \frac{k_f}{d_i} \times Re^{0.8}\ Pr^{0.4}$$

$$= \frac{0.023\times 0.59859}{0.0225} \times (44970)^{0.8} \times (6.97)^{0.4}$$

$$= 7020 \text{ W/m}^2\text{.K}$$

The overall heat transfer coefficient ;

$$U_o = \frac{1}{\frac{d_o}{d_i h_i} + \frac{1}{h_o}} = \frac{1}{\frac{0.025}{0.0225\times 7020} + \frac{1}{5000}}$$

or

$$U_o = 2791 \text{ W/m}^2\text{.K}$$

The heat transfer area of heat exchanger ;

$$Q = U_o\, A_o\, \Delta T_{lm}$$

or

$$A_o = \frac{23.08\times 10^6 \text{ W}}{(2791\text{ W/m}^2\text{.K})\times(29.72\text{ K})}$$

$$= \mathbf{278.23\ m^2.}\quad \text{Ans.}$$

(*iv*) The number of tube passes for each tube to be 2.5 m long

$$A_o = \pi d_o L\, p N$$

or

$$p = \frac{278.23\text{ m}^2}{(\pi\times 0.025\text{ m}\times 2.5\text{ m})\times 695\text{ tubes}}$$

$$= \mathbf{2\ passes.}\quad \text{Ans.}$$

The each tube will have two passes.

Example 14.15. *A two pass surface condenser is required to handle the exhaust steam 0.9 dry from a turbine, developing 15 MW with a steam consumption of 5 kg/kWh. The condenser maintains a vacuum of 660 mm of Hg, when barometric pressure is 760 mm of Hg. Condensate coming out the condenser as saturate water. The water enters tubes, 25 mm inner diameter and 33 mm outer diameter, at 24°C with a mean velocity of 3 m/s and leaves the tubes at a temperature which is 4°C less than the condensate temperature.*

The overall heat transfer coefficient based on outer surface of tubes is 4000 W/m².K. Calculate :

(*i*) *Mass of cooling water circulated in kg/min,*

(*ii*) *Condenser surface area,*

(*iii*) *Number of tubes required per pass,*

(*iv*) *Tube length per pass.*

Solution

Given : A two tube pass surface condenser with

$$U_o = 4000 \text{ W/m}^2\text{.K}$$

$$\text{Power} = 15 \text{ MW}$$

$$\dot{m}_s = 5 \text{ kg/kWh}$$

Steam side : $x = 0.9$

$$p_{\text{atm}} = 760 \text{ mm of Hg}$$

$$p_g = -\,660 \text{ mm of Hg}$$

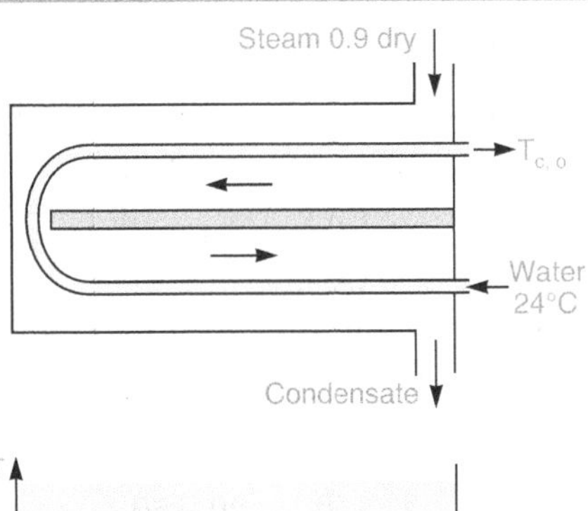

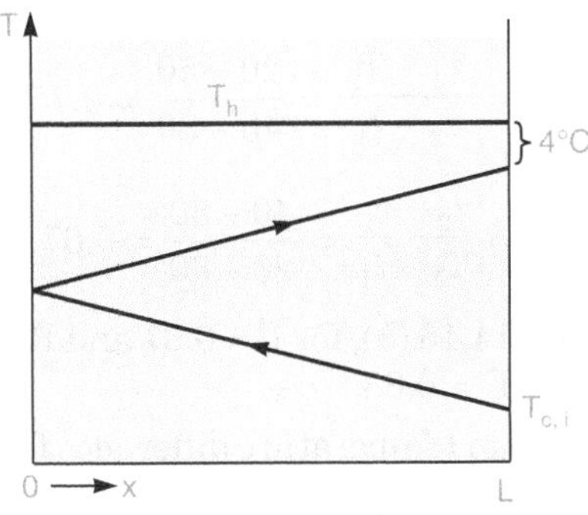

Fig. 14.26. Schematic and temperature distribution for two tube pass surface condenser

Water side : d_i = 25 mm

d_o = 33 mm

$T_{c,i}$ = 24°C

u_m = 3 m/s

$T_{c,o} = T_{h,o} - 4°C.$

To find :

(*i*) Mass flow rate of cooling water in kg/min,

(*ii*) Condenser surface area,

(*iii*) Number of tubes per pass,

(*iv*) Length of tubes per pass.

Assumptions :

(*i*) Shell of the condenser is well insulated, so that heat rejected during condensation is transferred to cooling water only.

(*ii*) No change in kinetic and potential energy.

(*iii*) Scale free surfaces.

(*iv*) Constant properties.

(*v*) Specific heat of water as 4.187 kJ/kg.K and density as 1000 kg/m³.

Analysis : The absolute pressure maintained in condenser is

$$p_{abs} = p_{atm} + p_g = 760 - 660 \text{ mm of Hg}$$
$$= 100 \text{ mm of Hg}$$
$$= \rho g h = 13.6 \times 10^3 \times 9.81 \times \frac{100}{1000}$$
$$= 13341.6 \text{ Pa}$$
$$= 0.133 \text{ bar or } 13.34 \text{ kPa}$$

From Table A-7 of Appendix A at p = 0.133 bar or 13.34 kPa

$$T_h = T_{sat} = 51°C \qquad h_{fg} = 2592 \text{ kJ/kg}$$

and $$T_{c,o} = 51 - 4 = 47°C$$

(*i*) Mass of steam condensed per minute

$$\dot{m}_h = 15 \times 10^3 \text{ kW} \times \frac{5 \text{ kg}}{\text{kWh}}$$
$$= 75000 \text{ kg/h}$$
$$= 1250 \text{ kg/min} = 20.83 \text{ kg/s}$$

Heat rejected by steam per minute

$$Q = \dot{m}_h \times (x h_{fg}) = 20.83 \times (0.9 \times 2592)$$
$$= 48,600 \text{ kW}$$

Mass of water circulated in the condenser can be determined by energy balance

$$Q = \dot{m}_c C_{p,c} (T_{c,o} - T_{c,i})$$

or $$48,600 = \dot{m}_c \times 4.187 \times (47 - 24)$$

or $$\dot{m}_c = 504.67 \text{ kg/s} = \mathbf{30,280 \text{ kg/min.}}$$ **Ans.**

(*ii*) The condenser surface area

The condenser operates in counter flow, therefore,

$$\Delta T_1 = T_h - T_{c,o} = 51 - 47°C = 4°C$$
$$\Delta T_2 = T_h - T_{c,i} = 51 - 24 = 27°C$$
$$\Delta T_{lm} = F\, \Delta T_{lm,\ \text{counter}}$$

(Here F = 1 for condenser)

LMTD

$$\Delta T_{lm} = \frac{\Delta T_1 - \Delta T_2}{\ln\left(\frac{\Delta T_1}{\Delta T_2}\right)} = \frac{4 - 27}{\ln\left(\frac{4}{27}\right)}$$
$$= 12.04°C \text{ (or } 12.04 \text{ K)}$$

The heat transfer rate is also expressed as

$$Q = U_o A_o \Delta T_{lm}$$
$$48,600 \times 10^3 \text{ W} = (4000 \text{ W/m}^2\cdot\text{K}) \times A_o \times (12.04 \text{ K})$$

or $$A_o = \mathbf{1009.1 \text{ m}^2.}$$ **Ans.**

(*iii*) Number of tubes required per pass :

These can be determined by continuity equation

$$\dot{m}_c = \left(\frac{\pi}{4}\right) d_i^2 \times \rho \times u_m \times \text{No. of tubes (N)}$$
$$504.67 \text{ kg/s} = \left(\frac{\pi}{4}\right) \times (0.025 \text{ m})^2 \times (1000 \text{ kg/m}^3) \times (3 \text{ m/s}) \times N$$

or $$N = \mathbf{342.7 \text{ say } 343 \text{ tubes.}}$$ **Ans.**

(*iv*) Length of tube per pass

$$A_o = \pi d_o L \times N \times 2 \text{ passes.}$$

$$L = \frac{1009.1 \text{ m}^2}{\pi \times (0.033 \text{ m}) \times 343 \times 2} = \mathbf{14.1 \text{ m.}} \quad \text{Ans.}$$

Example 14.16. *A two shell pass and four tube passes heat exchanger is used to heat glycerin from 20°C to 50°C by hot water, which enters thin walled 20 mm diameter tube at 80°C and leaves at 40°C. The total length of the tube in the heat exchanger is 60 m. The convection coefficient on shell side is 25 W/m².K and that on water (tube) side is 160 W/m².K. Calculate the rate of heat transfer in the heat exchanger*

(*i*) *For clean surfaces of tubes,*

(*ii*) *After fouling with fouling factor of 0.0006 m².K/W on outer surface of tubes.*

Solution

Given : Two shell pass and four tube passes heat exchanger as shown in Fig. 14.27

Shell side :

$$T_{c,i} = 20°C$$
$$T_{c,o} = 50°C$$
$$h_o = 25 \text{ W/m}^2.\text{K}$$
$$R_{f,o} = 0.0006 \text{ m}^2.\text{K/W}$$

Tube side : $d = 20$ mm,

$$L = 60 \text{ m}$$
$$T_{h,i} = 80°C,$$
$$T_{h,o} = 40°C$$
$$h_i = 160 \text{ W/m}^2.\text{K.}$$

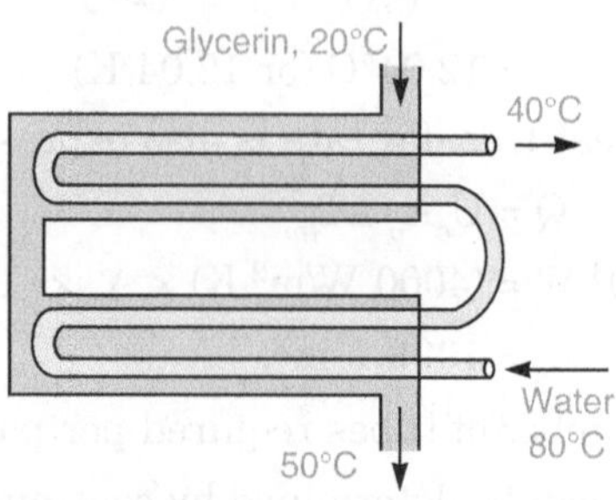

Fig. 14.27. Schematic

To find : Rate of heat transfer in exchanger,

(*i*) Before any fouling,

(*ii*) After fouling of outer tube surface.

Analysis : For thin walled tube surface area,

$$A = \pi dL$$
$$= \pi \times 0.02 \times 60 = 3.77 \text{ m}^2$$

Here $\Delta T_1 = T_{h,i} - T_{c,o} = 80 - 50 = 30°C$

$$\Delta T_2 = T_{h,o} - T_{c,i} = 40 - 20 = 20°C$$

$$\Delta T_{lm,\text{ counter}} = \frac{\Delta T_1 - \Delta T_2}{\ln\left(\frac{\Delta T_1}{\Delta T_2}\right)} = \frac{30 - 20}{\ln\left(\frac{30}{20}\right)} = 24.66°C$$

For shell side and tube heat exchanger

$$T_1 = T_{c,i} = 20°C, \qquad T_2 = T_{c,o} = 50°C$$
$$t_1 = T_{h,i} = 80°C \qquad t_2 = T_{h,o} = 40°C$$

$$R = \frac{T_1 - T_2}{t_2 - t_1} = \frac{20 - 50}{40 - 80} = 0.75$$

$$P = \frac{t_2 - t_1}{T_1 - t_1} = \frac{40 - 80}{20 - 80} = 0.67$$

From Fig. 14.15 (*b*), for P = 0.67 and R = 0.75, the correction factor F = 0.88.

The log mean temperature difference for shell and tube heat exchanger

$$\Delta T_{lm} = F\, \Delta T_{lm,\text{ counter}}$$
$$\Delta T_{lm} = 0.88 \times 24.66 = 21.7°C$$

(*i*) For clean tube surfaces (before fouling),

Overall heat transfer coefficient for thin walled tube ;

$$U = \frac{1}{\frac{1}{h_i} + \frac{1}{h_o}} = \frac{1}{\frac{1}{160} + \frac{1}{25}} = 21.62 \text{ W/m}^2.\text{K}$$

Heat transfer rate,

$$Q = UA\, \Delta T_{lm} = 21.62 \times 3.77 \times 21.7$$
$$= \mathbf{1768.7 \text{ W.}} \quad \text{Ans.}$$

(*ii*) When outer surface of tube is fouled with fouling factor of 0.0006 m².K/W

$$U = \frac{1}{\frac{1}{h_i} + \frac{1}{h_o} + R_{f,o}} = \frac{1}{\frac{1}{160} + \frac{1}{25} + 0.0006}$$
$$= 21.34 \text{ W/m}^2.\text{K}$$

$$Q = UA\, \Delta T_{lm} = 21.34 \times 3.77 \times 21.7$$
$$= \mathbf{1746.2 \text{ m}^2.} \quad \text{Ans.}$$

Example 14.17. *An automotive radiator has 40 tubes of inner diameter of 0.5 cm and 60 cm long in a closely spaced plate finned matrix, so both fluids unmixed. Hot water enters the tubes at 90°C at a rate of 0.6 kg/s and leaves at 65°C. Air flows across the radiator through the inter fin spaces and is heated from 20°C to 40°C. Calculate overall heat transfer coefficient based on inner surface for this radiator.*

Solution

Given : An automotive radiator as cross flow heat exchanger with unmixed fluid streams.

$$N = 40$$
$$d_i = 0.5 \text{ cm} = 0.005 \text{ m}$$
$$L = 60 \text{ cm} = 0.6 \text{ m}$$
$$T_{h,i} = 90°C,$$
$$T_{h,o} = 65°C$$
$$\dot{m}_h = 0.6 \text{ kg/s}$$
$$T_{c,i} = 20°C,$$
$$T_{c,o} = 40°C.$$

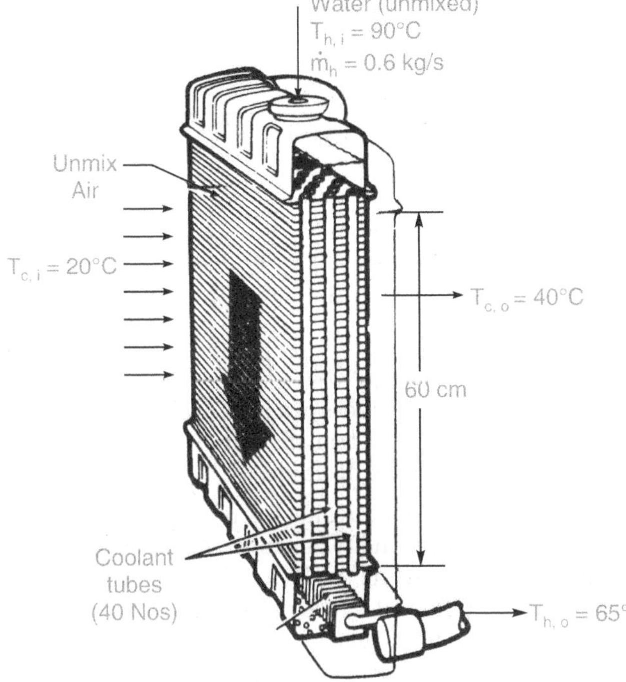

Fig. 14.28. Schematic of a radiator

To find : Overall heat transfer coefficient U_i based on inner surface of tube.

Assumptions :

(*i*) Specific heat of hot water as 4.18 kJ/kg.K.

(*ii*) Clean tube surfaces without any fouling.

Analysis : The heat transfer rate in the radiator

$$Q = \dot{m}_h C_{p,h}(T_{h,i} - T_{h,o})$$
$$= 0.6 \times 4.18 \times (90 - 65) = 62.7 \text{ kW}$$

Total heat transfer area on tube side

$$A_i = \pi d_i LN = \pi \times 0.005 \times 0.6 \times 40$$
$$= 0.377 \text{ m}^2$$

Here $\Delta T_1 = T_{h,i} - T_{c,o} = 90 - 40 = 50°C$

$\Delta T_2 = T_{h,o} - T_{c,i} = 65 - 20 = 45°C$

$$\Delta T_{lm,\text{ counter}} = \frac{\Delta T_1 - \Delta T_2}{\ln\left(\frac{\Delta T_1}{\Delta T_2}\right)} = \frac{50 - 45}{\ln\left(\frac{50}{45}\right)} = 47.45°C$$

For cross flow

$$t_1 = T_{h,i} = 90°C, \quad t_2 = T_{h,o} = 65°C$$
$$T_1 = T_{c,i} = 20°C, \quad T_2 = T_{c,o} = 40°C$$

$$\left.\begin{aligned} P &= \frac{t_2 - t_1}{T_1 - t_1} = \frac{65 - 90}{20 - 90} = 0.357 \\ R &= \frac{T_1 - T_2}{t_2 - t_1} = \frac{20 - 40}{65 - 90} = 0.8 \end{aligned}\right\}$$

F = 0.97 (From Fig. 14.15(*c*))

The log mean temperature difference for cross flow heat exchanger

$$\Delta T_{lm} = F\, \Delta T_{lm,\text{ counter}}$$
$$\Delta T_{lm} = 0.97 \times 47.45 = 46°C$$

The overall heat transfer coefficient U_i

$$U_i = \frac{Q}{A_i\, \Delta T_{lm}} = \frac{62.7 \times 10^3}{0.377 \times 46}$$
$$= \mathbf{3615.5 \text{ W/m}^2\text{.K.}} \quad \textbf{Ans.}$$

14.8 THE EFFECTIVENESS-NTU METHOD

It is very simple to use LMTD method of heat exchanger analysis when inlet and outlet temperatures, mass flow rates for hot and cold fluids and overall heat transfer coefficient are available or easily be determined from specified relations. The heat transfer surface area and thus size of heat exchanger can easily be determined from

$$Q = UA\Delta T_{lm}$$

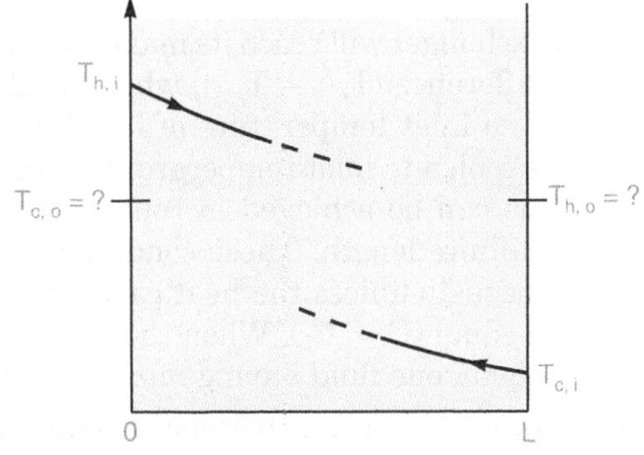

Fig. 14.29. Design problem encountered with LMTD method

However, if the type and size of heat exchanger are specified and for prescribed mass flow rates, only inlet temperatures of both fluids are known and outlet temperatures are to determine as shown in Fig. 14.29, then use of LMTD method becomes helpless. If it is used in such a case, then the procedure would require tedious iterations thus is not practical. In order to eliminate such complication from the solution, *Kays* and *London* came up with new method that is called the *effectiveness-NTU* method.

14.8.1. Heat Exchanger Effectiveness

It is a dimensionless parameter and defined as the ratio of actual heat transfer rate Q_{act} by heat exchanger to *maximum possible heat transfer rate* Q_{max}. It is denoted by ε and expressed as :

$$\varepsilon = \frac{Q_{act}}{Q_{max}} = \frac{\dot{m}_c C_{p,c}(T_{c,o} - T_{c,i})}{Q_{max}}$$

$$= \frac{\dot{m}_h C_{p,h}(T_{h,i} - T_{h,o})}{Q_{max}} \quad ...(14.40)$$

The actual heat transfer rate in a heat exchanger can be determined from an energy balance on hot and cold fluids and can be expressed as

$$Q = C_c(T_{c,o} - T_{c,i})$$
$$= C_h(T_{h,i} - T_{h,o}) \quad ...(14.41)$$

where, $C_c = \dot{m}_c C_{p,c}$ and $C_h = \dot{m}_h C_{p,h}$ are *heat capacity rates* of cold and hot fluids, respectively.

In any heat exchanger, the objective is to either maximization of heating or cooling *i.e.*, to gain the maximum temperature difference and hence the *maximum heat transfer rate* Q_{max} can be obtained with *maximum temperature difference* in a heat exchanger. The maximum temperature difference is the difference between inlet temperatures of hot and cold fluids. That is

$$\Delta T_{max} = T_{h,i} - T_{c,i} \quad ...(14.42)$$

A heat exchanger will reach its maximum possible temperature difference $(T_{h,i} - T_{c,i})$, when (1) the cold fluid is heated to inlet temperature of hot fluid, or (2) the hot fluid is cooled to inlet temperature of cold fluid. Such conditions can be achieved in counter flow heat exchanger of infinite length. These conditions will not reach simultaneously unless the heat capacity rates of both fluids are equal $(C_h = C_c)$. When the heat capacity rates are not equal, one fluid having *minimum value of heat capacity rate* $(\dot{m}\, C_p)$ can experience maximum temperature difference.

$$\therefore \quad Q_{max} = C_{min}(T_{h,i} - T_{c,i})$$
$$= (\dot{m}\, C_p)_{min}(T_{h,i} - T_{c,i}) \quad ...(14.43)$$

where C_{min} is the smaller value among $C_c = \dot{m}_c C_{p,c}$ and $C_h = \dot{m}_h C_{p,h}$ either for cold or hot fluid depending on product of mass flow rate and specific heat.

For cold fluid

Effectiveness,

$$\varepsilon = \frac{\dot{m}_c C_{p,c}(T_{c,o} - T_{c,i})}{(\dot{m} C_p)_{min}(T_{h,i} - T_{c,i})}$$

$$= \frac{C_c(T_{c,o} - T_{c,i})}{C_{min}(T_{h,i} - T_{c,i})} \quad ...(14.44)$$

and for hot fluid,

$$\varepsilon = \frac{\dot{m}_h C_{p,h}(T_{h,i} - T_{h,o})}{(\dot{m} C_p)_{min}(T_{h,i} - T_{c,i})}$$

$$= \frac{C_h(T_{h,i} - T_{h,o})}{C_{min}(T_{h,i} - T_{c,i})} \quad ...(14.45)$$

If the effectiveness ε, $T_{h,i}$, $T_{c,i}$ and minimum heat capacity C_{min} are known, the actual heat transfer rate can be obtained as

$$Q = \varepsilon\, C_{min}(T_{h,i} - T_{c,i}) \quad ...(14.46)$$

Thus the effectiveness of a heat exchanger enables us to determine the heat transfer rate without knowing the outlet temperatures of fluids.

The value of effectiveness, ε of a heat exchanger lies between 0 and 1 and it depends on the geometry of the heat exchanger as well as flow arrangement. Therefore, the different types of heat exchangers have different effectiveness relations.

14.8.2. NTU

This quantity is called the *number of transfer units.* It is a dimensionless parameter, which is expressed as ;

$$NTU = \frac{UA}{C_{min}} = \frac{UA}{(\dot{m}\, C_p)_{min}}$$

$$= \frac{\text{Heat capacity rate of exchanger W/K}}{\text{Heat capacity rate of fluid W/K}} \quad ...(14.47)$$

where U is the overall heat transfer coefficient and A is the heat transfer area of the heat exchanger and C_{min} is minimum heat capacity rate between the two fluids. The quantities U and C_{min} remain constant for given flow conditions, hence

$$NTU \propto A$$

Therefore, NTU is the measure of the physical size (heat transfer area) of the heat exchanger. Higher the value of NTU, larger the physical size.

14.8.3. Capacity Ratio

In heat exchanger analysis, it is convenient to define a dimensionless parameter called *capacity* ratio C as

$$C = \frac{\text{Minimum heat capacity rate}}{\text{Maximum heat capacity rate}} = \frac{C_{min}}{C_{max}} \quad ...(14.48)$$

For any heat exchanger, the effectiveness is the function of NTU and heat capacity ratio.

$$\varepsilon = f(\text{NTU, C}) \quad ...(14.49)$$

14.8.4. Effectiveness of a Parallel Flow Heat Exchanger

Rewriting eqn. (14.24), for a double pipe parallel flow heat exchanger,

$$\ln\left[\frac{\Delta T_2}{\Delta T_1}\right] = -UA\left\{\frac{1}{\dot{m}_h C_{p,h}} + \frac{1}{\dot{m}_c C_{p,c}}\right\}$$

where $\Delta T_1 = T_{h,i} - T_{c,i}$ and $\Delta T_2 = T_{h,o} - T_{c,o}$

It may be expressed in terms of inlet, outlet temperatures and heat capacity rates as

$$\ln\left[\frac{T_{h,o} - T_{c,o}}{T_{h,i} - T_{c,i}}\right] = -UA\left[\frac{1}{C_h} + \frac{1}{C_c}\right]$$

$$[C_h = \dot{m}_h C_{p,h} \text{ and } C_c = \dot{m}_c C_{p,c}]$$

Rearranging, we get

$$\frac{T_{h,o} - T_{c,o}}{T_{h,i} - T_{c,i}} = \exp\left[-\frac{UA}{C_c}\left(1 + \frac{C_c}{C_h}\right)\right] \quad ...(14.50)$$

Substituting for $T_{c,o}$ and $T_{h,o}$ from eqns. (14.44) and (14.45), respectively, we get

$$\frac{\left\{T_{h,i} - \varepsilon\frac{C_{min}}{C_h}(T_{h,i} - T_{c,i})\right\} - \left\{\varepsilon\frac{C_{min}}{C_c}(T_{h,i} - T_{c,i}) + T_{c,i}\right\}}{(T_{h,i} - T_{c,i})}$$

$$= \exp\left\{-\frac{UA}{C_c}\left(1 + \frac{C_c}{C_h}\right)\right\}$$

which is simplified to

$$1 - \varepsilon\frac{C_{min}}{C_c}\left\{1 + \frac{C_c}{C_h}\right\} = \exp\left\{-\frac{UA}{C_c}\left(1 + \frac{C_c}{C_h}\right)\right\}$$

$$\varepsilon\frac{C_{min}}{C_c}\left\{1 + \frac{C_c}{C_h}\right\} = 1 - \exp\left\{-\frac{UA}{C_c}\left(1 + \frac{C_c}{C_h}\right)\right\}$$

Rearranging, we get relation for effectiveness of a parallel flow heat exchanger,

$$\varepsilon_{\text{parallel flow}} = \frac{1 - \exp\left\{-\frac{UA}{C_c}\left(1 + \frac{C_c}{C_h}\right)\right\}}{\frac{C_{min}}{C_c}\left\{1 + \frac{C_c}{C_h}\right\}} \quad ...(14.51)$$

Taking either C_h or C_c as C_{min} (both approaches to same result), above relation yields to

$$\varepsilon_{\text{parallel flow}} = \frac{1 - \exp\left\{-\frac{UA}{C_{min}}\left(1 + \frac{C_{min}}{C_{max}}\right)\right\}}{1 + \frac{C_{min}}{C_{max}}} \quad ...(14.52)$$

Introducing the number of transfer units NTU and capacity ratio C as defined by eqns. (14.47) and (14.48), respectively, we get the relation for effectiveness in the form

$$\varepsilon_{\text{parallel flow}} = \frac{1 - \exp\{-\text{NTU}(1 + C)\}}{1 + C} \quad ...(14.53)$$

14.8.5. Effectiveness of a Counter Flow Heat Exchanger

Rewriting eqn. (14.35), for a double pipe counter flow heat exchanger,

$$\ln\left[\frac{\Delta T_2}{\Delta T_1}\right] = -UA\left\{\frac{1}{\dot{m}_h C_{p,h}} - \frac{1}{\dot{m}_c C_{p,c}}\right\}$$

where $\Delta T_1 = T_{h,i} - T_{c,o}$ and $\Delta T_2 = T_{h,o} - T_{c,i}$

It may be expressed in terms of inlet, outlet temperatures and heat capacity rates as

$$\ln\left[\frac{T_{h,o} - T_{c,i}}{T_{h,i} - T_{c,o}}\right] = -UA\left[\frac{1}{C_h} - \frac{1}{C_c}\right]$$

Rearranging, we get

$$\frac{T_{h,o} - T_{c,i}}{T_{h,i} - T_{c,o}} = \exp\left[-\frac{UA}{C_h}\left(1 - \frac{C_h}{C_c}\right)\right] \quad ...(14.54)$$

Substituting for $T_{c,o}$ and $T_{h,o}$ from eqns. (14.44) and (14.45), respectively, we get

$$\frac{\left\{T_{h,i} - \varepsilon\frac{C_{min}}{C_h}(T_{h,i} - T_{c,i}) - T_{c,i}\right\}}{\left\{T_{h,i} - \varepsilon\frac{C_{min}}{C_c}(T_{h,i} - T_{c,i}) - T_{c,i}\right\}}$$

$$= \exp\left\{-\frac{UA}{C_h}\left(1-\frac{C_h}{C_c}\right)\right\}$$

which is simplified to

$$\frac{(T_{h,i}-T_{c,i})\left\{1-\varepsilon\frac{C_{min}}{C_h}\right\}}{(T_{h,i}-T_{c,i})\left\{1-\varepsilon\frac{C_{min}}{C_c}\right\}} = \exp\left\{-\frac{UA}{C_h}\left(1-\frac{C_h}{C_c}\right)\right\}$$

or

$$\frac{1-\varepsilon\frac{C_{min}}{C_h}}{1-\varepsilon\frac{C_{min}}{C_c}} = \exp\left\{-\frac{UA}{C_h}\left(1-\frac{C_h}{C_c}\right)\right\}$$

Taking either C_h or C_c as C_{min}, both approaches to same result. Let $C_h = C_{min}$ and $C_c = C_{max}$, above relation yields to

$$\frac{1-\varepsilon}{1-\varepsilon\frac{C_{min}}{C_{max}}} = \exp\left\{-\frac{UA}{C_{min}}\left(1-\frac{C_{min}}{C_{max}}\right)\right\}$$

or

$$\frac{1-\varepsilon}{1-\varepsilon C} = \exp\{-NTU(1-C)\}$$

where

$$C = \frac{C_{min}}{C_{max}} = \frac{(\dot{m}C_p)_{min}}{(\dot{m}C_p)_{max}}$$

Rearranging, we get relation for effectiveness of a counter flow heat exchanger,

$$\varepsilon_{\text{counter flow}} = \frac{1-\exp[-NTU(1-C)]}{1-C\exp[-NTU(1-C)]} \quad \text{...(14.55)}$$

The similar expressions for variety of heat exchangers are presented in Table 14.3.

TABLE 14.3. Effectiveness relations for heat exchangers : $NTU = UA/C_{min}$ and $C = C_{min}/C_{max} = (\dot{m}C_p)_{min}/(\dot{m}C_p)_{max}$

	Flow geometries	*Relation for effectiveness, ε*
1.	Double pipe heat exchanger :	
	Parallel flow	$\varepsilon = \frac{1-\exp[-NTU(1+C)]}{1+C}$
	Counter flow	$\varepsilon = \frac{1-\exp[-NTU(1-C)]}{1-C\exp[-NTU(1-C)]}$
2.	Shell and tube heat exchanger : One shell pass	
	2, 4, ... tube passes	$\varepsilon = 2\left\{1+C+\sqrt{1+C^2}\,\frac{1+\exp[-NTU\sqrt{1+C^2}]}{1-C\exp[-NTU\sqrt{1+C^2}]}\right\}^{-1}$
3.	Cross flow	
	Both fluid unmixed	$\varepsilon = 1-\exp\left\{\frac{NTU^{0.22}}{C}[\exp(1-C\,NTU^{0.78})-1]\right\}$
	C_{max} mixed	
	C_{min} unmixed	$\varepsilon = \frac{1}{C}(1-\exp\{-C[1-\exp(-NTU)]\})$
	C_{min} mixed	
	C_{max} unmixed	$\varepsilon = 1-\exp\{-\frac{1}{C}[1-\exp(-C\,NTU)]\}$
4.	Boilers and condensers, $C = C_{min}/C_{max} \longrightarrow 0$	$\varepsilon = 1-\exp(-NTU)$

The *Kays* and *London* have presented effectiveness for various heat exchangers as function of NTU and ratio of heat capacities C (= C_{min}/C_{max}). Hence effectiveness can also be obtained directly from the charts [Fig. 14.30 to Fig. 14.35] for given NTU and C.

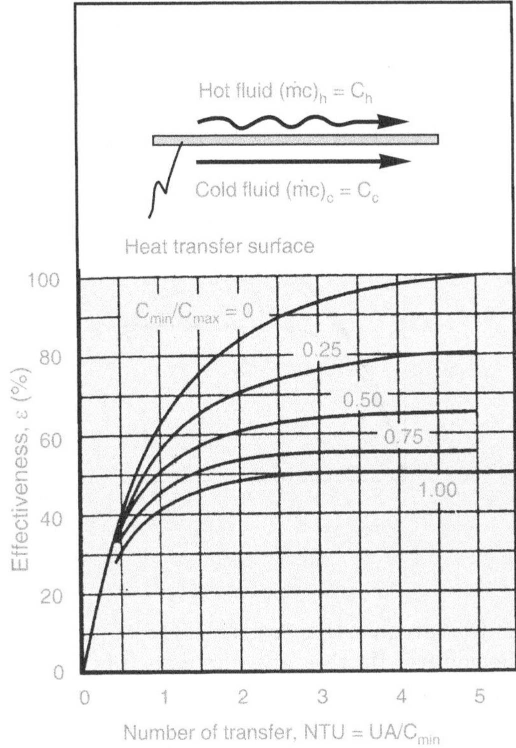

Fig. 14.30. Effectiveness for parallel flow heat exchanger

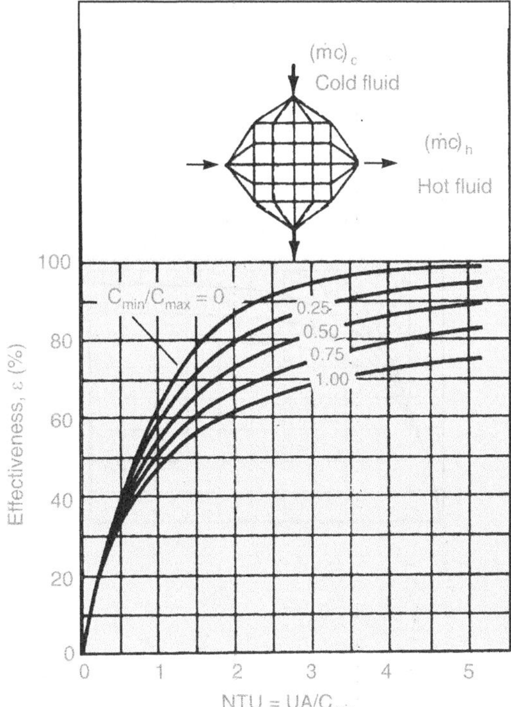

Fig 14.32. Effectiveness for cross-flow heat exchanger both fluids unmixed

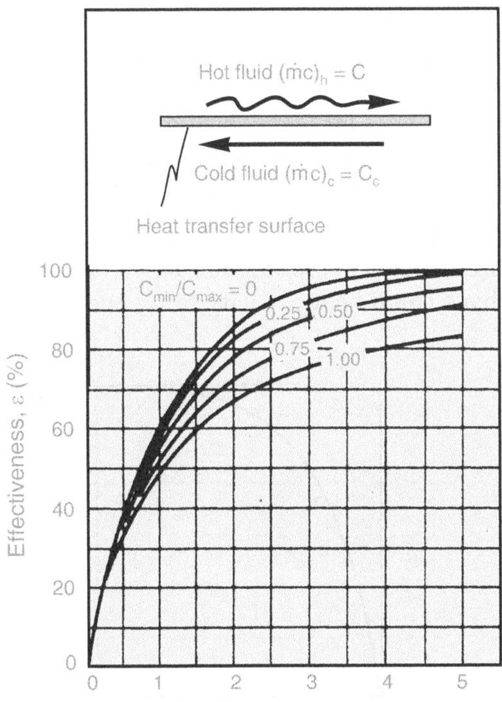

Fig. 14.31. Effectiveness for counter flow heat exchanger

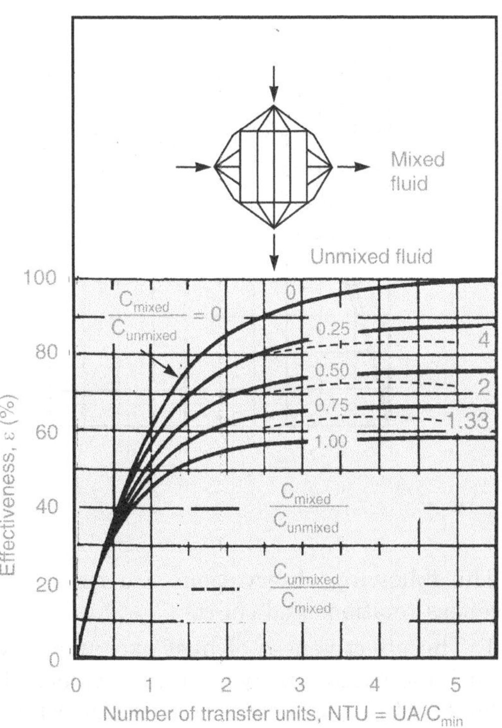

Fig. 14.33. Heat exchanger effectiveness for cross-flow with one fluid mixed

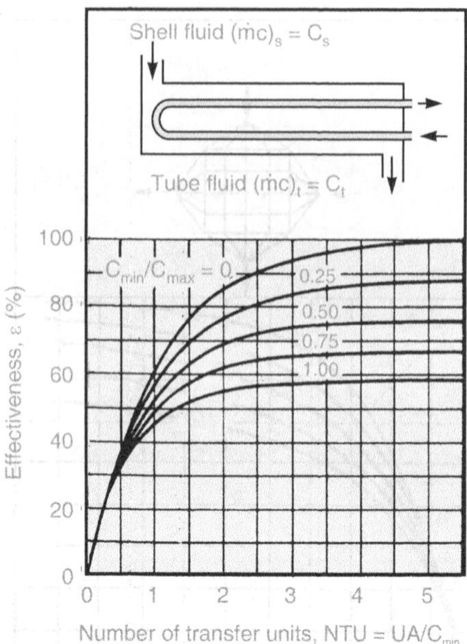

Fig. 14.34. Effectiveness for single shell pass with two, four etc. tube passes

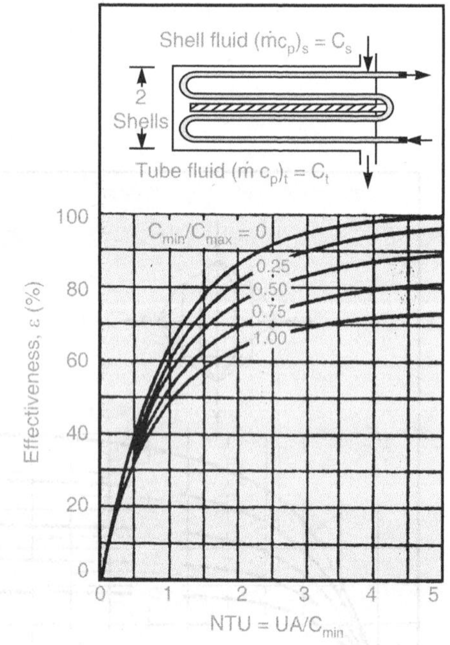

Fig. 14.35. Effectiveness for two shell pass with four, eight etc. heat exchanger

The following observations are revealed from effectiveness relations and charts :

1. The effectiveness of heat exchangers varies from 0 to 1, increases rapidly with small values of NTU up to about NTU = 1.5, and after rather slowly for larger values as shown in Fig. 14.36. Therefore, the use of a heat exchanger with a large NTU (usually more than 3) and thus a large size cannot be justified economically, since a large increase in NTU results into a small increase in effectiveness. Thus, a heat exchanger with a very high effectiveness may be highly desirable from a heat transfer point of view, but rather undesirable from an economical point of view.

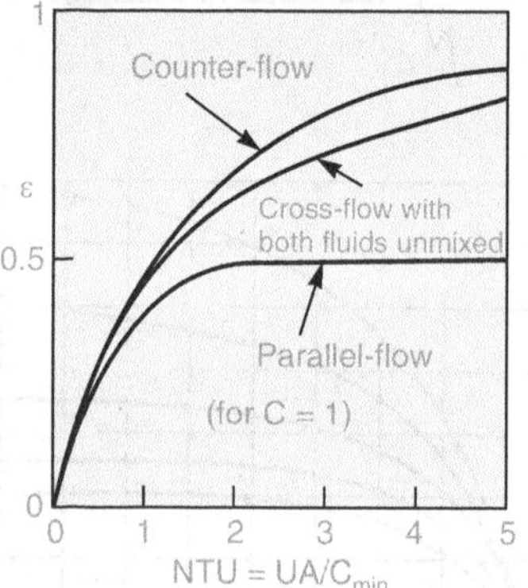

Fig. 14.36. Variation of effectiveness for a specified NTU and capacity ratio C

2. For given capacity ratio $C = C_{min}/C_{max}$ and NTU, the counter flow arrangement results into higher effectiveness, followed by closely by the cross flow heat exchanger with both unmixed fluids. A parallel flow arrangement gains a lowest value of effectiveness.

3. The effectiveness of all heat exchangers is more close for NTU values of less than about 0.3 and it is independent of capacity ratio C.

4. For given NTU, the effectiveness becomes a maximum for C = 0, and a minimum for C = 1. For the case C = 0 *i.e.,* for condensation and evaporation processes, the effectiveness relation reduced to

$$\varepsilon = \varepsilon_{max} = 1 - \exp(-\text{NTU})$$

Fig. 14.37 shows the effectiveness variation for condensers and evaporators. For case C = 1, when two fluids have equal heat capacity rates, the effectiveness is lowest.

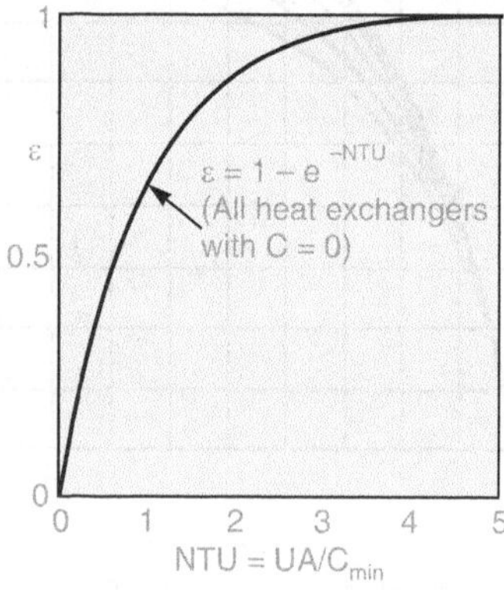

Fig. 14.37. Variation of effectiveness for condensers and evaporators for which capacity ratio C = 0

Once the quantities C and NTU are evaluated for a heat exchanger, its effectiveness ε can be determined either from the charts or from effectiveness relation for the specified type of heat exchanger. Then the rate of heat transfer Q and outlet temperature of two fluids can be obtained.

14.9. RATING OF HEAT EXCHANGERS

When the given data with flow arrangement are :

Inlet temperatures $= T_{h,i}$ and $T_{c,i}$

The mass flow rates $= \dot{m}_h$ and $\dot{m}_c$

The physical properties $= C_{p,c}$ and $C_{p,h}$

The overall heat transfer coeff. $= U$

The heat transfer area $= A$

The heat transfer rate $= Q$

The heat transfer rate and outlet temperatures are to be determined.

The following methodology should be adopted.

(1) Calculate $C = \dfrac{C_{min}}{C_{max}}$, $NTU = \dfrac{UA}{C_{min}}$

(2) Calculate effectiveness ε from chart or specific equation.

(3) Calculate $Q = \varepsilon C_{min}(T_{h,\,i} - T_{c,\,i})$

(4) Calculate the outlet temperatures

$$T_{h,\,o} = T_{h,\,i} - \frac{Q}{(\dot{m}C_p)_h} \text{ and } T_{c,\,o} = \frac{Q}{(\dot{m}\,C_p)_c} + T_{c,\,i}$$

14.10. SIZING OF HEAT EXCHANGERS

Given : When given data are with flow arrangement :

Inlet and outlet temperatures $= T_{h,\,i}, T_{c,\,i}, T_{h,\,o}$ and $T_{c,\,o}$

The mass flow rates $= \dot{m}_h$ and $\dot{m}_c$,

The physical properties $= C_{p,\,c}$ and $C_{p,\,h}$, k etc.

The heat transfer coefficient $= h_i$ and h_o.

To find : The heat transfer area or length of heat exchanger.

The procedure for the analysis follows as :

1. Calculate heat transfer rate as :

$$Q = \dot{m}_c C_{p,\,c}(T_{c,\,o} - T_{c,\,i}) = \dot{m}_h C_{p,\,h}(T_{h,\,i} - T_{h,\,o})$$

Any unknown data involved, can also be determined.

2. Calculate overall heat transfer coefficient using eqn. (14.9) ;

$$U_o = \frac{1}{\dfrac{d_o}{d_i h_i} + \dfrac{d_o}{d_i} R_{f,\,i} + \dfrac{d_o}{2k}\ln(r_o/r_i) + R_{f,\,o} + \dfrac{1}{h_o}}$$

(*i*) If $R_{f,\,i}$ and $R_{f,\,o}$ are not specified, assumed as zero.

(*ii*) If k is not specified, assume wall resistance negligible.

(*iii*) h_i and h_o are not specified, then calculate for turbulent flow (Re > 2300) as

$$h = 0.023\,\frac{k_f}{D_h}\,Re^{0.8}\,Pr^n$$

where, $D_h = \dfrac{4A_c}{P}$ = hydraulic diameter, m

A_c = cross-sectional area, m^2

P = perimeter, m.

3. Calculate the LMTD, ΔT_{lm}

4. Calculate the area of heat exchanger as

$$A_o = \frac{Q}{U_o \Delta T_{lm}}$$

Length of the heat exchanger

$$L = \frac{A_o}{\pi d_o p N}.$$

Alternative procedure :

1. Compute the heat capacity ratio,

$$C = \frac{C_{min}}{C_{max}}$$

2. With known temperatures, calculate effectiveness, ε.

3. Obtain NTU from chart or calculation.

4. Calculate area (if U_o available otherwise determine by procedure as given above)

$$NTU = \frac{U_o A_o}{C_{min}}.$$

Example 14.18. *Two fluids heat exchanger has inlet and outlet temperatures of 65 and 40°C for the hot fluid and 15 and 30°C for the cold fluid. Can you tell whether this exchanger is operating under counterflow or parallel flow conditions ? What is the effectiveness of the exchanger, if the cold fluid has the minimum capacity rate ?*

Solution

Given : $T_{h,\,i} = 65°C$, $T_{h,\,o} = 40°C$

$T_{c,\,i} = 15°C$, $T_{c,\,o} = 30°C$.

To find :

(*i*) Type of heat exchanger.

(*ii*) Effectiveness of heat exchanger.

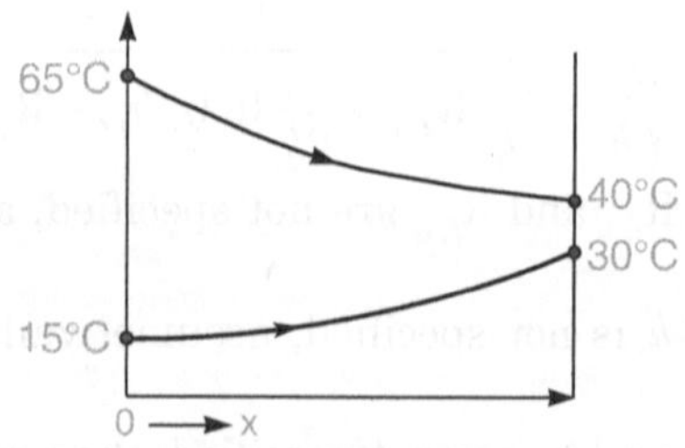

Fig. 14.38. Schematic for example 4.18

Analysis : (*i*) Since the cold fluid outlet temperature is less than hot fluid outlet temperature, hence the type of heat exchanger is parallel flow heat exchanger.

(*ii*) Effectiveness of the heat exchanger with minimum heat capacity of cold fluid :

$$\varepsilon = \frac{T_{c,o} - T_{c,i}}{T_{h,i} - T_{c,i}} = \frac{30 - 15}{65 - 15} = \mathbf{0.3.} \quad \textbf{Ans.}$$

Example 14.19. *Consider the following parallel flow heat exchanger specification*

cold flow enters at 40°C : $C_c = 20,000$ *W/K*

hot flow enters at 150°C : $C_h = 10,000$ *W/K*

$A = 30\ m^2$ $\quad U = 500\ W/m^2.K.$

Determine the heat transfer rate and the exit temperatures. (N.M.U., May 2002)

Solution

Given : A parallel flow heat exchanger with only inlet temperatures :

To find :

(*i*) Heat transfer rate,

(*ii*) Exit temperatures of fluid.

Analysis : Since the exit temperatures of both fluids are not known, so it is not possible to calculate ΔT_{lm}.

Heat capacity rates of two fluids are available and

$$C_{min} = C_h = 10,000 \text{ W/K}$$

and $$C_{max} = C_c = 20,000 \text{ W/K}$$

Therefore, using NTU method.

$$\text{NTU} = \frac{UA}{C_{min}} = \frac{500 \times 30}{10,000} = 1.5$$

$$\text{Capacity ratio } C = \frac{C_{min}}{C_{max}} = \frac{10,000}{20,000} = 0.5$$

For parallel flow heat exchanger, effectiveness is given by eqn. (14.53)

$$\varepsilon = \frac{1 - \exp[-\text{NTU}(1 + C)]}{1 + C}$$

$$= \frac{1 - \exp[-1.5 \times (1 + 0.5)]}{1 + 0.5} = 0.596$$

Maximum possible heat transfer rate,

$$Q_{max} = C_{min} \times (T_{h,i} - T_{c,i})$$

$$= 10,000 \times (150 - 40)$$

$$= 11 \times 10^5 \text{ W} = 1100 \text{ kW}$$

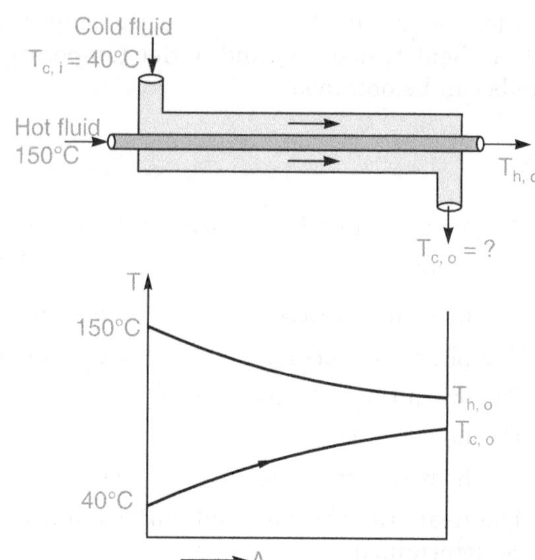

Fig. 14.39. Schematic and temperature profile for example 14.19

(*i*) Actual heat transfer rate in exchanger

$$Q = \varepsilon\, Q_{max} = 0.596 \times 1100$$

$$= \mathbf{655.6\ kW.} \quad \textbf{Ans.}$$

(*ii*) The exit temperatures of two fluids can be determined by energy balance

$$Q = C_c(T_{c,o} - T_{c,i}) = C_h(T_{h,i} - T_{h,o})$$

$$T_{h,o} = T_{h,i} - \frac{Q}{C_h} = 150 - \frac{655,600}{10000} = \mathbf{84.44°C}$$

$$T_{c,o} = T_{c,i} + \frac{Q}{C_c} = 40 + \frac{655,600}{20000}$$

$$= \mathbf{72.78°C.} \quad \textbf{Ans.}$$

Example 14.20. *A chemical having specific heat of 3.3 kJ/kg.K at a rate of 20,000 kg/h enters a parallel flow heat exchanger at 120°C. The flow rate of cooling water is 50,000 kg/h with an inlet temperature of 20°C. The heat transfer area is 10 m² and overall heat transfer coefficient is 1050 W/m².K. Find*

(*i*) *The effectiveness of the heat exchanger,*

(*ii*) *Outlet temperature of water and chemical.*

Take C_p of water as 4.186 kJ/kg.K.

(U.P.S.C., 1992)

Solution

Given : A parallel flow heat exchanger.

$A = 10\ m^2$, $\quad U = 1050\ W/m^2.K$

Hot fluid : Chemical

$$C_{p,h} = 3.3 \text{ kJ/kg.K}$$

$$\dot{m}_h = 20,000 \text{ kg/h}$$

$$T_{h,i} = 120°C$$

Cold fluid : Water

$$\dot{m}_c = 50{,}000 \text{ kg/h}$$
$$T_{c,\,i} = 20°C$$
$$C_{p,\,c} = 4.186 \text{ kJ/kg.K.}$$

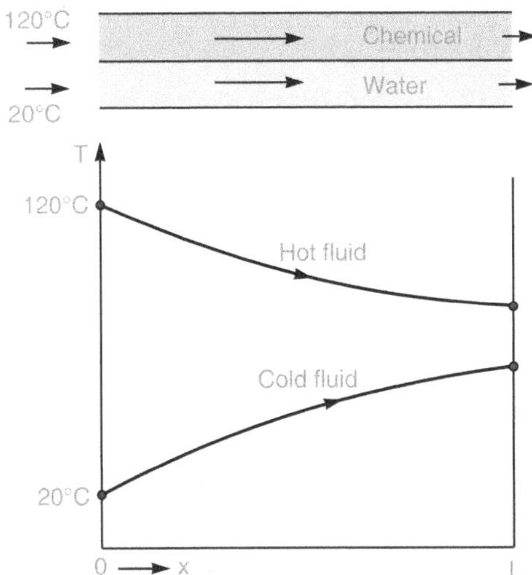

Fig. 14.40. Schematic and temperature distribution for example 14.20

To find :

(*i*) Effectiveness of heat exchanger, and

(*ii*) Outlet temperatures of water and chemical.

Analysis : (*i*) The heat capacity rates of two fluids

$$C_h = \dot{m}_h\, C_{p,\,h}$$
$$= \left(\frac{20{,}000}{3600}\text{ kg/s}\right) \times (3.3 \times 10^3 \text{ J/kg.K})$$
$$= 18{,}333.33 \text{ W/K}$$
$$C_c = \dot{m}_c\, C_{p,\,c}$$
$$= \left(\frac{50{,}000}{3600}\text{ kg/s}\right) \times (4186 \text{ J/kg.K})$$
$$= 58{,}138.89 \text{ W/K}$$

Therefore,

$$C_{min} = C_h = 18{,}333.33 \text{ W/K}$$
$$C_{max} = C_c = 58{,}138.89 \text{ W/K}$$

The capacity ratio,

$$C = \frac{C_{min}}{C_{max}} = \frac{18{,}333.33}{58{,}138.89} = 0.315$$

$$NTU = \frac{UA}{C_{min}} = \frac{1050 \times 10}{18{,}333.33} = 0.572$$

The effectiveness of heat exchanger

$$\varepsilon = \frac{1 - \exp\,[-\,NTU\,(1 + C)]}{1 + C}$$
$$= \frac{1 - \exp\,[-\,0.572 \times (1 + 0.315)]}{1 + 0.315}$$
$$= \mathbf{0.40}. \quad \textbf{Ans.}$$

(*ii*) Outlet temperature of hot and cold fluids can be obtained with the use of effectiveness as :

$$\varepsilon = \frac{C_h(T_{h,i} - T_{h,o})}{C_{min}(T_{h,i} - T_{c,i})} = \frac{C_c(T_{c,o} - T_{c,i})}{C_{min}(T_{h,i} - T_{c,i})}$$

On rearranging it leads to

$$T_{h,\,o} = T_{h,\,i} - \varepsilon\,\frac{C_{min}}{C_h}\,(T_{h,\,i} - T_{c,\,i})$$
$$= 120 - 0.4 \times (120 - 20) = \mathbf{80°C}$$
$$T_{c,\,o} = T_{c,\,i} + \varepsilon\,\frac{C_{min}}{C_c}\,(T_{h,\,i} - T_{c,\,i})$$
$$= 20 + 0.4 \times 0.315 \times (120 - 20) = \mathbf{32.6°C}$$

The outlet temperatures of chemical and water are 80°C and 32.6°C, respectively. **Ans.**

Example 14.21. *In an open heart surgery, under hypothermic conditions, the patient blood is cooled before the surgery and rewarmed afterwards. It is proposed that a concentric tube, counter flow heat exchanger of length 0.5 m be used for this purpose with the thin walled inner tube having a diameter of 55 mm. If the water at 60°C and 0.10 kg/s is used to heat the blood entering the exchanger at 18°C and 0.05 kg/s, what is the temperature of blood leaving the exchanger ? The overall heat transfer coefficient is 500 W/m²·K and specific heat of the blood is 3500 J/kg.K.* (N.M.U., Nov. 1998)

Solution

Given : A concentric tube counterflow heat exchanger :

$T_{c,\,i} = 18°C$, $\dot{m}_h = 0.1$ kg/s

$T_{h,\,i} = 60°C$, $\dot{m}_c = 0.05$ kg/s

$L = 0.5$ m, $d = 55$ mm $= 0.055$ m

$U = 500$ W/m^2.K, $C_{p,\,c} = 3500$ J/kg.K.

To find : Outlet temperature of blood leaving the exchanger.

Schematic :

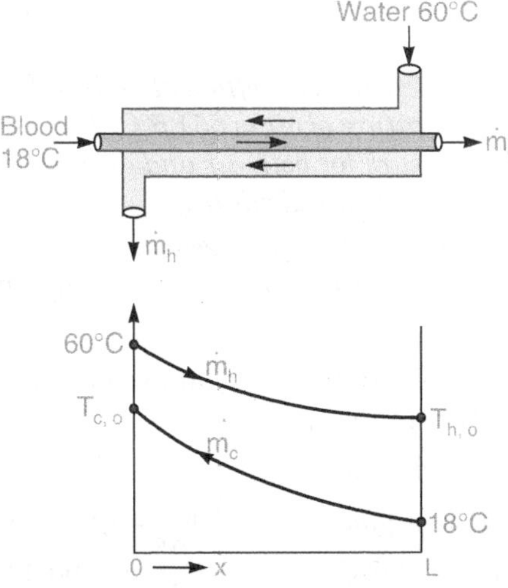

Fig. 14.41. Schematic and temperature distribution for counterflow heat exchanger for example 14.21

Assumptions :

1. Specific heat of water, $C_{p,h}$ as 4200 J/kg.K.
2. No heat exchange with the surroundings.

Analysis : The two heat capacities are :

$$C_h = \dot{m}_h C_{p,h} = (0.1 \text{ kg/s}) \times (4200 \text{ J/kg.K}) = 420 \text{ W/K}$$

$$C_c = \dot{m}_c C_{p,c} = (0.05 \text{ kg/s}) \times (3500 \text{ J/kg.K}) = 175 \text{ W/K}$$

Comparing the two heat capacities,

$$C_c = C_{min} = 175 \text{ W/K}$$

The ratio of two specific heats,

$$C = \frac{C_{min}}{C_{max}} = \frac{175}{420} = 0.4167$$

The number of transfer units,

$$NTU = \frac{UA}{C_{min}} = \frac{500 \times (\pi \times 0.055 \times 0.5)}{175} = 0.2468$$

The effectiveness of heat exchanger in counter flow ;

$$\varepsilon = \frac{1 - \exp\{-NTU(1-C)\}}{1 - C\exp\{-NTU(1-C)\}}$$

$$= \frac{1 - \exp\{-0.2468 \times (1 - 0.4167)\}}{1 - (0.4167) \times \exp\{-0.2468 \times (1 - 0.4167)\}} = 0.21$$

Outlet temperature of blood : (blood has C_{min})

$$\therefore \quad \varepsilon = \frac{T_{c,o} - T_{c,i}}{T_{h,i} - T_{c,i}} = \frac{T_{c,o} - 18}{60 - 18}$$

$$\Rightarrow \quad T_{c,o} = 18 + 0.21 \times (60 - 18) = \mathbf{26.83°C.} \text{ Ans.}$$

Example 14.22. *A counter flow heat exchanger is used to heat air entering at 400°C with a flow rate of 6 kg/s by exhaust gas entering at 800°C with a flow rate of 4 kg/s. The overall heat transfer coefficient is 100 W/m².K and the outlet temperature of air is 551.5°C. The specific heat at constant pressure for both air and exhaust gas can be taken as 1100 J/kg.K. Calculate :*

(i) Heat transfer area needed,

(ii) Number of transfer units. (GATE, 1995)

Solution

Given : A counterflow heat exchanger

$U = 100$ W/m².K

Cold fluid : Air

$\dot{m}_c = 6$ kg/s $\quad T_{c,o} = 551.5°C$

$T_{c,i} = 400°C, \quad C_{p,c} = 1100$ J/kg.K

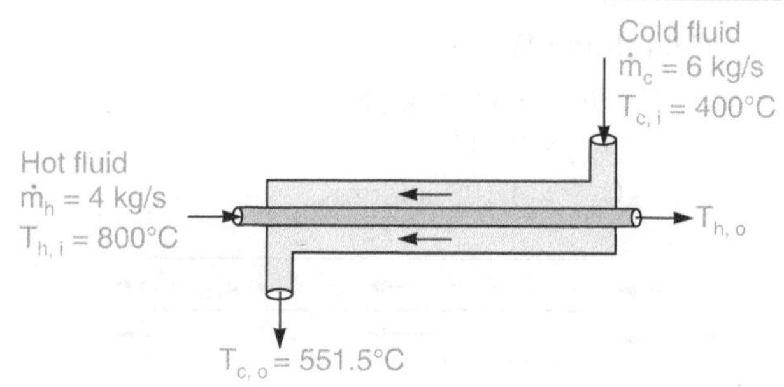

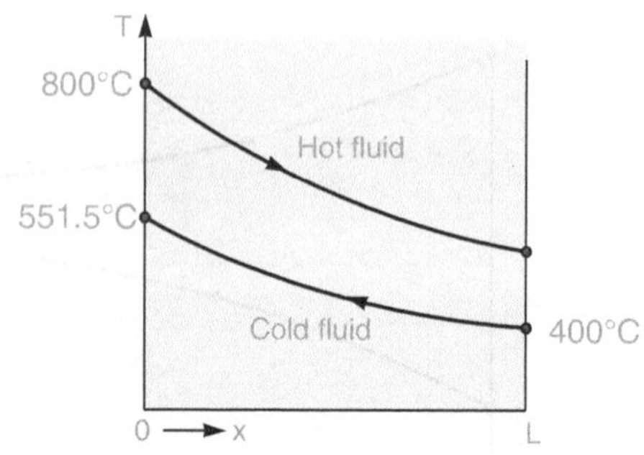

Fig. 14.42. Schematic and temperature distribution for example 14.22

Hot fluid : Exhaust gas

$\dot{m}_h = 4$ kg/s, $\quad T_{h,i} = 800°C$

$T_{h,o} = ?$ $\quad C_{p,h} = 1100$ J/kg.K.

To find :

(*i*) Heat transfer area, and

(*ii*) NTU.

Analysis : (*i*) The heat transfer rate in the exchanger

$$Q = \dot{m}_c C_{p,c} (T_{c,o} - T_{c,i}) = 6 \times 1100 \times (551.5 - 400) = 999{,}900 \text{ W}$$

The heat capacity rates of two fluids

$$C_h = \dot{m}_h C_{p,h} = 4 \times 1100 = 4400 \text{ W/K}$$

$$C_c = \dot{m}_c C_{p,c} = 6 \times 1100 = 6600 \text{ W/K}$$

Therefore,

$$C_{min} = 4400 \text{ W/K, and } C_{max} = 6600 \text{ W/K}$$

The heat capacity ratio

$$C = \frac{C_{min}}{C_{max}} = \frac{4400}{6600} = 0.667$$

The maximum possible heat transfer rate

$$Q_{max} = C_{min}(T_{h,i} - T_{c,i}) = 4400 \times (800 - 400) = 17{,}60{,}000 \text{ W}$$

The effectiveness of heat exchanger

$$\varepsilon = \frac{Q}{Q_{max}} = \frac{999{,}900}{17{,}60{,}000} = 0.568$$

From Fig. 14.31, for $\varepsilon = 0.568$ and $C = 0.667$

NTU = **1.1.** Ans.

(*ii*) Further, $\text{NTU} = \dfrac{\text{UA}}{\text{C}_{min}}$

or $$\text{A} = \frac{\text{NTU C}_{min}}{\text{U}} = \frac{1.1 \times 4400}{100}$$

$= \mathbf{48.4\ m^2}$. **Ans.**

Example 14.23. *The overall temperature rise of a cold fluid in a cross flow heat exchanger is 20°C and overall temperature drop of the hot fluid is 30°C. The effectiveness of the heat exchanger is 0.6. The heat transfer area is 1 m² and overall heat transfer coefficient is 60 W/m².K. Calculate the rate of heat transfer. Assume both fluids are unmixed.*

Solution

Given : A cross flow (both fluid unmixed) heat exchanger

$$\Delta\text{T}_c = \text{T}_{c,o} - \text{T}_{c,i} = 20°\text{C}$$
$$\Delta\text{T}_h = \text{T}_{h,i} - \text{T}_{h,o} = 30°\text{C}$$
$$\varepsilon = 0.6, \quad \text{A} = 1\ \text{m}^2$$
$$\text{U} = 60\ \text{W/m}^2.\text{K}.$$

To find : Rate of heat transfer in the heat exchanger.

Analysis : Making the energy balance between the two fluids

$$\dot{m}_c \text{C}_{p,c}\, \Delta\text{T}_c = \dot{m}_h \text{C}_{p,h}\, \Delta\text{T}_h$$

The capacity ratio

$$\text{C} = \frac{\dot{m}_h \text{C}_{p,h}}{\dot{m}_c \text{C}_{p,c}} = \frac{20}{30} = 0.667$$

Therefore, $\text{C}_{min} = \dot{m}_h \text{C}_{p,h}$

and $\text{C}_{max} = \dot{m}_c \text{C}_{p,c}$

From Fig. 14.32, for

$$\varepsilon = 0.6,\ \text{C} = 0.667,$$
$$\text{NTU} = 1.35$$

Further, $\text{NTU} = \dfrac{\text{UA}}{\text{C}_{min}}$

or $$\text{C}_{min} = \frac{60 \times 1}{1.35} = 44.44\ \text{W/K}$$

$$\text{C}_{min} = \text{C}_h = \dot{m}_h \text{C}_{p,h}$$

Heat transfer rate

$$\text{Q} = \text{C}_h\, \Delta\text{T}_h = 44.44 \times 30$$

$= \mathbf{1333.33\ W}$. **Ans.**

Example 14.24. *In a heat exchanger, hot fluid enters at 180°C and leaves at 118°C. The cold water enters at 99°C and leaves at 119°C. Find the LMTD, NTU and effectiveness in the following cases of heat exchanger :*

(*i*) *Counterflow,*

(*ii*) *One shell pass and multiple tube passes,*

(*iii*) *Two shell passes and multiple tube passes,*

(*iv*) *Cross flow both fluids unmixed, and*

(*v*) *Cross flow, the cold fluid unmixed.*

(J.N.T.U., May 2004)

Solution

Given : A heat exchanger in various modes with

$$\text{T}_{h,i} = 180°\text{C} \qquad \text{T}_{c,i} = 99°\text{C}$$
$$\text{T}_{h,0} = 118°\text{C} \qquad \text{T}_{c,o} = 119°\text{C}$$

To find :

(*a*) LMTD,

(*b*) effectiveness and

(*c*) NTU, for

(*i*) Counterflow heat exchanger,

(*ii*) One shell pass and multiple tube passes,

(*iii*) Two shell passes and multiple tube passes

(*iv*) Cross flow heat exchanger with both fluids unmixed, and

(*v*) Cross flow heat exchanger with cold fluid unmixed.

Assumption : Hot fluid flows through shell side and cold fluid through tubes.

Analysis : (*i*) Counterflow heat exchanger

(*a*) *LMTD* : $\Delta\text{T}_1 = \text{T}_{h,i} - \text{T}_{c,o} = 180 - 119 = 61°\text{C}$

$$\Delta\text{T}_2 = \text{T}_{h,o} - \text{T}_{c,i} = 118 - 99 = 19°\text{C}$$

$$\Delta\text{T}_{lm,\,counter} = \frac{\Delta\text{T}_1 - \Delta\text{T}_2}{\ln\left(\dfrac{\Delta\text{T}_1}{\Delta\text{T}_2}\right)} = \frac{61 - 19}{\ln\left(\dfrac{61}{19}\right)}$$

$= \mathbf{36°C}$. **Ans.**

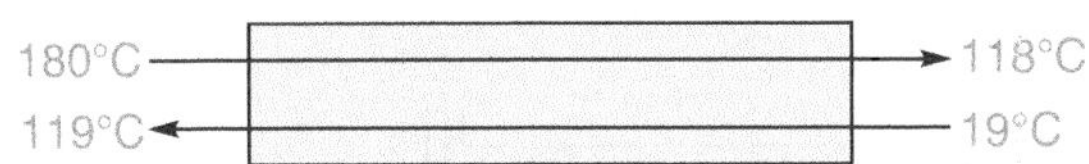

Fig. 14.43. (*a*) Counter flow exchanger

(*b*) *Effectiveness :* Making energy balance on heat exchanger

$$\dot{m}_c \text{C}_{p,c}(\text{T}_{c,o} - \text{T}_{c,i}) = \dot{m}_h \text{C}_{p,h}(\text{T}_{h,i} - \text{T}_{h,o})$$

or $$\text{C}_c(119 - 99) = \text{C}_h(180 - 118)$$

or $$\frac{\text{C}_h}{\text{C}_c} = \frac{20}{62} = 0.322$$

It indicates that the hot fluid has lower heat capacity rate, the effectiveness can be obtained as :

$$\varepsilon_{\text{counter}} = \frac{T_{h,i} - T_{h,o}}{T_{h,i} - T_{c,i}} = \frac{180 - 118}{180 - 99} = \mathbf{0.765.} \text{ Ans.}$$

(c) *NTU* :

For C = 0.322 and ε = 0.765 from Fig. 14.31

NTU = **1.72. Ans.**

(*ii*) One shell pass and multiple tube pass heat exchanger:

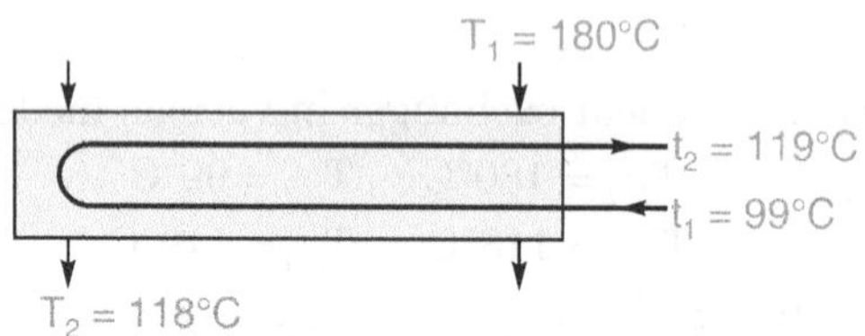

Fig. 14.43. (*b*) One shell and multipass tube exchanger

(*a*) *LMTD* : $\Delta T_{lm} = F\,\Delta T_{lm,\,\text{counter}}$

From Fig. 14.15(*a*)

$$P = \frac{t_2 - t_1}{T_1 - t_1} = \frac{119 - 99}{180 - 99} = 0.25$$

$$R = \frac{T_1 - T_2}{t_2 - t_1} = \frac{180 - 118}{119 - 99} = 3.1$$

It gives F = 0.88, thus

$$\Delta T_{lm} = 0.88 \times 36 = \mathbf{31.68°C.} \text{ Ans.}$$

(*b*) *Effectiveness* : Since hot fluid has lower heat capacity rate, thus the effectiveness can be obtained as :

$$\varepsilon = \frac{T_{h,i} - T_{h,o}}{T_{h,i} - T_{c,i}} = \frac{180 - 118}{180 - 99} = \mathbf{0.765.} \text{ Ans.}$$

(*c*) *NTU* :

From Fig. 14.34 , for C = 0.322 and ε = 0.765,

NTU = **2.3. Ans.**

(*iii*) Two shell passes and multiple tube passes heat exchanger :

(*a*) *LMTD* : $\Delta T_{lm} = F\,\Delta T_{lm,\,\text{counter}}$

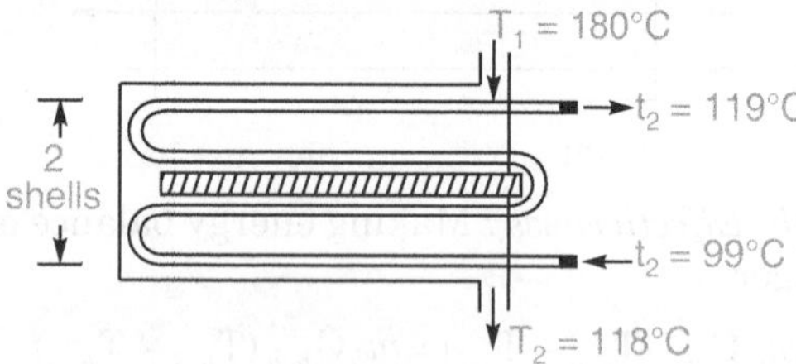

Fig. 14.43. (*c*) Two shell pass and multiple tube pass exchanger

For P = 0.25, and R = 3.1, from Fig. 14.15(*b*)

F = 0.97

$$\Delta T_{lm} = 0.97 \times 36 = \mathbf{34.92°C.} \text{ Ans.}$$

(*b*) *Effectiveness* : Since operating temperatures and heat capacity rates are same, thus effectiveness will not change, *i.e.*,

ε = **0.765. Ans.**

(*c*) *NTU* :

For C = 0.322 and ε = 0.765, from Fig. 14.35,

NTU = **1.7. Ans.**

(*iv*) Cross flow heat exchanger with both fluid unmixed :

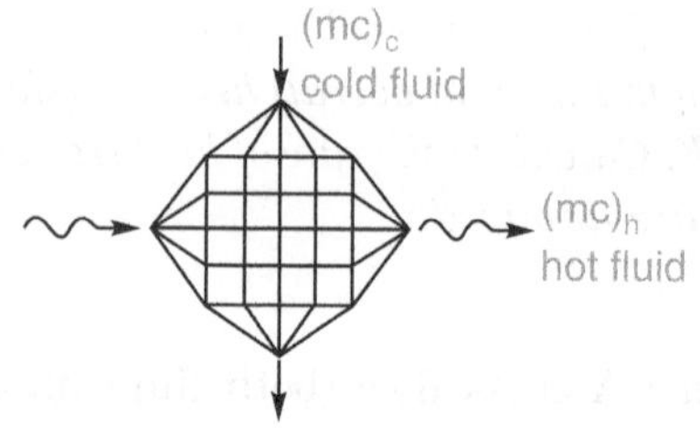

Fig. 14.43. (*d*) Cross flow exchanger with both fluid unmixed

(*a*) *LMTD* : $\Delta T_{lm} = F\,\Delta T_{lm,\,\text{counter}}$

For P = 0.25, and R = 3.1, from Fig. 14.15 (*c*), the correction factor

F = 0.93

$$\therefore \quad \Delta T_{lm} = 0.93 \times 36 = \mathbf{33.5°C.} \text{ Ans.}$$

(*b*) *Effectiveness* : It remains same

ε = **0.765. Ans.**

(*c*) *NTU* : For C = 0.322 and ε = 0.765, from Fig. 14.32

NTU = **2.0. Ans.**

(*v*) Cross flow heat exchanger with cold fluid unmixed :

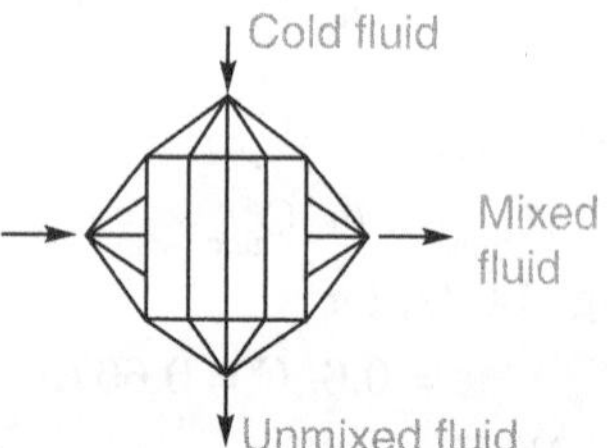

Fig. 14.43. (*e*) Cross flow exchanger with both fluid unmixed

(*a*) *LMTD* : $\Delta T_{lm} = F\,\Delta T_{lm,\,\text{counter}}$

For P = 0.25, R = 3.1, from Fig. 14.15(*d*), the correction factor

F = 0.92

$$\Delta T_{lm} = 0.92 \times 36 = \mathbf{33.12°C.} \text{ Ans.}$$

(*b*) *Effectiveness : It remains same*

i.e., ε = **0.765. Ans.**

(*c*) *NTU* : For C = 0.322 and ε = 0.765, from Fig. 14.33(*d*)

NTU = **2.4. Ans.**

Example 14.25. *Oil enters a heat exchanger at 100°C with a heat capacity rate of 3715 W/K. Water is available at 20°C with mass flow rate of 0.6 kg/s. Calculate the exit temperatures of both fluids in (a) counter flow, and (b) parallel flow arrangements for U = 500 W/m².K and surface area of 10 m². Take*

$$C_{p,h} = 1.88 \text{ kJ/kg.K and } C_{p,c} = 4.18 \text{ kJ/kg.K}$$

If the ratio of convection thermal resistance of oil to that of water is 1.2 and negligible wall and other resistances, calculate the wall temperature at each end of counter flow and parallel flow exchangers.

(NIT, Calicut 2008)

Solution

Given: Counter flow and parallel flow arrangement of heat exchanger with the following data

Hot fluid: Oil	Cold fluid: Water
$T_{h,i} = 100°C$	$\dot{m}_c = 0.6$ kg/s
$C_h = 3715$ W/K	$T_{c,i} = 20°C$
$C_{p,h} = 1.88$ kJ/kg.K	$C_{p,c} = 4.18$ kJ/kg.K
$U = 500$ W/m².K,	$A = 10$ m², $\dfrac{R_h}{R_c} = 1.2$

To find: (*i*) Exit temperatures of fluids in counter flow and parallel flow arrangements

(*ii*) Wall temperature at $x = 0$ and $x = L$ of counter flow and parallel heat exchangers.

Schematic :

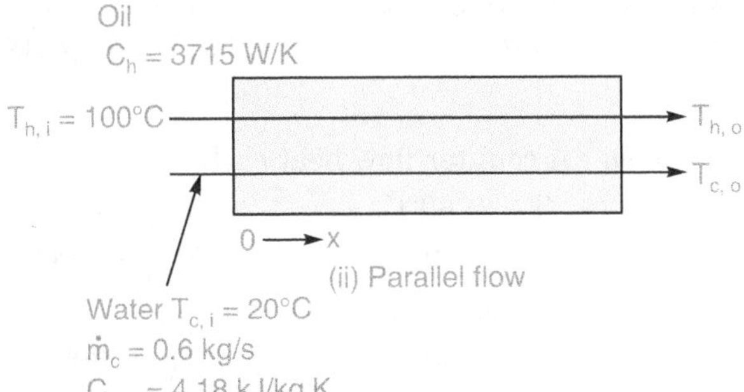

Fig. 14.44. (*a*) Counter flow and parallel flow arrangements of heat exchanger

Analysis: The heat capacity rates are

$$C_h = 3715 \text{ W/K}$$

$$C_c = \dot{m}_c C_{p,c} = 0.6 \times 4.18 \times 10^3 = 2508 \text{ W/K}$$

Ratio of heat capacity rates and NTU

$$C = \frac{C_{min}}{C_{max}} = \frac{C_c}{C_h} = \frac{2508}{3715} = 0.675$$

$$NTU = \frac{UA}{C_{min}} = \frac{500 \times 10}{2508} = 1.993$$

(*i*) Counter flow heat exchanger :

Effectiveness of heat exchanger, eqn. (14.55)

$$\varepsilon_{\text{counter flow}} = \frac{1 - \exp.[-NTU(1-C)]}{1 - C\exp[-NTU(1-C)]}$$

$$= \frac{1 - \exp[-1.993 \times (1 - 0.675)]}{1 - 0.675 \times \exp[-1.993 \times (1 - 0.675)]}$$

$$= 0.737$$

Further, effectiveness can be expressed in temperature terms and heat capacity rates as,

$$\varepsilon = \frac{C_c(T_{c,o} - T_{c,i})}{C_{min}(T_{h,i} - T_{c,i})} = \frac{C_h(T_{h,i} - T_{h,o})}{C_{min}(T_{h,i} - T_{c,i})}$$

It gives

$$T_{c,o} = 20 + 0.737 \times (100 - 20)$$

$$= \mathbf{78.97°C.} \quad \text{Ans.}$$

$$T_{h,o} = 100 - 0.737 \times \frac{2508}{3715} \times (100 - 20)$$

$$= \mathbf{60.20°C.} \quad \text{Ans.}$$

The energy balance at $x = 0$ of heat exchanger without wall and fouling resistances:

$$\frac{T_{h,i} - T_{w,x=0}}{R_h} = \frac{T_{w,x=0} - T_{c,o}}{R_c}$$

It leads to

$$T_{w,x=L} = \frac{T_{h,i} + \left(\frac{R_h}{R_c}\right) T_{c,o}}{1 + \left(\frac{R_h}{R_c}\right)}$$

$$= \frac{100 + 1.2 \times 78.97}{1 + 1.2}$$

$$= \mathbf{88.53°C.} \quad \text{Ans.}$$

Similarly at $x = L$

$$T_{w, x = L} = \frac{T_{h,o} + (R_h/R_c)\, T_{c,i}}{1 + (R_h/R_c)}$$

$$= \frac{60.2 + 1.2 \times 20}{1 + 1.2} = \mathbf{38.27°C.} \quad \text{Ans.}$$

(*ii*) Parallel flow heat exchanger:

Effective of heat exchanger, eqn. (14.53);

$$\varepsilon_{\text{parallel}} = \frac{1 - \exp[-\text{NTU}\,(1 + C)]}{1 + C}$$

$$= \frac{1 - \exp[-1.993 \times (1 + 0.675)]}{1 + 0.675}$$

$$= 0.576$$

Effectiveness in terms of temperatures and heat capacity rates leads to

$$T_{c,o} = T_{c,i} + \varepsilon(T_{h,i} - T_{c,i})$$

$$= 20 + 0.576 \times (100 - 20)$$

$$= \mathbf{66.0°C.} \quad \text{Ans.}$$

$$T_{h,o} = T_{h,i} - \varepsilon \times \frac{C_c}{C_h}(T_{h,i} - T_{c,i})$$

$$= 100 - 0.576 \times \frac{2508}{3715} \times (100 - 20)$$

$$= \mathbf{68.76°C.} \quad \text{Ans.}$$

The energy balance at $x = 0$ of heat exchanger without wall and fouling resistances:

$$\frac{T_{h,i} - T_{w,x=0}}{R_h} = \frac{T_{w,x=0} - T_{c,i}}{R_c}$$

It leads to

$$T_{w, x = 0} = \frac{T_{h,i} + \left(\frac{R_h}{R_c}\right) T_{c,i}}{1 + \left(\frac{R_h}{R_c}\right)}$$

$$= \frac{100 + 1.2 \times 20}{1 + 1.2} = \mathbf{56.36°C.} \quad \text{Ans.}$$

Similarly at $x = L$

$$T_{w, x = L} = \frac{T_{h,o} + \left(\frac{R_h}{R_c}\right) T_{c,o}}{1 + \left(\frac{R_h}{R_c}\right)}$$

$$= \frac{68.76 + 1.2 \times 66.0}{1 + 1.2}$$

$$= \mathbf{67.25°C.} \quad \text{Ans.}$$

Comment 1. The wall temperature across the exchanger varies from 88.53°C to 38.27°C in counter flow arrangement and 56.36°C to 67.25°C in parallel flow arrangement. Longitudinal wall temperature distribution is more uniform in parallel flow arrangement compared to counter flow arrangement.

2. Effectiveness of heat exchanger is better in counter flow mode.

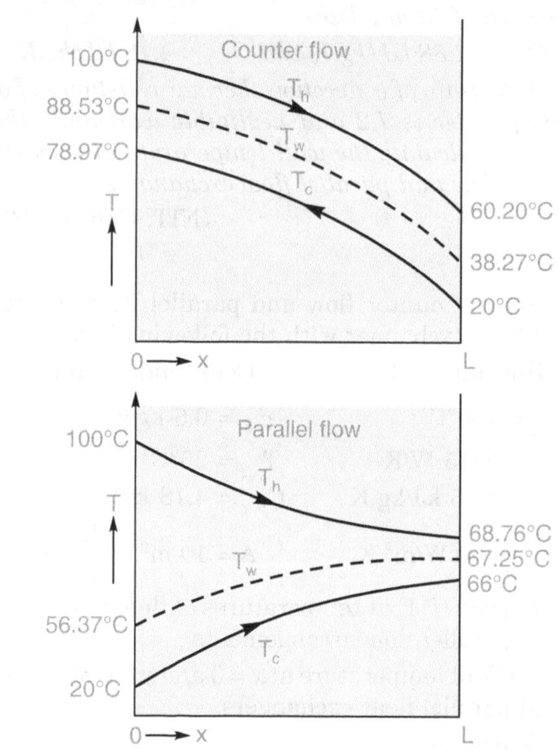

Fig. 14.44. (*b*) Temperature distribution along the length of counter flow and parallel flow heat exchanger.

Example 14.26. *A counter flow heat exchanger operates under the following conditions :*

Fluid A, inlet and outlet temperatures 80°C and 40°C

Fluid B, inlet and outlet temperatures 20°C and 40°C.

The exchanger is cleaned, causing an increase in overall heat transfer coefficient by 10% and the inlet temperature of fluid B is changed to 30°C, what would be the new outlet temperatures of fluid A and B ? Assume heat transfer coefficient and capacity rates are unchanged by temperature changes. (N.M.U., May 2002)

Solution

Given : A counter flow heat exchanger

$T_{h,i} = 80°C$, $T_{h,o} = 40°C$

$T_{c,i} = 20°C$, $T_{c,o} = 40°C$

After cleaning

$T_{c,i} = 30°C$, $T_{h,i} = 80°C$

$T_{c,o} = ?$ $T_{h,o} = ?$

$U_2 = 1.1\, U_1$.

To find : Exit temperatures of two fluids after cleaning of heat exchanger.

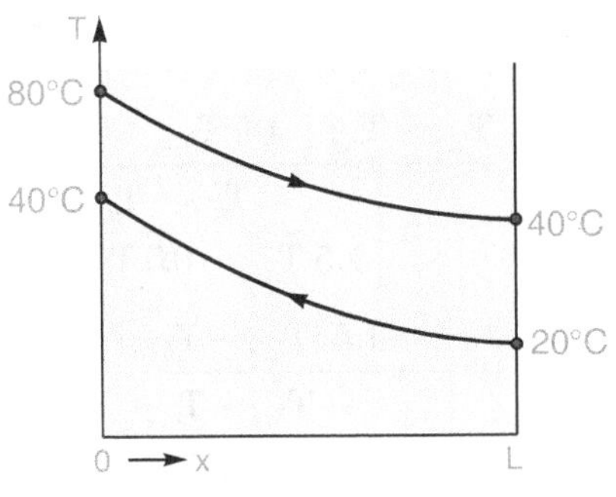

Before cleaning

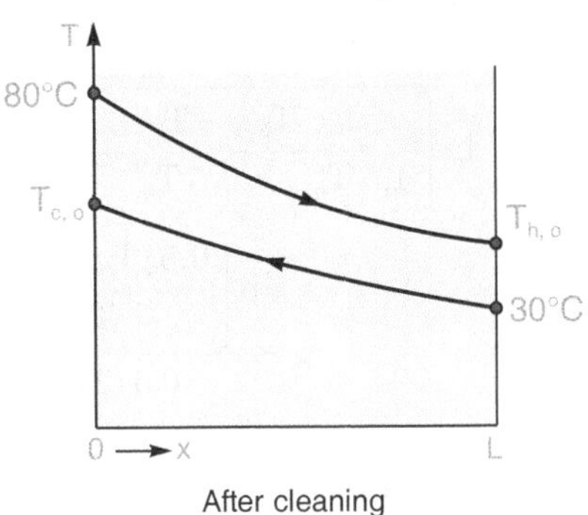

After cleaning

Fig. 14.45

Analysis : Only exit temperatures are to be changed after cleaning and other parameter will remain same using NTU method.

Case I. Before cleaning of heat exchanger

Energy balance between two fluid

$$\dot{m}_c C_{p,c}(T_{c,o} - T_{c,i}) = \dot{m}_h C_{p,h}(T_{h,i} - T_{h,o})$$

or $$C_c(40 - 20) = C_h(80 - 40)$$

Capacity ratio $$C = \frac{C_h}{C_c} = \frac{40 - 20}{80 - 40} = 0.5$$

and $$C_{min} = C_h$$
$$C_{max} = C_c$$

Effectiveness of heat exchanger

$$\varepsilon_1 = \frac{C_h(T_{h,i} - T_{h,o})}{C_{min}(T_{h,i} - T_{c,i})} = \frac{80 - 40}{80 - 20} = 0.667$$

For $\varepsilon_1 = 0.667$ and $C = 0.5$, from Fig. 14.31

$NTU_1 = 1.38.$

Case II. After cleaning of heat exchanger

$$NTU_2 = \frac{U_2 A}{C_{min}} = 1.1\frac{U_1 A}{C_{min}} = 1.1 \times 1.38 = 1.518$$

$C = 0.5$ (remains unchanged)

Again from Fig. 14.31.

$$\left.\begin{aligned} NTU_2 &= 1.518 \\ C &= 0.5 \end{aligned}\right\} \varepsilon_2 = 0.69$$

The effectiveness is also given by

$$\varepsilon_2 = \frac{C_h(T_{h,i} - T_{h,o})}{C_{min}(T_{h,i} - T_{c,i})} = \frac{C_c(T_{c,o} - T_{c,i})}{C_{min}(T_{h,i} - T_{c,i})}$$

$$T_{h,o} = T_{h,i} - \varepsilon \frac{C_{min}}{C_h}(T_{h,i} - T_{c,i})$$
$$= 80 - 0.69 \times (80 - 30) = \mathbf{45.5°C}$$

$$T_{c,o} = T_{c,i} + \varepsilon \frac{C_{min}}{C_c}(T_{h,i} - T_{c,i})$$
$$= 30 + 0.69 \times 0.5 \times (80 - 30) = \mathbf{47.25°C}$$

The outlet temperatures of hot and cold fluids are 45.5°C and 47.25°C, respectively. **Ans.**

Example 14.27. *In a double pipe heat exchanger, $\dot{m}_h C_{p,h} = 0.5\, \dot{m}_c C_{p,c}$. The inlet temperatures of hot and cold fluids are $T_{h,i}$ and $T_{c,i}$. Deduce an expression in terms of $T_{h,i}$, $T_{c,i}$ and $T_{h,o}$ for the ratio of area of counter flow heat exchanger to that of parallel flow heat exchanger, which will give same hot fluid outlet temperature $T_{h,o}$. Find the ratio, if $T_{h,i}$ = 150°C, $T_{c,i}$ = 30°C and $T_{h,o}$ = 90°C.* (U.P.S.C., 1994)

Solution

Given : A double pipe heat exchanger, with

$T_{h,i}$ = 150°C, $T_{h,o}$ = 90°C and $T_{c,i}$ = 30°C

$\dot{m}_h C_{p,h} = 0.5\, \dot{m}_c C_{p,c}$ and $T_{h,o,\,parallel} = T_{h,o,\,counter}$

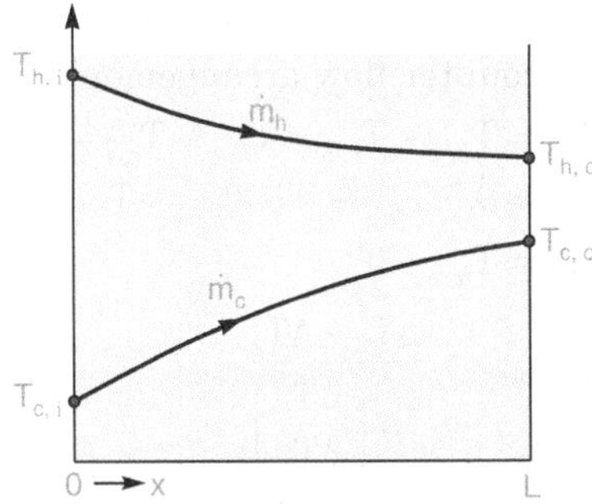

(a) Parallel flow mode

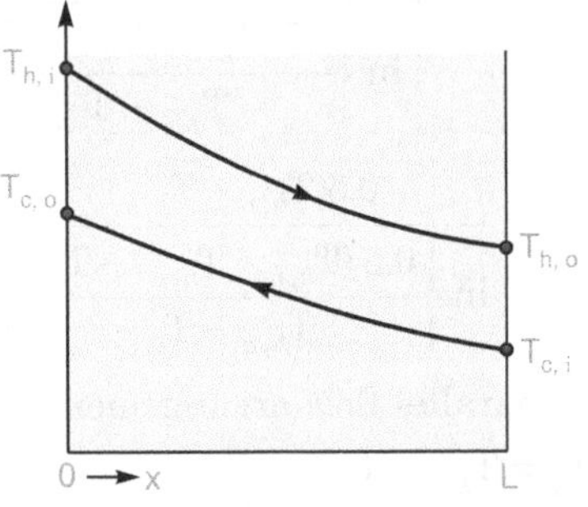

(b) Counter flow mode

Fig. 14.46

Analysis : The heat capacity rates for two fluids are related as :

$$\dot{m}_h C_{p,h} = 0.5\, \dot{m}_c C_{p,c}$$

which yields to

$$C_{min} = \dot{m}_h C_{p,h} = C_h$$

$$C_{max} = \dot{m}_c C_{p,c} = C_c$$

and capacity ratio C = 0.5

The effectiveness of heat exchanger

$$\varepsilon = \frac{C_h(T_{h,i} - T_{h,o})}{C_{min}(T_{h,i} - T_{c,i})} = \frac{C_c(T_{c,o} - T_{c,i})}{C_{min}(T_{h,i} - T_{c,i})}$$

Using $C_h = 0.5C_c$, or $C_c = 2C_h$, then

$$\varepsilon = \frac{T_{h,i} - T_{h,o}}{T_{h,i} - T_{c,i}} = \frac{2(T_{c,o} - T_{c,i})}{T_{h,i} - T_{h,o}}$$

or $$T_{c,o} = T_{c,i} + 0.5\,(T_{h,i} - T_{h,o}) \quad ...(i)$$

Let A_{PF} = area of parallel flow arrangement

A_{CF} = area of counter flow arrangement.

As, $T_{h,o,\,parallel} = T_{h,o,\,counter}$, therefore, the heat lost by hot fluid in both case be same.

Further, the heat transfer rate is also expressed as

$$Q = UA_{PF}\, \Delta T_{lm,\,parallel} = UA_{CF}\, \Delta T_{lm,\,counter}$$

or $$\frac{A_{CF}}{A_{PF}} = \frac{\Delta T_{lm,\,parallel}}{\Delta T_{lm,\,counter}} \quad ...(ii)$$

($\because$ U is constant in both cases)

(*a*) For counter flow arrangement

$$\Delta T_1 = T_{h,i} - T_{c,o} = T_{h,i} - T_{c,i} - 0.5(T_{h,i} - T_{h,o})$$

$$= 0.5\, T_{h,i} + 0.5\, T_{h,o} - T_{c,i}$$

$$\Delta T_2 = T_{h,o} - T_{c,i}$$

$$\Delta T_{lm,\,counter} = \frac{\Delta T_1 - \Delta T_2}{\ln\left(\dfrac{\Delta T_1}{\Delta T_2}\right)}$$

$$= \frac{0.5\, T_{h,i} + 0.5\, T_{h,o} - T_{c,i} - T_{h,o} + T_{c,i}}{\ln\left\{\dfrac{0.5\,(T_{h,i} + T_{h,o}) - T_{c,i}}{T_{h,o} - T_{c,i}}\right\}}$$

$$= \frac{0.5(T_{h,i} - T_{h,o})}{\ln\left\{\dfrac{0.5\,(T_{h,i} + T_{h,o}) - T_{c,i}}{T_{h,o} - T_{c,i}}\right\}} \quad ...(iii)$$

(*b*) For parallel flow arrangement

$$\Delta T_1 = T_{h,i} - T_{c,i}$$

$$\Delta T_2 = T_{h,o} - T_{c,o} = T_{h,o} - T_{c,i} - 0.5(T_{h,i} - T_{h,o})$$

$$= 1.5\, T_{h,o} - 0.5\, T_{h,i} - T_{c,i}$$

$$\Delta T_{lm,\,parallel} = \frac{\Delta T_1 - \Delta T_2}{\ln\left(\dfrac{\Delta T_1}{\Delta T_2}\right)}$$

$$= \frac{T_{h,i} - T_{c,i} - 1.5\, T_{h,o} + 0.5\, T_{h,i} + T_{c,i}}{\ln\left\{\dfrac{T_{h,i} - T_{c,i}}{1.5\, T_{h,o} - 0.5\, T_{h,i} - T_{c,i}}\right\}}$$

$$= \frac{1.5\,[T_{h,i} - T_{h,o}]}{\ln\left[\dfrac{T_{h,i} - T_{c,i}}{1.5\, T_{h,o} - 0.5\, T_{h,i} - T_{c,i}}\right]} \quad ...(iv)$$

Substituting eqns. (*iii*) and (*iv*) in eqn. (*ii*),

$$\frac{A_{CF}}{A_{PF}} = \frac{1.5\,(T_{h,i} - T_{h,o})}{\ln\left[\dfrac{T_{h,i} - T_{c,i}}{1.5\, T_{h,o} - 0.5\, T_{h,i} - T_{c,i}}\right]} \times \frac{\ln\left[\dfrac{0.5\,(T_{h,i} + T_{h,o}) - T_{c,i}}{T_{h,o} - T_{c,i}}\right]}{0.5\,(T_{h,i} - T_{h,o})}$$

$$= 3\left[\frac{\ln\left\{\dfrac{0.5\,(T_{h,i} + T_{h,o}) - T_{c,i}}{T_{h,o} - T_{c,i}}\right\}}{\ln\left\{\dfrac{T_{h,i} - T_{c,i}}{1.5\, T_{h,o} - 0.5\, T_{h,i} - T_{c,i}}\right\}}\right]$$

For given data :

$T_{h,i}$ = 150°C, $T_{c,i}$ = 30°C and $T_{h,o}$ = 90°C

$$\frac{A_{CF}}{A_{PF}} = 3\left[\frac{\ln\left\{\dfrac{0.5\,(150 + 90) - 30}{90 - 30}\right\}}{\ln\left\{\dfrac{150 - 30}{1.5 \times 90 - 0.5 \times 150 - 30}\right\}}\right]$$

$$= 3 \times \frac{\ln(1.5)}{\ln(4)} = \mathbf{0.877.} \quad \text{Ans.}$$

Example 14.28. *A counter flow heat exchanger is used to heat water from 20°C to 80°C at a rate of 1.2 kg/s. The heating is obtained by using geothermal water available at 160°C at a mass flow rate of 2 kg/s. The inner tube is thin walled, and has a diameter of 1.5 cm. If the overall heat transfer coefficient is 640 W/m²·K. Calculate the length of the heat exchanger required to achieve the desired heating by using effectiveness – NTU method.*

Take specific heat of geothermal water as 4.31 kJ/kg.K and that of ground water as 4.18 kJ/kg.K.

Solution

Given : A counter flow heat exchanger

$U = 640\ W/m^2.K$

Cold water :

$$\dot{m}_c = 1.2 \text{ kg/s}$$
$$T_{c,i} = 20°C$$
$$T_{c,o} = 80°C,\ C_{p,c} = 4.18 \text{ J/kg.K}$$
$$d = 1.5 \text{ cm} = 0.015 \text{ m}$$

Geothermal hot water :

$$\dot{m}_h = 2 \text{ kg/s}$$
$$T_{h,i} = 160°C$$
$$C_{p,h} = 4.31 \text{ kJ/kg.K.}$$

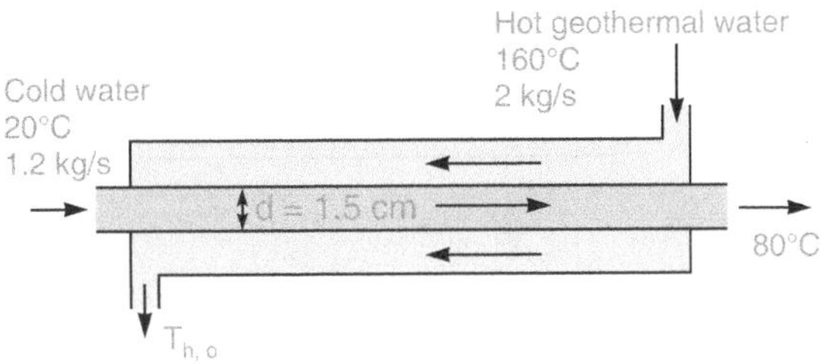

Fig. 14.47. Sechmatic for example 14.28

To find : Length of heat exchanger.

Analysis : This example can also be solved by LMTD method. Here we are using effectiveness – NTU method.

Heat capacity rates

$$C_h = \dot{m}_h C_{p,h} = 2 \times 4.31 = 8.62 \text{ kW/K}$$
$$C_c = \dot{m}_c C_{p,c} = 1.2 \times 4.18 = 5.02 \text{ kW/K}$$

Therefore,

$$C_{min} = C_c = 5.02 \text{ kW/K}$$
$$C_{max} = C_h = 8.62 \text{ kW/K}$$

Capacity ratio

$$C = \frac{C_{min}}{C_{max}} = \frac{5.02}{8.62} = 0.583$$

The actual heat transfer rate

$$Q = \dot{m}_c C_{p,c}(T_{c,o} - T_{c,i})$$
$$= 1.2 \times 4.18 \times (80 - 20) = 301.0 \text{ kW}$$

Maximum possible heat transfer rate

$$Q_{max} = C_{min}(T_{h,i} - T_{c,i})$$
$$= 5.02 \times (160 - 20) = 702.8 \text{ kW}$$

The effectiveness of heat exchanger is

$$\varepsilon = \frac{Q}{Q_{max}} = \frac{301.0}{702.8} = 0.428$$

For counter flow heat exchanger, the effectiveness is also expressed as

$$\varepsilon = \frac{1 - \exp[-NTU(1-C)]}{1 - C\exp[-NTU(1-C)]}$$

$$0.428 = \frac{1 - \exp[-NTU(1-0.583)]}{1 - 0.583 \times \exp[-NTU(1-0.583)]}$$

or $\quad 0.428 - 0.249 \exp[-0.417(NTU)]$
$$= 1 - \exp(-0.417\ NTU)$$

or $\quad \exp(-0.417\ NTU)(1 - 0.249) = 1 - 0.428$

or $\quad -0.417\ NTU = \ln\left(\frac{1-0.428}{1-0.249}\right) = -0.2722$

or $\quad NTU = 0.653$

Again $\quad NTU = \dfrac{UA}{C_{min}}$

or $\quad A = \dfrac{NTU\ C_{min}}{U} = \dfrac{0.653 \times 5.02 \times 10^3}{640}$
$$= 5.12 \text{ m}^2$$

and $\quad A = \pi d L$

or $\quad L = \dfrac{A}{\pi d} = \dfrac{5.12}{\pi \times 0.015} = \mathbf{108.6\ m.}$ **Ans.**

Example 14.29. *Consider a concentric tube heat exchanger with hot and cold water inlet temperatures of 200°C and 35°C, respectively. The flow rates of hot and cold water are 42 kg / h and 84 kg / h, respectively. Assume the overall heat transfer coefficient is 180 W / m² K.*

(i) What is the maximum possible heat transfer rate that could be achieved for the prescribed inlet conditions ?

(ii) If the exchanger is operated in counter flow arrangement with a heat transfer area of 0.33 m², determine the outlet fluid temperatures.

(iii) What is the largest possible heat transfer rate that could be achieved for the prescribed inlet conditions, if exchanger is operated in parallel flow mode and is very long ? What is the effectiveness of heat exchanger in this configuration ?

Solution

Given : Concentric tube heat exchanger

$$T_{h,i} = 200°C, \quad T_{c,i} = 35°C$$
$$\dot{m}_h = 42 \text{ kg/h}, \quad U = 180 \text{ W/m}^2\text{.K}$$
$$\dot{m}_c = 84 \text{ kg/h}, \quad A = 0.33 \text{ m}^3.$$

To find :

(*i*) Maximum possible heat transfer rate,

(*ii*) Outlet fluid temperatures in counter flow mode, and

(*iii*) Largest possible heat transfer rate in parallel flow mode and effectiveness of heat exchanger.

Assumptions :

1. Specific heat of water as C_p = 4200 J/kg.K.
2. No heat loss to the surroundings.

Analysis : (*i*) Maximum heat transfer rate :

Heat capacity of hot fluid

$$\dot{m}_h C_{p,h} = \frac{42}{3600} \text{ (kg/s)} \times (4200 \text{ J/kg.K}) = 49 \text{ W/K}$$

Heat capacity of cold fluid,

$$\dot{m}_c C_{p,c} = \frac{84}{3600} \times 4200 = 98 \text{ W/K}$$

Since, $\dot{m}_h C_{p,h} < \dot{m}_c C_{p,c}$

The maximum heat transfer rate

$$Q_{max} = (\dot{m}\, C_p)_{min}(T_{h,i} - T_{c,i})$$
$$= (49 \text{ W/K}) \times (200 - 35)(\text{K})$$
$$= \mathbf{8085\ W.} \quad \textbf{Ans.}$$

(*ii*) Outlet temperatures of fluid in counter flow mode :

$$NTU = \frac{UA}{C_{min}} = \frac{180 \times 0.33}{49} = 1.212$$

Heat capacity ratio

$$C = \frac{(\dot{m}\, C_p)_{min}}{(\dot{m}\, C_p)_{max}} = \frac{49}{98} = 0.5$$

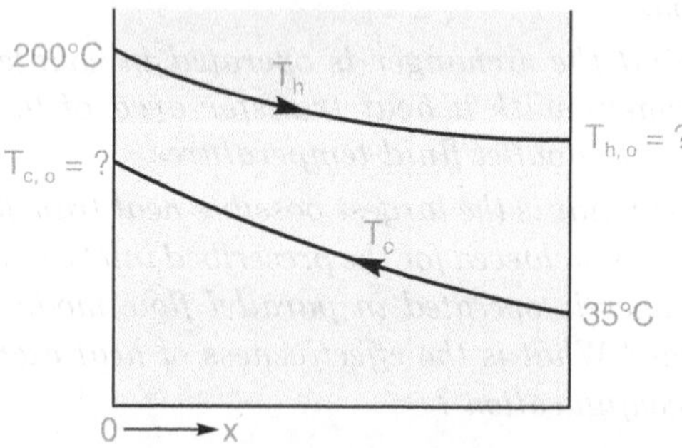

(a) Counter flow mode

The effectiveness

$$\varepsilon = \frac{1 - \exp\{-NTU(1-C)\}}{1 - C\exp\{-NTU(1-C)\}}$$

$$= \frac{1 - \exp\{-1.212(1-0.5)\}}{1 - 0.5\exp\{-1.212(1-0.5)\}} = 0.625$$

The outlet temperature of oil (fluid of minimum heat capacity) ;

(*a*) $T_{h,o} = T_{h,i} - \varepsilon\,(T_{h,i} - T_{c,i})$
$= 200 - 0.625 \times (200 - 35)$
$= \mathbf{97.88°C.}$ **Ans.**

(*b*) Cold fluid outlet temperature

$$T_{c,o} = T_{c,i} + \varepsilon\, C\,(T_{h,i} - T_{c,i})$$
$$= 35 + 0.625 \times 0.5 \times (200 - 35)$$
$$= \mathbf{86.568°C.} \quad \textbf{Ans.}$$

(*iii*) Very long parallel flow heat exchanger :

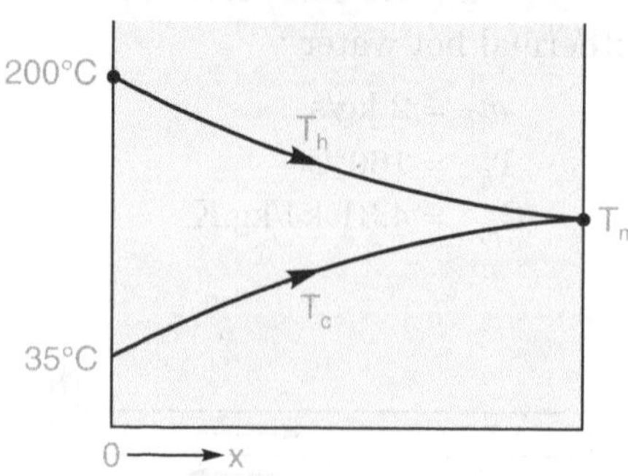

(b) Long parallel flow heat exchanger

Fig. 14.48. Temperature profile for example 14.29

Both fluid will have common outlet temperature. Let the common outlet temperature be T_m, hence

$$\dot{m}_c C_{p,c}(T_m - T_{c,i}) = \dot{m}_h C_{p,h}(T_{h,i} - T_m)$$

or $\quad 98 \times (T_m - 35) = 49 \times (200 - T_m)$

or $\quad 3T_m = 270 \quad \text{or} \quad T_m = 90°C$

The largest heat transfer rate, possible with parallel flow arrangement

$$Q_{max,\ parallel} = \dot{m}_c C_{p,c}(T_m - T_{c,i})$$
$$= (98 \text{ W/K}) \times (90 - 35)(\text{K})$$
$$= \mathbf{5390\ W.} \quad \textbf{Ans.}$$

Maximum possible effectiveness in parallel mode :

For the hot fluid,

$$\varepsilon = \frac{T_{h,i} - T_{h,o}}{T_{h,i} - T_{c,i}} = \frac{200 - 90}{200 - 35}$$
$$= \mathbf{0.667.} \quad \textbf{Ans.}$$

Example 14.30. *Oil is being cooled by water in a tube in tube parallel flow type heat exchanger. The water enters the centre tube at 15°C and is heated to 50°C. The oil flows in the annulus is cooled from 130°C to 60°C. Find :*

(*i*) *Exit temperature of each of the fluids, if the exiting heat exchanger was switched to counter flow operation.*

(*ii*) *The minimum temperature to which the oil may be cooled by increasing the tube length with parallel flow operations.*

(*iii*) *The maximum possible effectiveness in parallel flow operation.* (P.U., May 1994)

Solution

Given : A tube in tube heat exchanger,

$$T_{c,\,i} = 15°C, \qquad T_{h,\,i} = 130°C.$$

In parallel flow mode,

$$T_{c,\,o} = 50°C, \qquad T_{h,\,o} = 60°C$$

To find :

(*i*) Exit temperature of each of the fluids, if the existing heat exchanger was switched to counter flow operation.

(*ii*) The minimum temperature to which the oil may be cooled by increasing the tube length with parallel flow operations.

(*iii*) The maximum possible effectiveness in parallel flow operation.

Analysis : Making the energy balance for two fluids

$$(\dot{m}\,C_p)_{water}\,(\Delta T)_{water} = (\dot{m}\,C_p)_{oil}\,(\Delta T)_{oil}$$

or
$$C_{water} \times (50 - 15) = C_{oil} \times (130 - 60)$$

Hence, capacity ratio $C = \dfrac{C_{oil}}{C_{water}} = \dfrac{C_{min}}{C_{max}} = 0.5$

The temperature distribution in parallel flow heat exchanger, is shown in Fig. 14.49(*a*)

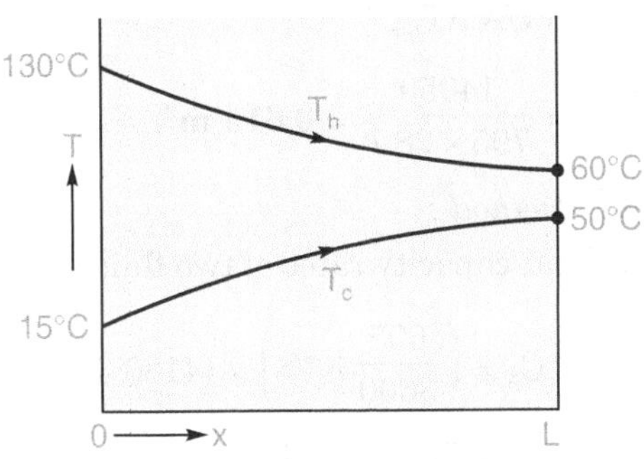

(*a*) Existing parallel flow exchanger

$$\varepsilon_{parallel} = \frac{130 - 60}{130 - 15} = 0.6086$$

Further, $$\varepsilon_{parallel} = \frac{1 - \exp\{-NTU(1 + C)\}}{1 + C}$$

or
$$0.6086 = \frac{1 - \exp\{-NTU(1 + 0.5)\}}{1 + 0.5}$$

It gives; NTU = 1.628

(*i*) If the existing heat exchanger was switched to counter flow mode Fig. 14.49(*b*), its NTU and heat capacity ratio remain unchanged. The effectiveness of exchanger in counter flow mode;

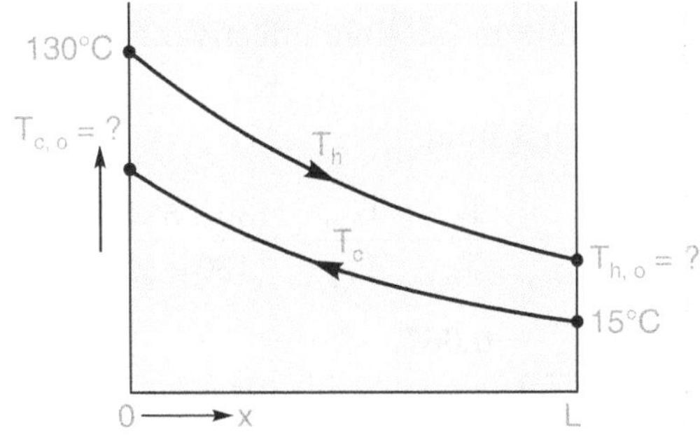

(*b*) Counter flow mode of exchanger

$$\varepsilon_{counter} = \frac{1 - \exp\{-NTU(1 - C)\}}{1 - C\exp\{-NTU(1 - C)\}}$$

or
$$\varepsilon_{counter} = \frac{1 - \exp\{-1.628(1 - 0.5)\}}{1 - 0.5 \times \exp\{-1.628(1 - 0.5)\}}$$
$$= 0.7154$$

The exit temperatures :

Hot fluid :

$$T_{h,\,o} = T_{h,\,i} - \varepsilon(T_{h,\,i} - T_{c,\,i})$$
$$= 130 - 0.7154 \times (130 - 15)$$
$$= \mathbf{47.72°C.} \quad \textbf{Ans.}$$

Cold fluid :

$$T_{c,\,o} = T_{c,\,i} + \varepsilon C(T_{h,\,i} - T_{c,\,i})$$
$$= 15 + 0.7154 \times 0.5 \times (130 - 15)$$
$$= \mathbf{56.135°C.} \quad \textbf{Ans.}$$

(*ii*) If the parallel flow heat exchanger is too long, then both fluids attain a common exit temperature (say T_m) as shown in Fig. 14.49 (*c*)

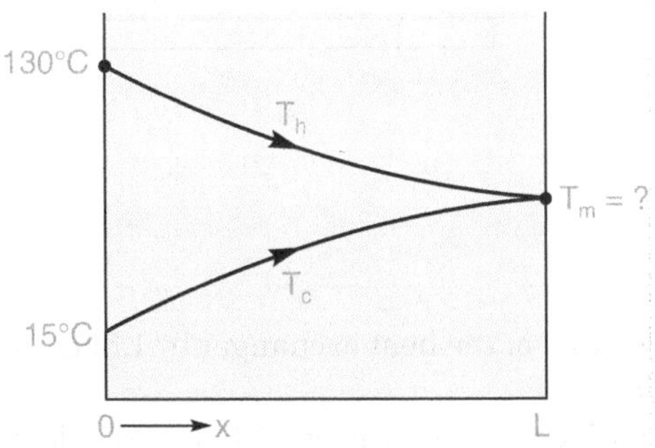

(*c*) Very long parallel flow heat exchanger

Fig. 14.49. Temperature profiles for example 14.30

$$(\dot{m}\,C_p)_{oil}\,(T_{h,\,i} - T_m) = (\dot{m}\,C_p)_{water}\,(T_m - T_{c,\,i})$$

or
$$130 - T_m = 2 \times (T_m - 15)$$

or
$$3T_m = 160$$

or
$$T_m = \mathbf{53.33°C.} \quad \textbf{Ans.}$$

It is minimum exit temperature of oil.

(*iii*) Maximum possible effectiveness in parallel flow mode :

For the hot fluid,

$$\varepsilon = \frac{T_{h,i} - T_{h,o}}{T_{h,i} - T_{c,i}} = \frac{130 - 53.33}{130 - 15}$$

$= \mathbf{0.667.}$ **Ans.**

Example 14.31. *Cold water at 1495 kg/h enters at 25°C through a parallel flow heat exchanger to cool 605 kg/h of hot water entering at 70°C and leaving at 50°C. Find the area of the heat exchanger. The individual heat transfer coefficients on both sides are 1590 W/m².K. Use LMTD and NTU methods. Find also the exit temperatures of cold and hot fluid streams, if the flow of hot water is doubled. Assume the individual heat transfer coefficients are proportional to 0.8th power of the flow rate. For water C_p = 4180 J/kg.K.* (J.N.T.U., May 2004)

Solution

Given : A parallel flow heat exchanger with

$\dot{m}_c$ = 1495 kg/h, $\dot{m}_h$ = 605 kg/h

$T_{c,i}$ = 25°C, $T_{h,i}$ = 70°C

C_p = 4180 J/kg.K, $T_{h,o}$ = 50°C

h_i = 1590 W/m².K, h_o = 1590 W/m².K

$h \propto \dot{m}^{0.8}$

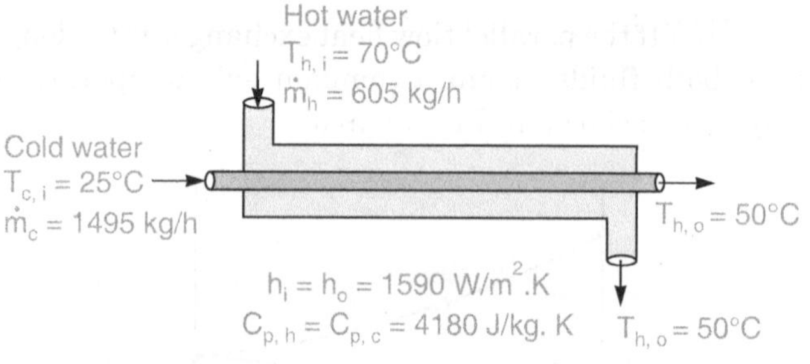

Fig. 14.50 (*a*) Parallel flow heat exchanger

To find :

(*i*) Area of the heat exchanger by LMTD and NTU methods

(*ii*) Exit temperatures of hot and cold fluids, if flow rate of hot fluid is doubled.

Assumptions :

(*i*) Thin walled heat transfer tube.

(*ii*) Clean and scale free surfaces.

(*iii*) Constant properties.

Analysis : (*i*) Making the energy balance on the heat exchanger

$$\dot{m}_c C_{p,c}(T_{c,o} - T_{c,i}) = \dot{m}_h C_{p,h}(T_{h,i} - T_{h,o})$$

with known numerical values

$$1495 \times 4180\,(T_{c,o} - 25) = 605 \times 4180\,(70 - 50)$$

It gives $T_{c,o}$ = 33.1°C

LMTD Method :

$$\Delta T_1 = T_{h,i} - T_{c,i} = 70 - 25 = 45°C$$

$$\Delta T_2 = T_{h,o} - T_{c,o} = 50 - 33.1 = 16.9°C$$

$$\Delta T_{lm} = \frac{\Delta T_1 - \Delta T_2}{\ln\left(\frac{\Delta T_1}{\Delta T_2}\right)} = \frac{45 - 16.9}{\ln\left(\frac{45}{16.9}\right)} = 28.7°C$$

The heat transfer rate

$$Q = \dot{m}_h C_{p,h}(T_{h,i} - T_{h,o})$$

$$= \left(\frac{605}{3600}\text{ kg/s}\right) \times (4180\text{ J/kg.K})\,(70 - 50)\,(\text{K})$$

$$= 14050\text{ W.}$$

The overall heat transfer coefficient

$$U = \frac{1}{\frac{1}{h_i} + \frac{1}{h_o}} = \frac{1}{\frac{1}{1590} + \frac{1}{1590}} = 795\text{ W/m}^2\text{.K}$$

The heat transfer rate is also expressed as :

$$Q = UA\,\Delta T_{lm}$$

or $$A = \frac{14050}{795 \times 28.7} = \mathbf{0.615\ m^2.}$$ **Ans.**

NTU Method :

The heat capacity rates of two fluid streams are :

$$C_h = \left(\frac{605}{3600}\text{ kg/s}\right) \times (4180\text{ J/kg.K})$$

$$= 702.47\text{ W/K}$$

$$C_c = \left(\frac{1495}{3600}\text{ kg/s}\right) \times (4180\text{ J/kg.K})$$

$$= 1736\text{ W/K}$$

$$C_{min} = C_h = 702.47\text{ W/m.K,}$$

$$C_{max} = C_c = 1736\text{ W/K}$$

The ratio of heat capacity rates

$$C = \frac{C_{min}}{C_{max}} = \frac{702.47}{1736} = 0.405$$

The effectiveness of heat exchanger

$$\varepsilon = \frac{C_h(T_{h,i} - T_{h,o})}{C_{min}(T_{h,i} - T_{c,i})} = \frac{70 - 50}{70 - 25} = 0.444$$

Further, the effectiveness of parallel flow heat exchanger is also given as :

$$\varepsilon = \frac{1-\exp[-\text{NTU}(1+C)]}{1+C}$$

or $$0.444 = \frac{1-\exp[-\text{NTU}(1+0.405)]}{1+0.405}$$

or $$\exp(-1.405\ \text{NTU}) = 0.376$$

or $$\text{NTU} = 0.696$$

The NTU is given as :

$$\text{NTU} = \frac{UA}{C_{min}}$$

or $$A = \frac{\text{NTU}\ C_{min}}{U} = \frac{0.696 \times 702.47}{795}$$

$$= \mathbf{0.615\ m^2} \quad \textbf{Ans.}$$

(*ii*) When flow rate of hot fluid is doubled

$$\dot{m}_{h,new} = 2 \times 605 = 1210 \text{ kg/h}$$

$$C_h = \left(\frac{1210}{3600}\text{ kg/s}\right) \times (4180 \text{ J/kg.K})$$

$$= 1404.94 \text{ W/K}$$

$$C_c = \left(\frac{1495}{3600}\text{ kg/s}\right) \times (4180 \text{ J/kg.K})$$

$$= 1736 \text{ W/K}$$

Thus $C_{min} = C_h = 1404.94$ W/K,

and $C_{max} = C_c = 1736$ W/K.

and $$C = \frac{C_{min}}{C_{max}} = \frac{1404.94}{1736} = 0.8093$$

Since the individual heat transfer coefficients are proportional to 0.8 power of flow rate, *i.e.*,

$$h_o \propto (\dot{m}_h)^{0.8}$$

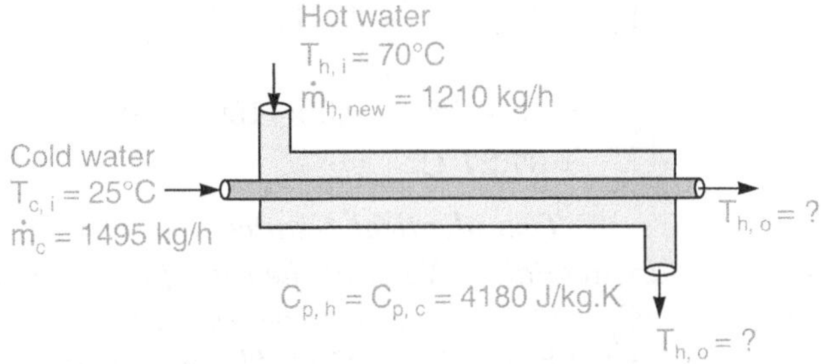

Fig. 14.50 (*b*) Parallel flow heat exchanger when flow rate of hot water is doubled

or $$1590 = K \times (605)^{0.8}$$

Constant $K = 9.4622$

Thus for changed flow condition

$$h_{o,new} = K(2\dot{m}_{h,new})^{0.8}$$
$$= 9.4622 \times (2 \times 605)^{0.8}$$
$$= 2768.35 \text{ W/m}^2\text{.K}$$

The overall heat transfer coefficient

$$U = \frac{1}{\frac{1}{h_i} + \frac{1}{h_{o,new}}} = \frac{1}{\frac{1}{1590} + \frac{1}{2768.35}}$$
$$= 1010 \text{ W/m}^2\text{.K}$$

$$\text{NTU} = \frac{UA}{C_{min}} = \frac{1010 \times 0.615}{1404.94} = 0.442$$

The effectiveness

$$\varepsilon = \frac{1-\exp[-\text{NTU}(1+C)]}{1+C}$$
$$= \frac{1-\exp[-0.442 \times (1+0.8093)]}{1+0.8093}$$
$$= 0.3043$$

The exit temperatures of two fluids.

$$T_{h,o} = T_{h,i} - \frac{\varepsilon C_{min}}{C_h}(T_{h,i} - T_{c,i})$$
$$= 70 - 0.3043 \times (70 - 25) = \mathbf{56.3°C}$$

$$T_{c,o} = T_{c,i} + \frac{\varepsilon C_{min}}{C_c}(T_{h,i} - T_{c,i})$$
$$= 25 + 0.3043 \times 0.8093 \times (70 - 25)$$
$$= \mathbf{36.08°C.} \quad \textbf{Ans.}$$

Example 14.32. *An oil is cooled to 100°C in a concurrent heat exchanger by transferring its heat to cooling water, that leaves the exchanger at 30°C. However, it is now required that the oil must be cooled down to 75°C by increasing the length of heat exchanger, while oil and water flow rates, their inlet temperatures and other dimensions of the exchanger keeping constant. The inlet temperatures of water and oil being 15°C and 150°C, respectively.*

If the original cooler was 1 m long, calculate

(i) Outlet temperature of water in new cooler,

(ii) Length of new cooler. (P.U., Nov. 2002)

Solution

Given : A concurrent (parallel) flow heat exchanger.

Case I. $L_1 = 1$ m

Hot fluid : Oil

$T_{h,i} = 150°C, \quad T_{h,o} = 100°C$

Cold fluid :Water

$T_{c,i} = 15°C, \quad T_{c,o} = 30°C$

Case II. $L_2 = ?$

Hot fluid : Oil

$T_{h,i} = 150°C, \quad T_{h,o} = 75°C$

Cold fluid $T_{c,i} = 15°C, \quad T_{c,o} = ?$

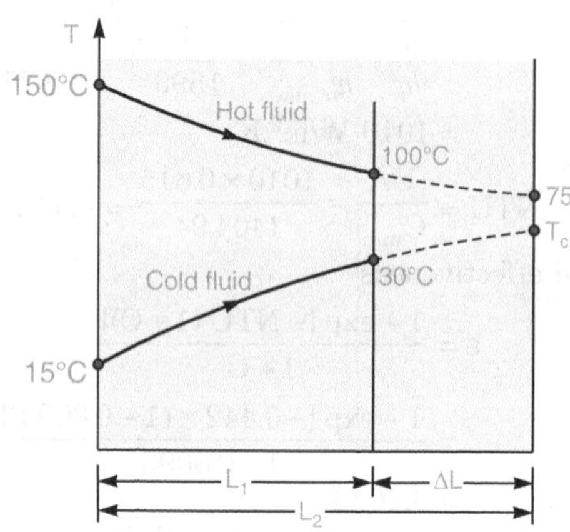

Fig. 14.51

To find :

(*i*) Outlet temperature of water in new cooler, and

(*ii*) Length of new cooler.

Analysis : **Case I.** Length of new heat exchanger :

Before increasing the length of cooler, the energy balance on exchanger

$$\dot{m}_h C_{p,h}(T_{h,i} - T_{h,o}) = \dot{m}_c C_{p,c}(T_{c,o} - T_{c,i})$$

$$\dot{m}_h C_{p,h}(150 - 100) = \dot{m}_c C_{p,c}(30 - 15)$$

or $$C = \frac{C_h}{C_c} = \frac{30-15}{150-100} = \frac{15}{50} = 0.3$$

and $C_{min} = C_h, \; C_{max} = C_c$

Effectiveness of heat exchanger

$$\varepsilon_1 = \frac{C_h(T_{h,i} - T_{h,o})}{C_{min}(T_{h,i} - T_{c,i})} = \frac{150-100}{150-15} = 0.37$$

Further, for parallel flow heat exchanger,

$$\varepsilon_1 = \frac{1 - \exp[-NTU_1(1+C)]}{1+C}$$

$$0.37 = \frac{1 - \exp[-NTU_1(1+0.3)]}{1+0.3}$$

$$\exp[-1.3\, NTU_1] = 1 - 1.3 \times 0.37 = 0.519$$

or $$NTU_1 = \frac{\ln(0.519)}{-1.3} = 0.50$$

Further, $$NTU_1 = \frac{UA_1}{C_{min}} = \frac{U(\pi d L_1)}{C_{min}}$$

or $$\frac{U\pi d}{C_{min}} = \frac{NTU_1}{L_1} = \frac{0.50}{1} = 0.5$$

The quantity $\frac{U\pi d}{C_{min}}$ is a constant quantity for heat exchanger.

Case II. After increasing the length of heat exchanger,

(*a*) Flow rates are same,

(*b*) Properties are unchanged.

(*c*) Dimension other than length are also unchanged.

Therefore,

$$C = 0.3, \quad C_{min} = C_h \text{ and } C_{max} = C_c$$

The effectiveness of heat exchanger in new case

$$\varepsilon_2 = \frac{C_h(T_{h,i} - T_{h,o})}{C_{min}(T_{h,i} - T_{c,i})} = \frac{150-75}{150-15} = \frac{75}{135} = 0.55$$

and $$0.55 = \frac{1 - \exp[-NTU_2(1+0.3)]}{1+0.3}$$

or $$NTU_2 = 0.965$$

Further,

$$NTU_2 = \frac{UA_2}{C_{min}} = \frac{U(\pi d L_2)}{C_{min}} = 0.5\, L_2$$

or $$L_2 = \frac{NTU_2}{0.5} = \frac{0.965}{0.5} = \mathbf{1.93\ m.}$$ **Ans.**

The outlet temperature of water :

$$\varepsilon_2 = \frac{C_c(T_{c,o} - T_{c,i})}{C_{min}(T_{h,i} - T_{c,i})}$$

or $$T_{c,o} = T_{c,i} + \varepsilon_2 \frac{C_{min}}{C_c}(T_{h,i} - T_{c,i})$$

$$= 15 + 0.55 \times 0.3 \times (150 - 15)$$

$$= \mathbf{37.27°C.}$$ **Ans.**

Example 14.33. *Two identical counter flow type heat exchangers are available. Water at the rate of 3600 kg/h and at 30°C ($C_p = 4.2$ kJ/kg.K) is to be heated by cooling of an oil ($C_p = 2.1$ kJ/kg.K) at 90°C. The oil flow rate is 2700 kg/h. The heat transfer area in each heat exchanger is 4 m².*

The heat exchangers are connected in series on the water side and in parallel on the oil side. The oil flow rate is split in the ratio of 2 to 1 i.e., 1800 kg/h in the first and 900 kg/h in the second heat exchanger. Water enters the first heat exchanger at 30°C. Find the final water and oil exit temperatures. Overall heat transfer coefficient in each heat exchanger is 300 W/m².K.

Solution

Given : Two counter flow heat exchangers

$C_{p,h}$ = 2100 J/kg.K, $C_{p,c}$ = 4200 J/kg.K

Heat exchanger 1	Heat exchanger 2
A = 4 m²,	A = 4 m²
U = 300 W/m².K,	U = 300 W/m².K.
$\dot{m}_{c_1}$ = 3600 kg/h	$\dot{m}_{c_2}$ = 3600 kg/h
$\dot{m}_{h_1}$ = 1800 kg/h	$\dot{m}_{h_2}$ = 900 kg/h
T_{c,i_1} = 30°C	$T_{c,i_2} = T_{c,o_1}$
T_{h,i_1} = 90°C	T_{h,i_2} = 90°C

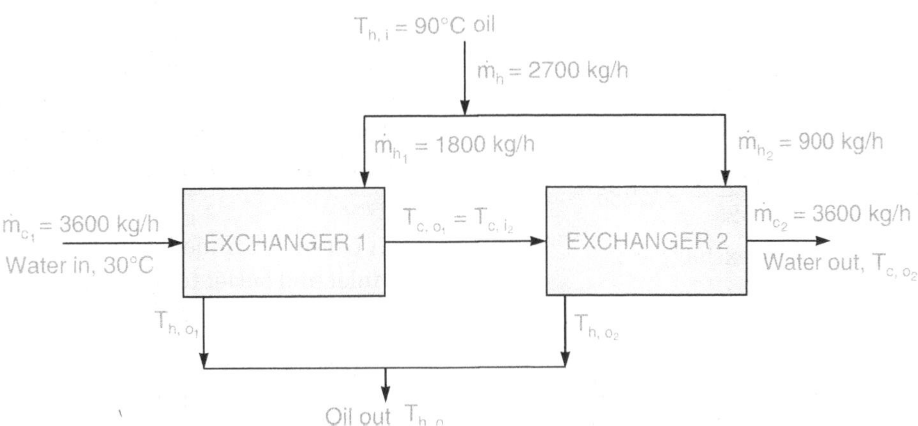

Fig. 14.52 Schematic for example 14.33

To find : The final exit temperature of oil and water.

Assumptions:

1. No heat loss to surroundings.
2. Steady state operation
3. Cold fluid leaving exchanger 1, enters exchanger 2 at same temperature.
4. Constant properties.
5. Scale free heat transfer surfaces.

Analysis : The exit temperatures are unknown, hence using NTU method for each heat exchanger.

Heat Exchanger 1 :

Heat capacity for cold fluid

$$(\dot{m}C_p)_{c_1} = (3600/3600)(\text{kg/s}) \times (4200 \text{ J/kg.K}) = 4200 \text{ W/K}$$

Heat capacity for hot fluid,

$$(\dot{m}C_p)_{h_1} = (1800/3600)(\text{kg/s}) \times (2100 \text{ J/kg.K}) = 1050 \text{ W/K}$$

Heat capacity ratio

$$C = \frac{C_{min}}{C_{max}} = \frac{1050}{4200} = 0.25$$

$$NTU = \frac{UA}{C_{min}} = \frac{300 \times 4}{1050} = 1.143$$

The effectiveness of heat exchanger 1

$$\varepsilon_1 = \frac{1 - \exp[-NTU(1-C)]}{1 - C\exp[-NTU(1-C)]}$$

$$= \frac{1 - \exp[-1.143 \times (1-0.25)]}{1-(0.25) \times \exp[-1.143 \times (1-0.25)]}$$

$$= 0.6434$$

Outlet temperatures :

For hot fluid; $(\dot{m}C_p)_{min}$

$$\varepsilon_1 = \frac{T_{h,i} - T_{h,o_1}}{T_{h,i} - T_{c,i_1}} = \frac{90 - T_{h,o_1}}{90 - 30}$$

or $$T_{h,o_1} = 90 - 0.6434 \times (90 - 30) = 51.36°C$$

For cold fluid;

$(\dot{m}C_p)_{max}$ = 4200 W/K

$$\varepsilon_1 = \frac{(\dot{m}C_p)_{max}(T_{c,o_1} - T_{c,i_1})}{(\dot{m}C_p)_{min}(T_{h,i} - T_{c,i_1})}$$

$$= \frac{4200 \times (T_{c,o_1} - 30)}{1050 \times (90 - 30)}$$

or $T_{c,o_1} = 30 + 0.6434 \times 0.25 \times (90 - 30)$

$= 39.65°C.$

Heat Exchanger 2 :

Heat capacity for cold fluid

$$(\dot{m}\, C_p)_{c_2} = 4200 \text{ W/K}$$

Heat capacity for hot fluid,

$$(\dot{m}\, C_p)_{h_2} = (900/3600)(\text{kg/s}) \times (2100 \text{ J/kg.K})$$

$$= 525 \text{ W/K}$$

Heat capacity ratio

$$C = \frac{C_{min}}{C_{max}} = \frac{525}{4200} = 0.125$$

$$NTU = \frac{UA}{C_{min}} = \frac{300 \times 4}{525} = 2.285$$

The effectiveness of heat exchanger 2

$$\varepsilon_2 = \frac{1 - \exp[-NTU(1-C)]}{1 - C\exp[-NTU(1-C)]}$$

$$= \frac{1 - \exp[-2.285 \times (1-0.125)]}{1 - 0.125 \times \exp[-2.285 \times (1-0.125)]}$$

$$= 0.8795$$

Outlet temperatures :

For hot fluid

$$(\dot{m}\, C_p)_{min} = 525 \text{ W/K}$$

$$\varepsilon_2 = \frac{T_{h,i} - T_{h,o_2}}{T_{h,i} - T_{c,i_2}} = \frac{90 - T_{h,o_2}}{90 - 39.65}$$

or $T_{h,o_2} = 90 - 0.8795 \times (90 - 39.65)$

$= 45.72°C$

For cold fluid

$$(\dot{m}C_p)_{max} = 4200 \text{ W/K}$$

$$\varepsilon_2 = \frac{(\dot{m}C_p)_{max}(T_{c,o_2} - T_{c,i_2})}{(\dot{m}C_p)_{min}(T_{h,i} - T_{c,i_2})}$$

$$= \frac{4200 \times (T_{c,o_2} - 39.65)}{525 \times (90 - 39.65)}$$

or $T_{c,o_2} = 39.65 + 0.8795 \times 0.125 \times (90 - 39.65)$

$= 45.18°C$

Final water outlet temperature = **45.18°C. Ans.**

Final outlet temperature of oil can be calculated as :

$$(\dot{m}C_p)_{h_1} T_{h,o_1} + (\dot{m}C_p)_{h_2} T_{h,o_2} = (\dot{m}_{h_1} + \dot{m}_{h_2})\, C_{p,h} T_o$$

or
$$T_o = \frac{1800 \times 51.36 + 900 \times 45.72}{2700}$$

$$= 49.48°C$$

Final outlet temperature of oil is **49.48°C. Ans.**

Example 14.34. *Hot gases enters a finned tube, cross flow heat exchanger at 300°C and leave at 100°C, are used to heat the water at a flow rate of 1 kg/s from 35°C to 125°C. The exhaust gas specific heat is approximately 1000 J/kg.K and overall heat transfer coefficient based on the gas side surface is 100 W/m²$.K. Calculate the required gas side area using NTU method.*

Solution

Given : A finned tube cross flow heat exchanger with inlet and outlet temperatures of hot and cold fluids.

$$U = 100 \text{ W/m}^2\text{.K}$$

Hot gas side :

$$T_{h,i} = 300°C$$
$$T_{h,o} = 100°C$$
$$C_{p,h} = 1000 \text{ J/kg.K}$$

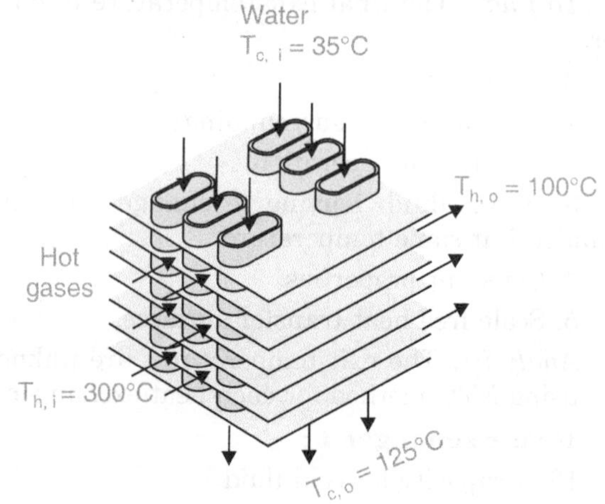

(a) Finned tube cross flow heat exchanger (cold fluid unmixed)

Cold water side :

$$\dot{m}_c = 1 \text{ kg/s}$$
$$T_{c,i} = 35°C$$
$$T_{c,o} = 125°C.$$

To find : Surface area of heat exchanger on gas side.

Assumptions :

(*i*) Finned tube cross flow heat exchanger with cold fluid unmixed.

(*ii*) Scale free surfaces.

(*iii*) Negligible heat loss to surroundings and negligible kinetic and potential energy changes.

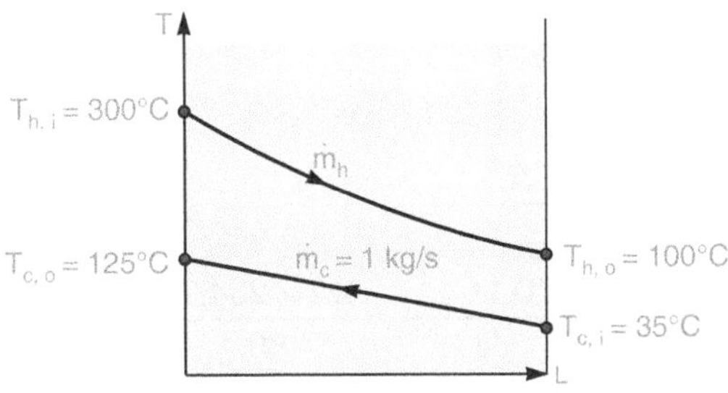

(b) Temperature distribution

Fig. 14.53. Schematic and temperature distribution for finned tube cross flow heat exchanger

Properties : The specific heat of water at its film temperature,

$$T_f = \frac{35 + 125}{2} = 80°C$$

From Table A-7,

$$C_{p,\,c} = 4197 \text{ J/kg.K}$$

The mass flow rate of hot gases can be obtained by energy balance

$$\dot{m}_h C_{p,h}(T_{h,i} - T_{h,o}) = \dot{m}_c C_{p,c}(T_{c,o} - T_{c,i})$$

or

$$\dot{m}_h = \frac{1 \times 4197 \times (125 - 35)}{1000 \times (300 - 100)}$$

$$= 1.88865 \text{ kg/s}$$

Heat transfer rate in the exchanger

$$Q = \dot{m}_c C_{p,c}(T_{c,o} - T_{c,i})$$

$$= 1 \times 4197 \times (125 - 35)$$

$$= 377{,}730 \text{ W}$$

The heat capacity rates of two fluids

$$C_c = \dot{m}_c C_{p,\,c} = 1 \times 4197$$
$$= 4197 \text{ W/K}$$

$$C_h = \dot{m}_h C_{p,\,h} = 1.88865 \times 1000$$
$$= 1888.65 \text{ W/K}$$

Therefore,

$$C_{min} = C_{mixed} = 1888.65 \text{ W/K}$$
$$C_{max} = C_{unmixed} = 4197 \text{ W/K}$$

Heat capacity ratio,

$$C = \frac{C_{min}}{C_{max}} = \frac{1888.65}{4197} = 0.45$$

Maximum possible heat transfer rate

$$Q_{max} = C_{min} \times (T_{h,\,i} - T_{c,\,i})$$
$$= 1888.65 \times (300 - 35)$$
$$= 500{,}492.25 \text{ W}$$

The effectiveness of heat exchanger

$$\varepsilon = \frac{Q}{Q_{max}} = \frac{377{,}730}{500{,}492.25} = 0.755$$

For $\varepsilon = 0.755$ and $\frac{C_{mixed}}{C_{unmixed}} = 0.45$, from Fig. 14.32

$$NTU = 2.8$$

and

$$NTU = \frac{UA}{C_{min}} \longrightarrow A = \frac{NTU\, C_{min}}{U}$$

$$= \frac{2.8 \times 1888.65}{100} = \mathbf{52.88\ m^2}.$$ **Ans.**

Example 14.35. *Air at 25°C flows over a cross flow car radiator and cools water from 99°C to 60°C. Water flows at the rate of 4 kg/min through a number of separate passages within the heat exchanger. The mass flow rate of air is 14 kg/min. If overall heat transfer coefficient is 80 W/m².K. Determine the required heat transfer area and heat exchanger effectiveness. Assume :*

(i) Clean and scale free heat transfer surfaces.

(ii) Constant specific heat of fluids over given temperatures ranges.

$$C_{p,\,water} = 4186 \ J/kg.K$$

and

$$C_{p,\,air} = 1006 \ J/kg.K$$

(iii) Overall heat transfer coefficient is constant.

(iv) Air is mixed fluid.

Given that for cross flow heat exchanger

$$\varepsilon = (1/C)\ [1 - exp\ \{- C\ (1 - exp\ (-NTU)\}]$$

$$R = \frac{C_{min}}{C_{max}}.$$ (P.U.P., May 1993)

Solution

Given : A cross flow heat exchanger

$T_{c,\,i} = 25°C$, $U = 80 \text{ W/m}^2\text{.K}$

$T_{h,\,o} = 60°C$, $T_{h,\,i} = 99°C$

$\dot{m}_h = 4$ kg/min, $\dot{m}_c = 14$ kg/min

$C_{p,h}$ = 4186 J/kg.K, $C_{p,c}$ = 1006 J/kg. K

and relation for effectiveness.

To find :

(*i*) Heat transfer area.

(*ii*) Heat exchanger effectiveness.

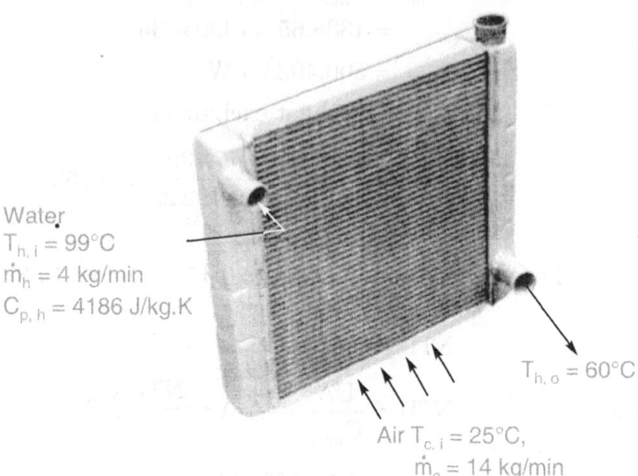

Fig. 14.54. Schematic of car radiator for example 14.35

Analysis : The exit temperature of air, by energy balance on heat exchanger;

$$\dot{m}_h C_{p,h} (T_{h,i} - T_{h,o}) = \dot{m}_c C_{p,c}(T_{c,o} - T_{c,i})$$

or $4 \times 4186 \times (99 - 60) = 14 \times 1006 \times (T_{c,o} - 25)$

or $T_{c,o} = 71.36°C$

Heat capacities :

For air $C_c = \frac{14}{60} \times 1006$

$= 234.7$ W/K

For water $C_h = \frac{4}{60} \times 4186$

$= 279$ W/K

Hence, $C_{min} = C_c$

and $C_{max} = C_h$

and capacity ratio, $C = \frac{C_{min}}{C_{max}} = \frac{234.7}{279} = 0.84$

Effectiveness $\varepsilon = \frac{C_c (T_{c,o} - T_{c,i})}{C_{min} (T_{h,i} - T_{c,i})}$

$= \frac{71.36 - 25}{99 - 25} = \mathbf{0.626.}$ **Ans.**

Now using given relation ;

$$\varepsilon = \frac{1}{C} [1 - \exp\{- C(1 - \exp(- NTU))\}]$$

$$0.626 = \frac{1}{0.84} \times [1 - \exp\{-0.84 \times (1 - \exp(-NTU))\}]$$

$0.474 = \exp[- 0.84 \times \{1 - \exp(- NTU)\}]$

or $0.888 = 1 - \exp(- NTU)$

or $NTU = 2.2$

Further,

$$NTU = \frac{UA}{C_{min}}$$

or $A = \frac{NTU\, C_{min}}{U} = \frac{2.2 \times 234.7}{80}$

$= \mathbf{6.45\ m^2}.$ **Ans.**

14.11 COMPACT HEAT EXCHANGERS

The compact heat exchangers are a special class of exchangers and they are the replacement to the conventional shell and tube heat exchangers. A compact heat exchanger accommodates a large heat transfer area within a small volume. A heat exchanger can be referred as a *compact heat exchanger* when its surface area density exceeds 700 m^2/m^3. These heat exchangers are typically used as gas to liquid and gas to gas heat exchange.

The gas has low convection heat transfer coefficient and thus for specified heat transfer rate, the large heat transfer area is required. The compact heat exchangers are light in weight and small in size. These are available in different tubular and plate configurations. *Kays* and *London* have studied these types of heat exchanger very extensively. Fig. 14.55 shows typical heat transfer matrices for compact heat exchanger.

Fig. 14.55 (*a*) and (*b*) show finned tube arrangement. Here the tubes are illustrated as rectangular, and circular. Fig. 14.55 (*c*) illustrates finned circular tube surface in which circular fins are attached to the outside of the circular tubes. Such arrangements are frequently used in gas to liquid heat exchangers, where optimum design demands maximum surface area on gas side.

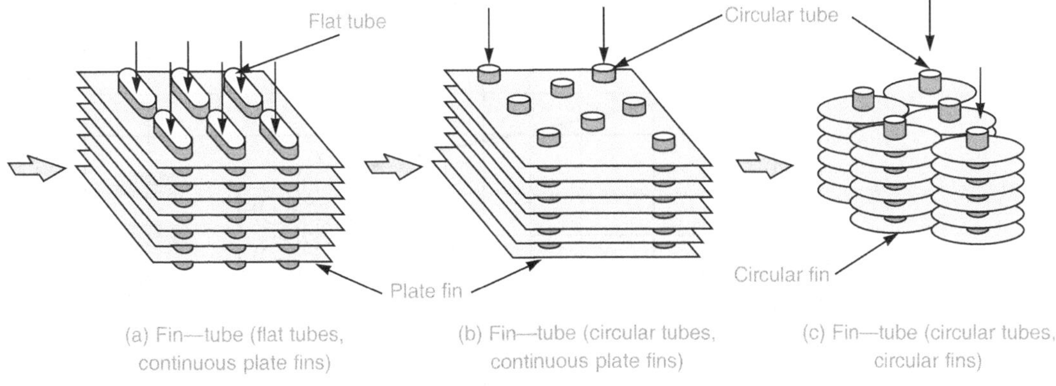

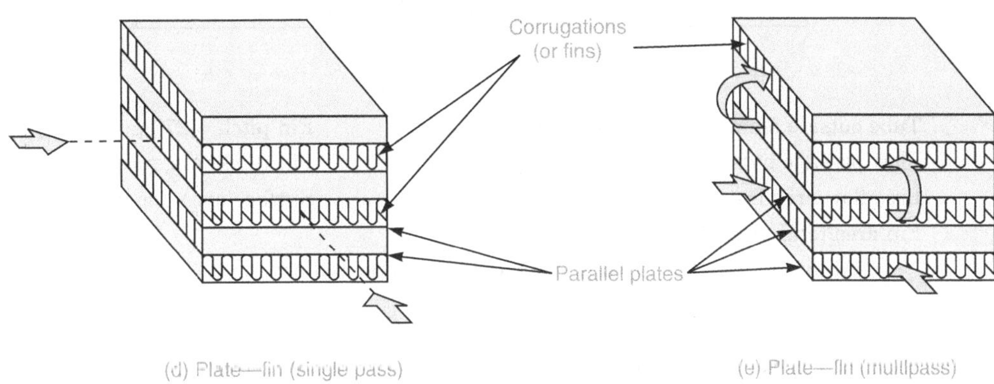

Fig. 14.55. Compact heat exchanger cores

In compact gas to gas heat exchanger, a large area density is required on both fluids side and flat fins arrangement of Fig. 14.55 (*d*) and (*e*) can accomplish this objective.

Heat transfer and flow characteristics have been determined experimentally for specific configurations and are typically presented in the format of Fig. 14.56 and Fig. 14.57. Heat transfer results are correlated in terms of Stanton, Prandtl and Reynolds numbers.

$$\text{St} = \frac{h}{\text{GC}_p},\ \text{Pr} = \frac{\mu\,\text{C}_p}{k_f}$$

and $$\text{Re} = \frac{\text{GD}_h}{\mu}$$

where G is the **mass velocity** defined as :

$$\text{G} = \frac{\dot{m}}{\text{A}_{\text{min}}} \qquad \text{...(14.56)}$$

where $\dot{m}$ = total mass flow rate of fluid, kg/s, and

A_{min} = minimum free flow cross-sectional area of finned passage, m^2

The hydraulic diameter, D_h is defined as

$$\text{D}_h = 4\,\frac{\text{LA}_{\text{min}}}{\text{A}} \qquad \text{...(14.57)}$$

where A is total heat transfer area and quantity LA_{min} becomes the minimum free flow passage volume.

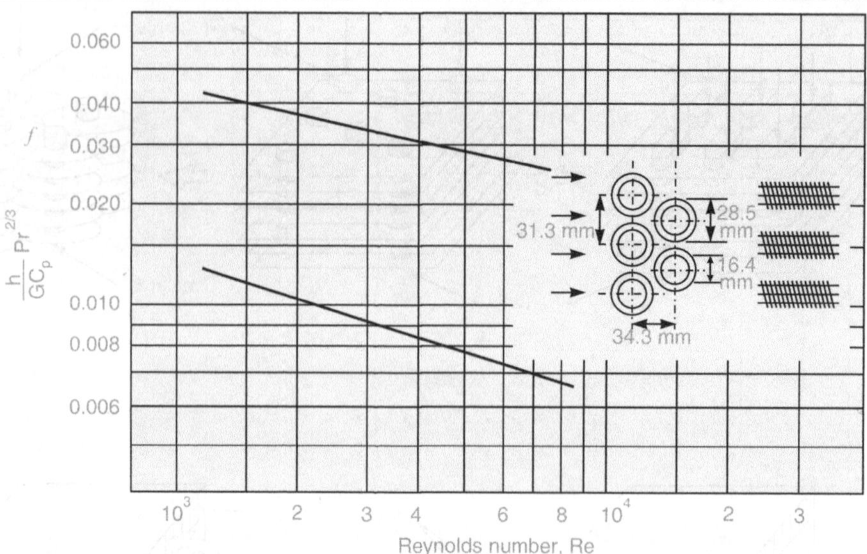

Tube outside diameter, D_o = 16.4 mm

Flow passage hydraulic diameter, D_h = 6.68 mm

Free-flow area/frontal area, σ = 0.449

Fin area/total area, A_{fin}/A_{total} = 0.830

Fin pitch = 275 per metre

Fin thickness, t = 0.254 mm

Surface area density, α = 269 m^2/m^3

Note. Minimum free-flow area is in spaces transverse to flow.

Fig. 14.56. Heat transfer and friction factor for a circular tube-circular fin heat

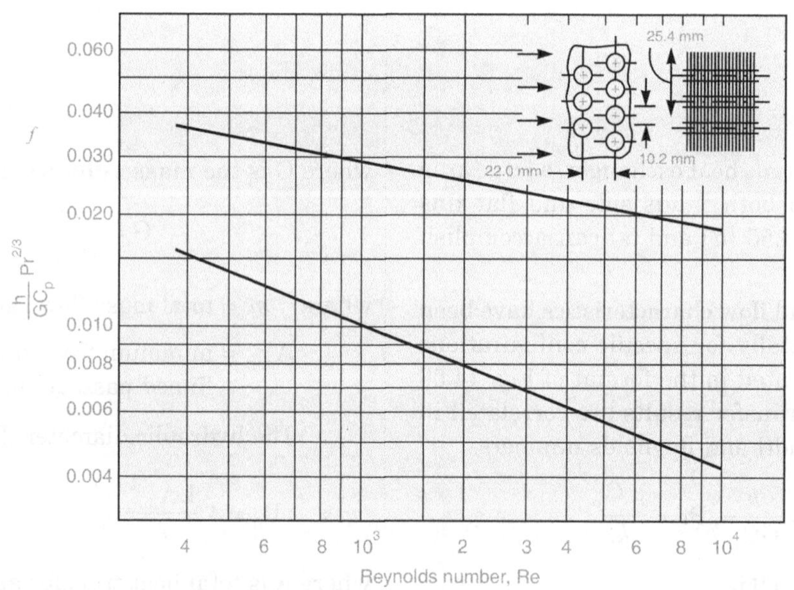

Tube outside diameter, D_o = 10.2 mm

Flow passage hydraulic diameter, D_h = 3.63 mm

Free-flow area/frontal area, σ = 0.534

Fin area/total area = 0.913

Fin pitch = 315 per metre

Fin thickness = 0.330 mm

Surface area density, α = 587 m^2/m^3

Note. Minimum free-flow area in spaces transverse to flow.

Fig. 14.57. Heat transfer and friction factor for a circular tube-continuous fin heat exchanger surface

The values of σ, D_h, α (ratio to total heat transfer area to total heat exchanger volume), with the A_{fin}/A_{total} (ratio of finned surface to total heat transfer area and other geometrical parameters are listed for each configuration with the charts.

Once the heat transfer and friction factor charts for a specified configuration are available and the Reynolds number Re, the heat transfer coefficient h and friction factor f can be evaluated. Then the rating and sizing problem associated with the heat exchanger matrix can be performed by using either LMTD or effectiveness method analysis.

The pressure drop associated with the flow through the heat exchanger may be obtained from

$$\Delta p = \frac{G^2}{2\rho_i}\left[\underbrace{(1+\sigma^2)\left(\frac{\rho_i}{\rho_o}-1\right)}_{\text{Flow acceleration}} + \underbrace{f\left(\frac{A_{total}}{A_{min}}\right)\left(\frac{\rho_i}{\rho_m}\right)}_{\text{Core friction}}\right] \quad ...(14.58)$$

where $$\sigma = \frac{\text{Minimum free flow area}}{\text{Frontal area}} = \frac{A_{min}}{A_{fr}} \quad ...(14.59)$$

$$\frac{A_{total}}{A_{min}} = \frac{\text{Total heat transfer area}}{\text{Minimum free flow area}} = \frac{4L}{D_h}$$

$$G = \text{mass velocity, (kg/m}^2\text{s)} = \frac{\rho u_\infty A_{fr}}{A_{min}} = \frac{\rho u_\infty}{\sigma} \quad ...(14.60)$$

$$\rho_m = \frac{2}{\dfrac{1}{\rho_i}+\dfrac{1}{\rho_o}}$$

ρ_i, ρ_o = fluid densities at inlet and outlet, respectively.

Example 14.36. *Consider a finned tube compact heat exchanger having the core configuration of Fig. 14.58. The core is fabricated from aluminium (k = 237 W/m.K) tubes have an inside diameter of 13.8 mm. In a waste heat recovery applications, the water flows through the tubes provides an inside convection coefficient of 1500 W/m^2.K, while the combustion gases at 1 atm and 700 K are in cross flow over the tubes. If the gas flow rate is 1.25 kg/s and the frontal area is 0.2 m^2, what is the overall heat transfer coefficient based on gas side. Take fin efficiency as 0.89.*

Solution

Given : Compact heat exchanger geometry, gas side flow rate and temperature and inside convection coefficient.

d_i = 13.8 mm, k = 237 W/m.K,

Schematic :

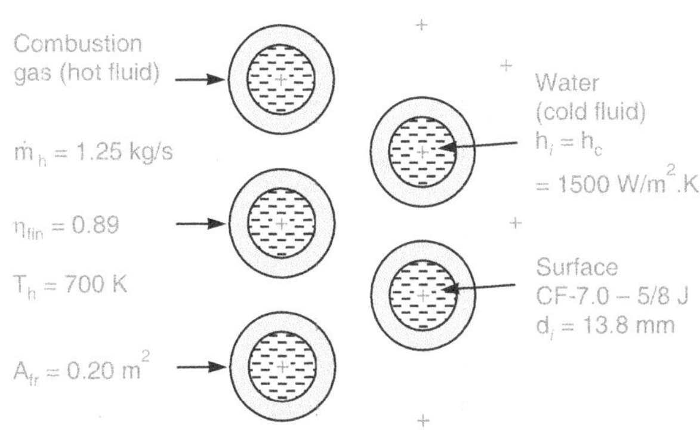

Fig. 14.58

To find : Gas side overall heat transfer coefficient.

Assumptions :

(*i*) Gas has properties of atmospheric air

(*ii*) Scale free surfaces.

Properties : Air (at 1 atm and 700 K), from Table A-4,

C_p = 1075 J/kg.K,

$\mu = 338.8 \times 10^{-7}$ kg/ms

Pr = 0.695.

Analysis : The overall heat transfer coefficient is given by

$$U_o = \frac{1}{\dfrac{A_o}{A_i h_i} + \dfrac{A_o}{2\pi k L}\ln\left(\dfrac{d_o}{d_i}\right) + \dfrac{1}{\eta_{total} h_o}} \quad ...(i)$$

where A_o (= A_{total}) is gas side and A_i is water side surface area, respectively.

If the fin thickness is assumed negligible, then

$$\frac{A_i}{A_o} = \frac{d_i}{d_o}\left(1 - \frac{A_{fin}}{A_{total}}\right)$$

A_{fin} = finned surface area

and $$\frac{A_{fin}}{A_{total}} = 0.830 \text{ from Fig. 14.56.}$$

Area ratio $$\frac{A_i}{A_o} = \frac{13.8}{16.4} \times (1 - 0.830) = 0.143$$

Thus $$\frac{A_o}{\pi L} = \frac{d_i}{0.143} = \frac{0.0138}{0.143}$$

The gas side convection coefficient :

$$G = \frac{\dot{m}}{\sigma A_{fr}} = \frac{1.25}{0.449 \times 0.2} = 13.9 \text{ kg/m}^2\text{s.}$$

$$\text{Re} = \frac{GD_h}{\mu} = \frac{13.9 \times 6.68 \times 10^{-3}}{338.8 \times 10^{-7}}$$

$$= 2740$$

From Fig. 14.56,

$$\frac{h}{GC_p} \text{Pr}^{2/3} = 0.0096$$

Hence $$h = \frac{0.0096 \times 13.9 \times 1075}{(0.695)^{2/3}}$$

$$= 183 \text{ W/m}^2\text{.K.}$$

Hot side temperature effectiveness or area weighted fin efficiency

$$\eta_{total} = 1 - \frac{A_{fin}}{A_{total}} (1 - \eta_{fin})$$

$$= 1 - 0.830 \times (1 - 0.89) = 0.91$$

Using values in eqn. (*i*),

$$U_o = \frac{1}{\frac{1}{1500 \times 0.143} + \frac{0.0138}{0.143} \times \frac{1}{2 \times 237} \times \ln\left(\frac{16.4}{13.8}\right) + \frac{1}{0.91 \times 183}}$$

$$= \mathbf{93.4 \text{ W/m}^2\text{.K.}}$$ **Ans.**

14.12. PLATE HEAT EXCHANGER (PHE)

The plate heat exchanger (PHE) is best alternative to shell and tube exchanger. It is widely used in food and drink processing industries. A plate heat exchanger consists of a large number of thin, cold pressed, corrugated metal plates or sheets fitted into a frame as shown in Fig. 14.59. These are provided with gaskets and corner portals. The gasket prevents the mixing of liquids and directs the fluid into their respective channel through the portals at the corner. Each plate has four corner holes. When plates corner holes are packed, these corner holes form four continuous flow lines for the two fluids : two of these corner hales serve as the inlet and outlet for the hot fluid and other two for the cold fluid. The Fig. 14.60 shows typical flow paths for two fluids. A plate separates the hot and cold fluid streams which are in counter flow on the two sides of the plate.

The corrugated design helps in inducing turbulence in the fluids, increasing the effective surface area and imparting additional strength and rigidity to the plate. Therefore, a plate heat exchanger offers greater compactness and accessibility of the heat transfer surfaces compared to shell and tube heat exchanger and ease of adding and removing plates for different applications of heat transfer.

The plates are made from thin sheets varying from 0.5 mm to 1 mm in thickness. The stainless steel, carbon steel, and some special alloys like Hestalloy, incoloy, aluminium, brass, titanium, etc. are used as plate material. The common gasket materials are lastomeric material like nitrile rubber, EPDM, viton, resin cured butyl rubber, silicon rubber etc.

A plate heat exchanger can withstand a maximum pressure of 20 atm and gasket material limits the operating temperature upto 190°C. A typical expression for calculation of Nusselt number is

$$\text{Nu} = 0.2536\,(\text{Re})^{0.65}(\text{Pr})^{0.4} \quad ...(14.61)$$

Fig. 14.59. Plate heat exchanger built up of corrugated plates assembled to optimize heat transfer

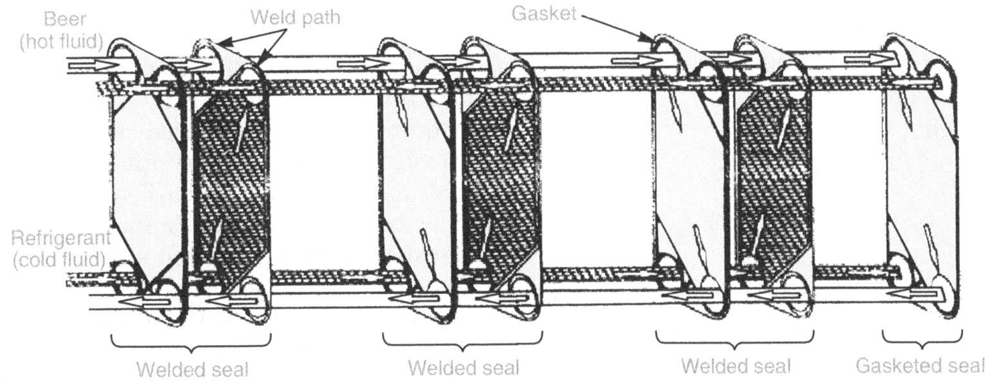

(a) Counter current flow in a PHE with welded plate pairs

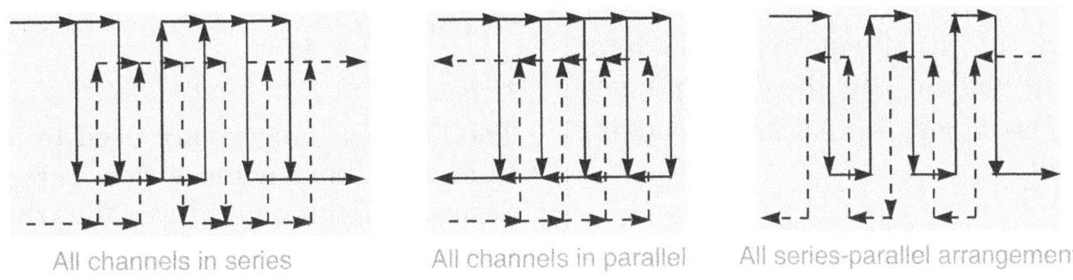

(b) Typical series-parallel flow arrangements

Fig. 14.60. Flow configuration in a plate heat exchanger

Applications of plate heat exchangers

The plate heat exchangers are widely used for

1. Liquid to liquid heat exchange,
2. Evaporation, boiling and condensing services,
3. As a practical use, these heat exchangers are extensively used in dairy and food processing industries.

Advantages of plate heat exchangers

The plate heat exchangers are preferred over shell and tube heat exchanger because of its greater compactness and accessibility. The major advantages of plate heat exchangers are :

1. It offers very high heat transfer coefficient on both sides of the plate. For clean water to water service, an overall heat transfer coefficient of about 7000 W/m^2.K is obtained as compared to 2300 W/m^2.K in shell and tube heat exchanger.

2. Plate heat exchangers are also suitable for close approach temperature as low as 2°C and for large temperature also.

3. It offers ease of inspection, service, maintenance and cleaning.

4. In plate heat exchanger, the heat transfer area can be increased or decreased by adding and removing some plates.

5. It requires much less floor space as compared to shell and tube heat exchanger.

6. These are cheaper than shell and tube heat exchangers.

7. It can perform multiple heat exchange duties in a simple heat exchanger.

14.13. REQUIREMENTS OF GOOD HEAT EXCHANGER

1. It should satisfy the purpose of
   (*a*) Maintaining specified temperature difference,
   (*b*) Cooling or heating duty of the fluid.
   (*c*) Transferring specified quantity of heat from hot fluid to cold one.
2. A heat exchanger can easily withstand plant environment for
   (*a*) Different temperatures and atmospheric conditions,
   (*b*) Vibrations produced as a result of fluids flow,
   (*c*) High fluid pressure,
   (*d*) Personal safety
   (*e*) Maintenance, servicing, cleaning and repairing.
3. Higher reliability *i.e.,* working without causing any problems or repairing.

4. It should be low cost.
5. A heat exchanger should be compact for
   (*a*) its dimensional area.
   (*b*) less weight
   (*c*) convenient to transport.

14.14. HEAT EXCHANGER DESIGN AND SELECTION

The heat exchangers are complicated devices and results obtained by simplified procedure presented in preceding sections 14.9 and 14.10 can work well in smaller size heat exchangers. Larger shell and tube heat exchangers consist of two practical difficulties in relation of overall heat transfer coefficient.

(*i*) Assumption of uniform overall heat transfer coefficient throughout the heat exchanger.

(*ii*) Prediction of heat transfer coefficient by using convection correlations.

Variation of overall heat transfer coefficient

Use of constant values of overall heat transfer coefficient U is more serious. The overall heat transfer coefficient U is never constant throughout the heat exchanger, it varies with change in flow configuration as well as with variation in fluid temperature. Fig. 14.61 shows a typical variation of overall heat transfer coefficients within a heat exchanger. In this case, the heat transfer mechanism is completely changed on liquid side, when finally liquid starts boiling.

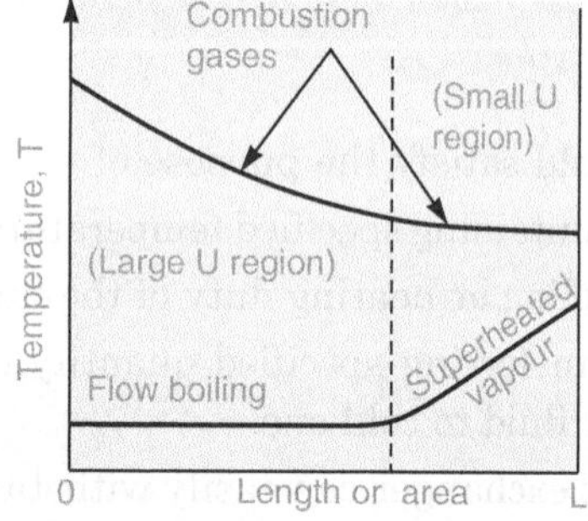

Fig. 14.61. A typical case of a heat exchanger in which U varies significantly

If overall heat transfer coefficients were uniform in each portion of heat exchanger, then we could treat two different heat exchangers in series.

However, the variation of overall heat transfer coefficient U with position within heat exchanger is common difficulty and this problem is more severe in large industrial shell and tube heat exchangers and less serious in compact heat exchangers with less surface area.

Prediction of heat transfer coefficients

The determination of heat transfer coefficient for fluid flow through the smooth tube of uniform cross section is quite simple. But prediction of convection heat transfer coefficients is very difficult for complicated flows that move through a baffled shell or twisted tape inserted tube. The determination of convection heat transfer coefficient in such situations remains a problem that cannot be solved analytically. It is normally estimated with the help of empirical relations or with the aid of computer programs. The prediction of convection heat transfer coefficient during the phase change, when the flow is boiling or condensing is even more complicated. It is a still chalange to the researchers.

Heat transfer rate

The heat exchangers are used to heat or cool a certain fluid with known mass flow rate and temperature to desired outlet temperature. Thus the rate of heat transfer in a heat exchanger can be determined by relation

$$Q = \dot{m}\, C_p(T_{in} - T_{out})$$

A heat exchanger must be capable of transferring heat at the specified rate in order to achieve desired temperature difference of the fluid under consideration at the specified mass flow rate.

Pumping power

The principle and operating cost of heat exchanger plays very important role in the selection of heat exchanger. In a heat exchanger, both fluids are usually forced to flow by pumps or fans which consume electrical power. The cost of electrical power

= pumping power, kW × hours of operation, *h* × price of electricity unit

where the pumping power input to motors and pumps

$$= \frac{\dot{m}\,\Delta p}{\rho}\ (\text{W}) \qquad \text{...(14.62)}$$

where $\dot{m}$ = the mass flow rate of fluid stream, kg/s

Δp = pressure drop of fluid stream as it passes through the heat exchanger, N/m².

ρ = fluid density, kg/m³.

The pressure drop in a straight run of pipe, is given by

$$\Delta p = f\left(\frac{L}{D_h}\right) \times \frac{\rho\, u_m^2}{2} \qquad \text{...(14.63)}$$

where L = length of the pipe

D_h = hydraulic diameter

u_m = mean velocity of fluid flow in tube

f = friction factor.

Minimizing the pressure drop and mass flow rate of fluids will minimise the operating cost of heat exchanger, but it will maximise the size of heat exchanger and thus initial cost. It is thumb rule by doubling the mass flow rate will reduce the initial cost by half, but it will increase pumping power by eight times.

Often, the heat exchangers are augmented by use of fins or roughning element in an exchanger. Such augmentation will invariable increase the pressure drop, but at the same time, it will reduce the fixed cost of an exchanger by increasing U and reducing the required area.

Size and weight

The smaller size and lighter weight heat exchanger are always preferred. In case of automotive and aerospace applications, size and weight of heat exchanger are very important. Larger heat exchanger size carries a higher price and space problem in some cases limits the tube length.

Type

Type of heat exchanger to be selected depends on types of fluids used, size and weight limitations and presence of any phase change process. For an example to cool a liquid by a gas, a heat exchanger is suitable, that has surface area on gas side many times that on the liquid side. On the other hand, a plate or shell and tube type heat exchanger is very suitable for heat exchanger between two liquids.

Material

Material of construction of the heat exchanger may play an important role in selection of heat exchangers. Thermal and structural effects need to be considered, if a heat exchanger operates above 70 atm or 550°C. A temperature difference of 50°C may cause uneven thermal expansion in the tube material. In case of chemical fluid, expensive corrosive resistance material such as stainless steel or titanium may be used in heat exchanger.

Other considerations

Other considerations involve ease of servicing, low maintenance cost, safety and reliability of heat exchanger. If a heat exchanger uses toxic or expensive fluid, then it must be leak tight.

A summary of design considerations for a large shell and tube exchanger is given below :

1. Decide the fluids on shell and tube side in order to minimise the pumping cost. For an example, for oil cooling by water, the oil (a viscous fluid) must flow on shell side and water on tube side. Corrosion behaviour, fouling, and cleaning problems of fouled tube are also some factors to be considered.

2. Compare the cost of operation and investment.

(*a*) Investment in the heat exchangers,

(*b*) Operating cost.

3. Make the rough estimation of size of heat exchanger for overall heat transfer coefficient value given in Table 14.1. It will help to size the flow rates and to anticipate the temperature variations.

4. Calculate the heat transfer rate, pressure drop and the cost of various exchanger configurations that reasonable suit the application.

Once the exchanger configuration is set, U will be approximately fixed and area becomes the design variable. If inlet and outlet temperatures are set, then use

$$Q = UAF\Delta T_{lm}$$

where $Q = C_h(T_{h,i} - T_{h,o}) = C_c(T_{c,o} - T_{c,i})$

Then A can be calculated and design completes.

14.15. PRACTICAL APPLICATIONS OF HEAT EXCHANGERS

(*i*) The *finned tube heat exchangers* are used in air preheating for evaporation of feed liquid in an ammonia nitrate concentration plant.

(*ii*) *Double pipe heat exchangers* are used for water cooling of distillation in an alcohol plant.

(*iii*) *Cooled oils* are used for water cooling of reaction mass in a reactor, oil cooling etc.

(*iv*) Plate heat exchangers are used for

(*a*) Water cooling of products (hydrogen, chlorine, caustic soda) in a caustic soda plant,

(*b*) Steam heating of an edible oil,

(*c*) Milk chilling and pasteurizing in a dairy plant,

(*d*) Process service water heat exchange in power plant,

(*e*) Low pressure steam preheating of process liquid in evaporating unit.

14.16. HEAT PIPES

A *heat pipe* is a novel device without any moving part that can transfer a large quantity of heat over a large distance at a constant temperature without any power input. A heat pipe is basically a sealed circular tube containing an annular layer of wicking material on the inside surface of the tube. The core of the system is hollow in the centre to permit working fluid to pass freely from one end to other as shown in Fig. 14.62. The wick transports the working medium in liquid form from one end to other end by capillary action.

A heat pipe is composed of three sections :

(*i*) An evaporator section at one end, where fluid absorbs heat of evaporation and turns into vapour ;

(*ii*) A condenser section at the other end where vapour is condensed and rejects its latent heat ;

(*iii*) The adiabatic section between evaporator and condenser sections, where liquid and vapour phases flow in opposite directions through the wick and core, respectively to complete the cycle without any heat transfer between fluid and its surroundings.

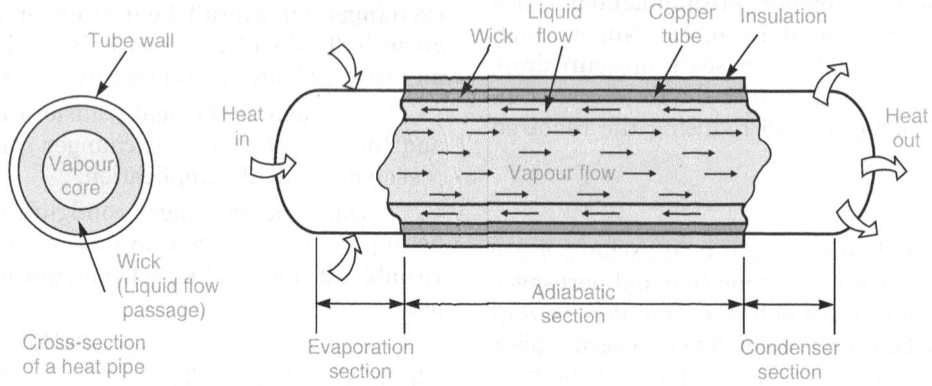

Fig. 14.62. Schematic and operation of a heat pipe

A heat pipe operates in the reversible manner and its operation is completely sailient. A heat pipe is very sensitive to the effect of gravity. If it is tilted so that the condenser is located at higher level than the evaporator, the action of gravity serves to speed up the flow of liquid down through the wick and vapour rises up because of its lower density. This is called a *faborable tilt.* In contrast, when the condenser is located at lower elevation than the evaporator, the action of gravity will obstruct the flow of liquid in the wick and heat pipe is said to have *adverse tilt.*

A variety of fluid and pipe materials have been used for heat pipe construction and some operating characteristics are summarised in Table 14.4. A number of wick materials have been used in practice, ranging from mesh screen, stainless steel fibers and cintered fibers and powders. The wick of a heat pipe provides the means for return of the liquid to evaporator. Therefore, the structure of the wick plays very important role on the performance of the heat pipe and thus the design and construction of the wick is most critical aspect of the manufacturing process.

TABLE 14.4. Some typical operating characteristics of heat pipes

Temperature range, K	*Working fluid*	*Pipe material*	*Measured axial heat flux, W/cm²*	*Measured surface heat flux, W/cm²*
230–400	Methanol	Copper, nickel, stainless steel	0.45 at 373 K	75.5 at 373 K
280–500	Water	Copper, nickel	0.67 at 473 K	146 at 443 K
360–850	Mercury	stainless steel	25.1 at 533 K	181 at 533 K
673–1073	Potassium	Nickel, stainless steel	5.6 at 1023 K	181 at 1023 K
773–1173	Sodium	Nickel, stainless steel	9.3 at 1123 K	224 at 1033 K

Adapted from reference [18]

Working of a heat pipe

Initially, the wick of the heat pipe is soaked with saturated liquid and core section is filled with saturated vapour. When the evaporator section of the heat pipe is placed in hot environment, the heat is transferred to saturate liquid, that starts vaporising and causing vapour pressure to rise. Thus resulting pressure difference drives the vapour through the core of the heat pipe from evaporator section to condenser section. The condenser section of the heat pipe is located in the cooling medium, thus its surface is slightly at lower temperature. The vapour comes in contact of the cooler surface, condenses, releasing its latent heat, to cooling medium. The condensed liquid then returns to the evaporator end of the heat pipe through the wick as a result of capillary action in the wick, thus completing the cycle. Thus, the heat is absorbed at one end of the heat pipe and is rejected at the other end, with the fluid inside serving as transport medium for heat.

A heat pipe involves evaporation and condensation processes, associated with very high heat transfer coefficient, thus a heat pipe is an extremely effective heat transfer device, with an effective thermal conductivity several hundred times that of copper or silver. A simple heat pipe with water as working fluid has an effective thermal conductivity of 100,000 W/m.K, which is 250 times that of copper.

The heat pipes are extensively used for waste gas heat recovery. In such applications, the condenser and evaporator sections of the heat pipe are attached with fins in order to increase the heat transfer area. Because the heat transfer coefficient for a gas side is always low.

The rating of a heat pipe is measured in terms of "*axial power rating*" (APR), which is equal to the energy transport axially along the pipe. Heat pipes in different APR like electrical heater are available in their standard rating (1 kW, 2 kW and so on).

Applications of heat pipe

The heat pipes are used for removing localised heat and waste heat recovery.

1. Electrical and Electronics system : the use of heat pipe can reduce size of most magnetic components such electrical transformer by 30%. The evaporator end of the heat pipe can be inserted into transformer core to dissipate more heat and to reduce its size significantly.

2. Air conditioning system. Exhaust air leaving the room can precool or preheat the incoming air with the use of a heat pipe.

3. VAPIPE is recent developed heat pipe used in a car engine to vaporise gasolene by exhaust gases before it enters the engine through carburettor. The vaporised fuel makes a homogeneous mixture of air and fuel, and thus combustion is improved.

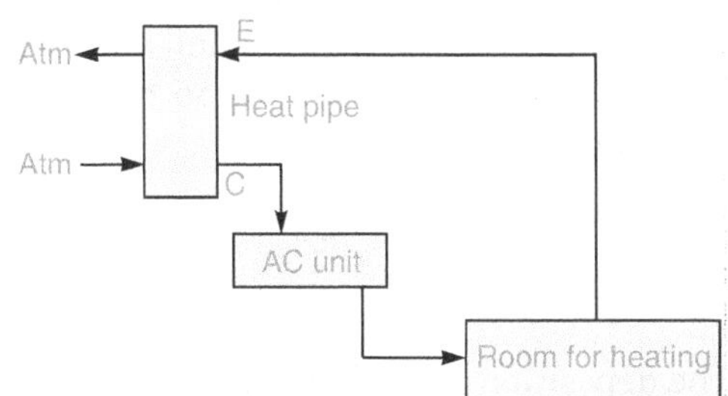

Fig. 14.63. Air-conditioning unit with heat pipe

4. In a gas turbine plant, the heat pipes are used to preheat the compressed air before it enters the combustion chamber, and thus the thermal efficiency of the plant improves.

5. Heat pipes can be used in extrusion of plastic materials to maintain the temperature gradient.

6. Heat pipes can also be used in solar applications, such as space heating, snow melting, kitchen cooking, spray drying, pollution control, pharmaceutical, laundaries, biscuit and bread ovens and so on.

14.17. SUMMARY

The heat exchangers are devices that are used for the heat exchange between the hot and cold fluids which are at different temperatures. These are manufactured in variety of types and simplest one is the double pipe heat exchanger. In a *parallel or concurrent flow* type, both the fluids enter the heat exchanger at the same end, move in same direction and leave the same other end. Whereas, in a *counter flow* type, the hot and cold fluids enter the heat exchanger at opposite ends, flow in opposite direction and leave the exchanger from opposite ends. The *compact* heat exchangers are usually cross flow type, in which two fluids flow perpendicular to each other.

The heat transfer in a heat exchanger usually involves convection in each fluid and conduction through the wall separating the two fluids. In the analysis of heat exchanger, it is convenient to use *overall heat transfer coefficient* U or *total thermal resistance* ΣR_{th} expressed as :

$$\Sigma R_{th} = R_{total} = R_{conv,\, i} + R_{wall} + R_{conv,\, o}$$

or
$$\Sigma R_{th} = \frac{1}{h_i A_i} + \frac{\ln (d_o / d_i)}{2\pi L k} + \frac{1}{h_o A_o}$$

and $$\Sigma R_{th} = \frac{1}{UA} = \frac{1}{U_i A_i} = \frac{1}{U_o A_o}$$

where A is the heat transfer area and d is diameter of tube and subscript i and o stand for *inner* and *outer surfaces* of the tube wall that separates the two fluids. When wall thickness of the tube is small and its thermal conductivity is high, then evaluation of overall heat transfer coefficient U simplifies to

$$U = \frac{1}{\frac{1}{h_i} + \frac{1}{h_o}}$$

The deposition of fluid impurities, rust formation and other reactions between the fluids and surfaces reduce the performance of heat exchanger. The effect of fouling on both surfaces of the heat exchanger can be accounted for by

$$\Sigma R_{th} = \frac{1}{A_i h_i} + \frac{R_{f,i}}{A_i} + \frac{\ln (d_o / d_i)}{2\pi L k} + \frac{R_{f,o}}{A_o} + \frac{1}{A_o h_o}$$

where $R_{f,i}$ and $R_{f,o}$ are the fouling factors at the inside and outside surfaces of the heat exchanger tube. The overall heat transfer coefficients based on outer and inner surfaces are :

$$U_o = \frac{1}{\frac{d_o}{d_i h_i} + \frac{d_o}{d_i} R_{f,i} + \frac{d_o}{2k} \ln (d_o/d_i) + R_{f,o} + \frac{1}{h_o}}$$

and $$U_i = \frac{1}{\frac{1}{h_i} + R_{f,i} + \frac{d_i}{2k} \ln (d_o/d_i) + \frac{d_i}{d_o} R_{f,o} + \frac{d_i}{d_o h_o}}$$

In a well insulated heat exchanger, the rate of heat transfer from the hot fluid is equal to the rate of heat transfer to cold fluid.

$$Q = \dot{m}_h C_{p,h}(T_{h,i} - T_{h,o}) = C_h(T_{h,i} - T_{h,o})$$

and $$Q = \dot{m}_c C_{p,c}(T_{c,o} - T_{c,i}) = C_c(T_{c,o} - T_{c,i})$$

where, subscripts c and h stand for cold and hot fluids, respectively and product of the mass flow rate and specific heat of a fluid $C = \dot{m} C_p$ is called the *heat capacity rate.*

The *log mean temperature difference* (or LMTD) is appropriate temperature difference between the hot and cold fluids and LMTD method of heat exchanger analysis is best suited for determining its size, when all the inlet and outlet temperatures are known. The log mean temperature difference for a double pipe heat exchanger is given by :

$$\Delta T_{lm} = \frac{\Delta T_1 - \Delta T_2}{\ln\left[\frac{\Delta T_1}{\Delta T_2}\right]}$$

where for a parallel flow arrangement :

$$\Delta T_1 = T_{h,i} - T_{c,i} \text{ and } \Delta T_2 = T_{h,o} - T_{c,o}$$

and for counter flow arrangement :

$$\Delta T_1 = T_{h,i} - T_{c,o} \text{ and } \Delta T_2 = T_{h,o} - T_{c,i}$$

For cross flow and multipass shell and tube heat exchangers, the logarithmic mean temperature difference is related to counter flow arrangement as :

$$\Delta T_{lm} = F\, \Delta T_{lm,\text{ counter flow}}$$

where F is the correction factor, which depends on the geometry of the heat exchanger and the inlet and outlet temperatures of the two fluids.

The *effectiveness* ε of a heat exchanger is the ratio of actual heat transfer to maximum possible heat transfer rate by a heat exchanger or

$$\varepsilon = \frac{\text{Actual heat transfer rate}}{\text{Maximum possible heat transfer rate}}$$

where $Q_{max} = C_{min}(T_{h,i} - T_{c,i})$

and C_{min} is the smaller of $C_h = \dot{m}_h C_{p,h}$ and $C_c = \dot{m}_c C_{p,c}$. The effectiveness of a heat exchangers can be determined from

$$\varepsilon = \frac{C_h(T_{h,i} - T_{h,o})}{C_{min}(T_{h,i} - T_{c,i})} = \frac{C_c(T_{c,o} - T_{c,i})}{C_{min}(T_{h,i} - T_{c,i})}$$

For particular type of heat exchanger, the effectiveness can be obtained from effectiveness relations or effectiveness charts.

The *Number of transfer units* or NTU is a dimensionless group. It is a measure of *heat transfer surface area* A in a heat exchanger and is expressed as:

$$NTU = \frac{UA}{C_{min}} = \frac{UA}{(\dot{m}\, C_p)_{min}}$$

$$= \frac{\text{Heat capacity of exchanger, W/K}}{\text{Heat capacity rate of flow, W/K}}$$

The design of heat exchanger depends on several factors such as heat transfer rate, pressure drop, pumping power, size, weight, construction type, material and operating environment.

The compact heat exchanger accommodates a large heat transfer area in a small volume. A heat exchanger having surface area density more than 700 is called the compact heat exchanger. The plate heat exchanger is a compact heat exchanger, consists of a large number thin, cold pressed and corrugated metal sheets fitted into a frame.

A heat pipe is a device without moving parts, that can transfer heat over a distance at constant temperature without any power input. It is a sealed circular tube, consists of an evaporator section, condenser section and adiabatic section between them.

REVIEW QUESTIONS

1. What is a heat exchanger ? Where are they used ?
2. Classify heat exchanger in three broad classes.
3. What is a direct contact type heat exchanger ? Give some examples.
4. Classify recuperative heat exchanger according to construction, flow arrangement.
5. What do you mean by fouling factor ? State the causes of fouling.
6. When a heat exchanger is referred as compact heat exchanger ? What are its applications ?
7. Explain shell and tube type heat exchanger. Why baffles are used ?
8. What are multipass heat exchanger ? When are they used ?
9. Compare parallel flow and counterflow heat exchanger.
10. For a balanced counterflow heat exchanger where $\dot{m}_h C_{ph} = \dot{m}_c C_{pc}$, prove that the temperature profiles of two fluids along the heat exchanger are linear and parallel.
11. Prove that as $\Delta T_1 \longrightarrow \Delta T_2$, $\Delta T_{lm} \longrightarrow \Delta T_{\text{arithmetic mean}}$.
12. Obtain an expression for overall heat transfer coefficient for tubular heat exchanger, subjected to fouling on its two sides of heat transfer surface.
13. Derive an expression for log mean temperature difference of parallel flow heat exchanger. How this expression can be modified for counter flow heat exchanger ?
14. Define effectiveness of heat exchanger. How is maximum heat transfer rate is obtained ?
15. Define NTU of a heat exchanger.
16. What are the limitations of LMTD method ? How is ε-NTU method superior to LMTD method ?
17. Derive for parallel flow heat exchanger.
$$\varepsilon = \frac{1 - \exp[-\text{NTU}(1 + C)]}{1 + C}$$
where $$C = \frac{C_{min}}{C_{max}} = \frac{(\dot{m}\,C_p)_{min}}{(\dot{m}\,C_p)_{max}}$$
18. Prove that for an evaporation process
$$\varepsilon = 1 - \exp(-\text{NTU})$$
19. If $(\dot{m}\,C_p)_{hot} = (\dot{m}\,C_p)_{cold}$, prove that
(*a*) for parallel flow, $\varepsilon = \dfrac{1 - \exp(-2\,\text{NTU})}{2}$
(*b*) for counter flow, $\varepsilon = \dfrac{\text{NTU}}{1 + \text{NTU}}$
20. Explain the operation in heat pipe. Why is a wick needed in a heat pipe ? Write its advantages.
21. Explain construction and operation of plate heat exchanger?
22. State advantages and applications of PHE.
23. What are the requirement of good heat exchanger?
24. What are design and selection criteria of heat exchangers?
25. What are applications of heat pipe?

PROBLEMS

1. In a counter flow double pipe heat exchanger, water is heated from 40°C to 80°C with an oil entering at 105°C and leaving at 70°C. Taking the overall heat transfer coefficient as 300 W/m^2.K and the water flow rate as 0.1 kg/s. Calculate the heat exchanger area. [**Ans.** A = 2.07 m^2]
2. A double pipe heat exchanger is constructed of 0.287 cm thick steel tubing (k = 35 W/m.K) with 2.09 cm inner tube and 2.66 cm outer tube. The inside and outside coefficients of heat transfer are 1135 W/m^2.K and 5677 W/m^2.K, respectively, and the inner fouling factor is 9.98 × 10^{-5} m^2.K/W. Calculate the overall coefficient of heat transfer. [**Ans.** U_i = 893.5 W/m^2.K]
3. Water at the rate of 4080 kg/h is heated from 35°C to 75°C by an oil having a specific heat of 1900 J/kg.K. The exchanger is of a counter flow double pipe design. The oil enters at 110°C and leaves at 75°C. Determine the area of the heat exchanger necessary to handle this load if the overall heat transfer coefficient is 320 W/m^2.K. [**Ans.** A = 15.82 m^2]
4. Hot oil having a specific heat of 2.09 kJ/kg.K flows through a counter flow heat exchanger at the rate of 2286 kg/h with an inlet temperature of 93°C and an outlet temperature of 65°C. Cold oil having a specific heat of 1.67 kJ/kg.K flows in at a rate of 3600 kg/h and leaves at 49°C. What area is required to handle this load, if the overall heat transfer coefficient based on the inside area is 0.7 kW/m^2.K ? [**Ans.** A = 8.5 m^2]
5. In a concentric tube counter flow heat exchanger water flows through the inner tube (d_i = 25 mm) at a rate of 12 kg/min while an oil flows through the outer annulus (D_i = 45 mm) at a rate of 6 kg/min. The inlet temperatures of oil and water are 100°C and 30°C, respectively. What should be the length of the exchanger for an oil outlet temperature of 60°C ? Take U_o = 37.8 W/m^2.K, C_p(oil) = 2130 J/kg.K and C_p (water) = 4178 J/kg.K. [**Ans.** L = 66.5 m]
6. An oil cooler, of the concentric tube type is used for cooling oil at 65.5°C to 54.4°C with water at 26.7°C with a temperature rise of 11.1°C. Assuming overall heat transfer coefficient of 738 W/m^2.K based on the outside area of the tubes. Determine the heat transfer surface area required for a design heat load of 190.5 kW for a single pass (*a*) parallel flow mode and (*b*) counter flow mode. [**Ans.** (*a*) A = 9.85 m^2, (*b*) 9.3 m^2]
7. A pipe (k = 59 W/m.K) with an I.D. of 3.175 cm and wall thickness of 0.318 cm is externally heated by steam at a temperature of 180°C. The water flows through the pipe with a velocity of 1.22 m/s. Calculate the length of the pipe required to heat water from

30°C to 90°C. Assuming the heat transfer coefficient on the steam side to be 11.3 kW/m^2.K.

[Ans. L = 4.82 m]

8. A refrigerator is designed to cool 250 kg/h of hot liquid of specific heat 3350 J/kg.K at 120°C using a parallel flow arrangement. 1000 kg/h of cooling water is available for cooling purposes at a temperature of 10°C. If the overall heat transfer coefficient is 1160 W/m^2.K and the surface area of the heat exchanger is 0.25 m^2, calculate the outlet temperatures of the cooled liquid and water and also the effectiveness of the heat exchanger.

[Ans. ε = 0.646, $T_{c,o}$ = 24.22°C, $T_{h,o}$ = 48.94°C]

9. A double pipe heat exchanger is used to heat water with a mass flow rate of 10 kg/s from 15°C to 33°C. The heating fluid enters at 75°C with a capacity rate of 25 kW/K and the mean overall coefficient of heat transfer is 1570 W/m^2.K. Determine the surface areas for counter flow and parallel flow arrangements.

[Ans. A = 13.5 m^2, 16.1 m^2]

10. In a parallel flow heat exchanger, engine oil (C_p = 2.29 kJ/kg.K), enters a heat exchanger at 150°C and leaves at 80°C. The cooling water enters at 30°C and leaves at 65°C. If the fluid rates and inlet conditions are unchanged, find the exit temperature of each fluid in the counter flow heat exchanger. Also find the lowest temperature to which the oil may be cooled in (*i*) parallel flow and (*ii*) counter flow by increasing the length of heat exchanger.

[Ans. 70°C, 70°C, 70°C, 30°C]

11. Fuel oil passing at the rate of 1.1 kg/s through the annulus of a counter flow double pipe heat exchanger is heated from 10°C to 20°C by using hot water available from the engine at 71°C. The water flows through the copper tube having an outside diameter of 2.13 cm and an inside diameter of 1.86 cm with a velocity of 0.76 m/s. The oil passes through the annulus formed by the inner copper tube and the outer steel pipe having an outside diameter of 3.34 cm and an inside diameter of 3 cm. Assume the fouling factor on water side as 0.0004 m^2.K/W and fouling factor on oil side as 0.0009 m^2.K/W. Properties of water and oil are tabulated as follows :

Property	*Water*	*Oil*
ρ (kg/m^3)	982	854
C_p (kJ/kg. K)	4.187	1.884
k(W/m.K)	0.657	0.138
ν (m^2/s)	4.81 × 10^{-7}	7.48 × 10^{-6}

Neglecting the resistance of the copper tube, determine the length of the tube. [Ans. L = 15 m long]

12. A shell-and-tube exchanger (two shells, four passes) is used to heat 10,000 kg/h of pressurised water from 35 to 120°C with 5000 kg/h water entering the exchanger at 300°C. If the over all heat transfer coefficient is 1500 W/m.K, determine the required heat exchanger area.

[Ans. A = 4.95 m^2]

13. A shell and tube type heat exchanger is designed to heat 2.5 kg/s of water from 15°C to 85°C. The heating is to be acomplished by passing hot engine oil, which is available at 160°C, through the shell side of the heat exchanger. The convection coefficient of oil on outer surface of the tubes is 400 W/m^2.K. Ten tubes pass the water through the shell. Each tube is thin walled, of diameter 25 mm and makes eight passes through the shell. If the oil leaves the exchanger at 100°C, what is its flow rate ? How long must the tubes be to accomplish the desired heating ?

[Ans. $\dot{m}_h$ = 5.19 kg/s, L = 37.9 m]

14. Water at a rate of 45,45,500 kg/h is heated from 80°C to 150°C in a heat exchanger having two shell passes and eight tube passes with a total surface area of 925 m^2. Hot exhaust gases having approximately the same thermophysical properties as air enter at 350°C and exit at 175°C. Determine the overall heat transfer coefficient.

[Ans. U = 28.5 W/m^2.K]

15. In a shell-and-tube type heat exchanger, 3000 kg/h of furnace oil is heated from 10°C to 90°C. The oil is to flow inside the tube, while the steam at 120°C is to flow through the shell. The tubes used are of 1.65 cm inside diameter and 1.90 cm outside diameter. The heat transfer coefficients of the oil and steam sides are 85 W/m^2.K and 7420 W/m^2.K, respectively. Find the number of passes and the number of tubes in each pass, if the length of each tube is limited to 2.85 m due to space limitations. The velocity of oil is limited to 5 cm/s. Given ρ = 900 kg/m^3 and C_p (oil) = 1970/J/kg.K.

[Ans. p = 2, N = 87]

16. Water at the rate of 3.783 kg/s is heated from 37.78°C to 54.44°C in a shell-and-tube heat exchanger. On a shell side one pass is used with water as the heating fluid, 1.893 kg/s of it entering the exchanger at 93.3°C. The overall heat transfer coefficient is 1419 W/m^2.K and the average water velocity in the 1.905 cm diameter tube is 0.366 m/s. Because of space limitations the tube length must not be longer than 2.438 m. Calculate the number of tube passes, the number of tubes per pass, and the length of the tubes.

[Ans. N = 36, p = 2, L = 1.646 m]

17. A shell and tube type heat exchanger consists of 135 thin walled tubes in a double pass arrangement, each of 12.5 mm diameter with total heat transfer area 47.5 m^2. The water (the tube side fluid) enters the heat exchanger at 15°C and 6.5 kg/s and is heated by exhaust gas entering at 200°C and 5 kg/s. The gas may be assumed to have the properties of atmospheric air, the overall heat transfer coefficient is approximated as 200 W/m^2.K. What are gas and water outlet temperatures ? [Ans. $T_{h,o}$ = 37.2°C, $T_{c,o}$ = 45°C]

18. In a chemical processing plant water at the rate of 13,620 kg/h is heated from 37.8°C to 54.4°C in tubes with 6810 kg/h hot water (on the shell side) entering

the exchanger at 93.33°C. The average velocity of water in the tubes is 36.6 cm/s. The tube diameter is 1.9 cm and the overall heat transfer coefficient is 1420 W/m^2.K. If the floor space available cannot accommodate tube longer than 2.44 m, calculate the number of tube passes, the number of tubes per pass, and the length of the tubes.

[Ans. p = 2, N = 36, L = 1.645 m]

19. In a food processing plant, water is to be cooled from 18°C to 6.5°C by using brine solution entering at an inlet temperature of – 1.1°C and leaving at 2.9°C. What area is required when using a shell-and-tube heat exchanger with the water making one shell pass and the brine making two tube passes ? Assume an average overall heat transfer coefficient of 850 W/m^2.K and a design heat load of 6000 W. [Ans. A = 0.67 m^2]

20. A steam condenser being cooled by water has been designed on the assumption of clean water and its overall heat transfer coefficient was assumed to be 4900 W/m^2. K. Neglecting the fouling resistance. Later on, it was discovered that the cooling water is not clean and its fouling factor is 0.002 m^2.K/W. Is there any need of redesigning ? Comment on the results.

[Ans. U = 435.7 W/m^2.K]

21. Calculate the overall heat transfer coefficient and the heat transfer rate possible from a shell and tube counterflow heat exchanger for a situation in which water flow through a copper pipe 2 cm I.D. and 2.3 cm O.D. while oil flows through the shell. Water enters at 25°C and comes out at 35°C while oil enters at 70°C and comes out at 64°C. The water and oil side film coefficients are 4650 W/m^2.K and 1280 W/m^2.K. The thermal conductivity of the tube wall is 350 W/m.K. The fouling factors on the water and oil sides may be taken to be 0.0004 and 0.001 respectively. The length of the tube is 2.5 m.

[Ans. U_o = 401 W/m^2.K, Q = 2677 W]

22. It is desired to condense 40,000 kg/h of dry and saturated steam at a pressure of 0.05 bar (saturation temperature = 32.76°C) in a single pass steam condenser. The cooling water enters the tube at 20°C and leaves at 26°C. The tubes are 2.5 cm I.D. and 2.75 cm O.D. and the thermal conductivity of tube material is 110 W/m^2.K. The film heat transfer coefficient on water and steam side are 7090 W/m^2.K and 5930 W/m^2.K respectively. Neglecting the scale resistance, calculate the surface area of the tubes required and the heat transfer rate. If the length of tube is 3 m, find the number of tubes.

[Ans. A_o = 927.23 m^2, Q = 2693.3 kW]

23. For experimental purpose, a steam surface condenser is to be designed to transfer 300 kW of thermal energy at a condensing temperature of 45°C. The cooling water enters the condenser at 20°C with a flow rate of 25,000 kg/h. Calculate the surface area required to handle this load, if the overall heat transfer coefficient for the condenser surface is 1400 W/m^2.K. If the outside tube diameter is 30 mm and the tube length is 1 m, calculate the number of tubes. Compare the result with the use of arithmetic mean temperature difference.

[Ans. A_o = 11.06 m^2, N = 118, Error = 2.3%]

24. A counter-flow tubular oil cooler is to be designed to cool 1500 kg/h of oil from temperature 90°C to 30°C by means of water entering the cooler at 20°C and leaving the cooler at 50°C. Calculate the amount of water flow rate required and the heat transfer area. Take C_p of oil as 3 kJ/kg.K and overall heat transfer coefficient equal to 1200 W/m^2.K.

[Ans. $\dot{m}_w$ = 0.598 kg/s, A = 2.88 m^2]

25. In a heat exchanger, it is desired to cool 50,000 kg/h of alcohol from 60°C to 35°C using 25,000 kg/h of water entering at 6°C. The overall heat transfer coefficient based on the outer tube is 600 W/m^2.K. The specific heat of alcohol is 3.768 kJ/kg.K. Calculate the surface area required for the following cases :
(*a*) Counter flow shell and tube type,
(*b*) Counter flow with 2 shell passes and 72 tube passes. Alcohol flows through the shell,
(*c*) Cross flow with one tube pass and one shell pass with water unmixed and alcohol mixed.
Comment on the results.

[Ans. (*a*) 86.42 m^2 (*b*) 107 m^2]

26. In an industry, 0.6 kg/s of oil (C_p = 2.5 kJ/kg.K) is to be cooled in a counter flow heat exchanger from 110°C to 30°C by the use of water entering at 20°C. The overall heat transfer coefficient is expected to be 1500 W/m^2.K. Assume that the exit temperature of water is not to exceed 80°C. Using NTU method calculate (*a*) the water flow rate, (*b*) the surface area required and (*c*) the effectiveness of the exchanger.

[Ans. (*a*) $\dot{m}_w$ = 0.4893 kg/s, (*b*) 2.74 m^2, (*c*) 0.882]

27. An industrial gas turbine plants employs a counter-flow concentric tube heat exchanger which cools the lubricating oil coming out from its bearings. The oil mass flow rate is 0.2 kg/s and its specific heat is 2.2 kJ/kg.K. The mass flow rate of the coolant (water) which flows in the opposite direction is 0.16 kg/s and its specific heat is 4.18 kJ/kg.K. The lubricating oil enters the cooler at 110°C and leaves at 40°C. The cooling water enters the cooler at 20°C. The heat transfer coefficient from the oil to the tube surface is 2300 W/m^2.K and from tube surfaces to water is 5700 W/m^2.K. The resistance of the tube wall may be neglected. If the mean diameter of the tube is 15 mm calculate the length of the tube. Neglect the resistance of the scale, if any formed. [Ans. L = 16.52 m]

28. Single pass surface condenser is to handle 5100 kg/h of dry and saturated steam at 50°C. The outer and inner diameters of the tubes are 20 mm and 17 mm, respectively. The length of the tube is 3 m. The cooling water enters the tube at 20°C. The velocity of water through the tube is 2 m/s and the temperature rise of the cooling water is 10°C. Calculate number of tubes. Take the following data :

Heat transfer coefficient of steam side
= 11700 W/m^2.K
Fouling factor of water side
= 0.0002 m^2.K/W
Thermal conductivity of tube material
= 93 W/m.K
Fouling factor of steam side
= 0.0009 m^2.W
[Ans. N = 1031]

29. In a chemical plant, 600 kg/h of sulphuric acid is to be cooled from 80°C to 40°C in a counter flow double pipe heat exchanger with acid flowing through inner tube and water surrounding it. Calculate the area of exchanger for the following data:
Mass flow rate of water = 500 kg/h
Water inlet temperature = 20°C
Specific heat of water = 4.18 kJ/kg.K
Heat transfer coefficient of water side = 580 W/m^2.K
Heat transfer coefficient of sulphuric acid side
= 465 W/m^2.K
Inside diameter of inner tube = 25 mm
Outside diameter of inner tube = 30 mm
Thermal conductivity of tube material = 93 W/m.K
Fouling factor of water side = 0.0002 m^2.K/W
Fouling factor of sulphuric acid side
= 0.0001 m^2.K/W
Neglect all other losses. [Ans. A = 17.684 m^2]

30. A multi stage reciprocating air compressor employs an intercooler between the first and second stage. The air at 6.5 bar and 160°C enters the intercooler and comes out at 30°C. The free air (15°C and 1.103 bar) delivered is 8 m^3/min. The cooling water flows parallel to the air over the tubes which are of 15 mm inside diameter and 18 mm outside diameter. The inlet and outlet water temperatures are 15°C and 25°C respectively. The velocity of air through the tube is 6 m/s.
Find (*a*) The number of tubes and (*b*) Length of each tube (*c*) Also calculate the saving in total tube length if the intercooler is made of counterflow having inlet and outlet temperatures same. Take the following data :
Heat transfer coefficient of water side = 2095 W/m^2.K.
Heat transfer coefficient of air side = 105 W/m^2.K.
C_p of air = 1.005 kJ/kg.K and R = 0.287 kJ/kg.K.
Neglect the resistance due to tube thickness and scale formation.
[Ans. (*a*) N = 25 tubes, (*b*) L = 32.88 m, (*c*) Saving length = 23.88%]

31. A cross-flow heat exchanger with both fluids unmixed is used to heat water (C_p = 4.18 kJ/kg.K) from 40°C to 80°C flowing at the rate of 1.0 kg/s. What is the overall heat transfer coefficient if hot engine oil (C_p = 1.9 kJ/kg.K), flowing at the rate of 2.6 kg/s, enters at 100°C ? The heat transfer area is 20 m^2.
[Ans. 425 W/m^2.K]

32. Air at 2 kg/s and 27°C and a stream of water at 1.5 kg/s and 60°C each enter a heat exchanger. Evaluate the exit temperatures, if A = 12 m^2, U = 185 W/m^2.K, and :
(*a*) The exchanger is parallel flow
(*b*) The exchanger is counter flow
(*c*) The exchanger is cross-flow, one stream mixed
(*d*) The exchanger is cross-flow, neither stream mixed.
[Ans. (*a*) [$T_{c,o}$ = 33.2°C and $T_{h,o}$ = 41.3°C]
(*b*) [$T_{c,o}$ = 33.6°C and $T_{h,o}$ = 40°C]
(*c*) [$T_{c,o}$ = 33.4°C, $T_{h,o}$ = 40.9°C]
(*d*) [$T_{c,o}$ = 33.3°C, $T_{h,o}$ = 41°C]

33. A particular two shell-pass, four tube-pass heat exchanger uses 20 kg/s of river water at 10°C on the shell side to cool 8 kg/s of processed water from 80°C to 25°C on the tube side. At what temperature will the coolant be returned to the river ? If U is 800 W/m^2.K, how large area must the exchanger be have? [Ans. 32°C, 88.21 m^2]

34. A particular cross-flow process heat exchanger operates with the fluid mixed on one side only. When U = 2000 W/m^2.K, $T_{c,i}$ = 25°C, $T_{c,o}$ = 80°C, $T_{h,i}$ = 160°C, and $T_{h,o}$ = 70°C. After 6 months of operation, the plant manager reports that the hot fluid is only being cooled to 90°C and that he is suffering a 30% reduction in total heat transfer. What is the fouling resistance after 6 months of use ? (Assume no reduction of cold-side flow rate by fouling.)
[Ans. 6.5 × 10^4 m^2.K/W]

35. Water at 0.8 kg/s enters the tubes of a two-shell-pass, four-tube-pass heat exchanger at 17°C and leaves at 37°C. It cools 0.5 kg/s of air entering the shell at 250°C with U = 432 W/m^2.K. Determine : (*a*) the exit air temperature ; (*b*) the area of the heat exchanger ; and (*c*) the exit temperature if, after some time, the tubes become fouled with R_f = 0.0005 m^2. K/W.
[Ans. (*a*) 118.6°C, (*b*) 1.03 m^2, (*c*) 140.5°C]

36. 0.72 kg/s of superheated steam enters a crossflow heat exchanger at 240°C and leaves at 120°C. It heats 0.6 kg/s of water entering at 17°C. U = 612 W/m^2.K. By what percentage will the area differ if a both-fluids-unmixed exchanger is used instead of a one-fluid-unmixed exchanger ? [Ans. – 1.8%]

37. The factors dictating the heat transfer coefficients in a certain two shell-pass, four tube-pass heat exchanger are such that U increases at $(\dot{m}_{shell})^{0.6}$. The exchanger cools 2 kg/s of air from 200°C to 40°C using 4.4 kg/s of water at 7°C, and U = 312 W/m^2.K under these circumstances. If we double the air flow, what will its temperature be leaving the exchanger ?
[Ans. $T_{air\ out}$ = 61°C]

38. A flow rate of 1.4 kg/s of water enters the tubes of a two-shell-pass, four-tube-pass heat exchanger at 7°C. A flow rate of 0.6 kg/s of liquid ammonia at 100°C is to be cooled to 30°C on the shell side : U = 573 W/m^2.K. (*a*) How large must the heat exchanger be ? (*b*) How large must it be if, after some months, a fouling factor of 0.0015 will build up in the tubes, and we still want to deliver ammonia at 30°C ? (*c*) If we make it large enough to accommodate fouling, to what temperature will it cool the ammonia when it is new ? (*d*) At what temperature does water leave the new, enlarged exchanger ? [Ans. (*d*) 49.9°C]

39. Both C's in a parallel-flow heat exchanger are equal to 156 W/K, U = 327 W/m^2.K and A = 2 m^2. The hot fluid enters at 140°C and leaves at 90°C. The cold fluid enters at 40°C. If both C's are halved, what will be the exit temperature of the hot fluid ? [Ans. 90°C]

40. A single-pass heat exchanger condenses steam at 1 atm on the shell side and heats water from 10°C to 30°C on the tube side with U = 2500 W/m^2.K. The tubing is thin-walled, 5 cm in diameter, and 2 m in length. (*a*) Your boss asks whether the exchanger should be counter flow or parallel-flow. How do you advise her ? Evaluate ; (*b*) the LMTD ; (*c*) $\dot{m}_c$; (*d*) ε .
[Ans. (*a*) Shell and tube exchangers are usually operated in counter flow, but operation of condenser can be in either mode, (*b*) 79.58°C, (*c*) 2691.52 kg/h, (*d*) 0.222]

41. Hot engine oil is to be cooled in a double pipe counter flow heat exchanger. The copper tube has a diameter of 2 cm with negligible thickness. The inner diameter of outer tube is 3 cm. The water flow through the inner tube at a rate of 0.5 kg/s and oil flows through the annular space at a rate of 0.8 kg/s. Taking the average temperature of water and oil as 47°C and 80°C, respectively. Assume fully developed flow, calculate overall heat transfer coefficient for flow conditions of the heat exchanger. [Ans. 75.1 W/m^2.K]

42. A concentric tube heat exchanger is used to cool the lubricating oil for a large marine diesel engine. The inner tube is constructed from 2 mm wall thickness stainless steel (*k* = 16 W/m.K). The flow rate of cooling water through the inner tube (r_i = 30 mm) is 0.3 kg/s. The flow rate of the oil through outer tube (R_i = 50 mm) is 0.15 kg/s. Assuming fully developed flow and taking fouling resistances on oil side as 0.0004 m^2.K/W and on water side as 0.0001 m^2.K/W. Calculate the overall heat transfer coefficient referred to inside surface. Use properties of oil at 80°C and water at 35°C. [Ans. 21.9 W/m^2.K]

43. Water enters the tubes of a small single-pass heat exchanger at 20°C and leaves at 40°C. On the shell side, 25 kg/min of steam condenses at 60°C. Calculate the overall heat transfer coefficient and the required flow rate of water, if the area of the exchanger is 12 m^2. (The latent heat, h_{fg} is 2358.7 kJ/kg at 60°C). Take specific heat of water as 4174 J/kg.K.
[Ans. 2838.5 W/m^2.K, 11.77 kg/s]

44. Find the surface area required in a counter flow steam superheater in which the steam enters at 180°C in a dry, saturated state and leaves at 250°C with an increase of enthalpy of 159 kJ/kg. The hot combustion gases (C_p = 1.05 kJ/kg.K) enter the superheater at 510°C. The steam flow rate is 1000 kg/h, the hot gas flow rate is 2000 kg/h and the overall heat transfer coefficient is 26 W/m^2.K. [Ans. 6.6 m^2]

45. 5.795 kg/s of oil flows through the shell side of a two shell pass, four tube-pass oil cooler. The oil enters at 101°C and leaves at 38°C. Water flows in the tubes, entering at 32°C and leaving at 49°C. In addition, $c_{p,\,oil}$ = 2282 J/kg.K and U = 416 W/m^2.K. Find how much area the heat exchanger must have ? [Ans. 106.85 m^2]

46. Hot oil with capacity rate of 2500 W/K flows through a double pipe heat exchanger. It enters at 360°C and leaves at 300°C. Cold fluid enters at 30°C and leaves at 200°C. If the overall heat transfer coefficient is 800 W/m^2.K, determine the heat exchanger area required for parallel and counter flow.
(*Anna Univ., April 1999, March 2000*)
[Ans. 0.971 m^2, 0.892 m^2]

47. A heat exchanger is used to cool the lubricating oil. Water flows at the rate of 0.3 kg/s through inner stainless steel tube (*k* = 16 W/m.K) of inner radius of 30 mm and wall thickness of 2 mm. The oil flows at a rate of 0.15 kg/s through outer tube of inner radius of 50 mm. The oil is cooled from 90°C to 50°C using water at 10°C. Calculate :
(*i*) length of tube required for parallel flow,
(*ii*) length of tube required for counter flow,
(*iii*) the area required for simple pass cross flow heat exchanger with both streams unmixed, operating at same temperatures and flow rates,
Take overall heat transfer coefficient based on inner radius as 21.9 W/m^2.K and assume fully developed flow.
Take specific heat of water as 4178 J/kg.K and that of oil as 2131 J/kg.K. (*J.N.T.U., May 2001*)
[Ans. (*i*) 61 m, (*ii*) 57.86 m, (*iii*) 11.13 m^2]

48. A rectangular tube 30 mm × 50 mm carries water at a rate of 2 kg/s. Calculate the length required to heat the water from 30°C to 50°C. If the wall temperature is maintained at 90°C.
Use the following properties of water at 40°C :
ρ = 992.2 kg/m^3, k_f = 0.634 W/m.K,
C_p = 4.174 kJ/kg.K
$\mu = 6.531 \times 10^{-4}$ Ns/m^2. (*P.U., Nov 2001*)
[Ans. 3.76 m]

49. A shell and tube type heat exchanger is designed for heating water from 25°C to 50°C with the help of

steam condensing at atmospheric pressure. The water flows through tubes (2.5 cm I.D., 2.9 cm O.D. and 2 m long) and the steam condenses on the outside. Calculate the number of tubes required if the water flow rate is 500 kg/min and the individual coefficients of heat transfer on the steam and water side are 8000 and 3000 W/m^2.K. Neglect all other resistances.

[Ans. 40 tubes]

50. A shell and tube type heat exchanger is to heat 10,000 kg/h of water from 16°C to 84°C by hot engine oil flowing through a shell. The oil makes a single shell pass, entering at 160°C and leaving at 94°C, with an average heat transfer coefficient of 400 W/m^2.K. The water flows through 11 brass (k = 100 W/m.K) tubes of 22.9 mm inside diameter and 25.4 mm outside diameter, with each tube making four passes through the shell. Assume fully developed flow for the water, determine the required tube length per pass. [Ans. 9.45 m]

51. The velocity of water flowing through a tube of 22 mm diameter is 2 m/s. The steam condenses at 150°C on outside surface of the tube, heats the water from 15°C to 60°C, over entire length of the tube. Neglect the tube and steam side film resistance, calculate the following :

(*i*) heat transfer coefficient, and

(*ii*) the length of the tube.

Take the following properties of water at mean temperature

ρ = 990 kg/m^3,

C_p = 4.2 kJ/kg.K,

k_f = 0.5418 W/m.K,

$\mu = 700 \times 10^{-6}$ kg/ms. (*P.U., Nov. 1994*)

[Ans. 2.43 m]

52. Water at 90°C enters a heat exchanger for heating oil and it leaves at 40°C. The oil enters at 25°C and leaves at 50°C. Find the effectiveness of heat exchanger. [Ans. 0.3846]

53. Consider a very long concentric tube heat exchanger having hot and cold water inlet temperatures of 85°C and 15°C, respectively. The flow rate of the hot water is twice that of the cold water. Assuming equivalent cold and hot water specific heats, determine the hot water outlet temperature for the following modes of operations.

(*a*) Counter flow

(*b*) Parallel flow. [Ans. (*a*) 50°C, (*b*) 61.67°C]

54. A concentric tube heat exchanger uses water which is available at 15°C to cool ethylene glycol from 100°C to 60°C. The water and glycol flow rates are each 0.5 kg/s. What are the maximum possible heat transfer rate and effectiveness of the heat exchanger ? Which is preferred, a parallel flow or a counter flow mode operation ?

Take C_p of glycol = 2637 J/kg.K,

C_p of water = 4180 J/kg.K.

[Ans. 112.07 kW, 0.47, and parallel flow mode]

55. Hot water at 2.5 kg/s and 100°C enters a concentric tube counter flow heat exchanger having a total area of 23 m^2. Cold water at 20°C enters at 5.0 kg/s and the overall heat transfer coefficient is 1000 W/m^2.K. Determine the total heat transfer rate and the outlet temperature of hot and cold fluids.

[Ans. 672 kW, 36°C and 52°C]

56. In a tube in tube type parallel flow heat exchanger, the hot water at 80°C is cooled to 65°C by cold water entering the heat exchanger at 20°C and leaving at 35°C. What would be the exit temperatures if the flow rates of water are doubled ? [Ans. 72.2°C, 28.8°C]

57. A parallel flow heat exchanger has hot and cold water streams running through it. The flow rates are 10 and 25 kg/min and inlet temperatures are 70 and 25°C on hot and cold sides, respectively. Calculate the area of heat exchanger, if the individual heat transfer coefficients on both sides are 700 W/m^2.K. The exit temperature on hot side is required to be 50°C. Calculate also the exit temperatures of hot and cold streams in this heat exchanger, if the hot water flow rate is doubled to 20 kg/min. It is known that the heat transfer coefficient is proportional 0.8th power of the flow rate. (*Anna Univ., Dec. 1999*)

[Ans. (*i*) 1.375 m^2, (*ii*) 56.3°C, 36°C]

58. Hot oil is to be cooled by water in a one shell and 8 tube passes heat exchanger. The tubes are thin walled and made of copper with an internal diameter of 1.4 cm. The length of each tube pass in the exchanger is 5 m, and the overall heat transfer coefficient is 310 W/m^2.K. The water flows through the tubes at the rate of 0.2 kg/s and oil through the shell at a rate of 0.3 kg/s. The water and oil enter at temperatures of 20°C and 150°C, respectively. Calculate the rate of heat transfer in the heat exchanger and the outlet temperatures of water and oil. Take specific heats of water and oil as 4.18 and 2.13 kJ/kg.K, respectively.

[Ans. 43.21 kW, 71.7°C, 82.4°C]

59. Exhaust gas from a furnace is used to preheat the combustion air supplied to the furnace burners. The gas, which has a flow rate 15 kg/s and an inlet temperature of 1100 K, passes through a bundle of tubes, while the air, which has a flow rate of 10 kg/s and an inlet temperature of 300 K, is in cross flow over the tubes. The tubes are unfinned, and the overall heat transfer coefficient is 100 W/m^2.K. Determine the total tube surface area required to achieve an air outlet temperature of 850 K. The exhaust gas and air may each be assumed to have a specific heat of 1075 J/kg.K.

Use the following relation

$$\varepsilon = (1/C)\,(1 - \exp\{-C\,[1 - \exp(-\text{NTU})]\})$$

where C = ratio of two specific heats = C_{min}/C_{max}.

NTU = number of transfer units.

ε = effectiveness of the heat exchanger.

(*P.U., Nov. 1998*) [Ans. 271.29 m^2]

60. A thin walled concentric tube heat exchanger is used to cool engine oil from 160°C to 60°C and water, which is available at 25°C acts as coolant. The oil and water flow rates are each 2 kg/s and the diameter of the inner tube is 0.5 m and corresponding value of overall heat transfer coefficient is 250 W/m^2.K. How long must the heat exchanger be to accomplish the desired cooling ?

Take, C_p of water = 4.187 kJ/kg.K

C_p of engine oil = 2.035 kJ/kg.K.

[Ans. 18.22 m]

61. A counter flow double pipe heat exchanger using superheated steam is used to heat the water at a rate of 3 kg/s. The steam enters the exchanger at 180°C and leaves at 130°C. The inlet and exit temperature of water are 30°C and 80°C, respectively. The overall heat transfer coefficient is 820 W/m^2.K. Calculate the heat transfer area required. What would be the increase in surface area, if fluids flow in parallel ? Take C_p of water as 4.187 kJ/kg.K.

[Ans. 7.66 m^2, 9.85%]

62. Water at 225 kg/h is to be heated from 35°C to 95°C by means of concentric tube heat exchanger. Oil at 225 kg/h and 210°C with a specific heat of 2095 J/kg.K is to be used as hot fluid. If the overall heat transfer coefficient based on the outer diameter of inner tube is 550 W/m^2.K. Determine the length of the heat exchanger, if the outer diameter is 100 mm.

[Ans. 1.11 m]

REFERENCES AND SUGGESTED READING

1. A. P. Fraas, *"Heat Exchanger Design"*, 2/e, John Wiley & Sons, 1989.
2. F. Krieth and M.S. Bohn, *"Principles of Heat Transfer"*, 5th ed, PWS Pub. Company, 1997.
3. J.P. Holman, *"Heat Transfer"*, 7th ed. McGraw Hill, New York, 1990.
4. F.P. Incropera and D.P. DeWitt, *"Introduction to Heat Transfer"*, 2/e, John Wiley & Sons, 1990.
5. M.N. Ozisik, *Heat Transfer—"A Basic Approach"*, McGraw Hill, New York, 1985.
6. Y. Bayazitoglu and M.N. Ozisik, *"Elements of Heat Transfer"*, McGraw Hill, New York, 1988.
7. W.M. Kays and A.L London, *"Compact Heat Exchangers"*, 3/e, McGraw Hill, New York, 1984.
8. W. M. Kays and H.C. Perkins, Chapter 7 in W.M. Rehsenow and J.P. Hartnett, *"Handbook of Heat Transfer"*, McGraw Hill, New York, 1972.
9. A. C. Mueller, Chapter 18, *"Heat Exchangers"* in W. M. Rehsenow and J.P. Hartnett, *"Handbook of Heat Transfer"*, McGraw Hill, New York, 1972.
10. E. U. Schlunder, *"Heat Transfer Design Hand Book"*, Hemisphere, Washington, 1982.
11. *"Standards of Tubular Exchanger Manufacturer Association"* 7th ed., Tubular Exchanger Manufacturer Association, New York, 1988.
12. L.C. Thomas, *"Heat Transfer"*, Prentice-Hall, Englewood Cliffs, NJ, 1982.
13. F. M. White, *"Heat and Mass Transfer"*, Addison Wesley, Reading, MA, 1988.
14. Yunus A Cenzel, *"Introduction to Thermodynamics and Heat Transfer"*, McGraw Hill, New York, 1997.
15. A. M. Hargis, A.T. Bechman and J. Lolacoma, *"The Plate Heat Exchangers"*, Trans. ASME, 1965.
16. P.D. Dunn and D.A. Reay, *"Heat Pipes"*, 3rd. ed., Pergaman Press, New York, 1982.
17. S.W. Chi, *"Heat Pipe Theory and Practice"*, Hemisphere, Washington DC, 1976.
18. C. H. Dutcher and M. R. Burke, *"Heat Pipes : A Cool Way to Cool Circuits, Electronics"*, February 16, 1970.
19. Sarit K. Das, *"Engineering Heat and Mass Transfer"*, Dhanpat Rai Pub. Co., New Delhi 1999.
20. Vinay K. Datta, *"Heat Transfer"*, Prentice hall of India, New Delhi, 2001.
21. B. V. Karlekar, and R.M. Desmond. *"Heat Transfer"*, Prentice hall of India, New Delhi, 1989.

Mass Transfer

15

15.1. Introduction. 15.2. Modes of Mass Transfer. 15.3. Comparison between Heat and Mass Transfer. 15.4. Concentrations, Velocities and Fluxes. 15.5. Fick's Law of Diffusion. 15.6. General Mass Diffusion Equation. 15.7. Boundary Conditions. 15.8. Mass Diffusion without Homogeneous Chemical Reactions—Steady state diffusion through a plane membrane—Water vapour migration—Equimolar counter diffusion—Diffusion through a stagnant gas: Stefan's flow. 15.9. Mass Diffusion with Homogeneous Chemical Reactions. 15.10. Convective Mass Transfer—Mass transfer coefficient dimensionless parameters in convective mass transfer—Analogy between heat and mass transfer—Correlation for convective mass transfer. 15.11. Dimensional Analysis of Convective Mass Transfer. 15.12. Evaporation of Water into Air. 15.13. Summary—*Review Questions—Problems—References and Suggested Reading.*

15.1. INTRODUCTION

We have so far dealt with conduction, convection and radiation modes of heat transfer, in which energy transfer takes place due to temperature difference in the medium(s). Similarly, if there is a concentration difference within two or more species (components) of a mixture, then mass transfer must occur in order to minimize the concentration difference within the system.

Mass transfer is mass in transit as a result of a species concentration difference in a mixture.

It should be noted that the transfer of mass with bulk fluid motion induced by mechanical work is not case of mass transfer; for example motion of air that is induced by fan or motion of fluid forces through a pipe. We do however; use the term mass transfer, when there is relative motion of species in a mixture as a result of diffusion from a region of high concentration to a region of lower concentration.

15.2. MODES OF MASS TRANSFER

The mechanism of mass transfer process can be classified as :

(*i*) Mass transfer by diffusion,

(*ii*) Mass transfer by convection,

(*i*) Mass transfer by diffusion (Molecular or eddy diffusion). The *molecular diffusion* is the transfer of mass on a microscopic level as a result of concentration gradient of one or more constituents (species) in the system. The diffusion of mass from a constituent occurs through a layer of stagnant fluid, and it may be occurred due to concentration gradients, temperature gradient, or pressure gradient.

The various examples of molecular diffusion can be taken from our day to day experiences. Some of these are :

(1) Dissolution of sugar or salt in the hot water and its uniform diffusion throughout, although their molecules are much heavier than water molecules.

(2) Evaporation of water from a pond to increase the humidity of passing air stream.

(3) Evaporation of petrol in engine carburettor.

(4) Evaporation of moisture during drying of clothes or wood.

(5) Humidification of air in air coolers, cooling towers etc.

(6) Spread of fragrance of perfume or flowers in surroundings.

(7) Diffusion of smoke through a tall chimney into atmosphere.

(8) Diffusion of small fraction of O_2 in the air into sea water and meets the oxygen need of marine animals.

When one of the diffusing fluids is in turbulent motion, the **eddy diffusion** takes place. Due to turbulence, the rate of diffusion increases, resulting in an increase in mass transfer. An example of eddy diffusion process is dissipation of smoke from a smoke stack.

(ii) **Mass transfer by convection.** The mass transfer between a surface and a moving fluid or between two immisible moving fluids is referred as *convective mass transfer*. The convective mass transfer depends on

(1) transport properties of fluid, and

(2) dynamic characteristics (laminar or turbulent flow) of flowing fluid.

The evaporation of water by moving air in a desert cooler is an example of convective mass transfer.

A change of phase of a fluid can also cause mass transfer, such mass transfer is a combination of processes of diffusion and convection. Some examples of mass transfer by phase change are :

(1) Hot flue gases leaving the chimney, rise by convection and then diffuse into atmospheric air,

(2) Mass transfer from boiling water into air.

15.3. COMPARISON BETWEEN HEAT AND MASS TRANSFER

The mass can be transferred by random molecular motion in fluids at the rest or it can be transferred from a surface into a moving fluid, aided by dynamic characteristics of flow. These two distinct modes of mass transport, molecular mass diffusion and convective mass transfer are analogous to conduction heat transfer and convection heat transfer respectively. There are following similarities between heat and mass transfer :

1. The heat conduction takes place, when a temperature gradient exists in a medium and it continues until the entire medium attains a uniform temperature. Similarly, the mass transfer takes place, when a concentration gradient of one or more chemical species (constituents) exists and it continues until the concentration of all species becomes uniform over the entire volume.

2. The conduction heat of transfer is a molecular phenomenon and the thermal conductivity is a material property. Similarly, the diffusion mass transfer is a molecular phenomenon and the diffusion coefficient is a property of particular combination of species and medium.

3. Convective heat transfer coefficient is a flow property depending on fluid flow properties and geometry. Similarly, the convective mass transfer coefficient also depends on flow and geometrical parameters and they can be described by similar governing equations.

There are some significant differences between heat and mass transfer. These are :

1. Heat transfer is a one way phenomenon. It always flows from high temperature region to low temperature region. But the mass transfer is a multiway phenomenon. When two locations P and Q are at different concentration within a system of species 1, 2, 3, etc. The different species diffuse simultaneously between P and Q, depending on the concentration gradient. Some species may diffuse from P to Q while some other may diffuse from Q to P and their diffusion rates will also be different.

2. Except at very low concentrations, the concentration change affects the momentum and continuity equations and they cannot be solved independently.

3. In heat transfer, the boundary conditions may involve specified temperature, specified heat flux, convection or radiation boundary, and their treatment is rather simple. On the other hand, in mass transfer, the boundary may involve different phases, interphase such as condensation, evaporation, adsorption, chemical reaction etc. This makes the boundary conditions much more complex.

4. The mass diffusion rate depends on molecular size of diffusing species compared to medium size in heat transfer.

5. The heat conduction in its pure state, does not deal with any medium velocity. But pure mass diffusion deals with bulk velocity of the medium.

15.4. CONCENTRATIONS, VELOCITIES AND FLUXES

Mass concentration or **mass density.** The mass concentration of the component A within a multicomponent mixture is defined as mass of species A per unit volume of the mixture under consideration. It is denoted by ρ_A and is expressed in kg/m^3.

Mass concentration ρ_A

$$= \frac{\text{Mass of component A}}{\text{Volume of mixture}} = \frac{m_A}{V} \quad ...(15.1)$$

Molar concentration. The molar concentration of the component A is defined as the number of moles of species A per unit volume of mixture. It is also called *molar density* and denoted by C_A and expressed in kg.mol/m^3.

The molar concentration,

$$C_A = \frac{\text{No. of moles of component A}}{\text{Volume of mixture}} = \frac{n_A}{V} \quad ...(15.2)$$

Number of moles of component

$$n_A = \frac{\text{Mass of component A}}{\text{Molecular weight of A}} = \frac{m_A}{M_A} \quad ...(15.3)$$

Therefore, molar concentration,

$$C_A = \frac{\text{Mass of component A}}{\text{Volume of mixture} \times M_A} = \frac{\rho_A}{M_A} \quad ...(15.4)$$

where M_A = molecular weight of component A.

Mass fraction. The mass fraction x_A is defined as the ratio of mass concentration of species A to the mass density ρ, of mixture,

$$x_A = \frac{\rho_A}{\rho} \quad ...(15.5)$$

Mole fraction. It is defined as the ratio of number of moles of component A to the total number of moles of mixture. It is denoted by γ_A and expressed as :

$$\gamma_A = \frac{C_A}{C} \quad ...(15.6)$$

Partial pressure. It is defined as the pressure exerted by a single component in a mixture, when it exists alone in the system at the temperature and volume of the mixture. The total pressure of a mixture is the summation of partial pressures of all components in the mixture *i.e.,*

$$p = p_1 + p_2 + p_3 + ... + p_n \quad ...(15.7)$$

For a binary mixture of component A and B, the following summation rules may be applied.

$$\rho_A + \rho_B = \rho \quad ...(15.8)$$

$$C_A + C_B = C \quad ...(15.9)$$

$$x_A + x_B = 1 \quad ...(15.10)$$

$$\gamma_A + \gamma_B = 1 \quad ...(15.11)$$

$$\frac{x_A}{M_A} + \frac{x_B}{M_B} = \frac{1}{M} \quad ...(15.12)$$

where ρ, C, M are the quantities pertaining to the mixture.

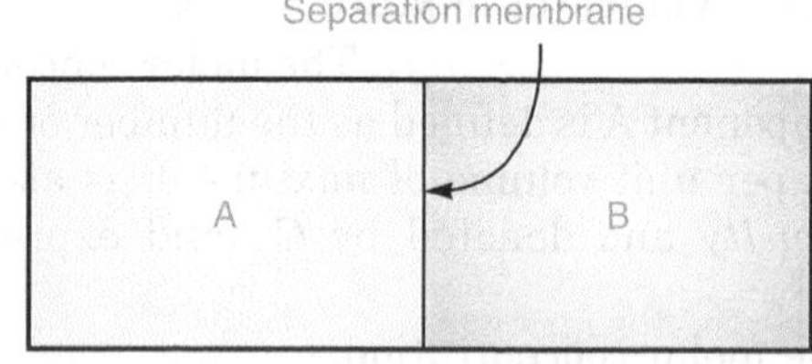

Fig. 15.1. Two components in a mixture

For a mixture of two perfect gases A, and B ;

$$p_A V = n_A R_u T$$

where R_u is the universal gas constant = MR = 8.314 kJ/kg–mole.K

and

$$C_A = \frac{n_A}{V} = \frac{p_A}{R_u T}$$

$$\gamma_A = \frac{C_A}{C} = \frac{p_A / R_u T}{p / R_u T} = \frac{p_A}{p} \quad ...(15.13)$$

$$C = C_A + C_B = \frac{p_A}{R_u T} + \frac{p_B}{R_u T} = \frac{p}{R_u T} \quad ...(15.14)$$

Velocities. In a multicomponent system, the various species normally move at different velocities. The **mass average velocity** u for a binary mixture of components A and B is defined in terms of mixture's mass flux ρu as :

$$u = \frac{\sum_{i=1}^{n} \rho_i u_i}{\sum_{i=1}^{n} \rho_i} = \frac{\rho_A u_A + \rho_B u_B}{\rho} \quad ...(15.15)$$

or

$$u = \frac{\rho_A}{\rho} u_A + \frac{\rho_B}{\rho} u_B = x_A u_A + x_B u_B \quad ...(15.16)$$

The **molar average velocity** for a binary mixture is defined in terms of molar concentrations of components A and B as :

$$v = \frac{\sum_{i=1}^{n} C_i u_i}{\sum_{i=1}^{n} C_i} = \frac{C_A u_A + C_B u_B}{C} \quad ...(15.17)$$

or

$$= \frac{C_A}{C} u_A + \frac{C_B}{C} u_B = \gamma_A u_A + \gamma_B u_B \quad ...(15.18)$$

The velocity of a particular component relative to mass average or molar average velocity is called *diffusion velocity*. Thus we can define mass diffusion velocity and molar diffusion velocity as

$u_A - u$ = mass diffusion velocity of component A,

$u_B - u$ = mass diffusion velocity of component B,

$u_A - v$ = molar diffusion velocity of component A,

$u_B - v$ = molar diffusion velocity of component B,

Fluxes. The **mass flux** of the species is defined as mass of a species that passes through a unit area per unit time relative to a stationary coordinate. For species A of a multicomponent mixture

Absolute mass flux $\mathbf{n}_A = \rho_A u_A$

Bulk motion flux $= \rho_A u$

$$\text{Diffusional mass flux } j_A = \frac{\dot{m}_A}{A}$$

where the quantity $j_A = \frac{\dot{m}_A}{A}$ is mass flow rate per unit area per unit time. The mass flux of any constituent seen by a stationary observer

= diffusional flux + bulk motion flux

$$\rho_A u_A = \frac{\dot{m}_A}{A} + \rho_A u$$

$$\text{or Diffusional flux, } j_A = \frac{\dot{m}_A}{A} = \rho_A (u_A - u) \quad \text{...(15.19)}$$

The **molar flux** can also be defined in similar way as number of moles of a species passing through a unit area per unit time with respect to stationary coordinate, as :

$$\text{Absolute molar flux, } N_A = C_A u_A$$
$$\text{Bulk motion flux} = C_A v$$
$$\text{Diffusional mole flux, } J_A = C_A (u_A - v) \quad \text{...(15.20)}$$

A summary of the definitions of the basic quantities used in mass transfer within a mixture is given in Table 15.1.

TABLE 15.1. Basic quantities in mass transfer

	Mass quantities	*Molar quantities*	*Interrelationship*
Concentrations :			
Definition	$\rho = \frac{m}{V}$	$C = \frac{n}{V}$	$m = nM$
Species	$\rho_i = \frac{m_i}{V}$	$C_i = \frac{n_i}{V}$	$\rho_i = C_i M_i$
Fraction	$x_i = \frac{\rho_i}{\rho}$	$\gamma_i = \frac{C_i}{C}$	$\frac{x_i}{\gamma_i} = \frac{M_i}{M}$
Mixture	$\Sigma x_i = 1$ $\rho = \Sigma \rho_i$ $\rho = \frac{p}{RT} = \sum \frac{p_i}{RT}$	$\Sigma \gamma_i = 1$ $C = \Sigma C_i$ $C = \frac{p}{R_u T} = \sum \frac{p_i}{R_u T}$	$M = \Sigma \gamma_i M_i$ $R = \frac{R_u}{M}$
Velocities :			
Species	u_i	u_i	
Average	$u = \Sigma x_i u_i$	$v = \Sigma \gamma_i u_i$	
Diffusional	$u_i - u$	$u_i - v$	
Fluxes :			
Absolute	$\mathbf{n}_i = \rho_i u_i$	$N_i = C_i u_i$	$n_i = N_i M_i = \rho_i u + j$
Bulk motion	$\rho_i u$	$C_i u$	
Diffusional	$J_i = \rho_i (u_i - u)$	$J_i = C_i (u_i - v)$	$N_i = C_i u + J_i$

Example 15.1. *A mixture of O_2 and CO_2 is kept in a container at 21°C and at 151.9 kPa. The mean velocity of O_2 is 0.08 m/s and that of CO_2 is –0.02 m/s. If the molar fraction of O_2 is 0.4, calculate :*

(i) The molar fraction of CO_2,
(ii) The molecular weight of mixture,
(iii) Molar concentration of CO_2 and O_2,
(iv) Mass concentration of CO_2 and O_2,
(v) Mass fraction of CO_2 and O_2,
(vi) Mass diffusion velocities of both components,
(vii) Molar diffusion velocities of both components,
(viii) Absolute molar fluxes of CO_2 and O_2,
(ix) Absolute mass fluxes of CO_2 and O_2.

Solution

Given : A mixture of CO_2 and O_2

$T = 21°C = 294$ K, $p = 151.9$ kPa

$u_{O_2} = 0.08$ m/s, $u_{CO_2} = -0.02$ m/s, $\gamma_{O_2} = 0.4$.

To find :

(*i*) The molar fraction of CO_2,
(*ii*) The molecular weight of mixture,
(*iii*) Molar concentration of CO_2 and O_2,

(*iv*) Mass concentration of CO_2 and O_2,

(*v*) Mass fraction of CO_2 and O_2,

(*vi*) Mass diffusion velocities of both components,

(*vii*) Molar diffusion velocities of both components,

(*viii*) Absolute molar fluxes of CO_2 and O_2,

(*ix*) Absolute mass fluxes of CO_2 and O_2.

Analysis: (*i*) Mole fraction of CO_2

We have $\gamma_A + \gamma_B = 1$

or $\gamma_B = 1 - \gamma_A = 1 - \gamma_{O_2} = 1 - 0.4 = \mathbf{0.6}$. **Ans.**

(*ii*) Molecular weight of mixture

$$M = \gamma_{O_2} M_{O_2} + \gamma_{CO_2} M_{CO_2}$$
$$= 0.4 \times 32 + 0.6 \times 44 = \mathbf{39.2}. \text{ Ans.}$$

(*iii*) Molar concentrations :

We have $\gamma_i = \dfrac{p_i}{p}$

Thus $p_{O_2} = 0.4 \times p = 0.4 \times 151.9 = 60.76$ kPa

and $p_{CO_2} = 0.6 \times 151.9 = 91.14$ kPa

Concentration of O_2

$$C_{O2} = \frac{p_{O_2}}{R_u T} = \frac{60.76}{8.314 \times 294}$$
$$= \mathbf{0.0248 \ kg\text{-}mol/m^3}. \text{ Ans.}$$

Concentration of CO_2

$$C_{CO_2} = \frac{p_{CO_2}}{R_u T} = \frac{91.14}{8.314 \times 294}$$
$$= \mathbf{0.0372 \ kg\text{-}mol/m^3}. \text{ Ans.}$$
$$C = C_{O_2} + C_{CO_2} = 0.0248 + 0.0372$$
$$= 0.062 \text{ kg-mol/m}^3.$$

(*iv*) Mass concentrations

$$\rho_{O_2} = C_{O_2} M_{O_2} = 0.0248 \times 32$$
$$= 0.793 \text{ kg/m}^3$$
$$\rho_{CO_2} = C_{CO_2} M_{CO_2} = 0.0372 \times 44$$
$$= 1.636 \text{ kg/m}^3$$

Thus $\rho = \rho_{O_2} + \rho_{CO_2} = 0.793 + 1.636$
$$= 2.429 \text{ kg/m}^3$$

(*v*) Mass fractions :

$$x_{O_2} = \frac{\rho_{O_2}}{\rho} = \frac{0.793}{2.429} = \mathbf{0.326}. \text{ Ans.}$$
$$x_{CO2} = \frac{\rho_{CO_2}}{\rho} = \frac{1.636}{2.429} = \mathbf{0.673}. \text{ Ans.}$$

(*vi*) Mass diffusion velocities :

$$u = \Sigma x_i u_i$$
$$= 0.326 \times 0.08 + 0.673 \times (-0.02)$$
$$= 0.0126 \text{ m/s}$$

The mass diffusion velocity of O_2

$$u_{O_2} - u = 0.08 - 0.0126 = \mathbf{0.0674 \ m/s}. \text{ Ans.}$$

The mass diffusion velocity of CO_2

$$u_{CO_2} - u = -0.02 - 0.0126 = \mathbf{-0.328 \ m/s}. \text{ Ans.}$$

(*vii*) Molar diffusion velocities :

$$v = \Sigma \gamma_i u_i = 0.4 \times 0.08 + 0.6 \times (-0.02)$$
$$= 0.02 \text{ m/s}$$

Molar diffusion velocity of O_2

$$= u_{O_2} - v = 0.08 - 0.02$$
$$= \mathbf{0.06 \ m/s}. \text{ Ans.}$$

Molar diffusion velocity of CO_2

$$= -0.02 - 0.02 = \mathbf{-0.04 \ m/s}. \text{ Ans.}$$

(*viii*) Absolute molar fluxes

$$N_{O_2} = C_{O_2} u_{O_2} = 0.0248 \times 0.08$$
$$= \mathbf{1.98 \times 10^{-3} \ kg\text{-}mol/m^2s}. \text{ Ans.}$$
$$N_{CO_2} = C_{CO_2} u_{CO2} = 0.0372 \times (-0.02)$$
$$= \mathbf{-7.44 \times 10^{-4} \ kg\text{-}mol/m^2s}. \text{ Ans.}$$

(*ix*) Absolute mass fluxes :

$$\mathbf{n}_{O_2} = \rho_{O_2} u_{O_2} = 0.793 \times 0.08$$
$$= 0.0634 \text{ kg/m}^3\text{s}$$
$$n_{CO_2} = \rho_{CO_2} u_{CO_2} = 1.636 \times (-0.02)$$
$$= \mathbf{-0.0327 \ kg/m^3s}. \text{ Ans.}$$

15.5. FICK'S LAW OF DIFFUSION

Fick's law relates the mass flux to concentration gradient of a species in a mixture. Consider a system with two components A and B as shown in Fig. 15.2. A thin partition which separates the components is removed at an instance, the two components diffuse through each other until they reach at an equilibrium state and concentration of components becomes uniform throughout.

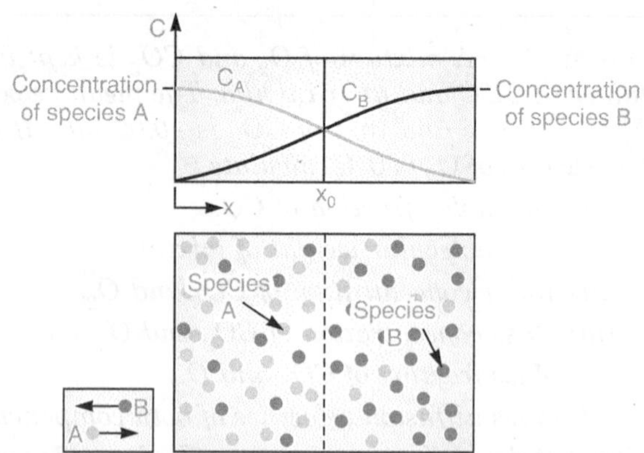

Fig. 15.2. Mass transfer by diffusion in a binary gas mixture.

The diffusion rate is governed by Fick's law. It states that in a gas mixture the diffusion mass flux of a constituent per unit area is directly proportional to the concentration gradient.

Thus $$j_A = \frac{\dot{m}_A}{A} = -D_{AB}\frac{d\rho_A}{dx} \quad ...(15.21)$$

where $\dot{m}_A$ = mass flow rate of component A, kg/s

A = area through which mass $\dot{m}_A$ is flowing, m^2

$j_A = \frac{\dot{m}_A}{A}$ = mass flux of component A, kg/sm^2

ρ_A = mass concentration of component A per unit volume, kg/m^3

D_{AB} = constant of proportionality, called the *diffusion coefficient* or mass diffusivity of component A in component B, m^2/s

$\frac{d\rho_A}{dx}$ = concentration gradient of component A, a driving potential for diffusion of species.

The negative sign indicates that the mass diffusion takes place in the direction of decreasing concentration.

The diffusion rate of component B in component A is given by

$$j_B = \frac{\dot{m}_B}{A} = -D_{BA}\frac{d\rho_B}{dx} \quad ...(15.22)$$

The diffusion coefficients D_{AB} and D_{BA} are relative diffusivities and they are equal in magnitude.

Fick's law in terms of diffusional molar flux J_A, molar concentration C_A and mass diffusivity D_{AB} can also be expressed as :

$$j_A = -D_{AB}\frac{dC_A}{dx} \quad ...(15.23)$$

Fick's law of diffusion is analogous to

(*i*) Fourier law of heat conduction,

$$\frac{Q}{A} = q = -k\frac{dT}{dx}, \text{ and}$$

(*ii*) Newton's law of viscosity, $\tau = -\mu\frac{du}{dy}$

Fourier law describes the transport of heat energy due to *temperature gradient*, Newton's law of viscosity describes the transport of momentum due to *velocity gradient* and Fick's law describes the mass transport due to *concentration gradient.*

Further, the units of mass diffusion coefficient

$$D_{AB} = \frac{\dot{m}_A/A}{\frac{d\rho_A}{dx}} = \frac{kg/s.m^2}{kg/m^4} = m^2/s$$

which is identical to units of thermal diffusivity α and momentum diffusivity ν.

The physical mechanism of mass diffusion can be understood by considering Fig. 15.3.

The concentration of component A is greater on the left of the imaginary plane than on the right side. The higher concentration means more molecules per unit volume, thus the more molecules moving, *cross the imaginary plane from left to right.* It results in net mass transfer from a region of high concentration to that of low concentration.

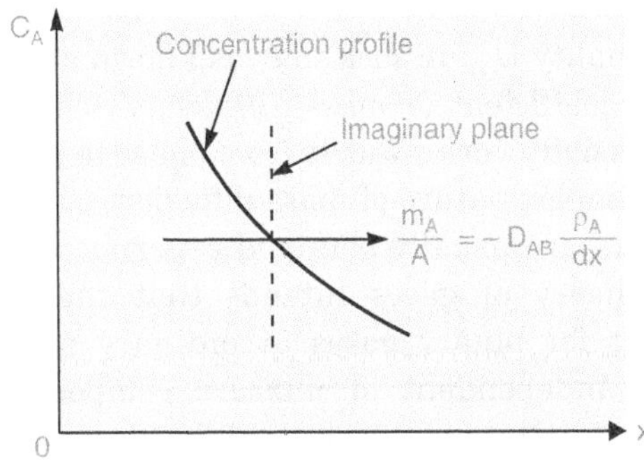

Fig. 15.3. Diffusion dependence on concentration profile

Some important aspects of Fick's law of diffusion are :

1. Fick's law is based on experimental evidence and hence, it cannot be derived from first principle.

2. Fick's law is valid in any state of all matters.

3. The mass diffusion rate depends on concentration gradient, caused by temperature gradient, a pressure gradient or an external force.

4. A component in a mixture diffuses in the direction of decreasing concentration.

5. Diffusion coefficient D_{AB}, in general depends upon temperature, pressure and the nature of system components.

When a small concentration of species A is moving through a large concentration of species B (such as air), then the diffusional mass flux j_A can be approximated as $\boldsymbol{n}_A$, and diffusional molar flux J_A as N_A.

When working with Fick's law, one may use mass flux $\dot{n}_A$ and mass concentration (ρ_A) as in eqn. (15.21) or Fick's law may be expressed in terms of molar concentration (C_A) and molar flux N_A. For gases, Fick's law may also be used in terms of partial pressures by using perfect gas equation of state.

$$p = \rho RT$$

where R is the specific constant. In terms of universal gas constant R_u(= MR) for gas component A

$$\rho_A = \frac{p_A}{RT} = \frac{M_A p_A}{R_u T}$$

The density ρ_A represents mass concentration of species A. Then Fick's law

$$\frac{\dot{m}_A}{A} = -D_{AB}\frac{M_A}{R_u T}\frac{dp_A}{dx} \quad \text{...(15.24(a))}$$

For gas component B in an isothermal diffusion

$$\frac{\dot{m}_B}{A} = -D_{BA}\frac{M_B}{R_u T}\frac{dp_B}{dx} \quad \text{...(15.24(b))}$$

Mass Diffusion Coefficient. The constant of proportionality D_{AB} in Fick's law is known as the mass diffusion coefficient or *mass diffusivity*. The value of mass diffusivity varies widely from phase to phase. Due to very complex nature of mass diffusion, the diffusion coefficients are usually obtained experimentally. The kinetic theory of gases intends that the diffusion coefficient for binary gases at ordinary pressure is basically independent of mixture composition and increases with temperature while decreases with pressure as :

$$D_{AB} \propto \frac{T^{3/2}}{p}$$

The binary diffusion coefficient at a specified temperature and pressure can be determined from equation given by *Gilliland;*

$$D_{AB} = 435.7\frac{T^{3/2}}{p(V_A^{1/3} + V_B^{1/3})} \times \sqrt{\frac{1}{M_A} + \frac{1}{M_B}} \quad \text{...(15.25)}$$

where D_{AB} = diffusion coefficient, cm^2/s
T = absolute temperature, K
p = total pressure of system, N/m^2 or Pa
V_A = molecular volume of component A
V_B = molecular volume of component B
M_A = molecular weight of component A
M_B = molecular weight of component B

Table 15.2 presents the molecular weights and molecular volumes of some selected gases.

The diffusion coefficient for gases at different temperatures and pressures can be related as;

$$\frac{D_1}{D_2} = \left(\frac{T_1}{T_2}\right)^{3/2}\left(\frac{p_2}{p_1}\right) \quad \text{...(15.26)}$$

More general, but complicated relations that account the effect of molecular collisions are also available in the literature. Table 15.3 lists some experimental values of the diffusion coefficient for binary gas mixtures, dilute liquid solutions and dilute solid solutions. For the diffusion coefficient, two important observations should be noted that

1. The diffusion coefficient is highest in gases and lowest in solids, and
2. Diffusion coefficient increases with temperatures and decreases with pressure.

TABLE 15.2. Molecular weight and molecular volume of gases

Gas	*Molecular weight*	*Molecular volume at normal boiling point*
Air	28.9	29.9 cm^3/gm mole
Ammonia NH_3	17	25.81
Carbon dioxide CO_2	44	34.00
Carbon monoxide	28	30.71
Hydrogen H_2	2	28.6
Nitrogen N_2	28	31.20
Oxygen O_2	32	14.8
Sulphur dioxide SO_2	64	44.78

TABLE 15.3. Typical diffusion coefficients for binary gas mixtures at 1 atm and dilute liquid solutions

(*a*) Diffusion through gases

Solute A	*Solvent B*	*T(K)*	$D_{AB}(m^2/s) \times 10^5$
Air	Carbon dioxide	276	1.42
Air	Ethanol	313	1.45
Air	Helium	276	6.24
Air	Naphthalene	303	0.86
Air	Water	313	2.88
Argon	Helium	295	8.3
		628	32.1

(*b*) Diffusion through liquids

(Dilute solute, A)	*(Liquid solvent B)*	*T(K)*	D_{AB} $(m^2/s) \times 10^9$
Ethanol	Benzene	288	2.25
Benzene	Ethanol	298	1.81
Water	Ethanol	298	1.24
Carbon dioxide	Water	298	2.00
Ethanol	Water	288	1.00
Methane	Water	275	0.85
		333	3.55
Pyridene	Water	288	0.58

(*c*) Diffusion through solids

(Solute, A)	*(Solid solvent, B)*	*T(K)*	D_{AB}, m^2/s
Carbon dioxide	Rubber	298	1.1×10^{-10}
Oxygen	Rubber	298	2.1×10^{-10}
Hydrogen	Iron	298	2.6×10^{-13}
Hydrogen	Nickel	358	1.2×10^{-12}
Zinc	Copper	773	4×10^{-18}
Antimony	Silver	293	3.5×10^{-25}
Copper	Aluminium	773	4×10^{-14}
Carbon	Iron	773	5×10^{-5}

Example 15.2. *The hydrogen gas diffuses through a steel wall of 2 mm thickness. The molar concentration of hydrogen at the interface is 1.5 kg-mol/m³ and it is zero on the outer face. Calculate the diffusion rate of hydrogen, if its diffusivity coefficient is 0.3×10^{-12} m²/s.*

Solution

Given : Diffusion of hydrogen through a steel wall.

$L = 2 \text{ mm} = 0.002 \text{ m}, \quad C_{A_1} = 1.5 \text{ kg-mol/m}^3,$

$D_{AB} = 0.3 \times 10^{-12} \text{ m}^2/\text{s}, \quad C_{A_2} = 0$

To find : Diffusion rate of hydrogen.

Assumptions :

(*i*) Steady state diffusion.

(*ii*) One dimensional diffusion without any chemical reaction.

(*iii*) Constant properties.

Analysis : The one dimensional molar diffusion is given by

$$N_A = D_{AB} \frac{C_{A_1} - C_{A_2}}{L}$$

$$= \frac{0.3 \times 10^{-12} \times (1.5 - 0)}{0.002}$$

$= \mathbf{1.5 \times 10^{-10}}$ **kg-mol/m²s.** Ans.

Example 15.3. *Calculate the diffusion coefficient for CO_2 in air at atmospheric pressure and 25°C.*

Solution

The diffusion coefficient in gases

$$D_{AB} = 435.7 \frac{T^{3/2}}{p(V_A^{1/3} + V_B^{1/3})} \sqrt{\frac{1}{M_A} + \frac{1}{M_B}}$$

From Table 15.1

$V_{CO_2} = 34.0, \quad V_{air} = 29.9$

$M_{CO_2} = 44, \quad M_{air} = 28.9$

$$D_{AB} = 435.7 \times \frac{(298)^{3/2}}{(1.01325 \times 10^5) \times (34^{1/3} + 29.9^{1/3})} \times \sqrt{\frac{1}{28.9} + \frac{1}{44}}$$

$= 0.132$ cm²/s

$= \mathbf{13.2 \times 10^{-6}}$ **m²/s.** Ans.

15.6. GENERAL MASS DIFFUSION EQUATION

Mass transfer by molecular diffusion is analogous to heat transfer by conduction or momentum transfer in laminar flow.

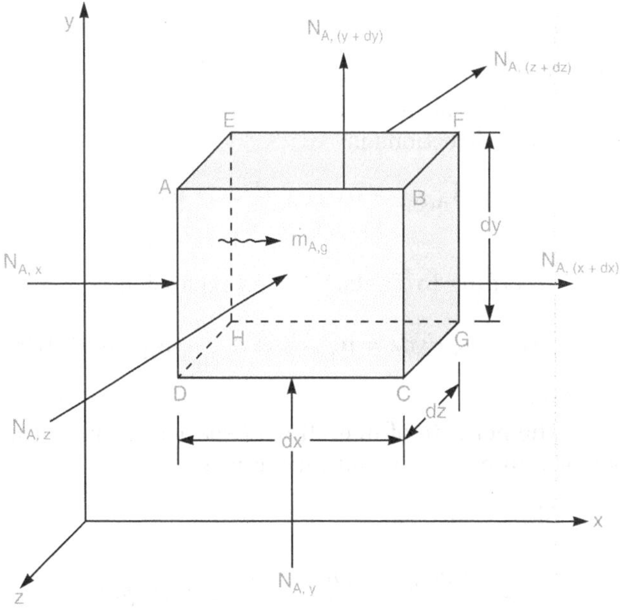

Fig. 15.4. Control volume for mass diffusion equation in Cartesian coordinates

Mass transfer by molecular diffusion may occur in a stagnant fluid or in a fluid in laminar flow. Like Fourier equation of heat conduction

$$\nabla^2 T + \frac{g(x)}{k} = \frac{1}{\alpha}\frac{\partial T}{\partial t},$$

The derivation for mass transfer equation can be taken.

Consider a differential control volume of sides dx, dy and dz as shown in Fig. 15.4. Any species A may enter and leave the control volume due to fluid motion and diffusion across the control volume. The some species A may also be generated and accumulated within the control volume. The mass balance equation of species A on a rate basis can be written as

$$\begin{Bmatrix}\text{Net rate of mass}\\ \text{flux of species A}\\ \text{from control volume}\end{Bmatrix} + \begin{Bmatrix}\text{Net rate of species A}\\ \text{generation within}\\ \text{control volume}\end{Bmatrix}$$

$$= \begin{Bmatrix}\text{Net rate of}\\ \text{accumulation of species}\\ \text{A within control volume}\end{Bmatrix} \quad \text{...(15.27)}$$

The individual terms for species A are evaluated as given below.

Total rate of mass flux of species A enters the control volume in x, y and z directions is

$$\dot{M}_{A,\text{in}} = \mathbf{n}_{A,x}dydz + \mathbf{n}_{A,y}dxdz + n_{A,z}dxdy$$

where $n_A = \rho_A u_A$ = mass flux

The rate of mass flux leaves the control volume in x, y and z directions is

$$\dot{M}_{A,\text{out}} = \mathbf{n}_{A,x+dx}\,dydz + \mathbf{n}_{A,y+dy}\,dxdz + \mathbf{n}_{A,z+dz}\,dxdy$$

According to Taylor's series expansion

$$\mathbf{n}_{A,x+dx}\,dydz = \mathbf{n}_{A,x}\,dydz + \frac{\partial}{\partial x}(\mathbf{n}_{A,x}\,dydz)\,dx$$

and so on.

The net rate of mass flux of species A enters the control volume in x, y and z directions.

$$\dot{M}_{A,\text{in}} - \dot{M}_{A,\text{out}}$$

$$= -\left[\frac{\partial \mathbf{n}_{A,x}}{\partial x} + \frac{\partial \mathbf{n}_{A,y}}{\partial y} + \frac{\partial \mathbf{n}_{A,z}}{\partial z}\right]dxdydz \quad \text{...}(i)$$

In addition, there may be uniform volumetric chemical reaction occurring within species A. This chemical reaction may generate or absorb a particular species. If $\dot{\mathbf{n}}_{A,g}$ is the volumetric species generation rate, then the net rate of species A generation within control volume.

$$\dot{M}_{A,g} = \dot{\mathbf{n}}_{A,g}\,dxdydz \quad \text{...}(ii)$$

Due to species diffusion and species generation process, the mass of species A may be accumulated or dispersed. The rate of change of mass of species A within the control volume is

$$\dot{M}_{A,st} = \frac{\partial \rho_A}{\partial t}\,dxdydz \quad \text{...}(iii)$$

Substituting eqns. (i) to (iii) in eqn. (15.27) to obtain

$$-\left[\frac{\partial \mathbf{n}_{A,x}}{\partial x} + \frac{\partial \mathbf{n}_{A,y}}{\partial y} + \frac{\partial \mathbf{n}_{A,z}}{\partial z}\right] + \mathbf{n}_{A.g} = \frac{\partial \rho_A}{\partial t} \quad \text{(15.28(}a\text{))}$$

It is the *equation of continuity* for species A. A similar equation of continuity may be developed for a second species B of mixture as

$$-\left[\frac{\partial}{\partial x}\mathbf{n}_{B,x} + \frac{\partial}{\partial y}\mathbf{n}_{B,y} + \frac{\partial}{\partial z}\mathbf{n}_{B,z}\right] + \dot{\mathbf{n}}_{B,g} = \frac{\partial \rho_B}{\partial t} \quad \text{(15.28(}b\text{))}$$

For a stationary medium the mass average velocity u is zero, then eqn. (15.19) leads to $\dot{\mathbf{n}}_A = j_A = \rho_A u_A$. Using Fick's law eqn. (15.21) in terms of mass concentration gradients in eqn. (15.28 (a)) to obtain

$$-\frac{\partial}{\partial x}\left[-D_{AB}\frac{\partial \rho_A}{\partial x}\right] - \frac{\partial}{\partial y}\left[-D_{AB}\frac{\partial \rho_A}{\partial y}\right] - \frac{\partial}{\partial z}\left[-D_{AB}\frac{\partial \rho_A}{\partial z}\right] + \dot{\mathbf{n}}_{A,g} = \frac{\partial \rho_A}{\partial t}$$

If D_{AB} is constant, then, above equation yields to

$$\frac{\partial^2 \rho_A}{\partial x^2} + \frac{\partial^2 \rho_A}{\partial y^2} + \frac{\partial^2 \rho_A}{\partial z^2} + \frac{\dot{\mathbf{n}}_{A,g}}{D_{AB}} = \frac{1}{D_{AB}}\frac{\partial \rho_A}{\partial t} \quad \text{...(15.29)}$$

It is governing differential mass diffusion equation in Cartesian coordinates.

The subsequent treatment of species diffusion phenomenon on mole basis in terms of molar concentration with volumetric species generation $\dot{\mathbf{N}}_{A,g}$ within the control volume yields to

$$\frac{\partial^2 C_A}{\partial x^2} + \frac{\partial^2 C_A}{\partial y^2} + \frac{\partial^2 C_A}{\partial z^2} + \frac{\dot{\mathbf{N}}_{A,g}}{D_{AB}} = \frac{1}{D_{AB}}\frac{\partial C_A}{\partial t} \quad \text{...(15.30)}$$

In steady state $\left(\frac{\partial C_A}{\partial t} = 0\right)$ and without chemical reaction ($\dot{N}_{A,g} = 0$), eqn. (15.30) takes the form

$$\frac{\partial^2 C_A}{\partial x^2} + \frac{\partial^2 C_A}{\partial y^2} + \frac{\partial^2 C_A}{\partial z^2} = 0$$

In x-direction, the above equation reduces to

$$\frac{d^2 C_A}{dx^2} = 0 \quad ...(15.31)$$

The forgoing species diffusion equations are analogous to heat conduction equations, thus they follow similar boundary conditions. Further, the species diffusion eqn. (15.30) may also be expressed in cylindrical and spherical coordinates.

Cylindrical coordinates:

$$\frac{1}{r}\frac{\partial}{\partial r}\left(r\frac{\partial C_A}{\partial r}\right) + \frac{1}{r^2}\frac{\partial}{\partial \theta}\left[\frac{\partial C_A}{\partial \theta}\right] + \frac{\partial}{\partial z}\left[\frac{\partial C_A}{\partial z}\right] + \frac{\dot{N}_{A,g}}{D_{AB}} = \frac{1}{D_{AB}}\frac{\partial C_A}{\partial t} \quad ...(15.32)$$

Spherical coordinates:

$$\frac{1}{r^2}\frac{\partial}{\partial r}\left(r^2\frac{\partial C_A}{\partial r}\right) + \frac{1}{r^2 \sin^2\theta}\frac{\partial}{\partial \phi}\left[\frac{\partial C_A}{\partial \phi}\right] + \frac{1}{r^2 \sin\theta}\frac{\partial}{\partial \theta}\left[\sin\theta\frac{\partial C_A}{\partial \theta}\right] + \frac{\dot{N}_{A,g}}{D_{AB}} = \frac{1}{D_{AB}}\frac{\partial C_A}{\partial t} \quad ...(15.33)$$

15.7. BOUNDARY CONDITIONS

The mass diffusion equation is analogous to heat conduction equation, the boundary conditions are required to determine the *species concentration distribution,* and *species flux.* These conditions are the physical situation existing at the boundary surfaces of the medium. For steady state diffusion of species, there are three types of boundary conditions:

1. Specified species concentration,
2. Specified species flux condition, and
3. Specified convective mass transfer flux condition.

1. Specified Species Concentration Condition. The first condition corresponds to a situation for which the species concentration at the surface is maintained constant value. Expressing this condition on molar basis:

$$\text{At } x = 0;\ C_A(0) = C_{A,s} \quad ...(15.34(a))$$

and on mass basis $\rho_A(0) = \rho_{A,s}$ ...(15.34(b))

2. Specified Species Flux Condition. If the mass or molar flux of the species A is provided at a surface, the appropriate form of the boundary condition yields to

$$j_{A,0} = -D_{AB}\left.\frac{d\rho_A}{dx}\right|_{x=0}$$

and

$$J_{A,0} = -D_{AB}\left.\frac{dC_A}{dx}\right|_{x=0} \quad ...(15.35)$$

where $j_{A,0}$ and $J_{A,0}$ are the specified diffusion mass and mole fluxes, respectively of species A at the boundary at $x = 0$.

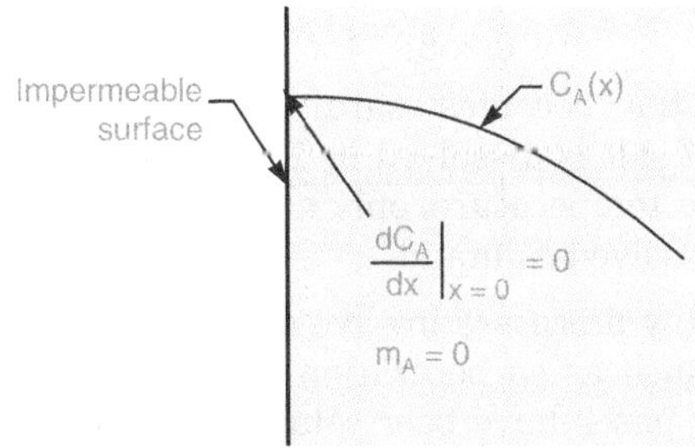

Fig. 15.5. Concentration distribution on an impermeable surface

A special case of specified species flux occurs for an impermeable surface, for which species flux is zero, *i.e.*,

$$J_{A,0} = 0 = -D_{AB}\left.\frac{dC_A}{dx}\right|_{x=0}$$

or

$$\left.\frac{dC_A}{dx}\right|_{x=0} = 0$$

3. Specified Convective Mass Transfer Flux Condition. The mass transfer convective coefficient h_m can be defined in similar way analogous with definition of heat transfer coefficient h.

$$j_{A,0} = h_m(\rho_{A,s} - \rho_{A,\infty})$$

Further, the convective mass transfer coefficient at the boundary can also be expressed as

$$j_{A,0} = h_m(\rho_{A,s} - \rho_{A,\infty}) = -D_{AB} \left.\frac{d\rho_A}{dx}\right|_x \quad ...(15.36)$$

And convective molar flux at the boundary

$$j_{A,0} = h_m (C_{A,s} - C_{A,\infty})$$

$$= -D_{AB} \left.\frac{dC_A}{dx}\right|_x \quad ...(15.37)$$

where h_m is the convective mass transfer coefficient in m/s.

Solubility and Permeability. The **solubility** $\mathcal{S}$ represents the *maximum amount of substance that can be dissolved in a medium at specified temperature.* For an example, the solubility of salt (NaCl) in water at 27°C is 36.2 kg per 100 kg of water. Therefore, the mass fraction of salt in brine solution at the interface is

$$x_{\text{salt, water}} = \frac{m_{\text{salt}}}{\text{Mass of solution}}$$

$$= \frac{36.2 \text{ kg}}{(100 + 36.2) \text{ kg}}$$

$$= 0.2658 \quad \text{(or 26.58 per cent)}$$

Water becomes saturated with salt, when 36.2 kg of salt is dissolved in 100 kg of water at 27°C. At a given temperature, only a certain amount of solid can be dissolved in liquid.

Many processes involve dissolution of gases into a liquid such as carbonization of soft drink with CO_2 gas. Most gases have poor solubility in liquid, such as air in the water. Further, the gases may dissolve in solids, but the dissolution process in this case is very complex. The dissolution of gas may be independent of the structure of the solid, but it may depend on the porosity of solid. The dissolution of hydrogen in titanium is reversible, while dissolution of oxygen gas in titanium to form TiO_2 on the surface is irreversible.

The concentration $C_{A,\text{ silid side}}$ of gas species A in the solid at interface is proportional to the partial pressure $p_{A,\text{ gas side}}$ of the species A in the gas on the gas side of the interface and is expressed as

$$C_{A,\text{ solid side}} = \mathcal{S}_{AB} \times p_{A,\text{ gas side}} \text{ (kg-mole/m}^3\text{)} \quad ...(15.38)$$

where $\mathcal{S}$ is the solubility. Expressing the pressure in bars and concentration in kg-mole/m^3, thus the unit of solubility is derived as kg-mole/(m^2.bar). The solubility data for selected gas-solid combination at 25°C is given in Table 15.4.

TABLE 15.4. Solubility of gases in solids at 25°C

Gas	*Solid*	$\mathcal{S}$, *kg-mole / (m^3.bar)*
O_2	Rubber	0.00312
N_2	Rubber	0.00156
CO_2	Rubber	0.04015
He	SiO_2	0.00045

The product of solubility of a gas and the diffusion coefficient of the gas in the solid is referred as the **permeability** $\mathcal{P}$, which is measure of the ability of the gas to penetrate a solid. That is

$$\mathcal{P} = \mathcal{S}_{AB} D_{AB} \quad ...(15.39)$$

where D_{AB} is the diffusivity of gas in the solid. The permeability is inversely proportional to thickness of solid and it has the unit of kg-mole/(s.bar).

If a process involves the *sublimation* of a pure solid or *evaporation* of a pure liquid, in a different medium such as air, the mass (or mole) fraction of the substance in the liquid or solid phase is simply taken as 1.0, and the partial pressure and the mole fraction of the substance in the gas phase can readily obtained from the saturation data table of the substance at specified temperature.

15.8. MASS DIFFUSION WITHOUT HOMOGENEOUS CHEMICAL REACTIONS

A *homogeneous* chemical reaction is a *volumetric phenomenon,* it takes place within the medium and generates species within the medium, thus it is analogous to internal source of heat generation. In contrast, the *heterogeneous* chemical reaction is *surface phenomenon,* and it results from contact between the reactants and a surface and thus analogous to the constant surface heat flux conditions.

There are many situations encountered in engineering, where the mass diffusion takes place in one direction in steady state conditions without chemical reaction. This situation is analogous to steady state heat conduction without heat generation. In this section, we do not consider the situations that do involve the homogeneous chemical reaction during mass transfer.

15.8.1. Steady State Diffusion Through a Plane Membrane

In the diffusion of small quantity of fluid A through a stationary gas B, or diffusion of dilute solution of A with a liquid or solid B, the mass or molar average velocity of the mixture is very small and such cases can be treated as the diffusion of A through a stationary medium or plane membrane B.

Consider the molar diffusion of a fluid A through a stationary medium B of thickness L as shown in Fig. 15.6. The molar concentrations of fluid A at two faces are $C_{A,1}$ and $C_{A,2}$, respectively. For steady state one dimensional, without any chemical reaction, the molar diffusion eqn. (15.31) is

$$\frac{d^2C_A}{dx^2} = 0$$

Subjected to boundary conditions

at $x = 0$, $C_A = C_{A,1}$

and at $x = L$, $C_A = C_{A,2}$

The solution of above equation after double integration

$$C_A = C_1x + C_2 \qquad ...(15.40)$$

where C_1 and C_2 are constants of integrations and are evaluated from boundary conditions ;

at $x = 0$, $C_2 = C_{A,1}$

at $x = L$, $C_1 = \frac{(C_{A,2} - C_{A,1})}{L}$

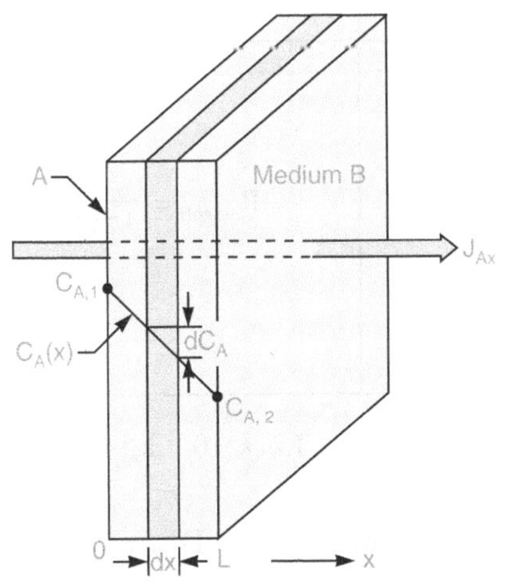

Fig. 15.6. One dimensional steady diffusion through a plane wall

Substituting in eqn. (15.40), we get

$$C_A = (C_{A,2} - C_{A,1})\frac{x}{L} + C_{A,1} \qquad ...(15.41)$$

The diffusional molar flux of A at any x

$$J_{Ax} = -D_{AB}\frac{dC_A}{dx}$$

$$= -D_{AB}\frac{d}{dx}\left[(C_{A,2} - C_{A,1})\frac{x}{L} + C_{A,1}\right]$$

$$= -D_{AB}\left(\frac{C_{A,2} - C_{A,1}}{L}\right)$$

Multiplying the surface area A, then the molar rate

or
$$\dot{N}_A = \frac{D_{AB}A}{L}(C_{A,1} - C_{A,2}) \qquad ...(15.42)$$

$$\dot{N}_A = \frac{C_{A,1} - C_{A,2}}{\frac{L}{AD_{AB}}}$$

$$= \frac{C_{A,1} - C_{A,2}}{R_{m,A}}$$

$$= \frac{\text{Concentration potential}}{\text{Diffusion resistance}} \qquad ...(15.43)$$

where
$$R_{m,A} = \frac{L}{AD_{AB}} \qquad ...(15.44)$$

$R_{m,A}$ is known as *diffusion resistance,* which is similar to the conduction resistance of a plane slab of thickness L and area A. The electrical analogy between thermal, electrical and molar diffusion is shown in Fig. 15.7.

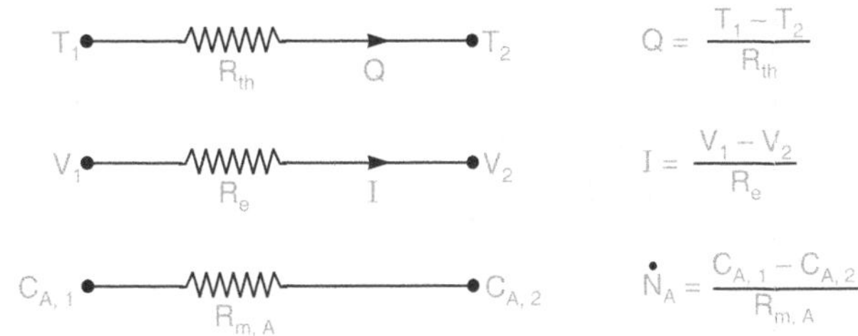

Fig. 15.7. Analogy between thermal, electrical and molar diffusion resistance concepts

The preceding analysis can also be repeated on the mass basis with a resulting equation for mass transfer rate in plane wall

$$\dot{m}_A = D_{AB}A\frac{\rho_{A,1} - \rho_{A,2}}{L}$$

$$= \rho D_{AB}A\frac{x_{A,1} - x_{A,2}}{L} \text{ (kg/s)} \qquad ...(15.45)$$

and concentration distribution on the mass basis as

$$\rho_A = (\rho_{A,2} - \rho_{A,1})\frac{x}{L} + \rho_{A,1} \qquad ...(15.46)$$

With eqn. (15.42) and eqn. (15.45), we can conclude that *the molar (or mass) diffusion rate through a plane wall is proportional to concentration gradient and wall area.* The molar and mass concentration at any location in the plane wall can be determined by using eqn. (15.41) and eqn. (15.46), respectively.

We can also obtain the concentration distribution and diffusion resistance for various other geometries as presented in Table 15.5.

TABLE 15.5. Summary of species concentration distribution and diffusion resistance

Geometry	*Species concentration distribution*	*Species diffusion resistance, $R_{m,A}$*
Plane wall: $C_{A,1}$ or $\rho_{A,1}$; $C_{A,2}$ or $\rho_{A,2}$; A; 0; L; x	$C_A(x) = (C_{A,2} - C_{A,1})\dfrac{x}{L} + C_{A,1}$ $\rho_A(x) = (\rho_{A,2} - \rho_{A,1})\dfrac{x}{L} + \rho_{A,1}$	$R_{m,\,wall} = \dfrac{L}{D_{AB}A}$
Cylinder: $C_{A,2}$; r_2; $C_{A,1}$; r; r_1; L	$C_A(r) = \dfrac{C_{A,2} - C_{A,1}}{\ln\left(\dfrac{r_2}{r_1}\right)} + C_{A,2}$ $\rho_A(r) = \dfrac{\rho_{A,2} - \rho_{A,1}}{\ln\left(\dfrac{r_2}{r_1}\right)} + \rho_{A,2}$	$R_{m,\,cyl} = \dfrac{\ln\left(\dfrac{r_2}{r_1}\right)}{2\pi LD_{AB}}$
Sphere: r_1; r_2; $C_{A,2}$; $C_{A,1}$; r	$C_A(r) = \dfrac{C_{A,2} - C_{A,1}}{\dfrac{1}{r_2} - \dfrac{1}{r_1}}\left(\dfrac{1}{r} - \dfrac{1}{r_2}\right) + C_{A,2}$ $\rho_A(r) = \dfrac{\rho_{A,2} - \rho_{A,1}}{\dfrac{1}{r_2} - \dfrac{1}{r_1}}\left(\dfrac{1}{r} - \dfrac{1}{r_2}\right) + \rho_{A,2}$	$R_{m,\,sph} = \dfrac{r_2 - r_1}{4\pi r_1 r_2 D_{AB}}$

Example 15.4. *The hydrogen gas is maintained at pressure of 2.4 bar and 1 bar on opposite of a plastic membrane 0.3 mm thick. The binary diffusion coefficient of hydrogen in the plastic is 8.7×10^{-8} m²/s. The solubility of hydrogen in the membrane is 1.5×10^{-3} kg-mole/m³ bar. Calculate under uniform condition of 25°C, the following:*

(*i*) *Molar concentration of hydrogen on opposite faces of membrane.*

(*ii*) *Molar and mass diffusion flux of hydrogen through membrane.*

Solution

Given : Diffusion of two hydrogen gas streams through a plastic membrane.

Solubility $\mathcal{S}_{AB} = 1.5 \times 10^{-3}$ kg-mole/(m³.bar)

$p_1 = 2.4$ bar, $\quad p_2 = 1$ bar

$L = 0.3$ mm, $\quad D_{AB} = 8.7 \times 10^{-8}$ m²/s

$T = 25°C = 298$ K, $\quad M_{H_2} = 2$ kg/kg-mole

To find : (*i*) Molar concentrations of hydrogen at opposite faces of membrane.

(*ii*) Molar and mass diffusion flux.

Analysis : (*i*) Molar concentrations : The molar concentration, pressure and solubility are related as :

$$C = \mathcal{S}_{AB}\, p$$

Therefore, surface concentration

$$C_{H_{2},1} = 1.5 \times 10^{-3} \times 2.4$$

$$= \mathbf{3.6 \times 10^{-3}\ kg\text{-}mole/m^3}$$

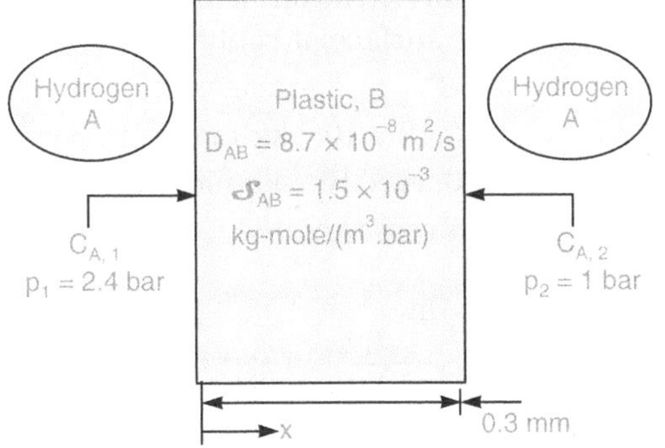

Fig. 15.8. Schematic for example 15.4

$$C_{H_{2,2}} = 1.5 \times 10^{-3} \times 1$$

$$= \mathbf{1.5 \times 10^{-3}\ kg\text{-}mole/m^3}. \quad \text{Ans.}$$

(*ii*) Molar and mass diffusion fluxes are calculated as :

$$N_A = \frac{D_{AB}(C_{H_{2,1}} - C_{H_{2,2}})}{L}$$

$$= \frac{(8.7 \times 10^{-8}\ \text{m}^2/\text{s}) \times (3.6 \times 10^{-3} - 1.5 \times 10^{-3})(\text{kg-mole/m}^3)}{(0.3 \times 10^{-3}\ \text{m})}$$

$$= \mathbf{6.09 \times 10^{-7}\ kg\text{-}mole/(m^2s)}. \quad \text{Ans.}$$

Mass diffusion flux,

$$j_A = M_{H_2} \times N_A = 2 \times 6.09 \times 10^{-7}$$

$$= \mathbf{1.218 \times 10^{-6}\ kg/m^2s}. \quad \text{Ans.}$$

Example 15.5. *The tyre tube of a vehicle has a surface area of 0.62 m^2 and wall thickness of 10 mm. The tube has air filled in it at a pressure of 2.2 bar. The air pressure drops to 2.18 bar in period of 6 days. The solubility of air in the rubber is 0.075 m^3 of air/m^3 of rubber at 1 bar. The volume of air in the tube is 0.034 m^3 and operating temperature is 300 K. Calculate diffusivity of air in the rubber.*

Solution

Given : Diffusion of air in a tyre tube system ;

A = 0.62 m^2, V = 0.034 m^3

L = 10 mm = 0.01 m, p_1 = 2.2 bar = 220 kPa

T = 300 K. p_2 = 2.18 bar = 218 kPa

t = 6 days = 518400 s, $\mathcal{S}$ = 0.075 m^3 of air/(m^3.bar)

To find : Diffusivity or diffusion coefficient of air.

Assumptions :

1. Steady diffusion of air in tube.
2. Gas constant of air as 0.287 kJ/kg.K.
3. Constant properties.
4. Atmospherie pressure as 1 bar.

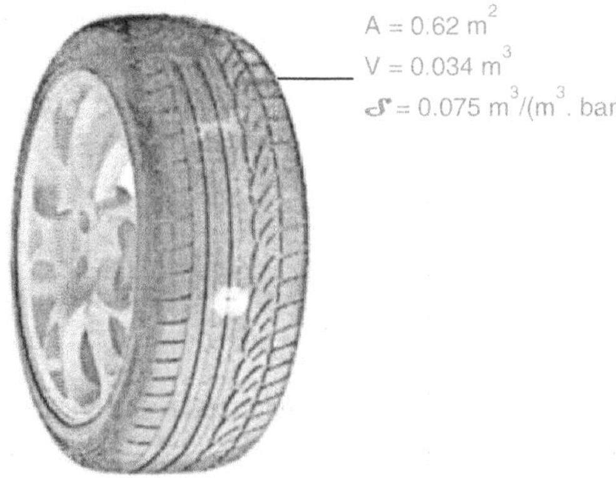

Fig. 15.9. Tyre and tube for example 15.5

Analysis : The initial mass of air in the tube

$$m_1 = \frac{p_1 V}{RT_1} = \frac{220 \times 0.034}{0.287 \times 300} = 0.0868 \text{ kg}$$

The mass of air in the tube after 6 days.

$$m_2 = \frac{p_2 V}{RT_2} = \frac{218 \times 0.034}{0.287 \times 300} = 0.0860 \text{ kg}$$

The mass of air escaped through tube

$$\Delta m = m_1 - m_2 = 0.0868 - 0.0860$$

$$= 0.0008 \text{ kg}$$

The mass flux of air escaped

$$\frac{\dot{m}_A}{A} = \frac{\text{Mass of air escaped}}{\text{Time elapsed} \times \text{Surface area}}$$

$$= \frac{0.0008 \text{ kg}}{518400 \text{ s} \times 0.62 \text{ m}^2}$$

$$= 2.49 \times 10^{-9} \text{ kg/m}^2\text{s}$$

Mean pressure of air at inner surface of tube

$$p_m = \frac{p_1 + p_2}{2} = \frac{2.2 + 2.18}{2} = 2.19 \text{ bar}$$

The solubility is the dissolved volume of air in the rubber, therefore,

$$V_i = \mathcal{S}_{AB}\, p_i \ (\text{m}^3/\text{m}^3.\text{bar})$$

V_1 = solubility at inside surface of rubber

$$= \mathcal{S}_{AB}\, p_m$$

$$= 0.075 \times 2.19$$

$$= 0.01642 \text{ m}^3/\text{m}^3 \text{ of rubber}$$

V_2 = solubility at outer surface of rubber tube

$$= \mathcal{S}_{AB}\, p_{atm}$$

$$= 0.075 \times 1 \text{ bar}$$

$$= 0.075 \text{ m}^3/\text{m}^3 \text{ of rubber tube}$$

Corresponding mass concentrations at inner and outer surfaces of rubber tube

$$\rho_{A_1} = \frac{p_m V_1}{RT_1} = \frac{219 \times 0.1642}{0.287 \times 300} = 0.4176 \text{ kg/m}^3$$

and $\rho_{A_2} = \frac{p_{atm} V_2}{RT_2} = \frac{100 \times 0.075}{0.287 \times 300} = 0.0871 \text{ kg/m}^3$

The diffusion flux rate of air through rubber tube

$$\frac{\dot{m}_A}{A} = \frac{D_{AB}(\rho_{A_1} - \rho_{A_2})}{L}$$

or $D_{AB} = \frac{\dot{m}_A}{A} \frac{L}{(\rho_{A_1} - \rho_{A_2})} = \frac{2.49 \times 10^{-9} \times 0.01}{(0.4176 - 0.0871)}$

$= \mathbf{7.509 \times 10^{-11}\ m^2/s.}$ **Ans.**

Example 15.6. *A rubber tube was used to transport hydrogen gas at 2.0 bar and 34°C. The inner and outer diameters of rubber tube are 25 mm and 40 mm, respectively. Diffusion coefficient of hydrogen in rubber is 2.2 × 10^{-8} m²/s and solubility of hydrogen in the rubber is 3.12 × 10^{-3} kg-mole of H_2 per m³ of rubber at 1 bar. Calculate the diffusion flux of hydrogen per metre length of tube. Assume concentration of hydrogen at the outer surface of the tube is negligible.*

Solution

Diffusion of hydrogen through a rubber tube

$d_1 = 25$ mm, or $r_1 = 12.5 \times 10^{-3}$ m

$d_2 = 40$ mm, or $r_2 = 20 \times 10^{-3}$ m

$p = 2$ bar, T = 34°C = 307 K

$D_{AB} = 2.2 \times 10^{-8}$ m²/s,

$\mathcal{S}_{AB} = 3.12 \times 10^{-3}$ kg-mole/(m³.bar)

$C_{A,2} = 0$,

To find : Diffusion flux of hydrogen per metre length of tube.

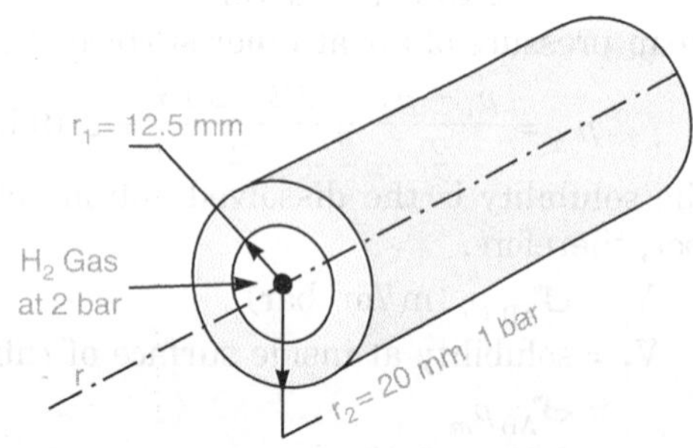

Fig. 15.10. Schematic for example 15.6

Assumptions :

1. Partial pressure of Hydrogen on outer surface tube is zero.
2. Steady diffusion through rubber tube.
3. Specific gas constant for hydrogen.

   $R_{H_2} = 4160$ J/kg.K
4. Constant properties.

Analysis: Concentration of hydrogen on inside surface and outside surface of rubber tube:

$C_{A,1} = \mathcal{S}_{AB}\, p_1$ kg-mole/m³

$= 3.12 \times 10^{-3}$ kg-mole/(m³. bar) × 2 bar

$= 6.24 \times 10^{-3}$ kg-mole/m³

The diffusion resistance

$$R_{m,\,cyl} = \frac{\ln\left(\frac{r_2}{r_1}\right)}{2\pi L D_{AB}} = \frac{\ln\left(\frac{20}{12.5}\right)}{2\pi \times 1 \times 2.2 \times 10^{-8}}$$

$$= 34 \times 10^5 \text{ s/m}^3$$

The diffusion flux through cylindrical rubber tubing is

$$N_A = \frac{C_{A,1} - C_{A,2}}{R_{m,\,cyl}} = \frac{6.24 \times 10^{-3} - 0}{34 \times 10^5}$$

$= \mathbf{1.835 \times 10^{-9}}$ **kg-mole/s.** **Ans.**

Example 15.7. *Pressurised hydrogen gas is stored at 85°C in a 4.5 m inner diameter spherical container made of nickel. The shell of the container is 15 cm thick. The molar concentration of hydrogen in the nickel at inner surface is 0.087 kg-mole/m³ and that at outer surface is negligible. Determine the mass diffusion rate of hydrogen through the nickel container. Take binary diffusion coefficient of hydrogen in the nickel as 1.2 × 10^{-12} m²/s.*

Solution

Given : Spherical container of nickel stores H_2 gas

$D_1 = 4.5$ m or $r_1 = 2.25$ m

$r_2 = r_1 + t = 2.25 + 0.15 = 2.4$ m

T = 85°C = 358 K

$C_{A,1} = 0.087$ kg-mole/m³

$C_{A,2} = 0$

$D_{AB} = 1.2 \times 10^{-12}$ m²/s

To find : Mass diffusion rate of H_2 gas through spherical nickel container.

Assumptions :

1. Mass diffusion is steady and one dimensional.
2. No chemical reaction in the nickel shell.
3. No accumulation of H_2 gas in the container.

Analysis : The molar diffusion resistance of spherical shell;

$$R_{m,\,sph} = \frac{r_2 - r_1}{4\pi r_1 r_2 D_{AB}}$$

$$= \frac{2.4 - 2.25}{4\pi \times 2.25 \times 2.4 \times 1.2 \times 10^{-12}}$$
$$= 1.842 \times 10^9 \text{ s/m}^3$$

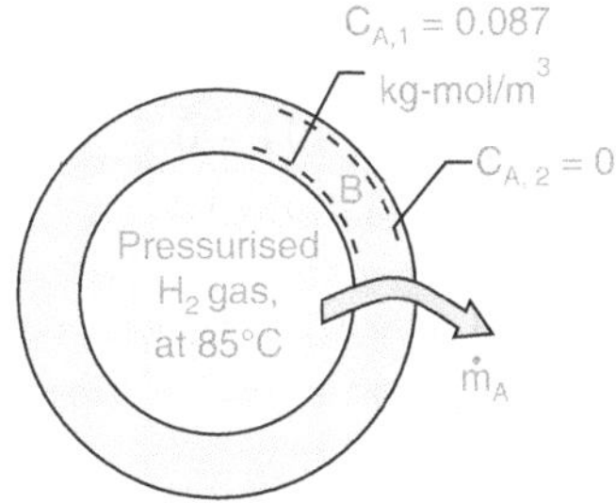

Fig. 15.11. Schematic for example 15.7

The molar diffusion rate of hydrogen gas through container is given by:

$$\dot{N}_A = \frac{C_{A,1} - C_{A,2}}{R_{m,\,sph}} = \frac{0.087 - 0}{1.842 \times 10^9}$$
$$= 47.23 \times 10^{-12} \text{ kg-mole/s}$$

The mass diffusion rate

$$\dot{m}_A = \dot{N}_A M_A$$
$$= (47.23 \times 10^{-12} \text{ kg-mole/s}) \times (2 \text{ kg/kg-mole})$$
$$= \mathbf{9.446 \text{ kg/s or } 2.98 \text{ gram/year.}}$$ Ans.

15.8.2. Water Vapour Migration

The presence of water vapour in material affects the performance, durability, effective thermal conductivity, specific heat and heat storage characteristic etc. During a hot and humid day, water vapour may transfer through a wall and condense on the inner side, releasing latent heat of vaporization, while the process may reverse during a cool night.

Moisture migration in the walls, floors, or ceiling is controlled by either *vapour barriers or vapour retarders*. **Vapour barriers** are materials, which are impermeable to moisture, and thus do not allow the transfer of moisture through them. These are sheet metals, metal foils, and thick plastic sheets. **Vapour retarders** are the materials, which retard or slow down the current of moisture flow through the materials, but do not completely eliminate it. Vapour retarders are available as solid, flexible and coating materials. The reinforced plastic, thin metal foils, plastic films, treated papers, and polymeric paints are some common vapour retarders. The vapour retarders has very low permeance of 5.75×10^{-10} kg/s per m^2.

Atmospheric air is the mixture of dry air and water vapour, and the atmospheric pressure is the sum of partial pressures of dry air and water vapour. The partial pressure of water vapour in air is called the *vapour pressure* p_v The *saturated air* contains maximum possible water vapour in moist air. The partial pressure of water vapour in saturated air is equal to *saturation pressure* p_{sat} of steam corresponding to the temperature of moist air.

The *ratio of actual mass of water vapour to mass of saturated vapour* produced in a mixture of air and water-vapour at same temperature and pressure is called the **relative humidity.** It is denoted by ϕ (phi).

The vapour pressure, saturation pressure and relative humidity are related as

$$p_v = \phi\, p_{sat} \qquad \text{...(15.47)}$$

The water vapour diffusion takes place from a region of high vapour pressure to that of lower vapour pressure. Using eqn. (15.38); and eqn. (15.39) in eqn. (15.42) then the mass flow rate of moisture through a plane layer of thickness L and normal area A can be expressed in terms of vapour pressures as

$$\dot{m}_v = \mathcal{P}A\, \frac{p_{A,1} - p_{A,2}}{L}$$
$$= \mathcal{P}A\, \frac{\phi_1 p_{sat,1} - \phi_2 p_{sat,2}}{L} \text{ (kg/s)} \qquad \text{...(15.48)}$$

where $\mathcal{P}$ is the vapour permeability of the material on the mass basis, expressed in kg/(s.m.Pa). The permeability of material per unit thickness is called the **permeance**, denoted by $\mathcal{M}$.

Permeance,

$$\mathcal{M} = \frac{\text{Permeability}}{\text{Thickness}} = \frac{\mathcal{P}}{L} \text{ kg/(s.m}^2\text{. Pa)} \qquad \text{...(15.49)}$$

The reciprocal of permeance is called **vapour resistance** R_v and is expressed as

Vapour resistance,

$$R_v = \frac{1}{\text{Permeance}}$$
$$= \frac{1}{\mathcal{M}} = \frac{L}{\mathcal{P}} \text{ (s.m}^2\text{.Pa/kg)} \qquad \text{...(15.50)}$$

The vapour resistance is the property of material, which offers the resistance to vapour transmission through the material. The overall vapour resistance to composite structure that consists of several layers in series is the sum of vapour resistances of individual layers and is given by

$$R_{v,\,total} = R_{v,1} + R_{v,2} + R_{v,3} + ... + R_{v,n}$$
$$= \sum_{i=1}^{i=n} R_{v,i} \quad ...(15.51)$$

Then the rate of vapour diffusion through a composite structure can be expressed in an analogous manner of heat transfer rate :

$$\dot{m}_v = A\frac{\phi_1 p_{sat,1} - \phi_2 p_{sat,2}}{R_{v,\,total}} \text{ (kg/s)} \quad ...(15.52)$$

Example 15.8. *A house is maintained at 20°C and 60% RH at a location where atmospheric pressure is 97 kPa. The inside of the wall is finished with 9.5 mm thick gypsum plaster. The vapour pressure at the outside of the plaster is zero. Calculate the maximum amount of water vapour that will diffuse through a 3 m × 8 m section of a wall during 24 hour period. The permeance of the 9.5 mm thick gypsum plaster to water vapour is 2.86 × 10⁻⁹ kg/(s.m².Pa).*

Solution

Given : A layer of gypsum plaster

$L = 9.5 \text{ mm} = 9.5 \times 10^{-3}$ m

$A = 3 \text{ m} \times 8 \text{ m} = 24 \text{ m}^2$

$\Delta t = 24$ h

$\phi_1 = 0.6,\quad p_1 = 97$ kPa

$T_1 = 20°C,\quad p_{v_2} = 0$

$\mathcal{M} = 2.86 \times 10^{-9}$ kg/(s.m².Pa)

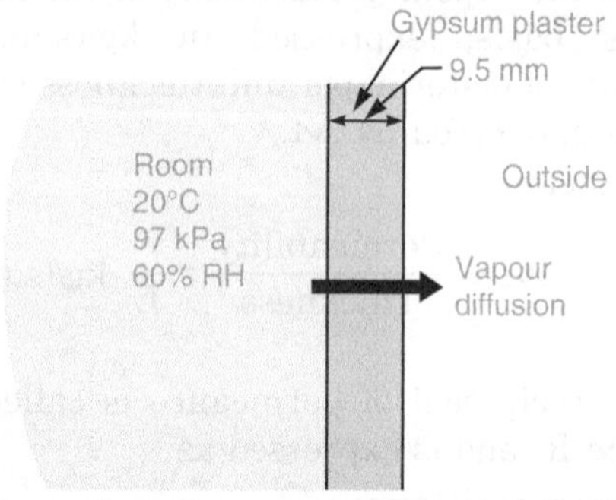

Fig. 15.12. Schematic for example 15.8

To find: Vapour diffusion rate through gypsum layer.

Analysis: The saturation pressure of water at 20°C (from steam tables) is 2340 Pa. Thus vapour pressure at inside of gypsum plaster

$$p_{v,1} = \phi_1 p_{sat,1} = 0.6 \times 2340 = 1404 \text{ Pa}$$
$$p_{v,2} = 0$$

Vapour resistance,

$$R_v = \frac{1}{\mathcal{M}} = \frac{1}{2.86 \times 10^{-9}}$$
$$= 3.496 \times 10^8 \text{(s.m}^2\text{.Pa/kg)}$$

The diffusion rate of vapour through gypsum plaster

$$\dot{m}_v = A\frac{\phi_1 p_{v,1} - \phi_2 p_{v,2}}{R_{v,\,total}}$$
$$= 24 \times \frac{1404 - 0}{3.496 \times 10^8}$$
$$= 9.64 \times 10^{-5} \text{ (kg/s)}$$

In 24 hours period

$$\dot{m}_v = 9.64 \times 10^{-5} \times 24 \times 3600$$
$$= \mathbf{8.32 \text{ kg/day.}} \quad \text{Ans.}$$

15.8.3. Equimolar Counter Diffusion

In the distillation of two components, whose molar latent heats of vaporisation are equal, a physical situation encountered, that the flux of one gaseous component is equal to but acting in opposite direction from other gaseous component. Such type of diffusion is termed as *equimolar counter diffusion.*

Consider two large chamber containing gas component A and gas component B are connected by a passage as shown in Fig. 15.13 in such a way that the each molecule of gas A is replaced by a molecule of B and *viceversa.*

The total pressure $p = p_A + p_B$ (uniform throughout)

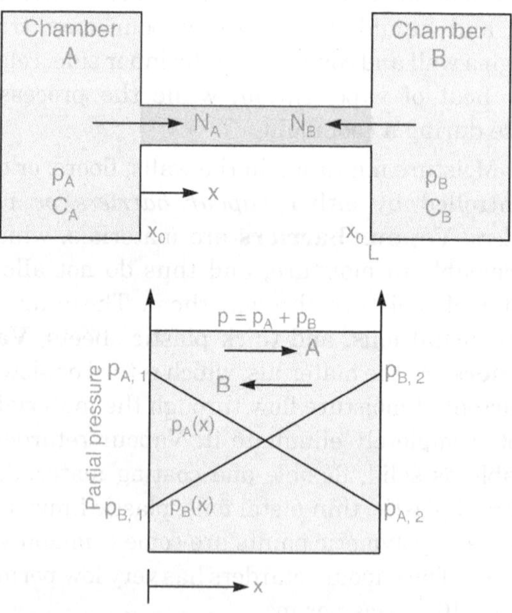

Fig. 15.13. Equimolar counter diffusion in a binary mixture

Using Fick's law, the molar diffusion rate of component A and B are given by

$$\dot{N}_A = -D_{AB} A \frac{dC_A}{dx} \quad ...(15.53)$$

and $$\dot{N}_B = -D_{BA} A \frac{dC_B}{dx} \quad ...(15.54)$$

For steady state conditions

$$\dot{N}_A = -\dot{N}_B \quad \text{or} \quad \dot{N}_A + \dot{N}_B = 0$$

and $$D_{AB} = D_{BA}$$

we get $$\frac{dC_A}{dx} = -\frac{dC_B}{dx}$$

Rearranging eqn. (15.53) and integrating

$$\dot{N}_A \int_0^L dx = -D_{AB} A \int_{C_{A,1}}^{C_{A,2}} dC_A$$

$$\dot{N}_A L = -D_{AB} A (C_{A,2} - C_{A,1})$$

or $$\dot{N}_A = \frac{D_{AB} A (C_{A,1} - C_{A,2})}{L} \quad ...(15.55)$$

For a perfect gas, the molar concentration of A is related to partial pressure

$$C_A = \frac{n_A}{V} = \frac{\rho_A}{M_A} = \frac{p_A}{R_u T}$$

For an ideal gas A,

$$\dot{N}_A = \frac{D_{AB} A}{R_u T} \frac{(p_{A,1} - p_{A,2})}{L} \text{ kg-mole/m}^2\text{s} \quad ...(15.56(a))$$

Similarly, for ideal gas B

$$\dot{N}_B = \frac{D_{BA} A}{R_u T} \frac{(p_{B,1} - p_{B,2})}{L} \text{ kg-mole/m}^2\text{s} \quad ...(15.56(b))$$

Equations (15.56(*a*)) and (15.56(*b*)) are equations of steady state equimolar counter diffusion.

It is noticed that the net molar flow rate through the channel is zero, but the net mass flow rate of the mixture through the channel is not zero and it is sum of two streams of mass flow.

$$\dot{m} = \dot{m}_A + \dot{m}_B = \dot{N}_A M_A + \dot{N}_B M_B = \dot{N}_A (M_A - M_B) \quad ...(15.57)$$

It is also observed that the net mass flows in the flow direction of the gas with the higher molar mass.

Example 15.9. *A tank contains a mixture of CO_2 and N_2 in the mole proportion of 0.2 and 0.8 at 1 bar and 290 K. It is connected by a duct of cross-sectional area 0.1 m², 0.5 m long to another tank containing mixture of CO_2 and N_2 in the molar proportion of 0.8 and 0.2 respectively. Calculate the diffusion rates of CO_2 and N_2. Assume diffusivity coefficient $D_{AB} = 0.17 \times 10^{-4}$ m²/s.*

Solution

Given : Equimolar diffusion of CO_2 and N_2

$L = 0.5$ m

$A_c = 0.1$ m²

$D_{AB} = 0.17 \times 10^{-4}$ m²/s

$p = 1$ bar

$T = 290$ K.

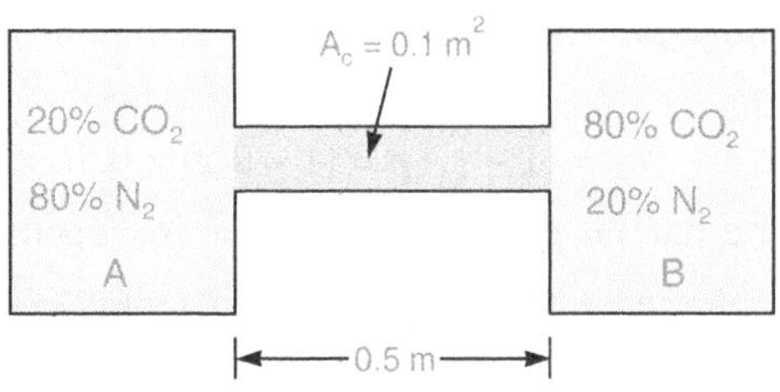

Fig. 15.14

To find : Diffusion rates of CO_2 and N_2.

Analysis : The partial pressures of two gases in two tanks are :

$$p = p_1 + p_2 \text{ and } p_i = x_i p$$

$$p_{A,1} = 0.2 \text{ bar}, \quad p_{A,2} = 0.8 \text{ bar},$$

and $$p_{B,1} = 0.8 \text{ bar}, \quad p_{B,2} = 0.2 \text{ bar}$$

For rate of molar diffusion of CO_2 using eqn. (15.56(*a*))

$$\dot{N}_A = \frac{D_{AB} A}{R_u T} \left[\frac{p_{A_2} - p_{A_1}}{L} \right]$$

$$= \frac{0.17 \times 10^{-4} \times 0.1 \times (0.8 - 0.2) \times 10^5}{8314 \times 290 \times 0.5}$$

$$= 8.46 \times 10^{-8} \text{ kg-mole/s}$$

and $$\dot{m}_A = \dot{N}_A M_A = 8.46 \times 10^{-8} \times 44$$

$$= \mathbf{3.73 \times 10^{-6} \text{ kg/s.}} \quad \text{Ans.}$$

Similarly for N_B

$$\dot{N}_B = -\dot{N}_A = 8.46 \times 10^{-8} \text{ kg-mole/s}$$

$$\dot{m}_B = \dot{N}_B M_B = 8.46 \times 10^{-8} \times 28$$

$$= \mathbf{2.37 \times 10^{-6} \text{ kg/s.}} \quad \text{Ans.}$$

Example 15.10. *The mixture of benzene and toluene is being distilled in a distillation column at 1 bar and 108°C. The liquid phase contains 20% mole of benzene and vapour phase contains 60% mole of benzene, respectively. At the operating conditions, the vapour pressure of toluene is 0.75 bar and its diffusivity is 5.2 × 10⁻⁶ m²/s. Assuming the equimolar diffusion, calculate molar diffusion flux of toluene, if the diffusion zone is 0.2 mm thick.*

Solution

Given : Equimolar diffusion of benzene and toluene.

$p = 1 \text{ bar} = 10^5 \text{ Pa}, \quad T = 108°C = 381 \text{ K}$

$\gamma_A = 0.2, \quad \gamma_B = 0.6$

$p_{v,\text{ toluene}} = 0.75 \text{ bar}, \quad D_{AB} = 5.2 \times 10^{-6} \text{ m}^2/\text{s}$

$L = 0.2 \text{ mm} = 2 \times 10^{-4} \text{ m}.$

To find : Molar diffusion flux of toluene.

Analysis : The partial pressure of toluene in liquid phase

$$p_{t_1} = (1 - \gamma_A)\, p = (1 - 0.2) \times 0.75 = 0.6 \text{ bar}$$

The partial pressure of toluene in vapour phase

$$p_{t_2} = (1 - \gamma_B)\, p = (1 - 0.6) \times 0.75 = 0.3 \text{ bar}$$

For the equimolar diffusion, the molar diffusion flux, eqn. (15.57)

$$N_{\text{toluene}} = \frac{D_{AB}}{R_u T} \frac{(p_{t_1} - p_{t_2})}{L}$$

$$= \frac{5.2 \times 10^{-6} \times (0.6 - 0.3) \times 10^5}{8314 \times 381 \times 2 \times 10^{-4}}$$

$$= \mathbf{2.46 \times 10^{-6} \text{ kg-mole/m}^2\text{s.}} \quad \text{Ans.}$$

Example 15.11. *The pressure of ammonia gas flowing at the rate of 2 kg/s through an industrial pipeline is maintained close to 1 atm by venting ammonia to atmosphere air through a 3 mm diameter tube that extends 15 m into the air. Assuming entire system is operating at 25°C, determine :*

(a) mass diffusion rate of ammonia lost to atmosphere through a tube,

(b) mass flow rate of air that infiltrates into the pipeline,

(c) mole and mass fractions of air in the pipe,

(d) the steady flow velocity at the bottom of the tube, where it is attached to the pipeline.

Take $D_{AB} = 2.8 \times 10^{-5} \text{ m}^2/\text{s}$.

Solution

Given : Venting of Ammonia pipeline to atmosphere

$p = 1 \text{ atm}, \quad T = 25°C$

$L = 15 \text{ m}, \quad d = 3 \text{ mm}$

$D_{AB} = 2.8 \times 10^{-5} \text{ m}^2/\text{s}$

To find :

(*a*) Mass rate of ammonia lost through the vent.

(*b*) Mass rate of air infiltration into pipeline.

(*c*) Mole and mass fraction of air in the pipeline.

(*d*) The steady flow velocity at the bottom of tube.

Assumptions :

1. Steady, one dimensional diffusion in the tube.
2. Ammonia and atmospheric air are ideal gases.
3. No chemical reaction occur in the tube.
4. Negligible mole fraction of air in the pipe line and negligible mole fraction of ammonia in the atmosphere.
5. The molecular mass of ammonia is 17 kg/kg-mole and that of air is 28.97 kg/kg-mole.

Analysis : This is a typical equimolar counter diffusion process between two large reservoirs.

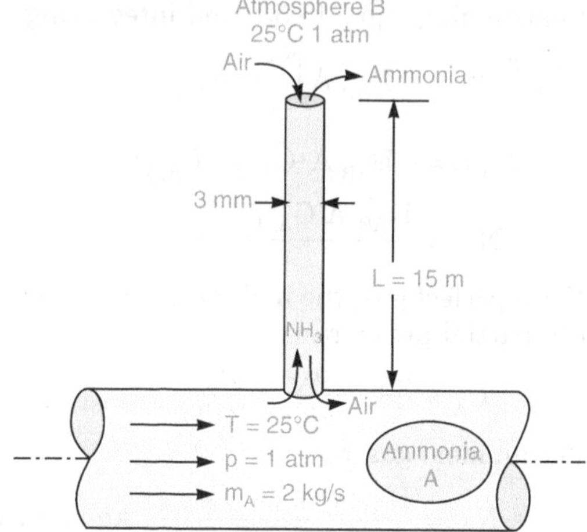

Fig. 15.15. Schematic for example 15.11

(*a*) Cross-sectional area of the vent tube:

$$A_c = \frac{\pi}{4} d^2 = \frac{\pi}{4} \times (3 \times 10^{-3})^2$$

$$= 7.068 \times 10^{-6} \text{ m}^2$$

For ammonia

$p_{A,\,0} = 1 \text{ atm}$

$= 1.01325 \times 10^2$ kPa at the bottom of vent tube

$p_{A,\,1} = 0$ at the top of vent tube,

From eqn. (15.56(*a*)), molar diffusion rate of ammonia:

$$\dot{N}_A = \frac{D_{AB} A}{R_u T} \times \left(\frac{p_{A,0} - p_{A,1}}{L} \right)$$

$$= \frac{(2.8 \times 10^{-5} \text{ m}^2/\text{s}) \times (7.068 \times 10^{-6} \text{ m}^2)}{8.314 \text{ (kJ/kg-mole K)} \times (298 \text{ K})} \times \frac{(1.01325 \times 10^2 - 0) \text{ (kPa)}}{(15 \text{ m})}$$

$$= \mathbf{5.395 \times 10^{-13} \text{ kg-mole/s.}} \quad \text{Ans.}$$

Using molecular mass of ammonia,

$$\dot{m}_A = \dot{N}_A M_A = 5.395 \times 10^{-13} \times 17$$
$$= \mathbf{9.171 \times 10^{-12}\ kg/s.} \quad \text{Ans.}$$

(*b*) Mass rate of air infiltration into pipeline is evaluated from equimolar counter-diffusion process

$$\dot{N}_B = -\dot{N}_A = -5.395 \times 10^{-13} \text{ kg-mole/s}$$

with molecular mass of air, M_B = 28.97 kg/kg-mole, the mass rate of air

$$\dot{m}_B = -\dot{N}_B M_B = -5.395 \times 10^{-13} \times 28.97$$
$$= \mathbf{-1.563 \times 10^{-11}\ kg/s.} \quad \text{Ans.}$$

(*c*) For ammonia flow rate of 2 kg/s, the mass fraction of air in the pipeline:

$$x_B = \frac{|\dot{m}_B|}{\dot{m}_{total}} = \frac{|\dot{m}_B|}{(\dot{m}_{NH_3} + \dot{m}_A - \dot{m}_B)}$$
$$= \frac{1.563 \times 10^{-11}}{(2 + 9.171 \times 10^{-12} - 1.563 \times 10^{-11})}$$
$$= 7.815 \times 10^{-12} \approx \mathbf{0.} \quad \text{Ans.}$$

The molar flow rate of ammonia

$$\dot{n} = \frac{\dot{m}_A}{M_A} = \frac{2 \text{ kg/s}}{(17 \text{ kg/kg-mole})}$$
$$= 0.1176 \text{ kg-mole/s}$$

Mole fraction of air in the pipeline

$$\gamma_B = \frac{|N_B|}{\dot{n}} = \frac{5.395 \times 10^{-13}}{0.1176}$$
$$= 4.585 \times 10^{-12} \approx \mathbf{0.} \quad \text{Ans.}$$

(*d*) Net mass flow rate through the tube is

$$\dot{m}_{net} = \dot{m}_A + \dot{m}_B$$
$$= 9.171 \times 10^{-12} - 1.563 \times 10^{-11}$$
$$= -6.46 \times 10^{-12} \text{ kg/s}$$

Density of ammonia

$$\rho_A = \frac{p_A M_A}{R_u T} = \frac{101.325 \times 17}{8.314 \times 298}$$
$$= 0.695 \text{ kg/m}^3$$

From continuity equation, the flow velocity

$$u = \frac{\dot{m}_{net}}{\rho_A A} = \frac{-6.46 \times 10^{-12}}{0.695 \times 7.068 \times 10^{-6}}$$
$$= \mathbf{-1.31 \times 10^{-6}\ m/s.} \quad \text{Ans.}$$

Flow in negative x direction *i.e.*, towards pipeline.

15.8.4. Diffusion Through a Stagnant Gas: Stefan's Flow

The evaporation of the pure liquids such as benzene, acetone, carbon tetrachloride, gasolene, water in air are some examples of diffusion through stagnant air.

Consider the ***isothermal evaporation*** of water from a surface and its diffusion upward through a stagnant air as shown in Fig. 15.16. The free surface of water is exposed to air in the tank and air flowing over the tank in perpendicular direction and it removes the water vapour which reaches to that level.

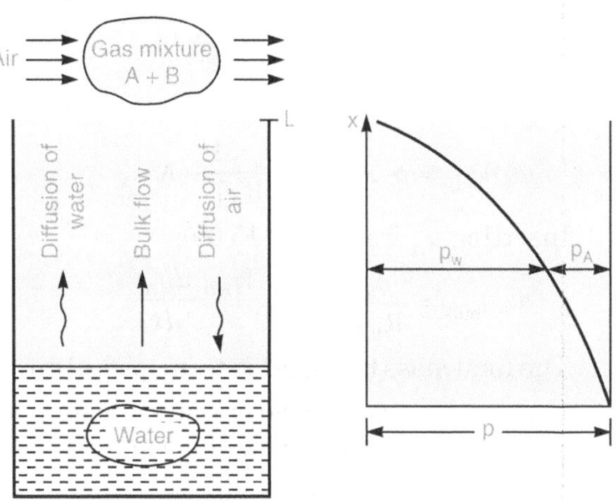

Fig. 15.16. Diffusion of water vapour into air

We assume that

1. System is isothermal and total pressure remains constant.

2. The system is in steady state and hence the concentration profile in air does not change.

3. Air and water vapour behave as an ideal gas.

As water evaporates, it diffuses upward through air and in steady state, the upward movement of water vapour is balanced by a downward diffusion of air.

Therefore, the concentration at any location x remains constant.

The diffusion rate of air downward is given by eqn. [15.24(*a*)]

$$\dot{m}_A = -\frac{D_{AB} A M_A}{R_u T} \frac{dp_A}{dx} \quad \text{...(15.58)}$$

where A = cross-section area of the tank

M_A = molecular weight of component A.

At the surface of water, the air cannot move downward ; hence there must be bulk mass movement upward with a velocity to compensate for the mass diffusion of air downward.

Upward mass transfer of air

$$= -\rho_A u_A A$$
$$= -\frac{p_A}{R_u T} M_A A u_A \quad \text{...(15.59)}$$

where u_A is bulk mass velocity in upward direction.

Equating the two eqns. (15.58) and (15.59)

$$-\frac{D_{AB} A M_A}{R_u T} \frac{dp_A}{dx} = -\frac{p_A}{R_u T} M_A A u_A$$

we get $$u_A = \frac{D_{AB}}{p_A}\frac{dp_A}{dx} \quad ...(15.60)$$

Diffusion rate of water upward is given as

$$\dot{m}_w = -\frac{D_{AB} A M_w}{R_u T}\frac{dp_w}{dx} \quad ...(15.61)$$

The bulk transport rate of water vapour upward;

$$(\dot{m}_w)_{bulk} = \rho_w A u_A = \frac{p_w M_w}{R_u T} A u_A$$

Inserting u_A from eqn. (15.60);

$$(\dot{m}_w)_{bulk} = \frac{p_w}{R_u T} M_w A \frac{D_{AB}}{p_A}\frac{dp_A}{dx} \quad ...(15.62)$$

The total mass transport rate in upward direction = upward mass diffusion of water + water vapour carried away by bulk movement of moving air

$$(\dot{m}_w)_{total} = -\frac{D_{AB} A M_w}{R_u T}\frac{dp_w}{dx} + \frac{p_w}{p_A}\frac{M_w A D_{AB}}{R_u T}\frac{dp_A}{dx}$$

or $$(\dot{m}_w)_{total} = -\frac{D_{AB} A M_w}{R_u T}\left[\frac{dp_w}{dx} - \frac{p_w}{p_A}\frac{dp_A}{dx}\right] \quad ...(15.63)$$

The partial pressure of water vapour p_w can be expressed in terms of total pressure, by using Dalton's law of partial pressures;

$$p_A + p_w = p \quad \text{(constant)}$$

or $$\frac{dp_A}{dx} = -\frac{dp_w}{dx}$$

Then total mass flow rate of water vapour,

$$(\dot{m}_w)_{total} = -\frac{D_{AB} A M_w}{R_u T}\frac{(p_A + p_w)}{p_A}\frac{dp_w}{dx}$$

or $$(\dot{m}_w)_{total} = -\frac{D_{AB} A M_w}{R_u T}\frac{p}{p - p_w}\frac{dp_w}{dx} \quad ...(15.64)$$

This relation is known as **Stefan's law**, and the induced convective flow that enhances the mass diffusion is called Stefan's flow. Integrating eqn. (15.64)

$$(\dot{m}_w)_{total}\int_0^L dx = -\frac{D_{AB} A M_w}{R_u T} p \int_{p_s}^{p_\infty}\frac{dp_w}{p - p_w}$$

$$(\dot{m}_w)_{total} L = \frac{D_{AB} A M_w}{R_u T} p \ln\left(\frac{p - p_\infty}{p - p_s}\right)$$

or $$\frac{(\dot{m}_w)_{total}}{A} = \frac{D_{AB} M_w p}{R_u T L}\ln\left(\frac{p - p_\infty}{p - p_s}\right)$$

$$= \frac{D_{AB} p M_w}{R_u T L}\ln\left(\frac{p_{A_2}}{p_{A_1}}\right) \quad ...(15.65)$$

The distribution of partial pressures of air and water vapour with distance x is shown in Fig. 15.17.

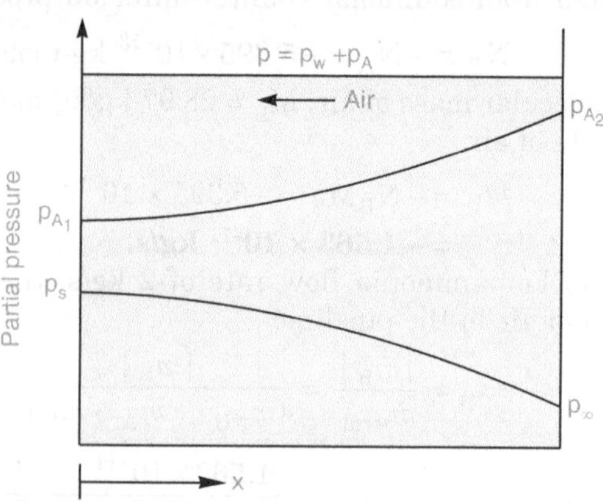

Fig. 15.17. Variation of partial pressures of air and water during evaporation

The diffusion of a substance through a stationary medium in terms of concentrations can be expressed as :

$$\frac{N_A}{A} = \frac{D_{AB} C}{L}\ln\left(\frac{C - C_{A_2}}{C - C_{A_1}}\right) \text{ mol/m}^2\text{s} \quad ...(15.66)$$

As some amount of water evaporates from the free surface of water, it absorbs the latent heat of vaporisation from the water thus the temperature of water decreases. In order to maintain the isothermal condition in the tank during evaporation, the heat must be supplied to the tank at the rate of

$$Q = \dot{m}_A h_{fg,A} = j_A A_s h_{fg,A}$$

$$= (J_A M_A) A_s h_{fg,A} \quad ...(15.67)$$

where J_A = diffusional molar flux,
M_A = molar mass,
j_A = diffusional mass flux,
$h_{fg,A}$ = latent of vaporisation,
A_s = surface area of liquid water interface of species A.

Example 15.12. *An open pan 20 cm in diameter 20 mm deep is filled with water to a level of 10 mm and is exposed to air at 25°C. Assuming mass diffusivity of 0.25 × 10⁻⁴ m²/s, calculate the time required for all the water to evaporate.*

Solution

Given : Evaporation of water from an open pan.
d = 20 cm = 0.2 m
L = 20 mm – 10 mm = 10 mm = 0.01 m

T = 25°C = 298 K,
$D_{AB} = 0.25 \times 10^{-4}$ m²/s.

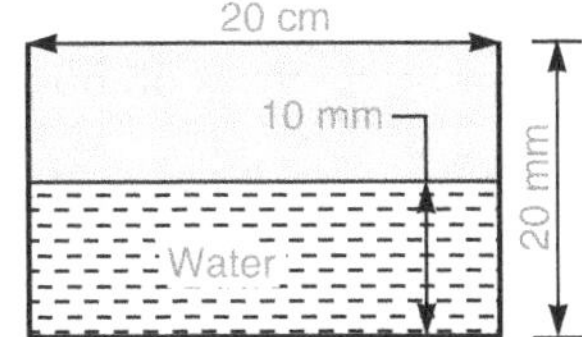

Fig. 15.18. Schematic for example 15.12

To find : Time required for all water to evaporate.

Assumption : Total pressure of 1 atm.

Analysis : The partial pressure of water vapour, corresponding to saturation temperature of 298 K.

$$p_s = 3.169 \text{ kPa (from steam table)}$$

$$p_\infty = 0 \text{ at the top of pan, dry air.}$$

For water diffusion in stagnant air

$$(\dot{m}_w)_{\text{total}} = \frac{D_{AB} A M_w}{R_u T} \frac{p}{L} \ln\left(\frac{p - p_\infty}{p - p_s}\right)$$

$$A = \frac{\pi}{4} d^2 = \frac{\pi}{4} \times (0.2)^2 = 0.0314 \text{ m}^2$$

$$M_w = 18,$$

$$p = 1.01325 \times 10^5 \text{ N/m}^2$$

$$\therefore \quad (\dot{m}_w)_{\text{total}} = \frac{0.25 \times 10^{-4} \times 0.0314 \times 18}{8314 \times 298} \times \frac{1.01325 \times 10^5}{0.01} \times \ln\left(\frac{101.32 - 0}{101.325 - 3.169}\right)$$

$$= 1.869 \times 10^{-6} \text{ kg/s}$$

Time for evaporation of total water

$$(\dot{m}_w)_{\text{total}} \times t = \text{Area} \times \text{height of water} \times \rho_w$$

$$1.869 \times 10^{-6} \times t = \frac{\pi}{4} \times (0.2)^2 \times 0.01 \text{ m} \times 1000$$

or $t = 168090 =$ **46.69 h. Ans.**

Example 15.13. *A well is 40 m deep and 9 m in diameter is exposed to atmosphere at 25°C. The air at the top has relative humidity of 50%. Calculate the rate of diffusion of water vapour through the wall.*

Take $D_{AB} = 2.58 \times 10^{-5}$ m²/s, $M_w = 18$ kg/kg-mole.

Solution

Given : The evaporation of water in stagnant air.

L = 40 m, $d = 9$ m

T = 25°C = 298 K, RH = 0.5

$D_{AB} = 2.58 \times 10^{-5}$ m²/s, $M_w = 18$ kg/kg-mole.

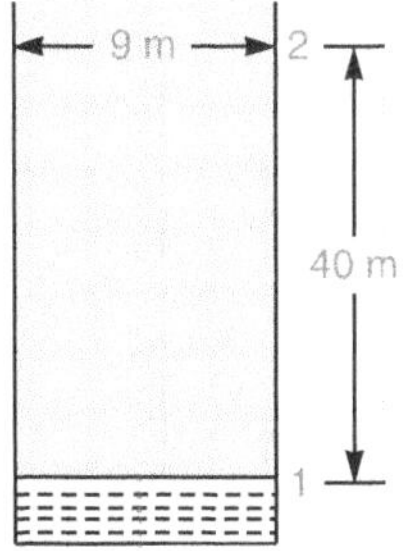

Fig. 15.19. Schematic for example 15.13

To find : Diffusion rate of water from the well.

Analysis : The saturation pressure p_s of water vapour at level 1 corresponds to 25°C

$$p_s = 3.169 \text{ kPa}$$

Partial pressure of air at top of well

$$p_\infty = \text{RH} \times p_s = 0.5 \times 3.169 = 1.585 \text{ kPa}$$

The diffusion rate of water into air

$$(\dot{m}_w)_{\text{total}} = \frac{D_{AB} A M_w}{R_u T} \frac{p}{L} \times \ln\left(\frac{p - p_\infty}{p - p_s}\right)$$

$$= \frac{2.58 \times 10^{-5} \times (\pi/4) \times 9^2 \times 18}{8314 \times 298} \times \frac{1.01325 \times 10^5}{40} \times \ln\left(\frac{101.325 - 1.585}{101.325 - 3.169}\right)$$

$$= 1.1924 \times 2533.125 \times 0.0160$$

$$= \mathbf{4.835 \times 10^{-7} \text{ kg/s.}} \quad \textbf{Ans.}$$

Example 15.14. *Estimate the diffusion rate of water from the bottom of the test tube 1.5 cm in diameter and 15 cm long into dry atmospheric air at 25°C.*

Take diffusion coefficient of 25.6×10^{-6} m²/s.

(V.T.U., July 2002 ; J.N.T.U., May 2004)

Solution

Given : Evaporation of water from a test tube.

$d = 1.5$ cm = 0.015 m

L = 15 cm = 0.15 m

T = 25°C = 298 K

$D_{AB} = 25.6 \times 10^{-6}$ m²/s.

To find : Diffusion rate of water.

Analysis : The saturation pressure of water vapour at level 1, corresponds to temperature of 25°C

$$p_s = 3.169 \text{ kPa}$$

The partial pressure of water at the top of tube, dry air

$$p_\infty = 0$$

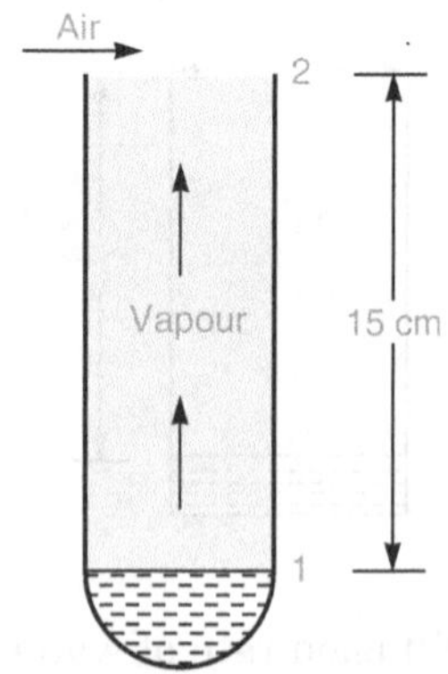

Fig. 15.20. Schematic for example 15.14

The diffusion rate of water from the tube

$$(\dot{m}_w)_{total} = \frac{D_{AB} A\, M_w}{R_u T} \frac{p}{L} \ln\left(\frac{p - p_\infty}{p - p_s}\right)$$

$$= \frac{25.6 \times 10^{-6} \times (\pi/4) \times (0.015)^2 \times 18}{8314 \times 298}$$

$$\times \frac{1.01325 \times 10^5}{0.15} \times \ln\left(\frac{101.32 - 0}{101.32 - 3.169}\right)$$

$$= 3.28 \times 10^{-14} \times 675500 \times 0.0317$$

$$= 7.05 \times 10^{-10} \text{ kg/s}$$

$$= \mathbf{0.0025 \text{ gm/h.}}$$ Ans.

Example 15.15. *In a solar pond, the salt is placed at the bottom of the pond 1.5 m deep. The surface is flushed constantly, so that the concentration of salt at the top layer is zero. The salt concentration at the bottom layer is 5 kg-mole/m³. Calculate the rate at which salt is washed off at the top at steady state condition per m². Take D_{AB} = 1.24 × 10⁻⁹ m²/s. Total concentration = 55.55 kg-mole/m³.*

Solution

Given : Diffusion of salt into a solar pond.

$L = 1.5$ m, $C_{A_2} = 0$,

$C_{A_1} = 5$ kg-mole/m³ $D_{AB} = 1.24 \times 10^{-9}$ m²/s,

$C = 55.55$ kg-mole/m³

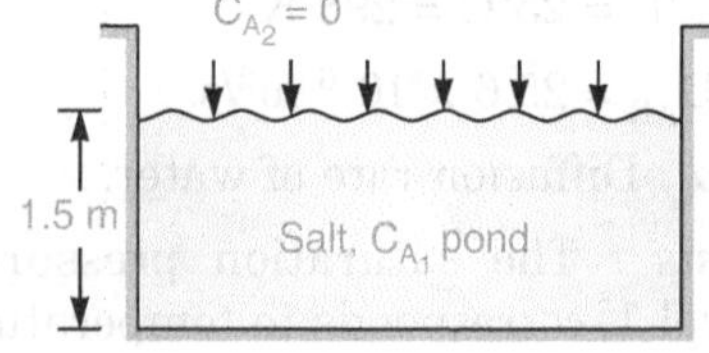

Fig. 15.21. Schematic for example 15.15

To find : Diffusion rate of salt.

Analysis : The diffusion rate of salt through solar pond is given by

$$\frac{N_A}{A} = \frac{D_{AB}}{L} C \ln\left(\frac{C - C_{A_2}}{C - C_{A_1}}\right)$$

$$= \frac{1.24 \times 10^{-9}}{1.5} \times 55.55 \times \ln\left(\frac{55.55 - 0}{55.55 - 5}\right)$$

$$= \mathbf{4.33 \times 10^{-9} \text{ kg-mole/m}^2\text{s.}}$$ Ans.

15.9. MASS DIFFUSION WITH HOMOGENEOUS CHEMICAL REACTIONS

When a homogeneous chemical reaction taken place, the species concentration may be influenced. The steady state, one dimension, x-directional molar diffusion equation with uniform species generation per unit volume is reduced to

$$D_{AB} \frac{d^2 C_A}{dx^2} + \dot{N}_{A,g} = 0 \qquad ...(15.68)$$

where $\dot{N}_{A,g}$ is volumetric species generation rate. If $\dot{N}_{A,g}$ is positive, the reaction results in species generation; if it is negative, it results in species consumption. When chemical reaction occurs at constant rate, then its reaction order is considered to be zero order; that is :

$$\dot{N}_{A,g} = k_0$$

and when the chemical reaction occurs that is proportional to local concentration, then it is first order reaction :

$$\dot{N}_{A,g} = -k_1 C_A$$

With first order chemical reaction, eqn. (15.68) yields to

$$D_{AB} \frac{d^2 C_A}{dx^2} - k_1 C_A = 0 \qquad ...(16.69)$$

Using $m = \sqrt{\frac{k_1}{D_{AB}}}$, then above equation modifies to

$$\frac{d^2 C_A}{dx^2} - m^2 C_A = 0$$

It is second order, linear homogeneous differential equation. Its generals solution is as in case of fins;

$$C_A(x) = C_1 e^{-mx} + C_2 e^{mx} \qquad ...(15.70)$$

where C_1 and C_2 are the constants of integrations, which are evaluated with the help of prescribed boundary conditions.

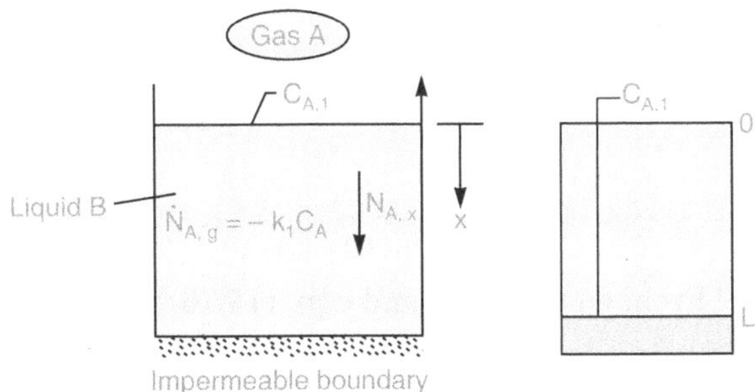

Fig. 15.22. Diffusion and homogeneous reaction of gas A in liquid B

Consider a situation shown in Fig. 15.22. The gas A is solute in liquid B, where it experiences diffusion process and first order chemical reaction. The concentration of A at the interface is known to be $C_{A,1}$. The bottom of the container is impermeable to A, then the boundary conditions are

$$x = 0 \qquad C_A(x) = C_{A,1}$$

and at $$x = L \qquad \left.\frac{dC_A}{dx}\right|_{x=L} = 0$$

using these boundary conditions in eqn. (15.70), we get

$$C_A(x) - C_{A,1}\frac{\cosh m(L-x)}{\cosh mL} \qquad \ldots(15.71)$$

At $x = L$, we get

$$C_{A,L} = \frac{C_{A,1}}{\cosh mL} \qquad \ldots(15.72)$$

Moreover, the molar flux at liquid-gas interface;

$$N_{A,x}(0) = -D_{AB}\left.\frac{dC_A}{dx}\right|_{x=0}$$

$$= -D_{AB}C_{A,1}\, m[\sinh mx - \tanh mL \cosh mx]_{x=0}$$

or $$N_{A,x}(0) = D_{AB}\, C_{A,1}\; m \tanh mL \qquad \ldots(15.73)$$

Example 15.16. *Consider the emission of NO_2 from automobiles. This NO_2 reacts chemically in atmosphere with unburned hydrocarbons (aided by sunlight) and produces smog. The reaction is first order with $k_1 = 0.03\ (s)^{-1}$. Assume steady state condition. Obtain the variation of NO_2 concentration in altitude in terms of surface flux N_A of NO_2, if the concentration of NO_2 well above the ground is zero. Take*

$$D_{AB} = 0.18 \times 10^{-4}\ m^2/s.$$

Solution

Given : First order reaction of NO_2 with atmospheric air to form smog.

$$k_1 = 0.035^{-1} \quad C_{A,1} = 0 \text{ as } x \to \infty$$
$$D_{AB} = 0.18 \times 10^{-4}\ m^2/s$$

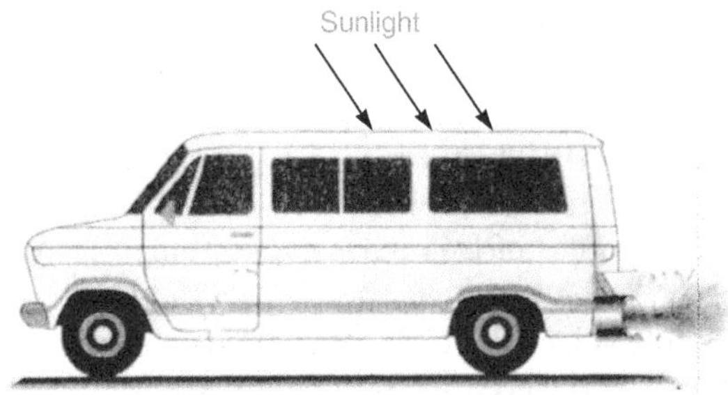

Fig. 15.23. Schematic for example 15.16

To find: Variation of NO_2 concentration with height.

Assumptions :

1. Uniform reaction rate.
2. Stagnant atmosphere.
3. Atmospheric temperature as 300 K.

Analysis : The molar concentration with first degree of reaction is given by eqn. (15.70)

$$C_A(x) = C_1\, e^{-mx} + C_2\, e^{mx}$$

where $$m = \sqrt{\frac{k_1}{D_{AB}}} = \sqrt{\frac{0.03}{0.18\times 10^{-4}}} = 40.82\ m^{-1}$$

using boundary condition well above ground $x \to \infty$

$$0 = C_1\, e^{-m\times\infty} + C_2 e^{m\times\infty}$$

It gives $C_2 = 0$, then

$$C_A(x) = C_1 e^{-mx}$$

The molar flux at ground level is given by

$$N_{A,0} = -D_{AB}\left.\frac{dC_A}{dx}\right|_{x=0}$$

$$= -D_{AB}\left[-C_1\, me^{-m\times 0}\right]$$

$$= 40.82 \times 0.18 \times 10^{-4}\, C_1$$

or $$C_1 = 1360.82\ N_A,$$

we get $$C_A(x) = \mathbf{1360.82\ N_A\ e^{-40.82\,x}}. \quad \text{Ans.}$$

15.10. CONVECTIVE MASS TRANSFER

Convective mass transfer involves the diffusion of a fluid by bulk motion of mixture. When a fluid moves over a surface, the transfer of mass occurs due to *mass diffusion* and *bulk movement of fluid.* The situation is analogous to convective heat transfer. The fluid movement augments the mass transfer considerably, by replacing the high concentration fluid near the surface by the low concentration fluid. In absence of bulk motion of fluid, the mass convection reduces to mass diffusion.

Convective mass transfer can also be classified as free mass convection and forced mass convection. In *free mass convection*, the concentration gradient changes the density of the fluid, which may be acted upon by the buoyancy force. The evaporation of alcohol or gasoline are the examples of free mass convection.

In *forced mass convection* one fluid moves with appreciable velocity over another fluid. The evaporation of water from the ocean, cooling towers, and the exhaust gases from an automobile that are forceed to pass over a catalytic converter are some examples of forced mass convection.

The rate of mass convection is influenced by fluid properties, flow geometry, flow velocity, flow regimes and composition. The mechanism of convection mass transfer is also intricated like convection heat transfer. Therefore, we rely on experimental relations to obtain mass transfer rate.

15.10.1. Mass Transfer Coefficient

Convective mass transfer coefficient h_m can be defined in similar manner that used for defining convective heat transfer coefficient h. The rate of mass convention across the boundary can be expressed as

$$\dot{m}_{\text{conv}} = h_m \, A_s(\rho_{A,s} - \rho_{A,\infty})$$

$$= h_m \, A_s\rho(x_{A,s} - x_{A,\infty}) \text{ (kg/s)} \quad ...(15.74)$$

Where h_m is the *average mass transfer coefficient*, measured in m/s, A_s is the surface area, $(\rho_{A,s} - \rho_{A,\infty})$ is the mass concentration difference of species A across the boundary layer, ρ is average density of the fluid in the boundary and product of $h_m\rho$ is the *mass transfer conductance* and measured in kg/m^2.s. The local mass transfer rate $\dot{m}_{conv}(x)$ is expressed in terms of local mass transfer coefficient h_{mx} as

$$\dot{m}_{conv}(x) = h_{mx} \, A_s \, (\rho_{A,s} - \rho_{A,\infty})$$

$$= h_{mx} \, A_s \, \rho(x_{A,s} - x_{A,\infty}) \quad ...(15.75)$$

Like heat convection at the boundary, there is no slip condition of species at the surface, therefore, the species transfer at the surface $(y = 0)$ is by diffusion only and the mass flux at the surface can be expressed by Fick's law:

$$j_A = \frac{\dot{m}_A}{A_s} = -D_{AB} \left.\frac{d\rho_A}{dy}\right|_{y=0}$$

$$= -\rho D_{AB} \left.\frac{dx_A}{dy}\right|_{y=0} \quad ...(15.76)$$

From eqn. (15.75) and eqn. (15.76), we get

$$h_{mx} = \frac{-D_{AB}\dfrac{d\rho_A}{dy}}{(\rho_{A,s} - \rho_{A,\infty})} = \frac{-D_{AB}\dfrac{dx_A}{dy}}{(x_{A,s} - x_{A,\infty})} \quad ...(15.77)$$

The convection mass transfer coefficient can also be defined in terms of molar flux and molar concentrations.

$$h_{mx} = \frac{J_A}{(C_{A,s} - C_{A,\infty})} = \frac{-D_{AB}\dfrac{dC_A}{dy}}{(C_{A,s} - C_{A,\infty})} \text{ (m/s)} \quad ...(15.78)$$

The average mass transfer coefficient can be evaluated from

$$h_m = \frac{1}{A_s}\int_{A_s} h_{mx} \, dA_s \quad ...(15.79)$$

15.10.2. Dimensionless Parameters in Convective Mass Transfer

The dimensionless parameters are often used to correlate convective transport data. In convection heat transfer, the Reynolds number. Prandtl number and Nusselt number were used. Some of these parameters along with some newly defined dimensionless quantities will be useful in correlation of convective mass transfer.

The molecular diffusivities of three transport phenomenon are :

(*i*) Momentum diffusivity $\nu = \dfrac{\mu}{\rho}$

(*ii*) Thermal diffusivity, $\alpha = \dfrac{k}{\rho C}$

(*iii*) Mass diffusivity, D_{AB}.

All these diffusivities are measured in m^2/s and thus ratio of any two diffusivities becomes dimensionless.

Schmidt number : It is the ratio of momentum diffusivity to the mass diffusivity. It is designated as Sc and expressed as :

$$Sc = \frac{\text{Momentum diffusivity}}{\text{Mass diffusivity}} = \frac{\nu}{D_{AB}} \quad ...(15.80)$$

The role of Schmidt number in convective mass transfer is analogous to role of Prandtl number in convection heat transfer. It compares the relative thicknesses of velocity and concentration boundary layers.

Lewis number: It is the ratio of thermal diffusivity to mass diffusivity. It is designated as Le and expressed as:

$$\text{Le} = \frac{\text{Thermal diffusivity}}{\text{Mass diffusivity}}$$

$$= \frac{\alpha}{D_{AB}} = \frac{k}{\rho C\, D_{AB}} \quad ...(15.81)$$

The Lewis number compares the relative thicknesses of thermal and concentration boundary layers. It comes in existance, when a process involves simultaneously convective transfer of mass and energy.

Sherwood number: It is analogous to Nusselt number in convection heat transfer and it may be expressed as ratio of concentration gradient at the surface to overall concentration gradient, it is also called *mass transfer Nusselt number* Nu_{AB}. It is designated as Sh.

$$\text{Sh} = \frac{\text{Concentration gradient}}{\text{Overall concentration gradient}}$$

$$= \frac{\text{Molecular mass transport resistance}}{\text{Convective mass transport resistance}}$$

$$= \frac{h_m L_c}{D_{AB}} = f(\text{Re, Sc}) \quad ...(15.82)$$

where L_c is characteristic length. The Sherwood number signifies the effectiveness of mass convection at the surface.

Mass Transfer Stanton number: Mathematically, it is interpreted

as $$St_m = \frac{h_m}{u_m} = \frac{\text{Sh}}{\text{Re Sc}} \quad ...(15.83)$$

where u_m is free stream velocity of fluid in external flow and bulk mean velocity in internal flow.

15.10.3. Analogy between Heat and Mass Transfer

The mechanism of heat, momentum and mass transfers are closely related and they are almost identical. Therefore, data of one transfer operation should be used to predict the rate of transfer in other operation.

The heat and momentum transfers are interrelated by Reynolds analogy for fluids having Pr = 1.

$$\frac{\text{Nu}}{\text{Re Pr}} = \frac{h}{\rho C_p u_m} = \text{St} = \frac{C_f}{2}$$

A similar analysis for mass and momentum transfer for Sc = 1 results

$$\frac{\text{Sh}}{\text{Re Sc}} = \frac{C_f}{2} \quad ...(15.84)$$

using $$\text{Sh} = \frac{h_m L}{D_{AB}} \quad \text{and} \quad \text{Sc} = \frac{\nu}{D_{AB}} = \frac{\mu}{\rho D_{AB}}$$

$$\therefore \quad \frac{h_m L}{D_{AB}\left(\frac{\rho u_m L}{\mu}\right)\left(\frac{\mu}{\rho D_{AB}}\right)} = \frac{C_f}{2}$$

or $$\frac{h_m}{u_m} = St_m = \frac{C_f}{2} \quad ...(15.85)$$

Reynolds Colburn analogy for tube flow may also be extended to mass transfer problems to express the friction factor as :

$$St_m\, Sc^{2/3} = \frac{f}{8} \quad ...(15.86)$$

which is analogous to

$$\text{St Pr}^{2/3} = \frac{f}{8}$$

When Pr and Sc are not unity, Reynolds Colburn equation for turbulent flow inside tubes is :

$$\text{Nu} = 0.023\, \text{Re}^{0.8}\, \text{Pr}^{1/3}$$

The analogous relation for mass transfer in wetted wall column is given as

$$\text{Sh} = 0.023\, \text{Re}^{0.8}\, \text{Sc}^{1/3} \quad ...(15.87)$$

A wetted column means a vertical tube with liquid flowing in a thin film down the inside wall of the tube and a gas flowing upward in the tube.

15.10.4. Correlation for Convective Mass Transfer

In convection heat transfer, it was observed that the energy and momentum equations for laminar boundary layer on a flat plate are :

$$u\frac{\partial u}{\partial x} + v\frac{\partial u}{\partial y} = \nu\frac{\partial^2 u}{\partial y^2}$$

and $$u\frac{\partial T}{\partial x} + v\frac{\partial T}{\partial y} = \alpha\frac{\partial^2 T}{\partial y^2}$$

If we consider laminar boundary layer on a flat plate in which mass diffusion occurs, the equation for concentration can be written as :

$$u\frac{\partial \rho_A}{\partial x} + v\frac{\partial \rho_A}{\partial y} = D_{AB}\frac{\partial^2 \rho_A}{\partial y^2} \quad ...(15.88)$$

where ρ_A = mass concentration of component A, diffusing through boundary layer.

The similarities between the governing equations for heat, mass and momentum transfer suggest that the empirical correlations for mass transfer coefficient would be similar to that of heat transfer coefficient. Thus

$$Sh_x = \frac{h_{mx} x}{D_{AB}} = f(Re_x, Sc) \quad ...(15.89)$$

For flow over a smooth flat plate, the local value of mass transfer coefficient can be obtained from relation

$$Sh_x = \frac{h_{mx} x}{D_{AB}} = 0.332 \, Re_x^{1/2} \, Sc^{1/3} \quad ...(15.90)$$

for laminar flow

$$Re < 5 \times 10^5$$

and

$$\frac{\delta}{\delta_c} = Sc^{1/3} \quad ...(15.91)$$

where δ = thickness of velocity boundary layer

δ_c = thickness of concentration boundary layer.

The corresponding average value

$$Sh = 0.664 \, Re_L^{1/2} \, Sc^{1/3} \quad ...(15.92)$$

or

$$h_m = 2h_{mx}$$

For mixed boundary condition, the average value of mass transfer coefficient, h_m can be obtained from

$$Sh = (0.037 \, Re_L^{0.8} - 870) \, Sc^{1/3} \quad ...(15.93)$$

The mass transfer coefficient from a turbulent boundary layer over a flat plate

$$Sh_x = 0.0292 \, Re_x^{0.8} \, Sc^{1/3} \quad \text{for } Re > 5 \times 10^5 \quad ...(15.94)$$

The coefficient of friction for flow over a smooth flat plate can be obtained from Reynolds analogy for mass transfer

$$\text{Laminar} \quad \frac{h_{mx}}{u_\infty} Sc^{2/3} = \frac{C_{fx}}{2} = 0.332 \, Re_x^{-1/2} \quad ...(15.95)$$

$$\text{Turbulent} \quad \frac{h_{mx}}{u_\infty} Sc^{2/3} = \frac{C_{fx}}{2} = 0.296 \, Re^{-1/5} \quad ...(15.96)$$

When both *heat and mass transfer* are occur *simultaneously*. The heat and mass transfer coefficient may be related as

$$\frac{h}{h_m} = \rho C_p \, Le^{2/3} \quad ...(15.97)$$

For *turbulent flow* through pipes, the mass transfer coefficient is correlated as

$$\frac{h_m d}{D_{AB}} = 0.023 \, Re^{0.83} \, Sc^{0.44} \quad ...(15.98)$$

For diffusion over a sphere

$$Sh = 2(1 + 0.276 \, Re^{1/2} \, Sc^{1/3}) \quad ...(15.99)$$

15.11. DIMENSIONAL ANALYSIS OF CONVECTIVE MASS TRANSFER

As discussed in chapter 7, the dimensional analysis predicts the various dimensionless parameters, correlating the experimental data.

Forced mass convection stream. Consider the mass transfer from a walls of a circular tube to a fluid. The mass transfer is the result of concentration gradient. The variables affecting the phenomenon are listed with their symbols and dimensions.

Variable	*Symbol, units*	*Dimensions*
Tube diameter	d, m	L
Fluid density	ρ, kg/m³	ML^{-3}
Fluid viscosity	μ kg/ms	$ML^{-1} t^{-1}$
Fluid velocity	u, m/s	$L t^{-1}$
Mass diffusivity	D_{AB}, m²/s	$L^2 t^{-1}$
Mass transfer coefficient	h_m m/s	$L t^{-1}$

There are six variables affecting the forced convection mass transfer phenomenon and they are expressed in three primary dimensions (M, L and t). According to Buckingham π theorem, the number of independent dimensionless group formed as = 6 – 3 = 3

or

$$f(\pi_1, \pi_2, \pi_3) = 0$$

The variables of core group are D_{AB}, ρ, d and h_m, u and μ are selected variables.

Thus

$$\pi_1 = D_{AB}^a \, \rho^b \, d^c \, h_m$$

$$\pi_2 = D_{AB}^d \, \rho^e \, d^f \, u$$

$$\pi_3 = D_{AB}^g \, \rho^h \, d^i \, \mu$$

Writing π_1 in dimensional form

$$M^0 L^0 t^0 = L^{2a} t^{-a} \, M^b L^{-3b} \, L^c \, L \, t^{-1}$$

Equating the exponents of dimensions on both sides

M : $0 = b$

L : $0 = 2a - 3b + c + 1$

t : $0 = -a - 1$

Solution of these simultaneous equations give

$$a = -1 \quad b = 0 \quad c = 1$$

Thus $\pi_1 = \frac{h_m d}{D_{AB}} = Sh$ (Sherwood number)

Similarly the π_2 and π_3 group yielding to

$$\pi_2 = \frac{u\,d}{D_{AB}}$$

and $\pi_3 = \frac{\mu}{\rho D_{AB}} = Sc$ (Schmidt number)

Dividing π_2 by π_3, we get

$$\frac{\pi_2}{\pi_3} = \frac{u\,d}{D_{AB}} \times \frac{\rho D_{AB}}{\mu} = \frac{\rho u d}{\mu} = Re.$$

The result of dimensional analysis for forced convection mass transfer indicates, that

$$Sh = f(Re, Sc)$$

which is analogous to heat transfer correlation

$$Nu = f(Re, Pr)$$

Natural convection mass transfer: The natural convection mass transfer phenomenon is affected by the following variables :

Variables	*Symbol units*	*Dimensions*
Characteristic length	L_c, m	L
Mass diffusivity	D_{AB}, m²/s	$L^2 t^{-1}$
Fluid density	ρ, kg/m³	ML^{-3}
Fluid viscosity	μ, kg/ms	$M L^{-1} t^{-1}$
Buoyancy force	$g\,\Delta\rho_A$, kg/m²s²	$M L^{-2} t^{-2}$
Mass transfer coefficient	h_m, m/s	$L t^{-1}$

There are six variables affecting the phenomenon and they are expressed in three primary dimensions thus, according to Buckingham π theorem, the independent dimensionless group formed are :

$$= 6 - 3 = 3, \pi_1, \pi_2, \text{ and } \pi_3 \text{ are :}$$

$$\pi_1 = D_{AB}^a\, L^b\, \mu^c\, h_m$$

$$\pi_2 = D_{AB}^d\, L^e\, \mu^f\, \rho$$

$$\pi_3 = D_{AB}^g\, L^h\, \mu^i\, g\, \Delta\rho_A$$

Solving for three π groups, we get

$$\pi_1 = \frac{h_m L_c}{D_{AB}} = Sh \quad \text{(Sherwood number)}$$

$$\pi_2 = \frac{\rho D_{AB}}{\mu} = \frac{1}{Sc}$$

(Reciprocal of Schmidt number)

and $\pi_3 = \frac{L_c^3\, g\, \Delta\rho_A}{\mu\, D_{AB}}$

Multiplying π_2 and π_3, we get

$$\pi_2\pi_3 = \left(\frac{\rho D_{AB}}{\mu}\right) \times \left(\frac{L_c^3\, g\, \Delta\rho_A}{\mu\, D_{AB}}\right)$$

$$= \frac{\rho L_c^3\, g \Delta\rho_A}{\mu^2} = Gr_{AB} \quad \text{...(15.100)}$$

The Gr_{AB} is called the *mass transfer Grashof number.*

The natural convection mass transfer suggests the correlation

$$Sh = f(Gr_{AB}, Sc) \quad \text{...(15.101)}$$

which is analogous to the relation for natural convection heat transfer.

15.12. EVAPORATION OF WATER INTO AIR

In the atmosphere, the continuous evaporation and condensation of water with moving wind from the soil, ocean, rivers and lakes influence every activity of life and provide variety of climate that governs the environment on the earth. There are also several engineering situations, in which evaporation of water along with heat transfer is important *e.g.*, humidifiers, dehumidifiers, absorbers, desert coolers, wet bulb thermometers etc.

During the evaporation of water by moving air over the water surface (evaporative cooling), the energy associated with the phase change is latent heat of vaporisation of liquid. The energy required for evaporation must come from water by lowering its temperature. However, under steady state conditions, the latent heat supplied by water during its evaporation is equal to heat supplied to water by surrounding air, which in turn also gets cooled.

Under steady state conditions (Refer Fig. 15.24), the energy balance at air water interface

$$q_{conv} = q_{evap}$$

or $h(T_\infty - T_s) = \frac{\dot{m}_w}{A} h_{fg}$ ...(15.102)

where $\frac{\dot{m}_w}{A} = h_m(\rho_s - \rho_\infty)$ from eqn. (15.74) and

$\dot{m}_w$ = mass rate of water that evaporates.

h and h_m = heat and mass transfer coefficients, respectively

ρ_s and ρ_∞ = mass concentrations at the surface and ambient, respectively

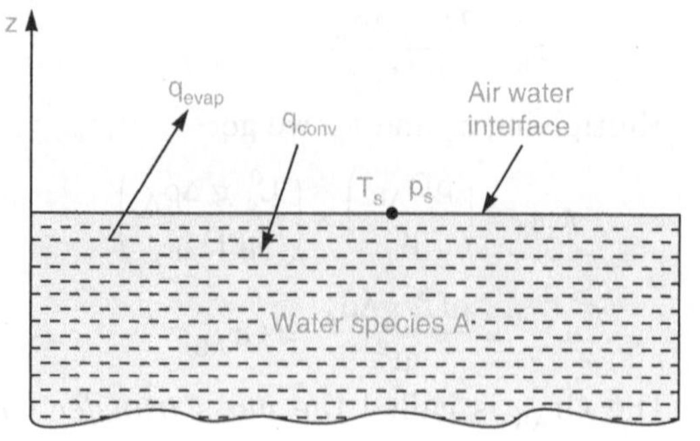

Fig. 15.24. Diffusion of water into air from a horizontal surface

Rearranging eqn. (15.102)

$$\frac{h}{h_m}(T_\infty - T_s) = h_{fg}(\rho_s - \rho_\infty)$$

Using eqn. (15.97), we get

$$\rho C_p \, Le^{2/3}(T_\infty - T_s) = h_{fg}(\rho_s - \rho_\infty) \quad ...(15.103)$$

This equation can be applied on wet bulb thermometer to obtain relative humidity or ambient temperature. In terms of partial pressures

$$\rho_s = \frac{M_w \, p_s}{R_u T_s} \text{ and } \rho_\infty = \frac{M_w \, p_\infty}{R_u T_\infty}$$

$$\therefore \quad (T_\infty - T_s) = \frac{M_w \, h_{fg}}{R_u \rho C_p \, Le^{2/3}}\left(\frac{p_s}{T_s} - \frac{p_\infty}{T_\infty}\right) \quad ...(15.104)$$

The air properties should be used at T_f

$$T_f = \frac{T_\infty + T_s}{2}$$

and T_f is approximately equal to T_s and T_∞

$$\text{Thus } (T_\infty - T_s) = \frac{h_{fg} \, M_w}{\rho C_p \, L_e^{2/3} \, R_u T_f}(p_s - p_\infty) \quad ...(15.105)$$

Replacing

$$\rho \, R_u T_f = M_{air} \, p$$

$$\text{Then } (T_\infty - T_s) = \frac{h_{fg}(M_w / M_{air})}{C_p \, Le^{2/3}} \times \left(\frac{p_s}{p} - \frac{p_\infty}{p}\right) \quad ...(15.106)$$

where p is total pressure.

Introducing specific humidity

$$\omega = \frac{m_v}{m_{air}} = \frac{M_w \, p_v}{M_{air} \, p_\infty}$$

$$\frac{p_{v_1}}{p_\infty} \simeq \frac{p_\infty}{p} = \omega_1 \frac{M_{air}}{M_w}$$

and $$\frac{p_{v_2}}{p_\infty} \approx \frac{p_s}{p} = \omega_2 \frac{M_{air}}{M_w}$$

Then eqn. (15.106) takes the form

$$\frac{\omega_2 - \omega_1}{T_\infty - T_s} = \frac{C_p}{h_{fg}} Le^{2/3} \quad ...(15.107)$$

Example 15.17. *Air at 35°C and 1 atm flows at a velocity of 60 m/s over*

(i) a flat plate 0.5 m long

(ii) a sphere 5 cm in diameter.

Calculate the mass transfer coefficient of water vapour in air. Neglect the concentration of vapour in air.

Solution

Given : Flow of air over the bodies with

$$p = 1 \text{ atm} = 1.01325 \times 10^5 \text{ N/m}^2$$

$$T_\infty = 35°C = 308 \text{ K},$$

$$u_\infty = 60 \text{ m/s}.$$

Air
T_∞ = 35°C
u_∞ = 60 m/s
p = 1 atm
0.5 m
(a)

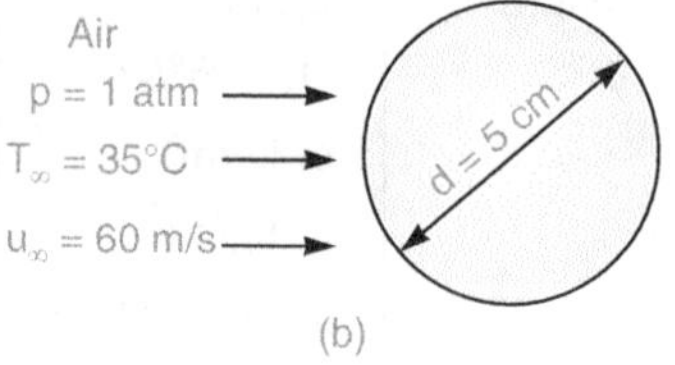

(b)

Fig. 15.25. Schematics for example 15.17

To find : Mass transfer coefficients.

Analysis : The diffusion coefficient of water vapour in air

$$D_{AB} = 0.256 \times 10^{-4} \text{ m}^2/\text{s}$$

(From Appendix Table A-11)

The properties of air at 35°C from Table A-4

Density $C_p = 1.006$ kJ/kg.K,

$\mu = 2 \times 10^{-5}$ kg/ms, Pr = 0.706

$$\rho = \frac{1.01325 \times 10^5}{287 \times 308} = 1.146 \text{ kg/m}^3$$

Schmidt number

$$Sc = \frac{\mu}{\rho D_{AB}}$$

$$= \frac{2 \times 10^{-5}}{1.146 \times 0.256 \times 10^{-4}} = 0.682$$

(*i*) Flow over flat plate,

$$L_c = 0.5 \text{ m}$$

$$\text{Re} = \frac{\rho u_\infty L_c}{\mu} = \frac{1.146 \times 60 \times 0.5}{2 \times 10^{-5}}$$

$$= 1.719 \times 10^6$$

greater than 5×10^5, thus flow is turbulent.

Using eqn. (15.94)

$$\text{Sh} = 0.0296 \text{ Re}^{0.8} \text{ Sc}^{1/3}$$

or $$\frac{h_m \times 0.5}{0.256 \times 10^{-4}} = 0.0296 \times (1.719 \times 10^6)^{0.8} \times (0.632)^{1/3}$$

$$= 2472.2$$

or $$h_m = \mathbf{0.1265 \text{ m/s.}}$$ **Ans.**

(*ii*) Flow over a sphere,

$$L_c = d = 0.05 \text{ m}$$

$$\text{Re} = \frac{\rho u_\infty d}{\mu} = \frac{1.146 \times 60 \times 0.05}{2 \times 10^{-5}}$$

$$= 171900$$

Using correlation, eqn. (15.99)

$$\text{Sh} = 2 [1 + 0.276 \text{ Re}^{1/2} \text{ Sc}^{1/3}]$$

$$\frac{h_m \times 0.05}{0.256 \times 10^{-4}} = 2 \times [1 + 0.276 \times (171900)^{1/2} \times (0.632)^{1/3}]$$

$$= 198.4$$

or $$h_m = \mathbf{0.101 \text{ m/s.}}$$ **Ans.**

Example 15.18. *Air at 50°C and 1 atm flow over the surface of a water reservoir at an average velocity of 2.3 m/s. The water surface is 0.65 m long and 0.65 m wide. The water surface temperature is estimated at 30°C. The relative humidity of air is 40%. The density of air is 1.105 kg/m³ and its viscosity is 1.943 × 10⁻⁵ kg/ms. Calculate the amount of water vapour evaporates per hour per sq.m of water surface and state the direction of diffusion.* (V.T.U., March 2001)

Solution

Given : Evaporation from the surface of a water reservoir.

$T_\infty = 50°\text{C}$,

$p = 1 \text{ atm} = 1.01325 \times 10^5 \text{ N/m}^2$,

$u_\infty = 2.3 \text{ m/s}$, $L = 0.65 \text{ m}$,

$A = 0.65 \times 0.65 = 0.4225 \text{ m}^2$,

$T_s = 30°\text{C} = 303 \text{ K}$, $\phi = 0.4$,

$\rho = 1.105 \text{ kg/m}^3$, $\mu = 1.943 \times 10^{-5} \text{ kg/ms}$.

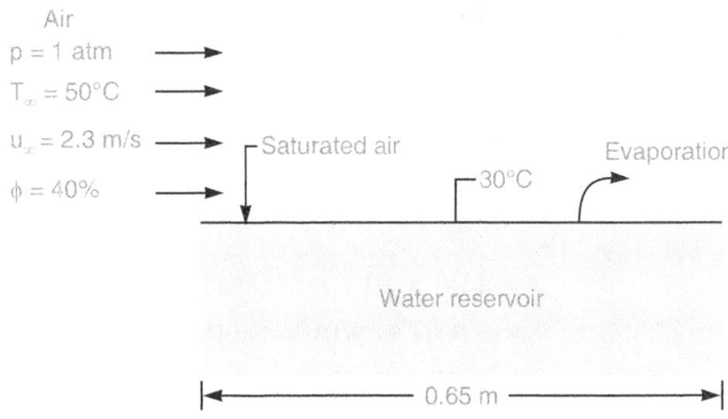

Fig. 15.26. Schematic for example 15.18

To find : Amount of water evaporation per hour per sq. m of the reservoir with direction of diffusion.

Assumption : Diffusion coefficient = 0.256×10^{-4} m²/s (from Appendix Table A-11)

Analysis : Reynolds number of fluid flow

$$\text{Re} = \frac{\rho u_\infty L}{\mu} = \frac{1.105 \times 2.3 \times 0.65}{1.943 \times 10^{-5}}$$

$$= 85.02 \times 10^3$$

It is less than the 5×10^5, thus flow is laminar.

Schmidt number

$$\text{Sc} = \frac{\mu}{\rho D_{AB}} = \frac{1.943 \times 10^{-5}}{1.105 \times 0.256 \times 10^{-4}}$$

$$= 0.686$$

Using correlation

$$\text{Sh} = 0.664 \text{ Re}_L^{1/2} \text{ Sc}^{1/3}$$

or $$\frac{h_m \times 0.65}{0.256 \times 10^{-4}} = 0.664 \times (85.02 \times 10^3)^{1/2} \times (0.686)^{1/3}$$

$$= 170.75$$

or $$h_m = 6.725 \times 10^{-3} \text{ m/s}$$

The partial pressure of water vapour at the air-water interface, corresponds to saturation temperature of 30°C

$$p_{s_1} = 4.246 \text{ kPa}$$

Saturation pressure of water vapour at 50°C

$$p_{s_2} = 12.35 \text{ kPa}$$

Partial pressure of water vapour in air at 50°C, 40% RH

$$p_\infty = \phi \times p_{s_2} = 0.4 \times 12.35 = 4.94 \text{ kPa.}$$

The concentrations

$$\rho_{w_1} = \rho_s = \frac{p_{s_1} M_w}{R_u T_s} = \frac{4.246 \times 18}{8.314 \times 303}$$

$$= 0.0303 \text{ kg/m}^3$$

$$\rho_{w_2} = \rho_\infty = \frac{p_\infty M_w}{R_u T_s} = \frac{4.94 \times 18}{8.314 \times 303} = 0.0353$$

The mass flux of water vapour, evaporated

$$\frac{\dot{m}_w}{A} = h_m(\rho_s - \rho_\infty)$$

$$= 9.03 \times 10^{-3} \times (0.0303 - 0.0353)$$

$$= -4.513 \text{ kg/m}^2\text{s} = -0.162 \text{ kg/m}^2\text{h}.$$

Since $\rho_\infty > \rho_s$, thus water vapour diffuses from air to water. **Ans.**

Example 15.19. (*i*) *The water in a 5 m × 15 m outdoor swimming pool is maintained at a temperature of 27°C. The average ambient temperature and relative humidity are 27°C and 40%, respectively. Assuming a wind speed of 2 m/s in the direction of long side of the pool, estimate the mass transfer coefficient for the evaporation of water from the pool surface.*

(*ii*) *If the air stream is at 50°C, while the wet bulb temperature is at 22°C. Calculate the relative humidity of air stream.* (Anna Univ., March 2000)

Solution

Given : (*i*) Evaporation of water from a swimming pool.

$L = 15$ m, $w = 5$ m,

$\phi = 40\%$,

$T_\infty = T = 27°C = 300$ K,

$u_\infty = 2$ m/s

(*ii*) $T_s = 22°C$, $T_\infty = 50°C$.

To find :

(*i*) Mass transfer coefficient for evaporation.

(*ii*) Relative humidity of air for change of state.

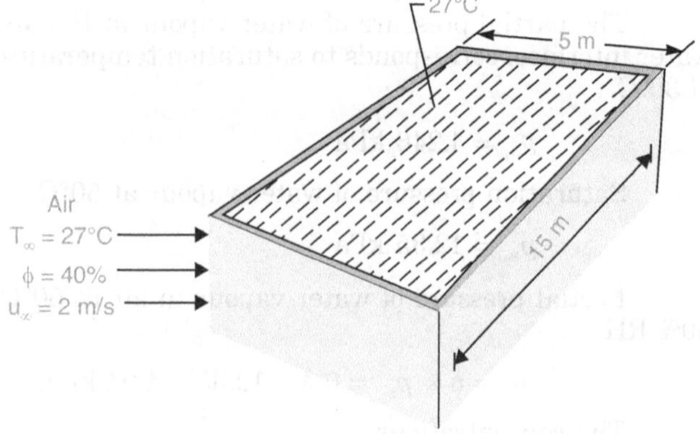

Fig. 15.27. Schematic of swimming pool for example 15.19

Analysis : (*i*) Thermophysical properties of air and water at 27°C (from appendix)

For air, $\nu = 15.89 \times 10^{-6}$ m²/s

For water,

$$v_g = 22.93 \text{ m}^3/\text{kg} \rightarrow \rho_A = \frac{1}{22.93}$$

$$= 0.0436 \text{ kg/m}^3$$

$$D_{AB} = 0.26 \times 10^{-4} \text{ m}^2/\text{s}$$

Schmidt number

$$Sc = \frac{\nu}{D_{AB}} = \frac{15.89 \times 10^{-6}}{0.26 \times 10^{-4}} = 0.611$$

Reynolds number

$$Re_L = \frac{u_\infty L}{\nu} = \frac{2 \times 15}{15.89 \times 10^{-6}} = 1887980$$

It is greater than 5×10^5, thus flow is turbulent.

For the mixed boundary conditions, using correlation, eqn. (15.93)

$$Sh_L = (0.037\, Re_L^{0.8} - 870)\, Sc^{1/3}$$

$$= [0.037 \times (1887980)^{0.8} - 870] \times (0.611)^{1/3}$$

$$= 2556$$

Further,

$$Sh_L = \frac{h_m L}{D_{AB}}$$

or

$$h_m = \frac{2556 \times 0.26 \times 10^{-4}}{15}$$

$$= \mathbf{4.43 \times 10^{-3}\ m/s.}$$ **Ans.**

(*ii*) The mean film temperature,

$$T_f = \frac{T_s + T_\infty}{2} = \frac{22 + 50}{2} = 36°C$$

From table at 36°C, the properties of air from Table A-4

$\rho = 1.14$ kg/m³, $C_p = 1.006$ kJ/kg.K

$Pr = 0.7$ $\mu = 2 \times 10^{-5}$ kg/ms

$D_{AB} = 0.26 \times 10^{-4}$ m²/s (from Table A-11)

Schmidt number,

$$Sc = \frac{\nu}{D_{AB}} = \frac{\mu}{\rho D_{AB}}$$

$$= \frac{2 \times 10^{-5}}{1.14 \times 0.26 \times 10^{-4}} = 0.675$$

Lewis number,

$$Le = \frac{\alpha}{D_{AB}} = \frac{Sc}{Pr} = \frac{0.675}{0.7} = 0.964$$

Properties of water at 22°C (from Table A-7)

$$v_g = 52.08 \text{ m}^3/\text{kg}, \qquad h_{fg} = 2449 \text{ kJ/kg.K}$$

$$\rho_s = \frac{1}{v_g} = 0.01920 \text{ kg/m}^3$$

Using eqn. (15.103) for energy transfer on wet bulb

$$\rho C_p \text{ Le}^{2/3} (T_\infty - T_s) = h_{fg} (\rho_s - \rho_\infty)$$

where ρ, ρ_s, ρ_∞ are density of air, water vapour at water surface and in air respectively,

Using values

$$1.14 \times 1.006 \times (0.964)^{2/3} \times (50 - 22)$$
$$= 2449 \times (0.01920 - \rho_\infty)$$

or $\rho_\infty = 0.01920 - 0.01295 = 0.00625 \text{ kg/m}^3$

The density of water vapour at 50°C,

$$\rho_s = \frac{1}{v_g} = \frac{1}{12.23} = 0.0817 \text{ kg/m}^3$$

Relative humidity,

$$\phi = \frac{\rho_\infty}{\rho_s} = \frac{0.00625}{0.0817}$$
$$= 0.0765 = \mathbf{7.65\%}. \quad \text{Ans.}$$

Example 15.20. *The dry bulb and wet bulb temperature recorded by a thermometer in moist air are 27°C and 17°C, respectively. Calculate the specific humidity of air, assuming following values:*

$Pr = 0.74$, $\qquad Sc = 0.6$,

$M_w = 18$, $\qquad M_{air} = 29$,

$C_p = 1.004$ *kJ/kg.K*, $\qquad p = 1.0132 \times 10^5$ *N/m²*.

Solution

Given : Temperature measured by dry bulb and wet bulb thermometers.

$$T_\infty = 27°C, \qquad T_s = 17°C.$$

To find : Specific humidity of air.

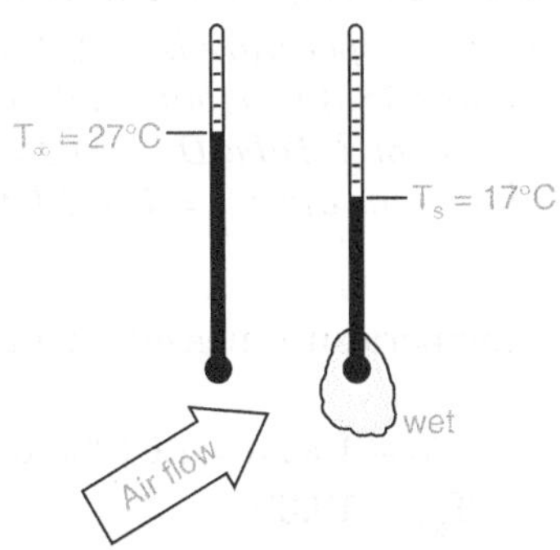

Fig. 15.28. Dry bulb and wet bulb thermomenters for example 15.20

Analysis : The specific humidity of air ω_1 can be obtained from eqn. (15.107)

$$\frac{\omega_2 - \omega_1}{T_\infty - T_s} = \frac{C_p}{h_{fg}} \text{Le}^{2/3}$$

where $\omega_2 = \dfrac{M_w}{M_{air}} \times \dfrac{p_s}{p - p_s}$,

during evaporation

At 17°C from Table A-7

$$p_s = 1.917 \text{ kPa},$$
$$h_{fg} = 2461 \text{ kJ/kg.K}$$

$$\therefore \qquad \omega_2 = \left(\frac{18}{28.9}\right) \times \frac{1.917}{101.32 - 1.917}$$
$$= 0.0120 \text{ kg/kg of dry air}$$

and Lewis number

$$\text{Le} = \frac{\text{Pr}}{\text{Sc}} = \frac{0.74}{0.6} = 1.233$$

Then

$$\frac{0.0120 - \omega_1}{27 - 17} = \frac{1.004}{2461} \times (1.233)^{2/3} = 0.00047$$

and $\omega_1 = \mathbf{0.0073 \text{ kg/kg of dry air.}}$ Ans.

Example 15.21. *Air at 1 atm, 25°C containing small quantity of iodine flows with a velocity of 4 m/s inside a 25 mm diameter tube. Calculate the mass transfer coefficient for iodine transfer from gas stream to the wall surface. If C_m is mean concentration of the iodine in kg-mole/m³ in air stream, calculate the rate of deposition of iodine on the tube surface, where the iodine concentration is zero.*

Solution

Given : Flow of air containing iodine through a tube.

$$p = 1 \text{ atm} = 1.01325 \text{ bar},$$
$$T = 25°C = 298 \text{ K},$$
$$u_\infty = 4 \text{ m/s},$$
$$d = 25 \text{ mm}.$$

To find :

(*i*) Mass transfer coefficient of iodine.

(*ii*) Rate of deposition of iodine on the tube surface.

Analysis : (*i*) For air at 25°C at 1 atm, (from Table A-4)

$$\nu = 1.58 \times 10^{-5} \text{ m}^2/\text{s}$$

For iodine

$$D_{AB} = 0.826 \times 10^{-5} \text{ m}^2/\text{s (from Table A-11)}$$

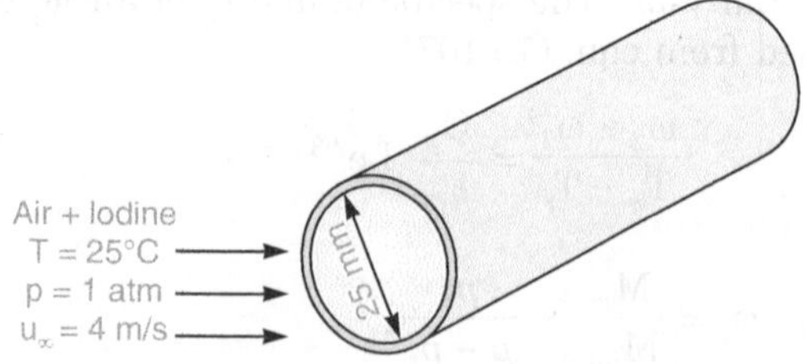

Fig. 15.29. Schematic for example 15.21

Reynolds number

$$\text{Re} = \frac{u_\infty d}{\nu} = \frac{4 \times 25 \times 10^{-3}}{1.58 \times 10^{-5}} = 6329$$

Schmidt number

$$\text{Sc} = \frac{\nu}{D_{AB}} = \frac{1.58 \times 10^{-5}}{0.826 \times 10^{-5}} = 1.913$$

Using correlation, eqn. (15.98)

$$\text{Sh} = 0.023\ \text{Re}^{0.83}\ (\text{Sc})^{0.44}$$

or $$\frac{h_m d}{D_{AB}} = 0.023 \times (6329)^{0.83} \times (1.913)^{0.44}$$

$$= 43.73$$

or $$h_m = \frac{43.73 \times 0.826 \times 10^{-5}}{25 \times 10^{-3}}$$

$$= \mathbf{0.0144\ m/s.}$$ **Ans.**

(*ii*) The rate of deposition of iodine

$$N_A = h_m (C_m - C_w) = 0.0144 \times (C_m - 0)$$

$$= \mathbf{0.0144\ C_m\ kg\text{-}mole/m^2s.}$$ **Ans.**

Example 15.22. *An earthenware pot cools water by keeping its outside surface wet with a volatile liquid of molecular weight 120 kg/kg-mole. If this pot is placed in dry air at 40°C with heat and mass transfer occurring simultaneously by forced convection. What is the steady temperature of cold water inside the pot ?*

Take $h_{fg} = 120$ *kJ/kg,* $p_s = 3530$ *N/m²,* $D_{AB} = 0.2 \times 10^{-4}$ *m²/s.*

Solution

Given : Evaporation from earthenware pot

$$M_w = 120 \text{ kg/kg-mole},$$
$$h_{fg} = 120 \text{ kJ/kg},$$
$$T_\infty = 40°C = 313 \text{ K},$$
$$D_{AB} = 0.2 \times 10^{-4} \text{ m}^2/\text{s},$$
$$p_s = 3530 \text{ N/m}^2.$$

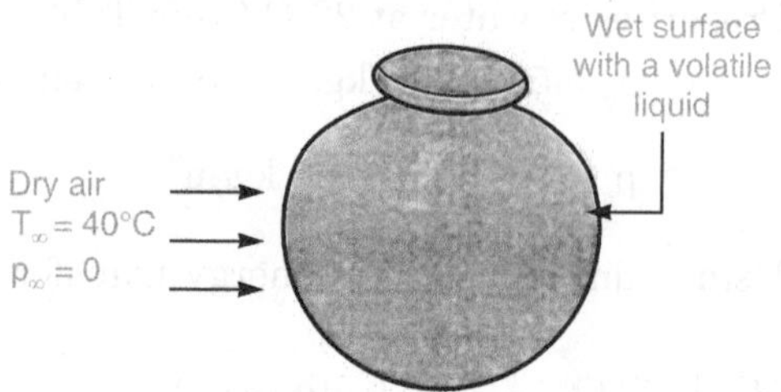

Fig. 15.30. Earthenware pot for example 15.22

To find : Water temperature inside the pot.

Analysis : Assuming mean film temperature as 300 K, from Table A-4, the properties of air

$$\rho = 1.16 \text{ kg/m}^3,$$
$$C_p = 1.007 \text{ kJ/kg.K},$$
$$\alpha = 22.5 \times 10^{-6} \text{ m}^2/\text{s}.$$

Then $$\text{Le} = \frac{\alpha}{D_{AB}} = \frac{22.5 \times 10^{-6}}{0.2 \times 10^{-4}} = 1.125$$

The evaporative cooling effect is given by eqn. (15.104);

$$T_\infty - T_s = \frac{M_w h_{fg}}{R_u \rho C_p \text{Le}^{2/3}} \left[\frac{p_s}{T_s} - \frac{p_\infty}{T_s}\right]$$

Putting $p_\infty = 0$ for dry air and rearranging, we get

$$T_s^2 - T_s T_\infty + \frac{h_{fg} M_w p_s}{R_u \rho C_p \text{Le}^{2/3}} = 0$$

or $$T_s^2 - 313\ T_s + \frac{120 \times 120 \times 3530 \times 10^{-3}}{8.314 \times 1.16 \times 1.007 \times (1.125)^{2/3}} = 0$$

or $$T_s^2 - 313\ T_s + 4838 = 0$$

Hence, $$T_s = \frac{+313 \pm \sqrt{313^2 - 4 \times 4838}}{2 \times 1}$$

$$= 296.7 \text{ K} = \mathbf{23.7°C.}$$ **Ans.**

Example 15.23. *Calculate the temperature of dry air at 1 atm, whose wet bulb temperature is 18.3°C.*

If air stream temperature is 32.2°C and wet bulb temperature remains 18.3°C, what would be the relative humidity of air stream ? Take $D_{AB} = 0.26 \times 10^{-4}$ *m²/s,* $\alpha = 0.221 \times 10^{-4}$ *m²/s for air,* $C_p = 1.004$ *kJ/kg.K.*

Solution

Given : Temperature measurement by wet bulb thermometer.

(*i*) $p = 1 \text{ atm} = 1.01325 \text{ bar}$

$T_{wb} = 18.3°C$

(*ii*) $T_\infty = 32.2°C,$

$T_s = 18.3°C,$

$$D_{AB} = 0.26 \times 10^{-4} \text{ m}^2/\text{s},$$
$$\alpha = 0.221 \times 10^{-4} \text{ m}^2/\text{s},$$
$$C_p = 1.004 \text{ kJ/kg.K.}$$

To find :

(*i*) Dry bulb temperature when T_{wb} = 18.3°C

(*ii*) R.H. when T_{db} = 32.2°C, and T_{wb} = 18.3°C.

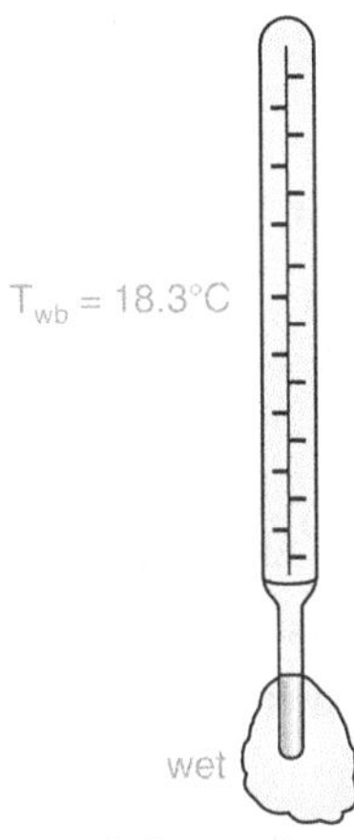

Fig. 15.31. (*a*) Schematice of wet bulb thermometer for example 15.23

Analysis : (*i*) From steam table at 18.3°C

$$p_g = p_s = 2.107 \text{ kPa}, \; v_g = 65.08 \text{ m}^3/\text{kg}$$
$$h_{fg} = 2456 \text{ kJ/kg}$$
$$\rho_s = \frac{1}{v_g} = \frac{1}{65.08} = 0.01536 \text{ kg/m}^3$$

For dry air $\rho_\infty = 0$

Density of air

$$\rho = \frac{p}{RT} = \frac{1.01325 \times 100}{0.287 \times (18.3 + 273)}$$
$$= 1.212 \text{ kg/m}^3$$

Lewis number

$$\text{Le} = \frac{\alpha}{D} = \frac{0.221 \times 10^{-4}}{0.26 \times 10^{-4}} = 0.85$$

Using eqn. (15.103)

$$\rho C_p (\text{Le})^{2/3} (T_\infty - T_s) = h_{fg} (\rho_s - \rho_\infty)$$

or $$T_\infty - 18.3 = \frac{2456 \times (0.01536 - 0)}{1.212 \times 1.004 \times (0.85)^{2/3}}$$
$$= 34.5°\text{C}$$

or Dry air temperature,

$$T_\infty = 34.5 + 18.3 = \mathbf{52.8°C}. \quad \text{Ans.}$$

(*ii*) Relative humidity of air stream:

Here T_∞ = 32.2°C, T_s = 18.3°C,

ρ = 1.212 kg/m³, h_{fg} = 2456 kJ/kg.K,

C_p = 1.004 kJ/kg.K, Le = 0.85 and $\rho_\infty \neq 0$

Using in eqn. (15.103), we get

$$(32.2 - 18.3) = \frac{2456 \times (0.01536 - \rho_\infty)}{1.212 \times 1.004 \times (0.85)^{2/3}}$$

or $$0.01536 - \rho_\infty = 0.0062 \quad \text{or} \quad \rho_\infty = 0.0092 \text{ kg/m}^3$$

From steam table at 32.2°C

$$v_g = 29.24 \text{ m}^3/\text{kg}$$

$$\therefore \quad \rho_s = \frac{1}{v_g} = \frac{1}{29.24} = 0.0342 \text{ kg/m}^3$$

The relative humidity

$$\phi = \frac{\rho_\infty}{\rho_s} = \frac{0.0092}{0.0342} = \mathbf{26.9\%}. \quad \text{Ans.}$$

Fig. 15.31. (*b*) Dry bulb and wet bulb thermometers for example 15.23

Example 15.24. *During an experimentation, the flow of dry air at 25°C and 1 atm at a free stream velocity of 2 m/s over a softdrink bottle covered with a layer of wet cloth. It is observed that 15 gram of water has evaporated in 15 minutes. The surface area of bottle is 0.3 m². Both the body and air are kept at 25°C during study. The vapour pressure of water at 25°C is 15 Pa and mass diffusivity of water in air at 25°C is 2.5 × 10⁻⁵ m²/s. Calculate heat transfer coefficient under the same flow conditions over same geometry.*

Solution

A softdrink bottle covered with layer of wet cloth.

T_∞ = 25°C, p = 1 atm,

u_∞ = 2 m/s, t = 15 min = 900 s

m_{ev} = 15 gram = 0.015 kg

A = 0.3 m², T_s = 25°C

$D_{AB} = 2.5 \times 10^{-5}$ m²/s, $p_v = 15 \; p_a$

To find : Heat transfer cofficient under identical conditions over same geometry.

Assumptions :

1. Both air and water vapour as an ideal gas.
2. Motar mass of water is 18.

Analysis : Properties of dry air at 25°C and 1 atm (from Table A-4)

$\rho = 1.184$ kg/m³, $C_p = 1007$ J/kg.K,
$\alpha = 2.141 \times 10^{-5}$ m²/s

The mass concentration of water vapour in air

$$\rho_s = \frac{p_v M_w}{R_u T} = \frac{(15 \times 10^{-3} \text{ kPa}) \times (18 \text{ kg/kg-mole.K})}{(8.314 \text{ kJ/kg-mole.K}) \times (298 \text{ K})}$$

$= 1.089 \times 10^{-4}$ kg/m³

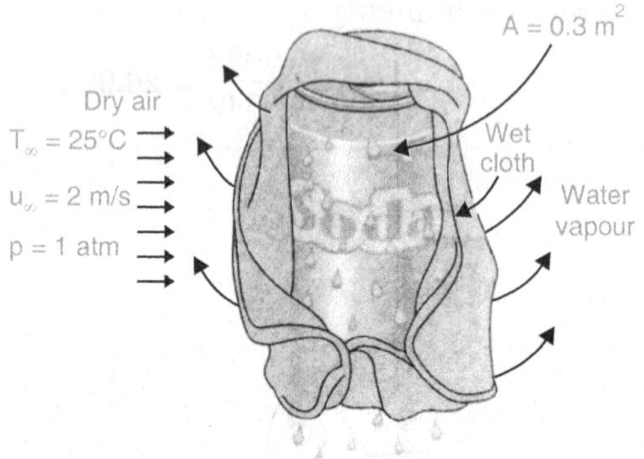

Fig. 15.32. Cooling of softdrink in desert conditions.

Mass concentration of water vapour in dry air

$\rho_\infty = 0$

Rate of evaporation of water

$$\dot{m}_{evap} = \frac{m_{ev}}{\Delta t} = \frac{0.015 \text{ kg}}{900 \text{ s}}$$

$= 1.667 \times 10^{-5}$ kg/s

The mass convection coefficient h_m using eqn (15.74)

$$h_m = \frac{\dot{m}_{evap}}{A(\rho_s - \rho_\infty)}$$

$$= \frac{(1.667 \times 10^{-5} \text{ kg/s})}{(0.3 \text{ m}^2) \times (1.089 \times 10^{-4} - 0)}$$

$= 0.510$ m/s

Lewis number

$$\text{Le} = \frac{\alpha}{D_{AB}} = \frac{(2.141 \times 10^{-4} \text{ m}^2/\text{s})}{(2.5 \times 10^{-5} \text{ m}^2/\text{s})}$$

$= 8.564$

Using analogy between heat and mass transfer, eqn. (15.97)

$h = \rho C_p h_m \text{Le}^{2/3}$

$= (1.184 \text{ kg/m}^3) \times (1007 \text{ J/kg.K}) \times (0.510 \text{ m/s}) \times (8.564)^{2/3}$

$=$ **2546.5 W/m².K. Ans.**

15.13. SUMMARY

The transport of mass on a microscopic level as a result of diffusion from a region of high concentration to that of a lower concentration in called mass transfer.

The mass can be transferred by molecular diffusion, convection and by change of phase. The diffusion of mass transfer is a molecular phenomenon and diffusion coefficient or mass diffusivity is a property.

The Fick's law states that the mass flux of a constituent per unit area is proportional to the concentration gradient.

$$j_A = \frac{\dot{m}_A}{A} = -D_{AB} \frac{d\rho_A}{dx}$$

Alternatively

$$J_A = -D_{AB} \frac{dC_A}{dx}$$

where $\dot{m}_A$ = mass flow rate of component A, kg/s
A = cross-sectional area, m²
ρ_A = mass concentration kg/m³
C_A = molar concentration, kg-mole/m³
N_A = molar flux
D_{AB} = mass diffusivity of component A in component B, m²/s.

The three dimensional mass diffusion equation is:

$$\nabla^2 C_A = \frac{1}{D_{AB}} \frac{\partial C_A}{\partial t}$$

The mass diffusion of water in stagnant gases is given as :

$$(\dot{m}_w)_{total} = \frac{D_{AB} A M_w p}{R_u T L} \ln\left(\frac{p - p_\infty}{p - p_s}\right)$$

where p_s and p_∞ are partial pressure of water at free water level and container level, respectively.

p is total pressure, R_u is universal gas constant, D_{AB} is diffusion coefficient and M_w is molecular weight of water and T is absolute temperature.

The convective mass transfer involves the diffusion of one fluid through the movement of the other fluid. In free mass convection, the concentration gradient changes the density of fluid, which may be acted upon by buoyancy force. In forced mass convection, one fluid is moving with an appreciable velocity over the other.

The steady state diffusion through a plane membrane is given as

$$J_A = \frac{C_{A,1} - C_{A,2}}{R_{m,A}}$$

where $R_{m,A}$ is known as diffusion resistance

$$= \frac{L}{AD_{AB}} \text{ for plane fluid column}$$

$$= \frac{\ln\left(\frac{r_2}{r_1}\right)}{2\pi LD_{AB}}$$

for cylindrical fluid column.

When diffusion fluxes of two components in a fluid mixture are equal but acting in opposite direction, then such diffusion is referred as equimolar counter diffusion and it is given as

$$\dot{N}_A = \frac{D_{ABA}(C_{A,1} - C_{A,2})}{L}$$

The mass transfer coefficient h_m is defined in terms of mass flux as

$$h_m = \frac{\dot{m}_A}{A(\rho_{A,1} - \rho_{A,2})}$$

$$= \frac{D_{AB}}{\Delta x} \text{ m/s}$$

In terms of molar flux

$$h_m = \frac{N_A}{(C_{A,1} - C_{A,2})} \text{ m/s}$$

The Reynolds Colburn analogy for mass transfer rate yields to

$$\frac{h_m}{u_m} Sc^{2/3} = \frac{f}{8}$$

and $\quad Sh_D = 0.023\, Re_D^{0.83}\, Sc^{0.44}$

where Sc is the Schmidt number, a ratio of momentum diffusivity to mass diffusivity. The other dimensionless numbers used in mass transfer are Lewis number, Le and Sherwood number, Sh, defined as :

$$Le = \frac{\text{Thermal diffusivity}}{\text{Mass diffusivity}} = \frac{\alpha}{D_{AB}}$$

and $$Sh = \frac{\text{Concentration gradient}}{\text{Overall concentration gradient}}$$

The continuous evaporation of water into air results into humidity increase in air and cooling of water simultaneously and yields into form

$$\frac{\omega_2 - \omega_1}{T_\infty - T_s} = \frac{C_p}{h_{fg}} Le^{2/3}$$

where ω_1 and ω_2 are specific humidities before and after evaporation.

REVIEW QUESTIONS

1. Why is mass transfer take place ?
2. State the modes of mass transfer with suitable examples.
3. Explain Fick's law of diffusion. What is mass diffusivity ? What is its dimension ?
4. Compare Newton's law of viscosity, Fourier law of heat conduction and Fick's law of diffusion.
5. Define mass fraction, mole fraction, molar concentration, mass flux and molar flux.
6. Derive the generalised mass diffusion equation.
7. Explain the molecular diffusion through a Stagnant gas.
8. Explain steady state diffusion through a plane membrane.
9. What is diffusion resistance ? What is convective mass transfer coefficient ? What is its dimension ?
10. Explain equimolar counter diffusion.
11. Define and explain the physical significance of
    (*i*) Schmidt number,
    (*ii*) Lewis number, and
    (*iii*) Sherwood number.
12. Explain the dimensional analysis for convective mass transfer.
13. Explain evaporation of water into air.
14. For simultaneous flow of heat and mass on a flat plate, prove that
$$\frac{h}{h_m} = \rho\, C_p\, Le^{0.67}$$
15. Derive an expression for diffusion of one gas through a stagnant gas in terms of logarithmic mean partial pressures. Consider the pressure and temperature of the system to be constant.
16. Show that the total mass of water vapour diffused from a water column to air passing over, it is given by
$$(\dot{m}_w)_{total} = \frac{D_{AB}\, A}{R_u T} \frac{M_w\, p}{(x_2 - x_1)} \ln\left(\frac{p - p_{w_2}}{p - p_{w_1}}\right)$$
where D_{AB} = diffusion coefficient,
A = cross-section area of water column,
R_u = universal gas constant,
M_w = molecular weight of water vapour,
$x_2 - x_1$ = height of container above water level,
p = total pressure
p_{w_1} = partial pressure of water vapour at water surface, and
p_{w_2} = partial pressure of water vapour at the top of container.
17. Derive the Stefan's equation for the rate of evaporation from the surface of a lake.

PROBLEMS

1. Estimate the diffusion rate of oxygen through 1 cm^2 area, when it is diffusing through stagnant carbon monoxide at 0°C and 1 atm under steady state conditions. The partial pressure of oxygen at two planes 0.3 cm apart is 100 mm of Hg and 25 mm of Hg, respectively.

   Assume $D_{AB} = 0.185\ cm^2/s$

   $R_u = 8.314$ kJ/kg-mole. K

   [Ans. 2.962×10^{-6} gm-mol/s]

2. An open tank, 6 mm in diameter, contains 1 mm deep layer of benzene (M = 78 kg/kg-mole) at its bottom. The vapour pressure of benzene in the tank is 13.15 kPa and its diffusion takes place through a stagnant air film 2.5 mm thick at the operating temperature of 20°C. The diffusivity of the benzene in the tank is $8.0 \times 10^{-6}\ m^2/s$. Calculate the diffusion rate of benzene.

   If the density of the benzene is 880 kg/m^3, calculate the time taken for the entire benzene to evaporate. Take atmospheric pressure as 101.3 kPa.

   [Ans. 0.04195 kg/s, 593 seconds]

3. At the bottom of a well 2.5 m in diameter, water is 5 m deep. Calculate the diffusion rate into dry atmospheric air at 25°C and 1.032 bar. Take diffusion coefficient $D_{AB} = 0.0925\ m^2/h$.

   [Ans. 2.122×10^{-3} kg/h]

4. Hydrogen gas is maintained at 5 bar and 1 bar on opposite of a plastic membrane, which is 0.3 mm thick. The temperature is 25°C and the binary diffusion coefficient of hydrogen in the plastic is $8.7 \times 10^{-8}\ m^2/s$. The solubility of hydrogen in the membrane is 1.5×10^{-3} kg-mole/m^3 bar. What is the mass flux of hydrogen by diffusion through membrane ?

   [Ans. 34.8×10^{-7} kg/m^2s]

5. Calculate the mass diffusivity of the binary gas mixture of air-water vapour at 273 K and 1 atm, and compare the result with that given in Table 10.2.

6. Calculate the diffusion coefficient for benzene in atmospheric air at 25°C.

7. Estimate the diffusion rate of water from the bottom of a test tube 10 mm in diameter and 15 cm long into dry atmospheric air at 25°C. Given: $D_{AB} = 0.256 \times 10^{-4}\ m^2/s$. [Ans. 1.131×10^{-10} kg/s]

8. Two large vessels contain uniform mixtures of nitrogen (component A) and carbon dioxide (component B) at 1 atm, T = 289 K, but at different concentrations. Vessel 1 contains 90% N_2 and 10% CO_2 by moles, whereas vessel 2 contains 20% N_2 and 80% CO_2 by moles. The two vessels are connected by a duct of 0.1524 m ID, and 1.22 m long. Determine the rate of transfer of nitrogen between the two vessels by assuming that steady-state transfer takes place in view of the large capacity of the two reservoirs. Take the mass diffusivity for N_2-CO_2 mixture as $0.16 \times 10^{-4}\ m^2/s$.

   [Ans. 0.71×10^{-8} kg-mole/s]

9. A deep narrow tube open at the top contains toluene at the bottom. Air inside the tube is motionless while at the top toluene concentration is zero. The entire system is at 1 atm, 18.7°C, when $D_{AB} = 0.826 \times 10^{-4}\ m^2/s$. The saturated vapour pressure of toluene at liquid surface is 0.026 atm. Determine the rate of evaporation of toluene per unit area, if the distance from the liquid surface to the top is 1.524 m.

   [Ans. 0.597×10^{-8} kg-mole/m^2s]

10. One method of measuring diffusion coefficients of vapours is to measure the rate of evaporation of a liquid in narrow tubes. In one such experiment, a glass tube 1 cm in diameter was filled with water at 20°C to within 4 cm of the top. Dry air at 20°C and 1.013 bar was blown across the top of the tube. At the end of 24 hours of steady-state operation, the level of the water dropped 0.1 cm. Calculate the diffusivity of the air-vapour system at 20°C.

    [Ans. $2.644 \times 10^{-5}\ m^2/s$]

11. Pure N_2 gas at 1 atm and 25°C is flowing through a 10 m long, 3 cm inner diameter pipe made of 1 mm thick rubber. Determine the rate at which N_2 leaks out of the pipe if the medium surrounding the pipe is (*a*) a vacuum and (*b*) atmospheric air at 1 atm and 25°C with 21 per cent O_2 and 79 per cent N_2.

    [Ans. (*a*) 4.48×10^{-10} kg-mole/s,
    (*b*) 9.4×10^{-11} kg-mole/s]

12. Dry air at atmospheric pressure and 10°C flows over a flat plate with a velocity of 1 m/s. The plate is covered with a film of water which evaporates into the air steam. Determine the average mass transfer coefficient for the transfer of water vapour over a distance of 0.6 m from the leading edge of the plate.

13. Atmospheric air at 30°C flows over a wet bulb thermometer which reads 20°C. Calculate the concentration of water vapour in the air stream and the relative humidity of the air.

    [Ans. 0.01 kg/kg of dry air]

14. Dry air at 25°C and atmospheric pressure blows over a 30 cm^2 surface of ice at a velocity of 1.5 m/s. Estimate the amount of moisture evaporated per hour, assuming that the block of ice is perfectly insulated except for the surface exposed to the air stream.

15. A thin plastic membrane is used to separate helium from a gas stream. Under steady-state conditions, the concentration of helium in the membrane is known to be 0.02 kg-mole/m^3 and 0.005 kg-mole/m^3 at the inner and outer surfaces, respectively. If the membrane is 1 mm thick and the binary diffusion coefficient of helium with respect to the plastic is $10^{-19}\ m^2/s$, what is the mass flux by diffusion ?

    [Ans. 6×10^{-8} kg/s.m^2]

16. Oxygen gas is maintained at pressure of 2 bar and 1 bar on opposite sides of a rubber membrane that is 0.5 mm thick, and the entire system is at 25°C. What is the molar diffusive flux of O_2 through the membrane ? What are the molar concentrations of O_2 on both sides of the membrane (outside the rubber) ? [Ans. $C_{A,1}$ = 0.0807 kg-mole/m^3
$C_{A,2}$ = 0.0404 kg-mole/m^3]

17. Helium gas is stored at 20°C in a spherical container of fused silica, which has an inner diameter of 0.20 m and a wall thickness of 2 mm. If the container is charged to an initial pressure of 4 bar, what is the rate at which this pressure decreases with time ? [Ans. 2.63 × 10^{-8} bar/s]

18. A tray 40 cm long and 20 cm wide is full of water. Air at 30°C flows over the tray along the length at 2 m/s. The moving air at 1.013 bar and partial pressure of water in the air is 0.007 bar. Calculate the rate of evaporation, if the temperature of the water is 25°C. Take for air ρ = 1.2 kg/m^3, ν = 15 × 10^{-6} m^2/s, D_{AB} = 0.145 m^2/h. *(Anna Univ., Dec. 1999)*
[Ans. 0.0827 kg/h]

19. Dry air at 15°C and 92 kPa flows over a 2 m long wet surface with free stream velocity of 4 m/s. Determine average mass transfer coefficient.
[Ans. 0.00514 m/s]

20. Pure N_2 gas at 1 atm and 25°C is flowing through a 10 m long, 3 cm inner diameter pipe made of 1 mm thick rubber. Determine the rate at which N_2 leaks out the pipe, if medium surrounding the pipe is (*a*) vacuum and (*b*) atmosphere air at 1 atm and 25°C.
[Ans. (*a*) 4.48 × 10^{-10} kg-mole/s
(*b*) 9.4 × 10^{-11} kg-mole/s]

21. During a hot summer day, a canned drink is to be cooled by wrapping it in a cloth that is kept wet continually, and blowing air to it by a fan. If the environment conditions are 1 atm, 30°C, and 40 per cent relative humidity, determine the temperature of the drink when steady conditions are reached.
[Ans. 19.4°C]

REFERENCES AND SUGGESTED READING

1. H. Perry, ed. *"Chemical Engineer's Handbook"*. 4th ed. McGraw-Hill, New York, 1963.
2. F. Kreith, R.F. Boehm, et. Al., *"Heat and Mass Transfer"*, *"Mechanical Engineering Handbook"*, CRC Press LLC, 1999.
3. J.P. Holman. *"Heat Transfer"*. 9th ed. McGraw-Hill, New York, 2010.
4. F.P. Incropera and D.P. De Witt. *"Fundamentals of Heat and Mass Transfer"*, 5th ed. John Wiley & Sons, New York, 2006.
5. W.M. Kays and M.E. Crawford. *"Convective Heat and Mass Transfer"*, 2nd ed. McGraw-Hill, New York, 1980.
6. A.F. Mills and V. Ganeshan. *"Heat Transfer"*. 2nd ed. Pearson Education, New Delhi, 2009.
7. M.M. Rathore. *"Thermal Engineering"*, McGraw-Hill Education, New Delhi, 2010.
8. Yunus A. Cengel, *"Heat Transfer"*, A Practical Approach, 2nd ed. McGraw-Hill, New York, 2002.
9. R.M. Barrer. *"Diffusion in and through Solids"*, MacMillan, New York, 1941.
10. R.B. Bird. *"Theory of Diffusion"*. *"Advances in Chemical Engineering"* vol. 1 (1956), p. 170.
11. A.H.P. Skelland. *"Diffusional Mass Transfer"*. John Wiley & Sons, New York, 1974.
12. W.F. Stoecker and J.W. Jones. *"Refrigeration and Air Conditioning"*. McGraw-Hill, New York, 1982.
13. L.C. Thomas. *"Mass Transfer Supplement—Heat Transfer"*. Prentice Hall, Englewood Cliffs, NJ, 1991.
14. C.J. Geankoplis. *"Mass Transport Phenomena"*. Holt, Rinehart, and Winston, New York, 1972.

Experiments in Engineering Heat Transfer

16

Engineering education has placed a great emphasis on the ability of an individual to perform experiments along with a theoretical analysis of the problems. The experimental methods have their own importance. They help in better understanding of the basic principles of the subject and to verify the result obtained analytically. Therefore, in engineering curiculla, the students are expected to devote one laboratory period a week for experimentation. The students are exposed to the basic instruments and get acquainted with the methods used for measuring the physical properties.

The experimentation in subject of Engineering Heat Transfer is for determination of thermal conductivities of conducting and non-conducting materials, heat transfer coefficient in natural and forced convection, emissivity of the surfaces, verification of value of Stefan Boltzmann constant, performance of heat exchangers, heat pipes etc.

An experimental instructional sheet consists of objective, prologue, experimental analysis *i.e.*, discription of set-up, schematic of apparatus, relevant theory, experimental procedure, specifications, and observation table. The students are required to prepare such instruction sheet before performing experiment. After performing the experiments, they have to complete the observation table and to make calculations, result, and post processing of result. They must mention the salient feature of experiment concentrating on the result and discussion. Each experimental write-up sheet must be headed with name and class of student, date of experimentation, and name of experiment.

This chapter provides a brief idea about write-up for some common experiments in subject of heat transfer. These write-ups are based on specific design of experimental setup. The design of experimental setup in some institute may differ from considered design. Therfore, observation tables, thus calculations may differ in some cases, but experiments run on same basic principles. Further, the list of experiments selected here may also be incomplete.

EXPT. 1 *THERMAL CONDUCTIVITY OF METALLIC ROD*

1.1. OBJECTIVE

To determine thermal conductivity of a metallic rod.

1.2. APPARATUS

Measuring flask, stop watch, thermocouples with temperature indicators.

1.3. PROLOGUE

Thermal conductivity is an important thermo-physical property of conducting materials, by virtue of which the material conducts the heat energy through it. From Fourier's law of heat conduction, the thermal conductivity is defined as,

$$k = -\frac{\mathrm{Q}}{\mathrm{A}}\frac{d\mathrm{T}}{dx} = -q\frac{d\mathrm{T}}{dx}\ (\mathrm{W/m.K})$$

where Q = heat transfer rate, watts,

q = heat flux, W/m^2,

A = area normal to heat transfer, m^2, and

$\frac{dT}{dx}$ = temperature gradient in the direction of heat flow.

The thermal conductivity for a given material depends on its state and it varies with direction, structure, humidity, pressure and temperature change. The following Table 16.1 shows the thermal conductivity of some commonly used materials.

TABLE 16.1 Thermal conductivities of some typical metals

Sr. No.	*Metals*	*Thermal Conductivity*	*Reference State*
1	Silver	419	20°C
2	Pure copper	386	20°C
3	Aluminium	204	20°C
4	Brass	111	20°C
5	Pure iron	73	20°C
6	Lead	35	20°C
7	Stainless steel	16.2	20°C

1.4. MECHANISM OF HEAT CONDUCTION IN METALS

Thermal energy can be transported in solids by two means.

(1) Lattice vibration,

(2) Transport of free electrons.

In good conducting materials, a large number of free electrons move about their lattice structure of metal. These electrons move from higher temperature region to lower temperature region, thus transport heat energy. Further, the increased temperature increases vibration energy of atoms in the lattice structure. Thus in hotter portion of the solid, the atoms which have larger vibration energy, transfer a part of its energy to the neighbouring low energy molecules and so on throughout the whole length of the body.

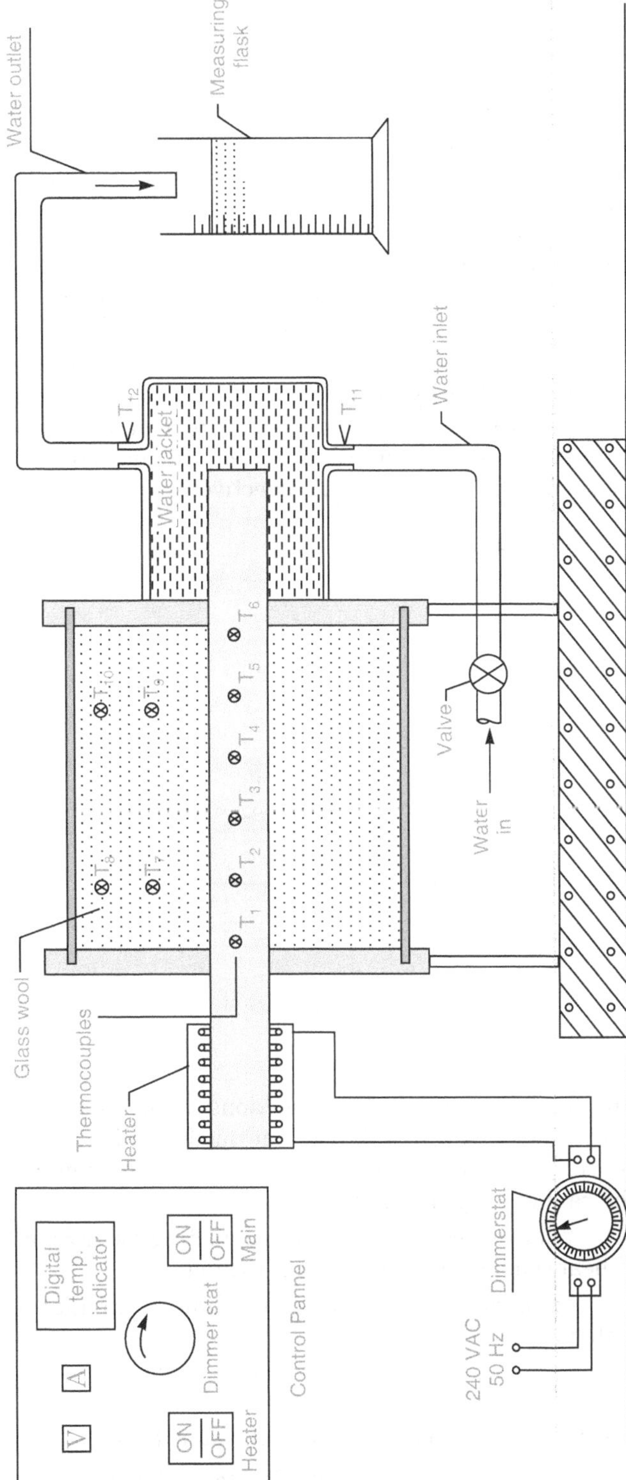

Fig.16.1. Experimental setup for determination of thermal conductivity of metal bar

1.5. EXPERIMENTAL ANALYSIS

1.5.1. Experimental Setup

The experimental setup consists of a brass metallic bar. The bar is horizontally placed and its test section is surrounded by a thick layer of insulating material. The bar is heated at its one end with the help of band type electrical heater and other end of the bar is projected in the cooling water jacket

Six thermocouples T_1 to T_6 are embedded at known distances along the length of the bar in test section and four thermocouples T_7 to T_{10} are equipped in the insulating cylinder in order to measure the heat loss in radial direction. The inlet and outlet temperatures of circulating water are measured by thermocouple T_{11} and T_{12}, respectively.

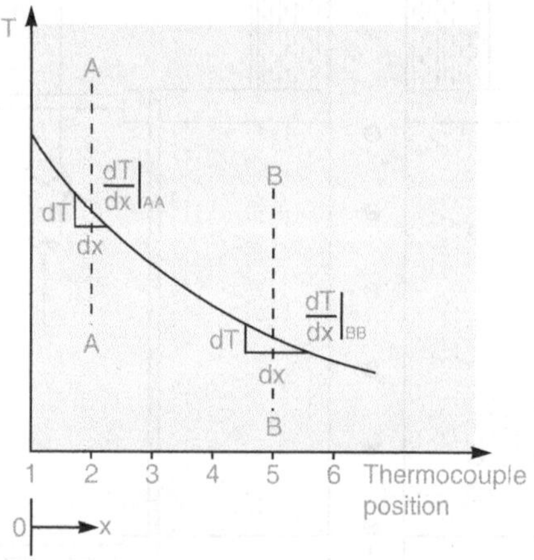

Fig. 16.2. Temperature distribution along the length of metal bar

1.5.2. Theory

The bar is analysed at two sections along its length. According to the first law of thermodynamics at any section, the rate of incoming energy must be equal to the rate of outgoing energy.

Rate of heat absorption by circulating water = Rate of heat conduction at free end of the rod

The rate of heat energy reaching to circulating water

$$Q_w = \dot{m}_w C_{pw}(T_{12} - T_{11}) \qquad ...(16.1)$$

where, $\dot{m}_w$ = mass flow rate of water, kg/s,

C_{pw} = specific heat of water in kJ/kg.K

$T_{12} - T_{11}$ = temperature rise of circulating water, °C

The rate of heat conduction through section B-B

$$Q_{BB} = Q_w + \frac{2\pi L_1 k_{ins}(T_9 - T_{10})}{\ln(r_o/r_i)} \qquad ...(16.2)$$

where L_1 = spacing between section B-B and cooling end,

k_{ins} = thermal conductivity of insulating material packed around the test section,

r_o = radius at which thermocouple 10 is embedded,

r_i = radius at which thermocouple 9 is attached,

T_{10} = temperature of insulation at radius r_o of section B-B,

T_9 = inner temperature of insulation at radius r_i of section B-B.

Further, the heat conduction rate at the section B-B is expressed as

$$Q_{BB} = -k_B A \left\{\frac{dT}{dx}\right\}_{BB} \qquad ...(16.3)$$

where k_B = thermal conductivity of bar at section B-B.

A = cross-section area of the bar, m^2.

$\left\{\frac{dT}{dx}\right\}_{BB}$ = temperature gradient at section B-B.

From eqns. (16.2) and (16.3), we get

$$k_B = \frac{Q_w + \dfrac{2\pi L_1 k_{ins}(T_9 - T_{10})}{\ln(r_o/r_i)}}{A\left\{\dfrac{dT}{dx}\right\}_{BB}} \qquad ...(16.4)$$

Similarly at section A-A heat conduction rate

$$Q_{AA} = Q_{BB} + \frac{2\pi L_2 k_{ins}(T_7 - T_8)}{\ln(r_o/r_i)} \qquad ...(16.5)$$

where, L_2 = length of bar between section A-A and B-B

The thermal conductivity of the bar material at section A-A in terms of heat conduction rate as :

$$Q_{AA} = -k_A A \left\{\frac{dT}{dx}\right\}_{AA} \qquad ...(16.6)$$

and

$$k_A = \frac{Q_{BB} + \dfrac{2\pi L_2 k_{ins}(T_7 - T_8)}{\ln(r_o/r_i)}}{A\left\{\dfrac{dT}{dx}\right\}_{AA}} \qquad ...(16.7)$$

1.5.3. Procedure

1. Start the water supply through the water jacket and regulate its uniform flow rate.
2. Put on the heater switch and adjust the heater input through dimmer-stat.
3. Wait until steady state condition is reached.
4. Note down the reading of all thermocouples through selector switch, voltmeter and ammeter.

5. Plot the temperature against the location of thermocouples along the test section of the bar.

1.6. SPECIFICATIONS

1. Total length of the bar = mm
2. Test length of the bar L = mm
3. Diameter of the bar d = mm
4. Temperature indicator = 0–300°C
5. Radius r_i of thermocouples T_7 and T_9 positions
6. Radius r_0 of thermocouples T_8 and T_{10} positions
7. Thermocouple type = Chromel Alumel
8. Measuring flask capacity = 0 to 1000 ml
9. Spacing between thermocouple = mm
10. Specific heat of water, C_{pw} = 4187 J/kg.K
11. Thermal conductivity of insulating material,

k_{ins} = 0.15 W/m.K for asbestos powder,

= 0.038 W/m.K for glass wool.

1.7. OBSERVATION TABLE

Sr. No.	*Mass flow (kg/s)*	*Heater input*		*Thermocouple readings, °C*											
		V	*I*	*1*	*2*	*3*	*4*	*5*	*6*	*7*	*8*	*9*	*10*	*11*	*12*

1.8. CALCULATION

Use eqn. (16.1) to eqn. (16.7) explained above in experiment theory to calculate the result.

1.9. RESULT

Take the average of two values of thermal conductivity and put it as outcome of the experiment.

The average thermal conductivity of the brass = ... W/m.K

1.10. EPILOGUE

1. Compare the value of thermal conductivity obtained by experimentation with standard value. State the reason for deviation, if any.
2. From the experiment it is concluded that the temperature goes on decreasing along the length of the rod.
3 The thermal conductivity of brass decreases with increase in temperature. This is due to obstruction in flow of free electrons caused due to increase in amplitude of lattice vibration.

1.11. PRECAUTIONS

1. Wait till perfect steady state is reached.
2. Input supply voltage must be constant.
3. Handle the change over switch of temperature gently.

EXPT. 2 *THERMAL CONDUCTIVITY OF INSULATING POWDER*

2.1. OBJECTIVE

To determine thermal conductivity of an insulating powder.

2.2. PROLOGUE

There are many heat exchanger equipments, where heat losses to the surroundings can be the subject of minimization. In such cases, the exterior surface of the equipment is covered by a material of lower thermal conductivity called as **insulator.** Because of demand of such materials, many industries have come up to produce such materials in different categories, shapes and sizes. One important category of insulating material is the powder form. The powder can take any complicated shape between any two containing surfaces. In addition, its thermal conductivity is much lower than that of its basic solid form, because of the larger number of air gaps present within powder.

2.3. EXPERIMENTAL ANALYSIS

2.3.1. Experimental Setup

The apparatus consists of two thin walled concentric copper spheres. The inner sphere houses the heating coil. The insulating powder filled in the annular gap between two copper spheres, takes the form of a hollow sphere. The power supply to the heating coil is adjusted by dimmerstat and it is measured by voltmeter and ammeter. The eight chromel alumel thermocouples are used to measure the temperature at inner and outer surface of insulating powder sphere. The four thermocouples, numbered as T_1 to T_4 are embedded on outer surface of the inner sphere and four thermocouples, numbered as T_5 to T_8 are embedded on inner surface of the outer sphere.

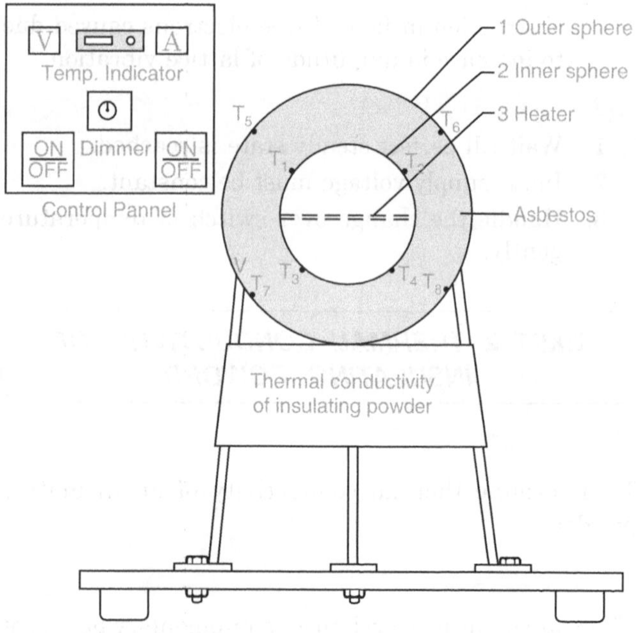

Fig.16.3. Setup for thermal conductivity of insulating powder

2.3.2. Theory

We assume that the insulating powder is an isotropic material and its thermal conductivity to be constant in all directions therefore, one-dimensional radial heat conduction across the powder is considered. Consider the heat transfer through hollow sphere formed by insulating powder layer packed between two copper spheres.

Let r_i = inner radius of the sphere, m,

r_o = outer radius of the sphere, m,

T_i = average temperature of inner surface of hollow insulating sphere, °C,

T_o = average temperature of outer surface of hollow insulating sphere, °C.

where $T_i = \frac{T_1 + T_2 + T_3 + T_4}{4}$ and

$$T_o = \frac{T_5 + T_6 + T_7 + T_8}{4}$$

According to Fourier law of heat conduction, the unknown thermal conductivity can be determined as:

$$k = \frac{Q(r_o - r_i)}{4\pi r_o r_i (T_i - T_o)} \quad ...(16.8)$$

where, Q = V.I

2.3.3. Experimental Procedure

1. Put on the heater switch and adjust the heater input through dimmer-stat.
2. Wait until steady state condition is reached.
3. Note down the reading of all thermocouples through selector switch, voltmeter and ammeter.

2.4. SPECIFICATIONS

1. Radius of inner copper sphere,

r_i = ... mm

2. Radius of outer copper sphere,

r_o = ... mm

3. Voltmeter = 0–100 / 200 Volt
4. Ammeter = 0–2 amp
5. Dimmer-stat = 0–260 Ω / 220 Volt
6. Heater capacity = W
7. Thermocouple type = Chromel Alumel

2.5. OBSERVATION TABLE

Sr. No.	*Heater input*		*Thermocouple readings, °C*							
	V	*I*	*1*	*2*	*3*	*4*	*5*	*6*	*7*	*8*

2.6. CALCULATION

Temperature of inner surface of hollow insulating sphere

$$T_i = \frac{T_1 + T_2 + T_3 + T_4}{4} = °C$$

Temperature of outer surface of hollow insulating sphere

$$T_o = \frac{T_5 + T_6 + T_7 + T_8}{4} = °C$$

The energy input to heater

Q = V.I (watt)

Thermal conductivity of insulating powder

$$k = \frac{Q(r_o - r_i)}{4\pi r_o r_i (T_i - T_o)} \text{ (W/m.K)}$$

2.7. RESULT

The thermal conductivity of given insulating powder

k = ... W/m.K.

2.8. EPILOGUE

The thermal conductivity for the given material depends upon its states and it may vary with structure, pressure and temperature changes. Thermal conductivity of insulating powder increases with temperature rise. Its variation with temperature is shown in Fig. 16.4.

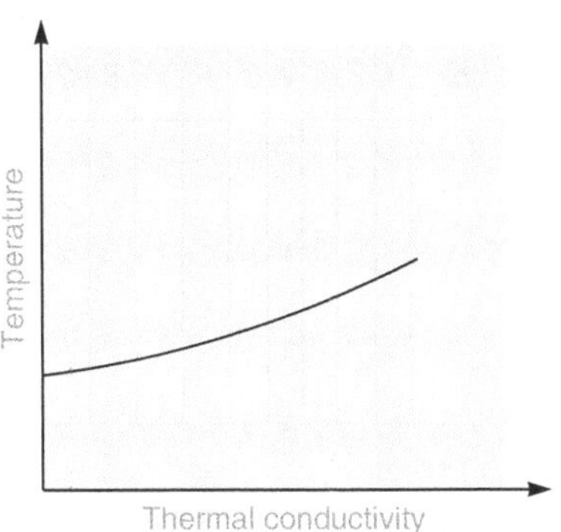

Fig.16.4. Variation of thermal conductivity of insulating material with temperature change

2.9. PRECAUTIONS

1. Wait till perfect steady state is reached.
2. Input supply must be constant.
3. Handle the change over switch of temperature gently.

EXPT. 3 *THERMAL CONDUCTIVITY OF COMPOSITE WALL*

3.1. OBJECTIVE

To determine thermal conductivity and temperature distribution across the thickness of the composite wall.

3.2. PROLOGUE

Many engineering situations involve the use of composite materials which consist of two or more materials of different thermal conductivities. Some situations involved are walls of a building, refrigerator, furnace, cold storage, etc. The knowledge of thermal conductivity of such composite medium helps in better design of equipments.

3.3. EXPERIMENTAL ANALYSIS

3.3.1. Experimental Setup

A typical experimental setup for composite wall is shown in Fig. 16.5. It consists of composite walls of three slabs of equal size and thickness, but of different thermal conductivities. Three materials are brass, asbestos and mild steel. These slabs are clamped together on both sides with the help of holding clamps, bolts and nuts. One side of the composite is exposed to heater while cooling water jacket is attached on other side. 14 thermocouples (T_1 to T_{14}) are embedded at different sections to record the temperature. Three thermocouples (T_1 to T_3) are embedded on heater side of mild steel slab, three (T_4 to T_6) on interface of mild steel and asbestos slabs, three (T_7 to T_9) on interface of asbestos and brass slabs and three (T_{10} to T_{12}) on water side of brass slab. Two thermocouples T_{13} and T_{14} for measuring water temperature at inlet and exit, respectively. In order to minimise the heat loss from the annular surface of the slabs, the glass wool insulation is provided. The temperature readings T_1 to T_{14} are displayed on a digital temperature indicator with the help of selector switch.

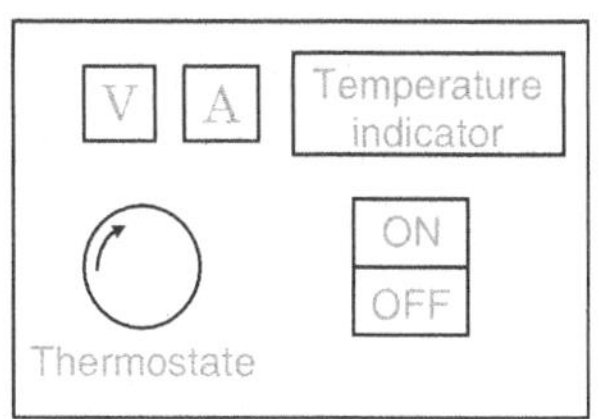

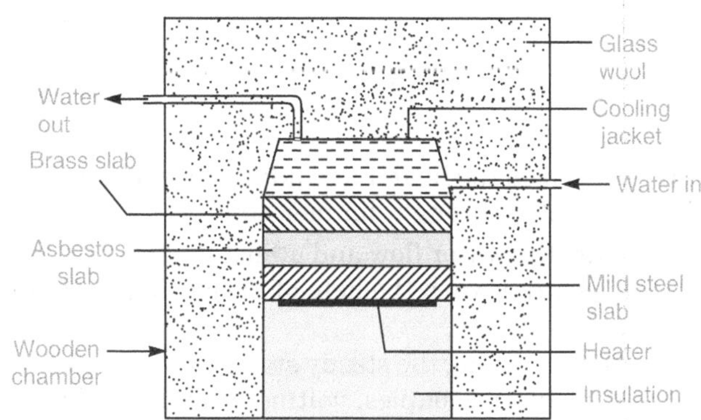

Fig. 16.5. Setup for composite wall

3.3.2 Theory

At steady state, the rate of heat input to heater

$$Q = VI$$

Rate of heat dissipation to water

$$Q_w = \dot{m}_w C_{pw} (T_{14} - T_{13}) \quad ...(16.9)$$

where $\dot{m}_w$ = mass flow rate of water,

C_{pw} = specific heat of water,

T_{13} = water inlet temperature to water jacket,

T_{14} = water exit temperature from water jacket.

The average temperature of heated surface of mild steel slab;

$$T_H = \frac{T_1 + T_2 + T_3}{3}$$

The average temperature of cooled surface of brass slab;

$$T_C = \frac{T_{10} + T_{11} + T_{12}}{3}$$

The thermal conductivity of composite slab may be expressed as :

$$k_{composite} = \frac{Q_w \, L_{composite}}{A(T_H - T_C)} \quad ...(16.10)$$

where $L_{composite} = L_{brass} + L_{steel} + L_{asbestos}$

The heat transfer area,

$$A = \frac{\pi}{4} D^2$$

where D is the diameter of plates, m.

Then total resistance of the composite wall per unit area can be expressed as :

$$\Sigma R_{th} = \frac{L_{brass}}{k_{brass}} + \frac{L_{steel}}{k_{steel}} + \frac{L_{asbestos}}{k_{asbestos}} \quad ...(16.11)$$

The overall heat transfer coefficient

$$U = \frac{1}{A \Sigma R_{th}} \quad ...(16.12)$$

3.3.3. Procedure

1. Switch on the heater and adjust the voltage input to heater through dimmer-stat.
2. Start the water flow and adjust its flow rate and measure it with the help of measuring flask and stop watch.
3. After reaching the steady state, take the readings of 14 thermocouples, voltmeter and ammeter.
4. Repeat the procedure two to three times with different power input to heater and water flow rate.

3.4. SPECIFICATIONS

1. Size of plates :
2. Width of plates :
3. Rating of heater :
4. Type of thermocouples : Alumel-chromel
5. Dimmerstat type :
6. Thermal conductivities,

k_{brass} = 110 W/m.K,

k_{steel} = 46 W/m.K,

$k_{asbestos}$ = 0.15 W/m.K

3.5. OBSERVATION TABLE

Sr. No.	*Heater input*		*Mass flow rate, $\dot{m}_w$*	*Thermocouple readings*													
	V	*I*		*1*	*2*	*3*	*4*	*5*	*6*	*7*	*8*	*9*	*10*	*11*	*12*	*13*	*14*

3.6. CALCULATION

At steady state, the rate of heat input to heater

$$Q = VI \text{ (watts)}$$

Under steady state conditions, the heat is dissipated to water at the same rate

$$Q_w = \dot{m}_w C_{pw}(T_{14} - T_{13}) = ... \text{ W}$$

The average temperature of heated surface of composite wall;

$$T_H = \frac{T_1 + T_2 + T_3}{3} = ... \text{°C}$$

The average temperature of cooled surface of composite wall;

$$T_C = \frac{T_{10} + T_{11} + T_{12}}{3} = ... \text{°C}$$

The heat transfer area,

$$A = \frac{\pi}{4} D^2 = ... \text{ m}^2$$

Then total resistance of the composite wall per unit area can be expressed as

$$\Sigma R_{th} = \frac{L_{brass}}{k_{brass}} + \frac{L_{steel}}{k_{steel}} + \frac{L_{asbestos}}{k_{asbestos}} = (\text{m}^2 \text{°C/W})$$

The overall heat transfer coefficient

$$U = \frac{1}{A \Sigma R_{th}} ... \text{ (W/m}^2\text{.K)}$$

The thermal conductivity of composite medium can be evaluated as

$$k_{composite} = \frac{Q_w \, L_{composite}}{A(T_H - T_C)} = ... \text{ (W/m.K)}$$

3.7. RESULT

The thermal conductivity of the composite slab is

k = ... W/m.K.

The overall heat transfer, U = ... W/m^2.K

3.8. EPILOGUE

The temperature distribution in three materials is shown in Fig. 16.6. The slope (temperature drop) in asbestos plate is large among three materials.

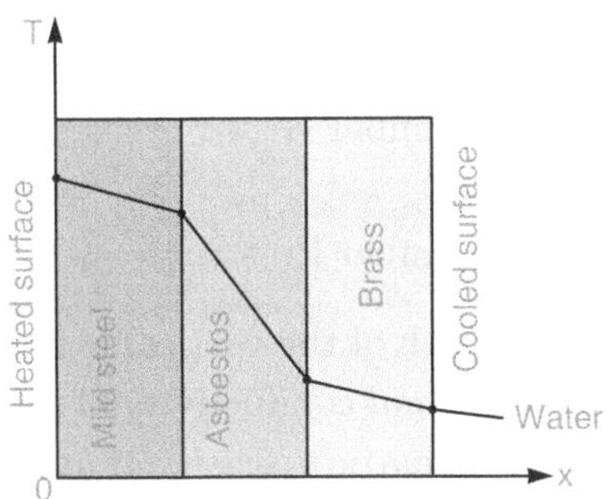

Fig. 16.6. Temperature distribution in three plates

The thermal conductivity for the composite medium is less than the avearge thermal conductivity of three materials. It is due to presence of air gaps and contact resistance at interfaces. The composite material provides insulating properties as well as good strength. Therefore, now a days its use is increasing.

3.9. PRECAUTIONS

1. Wait till perfect steady state is reached.
2. Input supply must be constant.
3. Handle the change over switch of temperature gently.

EXPT. 4 *NATURAL CONVECTION EXPERIMENT*

4.1. OBJECTIVE

To determine the heat transfer coefficient in natural convection.

4.2. PROLOGUE

Convection is a mode of heat transfer, which generally takes place in liquid and gases. Consider a fluid flow over a heated surface the molecules of fluid adjacent to the surface absorb heat and become hot. On heating, the molecules become lighter due to decrease in density, they rise up and the cold molecules of higher density come down in contact of heated surface. In this way, a motion of molecules (convection current) sets up in fluid due to developed density gradient.

4.3. EXPERIMENTAL ANALYSIS

4.3.1 Experimental Setup

The experimental setup consists of a cylindrical tube fitted in a rectangular duct, vertically as shown in Fig.16.7. The duct is open at the top and bottom. An electric heating element is kept in centre of vertical tube, which in turn heats the tube surface longitudinally. The heat is lost from the tube to the surrounding air by natural convection. The temperature of vertical tube is measured by six thermocouples at different locations and thermocouple T_7 measures the duct temperature. The heat energy input is measured by ammeter and voltmeter.

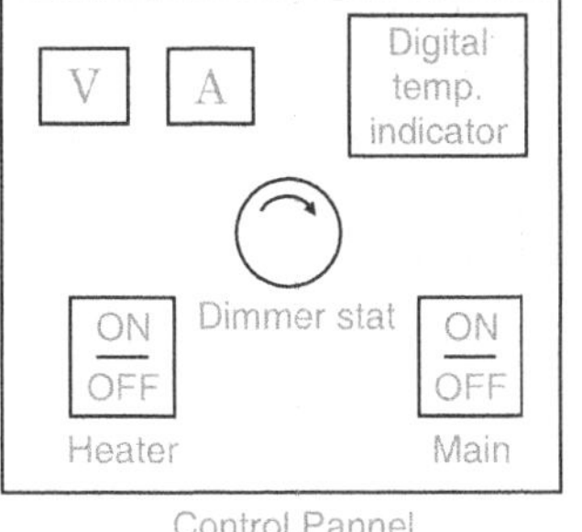

Fig.16.7. Experimental setup for determination of coefficient of heat-- transfer in natural convection

4.3.2. Theory

When a hot body is kept in still air, the heat is transferred to surrounding fluid adjacent to hot body. The adjacent fluid gets heated, it rises up due to decrease in its density and cold fluid rushes in to take place. Thus the fluid motion is setup and heat transfer takes place from the surface.

The heat transfer rate from the surface by natural convection is expressed by Newton's law as :

$$Q = A_s h(T_s - T_\infty) \quad ...(16.13)$$

where A_s = surface area,
h = heat transfer coefficient,
T_s = surface temperature,
T_∞ = ambient temperature.

Thus the convection coefficient can be evaluated as :

$$h = \frac{Q}{A_s(T_s - T_\infty)} \text{ W/m}^2\text{.K} \quad ...(16.14)$$

4.3.3. Experimental Procedure

1. Put on the heater switch and adjust the heater input through dimmerstat.
2. Wait until steady state condition is reached.
3. Note down the reading of all thermocouples through selector switch, voltmeter and ammeter.
4. Repeat above procedure for next readings.

4.4. SPECIFICATIONS

1. Diameter of tube, D = mm
2. Total length of the tube, L = mm
3. Capacity of heater = W
4. Temperature indicator = 0–300°C

(Multi-channel type calibrated for chromel, alumel, thermocouple)

4.5. OBSERVATION TABLE

Sr. No.	*Heater input*		*Thermocouple readings, °C*						
	V	*I*	*1*	*2*	*3*	*4*	*5*	*6*	*7*

4.6. CALCULATIONS

The rate of energy input to heat the cylindrical rod

$$Q = V I = ... \text{ W}$$

The average surface temperature of the brass rod,

$$T_s = \frac{T_1 + T_2 + T_3 + T_4 + T_5 + T_6}{6} = ... \text{ °C}$$

Ambient temperature

$$T_\infty = T_7 = ...\text{°C}$$

Thus the convection coefficient can be evaluated as :

$$h = \frac{Q}{A_s(T_s - T_\infty)} = ... \text{ W/m}^2\text{.K}$$

The heat transfer coefficient h can also be calculated from empirical relation:

$$\frac{hL}{k_{air}} = C\,(GrPr)^n$$

where, Grashof Number $Gr = \dfrac{g\,\beta\,L^3(\Delta T)}{\nu^2}$ ...(16.15)

Prandtl Number $Pr = \dfrac{\mu\,C_p}{k_{air}}$ ...(16.16)

where, L = length of the vertical cylinder, m.
k_{air} = thermal conductivity of air, W/m.K.
ν = kinematic viscosity of air, m²/s.
μ = dynamic viscosity of air, kg/ms.
C_p = specific heat of air, J/kg.K.
β = coefficient of volumetric expansion of air, K^{-1}
g = 9.81 m/s²
$\Delta T = (T_s - T_\infty)$°C.
$C = 0.56$ and $n = 0.25$ for $10^4 \le Gr\,Pr \le 10^8$
$C = 0.13$ and $n = 1/3$ for $10^8 \le Gr\,Pr \le 10^{12}$

The properties of fluid (air) are taken at film temperature at T_f, from properties of fluid Table A-4.

$$T_f = \frac{T_s + T_\infty}{2}$$

4.7. RESULT

1. Practical value of h = ...W/m².K
2. Theoretical value of h = ...W/m².K

4.8. GRAPH

The graph plotted for heat transfer coefficient *vs* location of thermocouples is shown in Fig.16.8.

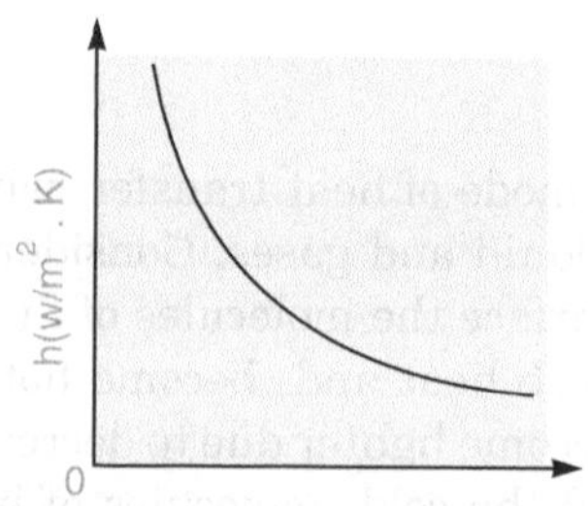

Fig.16.8. Variation of heat transfer coefficient along the height of a vertical pipe

4.9. EPILOGUE

The heat transfer rate in natural convection is much lower. Because, the heat is transferred due to density gradient only, thus the value of heat transfer coefficient is also small. The phenomenon of natural convection also depends upon viscosity, thermal conductivity, volumetric expansion coefficient β, characteristic length etc.

In situation, where the heat is transferred by natural convection then the fins are attached on heat transfer surface to enhance the heat transfer rate.

EXPT. 5 *FORCED CONVECTION EXPERIMENT*

5.1. OBJECTIVE

To determine the heat transfer coefficient in forced convection.

5.2. PROLOGUE

The heat transfer in forced convection occurs mechanically and depends on the motion of the fluid. When a fluid motion is caused by some external means such as pump or blower, then the convection is referred as *forced convection*. In the forced convection, the velocity gradients are more effective than density gradient. The forced convection heat transfer occurs in heat exchangers like automobile radiators, condensers, coolers etc.

5.3. EXPERIMENTAL ANALYSIS

5.3.1 Experimental Setup

The experimental setup consist of a blower unit fitted with test pipe as shown in Fig. 16.9. The four band type nichrome heaters surround the test section of pipe wall. The heat generated in heaters is conducted through the pipe wall of test section and then to flowing air by forced convection.

Three thermocouples (T_2 to T_4) are attached on test section. Thermocouple T_1 and T_5 record the temperature of incoming and outgoing air. Thermocouple T_6 measures temperature of atmospheric air. The test pipe is connected with the orifice to measure the flow rate of air through the pipe. A control valve is fitted in the delivery pipe in order to regulate the air flow rate. The pressure drop accros orifice is measured by U tube manometer in terms of water column. The energy input to the heater is measured by voltmeter and ammeter. Temperatures are displayed by digital temperature indicator with selection switch.

5.3.2. Theory

If Q is the rate of heat transfer to fluid

$$Q = \dot{m} C_p \Delta T = h A_s (T_s - T_\infty) \quad \text{...(16.17)}$$

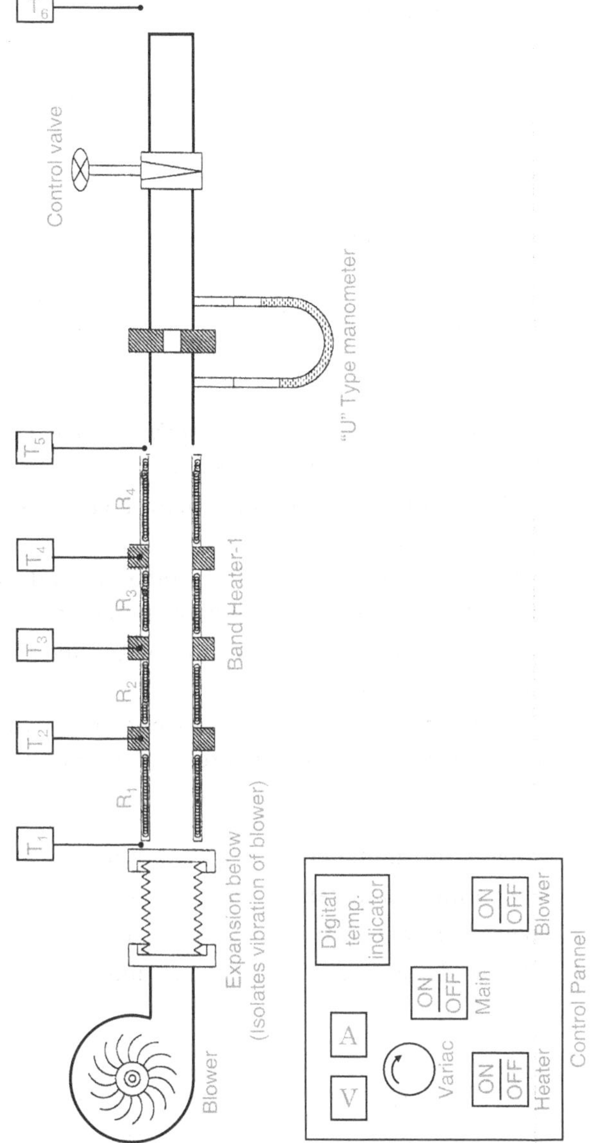

Fig. 16.9. Setup for coefficient of heat transfer in forced convection

where $\dot{m}$ = mass flow rate of air = $\dot{V} \rho_a$,

C_p = specific heat of air,

ΔT = temperature difference of air,

h = heat transfer coefficient,

A_s = surface area of the test pipe,

$(T_s - T_\infty)$ = temperature difference between surface and ambient.

The discharge (volume flow) rate, $\dot{V}$ of air through the orifice can be calculated as :

$$\dot{V} = \frac{\pi}{4} d_o^2 C_d \sqrt{2 g H_w \frac{\rho_w}{\rho_a}} \quad \text{...(16.18)}$$

where ρ_w = density of water,

ρ_a = density of air,

C_d = 0.64, coefficient of discharge for orifice,

d_o = diameter of orifice,

H_w = level difference of water in manometer arms.

Now heat transfer coefficient is given by

$$h = \frac{Q}{A_s(T_s - T_\infty)} \quad ...(16.19)$$

5.3.3. Procedure

1. Switch on the heater and adjust the voltage input to heater through dimmer-stat.
2. Adjust air flow rate with the help of valve fitted in delivery valve.
3. After reaching the steady state, take the readings of all thermocouples, voltmeter, ammeter.
4. Measure manometer deflection in U tube with the help of attached scale.
5. Repeat the procedure two to three times with different power input to heater and air flow rate.

5.4. SPECIFICATIONS

1. Pipe outer diameter, D = ... mm
2. Pipe inner diameter, d = ... mm
3. Length of test section, L = ... cm
4. Orifice diameter, d_o = ... mm

5.5. OBSERVATION TABLE

Sr. No.	*Heater input*		*Manometer reading,*	*Thermocouple readings, °C*					
	V	*I*	H_w	*1*	*2*	*3*	*4*	*5*	*6*

5.6. CALCULATIONS

The density of atmospheric air at mean temperature of

$$T_\infty = \frac{T_1 + T_5}{2}$$

$$\rho_a = \frac{p}{RT_\infty}$$

The discharge (volume flow) rate of air, $\dot{V}$ through the orifice can be calculated as :

$$\dot{V} = \frac{\pi}{4} d_o^2 \, C_d \sqrt{2gH_w \frac{\rho_w}{\rho_a}}$$

where density of water, ρ_w = 1000 kg/m³

the mass flow rate of air, $\dot{m} = \rho_w \dot{V}$

The heat transfer rate $Q = \dot{m} \, C_p \, \Delta T$ (W)

where C_p = 1005 J/kg.K for air,

$\Delta T = (T_s - T_\infty)$ = temperature difference of air in °C

Now heat transfer coefficient is given by

$$h = \frac{Q}{A_s(T_s - T_\infty)}$$

where $A_s = \pi d L$ (m²)

$$T_s = \frac{T_2 + T_3 + T_4}{3}$$

With use of empirical relations, the heat transfer coefficient can also be calculated as

$$\frac{hd}{k_{air}} = 0.023\,(Re)^{0.8}\,(Pr)^{0.4}$$

where

$$Re = \frac{\rho_a u_m d}{\mu}$$

$$Pr = \frac{\mu C_p}{k_{air}}$$

$$u_m = \frac{\dot{V}}{(\pi/4)d^2}$$

The properties of air are taken at $\frac{(T_s + T_\infty)}{2}$ from Table A-4 of Appendix A.

5.7. RESULT

The heat tansfer coefficient in forced convection = ... W/m².K.

5.8. GRAPH

Thermocouple location v/s heat transfer coefficient as shown in Fig.16.10.

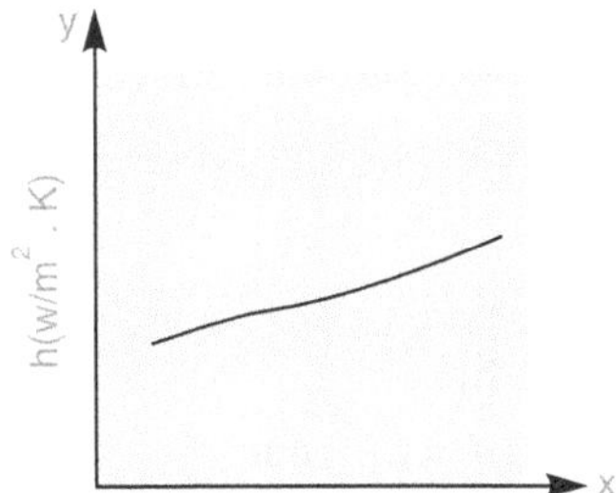

Fig. 16.10. Variation of heat transfer coefficient in forced convection

5.9. EPILOGUE

Heat transfer rate in forced convection is more than natural convection. As flow rate of fluid increases, the value of heat transfer coefficient also increases. Further, due to decrease in fluid density with the temperature rise, the turbulence increases in the flow, which causes an increase in heat transfer coefficient.

EXPT. 6 *HEAT TRANSFER FROM PIN FINS*

6.1. OBJECTIVE

To determine temperature distribution, heat transfer and fin efficiency of pin fins in natural and forced convection.

6.2. PROLOGUE

Extended surfaces or fins are used to increase the heat transfer rate from a surface to a fluid where it is not possible to increase the value of heat transfer coefficient or the temperature difference between the surface and fluid. The extended surface normally a thin strip of metal is called a fin.

The use of fins is very common and they are fabricated in variety of shapes. Circumferential fins are used around a cylinder of engine of the scooter and motorcycle and pin type fins are used on condenser tubes of a refrigerator.

6.3. EXPERIMENTAL ANALYSIS

6.3.1. Experimental Setup

Experimental setup consists of a mild steel duct in which a metallic body is heated by a band type heater. The input to this heater can be varied by dimmer-stat and measured by digital voltmeter and ammeter. Fig.16.11 shows a schematic of pin fin apparatus.

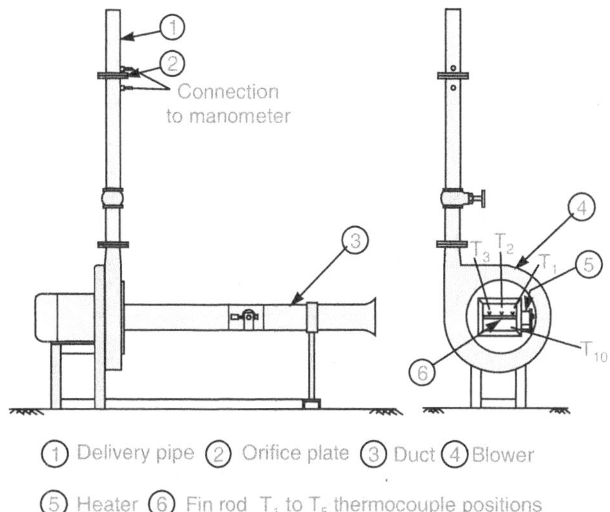

① Delivery pipe ② Orifice plate ③ Duct ④ Blower

⑤ Heater ⑥ Fin rod T_1 to T_5 thermocouple positions

Fig. 16.11. Schematic diagram of pin fin apparatus

Three identical cylindrical fins of materials : mild steel, brass and aluminium are fitted at their one end to the heated metallic body and other end of these fins are projected in the duct. The heat is conducted along the length of these fins and at the same time, the heat is also convected from the lateral surface of these fins. Three thermocouples are embedded along the length of each fin as shown in Fig. 16.12 (*a*). Thus the nine thermocouples show the temperature distribution of three fins. Two thermocouples are located in the duct to measure inlet and exit air temperature.

The duct is connected to delivery side of blower for arranging the forced convection over the fins. The air flow rate through the duct can be measured by using orifice which is fitted in duct with U tube manometer and scale.

6.3.2. Theory

Consider a fin connected at its base to a heated wall and transferring heat by convection to the surroundings. Other end of fin projecting inside the ducts is considered to be insulated.

Let, A = cross-section area of fin,

P = circumference of the fin,

L = length of the fin,

T_1 = temperature of the fin at heated end,

T_∞ = duct air temperature.

Temperature distribution at any location x in the fin takes the form of ;

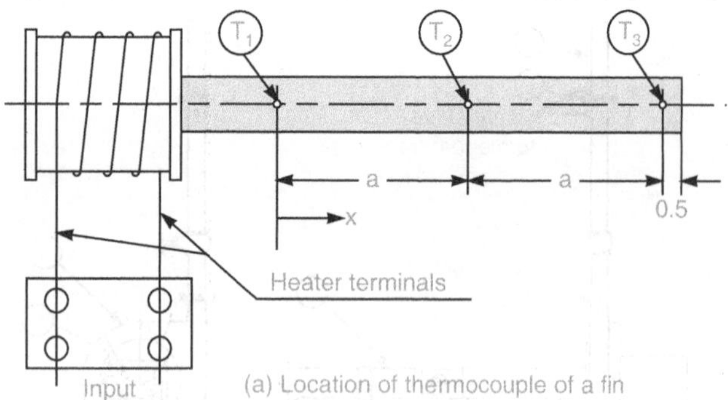

(a) Location of thermocouple of a fin

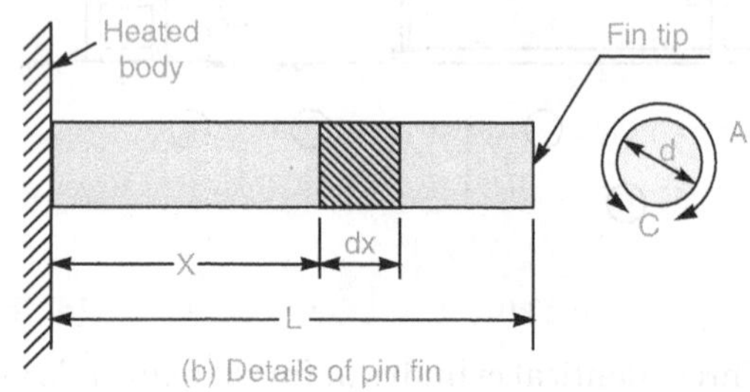

(b) Details of pin fin

Fig. 16.12

$$\frac{T(x) - T_\infty}{T_1 - T_\infty} = \frac{\text{Cosh } m\,(L - x)}{\text{Cosh } mL} \quad ...(16.20)$$

where $m = \sqrt{\frac{hP}{kA}}$ ...(16.21)

h = heat transfer coefficient between fin surface and surroundings, which can be evaluated from empirical relations.

The heat transfer from a fin with insulated tip fin can be obtained as :

$$Q = \sqrt{hPkA}\ (T_s - T_\infty)\tanh mL \quad ...(16.22)$$

The fin efficiency is given by

$$\eta_{fin} = \frac{\tanh mL}{mL} \quad ...(16.23)$$

6.3.3. Procedure

1. Electric supply is switched on and it is supplied to heater through dimmer-stat.
2. Steady state condition is waited, as it reaches the observations are made for:
   (*a*) Voltage from voltmeter,
   (*b*) Current from ammeter,
   (*c*) Temperatures indicated by eleven thermocouples through selector switch.
3. Repeat the procedure for next observation.
4. In case of forced convection, blower is run during the test and manometer deflection is also recorded.

6.4. SPECIFICATION

1. Duct size =
2. Diameter of fin
   d = ... cm
3. Diameter of orifice,
   d_o = ... cm
4. Coefficient of discharge,
   $C_d = 0.64$
5. Conductivity of brass fin,
   k = 110 W/m.K
6. Conductivity of aluminium fin,
   k = 210 W/m.K
7. Conductivity of mild steel fin,
   k = 46 W/m.K

6.5. OBSERVATION TABLE

Sr. No.	*Heater input*		*Manometer reading*	*Thermocouple readings, °C*										
	V	*I*	H_w	*1*	*2*	*3*	*4*	*5*	*6*	*7*	*8*	*9*	*10*	*11*

6.6. CALCULATION

(A) **Calculation of value of *h* in natural convection :**

(*i*) For brass fin

The average temperature of the fin

$$T_s = \frac{T_1 + T_2 + T_3}{3} = \text{ ... °C}$$

Air mean temperature

$$T_\infty = \frac{T_{10} + T_{11}}{2} = \text{ ... °C}$$

Bulk fluid mean temperature

$$T_f = \frac{T_s + T_\infty}{2} = \text{ ... °C}$$

The temperature difference

$$\Delta T = T_s - T_\infty = \ldots \; °C$$

Grashof number

$$Gr = \frac{g\,\beta\,\Delta T\, d^3}{\nu^2}$$

Prandtl number $\quad Pr = \dfrac{\mu\, C_p}{k_{air}}$

Nusselt number $\quad Nu = \dfrac{h\,d}{k_{air}}$

The properties of air μ, C_p, k_{air}, β, ν, and Pr are taken from Appendix Table A-4 at T_f.

The empirical relations for free convection are:

(*a*) $Nu = 1.1(Gr\,Pr)^{1/6}$ for $10^{-1} < Gr\,Pr < 10^4$

(*b*) $Nu = 0.53(Gr\,Pr)^{1/4}$ for $10^4 < Gr\,Pr < 10^9$

(*c*) $Nu = 0.13(Gr\,Pr)^{1/3}$ for $10^9 < Gr\,Pr < 10^{12}$

After calculation of Nusselt number, the quantity m is calculated by using eqn.(16.21), the fin heat transfer rate using eqn.(16.22) and fin efficiency by using eqn. (16.23).

Repeat the procedure for mild steel and aluminium fins.

(B) **Calculation of *h* in forced convection :**

(*a*) $Nu = 0.16(Re)^{0.466}$ When $40 < Re < 4000$

(*b*) $Nu = 0.174(Re)^{0.618}$ When $4000 < Re < 40000$

where $\quad Nu = \dfrac{hd}{k_{air}}$

$$Re = \frac{\rho u_m d}{\mu}$$

$$Pr = \frac{\mu C_p}{k_{air}}$$

The discharge rate of air through orifice

$$\dot{V} = \frac{\pi}{4} d_o^2\, C_d \sqrt{2gH_w \frac{\rho_w}{\rho_a}}$$

where, H_w = difference of level in manometer = ... m,

ρ_w = density of water = 1000 kg/m³

ρ_a = density of air $= \dfrac{p}{RT}$

orifice diameter d_o = ... m.

Again, $\dot{V} = u_m \times$ duct cross-section area

The velocity of air across the fins

$$u_m = \frac{\dot{V}}{\text{Duct C.S. Area}} \times \frac{T_f + 273}{T_\infty + 273}$$

After calculation of Nusselt number, the quantity m is calculated by using eqn.(16.21), the fin heat transfer rate using eqn.(16.22) and fin efficiency by using eqn. (16.23).

Repeat the procedure for mild steel and aluminium fins.

6.7. RESULT

(*i*) The efficiency of brass pin fin in natural convection =

The efficiency of brass pin fin in forced convection =

(*ii*) The efficiency of aluminium pin fin in natural convection =

The efficiency of aluminium pin fin in forced convection =

(*iii*) The efficiency of m.s. pin fin in natural convection =

The efficiency of m.s. pin fin in forced convection =

6.8. EPILOGUE

The temperature distribution along the length of three fins is shown in Fig.16.13. Observations are :

(*i*) It is observed that the temperature decreases along the length of each fin. But the temperature drop along the mild steel fin is large while along the aluminium fin is least.

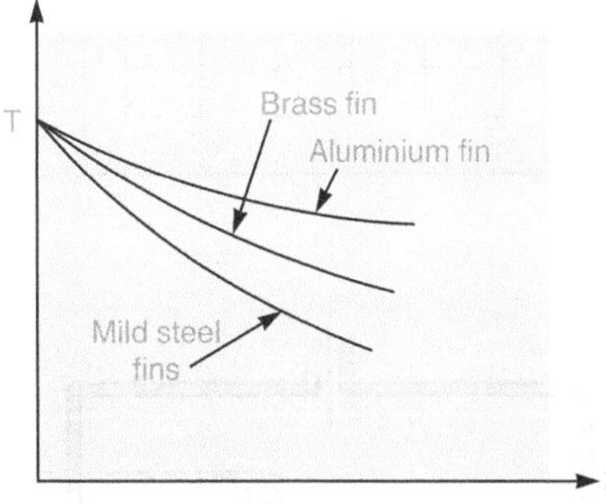

Fig. 16.13. Temperature distribution in three fins

(*ii*) The efficiency of each fin in natural convection is more than that in forced convection. It proves that the use of fins on a surface can only be justified in natural convection environment only.

(*iii*) It is also evident that the fin efficiency and heat transfer rate for aluminium fin are highest among the three material fins. It suggests that the fin should be made of light, conducting materials only.

EXPT. 7 STEFAN BOLTZMANN CONSTANT

7.1. OBJECTIVE

To determine Stefan Boltzmann constant of radiant heat transfer

7.2. PROLOGUE

Stefan Boltzmann law is a basic law of radiation heat transfer and it states that at thermal equilibrium, the heat flux or emissive power for a black surface is directly proportional to fourth power of absolute temperature of the surface and it is given by :

$$E_b = \sigma T^4$$

or

$$Q = A\,\sigma T^4 \quad \text{...(16.24)}$$

where σ is constant of proportionality and it is called Stefan Boltzmann constant and its standard value is :

$$\sigma = 5.667 \times 10^{-08} \ \text{W/m}^2.\text{K}^4$$

7.3. EXPERIMENTAL ANALYSIS

7.3.1. Experimental Setup

The experimental setup consists of a hemisphere having a black surface covered with a flat bakellite plate on the top of the tank as shown in Fig. 16.14. Five thermocouples (T_1 to T_5) are attached at different location in the hemisphere. A copper black coloured test disc fitted with a thermocouple T_6 is inserted in the bakellite plate at the centre of the hemisphere.

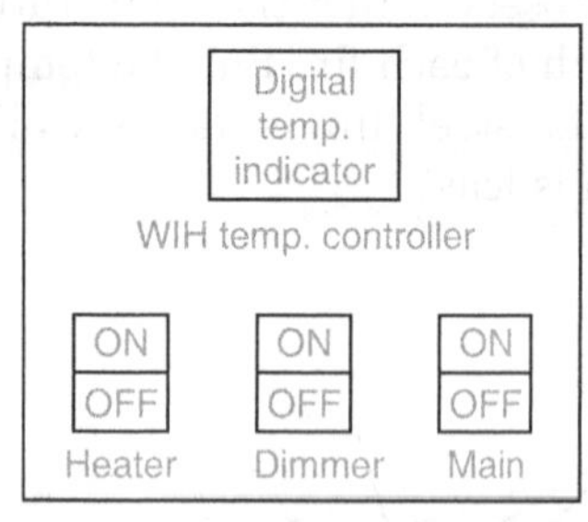

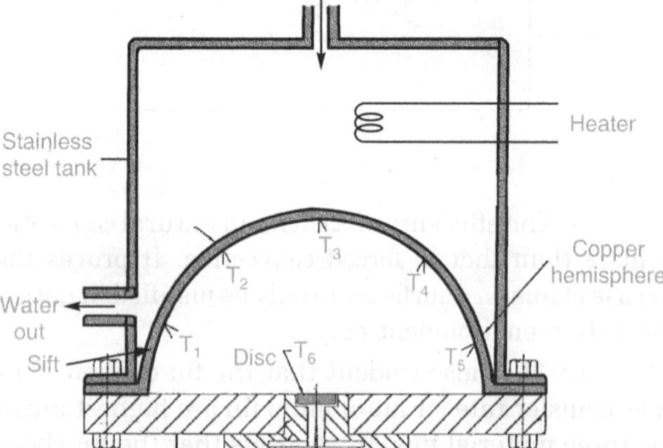

Fig. 16.14. Apparatus for Stefan Boltzmann constant

The hemisphere container is located in a cylindrical stainless steel tank equipped with an immersion heater. The water is filled in this tank. The heater is controlled by temperature controller to keep the inside temperature constant.

7.3.2. Theory

Initially the temperature of disc is lower than the hemisphere surface. Thus, when the test disc is inserted in its housing at the centre, the energy is radiated from hemisphere surface to test disc.

The rate of radiation energy is given by

$$Q = A_d\,(E_1 - E_2) = \sigma\, A_d\,(T_s^4 - T_d^4) \quad \text{...(16.25)}$$

where, $A_d = (\pi/4)\, d^2$

T_d = Initial temperature of test disk

$$T_s = \frac{T_1 + T_2 + T_3 + T_4 + T_5}{5} \ {}^\circ\text{C}$$

Due to this radiation heat exchange, the internal energy of disc increases and rate of internal energy increase is given by :

$$\Delta U = mC_p \left.\frac{dT}{dt}\right|_{t=0} \quad \text{...(16.26)}$$

where, m = mass of test disc,

C_p = specific heat of test disc,

$\left.\dfrac{dT}{dx}\right|_{t=0}$ = slope of curve at t = 0.0 second

The energy balance on the disc

$$\Delta U = mC_p \left.\frac{dT}{dt}\right|_{t=0} = A_d\, \sigma(T_s^4 - T_d^4)$$

Then Stefan Boltzmann constant

$$\sigma = \frac{mC_p \left.\dfrac{dT}{dt}\right|_{t=0}}{A_d(T_s^4 - T_d^4)} \quad \text{...(16.27)}$$

7.3.3. Procedure

The following steps should be followed during experimentation

(*i*) Check level of the water in the tank and fill it till it overflows.

(*ii*) Set the thermo-stat to desired temperature for water heating.

(*iii*) Switch on the heater and wait till the water is heated to set temperature.

(*iv*) Record the five thermocouple readings fitted to hemisphere surface.

(*v*) Record the temperature of the test disc and insert it in the bakellite plate.

(*vi*) Record the temperature of test disc at the interval of each 10 seconds.

(*vii*) Plot temperature time history for the test disc as shown in Fig. 16.15.

7.4. SPECIFICATIONS

1. Hemisphere enclosure diameter = ... mm
2. Test disc diameter, d = ... mm
3. Mass of the test disc, m = ... gm
4. Specific heat of the test disc, C_p = 381 J/kg °C

7.5. OBSERVATION TABLE

Bulk temperature of water = ... °C

Time	*Thermocouple readings, °C*					
(Seconds)	T_1	T_2	T_3	T_4	T_5	T_6

7.6. CALCULATIONS

The average surface temperature of the hemisphere

$$T_s = \frac{T_1 + T_2 + T_3 + T_4 + T_5}{5} = \ldots \text{°C}$$

The initial temperature of the test disc

$$T_d = \ldots \text{°C}$$

The initial slope of temperature rise of disc is

$$\left.\frac{dT}{dt}\right|_{t=0} = \ldots \text{°C/s}$$

(From temperature time history of the test disc, Fig. 16.15)

The Stefan Boltzmann constant

$$\sigma = \frac{mC_p \left.\frac{dT}{dt}\right|_{t=0}}{A_d(T_s^4 - T_d^4)} = \ldots \text{W/m}^2.\text{K}^4$$

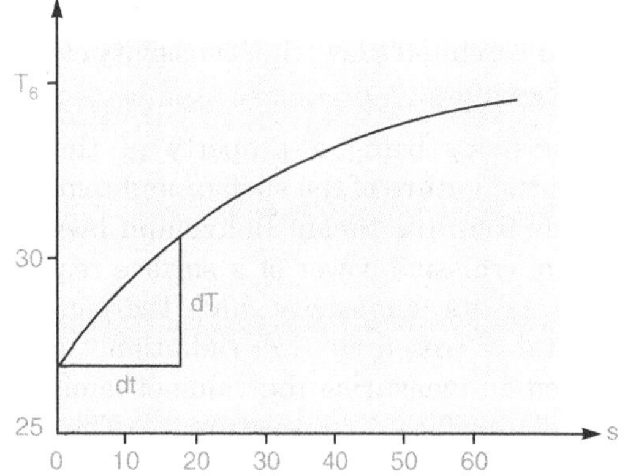

Fig. 16.15. Temperature history for test disc

7.7. RESULT

The value of Stefan Boltzmann constant, σ = ... W/m^2.K^4.

7.8. EPILOGUE

Explain if any variation is observed between calculated value and standard value.

EXPT. 8 *MEASUREMENT OF EMISSIVITY OF A TEST SURFACE*

8.1. OBJECTIVE

To measure the emissivity of the test surface in comparision to black surface.

8.2. PROLOGUE

All substances at all temperatures emit thermal radiation. Thermal radiation is an electromagnetic wave and does not require any material medium for propagation. All bodies can emit radiation and have the capacity to absorb all or a part of the radiation coming from the surrounding towards it.

An idealised black surface is one which absorbs all the incident radiation with reflectivity and transmissivity equal to zero. The radiant energy per unit time per unit area from the surface of body is called as the emissive power and is usually denoted by E. The emissivity of the surface is defined as the ratio of the emissive power of a surface to that of a black surface at the same temperature. It is denoted by ε.

Thus $$\varepsilon = \frac{E}{E_b} \quad \ldots(16.28)$$

where E_b = Emissive power of black body

For a black body absorptivity, $\alpha = 1$ and then according to Kirchhoff's law, the emissivity of the black body becomes unity.

Emissivity being a property of the surface depends on the nature of the surface and temperature. It is obvious from the Stefan Boltzmann law that the prediction of emissive power of a surface requires the knowledge of its emissivity and therefore, much experimental research in radiation has been concentrated on measuring the value of emissivity as function of surface temperature. The present experimental setup is designed and fabricated to measure the emissivity of the test plate surface at various temperature.

8.3. EXPERIMENTAL ANALYSIS

8.3.1. Experimental Setup

The experimental sets up consist of two similar circular aluminium discs and are provided with heating coil sandwiched. One disc is blackened by thick layer of lamp black to form the idealized black surface whereas the other disc having machined plane surface, whose emissivity is to be determined. The plates are mounted on brackets and are kept in enclosure so as to provide undisturbed natural convection surroundings. The schematic of experimental arrangement is shown in Fig. 16.16.

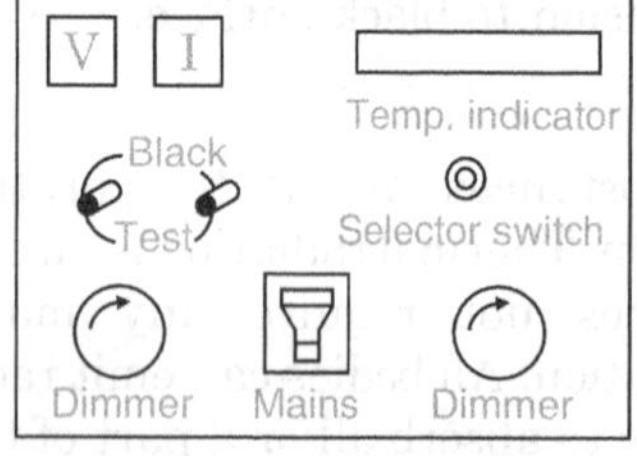

Fig. 16.16. Emissivity measurement apparatus

The power input to two plates can be varied by separate dimmer-stats and is measured by using an ammeter and a voltmeter with the help of double throw switch. The heat input to the black and test disc is to be adjusted in such a way that both discs should attain the same temperature. The each disc is equipped with three thermocouples to measure temperature at different locations. One thermocouple is kept in the enclosure.

8.3.2. Theory

The heat energy input to the two discs is dissipated by conduction, convection and radiation. Both discs are located in same environment and they cannot see each other, thus at identical temperature and in steady state conditions, the heat dissipation by conduction and convection will be same for both discs. The difference in the heater input readings is observed, it is because of the difference in the radiation characteristics of two surfaces.

Under steady state conditions, let

Q_1 = heater input to black plate $= V_1 I_1$

Q_2 = heater input to test plate $= V_2 I_2$

A = area of disc $= \frac{\pi}{4} d^2 + \pi dt$

t = thickness of discs

T_s = surface temperature of discs

T_∞ = ambient temperature of enclosure

ε_b = emissivity of black plate (To be assumed equal to unity)

ε = emissivity of non black (test) disc

σ = stefan Boltzmann constant

$= 5.667 \times 10^{-8}$ W/m^2.K^4

$$Q_1 - Q_2 = (\varepsilon_b - \varepsilon)\, \sigma A(T_s^4 - T_\infty^4) \quad ...(16.29)$$

8.3.3. Procedure

1. Switch on the heater and adjust the voltage input to both discs through dimmer-stat.
2. After reaching the steady state, take the readings of thermocouples, voltmeter, ammeter.
3. Repeat the procedure two to three times with different heat input rate.

8.4. SPECIFICATIONS

(*i*) Size of the enclosure =

(*ii*) Diameter of the discs, *d* = ... mm

(*iii*) Thickness of discs, *t* = ... mm

(*iv*) Capacity of heater attached to plates = ...W

8.5. OBSERVATION TABLES

Sr. No.	*Test disc*					*Black disc*					*Enclosure*
	V	I	T_1	T_2	T_3	V	I	T_4	T_5	T_6	T_7

8.6. CALCULATIONS

The heat input to black disc,

$$Q_1 = V_1 I_1$$

The heat input to test disc,

$$Q_2 = V_2 I_2$$

Area of discs,

$$A = \frac{\pi}{4} d^2 + \pi dt = \ldots \text{ m}^2$$

Surface temperature,

$$T_s = \frac{T_1 + T_2 + T_3}{3} = \ldots \text{ °C}$$

T_∞ = ambient temperature of enclosure, °C

$$\varepsilon = 1 - \frac{Q_1 - Q_2}{\sigma A (T_s^4 - T_\infty^4)}$$

8.7. RESULT

The emissivity of the test plate is ....

8.8. EPILOGUE

The emissivity of the test surface is lower than that of black surface and the emissivity of test surface increases with increase in temperature.

The emissivity of the test plate can be calculated at various surface temperatures of the discs. With increase in temperature, the test surface becomes somewhat dull and therefore, its emissivity increases with increase in surface temperature.

EXPT. 9 *HEAT EXCHANGER EXPERIMENT*

9.1. OBJECTIVE

To determine and compare LMTD, overall heat transfer coefficient and effectiveness of a heat exchanger in parallel flow and counter flow modes.

9.2. APPARATUS

Heat exchanger experimental setup, measuring flask, stop watch, and thermometers.

9.3. PROLOGUE

Heat exchanger is a device in which heat is transferred from a hot fluid to a cold fluid. Heat exchangers are used in various industries as well as in domestic use. Radiators and condensers are the common heat exchangers. The heat exchangers are classified in three categories:

1. Direct contact type,
2. Indirect contact type,
3. Storage type.

In *direct contact* type of heat exchanger, the hot and cold fluids are brought in direct contact. A good example is jet condenser, where steam jet and cooling water contact physically for heat exchange.

An *indirect contact* or *transfer* type heat exchanger is one in which hot and cold fluids do not make direct contact, but they are separated by a partition wall, acts as heat transfer medium. In practice, most of the heat exchangers now used are transfer type. One excellent example is shell and tube type heat exchanger. The transfer type heat exchangers are further classified as :

1. Parallel flow : Both fluids flow in same direction.

2. Counter flow : Hot and cold fluids flow in opposite direction to each other.

3. Cross flow : Hot and cold Fluids flow at right angle to each other.

A simple transfer type heat exchanger is tube in tube or double pipe (concentric tubes) type arrangement. One fluid flows inside the centre tube and another fluid flows in annular space formed between inner and outer tube.

In *storage* type or *regenerative* heat exchanger, hot and cold fluids pass alternately on the same heat transfer surface for their heat exchange.

9.4. EXPERIMENTAL ANALYSIS

9.4.1. Experimental Setup

Experimental setup consists of a concentric double pipe heat exchanger as shown in Fig. 16.17. The hot fluid is hot water, obtained from electric geyser and it flows through inner copper tube while cold water flows in annular space formed between inner and outer tube. The outer tube is provided with adequate insulation to minimise the heat loss to the surroundings.

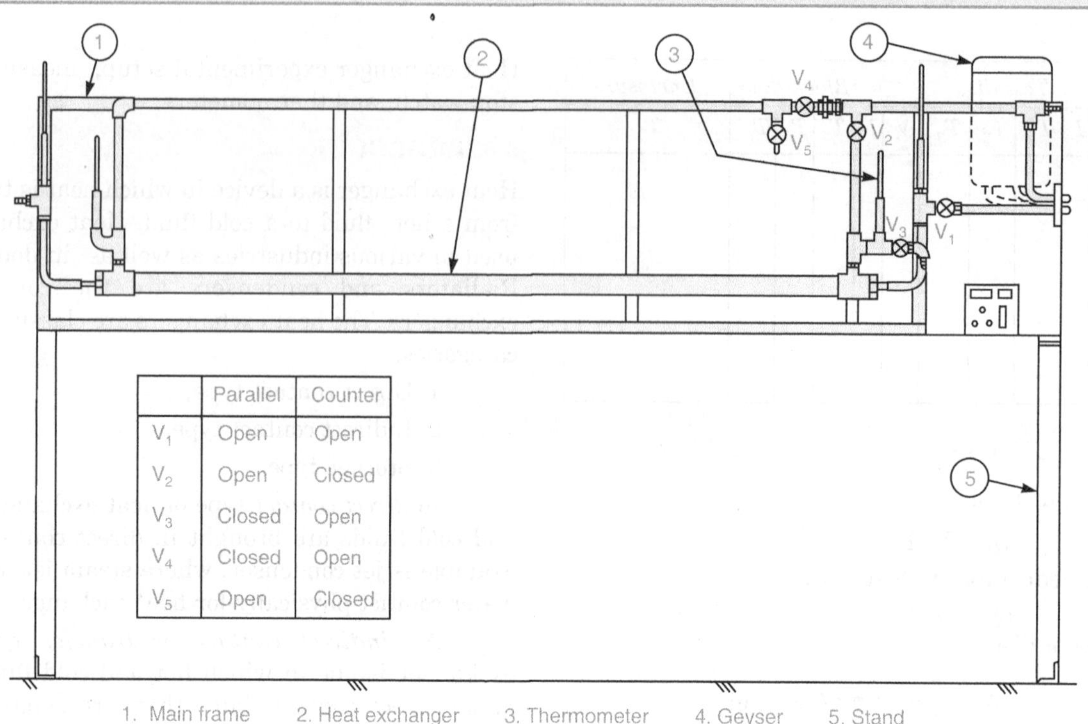

Fig. 16.17. Schematic of parallel flow counter flow heat exchanger

The hot water flows always in specified direction. A valve regulates the flow rate of hot water. With the valves arrangement and their proper operation, the cold water can be admitted at either end of the annular space to act as parallel flow or counter flow heat exchanger. Fig. 16.18 shows the temperature distribution in two cases.

The minimum objective of the experiment is to compare :

1. The temperature distribution in parallel flow and counter flow arrangement.

2. Heat transfer rate in two types of heat exchanger.

3. Overall heat transfer coefficient.

4. To compare the effectiveness of heat exchanger in both arrangements.

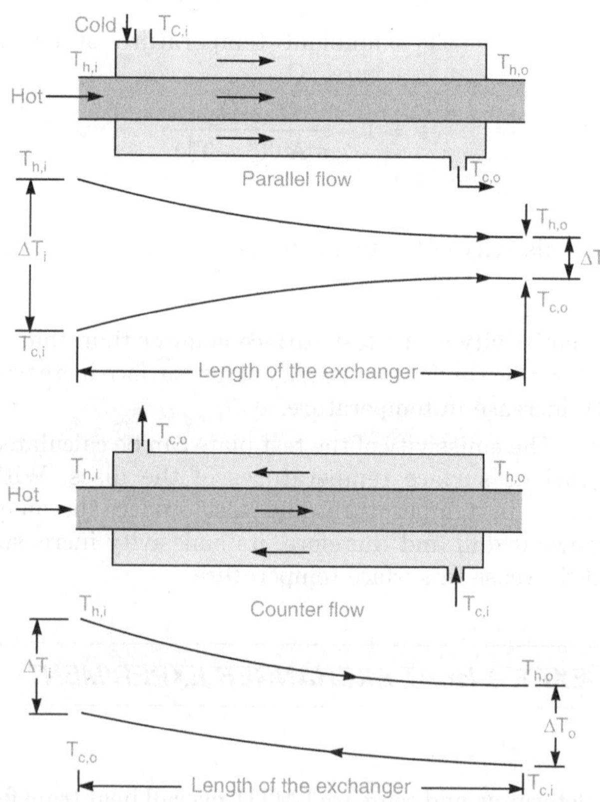

Fig. 16.18. Temperature distribution

9.4.2. Theory

1. The heat transfer rate is calculated as:

For hot fluid as,

$$Q_h = \dot{m}_h C_{ph} (T_{h,i} - T_{h,o}) \quad \text{...(16.30)}$$

For cold fluid,

$$Q_c = \dot{m}_c C_{pc} (T_{c,o} - T_{c,i}) \quad \text{...(16.31)}$$

where, suffix h for hot fluid and c for cold fluid.

The average heat transfer rate,

$$Q = \frac{Q_c + Q_h}{2} \quad ...(16.32)$$

2. Log mean temperature difference (LMTD)

$$\Delta T_{lm} = \frac{\Delta T_i - \Delta T_o}{\ln\left[\frac{\Delta T_i}{\Delta T_o}\right]} \quad ...(16.33)$$

where, $\Delta T_i = T_{h,i} - T_{c,i}$ For parallel flow

$= T_{h,i} - T_{c,o}$ For counter flow

$\Delta T_o = T_{h,o} - T_{c,o}$ For parallel flow

$= T_{h,o} - T_{c,i}$ For counter flow

3. Overall heat transfer coefficient

$$Q = U_i A_i(\Delta T)_{lm} \quad ...(16.34(a))$$

$$= U_o A_o(\Delta T)_{lm} \quad ...(16.34(b))$$

where, U_i = overall heat transfer coefficient based on inner surface,

U_o = overall heat transfer coefficient based on outer surface.

4. Effectiveness of heat exchanger

$$\varepsilon = \frac{\text{Actual Heat Transfer rate}}{\text{Max. Possible Heat Transfer rate}} \quad ...(16.35)$$

5. Overall heat transfer coefficient can also be calculated as :

$$\frac{1}{U_i} = \frac{1}{h_i} + \frac{r_i}{k} \times \ln\left[\frac{r_o}{r_i}\right] + \frac{r_i}{r_o h_o} \quad ...(16.36)$$

where, h_i and h_o are heat transfer coefficients based on inner and outer heat transfer surfaces, respectively, k = thermal conductivity of tube material (copper) = 380 W/m.K

The heat transfer coefficient h_i can be calculated by correlation;

$$Nu_{d_i} = 0.023 Re_{d_i}^{0.8} Pr^{0.3} \quad ...(16.37)$$

where,

$$Nu_{d_i} = \frac{h_i d_i}{k_{water}}$$

$$Re_{d_i} = \frac{4\dot{m}_h}{\pi d_i \mu}$$

The physical properties of water can be obtained from Table A-7 at mean hot water temperature

$$\frac{T_{h,i} + T_{h,o}}{2}$$

Similarly, the heat transfer coefficient h_o can be obtained from the relation,

$$Nu_{(D_i - d_o)} = 0.023\ Re_{(D_i - d_o)}^{0.8}\ Pr^{0.4} \quad ...(16.38)$$

where,

$$Nu_{(D_i - d_o)} = \frac{h_o(D_i - d_o)}{k_{water}}$$

$$Re_{(D_i - d_o)} = \frac{4\dot{m}_c}{\Pi(D_i - d_o)\mu}$$

The properties of fluid are obtained at mean temperature of cold water from Table A-7.

9.4.3. Procedure

1. With the help of valve arrangement shown on the apparatus, Fig. 16.17 it is made to act as parallel flow heat exchanger.
2. The heater is switch on and water flow rate for hot and cold streams are kept constant with the controlled opening of valves.
3. The flow rate for both the fluids is measured by measuring flask and stop watch.
4. Steady state condition is waited, as it prevails, the inlet and outlet temperatures of both fluids can be measured by thermometers.
5. Same procedure is repeated for counter flow arrangement.

9.5. SPECIFICATIONS

(*i*) Inner diameter of inner tube,

d_i = ... mm

(*ii*) Outer diameter of inner tube,

d_o = ... mm

(*iii*) Inner diameter of outer tube,

D_i = ... mm

(*iv*) Outer diameter of outer tube,

D_o = ... mm

(*v*) Length of the heat exchanger,

L = ... mm

(*vi*) Conductivity of copper,

k = 380 W/m.K

(*vii*) Specific heat of hot and cold fluids,

$C_{pc} = C_{ph}$ = 4180 J/kg.K

9.6. OBSERVATION TABLE

1. *Parallel flow run*

Readings	Hot water side			Cold water side		
	Flow rate $\dot{m}_h$ *(kg/s)*	*Inlet* $T_{h,i}$ *°C*	*Outlet* $T_{h,o}$ *°C*	*Flow rate* $\dot{m}_c$ *(kg/s)*	*Inlet* $T_{c,i}$ *°C*	*Outlet* $T_{c,o}$ *°C*
Initial reading Final reading Corrected reading						

2. *Counter flow run*

Readings	Hot water side			Cold water side		
	Flow rate $\dot{m}_h$ *(kg/s)*	*Inlet* $T_{h,i}$ *°C*	*Outlet* $T_{h,o}$ *°C*	*Flow rate* $\dot{m}_c$ *(kg/s)*	*Inlet* $T_{c,i}$ *°C*	*Outlet* $T_{c,o}$ *°C*
Initial reading Final reading Corrected reading						

9.7. CALCULATIONS

(A) *For parallel flow*

$$Q_h = \dot{m}_h C_{ph} (T_{h,i} - T_{h,o})$$

$$Q_c = \dot{m}_c C_{pc} (T_{c,o} - T_{c,i})$$

$$Q = \frac{Q_c + Q_h}{2}$$

$$\Delta T_{lm} = \frac{\Delta T_i - \Delta T_o}{\ln\left(\frac{\Delta T_i}{\Delta T_o}\right)}$$

$$\Delta T_i = T_{h,i} - T_{c,i} \quad \text{and} \quad \Delta T_o = T_{h,o} - T_{c,o}$$

Overall heat transfer coefficient

$$U_i = \frac{Q}{A_i (\Delta T)_{lm}} = \ldots \text{ W/m}^2\text{.K}$$

and $$U_o = \frac{Q}{A_o (\Delta T)_{lm}} = \ldots \text{ W/m}^2\text{.K}$$

For effectiveness

$$(\dot{m}C_p)_h = \ldots \text{ W/K}$$

$$(\dot{m}C_p)_c = \ldots \text{ W/K}$$

$$(\Delta T)_{max} = (T_{h,i} - T_{c,i}) = \ldots \text{ K}$$

$$\varepsilon = \frac{Q}{(\dot{m}C_p)_{min} (\Delta T)_{max}}$$

(B) *For counter flow*

Repeat above calculations as in similar way.

9.8. RESULTS

The following table shows computed values

Sr. No.	*Parameters*	*Parallel flow*	*Counter flow*
1	LMTD, °C		
2	U_i, W/m^2. K (practical)		
3	U_i, W/m^2. K (empirical)		
4	U_o, W/m^2. K (practical)		
5	U_o, W/m^2. K (empirical)		
6	Effectiveness		

9.9. EPILOGUE

The logmean temperature difference, heat transfer rate and effectiveness of heat exchanger in counter flow arrangement are more than that in parallel flow arrangement. Therefore, most of the tubular heat exchangers are operated in counter flow mode.

EXPT. 10 *CRITICAL HEAT FLUX*

10.1. OBJECTIVE

To study the pool boiling phenomenon and to determine critical heat flux.

10.2. PROLOGUE

When evaporation occurs at a solid liquid interface, it is termed to be *boiling*. The process occurs when the heated surface temperature T_s exceeds the saturation temperature T_{sat} corresponding to liquid pressure. The temperature difference $(T_s - T_{sat})$ is called *temperature excess*.

Depending upon the temperature excess, the physical mechanism of boiling may be studied by different modes, or regimes as shown in Fig. 16.19. The heat flux supplied to the surface is plotted against temperature excess $(T_s - T_{sat})$. It is seen that boiling curve obtained can be divided :

1. Natural convection boiling,
2. Nucleate boiling,
3. Film boiling.

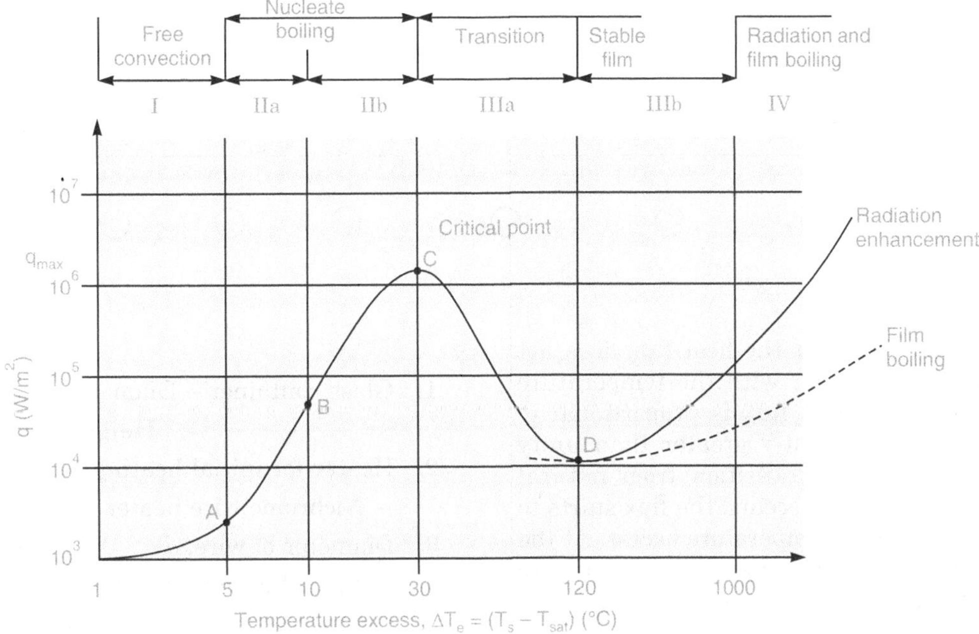

Fig. 16.19. Typical boiling curve for saturated water at 1 atmosphere

Natural Boiling

The natural convection boiling is said to be exist, if temperature excess ≅ 5°C. Heat transferred from the heated surface to liquid in this vicinity causes the liquid to be superheated. The superheated vapours rise to free liquid surface by natural convection, where insufficient vapour is produced by evaporation. As the excess temperature increases, the bubble formation starts. In this regime fluid motion depends on free convection effects.

Nucleate Boiling

As the temperature excess increases beyond 5°C, the nucleate boiling starts. In this region, it is observed that the bubbles starts to form at certain locations called *nucleation sites,* on the heated surface. In the nucleate boiling, two different flow regimes may be distinguished. In the IIa the bubbles formed are a few in numbers and small in size. They coalesce with liquid and do not reach the free surface. In the IIb part, the rate of bubble formation as well as the number of their locations, increase. At the end of nucleate boiling the heat flux reaches to its maximum value. This maximum heat flux is called *critical heat flux.*

Film Boiling

With increase in temperature excess, a stage is finally reached when the high bubble formation rate causes them to merge and blanket the surface with the vapour film. This is the beginning of region III, namely *film boiling*. In the first part of this region, IIIa, the vapour film is unstable, film boiling may occur on a portion on the heated surface area, while nucleate boiling may occur on the remaining area. In the second part, IIIb, a stable film covers the entire surface. The temperature excess in this region is approximately 100°C and consequently, the radiation heat transfer across the vapour film is also significant

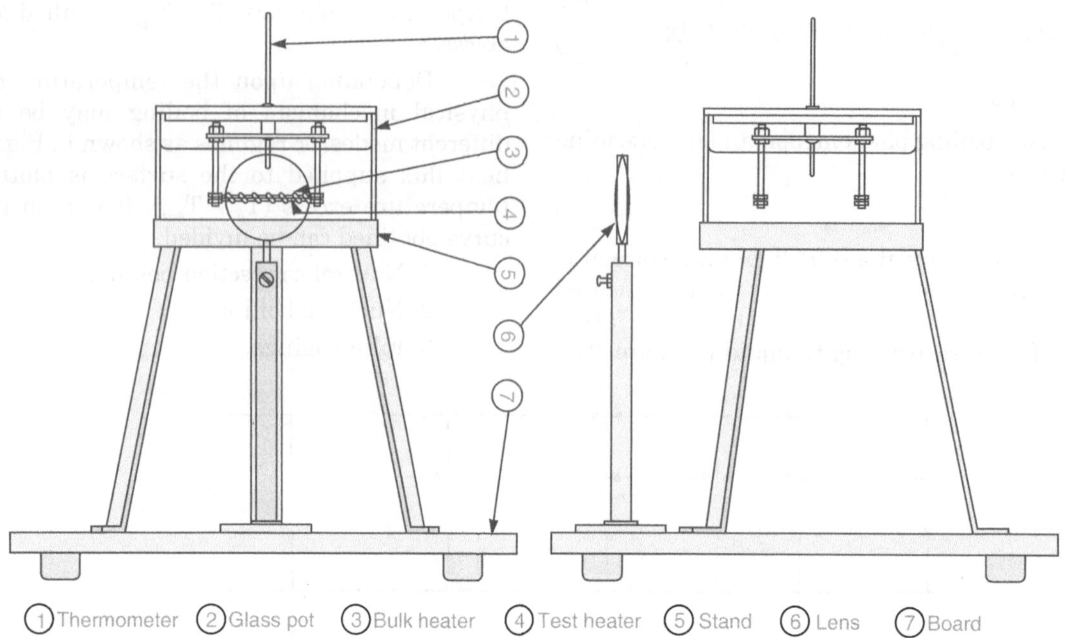

Fig. 16.20. Schematic diagram of critical heat flux apparatus

It will be observed that the heat flux does not increase in a regular manner with the temperature excess. In region I, the heat flux is proportional to $(T_s - T_{sat})^n$, where n is slightly greater than unity (approximately 1.3). When transition from natural convection to nucleate boiling occurs, the flux starts to increase more rapidly with temperature excess. At the end of region IIIa, inspite of increasing temperature excess, the heat flux starts to decrease, because the thermal resistance to heat flow increases with formation of vapour film. The heat flux passes through minimum value at point D at the end of region IIIa. It starts to increase again with temperature excess only when stable film boiling begins and radiation becomes increasingly significant, surface heat flux again shoots up to maximum. This peak heat flux is called critical heat flux.

10.3. EXPERIMENTAL APPARATUS

The apparatus consist of a cylindrical glass container housing, the test heater and a heater coil for initial heating of water. This heater coil is directly connected to mains and test heater (Nichrome wire) is also connected to mains via dimmer-stat, thus the energy input to test wire can be varied. An ammeter is connected in series while voltmeter across it to read. A stand, is fixed on wooden platform, which supports the apparatus.

10.4. SPECIFICATIONS

1. Glass container – Diameter = ... mm,
   Height = ... mm
2. Heater for initial heating
   – Nichrome wire heater = ... kW
3. Diameter of wire, d = ... mm
4. Length of Nichrome wire, L = ... mm
5. Dimmer-stat –10 A, 230 Volts
6. Voltmeter – 0 to 50/100 Volts
7. Ammeter – 0 to 10 A
8. Thermometer – 0 to 100°C

10.5. PROCEDURE

1. Switch the bulk heater to heat the water in the cylinder up to required temperature.
2. With the help of dimmer-stat, the heater input to nichrome wire is gradually increased.
3. Study the different types of boiling stages.
4. Observe the voltmeter and ammeter readings continuously.
5. Record the voltage and ammeter reading at the instance when nichrome wire burns out.
6. Repeat the procedure for different bulk temperature.

10.6. OBSERVATION

Sr. No.	*Bulk temperature*	*Ammeter*	*Voltmeter*

10.7. CALCULATION

The heat transfer area of the wire

$$A = \pi DL^2 = m^2$$

The energy input to the heater

$$Q = V\,I = \text{Watts} \quad \text{...(16.39)}$$

Critical Heat Flux

$$q = \frac{Q}{A} = \text{W/m}^2$$

The critical heat flux obtained practically can also be compared with peak heat flux in nucleate boiling using Zuber's analytical expression;

$$\left(\frac{Q}{A}\right)_{max} = \left(\frac{\pi}{4}\right)\rho_v^{1/2}\, h_{fg}\,[\sigma g(\rho - \rho_v)]^{1/4} \times \left(\frac{\rho + \rho_v}{\rho}\right)^{1/2} \quad \text{...(16.40)}$$

where ρ = density of saturated liquid kg/m^3

ρ_v = density of sturated vapour kg/m^3

σ = surface tension on liquid-vapour interface, N/m

h_{fg} = latent heat of vaporisation, J/kg

10.8. RESULT

The critical heat flux is ...... W/m^2

10.9. EPILOGUE

The heat transfer rate during pool boiling is maximum due to increase rate of bubbles formation.

EXPT. 11 *HEAT PIPE*

11.1. OBJECTIVE

Study the heat transfer phenomenon and to compare the performance of heat pipe with two geometrical similar pipes of copper and stainless steel.

11.2. PROLOGUE

A *heat pipe* is a very simple device without any moving part that can transfer heat energy over fairly large distance by using change of phase of working substance. It does not require any power input. It transfers large amount of heat energy through a small surface area with very little temperature difference from its one end to another.

A heat pipe is basically a sealed slender tube containing a wick structure lined on the inner surface and a small amount of fluid such as water at saturated state. The heat pipe comprises of three sections: *evaporator* section at one end, where heat is absorbed and fluid is vaporised, a *condenser* section at the other end, where vapour condeses and heat is rejected, and the *adiabatic* section in between, where the vapour and liquid phases of the working substance flow in opposite directions through the central core and the wick, respectively.

A variety of fluid and pipe materials have been used for heat pipe construction. The basic structure of a heat pipe is shown in Fig. 16.21. Typical copper heat pipe using water as the working fluid can transport an approximately 6500 kW/m^2 of heat flux in axial position.

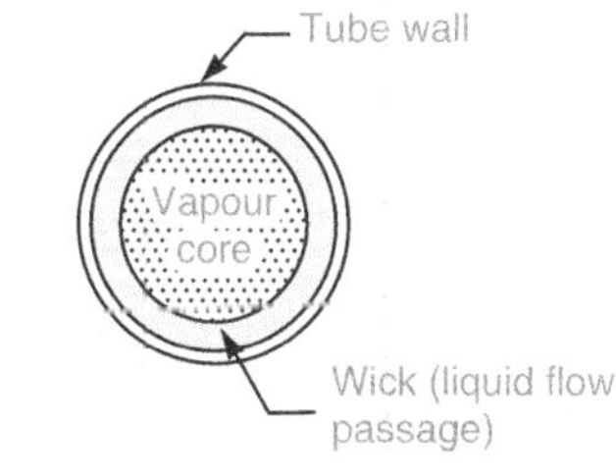

Cross-section of a heat pipe

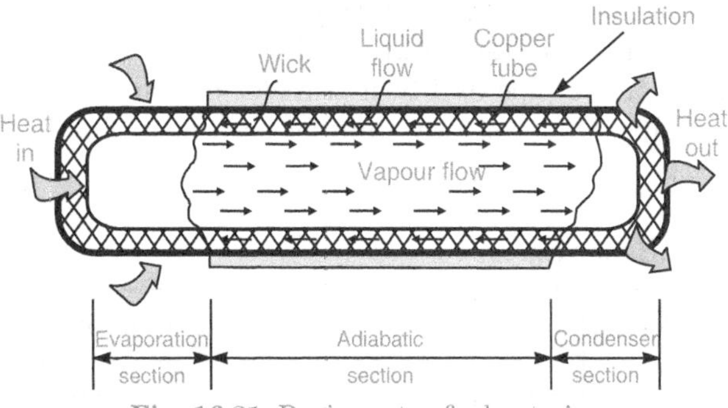

Fig. 16.21. Basic parts of a heat pipe

Presently, the heat pipes are used in several engineering applications *e.g.* cooling of microelectric circuit, power transisters, cryogenic targets in nuclear accelerators temperature control devices, etc.

11.3. EXPERIMENTAL ANALYSIS

11.3.1. Experimental Setup

The experimental setup as shown in Fig. 16.22, comprises of three identical (same diameter and same height) heat pipes. One of them is seamless pipe with copper perforated sheet used as wicking material with pure distilled water. The two pipes are made of copper and stainless steel. One end of these pipes is heated with help of electrical heaters. Other end is exposed to small capacity water tanks acting as heat sinks. Each tanks is provided with a thermometer to measure temperature of water. Four thermocouples are embedded along the length of each pipe to measure the temperature distribution. The power input to the heater can be varied through dimmer-stats and can be measured by voltmeter and ammeter.

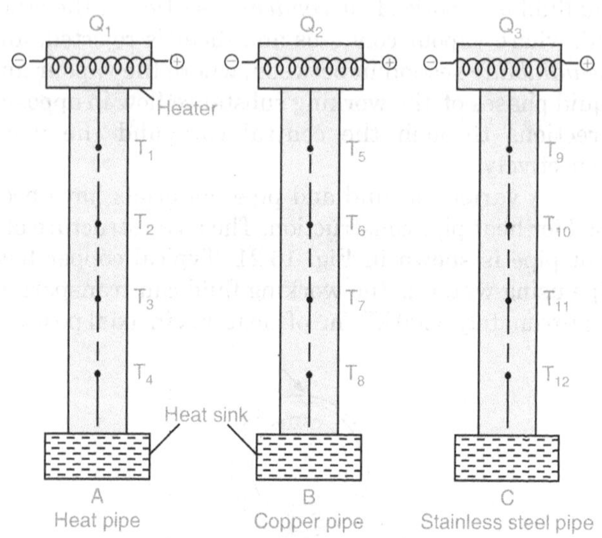

Fig. 16.22. Setup for heat pipe demonstration

11.3.2. Theory

The performance of heat pipes can be studied by measuring the temperature distribution along the length of the pipe and heat transfer characteristics of each pipe. Under steady state for each heat pipe

Energy input to heater in time Δt,

$$Q = V\,I\,\Delta t$$

Heat transferred to water,

$$Q_w = m_w C_{pw} (T_{final} - T_{initial})$$

11.3.3. Experimental Procedure

1. Fill the known quantity of water in three heat sinks and measure its initial temperature.
2. Switch on the mains and supply the same power input to each heater equipped with three pipes.
3. Wait for steady state conditions, and then note down readings of thermocouples connected to pipes.
4. Measure the final temperature of water in three heat sinks.
5. Repeat the experiment for different heat input.

11.4. SPECIFICATIONS

(A) Standard heat pipe, A

Diameter of heat pipe, d_A = ... mm,

Height of heat pipe, L_A = ... mm

Mass of water heat sink m_A = ... kg

(B) Copper pipe, B

Diameter of copper pipe, d_B = ... mm

Height of copper pipe, L_B = ... mm

Mass of water heat sink m_B = ... kg

(C) Stainless steel pipe, C

Diameter of stainless steel pipe, d_C = ... mm

Height of stainless steel pipe, L_C = ... mm

Mass of water heat sink, m_C = ... kg

11.5. OBSERVATION TABLES

(*i*) Heat pipe A

Sr. No.	*Heat input*		*Readings of thermocouple along pipe, °C*				*Heat sink temperature*	
	V	*I*	T_1	T_2	T_3	T_4	*Initial*	*Final*
1 2 3								

(*ii*) Copper pipe B

Sr. No.	*Heat input*		*Readings of thermocouple along pipe,°C*				*Heat sink temperature*	
	V	*I*	T_5	T_6	T_7	T_8	*Initial*	*Final*
1 2 3								

(*iii*) Stainless steel pipe C

Sr. No.	*Heat input*		*Readings of thermocouple along pipe*				*Heat sink temperature*	
	V	*I*	T_9	T_{10}	T_{11}	T_{12}	*Initial*	*Final*
1 2 3								

11.6. CALCULATIONS

(*i*) Standard heat pipe A

Energy input to heater in time Δt,

$$Q_A = V_A I_A \Delta t$$

Heat transferred to water,

$$Q_{wA} = m_A C_{pw} (T_{\text{final}} - T_{\text{initial}})_A$$

Temperatures distribution in heat pipe

T_1 = °C, T_2 = °C,

T_3 = °C, T_4 = °C,

(*ii*) Copper heat pipe B

Energy input to heater in time Δt ,

$$Q_B = V_B I_B \Delta t$$

Heat transferred to water ,

$$Q_{wB} = m_B C_{pw} (T_{\text{final}} - T_{\text{initial}})_B$$

Temperatures distribution in heat pipe

T_5 = °C, T_6 = °C,

T_7 = °C, T_8 = °C,

(*iii*) Stainless steel heat pipe C

Energy input to heater in time Δt ,

$$Q_C = V_C I_C \Delta t$$

Heat transferred to water,

$$Q_{wC} = m_C C_{pw} (T_{\text{final}} - T_{\text{initial}})_C$$

Temperatures distribution in heat pipe

T_9 = °C, T_{10} = °C,

T_{11} = °C, T_{12} = °C,

11.7. RESULT

1. Compare heat transfer rate to water.
2. Plot the temperature distribution along the length of three pipes.

11.8. EPILOGUE

Comment on the heat transfer characteristics and temperature distribution of three pipes.

EXPT. 12 *THERMOCOUPLES CALIBRATION TEST RIG*

Objective. To study and calibrate thermocouples

12.1. PROLOGUE

A thermocouple is a very simple temperature sensor, consisting essentially of two dissimilar wires in thermal contact. The operation of a thermocouple is based on the Seebeck effect, which results in the generation of a thermoelectric potential when two dissimilar metals are joined together to form a junction. The thermoelectric effect is produced by diffusion of electrons across the interface between the two materials. The electric potential of the material accepting electrons becomes negative at the interface, while the potential of the material providing the electrons becomes positive. Thus, an electric field is established by the flow of electrons across the interface. When this electric field becomes sufficient to balance the diffusion forces, a state of equilibrium with respect to electron migration is established. Since the magnitude of the diffusion force is controlled by the temperature of the thermocouple junction, the electric potential developed at the junction provides a measure of the temperature. The electric potential is usually measured by introducing a reference junction in an electric circuit and measuring the voltage E across one leg with a suitable voltmeter. The voltage E across terminals measuring the temperature difference $T_1 - T_2$ can be approximated by an empirical equation in the form

$$E = C_1(T_1 - T_2) + C_2(T_1 - T_2) \quad ...(16.41)$$

Where C_1 and C_2 are thermoelectric constants that depend on the materials, T_1 and T_2 are junction temperatures.

In practice, junction 1 is used to sense an unknown temperature T_1 while junction 2 is maintained at a known reference temperature T_2. The unknown temperature T_1 is determined with the help of

Temperature T_2 and measured voltage E. It is clear from above that the response of a thermocouple is a nonlinear function of the temperature. It is also observed that eqn. (16.41) is not a sufficiently accurate representation of the voltage-temperature relationship to be used with confidence when precise measurement of temperature is required. For this reason, the thermocouples are calibrated over the complete range of temperature for which they are useful and tables are obtained which can be used to relate temperature T_1 to the thermoelectric voltage E. Thermocouple tables are available for different make of thermocouples.

Reference Junction Temperature

Since a thermocouple circuit responds to a temperature difference ($T_1 - T_2$), it is essential that the reference junction be maintained at a constant and accurately known temperature T_2.

The simplest and most popular technique utilizes an ice and water bath, as illustrated in the reference junction is immersed in a mixutre of ice and water in a thermos bottle that is capped to prevent heat loss and constant reference temperature.

The ice bath method for maintaining a reference temperature at 0°C(32°F). Water (sufficient only to fill the voids) must be removed and ice must be replaced periodically to maintain a constant reference temperature. Such an ice bath can maintain the water temperature (and thus the reference temperature) to within 0.1°C (0.2°F) of the freezing point of water.

Principles of Thermocouple Behaviour

1. A thermocouple circuit must contain at least two dissimilar materials and at least two junctions. Typical situations encountered during use of thermocouples. (*a*) Basic thermocouple circuit. (*b*) Output depends on ($T_1 - T_2$) only. (*c*) Intermediate metal in circuit. (*d*) Intermediate metal in junction. (*e*) Voltage addition from identical thermocouples at different temperatures. (*f*) Voltage addition from different thermocouples at identical temperatures.
2. The output voltage E of a thermocouple circuit depends only on the difference between junction temperatures ($T_1 - T_2$) and is independent of the temperatures elsewhere in the circuit as shown in Fig. 16.23.

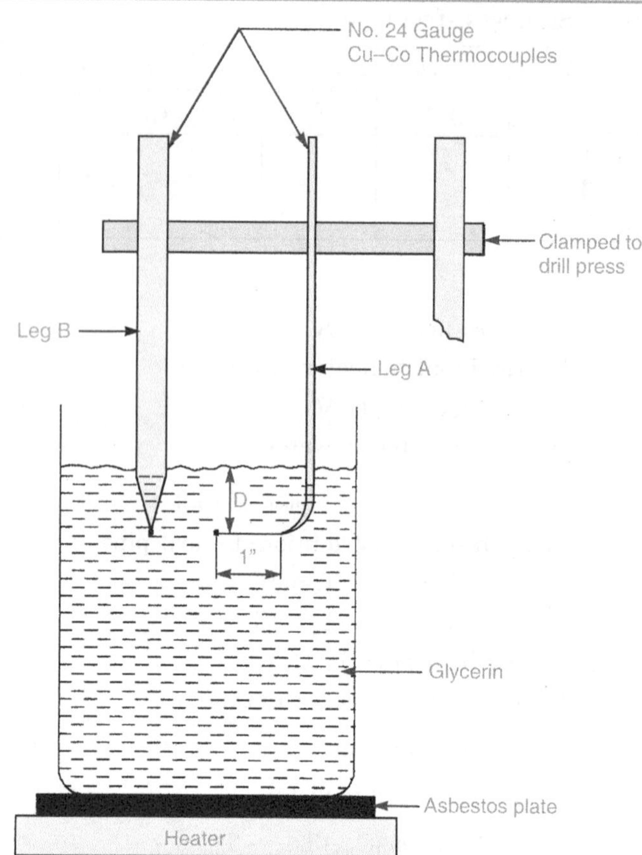

Fig. 16.23. Experimental set-up for calibration of thermocouples

3. If a third metal C is inserted into either leg (A or B) of a thermocouple circuit, the output voltage E is not affected, provided the two new junctions A/C and C/A are maintained at the same temperature (for example, temperature T_3).
4. The insertion of an intermediate metal C into junction 1 does not affect the output voltage E, provided the two junctions formed by the insertion (A/C and C/B) are maintained at the same temperature T_1.
5. A thermocouple circuit with temperatures T_1 and T_2 produces an output voltage $E_{1-2} = f(T_1 - T_2)$, and one exposed to temperatures T_2 and T_3 produces an output voltage $E_{2-3} = f(T_2 - T_3)$. If the same circuit is exposed to temperatures T_1 and T_3, the output voltage $E_{1-3} = f(T_1 - T_3) = E_{1-2} + E_{2-3}$
6. A thermocouple circuit fabricated from materials A and C generates an output voltage $E_{A/C}$ when exposed to temperatures T_1 and T_2, while a similar circuit fabricated from materials C and B

generates an output voltage $E_{C/B}$. Furthermore, a thermocouple fabricated from materials A and B generates an output voltage $E_{A/B} = E_{A/C} + E_{C/B}$.

REVIEW QUESTIONS

1. What is heat conduction ?
2. What is thermal conductivity ?
3. What is fundamental law of heat conduction ?
4. State Fourier law.
5. What is potential for heat conduction ?
6. What is effect of temperature on thermal conductivity of metals ?
7. What is effect of temperature on thermal conductivity of non-metals ?
8. What are uses of insulating powders ?
9. Why is thermal conductivity of insulating powders is less that their parent solids ?
10. What do you understand by isotropic and unisotropic materials ?
11. What is a fin ?
12. When a fin is called a pin fin ?
13. What do you understand by annular fin ?
14. What do you understand by longitudinal fin ?
15. What is profile area of fin ?
16. Why is fin tip considered insulated in most of cases ?
17. What is fin efficiency ?
18. When the use of fin can be justified ? In forced or natural convection.
19. What is emissive power ?
20. What is emissivity ?
21. Define black body.
22. What is gray body approximation? Why is it important ?
23. What is Kirchhoff's law of thermal radiation ?
24. Why the red signal can be seen from large distance in comparison to green signal ?
25. What is Stefan Boltzmann Law ?
26. What is heat convection ?
27. What is potential for heat convection ?
28. Defferentiate between forced convection and natural convection.
29. Define Reynolds number, Prandtl number and Grashof number.
30. What is pool boiling, nucleate boiling and film boiling ?
31. What is critical heat flux ?
32. What is the working principle of heat pipe ?
33. What are the main parts of heat pipe ?
34. What are the applications of heat pipe ?
35. What are the applications of longitudinal fins ?
36. What is the mechanism of heat conduction in solids ?
37. On what basic thermodynamic principle, the Stefan Boltzmann apparatus works ?
38. What are classifications of heat exchangers ?
39. Draw the temperature distribution for parallel flow and counter flow heat exchangers.
40. What is LMTD ? Why it is more for counterflow heat exchanger ?
41. Define for a heat exchanger: effectiveness and NTU.

REFERENCES

1. Holman J.P., "*Experimental Methods for Engineers*", fifth edition, McGraw Hill, New York, 1989.
2. Doolittle J.S. , "*Mechanical Engineering Laboratory*", McGraw Hill, New York, 1957.
3. Colclaser S.W., "*Heat Pipe Theory and Practice*", Hemisphere, Washington, DC, 1976.
4. Experimental manuals for Heat Transfer Laboratory, from Hytech Educational Equipments Pvt. Ltd. Pune, India.
5. Experimental manuals for Heat Transfer Laboratory, from ARE Educational Equipments Pvt. Ltd. Miraj, India.
6. Experimental manuals for Heat Transfer Laboratory, from Promark Equipments Pvt. Ltd. Banglore, India.

Appendix

A

Thermophysical Properties of Matter

APPENDIX A

TABLE A.1 Thermophysical properties of selected metallic solids

	Properties at 300 K					Properties at various temperatures (K)									
						k(W/mK)/C_p (J/kg K)									
Composition	Melting point (K)	ρ (kg/m³)	C_p (J/kg.K)	k (W/m.K)	$\alpha \times 10^6$ (m^2/s)	100	200	400	600	800	1000	1200	1500	2000	2500
Aluminium															
Pure	933	2702	903	237	97.1	302	237	240	231	218					
						482	798	949	1033	1146					
Alloy 2024-T6 (4.5% Cu, 1.5% Mg, 0.6% Mn)	775	2770	875	177	73.0	65	163	186	186						
						473	787	925	1042						
Alloy 195, Cast (4.5% Cu)		2790	883	168	68.2			174	185						
								—	—						
Beryllium	1550	1850	1825	200	59.2	990	301	161	126	106	90.8	78.7			
						203	1114	2191	2604	2823	3018	3227	3519		
Bismuth	545	9780	122	7.86	6.59	16.5	9.69	7.04							
						112	120	127							
Boron	2573	2500	1107	27.0	9.76	190	55.5	16.8	10.6	9.60	9.85				
						128	600	1463	1892	2160	2338				
Cadmium	594	8650	231	96.8	48.4	203	99.3	94.7							
						198	222	242							
Chromium	2118	7160	449	93.7	29.1	159	111	90.9	80.7	71.3	65.4	61.9	57.2	49.4	
						192	384	484	542	581	616	682	779	937	
Cobalt	1769	8862	421	99.2	26.6	167	122	85.4	67.4	58.2	52.1	49.3	42.5		
						236	379	450	503	550	628	733	674		
Copper															
Pure	1358	8933	385	401	117	482	413	393	379	366	352	339			
						252	356	397	417	433	451	480			
Commercial bronze (90% Cu, 10% Al)	1293	8800	420	52	14		42	52	59						
							785	460	545						
*Phosphor gear bronze (89% Cu, 11% Sn)	1104	8780	355	54	17		41	65	74						
Cartridge brass (70% Cu, 30% Zn)	1188	8530	380	110	33.9	75	95	137	149						
							360	395	425						
Constantan (55% Cu, 45% Ni)	1493	8920	384	23	6.71	17	19								
						237	362								
Germanium	1211	5360	322	59.9	34.7	232	96.8	43.2	27.3	19.8	17.4	17.4			
						190	290	337	348	357	375	395			

(Continued)

TABLE A.1 Continued

Composition	Melting point (K)	ρ (kg/m^3)	C_p ($J/kg.K$)	k ($W/m.K$)	$\alpha \times 10^6$ (m^2/s)	100	200	400	600	800	1000	1200	1500	2000	2500
	Properties at 300 K					Properties at various temperatures (K) $k(W/mK)/C_p\ (J/kg\ K)$									
Gold	1336	19300	129	317	127	327	323	311	298	284	270	255			
						109	124	131	135	140	145	155			
Iridium	2720	22500	130	147	50.3	172	153	144	138	132	126	120	111		
						90	122	133	138	144	153	161	172		
Iron															
Pure	1810	7870	447	80.2	23.1	134	94.0	69.5	54.7	43.3	32.8	28.3	32.1		
						216	384	490	574	680	975	609	654		
Armco															
(99.75% pure)		7870	447	72.7	20.7	95.6	80.6	65.7	53.1	42.2	32.3	28.7	31.4		
						215	384	490	574	680	975	609	654		
Carbon steels															
Plain carbon		7854	434	60.5	17.7			56.7	48.0	39.2	30.0				
(Mn ≤ 1%), Si ≤ 0.1%)								487	559	685	1169				
ALSL 1010		7832	434	63.9	18.8			58.7	48.8	39.2	31.3				
(Mn ≤ 1%								487	559	685	1168				
Carbon-silicon		7854	446	51.9	14.9			49.8	44.0	37.4	29.3				
(Mn ≤ 1% 0.1% < Si ≤ 0.6%)								501	582	699	971				
Carbon-manganese-		8131	434	41.0	11.6			42.2	39.7	35.0	27.6				
silicon (1% < Mn ≤ 1.65%, 0.1% < Si ≤ 0.6%)								487	559	685	1090				
Chromium (low) steels															
$\frac{1}{2}$Cr – $\frac{1}{4}$Mo – Si		7822	444	37.7	10.9			38.2	36.7	33.3	26.9				
(0.18% C, 0.65% Cr, 0.23% Mo, 0.6% Si)								492	575	688	969				
1 Cr – $\frac{1}{2}$ Mo		7858	442	42.3	12.2			42.0	39.1	34.5	27.4				
(0.16% C, 1% Cr, 0.54% Mo, 0.39% Si)								492	575	688	969				
1 Cr-V		7836	443	48.9	14.1			46.8	42.1	36.3	28.2				
(0.2% C, 1.02% Cr, 0.15% V)								492	575	688	969				
Stainless steels															

(Continued)

TABLE A.1 Continued

	Properties at 300 K					Properties at various temperatures (K) $k(W/mK)/C_p$ $(J/kg\ K)$									
Composition	Melting point (K)	ρ (kg/m^3)	C_p $(J/kg.K)$	k $(W/m.K)$	$\alpha \times 10^6$ (m^2/s)	100	200	400	600	800	1000	1200	1500	2000	2500
AISI 302		8055	480	15.1	3.91			17.3	20.0	22.8	25.4				
								512	559	585	606				
AISI 304	1670	7900	477	14.9	3.95	9.2	12.6	16.6	19.8	22.6	25.4	28.0	31.7		
						272	402	515	557	582	611	640	682		
AISI 316		8238	468	13.4	3.48			15.2	18.3	21.3	24.2				
								504	550	576	602				
AISI 347		7978	480	14.2	3.71			15.8	18.9	21.9	24.7				
								513	559	585	606				
Lead	601	11340	129	35.3	24.1	39.7	36.7	34.0	31.4						
						118	125	132	142						
Magnesium	923	1740	1024	156	87.6	169	159	153	149	146					
						649	934	1074	1170	1267					
Molybdenum	2894	10240	251	138	53.7	179	143	134	126	118	112	105	98	90	86
						141	224	261	275	285	295	308	330	380	459
Nickel															
Pure	1728	8900	444	90.7	23.0	164	107	80.2	65.6	67.6	71.8	76.2	82.6		
						232	383	485	592	530	562	594	616		
Nichrome	1672	8400	420	12	3.4			14	16	21					
(80% Ni, 20% Cr)								480	525	545					
Inconel X-750	1665	8510	439	11.7	3.1	8.7	10.3	13.5	17.0	20.5	24.0	27.6	33.0		
(73% Ni, 15% Cr,						—	372	473	510	546	626	—	—		
6.7% Fe)															
Niobium	2741	8570	265	53.7	23.6	55.2	52.6	55.2	58.2	61.3	64.4	67.5	72.1	79.1	
						188	249	274	283	292	301	310	324	347	
Palladium	1827	12020	244	71.8	24.5	76.5	71.6	73.6	79.7	86.9	94.2	102	110		
						168	227	251	261	271	281	291	307		
Platinum															
Pure	2045	21450	133	71.6	25.1	77.5	72.6	71.8	73.2	75.6	78.7	82.6	89.5	99.4	
						100	125	136	141	146	152	157	165	179	
Alloy 60Pt-40Rh	1800	16630	162	47	17.4			52	59	65	69	73	76		
(60% Pt, 40% Rh)								—	—	—	—	—	—		
Rhenium	3453	21100	136	47.9	16.7	58.9	51.0	46.1	44.2	44.1	44.6	45.7	47.8	51.9	
						97	127	139	145	151	156	162	171	186	
Rhodium	2236	12450	243	150	49.6	186	154	146	136	127	121	116	110	112	
						147	220	253	274	293	311	327	349	376	
Silicon	1685	2330	712	148	89.2	884	264	98.9	61.9	42.2	31.2	25.7	22.7		
						259	556	790	867	913	946	967	992		

(Continued)

TABLE A.1 Continued

	Properties at 300 K					Properties at various temperatures (K) $k(W/mK)/C_p\ (J/kg\ K)$									
Composition	*Melting point (K)*	ρ (kg/m^3)	C_p $(J/kg.K)$	k $(W/m.K)$	$\alpha \times 10^6$ (m^2/s)	*100*	*200*	*400*	*600*	*800*	*1000*	*1200*	*1500*	*2000*	*2500*
Silver	1235	10500	235	429	174	444	430	425	412	396	379	361			
					187		225	239	250	262	277	292			
Tantalum	3269	16600	140	57.5	24.7	59.2	57.5	57.8	58.6	59.4	60.2	61.0	62.2	64.1	65.6
					110		133	144	146	149	152	155	160	172	189
Thorium	2023	11700	118	54.0	39.1	59.8	54.6	54.5	55.8	56.9	56.9	58.7			
						99	112	124	134	145	156	167			
Tin	505	7310	227	66.6	40.1	85.2	73.3	62.2							
						188	215	243							
Titanium	1953	4500	522	21.9	9.32	30.5	24.5	20.4	19.4	19.7	20.7	22.0	24.5		
						300	465	551	591	633	675	620	686		
Tungsten	3660	19300	132	174	68.3	208	186	159	137	125	118	113	107	100	95
						87	122	137	142	145	148	152	157	167	176
Uranium	1406	19070	116	27.6	12.5	21.7	25.1	29.6	34.0	38.8	43.9	49.0			
						94	108	125	146	176	180	161			
Vanadium	2192	6100	489	30.7	10.3	35.8	31.3	31.3	33.3	35.7	38.2	40.8	44.6	50.9	
						258	430	515	540	563	597	645	714	867	
Zinc	693	7140	389	116	41.8	117	118	111	103						
						297	367	402	436						
Zirconium	2125	6570	278	22.7	12.4	33.2	25.2	21.6	20.7	21.6	23.7	26.0	28.8	33.0	
						205	264	300	322	342	362	344	344	344	

TABLE A.2 Thermophysical properties of selected non-metallic solids

	Properties at 300 K					Properties at various temperatures (K) $k(W/m.K)/C_p\ (J/kg.K)$									
Composition	Melting point (K)	ρ (kg/m^3)	C_p $(J/kg.K)$	k $(W/m.K)$	$\alpha \times 10^6$ (m^2/s)	100	200	400	600	800	1000	1200	1500	2000	2500
Alumium oxide,	2323	3970	765	46	15.1	450	82	32.4	18.9	13.0	10.5				
sapphire						—	—	940	1110	1180	1225				
Aluminum oxide,	2323	3970	765	36.0	11.9	133	55	26.4	15.8	10.4	7.85	6.55	5.66	6.00	
polycrystalline						—	—	940	1110	1180	1225	—	—	—	
Beryllium oxide	2725	3000	1030	272	88.0			196	111	70	47	33	21.5	15	
								1350	1690	1865	1975	2055	2145	2750	
Boron	2573	2500	1105	27.6	9.99	190	52.5	18.7	11.3	8.1	6.3	5.2			
						—	—	1490	1880	2135	2350	2555			
Boron fibre epoxy	590	2080													
(30% vol) composite															
k, ‖ to fibres				2.29		2.10	2.23	2.28							
k ⊥ to fibres				0.59		0.37	0.49	0.60							
C_p			1122			364	757	1431							
Carbon															
Amorphous	1500	1950	—	1.60	—	0.67	1.18	1.89	2.19	2.37	2.53	2.84	3.48		
Diamond,						—	—	—	—	—	—	—	—		
type IIa	—	3500	509	2300	—	10000	4000	1540							
insulator						21	194	853							
Graphite, pyrolytic	2273	2210													
k ‖ to layers				1950		4970	3230	1390	892	667	534	448	357	262	
k ⊥ to layers				5.70		16.8	9.23	4.09	2.68	2.01	1.60	1.34	1.08	0.81	
C_p			709			136	411	992	1406	1650	1793	1890	1974	2043	
Graphite fibre	450	1400													
epoxy (25% vol)															
composite															
k, heat flow															
‖ to fibres				11.1		5.7	8.7	13.0							
k,heat flow															
⊥ to fibres				0.87		0.46	0.68	1.1							
C_p			935			337	642	1216							
Pyroceram	1623	2600	808	3.98	1.89	5.25	4.78	3.64	3.28	3.08	2.96	2.87	2.79		
Corning 9606						—	—	908	1038	1122	1197	1264	1498		
Silicon carbide	3100	3160	675	490	230			—	—	—	87	58	30		
								880	1050	1135	1195	1243	1310		
Silicon dioxide,	1883	2650													
crystalline															
(quartz)															
k, ‖ to C axis				10.4		39	16.4	7.6	5.0	4.2					
k, ⊥ to C axis				6.21		20.8	9.5	4.70	3.4	3.1					
C_p			745			—	—	885	1075	1250					

(Continued)

TABLE A.2 Continued

Composition	Properties at 300 K					Properties at various temperatures (K) $k(W/m.K)/C_p\ (J/kg.\ K)$									
	Melting point (K)	ρ (kg/m^3)	C_p $(J/kg.K)$	k $(W/m.K)$	$\alpha \times 10^6$ (m^2/s)	100	200	400	600	800	1000	1200	1500	2000	2500
Silicon dioxide, polycrystalline (fused silicon)	1883	2220	745	1.38	0.834	0.69	1.14	1.51	1.75	2.17	2.87	4.00			
						—	—	905	1040	1105	1155	1195			
Silicon nitride	2173	2400	691	16.0	9.65	—	—	13.9	11.3	9.88	8.76	8.00	7.16	6.20	
						—	578	778	937	1063	1155	1226	1306	1377	
Sulphur	392	2070	708	0.206	0.141	0.165	0.185								
						403	606								
Thorium dioxide	3573	9110	235	13	6.1			10.2	6.6	4.7	3.68	3.12	2.73	2.5	
								255	274	285	295	303	315	330	
Titanium dioxide, polycrystalline	2133	4175	710	8.4	2.8			7.01	5.02	3.94	3.46	3.28			
								805	880	910	930	945			

TABLE A.3 Thermophysical properties of common materials

(a) Structural building materials

Description / composition	*Density, ρ (kg / m³)*	*Typical properties at 300 K* Thermal conductivity, k (W / m. K)	*Specific heat C_p (J / kg. K)*
Building boards			
Asbestos—cement board	1920	0.58	—
Gypsum or plaster board	800	0.17	—
Plywood	545	0.12	1215
Sheathing, regular density	290	0.055	1300
Acoustic tile	290	0.058	1340
Hardboard, siding	640	0.094	1170
Hardboard, high density	1010	0.15	1380
Particle board, low density	590	0.078	1300
Particle board, high density	1000	0.170	1300
Woods			
Hardwoods (oak, maple)	720	0.16	1255
Softwoods (fir, pine)	510	0.12	1380
Masonry materials			
Cement mortar	1860	0.72	780
Brick, common	1920	0.72	835
Brick, face	2083	1.3	—
Clay tile, hollow			
1 cell deep, 10 cm thick	—	0.52	—
3 cells deep, 30 cm thick	—	0.69	—
Concrete block, 3 oval cores			
Sand/gravel, 20 cm thick	—	1.0	—
Cinder aggregate, 20 cm thick	—	0.67	—
Concrete block rectangular core			
2 cores, 20 cm thick, 16 kg	—	1.1	—
Same with filled cores	—	0.60	—
Plastering materials			
Cement plaster, sand aggregate	1860	0.72	—
Gypsum plaster, sand aggregate	1680	0.22	1085
Gypsum plaster, vermiculite aggregate	720	0.25	—

(Continued)

TABLE A.3 Continued

(b) Insulating materials and systems

Description / composition	*Typical properties at 300 K* Density, ρ (kg / m³)	Thermal conductivity, k (W / m. K)	Specific heat C_p (J / kg. K)
Blanket and batt			
Glass fibre, paper faced	16	0.046	—
	28	0.038	—
	40	0.035	—
Glass fibre, coated ; duct liner	32	0.038	835
Board and slab			
Cellular glass	145	0.058	1000
Glass fibre, organic bonded	105	0.036	795
Polystyrene, expanded			
Extruded (R-12)	55	0.027	1210
Moulded beads	16	0.040	1210
Mineral fibreboard ; roofing material	265	0.049	—
Wood, shredded/cemented	350	0.087	1590
Cork	120	0.039	1800
Loose fill			
Cork, granulated	160	0.045	—
Diatomaceous silica, coarse	350	0.069	—
Powder	400	0.091	—
Diatomaceous silica, fine powder	200	0.052	—
	275	0.061	—
Glass fibre, poured or blown	16	0.043	835
Vermiculite, flakes	80	0.068	835
	160	0.063	1000
Formed/foamed-in-place			
Mineral wool granules with asbestos/ inorganic binders, sprayed	190	0.046	—
Polyvinyl acetate cork mastic ; sprayed or trowelled	—	0.100	—
Urethane, two-part mixture, rigid foam	70	0.026	1045
Reflective			
Aluminum foil separating fluffy glass mats ; 10–12 layers, evacuated ; for cyrogenic applications (150 K)	40	0.00016	—
Aluminum foil and glass paper laminate ; 75–150 layers ; evacuated ; for cyrogenic application (150 K)	120	0.000017	—
Typical silica powder, evacuated	160	0.0017	—

(Continued)

TABLE A.3 Continued

(c) Industrial insulation

Description / composition	*Maximum service temperature (K)*	*Typical density (kg/m^3)*	*Typical thermal conductivity, k (W/m.K), at various temperatures (K)* 200	215	230	240	255	270	285	300	310	365	420	530	645	750
Blankets																
Blanket, mineral	920	96–192									0.038	0.046	0.056	0.078		
fibre,																
metal reinforced	815	40–96									0.035	0.045	0.058	0.088		
Blanket, mineral	450	10				0.036	0.038	0.040	0.043	0.048	0.052	0.076				
fibre, glass ;																
fine fibre,																
organic bonded		12				0.035	0.036	0.039	0.042	0.046	0.049	0.069				
		16				0.033	0.035	0.036	0.039	0.042	0.046	0.062				
		24				0.030	0.032	0.033	0.036	0.039	0.040	0.053				
		32				0.029	0.030	0.032	0.033	0.036	0.038	0.048				
		48				0.027	0.029	0.030	0.032	0.033	0.035	0.045				
Blanket, alumina-																
silica fiber	1530	48												0.071	0.105	0.150
		64												0.059	0.087	0.125
		96												0.052	0.076	0.100
		128												0.049	0.068	0.091
Felt, semirigid ;	480	50–125						0.035	0.036	0.038	0.039	0.051	0.063			
organic bonded	730	50	0.023	0.025	0.026	0.027	0.029	0.030	0.032	0.033	0.035	0.051	0.079			
Felt, laminated ;																
no binder	920	120											0.051	0.065	0.087	
Blocks, boards and																
pipe insulations																
Asbestos paper,																
laminated and																
corrugated																
4-ply	420	190								0.078	0.082	0.098				
6-ply	420	255								0.071	0.074	0.085				
8-ply	420	300								0.068	0.071	0.082				

(Continued)

TABLE A.3 Continued

Industrial insulation (continued)

Description / composition	*Maximum service temperature (K)*	*Typical density (kg/m^3)*	*Typical thermal conductivity, k (W/m.K), at various temperatures (K)*													
			200	*215*	*230*	*240*	*255*	*270*	*285*	*300*	*310*	*365*	*420*	*530*	*645*	*750*
Magnesia, 85%	590	185									0.051	0.055	0.061			
Calcium silicate	920	190									0.055	0.059	0.063	0.075	0.089	0.104
Cellular glass	700	145			0.046	0.048	0.05	0.052	0.055	0.058	0.062	0.069	0.079			
Diatomaceous	1145	345												0.092	0.098	0.104
silica	1310	385												0.101	0.100	0.115
Polystyrene, rigid																
Extruded (R-12)	350	56	0.023	0.023	0.022	0.023	0.023	0.025	0.026	0.027	0.029					
Extruded (R-12)	350	35	0.023	0.023	0.023	0.025	0.025	0.026	0.027	0.029						
Moulded beads	350	16	0.026	0.029	0.030	0.033	0.035	0.036	0.038	0.040						
Rubber, rigid																
foamed	340	70						0.029	0.030	0.032	0.033					
Insulating cement																
Mineral fibre																
(rock, slag or glass)																
With clay binder	1255	430									0.071	0.079	0.088	0.105	0.123	
With hydraulic																
setting binder	922	560									0.108	0.115	0.123	0.137		
Loose fill																
Cellulose, wood																
or paper pulp	—	45							0.038	0.039	0.042					
Perlite, expanded	—	105	0.036	0.039	0.042	0.043	0.046	0.049	0.051	0.053	0.056					
Vermiculite,																
expanded	—	122			0.056	0.058	0.061	0.063	0.065	0.068	0.071					
		80			0.049	0.051	0.055	0.058	0.061	0.063	0.066					

(Continued)

TABLE A.3 Properties of common materials

(d) Other materials

Description / Composition	*Temperature* K	*Density* ρ kg/m^3	*Thermal conductivity k* $W/(m.K)$	*Specific heat* C_p $J/(kg.K)$
Asphalt	300	2115	0.062	920
Bakelite	300	1300	1.4	1465
Brick, refractory				
Carborundum	872	—	18.5	—
	1672	—	11.0	—
Chrome brick	473	3010	2.3	835
	823		2.5	
	1173		2.0	
Diatomaceous	478	—	0.25	—
silica, fired	1145	—	0.30	
Fire clay, burnt 1600 K	773	2050	1.0	960
	1073	—	1.1	
	1373	—	1.1	
Fire clay, burnt 1725 K	773	2325	1.3	960
	1073		1.4	
	1373		1.4	
Fire clay brick	478	2645	1.0	960
	922		1.5	
	1478		1.8	
Magnesite	478	—	3.8	1130
	922	—	2.8	
	1478		1.9	
Clay	300	1460	1.3	880
Coal, anthracite	300	1350	0.26	1260
Concrete (stone mix)	300	2300	1.4	880
Cotton	300	80	0.06	1300
Foodstuffs				
Banana (75.7% water content)	300	980	0.481	3350
Apple, red (75% water content)	300	840	0.513	3600
Cake, butter	300	720	0.223	—
Cake, fully baked	300	280	0.121	—
Chicken meat,	198	—	1.60	—
(74.4% water content)	233	—	1.49	
	253		1.35	
	263		1.20	
	273		0.476	
	283		0.480	
	293		0.489	
Glass				
Plate (soda lime)	300	2500	1.4	750
Pyrex	300	2225	1.4	835

(Continued)

TABLE A.3 Continued

(e) Properties of common materials

Description / Composition	*Temperature K*	*Density ρ kg / m³*	*Thermal conductivity k W / (m.K)*	*Specific heat C_p J / (kg.K)*
Ice	273	920	1.88	2040
	253	—	2.03	1945
Leather (sole)	300	998	0.159	—
Paper	300	930	0.180	1340
Paraffin	300	900	0.240	2890
Rock				
Granite, Barre	300	2630	2.79	775
Limestone, Salem	300	2320	2.15	810
Marble, Halston	300	2680	2.80	830
Quartzite, Sioux	300	2640	5.38	1105
Sandstone, Berea	300	2150	2.90	745
Rubber, vulcanized				
Soft	300	1100	0.13	2010
Hard	300	1190	0.16	—
Sand	300	1515	0.27	800
Soil	300	2050	0.52	1840
Snow	273	110	0.049	—
		500	0.190	—
Teflon	300	2200	0.35	—
	400		0.45	—
Tissue, human				
Skin	300	—	0.37	—
Fat layer (adipose)	300	—	0.2	—
Muscle	300	—	0.41	—
Wood, cross-grain				
Balsa	300	140	0.055	—
Cypress	300	465	0.097	—
Fir	300	415	0.11	2720
Oak	300	545	0.17	2385
Yellow pine	300	640	0.15	2805
White pine	300	435	0.11	—
Wood, radial				
Oak	300	545	0.19	2385
Fir	300	420	0.14	2720

TABLE A.4 Thermophysical properties of gases at atmospheric pressure

T (K)	ρ(kg/m³)	C_p (kJ/kg.K)	μ × 10⁷ (Ns/m²)	ν × 10⁶ (m²/s)	k × 10³ (W/m.K)	α × 10⁶ (m²/s)	Pr
Air							
100	3.5562	1.032	71.1	2.00	9.34	2.54	0.786
150	2.3364	1.012	103.4	4.426	13.8	5.84	0.758
200	1.7458	1.007	132.5	7.590	18.1	10.3	0.737
250	1.3947	1.006	159.6	11.44	22.3	15.9	0.720
300	1.1614	1.007	184.6	15.89	26.3	22.5	0.707
350	0.9950	1.009	208.2	20.92	30.0	29.9	0.700
400	0.8711	1.014	230.1	26.41	33.8	38.3	0.690
450	0.7740	1.021	250.7	32.39	37.3	47.2	0.686
500	0.6964	1.030	270.1	38.79	40.7	56.7	0.684
550	0.6329	1.040	288.4	45.57	43.9	66.7	0.683
600	0.5804	1.051	305.8	52.69	46.9	76.9	0.685
650	0.5356	1.063	322.5	60.21	49.7	87.3	0.690
700	0.4975	1.075	338.8	68.10	52.4	98.0	0.695
750	0.4643	1.087	354.6	76.37	54.9	109	0.702
800	0.4354	1.099	369.8	84.93	57.3	120	0.709
850	0.4097	1.110	384.3	93.80	59.6	131	0.716
900	0.3868	1.121	398.1	102.9	62.0	143	0.720
950	0.3666	1.131	411.3	112.2	64.3	155	0.723
1000	0.3482	1.141	424.4	121.9	66.7	168	0.726
1100	0.3166	1.159	449.0	141.8	71.5	195	0.728
1200	0.2902	1.175	473.0	162.9	76.3	224	0.728
1300	0.2679	1.189	496.0	185.1	82	238	0.719
1400	0.2488	1.207	530	213	91	303	0.703
1500	0.2322	1.230	557	240	100	350	0.685
1600	0.2177	1.248	584	268	106	390	0.688
1700	0.2049	1.267	611	298	113	435	0.685
1800	0.1935	1.286	637	329	120	482	0.683
1900	0.1833	1.307	663	362	128	534	0.677
2000	0.1741	1.337	689	396	137	589	0.672
2100	0.1658	1.372	715	431	147	646	0.667
2200	0.1582	1.417	740	468	160	714	0.655
2300	0.1513	1.478	766	506	175	783	0.647
2400	0.1448	1.558	792	547	196	869	0.630
2500	0.1389	1.665	818	589	222	960	0.613
3000	0.1135	2.726	955	841	486	1570	0.536
Ammonia (NH_3)							
300	0.6894	2.158	101.5	14.7	24.7	16.6	0.887
320	0.6448	2.170	109	16.9	27.2	19.4	0.870
340	0.6059	2.192	116.5	19.2	29.3	22.1	0.872
360	0.5716	2.221	124	21.7	31.6	24.9	0.872
380	0.5410	2.254	131	24.2	34.0	27.9	0.869
400	0.5136	2.287	138	26.9	37.0	31.5	0.853
420	0.4888	2.322	145	29.7	40.4	35.6	0.833

(Continued)

TABLE A.4 Continued

T (K)	ρ (kg/m^3)	C_p (kJ/kg.K)	$\mu \times 10^7$ (Ns/m^2)	$\nu \times 10^6$ (m^2/s)	$k \times 10^3$ (W/m.K)	$\alpha \times 10^6$ (m^2/s)	Pr
Ammonia (NH_3) (Continued)							
440	0.4664	2.357	152.5	32.7	43.5	39.6	0.826
460	0.4460	2.393	159	35.7	46.3	43.4	0.822
480	0.4273	2.430	166.5	39.0	49.2	47.4	0.822
500	0.4101	2.467	173	42.2	52.5	51.9	0.813
520	0.3942	2.504	180	45.7	54.5	55.2	0.827
540	0.3795	2.540	186.5	49.1	57.5	59.7	0.824
560	0.3708	2.577	193	52.0	60.6	63.4	0.827
580	0.3533	2.613	199.5	56.5	63.8	69.1	0.817
Carbon dioxide (CO_2)							
280	1.9022	0.830	140	7.36	15.20	9.63	0.765
300	1.7730	0.851	149	8.40	16.55	11.0	0.766
320	1.6609	0.872	156	9.39	18.05	12.5	0.754
340	1.5618	0.891	165	10.6	19.70	14.2	0.746
360	1.4743	0.908	173	11.7	21.2	15.8	0.741
380	1.3961	0.926	181	13.0	22.75	17.6	0.737
400	1.3257	0.942	190	14.3	24.3	19.5	0.737
450	1.1782	0.981	210	17.8	28.3	24.5	0.728
500	1.0594	1.02	231	21.8	32.5	30.1	0.725
550	0.9625	1.05	251	26.1	36.6	36.2	0.721
600	0.8826	1.08	270	30.6	40.7	42.7	0.717
650	0.8143	1.10	288	35.4	44.5	49.7	0.712
700	0.7564	1.13	305	40.3	48.1	56.3	0.717
750	0.7057	1.15	321	45.5	51.7	63.7	0.714
800	0.6614	1.17	337	51.0	55.1	71.2	0.716
Carbon monoxide (CO)							
200	1.6888	1.045	127	7.52	17.0	9.63	0.781
220	1.5341	1.044	137	8.93	19.0	11.9	0.753
240	1.4055	1.043	147	10.5	20.6	14.1	0.744
260	1.2967	1.043	157	12.1	22.1	16.3	0.741
280	1.2038	1.042	166	13.8	23.6	18.8	0.733
300	1.1233	1.043	175	15.6	25.0	21.3	0.730
320	1.0529	1.043	184	17.5	26.3	23.9	0.730
340	0.9909	1.044	193	19.5	27.8	26.9	0.725
360	0.9357	1.045	202	21.6	29.1	29.8	0.725
380	0.8864	1.047	210	23.7	30.5	32.9	0.729
400	0.8421	1.049	218	25.9	31.8	36.0	0.719
450	0.7483	1.055	237	31.7	35.0	44.3	0.714
500	0.67352	1.065	254	37.7	38.1	53.1	0.710
550	0.61226	1.076	271	44.3	41.1	62.4	0.710
600	0.56126	1.088	286	51.0	44.0	72.1	0.707
650	0.51806	1.101	301	58.1	47.0	82.4	0.705
700	0.48102	1.114	315	65.5	50.0	93.3	0.702
750	0.44899	1.127	329	73.3	52.8	104	0.702
800	0.42095	1.140	343	81.5	55.5	116	0.705

(Continued)

TABLE A.4 Continued

T (K)	ρ (kg/m^3)	C_p ($kJ/kg.K$)	$\mu \times 10^7$ (Ns/m^2)	$\nu \times 10^6$ (m^2/s)	$k \times 10^3$ ($W/m.K$)	$\alpha \times 10^6$ (m^2/s)	Pr
Helium (He)							
100	0.4871	5.193	96.3	19.8	73.0	28.9	0.686
120	0.4060	5.193	107	26.4	81.9	38.8	0.679
140	0.3481	5.193	118	33.9	90.7	50.2	0.676
160	—	5.193	129	—	99.2	—	—
180	0.2708	5.193	139	51.3	107.2	76.2	0.673
200	—	5.193	150	—	115.1	—	—
220	0.2216	5.193	160	72.2	123.1	107	0.675
240	—	5.193	170	—	130	—	—
260	0.1875	5.193	180	96.0	137	141	0.682
280	—	5.193	190	—	145	—	—
300	0.1625	5.193	199	122	152	180	0.680
350	—	5.193	221	—	170	—	—
400	0.1219	5.193	243	199	187	295	0.675
450	—	5.193	263	—	204	—	—
500	0.09754	5.193	283	290	220	434	0.668
550	—	5.193	—	—	—	—	—
600	—	5.193	320	—	252	—	—
650	—	5.193	332	—	264	—	—
700	0.06969	5.193	350	502	278	768	0.654
750	—	5.193	364	—	291	—	—
800	—	5.193	382	—	304	—	—
900	—	5.193	414	—	330	—	—
1000	0.04879	5.193	446	914	354	1400	0.654
Hydrogen (H_2)							
100	0.24255	11.23	42.1	17.4	67.0	24.6	0.707
150	0.16156	12.60	56.0	34.7	101	49.6	0.699
200	0.12115	13.54	68.1	56.2	131	79.9	0.704
250	0.09693	14.06	78.9	81.4	157	115	0.707
300	0.08078	14.31	89.6	111	183	158	0.701
350	0.06924	14.43	98.8	143	204	204	0.700
400	0.06059	14.48	108.2	179	226	258	0.695
450	0.05386	14.50	117.2	218	247	316	0.689
500	0.04848	14.52	126.4	261	266	378	0.691
550	0.04407	14.53	134.3	305	285	445	0.685
600	0.04040	14.55	142.4	352	305	519	0.678
700	0.03463	14.61	157.8	456	342	676	0.675
800	0.03030	14.70	172.4	569	378	849	0.670
900	0.02694	14.83	186.5	692	412	1030	0.671
1000	0.02424	14.99	201.3	830	448	1230	0.673
1100	0.02204	15.17	213.0	966	488	1460	0.662
1200	0.02020	15.37	226.2	1120	528	1700	0.659
1300	0.01865	15.59	238.5	1279	568	1955	0.655
1400	0.01732	15.81	250.7	1447	610	2230	0.650

(Continued)

TABLE A.4 Continued

T (K)	ρ (kg/m³)	C_p (kJ/kg.K)	μ × 10^7 (Ns/m²)	ν × 10^6 (m²/s)	k × 10^3 (W/m.K)	α × 10^6 (m²/s)	Pr
Hydrogen (H_2) (Continued)							
1500	0.01616	16.02	262.7	1626	655	2530	0.643
1600	0.0152	16.28	273.7	1801	697	2815	0.639
1700	0.0143	16.58	284.9	1992	742	3130	0.637
1800	0.0135	16.96	296.1	2193	786	3435	0.639
1900	0.0128	17.49	307.2	2400	835	3730	0.643
2000	0.0121	18.25	318.2	2630	878	3975	0.661
Nitrogen (N_2)							
100	3.4388	1.070	68.8	2.00	9.58	2.60	0.768
150	2.2594	1.050	100.6	4.45	13.9	5.86	0.759
200	1.6883	1.043	129.2	7.65	18.3	10.4	0.736
250	1.3488	1.042	154.9	11.48	22.2	15.8	0.727
300	1.1233	1.041	178.2	15.86	25.9	22.1	0.716
350	0.9625	1.042	200.0	20.78	29.3	29.2	0.711
400	0.8425	1.045	220.4	26.16	32.7	37.1	0.704
450	0.7485	1.050	239.6	32.01	35.8	45.6	0.703
500	0.6739	1.056	257.7	38.24	38.9	54.7	0.700
550	0.6124	1.065	274.7	44.86	41.7	63.9	0.702
600	0.5615	1.075	290.8	51.79	44.6	73.9	0.701
700	0.4812	1.098	321.0	66.71	49.9	94.4	0.706
800	0.4211	1.22	349.1	82.90	54.8	116	0.715
900	0.3743	1.146	375.3	100.3	59.7	139	0.721
1000	0.3368	1.167	399.9	118.7	64.7	165	0.721
1100	0.3062	1.187	423.2	138.2	70.0	193	0.718
1200	0.2807	1.204	445.3	158.6	75.8	224	0.707
1300	0.2591	1.219	466.2	179.9	81.0	256	0.701
Oxygen (O_2)							
100	3.945	0.962	76.4	1.94	9.25	2.44	0.796
150	2.585	0.921	114.8	4.44	13.8	5.80	0.766
200	1.930	0.915	147.5	7.64	18.3	10.4	0.737
250	1.542	0.915	178.6	11.58	22.6	16.0	0.723
300	1.284	0.920	207.2	16.14	26.8	22.7	0.711
350	1.100	0.929	233.5	21.23	29.6	29.0	0.733
400	0.9620	0.942	258.2	26.84	33.0	36.4	0.737
450	0.8554	0.956	281.4	32.90	36.3	44.4	0.741
500	0.7698	0.972	303.3	39.40	41.2	55.1	0.716
550	0.6998	0.988	324.0	46.30	44.1	63.8	0.726
600	0.6414	1.003	343.7	53.59	47.3	73.5	0.729
700	0.5498	1.031	380.8	69.26	52.8	93.1	0.744
800	0.4810	1.054	415.2	86.32	58.9	116	0.743
900	0.4275	1.074	447.2	104.6	64.9	141	0.740
1000	0.3848	1.090	477.0	124.0	71.0	169	0.733
1100	0.3498	1.103	505.5	144.5	75.8	196	0.736
1200	0.3206	1.115	532.5	166.1	81.9	229	0.725
1300	0.2960	1.125	588.4	188.6	87.1	262	0.721

TABLE A.5 Thermophysical properties of saturated liquids

T (K)	ρ (kg/m^3)	C_p (kJ/kg.K)	$\mu \times 10^2$ (Ns/m^2)	$\nu \times 10^6$ (m^2/s)	$k \times 10^3$ (W/m.K)	$\alpha \times 10^7$ (m^2/s)	Pr	$\beta \times 10^3$ (K^{-1})
Engine oil (unused)								
273	899.1	1.796	385	4280	147	0.910	47000	0.70
280	895.3	1.827	217	2430	144	0.880	27500	0.70
290	890.0	1.868	99.9	1120	145	0.872	12900	0.70
300	884.1	1.909	48.6	550	145	0.859	6400	0.70
310	877.9	1.951	25.3	288	145	0.847	3400	0.70
320	871.8	1.993	14.1	161	143	0.823	1965	0.70
330	865.8	2.035	8.36	96.6	141	0.800	1205	0.70
340	859.9	2.076	5.31	61.7	139	0.779	793	0.70
350	853.9	2.118	3.56	41.7	138	0.763	546	0.70
360	847.8	2.161	2.52	29.7	138	0.753	395	0.70
370	841.8	2.206	1.86	22.0	137	0.738	300	0.70
380	836.0	2.250	1.41	16.9	136	0.723	233	0.70
390	830.6	2.294	1.10	13.3	135	0.709	187	0.70
400	825.1	2.337	0.874	10.6	134	0.695	152	0.70
410	818.9	2.381	0.698	8.52	133	0.682	125	0.70
420	812.1	2.427	0.564	6.94	133	0.675	103	0.70
430	806.5	2.471	0.470	5.83	132	0.662	88	0.70
Ethylene glycol [$C_2H_4(OH)_2$]								
273	1130.8	2.294	6.51	57.6	242	0.933	617	0.65
280	1125.8	2.323	4.20	37.3	244	0.933	400	0.65
290	1118.8	2.368	2.47	22.1	248	0.936	236	0.65
300	1114.4	2.415	1.57	14.1	252	0.939	151	0.65
310	1103.7	2.460	1.07	9.65	255	0.939	103	0.65
320	1096.2	2.505	0.757	6.91	258	0.940	73.5	0.65
330	1089.5	2.549	0.561	5.15	260	0.936	55.0	0.65
340	1083.8	2.592	0.431	3.98	261	0.929	42.8	0.65
350	1079.0	2.637	0.342	3.17	261	0.917	34.6	0.65
360	1074.0	2.682	0.278	2.59	261	0.906	28.6	0.65
370	1066.7	2.728	0.228	2.14	262	0.900	23.7	0.65
373	1058.5	2.742	0.215	2.03	263	0.906	22.4	0.65
Glycerin [$C_3H_5(OH)_3$]								
273	1276.0	2.261	1060	8310	282	0.977	85000	0.47
280	1271.9	2.298	534	4200	284	0.972	43200	0.47
290	1265.8	2.367	185	1460	286	0.955	15300	0.48
300	1259.9	2.427	79.9	634	286	0.935	6780	0.48
310	1253.9	2.490	35.2	281	286	0.916	3060	0.49
320	1247.2	2.564	21.0	168	287	0.897	1870	0.50

(Continued)

TABLE A.5. Continued

T (K)	ρ (kg/m^3)	C_p (kJ/kg.K)	$\mu \times 10^2$ (Ns/m^2)	$\nu \times 10^6$ (m^2/s)	$k \times 10^3$ (W/m.K)	$\alpha \times 10^7$ (m^2/s)	Pr	$\beta \times 10^3$ (K^{-1})
Freon (Refrigerant-12) (CCl_2F_2)								
230	1528.4	0.8816	0.0457	0.299	68	0.505	5.9	1.85
240	1498.0	0.8923	0.0385	0.257	69	0.516	5.0	1.90
250	1469.5	0.9037	0.0354	0.241	70	0.527	4.6	2.00
260	1439.0	0.9163	0.0322	0.224	73	0.554	4.0	2.10
270	1407.2	0.9301	0.0304	0.216	73	0.558	3.9	2.25
280	1374.4	0.9450	0.0283	0.206	73	0.562	3.7	2.35
290	1340.5	0.9609	0.0265	0.198	73	0.567	3.5	2.55
300	1305.8	0.9781	0.0254	0.195	72	0.564	3.5	2.75
310	1268.9	0.9963	0.0244	0.192	69	0.546	3.4	3.05
320	1228.6	1.0155	0.0233	0.190	68	0.545	3.5	3.5
Mercury (Hg)								
273	13528	0.1404	0.1688	0.1240	8180	42.85	0.0290	0.181
300	13529	0.1393	0.1523	0.1125	8540	45.30	0.0248	0.181
350	13407	0.1377	0.1309	0.0976	9180	49.75	0.0196	0.181
400	13287	0.1365	0.1171	0.0882	9800	54.05	0.0163	0.181
450	13167	0.1357	0.1075	0.0816	10400	58.10	0.0140	0.181
500	13048	0.1353	0.1007	0.0771	10950	61.90	0.0125	0.182
550	12929	0.1352	0.0953	0.0737	11450	65.55	0.0112	0.184
600	12809	0.1355	0.0911	0.0711	11950	68.80	0.0103	0.187

TABLE A.6. Thermophysical properties of saturated liquid-vapour, 1 atm

Fluid	T_{sat} (K)	h_{fg} (kJ/kg)	ρ_f (kg/m^3)	ρ_g (kg/m^3)	$\sigma \times 10^3$ (N/m)
Ethanol	351	846	757	1.44	17.7
Ethylene glycol	470	812	1111[a]	—	32.7
Glycerin	563	974	1260[a]	—	63.0[a]
Mercury	630	301	12740	3.90	417
Refrigerant R-12	243	165	1488	6.32	15.8
Refrigerant R-113	321	147	1511	7.38	15.9

[a]Property value corresponding to 300 K

TABLE A.7 Thermophysical properties of saturated water

Temperature, T (K)	Pressure, p (bar)[a]	Specific volume (m^3/kg) $v_f \cdot 10^3$	v_g	Heat of vaporization, h_{fg} (kJ/kg)	Specific heat (kJ/kg.K) $C_{p,f}$	$C_{p,g}$	Viscosity ($N.s/m^2$) $\mu_f \cdot 10^6$	$\mu_g \cdot 10^6$	Thermal conductivity (W/m.K) $k_f \cdot 10^3$	$k_g \cdot 10^3$	Prandtl number Pr_f	Pr_g	Surface tension $\sigma_f \cdot 10^3$ (N/m)	Expansion coefficient $\beta_f \cdot 10^6$ (K^{-1})	Temperature T (K)
273.15	0.00611	1.000	206.3	2502	4.217	1.854	1750	8.02	569	18.2	12.99	0.815	75.5	– 68.05	273.15
275	0.00697	1.000	181.7	2497	4.211	1.855	1652	8.09	574	18.3	12.22	0.817	75.3	– 32.74	275
280	0.00990	1.000	130.4	2485	4.198	1.858	1422	8.29	582	18.6	10.26	0.825	74.8	46.04	280
285	0.01387	1.000	99.4	2473	4.189	1.861	1225	8.49	590	18.9	8.81	0.833	74.3	114.1	285
290	0.01917	1.001	69.7	2461	4.184	1.864	1080	8.69	598	19.3	7.56	0.841	73.7	174.0	290
295	0.02617	1.002	51.94	2449	4.181	1.868	959	8.89	606	19.5	6.62	0.849	72.7	227.5	295
300	0.03531	1.003	39.13	2438	4.179	1.872	855	9.09	613	19.6	5.83	0.857	71.7	276.1	300
305	0.04712	1.005	27.90	2426	4.178	1.877	769	9.29	620	20.1	5.20	0.865	70.9	320.6	305
310	0.06221	1.007	22.93	2414	4.178	1.882	695	9.49	628	20.4	4.62	0.873	70.0	361.9	310
315	0.08132	1.009	17.82	2402	4.179	1.888	631	9.69	634	20.7	4.16	0.883	69.2	400.4	315
320	0.1053	1.011	13.98	2390	4.180	1.895	577	9.89	640	21.0	3.77	0.894	68.3	436.7	320
325	0.1351	1.013	11.06	2378	4.182	1.903	528	10.09	645	21.3	3.42	0.901	67.5	471.2	325
330	0.1719	1.016	8.82	2366	4.184	1.911	489	10.29	650	21.7	3.15	0.908	66.6	504.0	330
335	0.2167	1.018	7.09	2354	4.186	1.920	453	10.49	656	22.0	2.88	0.916	65.8	535.5	335
340	0.2713	1.021	5.74	2342	4.188	1.930	420	10.69	660	22.3	2.66	0.925	64.9	566.0	340
345	0.3372	1.024	4.683	2329	4.191	1.941	389	10.89	668	22.6	2.45	0.933	64.1	595.4	345
350	0.4163	1.027	3.846	2317	4.195	1.954	365	11.09	668	23.0	2.29	0.942	63.2	624.2	350
355	0.5100	1.030	3.180	2304	4.199	1.968	343	11.29	671	23.3	2.14	0.951	62.3	652.3	355
360	0.6209	1.034	2.645	2291	4.203	1.983	324	11.49	674	23.7	2.02	0.960	61.4	697.9	360
365	0.7514	1.038	2.212	2278	4.209	1.999	306	11.69	677	24.1	1.91	0.969	60.5	707.1	365
370	0.9040	1.041	1.861	2265	4.214	2.017	289	11.89	679	24.5	1.80	0.978	59.5	728.7	370
373.15	1.0133	1.044	1.679	2257	4.217	2.029	279	12.02	680	24.8	1.76	0.984	58.9	750.1	373.15
375	1.0815	1.045	1.574	2252	4.220	2.036	274	12.09	681	24.9	1.70	0.987	58.6	761	375
380	1.2869	1.049	1.337	2239	4.226	2.057	260	12.29	683	25.4	1.61	0.999	57.6	788	380
385	1.5233	1.053	1.142	2225	4.232	2.080	248	12.49	685	25.8	1.53	1.004	56.6	814	385
390	1.794	1.058	0.980	2212	4.239	2.104	237	12.69	686	26.3	1.47	1.013	55.6	841	390
400	2.455	1.067	0.731	2183	4.256	2.158	217	13.05	688	27.2	1.34	1.033	53.6	896	400
410	3.302	1.077	0.553	2153	4.278	2.221	200	13.42	688	28.2	1.24	1.054	51.5	952	410
420	4.370	1.088	0.425	2123	4.302	2.291	185	13.79	688	29.8	1.16	1.075	49.4	1010	420
430	5.699	1.099	0.331	2091	4.331	2.369	173	14.14	685	30.4	1.09	1.10	47.2		430
440	7.333	1.110	0.261	2059	4.36	2.46	162	14.50	682	31.7	1.04	1.12	45.1		440
450	9.319	1.123	0.208	2024	4.40	2.56	152	14.85	678	33.1	0.99	1.14	42.9		450
460	11.71	1.137	0.167	1989	4.44	2.68	143	15.19	673	34.6	0.95	1.17	40.7		460
470	14.55	1.152	0.136	1951	4.48	2.79	136	15.54	667	36.3	0 .92	1.20	38.5		470
480	17.90	1.167	0.111	1912	4.53	2.94	129	15.88	660	38.1	0.89	1.23	36.2		480
490	21.83	1.184	0.0922	1870	4.59	3.10	124	16.23	651	40.1	0 .87	1.25	33.9	—	490
500	26.40	1.203	0.0766	1825	4.66	3.27	118	16.59	642	42.3	0.86	1.28	31.6	—	500
510	31.66	1.222	0.0631	1779	4.74	3.47	113	16.95	631	44.7	0.85	1.31	29.3	—	510
520	37.70	1.244	0.0525	1730	4.84	3.70	108	17.33	621	47.5	0.84	1.35	26.9	—	520
530	44.58	1.268	0.0445	1679	4.95	3.96	104	17.72	608	50.6	0.85	1.39	24.5	—	530
540	52.38	1.294	0.0375	1622	5.08	4.27	101	18.1	594	54.0	0.86	1.43	22.1	—	540
550	61.19	1.323	0 .0317	1564	5.24	4.64	97	18.6	580	58.3	0.87	1.47	19.7	—	550
560	71.08	1.355	0.0269	1499	5.43	5.09	94	19.1	563	63.7	0.90	1.52	17.3	—	560
570	82.16	1.392	0.0228	1429	5.68	5.67	91	19.7	548	76.7	0.94	1.59	15.0	—	570
580	94.51	1.433	0.0193	1353	6.00	6.40	88	20.4	528	76.7	0.99	1.68	12.8	—	580

TABLE A.7 Continued

Temperature, T (K)	Pressure, p (bar)[a]	Specific volume (m^3/kg) $v_f \cdot 10^3$	v_g	Heat of vaporization, h_{fg} (kJ/kg)	Specific heat (kJ/kg.K) $C_{p,f}$	$C_{p,g}$	Viscosity ($N.s/m^2$) $\mu_f \cdot 10^6$	$\mu_g \cdot 10^6$	Thermal conductivity (W/m.K) $k_f \cdot 10^3$	$k_g \cdot 10^3$	Prandtl number Pr_f	Pr_g	Surface tension $\sigma_f \cdot 10^3$ (N/m)	Expansion coefficient $\beta_f \cdot 10^6$ (K^{-1})	Temperature T (K)
590	108.3	1.482	0.0163	1274	6.41	7.35	84	21.5	513	84.1	1.05	1.84	10.5	—	590
600	123.5	1.541	0.0137	1176	7.00	8.75	81	22.7	497	92.9	1.14	2.15	8.4	—	600
610	137.3	1.612	0.0115	1068	7.85	11.1	77	24.1	467	103	1.30	2.60	6.3	—	610
620	159.1	1.705	0.0094	941	9.35	15.4	72	25.9	444	114	1.52	3.46	4.5	—	620
625	169.1	1.778	0.0085	858	10.6	18.3	70	27.0	430	121	1.65	4.20	3.5	—	625
630	179.7	1.856	0.0075	781	12.6	22.1	67	28.0	412	130	2.0	4.8	2.6	—	630
635	190.9	1.935	0.0066	683	16.4	27.6	64	30.0	392	141	2.7	6.0	1.5	—	635
640	202.7	2.075	0.0057	560	26	42	59	32.0	367	155	4.2	9.6	0.8	—	640
645	215.2	2.351	0.0045	361	90	—	54	37.0	331	178	12	26	0.1	—	645
647.3[a]	221.2	3.170	0.0032	0	∞	∞	45	45.0	238	238	∞	∞	0.0	—	647.3[b]

[a]1 bar = 10^5 N/m^2

[b]Critical temperature

TABLE A.8 Thermophysical properties of liquid metals

Composition	Melting point (K)	T (K)	ρ (kg/m^3)	C_p (kJ/kg.K)	$\nu \times 10^7$ (m^2/s)	k (W/m.K)	$\alpha \times 10^5$ (m^2/s)	Pr
Bismuth	544	589	10011	0.1444	1.617	16.4	0.138	0.0142
		811	9739	0.1545	1.133	15.6	1.035	0.0110
		1033	9467	0.1645	0.8343	15.6	1.001	0.0083
Lead	600	644	10540	0.159	2.276	16.1	1.084	0.024
		755	10412	0.155	1.849	15.6	1.223	0.017
		977	10140	—	1.347	14.9	—	—
Potassium	337	422	807.3	0.80	4.608	45.0	6.99	0.0066
		700	741.7	0.75	2.397	39.5	7.07	0.0034
		977	674.4	0.75	1.905	33.1	6.55	0.0029
Sodium	371	366	929.1	1.38	7.516	86.2	6.71	0.011
		644	860.2	1.30	3.270	72.3	6.48	0.0051
		977	778.5	1.26	2.285	59.7	6.12	0.0037
NaK, (45%/55%)	292	366	887.4	1.130	6.522	25.6	2.552	0.026
		644	821.7	1.055	2.871	27.5	3.17	0.0091
		977	740.1	1.043	2.174	28.9	3.74	0.0058
NaK, (22%/78%)	262	366	849.0	0.946	5.797	24.4	3.05	0.019
		672	775.3	0.879	2.666	26.7	3.92	0.0068
		1033	690.4	0.883	2.118	—	—	—
PbBi, (44.5%/55.5%)	398	422	10524	0.147	—	9.05	0.586	—
		644	10236	0.147	1.496	11.86	0.790	0.189
		922	9835	—	1.171	—	—	—
Mercury	234				See Table A.5			

TABLE A.9 Emissivities of some surfaces *(a) Metals*

Material	*Temperature K*	*Emissivity* ε
Aluminum		
Polished	300–900	0.04–0.06
Commercial sheet	400	0.09
Heavily oxidized	400–800	0.20–0.33
Anodized	300	0.8
Bismuth, bright	350	0.34
Brass		
Highly polished	500–650	0.03–0.04
Polished	350	0.09
Dull plate	300–600	0.22
Oxidized	450–800	0.6
Chromium, polished	300–1400	0.08–0.40
Copper		
Highly polished	300	0.02
Polished	300–500	0.04–0.05
Commercial sheet	300	0.15
Oxidized	600–1000	0.5–0.8
Black oxidized	300	0.78
Gold		
Highly polished	300–1000	0.03–0.06
Bright foil	300	0.07
Iron		
Highly polished	300–500	0.05–0.07
Cast iron	300	0.44
Wrought iron	300–500	0.28
Rusted	300	0.61
Oxidized	500–900	0.64–0.78
Lead		
Polished	300–500	0.06–0.08
Unoxidized, rough	300	0.43
Oxidized	300	0.63
Magnesium, polished	300–500	0.07–0.13
Mercury	300–400	0.09–0.12
Molybdenum		
Polished	300–2000	0.05–0.21
Oxidized	600–800	0.80–0.82
Nickel		
Polished	500–1200	0.07 0.17
Oxidized	450–1000	0.37–0.57
Platinum, polished	500–1500	0.06–0.18
Silver, polished	300–1000	0.02–0.07
Stainless steel		
Polished	300–1000	0.17–0.30
Lightly oxidized	600–1000	0.30–0.40
Highly oxidized	600–1000	0.70–0.80
Steel		
Polished sheet	300–500	0.08–014
Commercial sheet	500–1200	0.20–0.32
Heavily oxidized	300	0.81
Tin, polished	300	0.05
Tungsten		
Polished	300–2500	0.03–0.29
Filament	3500	0.39
Zinc		
Polished	300–800	0.02–0.05
Oxidized	300	0.25

TABLE A.9 Emissivities of some surfaces (*b*) *Non-metals*

Material	*Temperature* K	*Emissivity* ε
Ammonia	800–1400	0.65–0.45
Aluminum oxide	600–1500	0.69–0.41
Asbestos	300	0.96
Asphalt pavement	300	0.85–0.93
Brick		
Common	300	0.93–0.96
Fireclay	1200	0.75
Carbon filament	2000	0.75–0.90
Cloth	300	0.75–0.90
Concrete	300	0.88–0.94
Glass		
Window	300	0.90–0.95
Pyres	300–1200	0.82–0.62
Pyroceram	300–1500	0.85–0.57
Ice	273	0.95–0.99
Magnesium oxide	400–800	0.69–0.55
Masonry	300	0.80
Paints		
Aluminum	300	0.40–0.50
Black, Iacquer, shiny	300	0.88
Oils, all colors	300	0.92–0.96
White acrylic	300	0.90
White enamel	300	0.90
Red primer	300	0.93
Paper, white	300	0.90
Plaster, white	300	0.93
Porcelain, glazed	300	0.92
Quartz, rough, fused	300	0.93
Rubber		
Soft	300	0.86
Hard	300	0.93
Sand	300	0.90
Silicon carbide	600–1500	0.87–0.85
Skin, earth	300	0.95
Snow	273	0.80–0.90
Soil, earth	300	0.93–0.96
Soot 300–500	0.95	
Teflon	300–500	0.85–0.92
Water, deep	273–373	0.95–0.96
Wood		
Beech	300	0.94
Oak	300	0.90

TABLE A.10 Solar radiative properties for selected materials

Description / composition	α_s	ε^a	α_s/ε	τ_s
Aluminium				
Polished	0.09	0.03	3.0	
Anodised	0.14	0.84	0.17	
Quartz overcoated	0.11	0.37	0.30	
Foil	0.15	0.05	3.0	
Brick, red (Purdue)	0.63	0.93	0.68	
Concrete	0.60	0.88	0.68	
Galvanised sheet metal				
Clean, new	0.65	0.13	5.0	
Oxidised, weathered	0.80	0.28	2.9	
Glass, 3.2 mm thickness				
Float or tempered				0.79
Low iron oxide type				0.88
Metal, plated				
Black sulphide	0.92	0.10	9.2	
Black cobalt oxide	0.93	0.30	3.1	
Black nickel oxide	0.92	0.08	11	
Black chrome	0.87	0.09	9.7	
Mylar, 0.13 mm thickness				0.87
Paints				
Black (Parsons)	0.98	0.98	1.0	
White, acrylic	0.26	0.90	0.29	
White, zinc oxide	0.16	0.93	0.17	
Plexiglas, 3.2 mm thickness				0.90
Snow				
Fine particles fresh	0.13	0.82	0.16	
Ice granules	0.33	0.89	0.37	
Tedlar, 0.10 mm thickness				0.92
Teflon, 0.13 mm thickness				0.92

The emissivity values in this table correspond to a surface temperature of approximately 300 K

TABLE A.11 Diffusion coefficient of gases and vapours in air at 25°C and 100 kPa

Substance	D_{AB}, cm^2/s	$Sc = \frac{\nu}{D}$
Ammonia	0.28	0.78
Carbon dioxide	0.164	0.94
Hydrogen	0.410	0.22
Oxygen	0.206	0.75
Water	0.256	0.60
Ethyl ether	0.093	1.66
Methanol	0.159	0.97
Ethyl alcohol	0.119	1.30
Formic acid	0.159	0.97
Acetic acid	0.133	1.16
Aniline	0.073	2.14
Benzene	0.088	1.76
Toluene	0.084	1.84
Ethyl benzene	0.077	2.01
Propyl benzene	0.059	2.62

Table A.12 Molal specific volumes and latent heats of vaporization for selected liquids at their normal boiling points.

Substance	V_m *(m³/k-mol)*	$h_{fg} \times 10^3$ *(kJ/k-mol)*
Methanol	0.042	35.53
Ethanol	0.064	39.33
n-Propanol	0.081	41.97
Isopropanol	0.072	40.71
n-Butanol	0.103	43.76
tert-Butanol	0.103	40.63
n-Pentane	0.118	25.61
Cyclopentane	0.100	27.32
Isopentane	0.118	24.73
Neopentane	0.118	22.72
n-Hexane	0.141	28.85
Cyclohexane	0.117	33.03
n-Heptane	0.163	31.69
n-Octane	0.185	34.14
n-Nonane	0.207	36.53
n-Decane	0.229	39.33
Acetone	0.074	28.90
Benzene	0.096	30.76
Carbon tetrachloride	0.102	29.93
Ethyl bromide	0.075	27.41
Nitromethane	0.056	25.44
Water	0.0187	40.62

Index

D

E

F

G

H

I

J

K

L

M

N

O

P

R

S

T